Indian Insects

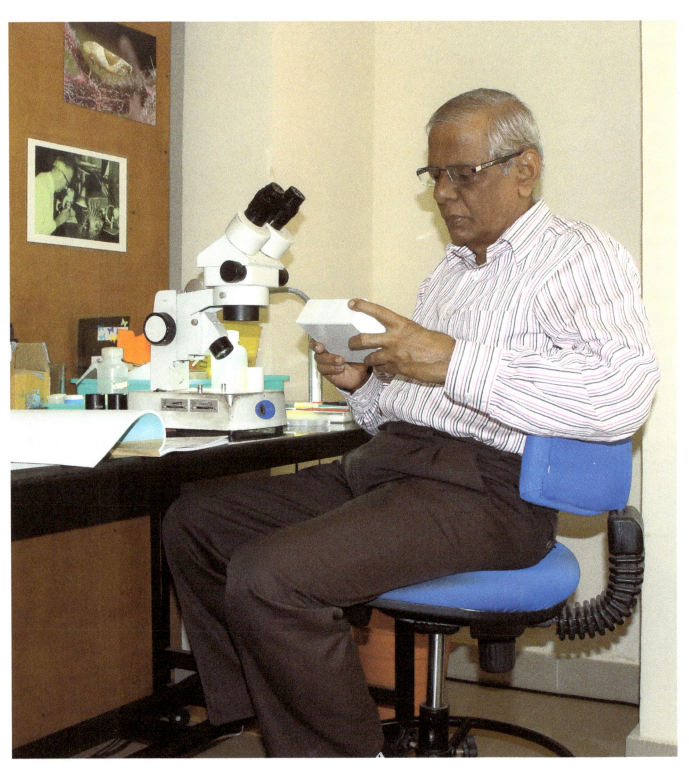

Vundlamath C.S

Indian Insects

Diversity and Science

A Festschrift for Professor C. A. Viraktamath's 75th Birthday

Edited by

S. Ramani

Prashanth Mohanraj

H. M. Yeshwanth

CRC Press
Taylor & Francis Group
Boca Raton London New York

CRC Press is an imprint of the
Taylor & Francis Group, an **informa** business

CRC Press
Taylor & Francis Group
6000 Broken Sound Parkway NW, Suite 300
Boca Raton, FL 33487-2742

© 2020 by Taylor & Francis Group, LLC
CRC Press is an imprint of Taylor & Francis Group, an Informa business

No claim to original U.S. Government works

Printed on acid-free paper

International Standard Book Number-13: 978-0-3671-8413-1 (Hardback)

This book contains information obtained from authentic and highly regarded sources. Reasonable efforts have been made to publish reliable data and information, but the author and publisher cannot assume responsibility for the validity of all materials or the consequences of their use. The authors and publishers have attempted to trace the copyright holders of all material reproduced in this publication and apologize to copyright holders if permission to publish in this form has not been obtained. If any copyright material has not been acknowledged please write and let us know so we may rectify in any future reprint.

Except as permitted under U.S. Copyright Law, no part of this book may be reprinted, reproduced, transmitted, or utilized in any form by any electronic, mechanical, or other means, now known or hereafter invented, including photocopying, microfilming, and recording, or in any information storage or retrieval system, without written permission from the publishers.

For permission to photocopy or use material electronically from this work, please access www.copyright.com (http://www.copyright.com/) or contact the Copyright Clearance Center, Inc. (CCC), 222 Rosewood Drive, Danvers, MA 01923, 978-750-8400. CCC is a not-for-profit organization that provides licenses and registration for a variety of users. For organizations that have been granted a photocopy license by the CCC, a separate system of payment has been arranged.

Trademark Notice: Product or corporate names may be trademarks or registered trademarks, and are used only for identification and explanation without intent to infringe.

Visit the Taylor & Francis Web site at
http://www.taylorandfrancis.com

and the CRC Press Web site at
http://www.crcpress.com

"My work enraptures but utterly exhausts me….To know that no one before you has seen an organ you are examining, to trace relationships that have occurred to no one before, to immerse yourself in the wondrous crystalline world of the microscope, where silence reigns, circumscribed by its own horizon, a blindingly white arena—all this is so enticing that I cannot describe it."

Letter to his sister Elena
Vladimir Nabakov (1945)

"Systematists will have only to decide (not that this will be easy) whether any form be sufficiently constant and distinct from other forms, to be capable of definition, and if definable whether the difference be sufficiently important to deserve a specific name."

Origin of Species
Charles Darwin (1858)

Alice explained, "…I can tell you the names of some of them."
"Of course they answer to their names?" the Gnat remarked carelessly.
"I never knew them do it."
"What's the use of their having names" the Gnat said, "if they won't answer to them?"
"No use to them," said Alice; "but it's useful to the people who name them, I suppose. If not, why do things have names at all?"

Through the Looking Glass
Lewis Carroll (1871)

Contents

Preface ix
Acknowledgments xi
Editors xiii
Chandrashekaraswami Adiveyya Viraktamath: The Prince of Indian Taxonomists xv
Contributors xxi

Chapter 1 Why Are Insects Abundant? Chance or Design? 1

    K. N. Ganeshaiah

Chapter 2 Mayflies (Insecta: Ephemeroptera) of India 7

    C. Selvakumar, K. A. Subramanian, and K. G. Sivaramakrishnan

Chapter 3 Dragonflies and Damselflies (Insecta: Odonata) of India 29

    K. A. Subramanian and R. Babu

Chapter 4 Stoneflies (Insecta: Plecoptera) of India 47

    R. Babu, K. G. Sivaramakrishnan, and K. A. Subramanian

Chapter 5 Taxonomy of Orthoptera with Emphasis on Acrididae 57

    Rajamani Swaminathan and Tatiana Swaminathan

Chapter 6 Taxonomy and Biodiversity of Soft Scales (Hemiptera: Coccidae) 69

    Sunil Joshi

Chapter 7 Whiteflies (Hemiptera: Aleyrodidae) of India 103

    R. Sundararaj, K. Selvaraj, D. Vimala, and T. Venkatesan

Chapter 8 Pentatomidae (Hemiptera: Heteroptera: Pentatomoidea) of India 121

    S. Salini

Chapter 9 Indian Mymaridae (Hymenoptera: Chalcidoidea) 147

    S. Manickavasagam and S. Palanivel

Chapter 10 Scelionidae and Platygastridae (Hymenoptera: Platygastroidea) of India 159

    K. Rajmohana and Sunita Patra

Chapter 11 Bumble Bees (Hymenoptera: Apidae: Apinae: Bombini) of India 173

    Martin Streinzer

Chapter 12 Potter Wasps (Hymenoptera: Vespidae: Eumeninae) of India 187

    P. Girish Kumar, Arati Pannure, and James M. Carpenter

Chapter 13 Tiger Beetles (Coleoptera: Cicindelidae): Their History and Future in Indian Biological Studies 201

David L. Pearson

Chapter 14 Coccinellidae of the Indian Subcontinent 223

J. Poorani

Chapter 15 Flea Beetles of South India (Coleoptera: Chrysomelidae: Galerucinae: Alticini) 247

K. D. Prathapan

Chapter 16 Taxonomy, Systematics, and Biology of Indian Butterflies in the 21st Century 275

Krushnamegh Kunte, Dipendra Nath Basu, and G. S. Girish Kumar

Chapter 17 Taxonomy and Diversity of Indian Fruit Flies (Diptera: Tephritidae) 305

K. J. David, S. Ramani, and S. K. Singh

Chapter 18 Hover-Flies (Diptera: Syrphidae) Recorded from "Dravidia," or Central and Peninsular India and Sri Lanka: An Annotated Checklist and Bibliography 325

Kumar Ghorpadé

Chapter 19 A Comparative Study of Antennal Mechanosensors in Insects 389

Harshada H. Sant and Sanjay P. Sane

Chapter 20 Cross-Kingdom Interactions in Natural Microcosms: The Worlds Within Fig Syconia and Ant-Plant Domatia 401

Renee M. Borges, Joyshree Chanam, Mahua Ghara, Anusha Krishnan, Yuvaraj Ranganathan, Megha Shenoy, Vignesh Venkateswaran, and Pratibha Yadav

Annexure I: Revisions and Reviews of Taxa by Professor C. A. Viraktamath 415

Annexure II: New Tribe and Genera Described by Professor C. A. Viraktamath 417

Annexure III: New Species Described by Professor C. A. Viraktamath 419

Annexure IV: Research Publications of Professor C. A. Viraktamath 427

Annexure V: Taxa Named in Honour of Professor C. A. Viraktamath 437

Annexure VI: Courses Offered by Professor C. A. Viraktamath 439

Annexure VII: Theses Submitted Under the Guidance of Professor C. A. Viraktamath (1980–2019) 441

Annexure VIII: Research Projects Operated by Professor C. A. Viraktamath 443

Index 445

Preface

> In the Agricultural Colleges the aim is to train farmers and fruit growers and not entomologists….. …they should have just a general knowledge of the external anatomy of insects so as to be able to place the insects at least in their orders. This amount of systematic work is quite sufficient for them.
>
> **C. C. Ghosh, Asst. to the Imperial Entomologist, Pusa (1919)**

> …During this meeting we have been constantly reminded of the very great difficulty of getting our specimens identified. … We do want … an institute such as the British Museum (Natural History) … where specialists can work and our insects be identified.…. Without pure science we cannot go very far with our applied science.
>
> **M. Afzal Husain, Supernumerary Entomologist, Pusa (1919)**

The last subject taken up for discussion at the Third Entomological Meeting at Pusa in February, 1919 was "The Organization of Entomological Work in India." When initiating the discussion on this subject, T. B. Fletcher, the then Imperial Entomologist to the Government of India, told the gathering of entomologists from across the country that after considerable thought, he felt that "entomological work, particularly entomological research had to be improved in the country." The antithetical views appearing as the epigraph above formed part of this discussion. Since then, Indian entomology has been perpetually plagued by this Janus-faced approach to insect taxonomy and systematics, the reverberations of which continue to haunt us to this day.

Linnaean taxonomy was introduced during Linnaeus's time to India by J. G. Koenig, a private student of his, who came to India in 1767. He was followed by I. K. Daldorff, a pupil of Fabricius, who came to India in 1790. This early start, however, failed to lend the necessary fillip to the development of indigenous expertise in insect taxonomy on Linnaean principles in India. Unfortunately, this situation did not improve either during the period of British rule or even after the country gained political independence in the mid twentieth century. Ironically, while a number of British entomologists worked on the taxonomy of Indian insects, only a handful of Indian entomologists developed the necessary expertise to work independently on taxonomy. This expertise was not, however, passed on and did not foster succeeding generations of home grown entomologists. Insect taxonomy remained "provincial" and dependent on metropolitan science. The British Museum (Natural History), London, was the main institution on which Indian entomologists largely depended for the determination of the taxonomic identity of insects. This situation was to change dramatically in the late twentieth century as the BM (NH), now the Natural History Museum, at London, changing policy, began to charge (exorbitantly for a developing economy) for their identification services. If the country was not to lose scarce foreign exchange reserves for the identification of large numbers of insect specimens, the necessary capabilities had to be built locally.

With the singular motive of serving this end, an ardent entomologist, Professor C. A. Viraktamath, with great foresight, had already set in motion a new approach to taxonomic pedagogy. The concept of a "practical record" (and all that it entailed) in taxonomy courses—till then the staple in any practical class—was dispensed with. He introduced students to taxonomic keys and trained them in the use of these keys to determine the identity of specimens. More students were allotted topics on taxonomy for their theses as part of the requirement for the completion of their post graduate programmes. As students thus became more independent and competent in the principles and practices of insect taxonomy, they continued to pursue this study in their places of employment. Thus developed centres of taxonomic excellence on selected insect taxa in various parts of the country.

In this *festschrift,* we have drawn together entomological expertise from those who have been associated with Professor Viraktamath and have drawn sustenance from his entomological knowledge. Insect taxonomy has of course been the focus of our attention as there is still no single book on the taxonomy of Indian insects that serious students of entomology can turn to. As Eliot Zimmermann, the American entomologist, put it "those who know most about this subject, know best how little they know." Acknowledging this sentiment to the hilt, we venture into the shallows knowing full well that one day the deep too will be conquered.

This book makes no pretensions to being a textbook on Indian insect taxonomy. It, however, takes the decisive preliminary step of being the nucleus of such a book. We have not been able to draw on the expertise of all the competent taxonomists for various reasons. Of the 20 chapters comprising this book, 17 deal with the taxonomy of different insect taxa. While four groups of insects (Ephemeroptera, Odonata, Plecoptera, and Orthoptera) have been dealt with at the order level, five others (Hemiptera, Hymenoptera, Coleoptera, Lepidoptera, and Diptera), which include the mega orders, have been treated in parts at various ranks below the order. A major endeavour has been to provide identification keys with appropriate illustrations in many cases, with the focus squarely on the Indian fauna. Depending on the nature of expertise of our contributors, these keys serve to identify Indian insects at different levels in the taxonomic hierarchy. The chapters focus on the present status and could form a starting point for students and young researchers to leverage further work on these groups. Fully aware of the incompleteness of a "book of even this limited scope," we hope that one that includes all Indian taxa will soon be forthcoming.

The three remaining chapters explore different facets of insect life by investigators who have been closely associated with Professor Viraktamath. The first is a novel explanation for why insects are so abundant, the other, an anatomical study comparing mechanosensors in the antennae of insects, and the third explores the exciting world of natural microcosms inhabited by insects.

Idiosyncrasies in presentation are a necessary concomitant of a compendium of this kind. While the authors were assigned taxa or topics of their expertise, they were given the leeway to present and organize the content of their chapters in any manner they chose to with the fidelity of the content being their sole responsibility.

Vive la Professor Viraktamath and may his illustrious legacy fill in the long standing requirement for a book on the taxonomy of all orders of Indian insects.

Editors

Acknowledgments

First and foremost, our greatest debt is to Professor Viraktamath, who instilled the love for entomology in each of us and taught us to observe and bring order into the bewildering diversity of insect life teeming around us. He also permitted us to peer unabashedly into and query his past, one that had undoubtedly dimmed with the steady, relentless passage of time.

It was with trepidation that the six of us—Ramani, Yeshwanth, Belavadi, Chandrashekar, Mallik, and Prashanth—embarked on this venture of filling a lacuna in the orbit of Indian entomological pedagogy. That there was a genuine requirement was never in doubt. But our concern of whether we could we pull it off, constrained by the bounds of a festschrift, was genuine too. Fortunately, the circle of those who were Professor Viraktamath's students or had been associated with him was so wide that we had little difficulty in getting together a band of the majority of the most respected taxonomists as well as those whose research depended on his taxonomic expertise in the country.

We thank all the contributors to this volume for having accepted our invitation to write with authority on taxa of their interest and take responsibility for the contents of their respective chapters. We also thank others who submitted chapters which could not be included in this book for a variety of reasons.

H. V. Ghate, A. K. Dubey, A. Ramesh Kumar, V. V. Belavadi, S. S. Anooj, and Professor C. A. Viraktamath graciously gave of their time to review the manuscripts given to them. We are grateful to all of them for this kind gesture.

Ms. Renu Upadhyay, assistant commissioning editor, Ms. Shikha Garg, and Ms. Jyotsna Jangra editorial assistants of the CRC Press, Taylor & Francis Group, eased the pressures of working on the manuscript of such an undertaking with great forbearance and understanding. Fully appreciative of the rationale of our request to increase the size of the book, they enabled it with alacrity. We remain indebted to them for answering our queries without rancour and for their near stoic patience in waiting for us to submit the completed manuscript even after it had crossed the deadline.

We wish to acknowledge Ms. Rachael Panthier, Production Editor, Taylor & Francis as well as the able team at Lumina Datamatics, Puducherry led by Mr. Sundaramoorthy Balasubramani for bringing in their expertise to ensure high production standards.

Editors

Editors

Dr. S. Ramani joined the Agricultural Research Service of the ICAR, New Delhi, India, in 1985 and after serving in different ICAR institutes retired as Project Coordinator, ICAR-AICRP on Honey Bees & Pollinators in 2011. His areas of specialization are pollination, biological control, and taxonomy. He has published more than 45 papers in peer-reviewed journals and several popular articles on insects. His special area of interest is taxonomy of Tephritidae. He collects stamps on insects as a hobby.

Dr. H. M. Yeshwanth is an assistant professor in the Department of Entomology at University of Agricultural Sciences, Bengaluru, India. His research interests are in the area of taxonomy, biodiversity, and natural history of insects. His special area of interest is taxonomy of Hemiptera, especially Miridae, and he has described about 25 new species of mirid bugs. He has published over a dozen papers in highly rated international journals. He has developed excellent techniques for photographing insects both in nature and from museum specimens.

Dr. Prashanth Mohanraj retired as a principal scientist from the ICAR-National Bureau of Agricultural Insect Resources, Bengaluru, India, in 2017. He joined the Agricultural Research Service of ICAR and worked at the ICAR-Central Islandm Agricultural Research Institute, Port Blair, Andaman & Nicobar Islands, for 14 years. He co-authored a book *Butterflies of Andaman & Nicobar Islands*, the Hindi translation of which won the best book award from the ICAR. His areas of specialization are biological control, natural history of insects, and taxonomy. He has specialized on the taxonomy of Trichogrammatidae. He has published over 65 research papers in peer-reviewed journals and more than a dozen popular articles on insects.

Chandrashekaraswami Adiveyya Viraktamath
The Prince of Indian Taxonomists

CAV, as professor C. A. Viraktamath is known to those in the country familiar with his work or have sought his advice or assistance on matters of insect taxonomy, are three letters that are as recognizable as if he were a brand like BMW, BEL, or HAL. Vakma, Mutt, and Chandra (with Viraktamath becoming Vakma or Mutt and Chandrashekaraswami—Chandra) are other sobriquets that he is known by to those who find his 38 letters long name—Chandrashekaraswami Adiveyya Viraktamath—containing over half the letters of the alphabet, more than a mouthful.

Seventy-five years ago, on January 31, 1944, when their second child, a boy was born to Adiveyya and Neelambika in Byadagi—in the heart of present-day Karnataka, famous for its deep red, spicy chillies—little did they imagine that he would turn out to be one of the most respected taxonomists not only in the country, but also internationally.

His father who worked in the Revenue Department was constantly transferred as part of government policy. This peripatetic existence saw CAV studying in different schools in the different places at which his father was posted. After school, he joined the College of Agriculture, Dharwad to pursue a BSc degree in Agriculture. As at school, he continued to be a brilliant student at college too and won the ASPEE Gold Medal for the highest marks in Plant Pathology. He then gained admission at the University of Agricultural Sciences, Bangalore for the MSc (Agricultural Entomology) program. Here, he did not fail to impress Dr. H. M. Harris (a taxonomist specializing on Nabidae, Hemiptera) and Dr. J. H. Lilly, visiting professors under the Ford Foundation program, who taught him insect systematics and insect physiology, respectively. It was at this time that the ragi (*Eleusine coracana*) crop was severely affected by the mosaic and streak diseases. Leafhoppers were known to transmit these diseases, but no one in India could identify the leafhoppers that were their vectors. The expertise of taxonomists in the USA and the UK had to be sought. At this juncture, Harris convinced "Vakma" (for this was how he abbreviated the, to him, unpronounceable name—Viraktamath) to take up the taxonomic study of leafhoppers. Recommended by Harris and Lilly, CAV gained admission to Oregon State University (OSU) as a Ford Foundation scholar to study leafhopper taxonomy under Dr. Paul W. Oman (PWO), the then most eminent leafhopper taxonomist in the world.

H. M. Harris, Visiting Professor, Ford Foundation at University of Agricultural Sciences, Bangalore

At OSU, he was a man in a hurry. He had to pick up everything from his mentor in a mere year's time. To do this, he reduced his course load (he'd complete the remaining courses on his return to UAS, Bangalore). Here, he had to devote all the available time to examining the large number of Old World agalliine leafhoppers (the group assigned to him for his PhD thesis) in Oman's laboratory as well as all the specimens arriving from museums and from PWO's fellow entomologists from all over the world. His day began early and went on into the wee hours of the night as he peered endlessly into the microscope to dissect and draw the intricate structures that served to distinguish one species of leafhopper from another. Most nights he'd be the last to leave the laboratory. Working at this frenetic, febrile pace, he not only studied the specimens, especially the types borrowed from different museums, but also found the time to travel to the California Academy of Sciences and sort their unsorted leafhopper material from the Old World so that he could borrow the agalliine leafhoppers from the collection for his research.

While G. P. Channabasavanna, the Head of the Department of Entomology, UAS, Bangalore, had taught him the fine art of camera lucida drawing, and Harris had introduced him to the rudiments of leafhopper taxonomy, he still had to learn the finer details under Oman, who patiently taught him the intricacies of the genital architecture of male leafhoppers for the first time and even made dissecting needles with minutens specifically for Chandra's (for that was the name he had been given in Oman's lab, and that was how he'd be addressed by his peers from Europe and the USA from now on) use. Oman also introduced him to the techniques of collecting leafhoppers in the field. Chandra could only look in wide-eyed disbelief as Oman took him to various locales and unerringly told him which species could be found on a specific plant or in a particular area. It was here that he picked up the importance of collecting a large series of specimens from the complete range of occurrence of each species.

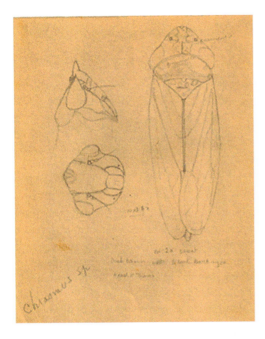

An early set of diagrams of a deltocephaline leafhopper by C. A. Viraktamath during his Dharwad days, identifies by Dr. H. M. Harris

After this short stint of a year and a half at OSU, Chandra returned a fledgling taxonomist to the College of Agriculture, Dharwad. Now began the trying task of locating for study the types of agalliine leafhoppers of the Indian subcontinent deposited in different museums, while simultaneously making fresh collections from all over India. These studies had to go hand in hand with the courses that he had to offer at the college. A few years after his return from the USA he got married to Lalitha (nee Shashikala Koranmath) at Dharwad on May 27, 1973. Persistent requests from him soon lead to his transfer to Bangalore. By then he had been promoted as an assistant professor.

It was in Bangalore that he began his research work in earnest. This had to be done in addition to teaching, which, as will be seen later, was a taxing schedule. To begin, he had to make his own collection, as leafhopper collections in Indian museums and colleges were modest or non-existent. In those days, it was not uncommon to see him wielding the outsized 5.5 feet long leafhopper net, designed by Oman—one that was over twice as long and as many times as heavy as the insect nets in common use—collecting leafhoppers on the university campus and elsewhere in the wilds of India. Since his thesis was to include the entire Old World Agalliinae, he had to source specimens of these leafhoppers from different museums and make collections, especially from the type localities from across the country. Lalitha, his wife, was not in the least resentful of his frequent absences from home in pursuit of leafhoppers. On the other hand, she cheerfully assisted him in processing specimens from the substantial catches from these expeditions. Till Sreedevi, their first daughter was born in 1976, they would walk each evening to the laboratory from home and she would lighten the hours by assisting him in the tedious task of sorting, mounting, and pinning leafhoppers. But when Smita, their second daughter was born a few years later, in 1983, she had no time at all for entomological work.

His first paper on the taxonomy of a leafhopper was published in 1972 (years before he completed his PhD), describing a new species of *Austroagallia* from the Galapagos. When he submitted his thesis in March 1982 titled "A Generic Revision of the Old World Agalliinae (Homoptera: Cicadellidae)," signalling the completion of his PhD, he had published 15 papers on leafhoppers. Recognition from fellow taxonomists came fairly early to him when he was invited to present a paper and chair the last session of the *First International Workshop on Biotaxonomy, Classification and Biology of Leafhoppers and Planthoppers (Auchenorrhyncha) of Economic Importance* held between 4th and 7th October 1982 under the auspices of the Commonwealth Institute of Entomology, Commonwealth Agricultural Bureau International held at London. His paper titled "Genera to be revised on a priority basis. The need for keys and illustrations of economic species of leafhoppers and preservation of voucher specimens in recognised institutions" raised issues of serious taxonomic concern particularly in the developing world.

Participants of the First International Workshop on Leafhoppers and Planthoppers of Economic Importance held between 4th–7th October, 1982 at Royal Entomological Society, London (Professor C. A. Viraktamath, standing fifth from right)

Subsequent studies on the taxonomy of leafhoppers saw him making repeated visits to the Natural History Museum, London and the National Museum of Natural History, Washington, DC. Some of these visits were at his own expense as official sources of funding for taxonomic studies are hard to come by.

The 114 papers on Cicadellidae so far published by him, some in association with national and international associates, include 46 revisionary studies and 11 reviews (Annexure I) of various taxa in addition to describing a new tribe, and many new genera and species. These also include a few papers on the biology and host plants of cicadellids as well as their role as vectors of plant pathogens. In all, 56 new genera (Annexure II) and 452 new species (Annexure III) have been described by him and his collaborators. Though he continued to teach and guide students after his retirement in 2004, he could from then onwards devote most of his time to taxonomic work. This is reflected in the spurt in the number of taxa described by him and the number of papers published since then. Over 50 percent of both genera and species were described in the 15 years after his retiring from the university than in the 32 years prior to that. His publications include 209 research papers and 6 edited books (Annexure IV). Recognition for his work came in many forms, but the most cherished for any taxonomist is recognition from his peers. That CAV got this in no small measure can be gauged from the number of species that have been named after him. Twenty-nine species of leafhoppers were named in his honour by 25 leafhopper taxonomists from laboratories across the world. Both admiration for his work and gratitude for his taxonomic assistance prompted 14 taxonomists, largely from India, to name new Lepidoptera (1 species), Hymenoptera (2 species), and Coleoptera (6 species) after him (Annexure V). In recognition of his pre-eminence as a leafhopper taxonomist, he was invited by Dr. Zhang Yalin of the Northwest A & F University, Yangling, Shaanxi Province, China, to assist post graduate students in their studies on the taxonomy of Cicadellidae. All four of these visits were funded by that university.

He had taught entomology for a few years at the undergraduate level at the College of Agriculture, Dharwad. It was at Bangalore, however, that he came into his own as a teacher. Here, he began by teaching the "Introductory Entomology" course to undergraduates in July 1974. While he continued to offer a few courses in entomology to the undergraduates till 1995, he was soon entrusted the more onerous task of teaching post graduate students. During the course of the next 30 years, he would offer a range of much sought after courses to those pursuing their MSc and PhD programmes. These courses included Insect Morphology, Internal Anatomy, Systematic Entomology I and II, Immature Insects, Bioecology of Crop Pests, Literature and Techniques in Entomology, and Insect Vectors of Plant Pathogens (some of them even for 12 years subsequent to his retirement) (Annexure VI). The most sought after, however, were the courses in systematics. Over the span of 35 years he offered over 90 courses.

He was among the most feared and revered members on the teaching faculty. His exacting standards and stern demeanour infused a sense of purpose and the determination to outperform oneself in every student. Over the years, however, he mellowed in his dealings with students with severity giving way to tact.

While one could complete most other courses by merely listening to the professor in class, he did not confine his teaching solely to lectures, but gave original scientific papers and chapters from the classics of taxonomy and evolutionary biology as reading assignments. The works of Borror and DeLong were required reading, so were those of R. F. Chapman, Ernst Mayr, Alvah Peterson, Theodosius Dobzhansky, S. L. Tuxen, Howard E. Hinton, Willi Hennig, and many others based on the course being offered. The International Code of Zoological Nomenclature had to be studied very closely. Practical classes also changed character under him. Having prevailed over a very reluctant Head of the Department, he dispensed with the need for the maintenance of practical record books in the course on Insect Systematics. Not only were these books the distinguishing feature of any practical class, they were inconceivable without them. For the first time at UAS—and perhaps in the country—students of insect systematics had to use keys and identify specimens during their practical classes. He began assigning "special topics" to motivate students to collect and study specified insect taxa. He removed the limit of 50 specimens that students were expected to collect for submission as part of the requirement for the course in Insect Systematics. The more diverse and the larger the collection submitted, the more marks were you likely to get. One more change he effected was in the pins used in pinning the insects collected. Insect pins were not given to students, as they had to be imported and were too expensive. The short stubby office pins then in use were replaced by fine sewing needles with plastic beads affixed to the blunt ends to serve as heads for ease of handling. This minimized damage to specimens, enabling the retrieval of valuable ones for retention in the departmental collection.

It was under him that the chaotic collection of insects at the Department of Entomology developed into a well-curated modern museum. The collection was so dear to him that once while a bomb scare made all others vacate the college building, he refused to move saying, "If there really is a bomb I would prefer to die along with the entire insect collection." Over the years, synoptic collections of taxa not represented in the collection were obtained from reputed taxonomists from across the country. The total number of insect specimens in the collection is currently estimated at about 350,000. The oldest specimen, recovered and preserved from the decrepit old collection at the department, is an *avare* weevil collected on October 29, 1908, by Leslie C. Coleman, the first director of agriculture, Mysore state. The collection also includes type specimens of all insects described by those working in the department. He set in place a free identification service for farmers, students, amateurs, and others desirous of ascertaining the identity of insects of their interest, which he continues to do even today. This gained greater significance when the Natural History Museum, London began levying a steep charge for the identification of insects sent to them, a service they had hitherto rendered free. Over 275 institutions from across the country, in addition to a very large number of interested individuals, have so far availed this facility.

Professor C. A. Viraktamath in the Insect Museum

He didn't have to wait too long after coming to Bangalore to be made a Major Advisor entrusted with guiding post graduate students (first MSc student allotted in 1975) for the completion of their research topics, a mandatory requirement for the completion of their MSc and PhD programmes. Forty-four students submitted their theses under his guidance (30 MSc students and the remaining, PhD students). Twenty-eight of these were on the taxonomy of different insect taxa, eight of which were on leafhoppers. Over half the remaining theses submitted under his guidance were on the bioecology and management of insect pests of crops and others in the areas of pollination biology, insecticide toxicology, studies on light traps, biological control, and forensic entomology (Annexure VII). He did not shy away from making his students take up taxonomic studies of taxa other than leafhoppers. Some of his students who have gone down these alternative paths have blossomed to become taxonomists of note in taxa as diverse as Orthoptera, Formicidae, Arctiidae, Chrysomelidae, Carabidae, Scarabaeidae, Cerambycidae, Coccinellidae, Tephritidae, Delphacidae, Pentatomidae, Miridae, Reduviidae, Aphididae, and Apidae.

In addition to three projects on insect taxonomy, he handled six others (with collaborators) on various aspects of economic entomology from "assessing the status of the invasive American leaf miner, *Liriomyza trifolii*" and the "utilization of non-edible oils in pest management," to the "management of insect pests of gherkins" (the last at the instance of the Gherkin Growers' Association). The pest management strategy for gherkins that he developed under this project continues to be in vogue. The results of the project on the "pest status of the Mexican bean beetle *Zygogramma bicolorata,* introduced for the biological control of *Parthenium hysterophorus* L.," helped stave off a major setback to the weed biocontrol programme in the country, as opinion in the higher echelons of the agricultural fraternity was dangerously veering towards a blanket ban on the import of all weed biocontrol agents. The "Network Project on Insect Biosystematics" which he handled with elan at one of the Bangalore centres enabled the growth of the departmental museum to become one of the finest in the country (Annexure VIII).

Though afflicted, since 2014, by Age-related Macular Degeneration (an affliction of the eyes that blurs vision), he continues to work unfazed. Anyone stepping into his laboratory these days is likely to find him with a hand lens clenched tightly in his right fist, poring intently, in Holmesian fashion, over a manuscript or scrutinizing a leafhopper from his collection. In spite of this impediment, however, there has been no let-up in the pace of his work. In the 4 years since the diagnosis of this ailment, he has published 17 papers on the taxonomy of leafhoppers, with at least four of them being 30 pages or more in length.

Professor C. A. Viraktamath Examining Leafhoppers

Hazarding the risk of being accused of hagiography, we assert that CAV remains an enviable brand in the international taxonomic landscape. Like any enduring brand, he remains unique and authentic, well known to the people who desire his services with no doubt having ever been cast on the consistency in the quality of his work. His brand equity and brand value can only wax in the years to come.

Editors

Contributors

R. Babu
Southern Regional Centre
Zoological Survey of India
Chennai, India

Dipendra Nath Basu
National Centre for Biological Sciences
Bengaluru, India

Renee M. Borges
Centre for Ecological Sciences
Indian Institute of Science
Bengaluru, India

James M. Carpenter
Division of Invertebrate Zoology
American Museum of Natural History
New York, New York

Joyshree Chanam
Centre for Ecological Sciences
Indian Institute of Science
Bengaluru, India

K. J. David
Division of Germplasm Collection and Characterisation
ICAR-National Bureau of Agricultural Insect Resources
Bengaluru, India

K. N. Ganeshaiah
School of Ecology and Conservation
University of Agricultural Sciences
Bengaluru, India

Mahua Ghara
Centre for Ecological Sciences
Indian Institute of Science
Bengaluru, India

Kumar Ghorpadé
Department of Entomology
University of Agricultural Sciences
Dharwad, India

G. S. Girish Kumar
Mangalore University
Mangalore, India

P. Girish Kumar
Western Ghats Regional Centre
Zoological Survey of India
Kozhikode, India

Sunil Joshi
Division of Germplasm Collection and Characterisation
ICAR-National Bureau of Agricultural Insect Resources
Bengaluru, India

Anusha Krishnan
Centre for Ecological Sciences
Indian Institute of Science
Bengaluru, India

Krushnamegh Kunte
National Centre for Biological Sciences
Bengaluru, India

S. Manickavasagam
Department of Entomology
Faculty of Agriculture
Annamalai University
Chidambaram, India

S. Palanivel
Department of Entomology
Faculty of Agriculture
Annamalai University
Chidambaram, India

Arati Pannure
Department of Entomology
College of Sericulture
University of Agricultural Sciences
Bengaluru, India

Sunita Patra
Zoological Survey of India
Kolkata, India

David L. Pearson
School of Life Sciences
Arizona State University
Tempe, Arizona

J. Poorani
ICAR-National Research Centre for Banana
Trichy, India

K. D. Prathapan
Department of Agricultural Entomology
Kerala Agricultural University
Trivandrum, India

K. Rajmohana
Zoological Survey of India
Kolkata, India

Yuvaraj Ranganathan
Centre for Ecological Sciences
Indian Institute of Science
Bengaluru, India

S. Ramani
ICAR, Department of Entomology
University of Agricultural Sciences
Bengaluru, India

S. Salini
Division of Germplasm Collection and Characterisation
ICAR-National Bureau of Agricultural Insect Resources
Bengaluru, India

Sanjay P. Sane
National Centre for Biological Sciences
Bengaluru, India

Harshada H. Sant
National Centre for Biological Sciences
Bengaluru, India

C. Selvakumar
Department of Zoology
The Madura College (Autonomous)
Madurai, India

K. Selvaraj
Division of Germplasm Conservation and Utilisation
ICAR-National Bureau of Agricultural Insect Resources
Bengaluru, India

Megha Shenoy
Centre for Ecological Sciences
Indian Institute of Science
Bengaluru, India

S. K. Singh
National Museum of Natural History
New Delhi, India

K. G. Sivaramakrishnan
Zoological Survey of India
Chennai, India

K. A. Subramanian
Zoological Survey of India
Southern Regional Centre
Chennai, India

R. Sundararaj
Forest and Wood Protection Division
Institute of Wood Science and Technology
Bengaluru, India

Martin Streinzer
Department of Neurobiology
Faculty of Life Sciences
University of Vienna
Vienna, Austria

Rajamani Swaminathan
Department of Entomology
Rajasthan College of Agriculture
Maharana Pratap University of Agriculture and Technology
Udaipur, India

Tatiana Swaminathan
Department of Entomology
Rajasthan College of Agriculture
Maharana Pratap University of Agriculture
  and Technology
Udaipur, India

T. Venkatesan
Division of Genomic Resources
ICAR-National Bureau of Agricultural Insect Resources
Bengaluru, India

Vignesh Venkateswaran
Centre for Ecological Sciences
Indian Institute of Science
Bengaluru, India

D. Vimala
Zoological Survey of India
Southern Regional Centre
Chennai, India

Pratibha Yadav
Centre for Ecological Sciences
Indian Institute of Science
Bengaluru, India

1 Why Are Insects Abundant? Chance or Design?

K. N. Ganeshaiah

CONTENTS

A Quip: God and the Beetles 1
Why Are Beetles Species Rich? 1
Are Beetles Uniquely Diverse Compared to Other Groups? 2
Species Radiation as a Positive Feedback Process: A Hypothesis 3
Simulating the Process 4
Results of the Simulation 4
Returning to the Quip and the Question 5
Acknowledgements 5
References 5

A QUIP: GOD AND THE BEETLES

Some quips live long; so much so that they come to be established almost as eternal truth. One such rare quip is by Haldane on God's fondness for beetles. Highly impressed by Haldane's deep insights on biological systems, an elite lady-theologian is believed to have asked him at a post dinner discussion, on any distinct feature Haldane would identify with, or attribute to "God." Haldane is supposed to have quipped immediately: *God must have an inordinate fondness for beetles* (Hutchinson 1959).

Though the exact line that Haldane is supposed to have uttered, the context in which he may have said so, and even the veracity of the event, have all been frequently questioned, discussed, and debated (Gould 1993). The implication of the quip *per se*, that beetles are the most diverse group of organisms seems to be a pervading message not only among the public, but even among biologists. Gould (1993), tracing the veracity of Haldane's quip stated, "Ultimate meaning must reside in the unparalleled diversity of that group that rarely rivets our attention." In fact, the quip has been so influential that several workers have repeatedly cited it in different contexts (e.g., Sagan [Anonymous 2002] while taking a jab at creationists), discussed it (Gould 1993; Hardy 2002), and derived inspiration to address its implication (Ganeshaiah 1998). In some sense, owing to this quip, "In evolutionary biologists' minds, beetles have been basking in the warm glow of God's favour ever since" (Hardy 2002).

Beetles have become a subject of special attention to biologists, not just because of Haldane's quip, but due to two of their features: (a) they are the most species rich group, and hence by derivation, (b) they are considered the most diverse group of organisms. Among the 1.82 million described species, beetles with about 400,000 species, constitute the largest group. Though these numbers may vary across the data sets compiled, it remains unquestionable that ~22 percent of all known species are beetles (Ghosh 1996; Ganeshaiah 1998); but see arguments for the greater abundance of Neoptera (Hardy 2002; Mayhew 2002) and Hymenoptera (Forbes et al. 2018) among insects. Thus, if we randomly sample a set of five species from the pool of all known species, almost always one among them would be a beetle. This unique, species rich feature of beetles remains equally strong when extinct groups from the fossils are also considered, suggesting that they have been the most predominant group in evolution.

WHY ARE BEETLES SPECIES RICH?

For Haldane, the God who created this species rich group is clearly natural selection and hence, for evolutionary biologists (or adaptationists), the most obvious fallout of the attention that beetles have gathered is, "why are there so many of them?" Not surprisingly, this has been the subject of several investigations. For instance, Farell (1998) addressed this question using a phylogenetic approach of the feeding behavior of beetles and showed that their high species richness is the result of the enhanced rates of speciation of the group when they shifted from feeding on lower plants to angiosperms. With this, Farell concluded that God's, "Inordinate fondness [had been] explained" (Farell 1998; Anonymous 2018a).

On the other hand, Smith and Marcot (2015) examined the rate of extinction of beetles in the fossil data and found that their richness is owing to their resistance to extinction. They then concluded that "focusing on ... factors that have inhibited beetle extinction...should be examined as important determinants of their great diversity today." And there are others who have suggested adaptive features such as wing flexion (Mayhew 2002), phytophagy, or complete metamorphosis (see Hardy 2002) as factors driving beetle species richness—none, however, being unequivocal determinants.

However, before addressing the factors causing the abundance of beetles, it may be important to ask whether they are indeed uniquely different in this feature of species loading. Because, only if they are found to be uniquely species rich, does it become meaningful to ask, "why (and how) are they species rich?" On the other hand, if their species richness could emerge merely from a general pattern through which biological diversity gets organized during evolution, then it would be futile and unnecessary to pursue the question. Alternatively, we may have to explore the processes shaping the organization of diversity. In other words, it may be worthwhile to assess whether the observed species richness of beetles is an inevitable consequence of the evolution of hierarchical organization of diversity even before assessing the biological and adaptive features that may have shaped their richness.

ARE BEETLES UNIQUELY DIVERSE COMPARED TO OTHER GROUPS?

Based purely on the numbers of species in different orders of insects, beetles (Coleoptera) do seem to be uniquely species rich. For instance, at the global level, Coleoptera, with 350,000 species, are almost two and a half times richer than their nearest competing order Lepidoptera (142,500 species), and three and a half times richer than the next rich order Hymenoptera (100,000 species) (Ghosh 1996). Thus, as long as we consider these groups as distinct taxonomic entities, it does seem to validate the general belief that beetles are uniquely rich compared to other orders. But when we attempt to assess if there exists a pattern among all the orders of insects in their species richness, the bubble of beetles' unique richness seems to collapse, as also God's "inordinate fondness" for them.

When the numbers of species in different orders are plotted against their size rank (on the basis of their species richness) an interesting and a very revealing pattern emerges. Species richness of orders exhibits a non-linear, exponential increase with the order size (Figure 1.1; $Y = 146.7 \times e^{0.238}$; $R^2 = 0.891$; $p < 0.001$) and Coleoptera, expectedly, occupies the top right corner of the space. This pattern is true for the data on Indian insects also (Figure 1.2; $Y = 6.28 \times e^{0.262}$; $R^2 = 0.956$; $p < 0.001$). Though, by the manner in which this relation is constructed, the species richness of the order has to increase, there is no immediate reason to expect it to be a nonlinear rise; it could have been a simple linear increase (with a constant number of species added for every size rank) or a power function (with decreasing numbers of species added with increase in the order size). The observed exponential increase suggests that the number of species added (rate of increase) increases with the order size; that means the higher the size rank of an order, there are disproportionately higher numbers of species in it. Further, the size rank of the order is not dictated by their phylogenetic relationship, nor by the

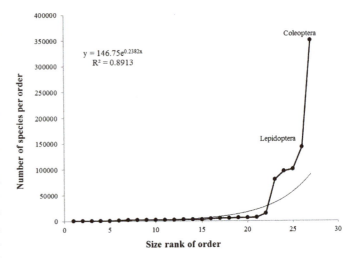

FIGURE 1.1 Relation between species richness of the insect orders with their rank based on the size for the global data. (Data from Ganeshaiah, K.N., *Curr. Sci.*, 74, 656–660, 1998; Ghosh, A.K., *Orient. Ins.*, 30, 1–10, 1996; Romoser, W.S. and Staffalano, J.G., *The Science of Entomology*, William C. Brown, Ottumwa, IA, 3rd ed., 1994.)

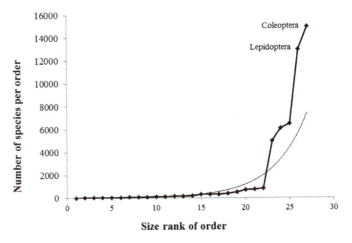

FIGURE 1.2 Relation between species richness of the insect orders with their rank based on the size for the Indian insect data. (Data from Ghosh, A.K., *Orient. Ins.*, 30, 1–10, 1996; and see Ganeshaiah, K.N., *Curr. Sci.*, 74, 656–660, 1998.)

antiquity of the orders. Rather, there appears to be an evolutionary process driving this relation across orders, a process that seems to be acting independent of the biological features of the orders. In this sense, the positioning of Coleoptera as a uniquely rich group could be merely a consequence of such hidden processes shaping this unique relation not necessarily due to any adaptive features of beetles.

Further, this relation does not seem to be unique to the class Insecta nor is it restricted to the ordinal level within a class. For example, plants (Figure 1.3), mammals (Figure 1.4), and birds (Figure 1.5) also show a similar exponential relation between the species loading of orders with their size rank. And in each such relation there is (and has to be) one order that always has disproportionately more species than its immediate predecessor. The species loading of families also shows a similar exponential increase with size rank (Figure 1.6).

Why Are Insects Abundant? Chance or Design?

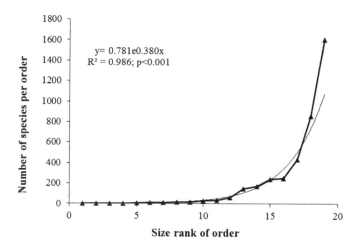

FIGURE 1.3 Relation between the species richness and rank order in mammals.

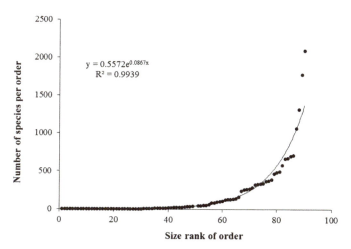

FIGURE 1.4 Relation between the species richness and rank order in angiosperms.

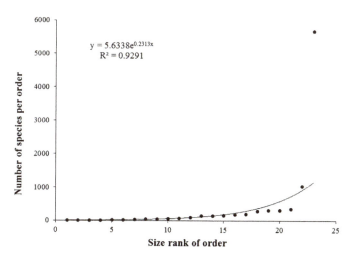

FIGURE 1.5 Relation between the species richness and rank order in birds.

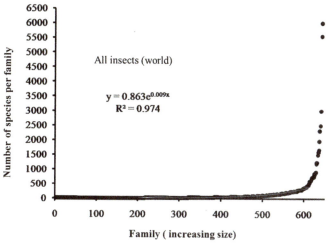

FIGURE 1.6 Relation between species load of families and their size rank in insects of the world.

Thus, there appears to be a repeating pattern of species loading of taxa with their size rank, and insects are no exception to this relation. Therefore, the apparent species richness of beetles could be explained if one understands the underlying process generating this relationship, found to be consistent across taxonomic groups and along different taxonomic hierarchies.

SPECIES RADIATION AS A POSITIVE FEEDBACK PROCESS: A HYPOTHESIS

Since the observed non-linear relation between species loading of groups and their rank size seems to be a pervading pattern across taxonomic groups and across different taxonomic hierarchies, it is not unlikely that the relation is shaped by an evolutionary process of radiation that operates across all levels, independent of the adaptive features of the groups. One simple mechanism that could generate such a pattern would be when larger groups speciate more frequently and hence add more species to themselves than do the smaller groups. In other words, a positive feedback process akin to "diversity begets diversity" in which species are accumulated in an order or a family, as a positive function of its size could shape such a relation.

The well-known phrase "diversity begets diversity" has generally been used to suggest that a diversity of hosts, or of resources, or heterogeneity of habitat features enhances diversity of organisms that use, or depend on, them (Janz et al. 2006; Palmer and Maurer 2007; Freckel et al. 2017; Maynard et al. 2017). But in this context, the phrase is used to specifically suggest that the higher the species diversity of a taxonomic group, the higher is the probability that it gives rise to new species. It may be argued to happen in several ways.

1. **Higher rates of hybridization and gene exchange**: In a group with a large number of species, there are greater opportunities for hybridization among related species and thence to generate fresh variations.

Natural selection can act on such variations and could lead to new species. Further, even horizontal gene transfer across related species, as has been noticed recently, is also highly possible in groups with more species facilitating the generation of new variation and hence speciation.

2. **Competitive swamping of habitats and resources**: Groups with higher species diversity are likely to occupy more and diverse habitats rendering them unavailable to other groups. Similarly, the larger groups could also usurp the resources making them less available to the species of smaller groups. Further, merely by richness of the species, they could competitively eliminate the species from smaller groups and remain an unassailable winner group.

3. **Providing a broader base for speciation**: New species emerge from the old ones. Thus, groups with more species provide a broader base for the evolution of new species purely by their numerical advantage.

In all these, and perhaps in more ways, larger groups are likely to add more species to their group than the smaller ones. What is also important is that, in this process, the smaller groups are rendered relatively weaker or less able to speciate owing to competition for the limited resources and habitats. This entire process was simulated in a computer using the famous Polya-Urn probability model.

SIMULATING THE PROCESS

The simulation began with a set of N groups (N = 50 here), all with equal numbers of species in them (i.e., one in each). They were also assumed to be equal in all respects with no differences in their features or in their propensity to speciate. In such a scenario, speciation was assumed to happen as a stochastic event dictated by the Polya-Urn probability that is proportional to the size of the group (Deneubourg et al. 1987, 1989; Ganeshaiah and Uma Shaanker 1992, 1994) as given by,

$$P_i = (n_i)^x / (\Sigma(n_i))^x \quad (1.1)$$

where:
  P_i is the probability that ith group speciates
  n_i is the number of species in ith group
  x is the power function that dictates the bias of the group as a function of its size (see a, b, and c in the previous sections). Note that this value is constant for all groups in a simulation.

At each stage, the group that speciates is chosen by a stochastic process (see Deneubourg et al. 1987, 1989; Ganeshaiah and Uma Shaanker 1992, 1994; Shaanker et al. 1995). Clearly, to begin with, since all groups are equal, the group that adds a fresh species is decided purely by a random process. Thus, some groups, purely by chance begin to grow and increase their species load. But the initial dominance of certain groups does not ensure that they eventually become the most dominant groups as the system reaches stability (see below). Because of the stochastic nature by which the probability is converted into the process of speciation, i.e., adding new species to groups that are initially larger may not always be favored at every step especially when the differences between them are smaller. With small differences in size among groups, there were a lot of fluctuations in the ranking of orders. In other words, there was enough flux in the system and groups began to move randomly along the size rank. However, as more and more species are added, certain groups begin to grow (accumulate more species in them) and irreversibly establish themselves as the dominant groups. The stage at which such irreversible dominance of some groups gets established is a function of relative differences among them, the absolute size of the dominant groups and the value of the parameter x. Thus as the process continues, and some groups begin to irreversibly dominate, the entire system moves towards a stable state with less fluctuations in the rank order.

RESULTS OF THE SIMULATION

After about 200 species were added, the programme began to continually monitor the stability of the system by checking the dominance hierarchy of all groups. If the rank sizes of the top ten groups were not disturbed over at least ten iterations, the simulation was stopped; else it was carried on until about 1,000 species were added. Thus, after a defined number of iterations, (1,000) or when the system was found to be attaining stability, simulation was halted and the species load in each group noted, groups were accordingly ranked, and the relation between the two plotted (Figure 1.7).

Clearly, the feedback process of speciation simulated here was able to generate the observed relationship between the species loads of the groups and their size ranking. Also note that in the simulation with x = 1.25, we do see a pattern almost similar to that seen among the orders of insects where

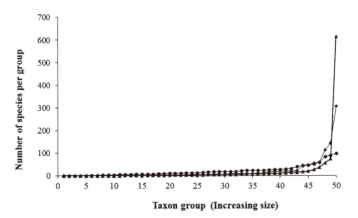

FIGURE 1.7 Relation between the number of species added to a group and the group size rank obtained from simulation. The three lines show the pattern of relation obtained for x = 1.00 (closed circle; $Y = 1.497e^{0.080x}$; $R^2 = 0.931$), x = 1.25 (closed diamond; $Y = 0.49e^{0.982x}$; $R^2 = 0.9383$), and x = 1.5 (closed triangle; $Y = 0.516e^{0.086x}$; $R^2 = 0.8728$). Points are joined for clarity.

the most dominant group is several times larger than the next smaller group. Further, it is interesting to see that even when x = 1, i.e., when the probability of a group speciating is a linear function of its relative size, dominance gets established, though to a lesser extent, and the non-linear relation between the group's size and size ranking does emerge. In other words, the wide range of exponential relations seen among different taxonomic groups could be reproduced by merely tweaking the value of x, the power function that decides the extent of feedback that the species load would have on the rate of speciation.

It is but expected that taxa differ in their abilities to exhibit dominance, in hybridizing and exchanging genes, and also in their competitive ability to exclude rival groups from habitats and resources (factors a, b, and c in the previous section), and these may determine the value of "x." Since these are factors determined by the adaptive and biological features of the group, it is likely that though the general relationship seen is a consequence of the feedback process of size dependent speciation of the groups, the rate at which the relation is built and the extent to which the dominance hierarchy is established could be attributed to differences in features among the groups. In other words, though the stochastic process of feedback growth of dominance hierarchy can generate the widely observed non-linear relation between species richness and rank size of groups, it is not unlikely that intrinsic biological features of the taxa could help in accelerating this process.

RETURNING TO THE QUIP AND THE QUESTION

It is argued here that the exceptionally high species load of Coleoptera need not be seen as a special feature of beetles, but as a consequence of the process of species diversification. The positive feedback model proposed here could explain the consistent relation observed between the species loads of taxonomic groups and their size ranks along any hierarchy and for any taxon. The relationship can be traced to even the level of the class ($Y = 889.7 \times e^{0.62x}$; $R^2 = 0.873$; data from Anonymous 2018b) with class Insecta occupying the top-most right corner (Figure 1.8). At a lower taxonomic level, within Coleoptera, weevils (family Curculionidae) dominate the order.

Thus, it is argued here that any group could become predominant purely by random drift in the initial stages of species radiation which in turn gets positively reinforced by the process of species radiation. However, one cannot rule out the possibility that certain intrinsic features of morphology, of behavior and of life cycle, or a combination of them, may favor certain groups to emerge as the predominant ones in the initial stages, which then gets established due to the positive feedback process proposed here. For instance, beetles may have become predominant in the early stages due to their shifting to feeding on angiosperms (Farell 1998), or due to their wing flexion (Mayhew 2002; Hardy 2002), or complete metamorphosis, etc. However, what is important to note is that their eventual emergence as the most successful group need not be because of such features, but purely due to the autocatalytic feedback process of species radiation. In other

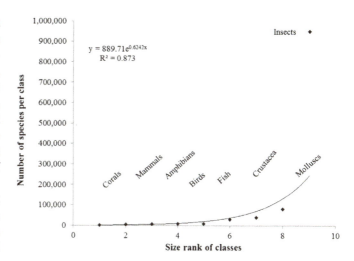

FIGURE 1.8 Relation between species load of different classes and their size rank.

words, God seems to have an inordinate fondness not for beetles, but to the autocatalytic process of speciation; beetles seem to have been the lucky ones that got the dice played to their luck in the initial stages.

ACKNOWLEDGEMENTS

This work and the argument therein have been the result of my frequent discussions on the issue with Drs. K. Chandrashekara, A. R. V. Kumar, R. Uma Shaanker, S. Ramani, and V. V. Belavadi. K. Chandrashekara and V. V. Belavadi also helped me improve the write-up and provided a good set of references relevant to the work.

REFERENCES

Anonymous. 2002. Was the Creator a Beetle Fan? http://creationsafaris.com/crev0402.htm

Anonymous. 2018a. Case study: Why so many beetles? Understanding Evolution. https://evolution.berkeley.edu/evolibrary/article/side_0_0/beetles_01 (accessed on November 15, 2018).

Anonymous. 2018b. Estimated number of animal and plant species on Earth. https://www.factmonster.com/science/animals/estimated-number-animal-and-plant-species-earth.

Deneubourg, J. L., S. S. Gross, J. M. Pasteels, D. Fresneau, and J. P. Lachahud. 1987. Self organization mechanisms in ant societies (II) In *Learning in Foraging and Division of Labor. From Individual to Collective Behavior in Insect Societies*, eds. J. M. Pasteels and J. L. Deneubourg, pp. 177–196. Basel, Switzerland: Birkhauser.

Deneubourg, J. L., S. S. Gross, N. Franks, and J. M. Pasteels. 1989. The blind leading the blind: Modeling chemically mediated army ant raid patterns. *Journal of Insect Behaviour* 2: 719–725.

Farell, B. D. 1998. Inordinate fondness explained: Why there are so many beetles? *Science* 281: 555–559.

Forbes, A. A., R. K. Bagley, M. A. Beer, A. C. Hippee, and H. A. Widmayer. 2018. Quantifying the unquantifiable: Why Hymenoptera, not Coleoptera, is the most speciose animal order. *BMC Ecology* 18: 21. doi:10.1186/s12898-018-0176-x.

Freckel, J., L. Theodosiou, and L. Becks. 2017. Rapid evolution of hosts begets species diversity at the cost of intraspecific diversity. *PNAS* 114(42): 11193–11198.

Ganeshaiah, K. N. 1998. Haldane's God and the honoured beetles: The cost of a quip. *Current Science* 74: 656–660.

Ganeshaiah, K. N., and R. Uma Shaanker. 1992. Frequency distribution of seed number per fruit in plants: A consequence of self-organizing process? *Current Science* 62(4): 359–365.

Ganeshaiah, K. N., and R. Uma Shaanker. 1994. Seed and fruit abortion as a process of self organization among developing sinks. *Physiologia Plantarum* 91: 81–89.

Ghosh, A. K. 1996. Insect biodiversity in India. *Oriental Insects* 30(1): 1–10.

Gould, S. J. 1993. A special fondness for beetles. *Natural History* 102(1): 4.

Hardy, I. C. W. 2002. A macroevolutionary fondness for Neoptera. *Trends in Ecology & Evolution* 17: 354.

Hutchinson, G. E. 1959. Homage to Santa Rosalia, or why are there so many kinds of animals. *American Naturalist* 93: 145–159.

Janz, N., N. Soren, and W. Niklas. 2006. Diversity begets diversity: Host expansions and the diversification of plant feeding insects. *Evolutionary Biology* 6: 4. doi:10.1186/1471-2148-6-4.

Maynard, D. S., M. A. Bradford, D. L. Lindner et al. 2017. Diversity begets diversity in competition for space. *Nature Ecology & Evolution* 151(6): 156. doi:10.1038/s41559-017-0156.

Mayhew, P. J. 2002. Shifts in hexapod diversification and what Haldane could have said. *Proceedings Royal Society London Series B* 269: 969–974.

Palmer, M. W., and T. A., Maurer. 2007. Does diversity beget diversity? A case study of crops and weeds. *Journal of Vegetation Science* 8: 235–240.

Romoser, W. S., and J. G. Staffalano. 1994. *The Science of Entomology*. 3rd ed. Ottumwa, IA: William C. Brown.

Shaanker, R. U., K. N. Ganeshaiah, and K. V. Krishnamurthy. 1995. Development of seeds as self organizing units: Testing the predictions. *International Journal of Plant Sciences* 156: 650–657.

Smith, D. M., and J. D. Marcot. 2015. The fossil record and macroevolutionary history of the beetles. *Proceedings Royal Society Series B* 282: 20150060. doi:10.1098/rspb.2015.0060.

2 Mayflies (Insecta: Ephemeroptera) of India

C. Selvakumar, K. A. Subramanian, and K. G. Sivaramakrishnan

CONTENTS

Introduction 7
Higher Classification 8
Key Diagnostic Characters 8
    Order Ephemeroptera 8
        Suborder Carapacea 9
        Suborder Furcatergalia 9
        Suborder Setisura 14
        Suborder Pisciforma 15
Key to Families of Ephemeroptera 16
    Adults 16
    Larvae 17
Major Work on Indian Fauna 18
Biodiversity and Species Richness 18
Distribution Patterns in the Indian Context 19
Immature Taxonomy 19
Molecular Characterization and Phylogeny 19
Integrative Taxonomy 19
Taxonomic Problems 20
Biology 20
    Life Cycle 20
    Ecology 21
Economic Importance 22
Conservation Status 22
Collection and Preservation 22
Useful Websites 23
Conclusion 23
Endnote 23
Acknowledgements 23
References 23

INTRODUCTION

Ephemeroptera, popularly known as mayflies, are the most primitive and ancient of the extant insect groups. Their evolutionary history dates back to the Carboniferous or Permian period about 290 million years ago, and they attained the highest diversity during the Mesozoic era. Ephemeroptera together with Odonata is traditionally considered as Palaeoptera, i.e., sister group of Neoptera or all other orders of insects. However, recent molecular phylogenetic studies suggest that Ephemeroptera is a sister group to Odonata and other Neopteran insect orders. Ephemeroptera are primarily aquatic insects. The larval stage is the dominant life history stage and is always aquatic. The larvae undergo a series of moults as they grow, the precise number being variable within a species, depending on external factors, such as temperature, food availability and current velocity (Brittain and Sartori 2003) and 10–50 instars have been reported (Ruffieux et al. 1996). Typically, larvae have up to seven pairs of abdominal gills, usually three caudal filaments, and mouthparts generally adapted for collecting/gathering and deposit feeding. A few species are predaceous and some are scrapers. Certain groups are burrowers and have variously developed mandibular tusks and frontal processes to loosen the substrate and flattened legs for digging. Burrowers usually have feathery gills, which are folded over the abdomen and used to create a current through their burrow. Mayfly larvae colonize all types of freshwaters, but are more diversified in running waters than in lakes or ponds. A couple of species can even be found in brackish waters.

Mayflies undergo hemimetabolous metamorphosis, having a unique maturation stage between the larva and adult, the subimago. Subimagos appear superficially similar to the adults, but are sexually immature. Their wings and abdomens are covered with small water-resistant microtrichia, which help them to leave the water after moulting from the final instar nymph (Edmunds and McCafferty 1988). Except for a few, such as female Polymitarcyidae and Palingeniidae (which are mature as subimagos), most adults have transparent wings and glossy abdomens, having shed the subimaginal cuticle. Males have extended forelegs for grasping the female during mating. Usually, mayfly adults live from a few hours to a few weeks depending on the species. Many species have male mating swarms forming at dawn or dusk. Females have various methods of oviposition, and the number of eggs laid varies according to species and size of female and eggs (Sartori and Sartori-Fausel 1991; Brittain and Sartori 2003). Length and number of life cycles per year depend largely on geographic locality and size of the species, with large burrowers in temperate climates taking over 2 years to mature, while tropical species may have several generations in a year.

HIGHER CLASSIFICATION

The original suborder classification of McCafferty and Edmunds (1979), based on thoracic morphology and wing pad position, comprised the holophyletic suborder Pannota (larvae with basally fused wingpads) and the paraphyletic suborder Schistonota (larvae with free wingpads) indicating the retention of plesiomorphic traits. Afterwards, McCafferty (1991a) proposed three different suborders viz., Pisciforma, Setisura, and Rectracheata and traced phylogenetic relationships within and among the suborders. Concurrent to McCafferty's work, Kluge (1988, 1998) independently proposed two suborders viz., Furcatergalia and Costatergalia. In contrast to previous hypotheses based on morphological observations by Ogden and Whiting (2005), the relationships inferred from the molecular data were congruent in some cases, but incongruent in others. In their investigation, the groups, Furcatergalia, Pannota, Carapacea, Ephemerelloidea, and Caenoidea, and 15 families were supported as monophyletic. On the other hand, Setisura, Pisciforma, Baetoidea, Siphlonuroidea, Ephemeroidea, Heptagenoidea, and five families were not supported as monophyletic (Sivaramakrishnan et al. 2011). Presently, four suborders are recognized in the order Ephemeroptera viz., Carapacea, Furcatergalia, Setisura, and Pisciforma (Sivaramakrishnan 2016).

Order Ephemeroptera
  Suborder: **Carapacea**
    Superfamily: **Prosopistomatoidea**
      Family: **PROSOPISTOMATIDAE**
  Suborder: **Furcatergalia**
    Superfamily: **Leptophlebioidea**
      Family: **LEPTOPHLEBIIDAE**
        Subfamily: Atalophlebiinae
        Subfamily: Leptophlebiinae
    Superfamily: **Ephemeroidea**
      Family: **EPHEMERIDAE**
        Subfamily: Ephemerinae
        Subfamily: Hexageniinae
        Subfamily: Palingeniinae
        Subfamily: Icthybotinae
      Family: **POLYMITARCYIDAE**
        Subfamily: Euthyplocynae
        Subfamily: Asthenopodinae
        Subfamily: Polymitarcyinae
      Family: **POTAMANTHIDAE**
    Superfamily: **Caenoidea**
      Family: **CAENIDAE**
        Subfamily: Caeninae
        Subfamily: Brachycercinae
      Family: **NEOEPHEMERIDAE**
    Superfamily: **Ephemerelloidea**
      Family: **EPHEMERELLIDAE**
        Subfamily: Ephemerellinae
      Family: **TELOGANODIDAE**
      Family: **TRICORYTHIDAE**
  Suborder: **Setisura**
    Superfamily: **Heptagenioidea**
      Family: **HEPTAGENIIDAE**
        Subfamily: Heptageniinae
  Suborder: **Pisciforma**
    Family: **AMELETIDAE**
    Family: **BAETIDAE**
      Subfamily: Baetinae
      Subfamily: Cloeoninae

KEY DIAGNOSTIC CHARACTERS

ORDER EPHEMEROPTERA (FIGURES 2.1 AND 2.2)

Mayflies are distinguished from other insect orders by the following characters:

Adult: (i) small-to medium-sized elongate fragile insects; (ii) antennae short and setaceous, mouthparts vestigial, compound eyes large, three ocelli present; (ii) generally two pairs of membranous wings (though hind pair greatly reduced) held vertically over body when at rest, with many cross veins; (iv) abdomen terminated with two very long cerci and frequently a median caudal filament; and (v) with subimaginal and imaginal winged stages.

Larvae: (i) aquatic; (ii) body campodeiform; (iii) antennae short, compound eyes well-developed, biting mouthparts; (iv) abdomen usually with long cerci and a median caudal filament; and (v) four to seven pairs of segmental tracheal gills (Gillott 2005).

Mayflies (Insecta: Ephemeroptera) of India

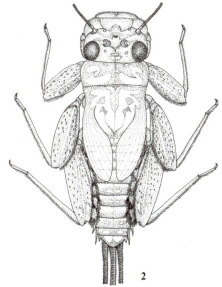

FIGURES 2.1–2.2 **1**. *Klugephlebia kodai* (adult) and **2**. *Petersula courtallensis* (larva).

Suborder Carapacea

Members of the Carapacea (an allusion to the carapace-like enlargement of the larval mesonotum) are included in a single superfamily Prosopistomatoidea (Gillott 2005).

Superfamily Prosopistomatoidea

The two small families in this group, the Baetiscidae and Prosopistomatidae, show considerable parallel evolution in the larval stage. Indeed, their larvae are remarkable in having an enormous, posteriorly projecting mesonotal shield that protects the gills so that they superficially resemble notostracan crustacea, into which group *Prosopistoma* was originally placed by the French biologist Latreille in 1833 (Berner and Pescador 1980). Larvae of most species live in moving water, from streams to large rivers, where the bottom has sand, fine gravel, or small stones. Adult baetiscids, which are medium sized insects, have an unusually large mesothorax; the eyes of males are large and almost contiguous, but not divided horizontally. Prosopistomatid adults of both sexes have small, widely separated eyes; males have relatively short forelegs; the legs of females are vestigial; and females do not have an adult moult (Gillott 2005).

Family Prosopistomatidae Lameere, 1917 (Figures 2.3 and 2.4)

Prosopistomatidae is a distinct, enigmatic family of mayflies presently confined to the Palaearctic, Afrotropical, Oriental, and Australasian realms and represented by a single genus viz. Prosopistoma Latreille, 1833. The larvae of this family have the following apomorphies: pronotum and mesonotum with rounded lateral margins without protuberances, with strongly derived mouthpart architecture, with symmetric mandibles with complete loss of molar, with first six ventral abdominal segments fused and with gills completely covered by carapace, hidden in a gill chamber (Bauernfeind and Soldán 2012).

Suborder Furcatergalia

Furcatergalia is the largest suborder of order Ephemeroptera. Its name derives from the forked nature of the larval gills. The group includes five superfamilies: Leptophlebioidea, Behningioidea, Ephemeroidea (burrowing mayflies), Ephemerelloidea, and Caenoidea. The last two superfamilies collectively form the pannote mayflies, so called because of the fused forewing pads of the larvae (Gillott 2005).

Superfamily Leptophlebioidea
Family Leptophlebiidae Banks, 1900 (Figures 2.5–2.8)

Leptophlebiidae is the only family of this primitive lineage. This has maximum diversity in the southern hemisphere (Edmunds 1972). The family is currently defined as monophyletic by the following apomorphies in the larvae: (i) maxilla broadened apically with a special arrangement of filtering setae and (ii) upper part of male compound eyes usually globular (superficially similar to turbinate eyes in Baetidae). In imagines and subimagines: (i) furcasternal protuberances separated; (ii) the metathoracic ganglion situated in between; (iii) cubital field of the forewing usually with two intercalaries or one bifurcate vein; and (iv) paracercus present in larval and imaginal stages (Bauernfeind and Soldán 2012). This represents one of the major stem groups within the Ephemeroptera consisting of relatively ancestral and highly derived components (McCafferty and Edmunds 1979). Leptophlebiidae is a basal lineage and a sister group to a relatively derived clade that includes a pair of sister groups viz., Scapphodonta and Pannota (McCafferty and Wang 2000), in addition to a more basal lineage represented by the Behningiidae (McCafferty 2004). Both the subfamilies of Leptophlebiidae viz., Atalophlebiinae and Leptophlebiinae are known from the Oriental region. Twelve genera of Leptophlebiidae, six of which are endemic to south India are distributed in Western Ghats, all belonging to the subfamily, Atalophlebiinae. The generic limits of Leptophlebiidae, especially those of the eastern hemisphere, are well defined.

FIGURES 2.3–2.8 Larvae: **3**. *Prosopistoma indicum*; **4**. *Prosopistoma someshwarensis*; **5**. *Edmundsula lotica*; **6**. *Nathanella indica*; **7**. *Notophlebia jobi*; and **8**. *Thraulus gopalani*.

Superfamily Ephemeroidea

Family Ephemeridae Latreille, 1810 (Figure 2.9)

Systematic and zoogeographical aspects of Asiatic Ephemeridae were dealt with extensively by McCafferty (1973, 2004). Diagnostic characters of this family are in larvae: mandibular tusks curved, without denticles and round in cross section; abdominal segments 7–9 elongated posteriorly, gills inserted in the middle of the segments; gill 1 vestigial and in imagines forewing with several veins from AA to hind margin of wing. The family is represented in India by the genera *Ephemera* (11 species), *Anagenesia* (3 species), and *Eatonigenia* (2 species). The genus *Ephemera* is the most speciose and widely distributed ephemerid genus, distributed all over the Holarctic, Ethiopian, and Oriental regions. The genus shows a relatively high diversity in the Oriental region from where it probably dispersed to the Holarctic and Afrotropical regions. In India, two subgenera of *Ephemera* viz., *Ephemera* and *Aethephemera* are represented by ten and one species, respectively (Sivaramakrishnan 2016).

Family Polymitarcyidae Banks, 1900 (Figure 2.10)

Polymitarcyidae is widely distributed, both in tropical and temperate areas with the exception of Australia and New Zealand (Kluge 2004; McCafferty 2004). They are one of the primitive, burrowing mayflies in the larval stage, making tunnels in submerged wood or living in aquatic plants and sponges and also inorganic sediment such as clay, mud, or sand (Hartland-Rowe 1958). Presently, three subfamilies are recognized viz., Polymitarcyinae in the Old World, with a northwestern extension into the Nearctic region, Asthenopodinae in

Afrotropical, Oriental and Nearctic regions, and Campsurinae in Nearctic and Neotropical regions (Needham et al. 1935). Three genera of this family viz., *Ephoron, Languidipes,* and *Povilla* are recorded from India.

Family Potamanthidae Albarda, 1888 (Figure 2.11) Potamanthidae is basically a Laurasian family comprising three genera viz., *Rhoenanthus, Anthopotamus,* and *Potamanthus,* of which *Anthopotamus* is Nearctic, while others occur in the Oriental region. A revision of the family by Bae and McCafferty (1991) presents a detailed account of the potamanthid phylogeny and biogeography. The larvae of this family are distinguished by somewhat dorsoventrally flattened bodies; small to large, somewhat convergent mandibular tusks projecting in front of the head; outspread legs; and posterolaterally oriented, fringed, and bilobed gills on abdominal segments 2–7. A distinct frontal process on the head and modifications of the legs which are associated with burrowing in other ephemeroids are not present in potamanthids. The adults and subimagos of Potamanthidae are distinguished primarily by wing characters. Besides a strongly arched MP_2 and CuA of the forewings, which is typical of all ephemeroids, the A_1 is distinctly forked and the hindwings each have an acute costal projection. Depending on the species, adults and subimagos can have three well-developed caudal filaments or the median terminal filament may be partially developed or rudimentary. The genus *Rhoenanthus* (*Rhoenanthus*) occurs in India, Vietnam, Thailand, Malaysia, and Cambodia (but does not reach Sulawesi), while *Rhoenanthus* (*Potamanthindus*) ranges from Korea through China to Vietnam and Thailand.

Superfamily Caenoidea

Family Caenidae Newman, 1853 (Figures 2.12 and 2.13) The family Caenidae is widespread in tropical and Palaearctic Asia. The family Caenidae is currently defined as monophyletic by the following apomorphies: in larvae second gill operculate, with a row of spiculae (Brachycercinae) or microtrichia (Caeninae) along the ventral outer margin. Imagines with fan-shaped forewings, veins iMP and MP_2 in forewings arise independently from one another, hindwings completely missing; forceps without segmentation; paracercus retained in the winged stages; caudal filaments in female imagines distinctly shorter than in male imagines (Bauernfeind and Soldán 2012). Three subfamilies are currently recognized, Brachycercinae, Caeninae and Madecocercinae. Five genera of this family are reported from the Oriental region of which two genera viz., *Caenis* and *Clypeocaenis* are recorded from India (Soldán 1978). Worldwide, about 15 genera including more than 100 species have been recorded. The genus *Caenis* is almost cosmopolitan, excluding Australia. *Caenis* is a large and apparently very old genus, and it could well be of Pangean origin (McCafferty and Wang 2000). The so-called brush-legged *Clypeocaenis* is an Oriental-Ethiopian genus which seems to have its centre of diversity in the Oriental region.

Family Neoephemeridae Traver, 1935 (Figures 2.14 and 2.15) Neoephemeridae is a small group of pannote mayflies presently confined to Holarctic and Oriental regions. The family is a distinct monophyletic group of mayflies (McCafferty and Edmunds 1979), but the taxonomy of its members has been problematic. Neoephemerid larvae are similar to caenid mayflies in that they have a pair of large, subquadrate, operculate

FIGURES 2.9–2.11 Larvae: **9**. *Ephemera* (*Aethephemera*) *nadinae*; **10**. *Languidipes* sp.; **11**. *Rhoenanthus* (*Potamanthindus*) sp.;

FIGURES 2.12–2.15 Larvae: **12**. *Caenis* sp.; **13**. *Clypeocaenis bisetosa*; **14**. *Potamanthellus caenoides*; and **15**. *Potamanthellus ganges*.

gills on abdominal segment 2. Adults, on the other hand, are similar to potamanthid mayflies, having similar wing venation (esp., basally arched MP2 and CuA, forked A1 in forewings). Bae and McCafferty (1991) clearly delimited both stages of Neoephemeridae and Potamanthidae. This family is represented by three genera viz., *Potamanthellus* Lestage, 1931 (seven species), *Neoephemera* McDunnough (six species), and *Ochernova* Bae and McCafferty (one species) (Bae and McCafferty 1998; Zhou and Zheng 2000; Nguyen and Bae 2003). Larvae of Neoephemeridae have unique operculate gills on the second abdominal segment that are fused medially. The larvae of *Potamanthellus* are distinguished from those of *Neoephemera* and *Ochernova* by their densely setate mouthparts, by their lack of well developed lateral expansions of the pronotum and mesonotum, and by their possession of rows of long setae on the caudal filaments (Bae and McCafferty 1998).

Superfamily Ephemerelloidea

Family Ephemerellidae Klapalek, 1909 (Figures 2.16–2.18)

The family comprises one of the abundant and widespread groups of known pannote mayfly genera mostly inhabiting pristine streams. They are popularly known as 'spiny crawlers'. The family is defined by the apomorphic loss of gills 2 from the larval abdomen (McCafferty and Wang 2000) and male genital forceps segment 1 that has its length much less than its width (Jacobus and McCafferty 2006). This family represents the Laurasian group of a monophyletic ephemerelloid clade differentiated conspicuously from the remaining seven Gondwanian families of Ephemerelloidea (Bauernfeind and Soldán 2012). The family comprises two subfamilies viz., Timpanoginae Allen and Ephemerellinae s.s. (McCafferty and Wang 2000; Jacobus and McCafferty 2006). It is predominantly Holarctic and Oriental, comprising 16–23 genera and 200–400 species globally (Bauernfeind and Soldán 2012). The ephemerelllid larvae like other mayfly larvae constitute vital links in the food chain of several insectivorous fishes and certain birds (Jenkins and Ormerod 1996; Feck and Hall 2004). The concept of Ephemerellidae is broadly outlined in Eaton's monumental monograph (1883–1888). However, Klapalek (1909) formalized it as a family. Traver (1935) recognized this group and Edmunds and Traver (1954), Edmunds et al. (1963), and Allen (1965, 1980, 1984) subsequently revised the concepts of the family. Further refinements in classification and phylogeny of the family reflecting Hennigian phylogenetic hypotheses were made mainly by Kluge (2000), Jacobus and McCafferty (2008), and Ogden et al. (2009a) among others. Though around ten genera of this family inhabit the Oriental region (Soldán 2001), according to Sivaramakrishnan (2016), only three genera viz., *Drunella*, *Ephemerella*, and *Torleya* are represented so far in India and he has attributed this to lack of extensive survey and investigation of ephemerellids in India, especially from the Himalayas. Five species under three genera have so far been reported from India viz., *Cincticostella indica* (Kapur and Kripalani) from Kulu Valley (N. W. Himalaya), *Drunella submontana* (Brodsky) from Kashmir, *Torleya coheri* (Allen and Edmunds) from Kashmir, *Torleya lacuna* (Jacobus, McCafferty, and Sites) from Kunjankhuzi (Tamil Nadu), and *Torleya nepalica* (Allen and Edmunds) from Karnataka (Kapur and Kripalani 1961; Jacobus and McCafferty 2004a, 2004b; Jacobus et al. 2007).

Family Teloganodidae Allen, 1965 (Figures 2.19–2.21)

Teloganodidae is an ancient group of mayflies of Gondwanan origin that currently are known from throughout the Oriental region and from the southern tip of Africa (McCafferty and Wang 2000; Jacobus and McCafferty 2006). This family can be separated from other families of Ephemerelloidea in larvae by the presence of gills on abdominal segment II, absence of gills on segment VII, glossae only

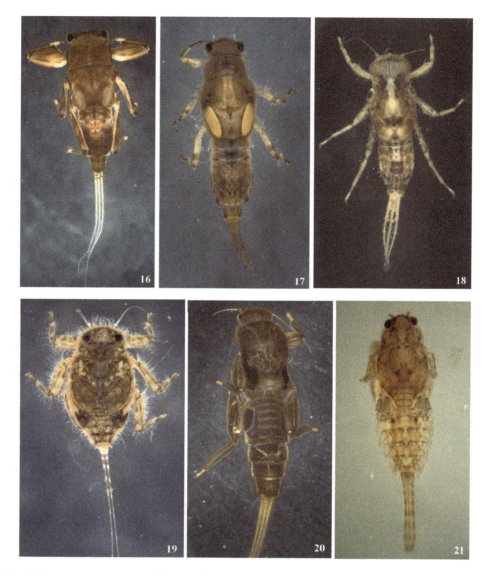

FIGURES 2.16–2.21 Ephemeropteran larvae: **16**. *Drunella submontana*; **17**. *Torleya lacuna*; **18**. *Torleya nepalica*; **19**. *Derlethina tamiraparaniae*; **20**. *Indoganodes jobini*; and **21**. *Teloganella indica*.

partially fused with paraglossae, and male eyes divided in two parts. A unique character shared by all teloganodid nymphs is the presence of stout spatulate setae on margins of coxal projections (Jacobus and McCafferty 2006). The Oriental lineage of Teloganodidae can be separated from the Afrotropical lineage by the absence of gills on abdominal segment I and the reduction of the median caudal filament giving the nymphs a two-tailed appearance (Sartori et al. 2008). Allen (1965) established the subfamily Teloganodinae within the Ephemerellidae. Teloganodinae was raised to family status by McCafferty and Wang (1997), and the composition of the family was refined by McCafferty and Wang (2000). Significant phylogenetic and biogeographic studies of teloganodid, and ephemerelloid mayflies in general, that have contributed to our current understanding of teloganodid systematics include works by McCafferty and Wang (1997, 2000), McCafferty and Benstead (2002), Jacobus and McCafferty (2006), and these works have incorporated various cladistic analyses of both Afrotropical and Oriental Teloganodidae. The recent landmark monograph on Oriental Teloganodidae by Sartori et al. (2008) distinguishes the Oriental lineages of Teloganodidae known at the time from the Afrotropical lineages and contributes to understanding patterns of distribution of the Oriental genera and species. The family currently includes the Afrotropical genera *Ephemerellina* Lestage, *Lestagella* Demoulin, *Lithoglea* Barnard, *Manohyphella* Allen, and *Nadinetella* McCafferty and Wang (Pereira-da-Conceicoa and Barber-James 2013) and the Oriental genera *Derlethina* Sartori, *Dudgeodes* Sartori, and *Teloganodes* Eaton (=*Macafertiella* Wang) (revised by Sartori et al. 2008). Recently, the genus *Indoganodes* Selvakumar, Sivaramakrishnan, and Jacobus was established and five new species were described from South India by Selvakumar et al. (2014a).

Family Tricorythidae Lestage, 1942 (Figure 2.22) The family Tricorythidae is Afrotropical and Oriental. Larval diagnostic characters: labium fused into single semicircular structure, palps with long setae; gills on abdominal segments II–V or

II–VII; gill II may overlay, partially conceal rest of series. All the subfamilies occur in Africa and Malagasy. Only the genus *Sparsorythus* is found in south India, Sri Lanka, and southeast Asia. Sroka and Soldán (2008) have established the genus *Sparsorythus*, based on *S. bifurcatus* from Vietnam. They described *S. gracilis* from northern Western Ghats. This advanced group of Oriental Tricorythidae with many apomorphies have been geographically isolated from African species for more than 100 million years. After splitting of Indian subcontinent from Africa, Indian species evolved further (Sroka and Soldán 2008).

Suborder Setisura

Included in this suborder are the families listed under the superfamily Heptagenoidea in older classifications. The major family is the Heptageniidae which ranks next to the Baetidae in terms of number of described species. Generally, dark coloured larvae are typically found clinging to the underside (occasionally the exposed face) of stones in fast-flowing streams and on wave-washed shores of large lakes. They are remarkably well adapted for this life. Their body is extremely flattened dorsoventrally; the femora are broad and flat; the tarsal claws have denticles on the lower side; the gills are strengthened on their anterior margin; and in some species the entire body takes on the shape (and function) of a sucking disc. Some larvae have fore tarsi with numerous setae that filter algae etc. from the water. Adults vary in size and colour. The eyes of male are large, but not contiguous (Gillott 2005).

Superfamily Heptagenoidea
Family Heptageniidae Needham, 1901 (Figures 2.23–2.25)

Heptageniidae is a family of mayflies with around 509 described species and distributed mainly in the Holarctic, Oriental, and Afrotropical regions (Barber-James et al. 2008). The family is currently defined as monophyletic by the following apomorphies: in larvae, the submentum vestigial, labium highly modified; in imagines, first tarsal segment distinctly articulated with tibia on all legs, cubital field of forewing with four intercalaries forming two pairs, in each pair the first intercalary concave (Bauernfeind and Soldán 2012). The phylogenetic origin of Heptageniidae was studied in detail by McCafferty (1991b), though Edmunds (1979) pioneered to highlight the biogeographic relationships of the Oriental and Ethiopian mayfly fauna. Excellent contributions regarding the understanding of the generic limits of the Heptageniidae were made by Jensen (1974) and Kluge (1988). Biogeography and evolution of the genera of Heptageniidae have been discussed at the global level in the context of a phylogenetic higher classification of the family by Wang and McCafferty (2004). They recognize three subfamilies viz., Ecdyonurinae, Heptageninae, and Rhithrogeninae. The family Heptageniidae is Laurasian in origin and is a conspicuous component of the benthic community in terms of abundance. The subfamily Heptageninae, though recorded from Himalayan streams, essentially as a Palaeactic spillover group, has not penetrated further south into the Western Ghats.

FIGURES 2.22–2.25 Larvae: **22.** *Sparsorythus gracilis*; **23.** *Afronurus kumbakkaraiensis*; **24.** *Epeorus petersi*; **25.** *Thalerosphyrus flowersi*.

FIGURES 2.26–2.30 Larvae: **26**. *Ameletus primitives*; **27**. *Baetis michaelohubbardi*; **28**. *Bungona* (*Chopralla*) *pusilla*; **29**. *Labiobaetis jacobusi*; and **30**. *Tenuibaetis frequentus*.

Suborder Pisciforma

McCafferty (1991) introduced the suborder Pisciforma (the name refers to the minnow-like body and actions of the larvae) for a group of families whose relationships remain unclear. For this reason, no arrangement into superfamilies is undertaken, though in earlier schemes the families were lumped in a single superfamily Baetoidea (Gillott 2005).

Family Ameletidae McCafferty, 1991 (Figure 2.26)

The family Ameletidae is present worldwide and about 50 species are distributed in the Holarctic, Central American, and Oriental region. The family is currently defined as monophyletic by the following apomorphies: in larvae, mouthparts highly specialized, e.g., maxilla apically broadened with pectinate, comb-shaped bristles; forceps base (styliger) of male imagines with dorsal membranous area (Bauernfeind and Soldán 2012). In India only one genus *Ameletus* Eaton has been recorded with two species.

Family Baetidae Leech, 1815 (Figures 2.27–2.35)

The homogenous family Baetidae commonly known as minnow mayflies encompasses around 100 genera and 900 species constituting one-quarter of the global Ephemeroptera diversity with a cosmopolitan distribution except for Antarctica and New Zealand (Gattolliat and Nieto 2009). They are one of the major components freshwater zoobenthos less diversified in standing water and mainly diversified in unimpacted lotic water, especially in the tropical belt. They are excellent bioindicators of water quality (Buss and Salles 2007; Kubendran et al. 2017a). Baetidae are distinguished by the presence of turbinate eyes in the male imago, detached MA_1 and MA_2 forewing veins, the presence of single or double free intercalary vein in the forewing, hind wings reduced or absent, three segmented mid and hind tarsi, and membranous penis (Edmunds et al. 1976). Larvae are pisciform, generally with long antennae and simple or double ovoid gills on segments I–VII or II–VII. They are unique in having the lateral branches of the epicranial suture anterior to the lateral ocellia ventral orientation of the dorsal lobe at the apex of femora (Wang and McCafferty 1996). Kazlauskas (1972) proposed dividing the family into two subfamilies viz., Baetinae and Cloeoninae. This is primarily based on the diversity of the Palaearctic representation of the Baetidae. Furthermore, two conflicting concepts have been proposed, the division of the Baetidae into different subfamilies (Gillies 1991) and the gathering of genera in several complexes (Waltz et al. 1994; Lugo-Ortiz and McCafferty 1996, 1998a, 1998b). Generic delimitation in Baetidae is being fine-tuned by taking into account larval characters rather than only imaginal ones and secondly by the use of phylogenetic methods and the splitting of paraphyletic and polyphyletic genera. Recent molecular studies reveal that the division into subfamilies is too simplistic and most of the complexes are not monophyletic (Gattolliat et al. 2008).

FIGURES 2.31–2.35 Adults and larvae: **31**. *Cloeon* sp. (larva); **32**. *Cloeon bicolor* (adult); **33**. *Cloeon bimaculatum* (adult); **34**. *Cloeon harveyi* (adult); and **35**. *Procloeon* sp. (larva).

KEY TO FAMILIES OF EPHEMEROPTERA

Adults

1. Venation in forewings considerably reduced, cross veins completely missing or missing at least between wing base and vein MA .... **Prosopistomatidae**
- Venation in forewings not considerably reduced, cross veins present also between wing base and vein MA 2
2. Forewings vein MP_2 (and CuA) strongly bent near the wing base, diverging from vein MP_1 at an angle of almost 80°; hindwings always well developed 3
- Forewings vein MP_2 (and CuA) not strongly bent near the wing base, almost parallel to vein MP_1; hindwings sometimes reduced or absent 6
3. Wings dull whitish, unicoloured; hindlegs short, weakly developed; paracercus strongly reduced (to several segments) in male, well developed in female .... **Polymitarcyidae**
- Wings clear, translucent with or without dark spots; hindlegs normally developed; paracercus either present or absent in both sexes 4
4. Wings with dark spots; vein AA bifurcate in forewings, costal cross veins distal to bulla simple (forming small quadrangular fields) .... **Ephemeridae**
- Wings without dark spots; vein AA simple in forewings, costal cross veins distal to bulla forked or branched (forming small polygonal fields) 5
5. Paracercus present; forewing length >10 mm; costal cross veins basal to bulla strong and well developed .... **Potamanthidae**
- Paracercus reduced, missing; forewing length <10 mm; costal cross veins basal to bulla weak, imperfect or atrophied .... **Neoephemeridae**

6. Paracercus missing (reduced to a few segments) 7
- Paracercus present (multisegmented) 9
7. Forewings with short basally detached single or double marginal intercalaries present in each interspace; forewing with MA_2 and MP_2 detached basally from their respective stems; hindwings small or absent; penes of male membranous; upper portion of eyes of male raised on a stalk-like structure .... **Baetidae**
- Forewings with marginal intercalaries attached basally to other veins; forewing with MA_2 and MP_2 attached basally; hindwings relatively large; penes of male well developed; eyes of male not raised on a stalk-like structure 8
8. Cubital field between veins CuA and CuP with two pairs of intercalaries .... **Heptageniidae**
- Cubital field between veins CuA and CuP with several sinuous intercalaries .... **Ameletidae**
9. Hindwings present and relatively large, with one or more veins forked; costal projection shorter than wing width... 10
- Hindwings absent 12
10. Forewing with short basally detached marginal intercalaries between veins along entire outer margin of wings; genital claspers of male with one short terminal segment... 11
- Forewing without short basally detached marginal intercalaries between veins along entire outer margin of wing; genital claspers of male with two or three short terminal segments ...... **Leptophlebiidae**
11. Basal part of forewing vein CuP very near to vein CuA; several free intercalaries present between veins MP_2 and CuA; subapical segment of forceps at least 5 times longer than apical segment .... **Ephemerellidae**
- Basal part of forewing vein CuP very near to vein AA; no free intercalaries present between veins MP_2 and CuA; apical and subapical segments of forceps equal or subequal in length .... **Teloganodidae**
12. Forewing with MA forming a more or less symmetrical fork and with MP_2 and 1MP extending less than ¾ distance to base of MP; genital claspers of male 2- or 3-segmented; thorax usually black or gray .... **Tricorythidae**
- Forewing with MA not forming a more or less symmetrical fork, but with MP_2 and 1 MP almost as long as MP, and extending nearly to base; genital claspers of male one segmented; thorax usually brown .... **Caenidae**

Larvae

1. Mesonotum forming a carapace; legs and abdominal segments 1–6 covered under carapace, not visible in dorsal view .... **Prosopistomatidae**
- Mesonotum not forming a carapace; legs and abdominal segments 1–10 visible in dorsal view 2

2. Mandibles with tusk-like projection; gills II–VII double and uniform in structure with fringed margin... 3
- Mandibles without tusk-like projection; gills II–VII not as above 5
3. Mandibular tusk long and sickle-shaped, bearing many long setae; maxillary palp more than twice as long as the galea-lacinia (the apical part of the maxilla).. 4
- Mandibular tusks, otherwise, bearing short bristles; maxillary palp as long as or slightly longer than the galea-lacinia .... **Potamanthidae**
4. Tusks curved outwards; abdomen yellowish with conspicuous black pattern .... **Ephemeridae**
- Tusks curved inwards; abdomen whitish without conspicuous black markings.......... **Polymitarcyidae**
5. Gills on abdominal segment II large, plate-like, touching or overlapping along dorsal midline, covering all or some of the gills arising posteriorly; gills III–VI with fringed margins.................................. 6
- Gills on abdominal segment II not as above 7
6. Gills on abdominal segment II meet along midline; terminal filament densely clothed with setae on both margins, lateral filaments with setae on inner margins only; mature larva has small hindwing pads beneath the fore wing pads on meta thorax...**Neoephemeridae**
- Gills on abdominal segment II overlap along the midline and covering all of the succeeding gills III–VI; lateral and terminal filament bearing rather short and sparse setae on the inner and outer margins, hind wing pads not present ..... **Caenidae**
7. Head flat, plate-like with dorsal eyes; body dorsoventrally flattened; gills plate-like, dorsal tuft of tracheae at base of lamellae .... **Heptageniidae**
- Head not plate-like; body not dorso-ventrally flattened; gill form various shapes without dorsal tracheal tuft at base of lamellae 8
8. Labium fused into single semicircular structure, palps with long setae; gills on abdominal segments II–V or II–VII; gill II may overlay, partially conceal rest of series ....**Tricorythidae**
- Mouthparts and gills not as above............................ 9
9. Gills on abdominal segment II absent and gills borne dorsally................................. **Ephemerellidae**
- Gills on abdominal segment II present and gills borne laterally 10
10. Lamellate gills on abdominal segment I–V or II–IV or II–V or II–VI................................ **Teloganodidae**
- Lamellate gills on abdominal segment I–VII 11
11. Head rectangular; gills similar, long, slender, and bifurcate form or first pair rudimentary (thread-like) and others plate-like and doubled; terminal filament well-developed and similar to cerci .... **Leptophlebiidae**
- Head round, antennae long and twice the width of the head; median terminal filament often much reduced and always shorter than the cerci 12

12. Maxilla apically broadened with pectinate, comb-shaped bristles; cerci and terminal filament densely beset with long bristles..........................**Ameletidae**
- Maxilla apically not broadened with pectinate, comb-shaped bristles; cerci and terminal filament not densely beset with long bristles .... **Baetidae**

MAJOR WORK ON INDIAN FAUNA

The first mayfly from India was described as early as 1843, when *Palingenia indica* (*Ephoron indicus*) was described by Pictet (1843). Subsequently, Walker (1853) described *Caenis perpusilla* and *Cloeon debilis* (*Procloeon debilis*) based on the specimens at the British Museum and on the collections of W. W. Saunders. During this period, Hagen (1858) worked on baetine mayflies of Sri Lanka. Up to 1900, two species of Ephemeridae and Palingenidae and one species of Heptageniidae were described. Needham (1909), Ulmer (1920), Chopra (1924, 1927), Navas (1931), Hafiz (1937), and Traver (1939) described many species. Needham worked on Ephemeroptera in the collection of the Indian Museum and Ulmer described *Ecdyonurus bengalensis* from Darjeeling, West Bengal. Chopra (1924, 1927) worked on Ephemeroptera of Chilka Lake and described four species. He also worked on Palingeniidae and Ploymitarcyidae. Hafiz (1937) and Traver (1939) worked on Ephemeroptera of the subcontinent.

Workers like Kimmins (1947), Gillies (1949, 1951, 1957), Kapur and Kripalani (1963), Dubey (1970, 1971), Kaul and Dubey (1970), Peters (1967, 1975), Peters and Edmunds (1970), McCafferty (1973), Hubbard and Peters (1978), Sivaramakrishnan (1984, 1985a, 1985b), Sivaramakrishnan and Hubbard (1984), Sivaramakrishnan and Peters (1984), Grant and Sivaramakrishnan (1985), Venkataraman and Sivaramakrishnan (1987, 1989), Sivaramakrishnan et al. (1996b), Dinakaran et al. (2009), Subramanian and Sivaramakrishnan (2009), Selvakumar et al. (2012, 2013, 2014a, 2015a, 2015b, 2016b, 2017b, 2017c, 2017d, 2017e), Sivaruban et al. (2013), Kluge et al. (2013), Kluge (2014), Kluge and Novikova (2014), Kluge et al. (2015), Anbalagan et al. (2014), Kubendran et al. (2014, 2015), Balachandran et al. (2016), and Ramya-Roopa et al. (2017), contributed substantially to the knowledge of Ephemeroptera of India. Significant discoveries of new genera were made during this period: *Petersula* Sivaramakrishnan 1984, *Edmundsula* Sivaramakrishnan 1985, *Indoganodes* Selvakumar, Sivaramakrishnan & Jacobus, 2014 and *Klugephlebia* Selvakumar, Subramanian & Sivaramakrishnan, 2016 from southern Western Ghats which are Gondwanian relicts.

BIODIVERSITY AND SPECIES RICHNESS

Ephemeroptera constitutes a small order of extant insects, with approximately 40 families, 440 genera, and 3330 species globally, and of these, 561 species in 84 genera and 20 families occur in the Oriental region (Sartori and Brittain 2015). The fauna of the Indian subregion (India, Sri Lanka, Pakistan, Nepal, Bhutan, and Bangladesh) is represented by 4 suborders, 15 families, 60 genera, and 204 species (Sivaramakrishnan et al. 2009). In India, we have 4 suborders, 15 families, 60 genera, and 152 species, and of these, 7 genera and 123 species are endemic (Table 2.1). The present diversity includes two

TABLE 2.1
Diversity and Endemism of Ephemeroptera

Suborder	Family	Diversity		Endemism	
		No. of Genera	No. of Species	No. of Genera	No. of Species
Carapacea					
	Prosopistomatidae	1	3	0	3
Furcatergalia					
	Leptophlebiidae	12	26	6	22
	Ephemeridae	4	16	0	15
	Polymitarcyidae	4	5	0	3
	Potamanthidae	2	2	0	2
	Caenidae	2	7	0	7
	Neoephemeridae	1	2	0	1
	Ephemerellidae	3	5	0	0
	Teloganodidae	5	8	1	8
	Tricorythidae	1	1	0	1
	Vietnamellidae	1	1	0	0
Setisura					
	Heptageniidae	10	26	0	22
	Isonychiidae	1	1	0	1
Pisciforma					
	Ameletidae	1	2	0	2
	Baetidae	12	47	0	36
Total		15	60	152	7	123

families viz., Vietnamellidae and Isonychiidae which were recently reported from India by Selvakumar et al. (2018) and Vasanth et al. (2019). Species rich families are Leptophlebiidae (12 genera, 26 species), Heptageniidae (10 genera, 26 species), Ephemeriidae (4 genera, 16 species), and Baetidae (12 genera, 47 species). Four families viz., Leptophlebiidae, Ephemeriidae, Heptageniidae, and Baetidae are represented by more than 10 species. Species rich genera with more than ten species each are *Ephemera* (Ephemeriidae), *Baetis,* and *Cloeon* (Baetidae).

DISTRIBUTION PATTERNS IN THE INDIAN CONTEXT

Mayflies are distributed in diverse inland freshwater habitats. Rich diversity is found in pristine hill streams. The following regions are reasonably well explored in India with regard to species diversity and distribution of Ephemeroptera viz., Western Ghats (76), Central Himalayas (29), and Gangetic plain (21). Going by the number of species in the Deccan peninsula (18), North East (10), Trans Himalaya (9), and Andaman and Nicobar Islands (3), they appear to be less explored. The remaining biogeographical regions (Coast, Desert, Semi-Arid, and Eastern Himalaya) have not been explored, and intensive survey and documentation of mayfly species diversity is needed in these regions.

IMMATURE TAXONOMY

Till the first half of the twentieth century, Ephemeroptera systematics mainly relied on imaginal characters. However, the crucial importance of ecology and phylogeny of larval stages of mayflies on freshwater ecosystem dynamics led to detailed taxonomic explorations of larval stages and associated adults in addition to the in-depth study of cryptic species complexes. Recently, considerable progress has been made in averting Wallacean shortfall through in-depth study of larval taxonomy in different families of Ephemeroptera in our country (Selvakumar et al. 2012, 2013, 2014a, 2015a, 2015b, 2016b, 2017b, 2017c, 2017d).

A recent publication by Ramya-Roopa et al. (2017) describing a new species and redescribing an already described species of *Prosopistoma* (Prosopistomatidae) from India has highlighted the importance of immature taxonomy in averting Wallacean shortfall besides contributing better understanding of the systematics and phylogeny of this genus especially in the context of the possession of several apomorphic features in the larval stage in contrast to the attenuated morphology of alate stages. This study has global significance in endorsing the hypothesis of Barber-James (2009) in explaining how the two clades of *Prosopistoma* viz., 'variegatum' clade and the 'African' clade were introduced into the Oriental region through drifting India and its collision with Asia some 34 million years back in geological history.

MOLECULAR CHARACTERIZATION AND PHYLOGENY

The first molecular phylogeny for the order Ephemeroptera was constructed by Ogden and Whiting (2005). Their analyses included 31 of the 37 families, representing more or less 24% of the genera. O'Donnell and Jockusch (2008) investigated the phylogenetic relationships of leptophlebiid mayflies as inferred by histone H3 and 28S ribosomal DNA from six continents. Gattolliat et al. (2008) reconstructed the first comprehensive molecular phylogeny of the Afrotropical Baetidae. The molecular reconstruction indicated the Afrotropical Baetidae require a global revision at a generic as well as supragenetic level. The investigation of Ogden et al. (2009b) represented the combined molecular and morphological analysis for the mayfly family Ephemerellidae, with a focus on the relationships of genera and species groups of the subfamily Ephemerellinae. Gattolliat and Monaghan (2010) have studied DNA-based association of adults and larvae in Baetidae. They used a general mixed Yule-coalescent (GMYC) model to combine population- and species-level sequence variation of mitochondrial deoxyribonucleic acid (mtDNA) to detect species boundaries in Baetidae.

DNA barcoding shows great potential for use by those studying the systematics including cryptic species of many Ephemeroptera species groups (Stahls and Savolainen 2008). A comprehensive DNA barcode library has been established for mayflies from Canada, Mexico, and the United States (Ball et al. 2005; Zhou et al. 2009, 2010; Webb et al. 2012; Gattolliat et al. 2015). Their study has demonstrated that DNA barcoding holds great promise as a tool for a rapid biodiversity assessment of unknown fauna. DNA barcodes of stream mayflies will improve descriptions of community structure and water quality for both ecological and bioassessment purposes (Sweeney et al. 2011). DNA barcodes have also had implications in studying the systematics, diversity, association of adults and larvae, ecology, biogeography, and conservation of aquatic insects (Sivaramakrishnan et al. 2014; Gattolliat et al. 2015). Williams et al. (2006) assessed the molecular diversity of this complex in one of the largest such studies of cryptic species in the order Ephemeroptera.

Molecular techniques have still not been used extensively in mayfly research in India. However, Sivaramakrishnan et al. (2011) reviewed the emerging trends in molecular systematics and molecular phylogeny of mayflies (Ephemeroptera) at the global level, and Selvakumar et al. (2016a) generated DNA barcodes for 40 species belonging to 32 genera under 10 families of Ephemeroptera from South India. Selvakumar et al. (2016a) calculated nucleotide sequence divergence using the Kimura two-parameter distance model and a neighbour-joining analysis was performed to provide a graphic display of the patterns of divergence among the species. Genetic diversity of south Indian endemic mayfly species *Petersula courtallensis* was investigated with wide geographic ranges using mitochondrial cytochrome oxidase gene sequences by Selvakumar et al. (2017a). Their results indicated a general pattern of high genetic diversity between the western and eastern streams and the presence of two genetically distinct populations.

INTEGRATIVE TAXONOMY

The primitive archaic order of aquatic insects viz., Ephemeroptera has received considerable attention globally in terms of referring species delimitation by incorporating 'integrative taxonomy' that integrates all available data sources and using species tree approaches (Yeates et al. 2011). However, application of integrative taxonomic approach to help in resolving taxonomic riddles vis-a-vis single method has not yet been initiated in studying mayflies of India. Recent integrated taxonomic revision of *Camplocia* belonging to the family Euthyplocidae by GonÇalves et al. (2017) has highlighted the advantages of this approach by including a morphological analysis of type and non-type material of species of *Camplocia* including their junior synonyms, ultrastructural analyses of the egg chorion, and neighbourhood joining based on Kimura 2 Parameter distances and Bayesian inference of 376 bp of the mitochondrial gene cytochrome oxidase (COI) of recently collected specimens to resolve problems in synonymy, fine tuning of species delimitations, precise descriptions of new species, from Amazon forest and Costa Rica, and exploring the possibility of existence of cryptic species.

TAXONOMIC PROBLEMS

Since most species described earlier were based on the imaginal or the larval stage alone, association of larvae with respective imagos by individual rearing is rather indispensable to arrive at precise taxonomic conclusions. Moreover, many regions of India especially the rivers, streams, and other wetlands of eastern and western Himalaya, central India, and Eastern Ghats are under explored. DNA barcodes were generated for 40 species belonging to 32 genera under 10 families of Ephemeroptera from South India by Selvakumar et al. (2016a), but no other aspects of molecular work were undertaken on mayflies in India so far. Future research should focus on correlating adult and larval stages and exploring under and unexplored regions as well using new tools like generation of DNA barcodes for mayflies (Selvakumar 2018).

BIOLOGY

Habitat: Mayflies inhabit all aquatic habitats except for marine environment, polluted, and underground waters. Some species are found in brackish waters also. Lotic-erosional habitats are more species rich than lotic-depositional and lentic-depositional habitats. In the higher altitudes (>3000 m), species diversity is poor. Species of lentic habitats are found in ponds, lakes, water tanks, paddy fields, etc. In lotic habitats, runs and riffles with bottom substrates such as boulders and cobbles have higher diversity than cascades or waterfalls. Species diversity is also reduced in habitats with bottom substrates such as sand or mud.

LIFE CYCLE

Emergence: In the last larval instar (nymph) food uptake stops, alimentary canal and malpighian tubules degenerate, the former fills up with water first and later with air to develop into an aerostatic organ. Spermatogenesis and oogenesis is already completed before moulting. Haemolymph, mouthparts, visceral muscles, and gonads undergo considerable changes. During this period, oxygen uptake and drift activity increases. Subimago leave the nymphal skin by rupturing the mesonotal cuticle along the midline, which is completed in 10–15 minutes. The subimago, depending upon the species, emerge either from the water surface, above water, or underwater. In some species, more than one type of emergence is observed. Temperature and light intensity influence the metamorphosis. In tropical regions, most of the species emerge within 2 hours after sunset.

Swarming: Conspicuous mating swarms of males are typical of mayflies. The mating swarm typically consists of several specimens to thousands of individuals. They swarm over land marks such as vegetation, rock, bush, tree, shore line, bridge, road, etc. The size, timing, height, and time of swarm depends on many factors such as weather, temperature, etc. Typical tropical species swarm during night. However, in the high altitudes of Western Ghats and Himalaya, swarming is also observed in the afternoon.

Mating: Mating usually takes place in flight which lasts from a few seconds to several minutes. Males grab the females from below using their forelegs curved around the wing roots. The male abdomen is turned up and the forceps grasps the apex of the female abdomen and the penis is inserted into oviduct opening or copulatory pouches.

Oviposition: Eggs are always deposited in water. However, sometimes females are attracted to oviposit in man-made objects such as car roof tops or smooth roads. Depending on species, several types of oviposition are observed: (i) females release a few eggs at a time by dipping the tip of abdomen on the water surface; (ii) releasing all eggs at one time on the water surface; (iii) females fall on the water surface and release the eggs by rupturing the abdominal wall; (iv) females approach the waterline from shore and release eggs; or (v) females crawl beneath the water surface to deposit eggs on stones or logs (Bauernfeind and Soldán 2012). Females typically lay 500–3000 eggs which is influenced by environmental degradation like eutrophication, xenochemicals, and acidification and other environmental variables (Sweeney 1978, 1984).

Larval stage: Species are morphologically adapted to current velocity which include hydrodynamic body shape (e.g., *Prosopistoma*), stabilizing and retention structures (e.g., *Epeorus*), friction discs, sclerotized gill margins with microtrichia, or suckers formed by gills. The mayfly larvae require high oxygen content in the water which is generally 3–4 times higher than in other aquatic insect groups such as Diptera. Apart from gills, cutaneous breathing is also important for mayfly larvae. The larvae are considered 'trophically generalized' or 'selectively omnivorous'. The feeding types are classified as: (i) grazers-scrappers feed on attached algae and mouth parts (maxillae) are scrape-like; (ii) shredders feed on coarse particulate organic matter (CPOM) and mouth parts are not particularly specialized; (iii) gatherers-collectors feed on fine particulate organic matter (FPOM) without specialized mouth parts; (iv) filter feeders use FPOM and seston (plankton, nekton, and detritus); and (v) predators feed on small benthic animals such as nematodes, oligochaetes, etc. Feeding is opportunistic and depends upon availability, substrate composition, and seasonality.

Adult stage: Adult mayflies do not feed, and reproduction and dispersal are the sole functions of adults. Adults do not move away from water, but some species are found far away from their emergence site. Females of most species exhibit 'upstream compensatory flight' to minimize downstream drift of eggs and larvae. This flight may vary from several metres to kilometres.

Longevity: The lifespan of typical adult mayflies usually lasts for 24 hours. However, depending on species, it varies from few hours to days.

ECOLOGY

Diversity profiles/trophic categorization: Pioneering attempts were made by Sivaramakrishnan and Job (1981) to study the mayfly populations of Courtallam streams. Sivaramakrishnan and Venkataraman (1990) have found that historical immigration, assured perennial flow of a stream, and its pollution-free nature appear to be factors mainly influencing the distribution of a few biogeographically significant genera of Leptophlebiidae in Palani hills, South India. Burton and Sivaramakrishnan (1993) conducted detailed investigations on the insect communities including mayflies and their trophic ecology in the streams of the Silent Valley National Park in Kerala part of the Western Ghats. Composition and zonation of mayfly nymphs were investigated along with other aquatic insects in the entire river basins of Kaveri and Gadananathi by Sivaramakrishnan et al. (1995) and Anbalagan and Dinakaran (2006), respectively.

Anbalagan et al. (2004) studied the diversity profiles and trophic categorization of aquatic insects including mayflies of Courtallam hills of the Western Ghats. There is paucity of information on the diversity of Eastern Ghats mayflies. Notable exceptions include studies on Karandamalai, Sirumalai, and Alagarmalai segments of the Eastern Ghats by Jahir-Hussain et al. (2006), Dinakaran and Anbalagan (2006), and Dinakaran and Krishnan-Kutty (1997), respectively. Kubendran et al. (2017b) have studied diversity and distribution of Baetidae larvae of streams and rivers of the southern Western Ghats. Kubendran et al. (2018) have investigated composition and trophic categorization of aquatic insects including mayflies in the three hill streams and rivers of the Western Ghats.

Emergence patterns/seasonal abundance: Sivaramakrishnan and Venkataraman (1990) have observed noon swarming of leptophlebiid and baetid mayflies at a few sites at higher altitudes (above 2000 msl in Palani hills, South India. Seasonal abundance and diet of *Cloeon* sp. were studied in a northeast Indian lake by Gupta et al. (1994). Selvakumar et al. (2016b) have observed noon emergence of *Klugephlebia kodai* in a stream near Pillar Rock of Palani hills, Tamil Nadu.

Life cycle/voltinism: Behavioural strategies of emergence, swarming, mating, and oviposition of some mayflies were investigated by Sivaramakrishnan and Venkataraman (1985). They also made observations on feeding propensities, growth rates, and fecundity in two baetine mayflies (Sivaramakrishnan and Venkataraman 1987). Fecundity of mayflies of the Western Ghats was studied by Sridhar and Venkataraman (1989). Sivaramakrishnan et al. (1990) studied life cycle patterns of mayflies of Cardamom hills of the Western Ghats. They found that five species viz., *Baetis frequentus*, *Caenis* sp., *Choroterpes* (*Euthraulus*) *alagarensis*, *Notophlebia jobi*, and *Afronurus kumbakkaraiensis* exhibited basically multivoltine pattern with overlapping generations and continuous emergence, whereas one species (*Epeorus* sp.) exhibited more than one univoltine brood. Bioecological studies on the burrowing mayfly, *Ephemera nadinae* (Ephemeridae) in Kurangani streams of the Western Ghats were done by Balasubramanian et al. (1992). Gupta (1993) investigated life histories of two species of *Baetis* in a small Northeast Indian stream. Life cycle and growth of *Cloeon* sp. were studied from Meghalaya by Gupta et al. (1993). Sivaruban et al. (2010) have investigated the life cycle of Heptageniidae in the Kumbbakarai stream of the Western Ghats, Tamil Nadu.

Biomonitoring tool: Utility of mayflies of peninsular Indian streams and rivers on biomonitoring water quality has been highlighted based on rapid bioassessment studies on the Kaveri river basin using macroinvertebrate assemblages by Sivaramakrishnan et al. (1996a). Sivaramakrishnan (2000) has generated a refined rapid bioassessment protocol for benthic macroinvertebrates for use in peninsular Indian streams and rivers in which the mayflies form an essential component of the macroinvertebrate assemblage on which the biotic indices are based. Kubendran et al. (2017) investigated Baetidae as biological indicators of environmental degradation in Tamiraparani and Vaigai River basins of southern Western Ghats.

Anthropogenic/global warming and climate change/land use impacts: Impact of riparian land use on stream insects of Kudremukh National Park, Karnataka was investigated by Subramanian et al. (2005). Anthropogenic impacts on aquatic insects including mayflies in six streams of southern Western Ghats were investigated by Dinakaran and Anbalagan (2007). Impact of global warming and climate change on aquatic insects especially mayflies has been reviewed by Sivaramakrishnan et al. (2008). Selvakumar et al. (2014b) studied the impact of riparian land use patterns on Ephemeroptera community structure in river basins of southern Western Ghats.

ECONOMIC IMPORTANCE

Mayflies are occupying freshwater and brackish water habitats across the world, with the exception of Antarctica. They constitute an important part of the food chain, mainly consuming primary producers such as algae and plants, and as a food source for vertebrate predators like fish. They are excellent biological indicators of water quality and habitat quality (Sivaramakrishnan et al. 1996a; Buffagni 1997; Selvakumar et al. 2014b). They are ideal objects for integrated phylogenetic, biogeographic, and phylogeographic studies, being endowed with several archaic traits in all life stages along with rather weak dispersal powers. Many of the montane mayflies, both nymphs and imagos are equally charismatic. Nymphs are important for freshwater ecological and biomonitoring studies.

CONSERVATION STATUS

Mayflies along with stoneflies and caddisflies have a significant role in the wetland food chain. Their species assemblages change primarily with levels of human disturbance in freshwaters, both lentic and lotic. Anthropogenic impacts result in habitat fragmentation, global warming, climate change, alien species invasion, as well as major land use changes, especially changes in the riparian zone. Mayfly larvae are microhabitat specialists and inhabit springs, habitats like sandy stretches of streams and rivers, impact-free erosional and depositional zones of pristine, montane headwater streams, as well reasonably clean lentic bodies. Conservation of habitats and microhabitats like leaf packs of leaf litter inhabiting species as well as the fragile microhabitats of phylogenetic relicts and macro and microendemics and pollution intolerant species is highly critical for protecting characteristic Ephemeroptera community assemblages in river catchments is of national importance. Significant Indian publications dealing with habitat diversity and land use impacts include Subramanian and Sivaramakrishnan (2005) on aquatic insects of the river basins of the Western Ghats and Selvakumar et al. (2014b) on Ephemeroptera communities in river basins of the southern Western Ghats. Recently, Sundar and Muralidharan (2017) have briefly reviewed the impact of climatic change on aquatic insects and habitats with particular reference to India. However, specific conservation measures to prioritize habitat and faunal conservation of mayflies within the overall framework of conservation of freshwater biota are yet to gain momentum in India. In this context, Massariol et al. (2014) recommend creation of conservation units of Ephemeroptera using biological information rather than economic, cultural, or political criteria.

COLLECTION AND PRESERVATION

Collection: Mayfly larvae can be collected by a large range of devices viz., kick net, hand screen, dip nets, drift net, etc. Adult mayflies are generally caught using a hand net with a long handle and large opening to catch swarming adults. Beating the vegetation with a stick can also be used to collect resting imagos and subimagos in a hand net. Tent traps, such as the Malaise traps or emergence traps can also be used, but they need to be checked regularly to remove subimagos, preventing them from drowning and enabling rearing to the adult stage. Light traps at dusk and dawn, especially in the tropics, give significant results (Sartori and Brittain 2015).

Rearing: Rearing is an important procedure because it allows the association between larval and adult characteristics. Several techniques exist, but the most useful is to select a single mature larva and put it in a rearing cage, either in situ or in a suitably equipped laboratory. Cages should be checked regularly for the emergence of the subimago. Once emerged, the larval skin must be placed in a vial with ethanol, and the subimago placed in another cage without direct sunlight and a relative humidity of more than 50%. Once the imago has moulted, it is necessary to wait for a couple of hours for the teguments to dry and the final colouration to be fixed. The specimen can then be placed in ethanol with its larval and subimaginal exuviae for further study (Sartori and Brittain 2015). Currently, molecular studies particularly DNA barcode are more and more frequently used to associate adult and larval stages (Monaghan and Sartori 2009).

Specimen preparation: Their soft cuticle and long appendices make mayflies fragile insects necessitating larvae and adults to be preserved in ethanol. The ethanol concentration should be approximately 80% for long-term preservation, but 100% if molecular studies are planned. If larvae are fixed in 80% ethanol in the field, the medium should be changed when arriving in the laboratory because of the high water and lipid concentration in the body. Preferably, specimens should be kept at low temperatures (<6°C) and definitely never in a warm place, which will rapidly fade the colours. Some small specimens (e.g., *Baetis*) can also be mounted entirely on a slide according to an appropriate protocol for slide preparation. For morphological examination and species identification, slide preparation of larval mouthparts and appendices as well as male genitalia and wings is often necessary (Sartori and Brittain 2015).

Preparation of slides: Mayfly parts must frequently be studied from slides. Most of the structures can be easily mounted on slides as outlined below. Some parts such as larval gills are usually best studied in alcohol, but for most of the structures, a better understanding can be achieved by studying both a slide specimen and an alcohol specimen. The wings of adult specimens can be mounted by floating them from clean alcohol onto the slide, arranging them properly in a thin film of alcohol, and covering them with a square cover slip. Narrow strips of white gummed paper are much better than any self-adhesive paper or tape for holding the cover slip tight on the slide. After the wings set in position on the slide, the alcohol is allowed to evaporate. Crumpled wings from dry specimens can be flattened by dropping them gently in boiling water and then quickly floating them onto a slide. Wings are best studied as dry mounts, wings mounted in Canada balsam generally show fewer details.

Male genitalia from dried specimens may need to be softened before mounting. A satisfactory mount results when they are placed for 1 or 2 hours in a solution of 10% potassium hydroxide or sodium hydroxide and then dehydrated and mounted in Canada balsam or Hoyer's medium. For male genitalia or other parts of adults or larvae preserved in alcohol, avoid hydroxide except for very large or dark specimens that otherwise would not clear. Structures can be mounted directly from 95% alcohol or cellosolve (ethylene glycol monoethyl ether) into specially prepared Canada balsam. Commercial neutral Canada balsam in xylene is allowed to dry until it is highly viscous. It is then returned to suitable consistency by replacing the evaporated xylene with cellosolve. Structures placed in this mixture may cloud temporarily, but seldom for more than an hour. The clouding can be reduced by passing the structures through pure cellosolve before mounting.

The cover slip for balsam mounts should be no larger than necessary. For small structures such as male genitalia and claws, use 8 mm round cover slips and larger structures use 12–18 mm round cover slips. An effective method for positioning mounts is to put the structures in a thin film of balsam on the cover slip or the slide. The parts can be repositioned periodically until the balsam has become quite firm. The cover slip can then be transferred to the slide with additional balsam. The structures must be completely covered with balsam and allowed to dry in a dust-free place or in a petri dish. The slides should also be fully labelled with the locality, the date, and the name of the collector or referenced to a specific specimen. When drawings are made for publication, the slide or the specimen used should be labelled (Edmunds et al. 1976).

USEFUL WEBSITES

Ephemeroptera of the World http://www.insecta.bio.spbu.ru/z/Eph-spp/Contents.htm from Russia

Ephemeroptera Galactica http://www.ephemeroptera-galactica.com/ from Florida A & M University, currently operating from Germany

Mayfly Central http://www.entm.purdue.edu/mayfly/ from Purdue University, U.S.A.

CONCLUSION

Mayfly fauna of India, a country endowed with two mega diversity hotspots, appears to be an assemblage of ancient Gondwanan derivatives, with a high percentage of endemism, a few Laurasian spillovers, along with some younger faunal elements that might have diversified in several spells at different periods in geological history by vicariant and dispersal events, through 'out of India and towards India' exchanges between Indian subcontinent on the one hand and Afrotropics including Madagascar, Oriental Southeast Asia, and Palearctic North on the other. Due to poor dispersal ability of adults, most of the mayfly species have restricted distribution. To advance our knowledge of Indian Ephemeroptera, the foremost need is a synthesis of the taxonomy of adults and larvae within a global systematic context for each family, precise delimitation of species, an understanding of intraspecific genetic diversity, as well as detection of cryptic species complexes.

ENDNOTE

After preparation of final manuscript two new families of mayflies*viz*., Vitenamellidae and Isonychidae were reported from India (Selvakumar et al. 2018 & Vasanth et al. 2019) increasing the total number of families reported from Indian region to 17.

ACKNOWLEDGEMENTS

The first author (CS) thanks Head, Department of Zoology, Principal, and Management, The Madura College (Autonomous), Madurai, for support and facilities. The second author (KAS) is grateful to Dr. Kailash Chandra, Director, Zoological Survey of India for encouragement and support.

REFERENCES

Allen, R. K. 1965. A review of the subfamilies of Ephemerellidae (Ephemeroptera). *Journal of the Kansas Entomological Society* 38: 262–266.

Allen, R. K. 1980. Geographic distribution and reclassification of the subfamily Ephemerellinae (Ephemeroptera: Ephemerellidae). In *Advances in Ephemeroptera Biology*, eds. J. F. Flannagan and K. E. Marshall, pp. 71–91, New York: Plenum.

Allen, R. K. 1984. A new classification of the subfamily Ephemerellinae and the description of a new genus. *Pan-Pacific Entomologist* 60: 245–247.

Anbalagan, S., and S. Dinakaran. 2006. Seasonal variation of diversity and habitat preferences of aquatic insects along the longitudinal gradient of the Gadana river basin, South-West Ghats, (India). *Acta Zoologica Bulgarica* 58: 253–264.

Anbalagan, S., C. Balachandran, M. Kannan, S. Dinakaran, and M. Krishnan. 2014. First record and a new species description of *Dudgeodes* (Ephemeroptera: Teloganodidae) from South India. *Turkish Journal of Zoology*. doi:10.3906/zoo-1401-74.

Anbalagan, S., B. Kaleeswaran, and C. Balasubramanian. 2004. Diversity and trophic categorization of aquatic insects of Courtallam hills of Western Ghats. *Entomom* 29(3): 215–220.

Bae, Y. J., and W. P. McCafferty. 1991. Phylogenetic systematics of the Potamanthidae (Ephemeroptera). *Transactions of the American Entomological Society* 117(3–4): 1–143.

Bae, Y. J., and W. P. McCafferty. 1998. Phylogenetic systematics and biogeography of the Neoephemeridae (Ephemeroptera: Pannota). *Aquatic Insects* 20: 35–68.

Balachandran, C., S. Anbalagan, M. Kannan, S. Dinakaran, and M. Krishnan. 2016. A new species of *Prosopistoma* Latreille, 1833 (Ephemeroptera: Prosopistomatidae) from South India. *Zootaxa* 4178: 289–294.

Balasubramanian, C., K. Venkataraman, and K. G. Sivaramakrishnan. 1992. Bioecological studies on the burrowing mayfly, *Ephemera nadinae* (Ephemeroptera: Ephemeridae) in Kurangani streams of Western Ghats. *Journal of Bombay Natural History Society* 89(1): 72–77.

Ball, S. L., P. D. N. Hebert, S. K. Burian, and J. M. Webb. 2005. Biological identifications of mayflies (Ephemeroptera) using DNA barcodes. *Journal of North American Benthological Society* 24: 508–524.

Barber-James, H. M. 2009. A preliminary phylogeny of Prosopistomatidae (Ephemeroptera) based on morphological characters of the larvae, and an assessment of their distribution. *Aquatic Insects* 31: 149–166.

Barber-James, H. M., J. L. Gattoliat, M. Sartori, and M. D. Hubbard. 2008. Global diversity of mayflies (Ephemeroptera, Insecta) in freshwater. *Hydrobiologia* 595: 339–350.

Bauernfeind, E., and T. Soldán. 2012. *The Mayflies of Europe (Ephemeroptera)*. Ollerup: Apollo Books.

Berner, L., and Pescador, M. L. 1980. The mayfly family Baetiscidae (Ephemeroptera). Part I. In *Advances in Ephemeroptera Biology*, eds. J. F. Flannagan and K. E. Marshall, pp. 511–524. New York: Plenum Press.

Brittain, J. E., and M. Sartori. 2003. Ephemeroptera (Mayflies). In *Encyclopedia of Insects*, eds. V. H. Resh and R. T. Card, pp. 373–380. Amsterdam, the Netherlands: Academic Press.

Buffagni, A. 1997. Mayfly community composition and the biological quality of streams. In *Ephemeroptera & Plecoptera: Biology-Ecology-Systematics*, eds. P. Landolt and M. Sartori, pp. 235–246. Fribourg, Switzerland: Mauron + Tinguely & Lachat.

Burton, T. M., and K. G. Sivaramakrishnan. 1993. Composition of the insect community in the streams of the Silent Valley National Park in Southern India. *Tropical Ecology* 34(1): 1–16.

Buss, D. F., and F. F. Salles. 2007. Using Baetidae species as biological indicators of environmental degradation in a Brazilian River Basin. *Environmental Monitoring and Assessment* 130: 365–372.

Chopra, B. 1924. The fauna of an island in the Chilka Lake, the Ephemeroptera of Barkuda Island. *Records of the Indian Museum* 26: 415–422.

Chopra, B. 1927. The Indian Ephemeroptera (Mayflies). Part I.—The sub-order Ephemeroidea: Families Palingeniidae and Polymitarcidae. *Records of the Indian Museum* 29: 91–138.

Dinakaran, S., and S. Anbalagan. 2006. Seasonal variation and substrate selection of aquatic insects in a small stream Sirumalai hills of southern Western Ghats. *Journal of Aquatic Biology* 21(1): 37–42.

Dinakaran, S., and S. Anbalagan. 2007. Anthropogenic impacts on aquatic insects in six streams of south Western Ghats. *Journal of Insect Science* 7(1). doi:10.1673/031.007.3701.

Dinakaran, S., and K. Krishnan-Kutty. 1997. A study on diversity of aquatic insects in a tropical stream of Alagar hill (Eastern Ghats). *Proceedings of Seminar on Biodiversity (SB'97)*, K. V. College, Tanjore, India pp. 111–112.

Dinakaran, S., C. Balachandran, and S. Anbalagan. 2009. A new species of *Choroterpes* (Ephemeroptera: Leptophlebiidae) from a tropical stream of south India. *Zootaxa* 2064: 21–26.

Dubey, O. P. 1970. Torrenticole insects of the Himalaya. III. Descriptions of two new species of Ephemerida from the Northwest Himalaya. *Oriental Insects* 4: 299–302.

Dubey, O. P. 1971. Torrenticole insects of the Himalaya. VI. Descriptions of nine new species of Ephemerida from the Northwest Himalaya. *Oriental Insects* 5(4): 521–548.

Eaton, A. E. 1883–1888. A revisional monograph of recent Ephemeridae or mayflies. *Transactions of the Linnean Society of London, Second Series, Zoology* 3: 1–352.

Edmunds, G. F. Jr. 1972. Biogeography and evolution of Ephemeroptera. *Annual Review of Entomology* 17: 21–42.

Edmunds, G. F. Jr. 1979. Biogeographical relationships of the Oriental and Ethiopian mayflies. In *Proceedings of the Second International Conference on Ephemeroptera*, eds. K. Pasternak and R. Sowa, pp. 11–14. Warszawa-Krakw: Panstwowe Wydawnictwo Naukowe.

Edmunds, G. F. Jr., and W. P. McCafferty. 1988. The mayfly subimago. *Annual Review of Entomology* 33: 509–529.

Edmunds, G. F. Jr., and J. R. Traver. 1954. An outline of a reclassification of the Ephemeroptera. *Proceedings of the Entomological Society of Washington* 56: 236–240.

Edmunds, G. F. Jr., R. K. Allen, and W. L. Peters. 1963. An annotated key to the nymphs of the families and subfamilies of mayflies (Ephemeroptera). *University of Utah Biological Series* 13: 1–49.

Edmunds, G. F. Jr., S. L. Jensen, and L. Berner. 1976. *The Mayflies of North and Central America*. Minneapolis, MN: University of Minnesota Press.

Feck, J., and R. O. Hall. 2004. Response of American dippers (*Cinclus mexicanus*) to variation in stream water quality. *Freshwater Biology* 49: 1123–1137.

Gattolliat, J. L., and M. T. Monaghan. 2010. DNA-based association of adults and larvae in Baetidae (Ephemeroptera) with the description of a new *Adnoptilum* in Madagascar. *Journal of the North American Benthological Society* 29(3): 1042–1057.

Gattolliat, J. L., and C. Nieto. 2009. The family Baetidae (Insecta: Ephemeroptera): Synthesis and future challenges. *Aquatic Insects* 31(1): 41–62.

Gattolliat, J. L., E. Cavallo, L. Vuataz, and M. Sartori. 2015. DNA barcoding of Corsican mayflies (Ephemeroptera) with implications on biogeography, systematics and biodiversity. *Arthropod Systematics & Phylogeny* 73: 3–18.

Gattolliat, J. L., M. T. Monaghan, M. Sartori et al. 2008. A molecular analysis of Afrotropical Baetidae. In *International Advances in the Ecology, Zoogeography and Systematics of Mayflies, and Stoneflies*, Vol. 128, eds. F. R. Hauer, J. A. Stanford, and R. L. Newell, pp. 219–232. Oakland, CA: University of California, Publication in Entomology.

Gillies, M. T. 1949. Notes on some Ephemeroptera Baëtidae from India and South-East Asia. *Transactions of the Royal Entomological Society London* 100: 161–177.

Gillies, M. T. 1951. Further notes on Ephemeroptera from India and South East Asia. *Proceedings of the Royal Entomological Society of London (B)* 20: 121–130.

Gillies, M. T. 1957. New records and species of *Euthraulus* Barnard (Ephemeroptera) from East Africa and the Oriental Region. *Proceedings of the Royal Entomological Society of London (B)* 26: 43–48.

Gillies, M. T. 1991. A diphyletic origin for the two-tailed baetid nymphs occurring in East African stony streams with a description of the new genus and species *Tanzaniella spinosa* gen. nov. sp. nov. In *Overview and Strategies of Ephemeroptera and Plecoptera*, eds. J. Alba-Tercedor and A. Sanchez-Ortega, pp. 175–187. Gainesville, Florida: Sandhill Crane Press.

Gillott, C. 2005. *Entomology*. Springer Science and Business Media.

Gonçalves, I. C., D. M. Takiya, F. F. Salles, J. G. Peters, and J. L. Nessimian. 2017. Integrative taxonomic revision of *Campylocia* (mayflies: Ephemeroptera, Euthyplociidae). *Systematics and Biodiversity* 15(6): 564–581.

Grant, P. M., and K. G. Sivaramakrishnan. 1985. A new species of *Thraulus* (Ephemeroptera: Leptophlebiidae) from southern India. *Florida Entomologist* 68(3): 424–432.

Gupta, A. 1993. Life histories of two species of *Baetis* (Ephemeroptera: Baetidae) in a small North-East Indian stream. *Archiv fur Hydrobiologie* 127(1): 105–114.

Gupta, A., S. Gupta, and R. G. Michael. 1994. Seasonal abundance and diet of *Cloeon* sp. (Ephemeroptera: Baetidae) in a northeast Indian lake. *Archiv fur Hydrobiologie* 130(3): 349–357.

Gupta, S., R. G. Michael, and A. Gupta. 1993. Laboratory studies on the life cycle and growth of *Cloeon* sp. (Ephemeroptera: Baetidae) in Meghalaya State, India. *Aquatic Insects* 15(1): 49–55.

Hafiz, H. A. 1937. The Indian Ephemeroptera (mayflies) of the sub-order Ephemeroidea. *Records of the Indian Museum* 39: 351–370.

Hagen, H. 1858. Synopsis der Neuroptera Ceylons. *Verhandelingen der Zoologisch Botanischen Gesellschaft in Wien* 8: 471–488.

Hartland-Rowe, R. 1958. The biology of a tropical mayfly *Povilla adusta* Navás (Ephemeroptera, Polymitarcidae) with special reference to the lunar rhythm of emergence. *Revue de Zoologie et de Botanique Africaines* 58(3–4): 185–202.

Hubbard, M. D., and W. L. Peters. 1978. A catalogue of the Ephemeroptera of the Indian Subregion. *Oriental Insects* 9(Supplement): 1–43.

Jacobus, L. M., and W. P. McCafferty. 2004a. Revisionary contributions to the genus *Drunella* (Ephemeroptera: Ephemerellidae). *Journal of the New York Entomological Society* 112: 127–147.

Jacobus, L. M., and W. P. McCafferty. 2004b. Revisionary contributions to the genus *Torleya* (Ephemeroptera: Ephemerellidae). *Journal of the New York Entomological Society* 112(2–3): 153–175.

Jacobus, L. M., and W. P. McCafferty. 2006. Reevaluation of the phylogeny of the Ephemeroptera Infraorder Pannota (Furcatergalia), with adjustments to higher classification. *Transactions of the American Entomological Society* 132: 81–90, 429–430.

Jacobus, L. M., and W. P. McCafferty. 2008. Revision of Ephemerellidae genera (Ephemeroptera). *Transactions of the American Entomological Society* 134(1–2): 185–274.

Jacobus, L. M., W. P. McCafferty, and R. W. Sites. 2007. A new species and first stage associations in *Crinitella* (Ephemeroptera: Ephemerellidae: Ephemerellinae). *Zootaxa* 1611: 45–53.

Jahir-Hussain, K., C. M. Jayachandra, and K. G. Sivaramakrishnan. 2006. Diversity, seasonability and trophic ecology of stream insects in a tropical dry deciduous forest of Karandamalai hills (Eastern Ghats), Tamil Nadu. *Indian Journal of Tropical Biodiversity* 14(2): 123–133.

Jenkins, R. K. B., and S. J. Ormerod. 1996. The influence of a river bird, the dipper (*Cinclus cinclus*), on the behaviour and drift of its invertebrate prey. *Freshwater Biology* 35: 45–56.

Jensen, S. L. 1974. A new genus of mayflies from North America (Ephemeroptera: Heptageniidae). *Proceedings of the Entomological Society of Washington* 76: 225–228.

Kapur, A. P., and M. B. Kripalani. 1961. The mayflies (Ephemeroptera) from the northwestern Himalaya. *Records of the Indian Museum* 59: 183–221.

Kapur, A. P., and M. B. Kripalani. 1963. The mayflies (Ephemeroptera) from the north-western Himalaya. *Records of the Indian Museum* (1961) 59(1–2): 183–221.

Kaul, B. K., and O. P. Dubey. 1970. Torrenticole insects of the Himalaya. I. Two new species of Ephemerida. *Oriental Insects* 4(2): 143–153.

Kazlauskas, R. S. 1972. Neues ber das System der Eintagsfliegen der Familie Baetidae (Ephemeroptera). *Proceedings of the 13th International Congress of Entomology, Moscow 1968* 3: 337–338.

Kimmins, D. E. 1947. New species of Indian Ephemeroptera. *Proceedings of the Royal Entomological Society of London (B)* 16: 92–100.

Klapalek, F. 1909. Ephemerida, Eintagsfliegen. *Die Susserwasserfauna Deutchlands* 8: 1–32.

Kluge, N. 1988. [Generic revision of the Heptageniidae (Ephemeroptera). 1. Diagnoses of tribes, genera, and subgenera of Heptageniinae.]. *Entomologicheskoye Obozreniye* 2: 291–313.

Kluge, N. J. 1998. Phylogeny and higher classification of Ephemeroptera. *Zoosystematica Rossica* 7(2): 255–269.

Kluge, N. J. 2000. *Modern Systematics of Insects. Part I. Principles of Systematics of Living Organisms and General System of Insects with Classification of Primary Wingless and Palepterous Insects*. Saint Petersburg, Russia: Lan' S-Petersburg.

Kluge, N. J. 2004. *The Phylogenetic System of Ephemeroptera*. Dordrecht, the Netherlands: Kluwer Academic Publishers.

Kluge, N. J. 2014. New Oriental tribe Iscini, new non-dilatognathan species of *Notophlebia* Peters & Edmunds 1970 and independent origin of *Dilatognathus*-type mouth apparatus in Atalophlebiinae (Ephemeroptera: Leptophlebiidae). *Zootaxa* 3760(4): 522–538.

Kluge, N. J., and E. A. Novikova. 2014. Systematics of *Indobaetis* Müller-Liebenau & Morihara 1982, and related implications for some other Baetidae genera (Ephemeroptera). *Zootaxa* 3835(2): 209–236.

Kluge, N. J., C. Selvakumar, K. G. Sivaramakrishnan, and L. M. Jacobus. 2015. Contribution to the knowledge of the mayfly genus *Teloganella* Ulmer, 1939 (Ephemeroptera: Ephemerelloidea). *Zootaxa* 4028(2): 287–295.

Kluge, N. J., K. G. Sivaramakrishnan, C. Selvakumar, and T. Kubendran. 2013. Notes about *Acentrella* (*Liebebiella*) *vera* (Müller-Liebenau, 1982) (= *Pseudocloeondifficilum* Müller-Liebenau, 1982 syn. n. = *Platybaetisarunachalae* Selvakumar, Sundar, and Sivaramakrishnan, 2012 syn. n.) (Ephemeroptera: Baetidae). *Aquatic Insects* 35(3–4): 63–70.

Kubendran, T., C. Balasubramanian, C. Selvakumar, J. L. Gattolliat, and K. G. Sivaramakrishnan. 2015. Contribution to the knowledge of *Tenuibaetis* Kang & Yang 1994, *Nigrobaetis* Novikova & Kluge 1987 and *Labiobaetis* Novikova & Kluge 1987 (Ephemeroptera: Baetidae) from the Western Ghats (India). *Zootaxa* 3957(2): 188–200.

Kubendran, T., S. Murali-Krishnan, C. Selvakumar, A. K. Sidhu, and A. Nair. 2018. Composition and trophic categorization of aquatic insects and biomonitoring potential of selected hill streams of Western Ghats, India. *International Journal of Ecology and Environmental Sciences* 44(2): 107–115.

Kubendran, T., T. Rathinakumar, C. Balasubramanian, C. Selvakumar, K. G. Sivaramakrishnan. 2014. A new species of *Labiobaetis* Novikova & Kluge, 1987 (Ephemeroptera: Baetidae) from the southern Western Ghats in India, with comments on the taxonomic status of *Labiobaetis*. *Journal of Insect Science* 14(86): 1–10.

Kubendran, T., C. Selvakumar, A. K. Sidhu, S. Murali-Krishnan, and A. Nair. 2017a. Baetidae (Ephemeroptera: Insecta) as biological indicators of environmental degradation in Tamiraparani and Vaigai river basins of southern Western Ghats, India. *International Journal of Current Microbiology and Applied Sciences* 6(6): 558–572.

Kubendran, T., C. Selvakumar, A. K. Sidhu, S. Murali-Krishnan, and A. Nair. 2017b. Diversity and distribution of Baetidae (Insecta: Ephemeroptera) larvae of streams and rivers of the southern Western Ghats, India. *Journal of Entomology and Zoology Studies* 5(3): 613–625.

Lugo-Ortiz, C. R., and W. P. McCafferty. 1996. Phylogeny and classification of the *Baetodes* complex (Ephemeroptera: Baetidae), with description of a new genus. *Journal of the North American Benthological Society* 15(3): 367–380.

Lugo-Ortiz, C. R., and W. P. McCafferty. 1998a. A new *Baetis*-complex genus (Ephemeroptera: Baetidae) from the Afrotropical Region. *African Entomology* 6(2): 297–301.

Lugo-Ortiz, C. R., and W. P. McCafferty. 1998b. The *Centroptiloides* complex of Afrotropical small minnow mayflies (Ephemeroptera: Baetidae). *Annals of the Entomological Society of America* 91(1): 1–26.

Massariol, F. C., E. D. G. Soares, and F. F. Salles. 2014. Conservation of mayflies (Insecta, Ephemeroptera) in Espírito Santo, southeastern Brazil. *Revista Brasileira de Entomologia* 58(4): 356–370.

McCafferty, W. P. 1973. Systematic and zoogeographic aspects of Asiatic Ephemeridae (Ephemeroptera). *Oriental Insects* 7: 49–67.

McCafferty, W. P. 1991a. Toward a phylogenetic classification of the Ephemeroptera (Insecta): A commentary on systematics. *Annals of the Entomological Society of America* 84(4): 343–360.

McCafferty, W. P. 1991b. The cladistics, classification and evolution of the Heptagenoidea. In *Overview and Strategies of Ephemeroptera and Plecoptera*, eds. J. Alba-Tercedor and A. Sanchez-Ortega, pp. 87–102. Gainesville, Florida: The Sandhill Crane Press.

McCafferty, W. P. 2004. Higher classification of the burrowing mayflies (Ephemeroptera: Scapphodonta). *Entomological News* 115(2): 84–92.

McCafferty, W. P., and J. P. Benstead. 2002. Cladistic resolution and ecology of the Madagascar genus *Manohyphella* Allen (Ephemeroptera: Teloganodidae). *Annales de Limnologie* 38: 41–52.

McCafferty, W. P., and G. F. Jr. Edmunds. 1979. The higher classification of the Ephemeroptera and its evolutionary basis. *Annals of the Entomological Society of America* 72(1): 5–12.

McCafferty, W. P., and T.-Q. Wang. 1997. Phylogenetic systematics of the family Teloganodidae (Ephemeroptera: Pannota). *Annals of the Cape Provincial Museums (Natural History)* 19(9): 387–437.

McCafferty, W. P., and T.-Q. Wang. 2000. Phylogenetic systematics of the major lineages of pannote mayflies (Ephemeroptera: Pannota). *Transactions of the American Entomological Society* 126(1): 9–101.

Monaghan, M. T., and M. Sartori. 2009. Genetic contributions to the study of taxonomy, ecology, and evolution of mayflies (Ephemeroptera): Review and future perspectives. *Aquatic Insects* 31: 19–39.

Navas, L. 1931. Communicaciones entomologicas. 14, Insectos de la India. 4a Serie. *Revista dela Real. Academia de Ciencias Exactas, F2sicas, Qu2micas y Naturales de Zaragoza* 15: 12–41.

Needham, J. G. 1909. Notes on the Neuroptera in the collection of the Indian Museum. *Records of the Indian Museum, Calcutta* 3: 185–210.

Needham, J. G., J. R. Traver, and Y.-C. Hsu. 1935. *The Biology of Mayflies*. New York: Comstock Publishing Co.

Nguyen, V. V., and Y. J. Bae. 2003. Taxonomic review of the Vietnamese Neoephemeridae (Ephemeroptera) with description of *Potamanthellus unicutibius*, new species. *Pan-Pacific Entomologist* 79: 230–236.

O'Donnell, B., and E. L. Jockusch. 2008. Phylogenetic relationships of leptophlebiid mayflies as inferred by histone H3 and 28S ribosomal DNA. *Systematic Entomology* 33: 651–667.

Ogden, T. H., and M. F. Whiting. 2005. Phylogeny of Ephemeroptera (mayflies) based on molecular evidence. *Molecular Phylogenetics and Evolution* 37: 625–643.

Ogden, T. H., J. L. Gattolliat, M. Sartori, A. H. Staniczek, T. Soldán, and M. F. Whiting. 2009a. Towards a new paradigm in mayfly phylogeny (Ephemeroptera): Combined analysis of morphological and molecular data. *Systematic Entomology* 34: 616–634.

Ogden, T. H., J. T. Osborne, L. M. Jacobus, and M. F. Whiting. 2009b. Combined molecular and morphological phylogeny of Ephemerellinae (Ephemerellidae: Ephemeroptera), with remarks about classification. *Zootaxa* 1991: 28–42.

Pereira-da-Conceicoa, L. L., and H. M. Barber-James. 2013. Redescription and lectotype designation of the endemic South African mayfly *Lestagellapenicillata* (Barnard, 1932) (Ephemeroptera: Teloganodidae). *Zootaxa* 3750(5): 450–464.

Peters, W. L. 1967. New species of *Prosopistoma* from the Oriental Region (Prosopistomatoidea: Ephemeroptera). *Tijdschrift voor Entomologie* 110: 207–222.

Peters, W. L. 1975. A new species of *Indialis* from India (Ephemeroptera: Leptophlebiidae). *Pan-Pacific Entomologist* 51(2): 159–161.

Peters, W. L., and G. F. Jr. Edmunds. 1970. Revision of the generic classification of Eastern Hemisphere Leptophlebiidae (Ephemeroptera). *Transactions of the Royal Entomological Society of London* 116: 225–253.

Pictet, F. J. 1843–1845. *Histoire naturelle gnerale et particuliere des insectes nevropteres. Famille des ephemerines.* Geneva, Switzerland: Chez J. Kessmann etAb.

Ramya-Roopa, S., C. Selvakumar, K. A. Subramanian, and K. G. Sivaramakrishnan. 2017. A new species of *Prosopistoma* Latreille, 1833 and redescription of *P. indicum* Peters, 1967 (Ephemeroptera: Prosopistomatidae) from the Western Ghats, India. *Zootaxa* 4242(3): 591–599.

Ruffieux, L., M. Sartori, and G. L'Eplattenier. 1996. Palmen body: A reliable structure to estimate the number of instars in *Siphlonurus aestivalis* (Eaton) (Ephemeroptera: Siphlonuridae). *International Journal of Insect Morphology & Embryology* 25(3): 341–344.

Sartori, M., and J. E. Brittain. 2015. Order Ephemeroptera. In *Thorp and Covich's Freshwater Invertebrates: Ecology and General Biology*, eds. J. Thorp and D. C. Rodgers, pp. 873–891. New York: Academic Press.

Sartori, M., and A. Sartori-Fausel. 1991. Variabilit de la dure du stade subimaginal et de la fcondit chez *Siphlonurus aestivalis* (Eaton). *Revue Suisse de Zoologie* 98(4): 717–723.

Sartori, M., J. G. Peters, and M. D. Hubbard. 2008. A revision of Oriental Teloganodidae (Insecta, Ephemeroptera, Ephemerelloidea). *Zootaxa* 1957: 1–51.

Selvakumar, C. 2018. Insecta: Ephemeroptera. In *Faunal Diversity of Indian Himalaya*, eds. K. Chandra, D. Gupta, K. C. Gopi, B. Tripathy, and V. Kumar, pp. 219–226. Kolkata, India: The Director, Zoological Survey of India.

Selvakumar, C., B. Sinha, M. Vasanth, K. A. Subramanian, and K. G. Sivaramakrishnan, 2018. A new record of monogeneric family Vietnamellidae (Insecta: Ephemeroptera) from India. *Journal of Asia-Pacific Entomology* 21: 994–998.

Selvakumar, C., M. Arunachalam, and K. G. Sivaramakrishnan. 2013. A new species of *Choroterpes* (Ephemeroptera: Leptophlebiidae) from Southern Western Ghats, India. *Oriental Insects* 47(2–3): 169–175.

Selvakumar, C., K. Chandra, K. G. Sivaramakrishnan, and E. E. Jehamalar. 2017b. A new species of *Thalerosphyrus* Eaton 1881 (Ephemeroptera: Heptageniidae: Ecdyonurinae) from India. *Zootaxa* 4350(1): 84–90.

Selvakumar, C., S. Janarthanan, and K. G. Sivaramakrishnan. 2015a. A new species of the *Choroterpes* Eaton, 1881 subgenus *Monophyllus* Kluge, 2012 and a new record of the subgenus *Choroterpes*, s.s. (Ephemeroptera: Leptophlebiidae) from southern Western Ghats, India. *Zootaxa* 3941(2): 284–288.

Selvakumar, C., T. Kubendran, K. Chandra, and A. K. Sidhu. 2017e. First record of the genus *Bungona* (Harker 1957) (Ephemeroptera: Baetidae) and range extension of two species belonging to *Bungona* in India. *Journal of Entomological Research* 41(4): 373–376.

Selvakumar, C., T. Kubendran, K. G. Sivaramakrishnan, and S. Janarthanan. 2017a. Genetic diversity and conservation of South Indian Mayfly, *Petersula courtallensis* Sivaramakrishnan, 1984 (Ephemeroptera: Leptophlebiidae). *Journal of Entomology and Zoology Studies* 5(3): 1110–1114.

Selvakumar, C., K. G. Sivaramakrishnan, L. M. Jacobus, S. Janarthanan, and M. Arumugam. 2014a. Two new genera and five new species of Teloganodidae (Ephemeroptera) from South India. *Zootaxa* 3846(1): 87–104.

Selvakumar, C., K. G. Sivaramakrishnan, and S. Janarthanan. 2015b. A new record of *Potamanthellus caenoides* Ulmer 1939 (Ephemeroptera: Neoephemeridae) from the southern Western Ghats of India. *Biodiversity Data Journal* 3: e5021: 1–5, 10.3897/BDJ.3.e5021.

Selvakumar, C., K. G. Sivaramakrishnan, S. Janarthanan, M. Arumugam, and M. Arunachalam. 2014b. Impact of riparian land use patterns on Ephemeroptera community structure in river basins of southern Western Ghats, India. *Knowledge and Management of Aquatic Ecosystems* 412(11): 1–15.

Selvakumar, C., T. Sivaruban, K. A. Subramanian, and K. G. Sivaramakrishnan. 2016b. A new genus and species of Atalophlebiinae (Insecta: Ephemeroptera: Leptophlebiidae) from Palani Hills of the southern Western Ghats, India. *Zootaxa* 4208(4): 381–391.

Selvakumar, C., K. A. Subramanian, K. Chandra, and E. E. Jehamalar. 2017c. A new species of *Choroterpes* Eaton, 1881 (Ephemeroptera: Leptophlebiidae) from India. *Zootaxa* 4338(1): 189–194.

Selvakumar, C., K. A. Subramanian, K. Chandra, K. G. Sivaramakrishnan, E. E. Jehamalar, and B. Sinha. 2017d. A new species and a new record of the subgenus *Dilatognathus* Kluge 2012 (Ephemeroptera: Leptophlebiidae: genus *Choroterpes* Eaton, 1881) from India. *Zootaxa* 4268(3): 439–447.

Selvakumar, C., S. Sundar, and K. G. Sivaramakrishnan. 2012. Two new mayfly species (Baetidae) from India. *Oriental Insects* 46(2): 116–129.

Sivaramakrishnan, K. G. 1984. New genus and species of Leptophlebiidae: Atalophlebiinae from southern India (Ephemeroptera). *International Journal of Entomology* 26(3): 194–203.

Sivaramakrishnan, K. G. 1985a. New genus and species of Atalophlebiinae (Ephemeroptera: Leptophlebiidae) from southern India. *Annals of the Entomological Society of America* 78: 235–239.

Sivaramakrishnan, K. G. 1985b. Description of the female imago and eggs of *Indialis badia* Peters & Edmunds (Ephemeroptera: Leptophlebiidae). *Oriental Insects* 18: 95–98.

Sivaramakrishnan, K. G. 2000. A refined rapid bioassessment protocol for benthic macroinvertebrates for use in peninsular Indian streams and rivers. In *Proceedings of Lake 2000 —Symposium on Restoration of Lakes and Wetlands*, eds. T. V. Ramachandra, C. Rajasekara Murthy, and N. Ahalya, pp. 302–314. Bangalore, India: Center for Ecological Sciences, Indian Institute of Science.

Sivaramakrishnan, K. G. 2016. Systematics of the Ephemeroptera of India: Present status and future prospects. *Zoosymposia* 11: 33–52.

Sivaramakrishnan, K. G., and M. D. Hubbard. 1984. A new species of *Petersula* from southern India (Ephemeroptera: Leptophlebiidae). *International Journal of Entomology* 26(3): 204–205.

Sivaramakrishnan, K. G., and S. V. Job. 1981. Studies on mayfly populations of Courtallam streams. *Proceedings of Symposium on Ecology, Animal Population, Zoological Survey of India* Pt.2: 105–116.

Sivaramakrishnan, K. G., and W. L. Peters. 1984. Description of a new species of *Notophlebia* from India and reassignment of the ascribed nymph of *Nathanella* (Ephemeroptera: Leptophlebiidae). *Aquatic Insects* 6(2): 115–121.

Sivaramakrishnan, K. G., and K. Venkataraman. 1985. Behavioural strategies of emergence, swarming, mating and oviposition in mayflies. *Proceedings of the Indian Academy of Sciences, Animal Sciences* 94(3): 351–357.

Sivaramakrishnan, K. G., and K. Venkataraman. 1987. Observations on feeding propensities, growth rate and fecundity in mayflies (Insecta: Ephemeroptera). *Proceedings of the Indian Academy of Sciences, Animal Sciences* 96(3): 305–309.

Sivaramakrishnan, K. G., and K. Venkataraman. 1990. Abundance, altitudinal distribution and swarming of Ephemeroptera in Palani Hills, South India. In *Mayflies and Stoneflies: Life histories and biology*, ed. I. C. Campbell, pp. 209–213. Dordrecht, the Netherlands: Kluwer Academic Publishers.

Sivaramakrishnan, K. G., M. Arunachalam, and M. Raja. 2008. Impact of global warming and climate change on aquatic insects including vector species—A global perspective. In *Proceedings of the 2nd Conference on Medical Arthropodology; Climate change & Vector Borne Diseases*, pp. 1–15. Madurai, India: Centre for Research in Medical Entomology (CRME).

Sivaramakrishnan, K. G., H. J. Morgan, and R. H. Vincent. 1996a. Biological assessment of the Kaveri river catchment, South India, and using benthic macroinvertebrates: Applicability of water quality monitoring approaches developed in other countries. *International Journal of Ecology & Environmental Science* 32: 113–132.

Sivaramakrishnan, K. G., S. Sridhar, and K. Venkataraman. 1990. Habitats, microdistribution, life cycle patterns and trophic relationships of mayflies of Cardamom Hills of Western Ghats. *Hexapoda* 2: 118–121.

Sivaramakrishnan, K. G., K. A. Subramanian, and V. V. Ramamoorthy. 2009. Annotated Checklist of Ephemeroptera of Indian subregion. *Oriental Insects* 43: 315–339.

Sivaramakrishnan, K.G., K. Venkataraman, and C. Balasubramanian. 1996b. Biosystematics of the genus *Nathanella* Demoulin (Ephemeroptera: Leptophlebiidae: Atalophlebiinae) from south India. *Aquatic Insects* 18(10): 19–28.

Sivaramakrishnan, K. G., S. Janarthanan, C. Selvakumar, and M. Arumugam. 2014. Aquatic insect conservation—A molecular genetic approach. *Conservation Genetic Resources* 6: 849–855.

Sivaramakrishnan, K. G., K. Venkataraman, S. Sridhar, and S. Marimuthu. 1995. Spatial patterns of benthic macroinvertebrate distributions along Kaveri River and its tributaries in south India. *International Journal of ecology and Environmental Sciences* 21: 141–161.

Sivaramakrishnan, K. G., K. A. Subramanian, M. Arunachalam, C. Selvakumar, and S. Sundar. 2011. Emerging trends in molecular systematics and molecular phylogeny of mayflies (Insecta: Ephemeroptera). *Journal of Threatened Taxa* 3(8): 1972–1980.

Sivaruban, T., S. Barathy, M. Arunachalam, K. Venkataraman, and K. G. Sivaramakrishnan. 2013. *Epeorus petersi*, a new species of Heptageniidae (Ephemeroptera) from the Western Ghats of southern India. *Zootaxa* 3731(3): 391–394.

Sivaruban, T., S. Barathy, K. Venkataraman, and M. Arunachalam. 2010. Life cycle studies of Heptageniidae (Insecta: Ephemeroptera) in Kumbbakarai Stream of Western Ghats, Tamil Nadu, India. *Journal of Threatened Taxa* 2(10): 1223–1226.

Soldán, T. 1978. New genera and species of Caenidae (Ephemeroptera) from Iran, India and Australia. *Acta Entomologica Bohemoslovaca* 75: 119–129.

Soldán, T. 2001. Status of the systematic knowledge and priorities in Ephemeroptera studies: The oriental region. In *Trends in Research in Ephemeroptera & Plecoptera*, ed. E. Domínguez, pp. 53–65. New York: Kluwer Academic/Plenum Publishers.

Sridhar, S., and K. Venkataraman. 1989. Fecundity of mayflies of Western Ghats of Peninsular India. *Current Science* 20: 1159–1160.

Sroka, P., and T. Soldán. 2008. The Tricorythidae of the oriental region. In *International Advances in the Ecology, Zoogeography and Systematics of Mayflies and Stoneflies*, eds. F. R. Hauer, J. A. Stanford, and R. L. Newell, pp. 313–354. Oakland, CA: University of California, Publications in Entomology.

Stahls, G., and E. Savolainen. 2008. MtDNA COI barcodes reveal cryptic diversity in the *Baetis vernus* group (Ephemeroptera, Baetidae). *Molecular Phylogenetics and Evolution* 46: 82–87.

Subramanian, K. A., and K. G. Sivaramakrishnan. 2005. Habitat and microhabitat distribution of stream insect communities of the Western Ghats. *Current Science* 89(6): 976–987.

Subramanian, K. A., and K. G. Sivaramakrishnan. 2009. A new species of *Symbiocloeon* (Ephemeroptera: Baetidae) associated with a freshwater mussel species from India. *Oriental Insects* 43: 71–76.

Subramanian, K. A., K. G. Sivaramakrishnan, and M. Gadgil. 2005. Impact of riparian land use on stream insects of Khudremukh National Park, Karnataka state. India. *Journal of Insect Science* 5(49): 1–10.

Sundar, S., and M. Muralidharan. 2017. Impacts of climatic change on aquatic insects and their habitats: A global perspective with particular reference to India. *Journal of Scientific Transactions in Environment and Technovation* 10(4): 157–165.

Sweeney, B. W. 1978. Bioenergetic and developmental response of a mayfly to thermal variation. *Limnology and Oceanography* 23: 461–477.

Sweeney, B. W. 1984. Factors influencing life-history patterns of aquatic insects. In *The Ecology of Aquatic Insects*, eds. V. H. Resh and D. M. Rosenberg, pp. 56–100. New York: Praeger Scientific Publishers.

Sweeney, B. W., J. M. Battle, J. K. Jackson, and T. Dapkey. 2011. Can DNA barcodes of stream macroinvertebrates improve descriptions of community structure and water quality? *Journal of the North American Benthological Society* 30(1): 195–216.

Traver, J. R. 1935. Part II: Systematic. North American mayflies order Ephemeroptera. In *The Biology of Mayflies*, eds. J. G. Needham, J. R. Traver, and Y.-C. Hsu, pp. 237–739. Ithaca, NY: Comstock.

Traver, J. R. 1939. Himalayan mayflies (Ephemeroptera). *Annals and Magazine of Natural History* 4: 32–56.

Ulmer, G. 1920. Neue Ephemeropteren. *Archiv für Naturgeschichte* (A) 85: 1–80.

Vasanth, M., C. Selvakumar, K. A. Subramanian, R. Babu, and K. G. Sivaramakrishnan, 2019. A new record of the family Isonychidae (Inescta: Ephemeroptera) from the Western Ghats, India with description of a new species. *Zootaxa* 4586 (1): 162–170.

Venkataraman, K., and K. G. Sivaramakrishnan. 1987. A new species of *Thalerospyrus* from India (Ephemeroptera: Hepatageniidae). *Current Science* 56 (21): 1126–1129.

Venkataraman, K., and K. G. Sivaramakrishnan. 1989. A new species of *Cinygmina* (Ephemeroptera) from south India and revaluation of genetic traits of *Cinygmina* Kimmins 1937. *Hexapoda* 1(1–2): 117–121.

Walker, F. 1853. Ephemerinae. In *List of the Specimens of Neuropterous Insects in the Collection of the British Museum. Part III.—(Termitidae-Ephemeridae)*, pp. 533–585. London, UK: Printed by order of the Trustees.

Waltz, R. D., W. P. McCafferty, and A. Thomas. 1994. Systematics of *Alainites* n. gen., *Diphetor*, *Indobaetis*, *Nigrobaetis* n. stat., and *Takobia* n. stat. (Ephemeroptera, Baetidae). *Bulletin de la Société d'Histoire Naturelle, Toulouse* 130: 33–36.

Wang, T.-Q., and W. P. McCafferty. 1996. New diagnostic characters for the mayfly family Baetidae (Ephemeroptera). *Entomological News* 107(4): 207–212.

Wang, T. Q. and W. P. McCafferty. 2004. Heptageniidae (Ephemeroptera) of the world. Part I: Phylogenetic higher classification. *Transactions of the American Entomological Society* 130: 11–45.

Webb, J. M., L. M. Jacobus, D. H. Funk et al. 2012. A DNA barcode library for North American Ephemeroptera: Progress and prospects. *PLoS One* 7: e38063.

Williams, H. C., S. J. Ormerod, and M. W. Bruford. 2006. Molecular systematics and phylogeography of the cryptic species complex (Ephemeroptera, Baetidae). *Molecular Phylogenetics and Evolution* 40: 370–382.

Yeates, D., A. Seago, L. Nelson, S. L. Cameon, L. Joseph, and J. W. H. Trueman. 2011. Integrative taxonomy, or iterative taxonomy? *Systematic Entomology* 36: 209–217.

Zhou, C.-F., and L.-Y. Zheng. 2000. A new species of the genus *Neoephemera* McDunnough from China (Ephemeroptera: Neoephemeridae). *Aquatic Insects* 23: 327–332.

Zhou, X., S. J. Adamowicz, L. M. Jacobus, R. E. DeWalt, and P. D. N. Hebert. 2009. Towards a comprehensive barcode library for arctic life—Ephemeroptera, Plecoptera, and Trichoptera of Churchill, Manitoba, Canada. *Frontiers in Zoology* 6: 30.

Zhou, X., L. M. Jacobus, R. E. DeWalt, S. J. Adamowicz, and P. D. N. Hebert. 2010. Ephemeroptera, Plecoptera, and Trichoptera fauna of Churchill (Manitoba, Canada): Insights into biodiversity patterns from DNA barcoding. *Journal of North American Benthological Society* 29: 814–837.

3 Dragonflies and Damselflies (Insecta: Odonata) of India

K. A. Subramanian and R. Babu

CONTENTS

Introduction 29
Classification of Indian Odonata 30
Key to Adults 30
    Key to Suborders 30
    Suborder ZYGOPTERA 30
        Key to Superfamilies 30
        Key to Families 30
    Suborder ANISOPTERA 31
        Key to Superfamilies 31
        Key to Families 31
Key to Larvae 31
    Key to Suborders 31
        Key to Zygoptera Families 31
        Key to Anisoptera Families 32
Odonatological Studies in India 32
Biodiversity and Species Richness 33
Molecular Phylogeny 33
Taxonomic Problems 36
Biology 37
    Habitat 37
    Life Cycle 37
Conservation Status and Economic Importance 39
Collection and Preservation 40
Useful Websites 40
Acknowledgements 40
References 40

INTRODUCTION

Order Odonata, popularly known as dragonflies and damselflies, are primarily associated with wetlands and surrounding landscape. The adults are terrestrial and the larvae aquatic. They are one of the ancient group of insects. Odonata fossils are available from the Permian (250 million years Before Present (BP)) era pointing to their remote evolutionary origin. The adults are characterized by a long slender abdomen, large globular eyes, long wings with nodus and pterostigma, and a unique mechanism of sperm transfer. Sperms are produced in the gonads situated in the last abdominal segment and transferred to the secondary genitalia at the second abdominal segment before copulation. Extant Odonata, based on morphology is divided into three groups, *viz.* damselflies (Zygoptera), Anisozygoptera, and dragonflies (Anisoptera). The suborder Anisozygoptera is a living fossil with three species of which *Epiophlebia laidlawi* Tillyard is known from Eastern and Central Himalaya (Tillyard 1921).

The oldest fossils of Odonata are from Upper Carboniferous (Pennsylvanian) sediments of Europe which are about 325 million years old, belonging to the Protodonata, a basal group which is now extinct. They were fast flying with a wingspan up to 75 centimetres and spiny legs, which presumably aided in capturing their prey. Protodonata went extinct in the Triassic when dinosaurs started appearing. The Protoanisoptera (family Meganeuridae), which have been found in Kansas, USA, and Commentry, France, were very large species with a wing span of over 50 cm and lacked a nodus and pterostigma. Fossil records of larvae are found from the Mesozoic, and some workers have suggested that aquatic larval stages of Odonata started during the Lower Permian. Recently, from mid Cretaceous Burmese amber, two species of Zygoptera, i.e., *Mesosticta electronica* (Platystictidae) and *Mesomegaloprepus magnificus* (Mesomegaloprepidae), were discovered (Zheng et al. 2016; Huang et al. 2017), pointing to the ancient origin of Odonata fauna of the Indian subcontinent.

CLASSIFICATION OF INDIAN ODONATA

Order Odonata Fabricius, 1793
  Suborder **Zygoptera** Selys, 1854
    Superfamily **Lestoidea** Calvert, 1901
      I. Family **Lestidae** Calvert, 1901
      II. Family **Synlestidae** Tillyard, 1917
    Superfamily **Platystictoidea** Kennedy, 1920
      III. Family **Platystictidae** Kennedy, 1920
    Superfamily **Calopterygoidea** Selys, 1850
      IV. Family **Calopterygidae** Selys, 1850
      V. Family **Chlorocyphidae** Cowley, 1937
      VI. Family **Euphaeidae** Yakobson & Bainchi, 1905
      VII. Family **Philogangidae** Kennedy, 1920
    Superfamily **Coenagrionoidea** Kirby, 1890
      VIII. Family **Coenagrionidae** Kirby, 1890
      IX. Family **Platycnemididae** Yakobson & Bainchi, 1905
      Genera *incertae sedis*
  Suborder **Anisozygoptera** Handlirsch, 1906
    Superfamily **Epiophlebioidea** Muttkowski, 1910
      X. Family **Epiophlebiidae** Muttkowski, 1910
  Suborder **Anisoptera** Selys, 1854
    Superfamily **Aeshnoidea** Leach, 1815
      XI. Family **Aeshnidae** Leach, 1815
    Superfamily **Gomphoidea** Rambur, 1842
      XII. Family **Gomphidae** Rambur, 1842
    Superfamily **Cordulegastroidea** Needham, 1903
      XIII. Family **Chlorogomphidae** Needham, 1903
      XIV. Family **Cordulegastridae** Hagen, 1875
    Superfamily **Libelluloidea** Leach, 1815
      XV. Family **Corduliidae** Selys, 1850
      XVI. Family **Libellulidae** Leach, 1815
      XVII. Family **Macromiidae** Needham, 1903
      XVIII. Family **Synthemistidae** Tillyard, 1917

KEY TO ADULTS

KEY TO SUBORDERS

- Fore and hind wings are petiolated and more or less similar in shape, the base of the wing not markedly broad, discoidal cell four sided; wings closed over the abdomen at rest (except Lestoidea)....... ....**ZYGOPTERA**
- Fore and hind wings are petiolated and similar in shape and size; discoidal cell four sided; wings open and perpendicular to body axis at rest; frons ridged and markedly raised ....
ANISOZYGOPTERA (Family EPIOPHLEBIDAE)
- Fore and hind wings are not petiolated and different in shape, hind wing broad at base; discoidal cell divided in to hypertriangle and triangle; wings open and perpendicular to body axis at rest .... **ANISOPTERA**

SUBORDER ZYGOPTERA

Key to Superfamilies

1. Wings with only two antenodal nervures; post-nodal nervures are in strict alignment with the cross veins below 2
- Wings with more than two antenodal nervures; post-nodals are not in line with the cross veins below ....**CALOPTERYGOIDEA**
2. The veins IR_3 and R_{4+5} nearer to the node than to arc 3
- The veins IR_3 and R_{4+5} nearer to the arc than to the node ....**LESTOIDEA**
3. A basal post-costal nervure always present and situated well proximal to *ac*....................................
PLATYSTICTOIDEA (Family PLATYSTICTIDAE)
- Absence of basal post-costal nervure in all the wings .... **COENAGRIONOIDEA**

Key to Families
Superfamily LESTOIDEA

- Cu_2 strongly arched towards the costa at origin; moderately large species................**SYNLESTIDAE**
- Cu_2 without any distinct arch towards the costa at origin; rather small species .... **LESTIDAE**

Superfamily COENAGRIONOIDEA

- Discoidal cell elongate, anterior and posterior sides subequal (the costal or anterior side slightly shorter than the basal), the distal end subacute; veins MA and IR_3 almost straight except the last end.............. ....**PLATYCNEMIDIDAE**
- Discoidal cell short, anterior portion shorter than the posterior (the costal or anterior side much shorter than the basal), the distal end very acute; veins MA and IR_3 zigzagged............... **COENAGRIONIDAE**

Superfamily CALOPTERYGOIDEA

1. Wings strongly petiolated basal to level of arculus; drab coloured; wings hyaline.....**PHILOGANGIDAE**
- Wings slightly petiolated or without petiolation 2
2. Cross veins distal to the level of arculus in costal and subcostal space not in alignment; ante- and post-clypeus produced into a long upturned horn-like structure; abdomen shorter than wings; wings with opaque metallic markings in males of some species ....**CHLOROCYPHIDAE**
- Clypeus not produced into a horn-like structure; abdomen longer than wings...................................... 2
3. Several cross veins in the cubital space; head, thorax, and abdomen metallic green; wings broad and similar in size; legs slender and long........................... ....**CALOPTERYGIDAE**
- One cross vein in cubital space; head, thorax, and abdomen not metallic green; hind wings shorter than forewings; and legs short ....**EUPHAEIDAE**

Suborder ANISOPTERA

Key to Superfamilies

1. Eyes meeting only at a point or slightly separated; discoidal cells of fore and hind wings equal in size and shape, or if dissimilar, the median space traversed by one or more veins...................... .... **CORDULEGASTEROIDEA**
- Eyes widely separated or more or less broadly confluent on vertex 2
2. Eyes widely separated; discoidal cells dissimilar in fore and hind wings.. .... **GOMPHOIDEA (Family GOMPHIDAE)**
- Eyes united; discoidal cells are almost of the same size in fore and hind wings....................................... 3
3. Discoidal cells nearly of the same size and shape in fore and hind wings and situated equally distant from the arculus; cross veins distal to the level of arculus in costal and subcostal space not in alignment; two robust primary antenodals present; middle lobe of labium large and fissured... .... **AESHNOIDEA (Family AESHNIDAE)**
- Discoidal cells differing in size and shape in fore and hind wings; and situated far distal of the arculus in forewings; cross veins distal to the level of arculus in costal and subcostal space in alignment, the robust primary antenodals absent; middle lobe of labium very small, not fissured, broadly overlapped by the lateral lobes... **LIBELLULOIDEA**

Key to Families

Superfamily CORDULEGASTEROIDEA

- Basal space traversed by 1 to 5 nervures; cubital space traversed by more than 5 nervures; tibiae of male with a long membranous keel on the flexor surface .... **CHLOROGOMPHIDAE**
- Basal space not traversed; cubital space traversed by 1 or 2 nervures; tibiae of male without a keel on the flexor surface .... **CORDULEGASTERIDAE**

Superfamily LIBELLULOIDEA

1. Eyes with a small sinous projection at the middle of posterior border; tibiae of male with a long membranous keel on the flexor surface; base of hind wing of male strongly angulated or rounded; thorax metallic green or blue.. 2
- Eyes without a projection at the middle of posterior border; tibiae of male without a membranous keel on the flexor surface; base of hind wing in both sexes always rounded; thorax rarely metallic colour .... **LIBELLULIDAE**
2. Base of hind wing rounded in both sexes; wings short, broad, and pointed at apices; anal triangle absent; discoidal field beginning with three rows of cells in fore wing.. **CORDULIIDAE** (*Hemicordulia, Somatochlora*)
- Base of hind wing of male angulated; apices variable; anal triangle present or absent 3
3. Base of hind wing strongly angulated and emarginated in the male; always rounded in the female; wings long and pointed at apices; anal triangle 2 celled; discoidal field commencing with two rows of cells in fore wing... **MACROMIIDAE** (*Epophthalmia, Macromia*)
- Base of hind wing shallowly notched in the male, broadly rounded in the female; apices rounded; anal triangle 2 celled or absent; discoidal field commencing with a single row of cells in fore wing .... **SYNTHEMISTIDAE** (*Idionyx, Macromidia*)

KEY TO LARVAE

Key to Suborders

- Slender, with tapering cylindrical abdomen, end of abdomen with 2 or 3 caudal appendages or gill lamellae...**ZYGOPTERA**
- Stout, abdomen variable and widened distally; end of abdomen without caudal appendages or lamellae, but with five short spinous projections ....**ANISOPTERA**
- Elongate; abdomen without caudal appendages or lamellae; slight petiolation at the base of the wing pad; antennae very minute and short with five segments; hard body covered with tubercles and without bristles... **ANISOZYGOPTERA (Family EPIOPHLEBIDAE)**

Key to Zygoptera Families

1. First segment of antennae as long as others combined.. 2
- First segment antennae much shorter 3
2. Larvae elongate with long legs; cylindrical body; triquetral laterals longer than lamellate central gill; labium with large premental cleft and small scattered premental setae...................... **CALOPTERYGIDAE**
- Larvae stout and depressed; flat labium without premental and palpal setae; lateral gills spiniform, sharply triquetral; central gill reduced to conical process .... **CHLOROCYPHIDAE**
3. Caudal gills saccoidal 4
- Caudal gills lamellate... 5
4. Abdomen with lateral gills............. **EUPHAEIDAE**
- Abdomen without lateral gills; caudal lamellae simple, elongate–oval sacs with long filamentous tips............. .... **PLATYSTICTIDAE**
- Caudal lamellae with scale-like setae, median lamella dissimilar in size and shape from laterals .... **AMPHIPTERYGIDAE**
5. Distal margin of labium with median cleft; movable hook of labial palps with setae; labial prementum narrowed proximally; secondary trachea of lamellae at right angles to main stem .... **LESTIDAE**

- Distal margins of labium deeply cleft; movable hook of labial palps without setae, but with two robust spines...**SYNLESTIDAE**
- Distal margin of labium entire 6
6. Lamellae not thickened basally and variable in shape, sometimes filamentous, nodate or subnodate; occiput rounded laterally....................**COENAGRIONIDAE**
- Lamellae denudate, variable in shape, ovoid, subquadrate or with filamentous fringe; if thickened, proximal half is thick and distal half is membranous; occiput variable or moderately angled lateral margins ....**PLATYCNEMIDIDAE**

Key to Anisoptera Families

1. Tarsi 3-segmented, antennae 5-7 segmented............ 2
- Pro- and mesotarsus 2-segmented; antennae 4-segmented, usually flattened; body shape variable .... **GOMPHIDAE**
2. Distal margin of labial palp crenulated or serrated, movable hook shorter than breadth; labium spoon-shaped... 3
- Distal margin of labial palp smooth, movable hook longer than breadth of palp; prementum flat; body elongated with hind leg shorter than abdomen .... **AESHNIDAE**
3. Distal margin of labial palp more evenly crenulated; body dorso-ventrally flattened, short and squat; hind legs longer than abdomen.. 4
- Distal margin of labial palp with deep, irregular, serrate dentations; prementum bifid; elongated body covered with fine hairs; tarsal claw of hind leg not reaching to the tip of abdomen .... **CHLOROGOMPHIDAE**
- Body elongate covered with bristles or tufts of setae; anterior margin of the mentum clefted; large irregular interlocking teeth in palpal lobes of labium .... **CORDULEGASTRIDAE**
4. Head with extended frontal shelf, horn, or ridge..... 5
- Head without extended frontal shelf, horn, or ridge 6
5. Legs very long; wing pads parallel; abdomen wide, with middorsal spines on segments 3–9 and lateral spines on segments 7–9 ....**MACROMIIDAE**
- Legs short, wing pads divergent, abdomen spinuliform, lacking middorsal and lateral spines, apex more or less sharply pointed; body covered with fine hairs ....**SYNTHEMISTIDAE**
6. Crenulations of distal margins of palpal lobes of labium separated by shallow notches; cerci not more than one-half as long as paraprocts................. ....**LIBELLULIDAE**
- Crenulations of distal margins of palpal lobes of labium separated by deep notches; cerci more than one-half as long as paraprocts....... **CORDULIIDAE**

ODONATOLOGICAL STUDIES IN INDIA

Dragonflies are very popular in folklore and stories from time immemorial in different Indian cultures. In southern India, it is widely believed among farmers and common people that the swarms of *Pantala flavescens* are harbingers of rains. The first scientific descriptions available on odonates found in India are that of *Neurobasis chinensis* Linnaeus, *Aeshna juncea* (Linnaeus), *Libellula quadrimaculata* Linnaeus, *Orthetrum cancellatum* (Linnaeus), and *Sympetrum vulgatum* (Linnaeus). These species descriptions were based on specimens collected beyond the biogeographic boundaries of the Indian subcontinent. However, *Rhyothemis variegata* (Linnaeus) was the first dragonfly to be described scientifically based on specimens from India. During the eighteenth century, Drury (1770, 1773) and Fabricius (1775–1798) described many species from India. Numerous species were later described by Selys–Longschamps (1840–1891) and Rambur (1842).

In the early half of the twentieth century, Laidlaw (1914–1932) and Fraser (1918–1953) contributed significantly to the knowledge on Indian Odonata. Between 1918 and 1935 Fraser published a series of papers which were eventually compiled into three volumes of *Fauna of British India–Odonata* (Fraser 1933, 1934, 1936). Post 1950s, Asahina (1960–1995), Bhasin (1953), Baijal (1955), Singh (1955), Singh and Baijal (1954), Singh et al. (1955), and Sahni (1964, 1965a, 1965b) studied Indian Odonata, especially of the Himalaya. Several new species described by Singh, Baijal, and Sahni from the Himalaya were later synonymized with widespread common species by Mitra (1973, 1992), Ram and Srivastava (1984), and Hämäläinen (1989). For example, *Prodasineura autumnalis* (Fraser, 1922) (= *Caconeura autumnalis gaudawricus* Sahni, 1964 syn.); *Ischnura aurora* (Brauer, 1865) (= *Ischnura bhimtalensis* Sahni, 1965 syn.); *Ischnura forcipata* Morton, 1907 (= *Agriocnemis nainitalensis* Sahni, 1964 syn.); *Coeliccia renifera* (Selys, 1886) (= *Coeliccia kumaonensis* Singh and Baijal, 1954 syn. and = *Calicnemia maheshi* Sahni, 1964 syn.); *Copera marginipes* (Rambur 1842) (= *Disparoneura bhatnagri* Sahni, 1965 syn.); *Indolestes cyaneus* (Selys, 1862) (= *Archibasis sushmae* Baijal, 1955 syn. and = *Lestes manaliensis* Singh, 1955 syn.); *Anisopleura lestoides* Selys, 1853 (= *Anisopleura kusumi* Sahni, 1965 syn.); *Onychogomphus bistrigatus* (Hagen in Selys, 1854) (= *Onychogomphus garhwalicus* Singh and Baijal, 1954 syn.); *Orthetrum taeniolatum* (Schneider, 1845) (= *Orthetrum garhwalicum* Singh and Baijal, 1954 syn.); and *Pantala flavescens* (Fabricius, 1798) (= *Orthetrum mathewi* Singh and Baijal, 1955 syn. and = *Sympetrum tandicola* Singh, 1955 syn.).

Zoological Survey of India surveyed and enlisted Odonata of many conservation areas, wetlands, and states (Kumar 1997, 2000, 2005; Kumar and Prasad 1981; Lahiri 1976, 1985, 1987, 2003; Lahiri et al. 2007; Mitra 1994, 1995, 2002a, 2002b, 2003a, 2003b, 2006a, 2006b; Mitra and Lahiri 1975; Mitra and Babu 2010; Mitra et al. 2006, 2013; Prasad 1996a, 1996b, 2004a, 2004b, 2007a, 2007b, 2007c; Prasad and Varshney 1988, 1995; Prasad and Kulkarni 2001, 2002; Prasad and Mishra 2008, 2009; Prasad and Mondal 2010; Prasad and Sinha 2010; Srivastava and Das 1987; Srivastava and Sinha 1993, 1995, 2000, 2004; Ram et al. 2000; Rao and Lahiri 1983; Kulkarni and Prasad 2002, 2005; Kulkarni et al. 2004, 2006a, 2006b, 2012; Mishra 2007; Emiliyamma and Radhakrishnan 2000, 2003, 2006, 2007a, 2007b; Emiliyamma and Palot 2016; Emiliyamma et al. 2007; Babu 2014; Babu and Mehta 2009; Babu and Srinivasan 2016; Babu et al. 2009; Nair and Subramanian 2014; Palot and Soniya 2000, 2004; Palot et al. 2005; Sharma 2010, 2013, 2015; Sharma and Kumar 2008; Nandy and Babu 2012; Rajeshkumar et al. 2017; Subramanian, 2007, 2009, 2012; Subramanian and Babu 2017a, 2018; Subramanian et al. 2013, 2018; Walia et al. 2016). Recently, researchers from other organizations *viz.,* D. B. Tembhare, Nagpur University, Nagpur; Raymond Andrew, Hislop College, Nagpur; B. K. Tyagi, Centre for Research in Medical Entomology (CRME), Madurai; B. Suri Babu, Regional Forensic Science Laboratory, Jagdalpur, Chhattisgarh; Gurinder K. Walia, Punjabi University, Patiala; Ashish Tiple, Vidhya Bharati College, Wardha; David Raju, Singinawa Jungle Lodges, Kanha National Park; Parag Rangnaker, Mrugaya, Goa; Pankaj Khoparde, Salim Ali Centre for Ornithology and Natural History (SACON), Coimbatore; Krushnamegh Kunte, National Centre for Biological Sciences (NCBS), Bangalore; and Shantanu Joshi, NCBS, Bangalore, also contributed to the knowledge on Indian Odonata (Subramanian and Babu 2017a, 2017b; Subramanian et al. 2018).

BIODIVERSITY AND SPECIES RICHNESS

Globally, 6383 species in 693 genera of odonates are known. In India, 493 species and 27 subspecies in 154 genera and 18 families are known (Schorr and Paulson 2019; Subramanian and Babu 2017b). Anisozygoptera with three relict species was earlier recognized as a third suborder of Odonata. However, recent studies group Anisozygoptera with Anisoptera, or some authors bring them together under a new name Epiprocta (Anisoptera + Anisozygoptera) (Trueman 1996, 2007; Lohmann 1996; Rehn 2003; Kalkman et al. 2008). In India, the suborder Zygoptera comprises of 213 species under 59 genera and 9 families; one species of Anisozygoptera under one genera and one family; 279 species of Anisoptera under 94 genera and 8 families (Subramanian and Babu 2017b). The checklist contains 58 nomenclatural changes, 30 new additions, 10 deletions, and 7 species which are under doubtful status. An overview of extant global and Indian Odonata diversity is provided in Table 3.1, and representative images of families are provided in Figures 3.1 through 3.12.

In India, high diversity and endemism are restricted to southern Western Ghats, Eastern Himalaya, Western Himalaya, and Andaman and Nicobar Islands. Western Ghats and Eastern Himalaya have 196 and 256 species, respectively (Subramanian et al. 2018; Subramanian and Babu 2018). High diversity is found in hill streams and forested riverine habitats and most of the endemic species are restricted to this habitat. Habitats like ponds, lakes, coastal marshes, irrigation canals, and paddy fields have common and wide spread species.

One hundred and ninety-five taxa, including subspecies belonging to 69 genera are endemic to India (Babu et al. 2013; Kiran et al. 2015; Emilyamma and Palot 2016; Joshi and Kunte 2017; Rajeshkumar et al. 2017; Rajeshkumar and Raghunathan 2018; Joshi and Sawant 2019; Babu and Subramanian 2019; Parag et al. 2019). Highest endemism is found in Western Ghats, especially in mountains south of Coorg in Karnataka. Here, the streams and rivers of Coorg, Wayanad, Nilgiris, Anamalais, Cardamom Hills, and Agasthyamalai are rich in endemic species (Subramanian 2007; Subramanian et al. 2018). High endemism is found in the family Gomphidae and genera such as *Protosticta* (Platystictidae), *Macromia* (Macromiidae), and *Idionyx* (Synthemistidae). In the Eastern Himalaya, high endemism is found in the Khasi Hills and Darjeeling–Sikkim Himalaya. Species of Platycnemididae and Gomphidae are highly diversified here with many endemics. The distribution of endemic species across different regions of India is provided in Table 3.2.

MOLECULAR PHYLOGENY

Recently, global level molecular phylogeny of damselflies (Zygoptera) and dragonflies (Anisoptera) has been carried out (Dijkstra et al. 2014; Carle et al. 2015). Phylogenetic relationships of 310 genera of all Zygoptera families except Hemiphlebiidae were investigated using mitochondrial (16S, COI) and nuclear (28S) data (Dijkstra et al. 2014). Carle et al. (2015) reconstructed the phylogeny of 184 Anisoptera genera using mitochondrial COI and COII and portions of the nuclear protein coding genes EF-1α and Histone H3. In Zygoptera, many traditional families were found to be monophyletic, and superfamily Coenagrionoidea is organized into three families *viz.,* Isostictidae, Platycnemididae, and Coenagrionidae. The study also proposes several new combinations at global level. Phylogeny of 11 families of Anisoptera reveal following lineages: (Austropetaliidae, Aeshnidae), (Gomphidae, Petaluridae), and [(Chlorogomphidae, (Neopetaliidae, Cordulegastridae)]. The taxonomic positions of Petaluridae, Chlorogomphidae, Neopetaliidae, and Cordulegastridae are equivocal.

TABLE 3.1
Diversity of Odonata in the World and in India

Suborder	Family	World		India	
		Genera	Species	Genera	Species
Zygoptera	Hemiphlebiidae	1	1	0	0
	Perilestidae	2	19	0	0
	Synlestidae	9	38	1	6
	Lestidae	9	153	5	25
	Platystictidae	10	269	3	15
	Amphipterygidae	1	5	0	0
	Argiolestidae	20	117	0	0
	Calopterygidae	21	184	6	9
	Chlorocyphidae	20	162	8	22
	Devadattidae	1	13	0	0
	Dicteriadidae	2	2	0	0
	Euphaeidae	9	77	6	19
	Heteragrionidae	2	58	0	0
	Hypolestidae	1	3	0	0
	Lestoideidae	2	9	0	0
	Megapodagrioniidae	3	29	0	0
	Pentaphlebiidae	1	3	0	0
	Philogangidae	1	4	1	1
	Philogeniidae	2	41	0	0
	Philosinidae	2	12	0	0
	Polythoridae	7	60	0	0
	Pseudolestidae	1	1	0	0
	Rimanellidae	1	1	0	0
	Thaumatoneuridae	2	5	0	0
	Isostictidae	12	47	0	0
	Platycnemididae	42	471	16	54
	Coenagrionidae	120	1356	12	61
	Genera *incertae sedis*	22	71	1	1
Anisozygoptera	Epiophlebiidae	1	3	1	1
Anisoptera	Austropetaliidae	4	11	0	0
	Aeshnidae	54	487	13	49
	Petaluridae	5	11	0	0
	Gomphidae	102	1015	29	88
	Chlorogomphidae	3	53	3	8
	Cordulegastridae	3	51	3	9
	Neopetaliidae	1	1	0	0
	Synthemistidae	28	149	2	15
	Macromiidae	4	124	2	17
	Corduliidae	22	166	2	2
	Libellulidae	140	1041	40	91
Total		**693**	**6323**	**154**	**493**

FIGURES 3.1–3.6 Dragonfly and damselfly adults: **1**. *Anax immaculifrons* (Aeshnidae); **2**. *Ictinogomphus rapax* (Gomphidae); **3**. *Orthetrum chrysis* (Libellulidae); **4**. *Epophthalmia vittata* (Macromiidae); **5**. *Macromedia donaldi* (Synthemistidae); and **6**. *Neurobasis chinensis* (Calopterygidae).

FIGURES 3.7–3.12 Damselfly adults: **7.** *Heliocypha bisignata* (Chlorocyphidae); **8.** *Agriocnemis pygmaea* (Coenagrionidae); **9.** *Euphaea fraseri* (Euphaeidae); **10.** *Indolestes davenporti* (Lestidae); **11.** *Disparoneura quadrimaculata* (Platycnemididae); and **12.** *Protosticta gravelyi* (Platystictidae).

TAXONOMIC PROBLEMS

Detailed taxonomic descriptions of adults of Indian Odonata are available (Fraser 1933, 1934, 1936; Lahiri 1987; Mitra 2002). However, many regions of Eastern Himalaya, Western Himalaya, Central Indian highlands, Eastern Ghats, and Andaman and Nicobar Islands are poorly explored. Several species such as *Indolestes assamicus* Fraser, *Lestes patricia* Fraser, *Orolestes durga* Lahiri (Lestidae); *Megalestes micans* Needham (Synlestidae); *Drepanosticta polychromatica* Fraser, *Protosticta rufostigma* Kimmins (Platystictidae); *Bayadera kali* Cowley (Euphaeidae); *Burmagomphus cauvericus* Fraser, *Gomphidia fletcheri* Fraser, *Gomphidia platyceps* Fraser, *Heliogomphus kalarensis* Fraser, *Onychogomphus malabarensis* (Fraser), *Nychogomphus striatus* (Fraser) (Gomphidae); *Gynacantha biharica* Fraser (Aeshnidae); and *Idionyx nadganiensis* Fraser and *Idionyx periyashola* Fraser (Synthemistidae) are known only from type specimens. Further, information gaps exist on description of larval stages and their ecology. Larval stages of only about 15 percent of Indian species are known, and full life history is worked out only for less than 10 percent of the species. A good understanding of larval ecology is crucial to assess wetland health. This paucity of ecological information is a serious lacuna when designing any biomonitoring tool. Future studies on dragonflies must be directed to enable a comprehensive understanding of their distribution ecology including that of larval stages and their value as a conservation tool.

TABLE 3.2
Distribution of Endemic Species of Odonata in India

Superfamily	Family	Eastern Himalaya	Western Himalaya	Western Ghats	Andaman & Nicobar	Peninsular India (Excluding Western Ghats)
Coenagrionoidea	Coenagrionidae	4	3	7	1	5
	Platycnemididae	10	1	14	3	7
Platystictoidea	Platystictidae	2	1	10	1	–
Lestoidea	Lestidae	4	–	3	–	1
	Synlestidae	2	–	–	–	–
Calopterygoidea	Calopterygidae	2	–	2	–	1
	Chlorocyphidae	3	–	3	3	1
	Euphaeidae	6	–	4	–	1
	Genera *incertae sedis*	1	–	–	–	–
Aeshnoidea	Aeshnidae	11	1	–	2	3
Gomphoidea	Gomphidae	17	4	29	2	7
Cordulegastroidea	Chlorogomphidae	3	–	3	–	–
	Cordulegastridae	–	1	–	–	–
	Corduliidae	1	–	–	–	–
Libelluloidea	Macromiidae	2	1	8	–	2
	Synthemistidae	3	–	10	–	–
	Libellulidae	3	2	2	–	–

BIOLOGY

HABITAT

Odonates use a wide range of flowing and stagnant water bodies, and their life history is closely associated with a particular wetland habitat. Though species are highly specific for a particular habitat, some have adapted to urbanization and use man-made water bodies. This habitat specificity has an important bearing on the distribution and ecology of odonates. Species using restricted habitats like hill streams tend to be narrowly distributed when compared to pool breeders, which are widespread.

LIFE CYCLE

Eggs: Odonates lay their eggs in a wide range of aquatic habitats from damp soil to waterfalls. Females select the egg-laying site mainly by physical characters such as length of shoreline, aquatic vegetation and bottom substrate. Species breeding in rivers prefer long straight shores (Corbet 1999). It is observed that long straight shores of lakes tend to be colonized by riverine species. Visual cues also play an important part in oviposition. It has been observed that many pool breeders are deceived by smooth shining surfaces, such as bonnets of cars and wet roads, and they try to lay their eggs in these unnatural sites. Some species like *Lyriothemis tricolor* Ris lay eggs and nymphs develop in phytotelmata, and *Tetrathemis platyptera* Selys lay eggs on dry twigs overhanging water bodies. The larvae emerge with the first rain and fall off to the underlying water body.

Dragonflies lay broad and elliptical eggs either in flight or by perching on an overhanging vegetation or rock. Damselflies with their elaborate serrated ovipositor insert elongate, cylindrical eggs into a plant body. Eggs are laid in successive batches: a damselfly lays about 100–400 eggs and dragonflies, usually about several hundreds to thousands per batch. Eggs hatch immediately in the tropics, usually in 5–40 days. In many stream dwelling dragonflies, the eggs are covered with a gelatinous substance which expands and becomes adhesive on contact with water. This helps the egg from being carried away far from its habitat by the water current.

Small parasitic Hymenoptera belonging to the families Eulophidae, Mymaridae, Trichogrammatidae, and Scelionidae parasitise eggs of odonates. Parasitizing females climb or swim beneath the water to search for the eggs in the submerged plants and lay their eggs (Corbet 1999). Water mites (Hydrachnida) of families Arrenuridae, Hydyphantidae, and Linocharidae are known to be ectoparasites of adult odonates (Corbet 1999). For example, *Arrenurus cupidator* is a common ectoparasite of coenagrionid damselflies. The mite larvae seek the final instar host larvae by random tactile search. The larvae briefly feed on host larva, and when the adult damselfly emerges, the mite larvae get attached to the adult. The mite larva pierces the host body and starts feeding. The larvae detach only when the host comes back to water for oviposition.

The detached larvae complete two more larval stages as predator before moulting into an adult.

Larval stages: Larvae are aquatic predators and feed on other invertebrates including odonate larvae, tadpoles, and small fish. Their cryptic colouration and keen eyesight makes them efficient ambush predators. They are voracious feeders and feed on any moving and any size prey including their own kind. Last instar larvae of bigger species are known to catch small fishes, tadpoles, and even freshly emerged adults of their own species (Corbet 1999).

In dragonflies, the inner surface of the rectum is foliate and richly supplied by trachae. These foliations or "rectal gills" are the respiratory organs. Pumping movements of the abdomen continually renew water in the rectum. In damselflies, there are foliaceous lamellae at the end of the abdomen. They are the supplementary respiratory devices in addition to rectum, general body surface, and wing sheaths where also gaseous exchange occurs.

A larva completes its development in 2 months. The number of larval instars is very variable within and between species and is usually 9–15. When they are ready to moult, they stop feeding and crawl up to emergent vegetation or rock. This usually happens after sunset, and the larvae moult into adults just before sunrise. The newly emerged adults are wet and delicate, and as the day warms up, they become dry and robust for their maiden flight. Species of the tropics and warm temperate latitudes complete one or more generations per year.

Adult stage: Newly emerged males and females leave their emergence site and occupy nearby landscapes. Generally, males travel farther than females. In a few species, the maturation period serves as a resting stage and lasts about 8–9 weeks. Damselflies complete their maturation period in about 3 weeks or less, whereas dragonflies take 2 weeks. During the maturation period, sequential changes occur in the colour of the body and wings.

Flight: Odonates surpass all other groups of insects in their flying skills. Odonates have uncoupled wings unlike moths, butterflies, wasps, and bees and fore and hind wings beat independently. The powerful thoracic muscles help them in long sustained flight and good manoeuvrability. Odonates can hover and turn 180° while in flight. Dragonflies are stronger fliers than damselflies, and they can reach a speed up to 25–30 km per hour. This difference in flying abilities influences their dispersal and geographic distribution. Big powerful fliers have wider geographic range than small weak fliers.

Migration: Like many other organisms, dragonflies also migrate. Common Indian species such as *Anax guttatus* (Burmeister), *Hemianax ephippiger* (Burmeister), *Pantala flavescens* (Fabricius), *Tholymis tillarga* (Fabricius), *Tramea basilaris* (Beauvois), *Tramea limbata* (Desjardins), and *Diplacodes trivialis* (Rambur) are known to migrate long distances. Among these, *Pantala flavescens* migrates immediately after the monsoon. Large swarms of these dragonflies move through prominent clearings in the landscapes such as highways and railway tracks. Recent studies suggest that *Pantala flavescens* migrate across the Arabian Sea to reach east African shores using winds of Intertropical Convergence Zone (ITCZ) covering nearly 3500 km (Anderson 2009). Stable isotope studies indicate that the migrating population of *Pantala flavescens* originates in northern India, especially from the Gangetic plains (Hobson et al. 2012).

Feeding: Adult dragonflies are aerial predators and catch small insects like mosquitoes, midges, small butterflies, moths, and bees on wing. Most of the dragonflies are day flying, but a few actively hunt during twilight hours. Dragonflies capture their prey by perching at a vantage point and making short sallying flights or by flying continuously. Large numbers of adults sometime congregate especially during dawn and dusk near tree canopies to feed on swarming insects. They feed in flight, using the legs to capture the prey and transfer it to the jaws. The legs are highly specialized for this purpose, particularly with regard to its position, relative length, articulation, and complement of spines. Their vision is well developed as in butterflies and as far as dragonflies are concerned, the whole head is an eye.

Reproduction: Sexually mature dragonflies return to the breeding habitat from their foraging or roosting site. Usually males mature earlier than females and reach the breeding habitat first. Mature males hold territory, but species may or may not show pronounced site fidelity. Resident males show aggressive behaviour towards conspecific males, which enter their territory. Aggressive behaviour may be simple "wing warning" by perched males and a display of abdomen. More elaborate aggressive encounters occur in flight, progressing from mutual threat display to physical fighting.

Odonates are sexually dimorphic. Newly emerged males and females are similarly coloured. Males acquire bright colouration as they become reproductively mature. Colours and patterns in the wings and body may play an important role in territoriality and courtship. Courtship is more evident in damselflies than in dragonflies. It ranges from simple submissive posture by males towards approaching females to elaborate displays where the male flies towards an egg laying site and allows itself to be carried by the water current for a short distance. Competition over sexually receptive females is very intense among male odonates.

A receptive female adopts a characteristic posture towards a potential male and pairing follows immediately. The last abdominal segments of the male have claspers, which are used to hold the female by the posterior side of head in the case of dragonflies and by prothorax in damselflies. During copulation or just before that, the male transfers his sperm into an accessory genital organ at the second abdominal segment. This has a complicated harpoon shaped structure, which removes sperm of previously mated male before insemination. Multiple mating in both males and females are common among odonates.

Egg laying: Egg laying commences immediately after copulation. The male continues to hold the female and flies with her to an egg-laying site or just accompanies her. It is usually observed that territory holding males accompany females and non-territory holding males maintain physical contact with the female while laying egg. Usually during this period, the female is very vulnerable to the attack by other males. Non-mated males attack the mated pair and try to hijack the female. Some damselflies lay eggs in submerged plants. In such cases, the hovering male anchors the egg-laying female.

Longevity: Most of the records of longevity in nature refer only to reproductive period. During this, most damselflies live up to 8 weeks and dragonflies up to 6 weeks. If the maturation period is included, it may extend up to 7–9 and 8–10 weeks, respectively. Dragonflies encounter a large number of predators throughout their life. Fishes are important predators during the larval stage. Birds such as falcons, hawks, bee eaters, kingfishers, herons, and terns have been observed to feed on odonates. Large dragonflies, robber flies (Asilidae), and spiders are important invertebrate predators.

CONSERVATION STATUS AND ECONOMIC IMPORTANCE

Odonata larvae live in freshwater habitats and only a few species can tolerate brackish waters. They are highly specific to particular aquatic habitat, and utilize both running and standing waters for breeding. This habitat specificity makes them an ideal model system to address questions in ecology, evolutionary biology, biogeography, and for monitoring health of freshwater ecosystems. Odonates have a significant role in the wetland food chain. Adult odonates feed on mosquitoes, blackflies, and other blood-sucking flies and act as an important biocontrol agent of these harmful insects. Recent studies documented 61 species of odonates from agro ecosystems of India (Subramanian et al. in press). Species such as *Orthetrum sabina*, *Orthetrum pruinosum*, *Neurothemis tullia*, *Crocothemis servilia*, *Lestes elatus*, *Coenagrion coramandalianum*, *Ischnura aurora*, and *Agriocnemis pygmaea* are commonly found in agro ecosystems and play a crucial role in controlling pest populations.

In addition to the role of odonates in ecosystem function, their value as indicators of quality of the biotope is now being increasingly recognized. Recent studies have shown how species assemblages of dragonflies change with levels of human disturbance. Dragonflies found at undisturbed habitats with good riparian vegetation were specialists with narrow distribution. On the other hand, species recorded at industrial land or urban areas with disturbed riparian vegetation were generalists with wide habitat preference and distribution (Kulkarni and Subramanian 2013; Koparde et al. 2015). These studies also show that dragonflies are sensitive not only to the quality of the wetland, but also to the major landscape changes, especially changes in the riparian zone.

Subramanian et al. (2018) studied the geographic distribution of 193 odonate species of the Western Ghats using GIS tools. Individual species distribution maps were prepared and areas of high diversity and endemism were mapped at a grid size of 0.125° (~14 km × 14 km). The study reported high diversity and endemism from the hills south of 13°N and also found that the highest diversity and endemism is at Nilgiri–Wayanad–Kodagu complex and Anaimalai Hills. In the Western Ghats, between 9°N and 13°N, there are nearly 130 species with 40 endemics. Further north, the diversity and endemism drastically drops. The work also identified geographic gap areas of the region for facilitating future surveys.

Recent studies conducted in Eastern Himalaya and Peninsular India (Mitra et al. 2010; Subramanian et al. 2011) demonstrate that odonate fauna of the subcontinent is threatened due to anthropogenic activities such as habitat destruction, agricultural expansion, and pesticide and industrial pollution. Current IUCN Red list Assessment (2016) categorizes 2 endangered, 15 vulnerable, and 8 near threatened species from India (Table 3.3).

TABLE 3.3
Threatened Species of Indian Odonata

Sl. No.	Species	IUCN Status
1.	*Idionyx galeata*	Endangered
2.	*Libellago balus*	Endangered
3.	*Anormogomphus kiritshenkoi*	Near threatened
4.	*Asiagomphus personatus*	Near threatened
5.	*Elattoneura atkinsoni*	Near threatened
6.	*Epiophlebia laidlawi*	Near threatened
7.	*Heliogomphus promelas*	Near threatened
8.	*Idionyx optata*	Near threatened
9.	*Indocypha vittata*	Near threatened
10.	*Indolestes indicus*	Near threatened
11.	*Indothemis carnatica*	Near threatened
12.	*Megalogomphus hannyngtoni*	Near threatened
13.	*Melanoneura bilineata*	Near threatened
14.	*Merogomphus martini*	Near threatened
15.	*Neallogaster ornata*	Near threatened
16.	*Phylloneura westermanni*	Near threatened
17.	*Planaeschna intersedens*	Near threatened
18.	*Bayadera hyalina*	Vulnerable
19.	*Chlorogomphus xanthoptera*	Vulnerable
20.	*Chloropetalia selysi*	Vulnerable
21.	*Coeliccia fraseri*	Vulnerable
22.	*Disparoneura apicalis*	Vulnerable
23.	*Libellago andamanensis*	Vulnerable
24.	*Platysticta deccanensis*	Vulnerable
25.	*Protosticta sanguinostigma*	Vulnerable

COLLECTION AND PRESERVATION

Adult Odonata can be collected using an aerial or butterfly net. Collected individuals should be kept alive for a day to remove faecal matter. They can be killed using ethyl acetate vapour. Killed specimens can be stretched in an insect stretching board and dried. However, dry specimens are very fragile, and body parts such as head, legs, and abdomen break off easily. Hence, for better dry preservation, large specimens need to be defatted by immersing in acetone for 5–8 hours and the acetone drained off. The specimen must be dried at room temperature and stored in paper packets with label. Specimens intended for molecular studies can be directly put into 50% ethyl alcohol and later transferred to absolute alcohol. Dry specimens after identification and labelling can be stored in insect cabinets with preservatives.

Larvae can be collected from rivers, streams, ponds, lakes, etc. using a kick net, D-Frame net, pond net, and by hand picking from aquatic vegetation. Collected larvae can be directly transferred to vials with 50% ethyl alcohol and later to absolute alcohol. Field labels written with alcohol proof ink with date, name of collector, locality, geo-coordinates, and habitat is to be placed with each specimen.

To study the larval mouth parts, the labium is dissected out in a watch glass and kept in 10% KOH overnight in room temperature. The cleared sample is thoroughly washed in distilled water and mounted in Hoyer's medium under a microscope.

USEFUL WEBSITES

1. https://www.pugetsound.edu/academics/academic-resources/slater-museum/biodiversity-resources/dragonflies/world-odonata-list2/
2. http://zsi.gov.in/WriteReadData/userfiles/file/Checklist/Odonata%20V3.pdf
3. www.indianodonata.org
4. https://indiabiodiversity.org/group/dragonflies_of_india/userGroup/show
5. http://www.allodonata.com
6. www.redlist.org
7. www.faunaofindia.nic.in

ACKNOWLEDGEMENTS

We are thankful to Dr. Kailash Chandra, Director, Zoological Survey of India, Kolkata and Prof. K. G. Sivaramakrishnan for critical inputs, encouragement, and support for the preparation of this manuscript.

REFERENCES

Anderson, R. 2009. Do dragonflies migrate across the western Indian Ocean? *Journal of Tropical Ecology* 25(4): 347–358. doi:10.1017/S0266467409006087.

Asahina, S. 1960. Notes on the relationship between the Himalayan and Japanese insect fauna. *Journal of Bengal Natural History Society* 31: 69–75.

Asahina, S. 1961a. Description of some dragonfly larvae from Darjeeling. *Kontyû* 29: 240–246.

Asahina, S. 1961b. Is *Epiophlebia laidlawi* Tillyard (Odonata: Anisozygoptera) a good species? *Internationale Revue der gesamten Hydrobiologie* 46: 441–446.

Asahina, S. 1961c. Taxonomic characteristics of Himalayan *Epiophlebia* larvae (Insecta: Odonata). *Proceedings of the Japan Academy Series B* 37: 42.

Asahina, S. 1962. *Anax nigrofasciatus* Oguma and *Anax nigrolineatus* Fraser (Odonata: Anisoptera). *Japanese Journal of Zoology* 13: 249–255.

Asahina, S. 1963a. Notes on three Indian dragonfly species. *Akitu* 11: 21–22.

Asahina, S. 1963b. Description of the possible adult dragonfly, *Epiophlebia laidlawi* from the Himalayas. *Tombo* 6: 17–20.

Asahina, S. 1967. A revision of the Asiatic species of the damselflies of the genus *Ceriagrion* (Odonata: Agrionidae). *Japanese Journal of Zoology* 15: 255–334.

Asahina, S. 1970. Burmese material collected by Dr. Arthur Svihla with supplementary notes on Asiatic *Ceriagrion* species. *Japanese Journal of Zoology* 16: 99–126.

Asahina, S. 1978. A new and some known species of Odonata from Kashmir (Insecta). *Senckenbergiana Biologica* 59(1/2): 115–120.

Asahina, S. 1981. A revision of the Himalayan dragonflies of the genus *Cephalaeschna* and its allies (Odonata: Aeshnidae). *Bulletin of the National Science Museum* (A) 7(2): 57–77.

Asahina, S. 1984a. Assamese and Burmese *Coeliccia* species in the collection of Ehrich Schmidt (Odonata: Platycnemididae). *Transactions of the Shikoku Entomological Society* 16: 1–9.

Asahina, S. 1984b. *Gynacantha arnaudi* sp. nov. and enigmatic *Gynacantha* from Assam (Odonata: Aeshnidae). *Chô chô* 7: 2–8.

Asahina, S. 1985a. Further contribution to the taxonomy of South Asiatic *Coeliccia* species (Odonata: Platycnemididae). *Chô chô* 8(2): 2–13.

Asahina, S. 1985b. Contributions to the taxonomic knowledge of the *Megalestes* species of Continental South Asia (Odonata: Synlestidae). *Chô chô* 8(10): 2–18.

Asahina, S. 1986. Revisional notes on Nepalese and Assamese dragonfly species of the genus *Chlorogomphus* (Odonata: Cordulegasteridae). *Chô chô* 9(1): 11–26.

Asahina, S. 1994. Records of Gomphid Dragonflies recently collected by Japanese Entomologists from Nepal and Darjeeling District. *Tombo* 37(1–4): 2–17.

Asahina, S. 1995. Records of Gomphid dragonflies recently collected by Japanese entomologists from Nepal and Darjeeling District, Part II. *Tombo* 38(1–4): 2–18.

Babu, R. 2014. Odonata. In *Faunal Diversity of Churdhar Wildlife Sanctuary, Himachal Pradesh, Conservation Area Series* No. 53, ed. Director, pp. 7–21. Kolkata, India: Zoological Survey of India.

Babu, R., and H. S. Mehta. 2009. Insecta: Odonata. In *Faunal Diversity of Simbalbara Wildlife Sanctuary, Conservation Area Series* No. 41, ed. Director, pp. 21–28. Kolkata, India: Zoological Survey of India.

Babu, R., and G. Srinivasan. 2016. Distribution of *Aeshna petalura* Martin, 1908 (Odonata: Anisoptera: Aeshnidae) in Indian Subcontinent. *Journal of Threatened Taxa* 8(7): 9034–9037.

Babu, R., H. S. Mehta, and S. Kamal. 2009. Insecta: Odonata. In *Faunal Diversity of Pong Dam, Wetland Ecosystem Series* No. 12, ed. Director, pp. 13–19. Kolkata, India: Zoological Survey of India.

Babu, R., and K. A. Subramanian. 2019. A new species of *Gomphidia* Selys, 1854 (Insecta: Odonata: Anisoptera: Gomphidae) from the Western Ghats of India. *Zootaxa* 4652: 155–164. doi: 10.11646/zootaxa.4652.1.9.

Babu, R., K. A. Subramanian, and S. Nandy. 2013. Endemic odonates of India. *Records of the Zoological Survey of India, Occasional Paper* No. 347: 1–60.

Baijal, H. N. 1955. Entomological Survey of the Himalayas part IV. Notes on some insects collected by the second entomological expedition to the North West Himalayas (1955) with description of three new species of Odonata. *Agra University Journal of Research (Science)* suppl. 4: 741–766.

Bhasin, G. D. 1953. Odonata. *Indian Forest Leafletter* No. 121(3): 63–69.

Carle, F. L., K. M. Kjer, and M. L. May. 2015. A molecular phylogeny and classification of Anisoptera (Odonata). *Arthropod Systematics & Phylogeny* 73(2): 281–301.

Corbet, P. S. 1999. *Dragonflies: Behavior and Ecology of Odonata*. Ithaca, NY: Cornell University Press.

de Selys Longchamps, E. 1840. *Monographie des Libellulidees D'Europe*. Paris, France: Librairie Encyclopedique de Roret.

de Selys Longchamps, E. 1853. Synopsis des Calopterygines. *Bulletin Academie Royale de Belgique, Classe des Sciences* (2) 1, 20: 1–73.

de Selys Longchamps, E. 1854a. Monographie des Calopterygines. *Memoires Societe Royale des Sciences de Liege* 9: 291 (with Hagen).

de Selys Longchamps, E. 1854b. Synopsis des Gomphines. *Bulletin Academie Royale de Belgique* (II) 21: 23–113.

de Selys Longchamps, E. 1859. Addition au Synopsis des Calopterygines. *Bulletin Academie Royale de Belgique, Classe des Sciences* (2) 7: 437–451.

de Selys Longchamps, E. 1860. Synopsis des Agrionines. Derniere Legion: *Protoneura*. *Bulletin Academie Royale de Belgique, Classe des Sciences* (2) 10: 431–462.

de Selys Longchamps, E. 1862. Synopsis des Agrionines. Seconde Legion: *Lestes*. *Bulletin Academie Royale de Belgique, Classe des Sciences* (2) 13: 288–338.

de Selys Longchamps, E. 1863. Synopsis des Agrionines. Quetrieme Legion: *Platycnemis*. *Bulletin Academie Royale de Belgique, Classe des Sciences* (2) 16: 150–176.

de Selys Longchamps, E. 1865. Synopsis des Agrionines. 5me Legion: *Agrion*. *Bulletin Academie Royale de Belgique, Classe des Sciences* (2) 20: 375–417.

de Selys Longchamps, E. 1869. Odonates recueillis a Madagascar, et. auxiles Mascareignes ec Comoros, Researches sur la Faune de Madagascar et de ses dependencea d après les decouvertes de F.P.L. Pollen et D.C. van dam ^{5}me parte, p. 24.

de Selys Longchamps, E. 1876. Synopsis des agrionines (suite). Le grand genre Agrion. *Bulletin Academie Royale de Belgique* (II) 41: 247–322, 496–539, 1233–1309; 42: 490–531, 952–989.

de Selys Longchamps, E. 1877. Synopsis des Agrionises. ^{5}me Legion, *Agrion* (suite et fin) les generes *Telebasis*, *Agriocnemis*, et *Hemiphlebia*. Bruxelles, 1–65.

de Selys Longchamps, E. 1883. Synopsis des Aeschnines. *Bulletin Academie Royale de Belgique, Classe des Sciences* (3) 5: 712–748.

de Selys Longchamps, E. 1891. Odonates in 'Viaggio' Di Leonardo Fea in Birmania e Regional Viccine, *Annali del Museo Civico di Storia Naturale "Giacomo Doria"* Serie 2, 10: 433–518.

Dijkstra, K.-D. B., V. J. Kalkman, R. A. Dow, F. R. Stokvis, and J. van Tol. 2014. Redefining the damselfly families: A comprehensive molecular phylogeny of Zygoptera (Odonata). *Systematic Entomology* 39: 68–96.

Drury, D. 1770. *Illustrations of Natural History–Exotic Insects* 1: 112–115.

Drury, D. 1773. Illustrations *of Natural History–Exotic Insects* 2: 84–85.

Emiliyamma, K. G., and C. Radhakrishnan. 2000. Odonata (Insecta) of Parambikulam Wildlife Sanctuary, Kerala, India. *Records of the Zoological Survey of India* 98(1): 157–167.

Emiliyamma, K. G., and C. Radhakrishnan. 2003. Odonata (Insecta) of Indira Gandhi Wildlife Sanctuary and National Park, Tamil Nadu. *Zoo's Print Journal* 18(11): 1264–1266.

Emiliyamma, K. G., and C. Radhakrishnan. 2006. Insecta: Odonata. In *Fauna of Biligiri Rangaswamy Wildlife Sanctuary, Conservation Area Series* No. 27, ed. Director, pp. 21–25. Kolkata, India: Zoological Survey of India.

Emiliyamma, K. G., and C. Radhakrishnan. 2007a. Insecta: Odonata. In *Fauna of Kudremukh National Park, Conservation Area Series* No. 32, ed. Director, pp. 27–48. Kolkata, India: Zoological Survey of India.

Emiliyamma, K. G., and C. Radhakrishnan. 2007b. Insecta: Odonata. In *Fauna of Bannerghatta National Park, Conservation Area Series* No. 33, ed. Director, pp. 39–41. Kolkata, India: Zoological Survey of India.

Emiliyamma, K. G., and M. J. Palot. 2016. A new species of *Protosticta* Selys, 1885 (Odonata: Zygoptera: Platystictidae) from Western Ghats, Kerala, India. *Journal of Threatened Taxa* 8(14): 9648–9652.

Emiliyamma, K. G., C. Radhakrishnan, and M. J. Palot. 2007. Odonata (Insecta) of Kerala. *Records of the Zoological Survey of India, Occasional Paper* No. 269: 1–195.

Fabricius, J. C. 1775. *Entomologica Systematica* 420–426.

Fabricius, J. C. 1787. Odonata. *Mantissa Insectorum* 1787: 336–340.

Fabricius, J. C. 1793. Libellula. *Entomologica Systematica* 2: 373–383.

Fabricius, J. C. 1798. Libellula. *Entomologica Systematica* Suppl. 283–286.

Fraser, F. C. 1918a. Indian dragonflies, part I. *Journal of the Bombay Natural History Society* 25: 454–471.

Fraser, F. C. 1918b. Indian dragonflies, part II. *Journal of the Bombay Natural History Society* 25: 608–627.

Fraser, F. C. 1918c. Indian dragonflies, part III. *Journal of the Bombay Natural History Society* 26: 141–171.

Fraser, F. C. 1919a. Indian dragonflies, part IV. *Journal of the Bombay Natural History Society* 26: 488–517.

Fraser, F. C. 1919b. Indian dragonflies, part V. *Journal of the Bombay Natural History Society* 26: 734–744.

Fraser, F. C. 1919c. Descriptions of four new Indian Odonata. *Records of the Indian Museum* 16(7): 451–455.

Fraser, F. C. 1920a. Indian dragonflies, part VI. *Journal of the Bombay Natural History Society* 26: 919–932.

Fraser, F. C. 1920b. Indian dragonflies, part VII. *Journal of the Bombay Natural History Society* 27: 48–56.

Fraser, F. C. 1920c. Indian dragonflies, part VIII. *Journal of the Bombay Natural History Society* 27: 253–269.

Fraser, F. C. 1921a. Indian dragonflies, part IX. *Journal of the Bombay Natural History Society* 27: 492–498.

Fraser, F. C. 1921b. Indian dragonflies, part X. *Journal of the Bombay Natural History Society* 27: 673–691.

Fraser, F. C. 1921c. Indian dragonflies, part XI. *Journal of the Bombay Natural History Society* 28: 107–122.

Fraser, F. C. 1922a. Indian dragonflies, part XII. *Journal of the Bombay Natural History Society* 28: 481–492.

Fraser, F. C. 1922b. Indian dragonflies, part XIII. *Journal of the Bombay Natural History Society* 28: 610–620.

Fraser, F. C. 1922c. Indian dragonflies, part XIV. *Journal of the Bombay Natural History Society* 28: 899–910.

Fraser, F. C. 1922d. New and rare Odonata from Nilgiri Hills. *Records of the Indian Museum* 24(1): 1–9.

Fraser, F. C. 1922e. New and rare Indian Odonata in the Pusa Collection. *Memoirs of the Department of Agriculture in India (Entomology)* 7(7): 39–77.

Fraser, F. C. 1923a. Indian dragonflies, part XV. *Journal of the Bombay Natural History Society* 29: 36–47.

Fraser, F. C. 1923b. Indian dragonflies, part XVI. *Journal of the Bombay Natural History Society* 29: 324–333.

Fraser, F. C. 1923c. Indian dragonflies, part XVII. *Journal of the Bombay Natural History Society* 29: 659–680.

Fraser, F. C. 1924. A survey of the Odonata fauna of western India with special remarks on the genera *Macromia* and *Idionyx* and descriptions of thirty new species, with Appendix I and II. *Records of the Indian Museum* 26(5): 423–522.

Fraser, F. C. 1926. A revision of the genus *Idionyx* Selys. *Records of the Indian Museum* 28(3): 195–207.

Fraser, F. C. 1931. Additions to the survey of the Odonata (dragonfly) fauna of Western India, with descriptions of nine new species. *Records of the Indian Museum* 33: 443–474.

Fraser, F. C. 1933. *Fauna of British India, Including Ceylon and Burma, Odonata,* Vol. I. London: Taylor & Francis Group.

Fraser, F. C. 1934. *Fauna of British India, Including Ceylon and Burma, Odonata,* Vol. II. London: Taylor & Francis Group.

Fraser, F. C. 1936. *Fauna of British India, Including Ceylon and Burma. Odonata,* Vol. III. London: Taylor & Francis Group.

Fraser, F. C. 1938. Two new species of Oriental Odonata. *Proceedings of the Royal Entomological Society of London* (B) 7(10): 197–198.

Fraser, F. C. 1939. Additions to the family Corduliidae including descriptions of two new species and a new genus (Order: Odonata). *Proceedings of the Royal Entomological Society of London* (B) 8: 91–94.

Fraser, F. C. 1946. *Hylaeothemis indica,* a new species of oriental libelluline (Order Odonata). *Proceedings of the Royal Entomological Society of London* (B) 15(7/8): 97–100.

Fraser, F. C. 1953. Notes on the family Gomphidae with descriptions of a new species and the females of another (Order: Odonata). *Proceedings of the Royal Entomological Society of London* (B) 22: 189–194.

Huang, D., C. Cai, and A. Nel. 2017. A new Burmese amber hawker dragonfly helps to redefine the position of the aeshnopteran family Burmaeshnidae (Odonata: Anisoptera: Aeshnoidea). *Cretaceous Research* 79: 153–158. doi:10.1016/j.cretres.2017.07.020.

Hämäläinen, M. 1989. Odonata from the Dehra Dun Valley (Uttar Pradesh, India) with notes on synonymy of some West Himalayan species. *Odonatologica* 18(1): 13–20.

Hobson, K. A., R. C. Anderson, D. X. Soto, and L. I. Wassenaar. 2012. Isotopic evidence that dragonflies (*Pantala flavescens*) migrating through the Maldives come from the Northern Indian subcontinent. *PLoS ONE* 7(12): e52594. doi:10.1371/journal.pone.0052594.

International Union for Conservation of Nature (IUCN) www.redlist.org. Version 2016-3 (accessed May 3, 2017).

Joshi, S., and K. Kunte. 2017. Two new dragonfly species (Odonata: Anisoptera: Aeshnidae) from north-eastern India. *Zootaxa* 4300(2): 259–268.

Kalkman, V. J., V. Clausnitzer, K.-D. B. Dijkstra, A. G. Orr, D. R. Paulson, and J. van Tol. 2008. Global diversity of dragonflies (Odonata) in freshwater. *Hydrobiologia* 595: 351–363.

Kiran, C. G., S. Kalesh, and K. Kunte. 2015. A new species of damselfly, *Protosticta ponmudiensis* (Odonata: Zygoptera: Platystictidae) from Ponmudi Hills in the Western Ghats of India. *Journal of Threatened Taxa* 7(5): 7146–7151.

Koparde, P., P. Mhaske, and A. Patwardhan. 2015. Habitat correlates of Odonata species diversity in the northern Western Ghats, India. *Odonatologica* 44(1/2): 21–43.

Kumar, A. 1997. Odonata: Imagos. In *Fauna of Delhi, State Fauna Series* No. 6, ed. Director, pp. 147–159. Kolkata, India: Zoological Survey of India.

Kumar, A. 2000. Odonata. In *Fauna of Renuka Wetland (Western Himalaya: Himachal Pradesh), Wetland Ecosystem Series* No. 2, ed. Director, pp. 45–53. Kolkata, India: Zoological Survey of India.

Kumar, A. 2005. Odonata. In *Fauna of Western Himalaya, Part–2 (Himachal Pradesh),* ed. Director, pp. 75–98. Kolkata, India: Zoological Survey of India.

Kumar, A., and M. Prasad. 1981. Field ecology, zoogeography and taxonomy of the Odonata of Western Himalaya, India. *Records of the Zoological Survey of India, Occasional Paper* No. 20: 1–118.

Kulkarni. A. S., and K. A. Subramanian. 2013. Habitat and seasonal distribution of Odonata (Insecta) of Mula and Mutha river basins, Maharashtra, India. *Journal of Threatened Taxa* 5(7): 4084–4095.

Kulkarni, P. P., and M. Prasad. 2002. Insecta: Odonata. In *Fauna of Ujani, Wetland Ecosystem Series* No. 3, ed. Director, pp. 91–104. Kolkata, India: Zoological Survey of India.

Kulkarni, P. P., and M. Prasad. 2005. Insecta: Odonata. In *Fauna of Melghat Tiger Reserve, Conservation Area Series* No. 24, ed. Director, pp. 297–316. Kolkata, India: Zoological Survey of India.

Kulkarni, P. P., M. Prasad, and S. S. Talmale. 2004. Insecta: Odonata. In *Fauna of Pench National Park, Conservation Area Series* No. 20, ed. Director, pp. 175–205. Kolkata, India: Zoological Survey of India.

Kulkarni, P. P., S. S. Talmale, and M. Prasad. 2006a. Insecta: Odonata. In *Fauna of Tadoba Andhari Tiger Reserve, Conservation Area Series* No. 25, ed. Director, pp. 197–226. Kolkata, India: Zoological Survey of India.

Kulkarni, P. P., S. S. Talmale, and M. Prasad. 2006b. Insecta: Odonata. In *Fauna of Sanjay Gandhi National Park (Invertebrates), Conservation Area Series* No. 26, ed. Director, pp. 19–40. Kolkata, India: Zoological Survey of India.

Kulkarni, P. P., R. Babu, S. S. Talmale, C. Sinha, and S. B. Mondal. 2012. Insecta: Odonata. In *Fauna of Maharashtra, State Fauna Series* No. 20 (Part–2), ed. Director, pp. 397–428. Kolkata, India: Zoological Survey of India.

Lahiri, A. R. 1976. *Calicnemia mukherjeei* spec. nov. from Khasi Hills, India (Zygoptera: Platycnemididae). *Odonatologica* 5: 273–276.

Lahiri, A. R. 1985. Odonata Fauna of Namdhapa, Arunachal Pradesh. *Records of the Zoological Survey of India* 82: 61–67.

Lahiri, A. R. 1987. Studies on the Odonata fauna of Meghalaya. *Records of the Zoological Survey of India, Occasional Paper* No. 99: 1–402.

Lahiri, A. R. 2003. On a new species of the genus *Bayadera* Selys (Odonata: Euphaeidae) from India with notes on its Indian representative. *Records of the Zoological Survey of India* 101(3–4): 39–42.

Lahiri, A. R., R. Sandhu, and G. K. Walia. 2007. *Gynacantha palampurica* sp. nov. from Northern Himachal Pradesh, India (Odonata: Aeshnidae). *Records of the Zoological Survey of India* 107(Part–3): 45–49.

Laidlaw, F. F. 1914. Zoological results of the Abor Expedition 1911–1912; XXV. Odonata. *Records of the Indian Museum* 8: 335–349.

Laidlaw, F. F. 1915. Notes on Oriental dragonflies in the Indian Museum No. 3. *Records of the Indian Museum* 11: 387–391.

Laidlaw, F. F. 1916a. Notes on Oriental dragonflies in the Indian Museum. No. 4. The genus *Pseudagrion*. *Records of the Indian Museum* 12: 21–25.

Laidlaw, F. F. 1916b. Notes on Indian Odonata. *Records of the Indian Museum* 12: 129–136.

Laidlaw, F. F. 1917a. A list of the dragonflies recorded from the Indian Empire with special reference to the collection of the Indian Museum Part–I. The family Calopterygidae. *Records of the Indian Museum* 13: 23–40.

Laidlaw, F. F. 1917b. A list of the dragonflies recorded from the Indian Empire with special reference to the collection of the Indian Museum Part–I. The family Agrioninae. *Records of the Indian Museum* 13: 321–348.

Laidlaw, F. F. 1919. A list of the dragonflies recorded from the Indian Empire with special reference to the collection of the Indian Museum Part–II. *Records of the Indian Museum* 16: 169–195.

Laidlaw, F. F. 1920. Description of a new species of the genus *Pseudophaea* (=*Euphaea*, Selys) from Western India with some remarks on the section *dispar* of the genus. *Records of the Indian Museum* 19: 23–27.

Laidlaw, F. F. 1922. A list of the dragonflies recorded from the Indian Empire with special reference to the collection of the Indian Museum Part–V. The subfamily Gomphinae. *Records of the Indian Museum* 24: 367–426.

Laidlaw, F. F. 1924. Notes on Oriental dragonflies of the genus *Aciagrion*. *Proceedings of the United States National Museum* 66(Art 10): 1–9.

Laidlaw, F. F. 1932. A revision of the genus *Coeliccia* (Order Odonata). *Records of the Indian Museum* 34: 7–42.

Linnaeus, C. 1758. *Systema Naturae. Regnum Animale*. Editio decima. 1. Holmiae, Sweden: Laurentii Salvii.

Linnaeus, C. 1763. *Amoenitates Academicae*. Holmiae, Sweden: Laurentii Salvii.

Lohmann, H. 1996. Das phylogenetische System der Anisoptera (Odonata). *Entomologische Zeitschrift* 106: 209–252.

Mishra, S. K. 2007. Insecta: Odonata. In *Fauna of Madhya Pradesh (Including Chhattisgarh), State Fauna Series* No. 15(Part–1), ed. Director, pp. 245–272. Kolkata, India: Zoological Survey of India.

Mitra, A., R. Dow, K. A. Subramanian, and G. Sharma. 2010. The status and distribution of dragonflies and damselflies (Odonata) of the Eastern Himalaya. In *Status and Distribution of Freshwater Biodiversity in the Eastern Himalaya*, comp. D. J. Allen, S. Molur, and B. A. Daniel, pp. 54–66. Cambridge, UK and Gland, Switzerland: IUCN, and Coimbatore, India: Zoo Outreach Organization.

Mitra, T. R. 1973. *Sympetrum tandicola* Singh, 1955, a synonym of *Pantala flavescens* (Fabr.) (Odonata, Libellulidae). *The Entomologist's Record and Journal of Variation* 85: 30–31.

Mitra, T. R. 1992. Note on taxonomic status of five Indian Odonata. *Journal of the Bengal Natural History Society* (NS) 11: 82–85.

Mitra, T. R. 1994. Observations on the habits and habitats of adult dragonflies of Eastern India with special reference to the fauna of West Bengal. *Records of the Zoological Survey of India, Occasional Paper* No. 166: 1–40.

Mitra, T. R. 1995. Odonata. In *Fauna of Indravati Tiger Reserve, Conservation Area Series* No. 6, ed. Director, pp. 31–44. Kolkata, India: Zoological Survey of India.

Mitra, T. R. 2002a. Endemic Odonata of India. *Records of the Zoological Survey of India* 100(3–4): 189–199.

Mitra, T. R. 2002b. Geographical distribution of Odonata (Insecta) of Eastern India. *Memoirs of the Zoological Survey of India* 19(9): 1–208.

Mitra, T. R. 2003a. Odonata. In *Fauna of Sikkim, State Fauna Series* No. 9 (Part–2), ed. Director, pp. 125–164. Kolkata, India: Zoological Survey of India.

Mitra, T. R. 2003b. Ecology and biogeography of Odonata with special reference to Indian fauna. *Records of the Zoological Survey of India, Occasional Paper* No. 202: 1–41.

Mitra, T. R. 2006a. *Handbook of Common Indian Dragonflies (Insecta: Odonata)*. Kolkata, India: Published by Director, Zoological Survey of India.

Mitra, T. R. 2006b. Insecta: Odonata. In *Fauna of Arunachal Pradesh, State Fauna Series* No. 13 (Part–2), ed. Director, pp. 67–149. Kolkata, India: Zoological Survey of India.

Mitra, T. R., and A. R. Lahiri. 1975. A new species of *Gynacantha* Rambur, 1842 (Odonata: Aeshnidae) from India. *The Entomologist's Record and Journal of Variation* 87: 148–149.

Mitra, T. R., and R. Babu. 2010. Revision of Indian species of the families Platycnemididae and Coenagrionidae (Insecta: Odonata: Zygoptera)–Taxonomy and Zoogeography. *Records of the Zoological Survey of India, Occasional Paper* No. 315: 1–103.

Mitra, T. R., M. Prasad, and C. Sinha. 2006. Odonata. In *Fauna of Nagaland, State Fauna Series* No. 12, ed. Director, pp. 75–87. Kolkata, India: Zoological Survey of India.

Mitra, T. R., R. Babu, and K. A. Subramanian. 2013. *Anax panybeus* Hagen, 1867: An addition to the Odonata (Aeshnidae) of India. *Journal of Threatened Taxa* 5(2): 3682–3683.

Nair, M. V., and K. A. Subramanian. 2014. A new species of *Agriocnemis* Selys, 1869 (Zygoptera: Coenagrionidae) from eastern India with redescription of *Agriocnemis keralensis* Peter, 1981. *Records of the Zoological Survey of India* 114(4): 669–679.

Nandy, S., and R. Babu. 2012. Insecta: Odonata. In *Fauna of Andaman and Nicobar Islands, State Fauna Series* No. 19(1), ed. Director, pp. 33–68. Kolkata, India: Zoological Survey of India.

Palot, M. J., and V. P. Soniya. 2000. Odonata from Courtallam, Tamil Nadu, Southern India. *Zoo's Print Journal* 15(7): 301–303.

Palot, M. J., and V. P. Soniya. 2004. Studies on the Odonata (Insecta) from a backwater swamp on Northern Kerala. *Journal of the Bombay Natural History Society* 101(1): 177–179.

Palot, M. J., C. Radhakrishnan, and V. P. Soniya. 2005. Odonata (Insecta) diversity of rice field habitat in Palakkad district, Kerala. *Records of the Zoological Survey of India* 104(Part 1–2): 71–77.

Parag, R., D. Omkar, S. Kalesh, and K. A. Subramanian. 2019. A new species of *Cyclogomphus* Selys, 1854 (Insecta: Odonata: Gomphidae) from the Western Ghats, India with comments on the status of *Cyclogomphus vesiculosus* Selys, 1873. *Zootaxa*. 4656: 515–524. doi:10.11646/zootaxa.4656.3.8.

Prasad, M. 1996a. An account of the Odonata of Maharashtra State, India. *Records of the Zoological Survey of India* 95: 305–327.

Prasad, M. 1996b. Studies on the Odonata fauna of Bastar, Madhya Pradesh, India. *Records of the Zoological Survey of India* 95: 385–428.

Prasad, M. 2004a. Insecta: Odonata. In *Fauna of Gujarat, State Fauna Series* No. 8(Part–3), ed. Director, pp. 19–40. Kolkata, India: Zoological Survey of India.

Prasad, M. 2004b. Insecta: Odonata. In *Fauna of Desert National Park, Conservation Area Series* No. 19, ed. Director, pp. 51–58. Kolkata, India: Zoological Survey of India.

Prasad, M. 2007a. Insecta: Odonata. In *Fauna of Mizoram, State Fauna Series* No. 14, ed. Director, pp. 143–186. Kolkata, India: Zoological Survey of India.

Prasad, M. 2007b. Insecta: Odonata. In *Fauna of Andhra Pradesh, State Fauna Series* No. 5(Part–3), ed. Director, pp. 115–181. Kolkata, India: Zoological Survey of India.

Prasad, M. 2007c. Insecta: Odonata. In *Fauna of Pichola Lake, Wetland Ecosystem Series* No. 8, ed. Director, pp. 79–83. Kolkata, India: Zoological Survey of India.

Prasad, M., and P. P. Kulkarni. 2001. Insecta: Odonata. In *Fauna of Nilgiri Biosphere Reserve, Conservation Area Series* No. 11, ed. Director, pp. 73–83. Kolkata, India: Zoological Survey of India.

Prasad, M., and P. P. Kulkarni. 2002. Insecta: Odonata. In *Fauna of Eravikulam National Park, Conservation Area Series* No. 13, ed. Director, pp. 7–9. Kolkata, India: Zoological Survey of India.

Prasad, M., and S. K. Mishra. 2008. Insecta: Odonata. In *Faunal Diversity of Jabalpur District (Madhya Pradesh)*, ed. Director, pp. 77–92. Kolkata, India: Zoological Survey of India.

Prasad, M., and S. K. Mishra. 2009. Insecta: Odonata. In *Fauna of Pachmarhi Biosphere Reserve, Conservation Area Series* No. 39, ed. Director, pp. 203–212. Kolkata, India: Zoological Survey of India.

Prasad, M., and S. B. Mondal. 2010. Insecta: Odonata: Zygoptera. In *Fauna of Uttarakhand, State Fauna Series* No. 18(Part–2), ed. Director, pp. 17–28. Kolkata, India: Zoological Survey of India.

Prasad, M., and C. Sinha. 2010. Insecta: Odonata: Anisoptera. In *Fauna of Uttarakhand, State Fauna Series* No. 18(Part–2), ed. Director, pp. 29–52. Kolkata, India: Zoological Survey of India.

Prasad, M., and R. K. Varshney. 1988. The Odonata of Bihar, India. *Records of the Zoological Survey of India, Occasional Paper* No. 110: 1–47.

Prasad, M., and R. K. Varshney. 1995. A check–list of the Odonata of India including data on larval studies. *Oriental Insects* 29: 385–428.

Rajeshkumar, S., C. Raghunathan, and K. Chandra. 2017. *Nososticta nicobarica* sp. nov. (Odonata: Platycnemididae: Disparoneurinae) from Great Nicobar Island, India. *Zootaxa* 4311(3): 426–434.

Rajeshkumar, S., and C. Ragunathan. 2018. Description of a new species of *Nososticta* Hagen (Odonata: Platycnemididae: Disparoneurinae) from Central Nicobar Islands, India. *Zootaxa* 4422(3): 431–441. doi:10.11646/zootaxa.4422.3.9.

Ram, R., and V. D. Srivastava. 1984. *Orthetrum mathewi* Singh and Baijal 1954, a synonym of *Pantala flavescens* (Fabr.) (Odonata: Libellulidae). *Bulletin of the Zoological Survey of India* 5: 181.

Ram, R., K. Chandra, and K. Yadav. 2000. Studies on the Odonata fauna of Andaman and Nicobar Islands. *Records of the Zoological Survey of India* 98(3): 25–60.

Rao, K. R., and A. R. Lahiri. 1983. First records of Odonata (Arthropoda: Insecta) from Silent Valley and New Amarambalam Reserved Forests. *Journal of the Bombay Natural History Society* 79: 557–562.

Rambur, M. P. 1842. *Histoire Naturelle des Insects. Névropteres.* Paris, France: Librairie Encyclopedique de Roret.

Rehn, A. C. 2003. Phylogenetic analysis of higher–level relationships of Odonata. *Systematic Entomology* 28: 181–239.

Sahni, D. N. 1964. Some new species of Odonata from Kumaon Hills (India). *Agra University Journal of Research (Science)* 13(3): 79–86.

Sahni, D. N. 1965a. Studies on the Odonata (Zygoptera) of Naini Tal. *Indian Journal of Entomology* 27: 205–216.

Sahni, D. N. 1965b. Studies on the Odonata (Ansioptera) of Naini Tal. *Indian Journal of Entomology* 27: 277–289.

Schorr, M., and D. Paulson. 2019. World Odonata List. https://www.pugetsound.edu/academics/academicresources/slater-museum/biodiversity-resources/dragonflies/worldodonata-list2/ (accessed 18 September, 2019).

Sharma, G. 2010. Insecta: Odonata. In *Fauna of Ranthambhore National Park, Rajasthan, Conservation Area Series* No. 43, ed. Director, pp. 67–74. Kolkata, India: Zoological Survey of India.

Sharma, G. 2013. Insecta: Odonata. In *Fauna of Kumbhalgargh Wildlife Sanctuary, Rajasthan, Conservation Area Series* No. 47, ed. Director, pp. 31–42. Kolkata, India: Zoological Survey of India.

Sharma, G. 2015. *Pictorial Handbook on Damselflies and Dragonflies (Odonata: Insecta) of Rajasthan.* Kolkata, India: Published by Director, Zoological Survey of India.

Sharma, G., and A. Kumar. 2008. Odonata diversity of Punjab Shivalik with their habitats and flight period. *Fraseria* (N.S.) 7: 29–33.

Singh, S. 1955. Entomological Survey of the Himalayas Part V–on two new species of Odonata. *Agra University Journal of Research (Science)* 4: 171–174.

Singh, S., and H. N. Baijal. 1954. Entomological Survey of the Himalayas II–on a collection of Odonata. *Agra University Journal of Research (Science)* 3: 385–400.

Singh, S., H. N. Baijal, V. K. Gupta, and K. Mathew. 1955. Entomological survey of the Himalayas, Part-XIV. Notes on some insects collected by the second entomological expedition to the North–West Himalayas (1955) with description of three new species of Odonata. *Agra University Journal of Research (Science)* 4: 741–766.

Srivastava, V. D., and S. Das. 1987. Insecta: Odonata. In *Fauna of Orissa, State Fauna Series* No. 1(Part–4), ed. Director, pp. 135–159. Kolkata, India: Zoological Survey of India.

Srivastava, V. D., and C. Sinha. 1993. Odonata. In *Fauna of West Bengal, State Fauna Series* No. 3(Part–4), ed. Director, pp. 51–168. Kolkata, India: Zoological Survey of India.

Srivastava, V. D., and C. Sinha. 1995. Insecta: Odonata. In *Fauna of Meghalaya, State Fauna Series* No. 4(Part–3), ed. Director, pp. 33–154. Kolkata, India: Zoological Survey of India.

Srivastava, V. D., and C. Sinha. 2000. Odonata. In *Fauna of Tripura, State Fauna Series* No. 7(Part–2), ed. Director, pp. 155–196. Kolkata, India: Zoological Survey of India.

Srivastava, V. D., and C. Sinha. 2004. Insecta: Odonata. In *Fauna of Manipur, State Fauna Series* No. 10(Part–2), ed. Director, pp. 75–110. Kolkata, India: Zoological Survey of India.

Subramanian, K. A. 2007. Endemic odonates of the Western Ghats: Habitat distribution and Conservation. In *Odonata–Biology of Dragonflies,* ed. B.K. Tyagi, pp. 257–271. Jodhpur, India: Scientific Publishers.

Subramanian, K. A. 2009. *Dragonflies of India–A field Guide.* New Delhi, India: Vigyan Prasar, Department of Science and Technology, Government of India.

Subramanian, K. A. 2012. Foraging and breeding behaviour of Peninsular Indian Odonata. In *Dynamics of Insect Behaviour,* ed. T. N. Ananthakrishnan and K. G. Sivaramakrishnan, pp. 158–171. India: Scientific Publishers.

Subramanian, K. A., and R. Babu. 2017a. Insecta: Odonata (damselflies and dragonflies). In *Current Status of Freshwater Faunal Diversity in India,* eds. K. Chandra, K.C. Gopi, D.V. Rao, K. Valarmathi, and J. R. B. Alfred, 401–418. Kolkata, India: Zoological Survey of India.

Subramanian, K. A., and R. Babu. 2017b. A checklist of Odonata (Insecta) of India, Version 3.0. pp. 1–51. www.zsi.gov.in.

Subramanian, K. A., and R. Babu. 2018. Insecta: Odonata. In *Faunal Diversity of Indian Himalaya,* eds. K. Chandra, D. Gupta, K. C. Gopi, B. Tripathy, and V. Kumar, 227– 240. Kolkata, India: Zoological Survey of India.

Subramanian, K. A., F. Kakkassery, and M. V. Nair. 2011. The status and distribution of dragonflies and damselflies (Odonata) of the Western Ghats. In *Status and Distribution of Freshwater Biodiversity in the Western Ghats,* comp. S. Molur, K. G. Smith, B. A. Daniel, and W. R. T. Darwall, 63–74. Cambridge, UK and Gland, Switzerland: IUCN, and Coimbatore, India: Zoo Outreach Organization.

Subramanian, K. A., P. Rangnekar, and R. Naik. 2013. *Idionyx* (Odonata: Corduliidae) of the Western Ghats with a description of a new species. *Zootaxa* 3652(2): 277–288.

Subramanian, K. A., K. G. Emiliyamma, R. Babu, C. Radhakrishnan, and S. S. Talmale. 2018. *Atlas of Odonata (Insecta) of the Western Ghats, India.* Kolkata: Published by Director, Zoological Survey of India.

Subramanian, K. A., R. Babu, M. Vasanth, C. Selvakumar, and K. G. Sivaramakrishnan. in press. Ephemeroptera (Mayflies) and Odonata (Dragonflies and Damselflies) of Agro ecosystems of India. In *Fauna of Agro Ecosystems of India.*

Tillyard, R. J. 1921. On an anisozygopterous larva from the Himalayas (Order Odonata). *Records of the Indian Museum* 22: 93–107.

Trueman, J. W. H. 1996. A preliminary cladistic analysis of odonate wing venation. *Odonatologica* 25: 59–72.

Trueman, J. W. H. 2007. A brief history of the classification and nomenclature of Odonata. *Zootaxa* 1668: 381–394.

Walia, G. K., S. S. Chahal, and R. Babu. 2016. Cytogenetic report on *Gynacanthaeschna sikkima* from India (Odonata: Aeshnidae). *Odonatologica* 45(1/2): 87–94.

Zheng, D., B. Wang, E. A. Jarzembowski, C. Su–Chin, and A. Nel. 2016. The first fossil Perilestidae (Odonata: Zygoptera) from mid–Cretaceous Burmese amber. *Cretaceous Research* 65: 199–205. doi:10.1016/j.cretres.2016.05.002.

4 Stoneflies (Insecta: Plecoptera) of India

R. Babu, K. G. Sivaramakrishnan, and K. A. Subramanian

CONTENTS

Introduction 47
Historical Review 48
Diversity and Classification 48
Plecoptera of India 48
Superfamily Nemouroidea 49
    Family Capniidae Banks, 1900 49
    Family Leuctridae Klapálek, 1905 49
    Family Nemouridae Newman, 1853 50
    Family Taeniopterygidae Klapálek, 1905 50
Superfamily Perloidea 50
    Family Chloroperlidae Okamoto, 1912 50
    Family Perlidae Latreille, 1802 50
    Family Perlodidae Klapálek, 1909 50
Superfamily Pteronarcyoidea 50
    Family Peltoperlidae Classen, 1931 50
Diagnostic Keys 51
    Key to Superfamilies in India 51
    Key to Families in India (nymphs) 51
    Key to Families in India (adults) 52
Phylogeny and Biogeography 52
Taxonomic Problems 52
Biology and Life Cycle 52
Biomonitoring Potential and Climate Change Impacts 53
Economic Importance 53
Conservation 53
Collection and Preservation 53
Conclusion 53
Acknowledgements 54
References 54

INTRODUCTION

The order Plecoptera, commonly called as stoneflies, is a small monophyletic Blattoid-Orthopteroid (Polyneoptera) order of hemimetabolous insects, distributed over all continents except Antarctica and constitute a significant ecological component of running water ecosystems (Fochetti and Tierno de Figueroa 2008). The order Plecoptera was erected by Burmeister (1839), "*plecto*" means "folded" and "*pteron*" means "wing". The pleated hind wings are folded under the front wings at rest. The terrestrial adults have long setaceous antennae; well developed compound eyes and two or three ocelli; non-functional biting type mouth parts; most species with two pairs of membranous wings folded over their back at rest, long and narrow forewings, slightly shorter hind wings, but wider than forewings, and the anal lobe of hind wings pleated which helps to fold beneath the forewings; wings are rudimentary (brachypterous) or absent (apterous) in a few species; each leg with two claws; tarsi three-segmented; abdomen terminates in a pair of long multi-segmented cerci or reduced (unsegmented) in some species. Nymphs are aquatic and are similar to adults, but without well developed wings and genitalia. They are characterized by flattened body; mandibulate mouth parts; well developed compound eyes and two or three ocelli; widely separated legs; three segmented tarsi;

distinctive paired claws in each leg; 10-segmented abdomen with apical segment ending in a pair of multi-segmented long cerci. Depending on the species, tracheal gills may be present or absent on the thorax, base of the legs, and sides and tip of the abdomen.

Stonefly larvae are generally found in high altitude cold temperate streams, though some genera have penetrated to the cooler areas of subtropics and tropics. The larvae are found on the surface and under the boulders, leaf litter, and decaying vegetation in hill streams. Nymphs of smaller species feed on a wide variety of small plants, algae, or decayed vegetation or wood, and larger species are predators feeding on a wide range of aquatic invertebrates, but in early instars they are herbivorous or detritivorous and become predatory in later instars. The adults are usually diurnal except for a few nocturnal species of superfamily Perloidea and are found near the streams or on tree trunks, stones, and bushes. Stoneflies are not conspicuous insects except the family Chloroperlidae which are bright green in colour. Adults are herbivores and feed on algae, lichens, pollen, rotting wood, etc. However, many species may not feed at all as adults and live only for a few weeks. They are weak fliers and prefer to run to elude predators. During winter season, brachypterous or wingless forms are usually found (Babu et al. 2017). The composition of stonefly fauna varies in different seasons and in different habitats.

HISTORICAL REVIEW

Historically, higher classifications of Plecoptera were proposed by different authors (Ricker 1952; Illies 1965; Zwick 1973). Currently, the classification scheme proposed by Zwick (2000) is widely accepted and followed in this paper. Order Plecoptera has received scanty attention in India. Several scientists, Klapálek (1909, 1916), Needham (1909), Banks (1914, 1920, 1939), Navas (1922), Kimmins (1946, 1950a, 1950b), Ricker (1952), Jewett (1958, 1960, 1970, 1975), Aubert (1967), Kawai (1968), Singh and Ghosh (1969), Harper (1974, 1977), Zwick (1981, 1982a, 1982b), Zwick and Sivec (1980), Stark (1989), Stark and Sivec (1991, 2007, 2008, 2014, 2015), Zwick et al. (2007), Muranyi and Li (2013, 2016), Muranyi et al. (2015), and Mason and Stark (2015) have contributed to the knowledge on Indian Plecoptera. Sivaramakrishnan et al. (2011) and recently Babu et al. (2017) have made an overview of the current status of this group in India.

DIVERSITY AND CLASSIFICATION

Plecoptera is a basal aquatic order of the lower Neoptera with 16 extant families in two suborders Antarctoperlaria, exclusively present in the Southern Hemisphere, and Arctoperlaria predominantly distributed in the Northern Hemisphere and a few in the Southern Hemisphere (Zwick 2000; De Walt et al. 2015). Currently, Antarctoperlaria includes two superfamilies Eusthenioidea (Diamphipnoidae and Eustheniidae) and Gripopterygoidea (Austroperlidae and Gripopterygidae) with two families each. The suborder Arctoperlaria comprises of two infraorders, Euholognatha and Systellognatha with six families each. In the Euholognatha group, the labial palps are short and blunt. Glossae and paraglossae are similar in size. The adults live for several weeks and feed on algae, lichens, pollen, rotting wood, etc. However, in Systellognatha, mandibles are reduced. Their labial palps are long and filiform. Glossae reduced than paraglossae. They do not feed, and they are short lived. Further, based on the morphological characters, Systellognatha is divided in to two superfamilies Perloidea (comprising families Perlidae, Perlodidae, and Chloroperlidae) and Pteronarcyoidea (families Pteronarcyidae, Peltoperlidae, and Styloperlidae). Euholognatha includes only one superfamily Nemouroidea and comprises of families Capniidae, Leuctridae, Nemouridae, Notonemouridae, Taeniopterygidae, and Scopuridae. The suborder Arctoperlaria comprises of 12 families mostly distributed in the Northern Hemisphere, of this, a few species of families Notonemouridae and Perlidae are distributed in the Southern Hemisphere. Globally, 3671 described species under 303 genera and 16 families are currently known (Table 4.1) (De Walt et al. 2018). The family Perlidae is the largest with 1101 known species followed by Nemouridae (683 species), Leuctridae (376 species), Perlodidae (340 species), and Capniidae (296 species).

Fochetti and Figueroa (2008) have given an exhaustive account of the global distribution of stoneflies which provides detailed information on the species of eight families so far recorded from India. The distribution of stoneflies drastically differs between Himalaya and Peninsular India. There are eight families known from the Himalaya, while only a single species of Nemouridae (Kimmins 1950a) and four genera of Perlidae occur in the southern peninsula (Stark and Sivec 2014; Zwick 1981, 1982a, 1982b; Zwick and Sivec 1980). So far, 128 species of stoneflies under 24 genera are known from India, of this, 93 species are endemic to the Indian region (Table 4.1). Among these, 56 species are endemic to the Eastern Himalaya, 12 species to the Western Himalaya, 17 species to the Western Ghats, and 1 species each to the Eastern Ghats and Central India. The families Perlidae and Nemouridae have high levels of endemism with 90% and 69%, respectively.

PLECOPTERA OF INDIA

Order **Plecoptera** Burmeister, 1839
  Suborder **Arctoperlaria**
  Infra Order **Euholognatha**
    Superfamily **Nemouroidea** Billberg, 1820
      I. Family **Capniidae** Banks, 1900
      II. Family **Leuctridae** Klapálek, 1905
      III. Family **Nemouridae** Billberg, 1820
      IV. Family **Taeniopterygidae** Klapálek, 1905
  Infra Order **Systellognatha**
    Superfamily **Perloidea** Latreille, 1802
      V. Family **Chloroperlidae** Okamoto, 1912
      VI. Family **Perlidae** Latreille, 1802
      VII. Family **Perlodidae** Klapálek, 1909
    Superfamily **Pteronarcyoidea** Newman, 1853
      VIII. Family **Peltoperlidae** Claassen, 1931

TABLE 4.1
Global Diversity of Stoneflies[a]

S. No.	Suborder	Superfamily	Family	World Genera	World Species	India Genera	India Species	Endemic Species
1	Antarctoperlaria	Eusthenioidea	Diamphipnoidae	2	6	—	—	
2	″	″	Eustheniidae	6	22	—	—	
3	″	Gripopterygoidea	Austroperlidae	10	15	—	—	
4	″	″	Gripopterygidae	53	313	—	—	
5	Arctoperlaria	Nemouroidea	Capniidae	21	296	1	7	5
6	″	″	Leuctridae	12	376	1	1	1
7	″	″	Nemouridae	21	683	7	67	46
8	″	″	Notonemouridae	23	120	—	—	
9	″	″	Taeniopterygidae	13	104	2	2	—
10	″	Not assigned	Scopuridae	1	8	—	—	
11	″	Perloidea	Chloroperlidae	20	197	1	2	—
12	″	″	Perlidae	51	1101	8	40	36
13	″	″	Perlodidae	55	340	2	3	1
14	″	Pteronarcyoidea	Peltoperlidae	10	68	2	6	4
15	″	″	Pteronarcyidae	2	12	—	—	
16	″	″	Styloperlidae	2	10	—	—	
			Total	**303**	**3671**	**24**	**128**	**93**

Source: De Walt, R. E. et al., *Plecoptera Species File Online*. Version 5.0/5.0 http://Plecoptera. SpeciesFile.org (accessed January 5, 2018).

Note: Number of genera and species adopted from Plecoptera Species File.

[a] *nomen nudum, nomen dubium, species inquirenda,* temporary names and subspecies are not included.

SUPERFAMILY NEMOUROIDEA

FAMILY CAPNIIDAE BANKS, 1900

The family Capniidae, commonly known as "Winter Stoneflies" is a remarkable group of mostly winter emerging species. Capniid stonefly larvae are small (5–10 mm), and they are shredder-detritivores. They are found in small streams and moderate rivers, but are most common in small streams, including temporary streams and springs. They are usually located in gravel or detritus in sections of fast or moderate flow. Capniids are characterized by wing pads separated by equal distance and abdominal segments 1–9 separated by a membranous fold. The body is slender and elongate with pronotum only slightly wider than abdomen. Labium is compact with three small notches; paraglossae do not extend beyond glossae; labial palps are robust. Wing pads are variously shaped and are not divergent from midline. Abdominal segments of larvae are also separated by membranous fold. They are often active on snow and can easily be seen (Bouchard 2004). The family has a Holarctic distribution with a few species known also from northern areas of the Oriental realm and nearly 300 recognized species classified in 21 valid genera.

Murányi et al. (2015) have recently established a new genus and species of Capniidae from Southwest China viz., *Sinocapnia kuankuoshui*, with a commented checklist of the family in the Oriental realm. While all other Oriental genera of Capniidae are restricted to high mountains, viz., all other *cordata* and *pedestris* group of species, *Sinocapnia kuankuoshui* is hitherto the only known Capniidae recorded from high and relatively warm areas of the Oriental realm. Seven species of genera *Capnia* are reported from India, and they are distributed in the Western and Eastern Himalaya (Murányi et al. 2015).

FAMILY LEUCTRIDAE KLAPÁLEK, 1905

They are popularly called "Rolled-Winged Stoneflies". They are shredders feeding on whole living or dead plant materials. Nymphs are typically found in the sand or gravel in streams and rivers. Their body length varies from 5–10 mm, rarely 20 mm. The larvae spend most of their time burrowing in the substrate and are not commonly collected except when they are close to emergence. Larval body is slender and elongate with pronotum only slightly wider than abdomen. Labium is compact with three small notches. Paraglossae do not extend beyond glossae. Labial palps are robust. Wing pads are not divergent from midline. Wing pads are similar in shape with mesothoracic wing pads, 2–3 times further apart than metathoracic wing pads. Abdominal segments 7–9 without membranous fold (Bouchard 2004). Their elongate body shapes allow them to move through gravel substrates. Globally, 376 species are known of which only 1 species (*Rhopalopsole magnicerca* Jewett) of this family is so far known from India and recorded from the Western and Eastern Himalaya (Zwick and Sivec 1980).

Family Nemouridae Newman, 1853 (Figure 4.1)

The family Nemouridae are popularly called "Spring or Brown Stoneflies". They are found in streams and rivers. Nymphs are shredder-detritivores, and their body length varies from 5 to 20 mm. Around 680 species are known worldwide, and 67 species are distributed mostly in the Himalayan ranges and are most diverse in terms of species richness in India (De Walt et al. 2018). Totally, 67 species are described from 7 genera of which 46 species are endemic to India. Both the subfamilies viz., Amphinemurinae and Nemourinae are represented in India. Twenty-nine species have been recorded from Arunachal Pradesh (Kimmins 1950b; Aubert 1967). Recently, while describing a new species of *Amphinemoura* from the Darjeeling district of West Bengal and another new species of *Sphaeronemoura* from the West Garo Hills district of Meghalaya, Murányi and Li (2013) have provided an exhaustive checklist of the family Nemouridae from the Indian subcontinent with zoogeographical comments.

Family Taeniopterygidae Klapálek, 1905

They are popularly known as "Winter Stoneflies." They are found in woody debris, leaf packs, or coarse sediments in streams and rivers. Nymphs are shredders and scrapers. Body length varies from small to medium (5–20 mm). Tarsal segments are more or less similar length. Glossae and paraglossae are identical in size and shape. Wing pads of nymphs are greatly divergent. Adults emerge in late winter to early spring. The family present mainly in the Nearctic and Palaearctic Regions, with more than 100 described species. Only two species, *Kyphopteryx dorsalis* Kimmins from Sikkim and *Mesyatsia karakorum* (Šámal) from Jammu and Kashmir, Himachal Pradesh, and Sikkim, have been reported from India (Tierno de Figueroa and Fochetti 2003).

SUPERFAMILY PERLOIDEA

Family Chloroperlidae Okamoto, 1912

Chloroperlids are small, slender, "Green or Brown Stoneflies." They are found in streams and rivers. Since they are smaller species with reduced body size and wing venation, they are adapted to burrow in sand and gravel. Nymphs of most species are carnivorous predators, and a few are detritivores or herbivores, and their size varies from 10 to 20 mm. Nymphs are without gills. Cerci are shorter than abdomen. Glossae are much reduced than paraglossae. Adults emerge in late spring and early summer. Globally, nearly 200 species are known from the Holarctic region and only 2 species of genera *Xanthoperla* are reported from the Western Himalaya of India so far.

Family Perlidae Latreille, 1802 (Figures 4.2 and 4.4)

The perlids are generally known as "Common Stoneflies," and they are carnivorous predators feeding on wide range of invertebrates. Larvae of this family are commonly found under logs, stones, and detritus leaf litters in streams and rivers. Their body length varies from 20 to 50 mm. Larvae bear finely branched gills in all the thoracic segments, but are absent in 1st & 2nd abdominal segments. Labium has well developed paraglossa. Adults of this family have very long cerci with more than 10 segments and larval gill remnants present in lateral and ventral side of thorax. About 1100 species are known worldwide, of which 40 species in 8 genera are distributed both north and south of Indo-Gangetic plains (Babu et al. 2017). The common genus *Neoperla* has 19 species recorded from India (Stark and Sivec 2015). The genus *Phanoperla* Banks is widely distributed over mainland Southeast Asia, the Indian subcontinent, and several Asian islands including Borneo, the Philippines, and Sri Lanka (Mason and Stark 2015). This genus is represented by eight species in India. *Tyloperla* is a small genus of Asian stoneflies which currently includes 15 species of which only 4 species are known from India (Stark and Sivec 2014). The tropical or subtropical stonefly genus, *Brahmana*, is distributed in insular and peninsular Southeast Asia to Indian subcontinent, including China, Nepal, and the Himalayas (Klapálek 1916). Five species are known from the region of which three are recorded from India. Eighteen species of this family are recorded from the Western Ghats of India of which 16 species are exclusively known from this region.

Family Perlodidae Klapálek, 1909

The family Perlodidae are popularly known as "Patterned Stoneflies or Springflies." Nymphs are typically found in slow flowing streams and rivers. They are carnivorous predators. Their body length varies from 10 to 55 mm. The perlodids live exclusively in crenel habitats, some usually wide-spread taxa are strictly potamal globally (Stewart and Stark 2002; Zwick 2004). The fossil records of this family extend from the Triassic period. The larvae have flattened bodies, often with patterns on their head and thorax, long tails, and divergent hind wing pads. They have no gills on thorax. Labium has a deep notch and labial palps are slender.

The majority of perlodid stoneflies are univoltine. Adults are similar to perlid stoneflies and emerge from April to June. Many species have an egg diapause during the warmer months; this allows them to inhabit otherwise unfriendly environments like temporary seeps or streams. Members of the subfamily Perlodinae are striking, usually large sized, and conspicuously coloured. Perlodinae are unique with having high genetic diversity, but with low species level diversification (Teslenko 2015). This family encompasses 55 genera and around 340 species. The subfamily is distributed in the Holarctic with a few species entering the Oriental region (Li and Muranyi 2015) and only three species are distributed in the Eastern Himalayan region of India.

SUPERFAMILY PTERONARCYOIDEA

Family Peltoperlidae Classen, 1931 (Figure 4.3)

The family Peltoperlidae is commonly known as "Roachlike Stoneflies." They are found among leaf packs and detritus accumulations in streams and rivers. They are shredders or detritus feeders. Their body length varies from 5 to 20 mm. Species are semivoltine, with life cycle lasting 1 to 2 years.

Stoneflies (Insecta: Plecoptera) of India

FIGURES 4.1–4.4 Mature nymphs and adult: **1**. Nemouridae (nymph); **2**. Perlidae (nymph) **3**. Peltoperlidae (nymph); and **4**. Perlidae (adult).

Adults emerge in late spring or early summer, April through June. Larvae are flattened and brown in colour, and they are roach-like in appearance because of the expanded thoracic plates covering the bases of their legs and abdomen (Voshell 2002). Peltoperlidae are generally found in lotic erosional and depositional habitats in streams. They inhabit in leaf litter and detritus piles trapped in either stream riffles or pools. The Peltoperlidae are classified in the feeding group shredders-detritivores. They chew and mine through leaf litter in their habitats. They are a significant contributor to leaf breakdown in streams. This family is very sensitive to disturbance in environmental conditions. They are intolerant to loss of coarse particulate organic matter for food and habitat. Given this low tolerance, peltoperlids make potential bio-indicators (Qin et al. 2014). The family Peltoperlidae comprises 68 species in 10 genera. Although the Peltoperlidae extend into the Oriental region, the distribution of the family as a whole is centred on Palaearctic East Asia and North America (Illies 1965). Stark (1989) placed Asian peltoperline stoneflies in five genera. Stark and Sivec (2007) provided a systematic checklist of 41 species of Asian Peltoperlidae. Six species of peltoperline stoneflies are known from India, and all the species are recorded from Eastern India, of this, one species is reported from Himachal Pradesh, Western Himalaya.

DIAGNOSTIC KEYS

Illustrations regarding salient diagnostic features of the nymphs of various families are available in Bouchard (2012).

KEY TO SUPERFAMILIES IN INDIA

1. Apical tarsal segment as long as or slightly shorter than basal and midtarsal segments; structure and length of two basal segments differ between families; glossae and paraglossae approximately same length............................... **NEMOUROIDEA**
- Apical tarsal segment twice as long as basal and midtarsal segments together; structure and length of two basal segments are equal size and very short; glossae similar or dissimilar........................2
2. Glossae and paraglossae approximately same length; maxillae well developed ... **PTERONARCYOIDEA**
- Glossae much shorter than paraglossae; maxillae reduced.................................... **PERLOIDEA**

KEY TO FAMILIES IN INDIA (NYMPHS)

1. Body wide and cockroach-like; flattened head; conical gills present behind coxae of middle and hind legs; thoracic sterna produced posteriorly and overlapping......................... **PELTOPERLIDAE**
- Body slender and not cockroach-like; head not flattened; conical gills not present behind coxae of middle and hind legs; sterna not produced posteriorly..2
2. Glossae and paraglossae of labium approximately same length…...….....................…..................3
- Glossae much shorter than paraglossae.................6
3. Hind legs extending beyond apex of abdomen.......4
- Hind legs not extending beyond apex of abdomen..5
4. First and second tarsal segments more or less equal in length; coxae with single telescoping gill.................................... **TAENIOPTERYGIDAE**
- Second tarsal segment wedge-shaped and much shorter than first tarsal segment; coxae lacking gills.................................... **NEMOURIDAE**
5. Abdominal segments 7–9 with lateral membranous fold dividing tergal and sternal region..**CAPNIIDAE**
- Abdominal segments 7–9 without lateral membranous fold................................. **LEUCTRIDAE**
6. Lateral and ventral sides of thoracic segments with highly branched gills........................ **PERLIDAE**
- Thoracic segments without highly branched gills...7
7. Cerci at least as long as abdomen; dorsum of body usually patterned........................ **PERLODIDAE**
- Cerci less than three-fourths the length of abdomen; dorsum of body rarely patterned... **CHLOROPERLIDAE**

Key to Families in India (adults)

1. Glossae and paraglossae approximately same length...2
- Glossae much shorter than paraglossae................6
2. Body roach-like appearance, thorax much wider than head or abdomen; basal tarsal segment short; gill remnants present in sides of thorax..**PELTOPERLIDAE**
- Body elongate and slender; thorax slightly wider than head or abdomen; basal tarsal segment longer; gill remnants absent in sides of thorax....................3
3. First and second tarsal segments almost same length........................**TAENIOPTERYGIDAE**
- Second tarsal segment much shorter than first.........4
4. Cerci multi-segmented; forewing with 1 or 2 intercubital cross veins.............................**CAPNIIDAE**
- Cerci with only one segment; forewing with 5 or more intercubital cross veins............................5
5. Wings keep flat over abdomen at rest; X-pattern oblique cross vein present in apical marginal space of forewing; cervical gill remnants present................................**NEMOURIDAE**
- Wings rolled down around sides of abdomen at rest; absence of X-pattern oblique cross vein in apical marginal space of forewing; cervical gill remnants absent..**LEUCTRIDAE**
6. Hindwing with 5–10 anal veins and folded anal area present..7
- Hindwing usually with less than 5 anal veins and absence of folded anal area...**CHLOROPERLIDAE**
7. Gill remnants usually present on sides or ventral side of thorax; arms of mesosternal Y-sulcus poorly developed...**PERLIDAE**
- Gill remnants absent on sides or ventral side of thorax; well developed arms of mesosternal Y-sulcus..**PERLODIDAE**

PHYLOGENY AND BIOGEOGRAPHY

According to Illies (1965), Banarescu (1990), and Zwick (1990, 2000), Arctoperlaria and Antarctoperlaria originated as independent lines at the splitting of Pangaea and the subsequent separation of Gondwanaland and Laurasia, at the end of the Triassic period. Antarctoperlaria possibly began their diversification before the continents separated, producing some sister-groups distributed in South America and New Zealand. The absence of Antarctoperlaria from South Africa and India may be interpreted as a later extinction event. These lands became warmer and drier during their northward journey after separation from Antarctica and Australia. The Oriental stonefly fauna inclusive of that of India was colonized from close Palearctic areas, as suggested by the decrease in species numbers towards the south. The first penetration was by members of three families viz., Nemouridae, Peltoperlidae, and Perlidae that extended down into the rain forests of Southeast Asia, thus having somewhat overcome the aversion of the order Plecoptera for warmer climates (Fochetti and Tierno de Figueroa 2008).

Zwick (2000) has given in a nutshell the phylogenetic origin and familial relationships of the order, duly selecting numerous characters supporting the currently valid phylogenetic system. He has highlighted, with ample evidence from past and present global distribution patterns, the origin of the extant suborder around the period of the breakup of Pangaea followed by extensive recent speciation all over the world. McCulloch (2010), using molecular sequence data estimated precisely the divergence of the suborders during the Jurassic period.

Molecular data are being generated to reconstruct precise phylogeny. Wang et al. (2018) have presented a complete mitochondrial genome of a chloroperlid stonefly and have highlighted its implications for the higher phylogeny of stoneflies. Their maximum likelihood (ML), and Bayesian inference (BI) analyses suggests that the Capniidae was monophyletic and the other five stonefly families form a monophyletic group (Wang et al. 2018).

TAXONOMIC PROBLEMS

Accurate identification of stoneflies and other aquatic insects is crucial for employing biotic indices as sensitive tools in bio-monitoring. Unfortunately, larvae and female specimens of stoneflies are very difficult to identify to species level, and this lacuna compromises on our ability to accurately assess water quality and to conduct ecological and conservation assessments of individual species (Gattolliat et al. 2016). Adults of some species groups, for instance, in the family Leuctridae exhibit little morphological differentiation and are difficult to distinguish (Ravizza 2002). In addition to traditional rearing in the laboratory, an integrative approach incorporating rearing and DNA barcoding of larvae and adults collected from the same habitat will be useful to associate life stages and to unravel cryptic diversity as highlighted by Vitecek et al. (2017).

BIOLOGY AND LIFE CYCLE

There is a dearth of information on seasonality, life cycles, egg incubation strategies, and general bionomics of tropical stonefly species in general and Indian species in particular. It is still a mystery whether the numerous morphospecies that are presently recorded from the tropics are true species or are there regional variation and morphoclines over vast tropical areas (Zwick 2003). In tropical Asia, stonefly life cycles appear to be aseasonal, with adult emergence throughout the year, while stoneflies in temperate regions typically have highly synchronized life cycles, with emergence of the entire population occurring within a few weeks. Life cycle studies on Indian stoneflies are virtually absent. Many stoneflies remain in the vicinity of the stream or lake, mate on or near the ground under stones or in vegetation, and therefore do not need the capacity of flight. This applies to the alpine Himalayan species where low air temperatures restrict flight activity. Sexual dimorphism occurs in several species where males are short-winged, while females have wings of normal length. This saves the energy of males, while normal-winged females once mated are able to function as the dispersal agents (Brittain 1990). Egg structure is mostly species specific.

BIOMONITORING POTENTIAL AND CLIMATE CHANGE IMPACTS

Regional biodiversity studies are crucial for establishing conservation priorities to determine the conservation status of species and for examining factors that govern diversity (Gomez de Silva and Medellin 2001). The resulting species lists help conservation biologists to know what species assemblages inhabit a region. This is especially important for aquatic entomologists studying diversity and distribution of stoneflies (Plecoptera), mayflies (Ephemeroptera), and caddis flies (Trichoptera), which constitute a significant component of bio-indicators of water quality in rapidly deteriorating fragile freshwater bodies all over the world with multifaceted, catastrophic human impacts. Conservation agencies can use periodically updated faunal checklists generated by taxonomic fine-tuning as well as the ecological relationships among the co-occurring taxa to help prioritize conservation initiatives including the rehabilitation of habitats and niches and establishment of imperilment risk for such environmentally and climatically sensitive taxa like Plecoptera or stoneflies.

Stonefly larvae show a complex response to habitat change, including loss of pollution-sensitive species and habitat specialists as well as common species which, in some cases, counterbalance by concurrent colonization of new sites (Bojkova et al. 2013). Li et al. (2015) analyzed the current geographical distribution pattern of Plecoptera, a thermally sensitive aquatic insect order. They evaluated its stability when coping with global change across both space and time throughout the Mediterranean region. Their results revealed that climate change caused the biodiversity of Plecoptera to slowly diminish in the past and will cause remarkably accelerated biodiversity loss in the future.

ECONOMIC IMPORTANCE

The larvae play a vital role in the freshwater food chain. Along with mayflies, they are the major source of food for fishes in cold water streams. They are used more extensively along with mayflies, caddis flies, and stoneflies as sensitive bio-indicators of aquatic pollution as well as indicators of climate change.

CONSERVATION

Plecoptera are one of the most endangered groups of insects in montane streams and in lowland rivers of India, most of which are heavily impacted due to anthropogenic activities. Due to their strict ecological needs many species have been reduced to small isolated populations and many others continue to become extinct. In developed countries of Europe, America, and Australasia, periodical revision of cataloguing of stonefly species and review of their present diversity and conservation status have been undertaken. The situation is particularly critical for several species, known from a small number of individuals and/or from restricted areas and that of relict and endemic species, especially in tropical countries with heavy human impacts.

Special efforts should be initiated to capture information from museum specimens and new collections to help answer important conservation questions about stoneflies (De Walt et al. 2012). Species distribution modelling may be useful in predicting historical ranges using these data. Predicting individual species ranges, biodiversity hotspots, and reactions by species to climate change (Tierno de Figueroa et al. 2010) can all result from this approach (Cao et al. 2013). Range shrinking of climatically vulnerable stonefly species is inevitable in this Anthropocene era, and future strategies should provide corridors for such vulnerable species to move to cooler areas (De Walt et al. 2015). Though holistic study of freshwater biota is far more desirable compared to taxon focused ecology, stoneflies warrant attention as vital components of lotic zoobenthos since they also provide different environmental information than other aquatic insect orders (Heino et al. 2003; Park et al. 2003). So far Indian Plecoptera species have not been assessed for threat and hence are not included in IUCN Red List.

COLLECTION AND PRESERVATION

The most reliable places to collect stoneflies are small rocky unpolluted streams, especially those in more mountainous regions. The larvae can be collected with a forceps or hand brush from under stones, decaying wood, and leaf litter packs in streams. They can also be collected with the help of a kick net. The adults can be collected from the vegetation along the side of the streams by using a beating tray, aerial, and sweeping nets. Light traps, Malaise traps, and emergence traps are useful for capturing adult stoneflies. The nymphs and adults can be preserved in 70–100 percent ethanol (De Walt et al. 2015). For molecular studies, the specimens should be collected in 100% ethyl alcohol. Body parts can be dissected and stored in micro vials with alcohol for detailed microscopic study.

CONCLUSION

This chapter will hopefully lay a foundation for planned future investigations, including natural range modelling of species within the larger framework of understanding the diversity and distribution of the Oriental Plecoptera in general and Plecoptera of India in particular. Ultimately, such baseline studies of stonefly distributions could be utilized to measure climate related changes in future patterns of distribution by modifying climate variables in light of predicted CO_2 emissions scenarios (De Walt et al. 2012).

ACKNOWLEDGEMENTS

We are much grateful to Dr. Kailash Chandra, Director, Zoological Survey of India, Kolkata for his constant encouragement and valuable support in terms of laboratory facilities to promote our taxonomic studies of Plecoptera of India. The second author (KGS) is indebted to Professors Peter Zwick, Robert De Walt, Bill Stark, B.C. Kondratieff, and Dr. D. Murányi for valuable academic inputs, guidance, and literature support.

REFERENCES

Aubert, J. 1967. Les Nemouridae de l'Assam (Plècoptères). *Mitteilungen der Schweizerischen Entomologischen Gesellschaft* 39: 209–253.

Banks, N. 1914. New neuropteroid insects, native and exotic. *Proceedings of the Academy of Natural Sciences of Philadelphia* 66: 608–632.

Banks, N. 1920. New Neuropteroid Insects. *Bulletin of the Harvard College Museum of Comparative Zoology* 64(3): 314–325.

Banks, N. 1939. New genera and species of neuropteroid insects. *Bulletin of the Harvard College Museum of Comparative Zoology* 85(7): 439–504 (Plecoptera 441–454).

Babu, R., K. G. Sivaramakrishnan, K. A. Subramanian, and C. Selvakumar. 2017. Insecta: Plecoptera (stoneflies). In *Current Status of Freshwater Faunal Diversity in India*, eds. K. Chandra, K. C. Gopi, D. V. Rao, K. Valarmathi, and J. R. B. Alfred, 429–444. Kolkata, India: Zoological Survey of India.

Banarescu, P. 1990. *Zoogeography of Fresh Water: General Distribution and Dispersal of Freshwater Animals* Vol. 1, 1–511. Wiesbaden, Germany: Aula-Verlag.

Bojkova, J., V. Radkova, T. Soldan, and S. Zahradkova. 2013. Trends in species diversity of lotic stoneflies (Plecoptera) in the Czech Republic over five decades. *Insect Conservation and Diversity* 7(3): 252–262.

Bouchard, R. W., Jr. 2004. *Guide to Aquatic Macroinvertebrates of the Upper Midwest*, 1–208. USA, St. Paul MN: Water Resources Center, University of Minnesota.

Bouchard, R. W. 2012. *Guide to Aquatic Invertebrate Families of Mongolia-Identification Manual for Students, Citizen Monitors, and Aquatic Resource Professionals*. Saint Paul, Minnesota.

Burmeister, H. C. C. 1839. *Handbuch der Entomologie*. I–XII, 757–1050. Berlin, Germany: Ensilin.

Brittain, J. E. 1990. Life history strategies in Ephemeroptera and Plecoptera. In *Mayflies and Stoneflies: Life Histories and Biology*, ed. I. C. Campbell, 1–12. Boston, MA: Kluwer Academic Publishers.

Cao, Y., R. E. De Walt, J. L. Robinson, T. Tweddale, L. Hinz, and M. Pessino. 2013. Using Maxent to model the historic distributions of stonefly species in Illinois streams and rivers: The effects of regularization and threshold selections. *Ecological Modelling* 259: 30–39. doi:10.1016/j.ecolmodel.2013.03.012.

De Walt, R. E., Y. Cao, T. Tweddale, S. A. Grubbs, L. Hinz, M. Pessino, and J. L. Robinson. 2012. Ohio stoneflies (Insecta, Plecoptera): Species richness estimation, distribution of functional niche traits, drainage affiliations and relationships to other states. *ZooKeys* 178: 1–26.

De Walt, R. E., B. C. Kondratieff, and J. B. Sandberg. 2015. Order Plecoptera. In *Ecology and General Biology: Thorp and Covich's Freshwater Invertebrates*, eds. J. Thorp and D.C. Rogers, 933–949. London, UK: Academic Press.

De Walt, R. E., M. D. Maehr, U. Neu–Becker, and G. Stueber. 2018. *Plecoptera Species File Online*. Version 5.0/5.0 http://Plecoptera.SpeciesFile.org (accessed January 5, 2018).

Fochetti, R., and J. M. Tierno de Figueroa. 2008. Global diversity of stoneflies (Plecoptera: Insecta) in freshwater. *Hydrobiologia* 595: 365–377.

Gattolliat, J. L., G. Vinson, S. Wyler, J. Pawlowski, and M. Sartori. 2016. Towards a comprehensive COI DNA barcode library for Swiss stoneflies (Insecta: Plecoptera) with special emphasis on the genus *Leuctra*. *Zoosymposia* 11: 135–155.

Gomez de Silva, H., and R. A. Medellin. 2001. Evaluating completeness of species lists for conservation and macroecology: A case study of Mexican land birds. *Conservation Biology* 15: 1384–1395.

Harper, P. P. 1974. New *Protonemura* (S.L.) from Nepal (Plecoptera: Nemouridae). *Psyche: A Journal of Entomology* 81(3–4): 367–376.

Harper, P. P. 1977. Capniidae, Leuctridae and Perlidae (Plecoptera) from Nepal. *Oriental Insects* 11: 53–62.

Heino, J., T. Muotka, H. Mykrä, R. Paavola, H. Hä Mä Lä Inen, and E. Koskenniemi. 2003. Defining macroinvertebrate assemblage types of headwater streams: Implications for bioassessment and conservation. *Ecological Applications* 13: 842–852.

Illies, J. 1965. Phylogeny and zoogeography of the Plecoptera. *Annual Review of Entomology* 10: 117–140.

Jewett, S. G. 1958. Entomological survey of the Himalaya. Part XXIII–Stoneflies (Plecoptera) from the North–West (Punjab) Himalaya. *Proceedings of the National Academy of Sciences (India)*, 28(4): 320–329.

Jewett, S. G. 1960. Entomological survey of the Himalaya. Part XXXI–New and little known stoneflies (Plecoptera) from the North–West (Punjab) Himalaya collected by Prof. Mani's third entomological expedition. *Agra University Journal of Research* 9: 229–232.

Jewett, S. G. 1970. Stonefly records from the Northwest (Punjab) Himalaya. *Oriental Insects* 4(4): 481–482. doi:10.1080/00305316.1970.10433984.

Jewett, S. G. 1975. Records and descriptions of stoneflies from Northwest (Punjab) Himalaya and Mt. Makalu, Nepal Himalaya. *Oriental Insects* 9(1): 1–7. doi:10.1080/00305316.1975.10434838.

Kawai, T. 1968. Stoneflies (Plecoptera) from Thailand and India with description of one new genus and two new species. *Oriental Insects* 2(2): 107–139.

Kimmins, D. E. 1946. New species of Himalayan Plecoptera. *The Annals and Magazine of Natural History* (Series 11) 13: 721–740. doi:10.1080/00222934608654596.

Kimmins, D. E. 1950a. Some new species of Asiatic Plecoptera. *The Annals and Magazine of Natural History* (Series 12) 3: 177–192. doi:10.1080/00222935008654705.

Kimmins, D. E. 1950b. Some Assamese Plecoptera, with descriptions of new species of Nemouridae. *The Annals and Magazine of Natural History* (Series 12) 3: 194–209. doi:10.1080/00222935008654707.

Klapálek, F. 1909. Vorlaufiger berhicht uber exotische Plecopteren. *Wiener Entomologische Zeitung* 28(7/8): 215–232.

Klapálek, F. 1916. Subfamilia Acroneuriinae Klp. *Acta Societatis Entomologicae Bohemiae (Časopis)* 13: 45–84.

Li, W., and D. Murányi. 2015. A remarkable new genus of Perlodinae (Plecoptera: Perlodidae) from China, with remarks on the Asian distribution of Perlodinae and questions about its tribal concept. *Zoologischer Anzeiger* 259: 41–53. doi:10.1016/j.jcz.2015.10.003.

Li, F., J. M. Tierno de Figueroa, S. Lek, and Y. S. Park. 2015. Continental drift and climate change drive instability in insect assemblages. *Scientific Reports* 17(5): 11343.

Mason, D., and B. P. Stark. 2015. Notes on the genus *Phanoperla* Banks from Sri Lanka and India (Plecoptera: Perlidae). *Illiesia* 11(04): 29–40.

McCulloch, G. A. 2010. Evolutionary Genetics of Southern Stoneflies. PhD diss., University of Otago, Dunedin.

Murányi, D., and W. Li. 2013. Two new species of stoneflies (Plecoptera: Nemouridae) from Northeastern India, with a checklist of the family in the Indian subcontinent. *Zootaxa* 3694(2): 167–177.

Murányi, D., and W. Li. 2016. On the identity of some Oriental Acroneuriinae taxa (Plecoptera: Perlidae), with an annotated checklist of the subfamily in the realm. *Opuscula Zoologica* (Budapest) 47(2): 173–196.

Murányi, D., W. Li, and D. Yang. 2015. A new genus and species of winter stoneflies (Plecoptera: Capniidae) from Southwest China, with a commented checklist of the family in the Oriental Realm. *Zootaxa* 4059(2): 371–382.

Navás, L. 1922. Insectos nuevos o poco conocidos I. *Memorias de la Real Academia de Ciencias y Artes de Barcelona* 3(17): 383–400.

Needham, J. G. 1909. Notes on the Neuroptera in the collection of the Indian Museum. *Records of the Indian Museum* 3: 185–210.

Park, Y. S., R. Cereghino, A. Compin, and S. Lek. 2003. Applications of artificial neural networks for patterning and predicting aquatic insect species richness in running waters. *Ecological Modelling* 160(3): 265–280.

Qin, C. Y., J. Zhou, Y. Cao, Y. Zhang, R. M. Hughes, and B. X. Wang. 2014. Quantitative tolerance values for common stream benthic macroinvertebrates in the Yangtze River Delta, Eastern China. *Environmental Monitoring and Assessment* 186(9): 5883–5895. doi:10.1007/s10661-014-3826-2.

Ravizza, C. 2002. Atlas of the Italian Leuctridae (Insecta, Plecoptera) with an appendix including Central European species. *Lauterbornia* 44: 1–42.

Ricker, W. E. 1952. *Systematic Studies in Plecoptera*, Science Series 18: 1–200. Bloomington, IN: Indiana University Publications.

Singh, R. K., and S. K. Ghosh. 1969. A new species of the genus *Perlodes* (Plecoptera: Perlodidae) from India. *Zoologischer Anzeiger* 182: 134–136.

Sivaramakrishnan, K. G., S. Arunachalam, S. Sundar, and C. Selvakumar. 2011. A brief conspectus and research priorities on Plecoptera (Stoneflies) of India. In *Entomology: Ecology and Biodiversity*, ed. B. K. Tyagi, 105–118. Jodhpur, India: Scientific Publishers.

Stark, B. P. 1989. Oriental Peltoperlinae (Plecoptera): A generic review and descriptions of a new genus and seven new species. *Entomologica scandinavica* 19: 503–525.

Stark, B. P., and I. Sivec. 1991. Descriptions of Oriental Perlini (Plecoptera: Perlidae). *Aquatic Insects. International Journal of Freshwater Entomology* 13(3): 151–160. doi:10.1080/01650429109361436.

Stark, B. P., and I. Sivec. 2007. New species and records of Asian Peltoperlidae (Insecta: Plecoptera). *Illiesia* 3(12): 104–126.

Stark, B. P., and I. Sivec. 2008. New stoneflies (Plecoptera) from Asia. *Illiesia* 4(1): 1–10.

Stark, B.P., and I. Sivec. 2014. Three new species of *Tyloperla* Sivec & Stark (Plecoptera: Perlidae) from India. *Illiesia* 10(04): 32–42.

Stark, B. P., and I. Sivec. 2015. New species and records of *Neoperla* (Plecoptera: Perlidae) from India. *Illiesia* 11(07): 75–91.

Stewart, K. W., and B. P. Stark. 2002. *Nymphs of North American Stonefly Genera (Plecoptera)*, 1–510. Columbus, OH: Caddis Press.

Teslenko, V. A. 2015. A new species of *Megaperlodes* Yokoyama et al., 1990 (Plecoptera: Perlodidae) from the South of the Russian Far East. *Zootaxa* 3904: 553–562.

Tierno de Figueroa, J. M., and R. Fochetti. 2003. *Mesyatsia karakorum* (Šámal, 1935) (Plecoptera: Taeniopterygidae): An addition to the stonefly-fauna of Sikkim (Himalaya, India). *Boletin de la Asociacion española de Entomología* 27(1–4): 237–238.

Tierno de Figueroa, J. M., M. J. López–Rodríguez, A. Lorenz, W. Graf, A. Schmidt–Kloiber, and D. Hering. 2010. Vulnerable taxa of European Plecoptera (Insecta) in the context of climate change. *Biodiversity and Conservation* 19: 1269–1277.

Vitecek, S., G. Vincon, W. Graf, and S. U. Pauls. 2017. High cryptic diversity in aquatic insects: An integrative approach to study the enigmatic *Leuctra inermis* species group (Plecoptera). *Arthropod Systematics and Phylogeny* 75(3): 497–521.

Voshell, J. R. 2002. *A Guide to Common Freshwater Invertebrates of North America*. Blacksburg, VA: McDonald & Woodward Publishing Company.

Wang, Y., J. J. Cao, and W. H. Li. 2018. Complete mitochondrial genome of *Suwallia teleckojensis* (Plecoptera: Chloroperlidae) and implications for the higher phylogeny of stoneflies. *International Journal of Molecular Sciences* 19(3): 680. doi:10.3390/ijms19030680.

Zwick P. 1973. *Insecta Plecoptera. Phylogenetisches System und Catalog. Das Tierreich* 94, XXXII + 465.

Zwick, P. 1981. The South Indian species of *Neoperla* (Plecoptera: Perlidae). *Oriental Insects* 15: 113–126.

Zwick, P. 1982a. A revision of the Oriental stonefly genus *Phanoperla* (Plecoptera: Perlidae). *Systematic Entomology* 7: 87–126.

Zwick, P. 1982b. Contribution to the knowledge of *Phanoperla* Banks, 1938 (Plecoptera: Perlidae). *Mittelungen der Schweizerischen Entomologischen Gesellschaft* 59: 151–158.

Zwick, P. 1990. Transantarctic relationships in the Plecoptera. In *Mayflies and Stoneflies, Life History and Biology*, ed. I. C. Campbell, 141–148. London, UK: Kluwer Academic Publishers.

Zwick, P. 2000. Phylogenetic system and zoogeography of the Plecoptera. *Annual Review of Entomology* 45: 709–746.

Zwick, P. 2003. Plecoptera research today: Questions to be asked in the new millennium. In *Research Updated on Ephemeroptera and Plecoptera*, ed. E. Gaino, 245–251. Perugia, Italy: University of Perugia.

Zwick, P. 2004. Key to the West Palaearctic genera of stoneflies (Plecoptera) in the larval stage. *Limnologia* 34: 315–348.

Zwick, P., and I. Sivec. 1980. Beiträge zur Kenntnis der Plecoptera des Himalaja. *Entomologica Basiliensia* 5: 59–138.

Zwick, P., S. Anbalagan, and S. Dinakaran. 2007. *Neoperla biseriata* sp. n., a new stonefly from Tamil Nadu, India (Plecoptera: Perlidae). *Aquatic Insects* 29: 241–245.

5 Taxonomy of Orthoptera with Emphasis on Acrididae

Rajamani Swaminathan and Tatiana Swaminathan

CONTENTS

Introduction 57
Orthoptera Classification 58
   Key to Families of Orthoptera in the Indian Subcontinent 59
Acrididae: The Locusts and Short-Horned Grasshoppers 60
   Diagnosis and Key Identification Characters 60
   Key Identification Characters for Subfamilies of Acrididae 61
   Classification of Acrididae Adopted during Different Periods 63
   Key to Subfamilies of Acrididae in the Indian Subcontinent 63
   Major Work on Indian Acrididae 64
   Taxonomic Problems 64
Economic Importance 65
Collection and Preservation 65
   Preservation for Molecular Studies 65
Useful Websites 66
Conclusion 66
References 66

INTRODUCTION

Orthoptera are a group of insects that include grasshoppers, locusts, crickets, bush-crickets, mole crickets, and camel-crickets that exist in terrestrial habitats throughout the world. Though often associated with fields and meadows, there are orthopteran species that prefer caves, deserts, bogs, and seashores. Body size varies from less than 5 mm to some of the world's largest insects, with body lengths up to 11.5 cm and wingspans of over 22 cm. Members of both suborders (Ensifera and Caelifera) are generally phytophagous, but many species are omnivores. They feed on a variety of agricultural crops, forests, fodder, and grasses. Females of most species lay clutches of eggs, either in the ground or in vegetation. Insects of this order undergo paurometabolous or incomplete metamorphosis. Many produce sound (known as "stridulation") by rubbing their wings against each other or their legs, the wings or legs containing rows of corrugated bumps. The tympanum or ear is located in the front tibia in crickets, mole crickets, and katydids and on the first abdominal segment in the grasshoppers and locusts. These organisms use vibrations to locate other individuals. With an ability to fold their wings, taxonomically, they have been placed in the group Neoptera (Howell et al. 1998). The first fossil Orthoptera appears in the Upper Carboniferous with the first Ensifera (Chopard 1920) appearing in the Permian and the first Caelifera (Ander 1939) in the Triassic period (Gorochov 1995; Kukalova-Peck 1991; Sharov 1968; Zeuner 1939).

Scientists have described many species in this group that are more diverse in the tropics. The 12th International Congress of Orthopterology "*Orthoptera in a Changing World*" was held during October 30–November 3, 2016, at Ilhéus, Brazil. As documented in the Newsletter of the Orthopterist's Society, Maria Kátia Matiotti da Costa (2016) quoted that with more than 27,213 identified species, Orthoptera has its maximum representation in the superfamily Acridoidea. There are 8,016 species in 1,711 genera of Acridoidea; however, the number of known species may be even higher, considering the wide variety of biomes. Representatives of Acrididae, popularly known as grasshoppers, have economic importance, especially in agriculture, given the severe damage they cause. Acrididae is the most numerous and widely distributed family of Acridoidea, with approximately 6,642 known valid species, and more than 1,401 genera. The taxonomic knowledge of Acridoidea has been obtained by examining numerous morphological materials coming from collections through surveys and examining those deposited in national and international institutions. The descriptions of genera and species of Acridoidea are performed by morphological analyses of the external and internal genitalia of males to assist in precise identification of the species. It is necessary that studies on taxonomic aspects are performed continuously the world over in order to have a better estimate of the total number of representatives of Acridoidea.

ORTHOPTERA CLASSIFICATION

Classification of the order Orthoptera has been dynamic, often with modifications and revisions. According to the Orthoptera Species File—OSF (*orthoptera.speciesfile.org*), the order Orthoptera has been placed under super order Orthopterida in the cohort Polyneoptera. Members of the insect order Orthoptera have traditionally been divided into two major lineages, presently recognized as suborders, Ensifera and Caelifera. The suborder Caelifera comprises two infra orders: Acrididea and Tridactylidea. The infra order Acrididea comprises one superfamily group, Acridomorpha, that in turn includes seven superfamilies *viz.*, Acridoidea, Eumastacoidea, Pneumoroidea, Proscopoidea, Pyrgomorphoidea, Tanaoceroidea, & Trigonopterygoidea within the group; and the superfamily Tetrigoidea. Among the superfamilies of the group Acridomorpha, Acridoidea is the most diverse with more than 8,000 species of grasshoppers and locusts grouped into 11 families: Acrididae, Dericorythidae, Lathiceridae, Lentulidae, Lithidiidae, Ommexechidae, Pamphagidae, Pamphagodidae (Syn. Charilaidae), Pyrgacrididae, Romaleidae, and Tristiridae. Romaleidae is a diverse family of Acridoidea, especially in the Neotropics, with large specimens that often possess colourful, flashy, and bright hind wings. Members of the family Ommexechidae are small grasshoppers with rough or tubercular integuments possessing varied and distinctive ornamentation. The infra order Tridactylidea includes one extant superfamily, Tridactyloidea. The suborder Caelifera in all has more than 11,000 species grouped into 27 families. The suborder Ensifera comprises two infra orders: Gryllidea and Tettigoniidea. Infra order Gryllidea includes two extant superfamilies, the Grylloidea and Gryllotalpoidea; while the infra order Tettigoniidea includes five extant superfamilies, Hagloidea, Stenopelmatoidea, Tettigonioidea, Rhaphidophoroidea, and Schizodactyloidea with the entire suborder Ensifera containing more than 10, 000 species placed into 14 families (Table 5.1).

The diversity of Orthoptera is considered to be more in the tropics and subtropics; and, of late, it has become increasingly clear that orthopteran fauna, the grasshoppers in particular, happen to be more diverse in the Neotropical region. However, in the Indian subcontinent, on account of the diverse terrestrial habitats, orthopteran faunal diversity is fairly high; besides, many more taxa are yet to be discovered. The compilation of the orthopteran fauna in the Indian subcontinent from the Orthoptera Species File indicate that there are 6 superfamilies and 10 families in the suborder Caelifera and 6 superfamilies and 13 families in the suborder Ensifera. The orthopteran families with their common names, present in the Indian subcontinent, are given in Table 5.2.

TABLE 5.1
Classification of Orthoptera

Suborder Ensifera

Infra order: Gryllidea
Superfamily: Grylloidea
Gryllidae, Mogoplistidae, Phalangopsidae, and Trigonidiidae
Superfamily: Gryllotalpoidea
Gryllotalpidae, and Myrmecophilidae
Infra order: Tettigoniidea
Superfamily: Hagloidea
Prophalangopsidae
Superfamily: Rhaphidophoroidea Rhaphidophoridae – camel crickets, cave crickets, cave wetas, and sand treaders
Superfamily: Schizodactyloidea Schizodactylidae – dune crickets, splay-footed crickets
Superfamily: Stenopelmatoidea Anostostomatidae – wetas, king crickets; Cooloolidae, Gryllacrididae – leaf-rolling crickets, camel crickets
Stenopelmatidae – Jerusalem crickets, potato bugs, earth baby
Superfamily: Tettigonioidea Tettigoniidae – katydids/bush crickets

Suborder Caelifera

Infra order: Acrididea
Superfamily: Acridoidea
Acrididae – grasshoppers, locusts, Charilaidae, Dericorythidae, Lathiceridae, Lentulidae, Lithidiidae, Ommexechidae, Pamphagidae – toad grasshoppers, Pyrgacrididae, Romaleidae – lubber grasshoppers, and Tristiridae
Superfamily: Eumastacoidea Chorotypidae, Episactidae, Eumastacidae, Euschmidtiidae, Mastacideidae, Morabidae, and Thericleidae
Superfamily: Proscopioidea Proscopiidae
Superfamily: Pneumoroidea Pneumoridae – bladder grasshoppers
Superfamily: Pyrgomorphoidea Pyrgomorphidae – gaudy grasshoppers
Superfamily: Tanaoceroidea Tanaoceridae
Superfamily: Tetrigoidea Tetrigidae – grouse locusts, pygmy locusts, groundhoppers, and pygmy grasshoppers
Superfamily: Trigonopterygoidea Trigonopterygidae, Xyronotidae
Infra order: Tridactylidea
Superfamily: Tridactyloidea Cylindrachetidae (sandgropers) Rhipipterygidae, Tridactylidae – pygmy mole crickets

Source: Cigliano, M.M. et al., *Orthoptera Species File*. Version 5.0/5.0, http://Orthoptera. SpeciesFile.org, November 7, 2018.

TABLE 5.2
Orthoptera in the Indian Subcontinent

Orthoptera			Common Names
Suborder CAELIFERA			
Super family **Acridoidea**			
	Family	Acrididae	Grasshoppers
		Dericorythidae	
		Pamphagidae	Rugged earth hoppers
Super family **Eumastacoidea**			
	Family	Chorotypidae	Crested monkey hoppers
		Eumastacidae	Monkey grasshoppers
		Mastacideidae	South Indian wingless monkey hoppers
Super family **Pyrgomorphoidea**			
	Family	Pyrgomorphidae	True bush hoppers
Super family **Trigonopterygoidea**			
	Family	Trigonopterygidae	Broad-leaf bush hoppers
Super family **Tetrigoidea**			
	Family	Tetrigidae	Grouse locusts, pygmy grasshoppers
Super family **Tridactyloidea**			
	Family	Tridactylidae	Pygmy mole crickets
Suborder ENSIFERA			
Super family **Hagloidea**			
	Family	Prophalangopsidae	Hump-winged crickets
Super family **Stenopelmatoidea**			
	Family	Anostostomatidae	Silk-spinning crickets
		Gryllacrididae	Leaf-rolling grasshoppers
		Stenopelmatidae	Sand or stone crickets
Super family **Tettigonioidea**			
	Family	Tettigoniidae	Bush katydids, katydids, and long-horned grasshoppers
Super family **Grylloidea**			
	Family	Gryllidae	True crickets
		Mogoplistidae	Scaly crickets
		Phalangopsidae	
		Trigonidiidae	Sword-tailed crickets
		Gryllotalpidae	Mole crickets
		Myrmecophilidae	Ant loving crickets
Super family **Rhaphidophoroidea**			
	Family	Rhaphidophoridae	Cave or camel crickets
Super family **Schizodactyloidea**			
	Family	Schizodactylidae	Splay-footed crickets

Source: Cigliano, M.M. et al., *Orthoptera Species File*. Version 5.0/5.0., http://Orthoptera.SpeciesFile.org, November 7, 2018.

KEY TO FAMILIES OF ORTHOPTERA IN THE INDIAN SUBCONTINENT

1. Antennae as long as or longer than body, if shorter consisting of more than 30 segments; tympana (ear drum) may be present or absent, when present at base of fore tibiae; ovipositor when present sword-like or needle-like................suborder **Ensifera** 2
- Antennae at most half as long as body or shorter, consisting of less than 30 segments; tympana (ear drum) when present at base of first abdominal segment; ovipositor when present short, lobes triangular......................suborder **Caelifera** 13
2. Legs with tarsi 3-segmented (tarsal formula 3-3-3); ovipositor when developed needle-like or cylindrical...3
- Legs with tarsi 4-segmented (tarsal formula 4-4-4); ovipositor when developed sword-like. 6
3. Front legs modified for digging, with broad, flat femur and tibia, and with powerful teeth on both tibia and tarsus................................. **Gryllotalpidae**
- Front legs normal, not modified for digging 4

4. Body broadly oval; eyes greatly reduced; hind coxae closely approximated ventrally; wingless; living in ant nests ....**Myrmecophilidae**
- Body elongate; eyes not reduced; hind coxae well separated ventrally; winged or wingless; free living 5
5. Wings very short or absent; hind tibiae without strong spines, but with apical spurs; body covered with scales; hind femora stout................................**Mogoplistidae**
- Wings usually well developed; hind tibiae nearly always with long spines; body not covered with scales; hind femora only moderately developed....... ....**Gryllidae**
6. First tarsal segment of legs with lateral wing-like projections; wings when present spirally coiled at tips .... **Schizodactylidae**
- First tarsal segment of legs without wing-like lateral projections; wings when present not coiled.......... 7
7. Wings usually absent, if present, then with 8 or more principal longitudinal veins; front tibiae with or without tympana.. 8
- Wings present and with fewer than 8 principal longitudinal veins; front tibia with tympanum 11
8. Antennae contiguous at base or nearly so................ ....**Rhaphidophoridae**
- Antennae separated at base by a distance equal to or greater than length of first antennal segment........... 9
9. Tarsi lobed, and more or less flattened dorsoventrally; hind femora extending beyond apex of abdomen .... **Gryllacrididae**
- Tarsi not lobed and more or less flattened laterally; hind femora not extending beyond apex of abdomen 10
10. Base of abdomen and often head much wider than thorax, tibiae short and thick.........**Stenopelmatidae**
- Base of abdomen and head almost the same width as thorax; tibiae thin .... **Anostostomatidae**
11. Antennal sockets located about half way between epistomal suture and top of head; wings produced, broad in male, minute in female; ovipositor extremely short; hind femora extending to about apex of abdomen .... **Prophalangopsidae**
- Antennal sockets located near top of head; wings and ovipositor variable; hind femora usually extending beyond apex of abdomen................................. 12
12. Dorsal surface of first tarsal segment smoothly rounded; hind wings usually longer than front wings ....**Tettigoniidae** (Phaneropterinae)
- Dorsal surface of first tarsal segment grooved laterally; prosternal spines usually present; front wing about as long as hind wings................ **Tettigoniidae**
13. All tarsi 3-segmented, tarsal formula 3-3-3.............5
- All tarsi not 3-segmented, tarsal formula 2-2-3 or 2-2-1 or 2-2-0... 14
14. Pronotum conspicuously prolonged backwards over abdomen and tapering; tarsi of hind legs clearly visible, tarsal formula 2-2-3 .... **Tetrigidae**
- Pronotum not prolonged over abdomen; tarsi of hind legs concealed by long spurs; tarsal formula 2-2-1 or 2-2-0 .... **Tridactylidae**
15. Antennae shorter than front femora, if longer metabasitarsomere toothed dorsally............................. 16
- Antennae longer than front femora; metabasotarsomere not toothed dorsally...................................... 18
16. Face with weak frontal ridge; head raised above level of thorax .... **Chorotypidae**
- Face with distinct frontal ridge; head not raised above level of thorax 17
17. Head not pointed above; wingless; body comparatively short; distal antennal organ on antepenultimate (third from apex) segment; aedeagus very long .... **Mastacideidae**
- Head usually pointed, rarely not so; winged or wingless; body elongate; distal antennal organ on 2nd to 4th segment from apex; aedeagus not so long .... **Eumastacidae**
18. Fastigial furrow present.. 19
- Fastigial furrow absent 21
19. Front wing widened apically, often leaf-like, tympanum absent; phallic complex reversed so that epiphallus ventral .... **Trigonopterygidae**
- Fore wing narrowed apically or wingless; tympanum present; phallic complex not reversed, epiphallus dorsal 20
20. Front of head forming rostrum which is often triangular, with apical furrow and one pair of prominent concavities; first abdominal segment with a diagnostic granular or ridged convex patch called Krauss' organ..**Pamphagidae**
- Front of head not forming rostrum, without apical furrow and prominent concavities; first abdominal segment without Krauss' organ .... **Pyrgomorphidae**
21. Pronotum with crest or hump in prozona only; hind tibiae curved .... **Dericorythidae**
- Pronotum without crest, if present crest occupying both prozona and metazona; hind tibia straight .... **Acrididae**

ACRIDIDAE: THE LOCUSTS AND SHORT-HORNED GRASSHOPPERS

Diagnosis and Key Identification Characters

Members of the family Acrididae are characterized by the following traits: body slender or often thickset, having different dimensions. Head oval or conical; frons vertical or sloping; foveolae, when developed, of different forms, but may be sometimes wanting (Figure 5.1); antennae usually longer than fore-femora with 8–30 segments and can be differently shaped: filiform, club-shaped, swordshaped, or leaf-shaped, and without any special apical antennal organ. Pronotum short, does not overlap the abdomen dorsally, distinctly indicating the pro-thoracic

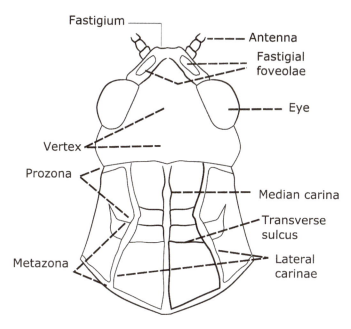

FIGURE 5.1 Head and pronotum (dorsal view). (Adapted and redrawn from Bei-Bienko, G.Ya., and Mishchenko, L.L., *Akademii Nauka, USSR, Moscow* Part I no. 38: xxi + 291 p, 1951.)

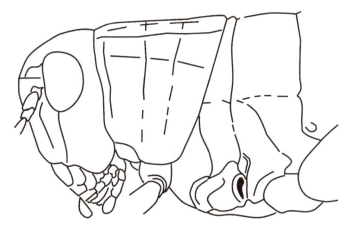

FIGURE 5.2 Head and pronotum (lateral view).

zones: prozona and metazona, which may or may not have lateral carinae (Figures 5.1 and 5.2). Tegmina often well developed, if so, almost always longer than wings, narrowing apically; however, some species are brachypterous or apterous. The wing pads of developing nymph (IV & V-instars) (Figure 5.3a) and the rudimentary wing of an adult (Figure 5.3b) can be easily differentiated. The tegmina of the fourth and fifth instar nymphs rise and arch in such a way that the wing overlaps tegmina. The typical acridid wing venation (Figure 5.4) has great taxonomic significance with the longitudinal veins and enclosed fields having typical nomenclature. The degree of development of fore and hind wings may vary between the sexes within species. Forelegs and middle legs are employed for walking, while the hind legs are modified for jumping. The lower lobe of the hind femora is shorter than the upper lobe in Acrididae (Figure 5.5); whereas, the reverse is true for the members of family Pyrgomorphidae. All tarsi of the acridids are typically 3-segmented, and the hind tarsi (Figure 5.6) have much taxonomical significance. Members of Acrididae have the tympanum on the first abdominal segment (Figure 5.7). The lateral and dorsal views of the external genitalia of the males (Figure 5.8) and females (Figure 5.9) of Acrididae decipher typical taxonomic characters for identification of the fauna at the subfamily level; besides, often assisting at the generic-level identification for many grasshoppers.

Family Acrididae is typically characterized by the following:

Pronotum not prolonged backward over abdomen; antennae longer than fore femora; wings and tympana nearly always present; arolia present; all tarsi 3-segmented; males without a file on third abdominal tergum; size variable, but usually over 15 mm in length; hind tibia with only inner immovable spine at apex, outer spine wanting; prosternum with or without median spine or tubercle; widely distributed.

Key Identification Characters for Subfamilies of Acrididae

The family Acrididae can be classified into different subfamilies based on the presence or absence of the following morphological characters or a combination of them:

- Head orientation with respect to thorax: slant faced/forming an obtuse angle
- Presence or absence of fastigial foveolae, its shape
- Presence or absence of prosternal process, its shape and direction in relation to thorax

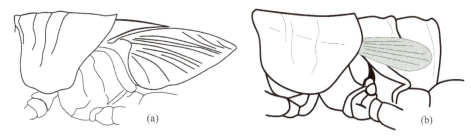

FIGURE 5.3 (a) Fifth instar nymph with wing pads. (b) Adult with rudimentary wing.

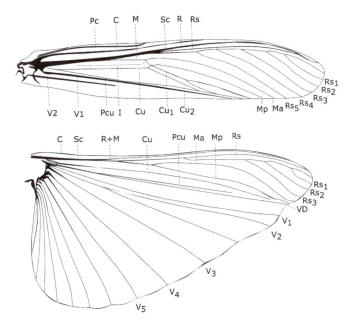

FIGURE 5.4 Developed wings with venation. C, costa; M, median; Sc, subcosta; R, radial; Rs, radial sector; R_1 to R_5, radial veins; Cu, cubitus; Cu_1, Cu_2, cubitus 1 and 2; Ma, median anterior; Mp, median posterior; V_1 and V_2, anal veins; and V_3, V_4, and V_5, jugal veins.

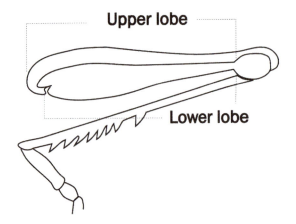

FIGURE 5.5 Hind femur depicting lower and upper lobes.

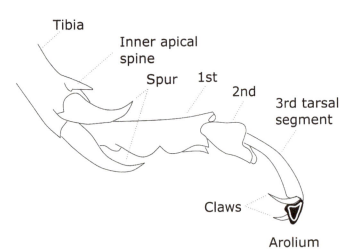

FIGURE 5.6 Tarsal details of hind leg.

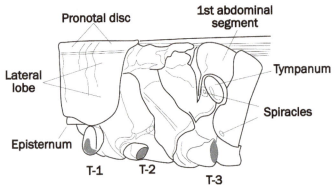

FIGURE 5.7 Prothorax, pterothorax, and first abdominal segment showing tympanum.

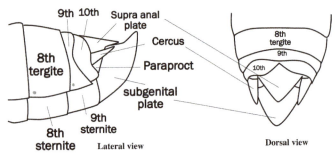

FIGURE 5.8 External male genitalia.

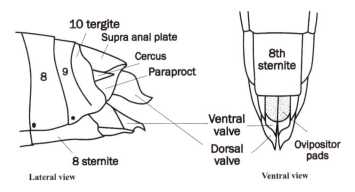

FIGURE 5.9 Female genitalia.

- Shape of pronotum, saddle-shaped or flat; raised or tectiform; with or without distinct carinae (lateral and dorsal)
- Tegmina and wings fully developed or reduced; radial area of tegmina with or without a series of parallel stridulatory veinlets
- Hind wings: colour at base, presence or absence of band or fasciae
- Hind femur shape, structure, presence or absence of band or fasciae
- Stridulatory mechanism type: its presence or absence; if present, then form/type
- Presence or absence of external apical spine of hind tibia
- Lower external lobe of hind knee (genicular lobe): rounded/sub-acute/spine-like

- Meso-sternal interspace closed or open/shape of lobes, acute angled or rounded
- Male cerci shape and size
- Variation in shape and size of epiphallus.

CLASSIFICATION OF ACRIDIDAE ADOPTED DURING DIFFERENT PERIODS

The family name Acrididae was proposed by Latreille (1802) and Krauss (1877) based on the genus *Acrida* Linnaeus (1758) in both cases. Thereafter, as a family name, Acrididae was adopted by a greater majority of taxonomists; however, significant variation can be observed in further subdivision up to subfamily and tribe levels. Thomas (1880), Brunner von (1893), and Lefroy (1909) adopted the family name Acrididae. Later, Lucas (1920) raised the family Acrididae to the rank of suborder Acridoidea, but maintained Acrididae as one of the families. Similarly, Uvarov (1921), Chopard (1943), Willemse (1951), Bei-Bienko and Mishchenko (1952), Rehn (1953), Brues et al. (1954), Mason (1954), Johnston (1956), Dirsh (1956), Shumakov (1963), Jago (1968), Avakyan (1968), and Amedegnato (1974) adopted Acrididae as one of the families of Acridoidea.

Harz (1975) followed the classification given by Willemse (1951); later, Willemse and Kruseman (1976) and Herrera and Schnidrig (1983) followed the system of classification of Acridomorphoid insects given by Dirsh (1975). Mason (1979) followed Dirsh (1975) in arranging the taxa, but retained the superfamily name Acridoidea. Leorente (1980), Garcia and Presa (1984), and Clemente et al. (1987) followed Harz (1975), but did not make any attempt to divide the subfamilies into tribes. Johnsen (1982–1987, 1990) upheld Dirsh's (1965) system of dividing the superfamily Acridoidea into families and subfamilies and followed the concept of Uvarov (1966) in treating the Truxalinae *sensu* Dirsh (1965). Herrera and Schnidrig (1983) divided the suborder Caelifera into two superfamilies: Pamphagoidea with one family Pamphagidae and Acridoidea with two families: Catantopidae (with two subfamilies: Catantopinae and Calliptaminae) and Acrididae (with two subfamilies: Oedipodinae and Gomphocerinae). Gracia and Presa (1984) recognized four distinct families: Pamphagidae, Catantopidae, Pyrgomorphidae, and Acrididae (with three subfamilies: Truxalinae, Locustinae, and Gomphocerinae).

Key (1986) classified Australian Acridoidea and recognized two distinct families: Pyrgomorphidae (with one subfamily Pyrgomorphinae) and Acrididae (with four subfamilies: Oxyinae, Catantopinae, Cyrtacanthacridinae, and Acridinae). Clemente et al. (1987) followed Garcia and Presa (1984). Balderson and Yin (1991), in their work on Acridoidea of Nepal, divided the superfamily Acridoidea into four families: Chrotogonidae (with two subfamilies: Chrotogoninae and Tagastiriae), Pyrgomorphidae (with one family Atractomorphinae), Oedipodidae (with eight subfamilies: Spathosterninae, Catantopinae, Habrocneminae, Oxyinae, Oedipodinae, Locustinae, Ceracrinae, and Arcypterinae), and Acrididae (with two subfamilies: Phlaeobinae and Acridinae). Gomez et al. (1991) considered three distinct families: Pyrgomorphidae, Pamphagidae (with two subfamilies: Akicerinae and Pamphaginae), and Acrididae (with five subfamilies: Catantopinae, Cyrtacanthacridinae, Acridinae, Oedipodinae, and Gomphocerinae) in Acridoidea. Herrera (1982) recognized two superfamilies under the suborder Caelifera: Tetrigoidea with one family Tetrigidae and Acridoidea with two families: Catantopidae (with three superfamilies: Cyrtacanthacridinae, Catantopinae, and Calliptaminae) and Acrididae (with three subfamilies: Oedipodinae, Acridinae, and Gomphocerinae); later followed by Herrera and Larumbe (1996). Lachininsky et al. (2002), in their work on short-horned grasshoppers, detailed out the classification systems followed by different acridologists, in a tabulated form, covering 8 genera of Pamphigidae and 84 genera of Acrididae.

The other names proposed independently were Truxalidae, Oedopodidae (Serville 1838; Walker 1870; Scudder 1875; and Saussure 1884), and Locustidae Leach (Westwood 1840, 1845; Zanon 1924; Comstock 1954; and Essig 1958); however, only a minor group of taxonomists upheld this system of classification.

As per OSF, the family Acrididae has been subdivided into 26 subfamilies, which are: *Acridinae*, Calliptaminae, Catantopinae, Copiocerinae, Coptacrinae, *Cyrtacanthacridinae*, Egnatiinae, Eremogryllinae, Euryphyminae, Eyprepocnemidinae, *Gomphocerinae*, Habrocneminae, Hemiacridinae, Leptysminae, Marelliinae, Melanoplinae, *Oedipodinae*, Ommatolampinae, Oxyinae, Pauliniinae, Proctolabinae, Rhytidochrotinae, Pezotettiginae, Spathosterninae, Teratodinae, and Tropidopolinae. The subfamily Romaleinae, formerly in Acrididae, has been upgraded to family status. The members of subfamily Pezotettiginae, a recent addition, are known to occur in Europe, West Asia, and the Middle East. In the Indian subcontinent, so far, the acridid fauna of 14 subfamilies have been reported; besides, the subfamily Habrocneminae is also known to occur in the Indo-China region. Nevertheless, keeping in view the diverse ecosystems, it is quite possible that a few more subfamilies do exist that need to be discovered. Consequent to the fact that the common ancestor of Acrididae must have originated from Latin America, rather than Africa as was earlier presumed, some of the subfamilies like Copiocerinae, Egnatiinae, Eremogryllinae, Euryphyminae, Leptysminae, Marelliinae, Ommatolampinae, Pauliniinae, Proctolabinae, and Rhytidochrotinae are restricted to Central and South America justifying the bio-geographical analysis. All the same, fauna of the subfamilies Acridinae, Cyrtacanthacridinae, Gomphocerinae, and Oedipodinae are known to have worldwide distribution. In addition to this, the subfamily status of the tribe Eucopiocerini, that is known to occur in Central America, has not been determined.

KEY TO SUBFAMILIES OF ACRIDIDAE IN THE INDIAN SUBCONTINENT (MODIFIED FROM DIRSH 1961)

1. Pronotum very high, sublaminately compressed, crest shaped; lower basal lobe of hind femur as long as upper one................................ **Teratodinae**

- Pronotum from saddle-shaped to tectiform, but not crest shaped; lower basal lobe of hind femur distinctly shorter than upper one 2
2. Prosternum without process, if present, antennae ensiform and body elongate 3
- Prosternum with well developed process; antennae variable, body not elongate....................... 5
3. Frons vertical or nearly so; integument rugulose; intercalary veins well developed, serrate, forming file for stridulatory mechanism................... **Oedipodinae**
- Frons oblique; integument not rugulose; intercalary veins neither well developed nor serrate; stridulatory structures when present on hind femur 4
4. Eyes nearer apex than to base of head; stridulatory file on hind femur when present, represented by closely set rigid tubercles and articulated bristles .... **Acridinae**
- Eyes not nearer to apex than to base of head; stridulatory file on hind femur represented by articulated pegs..................................**Gomphocerinae**
5. Lower external lobe of hind knee with spine-like apex ....**Oxyinae**
- Lower external lobe of hind knee with rounded apex, angular or subacute, but not spine-like 6
6. Radial area of tegmina with a series of regular parallel stridulatory veinlets............................ 7
- Radial area of tegmina without a series of regular, parallel stridulatory veinlets................. 8
7. Prosternal process transverse, lamellate or subquadrate; apical abdominal tergite with furcular lobes .... **Spathosterninae**
- Prosternal process conical; apical abdominal tergite without furcular lobes .... **Hemiacridinae**
8. Mesosternal interspace closed; male subgenital plate folded..**Tropidopolinae**
- Mesosternal interspace open; male subgenital plate not folded... 9
9. Last abdominal tergite in male with well developed furcula; supra anal plate with attenuated apex ....**Coptacrinae**
- Last abdominal tergite in male without well developed furcula; supra anal plate variable 10
10. Mesosternal lobes rectangular .... **Cyrtacanthacridinae and Melanoplinae**
- Mesosternal lobes rounded or obtusely angular or acutely angular 11
11. Male cerci pincer-like, strong, regularly incurved with small apical lobes ....**Calliptaminae**
- Male cerci variable, but not pincer-like.................. 12
12. Dorsum of pronotum flat or weakly tectiform, with linear, median and lateral carinae; male cerci with strongly compressed or subacute apex........................ ....**Eyprepocnemidinae**
- Dorsum of pronotum of variable shape; lateral carinae absent; male cerci variable, but not with strongly compressed lobiform or subacute apex........................ ....**Catantopinae**

Major Work on Indian Acrididae

Probably Stål (1860, 1873) initiated the study of Indian Acrididae. Walker (1870, 1871) and Saussure (1884, 1888) also studied some Indian fauna. From 1900 onwards, major contributions were made by Bolivar (1902, 1909, 1917, 1918). A notable taxonomical work on Acrididae was made by Kirby (1914) in the series "Fauna of British India". Uvarov (1921, 1924, 1927, 1942) studied Indian Acrididae in detail. Roonwal (1936, 1945, 1946, 1958, 1976) contributed some studies on the nymphal structures and ecology of Acrididae. Bhowmik (1964, 1965, 1985, 1986), Tandon and Shishodia (1969, 1989) have contributed to the taxonomy of this group working under the aegis of Zoological Survey of India. Kushwaha and Bhardwaj (1977) studied the acridid fauna associated with forage and pasture lands of Rajasthan. Kumar and Viraktamath (1989) have studied the taxonomy and diversity of short-horned grasshoppers of Bangalore district of Karnataka. Shishodia and Hazra (1985) and Tandon and Hazra (1988) have done work on taxonomy as well as on ecology of this group. Usmani and Shafee (1984) added a new tribe Gesonulini to the subfamily Oxyinae. Shishodia et al. (2010) divided the Indian Acridoidea into three families: Acrididae, Dericorythidae, and Pamphagidae and further divided the family Acrididae into 15 subfamilies: Acridinae, Calliptaminae, Catantopinae, Coptacridinae, Cyrtacanthacridinae, Eyprepocnemidinae, Gomphocerinae, Hemiacridinae, Melanoplinae, Oedipodinae, Ommatolamphinae, Oxyinae, Spathosterninae, Teratodinae, and Tropidopolinae. Studies on the diversity and abundance of Orthopteran fauna in rice ecosystems of south India were made by Chitra et al. (2001) and Kandibane et al. (2004). Rathore (2009) carried out bio-systematic investigations on *H. nigrorepletus* in the maize based agro-ecosystems of southwest Rajasthan; Dhakad (2013) studied the Orthopteran fauna of sugarcane in the subhumid Aravali agro-ecosystem of Rajasthan; and Swaminathan et al. (2018a) described a new species of the genus *Hieroglyphus*, *Hieroglyphus kolhapurensis* from Maharashtra. Some representative species of the tribe Catantopini (Acrididae) from India have been reported by Swaminathan et al. (2018b).

Taxonomic Problems

Within the superfamily Acridoidea, the Acrididae is the most numerous and widely distributed family. The other families vary in their diversity world over and within the Indian subcontinent. The number of families and/or subfamilies is often broadly debated owing to the fact that historically, several taxonomical schools studied the group, adopting distinct taxonomic classifications and different morphological interpretations. Opinions of specialists differ over the subfamilies within Acrididae. The descriptions of genera and species of Acridoidea, following morphological analyses of the external and internal genitalia of males, have stabilized and greatly aided in the precise identification of species.

ECONOMIC IMPORTANCE

Grasshoppers are noted in the list of destructive crop pests, and the family Acrididae alone has more than 100 species that are pests of agricultural crops and pastures. Further, life economy, long span of life, discrimination of micro climatic condition or wild flora besides divergent biology, and colour patterns have provided the grasshopper special status and superiority among the agricultural insect pests. Orthoptera species and assemblages vary enormously in biology, abundance, population variability, and geographic range. Some are major pests, but others are threatened with extinction or have become extinct due to human activities. Most pest species are in the Acrididae, yet proportionately more threatened species are in the less speciose families. Pest Orthoptera species are unusual on islands, which nevertheless support several threatened non-acridid species. In contrast, continental species of Acrididae and Tettigoniidae are the ones principally threatened. Many of the threatened Orthoptera species are confined to a small geographical area especially due to anthropogenic impacts that coincide with their small ranges (Samways and Lockwood 1998).

Being familiar insects in terrestrial habitats, grasshoppers are dominant herbivores especially in the major grasslands around the world (Uvarov 1966; Mitchell & Pfadt 1974; Gangwere et al. 1997; Cigliano et al. 2000; Guo et al. 2006). Several species of grasshoppers are considered major pests, especially when they periodically develop into local and large-scale outbreaks, causing enormous economic damage (COPR 1982). Some of the most important acridid pests around the world are locusts, which are grasshoppers that can form dense migrating swarms and exhibit density-dependent phase polyphenism (Uvarov 1966; Pener 1983; Pener & Simpson 2009; Cullen et al. 2017).

In many tropical and subtropical agro-ecosystems, short-horned grasshopper species that belong to the subfamilies Hemiacridinae (including the *Hieroglyphus* complex: *Hieroglyphus nigrorepletus* Bolivar; *Hieroglyphus banian* (Fabricius); *Hieroglyphus oryzivorous* Carl; and *Parahieroglyphus bilineatus* Bolivar) and Oxyinae (*Oxya* spp.) are serious pests, often attaining major status under favourable environmental conditions; while elsewhere, some species are beneficial, such as *Cornops aquaticum* (Bruner) (Leptysiminae), which has been used as a successful biocontrol agent of water hyacinth in South Africa (Bownes et al. 2011, Coetzee et al. 2011), and *Hesperotettix viridis* (Thomas) (Melanoplinae), which prefers to feed on noxious snakeweeds that can harm cattle and other livestock (Thompson & Richman 1993). In the Indian context, agriculturally important species of Acridoidea have been enlisted by the Zoological Society of India (Mandal et al. 2007).

COLLECTION AND PRESERVATION (MODIFIED FROM SCHAUFF 1986)

Orthopterans are generally collected by hand, net, suction trap, and light trap. If collected specimens cannot be mounted immediately, they can be kept for a few days in a refrigerator in a sealed jar and then pinned out afterwards. Alternatively, they can simply be put in paper envelopes and kept dry. They will need to be moistened before you can pin them out.

- **Processing specimens:** Pin the specimens as soon as you can after killing.
- **Evisceration:** It is essential to eviscerate large species and recommended for small species. Fill the cavity with cotton wool or filter paper and boric acid.
- **Softening:** Put some damp sand into a hermetically sealed container with dried naphthalene or carbolic acid to prevent the development of mould. Leave specimens for 1–2 days. The process can be speeded up by putting the container over boiling water.
- **Pinning out:** Spread the legs of the insect symmetrically and open out the wings on one side to study major taxonomic characters. In the tettigonids and crickets, the male fore wing is of specific importance; hence, sometimes wings on both sides need to be spread out.
- **Drying:** Leave the pinned specimens to dry for 1–2 weeks in an air-conditioned room or dry place and ensure protection from ants.
- **Labelling:** Source/host from which collected; name of collector; date collected; and place from where collected. Place the label on the body pin, under the insect.

Use a glass-topped box or entomological cabinet containing paradichlorbenzene or dichlorvos in a glass or cardboard container and arrange the insects systematically. Always preserve at least one male and one female of each species, preferably a series.

PRESERVATION FOR MOLECULAR STUDIES (UPTON AND MANTLE 2010)

Molecular studies are dependent on good-quality DNA, the successful extraction of which is based on appropriate preservation of insect samples. The underlying objective of tissue preservation is to prevent the damage and degradation of nucleic acids. Of late, natural history collections lay more emphasis on establishing tissue and/or DNA collections. As the enzymes involved in the degradation of DNA are most active in water; hence, the most important step in preserving tissue is to eliminate water. The most frequently used method of DNA preservation is immersion of the live specimen in 95%–99% ethanol, as it is reliable and practical for use in the field. Freezing specimens in liquid nitrogen or on dry ice produces excellent samples that can be stored long-term and freezing is recommended for archival storage. This method is equally suitable for DNA, RNA, and protein samples, as well as living tissue. However, it is expensive for long-term storage and using liquid nitrogen in the field can be complicated. RNAlater® (Ambion) and Allprotect Tissue Reagent (Qiagen) are newly developed fluids that preserve DNA and RNA at room temperature, for a short duration. Extraction kits for processing samples stored in either of these fluids are available from the manufacturers. These products are relatively expensive compared to the other methods. It is of utmost importance to keep instruments and containers clean and store the specimens separately. Polymerase Chain Reaction (PCR) amplifies very small quantities of DNA so any contamination by a foreign organism will cause spurious results.

USEFUL WEBSITES

Orthoptera Species File—OSF (*orthoptera.speciesfile.org*)

CONCLUSION

The taxonomic knowledge of Acridoidea in India is yet in a developing state. As a result of different taxonomical classifications adopted by specialists, the progress made has been rather slow; however, with the availability of the globally common organization, the *Orthoptera Species File*, we can strengthen our current knowledge on this group. Nevertheless, descriptions, identification keys, catalogues, and taxonomic revisions, at all levels, are still very needed, in order to increase the knowledge of its diversity. The systematic knowledge of Acridoidea is incipient, with a lack of phylogenetic hypotheses and cladistic studies. Further, Song et al. (2018) have inferred that the common ancestor of modern grasshoppers originated in Latin America, contrary to the popular belief that they originated in Africa, based on a biogeographical analysis. It becomes increasingly necessary that studies on taxonomic aspects be carried out on a regular basis in different parts of the country to have a better estimate of the total number of Indian representatives of Acridoidea. Thus, for the beginners and those who are already ahead into the taxonomy of this group, updating our knowledge vis-a-vis OSF is essential. Molecular approaches can assist the traditional morphological taxonomy in a great way, but cannot replace it.

REFERENCES

Amedegnato, C. 1974. Les genres d'acridiens neotropicaux, leur classification par familles, sous-familles et tribus. *Acrida* 3: 193–203.

Ander, K. 1939. *Comparative anatomical and phylogenetic studies on the Ensifera (Saltatoria)*. Opuscula Entomologica, Suppl. II. Lund, c.f.: Gwynne, Darryl T., Laure DeSutter, Paul Flook and Hugh Rowell 1996. Orthoptera. Crickets, katydids, grasshoppers, etc., http://tolweb.org/Orthoptera/8250/1996.01.01 in the Tree of Life Web Project, http://tolweb.org/.

Avakyan, G. D. 1968. Fauna of the Armenian SSR, (Orthoptera: Acridoidea). *Erevan Akademii Nauka, Armenia SSR* 259 p.

Balderson, J., and X. Yin. 1991. Grasshoppers (Orthoptera): Eumastacoidea and Acridoidea collected in Kashmir. *Entomologist's Gazette* 42: 189–205.

Bei-Bienko, G. Ya., and L. L. Mishchenko. 1951. *Locusts and Grasshoppers of the U. S. S. R. and Adjacent Countries*. Akademii Nauka, USSR, Moscow Part I no. 38: xxi + 291 p.

Bhowmik, H. K. 1964. On a new species of grasshopper, Euprepocnemis (Insecta: Orthoptera: Acrididae), from India. *Journal of the Bengal Natural History Society* 32: 89–92.

Bhowmik, H. K. 1965. A new species of the grasshopper Genus Sjostedtia Bolivar, from India. *Proceedings of the Zoological Society of Calcutta* 18: 59–62.

Bhowmik, H. K. 1985. Outline of distribution with an index catalogue of Indian grasshopper (Orthoptera: Acrodoidea). *Records of Zoological Survey of India, Miscellaneous Publications, Occasional Paper* 78: 1–51.

Bhowmik, H. K. 1986. Grasshopper fauna of West Bengal, India (Orthoptera: Acrididae). *Zoological Survey of India, Technical Monograph* No. 14: 1–180.

Bolivar, I. 1902. Les Orthopteres de St. Joseph's College a Trichinopoly (Sud de rinde). *Annales de la Société Entomologique de France* 70: 580–635.

Bolivar, I. 1909. Orthoptera, Fam. Acrididae. Subfamily Pyrgomorphinae. *Boletín de la Sociedad Española de Historia Natural* 9: 285–296.

Bolivar, I. 1917. Contribution alia conocimento de la fauna India. *Revista Acad. Ciene. Madr.* 16: 278–412.

Bolivar, I. 1918. Contribution al concocimento de la fauna India. *Revista de la Real Academia de Ciencias Exactas.* 16: 278–412.

Bownes, A., A. King, and A. Nongogo. 2011. Prerelease studies and release of the grasshopper, Cornops aquaticum in South Africa—A new biological control agent for water hyacinth, Eichhornia crassipes, pp. 3–13. In *XIII International Symposium on Biological Control of Weeds*, September 11–16, 2011, Waikoloa, Hawaii.

Brues, C. T., A. L. Melander, and F. M. Carpenter. 1954. *Classification of insects* (Revised ed.). *Bulletin of the Museum of Comparative Zoology* 108: 1–917 p.

Brunner, von W. C. 1893. Revision du systeme des Orthopteres et description des especes rapportees par M. Leonardo Fea de Birmanie. *Annali del Museo Civico di Storia Naturale di Genova* 13(2): 1–230.

Chitra, N., K. Gunathilagaraj, and R. P. Soundararajan. 2001. Orthopteran fauna in a rice ecosystem of South India. *Bionotes* 3: 8.

Chopard, L. 1920. Recherches sur la conformation et la développement des derniers segmentes abdominaux des Orthoptères. Thèse, Faculté des Sciences de Paris, Oberthur, Rennes, c.f., Chopard, L. 1969. The Fauna of India and Adjacent Countries. Orthoptera, 2, Grylloidea. The Manager of Publication Delhi, xviii+ 421.

Chopard, L. 1943. *Orthopteroïdes de L'Afrique du Nord, Faune de l'Empire Francais*, Lib. Larose, Paris 1: 450 p.

Cigliano, M. M., H. Braun, D. C. Eades, and D. Otte. 2018. *Orthoptera Species File*. Version 5.0/5.0. November 7, 2018. http://Orthoptera.SpeciesFile.org.

Cigliano, M. M., M. L. de Wysiecki, and C. E. Lange. 2000. Grasshopper (Orthoptera: Acridoidea) species diversity in the Pampas, Argentina. *Diversity and Distribution* 6: 81–91.

Clemente, M. E., M. D. García, and J. J. Presa. 1987. Clave de los géneros de saltamontes ibéricos (Orthoptera; Caelifera). Secretariado de Publicaciones e intercambio científico. Universidad de Murcia, Murcia, Spain, 64 p.

Coetzee, J. A., M. P. Hill, M. J. Byrne, and A. Bownes. 2011. A review on the biological control programmes on Eichhornia crassipes (C. Mart.) Solms (Pontederiaceae), Salvinia molesta D.S. Mitch (Salviniaceae), Pistia stratiotes L. (Araceae), Myriophyllum aquaticum (Vell.) Verdc. (Halorogaceae) and Azolla filiculoides Lam. (Azollaceae) in South Africa. *African Entomology*, 19: 451–468.

Comstock, J. H. 1954. *An Introduction to Entomology*, revised ed., Conmstock Publ. Assoc., Ithaca, NY. 1064 p.

COPR. 1982. *The Locust and Grasshopper Agricultural Manual*. Centre for Overseas Pest Research, London, UK.

Cullen, D. A., A. Cease, A. V. Latchininsky et al. 2017. From molecules to management: Mechanisms and consequences of locust phase polyphenism. *Advances in Insect Physiology*, 53: 167–285.

da Costa, Maria Kátia Matiotti. 2016. Taxonomy of orthoptera with emphasis in Acridoidea. *Metaleptea* 36 (2): 7–8.

Dhakad, D. 2013. Diversity of orthopteran fauna in sugarcane. M. Sc. (Ag.) Thesis submitted to Maharana Pratap University of Agriculture and Technology, Udaipur, India.

Dirsh, V. M. 1956. The phallic complex in Acridoidea (Orthoptera) in relation to taxonomy. *Transactions of the Royal Entomological Society, London,* 108: 223–356.

Dirsh, V. M. 1961. A preliminary revision of the families and subfamilies of Acridoidea (Orthoptera: Insecta). *Bulletin of the British Museum (NH) (Entomology),* 10: 351–419.

Dirsh, V. M. 1975. *Classification of the Acridomorphoid Insects,* E. W. Classey, Farringdon, UK, 171 p.

Essig, E.O. 1958. *Insects and Mites of Western North America.* Macmillan, New York, 1050 p.

Gangwere, S. K., M. C. Muralirangan, and M. Muralirangan (eds.). 1997. *The Bionomics of Grasshoppers, Katydids, and Their Kin.* CAB International, New York.

Garcia, M. D., and J. J. Presa. 1984. *Dociostaurus (Kazakia) monserrati* un Nuevo Gomphocerini de la fauna iberica (Orth. Acrididae). *Boletin de la Asociacion Espanola de Entomologia* 8: 21–24.

Gomez, R., J. J. Presa, and M. D. Garcia. 1991. Orthopteroidea del sur de la provincia de Albacete (España) Ensifera, Mantodea, Phasmoptera, Blattoptera, Dermaptera. *Anales de Biologia. Biologia Animal. Universidad de Murcia* 17 (6): 7–21.

Gorochov, A. V. 1995. Contribution to the system and evolution of the order Orthoptera. *Zoologichesky Zhurnal* 74: 39–45.

Guo, Z.-W., H.-C. Li, and Y.-L. Gan. 2006. Grasshopper (Orthoptera: Acrididae) biodiversity and grassland ecosystems. *Insect Science* 13: 221–227.

Harz, K. 1975. Die Orthopteren Europes. *The Orthoptera of Europe.* II - Dr. W. Junk B. V., The Hague, the Netherlands, 939 p.

Herrera, L. 1982. *Catalogue of the Orthoptera of Spain. Series Entomologica.* Dr. W. Junk. The Hague, the Netherlands, Vol. 22, pp. 1–162.

Herrera, L., and J. A. Larumbe. 1996. Distribución de Grylloidea y Caelifera (Orthoptera) en Cantabria (España). *Boletín de la Real Sociedad Española de Historia Natural* 92 (1–4): 101–111.

Herrera, L., and S. Schnidrig. 1983. *Andropigios de los Ortopteros de Navarra (Orthoptera), Publicationes Biologia,* Universidad De Navarra, 5. *Zoologia* 10: 1–52.

Howell, H. V., J. T. Doyen, and A. H. Purcell. 1998. *Introduction to Insect Biology and Diversity,* 2nd ed. Oxford University Press, Oxford, UK, pp. 392–394.

Jago, N. D. 1968. A checklist of grasshoppers (Orthoptera: Acrida) recorded from Ghana, with biological notes and extracts from the recent literature. *Transactions of American Entomological Society, London* 94: 209–353.

Johnston, H. B. 1956. *Annotated Catalogue of African Grasshoppers.* Cambridge, UK, 833 p.

Johnsen, P. 1982–1987. Acridoidea of Zambia. 1–7, *Zool. lab.,* 1: 1–81, 2: 82–162, 3: 163–241, 4: 242–266, 5: 267–354, 6: 355–442, 7: 443–505.

Johnsen, P. 1990. Acridoidea of Botswana. *Zool. lab.,* 1: 1–129.

Kandibane, M., S. Raguraman, and D. N. Ganapathy. 2004. Diversity and abundance of Orthoptera in the irrigated rice ecosystem of Tamil Nadu, India. *Pest Management and Economic Zoology* 12: 71–76.

Key, K. H. L. 1986. *A provisional synonymic list of the Australian Acridoidea (Orthoptera).* CSIRO. Division of Entomology, Melbourne, Australia, 47 p.

Kirby, W. F. 1914. *Fauna of British India, Including Ceylon and Burma,* Orthoptera (Acrididae), London, UK, 97 p.

Kukalova-Peck, J. 1991. Fossil history and the evolution of hexapod structures. In: *The Insects of Australia* (edited by CSIRO). Melbourne University Press, Melbourne, Australia, pp. 141–179.

Kumar, P., and C. A. Viraktamath. 1989. A list of short-horned grasshoppers (Orthoptera: Acridoidea) of Bangalore district and their host plants. *Mysore Journal of Agricultural Sciences* 23: 320–326.

Kushwaha, K. S., and Bhardwaj, S. C. 1977. *Forage and Pasture Insect Pests of Rajasthan.* ICAR Publication, New Delhi, India, 186 p.

Lachininsky A. V., M. G. Sergeev, M. K. Childebayev et al. 2002. *Locusts of Kazakhstan, Central Asia and Adjacent Territories—* Association for Applied Acridology International/University of Wyoming, Laramie, WY, p. 387.

Latreille, P. A. 1802. Families naturelles du regne animal, exposees succinctement et dans un ordre analytique, avec Vindication de leurs genres. J.B. Bailliere, Paris, 1–570.

Lefroy, H. M. 1909. *Indian Insect Life. Manual of the Insects of Plains* (Tropical India): xii + 786 pp, Thacker, Spink & Co., Calcutta and Simla, W. Thacker & Co., London.

Leorente del Moral V. 1980. Los Orthopteroidea del coto de Donana (Huvela). *Eos,* 54: 117–165.

Lucas, W. J. 1920. *A monograph of the British Orthoptera.* Ray Society Publication, London, UK.

Mandal, S. K., A. Dey, and A. K. Hazra. 2007. *Pictorial Handbook on Indian Short-horned Grasshopper Pests (Acridoidea: Orthoptera):* pp. 1–57. Zoological Survey of India, Kolkata, India.

Mason, J. B. 1954. Number of antennal segments in adult Acrididae (Orthoptera). *Proceedings of the Royal Entomological Society of London* 23: 228–238.

Mason, J. B. 1979. Acridoidea of south-west Angola. *Eos: Revista española de entomología, Madrid* 53 (1977): 91–132.

Mishchenko, L. L. 1952. Fauna of the U. S. S. R., orthoptera, locusts and grasshoppers, Catantopinae. *Akademii Nauka, USSR, Moscow* 4: 560 p.

Mitchell, J. E. and R. E. Pfadt. 1974. The role of grasshoppers in a shortgrass prairie ecosystem. *Environmental Entomology* 3: 358–360.

Pener, M. P. 1983. Endocrine aspects of phase polymorphism in locusts, pp. 379–394. In R.G.H. Downer and H. Laufer (eds.), *Invertebrate Endocrinology,* vol. 1, *Endocrinology of Insects.* Alan R. Liss, New York.

Pener, M. P., and S. J. Simpson. 2009. Locust phase polyphenism an update. *Advances in Insect Physiology* 36: 1–286.

Rathore, P. S. 2009. Bio-systematic investigations of the acridid, *Hieroglyphus nigrorepletus* Bolivar in South Western Rajasthan. Ph. D. Thesis submitted to Maharana Pratap University of Agriculture and Technology, Udaipur, India.

Rehn, J. A. G. 1953. *The grasshoppers and Locusts (Acridoidea) of Australia. Vol. II. Family Acrididae (Subfamily Pyrgomorphinae).* CSIRO: Melbourne, Australia, 270 p.

Roonwal, M. L. 1936. Studies on the embryology of the African migratory locust, *Locusta migratoria migratorioides* R. and F. I. The early development, with a new theory of multi-phased gastrulation among insects. *Philosophical Transactions of Royal Society, London* (B) 226: 391–421.

Roonwal, M. L. 1945. Notes on the bionomics of *Hieroglyphus nigrorepletus* Bolivar (Orthoptera: Acrididae) at Benaras, United Provinces, India. *Bulletin of Entomological Research* 36 (3): 339–341.

Roonwal, M.L. 1946. Studies in intraspecific variation. I. on the existence of two colour types in the adults and hoppers of the *solitaria* phase in the Desert locust, *Schistocerca gregaria* (Forsskal) [Orthoptera, Acrididae]. *Rec. Indian Mus.,* 44: 369–374.

Roonwal, M.L. 1958. Bibliographia Acrididiorum. *Rec. Indian Mus.,* 56: 611.

Roonwal, M. L. 1976. Ecology and biology of the grasshopper, *Hieroglyphus nigrorepletus* Bolivar (Acrididae). Distribution, economic importance, Hfe-history, colour forms, and problems of control. *Zeitschrift fur angewandte Zoologie, Berlin* 63: 307–332.

Samways, M. J., and J. A. Lockwood. 1998. The conservation of Orthoptera: Pests and paradoxes. *Journal of Insect Conservation* 2: 143–149.

Saussure, H. De. 1884. Prodromus Oedipodiorum, insectorum ex ordine Orthopterorum. *Memoires de la Societe de physique d'histoire naturelle de Geneve* 28: 1–256.

Saussure, H. De. 1888. Addimenta ad Prodromum Oedipodiorum, Insectorum ex ordine Orthopterorum. *Mémoires de la Société de Physique et d'Histoire Naturelle de Genève* 30: 1–180.

Schauff, M. E. 1986. *Collecting and Preserving Insects and Mites: Techniques and Tool.* Updated and modified version of the USDA Misc. Publication no. 1443 published by the Agricultural Research service in 1986 and edited by George C. Steyskal, William L. Murphy, and Edna M. Hoover.

Scudder, S. H. 1875. A century of Orthoptera. Decade IV. Acrydii. *Proceedings of the Boston Society of Natural History* 17: 274.

Serville, J. G. A. 1838 [1839]. *Histoire Naturelle des Insectes.* Orthoptères. Libraire Encyclopèdique de Roret. Paris, France. 776 p.

Sharov, A. G. 1968. Phylogeny of the Orthopteroidea. *Transactions of Paleontological Institute*, Russian Academy of Sciences, 118: 1–216.

Shishodia, M. S., and A. K. Hazra. 1985. Records of the Zoological Survey of India, Orthoptera Fauna of Silent Valley, Kerala *Records of the Zoological Survey of India* 82 (1–4): 17.

Shishodia, M. S., K. Chandra, and S. K. Gupta. 2010. An annotated checklist of Orthoptera (Insecta) from India. *Records of the Zoological Survey of India* 314: 1–366.

Shumakov, E. M. 1963. Acridoidea of Afghanistan and Iran. *Horae Societatis Entomologicae Unionis Soveticae* 49: 3–248.

Song, H., R. Marino-Perez, D. A. Woller, and M. M. Cigliano. 2018. Evolution, diversification, and biogeography of grasshoppers (Orthoptera: Acrididae). *Insect Systematics and Diversity* 2(4): 1–25. doi: 10.1093/isd/ixy008.

Stål, C. 1861 [1860]. Kongliga Svenska fregatten Eugenies Resa omkring jorden under befäl af C.A. *Virgin åren 1851–1853* (Zoologi) 2 (1): 324–336.

Stål, C. 1873. Recencio Orthopterorum. *Revue critique des Orthoptères décrits par Linné, De Geer et Thunberg* 1: 81.

Swaminathan, R., T. Swaminathan, and Rajendra Nagar 2018a. A new species of *Hieroglyphus* Krauss, 1877 (Orthoptera: Hemiacridinae: Acrididae) from India. *Transactions of the American Society of Entomology*, 143: 625–632.

Swaminathan, R., R. Nagar, and T. Swaminathan. 2018b. Representative species of the tribe Catantopini (Orthoptera: Acrididae) from India. *Transactions of the American Society of Entomology*, 144: 239–261.

Tandon, S. K., and M. S. Shishodia. 1969. Acridoidea (Insecta: Orthoptera) collected along the bank of River Tawi (Jammu and Kashmir), India. *Newsletter Zoological Survey of India*, 2(6): 269–271.

Tandon, S. K. and M. S. Shishodia. 1976. On a collection of Orthoptera from Rajasthan, India. *Newsletter Zoological Survey of India*, 2(1): 7–11.

Tandon, S. K., and M. S. Shishodia. 1989. Insecta: Orthoptera: Acridoidea, *Fauna of Orissa: Zoological Survey of India State Fauna* Series, 1(2): 93–145.

Tandon, S. K., A. K. Hazra, and S. K. Mandal. 1988. Observation on the field biology and ecology of some grasshoppers (Orthoptera: Acridoidea) near Calcutta. *Records of the Zoological Survey of India*, 85(2): 301–318.

Thomas, C. 1880. Influence of meteorological conditions on the development and migrations of locusts. In Riley, C.V., Packard, A.S., and Thomas, C. eds., *Second Report of the United States Entomological Commission for the years 1878 and 1879, relating to the Rocky Mountain Locust and the Western Cricket*, pp. 109–155. Washington, DC, Government Printing Office.

Thompson, D. C., and D. B. Richman. 1993. A grasshopper that only eats snakeweed? pp. 18–19. In Sterling, T.M. and Thompson, D.C. (ed.) *Research Report 674, Agricultural Experiment Station Cooperative Extension Service New Mexico State University, Snakeweed Research Updates and Highlights*. New Mexico Agricultural Experimental Station Report, Ames, IA.

Upton, M. S., and B. L. Mantle. 2010. *Methods for Collecting, Preserving and Studying Insects and Other Terrestrial Arthropods*. The Australian Entomological Society, Paragon Printers Australasia, Canberra, Australia.

Usmani, M. K., and S. A. Shafee. 1984. A new tribe of Oxyinae (Orthoptera: Acrididae). *Bulletin de la Société Entomologique Suisse* 57: 295–296.

Uvarov, B. P. 1921. Notes on Orthoptera in the British Museum. I. The group Euprepocnemini. *Transactions of the Entomological Society of London* 7: 106–144.

Uvarov, B.P. 1927. Distributional records of Indian Acrididae. *Rec. Indian Mus.*, 29: 233–239.

Uvarov, B. P. 1924. A revision of the Old World Cyrtacanthacridinae (Orthoptera, Acrididae) IV, V. *Annals and Magazine of Natural History, London, Series* 13: 1–19; 14: 96–113.

Uvarov, B.P. 1942. Differentiating characters of Oedipodinae and Acridinae. *Trans. Am. Ent. Soc.*, 67: 303–361.

Uvarov, B. P. 1966. *Grasshoppers and Locusts*, Vol. 1. Cambridge University Press, Cambridge, UK.

Walker, F. 1870. *Catalogue of the Specimens of Dermaptera Saltatoria in the Collection of the British Museum, Part I. Locustidae (concluded) and Acrididae (part) pp. 1–117; Part II, pp. 154–224; Part III, pp. 425–604; Part IV. Acrididae* (concluded), E. Newman, Printers, London, UK, pp. 605–801.

Walker, F. 1871. *Catalogue of the specimens of Dermaptera Saltatoria in the Collection of the British Museum*, Supplement, Part V, E. Newman, Printers, London, UK, pp. 49–89.

Westwood, J.O. 1840. Order XIII Diptera Aristotle. (Antliata Fabricius. Halteriptera Clairv.). In: *Synopsis of the genera of British insects*. [1838 – 1840: 158 pp., separately paginated]. – pp. 125–154 in: *An Introduction to the Modern Classification of Insects; founded on the natural habits and corresponding organization of the different families*, Vol. II (1838 – 1840). Longman, Orme, Brown, Green and Longmans, London, xi + 587 + 158 pp.

Westwood, J. O. 1845 [1843–1945] 1. Arcana Entomologica or illustrations or new, rare and interesting insects. *Arcana Entomologica* 2: 1–190 pls. 49–95.

Willemse, C. 1951. Synopsis of the Acridoidea of the Indo-Malayan and adjacent regions (Insecta, Orthoptera) Part I. Fam. Acrididae, subfam. Acridinae. *Publicaties van het Natuurhistorisch Genootschap in Limburg* 4: 41–114.

Willemse, F., and G. Kruseman. 1976. Orthopteroides of Crete. *Tijdschrift voor entomologie* 119: 123–164.

Zanon, V. 1924. Contributo alla conoscenza della fauna entomologica di Bengasi. Ortotteri di Bengasi. *Memorie Ponticia Accademia Scienze Nuovi Lincei* (2a serie) 7: 229–249.

Zeuner, F. E. 1939. *Fossil Orthoptera Ensifera*. British Museum Natural History, London, UK.

6 Taxonomy and Biodiversity of Soft Scales (Hemiptera: Coccidae)

Sunil Joshi

CONTENTS

Introduction	70
Family Diagnosis	70
Diagnostic Characters of Female	70
Diagnostic Characters of Mounted Female	71
Taxonomic Characters of Adult Male	71
Head	71
Thorax	71
Wings	71
Legs	71
Abdomen	72
Dermal Structure	72
Taxonomic Characters of Adult Female	72
Dorsal Structures	73
Marginal Structures	78
Ventral Structures	79
Ventral Pores	83
Classification of Family Coccidae	86
Characteristics of the Subfamilies and Tribes	86
Cardiococcinae Hodgson	86
Ceroplastinae Atkinson	86
Cissococcinae Brain	86
Coccinae Fallen	86
Cyphococcinae Hodgson	87
Eulecaniinae Koteja	87
Eriopeltinae Sulc	87
Filippiinae Bodenheimer	87
Myzolecaniinae Hodgson	88
Pseudopulvinariinae Tang, Hao, Xie, and Tang	88
Major Work on Indian Fauna	88
Distribution Patterns in Indian Context	88
Checklist of Indian Soft Scales	89
Phylogeny	91
Zoogeography of the Coccidae	92
Immature Taxonomy	93
Molecular Characterisation	93
Taxonomic Problems	94
Biology	94
Host Plant Association	95
Economic Importance	96
Collection, Preservation, and Mounting Techniques	96
Collection	96
Preservation and Storage	96
Slide Preparation	97

Useful Websites..97
Conclusion 98
References..98

INTRODUCTION

The branch of entomology that deals with the study of hemipterous insects of the superfamily Coccoidea, particularly on areas related to systematics is known as coccidology. The starting point of coccidology is considered as 1758, beginning with Carl Linnaeus' 10th edition of the *Systema Naturae* (Linnaeus 1758). During this 250-year period, the number of described scale insects have increased from 24 species described by Linnaeus (Williams 2007) to 7,700 species in more than 1,050 genera (Ben-Dov et al. 2009). The word "coccidology" is derived from the word "*Coccus,*" the genus in which Linnaeus included many species of scale insects. Most scale insects were not recognisable as insects, but were considered as seeds or berries and were given the ancient Greek word "Kokkos" and then the later Latin word "Coccus," meaning a berry. The word "coccidology," as a branch of entomology, was probably coined for the first time by Tinsley (1899) in his article "Contributions to coccidology I." Ferris (1957) provided a brief history of Coccoidea.

Scale insects are known by various names depending on the family to which they belong, e.g., the armoured scales (Diaspididae), the mealybugs (Pseudococcidae), the putoids (Putoidae), the soft scales (Coccidae), the felt scales (Eriococcidae), ground pearls (Margarodidae), lac insects (Kerriidae), cochineal insects (Dactylopiidae), and ensign scales (Ortheziidae). The most commonly encountered families are those with the most species, namely, the Diaspididae, Pseudococcidae, and Coccidae. The family Coccidae or "soft scales" is the third most species-rich family within Coccoidea. It is estimated to include over 1,150 described species that occur on more than 200 families of host plants (Ben-Dov et al. 2009), especially on trees and woody shrubs (Ben-Dov 1993).

Scale insects are sap sucking hemipterous insects usually less than 5 mm in length and include all members of the superfamily Coccoidea. These are closely related to aphids (Aphidoidea), whiteflies (Aleyrodoidea), and jumping plant lice (Psylloidea), in the suborder Sternorrhyncha (Gullan & Martin 2003). The taxonomy of scales is based mainly on the microscopic cuticular features of the adult female. The adult female is paedomorphic, maturing in a juvenile form, whereas the adult male (when present), after going through a prepupal and pupal stage, turns into an alate with non-functional mouthparts.

The Coccoidea form a rather small group of insects in terms of species richness with about 7,700 species described. At present, 46 scale insect families are known, of which, 32 are extant and 14 are known only as fossils. Scale insects are generally divided into two informal groups, the archaeococcoids and the neococcoids. The archaeococcoids are defined by the presence of 2–8 pairs of abdominal spiracles, which are absent in the neococcoids. The archaeococcoids consist of 27 families, i.e., 15 extant families (Callipappidae, Carayonemidae, Coelostomidiidae, Kuwaniidae, Marchalinidae, Margarodidae, Matsucoccidae, Monophlebidae, Ortheziidae, Phenacoleachiidae, Pityococcidae, Putoidae, Steingeliidae, Stigmacoccidae, and Xylococcidae) and 12 fossil families [Electrococcidae, Jersicoccidae, Kukaspididae, Labiococcidae, Naibiidae and seven recently described families, namely, Arnoldidae, Lithuanicoccidae, Weitschatidae, Grohnidae, Serafinidae (Koteja 2008), and Hammanococcidae and Lebanococcidae (Koteja & Azar 2008)]. The neococcoids are composed of 17 extant families, i.e., Aclerdidae, Asterolecaniidae, Beesoniidae, Cerococcidae, Coccidae, Conchaspididae, Dactylopiidae, Diaspididae, Eriococcidae, Halimococcidae, Kermesidae, Kerriidae, Lecanodiaspididae, Micrococcidae, Phoenicococcidae, Pseudococcidae, and Stictococcidae; and two extinct families, namely, Inkaidae and the recently described Pennygullaniidae (Koteja & Azar 2008), a neococcoid in the family Putoidae, however, Kondo et al. (2008) placed this family in the archaecoccoids.

FAMILY DIAGNOSIS

Taxonomists working on soft scales have provided diagnostic characters of both live as well as mounted females (Gill 1988; Kosztarab and Kozár 1988; Hodgson 1994; Kosztarab 1996; Hodgson & Henderson 2000) which is summarised below.

DIAGNOSTIC CHARACTERS OF FEMALE

Field diagnostic characters: The appearance of the soft scales in the field is highly variable. Generally, the body shape is round or broadly oval, but, in some species, females are elongate, especially the species that feed on plant family Poaceae. They can be nearly flat (*Eucalymnatus, Pulvinaria, Trijuba*) or highly convex in lateral view (*Saissetia*, some species of *Coccus*). Wax coverings are thin and transparent, filamentous (*Metaceronema*) or powdery (*Hemilecanium*), thick and opaque (*Ceroplastes*), thin, opaque, and brittle (*Drepanococcus*), or even thin and glassy. Soft scales colonise on all parts of the plant, but are predominantly found on leaves and stems. Few species are subterranean. Some species produce ovisacs which generally are filamentous and white (*Pulvinaria, Protopulvinaria*). Newly matured females vary in colour forms from green to brown, mottled to checkered, white to nearly transparent. Older females are usually brown or black. Some species of *Coccus* are so clear that it is possible to watch the malphigian tubules move inside the body (*Coccus viridis*).

Diagnostic Characters of Mounted Female

1. Posterior apex of body usually with conspicuous anal cleft
2. Anal area with two anal plates
3. Eversible anal ring set at end of anal tube
4. Spiracular atrium connected to body margin by furrow containing wax pores
5. Differentiated spiracular setae set at end of spiracular furrow
6. Tarsus without campaniform sensilla near juncture of tibia and tarsus.

Taxonomic Characters of Adult Male

The morphology of the male soft scale insect has been studied by Giliomee (1967) and Miller (1991), and they have described males of about 50 species. Their studies were based mostly on the detailed morphological study of just one species (Theron 1958). Based on the terminology of Giliomee (1967), a few more species have been described by Gimpel et al. (1974), Ray and Williams (1980; 1983), Manawadu (1986), Farrel (1990), and Hodgson (1991a; 1993).

The taxonomic characters of adult male soft scales given below follow Giliomee (1997).

Head

The shape of head: This character separates the *Eriopeltis*-group from the other groups of species. In lateral view, it appears flattened in the *Eriopeltis*-group, but dorsoventrally elongated or rounded in other groups.

The midcranial ridge: The degree of its development varies considerably within the family and the absence of the dorsal part; the lateral arms and the posterior extension of the ventral part can be used to separate genera.

The preocular ridge: Its length also varies considerably. In some genera and species-groups, it reaches the midcranial ridge ventrally, but in others, it does not extend much below the articulation with the scape.

The postocular ridge: The development of this ridge is used to separate genera or species-groups. It is absent dorsally in the *Protopulvinaria*-group and vestigial ventrally in the *Philephedra*-group, while it forks below the ocellus in all species examined by Giliomee (1967), except those of the *Eriopeltis*-group.

The preoral ridge: It shows little variation, but is absent in some species.

The cranial apophysis: This varies in size and in the shape of its apex (truncate, bifurcate, or trifurcate), and this character can be used to separate species and genera.

The simple eyes: These are very conspicuous and, since their number varies between four and ten, it is a very useful character. It is not constant for groups, but can be used to separate closely related genera or species. The large size of the dorsal and lateral eyes is characteristic of the *Philephedra*- and *Inglisia*-groups, respectively.

The antenna: The number of antennal segments is used to separate the *Toumeyella*-group from all other species-groups: in this group, the antennae have nine segments (eight in some apterous forms) compared to ten in all other groups (Ray and Williams 1983; Miller 1991). Other structural characters of the antennae that can be used to separate genera are its length in relation to some other body structures, the length and width of the first three segments, and the shape of the terminal segment.

Thorax

The mesoprephragma: The emargination of mesophragma can be used to separate genera and species.

The mesothoracic scutum: Its length to width ratio can be used to separate genera and species.

The scutellar foramen: The size of foramen can serve as a generic character or to separate some species-groups from other groups.

The basalare: Pleurally, a well-developed basalare (joining the pleural wing process with the mesepisternum) separates the *Eulecanium*-, *Philephedra*-, *Eriopeltis*-, and *Sphaerolecanium*-groups from the *Coccus*-, *Protopulvinaria*-, and *Toumeyella*-groups, where it is vestigial (Miller 1991).

The basisternum: Ventrally, the development of the median ridge of the basisternum can be used to separate genera where it is complete, from those where it is absent or incomplete.

Wings

The shape of the fore wings: An important character and is used to separate the *Eriopeltis*-group, where the wing is more than 2.75 times longer than wide, from the other groups. Within other groups, the length to width ratio can be used to separate genera.

The halters: These are absent in the winged males of all species-groups, except the *Eulecanium*-group as redefined by Miller (1991), where they are present and serve as an exclusive species-group character.

Legs

The appearance of the legs is fairly constant within the family, but the relative length of the legs, and the length to width ratio of the hind femur and hind tarsus can be used to differentiate amongst species-groups, genera, and species.

Abdomen

The IXth tergite: The absence or presence of some of the abdominal tergites can be used at the generic and species levels. However, care should be taken in using the sclerites as diagnostic characters as their appearance is affected by the degree to which the specimens are stained.

Segment VII: A prominent caudal extension of this segment is characteristic of the *Coccus-* and *Toumeyella-*groups. The shape of the caudal extension on segment VIII (cylindrical, rounded, geniculate, mammillate, etc.) is characteristic for species, genera, or species-groups.

Segment VIII: Conspicuous structure on segment VIII is the glandular pouch with its multilocular pores and setae. It is absent in all the genera so far studied in certain species-groups (*Protopulvinaria* and *Toumeyella*), but in the *Coccus-*group, it is only absent in one species of the genus *Ceroplastes*.

The Cicatrices: The presence of a cicatrix on the distal part of this extension or lobe is the only exclusive character of the *Coccus-*group (Miller 1991).

The genital segment: The structures of the genital segment vary in size at the generic and specific level, and this can be expressed in ratios such as length of penial sheath to body length, length of basal rod to length of aedeagus, length of aedeagus to length of penial sheath, and length of aedeagus to length of basisternum. The absence of the basal rod separates *Mesolecanium nigrofasciatum* Pergande from all other species.

Dermal Structure

Fleshy dorsal and ventral setae on head: These are absent in 11 species of the *Eulecanium-*group studied by Giliomee (1967) and present in most other species-groups. However, Miller (1991) found a single fleshy seta on both locations in one of the three *Eulecanium-*species studied by him, indicating that these characters can even vary within a genus.

Dorsal ocular setae: The presence of dorsal ocular setae seemed to be an exclusive character of the *Coccus* group, but were then found to be present in the genus *Ctenochiton* of the *Eulecanium-*group (Giliomee 1967). More recently, Miller (1991) found that they were absent in one of the four species of *Pulvinaria*, but present in all the other species of the *Coccus-*group to which *Pulvinaria* belongs.

The genal setae: These are absent in all the species of some species-groups studied, *Eulecanium, Philephedra, Sphaerolecanium,* and *Eriopeltis* and present in others, *Coccus* and *Protopulvinaria*. In the *Toumeyella-*group, however, they are present in all the species, but in *Mesolecanium nigrofasciatum* may be either absent or present (Miller and Williams 1995).

Setae on antenna: On the antennae, the absence of the capitate subapical setae is characteristic of the *Philephedra-*group, while in other groups deviation from the regular number of three can be used to separate genera. The length of the fleshy antennal setae relative to the length or width of the associated segments appears to be a species-specific character in some genera.

Basisternal setae: Miller (1991) regarded the presence of more than nine fleshy basisternal setae as unique to the monospecific *Protopulvinaria-*group. Hair-like basisternal setae have so far only been found in two genera, but unfortunately, they are absent in some specimens and can therefore only be used as a supporting character.

The scutal and tegular setae: The presence or absence of scutal and tegular setae was used to separate genera by Giliomee (1967), while Miller (1991) used the presence of hair-like scutellar setae to distinguish *Toumeyella cerifera* Ferris from the other species in this genus. The latter also used the presence of circular pores on the posttergite and scutellum to separate two species of the genera *Toumeyella* and *Philephedra*, respectively.

Setae on wings: On the wings, the presence of alar setae and the number of haltere setae are important.

Setae on legs: The presence and shape (capitate or not) of the coxal bristles, the length of the apical seta of the front coxa, the length of the apical seta on the front trochanter, the length of the tibial spur in relation to the length of the tarsus, the length of the fleshy setae in relation to the length of the tibia, and the ratio of fleshy to hair-like setae on specific segments are important taxonomic characters.

The setae of the abdomen: These can also be used to separate groups, genera, and species, but because of individual variation, they serve mostly as supplementary characters. Thus, the absence of fleshy abdominal setae is characteristic of the *Eulecanium-* and *Philephedra-*groups, as defined by Miller (1991), while none of these setae are found on the IVth to VIIIth abdominal segments in the *Sphaerolecanium-*group. The presence of fleshy setae on the penial sheath distinguishes *Ceroplastes ceriferus* (Fabricius) from the other *Ceroplastes* species.

The fleshy setae lateral to the glandular pouch on the VIIIth abdominal segment appeared to be characteristic for the genus *Ceroplastes* (Giliomee 1967), but was found to be absent in *C. ceriferus* (Miller 1991).

Taxonomic Characters of Adult Female

Adult female morphology and important taxonomic characters have been described in detail by Hamon and Williams (1984), Gill (1988), Matile-Ferrero (1997), and Hodgson and Henderson (2000). The taxonomic characters described below have been collated from their publications.

Basic shape: young teneral females of most Coccidae are broadly oval and flat, but a few tend to be long and narrow (e.g., *Prococcus acutissimus*). A few remain more or less flat as they age (e.g., *Coccus hesperidum*), but some expand laterally, becoming nearly round in outline, while others often become highly convex (e.g., *Saissetia* spp.).

General morphological characters on dorsal and ventral aspects of completely grown up typical female are given in Figure 6.1.

Size: most adult female Coccidae are between about 2 and 6 mm long, but some species may grow to a large size up to 10 mm in length, as with the *Poropeza* species, which are not found in India.

Ovisacs and egg protection: a separate cover secreted to protect the eggs is produced only by the Pulvinariini and is probably secreted mainly by the ventral tubular ducts. It takes the form of a white woolly "sac," which protrudes from under the posterior end of the adult female and is often characteristically ridged, into which the eggs are laid. It can be quite long, up to several times the length of the female and sticks to the substrate. It is likely that females of Pulvinariini withdraw their stylets once they start egg-laying and move forwards at the speed the ovisac is secreted; once egg laying is complete, the females die and may fall away, leaving the ovisac behind.

Dorsal Structures

Anal cleft (Figure 6.2): It is the cleft between the anal lobes at the posterior end of the body and is usually about 1/7th to 1/10th of the total body length, but may be very short in some species.

Anal plates (Figure 6.3): These are the paired, approximately triangular plates, which lie at the anterior end of the anal cleft and are sometimes referred to as anal opercula. They are a major recognition character of Coccidae as similar plates are otherwise found only on a few species of Eriococcidae. These plates are "hinged" along their anterior margin and open anterolaterally to allow the anal tube to evert, aiding the elimination of honeydew from the anal ring. On most adventive soft scales, they are approximately quadrate when combined, with the two inner margins lying more or less parallel.

Anal ring (Figure 6.4): It is a sclerotised ring, composed of two lateral crescents, that surrounds the anal opening. Typically, it has numerous wax-exuding pores and three, four, or more pairs of long setae, the anal ring setae. The anal ring is at

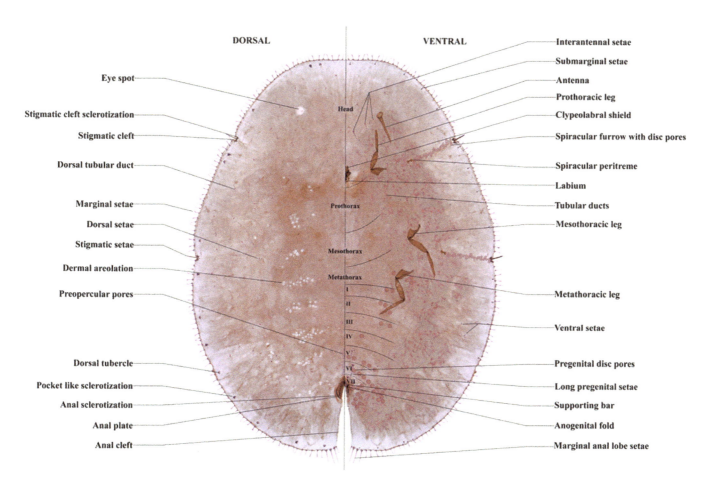

FIGURE 6.1 General morphological characters on dorsal and ventral aspect of typical female.

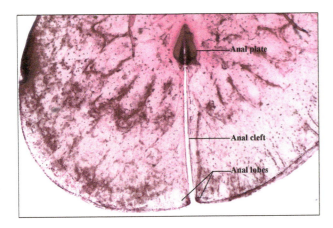
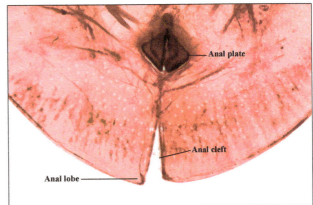

FIGURE 6.2 Anal cleft.

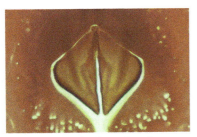
Anterolateral margin and posterolateral margin equal plates together forming square

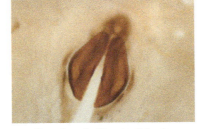
Pear shaped with anterolateral marging longer than posterolateral margin with rounded corners

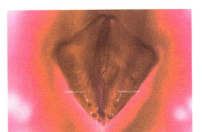
With posterolateral margin longer than anterolateral margin

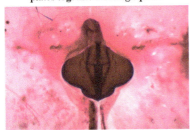
Anterolateral margin slightly longer than immarginate posterolateral margin

Anal plate together oval and surrounded by sclerotized area, inner margin bearing setae

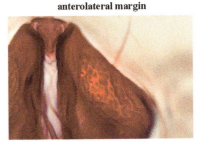
Anal plates with dorsal reticulation

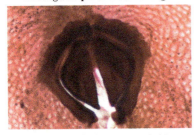
Quadrate anal plate surrounded by sclerotized area

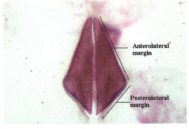
Anterolateral margin 2 - 2.5 times longer than posterolateral margin

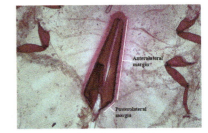
Anterolateral margin 5 time longer than posterolateral margin

FIGURE 6.3 Anal plates.

the inner end of the anal tube. When withdrawn and inverted, this tube lies within the body cavity and usually extends well anterior to the anal plates; rarely the tube is short.

Anal sclerotisation (Figure 6.5): This is a horseshoe-shaped sclerotisation around the anterior margin of the anal plates. On members of the Ceroplastinae, the anal sclerotisations is greatly enlarged to form the caudal or anal process, which carries the anal plates above the thick wax test.

Anogenital fold (Figure 6.6): This fold lies across the anterior end of the anal cleft, more or less at right

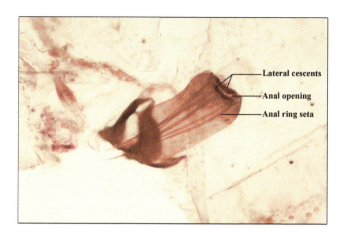
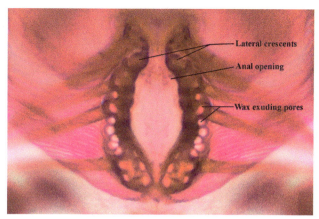

FIGURE 6.4 Anal ring.

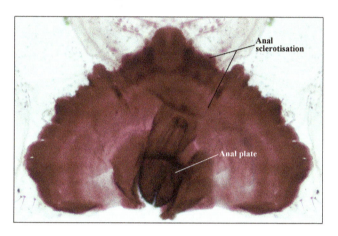
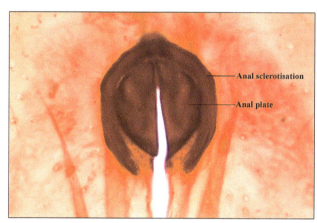

FIGURE 6.5 Anal sclerotisation.

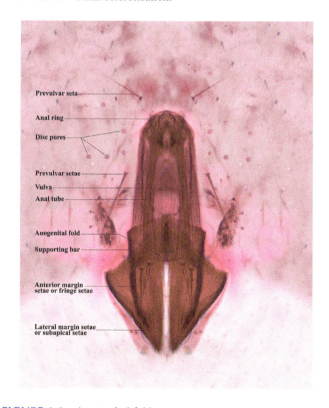

FIGURE 6.6 Anogenital fold.

angles to the long axis of the body, beneath the anal plates. It separates the anus on the dorsum from the vulva on the venter. The anogenital fold usually has setae at either corner and along the anterior margin, and these are referred to as anterior margin setae (also called fringe setae). Setae are also generally present along the lateral margins, and these are referred to as lateral margin setae. On either side of the anogenital fold, there is frequently a sclerotised supporting bar that appears to be part of the underside of each anal plate, and which is probably for the attachment of the muscles used in opening the anal plates. These bars extend anteriorly beneath the derm, where they frequently expand and may even meet medially.

Derm (Figure 6.7): Derm of teneral females tends to be thin and unsclerotised, and this is the best stage for making slide preparations for identification. Species with dense areolations (small clear areas in the derm), such as *Saissetia coffeae*, may have quite a thick derm, even just after the final moult, but, even then, the derm becomes much thicker with age, with the area within each areolation remaining quite thin, usually with a microduct. The size and distribution of the dorsal areolations may be useful in separating species. In the Ceroplastinae, the derm becomes heavily

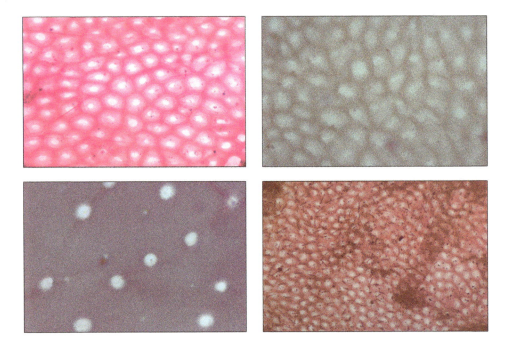

FIGURE 6.7 Derm areolations.

sclerotised quite soon after the final moult and only young females can be easily identified to species.

Dorsal setae (Figure 6.8): These are absent from all known indigenous species, but present on all adventive species. They are short (4–10 μm long) and spinose, arising from a distinct basal socket and are randomly distributed throughout the dorsum.

Dorsal pores: Most species have at least three or four different types of dorsal pores, whose function is probably to secrete different types of wax.

1. Dorsal microductules: each consists of a small sclerotised pore, round or slightly oval in shape, usually about 2–3 μm in diameter, set at the base of a short ductule and with a minute pore or slit-like opening, when viewed from above. Each microductule has a non-staining membranous inner ductule or filament, that is often balloon-shaped proximally and quite long distally
2. Simple pores: small pores without an inner filament; represented by at least two types, although they may appear similar under the light microscope:
    a. "open" pores that have a distinct pore opening and are flat (and should not be confused with closed pores that have lost their structure through being over cleared during slide preparation)
    b. "closed" pores that have no apparent aperture under the light microscope, but generally have a granulate surface (through which minute ducts emerge) and may be either flat or convex. Most are about 2–4 μm wide, but they can be quite a lot larger (<10 μm), while some have slightly thicker margins and may then appear "dark-rimmed." They are usually round, but may be oval. They have been referred to by previous workers as "dark-rimmed," "disc," or "discoidal" pores.
3. Macropores: this term is introduced here to describe some large pores, with a glandular internal structure. They are usually at least twice the size of simple pores, but are highly variable between species, both in size and shape, and are therefore very useful diagnostic characters
4. Multilocular disc-pores: these are rare on the dorsum
5. Preopercular pores (Figure 6.9): these pores are found in loose groups, typically just anterior to the anal plates, although they may be much more widespread in some genera and species. They are "closed" pores and are rather variable in shape. They may be small, flat, round to oval, relatively unsclerotised, and may have a granular surface, as on *Coccus hesperidum*, when they look very similar to closed simple pores. Their function is unknown
6. *Ceroplastes*-type pores: these are possibly the only sclerotised pores present on the dorsum of the Ceroplastinae, to which they are restricted. They are heavily sclerotised, with a large central pore and 0–4 smaller (satellite) pores. Ceroplastes-type pores generally have a long inner filament arising from the base of the central pore, which is much branched distally. *Ceroplastes*-type pores are abundant throughout the dorsum except on the lateral lobes or clear areas (areas on the dorsum without visible pores) and each opening is 2–5 μm wide. They are almost certainly involved in the production of the thick, soft, waxy test typical of the Ceroplastinae.

Taxonomy and Biodiversity of Soft Scales (Hemiptera: Coccidae)

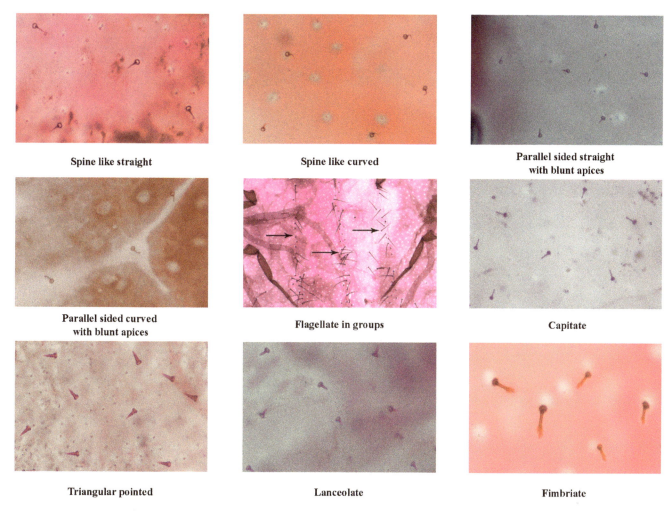

FIGURE 6.8 Dorsal setae.

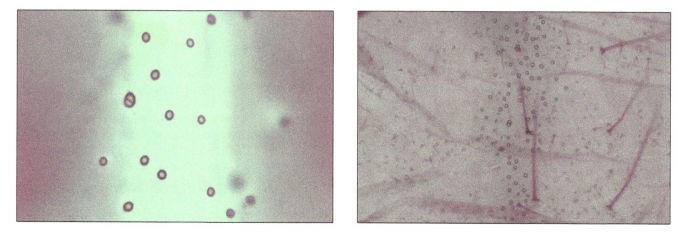

FIGURE 6.9 Preopercular pores.

Dorsal tubercles (Figure 6.10): These are also called submarginal tubercles. They are rather variable in structure, but those on species currently known from New Zealand are convex, usually wider than tall and rather sclerotised, with a central duct which has a small swelling/thickening at its inner end, with an inner filamentous ductule on one side, rather similar in structure to the cup-shaped invagination and inner ductule of a tubular duct.

Pocket-like sclerotisations (Figure 6.11): These are sclerotisations which have a pocket-like invagination and mark the site of the dorsal tubercles on the previous instar. Pocket-like sclerotisations are found on some species of *Saissetia* and

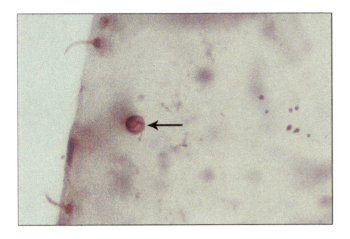
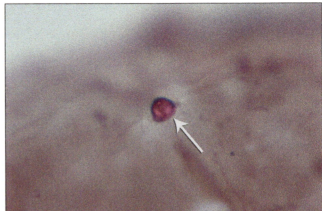

FIGURE 6.10 Dorsal tubercle.

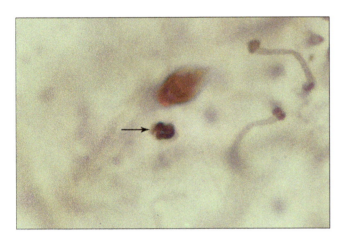
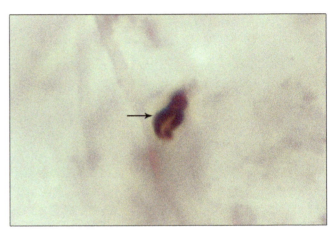

FIGURE 6.11 Pocket like sclerotisation.

Parthenolecanium. When present, pocket-like sclerotisations are generally found in association with a dorsal tubercle, either between it and the margin or close by, and only in a submarginal ring. Their function is unknown. Each sclerotisation is usually about 5 μm wide.

Tubular ducts: Dorsal tubular ducts consist of four parts-

1. An outer ductule, thin-walled, barely sclerotised, round in cross section and generally at least 10 μm long, which opens through the dorsum by a small inconspicuous pore that may occasionally be mildly sclerotised at its inner end is
2. A characteristic structure, here referred to as the cup-shaped invagination because the outer ductule terminates in a thick-walled structure that is bowl- or cup-shaped, and usually slightly asymmetrical; from one side of the cup arises
3. The inner ductule which is usually narrower and shorter than the outer ductule; this terminates in
4. A "flower-head"-like structure, here referred to as the terminal gland. Tubular ducts vary in the relative lengths and widths of the inner and outer ductules, in the form (particularly the depth) of the cup-shaped invagination, and in the size of the terminal gland. Including those types of tubular duct found on the venter, some *Pulvinaria* species have four or five types, and the structure of each type and their distribution within a species are good diagnostic characters. Usually the dorsum has only one type tubular duct.

Marginal Structures

Margin: on most Coccidae, the margin is distinct and marked by the presence of marginal setae.

Marginal setae (Figure 6.12): These form a marginal line. They are usually of one shape, are distinctly differentiated from other setae, and are frequently abundant. The shape and structure of these setae are highly variable, but are

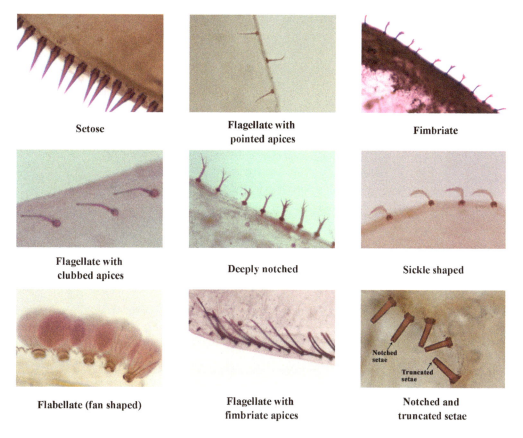

FIGURE 6.12 Marginal setae.

usually constant for a given species. As a result, marginal setae are significant taxonomic features at all levels, and the number of setae laterally between the anterior and posterior stigmatic clefts (or occasionally round the head between the anterior stigmatic clefts) is given in most descriptions as an indication of their frequency. On most genera, marginal setae are absent from the margins of the anal cleft. Also, as on *Saissetia coffeae*, one or more marginal setae on the anal lobes may be significantly longer than normal marginal setae (marginal anal lobe setae). These are in addition to the (often long) pair of ventral anal lobe setae. Each seta is set in a basal socket that is usually well developed and may be narrow or shallow.

Stigmatic clefts: Clefts may be distinct clefts, with parallel sides, or only shallow indentations at the point where the spiracular disc pore band meets the margin, or they may be absent.

Stigmatic spines (Figure 6.13): These are one or more marginal setae that are usually differentiated from the other setae in each stigmatic cleft or stigmatic area. There are usually three or more stigmatic spines and, when three, the middle (median) spine is generally longer than the laterals. On the Ceroplastinae, the stigmatic spines are in a large group in each stigmatic area and may extend medially onto the dorsum. When stigmatic clefts are present, the stigmatic spines are located at their base. The number, shape, and relative lengths of stigmatic spines are of taxonomic importance.

Eyespots (Figure 6.14): In the Coccidae, the eyespots are placed on the dorsum, very close to the margin, each consisting of a single lens. These can be hard to detect on slide-mounted specimens, but are often clear on fresh specimens.

Ventral Structures

Antennae (Figures 6.15 and 6.16): The basic structure of most normally developed antennae is as follows. The basal segment or scape (segment I) is well developed and has three setae. Segment II (pedicel) has one long and one short seta on its ventral surface and a campaniform sensillum on the dorsal surface. Between the basal two segments and the terminal three segments, there are normally one to four further segments, although usually it is only the segment nearest the apical three segments (or the distal end of segment III) which has setae, and then there are three flagellate setae. The apical segment has about five normal setae and probably three or four fleshy setae (these can appear rather flagellate on some species), two or three on the ventral surface and one on the dorsal; on most species, the terminal seta (apical seta) is rather straight, non-flagellate, and its length appears to be of taxonomic significance; in addition, one of the flagellate setae on the dorsal side is usually very long, but is less useful taxonomically as it is often broken.

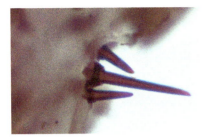
Group of three with setae having pointed apices

Group of three with middle setae curved and lateral setae straight

Group of three with all setae pointed and curved

Group of three with middle setae cylindrical and long and lateral setae small stubb like

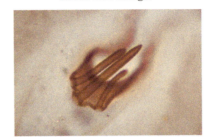
Group of three, cylindrical setae seated in deep sclerotized furrow

Multiple cylindrical setae in sclerotized spiracular furrow

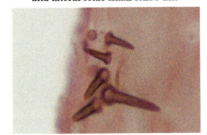
Multiple setose setae of different sizes

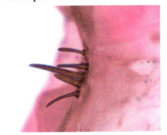
Multiple setae with variable length and width, some straight and some curved

Multiple cylindrical curved and straight setae with variable length and width, situated in deep furrow

FIGURE 6.13 Stigmatic spines.

Labium and mouthparts (Figure 6.17): The structure of the mouthparts is reasonably constant throughout the Coccidae, varying mainly in size—here usually indicated by the length of the clypeolabral shield. The labium is 1-segmented (occasionally indistinctly 2-segmented on some nymphs) and cone-shaped, usually with four pairs of setae, although these are difficult to see on some specimens.

Legs (Figure 6.18): When present, the legs are normal insect legs, each with five segments, although they are generally slightly small in proportion to the rest of the body. The coxae are attached to the venter along their width and articulate with a sclerotisation at the lateral corner; this is often most obvious on legs that are much reduced. The structure of the coxae appears to be very similar throughout the Coccidae; nor is there much variation in the trochanter and femur. Each trochanter has a pair of large pores on each side (whose function is unknown) and one or two long setae on its ventral surface. The setae on the femur have not been found to vary much. The tibia is always longer than the single tarsus, and usually the proportions are similar on all three pairs of legs. The tibia generally lacks the campaniform sensillum typically present at its base on most other coccoid families, the absence of this pore is otherwise a major taxonomic character of the Coccidae. The tibia and tarsus are usually separate, but without any articulation on most indigenous species, however, they are fused in several species. On other coccids, there is clearly a true tibio-tarsal articulation with an articulatory sclerosis (as on *Pulvinaria* species); the presence or absence of this sclerosis is of taxonomic importance. The frequency, distribution, and length of the setae on the tibia and tarsus may also be of some significance. At the distal end of the tarsus is a pair of thin digitules, the tarsal digitules; these tend to be slightly dissimilar, one being slightly shorter and slimmer than the

Taxonomy and Biodiversity of Soft Scales (Hemiptera: Coccidae)

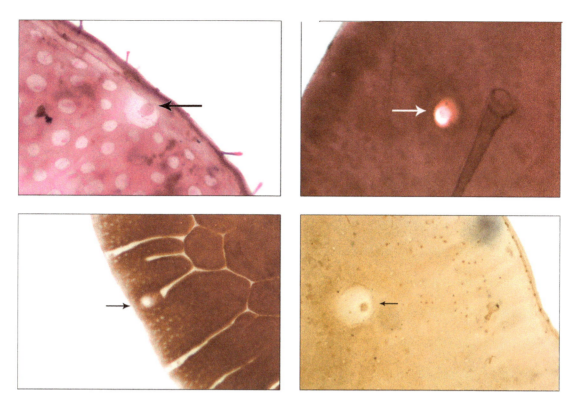

FIGURE 6.14 Eyespots.

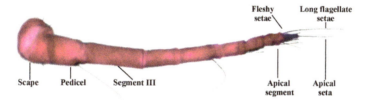

FIGURE 6.15 Antenna structure.

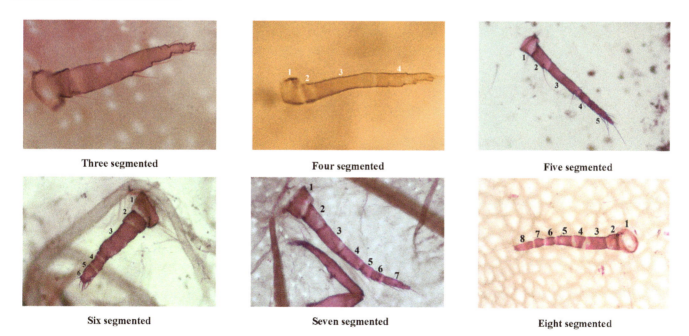

FIGURE 6.16 Types of antenna.

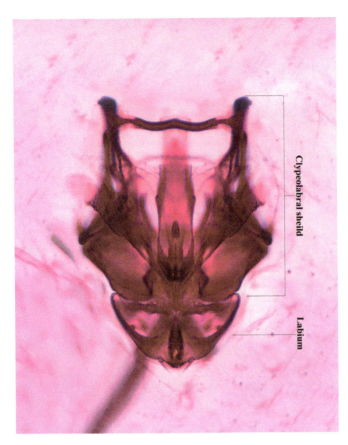

FIGURE 6.17 Labium and mouthparts.

other; they tend to be shortest when the leg shows signs of reduction. The structure of the claw and claw digitules show several features of taxonomic importance. The claw may be short and broad or long and thin and may or may not have a small denticle near the apex. The claw digitules may both be broad, both be narrow, or of distinctly dissimilar width. Fine claw digitules are generally associated with a reduction in the size of the legs.

Segmentation (can be seen in Figure 6.1): This is usually reasonably obvious medially on the abdomen and thorax. There are six visible segments between the vulva and the metathoracic coxa, and these are here numbered segments II to VII, following the system of previous authors who considered that the first visible segment ventrally on the abdomen is the 2nd (segment II), the 1st being represented by an area laterad to the metathoracic legs; thus, the pregenital segment is the 7th (segment VII). The segmentation on the thorax also usually can be seen, including the demarcation between the thorax and head, where the line runs posteriorly to the labium from near each procoxa. No segmentation is visible on the head.

Spiracles (Figure 6.19): Each spiracle is composed of a sclerotised, funnel-shaped outer peritreme, which opens through the spiracular opening or atrium into the tracheae. The size of each peritreme is important and can be useful in placing a species at the generic level. On some species, there is a sclerotisation around each spiracle mesad to the peritreme and, this is here referred to as the sclerotised spiracular plate.

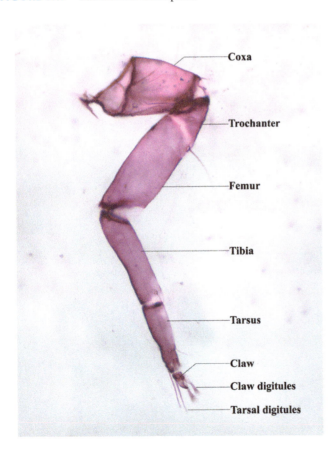

FIGURE 6.18 Leg and its structure.

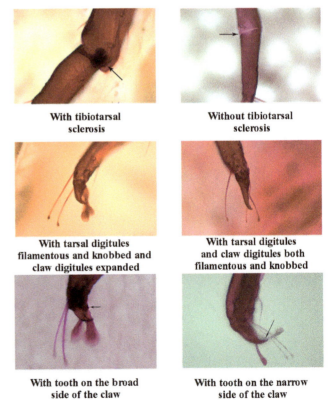

Taxonomy and Biodiversity of Soft Scales (Hemiptera: Coccidae)

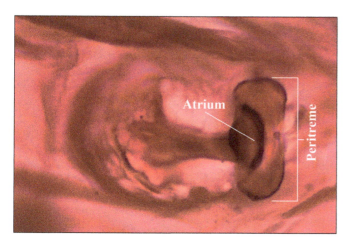
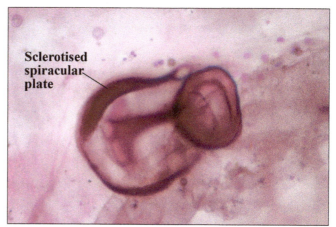

FIGURE 6.19 Spiracle.

Ventral Pores

1. Disc-pores: each disc-pore has a central loculus which is usually round, but may be oval, surrounded by a number of similarly shaped loculi or pores, the complete disc-pore looking rather like a wheel with spokes. Each disc-pore usually has 5–10 loculi in the outer ring, and the pores are therefore known as multilocular disc-pores. Multilocular disc-pores can be divided into two groups: the pregenital disc-pores and the spiracular disc-pores

   a. Pregenital disc-pores (Figure 6.20): are primarily located on the pregenital segment VII, thus the name. However, on most genera, they are also found across some of the more anterior abdominal segments and, less frequently, medially on

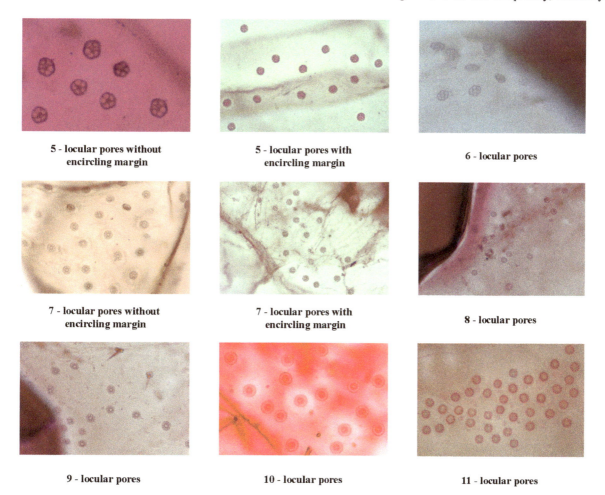

FIGURE 6.20 Pregenital disc pores.

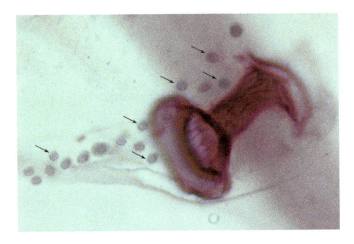

FIGURE 6.21 Spiracular disc pores.

the thorax and head (where they are referred to as multilocular disc-pores)
  b. Spiracular disc-pores (Figure 6.21): are in bands in the stigmatic furrows between the stigmatic area on the margin and the spiracles. On most species, each pore has five loculi and most authors refer to them as quinquelocular pores.
2. Ventral microducts: the oval, sclerotised pore of each ventral microduct is located at the base of a short, outer ductule (longer than that of the dorsal microductules). The pore opening is across the widest part of the pore, and the non-staining inner ductule is usually broad and skirt-like, not filamentous. Each pore is usually about 2–3 µm wide. Generally, ventral microducts are present more or less throughout the venter, although they may be more frequent submarginally; occasionally they have distinctive distributions. The function of the ventral microducts is uncertain
3. Pre-antennal pores: these are small, convex pores that are present just anterior to each scape. Their function is unknown.
4. Other ventral pores: other types of pores are infrequent on the venter and are therefore good taxonomic characters. Simple pores, similar to those on the dorsum, are occasionally present; they are most frequently associated with the margin, but may have a wider distribution. Ventral setae: most ventral setae are short and flagellate, but a few are longer, and these and their frequency can be of some taxonomic significance. The most common distribution of the longer setae is:

A pair medially on the pregenital segment (segment VII) (the pregenital setae) (Figure 6.22) and often with additional pairs on the preceding two segments (segments VI and V);
There may be long setae elsewhere, such as just mesad to each coxa and between the antennae (interantennal setae) (Figure 6.23).

Some other groupings of setae are now thought to be of significance and are here referred to as follows:

The sub-marginal setae are a single row of setae just mesad to the margin, and their frequency is given as the number laterally between the stigmatic clefts;
The anterior anal cleft setae are a group of 1 to several setae that occur on either side towards the anterior end of the anal cleft;
A pair of ventral anal lobe setae occur on the anal lobes near the margin, and their length appears to be taxonomically significant;
The abdominal setae are those found medially on each abdominal segment, and their frequency are also thought to be significant.

Ventral tubular ducts (Figure 6.24): The structure of the tubular ducts on the venter is similar to that of those on the dorsum and, like them, their structure and distribution are important taxonomic characters. When present, most species have only one type of dorsal and one type of ventral tubular duct, and then they are usually identical, but some species may have several types, as on members of the Pulvinariini, which have

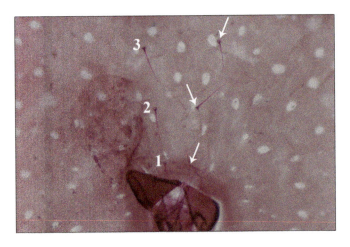

FIGURE 6.22 Pregenital setae.

Taxonomy and Biodiversity of Soft Scales (Hemiptera: Coccidae)

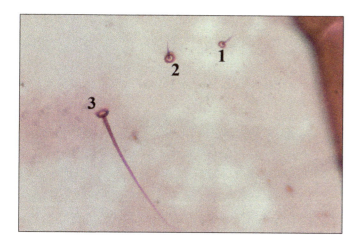
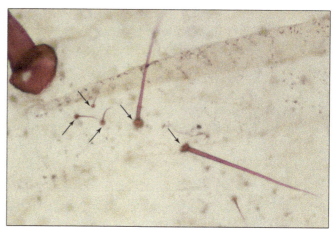

FIGURE 6.23 Interantennal setae.

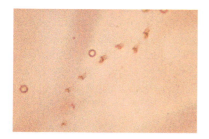

With small outer duct, sclerotised cup shaped invagination and invisible inner duct

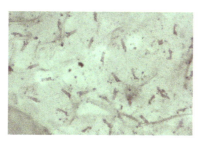

With medium sized outer duct, sclerotized cup shaped invagination and small inner duct

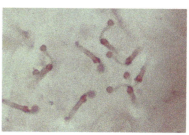

With long outer duct having sclerotized anterior end, sclerotized cup shaped invagination and narrow medium sized inner duct with flower shaped terminal gland

With long outer duct, sclerotized cup shaped invagination and broad (as broad as outer duct) medium sized inner duct with flower shaped terminal gland

With long outer duct, sclerotized cup shaped invagination and medium sized inner duct having narrow proxinmal end and broader distal end and flower shaped terminal gland

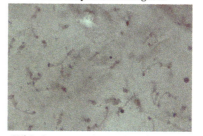

With short outer duct, sclerotized cup shaped invagination and long inner duct (longer than outer duct) with flower shaped terminal gland

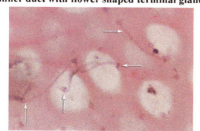

With short outer duct, sclerotized cup shaped invagination and very long inner duct with flower shaped terminal gland

With medium outer duct, sclerotized cup shaped invagination and inflated inner duct having narrow proximal end and wide distal end and flower shaped terminal gland

With medium outer duct, sclerotized cup shaped invagination and inflated inner duct having similar sized proximal and distal end and flower shaped terminal gland

FIGURE 6.24 Ventral tubular ducts.

three or four different types of ventral tubular duct, each with its own distinctive distribution. These ducts are most frequently distributed in one of three patterns:

1. Restricted to a more or less complete sub-marginal band, as on *Saissetia oleae*
2. More or less throughout, as on species of *Crystallotesta* and *Poropeza* (neither of these genera are found in India)
3. In a group medially on the abdomen.

Vulva: This is the female genital opening and is found on abdominal segment VII. Whilst the derm surrounding the opening of the vulva is thin and membranous and obscured by the anal tube, it can occasionally be found just anterior to the anogenital fold.

CLASSIFICATION OF FAMILY COCCIDAE

The classification suggested by Hodgson (1994) when he redescribed the adult females of the type species of all the known soft scale genera is followed here. This classification is based on the structure of the adult females and on male morphology studied since 1960. He divided the Coccidae into ten subfamilies, namely, CARDIOCOCCINAE Hodgson, CEROPLASTINAE Atkinson, CISSOCOCCINAE Brain, COCCINAE Fallen, CYPHOCOCCINAE Hodgson, EULECANIINAE Koteja, ERIOPELTINA Šulc, FILIPPIINAE Bodenheimer, MYZOLECANIINAE Hodgson, and PSEUDOPULVINARIINAE Tang, with the Coccinae divided into four tribes, viz., Coccini Fallén, Paralecaniini Williams, Pulvinariini Targoni Tozzetti, and Saissetini Hodgson. He considered that the status of these groupings needed further study. The family Coccidae is the third largest family within the Coccoidea, with approximately 1,100 species in about 160 genera (Ben-Dov 1993; Hodgson 1994).

CHARACTERISTICS OF THE SUBFAMILIES AND TRIBES

Cardiococcinae Hodgson

This subfamily, which might be loosely called the glassy scales, includes 16 genera.

1. The presence of a glassy test
2. The absence of dorsal setae, dorsal tubular ducts, and dorsal tubercles
3. Presence of spinose marginal setae
4. A distinctive distribution of dorsal pores, which occur in a distinctive pattern, often in a mid-dorsal line from the anal plates to the anterior margin of the head or forming a large reticulate pattern
5. Presence of pregenital disc-pores with typically five loculi
6. Presence of a sub-marginal band of ventral tubular ducts
7. The inner margins of the anal plates often diverging posteriorly, with spinose setae along the inner margin in many genera
8. Absence of pairs of long setae medially on the pregenital segments, these replaced by bands of rather spinose setae.

Ceroplastinae Atkinson

This subfamily includes all the genera related to *Ceroplastes* Gray.

1. A thick waxy test covering the dorsum
2. The distinctive dorsal *Ceroplastes*-type pores
3. A sclerotised caudal process, which lifts the anal plates above the thick wax cover
4. The presence of dorsal lobes or clear areas free from pores
5. The form of the ventral microducts, which appear to have a cruciform opening
6. The characteristics of the stigmatic areas.

Cissococcinae Brain

1. The reduction of the dorsum to a small area around the anal plates and their associated sclerotisations, the median areas of the venter becoming expanded, so that the legs, spiracles, and mouthparts lie dorsally
2. Possible complete absence of antennae
3. Reduced legs
4. Large spiracles, placed on the apparent dorsum with their atria facing medially, each with sparse bands of five-locular disc-pores also extending medially
5. Mouthparts also on the apparent dorsal surface, with the labium pointing anteriorly
6. Presence of anal plates typical of the Coccidae, with numerous setae on their dorsal surface (as in the Myzolecaniinae)
7. Anal plates surrounded by a very large area of sclerotisation, which appears to be structurally quite different to that of other Coccidae
8. True venter covered in numerous long setae and 10-locular disc-pores
9. Absence of the marginal setae and stigmatic spines.

Coccinae Fallen

Hodgson (1994) included 55 genera in four tribes in this subfamily. The tribes with their characteristic features are provided below.

1. **Coccini Fallen**
   a. Lack of dorsal tubular ducts (except very sparsely sub-marginally in *Coccus*)
   b. Absence of ventral tubular ducts or their restriction to medially in the thorax
   c. Lack of pocket-like sclerotisations
   d. Presence of eyespots, usually close to the margin

    e. Stigmatic areas unsclerotised
    f. Presence of stigmatic spines differentiated from the marginal setae
    g. Presence of pregenital disc-pores concentrated on the pregenital segment, never present medially on the thorax or head.

2. **Paralecaniini Williams**
    a. A distinct cleft which is sclerotised on the dorsum at its base
    b. No dorsal tubular ducts
    c. Ventral tubular ducts, when present, restricted to a group on either side of the genital opening
    d. Pregenital disc-pores restricted to the abdominal segments immediately anterior to the genital opening
    e. Eyespots displaced onto the dorsum, typically nearly dorsal to the base of each antenna.

3. **Pulvinariini Targioni Tozzetti**
    a. A woolly ovisac secreted by the reproducing female from beneath the posterior end of the abdomen, often lifting the insect so that its body is held almost vertically above the head by the ovisac
    b. Ventral tubular ducts of generally three or four types (rarely two), including (a) a small duct with a fine inner ductile, generally occurring in a sub-marginal band and (b) a larger duct with the inner and outer ductules of subequal width, these typically present medially in the head and thorax, but occasionally elsewhere
    c. No woolly test covering the dorsum, (or, if a mealy covering is present, this is very sparse)
    d. No dorsal tubular ducts, or if present, of one type only and typically similar to the smallest ventral ducts
    e. Spinose dorsal setae
    f. Each leg with a tibio-tarsal articulation
    g. No pocket-like sclerotisations
    h. Eyespots present near the margin
    i. Shallow, unsclerotised stigmatic clefts.

4. **Saissetiini Hodgson**
    Members of this tribe differ from other members of the Coccinae in:
    a. Presence of a broad sub-marginal band of ventral tubular ducts of one or two types
    b. Absence (typically) of dorsal tubular ducts
    c. Typically with dorsal tubercles and, often, also pocket-like sclerotisations, though both may be absent
    d. Presence of pregenital disc-pores, each usually with ten loculi, extending medially onto thorax
    e. Presence of eyespots near the margin
    f. Presence of unsclerotised, shallow stigmatic clefts.

Cyphococcinae Hodgson
1. Presence of microtubular ducts on the dorsum, otherwise unknown in the Coccidae
2. Dorsum divided into two areas, the median area with few pores and no dorsal setae, the lateral areas with abundant pores and setae
3. These two areas separated by a sinuous line of strongly spinose setae
4. Presence of pregenital disc-pores, each with six or seven loculi, frequent medially in all the abdominal and thoracic segments
5. Absence of preopercular pores
6. Spiracular disc-pores present in broad bands
7. Presence of spinose marginal setae
8. Presence of tibio-tarsal pseudo-articulations
9. Claw digitules both fine, (x) absence of eyespots
10. Presence of a glassy test covering the median area of the dorsum.

Eulecaniinae Koteja
1. Pregenital disc-pores, each with ten loculi, present medially on all the abdominal and thoracic segments and usually also on the head
2. Presence of spinose or setose marginal setae, which are never fimbriate
3. Absence of dorsal tubercles and pocket-like sclerotisations
4. Typically with a complete ring of ventral tubular ducts present
5. Legs without a tibio-tarsal articulatory sclerosis
6. Even though the legs are well developed, the claw digitules are either both fine or dissimilar, never both broad.

Eriopeltinae Sulc
1. Body elongate
2. Production of a felted ovisac over the whole or part of the dorsum, which is secreted by large tubular ducts on the dorsum, which are similar to the tubular ducts found sub-marginally on the venter
3. A membranous dorsum, without areas of dense sclerotisation
4. Each anal plate frequently with one or two setose or spinose setae along the inner margin
5. Lack of stigmatic clefts
6. Pregenital disc-pores each with 7–10 loculi
7. Presence typically of two types of ventral tubular ducts, the larger sub-marginally
8. Legs and antennae well developed
9. Presence of either 0 or two stigmatic spines in each stigmatic area
10. Lack of dorsal tubercles and pocket-like sclerotisations. In addition, members of this subfamily are usually restricted to monocotyledonous plants.

Filippiinae Bodenheimer
1. The Filippiinae mainly occur on dicotyledonous plants
2. The adult females are roundly oval in shape, sometimes have dorsal tubercles and pocket-like sclerotisations

3. They have 0, one, two, or three stigmatic spines in each stigmatic cleft.
4. Otherwise, they share most of the characters given above for the Eriopeltinae.

Myzolecaniinae Hodgson

1. The lack of dorsal tubular ducts
2. Absence of eyespots
3. Presence of anal plates with typically numerous setae on the dorsal surface
4. Particularly large and often somewhat modified spiracles (Hodgson 1995) with broad bands of spiracular disc-pores between the margin and the spiracle
5. Ventral tubular ducts of one type, frequently restricted to a group on either side of the genital opening
6. Without median pairs of long pregenital setae, but with segmental bands of short spinose setae
7. Legs reduced, with both claw digitules fine
8. Reduced antennae
9. A short anal tube.

Pseudopulvinariinae Tang, Hao, Xie, and Tang

1. The production of a dense, woolly test which covers the entire insect
2. Abundant sclerotised, five-locular disc-pores or cribriform plates which cover the entire dorsum and also form a narrow sub-marginal band ventrally
3. Anal plates which appear to be joined along both the dorsal and ventral margins, each plate rather triangular in shape, with long spinose setae, appearing rather like a crown
4. Tube either very short or absent
5. Extremely large spiracles
6. Lacking tubular ducts, but
7. With one or two types of tubular duct ventrally
8. Presence of strongly spinose marginal setae, but
9. With the stigmatic spines undifferentiated and
10. Legs and antennae more or less well developed.

Hodgson (1994) has provided a key to the world genera of family Coccidae. Coccidologists from India have provided keys to the species of different genera occurring in India (please refer to "Major Work on Indian Fauna" part of this chapter).

MAJOR WORK ON INDIAN FAUNA

Work on Indian soft scales was initiated by T.V. Ramakrishna Ayyar in 1919 contributing mainly to the knowledge on coccids of south India through publishing new records from different parts of the country (mainly Tamil Nadu) and finally by publishing a checklist of the Coccidae from the Indian region and Indo-Ceylonese fauna (Ayyar 1919a; 1919b; 1919c; 1921; 1924; 1926; 1930; and 1936). Sankaran (1954; 1955; 1959; 1962; 1965; Sankaran and Pattar 1977) worked on morphology and anatomy of species of scale insects belonging to the genus *Ceroplastes* and parasitoids of *Saissetia*, *Parasaissetia*. Borchsenius (1964; 1967) examined scale insect material collected from India and published their taxonomic account, but it mainly had scale insect species belonging to armoured scales. Ali (1962; 1964; 1967a,b; 1968a,b; 1969a,b; 1970a,b; 1971; 1973) worked for about a decade on fauna of soft scales of Bihar and published descriptions of few species and wrote a catalogue of the Oriental Coccoidea in five volumes. Das (1959) and Das and Ganguli (1961) worked on coccids of North-East India, mainly those infesting tea plantations. Ghose (1961) and Ganguli and Ghose (1964) worked on fauna of soft scales of West Bengal and Tripura, respectively. Avasthi and Shafee (1979; 1984a,b,c; 1985; 1986) and Yousuf and Shafee (1988) surveyed Uttar Pradesh and Andaman Islands and published several new species of soft scales and reviewed Indian species belonging to the genera *Coccus* and *Ceroplastes*. Shafee et al. (1989) compiled a list of host plants and distribution of soft scale pests in India. Varshney (1984; 1985) published new records of coccids and their host plants from India and published a review on Indian coccids. Varshney and Moharana (1987) studied fauna of soft scales of Orissa. Varshney (1992) published a checklist of the soft scale insect of India. Suresh and Mohanasundaram (1996) studied fauna of Coccidae of Tamil Nadu. Joshi (2017) and Joshi and Rameshkumar (2017) published four new records of soft scales from India.

DISTRIBUTION PATTERNS IN INDIAN CONTEXT

A total of 73 species under 32 genera have so far been recorded from India (Table 6.1). The major genera are *Coccus*, *Ceroplastes*, and *Pulvinaria* with 14, 12, and 11 species under

TABLE 6.1
Distribution of Coccidae in Different States of India

State	Number of Species	Number of Genera
Andaman and Nicobar Islands	2	2
Andhra Pradesh	9	2
Assam	8	6
Bihar	11	7
Goa	2	3
Gujarat	3	3
Himachal Pradesh	6	3
Jammu and Kashmir	5	1
Karnataka	18	9
Kerala	8	4
Madhya Pradesh	1	1
Maharashtra	3	4
Orissa	1	1
Punjab	1	1
Rajasthan	1	1
Tamil Nadu	34	19
Tripura	1	1
Uttar Pradesh	11	4
West Bengal	19	10

them, respectively. Indian states of Tamil Nadu, West Bengal, and Karnataka have records of 34, 19, and 18 species, respectively, while Bihar and Uttar Pradesh have 11 species records from each. This also indirectly reflects the number of taxonomists working in those particular areas and the intensity of surveys done in these states by them. The data presented in Table 6.1 indicate that there is a need for exploring Indus plains, North-West India, and Gangetic plains as very few scales have been recorded from these areas.

CHECKLIST OF INDIAN SOFT SCALES

This checklist (Table 6.2) has been generated by using distribution records given in ScaleNet (García et al. 2016).

TABLE 6.2
Checklist of Indian Coccidae

Subfamily, tribe, and species	Distribution
Subfamily: Myzolecaniinae	
1. *Akermes montanus* (Green) 1908	Himachal Pradesh
2. *Cribrolecanium radicicola* Green 1921	Tamil Nadu
Subfamily: Cardiococcinae	
1. *Cardiococcus bivalvatus* (Green) 1903	Karnataka, Tamil Nadu
2. *Dicyphococcus castilloae* (Green) 1911	Tamil Nadu
3. *Drepanococcus cajani* (Maskell) 1891	Bihar, Maharashtra, Tamil Nadu, and West Bengal
4. *Drepanococcus chiton* (Green) 1909	Himachal Pradesh, West Bengal
5. *Inglisia chelonioides* Green 1909	Maharashtra, Tamil Nadu
Subfamily: Filippiinae	
1. *Ceronema koebeli* Green 1909	Tamil Nadu
2. *Metaceronema japonica* (Maskell) 1897	West Bengal
3. *Ceronema fryeri* Green 1922	Karnataka
Subfamily: Ceroplastinae	
Tribe: Ceroplastini	
1. *Ceroplastes actiniformis* Green 1896	Bihar, Goa, Tamil Nadu, and West Bengal
2. *Ceroplastes ajmerensis* (Avasthi & Shafee) 1979	Himachal Pradesh, Rajasthan
3. *Ceroplastes alami* Avasthi & Shafee 1986	Tamil Nadu, Uttar Pradesh
4. *Ceroplastes ceriferus* (Fabricius) 1798	Assam, Bihar, Madhya Pradesh, Tamil Nadu, and West Bengal
5. *Ceroplastes destructor* (Newstead) 1917	Tamil Nadu
6. *Ceroplastes floridensis* (Comstock) 1881	Kerala, Uttar Pradesh, and West Bengal
7. *Ceroplastes neoceriferrus* (Avasthi & Shafee) 1979	Uttar Pradesh
8. *Ceroplastes pseudoceriferus* Green 1935	Uttar Pradesh
9. *Ceroplastes rubens* Maskell 1893	West Bengal
10. *Ceroplastes rusci* (Linnaeus) 1758	Uttar Pradesh
11. *Ceroplastes stellifer* (Westwood) 1871	Karnataka, Kerala, and West Bengal
12. *Ceroplastes stipulaeformis* (Haworth) 1812	INDIA
13. *Ceropastodes wandoorensis* (Yousuf & Shafee) 1988	Andaman Islands
Subfamily: Coccinae	
Tribe: *Coccini*	
1. *Prococcus acutissimus* (Green) 1896	Kerala, Tamil Nadu
2. *Ctenochiton olivaceus* (Green) 1922	Tamil Nadu
3. *Coccus almoraensis* Avasthi & Shafee 1984	Bihar, Uttar Pradesh
4. *Coccus capparidis* (Green) 1904	West Bengal
5. *Coccus colemani* Kannan 1918	Karnataka
6. *Coccus discrepans* (Green) 1904	Andhra Pradesh, Assam, Bihar, and Kerala
7. *Coccus formicarii* (Green) 1896	Karnataka
8. *Coccus gymnospori* (Green) 1908	Andhra Pradesh, Maharashtra, and Tamil Nadu
9. *Coccus hesperidum* (Linnaeus) 1758	Andhra Pradesh, Bihar, Goa, Gujarat, Jammu & Kashmir, Karnataka, Kerala, Tamil Nadu, Tripura, Uttar Pradesh, and West Bengal
10. *Coccus kosztarabi* Avasthi & Shafee 1984	Karnataka
11. *Coccus latioperculatum* (Green) 1922	Tamil Nadu

(Continued)

TABLE 6.2 (Continued)
Checklist of Indian Coccidae

Subfamily, tribe, and species	Distribution
12. *Coccus longulus* (Douglas) 1887	Andhra Pradesh, Karnataka, and Tamil Nadu
13. *Coccus ophiorrhizae* (Green) 1896	Andhra Pradesh, Tamil Nadu
14. *Coccus ramakrishnai* (Ramakrishna Ayyar) 1919	Andhra Pradesh, Himachal Pradesh
15. *Coccus trichodes* Anderson 1787	Tamil Nadu
16. *Coccus viridis* (Green) 1889	Andhra Pradesh, Assam, Karnataka, and Tamil Nadu
17. *Eucalymnatus tessellatus* (Signoret) 1873	Gujarat, Tamil Nadu
18. *Kilifia acuminate* (Signoret) 1873	Karnataka
19. *Trijuba oculata* (Brain) 1920	Karnataka
Tribe: Saissetiini	
20. *Hemilecanium imbricans* (Green) 1903	Karnataka, Tamil Nadu
21. *Parasaissetia nigra* (Nietner) 1861	Assam, Bihar, Tamil Nadu, and West Bengal
22. *Parthenolecanium corni* (Nuzzaei) 1969	Himachal Pradesh
23. *Parthenolecanium persicae* (Fabricius) 1776	Punjab
24. *Saissetia coffeae* (Walker) 1852	Assam, Bihar, Tamil Nadu, and West Bengal
25. *Saissetia miranda* (Cockerell & Parrott in Cockerell) 1899	Tamil Nadu
26. *Saissetia oleae* (Olivier) 1791	Assam
27. *Saissetia privigna* De Lotto 1965	Andhra Pradesh
Tribe: Paralecaniini	
28. *Maacoccus bicruciatus* (Green) 1904	Tamil Nadu
29. *Maacoccus watti* (Green) 1900	Assam, West Bengal
30. *Maacoccus piperis* (Green) 1922	Karnataka, Kerala
31. *Marsipococcus marsupialis* (Green) 1904	Tamil Nadu
32. *Neoplatylecanium adersi* (Newstead) 1917	Tamil Nadu
33. *Paralecanium expansum* (Green) 1896	Karnataka, Tamil Nadu
34. *Paralecanium maritimum* (Green) 1896	Ceylon
35. *Saccharolecanium krugeri* (Zehntner) 1897	Ceylon
Tribe: Pulvinariini	
36. *Megapulvinaria burkilli* (Green) 1908	Tamil Nadu, West Bengal
37. *Megapulvinaria maxima* (Green) 1904	Kerala, Tamil Nadu, Uttar Pradesh, and West Bengal
38. *Milviscutulus mangiferae* (Green) 1889	Bihar, Tamil Nadu, and West Bengal
39. *Protopulvinaria longivalvata* Green 1909	Karnataka
40. *Pulvinaria aligarhensis* Avasthi & Shafee 1985	Uttar Pradesh
41. *Pulvinaria durantae* Takahashi 1931	Tamil Nadu
42. *Pulvinaria floccifera* (Westwood) 1870	West Bengal
43. *Pulvinaria iceryi* (Signoret) 1869	Uttar Pradesh
44. *Pulvinaria indica* Avasthi & Shafee 1985	Andhra Pradesh
45. *Pulvinaria ixorae* Green 1909	Ceylon
46. *Pulvinaria obscura* Newstead 1894	Tamil Nadu
47. *Pulvinaria polygonata* Cockerell 1905	Bihar, West Bengal
48. *Pulvinaria portblairensis* (Yousuf & Shafee) 1988	Andaman Islands
49. *Pulvinaria psidii* Maskell 1893	Bihar, Gujarat, Karnataka, Kerala, Maharashtra, Odisha, Tamil Nadu, Uttar Pradesh, and West Bengal
50. *Pulvinaria urbicola*	Karnataka
Subfamily: Eulecaniinae	
1. *Eulecanium tiliae* (Linnaeus) 1758	Himachal Pradesh
2. *Lecanium mercarae* Ramakrishna Ayyar 1919	Karnataka
Subfamily: Eriopeltinae	
1. *Membranaria ceylonica* (Green) 1922	Ceylon
Subfamily: Pseudopulvinariinae	
Tribe: Pseudopulvinariini	
1. *Pseudopulvinaria sikkimensis* Atkinson 1889	Assam, Sikkim, Meghalaya

PHYLOGENY (FIGURE 6.25)

The first phylogenetic analysis of soft scales was done by Qin (1993) using several characters of several morphs, but the phylogenetic portion was never published. Miller and Hodgson (1997) chose taxa for each of the higher-level taxa in the Coccidae classification provided by Hodgson (1994a) for their phylogenetic analysis and also considering the results of the study of Qin (1993). The primary thrust of this analysis was to study the monophyly of the family Coccidae, to understand the relationships of the higher-level taxa within the Coccidae and to examine the relationships of families that have been hypothesised as being closely related to the Coccidae.

The family Pseudococcidae was chosen as an outgroup since there is general agreement that it is outside of the Coccidae. To score characters for an exemplar taxon, adult males, adult females, and first instars of two or more different species were taken into consideration.

The monophyly of the Coccidae is well supported by this analysis (Miller & Hodgson 1997). There is also support for the transfer of Eriochiton to the Eriococcidae (Hodgson 1994b).

It is of interest that the clade comprising the Coccidae plus the Aclerdidae is strongly supported as well, suggesting the possibility that aclerdids are simply basal coccids that have specialised on monocotyledonous plants, especially grasses. Since both groups are easily diagnosed as separate entities, and since there is some question about whether the tachardiids are more closely related to the coccids than are the aclerdids and because the current literature consistently treats the Coccidae, Aclerdidae, and Tachardiidae as distinct families, Miller and Hodgson (1997) accepted them as separate but closely related. The relationships of the higher-level groups within the family Coccidae are fairly consistent. One of the most strongly defined groups is comprised of the Ceroplastinae, Saissetiini, Pulvinariini, and Coccini. The placement of the Paralecaniini as most closely related to the Cardiococcinae suggests that it is not part of the Coccinae as previously suggested (Hodgson 1994a). This hypothesis needs to be tested more fully, since the lineage containing the Cardiococcinae and Paralecaniini is weakly defined. It is also clear that the Cissococcinae is basal within the Coccidae and might even be more basal than the Aclerdidae. Relationships within the Coccidae suggest that the Eulecaniinae and Filippiinae are basal and the Myzolecaniinae, Eriopeltinae, Pseudopulvmariinae, and Cyphococcinae are intermediate in their position.

There have been about ten different non-cladistic phylograms proposed for the Coccoidea in the last 35 years by Borchsenius (1958) (based mainly on adult female characters); Boratyński and Davies (1971) (based on adult male characters); Koteja (1974) (based on the structure of the adult female mouthparts); Miller and Kosztarab (1979) (modified on the basis of more recent descriptions of males); Danzig (1986) (using the morphology of the adult female, adult male, and crawler and also life history characters); Miller (1984) (based on a wide range of characters), Koteja (in Kosztarab and Kozár 1988), Miller and Miller (1993a; 1993b) (which considered the affinity of the Kermesidae and *Puto* in relation to a range of other taxa), and Miller and Williams (1995) (which considered the affinity of the Micrococcidae in relation to 11 other taxa). There is broad agreement in these phylograms as to which families fall within the lecanoid group, but there is less agreement as to their relationships. Most consider that the Pseudococcidae are the most primitive and that they and the Eriococcidae and Kermesidae probably arose from a mutual ancestor [although Boratyński and Davies (1971), on the basis of male structure, and Brown (1977), on the basis of chromosome structure, postulated that the neococcids are polyphyletic, and considered that the pseudococcids, developed separately]. All but Miller and Kosztarab

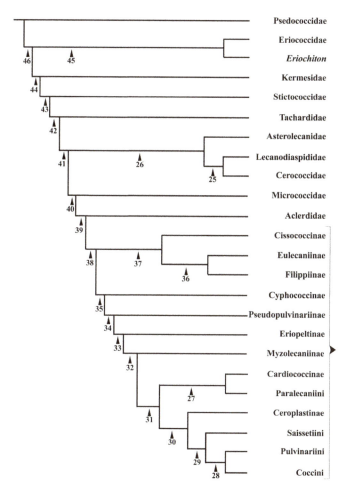

FIGURE 6.25 Cladogram from 14 equally parsimonious trees by successive weighting, using 105 characters from the adult female, adult male, and first instar-nymphs (character-states unordered); with Aclerdidae, Asterolecaniidae, Cerococcidae, Eriococcidae, Eriochiton, Kermesidae, Lecanodiaspididae, Micrococcidae, Pseudococcidae, Stictococcidae, and Tachardiidae as outgroups and higher taxa within Coccidae (bracketed) as ingroup (Length = 479; CI = 0.33; RI = 0.52). Note that one of the 14 trees was identical to the successively weighted tree. (With permission from Miller, D. R., and C. J. Hodgson. Phylogeny, In: *Soft Scale Insects: Their Biology, Natural Enemies and Control*, Vol. 7A. ed. Y. Ben-Dov and C. J. Hodgson, Amsterdam, the Netherlands: Elsevier, pp. 229–250, 1997.)

(1979) considered the Coccidae (along with the Aclerdidae) to be the most advanced lecanoids, and many workers also consider that the Tachardiidae belong to this group. The remaining families, namely, the Lecanodiaspididae, Cerococcidae, Asterolecaniidae, and Cryptococcidae, are believed to have evolved from ancestors which were extant before those from which the advanced lecaniids evolved, but after the appearance of the ancestors to the Pseudococcidae, Eriococcidae, and Kermesidae. Thus, there appears to be general agreement that the Aclerdidae, Asterolecaniidae, Cerococcidae, Cryptococcidae, Eriococcidae, Kermesidae, Lecanodiaspididae, Micrococcidae, and Tachardiidae are probably the families most closely related to the Coccidae. All of them belong to the lecanoid group, and the adult females can be separated by the keys in Balachowsky (1948), Howell and Williams (1976), Ben-Dov (1985), Danzig (1986), and Kosztarab and Kozár (1988).

ZOOGEOGRAPHY OF THE COCCIDAE

The zoogeography of this group has been discussed by Ben-Dov (1993) and Kozar and Ben-Dov (1997), which is summarised here. Distribution of Coccidae in different zoogeographical regions of the world is given in Table 6.3.

In his systematic catalogue of the soft scale insects of the World, Ben-Dov (1993) suggested that the family Coccidae contained about 1,088 species in 144 genera, whereas Hodgson (1994) divided them among 160 genera. Borchsenius (1957) was the first to list world scale insects, and he enumerated about 900 species and 100 genera, this was followed by Eastop (1978), who listed 1,162 species based on data in the Natural History Museum, London. Kosztarab and Kozár (1988) listed about 1,000 species and 100 genera and Tang (1991) 928 species under 139 genera. The species and genetic richness for each region are shown in Table 6.1, based on Ben-Dov's (1993) Systematic Catalogue of the Coccidae of the World. These show that the richest regions are the Palaearctic and Neotropical, containing about 50% of known Coccidae, many of them endemic to these regions, and the Ethiopian region harbours only about 20% of known species, followed by the Oriental region with only 8.1%. The ratio of genera to species differs between regions, with the highest ratios (maximum species per genus) being found in the Neotropical, Palaearctic, and Ethiopian regions. The highest percentage (31%) of endemic genera is found in the Palaearctic, where the number of endemic species is higher than 80%. Comparing zoogeographic regions individually, the highest species:genus ratios are in the Neotropical, Palaearctic, and Ethiopian regions, suggesting a high rate of speciation in these areas, especially on widely distributed woody plants. Species-rich soft scale genera include *Ceroplastes, Coccus, Eulecanium, Pulvinaria,* and *Saissetia.*

The number of genera and species found in different zoogeographical areas (Table 6.1) is discussed below along with the predominant genera in each region.

Palaearctic: This region, has the richest fauna, but this may be because it has been best explored. The total species richness of this region, including cosmopolitan species, is shown by a number of genera—35 spp. of *Eulecanium* on woody plants, 14 of *Lecanopsis* on grasses, 12 of *Luzulaspis* on sedges, 63 of *Pulvinaria* on trees and shrubs, and 32 of *Rhizopulvinaria* on the roots of herbs.

TABLE 6.3
Distribution of Coccidae in Different Zoogeographical Regions

Geographic Regions	Number of		Ratio of Genera to Species	Number of Genera Endemic to Region		Total Endemic Species[a]
	Genera	Species		Number of Genera	Number of Species in these Genera	
Palaearctic	58	299	1:5.2	28	78	244
Nearctic	27	105	1:3.9	1	1	48
Neotropical	51	298	1:5.8	20	26	247
Ethiopian	48	251	1:5.2	18	47	211
Oriental	44	126	1:2.9	6	6	80
Australian	23	73	1:3.2	5	18	54
New Zealand	18	60	1:3.3	1	2	23
and Pacific	28	99	1:3.5	5	7	61
Austro-Oriental Madagasian	20	43	1:2.2	2	3	17
Total (World)	**144**	**1088**	**1:7.6**	**86**	**188**	**985**

Source: Kozar, F., and Y. Ben-Dov., Zoogeographical considerations and status of knowledge of the family, In: *Soft Scale Insects, Their Biology, Natural Enemies and Control,* Vol. 7A. eds. Y. Ben-Dov and C. J. Hodgson, Elsevier, Amsterdam, the Netherlands, pp. 213–228, 1997.

[a] Total endemic species includes species not found in endemic genera.

Nearctic: The soft scale fauna of this region is poor. The majority of species of Nearctic soft scales are typically found on trees and shrubs. Particularly characteristic genera are *Toumeylla* and *Neolecanium*, living on pine and magnolias, respectively, while the only monotypic genus appears to be *Pseudophilippia*. The fauna also includes some widely distributed genera with numerous species, such as *Ceroplastes* with 11 spp., *Coccus* with 7 spp., and *Pulvinaria* with 23 spp., all typical of woodland and shrubs (Gill 1988).

Neotropics: The Neotropics are also species rich, particularly in some of the more cosmopolitan genera, e.g., *Ceroplastes* (73 spp.), *Coccus* (12 spp.), *Eucalymnatus* (13 spp.), and *Saissetia* (17 spp.), most of which are typical of woody plants. Few species are known from grassland and semidesert environments.

Ethiopian: This region is species rich with many widespread genera, such as *Ceroplastes* (40 spp.), *Coccus* (33 spp.), *Etiennea* (20 spp., but only in Africa), *Pulvinaria* (22 spp.), and *Saissetia* (22 spp.) (Ben-Dov 1986; 1993; De Lotto 1959; 1965; 1978; Hodgson 1968; 1969a,b; 1991b).

Oriental: The region has many monotypic genera, e.g., *Ericeroides*, *Metaceronema*, *Paracardiococcus*, *Paractenochiton*, *Podoparalecanium*, and *Taiwansaissetia*, some of which are considered to be endemic (Borchsenius, 1957). In addition, the Oriental region has other species-rich genera—*Ceroplastes* (8 spp.), *Coccus* (23 spp.), *Paralecanium* (13 spp.), *Pulvinaria* (16 spp.), and *Saissetia* (5 spp.) (Ben-Dov 1993). However, only China, India, and Sri Lanka have been more or less explored (Varshney 1985; 1992; Tang 1991). Varshney (1985; 1992) suggested that there might have been two centres of genetic evolution, one in the humid north-east of India, where Chinese, Indian, and Malaysian elements meet, and the other in the dry lowland forest, plains, and ephemeral habitats of southern India. The other areas within this region all need further exploration.

Australian: There is no comprehensive work on the Coccidae for this region. Several endemic genera, such as *Alecanopsis* (8 spp.), *Austrolichtensia* (1 sp.), *Cryptes* (2 spp.), *Symonicoccus* (6 spp.), and *Waricoccus* (1 spp.) have been recorded.

Pacific: According to Williams and Watson (1990), this region has 46 species belonging to 19 genera, most of which are probably of tropical origin, but also including one genus, *Anthococcus*, known only from this area. The most species rich genera are *Ceroplastes* (9 spp.), *Coccus* (7 spp.), *Milviscutulus* (4 spp.), *Pulvinaria* (5 spp.), and *Saissetia* (5 spp.) (Williams and Watson 1990).

New Zealand: The soft scale fauna of this region appears to be poor, with only 5 genera and 22 species (Eastop 1978).

Madagasian: Very meagre information is available for Madagascar, with only one monotypic genus, *Suareziella*, and the genus *Antandroya*, which resembles the Eriococcidae and perhaps represents the most primitive species of Coccidae (Williams 1984).

Austro-Oriental: This region appears to be characterised by several monotypic genera, such as *Alecanium*, *Anthococcus*, *Halococcus*, and *Myzolecanium*. This region includes large number of species in following genera: *Ceroplastes* (12 spp.), *Coccus* (21 spp.), and *Paralecanium* (15 spp., of which 14 appear to be restricted to this region).

IMMATURE TAXONOMY

Taxonomic characters of nymphs of soft scales have been studied and reviewed in detail by Williams and Hodges (1997). The fact that the taxonomy of the immature stages has proven to be very helpful in determining relationships within the armoured scales (Diaspididae) (Howell and Tippins 1990) has not greatly helped in the study of the immature stages of the Coccidae in fine detail. Most of the research has been conducted on the first-instar nymphs, with little attention being paid to other immature stages. All the taxonomic characters used in adult female are taken into consideration while studying first instar nymphs. These are dorsal setae, dorsal pores, dorsal tubular ducts, dorsal microductules, anal plate, and anal ring as dorsal structures; eye spots, marginal setae, stigmatic setae, and spiracular cleft as marginal structures; ventral setae, ventral pores, ventral microducts, ventral tubular ducts, antenna, mouthparts, spiracles, and legs as ventral structures. Their presence and absence, their types, and their distribution are important morphological characters in first instar nymph as in completely grown up female. Dorsal setae are absent in first nymphal instar in most of the species, similarly dorsal tubular ducts are rarely found. Ventral tubular ducts are absent in first instars of almost all species of soft scales.

MOLECULAR CHARACTERISATION

DNA barcoding has become a popular tool for species delimitation in vertebrates (Hebert et al. 2004; Wong et al. 2009) and invertebrates (Hajibabaei et al. 2006; Costa et al. 2007; Mikkelsen et al. 2007) and might prove ideal for accurate and rapid identification of scale insects. However, universal primers fail to amplify the standard barcode region of the COI for any but a few taxa (Kondo et al. 2008; Park et al. 2011). Some recent studies on barcoding of scale insects have gradually expanded the range of application to several families, including Diaspididae, Pseudococcidae, Coccidae, and Margarodidae (Ball & Armstrong 2007; Malausa et al. 2011; Park et al. 2011; Abd-Rabou et al. 2012; Beltrà et al. 2012; Deng et al. 2012; Sethusa et al. 2014). However, among the 1,140 coccid species, only the COI barcode regions of 41 (with complete scientific names) have been submitted to GenBank. This limited information calls for further investigation of the performance of DNA barcoding on a broader scale and development of a more efficient means of DNA barcoding identification in soft scales. The 28S nuclear gene can identify species in various insect taxa (Campbell et al. 1994; Smith et al. 2008; Monaghan et al. 2009). Although the 28S rDNA lacks sufficient variation to delimitate some species (Park et al. 2011; Deng et al. 2012), it is presently being proposed as a complementary marker to COI in scale insects (Sethusa et al. 2014). They have sequenced the COI and 28S genes of 340 individuals belonging to 36 common soft scale species in China with the aim of: (1) exploring the efficacy of DNA barcoding in Coccidae using multiple methods and (2) providing a comprehensive barcode library of common soft scales in China.

TAXONOMIC PROBLEMS

Soft scales have to be directly collected from the plant parts and cannot be collected using sweep nets or be trapped in any insect traps. The collector has to go to an individual plant/tree to look for the presence of soft scales. This makes it time consuming and more labour intensive.

During collection, getting right the stage of scale insect is very important, as only adult females are used for taxonomic studies and most of the keys are based on adult female characters. It is not always possible to grow nymphs obtained in the field to adult stage as many of them do not develop on factitious laboratory hosts.

Many coccids are too large or are covered with large quantities of wax. These scale insects are extremely difficult to process and often break during slide preparation. Similarly, old specimens often become hard and absorb too much of the stain making it almost impossible to look into the difficult-to-observe taxonomic characters.

Many scale insects look very different in live condition because of the geographical areas in which they are collected, and there are host plant induced variations in their live appearance, making it difficult to assign identity in live condition. In mounted conditions, they can be identified confidently, only if the literature on that particular species is accessible.

The above difficulties are further complicated due to major intraspecies variation. Intraspecific variation of taxonomic characters is widespread in the Coccoidea in general and in the soft scales in particular. This variation in the soft scale species so far studied is favoured by their wide geographic distribution, by their polyphagy, and because they reproduce by parthenogenesis. This intraspecific variation is shown by both morphological and biological characters. The greatest variation in morphology is shown by the number of dorsal sub-marginal tubercles. Some of these characters depend on the nutritive condition of the host plant, but some inter-population variation is also occasionally caused by parasitoids.

Intraspecific variation has been seen in two species viz., *Parthenolecanium corni* Bouche and *Coccus hesperidum* (Linnaeus). Variability of morphological characters in *P. corni* is widespread in both Old and New World, and it is also highly polyphagous, with about 350 known host plant species (Kawecki 1958; Ben-Dov 1993). Intraspecific variation in *P. corni* has been studied in detail by various authors from different zoogeographical areas (Sulc 1932; Sivescu 1943; 1944; Borchsenius 1957; Marchal 1908; Sanders 1909; Voukassovitch 1930; Ebeling 1938; Habib 1957; Borchsenius 1957; Dziedzicka and Sermak 1967; Danzig 1980 and 1986; Richards 1958; Phillips 1965; Williams and Kosztarab 1972; Hamon and Williams 1984; Gill 1988; Nakahara 1981; Saakyan-Baranova et al. 1971; Nuzzaci 1969). Intraspecific variation in *C. hesperidum* has been studied in detail by Hodgson (1967), and he described considerable variation in a single population of this species that had relatively consistent differences between specimens on the stems and leaves.

BIOLOGY

The biology of soft scales has been dealt in detail by Hamon and Williams (1984). Most species of soft scales have one generation per year, but some species have multiple generations in the warmer southern areas. Under greenhouse conditions, overlapping generations often occur and several developmental stages may be found at any one time. Soft scales reproduce sexually or parthenogenetically, and some species have both bisexual and parthenogenetic strains. Nur (1971) studied the cytology of 33 parthenogenetic scale insects and identified and characterised 7 different types of parthenogenesis. Most species of soft scale are capable of reproducing parthenogenetically.

The females have a simple metamorphosis, whereas the males have a complete metamorphosis. There are three or four instars in the female and five instars in the male. Females are either oviparous or ovoviviparous.

The first instar or "crawler" stage is the most active developmental stage in soft scales. Most soft scale crawlers prefer to feed on the leaves of their host and will usually settle there after hatching. Crawlers feed by inserting their stylet-like mouthparts directly into cells and suck out the cell contents. For those species which feed on deciduous hosts, this requires them to move from the leaves to the twigs or stems of their host before leaf drop in the fall. Some species of *Pulvinaria* begin their development on the leaves, move to the twigs to overwinter, and return to the leaves as adult females in the spring. Most soft scales overwinter as second instars, but some of the wax scales (*Ceroplastes* spp.) overwinter as fertilised adult females which produce eggs very early in the spring. Characters of different instars of male and females and adults of both the sexes are given in the Table 6.4.

TABLE 6.4
Characters of Different Instars of Males and Females

Character	First Instar Male & Female	Second Instar Female	Third Instar Female	Adult Female
Eyes	Present	Present	Present	Present
Anal plates	Present, each plate with a long apical seta	Anal plates are without the long apical seta	Present	Present
Anal cleft		Present	Present	Present
Legs	Well developed and 5-segmented	Reduced or well developed	Present	Present
Antenna	Well-developed 5- or 6-segmented	Reduced or well developed	Segments lesser than adult	Segments more than all other instars
Spiracular setae	Usually differentiated from marginal setae	Usually differentiated from the marginal setae	Usually differentiated from the marginal setae	Usually differentiated from the marginal setae
Dorsal & Ventral setae	Absent	Fewer than completely grown up female	Present	More than all other stages
Multilocular pores	Absent	Always absent from the venter of the abdomen	Absent from the venter of the abdomen	Present on the venter of the abdomen
Tubular ducts	Absent	Always absent from the venter of the abdomen		Present

Character	Second Instar Male	Third Instar Male	Fourth Instar Male (pupa)	Adult Male
Eyes	Present	Absent	Absent	2–5 pairs of simple eyes & a pair of lateral ocelli present
Anal plates	Present, but long apical setae are absent	Anal plate replaced by 2 quadrate sclerotised lobes	Absent	Absent
Legs	Present	Present	Present	Well developed, 5-segmented
Antenna	Present	Present	Present	Present
Spiracular setae	Absent	Absent	Absent	Absent
Multilocular pores	Absent	Absent	Absent	Absent
Tubular ducts	Present	Absent	Absent	Absent
Anal ring	Absent	Absent	Absent	Absent
Penial sheath		Short and rounded	Elongate triangular	Long and about ¼ of the body
Wing buds		Present	Present	Wings completely developed

HOST PLANT ASSOCIATION

Host plant relationships of Coccidae and their hosts recorded throughout the world was analysed by Lin et al. (2010). They collected host-plant data for 1,035 described species (excluding subspecies and species of unknown host use) of Coccidae from the scale insect database, ScaleNet (Ben-Dov et al. 2009). They scored coccid species as being either family-level monophagous (occurring on only a single host-plant family) or polyphagous (occurring on two or more host-plant families). Coccids were recorded from 200 plant families—171 angiosperms, 10 gymnosperms, and 19 non-seeded land plants. On angiosperms, approximately 63% of coccid species were restricted to only one plant family, and most of these (about 90%) were recorded from only one plant genus. About 37% of coccid species are polyphagous on angiosperms, including notorious agricultural pests such as *Ceroplastes rubens*, *Parasaissetia nigra*, *Saissetia coffeae*, and *Saissetia oleae*, which have broad host ranges of more than 20 plant families. Fewer coccids feeding on gymnosperms are specialists and approximately 48% are polyphagous. There are few coccid species (30) reported on non-seeded land plants, and the majority of these are polyphagous (about 83%). Of the five species of coccid on non-seeded land plants that exhibit family-level monophagy, two have only a single host record (*Alecanopsis filicum* and *Pulvinaria satoi*). The other three, *Kilifia diversipes*, *Pounamococcus cuneatas*, and *Saissetia carnosa*, have at least two collection records, and so their inferred host-use associations might be more reliable. Analysis of the relationship between species richness of host-plant families and the number of species of coccids recorded on these plants showed a significant positive correlation between host-plant species richness per angiosperm plant family and coccid species richness.

ECONOMIC IMPORTANCE

All soft scales feed on sap of economically important plants which are valuable to man. The continuous feeding results in loss of plant vigour, poor growth, die back of twigs and branches, early leaf drop, and even, sometimes, death of the entire plant. They cause indirect damage by excreting honeydew, which acts as a growth medium for sooty moulds, which produce a black coating over the leaf surface and other plant parts. This coating interferes with photosynthesis and may cause poor growth, a reduction in fruit size, and generally give an unsightly appearance to the crop. A substantial number of soft scale species have been proven to be especially troublesome to the plants cultivated or used by man. In some cases, the monetary loss caused by some of these species has devastated the agricultural economy of several nations. The economic loss worldwide attributed to all scale insects, including the cost of control, has been estimated to be 5 billion US$ annually (Kosztarab and Kozár 1988) and probably a quarter of that loss was due to species of Coccidae.

Although most of the species are damaging, a few are beneficial to man. Some species of scales have been found to produce useful by-products. *Ericerus pela* (Chavannes) has been used for more than 1,000 years to produce "China wax," a high quality, high melting-point wax with many uses in industry (Li 1985). The American Indians in the southwestern United States of America have used the wax of the irregular wax scale, *Ceroplastes irregularis* Cockerell, to water-proof or seal baskets and pottery. Soft scales produce large quantities of honeydew, which can form an important part of the diet of many ants, wasps, flies, and other animals. For example, Krombein (1951) recorded 176 species of wasps, belonging to 5 families, visiting the tulip tree scale, *Toumeyella liriodendri* (Gmelin). Honeydew is also an important component of many honeys.

COLLECTION, PRESERVATION, AND MOUNTING TECHNIQUES

Very few species of soft scales can be identified by their live appearance, but many cannot even be identified up to a genus level. Therefore, the majority of species have to be identified by microscopic examination of slide-mounted specimens. This need for adequately prepared specimens was recognised by workers, such as Green (1896), Newstead (1903), and Steinweden (1929) who proposed specific methods for the preparation and mounting of scale insects for microscopic study and which have been refined and improved by Cilliers (1967), Kozarzhevskaya, (1968), Williams and Kosztarab (1972), and Wilkey (1990).

COLLECTION

Soft scale insects infest different parts of their host plants, from the roots to the fruit. Presence of some species can be easily made out due to the presence of white ovisacs and sometimes because of presence of brightly coloured thick wax covers, but many cannot be easily detected, either because they blend with the environment or because they are concealed. A collector may increase his success by looking for intensive ant activity, honeydew droplets, and/or sooty mould. On trees, soft scales are commonly present on the branches (particularly in protected and shaded parts of the trunk), leaves, in bark crevices, forks between twigs, or other sheltered areas of the plants. On all plants, it is important to pay special attention to such neglected areas as roots, root crowns, leaf sheaths, both sides of grass leaves, fruits, galls, and areas under bark flakes. In order to secure high-standard, slide-mounted specimens, it is almost essential to collect the young adult female within a short period following its final moult because, in most species, the body then expands and the dorsum becomes sclerotised. Also, it is essential to make slides immediately after collection as the quality of the slide depends on the period for which it has been preserved in ethanol.

PRESERVATION AND STORAGE

Wet preservation: It is recommended that the best storage media is acid alcohol, either acetic acid alcohol or lactic acid alcohol. The problem with these media is that the alcohol can quickly evaporate and the specimens then dry out. Because of this, the liquid needs to be topped-up from time to time. The best storage containers are glass bottles with a metal screw-on cap and a thick rubber washer. Alternatively, the small bottles can themselves be stored in large bottles full of alcohol, which can be easily topped-up when necessary. Air bubbles in the bottle can damage the specimens. The bottles should be individually labelled, preferably with full collection and other data.

Dry preservation: Specimens collected even one hundred years ago and kept in a dried state can be made into excellent slides (Ben-Dov and Hodgson 1997). Storage of material in a dry state is, therefore, an option, but needs care because many Coccidae contain much water and so need to be dried out before storage, either by gentle heating or by keeping them in a dry atmosphere (perhaps with silica gel). However, as with storage in alcohol, there are problems with the storage of dry material. Firstly, the specimens may be damaged by museum beetle, which can be devastating. The stored scales, therefore, need to be kept in air-tight containers, preferably with naphthalene. Secondly, the dry insects become extremely brittle and can easily lose their setae, legs, and antennae. This can be a major problem if the material is handled or moved around much. On the other hand, they do not need to be checked at such regular intervals as material stored in liquids if they are stored in solid, reasonable insect-proof cabinets. Dried stored specimens should be kept in small cardboard, glass, or plastic containers, preferably wrapped in tissue paper rather than cotton wool to prevent the specimens from moving. They should then be stored in strong, insect-proof cabinets. Sending dried material through post is not recommended as jerks during transit will knock the specimens together and cause much damage.

SLIDE PREPARATION

There are a number of publications outlining various staining and mounting techniques for scale insects, such as Cilliers (1967) and Williams and Kosztarab (1972) for the Coccidae.

The best slides can only be made from freshly collected material. Whatever the age of the material, the methodology usually involves the same five or six stages. The proportion of the chemicals to be used is given in Table 6.5. The following steps have been developed by Ben-Dov and Hodgson (1997).

Initial fixation: Live specimens should be fixed in a mixture of acetic acid alcohol (see Table 6.5) for 1–2 minutes. All specimens should be fixed, even if they are going to be mounted immediately. If necessary, material can also be stored in acid alcohol for long periods.

Maceration: Make an incision (probably best mediolaterally in the abdomen) and transfer to 10% potassium hydroxide (KOH) at room temperature for 12–24 hours. Clearing specimens with KOH can be quickened by slight warming, but it is strongly recommended that this should never exceed 40°C as specimens which have been boiled or over-heated can be unstainable. Heavily sclerotised specimens can be left in cold KOH for at least a week.

Dehydration: Transfer, by means of a small spatula, to water and gently pump or press out body contents. The pumping out of body contents and the general handling of specimens needs to be done very gently and special care should be taken not to break off the various appendages and setae, which are almost certainly needed for identification. Nonetheless, clearing the body of all contents is vital, and the procedures towards staining should not be continued until as much contents as possible have been removed. Transfer to clean water and then gradually add glacial acetic acid (GAA) to replace the water until the concentration of GAA = about 70%, and then transfer to 100% GAA.

Staining: Add to the 100% GAA solution (in which the specimens are contained) acid fuchsin, at the rate of one drop to every 1–2 cc GAA; retain in staining solution for 1–24 hours.

Washing: Transfer to GAA to wash out excess stain.

Clearing: If any wax or fatty material is still present on or in the specimens, add a drop of xylene to the GAA solution to dissolve the wax; retain for 15–30 minutes.

Final fixation: Transfer to oil of cloves for 24 hours. Although 24 hours is not essential, this should ensure the full displacement of water (dehydration) from the specimens, which otherwise could cloud the Canada balsam.

Mounting: Place a small drop of Canada balsam on the slide and transfer the specimen to it. Gently push the specimen so that it is at the bottom of the drop of Canada balsam, and then cover with a cover-glass. Except with very convex specimens, the cover-glass should not be pressed down. It is recommended to use No. 0 or thinner cover-glasses to allow their use at oil-immersion magnifications. Placing the cover-glass over a freshly prepared slide may result in the specimen drifting away from under the cover-glass. To avoid this, place the specimen in a minute drop of Canada balsam and arrange it on the slide; then place the slide in a covered petri-dish (to prevent dust particles from sticking to the Canada balsam) and allow the surface of the drop of balsam to dry between 15 minutes and 24 hours. Then place a drop of Canada balsam on the cover-glass and cover.

Label slide: The future scientific value of the specimen will depend on the data on the slide. Unless the accession book is kept with the slides, no collection data will be available and the slides are almost useless. An unlabelled slide is useless.

Keep slides to cure at 40°C for 4–6 weeks. Slides should not be cured for longer than about 6 weeks, as this may cause darkening of the Canada balsam.

USEFUL WEBSITES

ScaleNet: Scale Insects (Coccoidea) Database (*http://scalenet.info*).

This database aims to provide comprehensive information on the scale insects (Coccoidea) of the world, of which there are about 7,800 species. Through keyword searches and other queries, ScaleNet provides comprehensive information including the insects' biology, classification, naming history, distribution, plant hosts, economic importance, controls, and scientific literature about them. Currently, information can be retrieved for 49 families including Coccidae.

TABLE 6.5
Chemicals and Their Proportions for Slide Preparation

Acid alcohols:	
a. acetic acid alcohol	4 parts 96% ethyl alcohol to 1 part glacial acetic acid.
b. lactic acid alcohol	2 parts 96% ethyl alcohol to 1 part lactic acid.
Acid fuchsin stain	either dissolve about 0.7 g acid fuchsin in 300°cc lactophenol or 0.5 g acid fuchsin + 25°cc 10% HCL + 300 cc water.
Lactophenol	Composed of phenol 100 g + 10 cc lactic acid + 100°cc glycerine + 200 cc water.

The reference database of ScaleNet includes about 25,000 references.

Scale Insects: Identification Tool for Species of Quarantine Significance

(*http://www.idtools.org/id/scales/index.php*).

This website provides information under the following heads.

Fact sheets: Detailed diagnostic information and images are presented in fact sheets for each family and species represented. The fact sheets have been updated with the latest taxonomic and quarantine information.

Identification: There are four separate keys to identify slide-mounted adult females: one for scale families, one for mealybugs and mealybug-like scales, one for soft scales, and one for pest species in other families.

Image gallery: A comprehensive image gallery is included. The gallery is filterable, offering an image-based option to find the pest fact sheet you're looking for. Each image links directly to the fact sheet in which it's found.

Online Lucid key for identification of Australian scale insects (https://keyserver.lucidcentral.org/key-server/player.jsp?keyId=30).

This website provides Lucid keys for identification of scale insects belonging to 27 genera and has species reference keys to scales that are found in Australia.

Common Soft Scales of India (http://www.nbair.res.in/Coccidae/index.html).

This website provides an account of 32 species of common soft scales of India. This site has been developed as an identification aid to common coccids of India, many of which also occur in other parts of the world. Brief diagnostic accounts, based on external characters to aid in field identification supported by taxonomic characters aided by microphotographs of mounted female, have been given for the species included. It gives information about classification of Coccidae, checklist of Indian Coccidae, and fact sheet that has descriptions of live and mounted characters of 32 species of common soft scales of India.

CONCLUSION

The advances in coccidology during the last 250 years have been remarkable. The number of scale insect species described has gone up from just 24 species in 1758 to nearly 8,000 species today. Our knowledge about the taxonomy of Coccoidea has been substantial; the advent of high-resolution microscopes, electron microscopy, digital photography, computer illustration and graphic editing programs, molecular biology, and phylogenetic methods have contributed to high standard descriptions of scale insects. The worldwide web has made possible the creation of online scale insect databases, which allow users to obtain information on numerous aspects of most described species. Advances in molecular genetics are helping to resolve the phylogenetic relationships of this morphologically highly derived group, and new techniques such as barcoding are being contemplated as a tool for identification of common pest species. Despite all of these advances in technology, and our accumulated knowledge about scale insects, the field of coccidology still faces many challenges. The higher-level relationships of scale insects are far from being resolved. Every year there are new species of scale insects being added to the list of agricultural pests. Perhaps the greatest problem that the field of coccidology faces today is the decline of scale insect specialists worldwide. Museums that used to employ coccidology experts no longer replace the retirees. There are over 1,150 species of soft scales known to occur throughout the world, while from India only 73 species have been recorded. DNA barcode studies have not even been initiated in India (only one species), while barcodes of 339 soft scale species have been generated from other parts of the world (*http://www.boldsystems.org*). This shows that taxonomic studies on soft scales are still in the early stages in India. More intensive exploratory surveys in different ecological areas need to be conducted to collect and identify soft scales. Molecular studies need to be initiated and collaborations developed with coccidologists working in other countries.

REFERENCES

Abd-Rabou, S., H. Shalaby, J. F. Germain, N. Ris, P. Kreiter, and T. Malausa. 2012. Identification of mealybug pest species (Hemiptera: Pseudococcidae) in Egypt and France, using a DNA barcoding approach. *Bulletin of Entomological Research* 102: 515–523.

Ali, S. M. 1962. Coccids affecting sugarcane in Bihar (Coccidae: Hemiptera). *Indian Journal of Sugarcane Research and Development* 6: 72–75.

Ali, S. M. 1964. Some studies on *Pulvinaria cellulosa* Green a mealy scale of mango in Bihar, India. *Indian Journal of Entomology* 26: 361–362.

Ali, S. M. 1967a. Current trends in the taxonomy of coccids (Insecta). *Bulletin of the National Institute of Sciences of India* 34: 212–218.

Ali, S. M. 1967b. Description of a new and records of some known coccids (Homoptera) from Bihar, India. *Oriental Insects* 1: 29–43.

Ali, S. M. 1968a. Coccids (Coccoidea: Hemiptera: Insecta) affecting fruit plants in Bihar (India). *Journal of the Bombay Natural History Society* 65: 120–137.

Ali, S. M. 1968b. Description of a new and records of some known coccids (Homoptera) from Bihar, India. *Oriental Insects* 1: 29–43.

Ali, S. M. 1969a. A catalogue of the oriental Coccoidea. Part II. (Insecta: Coccoidea: Diaspididae). *Indian Museum Bulletin* 4: 38–73.

Ali, S. M. 1969b. A catalogue of the Oriental Coccoidea. Part I. (Insecta: Coccoidea: Diaspididae). *Indian Museum Bulletin* 4: 67–83.

Ali, S. M. 1970a. A catalogue of the Oriental Coccoidea. (Part III.) (Insecta: Homoptera: Coccoidea). *Indian Museum Bulletin* 5: 9–94.

Ali, S. M. 1970b. A catalogue of the Oriental Coccoidea. (Part IV.) (Insecta: Homoptera: Coccoideae). *Indian Museum Bulletin* 5: 71–150.

Ali, S. M. 1971. A catalogue of the Oriental Coccoidea (Part V) (Insecta: Homoptera: Coccoidea) (with an index). *Indian Museum Bulletin* 6: 7–82.

Ali, S. M. 1973. Some coccids from Goa. *Journal of the Bombay Natural History Society* 69: 669–671.

Avasthi, R. K., and S. A. Shafee. 1979. A new species of *Cerostegia* De Lotto (Homoptera: Coccidae) from Ajmer (India). *Current Science* 48: 36–37.

Avasthi, R. K., and S. A. Shafee. 1984a. A new genus of Coccidae (Homoptera). *Indian Journal of Systematic Entomology* 1: 7–9.

Avasthi, R. K., and S. A. Shafee. 1984b. *Kozaricoccus* gen. n. for *Ceroplastodes bituberculatus* Brain (Homoptera: Coccidae). *Indian Journal of Systematic Entomology* 1: 31–33.

Avasthi, R. K., and S. A. Shafee. 1984c. Two new species of *Coccus* Linnaeus (Homoptera: Coccidae) from India. *Proceedings of the 10th International Central European Entomofaunistic Symposium*, Budapest, August 15–20, 1983: 389–392.

Avasthi, R. K., and S. A. Shafee. 1985. Two new species of *Pulvinaria* Targ.-Tozz. (Homoptera: Coccidae) from India. *Current Science*, 54: 1289–1291.

Avasthi, R. K., and S. A. Shafee. 1986. Species of Ceroplastinae (Homoptera: Coccidae) from India. *Journal of the Bombay Natural History Society* 83: 327–338.

Ayyar, T. V. R. 1919a. A contribution to our knowledge of South Indian Coccidae. *Bulletin of the Agricultural Research Institute, Pusa, India* 87: 1–50.

Ayyar, T. V. R. 1919b. Notes on new and unrecorded species of Indian Coccidae. *Bulletin of the Agricultural Research Institute, Pusa* 89: 91–99.

Ayyar, T. V. R. 1919c. Some south Indian coccids of economic importance. *Journal of the Bombay Natural History Society* 26: 621–628.

Ayyar, T. V. R. 1921. A check list of the Coccidae of the Indian region. *Proceedings of the Entomology Meetings. India* 4: 336–362.

Ayyar, T. V. R. 1924 A further contribution to our knowledge of south Indian Coccidae. *Proceedings of the Entomology Meetings. India* 5: 339–344.

Ayyar, T. V. R. 1926. Recent additions to the Indo-Ceylonese coccid fauna, with notes on known and new forms. *Journal of the Bombay Natural History Society* 31: 450–457.

Ayyar, T. V. R. 1930. A contribution to our knowledge of South Indian Coccidae (Scales and Mealybugs). *Bulletin of the Imperial Institute of Agricultural Research, Pusa, India* 197: 1–73.

Ayyar, T. V. R. 1936. Recent records of south Indian Coccidae. *Proceedings of the Indian Science Congress* 23: 351.

Balachowsky, A. 1948. Les cochenilles de France, d'Europe, du Nord de l'Afrique, et du bassin Méditerranéen. IV. Monographie des Coccoidea; Classification-Diaspididae (Premiere pattie). *Actualites Science et Industrie Entomologie Appliquce* 1054: 243–394.

Ball, S. L., and K. F. Armstrong. 2007. *Using DNA Barcodes to Investigate the Taxonomy of the New Zealand Sooty Beech Scale Insect*. Science and Technical Publishing, Department of Conservation. Wellington, New Zealand, 287, 1–14.

Beltrà, A., A. Soto, and T. Malausa. 2012. Molecular and morphological characterisation of Pseudococcidae surveyed on crops and ornamental plants in Spain. *Bulletin of Entomological Research* 102: 165–172.

Ben-Dov, Y. 1985. Coccoidea. In: *Insects of Southern Africa*, eds. C. H. Scholtz and E. Holm, Butterworths, Durban, South Africa, pp. 168–175.

Ben-Dov, Y. 1986. Taxonomy of two described and one new species of *Waxiella* De Lotto (Homoptera: Coccoidea: Coccidae). *Systematic Entomology* 11: 165–174.

Ben-Dov, Y. 1993. A systematic catalogue of the soft scale insects of the world (Homoptera: Coccoidea: Coccidae) with data on geographical distribution, host plants, biology and economic importance. *Flora and Fauna Handbook No. 9*. Sandhill Crane Press, Inc., Gainesville, FL, xxviii + 536 pp.

Ben-Dov, Y. D., R. Miller, and G. A. P. Gibson. 2009. ScaleNet: A data base of the scale insects of the world. http://www.sel.barc.usda.gov/scale net/scalenet.html (Accessed November 10, 2018).

Ben-Dov, Y., and C. J. Hodgson. 1997. 1.4 Techniques. 1.4.1 Collecting and mounting. In *Soft Scale Insects: Their Biology, Natural Enemies and Control* (Vol. 7A), eds. Y. Ben-Dov and C. J. Hodgson, 389–395. Elsevier, Amsterdam, the Netherlands & New York.

Boratyński, K., and R. G. Davies.1971. The taxonomic value of male Coccoidea (Homoptera) with an evaluation of some numerical techniques. *Biological Journal of the Linnean Society* 3: 57–102.

Borchsenius, N. S. 1957. *Sucking insects, Vol. IX. Suborder mealybugs and scale insects (Coccoidea). Family cushion and false scale insects (Coccidae)*. Fauna USSR, Novaya Seriya 66:493 pp (in Russian).

Borchsenius, N. S. 1958. On the evolution and phylogenetic interrelations of the Coccoidea. *Zoologicheskii Zhurnal* 37: 765–780.

Borchsenius, N. S. 1964. Notes on the Coccoidea of India. 1. A new genus and three new species of Leucaspidini (Diaspidae). *Entomologicheskoe Obozrenye* 43: 864–872.

Borchsenius, N. S. 1967. Materials on the fauna of scale insects (Homoptera, Coccoidea) from India. 2. *Andaspis* Macg. with three new allied genera (Diaspididae). *Entomologicheskoe Obozrenye* 46: 724–734.

Brown, S. W. 1977. Adaptive status and genetic regulation in major evolutionary changes of coccid chromosome systems. *Nucleus* 20: 145–157.

Campbell, B. C., J. D. Steffen-Campbell, and J. H. Werren. 1994. Phylogeny of the Nasonia species complex (Hymenoptera: Pteromalidae) inferred from an internal transcribed spacer (ITS2) and 28S rDNA sequences. *Insect Molecular Biology* 2: 225–237.

Cilliers, C. J. 1967. A comparative biological study of three *Ceroplastes* species (Hem., Coccidae) and their natural enemies. *Entomology Memoirs*, 13: 1–59.

Costa, F. O., J. R. DeWaard, J. Boutillier et al. 2007. Biological identifications through DNA barcodes: The case of the Crustacea. *Canadian Journal of Fisheries and Aquatic Sciences* 64: 272–295.

Danzig, E. M. 1980. *Scale Insects of the Far East of the USSR (Homoptera, Coccinea) with Phylogenetic Analysis of Scale Insect Fauna of the World*. Nauka Publishers, Leningrad, Russian, 363 pp.

Danzig, E. M. 1986. *Coccids of the Far-Eastern USSR (Homoptera, Coccinea). Phylogenetic Analysis of Coccids in the World Fauna*. Oxonian press, New Delhi, India, 450 pp.

Das, G. M. 1959. Observations on the association of ants with coccids of tea. *Bulletin of Entomological Research* 50: 437–448.

Das, G. M., and R. N. Ganguli. 1961. Coccoids on tea in North-East India. *Indian Journal of Entomology* 23: 245–256.

De Lotto, G. 1959. Further notes on Ethiopian species of the genus *Coccus* (Homoptera: Coccoidea: Coccidae). *Journal of the Entomological Society of Southern Africa* 22: 150–173.

De Lotto, G. 1965. On some Coccidae (Homoptera), chiefly from Africa. *Bulletin of the British Museum (Natural History) Entomology* 16: 177–239.

De Lotto, G. 1978. The soft scales (Homoptera: Coccidae) of South Africa, HI. *Journal of the Entomological Society of Southern Africa* 41: 135–147.

Deng, J., F. Yu, T. X. Zhang et al. 2012. DNA barcoding of six *Ceroplastes* species (Hemiptera: Coccoidea: Coccidae) from China. *Molecular Ecology Resources* 12: 791–796.

Dziedzicka, A., and W. Sermak 1967. The variability of the dorsomarginal glands in larvae H and females in *Lecanium corni* Bouch of *Taxus baccata*. 1. *Rosznik Nauk-Dydactyczny WSP w Krakowie*, 29: 25–31.

Eastop, V. F. 1978. Diversity of the Sternorrhyncha within major climatic zones. *Symposia of the Royal Entomological Society of London* 9: 71–88.

Ebeling, W. 1938. Host-determined morphological variations in Lecanium corni. *Hilgardia* 11: 613–631.

Farrel, G. S. 1990. Redescription of *Cryptes baccatus* (Maskell) (Coccoidea: Coccidae), an Australian species of soft scale. *Memoirs of the Museum of Victoria* 51: 65–82.

Ferris, G. F. 1957. A brief history of the study of the Coccoidea. *Microentomology* 22: 39–57.

Ganguli, R.N., and M. R. Ghosh. 1964. Coccoids of Tripura. *Indian Journal of Entomology* 26: 358–359.

García Morales, M., B. D. Denno, D. R. Miller, G. L. Miller, Y. Ben-Dov, and N. B. Hardy. 2016. ScaleNet: A literature-based model of scale insect biology and systematics. Database. doi: 10.1093/database/bav118. http://scalenet.info. (Accessed November 10, 2018).

Ghose, S. K. 1961. Studies on some coccids (Coccoidea: Hemiptera) of economic importance of West Bengal, India. *Indian Agriculturist* 5: 57–78.

Giliomee, J. H. 1967. Morphology and taxonomy of adult males of the family Coccidae (Homoptera: Coccidae). *Bulletin of the British Museum (Natural History), Entomology Supplement* 7: 1–168.

Giliomee, J. H. 1997. The adult male. In: *Soft Scale Insects: Their Biology, Natural Enemies and Control*, eds. Y. Ben-Dov and C. J. Hodgson, Elsevier Science B.V, pp. 23–30.

Gill, R. J. 1988. *The Scale Insects of California. Part 1. The Soft Scales (Homoptera" Coccoidea: Coccidae)*. California Department of Food and Agriculture: Technical Series in Agricultural Biosystematics and Plant Pathology, Sacramento, CA, 1: 132 pp.

Gimpel, W. F., D. R. Miller, and J. A. Davidson. 1974. *A Systematic Revision of the Wax Scales, Genus Ceroplastes, in the United States (Homoptera: Coccoidea: Coccidae)*. Miscellaneous Publication Agricultural Experiment Station, University of Maryland, 841: 1–85.

Green, R. E. 1896. *The Coccidae of Ceylon. Part 1*. Dulau & Co., London, UK. 101 pp.

Gullan, P. J., and J. H. Martin. 2003. Sternorrhyncha (jumping plant-lice, whiteflies, aphids and scale insects). In *Encyclopedia of Insects*, eds. V. H. Resh and R. T. Cardé, 1079–1089. Academic Press, Amsterdam, the Netherlands.

Habib, A. 1957. The morphology and biometry of the *Eulecanium corni*-grop, and its relation to host-plants (Hemiptera, Homoptera: Coccoidea). *Bulletin de la Socirt Entomologique d'Egypte* 50: 381–410.

Hajibabaei, M., D. H. Janzen, J. M. Burns, W. Hallwachs, and P. D. N. Hebert. 2006. DNA barcodes distinguish species of tropical Lepidoptera. *Proceedings of the National Academy of Sciences of the United States of America* 103: 968–971.

Hamon, A. B., and M. L. Williams. 1984. *The Soft Scale Insects of Florida (Homoptera: Coccoidea: Coccidae). Arthropods of Florida and Native Land Areas, 11*, Florida Department of Agriculture and Consumer Services, Gainesville, FL, pp. 1–194.

Hebert, P. D., M. Y. Stoeckle, T. S. Zemlak, and C. M. Francis. 2004. Identification of birds through DNA barcodes. *PLoS Biology* 2: 312.

Hodgson, C. J. 1967. Notes on Rhodesian Coccidae (Homoptera: Coccoidea): Part 1: The genera *Coccus, Parasaissetia, Saissetia* and a new genus *Mashona*. *Arnoldia* (Rhodesia) 3: 1–22.

Hodgson, C. J. 1968. Further notes on the genus *Pulvinaria* Targ. (Homoptera: Coccoidea) from the Ethiopian region. *Journal of the Entomological Society of Southern Africa* 31: 141–174.

Hodgson, C. J. 1969a. Notes on Rhodesian Coccidae (Homoptera: Coccoidea), II. *Arnoldia* (Rhodesia) 40: 1–43.

Hodgson, C. J. 1969b. Notes on Rhodesian Coccidae (Homoptera. Coccoidea), III. *Arnoldia* (Rhodesia) 4: 1–42.

Hodgson, C. J. 1991a. A redescription of *Pseudopulvinariasildamensis* Atkinson (Homoptera, Coccoidea), with a discussion of its affinities. *Journal of Natural History*, 25: 1513–1529.

Hodgson, C. J. 1993. The immature instars and adult male of *Etiennea* (Homoptera: Coccidae) with a discussion of its affinities. *Journal of African Zoology* 107: 193–215.

Hodgson, C. J. 1994a. *The Scale Insect Family Coccidae: An Identification Manual to Genera*. CAB International Wallingford, Oxon, UK. 639 pp.

Hodgson, C. J. 1995. A brief review of the structure of the spiracle in the family Coccidae. *Israel Journal of Entomology* 29: 47–55.

Hodgson, C. J. 1997. Classification of Coccidae and Coccid families. In *Soft Scale Insects: Their Biology, Natural Enemies and Control* (Vol. 7A), eds. Y. Ben-Dov and C. J. Hodgson, 157–201. Elsevier, Amsterdam, the Netherlands & New York.

Hodgson, C. J., 1991b. A revision of the scale insect genera *Etiennea* and *Platysaissetia* (Homoptera: Coccidae) with particular reference to Africa. *Systematic Entomology* 16: 173–221.

Hodgson, C. J., 1994b. *Eriochiton* and a new genus of the scale insect family Eriococcidae (Homoptera: Coccoidea). *Journal of the Royal Society of New Zealand* 24: 171–208.

Hodgson, C. J., and R. C. Henderson. 2000. *Coccidae (Insecta: Hemiptera: Coccoidea)*. Manaaki Whenua Press Lincoln, Canterbury, New Zealand. 259 pp.

Howell, J. O., and H. H. Tippins, 1990. The immature stages. In: *Armored Scale Insects, Their Biology, Natural Enemies and Control* ed. D. Rosen, World Crop Pests 4A. Elsevier, Amsterdam, the Netherlands, pp. 29–42.

Howell, J. O., and M. L. Williams. 1976. An annotated key to the families of scale insects (Homoptera: Coccoidea) of America, North of Mexico, based on characters of the adult female. *Annals of the Entomological Society of America* 69: 181–189.

Joshi, S. 2017. First record of *Pulvinaria urbicola* Cockerell (Hemiptera: Coccidae) from India, with a key to the Indian species of *Pulvinaria* Targioni Tozzetti. *Zootaxa* 4236: 533–542.

Joshi, S., and A. Rameshkumar. 2017. Silent foray of three soft scale insects in India. *Current Science* 112: 629–635.

Kawecki, Z. 1958. Studia nad rodzajem Lecanium Burm. 4. Materialy do monographii misecnica sliwowego, *Lecanium corni* Bouchr, Marchal (9 nee d') (Homoptera, Coccoidea, Lecaniidae). *Annales Zoologici Warszawa* 17: 135–246.

Kondo, T. P. J. Gullan, and D. J. Williams. 2008. Coccidology. The study of scale insects (Hemiptera: Sternorrhyncha: Coccoidea). *Revista Corpoica–Ciencia Tecnología Agropecuaria* 9: 55–61.

Kosztarab, M. P. 1996. *Scale insects of Northeastern North America. Identification, biology, and distribution*. Virginia Museum of Natural History Martinsburg, Martinsburg, Virginia 650 pp.

Kosztarab, M. P., and F. Kozár. 1988. *Scale Insects of Central Europe. Series Entomologica*, Vol. 41. Dordrecht, W. Junk, 456 pp.

Koteja, J. 1974. Comparative studies on the labium in the Coccinea (Homoptera). *Zeszyty Naukowe Akademii Rolniczej w Krakowie* 27: 1–162.

Koteja, J. 2008. Xylococcidae and related groups (Hemiptera: Coccinea) from Baltic amber (In English; Summary in Polish). *Prace Muzeum Ziemi* 49: 19–56.

Koteja, J., and D. Azar. 2008. Scale insects from lower cretaceous amber of Lebanon (Hemiptera: Sternorrhyncha: Coccinea). *Alavesia* 2:133–167.

Kozar, F., and Y. Ben-Dov. 1997. Zoogeographical considerations and status of knowledge of the family. In: *Soft Scale Insects, Their Biology, Natural Enemies and Control*, Vol. 7A, eds. Y. Ben-Dov and C. J. Hodgson, Elsevier, Amsterdam, the Netherlands, pp. 213–228.

Kozarzhevskaya, E. F. 1968. Methods of preparing slides for Coccid (Homoptera, Coccoidea) determination. *Entomologicheskoe Obozrenie* 47: 248–253.

Krombein, K. V. 1951. Wasp visitors of tulip-tree honeydew at Dunn Loring, Virginia. *Annals of the Entomological Society of America* 44: 141–143.

Li, C. 1985. China wax and the China wax scale insect. *World Animal Review* 55: 26–33.

Lin, Y. P., P. J. Gullan, and L. G. Cook. 2010. Species richness and host-plant diversity are positively correlated in Coccidae. *Entomologia Hellenica* 19: 90–98.

Linnaeus, C. 1758. *Systema Naturae per Regna Tria Naturae, Secundum Classes, Ordines, Genera, Species, cum Characteribus, Differentiis, Synonymis, Locis. Tomis I. Edirio Decima, Reformata. Cum Privilegio Siae Riae Mitis Sveciae*. Impensis Direct. Laurentii Salvii, Holmiae, Stockholm, 823 p.

Malausa, T., A. Fenis, S. Warot et al. 2011. DNA markers to disentangle complexes of cryptic taxa in mealybugs (Hemiptera: Pseudococcidae). *Journal of Applied Entomology* 135: 142–155.

Manawadu, D. 1986. A new species of *Eriopeltis* Signoret (Homoptera: Coccidae) from Britain. *Systematic Entomology* 11: 317–326.

Marchal, P. 1908. Notes sur les cochenilles de l'Europe et du Nord de l'Afrique. *Annales de la Socirts Entomologique de France* 77: 221–309.

Matile-Ferrero, D. 1997. 1.1.2 Morphology. 1.1.2.1 The adult female. *Soft Scale Insects: Their Biology, Natural Enemies and Control* (Vol. 7A), eds. Y. Ben-Dov and C. J. Hodgson, 5–21. Elsevier, Amsterdam, the Netherlands.

Mikkelsen, N. T., C. Schander, and E. Willassen. 2007. Local scale DNA barcoding of bivalves (Mollusca): A case study. *Zoologica Scripta* 36: 455–463.

Miller, D. R. 1984. Phylogeny and classification of the Margarodidae and related groups (Homoptera: Coccoidea). Verhandlungen des Zehnten Internationalen Symposiums fiber Entomofaunistik Mitteleuropas (SIEEC X), 15–20 Aug. 1983, Budapest, pp. 321–324.

Miller, D. R., and C. J. Hodgson. 1997. Phylogeny. In: *Soft Scale Insects: Their Biology, Natural Enemies and Control*, Vol. 7A. eds. Y. Ben-Dov and C. J. Hodgson, Amsterdam, the Netherlands: Elsevier, pp. 229–250.

Miller, D. R., and D. J. Williams. 1995. Systematic revision of the family Micrococcidae (Homoptera: Coccoidea), with a discussion of its relationships, and a description of a gynandromorph. *Bollettino del Laboratorio de Entomologia Agraria "Filippo Silvestri," Portici* 50: 199–247.

Miller, D. R., and G. L. Miller. 1993a. A new species of *Puto* and a preliminary analysis of the phylogenetic position of the *Puto* group within the Coccoidea (Homoptera: Pseudococcidae). *Jeffersoniana* 4: 1–35.

Miller, D. R., and G. L. Miller. 1993b. Description of a new genus of scale insect with a discussion of relationships among families related to the Kermesidae (Homoptera: Coccoidea). *Systematic Entomology* 18: 237–251.

Miller, D. R., and M. Kosztarab. 1979. Recent advances in the study of scale insects. *Annual Review of Entomology* 24: 1–27.

Miller, G. L. 1991. Morphology and Systematics of the Male Tests and Adult Males of the family Coccidae (Homoptera: Coccoidea) from America North of Mexico. PhD diss., Auburn University, Auburn, AL.

Miller, G. L., and M. L. Williams. 1995. Systematic analysis of the adult males of *Toumeylla* group, including Mesolecanium nigrofasciatum, Neolecanium cornuparvum, Psedophilippia quaintancii, and Toumeyella spp. (Homoptera: Coccidae) from America North of Mexico. *Contributions of the American Entomological Institute* 28: 1–68.

Monaghan, M.T., R. Wild, M. Elliot et al. 2009. Accelerated species inventory on Madagascar using coalescent-based models of species delineation. *Systematic Biology* 58: 298–311.

Nakahara, S. 1981. The proper placements of the Nearctic soft scale species assigned to the genus *Lecanium* Burmeister (Homoptera: Coccidae). *Proceedings of the Entomological Society of Washington* 83: 283–286.

Newstead, R. 1903. *Monograph of the Coccidae of the British Isles*. Vol. 2. Ray Society, London, UK. 270 pp.

Nur, U. 1971. Parthenogenesis in coccids (Homoptera). *American Zool* 2: 301–308.

Nuzzaci, G. 1969. Nots morfo-biologica sull *Eulecanium corni* (Bouche) ssp. *apuliae* nov. *Entomologica* 5: 9–36.

Park, D. S., S. J. Suh, P. D. N. Hebert, H. W. Oh, and K. J. Hong, 2011. DNA barcodes for two scale insect families, mealybugs (Hemiptera: Pseudococcidae) and armored scales (Hemiptera: Diaspididae). *Bulletin of Entomological Research* 101: 429–434.

Phillips, J. H. H. 1965. Notes on species of *Lecanium* Burmeister (Homoptera: Coccoidea) in the Niagara Peninsula, Ontario, with a description of a new species. *The Canadian Entomologist* 97: 231–238.

Qin, T. K. 1993. Phylogeny and Biogeography of the Wax Scales (Hemiptera, Coccoidea: Coccidae) with Special Reference to Ceroplastes sinensis Del Guercio. PhD diss., Australian National University, Canberra, Australia.

Ray, C. H., and M. L. Williams. 1983. Description of the immature stages and adult male of Neolecanium cornuparvum (Homoptera: Coccidae). *Proceedings of the Entomological Society of Washington* 85: 161–173.

Ray, C. H., and M. L. Williams.1980. Description of the immature stages and adult male of *Pseudophilippia quaintancii* (Homoptera: Coccoidea: Coccidae). *Annals of the Entomological Society of America* 73: 437–447.

Richards, W. R. 1958. Identities of species of *Lecanium* Burmeister in Canada (Homoptera: Coccoidea). *The Canadian Entomologist* 90: 305–313.

Saakyan-Baranova, A. A., E. S. Sugonyaev, and G. G. Sheldeshova. 1971. Brown fruit scale (*Parthenolecanium corni* Bouche) and its parasites (Chalcidoidea). The essay of the complex investigation of host-parasite relations. Nauka Publishers, Leningrad, Russian, 166 pp.

Sanders, J. G. 1909. The identity and synonymy of our soft scale insects. *Journal of Economic Entomology* 2: 428–448.

Sankaran, T. 1954. The natural enemies of *Ceroplastes pseudoceriferus* Green (Hemiptera—Coccidae). *Journal of Scientific Research of Benares Hindu University* 5: 100–120.

Sankaran, T. 1959. The life-history and biology of the wax-scale, *Ceroplastes pseudoceriferus* Green (Coccidae: Homoptera). *Journal of the Bombay Natural History Society* 56: 39–59.

Sankaran, T. 1962. The external characters of the post-larval stages of the wax scale, Ceroplastes pseudoceriferus Green (Hemiptera: Coccidae). *Indian Journal of Entomology* 24: 1–18.

Sankaran, T. 1965. Anatomical studies of the wax scale (*Ceroplastes pseudoceriferus* Green. *Zootomical Indian Journal* 6: 1–14.

Sankaran, T. and G. L. Pattar. 1977. Some parasites of *Parasaissetia nigra* and *Saissetia* spp. (Homoptera: Coccideae) from South India. *Technical Bulletin of the Commonwealth Institute of Biological Control* 18: 145–150.

Sethusa, M. T., I. M. Millar, K. Yessoufou, A. Jacobs, M. Van der Bank, and H. Van der Bank. 2014. DNA barcode efficacy for the identification of economically important scale insects (Hemiptera: Coccoidea) in South Africa. *African Entomology* 22: 257–266.

Shafee, S. A., M. Yousuf, and M. Y. Khan. 1989. Host plants and distribution of coccid pests (Homoptera: Coccoidea) in India. *Indian Journal of Systematic Entomology* 6: 47–55.

Sivescu, A. D. 1943. Oecoarten bei *Lecanium*. *Bulletin de la Section Scientifique de l'Academic Roumaine* 25: 212–223.

Sivescu, A. D. 1944. Formes dcologiques des ldcanides de la faune Roumaine. *Bulletin de la Section Scientifique de l'Acadrmie Roumaine* 27: 230–246.

Smith, M. A., J. J. Rodriguez, J. B. Whitfield et al. 2008. Extreme diversity of tropical parasitoid wasps exposed by iterative integration of natural history, DNA barcoding, morphology, and collections. *Proceedings of the National Academy of Sciences of the United States of America* 105: 12359–12364.

Steinweden, J. B. 1929. Bases for the generic classification of the coccoid family Coccidae. *Annals of the Entomological Society of America* 22: 197–245.

Sulc, K. 1932. Cskoslovenske druny rodu puclice (gn. *Lecanium*, Coccidae, Homoptera). *Acta Societatis Scientiarum Naturalium Moravicae* 7: 134.

Suresh, S. and M. Mohanasundaram. 1996. Coccoid (Coccoidea: Homoptera) fauna of Tamil Nadu, India. *Journal of Entomological Research* 20: 233–274.

Tang, Fang-tch 1991. *The Coccidae of China*. Shanxi United Universities Press, Shanxi, 377 pp.

Theron, J. G. 1958. Comparative studies on the morphology of male scale insects (Hemiptera: Coccoidea). *Annals of the University of Stellenbosch* 34: 1–71.

Tinsley, J. D. 1899. Contributions to coccidology. I. *Canadian Entomologist* 31: 45–47.

Varshney, R. K. 1984 New records of host-plants and distribution of some coccids from India (Homoptera: Coccoidea). *Bulletin of the Zoological Survey of India* 6: 137–142.

Varshney, R. K. 1985. A review of Indian coccids (Homoptera: Coccoidea). *Oriental Insects* 19: 1–101.

Varshney, R. K. 1992. A check list of the scale insects and mealy bugs of South Asia. Part-1. *Records of the Zoological Survey of India, Occasional Paper (No. 139)*: 1–152.

Varshney, R. K., and D. Moharana. 1987. Insecta: Homoptera: Coccoidea. *Fauna of Orissa: State Fauna Series* 1: 161–181.

Voukassovitch, P. 1930. Sur la polyphagie de la cochenille *Lecanium corni* L. Compte Rendu des Seances de la Socirt6 de Biologie, Paris 104: 1065–1067.

Wilkey, R. F. 1990. Collection, preservation and microslide mounting. In: *Armoured Scale Insects, their Biology, Natural Enemies and Control*. Vol. 4A, ed. D. Rosen, Elsevier, Amsterdam, the Netherlands, pp. 345–352.

Williams, D. J. 1984. Some aspects of the zoogeography of scale insects (Homoptera: Coccoidea). In: *Verhandlungen des Zehntcn Internationalen Symposiums über Entomofaunistik Mitteleuropas (SIEEC)*, Z. Kaszab ed. 15–20 August 1983, Muzsak Kozmuvelodesi Kiado, Budapest, pp. 331–333.

Williams, D. J. 2007. Carl Linnaeus and his scale insects (Hemiptera: Coccoidea). Zootaxa 1668: 427–490. In: *Linnaeus Tercentenary: Progress in Invertebrate Taxonomy. Zootaxa*. eds. Z. Q. Zhang & W. A. Shear, Magnolia Press, Auckland, NZ, 1668, pp. 1–766.

Williams, D. J., and G. W. Watson. 1990. *The Scale Insects of the Tropical South Pacific Region. Part 3, The Soft Scales (Coccidae) and Other Families*. CAB International, Wallingford, UK, 267 pp.

Williams, M. L., and G. S. Hodges. 1997. 1.1.3.3 Taxonomic characters: Nymphs. In *Soft Scale Insects: Their Biology, Natural Enemies and Control*. Vol. 7A, eds. Y. Ben-Dov and C. J. Hodgson, Elsevier, Amsterdam, the Netherlands, pp. 143–156.

Williams, M. L., and M. Kosztarab. 1972. *Morphology and Systematics of the Coccidae of Virginia with Notes on Their Biology* (Homoptera: Coccoidea). Research Division Bulletin, Virginia Polytechnic Institute and State University, Blacksburg, VA, 74, pp. 1–215.

Wong, E. H. K., M. S. Shivji, and R. H. Hanner. 2009. Identifying sharks with DNA barcodes: Assessing the utility of a nucleotide diagnostic approach. *Molecular Ecology Resources* 9: 243–256.

Yousuf, M., and S. A. Shafee. 1988. Four new species of Coccidae (Homoptera) from Andaman Islands. *Indian Journal of Systematic Entomology* 5: 57–63.

7 Whiteflies (Hemiptera: Aleyrodidae) of India

R. Sundararaj, K. Selvaraj, D. Vimala, and T. Venkatesan

CONTENTS

Introduction 103
Family Diagnosis and Key Characters 103
    Family Diagnosis 103
    Key Characters 104
    Subfamilies and Tribes 105
    Key to Subfamilies of Aleyrodidae 105
    Key to the Genera of the Subfamily Aleurodicinae of India 105
    Key to Genera of Subfamily Aleyrodinae from India 105
Major Work on Indian Fauna 108
Species Diversity of Whiteflies in India 108
Molecular Characterization and Phylogeny 108
Economic Importance 114
Conclusion 115
References 115

INTRODUCTION

Whiteflies are an economically important group of insects infesting a wide range of host plants. They are small, inconspicuous, phytophagous insects belonging to the family Aleyrodidae (Hemiptera: Sternorrhyncha) and are often overlooked despite their abundance on the surfaces of leaves. The name "whitefly" is derived from the white appearance of adults of most species due to the deposition of wax on the body and wings and from their tendency to fly when disturbed.

The biology of whiteflies is unusual. Many species of Aleyrodidae lay eggs in partial or complete circles, and members of Aleurodicinae lay their eggs in spirals on leaves, and some aleurodicines oviposit on non-foliar surfaces, such as fruits. A very few species, in both the families, habitually develop on the upper surfaces of leaves, whilst others develop on both leaf surfaces. Immature stages of some species develop on the petioles or woody stems of their hosts (Martin 2004). The whitefly egg has a stalk or pedicel at one end, with which it attaches itself to the leaf. The first nymphal instar is called the crawler because of its habit of crawling on the leaf surface after eclosion until it finds a suitable place to settle down and start feeding. The legs and antennae of the second, third, and fourth nymphal instars are atrophied. Hence these instars are sessile. The adult develops within the quiescent fourth instar which is usually referred to as the "puparium."

FAMILY DIAGNOSIS AND KEY CHARACTERS

FAMILY DIAGNOSIS

Although species and genera of whiteflies are usually distinguishable by characters of the puparium, the higher taxa are defined based on adult characters. The family Aleyrodidae is most closely related to the Psyllidae. In both groups, the adults have two tarsal segments that are almost equal in size. In contrast, aphids and coccids have the first tarsal segment reduced or even absent. These four groups together comprise the Sternorrhyncha in the Homoptera (Bink-Moenen and Mound 1990). Adult Aleyrodidae have well-developed, seven-segmented antennae, and two ocelli, generally placed at the anterior margins of the compound eyes. The upper and lower regions of the compound eyes differ both functionally and anatomically (Gill 1990). In some species like *Trialeurodes vaporariorum,* there is a complete separation of each eye. The degree of separation is useful in recognizing the species (Hahbazvar 2011). Both sexes have functional mouthparts and two pairs of membranous, functional wings; the rear wings are neither much reduced nor modified into hooked or haltere-like structures as occur in some other Hemiptera as in many Coccoidea. The wing venation is reduced, like that of the Psyllidae, though generally much more so. In many genera, only one conspicuous and unbranched vein is found in each wing; however, wings of larger species such as *Udamoselis* have less reduced venation, though their veins are still simple and few (Martin 2007). The wings of aleyrodids are variously marked or mottled according to species, and many species are covered with fine wax powder, giving most species a floury, dusted appearance. The legs of are well developed and fairly long, but gracile, and in contrast to Psyllidae, not adapted to leaping. The tarsi have two segments of roughly equal length. The pretarsus has paired claws with an empodium. In some species, the empodium is a bristle while in others it is a pad.

KEY CHARACTERS

The classification of aleyrodids is not based on the structure of the adults, but on the structure of the fourth larval instar, called the "puparium." Russell (1947) pointed out that the "accurate definition of tribes in the Aleyrodidae is a perplexing problem, owing to paucity of morphological characteristics indicative of tribal limits," and "tribal divisions must be principally recognized from the pupae, the stage that is most often collected and indeed the only one known for the majority of species." Frohlich et al. (1999) commented that the morphological traits of adults are currently poorly understood and so do not readily permit differentiation between genera or species. Some polyphagous whitefly species vary in their puparia depending on the host plant cuticle on which they develop, and this has caused a considerable amount of misidentification (Mound 1963). Hence, deduction from host plant associations must always be approached with caution (Mound and Halsey 1978). Martin (2003) commented that "it seems almost certain that puparia will continue to be dominant in whitefly systematics and there is no particular reason for larval characters to be regarded as second-rate." The accompanying diagram (Figure 7.1) illustrates the structural features of a generalized whitefly puparium with taxonomic characteristics considered in the classification of whiteflies. Vasiform orifice on the dorsal surface of the caudal segment of the abdomen is characteristic of the Aleyrodidae, and within the family, it is taxonomically diagnostic because it varies in shape according to the species. This orifice is large and is covered by an operculum. Within the orifice beneath the operculum there is a tongue-like lingula which in some species protrudes from beneath the operculum, while in others it is hidden (Gill 1990).

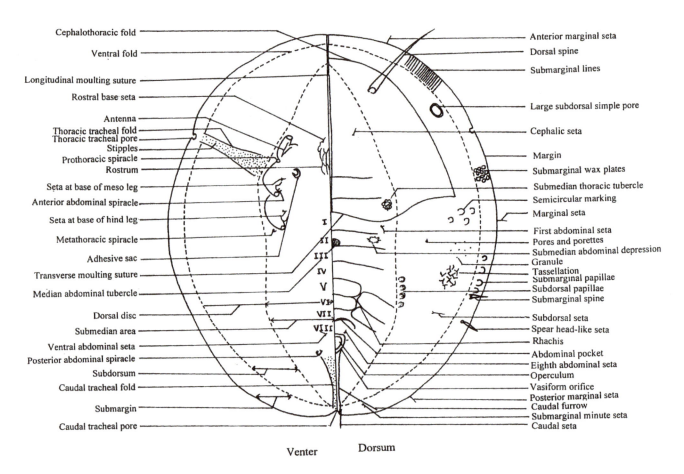

FIGURE 7.1 Whitefly puparium showing morphological characters. (From Dubey, A.K., and David, B.V, *Zootaxa*, 3303, 50–58, 2012a. With permission; Dubey, A.K., and David, B.V. 2012b. With permission.) Indian whiteflies (Hemiptera: Aleyrodidae) with their host plants, In: *The Whitefly or Mealywing Bugs: Bioecology, Host Specificity and Management*, ed. V. David, Saarbrucken, Germany, Lambert Academic Publishing GMbH & Co KG, 411 pp.)

Subfamilies and Tribes

In India, the family Aleyrodidae is represented by two subfamilies, *viz.*, Aleurodicinae Quaintance & Baker 1913 and Aleyrodinae Westwood 1840. The subfamily Aleurodicinae is represented by two genera viz., *Aleurodicus* and *Palaealeurodicus,* and Aleyrodinae is represented by 63 genera. It should be noted in this context that while no tribes have been recognized in the subfamily Aleurodicinae, Indian species of Aleyrodinae have been classified into 12 tribes (David 1990). A recent study (Manzari and Quicke 2006), however, found no support for any of the presently proposed tribes in Indian Aleyrodinae.

Key to Subfamilies of Aleyrodidae

1. Puparium with compound or agglomerate pores, if with simple pores, thin and flat; operculum transversely rectangular; tracheal folds rarely present; margin smooth or with one row of teeth; lingula very long, extending past the vasiform orifice, with two or more pairs of setae at its apex..**Aleurodicinae Quaintance & Baker**
- Puparium without compound or agglomerate pores; tracheal folds often present; margin often with rows of teeth; lingula usually not long and not extending past the vasiform orifice, generally with 1 pair of setae.............................**Aleyrodinae Westwood**

Key to the Genera of the Subfamily Aleurodicinae of India

1. Sub-margin with 0–2 pairs of setae; sub-median cephalothoracic setae absent; sub-margin and or dorsal disc usually with no evident simple pores, or with a few of one type only; ventrally, each leg rounded apically, without an apical claw.......***Palaealeurodicus* Martin**
- Sub-margin with 12 pairs of setae; sub-median cephalothoracic setae present; sub-margin and/or dorsal disc usually punctuated by pores of several types, but loculate pores absent; ventrally, each leg bears a large and distinct claw.......................***Aleurodicus* Douglas**

Key to Genera of Subfamily Aleyrodinae from India

1. Puparium with dorsal spines/elongate glandular spines or siphons..2
- Puparium without dorsal spines/elongate glandular spines or siphons..5
2. Puparium with dorsal spines; sub-margin without a row of setae; caudal tracheal comb absent 3
- Puparium with glandular siphons; sub-margin with a paired row of setae; caudal tracheal comb present..............................***Siphoninus* Silvestri**
3. Vasiform orifice not on an eminent protuberance, median tubercles on abdominal segments absent.......4
- Vasiform orifice on an eminent protuberance, median tubercles on abdominal segments often present..................................***Acanthaleyrodes* Takahashi**
4. Vasiform orifice not elevated, almost triangular; lingula with a pair of long distal setae...***Acanthobemisia* Takahashi**
- Vasiform orifice elevated and sub-circular or sub-cordate; lingula without a pair of long distal setae.........***Aleurocanthus* Quaintance & Baker**
5. Sub-margin distinct separated by complete or incomplete sub-marginal furrow.................................6
- Sub-margin not distinct, sub-marginal furrow absent..22
6. Sub-marginal furrow complete...........................7
- Sub-marginal furrow incomplete.......................10
7. First abdominal setae present…........................8
   First abdominal setae absent............................9
8. Sub-margin broad; pro, meso, and metathorax without a pair of setae.........***Icfrealeyrodes* Dubey & Sundararaj**
- Sub-margin narrow; pro, meso, and metathorax each with a pair of setae…***Vanaleyrodes* Pushpa & Sundararaj**
9. Margin lobulate; median area of abdominal segments tuberculate; operculum filling half the orifice.............................***Vasantharajiella* David**
- Margin crenulate; median area of abdominal segments not tuberculate; operculum filling the orifice......................***Orientaleyrodes* Regu & David**
10. Sub-margin with distinct crescent-shaped or circular-shaped wax secreting pores..............................11
- Sub-margin without distinct crescent-shaped or circular-shaped wax secreting pores..........................12
11. Sub-margin with two pairs of circular-shaped wax secreting pores, without minute setae; lingula tip not exposed..........................***Aleuroparvus* Dubey**
- Sub-margin with distinct crescent-shaped pores arranged equidistantly on sub-margin, with minute setae; lingula tip exposed......***Crescentaleyrodes* David & Jesudasan**
12. Thoracic and caudal tracheal combs, clefts, or pores, or folds absent...13
- Thoracic and caudal tracheal combs, clefts, or pores, or folds present...15
13. Sub-margin usually with a row of setae; vasiform orifice not elevated..14
- Sub-margin usually without a row of setae; vasiform orifice elevated................***Tetraleyrodes* Cockerell**
14. Sub-margin with a distinct row of wax secreting papillae; operculum cordate and not recessed posteriorly ....***Aleuropapillatus* Regu & David**
- Sub-margin without a distinct row of wax secreting papillae; operculum transversely rectangular and recessed posteriorly............***Aleuroputeus* Corbett**
15. Vasiform orifice closed; tracheal pore areas differentiated from margin as comb.............................16
- Vasiform orifice open, tracheal pore areas differentiated form margin as pores or invaginated clefts......21
16. Sub-marginal furrow present only on cephalothorax and merging with the transverse moulting suture…17
- Sub-marginal furrow extending beyond cephalothorax and not merging with the transverse moulting suture 18

17. Dorsum tessellated; sub-margin with a row of setae ....***Davidiella*** **Dubey & Sundararaj**
- Dorsum smooth; sub-margin without a row of setae....................***Pseudcockerelliella*** **Sundararaj**
18. Puparium not in spherical shape, paired setae on meso- and metathorax absent; vasiform orifice not sphere-shaped.. **19**
- Puparium in spherical shape, paired setae on meso- and metathorax present; vasiform orifice sphere-shaped....***Sphericaleyrodes*** **Selvakumaran & David**
19. Thoracic and caudal tracheal pores distinct; dorsum with papillae; dorsal setae not fixed in large cup-shaped "setal alveoli".....................................**20**
- Only caudal tracheal pore indicated; dorsum without papillae; dorsal setae fixed in large cup-shaped "setal alveoli"......................***Aleurocryptus*** **Dubey**
20. Sub-marginal furrow long almost reaching near vasiform orifice; vasiform orifice not notched at caudal end............................***Asialeyrodes*** **Corbett**
- Sub-marginal furrow short extending little beyond cephalothorax; vasiform orifice notched at caudal end................***Cockerelliella*** **Sundararaj & David**
21. Puparium lighter in colour, sub-marginal furrow complete, first abdominal setae generally absent, vasiform orifice triangular and lingula exposed........ ....***Africaleurodes*** **Dozier**
- Puparium black, sub-marginal furrow incomplete, first abdominal setae present, abdominal segment VIII often trilobed, vasiform orifice cordate and lingula concealed..............***Aleurolobus*** **Quaintance & Baker**
22. Dorsum without two pairs of long two-segmented setae on cephalic and first abdominal segments.... ....**23**
- Dorsum with two pairs of long two-segmented setae on cephalic and first abdominal segments...... ....***Martiniella*** **Jesudasan & David**
23. Thoracic and caudal tracheal combs, clefts, or pores, or folds absent...**24**
- Thoracic and caudal tracheal combs, or clefts, or pores, or folds present....................................**37**
24. Vasiform orifice not elevated...........................**25**
- Vasiform orifice elevated...............................**31**
25. First abdominal setae present; lingula exposed......**26**
- First abdominal setae absent; lingula concealed....**30**
26. Dorsum not elevated from sub-margin by two longitudinal ridges; usual anterior and posterior marginal setae only present; caudal furrow indistinc **27**
- Dorsum elevated from sub-margin by two longitudinal ridges which connect anterior and posterior regions; usual anterior and posterior marginal setae absent, but with five pairs of marginal setae; caudal furrow distinct with granular markings..........***Editaaleyrodes*** **David**
27. Dorsal setae less than 19 pairs; transverse moulting suture not reaching margin.............................**28**
- Dorsal setae 19–21 pairs arranged on cephalothoracic and abdominal segments sub-medially and sub-dorsally; transverse moulting suture reaching margin ....***Aleuromarginatus*** **Corbett**
28. Cephalic and sub-dorsal setae present...................**29**
- Cephalic and sub-dorsal setae absent...................... ....***Aleyrodes*** **Latreille**
29. Puparium narrowly elongate; sub-median setae on thoracic segments absent........***Vasdavidius*** **Russell**
- Puparium elliptical to slightly oval; sub-median setae on thoracic segments present................................... ....***Aleurotulus*** **Quaintance & Baker**
30. Puparium narrowly elongate with raised longitudinal ridges on dorsum ....***Agrostaleyrodes*** **Ko**
- Puparium oval with scallop-shaped thickening in inner sub-dorsal area............***Crenidorsum*** **Russell**
31. Marginal teeth have no glands at their bases, rhachis absent...**32**
- Marginal teeth have glands at their bases, often appearing double ranked; rhachis present, often without lateral arms......***Aleurotrachelus*** **Quaintance & Baker**
32. Margin with two rows of teeth **33**
- Margin with single row of teeth or smooth...........**34**
33. Meso- and metathoracic setae absent........................ ....***Zaphanera*** **Corbett**
- Meso- and metathoracic setae present.................. ....***Keralaleyrodes*** **Meganathan & David**
34. Dorsal disc without a rhachis or pair of longitudinal fold or furrows...**35**
- Dorsal disc mostly with a rhachis and a pair of longitudinal fold or furrows **36**
35. Puparium ovoid; outer sub-dorsum without a row of setae; dorsal setae (except caudal setae) apically rounded ensiform...................***Arunaleyrodes*** **Dubey**
- Puparia elongately to broadly oval; outer sub-dorsum with a row of setae; dorsal setae not apically rounded..........***Neomaskellia*** **Quaintance & Baker**
36. Sub-margin of cephalothorax with six pairs of small setae; vasiform orifice cordate, operculum almost filling vasiform orifice, concealing lingula tip.....................***Cohicaleyrodes*** **Bink–Moenen**
- Sub-margin of cephalothorax without six pairs of small setae; vasiform orifice elongately elliptical with extremely short lingula; operculum transversely rectangular ....***Acaudaleyrodes*** **Takahashi**
37. Thoracic and caudal tracheal pores/clefts indicated or combs sometimes developed.............................**38**
- Only caudal tracheal pore/cleft indicated............... ....***Fippataleyrodes*** **Sundararaj & David**
38. Tracheal pores, clefts, folds, furrows present, sometimes combs present; rhachis present or absent; dorsum generally with setae/tubercles **39**
- Tracheal pores, clefts, folds, furrows absent, but combs sometimes developed **61**
39. Meso- and metathoracic setae present; sub-median setae present on abdominal segments III–VII or

prothoracic setae present; median area of abdominal segment I–VII with minute setae **40**
- Meso- and metathoracic setae absent; sub-median setae or median setae absent on abdominal segments.... .... **41**
40. Meso- and metathoracic setae present; sub-median setae present on abdominal segments III–VII; major dorsal setae in grossly swollen form; vasiform orifice elevated......*Distinctaleyrodes* **Dubey & Sundararaj**
- Prothoracic setae present; median area of abdominal segment I–VII with minute setae; major dorsal setae not in grossly swollen form; vasiform orifice not elevated......*Milleraleurodes* **Phillips & Jesudasan**
41. Vasiform orifice cordate or sub-cordate; lingula usually small, concealed, and wanting setae............**42**
- Vasiform orifice subquadrate or triangular or sub-rectangular, sometimes in a ribbed pyriform pit; lingula long or spatulate, setose and exposed.............**55**
42. Thoracic tracheal pores indicated only as smooth emargination or slight invagination...................**43**
- Thoracic tracheal pores/clefts distinct..................**45**
43. 10–15 pairs of sub-marginal setae present.............**44**
- Sub-marginal row of setae absent...*Aleuroclava* **Singh**
44. Vasiform orifice not in a trilobed area; 13–15 pairs of sub-marginal setae present.......*Massileurodes* **Goux**
- Vasiform orifice in a trilobed area; 10 pairs of sub-marginal setae present......*Dialeurolobus* **Danzig**
45. Median length of cephalothorax shorter than abdomen; transverse moulting suture does not bend sharply cephalad on sub-dorsum and not reaching margin **46**
- Median length of cephalothorax longer than abdomen; transverse moulting suture bends sharply cephalad on sub-dorsum and reaching margin................ ....*Minutaleyrodes* **Jesudasan & David**
46. Vasiform orifice with comb of teeth on inner caudal and lateral margins................................. **47**
- Vasiform orifice without comb of teeth on inner caudal or lateral margins **49**
47. Series of sub-dorsal/sub-marginal setae present; vasiform orifice not distinctly longer than wide, sub-cordate, operculum entirely filling the orifice and lingula tip not exposed **48**
- Series of sub-dorsal/sub-marginal setae absent; vasiform orifice cordate, distinctly longer than wide, operculum not entirely filling the orifice and lingula tip partly exposed .... *Dialeurolonga* **Dozier**
48. Sub-dorsum with a row of distinct papillae; margin crenulate or smooth with a series of setae; sub-marginal/sub-dorsal setae absent; vasiform orifice broadly sub-cordate..........*Dialeuronomada* **Quaintance & Baker**
- Sub-dorsum without a row of distinct papillae; margin toothed or crenulated without a series of setae; sub-dorsum/sub-margin generally with setae; vasiform orifice sub-cordate to sub-circular......... .... *Dialeurodes* **Cockerell**
49. Tracheal pores without distinct teeth or fimbriae.....**50**
- Tracheal pores with distinct teeth or fimbriae.......**53**

50. Puparium symmetrical; first abdominal setae present **51**
- Puparium asymmetrical; first abdominal setae absent ....*Aleuropositus* **Dubey**
51. Vasiform orifice sub-circular to cordate, operculum sub-circular; caudal furrow indistinct or indicated without granules or dots, not merging with posterior end of vasiform orifice **52**
- Vasiform orifice rectangular shaped; operculum sub-cordate, setose at its distal end; caudal furrow distinct with granulations or dots, merging with posterior end of vasiform orifice......................*Singhius* **Takahashi**
52. Sub-margin without a row of large circular pores; caudal furrow distinct....*Kanakarajiella* **David & Sundararaj**
- Sub-margin with a row of large circular pores; caudal furrow indistinct.........*Dialeuropora* **Quaintance & Baker**
53. Sub-margin without a row of spines; dorsal disc without prominent rhachis; vasiform orifice cordate or sub-cordate; eye spots lacking **54**
- Sub-margin with a row of spines; dorsal disc with prominent rhachis and thickened ridges radiating from it; vasiform orifice cordate; operculum similarly shaped obscuring the lingula; eye spots often present ....*Rhachisphora* **Quaintance & Baker**
54. Tracheal folds covered with linear ridges; vasiform orifice roundly sub-cordate to subcircular............... .... *Rabdostigma* **Quaintance & Baker**
- Tracheal folds covered with polygonal markings; vasiform orifice broadly sub-cordate................... .... *Rusostigma* **Quaintance & Baker**
55. Sub-margin without a row of setae, if present not typically in 13 pairs; dorsal disc lacking small geminate pores **56**
- Sub-margin typically with a row of 13 pairs of setae; dorsal disc covered by small geminate pores .... *Singhiella* **Sampson**
56. Thoracic and caudal tracheal pore areas not deeply invaginated and inset.. **57**
- Thoracic and caudal tracheal pore areas deeply invaginated and inset ....*Indoaleyrodes* **David & Subramaniam**
57. Vasiform orifice elongate triangular with regular lateral craniae (ridges) or without craniae; lingula tip typically elongate spatulate with a pair of setae **58**
- Vasiform orifice subtriangular to sub-cordate with or without irregular lateral ridges; lingula tip short, knobbed, rounded, or D-shaped............................**60**
58. First abdominal setae absent; a row of setae along the body margin present......................................**59**
- First abdominal setae present; a row of setae along the body margin absent........*Bemisia* **Quaintance & Baker**
59. Sub-median area of puparium with a band of wax secreting glands...............*Himalayaleyrodes* **Dubey**
- Sub-median area of puparium without a band of wax secreting glands.................*Parabemisia* **Takahashi**
60. Puparium often oval; vasiform orifice situated in a ribbed pyriform pit; abdominal segments without

median tubercles; sub-median depressions normal, not pouch-like.............. ***Pealius* Quaintance & Baker**
- Puparium elongate; vasiform orifice not situated in a ribbed pyriform pit; abdominal segments with distinct median tubercles; sub-median depressions prominent, characteristically pouch-like .... ***Setaleyrodes* Takahashi**
61. Margin toothed; sub-margin without a row of glandular papillae; lingula without lobulate head............... **62**
- Margin smooth or crenulate; sub-margin with a row of glandular papillae; lingula with characteristic lobulate head .... ***Trialeurodes* Cockerell**
62. Tracheal combs indicated by shortened marginal teeth with no distinct incision in between tooth; dorsal disc separated from margin with a row of spherical granules; first abdominal setae present **63**
- Thoracic and caudal tracheal combs distinct with one to several finger-like teeth with incision between each tooth rather deep; dorsal disc not separated from margin; first abdominal setae absent............***Aleuroplatus* Quaintance & Baker**
63. Caudal tracheal pore and furrow indicated; thoracic tracheal furrows tuberculate, pouch-like, meso- and metathoracic setae absent; sub-margin with a row of setae; vasiform orifice trilobed......................***Rutaleyrodes* Dubey & Ko**
- Caudal tracheal pore and furrow not indicated; thoracic tracheal furrows not tuberculate, not pouch-like, meso- and metathoracic setae present; sub-margin with a row of raised rays or tubercles; vasiform orifice not trilobed..........................***Viennotaleyrodes* Cohic**

MAJOR WORK ON INDIAN FAUNA

Maskell (1896) pioneered the study of Indian aleyrodids and described five new species. Buckton (1903) and Peal (1903) contributed one and seven new species, respectively. Quaintance and Baker (1917) added a number of new species from the materials collected by Woglum from the Orient in 1910. Misra (1924) and Dozier (1928) described one new species each. Singh (1931) made the first detailed study of North Indian whiteflies and recorded 44 species, of which 25 were new. Singh (1938, 1940, 1945) later contributed another five new species. Husain and Khan (1945) described the whiteflies on *Citrus* sp. in Punjab. Takahashi (1950) added one new species, *Tetraleurodes pusana*. In 1955, eight species from Karnataka (Mysore) were listed by Usman and Puttarudriah. *T. semilunaris* Corbett had been reported by Abraham and Joy (1978) on lemon grass (*Cymbopogon flexosus*). Rao (1958) furnished a note on the whiteflies of Hyderabad. Nath (1970) reported the occurrence of *Aleurocanthus citriperdus* Quaintance & Baker on citrus plants in West Bengal. David and Subramaniam (1976) reported 60 species, representing 24 genera of Indian Aleyrodidae with special reference to South Indian species. David (1972, 1976a, 1976b, 1977, 1978, 1981, 1987) described seven new species of the subfamily Aleyrodinae and one species of *Aleurodicus* Douglas representing the first record of the subfamily Aleurodicinae from India. Following this, David (1988, 1990, 1994, 2005a, 2005b) and his students (David et al.1988, 1991, 1994, 2006, 2010; David and Jesudasan 1987, 1988, 1989a, 1989b, 2002; Jesudasan and David 1990, 1991; David and Selvakumaran, 1987; David and Augustine 1988; David and Regu 1989, 1990, 1991, 1995; Regu and David 1990, 1991, 1992a, 1992b, 1992c, 1992d, 1993a, 1993b, 199c, 199d; Sundararaj and David, 1990, 1991a, 1991b, 1992a, 1992b, 1992c, 1993a, 1993b, 1994, 1995, 2003; David and Sundararaj 1991a, 1991b, 1992, 1993; Meganathan and David 1994; David and Thenmozhi 1995; Selvakumaran and David 1996; David and David 2000, 2001, 2007a, 2007b; David and Manjunatha 2003; Jesudasan et al. 2003; David and Dubey 2006, 2009; Dubey and David 2012a, 2012b) contributed significantly towards documenting the whitefly fauna of India. Subsequently, notable contributions, particularly on the whitefly fauna of South India were made by Sundararaj (1999, 2000, 2001, 2007) and his students (Sundararaj and Dubey 2003, 2004, 2005, 2006a, 2006b, 2007; Dubey and Sundararaj, 2004a, 2004b, 2004c, 2005a, 2005b, 2005c, 2005d, 2005e, 2005f, 2006a, 2006b, 2006c, 2006d, 2006e, 2006f, 2006g; Dubey et al. 2004; Pushpa and Sundararaj 2008, 2009a, 2009b, 2010a, 2010b, 2010c, 2010d, 2010e, 2010f, 2010g, 2010h, 2011a, 2011b, 2011c, 2012; Sundararaj and Pushpa 2011a, 2011b, 2014; Vimala and Sundararaj 2015, 2018a, 2018b, 2018c; Revathi and Sundararaj 2016). Other contributions on Indian whiteflies include David (2000), Dubey and Ko (2007), Lalneihpuia and William (2011), Dubey and Ramamurthy (2013a, 2013b), Phillips and Jesudasan (2013), Dooley and Smith-Pardo (2013), Mohan and Jesudasan (2014), Dubey et al. (2014), Dubey and Ramamurthy (2013, 2015), Baig et al. (2016), Dubey (2016, 2017, 2018), Dubey and Singh (2016a, 2016b, 2017), Sundararaj and Selvaraj (2017), Vimala et al. (2017), Phillips et al. (2018), and Sundararaj and Vimala (2018).

SPECIES DIVERSITY OF WHITEFLIES IN INDIA

The Indian whitefly fauna comprises 454 species under 66 genera. The checklist of Indian whiteflies is presented in Table 7.1.

MOLECULAR CHARACTERIZATION AND PHYLOGENY

A considerable effort has been invested in attempting to resolve the systematics of the cryptic species complex of *Bemisia tabaci* and its population's classified into biotypes and host races based on various biological and biochemical markers. A genetic distance threshold of 3.5% was identified based on a gap in the distribution of pairwise sequence divergences amongst unique mitochondrial cytochrome oxidase subunit I (mtCOI) partial sequences of *B. tabaci*. In subsequent analyses, patterns of clusters of putative species were recognized which could be defined by sequence divergences equal to or higher than 3.5%. Mating experiments among some of these phylogenetic species have shown either complete or partial reproductive isolation (Maruthi et al. 2004). At least 39 putative species have now been proposed through

TABLE 7.1
Checklist of Indian Whiteflies

S. No	Whitefly Species

Genus I: *Aleurodicus* Douglas, 1892
1. *dispersus* Russell
2. *rugioperculatus* Martin

Genus II: *Palaealeurodicus* Martin, 2008
3. *holmesii* (Maskell)
4. *indicus* (Regu & David)
5. *machili* (Takahashi)

Genus III: *Acanthaleyrodes* Takahashi, 1931
6. *elevates* Dubey, Sudhirsingh & Martin

Genus IV: *Acanthobemisia* Takahashi, 1934
7. *indicus* Meganathan & David

Genus V: *Acaudaleyrodes* Takahashi, 1951
8. *rachipora* (Singh)

Genus VI: *Africaleurodes* Dozier, 1934
9. *ananthakrishnani* Dubey & Sundararaj
10. *citri* (Takahashi)
11. *indicus* Regu & David
12. *karwarensis* Dubey & Sundararaj
13. *orientalis* Dubey & David
14. *simula* (Peal)

Genus VII: *Agrostaleyrodes* Ko, 2001
15. *arcanua* Ko

Genus VIII: *Aleurocanthus* Quaintance & Baker, 1914
16. *arecae* David & Manjunatha
17. *ayyari* Regu & David
18. *bambusae* (Peal)
19. *bangalorensis* Dubey & Sundararaj
20. *bucktoni* Sundararaj & Pushpa
21. *citriperdus* Quaintance & Baker
22. *clitoriae* Jesudasan & David
23. *davidi* David & Subramaniam
24. *euphorbiae* Jesudasan & David
25. *firmianae* Dubey & Sundararaj
26. *gateri* Corbett
27. *goaensis* Dubey & Sundararaj
28. *gymnosporiae* Jesudasan & David
29. *husaini* Corbett
30. *icfreae* Sundararaj & Pushpa
31. *indicus* David & Regu
32. *ixorae* Jesudasan & David
33. *lobulatus* Jesudasan & David
34. *longispinus* Quaintance & Baker
35. *loyolae* David & Subramaniam
36. *mangiferae* Quaintance & Baker
37. *martini* David
38. *marudamalaiensis* David & Subramaniam
39. *mizoramensis* Chhakchhuak & Sundararaj
40. *musae* David & Jesudasan
41. *perseae* Chhakchhuak & Sundararaj
42. *rugosa* Singh
43. *russellae* Jesudasan & David
44. *satyanarayani* Dubey & Sundararaj
45. *seshadrii* David & Subramaniam
46. *shillongensis* Jesudasan & David
47. *singhi* Jesudasan & David
48. *spiniferus* (Quaintance)
49. *splendens* David & Subramaniam
50. *terminaliae* Dubey & Sundararaj
51. *valparaiensis* David & Subramaniam
52. *vindhyachali* Dubey & Sundararaj
53. *woglumi* Ashby

Genus IX: *Aleuroclava* Singh, 1931
54. *afriae* Sundararaj & David
55. *bilineata* Sundararaj & David
56. *calicutensis* Dubey & Sundararaj
57. *calycopteriseae* Dubey & Sundararaj
58. *cardamomi* (David & Subramaniam)
59. *celtise* Pushpa & Sundararaj
60. *cinnamomi* Jesudasan & David
61. *citri* Jesudasan & David
62. *citrifolii* (Corbett)
63. *combiformis* Pushpa & Sundararaj
64. *complex* Singh
65. *dehradunensis* Jesudasan & David
66. *vadipterocarpi* Pushpa & Sundararaj
67. *doddabettaensis* Dubey & Sundararaj
68. *ehretiae* Jesudasan & David
69. *evanantiae* Jesudasan & David
70. *goaensis* Jesudasan & David
71. *grewiae* Sundararaj & David
72. *hexcantha* (Singh)
73. *hindustanicus* (Meganathan & David)
74. *indicus* (Singh)
75. *jasmini* (Takahashi)
76. *kanyakumariensis* Sundararaj & David
77. *kavalurensis* Jesudasan & David
78. *kudremukhensis* Dubey & Sundararaj
79. *longisetosus* Jesudasan & David
80. *louiseae* Sundararaj & David
81. *madhucae* Jesudasan & David
82. *manii* (David)
83. *martini* Dubey & Sundararaj
84. *mizoramensis* Chhakchhuak & Sundararaj
85. *multituberculata* Sundararaj & David
86. *murrayae* (Singh)
87. *mysorensis* Jesudasan & David
88. *nagercoilensis* Sundararaj & David
89. *nanjangudensis* Jesudasan & David
90. *nigrus* Sundararaj & Pushpa
91. *nothapodytese* Pushpa & Sundararaj
92. *orientalis* Jesudasan & David
93. *palakkadensis* Pushpa & Sundararaj

(Continued)

TABLE 7.1 (Continued)
Checklist of Indian Whiteflies

S. No	Whitefly Species
94.	*pambaensis* Pushpa & Sundararaj
95.	*paracrotone* Pushpa & Sundararaj
96.	*papillata* Dubey & Sundararaj
97.	*pentatuberculata* Sundararaj & David
98.	*philomenae* Jesudasan & David
99.	*pongamiae* Jesudasan & David
100.	*premnae* Pushpa & Sundararaj
101.	*psidii* (Singh)
102.	*ramachandrani* Dubey & Sundararaj
103.	*regui* Sundararaj & David
104.	*rhamnacei* Pushpa & Sundararaj
105.	*saputarensis* Sundararaj & David
106.	*schimea* Chhakchhuak & Sundararaj
107.	*scolopiae* Pushpa & Sundararaj
108.	*selvakumarani* Sundararaj & David
109.	*serchhipensis* Chhakchhuak & Sundararaj
110.	*sindhuiae* Sundararaj & Pushpa
111.	*singhi* Jesudasan & David
112.	*sivakasiensis* Sundararaj & David
113.	*stereospermi* (Corbett)
114.	*takahashii* (David & Subramaniam)
115.	*tarennae* Martin & Mound
116.	*terminaliae* Sundararaj & David
117.	*trilineata* Sundararaj & David
118.	*trivandricus* Dubey & Sundararaj
119.	*vernoniae* Meganathan & David
120.	*viraktamathi* Pushpa & Sundararaj
121.	*vitexae* Sundararaj & David
122.	*wrightiae* Jesudasan & David

Genus X: *Aleurocryptus* Dubey, 2016

123.	*rhynchosiae* Dubey

Genus XI: *Aleurolobus* Quaintance & Baker, 1914

124.	*andhraensis* Pushpa & Sundararaj
125.	*antennata* Regu & David
126.	*azadirachtae* Regu & David
127.	*azimae* Jesudasan & David
128.	*barleriae* Jesudasan & David
129.	*barodensis* (Maskell)
130.	*bidentatus* Singh
131.	*burliarensis* Jesudasan & David
132.	*cassiae* Jesudasan & David
133.	*cephalidistinctus* Regu & David
134.	*cissampelosae* Regu & David
135.	*cohici* Regu & David
136.	*confusus* David & Subramaniam
137.	*curvata* Pushpa & Sundararaj
138.	*dalbergiae* Dubey & Sundararaj
139.	*delhiensis* Regu & David
140.	*diacritica* Regu & David
141.	*distinctus* Regu & David
142.	*exceptionalis* Regu & David
143.	*fluggeae* Pushpa & Sundararaj
144.	*gmelinae* Vimala & Sundararaj

S. No	Whitefly Species
145.	*gruveli* Cohic
146.	*hosurensis* Regu & David
147.	*indigoferae* Regu & David
148.	*janagarajani* David & David
149.	*karunkuliensis* Jesudasan & David
150.	*killikulamensis* David & David
151.	*lagerstroemiae* Regu & David
152.	*longisetosus* Dubey & Sundararaj
153.	*macarangae* Regu & David
154.	*madrasensis* Regu & David
155.	*marlatti* (Quaintance)
156.	*moundi* David & Subramaniam
157.	*mundanthuraiensis* David & David
158.	*musae* Corbett
159.	*nagercoilensis* Regu & David
160.	*onitshae* Mound
161.	*oplismeni* Takahashi
162.	*orientalis* David & Jesudasan
163.	*ovalis* Regu & David
164.	*pachamalaiensis* Kandasamy mohan & Jesudasan
165.	*padappaiensis* Regu & David
166.	*panvelensis* Regu & David
167.	*patchily* Regu & David
168.	*psidii* Jesudasan & David
169.	*rhachisphora* Regu & David
170.	*riveae* Regu & David
171.	*russellae* Regu & David
172.	*sairandhryensis* Meganathan & David
173.	*saklespurensis* Regu & David
174.	*saputarensis* Regu & David
175.	*scutiae* Vimala & Sundararaj
176.	*singhi* Regu & David
177.	*spinosus* Jesudasan & David
178.	*sterculiae* Jesudasan & David
179.	*sundararaji* Regu & David
180.	*tassellatus* Regu & David
181.	*thomasi* Philips, Jesudasan & David
182.	*tomkinsae* Dooley & Smith-Pardo
183.	*tuberculatus* Regu & David
184.	*vallanadensis* David & David
185.	*valparaiensis* Jesudasan & David
186.	*walayarensis* Jesudasan & David

Genus XII: *Aleuromarginatus* Corbett, 1935

187.	*kalakkadensis* David & David
188.	*kallarensis* David & Subramaniam
189.	*pseudokallarensis* David & David
190.	*tephrosiae* Corbett
191.	*thirumurthiensis* David

Genus XIII: *Aleuropapillatus* Regu & David, 1993c

192.	*gmelinae* (David, Jesudasan & Mathew)
193.	*kumariensis* Regu & David

Genus XIV: *Aleuroparvus* Dubey, 2018

194.	*theae* Dubey

(Continued)

TABLE 7.1 (Continued)
Checklist of Indian Whiteflies

S. No	Whitefly Species

Genus XV: *Aleuroplatus* Quaintance & Baker, 1914
- 195. *alcocki* (Peal)
- 196. *cinnamomi* Jesudasan & David
- 197. *ficusrugosae* Quaintance & Baker
- 198. *hoyae* (Peal)
- 199. *incisus* Quaintance & Baker
- 200. *keralica* David & David
- 201. *lepidoformis* David & David
- 202. *mysorensis* David & Subramaniam
- 203. *pectiniferus* Quaintance & Baker
- 204. *quaintancei* (Peal)
- 205. *spina* (Singh)

Genus XVI: *Aleuropositus* Dubey, 2013b
- 206. *sinus* Dubey

Genus XVII: *Aleuroputeus* Corbett, 1935
- 207. *baccaureae* Corbett

Genus XVIII: *Aleurotrachelus* Quaintance & Baker, 1914
- 208. *corbetti* Takahashi
- 209. *longispinus* Corbett
- 210. *multipapillus* Singh
- 211. *tuberculatus* Singh

Genus XIX: *Aleurotulus* Quaintance & Baker, 1914
- 212. *arundinacea* Singh
- 213. *kalakkadensis* David & David

Genus XX: *Aleyrodes* Latreille, 1796
- 214. *shizuokensis* Kuwana

Genus XXI: *Arunaleyrodes* Dubey, 2016
- 215. *geminus* Dubey

Genus XXII: *Asialeyrodes* Corbett, 1935
- 216. *elegans* Meganathan & David
- 217. *indicus* Sundararaj & David
- 218. *meghalayensis* Regu & David
- 219. *menoni* Meganathan & David
- 220. *papillatus* Regu & David
- 221. *spherica* (Sundararaj & Dubey)
- 222. *splendens* Meganathan & David
- 223. *tuberculata* Pushpa & Sundararaj

Genus XXIII: *Bemisia* Quaintance & Baker, 1914
- 224. *breyniae* (Singh)
- 225. *capitata* Regu & David
- 226. *cordiae* (David & Subramaniam)
- 227. *crossandrae* David & Subramaniam
- 228. *elongata* Sundararaj & Pushpa
- 229. *euphorbiae* David & Subramaniam
- 230. *formosana* Takahashi
- 231. *giffardi* (Kotinsky)
- 232. *grossa* Singh
- 233. *leakii* (Kotinsky)
- 234. *pongamiae* Takahashi
- 235. *moringae* (David & Subramaniam)
- 236. *multituberculata* Sundararaj & David
- 237. *religiosa* (Peal)
- 238. *tabaci* (Gennadius)
- 239. *twista* Sundararaj & Pushpa
- 240. *vernoniae* David & Thenmozhi
- 241. *vitexae* David & David

Genus XXIV: *Cockerelliella* Sundararaj & David, 1992c
- 242. *cinnamomi* Pushpa & Sundararaj
- 243. *dehradunensis* (Jesudasan & David)
- 244. *dioscoreae* Sundararaj & David
- 245. *indica* Sundararaj & David
- 246. *kudremukhensis* Sundararaj
- 247. *meghalayensis* Sundararaj & David
- 248. *papillata* Sundararaj & Pushpa
- 249. *quaintancei* Sundararaj & David
- 250. *rotunda* Regu & David
- 251. *schimae* Chhakchhuak & Sundararaj
- 252. *somnathensis* Sundararaj
- 253. *splendens* Meganathan & David
- 254. *vijendrai* Pushpa & Sundararaj
- 255. *williamsi* Chhakchhuak
- 256. *zingiberae* Sundararaj & David

Genus XXV: *Cohicaleyrodes* Bink-Moenen, 1983
- 257. *elongatus* (Meganathan & David)
- 258. *indicus* (David & Selvakumaran)
- 259. *jesudasani* David
- 260. *maduraiensis* David & David
- 261. *mappiae* Selvakumaran & David
- 262. *padminiae* Phillips & Jesudasan
- 263. *saklespurensis* (Regu & David)

Genus XXVI: *Crenidorsum* Russell, 1945
- 264. *binkae* (Jesudasan & David)
- 265. *cinnamomi* (Jesudasan & David)
- 266. *coimbatorensis* (David & Subramaniam)
- 267. *goaensis* (Jesudasan & David)
- 268. *pykarae* (Jesudasan & David)
- 269. *rubiae* (David)
- 270. *russellae* David & David
- 271. *wendlandiae* (Jesudasan & David)

Genus XXVII: *Crescentaleyrodes* David & Jesudasan, 1987
- 272. *semilunaris* (Corbett)
- 273. *vetiveriae* Dubey & Ko

Genus XXVIII: *Davidiella* Dubey & Sundararaj, 2005e
- 274. *cinnamomi* Dubey & Sundararaj

Genus XXIX: *Dialeurodes* Cockerell, 1902
- 275. *abbotabadiensis* Qureshi
- 276. *armatus* David & Subramaniam
- 277. *atalantiae* Dubey & Ko
- 278. *cinnamomi* Takahashi
- 279. *citri* (Ashmead)

(Continued)

TABLE 7.1 (Continued)
Checklist of Indian Whiteflies

S. No	Whitefly Species
280.	*davidi* Mound & Halsey
281.	*delhiensis* David & Sundararaj
282.	*icfreae* Sundararaj & Dubey
283.	*ichnocarpae* David & David
284.	*indicus* David & Subramaniam
285.	*keralica* Sundaraj & Pushpa
286.	*kirkaldyi* (Kotinsky)
287.	*kumargiriensis* Sundararaj & Dubey
288.	*loganiacei* Pushpa & Sundararaj
289.	*megaspina* Phillips & Jesudasan
290.	*martini* Jesudasan & David
291.	*radiipuncta* Quaintance & Baker
292.	*rotunda* Singh
293.	*sheryli* (David)
294.	*sundararajani* Sundararaj & Dubey
295.	*tamilica* Sundararaj & Pushpa
296.	*trilobiata* Pushpa & Sundararaj
297.	*wendlandiae* Meganathan & David

Genus XXX: *Dialeurolobus* Danzig, 1964

298.	*erythrinae* Selvakumaran & David

Genus XXXI: *Dialeurolonga* Dozier, 1928

299.	*cephalidistincta* David & David
300.	*connari* Pushpa & Sundararaj
301.	*davidi* Dubey & Sundararaj
302.	*elongata* Dozier
303.	*kumargiriensis* Dubey & Sundararaj
304.	*lagerstroemiae* Jesudasan & David
305.	*maculata* (Singh)
306.	*malleshwaramensis* Sundararaj
307.	*multipori* Dubey & Sundararaj
308.	*multituberculata* Dubey & Sundararaj
309.	*pseudocephalidistincta* Dubey & Sundararaj

Genus XXXII: *Dialeuronomada* Quaintance & Baker, 1917

310.	*ayyanarensis* Sundararaj & David
311.	*binkae* Sundararaj & David
312.	*biventralis* Sundararaj & David
313.	*canthiae* Sundararaj & David
314.	*dissimilis* (Quaintance & Baker)
315.	*gigantica* Sundararaj & David
316.	*granulata* Sundararaj & David
317.	*ixorae* (Singh)
318.	*keralaensis* Meganathan & David
319.	*martini* Sundararaj & David
320.	*multitubercuta* Sundararaj & Pushpa
321.	*nagpurensis* Sundararaj & David
322.	*nellaiensis* David & David
323.	*palmata* Sundararaj & David
324.	*papanasamensis* David & David
325.	*papillata* Dubey & Ko
326.	*remadeviae* Dubey & Sundararaj
327.	*rubiphaga* Dubey & Sundararaj
328.	*russellae* Sundararaj & David
329.	*saklespurensis* Regu & David
330.	*vedakani* Sundararaj & Pushpa

Genus XXXIII: *Dialeuropora* Quaintance & Baker, 1917

331.	*decempuncta* (Quaintance & Baker)
332.	*heptapora* Regu & David
333.	*hexapunctata* Chhakchhuak & David
334.	*murrayae* (Takahashi)
335.	*pterolobiae* David & Subramaniam

Genus XXXIV: *Distinctaleyrodes* Dubey & Sundararaj, 2006a

336.	*setosus* Dubey & Sundararaj

Genus XXXV: *Editaaleyrodes* David, 2005b

337.	*indicus* David

Genus XXXVI: *Fippataleyrodes* Sundararaj & David, 1992a

338.	*bituberculata* Pushpa & Sundararaj
339.	*cinnamomi* Dubey & Sundararaj
340.	*indica* Sundararaj & David
341.	*litseae* Sundararaj & David
342.	*multipori* Dubey & Sundararaj
343.	*rajmohani* Pushpa & Sundararaj
344.	*yellapurensis* Dubey & Sundararaj

Genus XXXVII: *Himalayaleyrodes* Dubey, 2017

345.	*sarcococcae* Dubey

Genus XXXVIII: *Icfrealeyrodes* Dubey & Sundararaj, 2006e

346.	*indica* Dubey & Sundararaj
347.	*radiata* Pushpa & Sudararaj

Genus XXXIXI: *Indoaleyrodes* David & Subramaniam, 1976

348.	*geminata* Pushpa & Sundararaj
349.	*laos* (Takahashi)
350.	*strychnosae* Pushpa & Sundararaj

Genus XL: *Kanakarajiella* David & Sundararaj, 1993

351.	*rotunda* Sundararaj & Pushpa
352.	*turpiniae* Meganathan & David
353.	*vulgaris* (Singh)

Genus XLI: *Keralaleyrodes* Meganathan & David, 1994

354.	*indicus* Meganathan & David

Genus XLII: *Martiniella* Jesudasan & David, 1990

355.	*ayyari* Sundararaj & David
356.	*fletcheri* (Sundararaj & David)
357.	*indica* (Singh)
358.	*lefroyi* Sundararaj & David
359.	*multituberculata* Vimala & Sundararaj
360.	*papillata* Sundararaj & Dubey

(Continued)

TABLE 7.1 (Continued)
Checklist of Indian Whiteflies

S. No	Whitefly Species
361.	*sepangensis* (Martin & Mound)
362.	*tripori* Dubey & Sundararaj

Genus XLIII: *Massilieurodes* Goux, 1949
363.	*formosensis* (Takahashi)
364.	*homonoiae* (Jesudasan & David)
365.	*multipori* (Takahashi)

Genus XLIV: *Milleraleurodes* Phillips & Jesudasan, 2013
366.	*illuminata* Phillips & Jesudasan

Genus XLV: *Minutaleyrodes* Jesudasan & David, 1990
367.	*indica* Meganathan & David
368.	*kolliensis* (David)
369.	*minuta* (Singh)
370.	*pearlis* Pushpa & Sundararaj
371.	*tricolorata* Pushpa & Sundararaj

Genus XLVI: *Neomaskellia* Quaintance & Baker, 1913
372.	*andropogonis* Corbett
373.	*bergii* (Signoret)

Genus XLVII: *Orientaleyrodes* Regu & David, 1993c
374.	*indicus* Regu & David

Genus XLVIII: *Parabemisia* Takahashi, 1952
375.	*indica* Meganathan & David
376.	*myricae* (Kuwana)

Genus XLIX: *Pealius* Quaintance & Baker, 1914
377.	*azalea* (Baker & Moles)
378.	*bengalensis* (Peal)
379.	*cinnamomi* David & Sundararaj
380.	*durairaji* David & David
381.	*elongatus* (David, Sundararaj, & Regu)
382.	*indicus* (David)
383.	*misrae* Singh
384.	*mori* Takahashi
385.	*nagerkoilensis* Jesudasan & David
386.	*nelson* David & Dubey
387.	*nilgiriensis* (David)
388.	*sairandhryensis* Meganathan & David
389.	*schimae* Takahashi
390.	*spinosus* Jesudasan & David
391.	*splendens* (David, Sundararaj, & Regu)
392.	*walayarensis* Jesudasan & David

Genus L: *Pseudocockerelliella* Sundararaj, 2007
393.	*curvata* Sundararaj

Genus LI: *Rabdostigma* Quaintance & Baker, 1917
394.	*atalantiae* David & David
395.	*mahableshwarensis* Sundararaj & David
396.	*saklaspurensis* (David)

Genus LII: *Rhachisphora* Quaintance & Baker, 1917
397.	*combiformis* Pushpa & Sundararaj
398.	*elongatus* Regu & David
399.	*indica* Sundararaj & David
400.	*ixorae* Sundararaj & David
401.	*kallarensis* Jesudasan & David
402.	*pechipparaiensis* David & David
403.	*rutherfordi* (Quaintance & Baker)
404.	*trilobitoides* (Quaintance & Baker)

Genus LIII: *Rusostigma* Quaintance & Baker, 1917
405.	*eugeniae* (Maskell)

Genus LIV: *Rutaleyrodes* Dubey & Ko, 2007
406.	*atalantiae* Dubey & Ko

Genus LV: *Setaleyrodes* Takahashi, 1931
407.	*litseae* David & Sundararaj
408.	*machili* Dubey
409.	*thretaonai* David

Genus LVI: *Singhiella* Sampson, 1943
410.	*aizawlensis* Chhakchhuak & David
411.	*bassiae* (David & Subramaniam)
412.	*bauhiniae* Dubey & Sundararaj
413.	*bicolor* (Singh)
414.	*brideliae* (Jesudasan & David)
415.	*cardamomi* (David & Subramaniam)
416.	*crenulata* Qureshi & Qayyum
417.	*globulata* Sundararaj & Pushpa
418.	*keralica* Pushpa & Sundararaj
419.	*malabaricus* Jesudasan & David
420.	*pallida* (Singh)
421.	*pterygotae* Pushpa & Sundararaj
422.	*simplex* (Singh)

Genus LVII: *Singhius* Takahashi, 1932
423.	*hibisci* (Kotinsky)
424.	*meiogynea* Pushpa & Sundararaj
425.	*morindae* Sundararaj & David
426.	*pandalamensis* Pushpa & Sundararaj
427.	*philomenae* Pushpa & Sundararaj
428.	*russellae* (David & Subramaniam)

Genus LVIII: *Siphoninus* Silvestri, 1915
429.	*phillyreae* (Haliday)

Genus LVIX: *Sphericaleyrodes* Selvakumaran & David, 1996
430.	*bambusae* Selvakumaran & David
431.	*regui* Dubey & Sundararaj

Genus LX: *Tetraleurodes* Cockerell, 1902
432.	*acaciae* (Quaintance)

(Continued)

TABLE 7.1 (Continued)
Checklist of Indian Whiteflies

S. No	Whitefly Species
433.	*bambusae* Jesudasan & David
434.	*burliarensis* Jesudasan & David
435.	*champaiensis* Dubey
436.	*dendrocalamae* Dubey & Sundararaj
437.	*kunnathoorensis* Regu & David
438.	*pusana* Takahashi
439.	*rubiphagus* David & David
440.	*thenmozhiae* Jesudasan & David
441.	*thassammaiae* Sundararaj & Pushpa
Genus LXI : *Trialeurodes* Cockerell, 1902	
442.	*ricini* (Mishra)
443.	*vaporariorum* (Westwood)
Genus LXII: *Vanaleyrodes* Pushpa & Sundararaj, 2011	
444.	*myristicae* Pushpa & Sundararaj
445.	*pseudopteriae* Pushpa & Sundararaj
Genus LXIII: *Vasantharajiella* David, 2000	
446.	*kalakadensis* David
Genus LXIV: *Vasdavidius* Russell, 2000	
447.	*indicus* (David & Subramaniam)
448.	*setiferus* (Quaintance & Baker)
Genus LXV: *Viennotaleyrodes* Cohic, 1968	
449.	*megapapillae* (Singh)
450.	*nilagiriensis* David, Krishnan & Thenmozhi
Genus LXVI: *Zaphanera* Corbett, 1926	
451.	*alysicarpae* David & David
452.	*indicus* Jesudasan & David
453.	*publicus* (Singh)
454.	*splendens* David & David

Bayesian and Maximum Likelihood phylogenetic analyses of the mtCOI sequenced from *B. tabaci* populations collected worldwide. Of these, Mediterranean (MED) and Middle East-Asia Minor 1 (MEAM 1) are found to be of significant economic importance as highly invasive pests (De barro et al. 2011; Vyskocilova et al. 2018).

In India, 454 species of whiteflies under 66 genera have been identified, of which, DNA barcodes have been generated for 11 species *viz.*, *Bemisia tabaci*, *Aleurothrixus trachoides*, *Aleurodicus rugioperculatus*, *A. dispersus*, *Aleurocanthus bangalorensis*, *A. woglumi*, *Trialeurodes vaporariorum*, *Neomaskellia bergii*, *Dialeurolonga malleshwaransis*, *Aleurolobus barodensis*, and *A. musae* (NCBI database 2018). Among whiteflies, occurrence of genetic groups in *Bemisia tabaci* (Gennadius) have been reported (Dinsdale et al. 2010; Ellango et al. 2015). Various populations of *B. tabaci* are morphologically indistinguishable; however, they differ with respect to their genetic, biological, and physiological characteristics, and so are designated as a cryptic species complex (Boykin et al. 2007; Dinsdale et al. 2010; De Barro et al. 2011; Boykin et al. 2012; Tay et al. 2012). Further, several workers reported that *B. tabaci* (Genetic group MED) are highly resistant to many insecticides and the MEAM 1 genetic group has high fecundity (Dalton 2006; Horowitz et al. 2005). Ellango et al. (2015) mentioned that populations of *B. tabaci* are highly invasive and may replace the already existing local populations to establish themselves quickly in their new locations. Populations of *B. tabaci* even differ with respect to virus transmission and host range. Different molecular markers, namely, mtCO1, ITS-2 have been used to identify the populations of *B. tabaci*. However, Btab (850 bp) is used to identify the genetic groups in *B. tabaci* populations (Simon et al. 1994). Recently, microsatellite-based markers have been used to identify the genetic groups in *B. tabaci*.

Worldwide, 34 genetic groups in *B. tabaci* have been reported. However, nine genetic groups *viz.*, Asia I, Asia I-India, Asia II-1, Asia II-5, Asia II-7, Asia II-8, Asia II-11, China-3, and the invasive group MEAM 1 have been recognized in India (Ellango et al. 2015). Subsequently, Roopa et al. (2015) reported the prevalence of the four previously existing genetic groups, namely, Asia-I, Asia-II-7, Asia-II-8, and MEAM 1, and of a new group called Middle East Asia Minor-K, which is genetically close (92.6%) to MEAM 1. Selvaraj et al. (2017) reported that Asia-1 and Asia-II-1 are found to be the commonly occurring genetic groups in different geographical locations and crops in India. Therefore, there is a need for documenting the genetic groups in *B. tabaci* occurring in India with support from morphological taxonomy.

ECONOMIC IMPORTANCE

Aleyrodids rank among the most noxious insects attacking field crops, green house crops, and trees around the world. Economic losses result not only as a consequence of their sucking plant sap, but also because, as vectors, they transmit plant diseases. The copious quantities of honey dew that they excrete encourages the growth of sooty mould which adversely impacts photosynthesis. Although there are approximately 1,600 species of whiteflies worldwide, only a few are of economic importance. The species that cause economic damage in India are the greenhouse whitefly *Trialeurodes vaporariorum* (Westwood), the sugarcane whitefly *Aleurolobus barodensis* (Maskell), the jasmine aleyrodids *Dialeurodes kirkaldyi* (Kotinsky) and *Kanakarajiella vulgaris* (Singh), the cardamom whitefly *Singhiella cardamomi* (David & Subramaniam), the betelvine whiteflies *Singhiella pallida*

(Singh) and *Aleurocanthus rugosa* Singh, the citrus whiteflies *Aleurocanthus woglumi* Ashby and *Dialeurodes citri* (Ashmead), the cotton whitefly *Bemisia tabaci* (Gennadius), the babul whitefly *Acaudaleyrodes rachipora* (Singh), and the spiralling whitefly *Aleurodicus dispersus* Russell. Most whitefly species have a narrow range of host plants, but the ones that are considered pests may feed on and damage many vegetable and field crops, greenhouse, and nursery crops and house plants. The cotton whitefly (*B. tabaci*) and silver leaf whitefly (*B. argentifolii* or *B. tabaci* biotype B) are common pests of various crops and ornamentals throughout India. In recent times, severe outbreaks of *B. tabaci* in North India, especially in Haryana, Rajasthan, and Punjab were reported leading to complete failure of the crop. Besides *B. tabaci*, two invasive whiteflies, the Solanum whitefly *Aleurothrixus trachoides* that attacks many solanaceous crops like chilli, brinjal, tomato, tobacco, capsicum, and *Duranta* in Tamil Nadu, Karnataka, and Kerala, and the rugose spiralling whitefly *Aleurodicus rugioperculatus* Martin on coconut, banana, sapota, custard apple, maize, oil palm, guava, mango, cashew, and many ornamental plants in Tamil Nadu, Karnataka, Kerala, and Andhra Pradesh, are emerging as potential pests. Generally, aleyrodid populations are kept in check by natural parasites and predators, but in agricultural crops or on ornamentals, where humans have upset the natural balance, consistently high and often damaging populations may occur.

CONCLUSION

Morphological systematics play a vital role even in the postgenomic era of insect systematics. Puparial taxonomy in whitefly systematics will continue to play a major role, and studies on adult whiteflies will play a supportive role. Molecular studies will provide an independent data set for a critical evaluation of morphological inferences. Advances in molecular biology as well as new techniques enabling the discovery of morphological characters will further assist in the resolution of problems in whitefly systematics.

REFERENCES

Abraham, C. C., and P. J. Joy. 1978. New record of *Tetraleurodes semilunaris* Corbett (Aleyrodidae: Hemiptera) as a pest of lemon grass *Cymbopogon flexuosus* (Steud.). *Entomon* 3 (2): 313–314.

Baig, M. M., A. K. Dubey, and V. V. Ramamurthy. 2016. Determination of sexual dimorphism in the puparia of four whitefly pest species from India (Hemiptera: Aleyrodidae). *Acta Entomologica Musei Nationalis Pragae* 56(2): 447–460.

Bink-Moenen, R. M., and L. A. Mound. 1990. Whiteflies. Diversity, biosystematics and evolutionary patterns. In. *Whiteflies: Their Bionomics, Pest Status and Management*, ed. Gerling, D. Andover, UK: Intercept Ltd., pp. 1–11.

Boykin, L. M., K. F. Armstrong, L. Kubatko, and P. De Barro. 2012. Species delimitation and global biosecurity. *Evolutionary Bioinformatics* 8: 1–37.

Boykin, L. M., R. G. J. Shatters, R. C. Rosell. et al. 2007. Global relationships of *Bemisia tabaci* (Hemiptera: Aleyrodidae) revealed using Bayesian analysis of mitochondrial COI DNA sequences. *Molecular Phylogenetics and Evolution* 44: 1306–1319.

Buckton, G. B. 1903. Description of a new species of *Aleurodes*. Destructive to beetle. *Indian Museum Notes* 5(2): 36.

Dalton, R. 2006. The Christmas invasion. *Nature* 443: 898–900.

David, B. V. 1972. Two new species of *Odontaleyrodes* Takahashi (Homoptera: Aleyrodidae) from India. *Oriental Insects* 6(3): 309–312.

David, B. V. 1976a. A new species of the genus *Aleuromarginatus* Corbett (Aleyrodidae: Hemiptera) from India. *Entomon* 1(1): 85–86.

David, B. V. 1976b. On a new species of leaf pit-gall forming aleyrodid and *Aleurotuberculatus hexcantha* (Singh) comb. nov. (Aleyrodidae: Homoptera). *Record of Zoological Survey of India* 69: 261–265.

David, B. V. 1977. A new species of *Aleurotuberculatus* Takahashi and redescription of *Aleurotuberculatus minutus* (Singh) (Aleyrodidae, Hemiptera). *Entomon* 2(1): 89–92.

David, B. V. 1978. On a new species of *Aleurotuberculatus* (Hemiptera: Aleyrodidae) from India with a key to Indian species. *Oriental Insects* 12(1): 133–135.

David, B. V. 1981. A new species of *Setaleyrodes* (Homoptera: Aleyrodidae) from India. *Colemania* 1(1): 37–38.

David, B. V. 1987. First record of the whitefly subfamily Aleurodicinae (Aleyrodidae: Homoptera) from India. *Current Science* 56(23): 1247–1248.

David, B. V. 1988. *Aleuromarginatus bauhiniae* (Corbett) comb. nov. and *A. thirumurthiensis* nom. nov. (Aleyrodidae: Homoptera). *Journal of the Bombay Natural History Society* 85(2): 445.

David, B. V. 1990. Key to tribes of whiteflies (Aleyrodidae: Homoptera) of India. *Journal of Insect Science* 3(1): 13–17.

David, B. V. 1994. A new species of *Viennotaleyrodes* Cohic (Aleyrodidae: Homoptera) from India. *Hexapoda* 6: 33–38.

David, B. V. 2005a. *Editaaleyrodesindicus*, a new genus and species of whitefly (Hemiptera: Aleyrodidae) from India. *Entomon* 30(4): 317–320.

David, B. V. 2005b. *Editaaleyrodes indicus*, a new genus and species of whitefly (Hemiptera: Aleyrodidae) from India. *Entomon* 30(4): 317–320.

David, B. V., and A. W. Augustine. 1988. A new whitefly *Bemisia graminis* sp. nov. (Aleyrodidae: Homoptera) from India. *Entomon* 13(1): 33–35.

David, B. V., and P. M. M. David. 2000. Occurrence of a new whitefly species of the Neotropical genus *Crenidorsum* Russell (Homoptera: Aleyrodidae). *Entomon* 25(2): 155–158.

David, B. V., and A. K. Dubey. 2006. Whitefly (Hemiptera: Aleyrodidae) fauna of Andaman and Nicobar Islands, India with description of a new species. *Entomon* 31(3): 191–205.

David, B. V., and A. K. Dubey. 2009. New synonymies and combinations in *Bemisia* (Aleyrodidae: Hemiptera). *Oriental Insects* 43: 1–6.

David, B. V., and R. W. A. Jesudasan. 1987. Description of a new genus *Crescentaleyrodes* for *Aleurolobus semilunaris* (Corbett) (Aleyrodidae: Homoptera) and two new combinations. *Current Science* 56(1): 42–44.

David, B. V., and R. W. A. Jesudasan. 1988. On two new species of whiteflies (Aleyrodidae: Homoptera) from India and Sri Lanka. *Entomon* 13(1): 29–32.

David, B. V., and R. W. A. Jesudasan. 1989a. *Dialeurolonga maculata* (Singh) comb. nov. and *Dialeurolonga takahashi* nom. nov. for *Dialeurolonga maculata* Takahashi (Aleyrodidae: Homoptera) from Madagascar. *Entomon* 14(3–4): 371.

David, B. V., and R. W. A. Jesudasan. 1989b. Redescription of the whitefly *Aleyrodes shizuokensis* Kuwana (Aleyrodidae: Homoptera). *Journal of the Bombay Natural History Society* 86: 260–261.

David, B. V., and R. W. A. Jesudasan. 2002. A new species of *Aleurocanthus* Quaintance & Baker and *Asialeyrodes indicus* Sundararaj & David (Aleyrodidae: Homoptera) from Andaman & Nicobar Islands. *Entomon* 27(3): 323–325.

David, B. V., and M. Manjunatha. 2003. A new species of *Aleurocanthus* Quaintance & Baker (Homoptera: Aleyrodidae) from *Arecacatechu* in India, with comments on the status of *Aleurodesnubilans* Buckton. *Zootaxa* 173: 1–4.

David, B. V., and K. Regu. 1989. A new whitefly *Aleurocanthus indicus* sp. nov. (Aleyrodidae: Homoptera) from India. *Entomon* 14(3–4): 275–276.

David, B. V., and K. Regu. 1991. A new record and Redescription of *Rhachisphora rutherfordi* (Quaintance & Baker) (Homoptera: Aleyrodidae) from India. *Journal of Insect Science* 4(1): 69–70.

David, B. V., and K. Regu. 1995. Aleurodicus dispersus Russell (Aleyrodidae: Homoptera) a whitefly pest, new to India. *Pestology* 19(3): 5–7.

David, B.V., and S. Selvakumaran. 1987. A new species of whitefly *Mixaleyrodes indicus* sp. nov. (Aleyrodidae: Homoptera) from India. *Journal of the Bombay Natural History Society* 84(3): 654–656.

David, B. V., and T. R. Subramaniam. 1976. Studies on some Indian Aleyrodidae. *Records of the Zoological Survey of India* 70: 133–233.

David, B. V., and R. Sundararaj. 1991a. A new species of whitefly genus *Pealius* Quaintance and Baker (Aleyrodidae: Homoptera) from India. *Journal of Insect Science* 4(1): 67–68.

David, B. V., and R. Sundararaj. 1991b. A new species of *Setaleyrodes* Takahashi (Aleyrodidae: Homoptera) from India. *Entomon* 16(4): 317–318.

David, B. V., and R. Sundararaj. 1992. *Dialeurodes delhiensis* sp. nov. (Aleyrodidae: Homoptera)—A new species of whitefly from India. *Journal of Insect Science* 5(1): 62–63.

David, B. V., and R. Sundararaj. 1993. Studies on Dialeurodini (Aleyrodidae: Homoptera) of India: *Kanakarajiella* gen. nov. *Journal of Entomological Research* 17(4): 289–295.

David, B. V., and K. Thenmozhi. 1995. On the characteristics of pupal case, adult and egg of Indian species of *Libaleyrodes* Takahashi (Aleyrodidae: Homoptera) with description of a new species. *Journal of the Bombay Natural History Society* 92: 339–349.

David, B. V., R. W. A. Jesudasan, and A. Phillips. 2006. A review of *Aleurotrachelus* Quaintance & Baker (Hemiptera: Aleyrodidae) and related genera in India, description of two new species of the genus *Cohicaleyrodes* Bink. *Hexapoda* 13(1&2): 16–27.

David, B. V., R. W. A. Jesudasan, and G. Mathew. 1988. Description of a new species of the genus *Aleurolobus* Quaintance & Baker (1914) (Aleyrodidae: Homoptera). *Journal of the Bombay Natural History Society* 85: 165–167.

David, B. V., B. Krishnan, and K. Thenmozhi. 1994. A new species of *Viennotaleyrodes* Cohic (Aleyrodidae: Homoptera) from India. *Hexapoda* 6(1): 33–38.

David, B. V., R. Sundararaj, and K. Regu. 1991 On the four species of *Odontaleyrodes* Takahashi (Aleyrodidae: Homoptera) with a key to Indian species. *Journal of Insect Science* 4(2): 117–119.

David, P. M. M. 2000. Three new genera of whiteflies *Mohanasundaramiella*, *Shanthiniae* and *Vasantharajiella* (Aleyrodidae: Homoptera) from India. *Journal of the Bombay Natural History Society* 97(1): 123–130.

David, P. M. M., and B. V. David. 2001. Revision of whiteflies (Aleyrodidae) infesting rice in India. *Entomon, Kariavattom* (Special issue): 353–356.

David, P. M. M., and B. V. David. 2007b. Descriptions of new species of whiteflies (Hemiptera: Aleyrodidae) from south India. *Oriental Insects* 41: 391–426.

David, P. M. M., and B. V. David. 2007a. The Indian species of *Zaphanera* Corbett (Hemiptera: Aleyrodidae) with description of a new species. *Biosystematica* 1(1): 41–44.

David, P. M. M., B. V. David, and A. K. Dubey. 2010. Two new species and one new record of whiteflies (Hemiptera: Aleyrodidae) from India. *Hexapoda* 17(1): 6–11.

De Barro, P. J., S. S. Liu, L. M. Boykin, and A. B. Dinsdale. 2011. *Bemisia tabaci*: A statement of species status. *Annual Review of Entomology* 56: 1–19.

Dinsdale, A., L. Cook, C. Riginos, Y. M. Buckley, and P. J. De Barro. 2010. Refined global analysis of *Bemisia tabaci* (Gennadius) (Hemiptera: Sternorrhyncha: Aleyrodoidea) mitochondrial CO1 to identify species level genetic boundaries. *Annals of the Entomological Society of America* 103:196–208.

Dooley, J. W., and A. Smith-Pardo. 2013. Two new species of whiteflies (Hemiptera: Sternorrhyncha: Aleyrodidae: Aleyrodinae) intercepted in quarantine on plants from Asia. *The Pan-Pacific Entomologist* 89(2): 84–101.

Douglas, J. W. 1892. Footnote in p.32. In: Morgan, A.C.F., A new genus and species of Aleurodidae. *Entomologist's Monthly Magazine* 28: 29–33.

Dozier, H. L. 1928. Two new aleyrodid (Citrus) pests from India and the South Pacific. *Journal of Agricultural Research* 36: 1001–1005.

Dubey, A. K. 2016. New genus and species of Aleyrodidae (Hemiptera: Sternorrhyncha) from North-Eastern India, with remarks on its relationships with allied genera, *Entomological Science* 19: 161–173.

Dubey, A. K. 2017. Description of a new species, *Setaleyrodes machili* Dubey, sp. nov. (Hemiptera: Aleyrodidae) infesting *Machilus odoratissima* Nees (Lauraceae) in Western Himalaya, India. *Zootaxa* 4363(2): 291–300.

Dubey, A. K. 2018. A new whitefly genus and species, *Aleuroparvus theae* Dubey (Hemiptera: Aleyrodidae) colonising Assam tea (*Camellia sinensis*) and *Cinnamomum bejolghota*, in North-East India. *Zootaxa* 4486(2): 169–179.

Dubey, A. K., and B. V. David. 2012a. Studies on the Indian species of *Aleuroplatus* (Hemiptera: Aleyrodidae), with designation of a neotype puparium for *Aleurodes alcocki* Peal. *Zootaxa* 3303: 50–58.

Dubey, A. K., and B. V. David. 2012b. Indian whiteflies (Hemiptera: Aleyrodidae) with their host plants. In: *The Whitefly or Mealywing Bugs: Bioecology, Host Specificity and Management*, ed. V. David. Saarbrucken, Germany: Lambert Academic Publishing GMBH & Co KG, 411 pp.

Dubey, A. K., and C. C. Ko. 2007. *Rutaleyrodes atalantiae*, a new genus and species (Hemiptera: Aleyrodidae) from India. *Current Science*, 92(12): 1685–1687.

Dubey, A. K., and V. V. Ramamurthy. 2015. Description of a grass feeding whitefly of the genus *Tetraleurodes* (Hemiptera: Aleyrodidae) from the Indo-Myanmar border. *Florida Entomologist* 98(1): 32–36.

Dubey, A. K., and V. V. Ramamurthy. 2013. *Dialeurolonga* re-defined (Hemiptera: Aleyrodidae): With a new genus and species from India, two new genera from Australia, and discussion of host-correlated puparial variation. *Zootaxa* 3616(6): 548–562.

Dubey, A. K., and V. V. Ramamurthy. 2013a. *Icfrealeyrodes* Stat. Rev. (Hemiptera: Aleyrodidae) with description of a new species from Myanmar. *Florida Entomologist* 96(2): 463–468.

Dubey, A. K., and V. V. Ramamurthy. 2013b. *Dialeurolonga* redefined (Hemiptera: Aleyrodidae): With a new genus and species from India, two new genera from Australia, and discussion of host-correlated puparial variation. *Zootaxa* 3616(6): 548–562.

Dubey, A. K., and S. Singh, 2016a. Description of a new genus and species of whitefly (Hemiptera: Sternorrhyncha) infesting *Rhynchosia minima* (Fabaceae) in Karnataka, India. *Entomological Science* 19: 367–375.

Dubey, A. K., and S. Singh. 2016b. New record of the genus and species, *Agrostaleyrodes arcanus* Ko (Hemiptera: Aleyrodidae) from India, now colonizing on sugarcane. In: *Recent Advances in Life Sciences: Proceedings of XV AZRA International Conference*, eds. A. Prakash, J. Rao, and K. Revathi. Chennai, India: Ethiraj College for Women, pp. 37–40.

Dubey, A. K., and S. Singh. 2017. A new whitefly genus and species, *Himalayaleyrodes sarcococcae* Dubey (Hemiptera: Aleyrodidae) infesting Christmas box (Buxaceae) in Western Himalaya, India. *Zootaxa* 4269(4): 531–544.

Dubey, A. K., and R. Sundararaj. 2004a. Host range of the spiralling whitefly *Aleurodicus dispersus* Russell (Aleyrodidae: Homoptera) in Western Ghats of south India. *Indian Journal Forestry* 27(1): 63–65.

Dubey, A. K., and R. Sundararaj. 2004b. A review of the genus *Dialeuronomada* Quaintance & Baker (Hemiptera: Aleyrodidae) with descriptions of two new species. *Formosan Entomology* 24: 147–157.

Dubey, A. K., and R. Sundararaj. 2004c. Whitefly species of the genus *Aleurocanthus* Quaintance & Baker (Hemiptera: Aleyrodidae) from India, with description of six new species. *Oriental Insects* 39: 295–321.

Dubey, A. K., and R. Sundararaj. 2005a. Description of a new species of the genus *Tetraleurodes* Cockerell (Hemiptera: Aleyrodidae) with a key to Indian species. *Zoos' Print Journal* 20(7): 1924–1926.

Dubey, A. K., and R. Sundararaj. 2005b. A review of the genus *Aleuroclava* Singh (Hemiptera: Aleyrodidae) with descriptions of eight new species from India. *Oriental Insects* 39: 241–272.

Dubey, A. K., and R. Sundararaj. 2005c. New record of *Aleurocanthus martini* David (Homoptera: Aleyrodidae) from India. *Journal of Bombay Natural History Society* 102(1): 131.

Dubey, A. K., and R. Sundararaj. 2005d. A taxonomic study of the genus *Pealius* Quaintance & Baker (Homoptera: Aleyrodidae) in India. *Journal of Bombay Natural History Society* 102(2): 158–163.

Dubey, A. K., and R. Sundararaj. 2005e. *Davidiella cinnamomi*, a new genus and species of whitefly (Hemiptera: Aleyrodidae) from India. *Entomon* 30(4): 351–354.

Dubey, A. K., and R. Sundararaj. 2005f. Three new species of *Fippataleyrodes* Sundararaj & David (Aleyrodidae: Hemiptera) from Western Ghats of south India. *Journal of Bombay Natural History Society* 102(2): 204–207.

Dubey, A. K., and R. Sundararaj. 2006a. *Distinctaleyrodes setosus* Dubey & Sundararaj (Sternorrhyncha: Aleyrodidae), a new whitefly genus and species from India. *Zootaxa* 1154: 35–39.

Dubey, A. K., and R. Sundararaj. 2006b. Key to whiteflies of the tribe *Aleurolobini* (Hemiptera: Aleyrodidae) of India with description of five new species and host records. *Oriental Insects* 40: 33–60.

Dubey, A. K., and R. Sundararaj. 2006c. Descriptions of five new species of the whitefly genus *Dialeurolonga*, Dozier (Hemiptera: Aleyrodidae) from India. *Oriental Insects* 40: 159–170.

Dubey, A. K., and R. Sundararaj. 2006d. A new whitefly species of the genus *Taiwanaleyrodes* Takahashi (Homoptera: Aleyrodidae) from India. *Entomon* 31(1): 73–76.

Dubey, A. K., and R. Sundararaj. 2006e *Icfrealeyrodes indica*, a new genus and species of whitefly (Hemiptera: Aleyrodidae) from India. *Entomon* 31(2): 125–128.

Dubey, A. K., and R. Sundararaj. 2006f. Two new aleyrodids (Hemiptera: Aleyrodidae) from India. *Entomon* 31(3): 229–235.

Dubey, A. K., and R. Sundararaj. 2006g. On the genus *Asialeyrodes* Corbett (Hemiptera: Aleyrodidae) of India with a key to Indian species. *Journal of Bombay Natural History Society* 103(1): 117–119.

Dubey, A. K., S. Singh, and J. H. Martin. 2014. *Acanthaleyrodes elevatus* sp. n. (Hemiptera: Aleyrodidae) from India, with key to species and discussion of tuberculate setae. *Zootaxa* 3881(1): 33–48.

Dubey, A. K., R. Sundararaj, and K. Regu. 2004. Aleyrodidae (Hemiptera: Aleyrodidae) Fauna of the Lakshadweep, India. *Entomon* 29(3): 279–286.

Ellango, R., S. T. Singh, V. S. Rana et al. 2015. Distribution of *Bemisia tabaci* genetic groups in India. *Environmental Entomology* 1–7: 1–5. doi:10.1093/ee/nvv062.

Frohlich, D. R., I. Torres-Jerez, I. D. Bedford, P. G. Markham, and J. K. Brown. 1999. A phylogeographical analysis of the *Bemisia tabaci* species complex based on mitochondrial DNA markers. *Molecular Ecology* 8: 1683–1691.

Gill, R. J. 1990. The morphology of whiteflies. In *Whiteflies, Their Bionomics, Pest status and Management*, ed. D. Gerling. Andover, UK: Intercept Ltd, pp. 13–46.

Hahbazvar, N. S., A. S. Ahragard, R. H. Osseini, and J. H. Ajizadeh. 2011. A preliminarily study on adult characters of whiteflies. *Entomofauna Zeitschrift Für Entomologie* 32: 413–420.

Horowitz, A. R., S. Kontsedalov, V. Khasdan, and I. Ishaaya. 2005. Biotypes B and Q of *Bemisia tabaci* and their relevance to neonicotinoid and pyriproxyfen resistance. *Archive of Insect Biochemistry and Physiology* 58: 216–225.

Husain, M. A., and A. W. Khan. 1945. The citrus Aleyrodidae (Homoptera) in Punjab and their control. *Memoirs of Entomological Society of India* 1: 1–41.

Jesudasan, R. W. A., and B. V. David. 1991. Taxonomic studies on Indian Aleyrodidae (Insecta: Homoptera). *Oriental Insects* 25: 231–434.

Jesudasan, R. W. A., and B. V. David. 1990. Revision of two whitefly genera, *Aleuroclava* Singh and *Aleurotuberculatus* Takahashi (Homoptera: Aleyrodidae). *FIPPAT Entomology Series* 2: 1–13.

Jesudasan, R. W. A., E. Ragupathy, and A. Joshy. 2003. Whiteflies of Jawadhi and Yelagiri hills (Eastern Ghats). *Insect Environment* 9(3): 135–137.

Lalneihpuia, C., and S. J. William. 2011. Taxonomic studies on the whitefly (Aleyrodidae: Hemiptera: Insecta) fauna of Mizoram. *Memoirs of the Entomological Society of India* 16: 111 pp.

Manzari, S., and D. L. J. Quicke. 2006. A cladistic analysis of whiteflies, subfamily Aleyrodinae (Hemiptera: Sternorrhyncha: Aleyrididae). *Journal of Natural History* 40(44–46): 2423–2554.

Martin, J. H. 2003. Whiteflies (Hemiptera: Aleyrodidae)—their systematic history and the resulting problems of conventional taxonomy, with special reference to descriptions of *Aleyrodes proletella* (Linnaeus, 1758) and *Bemisia tabaci* (Gennadius, 1889). *Entomologist's Gazette* 54: 125–136.

Martin, J. H. 2007. Giant whiteflies (Sternorrhyncha, Aleyrodidae), a discussion of their taxonomic and evolutionary significance, with the description of a new species of *Udamoselis* Enderlein from Ecuador. *Tijdschrift voor Entomologie* 150: 13–29.

Martin, J. H. 2004. The whiteflies of Belize (Hemiptera: Aleyrodidae) Part 1 - introduction and account of the subfamily Aleurodicinae Quaintance & Baker. *Zootaxa* 681: 1–119.

Maruthi, M. N., J. Colvin, R. M. Thwaites, G. K. Banks, G. Gibson, and S. E. Seal. 2004. Reproductive incompatibility and cytochrome oxidase I gene sequence variability amongst host-adapted and geographically separate *Bemesia tabaci* populations (Hemiptera: Aleyrodidae). *Systematic Entomology* 29: 560–568. doi:10.1111/j.0307-6970.2004.00272x.

Maskell, W. M. 1896. Contributions towards a monograph of the Aleurodidae, a family of Hemiptera–Homoptera. *Transactions and Proceedings of the New Zealand Institute* 28: 411–449.

Meganathan, P., and B. V. David. 1994. Aleyrodidae fauna (Aleyrodidae: Homoptera) of Silent Valley, A tropical evergreen rain-forest, in Kerala, India. *FIPPAT Entomology Series* 5: 1–66.

Misra, C. C. 1924. The citrus whitefly, *Dialeurodes* in India and its parasite, together with the life history of *Aleurodes ricini*, n. sp. *Report of Proceedings of Vth Entomological Meetings at Pusa* 1923: 129–135.

Mohan, K., and R. W. A. Jesudasan. 2014. *Aleurolobus pachamalaiensis*, a new species of whitefly (Aleyrodiae: Hemiptera) from India. *Pestology* 38(11): 15–16.

Mound, L. A. 1963. Host-correlated variation in *Bemisia tabaci* (Gennadius) (Homoptera: Aleyrodidae). *Proceeding of Royal Entomological Society Lonon* (A) 38: 171–180.

Mound, L.A. and Halscy, S.H. 1978. *Whitefly of the World*. A systematic catalogus of the Aleyrodidae (Homoptera) with host plant and natural enemy data. British Museum (Natural History) and John Wiley and Sons. *Chichester* 340 pp.

Nath, D. K. 1970. Occurrence of *Aleurocanthus citriperdus* Quaintance & Baker (Aleyrodidae: Hemiptera) on citrus plants in Darjeeling district, West Bengal. *Indian Journal of Entomology* 32(3): 268.

Peal, H. W. 1903. Contribution towards a monograph of the oriental Aleurodidae. *Journal of Asiatic Society of Bengal* 72: 61–93.

Phillips, A., and R. W. A. Jesudasan. 2013. A new genus, two new species and two new records of whiteflies (Aleyrodidae: Hemiptera) from India. *The Bioscan* 8(1): 343–347.

Phillips, A., R. W. A. Jesudasan and B. V. David. 2018. *Aleuolobus thomasi* sp. nov. (Aleyrodidae: Insecta) from India. *Pestology* 42(10): 29–31.

Pushpa, R., and R. Sundararaj. 2008. Description of two new species of the whitefly genus *Fippataleyrodes* Sundararaj & David (Hemiptera: Aleyrodidae) from India. *Hexapoda* 15(2): 83–87.

Pushpa, R., and R. Sundararaj. 2009a. Whiteflies of the genus *Cockerelliella* Sundararaj and David (Hemiptera: Aleyrodidae) of India with descriptions of two new species. *Biosystematica* 3(2): 29–36.

Pushpa, R., and R. Sundararaj. 2009b. Description of a new species of the genus *Rhachisphora* Quaintance & Baker (Hemiptera: Aleyrodidae) with a key to Indian species. *Entomon* 34(3): 167–174.

Pushpa, R., and R. Sundararaj. 2010a. Description of two new species of the genus Indoaleyrodes David & Subramaniam (Hemiptera: Aleyrodidae) from India. *Hexapoda* 17(1): 1–5.

Pushpa, R., and R. Sundararaj. 2010b. A key to the Indian species of the genus *Asialeyrodes* Corbett (Hemiptera: Aleyrodidae), with description of two new species. *Entomon* 35(1): 31–41.

Pushpa, R., and R. Sundararaj. 2010c. The genus *Aleuroclava* Singh (Hemiptera: Aleyrodidae) from India. *Oriental Insects* 44: 95–146.

Pushpa, R., and R. Sundararaj. 2010d. On the genus *Singhius* Takahashi (Hemiptera: Aleyrodidae) with descriptions of three new species from India. *Oriental Insects* 44: 147–156.

Pushpa, R., and R. Sundararaj. 2010e. Description of three new and redescriptions of two aleyrodids of the genus *Aleurolobus* Quaintance & Baker from India. *Annals of Entomology* 28(1): 1–8.

Pushpa, R., and R. Sundararaj. 2010f. Description of two new species of the genus *Minutaleyrodes* Jesudasan and David (Hemiptera: Aleyrodidae) with the list of species from India. *Biosystematica* 4(2): 45–50.

Pushpa, R., and R. Sundararaj. 2010g. Description of two new species of the genus *Minutaleyrodes* Jesudasan and David (Hemiptera: Aleyrodidae) with the list of species from India. *Biosystematica* 4(2): 45–50.

Pushpa, R., and R. Sundararaj. 2010h. On the genus *Kanakarajiella* Sundararaj & David (Hemiptera: Aleyrodidae) with description of a new species. *Journal of Bombay Natural History Society* 107(2): 159–162.

Pushpa, R., and R. Sundararaj. 2011a. *Vanaleyrodes* (Hemiptera: Aleyrodidae) A new genus with two new species of whiteflies from Western Ghats of India. *Hexapoda* 18(1): 13–16.

Pushpa, R., and R. Sundararaj. 2011b. A review of the genus *Dialeurolonga* Dozier (Hemiptera: Aleyrodidae) with description of a new species from India. *Journal of Bombay Natural History Society,* 108(1): 47–50.

Pushpa, R., and R. Sundararaj. 2011c. Description of two new aleyrodids of the genus *Dialeurodes* Cockerell (Aleyrodidae: Hemiptera) from India. *Entomon* 36(1–4): 71–76.

Pushpa, R., and R. Sundararaj. 2012. The genus *Singhiella* Sampson from India (Aleyrodidae: Hemiptera) with description of two new species. *Oriental Insects* 46: 19–29.

Quaintance, A. L., and A. C. Baker. 1913. Classification of the Aleyrodidae Part I. *Technical Series Bureau of Entomology U. S.* 27: 1–94.

Quaintance, A. L., and A. C. Baker. 1917. A contribution to our knowledge of the whiteflies of the subfamily Aleyrodinae (Aleyrodidae). *Proceedings of the United States National Museum* 51: 335–445.

Rao, A. S. 1958. Notes on Indian Aleurodidae (Whiteflies) with special reference to Hyderabad. *Proceedings 10th International Congress of Entomology*. Montreal, Canada, 1956 1: 331–336.

Regu, K., and B. V. David. 1990. *Rhachisphora elongatus* sp. nov. (Aleyrodidae Homoptera)—A new species of whitefly from India. *Entomon* 15: 277–279.

Regu, K., and B. V. David. 1991. A new species of *Bemisia* (Homoptera: Aleyrodidae) from India with a key to Indian species. *Entomon* 16: 77–81.

Regu, K., and B. V. David. 1992a. On two species of *Aleurodicus* Douglas (Aleurodicinae: Aleyrodidae: Homoptera) from India with a key to Indian species. *Entomon* 17: 99–102.

Regu, K., and B. V. David. 1992b. A new whitefly *Aleurotrachelus saklespurensis* sp. nov. (Aleyrodidae: Homoptera) from India. *Entomon* 17: 135–136.

Regu, K., and B. V. David. 1992c. Two new species of *Asialeyrodes* Corbett (Aleyrodidae: Homoptera) from India. *Journal of the Bombay Natural History Society* 88: 256–258.

Regu, K., and B. V. David. 1992d. A new species of whitefly *Dialeuropora heptapora* sp. nov. (Aleyrodidae: Homoptera) from India. *Journal of the Bombay Natural History Society* 88: 413–414.

Regu, K., and B. V. David. 1993a. On three new species of whiteflies of the tribe Dialeurodine Sampson, 1943 (Aleyrodidae: Homoptera) from India. *Journal of Bombay Natural History Society* 89: 82–87.

Regu, K., and B. V. David. 1993b. Two new species of whiteflies (Aleyrodidae: Homoptera) from India. *Hexapoda* 5: 53–56.

Regu, K., and B. V. David. 1993c. Taxonomic studies on Indian Aleyrodids of the tribe *Aleurolobini* (Aleyrodinae: Aleyrodidae: Homoptera). *FIPPAT Entomology Series* 4: 1–79.

Regu, K., and B. V. David. 1993d. *Asialeyrodes saklespurensis* sp. nov. (Aleyrodidae: Homoptera) from India. *Entomon* 18: 91–93.

Revathi, T. G., and R. Sundararaj. 2016. New record of a genus and two species of whiteflies (Hemiptera: Aleyrodidae) from India. *Entomon* 41(2): 121–124.

Roopa, H. K., R. Asokan, K. B. Rebijith, H. Ranjitha, R, Mahmood, and N. K. Kumar. 2015. Prevalence of a new genetic group, MEAM-K, of the whitefly *Bemisia tabaci* (Hemiptera: Aleyrodidae) in Karnataka, India, as evident from mtCOI sequences. *Florida Entomologist* 98(4): 1062–1071.

Russell, L. M. 1947. A classification of the whiteflies of the new tribe *Trialeurodini* (Homoptera: Aleyrodidae). *Revista de Entomologia, Rio de Janeiro* 18: 1–44.

Selvakumaran, S., and B. V. David. 1996. A new genus of whiteflies (Aleyrodidae: Homoptera) from the cardamom ecosystem, South India. *Journal of Spices and Aromatic Crops* 5: 58–63.

Selvaraj, K., T. Venkatesan, S. K. Jalali et al. 2017. Molecular characterization of different populations of whitefly *Bemisia tabaci* (Gennadius) occurring in India and their phylogenetic relationship. Abstract of Indo-US symposium on curbing whitefly-plant virus pandemics-The departure from pesticides to genomic solutions held on 4–5 December, 2017 at PAU. Ludhiana, India, P. 45.

Simon, C., F. Frati, A. Beckenbach, B. Crespi, H. Liu, and P. Flook. 1994. Evolution, weighting, and phylogenetic utility of mitochondrial gene sequences and a compilation of conserved polymerase chain reaction primers. *Annals of the Entomological Society of America* 87:651–701.

Singh, K. 1931. A contribution towards our knowledge of the Aleyrodidae (whiteflies) of India. *Memoires of Department of Agriculture, India Entomology Series* 12: 1–98.

Singh, K. 1938. Notes on Aleurodidae (Rhynchota) from India I. *Records of the Indian Museum* 40: 189–192.

Singh, K. 1940. Notes on Aleurodidae (Rhynchota) from India II. *Records of the Indian Museum* 42: 453–456.

Singh, K. 1945. Notes on Aleurodidae from India III. *Indian Journal of Entomology* 6: 75–78.

Sundararaj, R. 1999. Redescription and a new record of *Aleuroclava jasmini* (Takahashi) from India. *Indian Journal of Entomology* 61(2): 192–194.

Sundararaj, R. 2000. A new whitefly *Cockerelliella somnathensis* sp. n. from India (Insecta: Hemiptera: Aleyrodidae). *Entomon* 36: 313–316.

Sundararaj, R. 2001. Description of a new species of *Dialeurolonga* Dozier (Hemiptera: Aleyrodidae) breeding on *Polyalthia longifolia* Hook with key to Indian species. *Entomon* 26(2): 191–194.

Sundararaj, R. 2007. On the genera *Cockerelliella* Sundararaj & David and *Pseudcockrelliella* Sundararaj gen. nov. with a key to the Indian genera of Aleyrodidae (Hemiptera). *Oriental Insects* 41: 243–257.

Sundararaj, R., and B. V. David. 1990. A new whitefly *Bemisia multituberculata* sp. nov. (Aleyrodidae: Homoptera) from India. *Entomon* 15(1& 2): 113–115.

Sundararaj, R., and B. V. David. 1991a. On the whiteflies of the genus *Rhachisphora* Quaintance & Baker (Aleyrodidae: Homoptera) from India. *Entomon* 16(4): 311–315.

Sundararaj, R., and B. V. David. 1991b. Ten new species of *Dialeuronomada* Quaintance & Baker from Indian Subcontinent. *Hexapoda* 3: 27–47.

Sundararaj, R., and B. V. David. 1992 On the genera *Fippataleyrodes* n. gen. and *Taiwanaleyrodes* Takahashi from India (Insecta, Homoptera, Sternorrhyncha: Aleyrodidae). *Reichenbachia* 29(40): 15–18.

Sundararaj, R., and B. V. David.1993a. New species of *Aleuroclava* Singh from India (Homoptera: Aleyrodidae). *Oriental Insects* 27: 233–270.

Sundararaj, R., and B. V. David. 1993b. First record of whitefly genus *Martiniella* Jesudasan & David (Aleyrodidae: Homoptera) from India. *Entomon* 18(1&2): 95–99.

Sundararaj, R., and B. V. David. 1994. On the Indian species of the genus *Gigaleurodes* Quaintance & Baker (Aleyrodidae: Homoptera) from India. *Journal of Bombay Natural History Society* 91: 328–330.

Sundararaj, R., and B. V. David. 1995. *Aleuroclava afriae,* a new species of whitefly from India (Insecta, Homoptera, Sternorrhyncha: Aleyrodidae). *Reichenbechia* 31(4): 17–18.

Sundararaj, R., and A. K. Dubey. 2003. Whiteflies (Hemiptera: Aleyrodidae) associated with sandal (*Santalum album* L.) in south India. *Entomon* 28(4): 293–298.

Sundararaj, R., and A. K. Dubey. 2004. The Whitefly genus *Martiniella* Jesudasan & David (Aleyrodidae: Hemiptera) of India with description of one new species. *Entomon* 29(4): 357–360.

Sundararaj, R., and A. K. Dubey. 2005. *Parabemisiamyricae* (Kuwana) (Hemiptera: Aleyrodidae), a new record from India. *Zoos' Print Journal* 20(7): 1933.

Sundararaj, R., and A. K. Dubey. 2006a. A review of the whitefly genus *Dialeurodes* Cockerell (Aleyrodidae: Hemiptera) with descriptions of two new species from India. *Journal of Bombay Natural History Society* 103(1): 62–67.

Sundararaj, R., and A. K. Dubey. 2006b. On the genus *Rhachisphora* Quaintance & Baker (Hemiptera: Aleyrodidae) with descriptions of a species. *Journal of Bombay Natural History Society* 103(1): 68–70.

Sundararaj, R., and A. K. Dubey. 2007. Identification of Indian species of whitefly genus *Dialeuropora* Quaintance & Baker (Hemiptera: Aleyrodidae) and their host plants. *Indian Journal of Forestry* 30(2): 185–188.

Sundararaj, R., and R. Pushpa. 2011a. Aleyrodids (Aleyrodidae: Hemiptera) of India with description of some new species and new host records. 407–534. In: *Advancements in Invertebrate Taxonomy and Biodiversity*, ed. R. K. Gupta. AgroBios (International), Jodhpur, Rajasthan, India, viii + 534 pp.

Sundararaj, R., and R. Pushpa, 2011b. Whiteflies (Hemiptera-Aleyrodidae) breeding on Teak (*Tectona grandis* L. f) in India with description of a new species. *Journal of Biodiversity and Ecological Sciences* 1(2): 143–150.

Sundararaj, R., and R. Pushpa. 2014. Two new species of the whitefly genus *Bemisia* Quaintance and Baker (Hemiptera: Aleyrodidae) from India. *Biosystematica* 8(1&2): 19–23.

Sundararaj, R., and K. Selvaraj. 2017. Invasion of rugose spiraling whitefly, *Aleurodicus rugioperculatus* Martin (Hemiptera: Aleyrodidae): A potential threat to coconut in India. *Phytoparasitica* 45: 71–74.

Sundararaj, R., and D. Vimala. 2018. New record of the legume feeding whitefly *Tetraleyrodes acaicae* (Quaintance) (Hemiptera: Aleyrodidae) from India. *Indian Journal of Entomology* 80(3): 1168–1169.

Takahashi, R. 1950. Four new species of Aleyrodidae (Homoptera) from Australia, India and Borneo. *Annotationes Zoologicae Japonenses* 23: 85–88.

Tay, W. T., G. A. Evans, L. M. Boykin, and P. J. De Barro. 2012. Will the real *Bemisia tabaci* please stand up? *PLoS ONE* 7: e50550.

Vimala, D., and R. Sundararaj. 2015. Revision of the whitefly genus *Martiniella* Jesudasan and David (Hemiptera: Aleyrodidae) with a new record of *Martiniella sepangensis* (Martin and Mound) from India. *Entomon* 40(4): 221–234.

Vimala, D., and R. Sundararaj. 2018a. Key to the Indian species of the whitefly genus *Martiniella* Jesudasan and David (Hemiptera: Aleyrodidae) with description of a new species. *Entomon* 43(1): 49–52.

Vimala, D., and R. Sundararaj. 2018b. Two new species of the genus *Aleurolobus* Quaintance & Baker, 1914 (Hemiptera: Aleyrodidae) from India. *Journal of Insect Biodiversity* 007(2): 024–032. (IF 1.04)

Vimala, D., and R. Sundararaj. 2018c. Redescription and new record of *Bemisia pongamiae* Takahashi (Hemiptera: Aleyrodidae) from India. *Indian Journal of Entomology* 80(3): 1159–1160

Vimala, D., R. Sundararaj, and S. Prabakaran. 2017. New record of *Aleuroclava citrifolii* (Corbett) (Hemiptera: Aleyrodidae) from India. *Entomon* 42(3): 245–246.

Vyskocilova, S., W. T. Tay, S. V. Brunschot, and J. Colvin. 2018. An integrative approach to discovering cryptic species within the *Bemisia tabaci* whitefly species complex. *Scientific Reports* 8: 10886. doi: 10.1038/s41598-018-29305-w.

Westwood, J.O. 1840. *An introduction to the modern classification of insects*; founded on the natural habits and corresponding organization of different families. Longman, Orme, Brown and Green. London, 587 pp.

8 Pentatomidae (Hemiptera: Heteroptera: Pentatomoidea) of India

S. Salini

CONTENTS

Introduction 121
Diagnosis of the Family 122
Classification of Pentatomidae 122
    Key to Subfamilies of Pentatomidae 122
Subfamily Asopinae 123
    Key to Genera of Asopinae from India 124
Subfamily Pentatominae 125
    Key to Economically Important Genera of Pentatominae from India 128
Subfamily Phyllocephalinae 129
    Key to Commonly Occurring Genera of Phyllocephalinae from India 130
Subfamily Podopinae 130
    Key to Genera of Podopinae from India 130
Major Work on Indian Fauna 131
Biodiversity and Species Richness 132
Molecular Characterization and Phylogeny 132
Taxonomic Problems 134
Biology 135
    Predators 137
    Host Plants of Phytophagous Pentatomidae in India 137
Economic Importance 141
Collection and Preservation 141
    Procedure for Dissection of Male and Female Genitalia 142
Websites 142
Conclusion 142
References 142

INTRODUCTION

Pentatomidae, commonly known as stink bugs, are the most diverse family of pentatomomorphan bugs, found in all major zoogeographic regions of the world. They are very variable in colour, size, and shape of the body. Some of them are brilliantly coloured, whereas a few blend with their surroundings. A few of the smaller bugs, such as *Sepontia* Stål and *Spermatodes* Bergroth, measure only a couple of millimetres, while quite a few large ones like *Catacanthus* Spinola and *Placosternum* Amyot & Serville, range from 22 to 28 mm or more in size. They feed on all kinds of vegetation from herbs to trees. New flush and reproductive parts of plants like flower buds or seeds are the most preferred plant parts. They feed on plants by injecting watery saliva containing digestive enzymes and suck the liquefied contents with their maxillary stylets. Though they are considered minor, occasional pests, damage by them could be appreciable when they infest the economic parts of plants such as pods or grains.

Pentatomidae are confirmed vectors of plant pathogens like phytoplasmas. Adults and nymphs of the brown marmorated stink bug *Halyomorpha halys* (Stål) transmit Witches' broom phytoplasma to trees of *Paulownia* spp. in Asia (Weintraub and Beanland 2006); the spined fruit bug of citrus (*Rhynchocoris poseiden* Kirkaldy) transmits fungal spores of the yeast spot disease caused by *Nematospora coryli* (Kalshoven 1981) to soybean and the flagellate protozoan *Phytomonas* spp. is transmitted to palm trees (Panizzi 1997). A few species of Pentatomidae such as *Catacanthus* Spinola and *Udonga* Distant are known for their mass emergence behaviour which causes public nuisance. Apart from these, a few members of Pentatomidae like *Udonga montana* (Distant) are edible and serve as a dietary supplement in parts of India (Thakur and Firake 2012).

Among Pentatomidae, the subfamily Asopinae are exclusively predatory, at least in India. They feed on a variety of insects from among the Coleoptera and Lepidoptera, especially during their larval stages, as well as on other small and soft bodied arthropods (Lefroy and Howlett 1909; Fletcher 1914;

Kalshoven 1981). Many asopines are known for their brilliant and aposematic colouration like *Amyotea* Ellenreider, *Zicrona* Amyot, & Servile, etc. A few of them like *Cazira* Amyot & Servile are well equipped with varying armature giving them a bizarre appearance. Pentatominae is the most diverse and abundant group, which alone constitutes more than 80 percent of the Pentatomidae known from India. Phyllocephaline bugs are known to be associated with grass roots and also to feed on the roots of several grass species. Podopinae, commonly called black bugs, including those which feed on the roots of rice plants, are worldwide in distribution. Among Pentatomidae, Podopinae are the least represented in India.

Pentatomids lay eggs in clusters on plants or elsewhere in the open. Eggs (Figure 8.65) are barrel-shaped with an operculum. Soon after hatching, the nymphs cluster over the egg shell for sometime and later disperse and begin to suck sap from tender parts of host plants. Nymphs (Figure 8.67) usually undergo five moults to become adults. They resemble adults except in the absence of wings and reproductive structures. Adult pentatomids emit a characteristic stinking odour whenever disturbed by producing an irksome oily fluid from the external scent efferent system (Figures 8.58–8.63) located on the metapleura. In nymphs, the scent efferent system, also referred to as dorsal abdominal glands (DAG), is usually located between abdominal tergites III–IV (DAG 1), IV–V (DAG 2), and V–VI (DAG 3). The secretion of the external scent efferent system in Pentatomidae is assumed to have some defence function which helps ward off natural enemies. Adults exhibit thanatosis and quite a few species are attracted to light.

The first and the earliest comprehensive work on this group was by Atkinson (1884, 1887, 1888) and later by Distant (1902, 1908, 1918). Atkinson (1884) classified the present Pentatomidae into various families such as Asopidae, Halydidae, Pentatomidae, Podopidae, Phyllocephalidae, and Sciocoridae. He placed a few genera presently recognized as Acanthosomatidae under Pentatomidae. Atkinson (1887, 1888) further subdivided the Pentatomidae into subfamilies such as Plataspina, Cydnina, Scutellerina, Pentatomina, and Asopina. Distant (1902, 1908, 1918) in the Fauna of British India Series considered Pentatomidae as a subfamily and arranged them in 19 separate divisions.

Studies pertaining to the fauna of Pentatomidae from southern India were undertaken by various workers (Chatterjee 1934; Usman and Puttarudriah 1955; Hegde 1986a, b, 1995; Mathew 1986). Salini (2006) described and illustrated 66 species belonging to 43 genera from Karnataka. Later, Salini and Viraktamath (2015) provided an illustrated key to 86 genera and a checklist of 164 species of Pentatomidae from South India. Apart from these, new species descriptions, redescriptions, or new records of various taxa of Pentatomidae were recently undertaken by Salini (2016a, 2016b, 2017a, 2017b) and Salini and Schmidt (2018).

DIAGNOSIS OF THE FAMILY

Pentatomidae are characterized by a well-developed scutellum that is triangular or subtriangular to semi-elliptical in shape, sometimes even U-shaped (as in the case of *Alcimocoris* Bergroth, *Brachycoris* Stål, *Eurysaspis* Signoret, *Paracritheus* Bergroth, *Saceseurus* Breddin, and in many podopines) or completely covering the abdomen (as in *Spermatodes* Bergroth, and some podopines); claval commissure reduced or absent; paired abdominal trichobothria usually positioned laterally and posteriad of spiracles; tarsi three segmented (except in *Rolstoniellus* Rider, where it is two segmented); and frena present, usually extending beyond the middle of scutellar margins, spiracles of second abdominal segment or first visible abdominal segment concealed by metapleura except in a few genera of large body size, where it is at times partially exposed (Schuh and Slater 1995). The female spermathecal duct has a dilation with a long, slender, sclerotized rod evaginated from the distal orifice; the spermathecal pump is well-developed with both a proximal and a distal flange; and the spermathecal bulb may be simple, digitoid, or ball-shaped, often with one to three tubular diverticula (Rider et al. 2018).

CLASSIFICATION OF PENTATOMIDAE

Classification of the Pentatomidae remains conjectural (Schuh & Slater 1995) and requires urgent attention. Schuh and Slater (1995) recognized eight subfamilies in Pentatomidae with the subsequent addition of two more subfamilies (Rider 2000; Grazia et al. 2008). Later, Grazia et al. (2008) transferred Serbaninae, earlier placed in Pentatomidae to Phloeidae. Rider et al. (2018) recognized ten subfamilies such as Aphylinae, Asopinae, Cyrtocorinae, Discocephalinae, Edessinae, Pentatominae, Phyllocephalinae, Podopinae, Serbaninae, and Stirotarsinae. Of these only four subfamilies, Asopinae, Pentatominae, Phyllocephalinae, and Podopinae are known to occur in India (Salini and Viraktamath 2015). Pentatominae is the largest subfamily, comprising 43 tribes including the recently erected tribe Pentamyrmecini. A few genera are still not formally classified into tribes (unplaced) (Rider et al. 2018). Only 29 tribes are known to occur in South India (Salini and Viraktamath 2015). There is no generally accepted tribal level arrangement in Asopinae. Cressonini, Megarrhamphini, Phyllocephalini, and Tetrodini are the tribes recognized in Phyllocephalinae, whereas in Podopinae, five groups are recognized such as *Brachycerocoris* group, *Deroploa* group, *Graphosoma* group, *Podops* group, and *Tarisa* group (Rider et al. 2018).

KEY TO SUBFAMILIES OF PENTATOMIDAE (FOLLOWING SALINI AND VIRAKTAMATH, 2015)

1. Labium short, not extending beyond posterior margin of fore coxae (Figure 17 in Salini & Viraktamath 2015)..**Phyllocephalinae**
1'. Labium variable in length, but considerably extending beyond fore coxae...**2**
2. First segment of labium incrassate not concealed between bucculae (Figures 18 & 19 in Salini & Viraktamath 2015) ....**Asopinae**

2'. First segment of labium not considerably incrassate; usually concealed between bucculae..................... **3**
3. Anterolateral angles of pronotum with well-developed tooth or scutellum with tubercle (Figs 20, 237, 238, 239 in Salini & Viraktamath 2015); scutellum U-shaped, usually extending nearly to tip of abdomen or sometimes posterior part of membrane of hemelytra exposed .... **Podopinae**
3'. Anterolateral angles of pronotum with minute tooth and scutellum without tubercle; scutellum triangular or subtriangular, rarely U-shaped as in *Alcimocoris*, *Saceseurus*, *Eurysaspis*, and *Paracritheus* that lack tooth on anterolateral angles of pronotum .... **Pentatominae**

SUBFAMILY ASOPINAE (FIGURES 8.1–8.9)

Asopinae are characterized by having a crassate labium (Figure 8.64), the first segment of labium being markedly thickened and free, facilitating the full forward extension of the labium to catch active prey. Unlike in other Pentatomidae, the labium is usually not adpressed to the thoracic region.

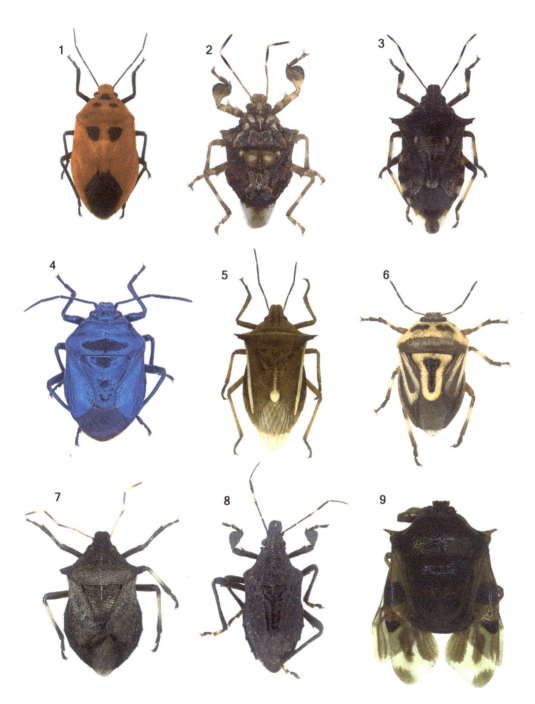

FIGURES 8.1–8.9 Asopinae (dorsal habitus) **1**. *Amyotea malabarica*; **2**. *Cazira verrucosa*; **3**. *Eocanthecona furcellata*; **4**. *Zicrona caerulea*; **5**. *Andrallus spinidens*; **6**. *Perillus bioculatus*; **7**. *Picromerus* sp.; **8**. *Cecyrina platyrhinoides*; and **9**. *Blanchia ducalis*.

They are distributed all over the world. This subfamily is represented by nearly 30 species in 17 genera from India. They are predatory in habit and feed on several soft bodied arthropods, mainly larvae of Lepidoptera, Diptera, and Coleoptera (Lefroy and Howlett 1909; Fletcher 1914; Kalshoven 1981; Schuh and Slater 1995; De Clercq 2000). A few of them are brilliantly coloured like *Amyotea* (Figure 8.1), *Zicrona* (Figure 8.4), and species of *Cazira* (Figure 8.2) which are bizarre in appearance. Thomas (1994) revised Asopinae of the Old world and reviewed 187 species. He proposed several synonymies, new combinations, and other nomenclatural changes. Gapon and Konstantinov (2006) studied the structure of the aedeagus in Asopinae. Asopinae of India is poorly studied. The following key is useful in identifying members of this group to the generic level.

Key to Genera of Asopinae from India

1. Forefemora with preapical spine (Figure 8.57) 2
1'. Forefemora without preapical spine 12
2. Jugae much longer than tylus and usually meeting in front of tylus, head as long as or longer than pronotum; weevil-like insects (Figure 8.8) .... ***Cecyrina* Walker**
2'. Jugae as long as or subequal or slightly longer than tylus, sometimes convergent, meeting or not meeting in front of tylus; variably looking insects) 3
3. Apex of basal abdominal tubercle bifid .... ***Glypsus* Dallas**
3'. Apex of basal abdominal tubercle, if present, not bifid 4
4. Base of scutellum with large tubercular gibbosities (Figure 8.2); fore femora with two preapical spines .... ***Cazira* Amyot & Serville**
4'. Base of scutellum without gibbosities; fore femora with one preapical spine 5
5. Humeri prominent, either produced elongate into spinous projection or angular 6
5'. Humeri not produced, rather rounded (Figure 8.6) .... ***Perillus* Stål**
6. Scutellar apex wider than corium (Figure 8.9) .... ***Blachia* Walker**
6'. Scutellar apex narrower than corium 7
7. Basal abdominal sternite without a distinct spine or tubercle or sometimes with weak tubercle or prominence, not protruding between metacoxae 8
7'. Basal abdominal sternite with a distinct spine or tubercle apposing metasternum and protruding between metacoxae 10
8. Humeri strongly spinose, usually bispinose at apex; peritreme flat and spatulate extending half way to metapleural margin 9
8'. Humeri prominent, angular, but not spinose; peritreme narrow, strongly curved, elongate, reaching more than halfway to metapleural margin .... ***Pseudanasida* Schouteden**
9. Posterior angles of pronotum with hook-like structure; membrane usually not extending much beyond apex of abdomen; pronotum, scutellum, and corium concolourous (Figure 8.7) .... ***Picromerus* Amyot & Serville**
9'. Posterior angles of pronotum without hook-like structure, rather angulate; membrane usually extending well beyond apex of abdomen; pronotum, scutellum, and corium dark brown except a narrow, median transverse line on prontoum connecting humeri, lateral margins of corium (embolium), and apex of scutellum ochraceous (Figure 8.5) .... ***Andrallus* Bergroth**
10. Metasternum with lateral margins strongly elevated or bicarinate and embracing apex of rostrum in repose; protibiae prismatic .... ***Cantheconidea* Schouteden**
10'. Metasternum sub-elevated, margins not or feebly carinate; protibiae dilated foliate, dilate or not 11
11. Protibiae dilated foliate; humeral angles produced into elongate stout process apex of which usually sinuate .... ***Platynopus* Amyot & Serville**
11'. Protibiae dilate or not; humeral angles produced into spinous process, apex of which sometimes bidentate .... ***Eocanthecona* Bergroth**
12. Reduction of Peritremal disc into peritremal surface only, merges with pleural surface and become indistinguishable; anterior and posterior margins of orificial peritreme diverging and evanescent; metallic blue coloured bugs (Figure 8.4) .... ***Zicrona* Amyot & Serville**
12'. Peritreme well developed and forms peritremal disc which reaches or extends beyond middle of metapleuron, sometime reduced into long, curved peritremal ruga extends much beyond middle of metapleuron; variously coloured bugs 13
13. Humeri not produced, rather rounded; anterolateral margins of pronotum smooth without any crenulations or serrations; reddish coloured bugs (Figure 8.1) .... ***Amyotea* Ellenrieder**
13'. Humeri broadly or narrowly angulate, sometimes produced into moderately elongate process with truncate apex; anterolateral margins of pronotum with serrations or crenulations; variously coloured insects 14
14. Peritreme long, curved, and ruga-like, extending about two-thirds distance to metapleural margin; evaporatorium obsolescent; labium short reaching only to mesocoxae .... ***Anasida* Karsch**
14'. Peritreme modified into peritremal disc, reaching or slightly extending beyond middle of metapleron; evaporatorium well developed; labium reaching or passing metacoxae 15
15. Humeri produced into moderately elongate, stout process, truncate apically .... ***Martinina* Schouteden**
15'. Humeri produced into either broad, laminate, and angulate process or into stout, acute spines 16

16. Humeri produced into broad, laminate, and angulate process; base of abdomen usually with short cylindrical spine... ***Troilus* Stål**
16'. Humeri produced into stout, acute process; basal abdominal sternite usually without spine or tubercle........***Arma* Hahn**

SUBFAMILY PENTATOMINAE (FIGURES 8.10–8.36)

It is one of the most diverse subfamilies of heteropteran bugs, and the tribal classification is in a state of considerable confusion (Cassis and Gross 2002). Pentatominae, referred to as a "catch all" taxon by Rider et al. (2018), includes several genera and groups of genera not recognized in any of the other subfamilies. There is no unique diagnostic feature to distinctly identify this subfamily, and this hampers the proposal of a stable classification for this group. Pentatominae is the most diverse subfamily of Pentatomidae, containing 660 genera and 3484 species (Rider et al. 2018). Nearly 300 species belonging to 120 genera of 30 tribes (a few genera are still not formally included in any tribe) of Pentatominae are reported from India. Members of this subfamily are quite variable in size and in colour. Most species have five segmented antennae, while in some they are four segmented (e.g., *Degonetus* Distant, *Phricodus* Spinola). Tarsi usually

FIGURES 8.10–8.18 Pentatominae (dorsal habitus) **10**. *Agonoscelis nubilis*; **11**. *Antestiopsis cruciata*; **12**. *Bagrada hilaris*; **13**. *Brachycoris tralucidus*; **14**. *Cappaea taprobanensis*; **15**. *Carbula biguttata*; **16**. *Catacanthus incarnates*; **17**. *Cresphontes monsoni*; and **18**. *Degonetus serratus*.

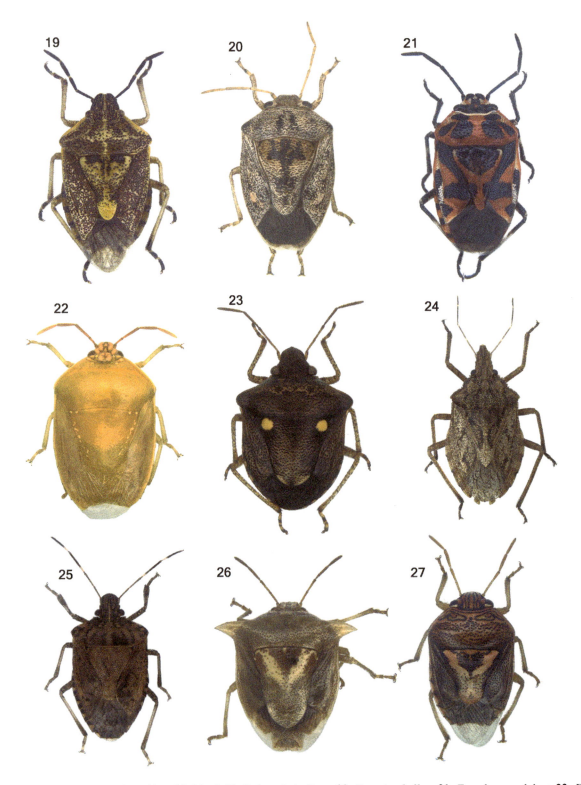

FIGURES 8.19–8.27 Pentatominae (dorsal habitus) **19**. *Dolycoris indicus*; **20**. *Dunnius bellus*; **21**. *Eurydema pulchra*; **22**. *Eurysaspis flavescens*; **23**. *Eysarcoris montivagus*; **24**. *Halys serrigera*; **25**. *Halyomorpha picus*; **26**. *Hoplistodera recurva*; and **27**. *Menida versicolor*.

three segmented except *Rolstoniellus* Rider. Humeral angles are highly variable from rounded to spine-like or produced into broad lobe-like processes. Scutellum triangular or subtriangular, sometimes U-shaped, usually not reaching the apex of abdomen. Frena present and usually extend at least two fifths the length of the scutellum. Other than a few regional studies, particularly for S. India (Salini 2015; Salini and Viraktamath 2015), a holistic study of the subfamily is lacking from India. Apart from these, new species descriptions, redescriptions, and new records of various taxa were undertaken recently by Salini (2016a, 2016b, 2017a, 2017b) and Salini and Schmidt (2018).

FIGURES 8.28–8.36 Pentatominae (dorsal habitus) **28**. *Neohalys serricollis*; **29**. *Nezara viridula*; **30**. *Phricodus hystrix*; **31**. *Piezodorus hybneri*; **32**. *Placosternum* sp.; **33**. *Plautia crossota*; **34**. *Rhynchocoris humeralis*; **35**. *Sciocoris indicus*; and **36**. *Udonga montana*.

The classification within this subfamily is in a chaotic state. The number of tribes or generic groups recognized in this subfamily varies between workers (Table 8.1).

All Pentatominae known from India are phytophagous. Some species like *Nezara viridula* (Linnaeus), *Bagrada hilaris* (Burmeister), *Antestiopsis cruciata* (Fabricius), *Plautia crossota* (Dallas), *Halyomorpha picus* (Fabricius), *Dolycoris indicus* Stål, *Menida versicolor* (Gmelin), *Piezodorus hybneri* (Gmelin), and *Rhynchocoris plagiatus* (Walker) are at times known to attain pest status on various cultivated plants. A key to Indian genera of economic importance is furnished.

TABLE 8.1
Major Suprageneric Classification of Pentatominae

Kirkaldy (1909)	Gross (1975) Genus Group	Hasan & Kitching (1993)	Schuh & Slater (1995)	ªCassis & Gross (2002)	Rider et al. (2018)
Acanthosomini	*Antestia*	Aeliini	Aeptini	Aeptini	Aeliini
Aeptini	*Asopus*	Agonoscelini	Diemeniini	Agonoscelidini	Aeptini
Diemeniini	*Carpocoris*	Antestiini	Halyini	Antestiini	Aeschrocorini
Discocephalini	*Cephaloplatus*	Asopini	Lestonocorini	Cappaeini	Agaeini
Edessini	*Dictyotus*	Catacanthini	Mecideini	Carpocorini	Agonoscelidini
Graphosomini	*Diemenia*	Carpocorini	Myrocheini	Catacanthini	Amyntorini
Halyini	*Eysarcoris*	Caystrini	Pentatomini	Caystrini	Antestiini
Myrocheini	*Halyini*	Eysarcorini	Sciocorini	Diemeniini	Aulacentrini
Pentatomini	*Ippatha*	Halyini		Eysarcorini	Axiagastini
Sciocorini	*Kapunda*	Lestonocorini		Halyini	Bathycoeliini
	Kitsonia	Mecideini		Menidini	Cappaeini
	Kumbutha	Megarrhamphini		Myrocheini	Carpocorini
	Macrocarenus	Menidini		Nezarini	Catacanthini
	Mecidea	Myrocheini		Pentatomini	Caystrini
	Menida	Pentatomini		Piezodorini	Coquereliini
	Menestheus	Phyllocephalini		Rhynchocorini	Degonetini
	Mycoolona	Rhynchocorini		Sciocorini	Diemeniini
	Ochisme	Sciocorini		Strachiini	Diplostirini
	Pentatoma	Strachini			Diploxyini
	Phyllocephala	Tetrodini			Eurysaspidini
	Piezodorus	Tropicorini			Eysarcorini
	Poecilotoma				Halyini
	Podops				Hoplistoderini
	Rhyncocoris				Lestonocorini
	Strachia				Mecideini
	Tarisa				Memmiini
	Tholosanus				Menidini
					Myrocheini
					Nealeriini
					Nezarini
					Opsitomini
					Pentamyrmecini
					Pentatomini
					Phricodini
					Piezodorini
					Procleticini
					Rhynchocorini
					Rolstoniellini
					Sciocorini
					Sephalini
					Strachiini
					Triplatyxini

Source: Modified from Cassis, G., and Gross, G. F., Hemiptera: Heteroptera (Pentatomomorpha), In: Houston, W. W. K. and Wells, A. (Eds.), *Zoological Catalogue of Australia*. 27(3B), CSIRO Publishing, Melbourne, Australia. xiv + 737, 2002.

ª Cassis & Gross listed only those tribes that are present in Australia.

KEY TO ECONOMICALLY IMPORTANT GENERA OF PENTATOMINAE FROM INDIA

1. Antennae with four segments 2
1'. Antennae with five segments 3
2. Lateral margins of pronotum and abdomen laminate (Figure 8.30); posterolateral angles of abdominal sternites laminate; basal abdominal sternite without tubercle .... ***Phricodus* Spinola**
2'. Lateral margins of pronotum and abdomen not laminate (Figure 8.18); posterolateral angles of abdominal sternites not laminate; basal abdominal sternite with tubercle .... ***Degonetus* Distant**
3. Basal abdominal sternite with spine or tubercle 4

3'. Basal abdominal sternite without spine or tubercle 11
4. Humeral processes well developed into elongate spine (Figures 8.34, 8.54) or broad apically with single spine-like process or into broad, apically truncate processes (Figure 8.32) 5
4'. Humeral processes not prominently produced rather rounded, triangular, or subtriangular 6
5. Humeral processes produced into elongate spine (Figure 8.54) or broad, but with spine-like process apically (Figure 8.34); posterior apex of metasternal carina with V-notch holding apex of basal abdominal spine (Figure 8.54) .... ***Rhynchocoris* Westwood**
5'. Humeral processes produced into broad, apically truncate processes (Figure 8.32); posterior apex of metasternal carina without V-notch rather contiguous with basal abdominal tubercle (Figure 8.55) .... ***Placosternum* Amyot & Serville**
6. Mesosternum with median, longitudinal carina well developed anteriorly into angular process, terminating between forecoxae .... ***Piezodorus* Fieber**
6'. Mesosternum with or without median, longitudinal carina, if present not produced into angular process 7
7. Peritreme short, not extending beyond middle of metapleuron (Figure 8.62) 8
7'. Peritreme elongate, distinctly extending beyond middle of metapleuron (Figure 8.61) 9
8. Anterolateral margins of pronotum crenulated, sometimes with forwardly directed short spine at humeral angle; lateral margin of head not smooth, rather with one short tooth .... ***Udonga* Distant**
8'. Anterolateral margins of pronotum smooth, humeri without spine; lateral margin of head smooth .... ***Nezara* Amyot & Serville**
9. Basal abdominal sternite with short tubercle not reaching anterior margin of forecoxae .... ***Plautia* Stål**
9'. Basal abdominal sternite with elongate spine, reaching or extending beyond anterior margin of forecoxae 10
10. Foretibiae dilated; clypeus slightly longer than mandibular plates (Figure 8.48) .... ***Catacanthus* Spinola**
10'. Foretibiae not dilated; clypeus as long as mandibular plates .... ***Menida* Motschulsky**
11. Body with conspicuous, long, erect pilosity 12
11'. Body glabrous 13
12. Clypeus shorter than mandibular plates; apex of mandibular plates slightly reflexed upwards (Figures 8.19, 8.46) .... ***Dolycoris* Mulsant & Rey**
12'. Clypeus as long as mandibular plates; apex of mandibular plates not reflexed upwards (Figure 8.10) .... ***Agonoscelis* Spinola**
13. Abdominal sternites with median longitudinal furrow (Figure 8.56) 14
13'. Abdominal sternites devoid of median longitudinal furrow 15
14. Lateral margins of mandibular plates with two pairs of teeth; sometimes the one towards the apex more prominent compared to the one just in front of eyes .... ***Halys* Fabricius**
14'. Lateral margins of mandibular plates smooth, without tooth; sometimes with indistinct tooth just in front of compound eyes .... ***Neohalys* Ahamd & Perveen**
15. External scent efferent system well developed, peritreme elongate or spout shaped... 17
15'. External scent efferent system rudimentary, sometimes ostiole visible 16
16. Head coarsely punctuate dorsally; anterior collar of pronotum well developed (Figure 8.21) .... ***Eurydema* Laporte**
16'. Head impunctate dorsally; anterior collar of pronotum not well developed (Figure 8.12) .... ***Bagrada* Stål**
17. Peritreme elongate, distinctly extending beyond middle of metapleuron (Figures 8.58, 8.59) 18
17'. Peritreme short, not extending beyond middle of metapleuron (Figure 8.60) 20
18. Humeri with elongate, stout spine-like processes (Figure 8.26) .... ***Hoplistodera* Westwood**
18'. Humeri without elongate processes, rather rounded (Figures 8.11, 8.25) 19
19. Anterior margin of pronotum with well-developed collar-like process; yellowish green medium-sized bugs with black spots dorsally (Figure 8.11) .... ***Antestiopsis* Leston**
19'. Anterior margin of pronotum flat without collar-like process; brownish yellow large-sized bugs with mosaics of black spots or streaks dorsally formed by coarse punctae (Figure 8.25) .... ***Halyomorpha* Mayr**
20. Apex of head dome-shaped (Figure 8.47); lateral margins of pronotum and abdomen laminate (Figure 8.35) .... ***Sciocoris* Fallén**
20'. Apex of head triangular or subtriangular (Figure 8.49); lateral margins of pronotum and abdomen not laminate (Figures 8.15, 8.23) 21
21. Head or propleura or both covered with short hairs associated with punctation (Figure 8.60) .... ***Eysarcoris* Hahn**
21'. Punctation on head or propleura without short hairs .... ***Carbula* Stål**

SUBFAMILY PHYLLOCEPHALINAE (FIGURES 8.37–8.42)

Members of this subfamily are large and flattened bugs where the labium does not extend beyond the posterior margin of the forecoxae. Presently, it contains 247 species belonging to 45 genera in the world (Rider et al. 2018). Twenty seven species belonging to 13 genera are known from India. Colouration varies from yellowish brown to dark brown or reddish enabling them to blend with their surroundings. The juga are usually longer than the tylus, sometimes meeting in front of the tylus, and may be expanded or foliate. The single most diagnostic character is the distinctively short labium, which does not or only barely surpasses the procoxae (Rider et al. 2018). First segment of labium as well as half of the second segment lie between the bucculae. Several studies have been undertaken on the phyllocephaline fauna of India and Pakistan with the discovery of new species and the erection of new tribes (Ahmad and Kamaluddin 1978, 1988, 1990, 1992; Ahmad et al. 1994; Kamaluddin 1982; Kamaluddin and Ahmad 1988, 1995). Besides this, Konstantinov and Gapon (2005) studied the structure of the aedeagus in Phyllocephalinae. There are four recognized tribes—Cressonini, Phyllocephalini, Megarrhamphini, and Tetrodini—in Phyllocephalinae (Rider et al. 2018).

KEY TO COMMONLY OCCURRING GENERA OF PHYLLOCEPHALINAE FROM INDIA

1. First segment of antennae extending beyond apex of head (Figure 8.37). ....***Cressona* Dallas**
1'. First segment of antennae not extending beyond apex of head..2
2. Anterolateral angles of pronotum produced forward (Figures 8.50, 8.51)...3
2'. Anterolateral angles of pronotum not produced forward 4
3. Second segment of antennae much shorter than segment III and not extending beyond apex of head; tooth on lateral margins of head in front of compound eyes, well developed with acute apex (Figure 8.50) .... ***Gellia* Stål**
3'. Second segment of antennae subequal to segment III and extends beyond apex of head; tooth on lateral margins of head in front of compound eyes, much shorter and blunt (Figure 8.51) ....***Tetroda* Amyot & Serville**
4. Humeri rounded and not produced angularly (Figure 8.40) ....***Megarrhamphus* Bergroth**
4'. Humeri not rounded, rather produced angularly (Figures 8.39, 8.41)..5
5. Humeral angles short, with sharply angulate apex (Figure 8.39); lateral margins of head straight not sinuate in front of compound eyes ....***Gonopsis* Amyot & Serville**
5'. Humeral angles much elongate, with rounded finger-like apex (Figure 8.41); lateral margins of head sinuate in front of compound eyes ....***Salvianus* Distant**

SUBFAMILY PODOPINAE (FIGURES 8.43–8.45)

Podopinae are cosmopolitan in distribution (269 species belonging to 68 genera are known worldwide). They are a heterogenous group, primarily because of the difficulty in classifying some of the taxa (Rider et al. 2018). India is represented by 30 species of Podopinae belonging to 10 genera. Podopinae possess the following characters. Dark yellowish brown to dark brown or sometimes completely black bugs with well-developed antenniferous tubercles, usually visible from above. Anterolateral angles usually with well-developed tooth (Figures 8.44, 8.45); trichobothria paired; scutellum enlarged, always attaining membrane of forewing and often reaching apex of abdomen, covering most of the forewings; and frena well developed, but extending less than one third length of scutellum.

A large portion of the literature and references pertaining to this subfamily are found under the name Graphosomatinae, a polyphyletic group with some of its members belonging to the Podopinae and most others belonging to Pentatominae (Schuh and Slater 1995). Gapon (2008) accepted Podopinae as a holophyletic taxon with two sister groups, Graphosomatini *s. l.* and Podopini, with the tribe Podopini being divided into three subtribes, namely, the Kayesiina, the Podopina, and the Scotinopharina. Recently, Rider et al. (2018) detailed the morphological characters to recognize members of Podopinae along with brief descriptions, a key, and phylogenetic notes on various included genus—groups. A prime autapomorphy in this subfamily is the frena, one on each side on the ventral surface of the scutellum, as an oblique ledge-shaped structure reaching half its length.

Podopinae are phytophagous, feeding on vegetative as well as reproductive parts of plants belonging to different families. As far as the podopine fauna of India is concerned, *Scotinophara* Stål, commonly known as rice black bugs, is pestiferous being found infesting the roots of rice. Joshi et al. (2007) worked extensively on the taxonomy, ecology, and management of invasive species of rice black bugs.

KEY TO GENERA OF PODOPINAE FROM INDIA

1. Disc of pronotum with tubercles 2
1'. Disc of pronotum without tubercles 4
2. Body convex, more or less gibbous, lateral margins of pronotum without tubercles (Figure 8.43)........................ ....***Brachycerocoris* Costa**
2'. Body neither gibbous, nor prominently convex; lateral margins of pronotum with tubercles..............................3
3. Scutellum conspicuously wide basally, completely covering clavus and membrane of hemelytra.......................... ....***Tarisa* Amyot & Serville**
3'. Scutellum broad, reaching apex of abdomen, but not completely covering clavus and membrane, corium and membrane partially exposed................... ***Burrus* Distant**
4. Lateral lobes of head not meeting in front of tylus 5
4'. Lateral lobes of head meeting in front of tylus 9
5. Lateral lobes of head dilated and divergent (Figure 8.45) .... ***Storthecoris* Horváth**
5'. Lateral lobes of head not dilated (Figure 8.44) 6
6. Pronotum with processes on anterolateral angles flattened and spatulate ....***Podops* Laporte**
6'. Pronotum with processes on anterolateral angles not flattened, and rather spinous 7
7. Lateral margins of pronotum uniformly rounded and denticulate ....***Amauropepla* Stål**

FIGURES 8.37–8.45 Phyllocephalinae (dorsal habitus) **37**. *Cressona valida*; **38**. *Gellia nigripennis*; **39**. *Gonopsis pallescens*; **40**. *Megarrhamphus* sp.; **41**. *Salvianus* sp.; and **42**. *Tetroda histeroides*. Podopinae (habitus) **43**. *Brachycerocoris camelus* (lateral); **44**. *Scotinophara* sp. (dorsal); and **45**. *Storthecoris nigriceps* (dorsal).

7'. Lateral margins of pronotum more or less straight or slightly concave or sometimes only posterolateral margin slightly convex **8**
8. Paramere with large pyramidal lateral parandria......... .... ***Stysiellus* Gapon**
8'. Paramere without lateral parandria .... ***Scotinophara* Stål**
9. Lateral margins of pronotum uniformly convex with several strong, coarse tooth-like structures.................................. .... ***Aspidestrophus* Stål**
9' Lateral margins of pronotum anteriorly convex with sinuation adjacent to humeri; convex anterolateral margins of pronotum with four small teeth...........***Melanophara* Stål**

MAJOR WORK ON INDIAN FAUNA

The earliest works on Indian Pentatomidae are by Atkinson (1887, 1888), published in the *Journal of Asiatic Society of Bengal* followed by Distant (1902, 1908, 1918) in the *Fauna of British*

India series. One of the earliest works on the Pentatomidae of South India was by Chatterjee (1934), wherein he dealt with 77 species belonging to 48 genera of Pentatomidae and described *Platynopus indicus* Chatterjee. Later, Usman and Puttarudriah (1955) reported a total of 46 genera and 72 species from the erstwhile Mysore state (now Karnataka) along with distribution and host plants. Datta et al. (1985) dealt with ten families of Indian Pentatomoidea and provided information on genitalia of 61 species belonging to 44 genera. Memon (2002) revised Halyini from the Indian subcontinent with a key to 106 species. Salini (2006, 2015) studied and illustrated the pentatomid and pentatomine fauna of Karnataka and South India. Salini and Viraktamath (2015) gave an illustrated key to 86 genera and a checklist of 164 species of Pentatomidae from South India.

Apart from this, several regional studies based on one or several taxa of Pentatomidae in India were undertaken by various workers (Distant 1887; Ahmad and Kamaluddin 1978, 1988, 1990, 1992; Kamaluddin 1982; Kamaluddin and Ahamd 1988, 1994; Ahmad et al. 1993; Mathew 1969, 1977, 1980, 1986; Chopra 1972, 1974; Pawar 1973; Azim and Shafee 1979, 1982, 1983, 1984a, 1984b, 1984c, 1985, 1986a, 1986b, 1987a, 1987b; Vidyasagar and Bhat 1986; Ghauri 1963, 1975a, 1975b, 1977a, 1977b, 1978, 1980, 1982a, 1982b, 1988; Ahmad and Rana 1988; Rana and Ahmad 1988; Hegde 1986a, 1986b, 1995; Ahmad and Afzal 1984a, 1984b; Chakrabarty et al. 1994; Azim 2002; Azim and Bhat 2010; Azim 2011; Salini 2006, 2011, 2015, 2016a, 2016b, 2017a, 2017b; Salini and Belavadi 2015; Salini and Schmidt 2018; Srikumar et al. 2018; Kaur et al. 2012, 2013; Parveen and Gaur 2012; Ghate et al. 2012; Ghate 2013, 2015; Waghmare et al. 2015; Waghmare and Gaikwad 2017).

BIODIVERSITY AND SPECIES RICHNESS (TABLE 8.2)

Pentatomidae are cosmopolitan in distribution, found in all major zoogeographical regions and are particularly diverse in tropical and subtropical regions of the Southern Hemisphere. The pentatomid fauna of the Neotropical region has been studied by Grazia et al. (2015). The subfamilies Asopinae, Cyrtocorinae, Discocephalinae, Edessinae, and Pentatominae are widely represented in the Neotropical region, while Podopinae and Stirotarsinae are poorly represented. The knowledge of pentatomid fauna from India is fragmentary and incomplete, though it is one of the most diverse families of Heteroptera. Among the various subfamilies of Pentatomidae, the Pentatominae are the most abundant and richest, which includes 80 percent of pentatomid species reported from India. Asopinae is the next species rich subfamily after Pentatominae, and the subfamilies Phyllocephalinae and Podopinae are poorly represented. South India is rich in diversity of Pentatomidae (Salini and Viraktamath 2015) with 57 percent of the genera and 43 percent of the species of the total number of genera and species reported from India.

Asopinae, Pentatominae, Phyllocephalinae, and Podopinae are the only subfamilies known from India and are distributed throughout the country. Eighty percent (approximately 300 species) of the pentatomid fauna known from India belong to subfamily Pentatominae, followed by Asopinae (approximately 30 species). In India, the subfamilies Podopinae and Phyllocephalinae are almost equal in species numbers.

MOLECULAR CHARACTERIZATION AND PHYLOGENY

The phylogenetic relationships among pentatomid lineages remain to be studied rigorously (Rider et al. 2018). McDonald (1966), Gross (1975), and Linnavuori (1982) suggested good apomorphic characters for the recognition of several monophyletic groups within the family. The possible phylogenetic relationships among pentatomoid lineages were discussed by Gross (1975) and Linnavuori (1982). Gapud (1991) and Hasan and Kitching (1993) were the first to discuss the phylogenetic relationships within the Pentatomidae. Roell and Campos (2015) conducted a cladistic analysis to investigate the possible relationships among newly described species within Discocephalinae as well as to determine the legitimacy of erecting a new genus in Ochlerini.

Phylogenetic relationships within the Pentatomoidea were also investigated by Grazia et al. (2008) through the analysis of characters derived from both morphology and DNA sequences. They investigated a total of 135 terminal taxa representing most of the major family groups; 84 ingroup taxa were analysed for 57 characters in a morphological matrix. As many as 3500 bp of DNA data were adduced for each of 52 terminal taxa, including 44 ingroup taxa, comprising 18S rRNA, 16S rRNA, 28S rRNA, and COI gene regions. They declared that monophyly of the Pentatomidae (inclusive of the Aphylinae and Cyrtocorinae) appeared unequivocal.

Barcoding of pentatomorphan bugs from Western Ghats, India was carried out by Tembe et al. (2014). They expanded the database by adding mitochondrial Cytochrome c Oxidase I (mtCOI) sequences from 43 species of indigenous true bugs from India. Though they could not adduce the monophyly of family Pentatomidae through maximum likelihood and maximum parsimony trees, they came to similar conclusions as that of Grazia et al. (2008) for Dinidoridae, Scutelleridae, and Tessaratomidae.

TABLE 8.2
Number of Genera and Species in the Subfamilies of Pentatomidae

Subfamily	World[a]		India	
	Genera	Species	Genera	Species
Asopinae	63	303	17	30
Pentatominae	660	3484	120	300
Phyllocephalinae	45	214	13	27
Podopinae	68	269	10	30

[a] From Rider, D. A., C. F. Schwertner, J. Vilímová, et al. Higher systematics of the pentatomoidea, In: McPherson, J. E. (Ed.), *Invasive Stink Bugs and Related Species (Pentatomoidea) Biology, Higher Systematics, Semiochemistry and Management*, CRC Press/Taylor & Francis Group, Boca Raton, FL, pp. 25–201, 2018.

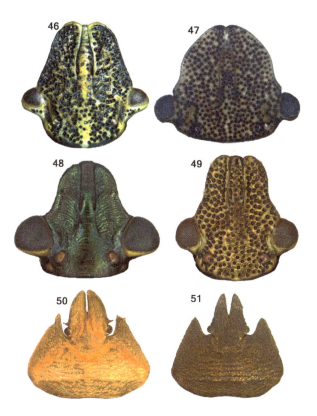

FIGURES 8.46–8.51 Pentatominae (head-dorsal) **46**. *Dolycoris indicus*; **47**. *Sciocoris indicus*; **48**. *Catacanthus incarnatus*; **49**. *Carbula scutellata*. Phyllocephalinae (head-dorsal) **50**. *Gellia nigripennis*; and **51**. *Tetroda histeroides* (Figures 46–51 from Salini and Viraktamath 2015 with permission from Zootaxa www.mapress.com/j/zt).

FIGURES 8.52–8.57 Pentatominae **52**. *Antestiopsis cruciata* (pronotum); **53**. *Halyomorpha picus* (pronotum); **54**. *Rhynchocoris plagiatus* [ventral side of head, thorax and abdomen (arrow showing U-shaped notch on posterior end of metasternal carina)]; **55**. *Placosternum* sp. (ventral side of head, thorax and part of abdomen); and **56**. *Halys serrigera* (abdominal venter showing median longitudinal groove). **57**. Asopinae, *Eocanthecona furcellata* (foreleg with preapical spine) (Figure 57 from Salini and Viraktamath 2015 with permission from Zootaxa www.mapress.com/j/zt).

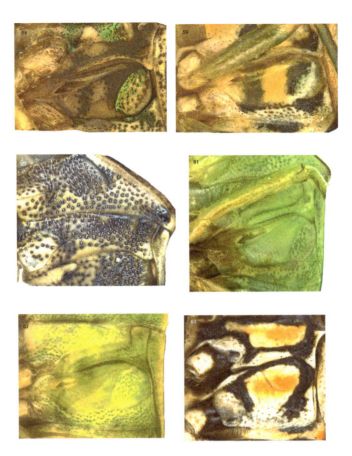

FIGURES 8.58–8.63 Pentatominae (external scent efferent system) **58**. *Halyomorpha picus*; **59**. *Antestiopsis cruciata*; **60**. *Eysarcoris ventralis*; **61**. *Plautia crossota*; **62**. *Nezara viridula*; and **63**. *Bagrada hilaris*.

Almost all subfamilies of Pentatomidae are well-defined by unique apomorphies, supporting the monophyly of these taxa (Rolston and McDonald 1979; Gapud 1991; Konstantinov and Gapon 2005; Campos and Grazia 2006; Gapon and Konstantinov 2006). The only exception being the subfamily Pentatominae, the "catch-all" taxon, that includes several genera and groups of genera not recognized in any of the other subfamilies (Cassis and Gross 2002). Phylogenetic relationships at the tribal level within Pentatominae was carried out by various workers (Schaefer and Ahmad 1987; Memon et al. 2011; Schwertner and Grazia 2012) and at the genus level (*Euschistus* Dallas) by Bianchi et al. (2017). Salini (2015) undertook a phylogenetic analysis of 96 species belonging to 63 genera under 29 tribes of Pentatominae from South India using 64 characters derived from both external morphology as well as male and female genitalia. She concluded that the present tribe - level classification of Pentatominae is unstable and paraphyletic.

The Pentatomidae is strongly supported as a monophyletic group on the basis of morphological and molecular evidence. The Pentatomidae *sensu lato*, with the inclusion of Aphylinae and Crytocorinae, were strongly supported by available morphological data. Many of the characters are difficult to interpret as certain characteristics have arisen multiple times and some have been lost secondarily. Spermathecal duct with large, membranous dilation around its middle and median penial plates (mesal, sclerotized portions of the second pair of conjunctival processes fused along the midline and closely associated with the distal portion of the vesica) are among a few characters of phylogenetic significance to support the monophyly of the family Pentatomidae. The structure of the thoracic sterna, presence or absence of structures at base of abdomen (rounded or produced), structure of ostiole, and its associated structures are useful for the classification within Pentatomidae (Rider et al. 2018).

TAXONOMIC PROBLEMS

Phenotypic plasticity is a major hurdle to studies on the taxonomy of Pentatomidae. This is especially true of those species inhabiting tree trunks such as *Agaeus* Dallas, *Ameridalpa* Ghauri, *Cappaea* Ellenrieder, *Dalpada* Amyot & Serville, *Eupaleopada* Ghauri, *Halys* Fabricius, *Meridalpa* Ghauri, *Meridindia* Ghauri, *Neohalys* Ahmad and Perveen, *Placosternum* Amyot & Serville, and *Tipulparra* Ghauri. The "Dalpada complex" which was extensively worked by several researchers (Ghauri 1975a, 1975b, 1977a, 1977b, 1978, 1980, 1982a, 1982b, 1988; Afzal & Ahmad 1981; Ahmad & Afzal 1986; Memon & Ahmad 2002; etc.) exhibits morphological plasticity such as variation in length of mandibular plates with respect to clypeus, tooth on lateral margins of head (present/absent/transition stages), length of antennal segments, length of labium, serration or tooth-like structures

FIGURES 8.64–8.69 64. *Amyotea malabarica* (arrow indicating incrassate labium). 65. *Halys serrigera* (eggs). 66. *Halyomorpha picus* (first instar nymphs clustering over egg shell). 67. *Degonetus serratus* nymph). 68, 69. Pentatomidae male genital capsule.

on anterolateral margins of pronotum, length of head with respect to first antennal segment, etc. *Placosternum* Amyot & Serville is one of the most problematic Oriental genera because of structural variation in both its external and internal morphology (characters of genitalia). This genus probably contains cryptic species where integrative taxonomy is essential to delineate species accurately.

Another important aspect is the significance of colour in pentatomid taxonomy. Several workers (Atkinson 1888; Distant 1902, 1908, 1918) relied heavily on colouration at the expense of structural characters. Use of colour as a diagnostic character to delineate taxa in Pentatomidae is not advisable as colour polymorphism is quite common in several species. Salini (2011) studied polymorphism in southern green stink bug, *Nezara viridula* Linnaeus and reported three colour forms. *Menida* Motschulsky, is another such example. Since colour is variable, taxonomic studies based on colour need validation through other characters.

BIOLOGY

Pentatomids lay eggs in clusters on plants or elsewhere in the open. Eggs are barrel-shaped with an operculum (Figure 8.65). Soon after hatching, the nymphs cluster over the egg shell for some time (Figure 8.66), to later disperse and begin sucking sap from tender parts of host plants. Nymphs (Figure 8.67) undergo five moults to become adults. They resemble adults except for the absence of fully developed wings and reproductive structures. Unlike many insects, a few species of Pentatomidae show parental care, guarding their young from predators. Whenever

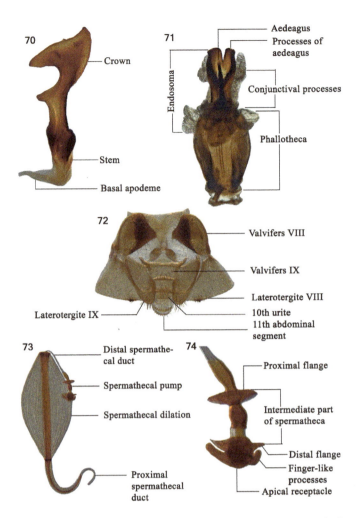

FIGURES 8.70–8.74 Male genitalia of Pentatomidae **70**. Paramere; **71**. Phallus. Female genitalia of Pentatomidae **72**. Terminalia; **73**. Spermatheca; and **74**. Spermathecal pump.

disturbed, adults secrete a stinking, irksome oily fluid from the external scent efferent system, which is located on the metapleura. In nymphs, the scent efferent system, also referred to as the DAG, is usually located between abdominal tergites III–IV (DAG 1), IV–V (DAG 2), and V–VI (DAG 3). The secretion of the external scent efferent system in Pentatomidae is reported to have some defensive function in warding off natural enemies.

Studies on the biology of Pentatomidae are inadequate. The biology of *Bagrada hilaris* (Burm) was studied by several workers (Rakshpal 1949; Azim and Shafee 1986b; Verma et al. 1993; Singh and Malik, 1993) and the internal anatomy by Rai and Trehan (1964). Sexual dimorphism is usually absent except in a few genera like *Aeliomorpha* Stål or *Mecidea* Dallas, where the antennae show sexual dimorphism and a slight variation in size between sexes in a few species. A few species exhibit thanatosis or feign death. Adults of quite a few species of Pentatomidae are attracted to light.

Pentatomidae are known for their cryptic and warning colouration. Asopinae, the predatory bugs, are usually brilliantly coloured and known for their aposematic colouration. Though colouration of Pentatomidae varies, some are known to merge with their surroundings, especially those genera which are specialized to inhabit tree trunks. They have the ability to camouflage themselves with the colouration of bark. The majority of these genera have been found to be nocturnal and are confined to concealed habitats like trunk holes or cracks or crevices of bark during the day. The genera of tree trunk dwellers include *Agaeus* Dallas, *Ameridalpa* Ghauri, *Cappaea* Ellenrieder, *Dalpada* Amyot & Serville, *Eupaleopada* Ghauri, *Halys* Fabricius, *Meridalpa* Ghauri, *Meridindia* Ghauri, *Neohalys* Ahmad and Perveen, *Placosternum* Amyot & Serville, and *Tipulparra* Ghauri. These genera exhibit morphological plasticity such as variation in lengths of mandibular plates with respect to clypeus, tooth on lateral margins of head (present/absent/transitory stages), length of antennal segments, length of labium, serration or tooth-like structures on anterolateral margins of pronotum, length of head with respect to first antennal segment, etc. The degree of phenotypic plasticity in this group is to such an extent that dissection of male genitalia is a must to delineate species or at times even of genera. The gradation in the shape of genital capsule (depth of concavity in ventral rim or dorsal rim, shape of caudal lobes), paramere (position of tooth or any sclerotized ridge), and overall shape of crown or phallus adds to the complexity. For example, in the case of species of *Placosternum* (a most problematic Oriental genus), an integrative taxonomic approach is perhaps the only means to delineate species.

Predators

Members of Asopinae, commonly known as predatory stink bugs or soldier bugs, are predaceous as both nymphs (with the exception of first instars) and adults. First instars of Asopinae are not predaceous, feeding on only water or plant juices (De Clercq 2000). The stylets are inserted into any soft area of the body of the prey and a salivary toxin is injected which quickly immobilizes it. While feeding on the prey, the only contact between predator and prey is by the labium and stylets (De Clercq 2000). Members are often observed to feed on plants that provide them with moisture and perhaps some supplemental nutrients at critical times, though the damage inflicted to plants by sucking plant sap is negligible (De Clercq 2000, 2002). Though they feed on plants at times, in contrast to other predatory heteropterans (e.g., *Geocoris*, *Dicyphus*, *Macrolophus*), plant feeding asopines have not been reported to injure crops. They are usually generalist predators attacking prey belonging to various taxa. It seems that none of the Asopinae is truly host specific, although there appears to be some degree of specialization in a number of species (De Clercq 2000). For example, the spined soldier bug, *Podisus maculiventris* (Say), which is being commercialized in North America and Europe mainly for the biological control of the Colorado potato beetle in potatoes and noctuid caterpillars in vegetables, shows a particular preference for lepidopteran larva. *P. maculiventris* was found feeding on more than 90 insect species from eight orders, including Coleoptera, Diptera, Ephemeroptera, Heteroptera, Homoptera, Hymenoptera, Lepidoptera, and Orthoptera (De Clercq 2000). Another popular predatory bug, the two-spotted stink bug, *Perillus bioculatus* (Fabricius), a native of North America, is recorded as a predator of grubs of *Zygogramma bicolorata* (Prasad and Pal 2015; Kaur et al. 2012) and is restricted to the northern parts of India. *Eocanthecona furcellata* (Wolff.), a predominant species, is found throughout India. It has increasingly been studied as a biocontrol agent against agricultural insect pests in southeastern Asia and India. It is documented as a predator, of mostly lepidopteran pests, of agricultural crops and forest plantations in the Indian subcontinent (De Clercq 2000).

Lenin and Rajan (2016) found *Corcyra cepahlonica* as a suitable host for rearing *E. furcellata*. Shophiya and Sahayaraj (2014) tested the biocontrol potential of *E. furcellata* against *Pericallia ricini* (Fab.) larva.

There are about 30 species of Asopinae in India, but the knowledge of their prey or feeding potential of the majority of these asopines remain poorly understood. Only a small proportion of the species belonging to this subfamily have been studied so far. Further studies on the biology of this group of bugs are essential if they are to be exploited for the biological control of crop pests.

Host Plants of Phytophagous Pentatomidae in India

Phytophagous Pentatomidae feed on several host plants belonging to diverse plant families. An exhaustive host plant list of various species of Pentatomidae is available at the Pentatomoidea Home Page (maintained by Rider 2018). A few species of Pentatomidae emerge occasionally to congregate in large numbers on various living and non-living structures. For example, *Udonga montana* (Distant) emerges in large numbers when bamboo blooms; *Catacanthus incarnatus* (Drury) aggregates on trees of *Pavetta indica* Linn.; and *Agonoscelis nubilis* (Fabricius) were found to aggregate on the bark of cashew (Anacardiaceae) as well as on the bark and fruits of citrus (Rutaceae). The following list of host plants in Table 8.3, though not exhaustive, has been compiled from various sources (Kalshoven 1981; Distant, 1902; Lefroy and Howlett 1909; Fletcher 1914; Chatterjee 1934; Usman and Puttarudriah 1955; apart from Salini 2006, 2015).

TABLE 8.3
Host Plants of Phytophagous Pentatomidae

Sl. No.	Subfamily/Species	Host Plants	Family
I. Pentatominae			
1.	*Acesines bambusana* Distant	*Bambusa vulgaris* Schrad. *ex* J.C. Wendl.	Poaceae
2.	*Acrosternum gramineum* (Fabricius)	*Triticum aestivum* Linn.	Poaceae
		Dodonaea viscosa (Linn.) Jacq.	Sapindaceae
		Leucas aspera (Willd) Spreng.	Lamiaceae
		Abelmoschus esculentus (Linn.) Moench	Malvaceae
		Cynadon dactylon (Linn.) Pers.	Poaceae
		Ocimum sanctum Linn.	Lamiaceae
		Ravolfia tetraphylla Linn.	Apocyanaceae
		Ruta graveolens Linn.	Rutaceae
		Glycine max (Linn.) Merr.	Fabaceae
		Spinacea oleracea Linn.	Amaranthaceae
		Crotalaria juncea Linn.	Fabaceae
		Catharanthus roseus (Linn.) G. Don.	Apocyanaceae
		Solanum torvum Sw.	Solanaceae

(Continued)

TABLE 8.3 (Continued)
Host Plants of Phytophagous Pentatomidae

Sl. No.	Subfamily/Species	Host Plants	Family
3.	*Adria parvula* (Dallas)	*Eleusine coracana* (Linn.) Gaertn.	Poaceae
		Sorghum bicolour (Linn.) Moench.	Poaceae
		Crotalaria juncea Linn.	Fabaceae
4.	*Aeliomorpha lineatocollis* (Westwood)	*Cynadon dactylon* (Linn.) Pers.	Poaceae
		Ocimum sp.	Lamiaceae
		Abelmoschus esculentus (Linn.) Moench	Malvaceae
		Erythroxylon monogynum Roxb.	Erythroxylaceae
5.	*Aeliomorpha viridescens* Rider & Rolston	*Oryza sativa* Linn.	Poaceae
6.	*Agaeus tessellatus* (Dallas)	*Lavendula officianalis* Chaix	Lamiaceae
		Markhamia lutea (Benth.) K. Schum.	Bignoniaceae
7.	*Agonoscelis nubila* (Fabricius)	*Leucas aspera* (Willd) Spreng.	Lamiaceae
8.	*Anaca florens* (Walker)	*Murrya paniculata* Jacq.	Rutaceae
9.	*Antestiopsis cruciata* Leston	*Santalum album* Linn.	Santalaceae
		Jasminum officinale Linn.	Oleaceae
10.	*Bagrada hilaris* Burmeister	*Brassica oleracea* Linn var. *Botrytis*	Brassicaceae
		Raphanus sativus Linn.	Brassicaceae
		Brassica rapa Linn.	Brassicaceae
		Mangifera indica Linn.	Anacardiaceae
		Brassica nigra Linn.	Brassicaceae
11.	*Brachycoris tralucidus* Salini	*Cajanus cajan* (Linn.) Millsp.	Fabaceae
12.	*Cappaea taprobanensis* (Dallas)	*Citrus aurantium* Linn.	Rutaceae
13.	*Carbula biguttata* (Fabricius)	*Glycine max* (Linn.) Merr.	Fabaceae
14.	*Carbula scutellata* Distant	*Barleria prionitis* Linn.	Acanthaceae
15.	*Carbula socia* (Walker)	*Barleria prionitis* Linn.	Acanthaceae
16.	*Catacanthus incarnatus* (Drury)	*Pavetta indica* Linn.	Rubiaceae
17.	*Cresphontes monsoni* (Westwood)	*Magnolia champaca* (L.) Baill. Ex Pierre	Magnoliaceae
		Lantana camara Linn.	Verbenaceae
18.	*Critheus lineatifrons* Stål	*Bambusa vulgaris* Schrad. *ex* J. C. Wendl.	Poaceae
19.	*Degonetus serratus* (Distant)	*Tectona grandis* L. F.	Lamiaceae
20.	*Dolycoris indicus* Stål	*Cajanus cajan* (Linn.) Millsp.	Fabaceae
		Arachis hypogaea Linn.	Fabaceae
		Solanum tuberosum Linn.	Solanaceae
		Zea mays Linn.	Poaceae
		Eleusine coracana (Linn.) Gaertn.	Poaceae
		Solanum melongena Linn.	Solanaceae
		Gossypium hirsutum Linn.	Malvaceae
		Hibiscus sabdariffa Linn.	Malvaceae
		Morus alba Linn.	Moraceae
		Triticum aestivum Linn.	Poaceae
		Oryza sativa Linn.	Poaceae
		Rosmarinus officinalis Linn.	Lamiaceae
		Glycine max (Linn.) Merr.	Fabaceae
		Spinacea oleracea Linn.	Amaranthaceae
21.	*Erthesina acuminata* Dallas	*Anacardium occidentale* Linn.	Anacardiaceae
		Cassine glauca Rottb. Kuntze.	Celastraceae
		Linum usitatissimum Linn.	Linaceae
22.	*Erthesina fullo* (Thunberg)	*Cassine glauca* (Rottboell) Kuntze	Celastraceae
		Hibiscus rosa sinensis Linn.	Malvaceae
		Anacardium occidentale Linn.	Anacardiaceae
23.	*Eupaleopada concinna* (Westwood)	*Artocarpus heterophyllus* Lam.	Moraceae
		Oryza sativa Linn.	Poaceae
		Croton sp.	Euphorbiaceae
		Murraya koenigii (Linn.) Sprengel	Rutaceae

(Continued)

TABLE 8.3 (Continued)
Host Plants of Phytophagous Pentatomidae

Sl. No.	Subfamily/Species	Host Plants	Family
24.	*Eusarcoris montivagus* Distant	*Phaseolus vulgaris* Linn.	Fabaceae
		Mangifera indica Linn.	Anacardiaceae
25.	*Eusarcoris ventralis* (Westwood)	*Cajanus cajan* (Linn.) Millsp.	Fabaceae
		Luffa acutangula (Linn.) Roxb.	Cucurbitaceae
		Richardia scabra Linn.	Rubiaceae
26.	*Glaucias albomaculatus* (Distant)	*Morinda tinctoria* Roxb.	Rubiaceae
27.	*Glaucias nigromarginatus* (Stål)	*Duranta* sp.	Verbenaceae
28.	*Gynenica affinis* Distant	*Barleria prionitis* Linn.	Acanthaceae
29.	*Gynenica alami* Shafee and Azim	*Barleria prionitis* Linn.	Acanthaceae
30.	*Halyomorpha picus* (Fabricius)	*Phaseolus vulgaris* Linn.	Fabaceae
		Cyamopsis tetragonoloba (Linn.) Taub.	Fabaceae
		Vanilla spp.	Orchidaceae
		Artocarpus heterophyllus Lam.	Moraceae
		Ageratum conyzoides Linn.	Asteraceae
		Areca catechu Linn.	Arecaceae
		Luffa acutangula (Linn.) Roxb.	Cucurbitaceae
		Sorghum bicolour (Linn.) Moench.	Poaceae
		Tabebuia cassinoides A. P. De. Candolle	Bignoniaceae
31.	*Halys serrigera* Westwood	*Tamarindus indica* Linn.	Caesalpiniaceae
		Tabebuia rosea Zhang.	Bignoniaceae
		Spathodea campanulata P. Beauv.	Bignoniaceae
		Swietenia mahagoni (Linn.) Jacq.	Meliaceae
		Syzygium cumini (Linn.) Skeels.	Myrtaceae
		Phyllanthus emblica Linn.	Phyllanthaceae
		Magnolia champaca (L.) Baill.ex Pierre	Magnoliaceae
32.	*Halys shaista* Ghauri	*Murraya* sp.	Rutaceae
		Magnolia champaca (L.) Baill. ex Pierre	Magnoliaceae
		Lagerstroemia indica (Linn.) Pers.	Lythraceae
		Butea monosperma (Lam.) Taub.	Fabaceae
33.	*Hermolaus typicus* Distant	*Sesamum indicum* Linn.	Pedaliaceae
		Richardia scabra L.	Rubiaceae
34.	*Hermolaus rolstoni* Azim & Shafee	*Eleusine coracana* (L.). Gaertn.	Poaceae
		Euphorbia sp.	Euphorbiaceae
		Lantana camara Linn.	Verbenaceae
		Ocimum sanctum Linn.	Lamiaceae
		Ocimum tenuiflorum Linn.	Lamiaceae
		Ocimum sp.	Lamiaceae
35.	*Madates parva* Rider	*Brassica nigra* Linn.	Brassicaceae
36.	*Menida varipennis* (Westwood)	*Dodonaea viscosa* (Linn.) Jacq.	Sapindaceae
		Phaseolus vulgaris Linn.	Fabaceae
		Santalum album Linn.	Santalaceae
		Ocimum sanctum Linn.	Lamiaceae
37.	*Menida versicolor* (Gmelin)	*Oryza sativa* Linn.	Poaceae
		Sorghum bicolor Linn.	Poaceae
		Cyanodon dactylon (Linn.) Pers.	Poaceae
		Panicum sp.	Poaceae
		Eleusine coracana (L.). Gaertn.	Poaceae
38.	*Neohalys sericollis* (Westwood)	*Ipomoea purpurea* (Linn.) Roth.	Convolvulaceae
		Ficus carica Linn.	Moraceae

(Continued)

TABLE 8.3 (Continued)
Host Plants of Phytophagous Pentatomidae

Sl. No.	Subfamily/Species	Host Plants	Family
39.	*Nezara viridula* (Linnaeus)	*Oryza sativa* Linn.	Poaceae
		Zea mays Linn.	Poaceae
		Sorghum bicolor Linn.	Poaceae
		Vigna radiata (Linn.) Wilz.	Fabaceae
		Glycine max (Linn.) Merr.	Fabaceae
		Vigna unguiculata (Linn.) Walp.	Fabaceae
		Cajanus cajan (Linn.) Millsp.	Fabaceae
		Gossypium hirsutum Linn.	Malvaceae
		Lycopersicon esculentum Miller.	Solanaceae
		Crotalaria juncea Linn.	Fabaceae
		Santalum album Linn.	Santalaceae
		Abelmoschus esculentus (Linn.) Moench	Malvaceae
		Anacardium occidentale Linn.	Anacardiaceae
		Syzygium cumini (Linn.) Skeels.	Myrtaceae
		Phaseolus vulgaris Linn.	Fabaceae
		Ocimum sp.	Lamiaceae
		Vanilla planifolia Andrews	Orchidaceae
		Triticum aestivum Linn.	Poaceae
40.	*Phricodus hystrix* (Germar)	*Ocimum sanctum* Linn.	Lamiaceae
	Piezodorus hybneri (Gmelin)	*Phaseolus vulgaris* Linn.	Fabaceae
		Luffa acutangula (Linn.) Roxb.	Cucurbitaceae
		Sesbania sesban (Linn.) Merr.	Fabaceae
		Eleusine coracana (Linn.) Gaertns	Poaceae
		Santalum album Linn.	Santalaceae
		Sesamum indicum Linn.	Pedaliaceae
		Cajanus cajan (Linn.) Millsp.	Fabaceae
		Vigna radiata (L.) R. Wilczek	Fabaceae
		Gossypium hirsutum Linn.	Malvaceae
41.	*Placosternum alces* Stål	*Tectona grandis* L. F.	Lamiaceae
		Santalum album Linn.	Santalaceae
42.	*Placosternum obtusum* Montandon	*Ficus benghalensis* Linn.	Moraceae
43.	*Placosternum taurus* (Fabricius)	*Ixora coccinea* Linn.	Rubiaceae
44.	*Plautia crossota* (Dallas)	*Santalum album* Linn.	Santalaceae
		Rauvolfia tetraphylla Linn.	Apocyanaceae
		Chromolaena odorata (Linn.) King and H. Robinson	Asteraceae
		Lantana camera Linn.	Verbenaceae
		Dodonaea viscosa (Linn.) Jacq.	Sapindaceae
		Eugenia uniflora Linn.	Myrtaceae
		Vicia faba Linn.	Fabaceae
		Gossypium hirsutum Linn.	Malvaceae
		Barleria prionitis Linn.	Acanthaceae
		Eleusine coracana (Linn.) *Erythroxylon monogynum* Roxb	Poaceae Erythroxylaceae
		Cajanus cajan (Linn.) Millsp.	Fabaceae
		Glycine max (Linn.) Merr.	Fabaceae
45.	*Rhynchocoris plagiatus* (Walker)	*Coffea arabica* Linn.	Rubiaceae
		Santalum album Linn.	Santalaceae
		Erythroxylon monogynum Roxb.	Erythroxylaceae
		Sapindus laurifolius Vahl.	Sapindaceae
		Murraya paniculata (Linn.) Jack.	Rutaceae
46.	*Sabaeus humeralis* (Dallas)	*Tabernaemontana dichotona* Roxb. Ex Wall.	Apocyanaceae
		Murraya paniculata (Linn.) Jack.	Rutaceae
47.	*Sciocoris indicus* Dallas	*Leucas aspera* (Willd.)	Lamiaceae
		Sorghum bicolor Linn.	Poaceae

(Continued)

TABLE 8.3 (Continued)
Host Plants of Phytophagous Pentatomidae

Sl. No.	Subfamily/Species	Host Plants	Family
48.	*Spermatodes variolosus* (Walker)	*Richardia scabra* Linn.	Rubiaceae
49.	*Stenozygum speciosum* (Dallas)	*Lantana camera* Linn.	Verbenaceae
		Oryza sativa Linn.	Poaceae
50.	*Tolumnia basalis* (Dallas)	*Glycine max* (Linn.) Merr.	Fabaceae
51.	*Udonga montana* (Distant)	*Bambusa vulgaris* Schrader ex Wendl.	Poaceae
		Coffea arabica Linn.	Rubiaceae
		Areca catechu Linn.	Arecaceae
		Bambusa vulgaris Schrader *ex* Wendl	Poaceae
		Coffea arabica Linn.	Rubiaceae
II. Phyllocephalinae			
1.	*Gonopsis rubescens* Distant	*Cymbopogon flexuous* (Nees ex Steud.) W. Watson	Poaceae
2.	*Tetroda histeroides* (Fabricius)	*Oryza sativa* Linn.	Poaceae
III. Podopinae			
1.	*Brachycerocoris camelus* Costa	*Santalum album* Linn.	Santalaceae

ECONOMIC IMPORTANCE

The economic importance of these bugs due to their destructive feeding habit has been extensively reported by Panizzi (1997) and Panizzi et al. (2000). Aggregated feeding by some species aggravates the injury to crops. The feeding injuries caused by these bugs vary across plant species from minute dark spots on seeds due to feeding punctures in soybean, formation of chaffy grains and chalky discolouration around the feeding sites resulting in pecky rice in paddy, premature fruit ripening, and the formation of cocoa beans without a sugary mucilage cover in cocoa, to leaf wilt and plant death in potato, etc. (Panizzi 1997). Phytophagous species feed on plants by injecting watery saliva containing digestive enzymes and suck the liquefied contents through the food canal formed by the maxillary stylets. Pentatomidae also act or serve as vectors of plant pathogens like phytoplasmas. Adults and nymphs of the brown marmorated stink bug, *Halyomorpha halys* (Stål) have been shown to transmit Witches' broom phytoplasma to *Paulownia* spp. trees in Asia (Weintraub and Beanland 2006). *Cappaea taprobanensis* (Dallas) was reported to suck the juice of twigs and shoots of orange and *Dolycoris indicus* Stål on various cereals like wheat, tenai, etc. and oilseeds like safflower, sunflower, etc. *Eysarcoris ventralis* (Westwood) is reported as a serious pest on sesame occasionally. *Antestiopsis cruciata* (Fabricius) was reported as a pest of jasmine and coffee; *Agonoscelis nubilis* (Fabricius) was reported as a minor pest in various cereals and pulses; *Bagrada hilaris* (Burmeister) was reported as a pest of cruciferous plants such as cabbage, cauliflower, mustard, turnip, etc.; *Nezara viridula* (Linnaeus) on various cereals and pulses; and *Menida versicolor* (Gmelin) on paddy (Fletcher 1914). *Rhynchocoris poseiden* Kirkaldy, the spined fruit bug of citrus, transmits fungal spores of *Nematospora* (Kalshoven 1981). Transmission of yeast spot disease, *Nematospora coryli*, by *Neomegalotomus parvus* (Westwood) to soybean and the flagellate protozoan *Phytomonas* spp. to palm trees (Panizzi 1997) is a problem. *Bagrada hilaris* (Burmeister) is a pest of oilseeds and vegetables in India. *Plautia* spp. infests orchard plants in the Oriental region (Panizzi et al. 2000). However, species of the subfamily Asopinae (both nymphs and adults) are predaceous on a variety of insects such as Coleoptera, Lepidoptera, especially in their larval stages, and other small and soft bodied arthropods (Lefroy and Howlett 1909; Fletcher 1914; Kalshoven 1981). The members of the subfamily Phyllocephalinae were found mainly associated with grass roots, and Podopinae were reported as pests of rice from various Asian countries (Panizzi et al. 2000). *Tetrodahisteroides* was recorded as a pest of paddy in Salem and Coimbatore (Fletcher 1914). A few species of Pentatomidae are invasive in nature, and their entry needs to be monitored. For example, *Halyomorpha halys* (Stål) (brown marmorated stink bug) is an invasive stink bug affecting various fruit crops in different countries such as China, Japan, Korea, Taiwan, etc. Apart from its direct injury, it is also known to transmit Witches' broom phytoplasma.

COLLECTION AND PRESERVATION

Pentatomidae can be collected by a variety of methods including the sweeping of vegetation with an insect net; hand picking from host plants (ideal for the bigger species and those inhabiting tree trunks); and use of light traps (both mercury as well as sodium vapour lamps).

Freshly collected bugs are killed using ethyl acetate. Samples for DNA extraction should be preserved directly in 100 percent ethanol immediately after field collection.

Procedure for Dissection of Male and Female Genitalia

Ahmad (1986) developed a fool-proof technique for inflation of male genitalia which was modified for Pentatomidae (Salini 2016b as detailed below).

Gapon (2001) developed and described the following technique for aedeagi inflation of heteropteran bugs using microcapillaries.

Place the pinned, dry specimens (with pin still inserted, but with labels removed) in boiling water in a 500 mL beaker for 5 minutes. Remove the specimens from the boiling water with forceps and slip off the pin. Detach the genital capsule (pygophore) (Figures 8.68, 8.69) after placing the specimen in distilled water in a cavity block. Apply gentle pressure posteriorly with the help of forefinger, which is sufficient to pop the genital capsule out. Remove the genital capsule carefully from the abdomen. Reinsert the pin into the specimen and replace the labels. Place the genital capsule in 5–10 mL of 10 percent KOH in a test tube. The test tube along with genital capsule is to be placed in a hot water bath at about 98°C. After boiling for about 10–15 minutes, remove the genital capsule and wash thoroughly with distilled water. Clean the anterior opening of the genital capsule with forceps under a Stereozoom microscope. Dissect the genital capsule in water in a cavity block. Separate the parameres (a pair are usually found) (Figure 8.70) and phallus (Figure 8.71). After the separation of phallus, it should be inflated to study the details. Sometimes the phallus will be naturally inflated. If not, the inflation of phallus can be facilitated by inserting a sharp, pointed tip of a needle carefully through the posterior part of the phallus. This will help the endosoma (Figure 8.71) to come out of the phallotheca and transfer the dissected parts to glycerine in cavity block and preserve in a drop of glycerin held in Arthropod Microvial®. The vial should be stoppered and transfixed to the pin holding the rest of the specimen.

Female specimens should be dissected by detaching and boiling the whole abdomen in 5–10 mL of 10% KOH taken in a test tube. Take out the digested abdomen and wash it in distilled water. After removing the fatty tissues, detach the terminalia (Figure 8.72) and spermatheca (Figure 8.73) using a fine forceps. Pentatomid spermatheca mainly possesses a spermathecal dilation and a spermathecal pump (Figure 8.74). Preserve the dissected parts as explained in the case of male genitalia.

WEBSITES

Pentatomoidea Home Page: https://www.ndsu.edu/pubweb/~rider/Pentatomoidea/
Includes information on taxa, details of types, literature (with PDFs), host plant information, biographies of research workers, etc.

BugGuide: https://bugguide.net/node/view/182
A good preliminary guide to insect identification including Pentatomidae.

Biodiversity Heritage Library: https://www.biodiversitylibrary.org
An open access digitized library of legacy literature on all aspects of natural history.

CONCLUSION

Pentatomidae has not received adequate attention although it is the largest family of Pentatomoidea and the second largest in the Pentatomorpha. Till recently, the catalogue of Hemiptera by Kirkaldy (1909) was the only available catalogue for this group. Aukema and Rieger (2006) published a recent catalogue of the Heteroptera of the Palaearctic region. Though several species of Pentatomidae cause significant damage to various crop plants, their damage is underestimated. Our country has a rich fauna of Pentatomidae, the majority of which are poorly studied with several taxa awaiting description as new. DNA barcodes are available for only a few taxa. Phylogenetic studies in this group are inadequate with nothing conclusive to prove the monophyly of many of the higher taxa. Most taxonomic studies undertaken in this group are based largely on morphological data. An integrative taxonomic approach using morphological, biochemical, and molecular tools may clarify the taxonomic confusion associated with several taxa in Pentatomidae.

REFERENCES

Afzal, M., and I. Ahmad. 1981. A new genus and three new species of Halyini Stål (Heteroptera: Pentatominae) from Pakistan. *Pakistan Journal of Zoology* 13(1–2): 63–72.

Ahmad, I., and M. Afzal. 1984a. A revision of the genus *Sarju* Ghauri (Hemiptera: Pentatomidae: Pentatominae: Halyini) with description of a new species from Pakistan. *Türkiye Bitki Koruma Dergisi* 8: 131–142.

Ahmad, I., and M. Afzal. 1984b. Revision of the Indo-Malayan genus *Dalpada* (Halyini, Pentatomidae, Heteroptera). *Zoologischer Anzeiger* 213(3–4): 170–176.

Ahmad, I., and M. Afzal. 1986. A new genus and a new species of Halyini Stål (Hemiptera: Pentatomidae: Pentatominae) with a note on its relationships. *Türkiye Bitki Koruma Dergisi* 10(4): 199–202.

Ahmad, I., and S. Kamaluddin. 1978. A new genus and a new species of Phyllocephalinae (Hemiptera: Pentatomidae) with phylogenetic considerations. *Transactions of the Shikoku Entomological Society* 14(1–2): 1–6.

Ahmad, I., and S. Kamaluddin. 1988. A new tribe and a new species of the subfamily Phyllocephalinae (Hemiptera: Pentatomidae) from the Indo-Pakistan subcontinent. *Oriental Insects* 22: 241–258.

Ahmad, I., and S. Kamaluddin. 1990. A new tribe for Phyllocephaline genera *Gellia* Stål and *Tetroda* Amyot and Serville (Hemiptera: Pentatomidae) and their revision. *Annotationes Zoologicae et Botanicae*, 195: 20 pp.

Ahmad, I., and S. Kamaluddin. 1992. New generic status of a rice-feeding tetrodine subgenus *Tetrodias* Kirkaldy and redescription of *Tetroda* Amyot and Serville (Hemiptera: Pentatomidae: Phyllocephalinae) and their cladistic analysis. *Pakistan Journal of Zoology*, 24(2): 123–127.

Ahmad, I., and N. A. Rana. 1988. A revision of the genus *Canthecona* Amyot et Serville (Hemiptera: Pentatomidae: Pentatominae: Asopini) from Indo-Pakistan subcontinent with description of two new species from Pakistan. *Türkiye Bitki Koruma Dergisi* 12(2): 75–84.

Ahmad, I. 1986. A fool-proof technique for inflation of male genitalia in Hemiptera (Insecta). *Pakistan Journal of Entomology* 1(2): 111–112.

Ahmad, I., and A. S. Siddiqui. 1993. Two new species of the Palaearctic subgenus *Parasciocoris* Wagner of *Sciocoris* Fallen (Heteroptera: Pentatomidae) from Pakistan. *Mitteilungen der Schweizerischen Entomologischen Gesellschaft* 66(3–4): 233–242.

Ahmad, I., S. S. Shaukat, and S. Kamaluddin. 1994. Taxometric studies on the tribe Phyllocephalini Amyot and Serville (Hemiptera: Pentatomidae: Phyllocephalinae) from Indo-Pakistan subcontinent. *Pakistan Journal of Zoology* 26(3): 265–268.

Atkinson, E. T. 1884. *The Himálayan Districts of the North-Western Provinces of India.* Vol. II. [Volume X of Statistical, descriptive and historical account of the North-Western Provinces of India.]. North-Western Provinces and Oudh Government Press, Allahabad, India, xviii + 964 pp.

Atkinson, E. T. 1887. Notes on Indian Rhynchota, Heteroptera No. 2. *Journal of Asiatic Society of Bengal* 56: 145–205.

Atkinson, E. T. 1888. Notes on Indian Rhynchota, Heteroptera No. 3 & 4. *Journal of Asiatic Society of Bengal* 57: 1–72, 118–184, 333–345.

Aukema, B., and C. Rieger, 2006. *Catalogue of the Heteroptera of Palaearctic Region.* The Netherlands Entomological Society, c/o Plantage Middenlaan 64, NL-1018, DH Amsterdam, the Netherlands.

Azim, M. N., and M. S. Bhat. 2010. A preliminary survey for pentatomid bugs (Heteroptera: Pentatomidae) in Kashmir Himalaya. *Journal of Entomological Research* 34(2): 165–170.

Azim, M. N., and S. A. Shafee, 1979. Indian species of the genus *Nezara* Amyot and Serville (Hemiptera: Pentatomidae). *Journal of the Bombay Natural History Society* 75(2): 507–511.

Azim, M. N., and S. A. Shafee. 1982. New descriptions, a new species of the genus *Asopus* (Heteroptera: Pentatomidae). *Journal of the Bombay Natural History Society* 79(2): 361–362.

Azim, M. N., and S. A. Shafee. 1983. A new species of the genus *Dalpada* Amyot and Serville (Heteroptera: Pentatomidae) from India. *Mitteilungen der Schweizerischen Entomologischen Gesellschaft* 56: 191–194.

Azim, M. N., and S. A. Shafee. 1984a. New descriptions. A new species of *Cresphontes* Stål, (Heteroptera: Pentatomidae) from India. *Journal of the Bombay Natural History Society* 81(2): 428–430.

Azim, M. N., and S. A. Shafee. 1984b. Indian species of the genus *Stollia* Ellenrieder (Heteroptera: Pentatomidae). *Mitteilungen der Schweizerischen Entomologischen Gesellschaft* 57: 291–293.

Azim, M. N., and S. A. Shafee. 1984c. Degonetini trib. n. (Heteroptera: Pentatomidae). *Current Science* 53(20): 1094–1095.

Azim, M. N., and S. A. Shafee. 1985. Studies on Indian species of *Hermolaus* (Heteroptera: Pentatomidae). *International Journal of Entomology* 27(4): 394–397.

Azim, M. N., and S. A. Shafee. 1986a. Studies on the Indian Strachiini (Pentatomidae: Pentatominae). *Journal of the Bombay Natural History Society* 82(3): 586–593.

Azim, M. N., and S. A. Shafee. 1986b. The life cycle of *Bagrada picta* (Fabricius) (Hemiptera: Pentatomidae). *Articulata* 2(8): 261–265.

Azim, M. N., and S. A. Shafee. 1987a. Studies on Indian species of the genus *Aeliomorpha* Stål. (Heteroptera: Pentatomidae). *Beiträge zur Entomologie* 37(2): 421–424.

Azim, M. N., and S. A. Shafee. 1987b. Studies on Indian species of *Sciocoris* Fallen (Heteroptera: Pentatomidae). *Articulata* 2(10): 367–370.

Azim, M. N. 2002. Studies on Indian genera of the tribe Halyini (Pentatomidae: Pentatominae). *Oriental Science* 7: 41–52.

Azim, M. N. 2011. Taxonomic survey of stink bugs (Heteroptera: Pentatomidae) of India. *Halteres* 3: 1–10.

Bianchi, F. M., M. Deprá, A. Ferrari et al. 2017. Total evidence phylogenetic analysis and reclassification of *Euschistus* Dallas within Carpocorini (Hemiptera: Pentatomidae: Pentatominae). *Systematic Entomology* 42: 399–409.

Campos, L. A., and J. Grazia. 2006. Análise cladística e biogeografia de Ochlerini (Heteroptera, Pentatomidae, Discocephalinae). *Iheringia Série Zoologia* 96(2): 147–163.

Cassis, G., and G. F. Gross. 2002. Hemiptera: Heteroptera (Pentatomomorpha). *In*: Houston, W. W. K. and A. Wells (eds.), *Zoological Catalogue of Australia.* 27(3B). CSIRO Publishing, Melbourne, Australia. xiv + 737 pp.

Chakrabarty, S. P., L. K. Ghosh, and R. C. Basu. 1994. On a collection of Hemiptera from Namdapha Biosphere Reserve in Arunachal Pradesh, India. *Zoological Survey of India, Occasional Paper No.* 161: 25–33.

Chatterjee, N. C. 1934. Entomological investigations on the spike disease of sandal (24). Pentatomidae (Hemipt.). *Indian Forest Records* 20: 1–31.

Chopra, N. P. 1972. A new species of the genus *Agaeus* Dallas (Hemiptera: Pentatomidae). *Oriental Insects* 6(3): 305–307.

Chopra, N. P. 1974. Studies on the genus *Halys* (Hemiptera: Pentatomidae). *Oriental Insects* 8(4): 473–479.

Datta, B., L. K. Ghosh, and M. Dhar. 1985. Study on Indian Pentatomoidea (Heteroptera: Insecta). *Records of Zoological Survey of India* 80: 1–43.

De Clercq, P. 2000. Predaceous bugs (Pentatomidae: Asopinae). *In*: C. W. Schaefer & Panizzi, A. R (Eds), *Heteroptera of Economic Importance.* CRC Press, Boca Raton, FL, pp. 737–789.

De Clercq, P. 2002. Dark clouds and their silver linings: Exotic generalist predators in augmentative biological control. *Neotropical Entomology* 31(2): 169–176.

Distant, W. L. 1887. Contributions to knowledge of Oriental Rhynchota. Part I. Fam. Pentatomidae. *Transactions of the Entomological Society of London* 3: 341–359.

Distant, W. L. 1902. Rhynchota Vol. I, Heteroptera. *In*: Blanford, W. T. (Ed). *The Fauna of British India Including Ceylon and Burma.* Taylor and Francis, London, UK, 307 + 438 pp.

Distant, W. L. 1908. Rhynchota Vol. IV. Homoptera and Appendix (Pt). *In*: Bingham, C. T (Ed). *The Fauna of British India Including Ceylon and Burma.* Taylor and Francis, London, UK, 15 + 501 pp.

Distant, W. L. 1918. Rhynchota Vol. VII. Homoptera: Appendix. Heteroptera: Addenda. *In*: Shipley, A. E & Marshall, A. K. (eds). *The Fauna of British India Including Ceylon and Burma.* Taylor & Francis Group, London, UK, 7 + 210 pp.

Fletcher, T. B. 1914. *Some South Indian Insects and Other Animals of Importance.* Bishen Singh Mahendra Pal Singh, Dehra Dun (publishers), The Superintendent, Government Press. 22 + 565 pp.

Gapon, D. A., and F. V. Konstantinov. 2006. On the structure of the aedeagus in shield bugs (Heteroptera, Pentatomidae): III. Subfamily Asopinae. Entomological Review 86(7), 806–819 (Published in *Entomologicheskoe Obozrenie* 85(3): 491–507.

Gapon, D. A. 2001. Inflation of heteropteran aedeagi using microcapillaries (Heteroptera: Pentatomidae). *Zoosystematica Rossica* 9(1): 157–160.

Gapon, D. A. 2008. New subtribes and a new genus of Podopini (Heteroptera: Pentatomidae: Podopinae). *Acta Entomologica Musei Nationalis Pragae* 48(2): 523–532.

Gapud, V. P. 1991. A generic revision of the subfamily Asopinae, with consideration of its phylogenetic position in the family Pentatomidae and superfamily Pentatomoidea (Hemiptera-Heteroptera). Parts. I and II. *Philippines Entomology* 8(3): 865–961.

Ghate, H. V. 2013. *Eurysaspis flavescens* Distant, a new Pentatomidae member for Indian Territory. *Prommalia* 1: 45–48.

Ghate, H. V. 2015. Rediscovery of the shieldbug *Menedemus vittatus*, with notes on *M. hieroglyphicus* (Heteroptera: Pentatomidae: Pentatominae: Sciocorini), from Pune, Maharashtra, India. *Entomon* 40(4): 243–248.

Ghate, H. V., G. P. Pathak, Y. Koli, and G. P. Bhawane. 2012. First record of two Pentatomidae bugs from Chandoli area, Kolhapur. Maharashtra, India. *Journal of Threatened Taxa* 4(4): 2524–2528.

Ghauri, M. S. K. 1963. A preliminary revision of the little-known genus *Critheus* Stål (Pentatomidae, Heteroptera). *Annals and Magazine of Natural History* 5(13): 407–415.

Ghauri, M. S. K. 1975a. Revision of the Himalayan genus *Paranevisanus* Distant (Halyini, Pentatominae, Pentatomidae, Heteroptera). *Zoologischer Anzeiger* 195(5–6): 407–416.

Ghauri, M. S. K. 1975b. *Jugalpada*, a new genus of Halyini (Pentatomidae, Heteroptera). *Journal of Natural History* 9(6): 629–632.

Ghauri, M. S. K. 1977a. A revision of *Apodiphus* Spinola (Heteroptera: Pentatomidae). *Bulletin of Entomological Research* 67: 97–106.

Ghauri, M. S. K. 1977b. *Sarju*—A new genus of Halyini (Heteroptera, Pentatomidae, Pentatominae) with new species. *Turkiye Bitki Koruma Dergisi* 1(1): 9–27.

Ghauri, M. S. K. 1978. *Cahara*—A new genus of Halyini (Heteroptera, Pentatomidae, Pentatominae) with new species on fruit and forest trees in the Sub-Himalayan region. *Journal of Natural History* 12(2): 163–176.

Ghauri, M. S. K. 1980. *Tipulparra*—A new genus of Halyini with new species (Heteroptera, Pentatomidae, Pentatominae). *Reichenbachia Staatliches Museum für Tierkunde in Dresden* 18(21): 129–146.

Ghauri, M. S. K. 1982a. New genera and new species of Halyini, mainly from south India (Heteroptera, Pentatomidae, Pentatominae). *Reichenbachia Staatliches Museum für Tierkunde in Dresden* 20(1): 1–24.

Ghauri, M. S. K. 1982b. The identity of *Halys versicolor* Herrich-Schäffer, 1840 (Heteroptera, Pentatomidae). *Reichenbachia Staatliches Museum für Tierkunde in Dresden* 20(21): 175–181.

Ghauri, M. S. K. 1988. A revision of Asian species of the genus *Halys* Fabricius based on the type material (Insecta, Heteroptera, Pentatomidae, Pentatominae). *Entomologische Abhandlungen Staatliches Museum für Tierkunde Dresden* 51(6): 77–92.

Ghauri, M. S. K. 1980. *Tipulparra*—a new genus of Halyini (Heteroptera: Pentatomidae) with new species (Heteroptera, Pentatomidae, Pentatominae). *Reichenbachia Staatliches Museum für Tierkunde in Dresden* 18(21): 129–146.

Grazia, J., A. R. Panizzi, C. Greve et al. 2015. Stink bugs (Pentatomidae). *In*: Panizzi, A. R. & Grazia, J. (Eds), *True Bugs (Heteroptera) of the Neotropics*, Entomology in Focus 2. Springer Science + Business Media Dordrecht, pp. 681–756.

Grazia, J., R. T. Schuh, and W. C. Wheeler. 2008. Phylogenetic relationship of family groups in Pentatomoidea based on morphology and DNA sequences (Insecta: Heteroptera). *Cladistics* 24: 932–976.

Gross, G. F. 1975. Handbook of the flora and fauna of South Australia. *Plant-feeding and other bugs (Hemiptera) of South Australia.* Heteroptera—Part 1. Handbooks Committee, South Australian Government, Adelaide, Australia, 250 pp.

Hasan, S. A., and I. J. Kitching. 1993. A cladistic analysis of the tribes of Pentatomidae (Heteroptera). *Japanese Journal of Entomology* 61: 651–669.

Hegde, V. 1986a. On a collection of bugs (Heteroptera) from the Tamil Nadu uplands, Eastern Ghats. *Zoological Survey of India* Part-I: 24–29.

Hegde, V. 1986b. Additional notes on bugs from Tamil Nadu uplands, Eastern Ghats. *Zoological Survey of India* Part-II: 26–29.

Hegde, V. 1995. Heteroptera (Insecta) from the Eastern Ghats, India. *Occasional Paper on Records of Zoological Survey of India* 168: 14–39.

Joshi, E. E., R. C. Joshi, and M. A. Florague. 2007. Selected knowledge-based materials on rice black bugs available on the world wide web. *In*: Joshi, R. C., A. T. Barrion & L. S. Sebastian (Eds), *Rice Black Bugs. Taxonomy, Ecology, and Management of Invasive Species*. Philippine Rice Research Institute, Science City of MuZoz, Philippines. pp.751–758.

Kalshoven, L. G. E. 1981. *Pests of Crops in Indonesia*. Revised and translated by P. A. Van Der Laan., P.T. Ichtiar Basu-Van Hoeve, Jakarta, xix + 701 pp.

Kamaluddin, S. 1982. Morphotaxonomy and larval systematics of Phyllocephalinae Amyot et Serville (Pentatomomorpha: Pentatomidae) of Indo-Pakistan subcontinent with special reference to phylogeny. PhD Dissertation (Unpublished), University of Karachi, 292 pp.

Kamaluddin, S., and I. Ahmad. 1988. A revision of the tribe Phyllocephalini (Hemiptera: Pentatomidae: Phyllocephalinae) from Indo-Pakistan subcontinent with description of five new species. *Oriental Insects* 22, 185–240.

Kamaluddin, S., and I. Ahmad. 1994. Cladistic analysis of the tribe Phyllocephalini Amyot and Serville (Hemiptera: Pentatomidae: Phyllocephalinae) from Indo-Pakistan subcontinent. *Proceedings of the Pakistan Congress of Zoology* 14: 89–96.

Kamaluddin, S., and I. Ahmad. 1995. A new tribe of the subfamily Phyllocephalinae Amyot et Serville (Hemiptera: Pentatomidae) from Indo-Pakistan subcontinent and their relationships. *Acta Entomologica Musei Nationalis Pragae* 44: 321–326.

Kaur, H., S. Devinder, and S. Vikas. 2012. Faunal diversity of terrestrial Heteroptera (Insecta: Hemiptera) in Punjab, India. *Journal of Entomological Research* 36(2): 177–181.

Kaur, R., D. Singh, and H. Kaur, 2013. Taxonomic significance of external genitalia in differentiating four species of genus *Carbula* Stål (Heteroptera: Pentatomidae) from North India. *Journal of Entomology and Zoology Studies* 1(3): 33–42.

Kirkaldy, G. W. 1909. *Catalogue of the Hemiptera (Heteroptera) with Biological and Anatomical References, Lists of Food Plants and Parasites etc.* Vol. I Cimicidae (M). Berlin, Germany, XI + 392 pp.

Konstantinov, F. V., and D. A. Gapon. 2005. On the structure of the aedeagus in shield bugs (Heteroptera, Pentatomidae): 1. Subfamilies Discocephalinae and Phyllocephalinae. *Entomological Review* 85(3), 221–235. Translated from *Entomologicheskoe Obozrenie,* 84(2): 334–352.

Lefroy, M. H., and Howlett, F. M. 1909. *Indian Insect Life, A Manual of the Insects of the Plains (Tropical India)*. Today and Tomorrow's Printers and Publishers, New Delhi, India. 786 pp. [reprinted edition].

Lenin, L. A., and S. J. Rajan. 2016. Biology of predatory bug *Eocanthecona furcellata* Wolff (Hemiptera: Pentatomidae) on *Corcyra cephalonica* Stainton. *Journal of Entomology and Zoology Studies* 4(3): 338–340.

Linnavuori, R. E. 1982. Pentatomidae and Acanthosomatidae (Heteroptera) of Nigeria and the Ivory Coast, with remarks on species of the adjacent countries in West and Central Africa. *Acta Zoologica Fennica* no. 163, 176 pp.

Mathew, K. 1969. A new species of *Degonetus* Distant (Hemiptera: Heteroptera, Pentatomidae) from Sikkim, India. *Oriental Insects* 3(2): 197–198.

Mathew, K. 1977. Studies on *Agonoscelis* from Southern India (Hemiptera: Heteroptera: Pentatomidae). *Oriental Insects* 11(4): 521–530.

Mathew, K. 1980. A new species of *Gynenica* from south India (Heteroptera: Pentatomidae). *Oriental Insects* 14(3): 379–382.

Mathew, K. 1986. On a collection of Pentatomidae (Hemiptera) from Silent Valley, Kerala, India. *Records of Zoological Survey of India* 84(1–4): 35–47.

McDonald, F. J. D. 1966. The genitalia of North American Pentatomoidea (Hemiptera: Heteroptera). *Quaestiones Entomologicae* 2: 7–150.

Memon, N. 2002. A revision of the berry bugs (Heteroptera: Pentatomoidea: Halyini) of Indo-Pakistan subcontinent with special reference to cladistic analysis of halyine genera. PhD thesis (unpublished), University of Karachi, Pakistan, 409 pp.

Memon, N., and I. Ahmad. 2002. A new genus and a new species of Halyini Stal (Hemiptera: Heteroptera: Pentatominae). *Pakistan Journal of Zoology* 34(3): 189–192.

Memon, N., F. Gilbert, and I. Ahmad. 2011. Phylogeny of the South Asian Halyine stink bugs (Hemiptera: Pentatomidae: Halyini) based on morphological characters. *Annals of Entomological Society of America* 104(6): 1149–1169.

Panizzi, A. R. 1997. Wild hosts of pentatomids: Ecological significance and role in their pest status on crops. *Annual Review of Entomology* 42: 99–122.

Panizzi, A. R., J. E. Mcpherson, D. G. James, M. Javahery, and R. M. Mcpherson. 2000. Stink bugs (Pentatomidae). *In*: C. W. Schaefer & Panizzi, A. R. (Eds), *Heteroptera of Economic Importance*, CRC Press, Boca Raton, FL, pp. 421–474.

Parveen, S., and A. Gaur. 2012. Redescription of Pentatomid *Spermatodes variolosa* (Walker) from India. *Indian Journal of Entomology* 74(2): 154–158.

Pawar, A. D. 1973. A new record of Pentatomoidea (Hemiptera) from India. *Current Science* 42(11): 392–393.

Prasad, C. S., and R. Pal. 2015. First record of two spotted stink bug, *Perillus bioculatus* (Fab.) from Meerut (U.P.) North India. *International Journal of Environmental and Agriculture Research (IJOER)* 1(3): 9–11.

Rai, K., and K. N. Trehan. 1964. The internal anatomy of *Bagrada cruciferarum* Kirk. (Heteroptera: Pentatomidae). *Research Bulletin of the Panjab University (ns)* 15(3–4): 339–343.

Rakshpal, R. 1949. Notes on the biology of *Bagrada cruciferarum* Kirk. *Indian Journal of Entomology* 11(1): 11–16.

Rana, N. A., and I. Ahmad. 1988. A revision of the genus *Zicrona* Amyot et Serville (Hemiptera: Pentatomidae: Pentatominae: Asopini) from Indo-Pakistan sub-continent with description of a new species from Pakistan, and their distribution and relationships. *Sarhad Journal of Agriculture* 4(5): 645–657.

Rider, D. A. 2000. Stirotarsinae, new subfamily for *Stirotarsus abnormis* Bergroth (Heteroptera: Pentatomidae). *Annals of Entomological Society America* 93(4): 802–806.

Rider, D. A. 2018. Pentatomoidea Home Page. North Dakota State University, Fargo, N. D., Available online at http://www.ndsu.nodak.edu/ndsu/rider/Pentatomoidea/ (accessed on May 30, 2018).

Rider, D. A., C. F. Schwertner, J. Vilímová et al. 2018. Higher systematics of the Pentatomoidea. *In*: McPherson, J. E. (Ed.), *Invasive Stink Bugs and Related Species (Pentatomoidea) Biology, Higher Systematics, Semiochemistry and Management*. CRC Press, Tayler & Francis Group, Boca Raton, FL, pp. 25–201.

Roell, T., and L. A. Campos. 2015. *Candeocoris bistillatus*, new genus and new species of Ochlerini from Ecuador (Hemiptera: Heteroptera: Pentatomidae). *Zootaxa* 4018(4): 573–583.

Rolston, L. H., and F. J. D. McDonald. 1979. Keys and diagnoses for the families of Western Hemisphere Pentatomoidea, subfamilies of Pentatomidae and tribes of Pentatominae (Hemiptera). *Journal of the New York Entomological Society* 87(3): 189–207.

Salini, S. 2006. Faunistic Studies on Pentatomidae (Hemiptera: Pentatomoidea) in Karnataka. MSc Thesis (unpublished), Department of Agricultural Entomology, University of Agricultural sciences, Bangalore, India, pp. 152.

Salini, S. 2011. Polymorphism in southern green stink bug, *Nezara viridula* (L.) (Hemiptera: Pentatomidae). *Current biotica* 4(4): 482–485.

Salini, S. 2015. Systematic studies on Pentatomidae (Hemiptera: Pentatomoidea) of South India. Department of Agricultural Entomology, University of Agricultural Sciences, Bangalore, India, PhD Dissertation, (unpublished) pp. 385.

Salini, S. 2016a. Redescription of *Dardjilingia* Yang (Hemiptera: Pentatomidae) from India. *Zootaxa* 4144(1): 131–137.

Salini, S. 2016b. Redescription of predatory stink bug, *Amyotea malabarica* (Fabricius, 1775) (Hemiptera: Pentatomidae: Asopinae). *Journal of Biological Control* 30(4): 240–247.

Salini, S. 2017a. First record of *Brachycoris* Stål (Hemiptera: Heteroptera: Pentatomidae) in India, with description of a new species. *Zootaxa* 4236(3): 563–572.

Salini, S. 2017b. First record of *Neojurtina typica* from India (Hemiptera: Heteroptera: Pentatomidae). *Journal of Threatened Taxa* 9(4): 10133–10137.

Salini, S., and V. V. Belavadi. 2015. External scent efferent system of Pentatomidae (Hemiptera: Pentatomoidea). *Mysore Journal of Agricultural Sciences* 49(2): 324–328.

Salini, S., and C. Schmidt. 2018. Revision of genus *Acrozangis* Breddin (Hemiptera: Pentatomidae: Pentatominae) with description of a new species from India. *Zootaxa* 4413(3): 507–523.

Salini, S., and C. A. Viraktamath. 2015. Genera of Pentatomidae (Hemiptera: Pentatomoidea) from south India—An illustrated key to genera and checklist of species. *Zootaxa* 3924(1): 1–76.

Schaefer, C. W., and I. Ahmad. 1987. A cladistic analysis of the genera of the Lestonocorini (Hemiptera: Pentatomidae: Pentatominae). *Proceedings of the Entomological Society of Washington* 89(3): 444–447.

Schuh, R. T., and J. A. Slater. 1995. *True Bugs of the World (Hemiptera: Heteroptera) Classification and Natural History*. Cornell University Press, Ithaca, NY, Xii + 336 pp.

Schwertner, C. F., and J. Grazia. 2012. Review of the neotropical genus *Aleixus* McDonald (Hemiptera: Heteroptera: Pentatomidae: Procleticini), with description of a new species and cladistic analysis of the tribe Procleticini. *Entomologica Americana* 118(1–4): 252–262.

Shophiya, N. J., and K. Sahayaraj. 2014. Biocontrol potential of entomophagous predator, *Eocanthecona furcellata* (Wolff.) against *Pericallia ricini* (Fab.) larvae. *International Journal of Current Research* 6(10): 9052–9056.

Singh, H., and V. S. Malik. 1993. Biology of painted bug (*Bagrada cruciferarum*). *Indian Journal of Agricultural Sciences* 63(10): 672–674.

Srikumar, K. K., S. Salini, S. Smitha, B. S. Kumar, and B. Radhakrishnan. 2018. Taxonomy, bionomics and predatory potential of *Eocanthecona concinna* (Walker) (Hemiptera: Pentatomidae: Asopinae). *Journal of Biological Control* 32(2): 81–86.

Tembe, S., Y. Shouche, and H. V. Ghate. 2014. DNA barcoding of Pentatomomorpha bugs (Hemiptera: Heteroptera) from Western Ghats of India. *Meta Gene* 2: 737–745.

Thakur, N. S. A., and D. M. Firake. 2012. *Ochrophora montana* (Distant): A precious dietary supplement during famine in northeastern Himalaya. Scientific correspondence. *Current Science* 102(6): 845–846.

Thomas, D. B. 1994. Taxonomic synopsis of the Old World Asopine genera (Heteroptera: Pentstomidae). *Insecta Mundi* 8(3–4): 145–212.

Usman, S., and M. Puttarudriah. 1955. *A List of the Insects of Mysore Including Mites,* Entomology Series Bulletin. Department of Agriculture, Government of Mysore state, 16: vi + 194 pp.

Verma, A. K., S. K. Patyal, O. P. Bhalla, and K. C. Sharma. 1993. Bioecology of painted bug (*Bagrada cruciferarum*) (Hemiptera, Pentatomidae) on seed crop of cauliflower (*Brassica oleracea* var. *botrytis* subvar. *cauliflora*). *Indian Journal of Agricultural Sciences* 63(10): 676–678.

Vidyasagar, P. S. P. V., and K. S. Bhat. 1986. A pentatomid bug causes tender nut drop in arecanut. *Current Science* 55(21): 1096–1097.

Waghmare, S. H., G. P. Bhawane, Y. J. Koli, and S. M. Gaikwad. 2015. A case of extensive congregation of man-faced stink bug *Catacanthus incarnatus* (Drury) (Hemiptera: Pentatomidae) together with new host records from western Maharashtra, India. *Journal of Threatened Taxa* 7(8): 7490–7492.

Waghmare, S. H., and S. M. Gaikwad. 2017. First record of the predatory stinkbug *Eocanthecona concinna* (Walker, 1867) (Pentatomidae: Asopinae) from India. *Journal of Threatened Taxa* 9(2): 9870–9873.

Weintraub, P. G., and L. Beanland. 2006. Insect vectors of phytoplasma. *Annual Review of Entomology* 51: 91–111.

9 Indian Mymaridae (Hymenoptera: Chalcidoidea)

S. Manickavasagam and S. Palanivel

CONTENTS

Introduction 147
Family Diagnosis 148
Subfamilies and Tribes 148
    Subfamilies and Tribes Proposed by Earlier Workers 149
Key to Indian Genera of Mymaridae (Females) 149
Major Work on Indian Fauna 151
Biodiversity and Distribution Patterns in India 152
Immature Taxonomy 152
Molecular Characterization and Phylogeny 152
Taxonomic Problems 152
Biology 152
Economic Importance 152
Conservation Status 153
Collection and Preservation 154
    Collection Methods 154
    Rearing of Host Eggs 154
    Preservation 154
Useful Websites 154
Conclusion 154
References 154

INTRODUCTION

Among the parasitic Hymenoptera, the chalcidoid family Mymaridae, commonly called fairyflies, are important in the natural management of many insects. They are internal, primary parasitoids on insect eggs, especially Auchenorrhyncha. Eggs of Hemiptera (Cicadellidae, Membracidae, Coccidae, Miridae, etc.), Coleoptera (Curculionidae and Chrysomelidae), Diptera (Tephritidae), Odonata (dragonflies and damselflies), Orthoptera (Tettigoniidae), Psocoptera (booklice), and Thysanoptera (thrips) are also parasitized by them. When reviewing the host range of Mymaridae, Huber (1986) made it clear that mymarids preferred concealed to naked eggs.

Some mymarids can swim in water to parasitize the eggs of aquatic insects (Matheson and Crosby 1912; Jackson 1966). Among the aquatic mymarids, *Caraphractus cinctus* Walker parasitizes the submerged eggs of dytiscids (Coleoptera) (Hagen 1996), while *Anagrus amazonensis* Triapitsyn, Querino, and Feitosa parasitize the eggs of Odonata (Zygoptera) deposited on *Rhynchospora pubera* Bockeler (Cyperaceae) (Triapitsyn et al. 2008). Feitosa et al. 2016 found *A. amazonensis* parasitizing the eggs of Hemiptera, Odonata, and Lepidoptera and listed the 12 associated aquatic plant species. No mymarid has so far been reported from the eggs aquatic hosts in India. Mymaridae were considered as exclusively oophagous parasitoids till Huber et al. (2006) reported two new species of Mymaridae, *Stethynium ophelimi* Huber and *S. breviovipositor* Huber, from the larvae of *Ophelimus maskelli* (Ashmead), a gall inducing pest of *Eucalyptus camalduensis* Dehnh. Similarly, *Camptoptera dryophantae* was recorded from cynipid (*Dryophanta folii* Linn.) galls in Europe (Kieffer 1902) and *Narayanella pilipes* (Subba Rao) from galls of *Lagerstroemia speciosa* Linn. (Lythraceae) in India and Myanmar (Subba Rao 1976).

Mymarids are usually solitary parasitoids with a few being gregarious in habit. They are very tiny insects, with an average size of only 0.5–1.00 mm and are black, brown, or yellow in colour. The diversity of mymarids is yet to be ascertained as the mymarid fauna from all parts of the world remains incompletely known. The classification of this group is in an elementary stage, and their abundance and economic importance is yet to be understood (Noyes and Valentine 1989). Huber (1986) reviewed the systematics and status of biological research of mymarids. Keys to the world genera of mymarids were first given by Annecke and Doutt (1961). Other major regional taxonomic contributions were by DeBauche (1949) for African genera, Kryger (1950) for Europe, Subba Rao and Hayat (1983) for the Oriental region, Schauff (1984) for Holarctic genera, Noyes and Valentine (1989) for New Zealand fauna, Yoshimoto (1990) for New World genera, Huber (1997) for Nearctic genera, Triapitsyn and Huber (2000) for Palaearctic genera, Lin et al. (2007) for Australian genera, and Huber et al. (2009) for genera of the Arabian Peninsula.

FAMILY DIAGNOSIS (MODIFIED FROM LIN ET AL. 2007)

The body is usually 1.5 mm or less in length. The head has an H-like pattern [transverse membranous suture or trabecula (Figure 9.1[**1**]) or carina below anterior ocellus and also along inner eye margins, rarely with additional sutures behind ocelli]. The antennal torulus is usually only about its own diameter from the eye, being conspicuously closer to the eye than to the other torulus (Figure 9.1[**1**]). The antenna of a mymarid is depicted in Figure 9.1(**3**). The scutellum is normally divided transversely into an anterior and posterior part (Figure 9.1[**2**]) and fore wing venation is greatly reduced. Hind wing is stalked or petiolate, the membrane of the disc usually originates apically from stalk; rarely it is a stalk only. Tarsi are 4- or 5-segmented (3-segmented in *Kikiki*).

SUBFAMILIES AND TRIBES

Ashmead (1904) classified Mymaridae into subfamilies based on the number of tarsal segments and further to tribes based on attachment of gaster with propodeum and projection of mesopostphragma into gaster. Girault (1911, 1912) followed Ashmead's classification, but later (Ashmead 1904) gave more importance to mode of attachment of gaster with propodeum for classifying subfamilies and used number of tarsal segments to denote tribes. Annecke and Doutt (1961) also followed Ashmead's (1929) system except that they added Anaphini in tetramerous Mymaridae. Yoshimoto et al. (1972) added a third subfamily Eubroncinae mainly because of the peculiar shaped head and enormously developed mandibles. Subba Rao and Hayat (1983) followed Ashmead's (1929) system of classification while dealing with Oriental Mymaridae. However, they did not accept Anaphini and Eubroncinae proposed by the earlier authors, as these tribes fail to reflect natural affinities of the included genera.

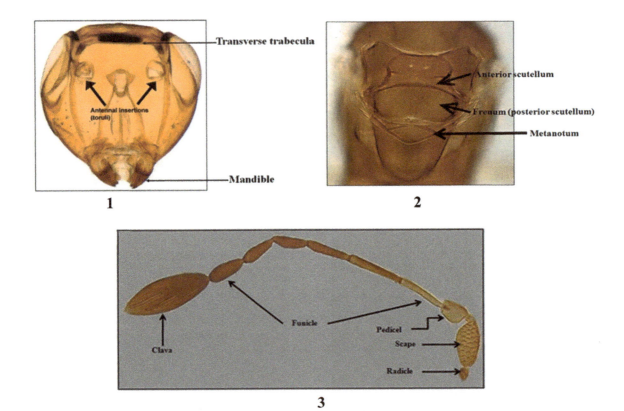

FIGURE 9.1 1. Head (frontal view) 2. Mesosoma 3. Antenna.

Subfamilies and Tribes Proposed by Earlier Workers

Ashmead (1904)
1. Tarsi 5-segmented (Gonatocerinae) 2
- Tarsi 4-segmented (Mymarinae) 3
2. Gaster sessile or subsessile .... Gonatocerini
- Gaster petiolate .... Ooctonini
3. Gaster sessile or subsessile .... Anaphini
- Gaster petiolate .... Mymarini

Girault (1929)
1. Convexly rounded, subsessile or petiolate gaster (Mymarinae) 2
- Truncately sessile gaster (Alaptinae) 3
2. Tarsi 4-segmented .... Mymarini
- Tarsi 5-segmented .... Ooctonini
3. Tarsi 4-segmented .... Anagrini
- Tarsi 5-segmented .... Alaptini

Subba Rao and Hayat (1983)
1. Gaster broadly attached to propodeum; mesopostphragma plainly projecting into gaster (Alaptinae) 2
- Gaster convexly rounded at base, with a more or less distinct petiole; mesopostphragma not or hardly projecting into gaster (Mymarinae) 3
2. Tarsi 4-segmented .... Anagrini
- Tarsi 5-segmented .... Alaptini
3. Tarsi 4-segmented .... Mymarini
- Tarsi 5-segmented .... Ooctonini

Viggiani (1989)
He classified subfamilies based on male genitalia and proposed three subfamilies *viz.*, Lymaenoninae (type 1 male genitalia), Mymarinae (type 2 male genitalia), Camptopterinae (type 3 male genitalia). He recognized two tribes under Lymaenoninae (Lymaenonini, Stethynini), eight tribes under Mymarinae (Aresconini, Ooctonini, Erythmelini, Alaptini, Anagrini, Mymarini, Anaphini, and Ptilomymarini), and did not recognize tribes under Camptopterinae.

KEY TO INDIAN GENERA OF MYMARIDAE (FEMALES)

(Modified from Lin et al. 2007)
1. Tarsi 3-segmented .... ***Kikiki* Huber**
- Tarsi 4-5-segmented 2
2(1) Tarsi 5-segmented 3
- Tarsi 4-segmented 16
3(2) Gaster with petiole distinct, narrow, at most about one third width of propodeal apex and varying in length from shorter than wide to distinctly longer than wide; mesophragma not projecting into gaster 4
- Gaster broadly joined to propodeum, the petiole indistinct and apparently almost as wide as propodeal apex or gaster (actually only about half as wide), much shorter than wide, ring-like; mesophragma projecting into gaster 12
4(3) Funicle 8-segmented 5
- Funicle 5- or 7-segmented or less than 8-segmented ... 10
5(4) Petiole short, wider than long; hypochaeta about midway between proximal and distal macrochaeta; propodeum smooth or with longitudinal sub-median carinae 6
- Petiole at least as long as wide; hypochaeta much closer to proximal macrochaeta than to distal macrochaeta; propodeum with well developed carinae .... ***Ooctonus* Haliday**
6(5) Dorsellum strap shaped, narrow, at least 5 × as wide as long, with anterior and posterior margins parallel; pronotum longitudinally divided, with two or three lobes; propodeum almost always with two longitudinal converging sutures; ocellar triangle with three or four setae .... ***Lymaenon* Walker**
- Dorsellum rhomboidal, triangular, or biconvex, usually much less than 5 × as wide as long and the anterior margin not parallel with posterior margin; ocellar triangle with two setae 7
7(6) Face with sub-antennal sulcus extending from ventral margin of each torulus to mouth margin; ovipositor slightly exserted, less than 0.33 × of gaster 8
- Face without sub-antennal sulcus; ovipositor exserted, at least 0.33 × of gaster .... ***Zeyanus* Huber**
8(7) Pronotum longitudinally divided 9
- Pronotum entire .... ***Tanyxiphium* Huber**
9(8) Fore wing with microtrichia uniformly distributed to base of parastigma and usually fairly narrow; dorsellum triangular to rhomboidal and sometimes margined with lighter colour; funicle of female often with F_1 obliquely truncate dorsoapically, and F_2 and F_3 almost longer than following funicle segments .... ***Gonatocerus* Nees**
- Fore wing almost always bare behind venation, then microtrichia usually not uniformly distributed and less densely spaced than those beyond venation; other features not as above .... ***Cosmocomoidea* Howard**
10(4) Funicle 5-segmented; fore wing relatively broad; fore wing venation almost two-third wing length .... ***Arescon* Walker**
- Funicle 7-segmented or, apparently, 6-segmented (segment 2 ring-like and easily overlooked); fore wing narrow and distinctly curved at apex; fore wing venation much less than half wing length 11
11(10) Propodeum with translucent, mesh-like lamellae; fore wing relatively broad, with numerous, scattered microtrichia .... ***Stephanocampta* Mathot**
- Propodeum without such lamellae; forewing relatively narrow, with few microtrichia in one or two rows .... ***Camptoptera* Forster**
12(3) Funicle 5-segmented; hind margin of fore wing excised beneath venation (scutellum and Frenum (postscutellum) not strongly sculptured) .... ***Alaptus* Westwood**
- Funicle 6- or 7-segmented; hind margin of fore wing convexly rounded, not excised 13

13(12) Funicle 6-segmented; mesoscutum, frenum and propodeum strongly sculptured.........***Litus* Haliday**
- Funicle 7-segmented; mesoscutum, frenum and propodeum not strongly sculptured..................... **14**
14(13) Petiole apparently absent (F2 either ring-like or not, fore wing with posterior margin either straight or slightly curved at apex, wing widening gradually towards apex).................... ***Callodicopus* Ogloblin**
- Petiole indistinct and apparently almost as wide as propodeal apex or gaster much shorter than wide, ring-like (mesophragma projecting into gaster)........ **15**
15(14) Head in anterior view relatively long and narrow below eyes, together with ventrally pointing mandibles giving head a beak-like appearance; mandible with one long and one short tooth; scape either relatively short, widest medially and with two ventral, setate denticles or very long swollen apically and without denticles; fore wing very narrow medially; with posterior margin evenly curved medially and almost straight towards apex...........................
....***Dicopus* Enock**
- Head in anterior view shorter and wider below eyes, not beak-like; mandibles pointing towards each other and with two subequal teeth; scape with ventral setae not on denticles; fore wing slightly wider medially than above, with posterior margin straight medially and more abruptly curved towards apex....
....***Dicopomorpha* Ogloblin**
16(2) Funicle 8-segmented (except *Ptilomymar heptafuniculata*, 7-segmented); propodeum with sub-median longitudinal pair of tall, translucent, areolate carinae and propodeal seta branched; gastral tergum 1 sub-lateral pair of translucent, areolate carinae ....***Ptilomymar* Annecke and Doutt**
- Funicle at most 6-segmented; propodeum and gaster without such carinae and propodeal seta not branched; gastral tergum 2 without carinae **17**
17(16) Gaster petiolate, the petiole tube-like and slightly to considerably longer than wide, rarely somewhat wider than long; phragma not projecting into gaster **18**
- Gaster appearing sessile or subsessile, the petiole ring-like, wider than long and barely recognizable or narrower and distinguishable; phragma usually projecting at least slightly into gaster................ **27**
18(17) Body minute (about 0.3 mm); fore wing very narrow and slightly, but distinctly curved apically as in *Camptoptera*; mandible apparently with only one tooth, sharply pointed; tarsi apparently 4-segmented, but actually 5-segmented with the apical two segments broadly fused…............. ***Eofoersteria* Mathot**
- Body longer (usually more than 0.5 mm), fore wing wider and not curved apically; tarsi distinctly 4-segmented... **19**
19(18) Mandible pointing ventrally, not crossing each other medially, with several small teeth on ventral surface; hind wing relatively wide, with rounded apex.........**20**
- Mandibles pointing towards each other, crossing each other medially, with usually three equal, normal-sized teeth on inner surface; hind wing with relatively narrow apex **21**
20(19) Head in lateral view only slightly triangular, not longer than high with small projection between toruli; mandibles not longer than width of mouth opening; antenna double geniculate and first funicle segment longer than any other segment and about as long as pedicel.....................***Anagroidea* Girault**
- Head in lateral view strongly and sharply triangular, much longer than high and with large, distinct shelf projecting between toruli; mandible at least as long as width of mouth opening; antenna not double geniculate and F1 shortest of all segments and shorter than pedicel...
......***Eubroncus* Yoshimoto, Kozlov and Trjapitsyn**
21(19) Fore wing very narrow, oar-shaped, with a long narrow petiolate basal half or more, and short, oval, partly infuscate blade; hind wing filamentous; antennal scape constricted medially.....***Mymar* Curtis**
- Fore wing not oar-shaped; hind wing not filamentous (sometimes membrane very narrow); antennal scape not constricted medially **22**
22(21) Hind leg with very long spine-like setae; hind coxa longer than petiole; fore wing with the discal setae arranged in curved and alternating strong and weaker rows; last segment of funicle like a segment of clava ....***Narayanella* Subba Rao**
- Hind leg without such long setae; hind coxa shorter than petiole; fore wing with the discal setae not arranged in curved rows; clava clearly differentiated from last funicle segment **23**
23(22) Petiole attached to gastral tergum **24**
- Petiole attached to gastral sternum..................... **26**
24(23) Face with small pit sub-medially next to each torulus ....***Himopolynema* Taguchi**
- Face without a pit next to each torulus................ **25**
25(24) Propleura abutting each other anteriorly along midline, the prosternum thus closed anteriorly; fore wing usually narrow and slightly narrowing just beyond apex of venation; propodeum smooth, without carinae; pro-and mesothorax with enlarged and blunt or cuspidate setae ....***Palaeoneura* Waterhouse**
- Propleura not abutting anteriorly along midline, the prosternum thus open anteriorly; fore wing usually wider just beyond apex of venation than at marginal vein; propodeum with at least an incomplete median carina; pro- and mesothoracic setae usually normal, neither blunt nor cuspidate at apices...........
.... ***Polynema* Haliday**
26(23) Propodeum with V-shaped sub-median carinae (except in *Acmopolynema campylurum*, propodeum with a medial groove extending from anterior margin to base of sub-medial carinae at posterior margin); antennal scape without imbricate,

rasp-like sculpture, but with cross-ridges on inner surface; vertex usually without a depression outside each ocellus; prothoracic spiracle at posterolateral angle of pronotum ....*Acmopolynema* **Ogloblin**

- Propodeum without median carinae; antennal scape with imbricate, rasp-like sculpture on inner surface; vertex with a wide, shallow depression outside each ocellus; prothoracic spiracle advanced forward, near anterior apex of notauli ....*Stephanodes* **Enock**

27(17) Frenum usually divided medially by a longitudinal sulcus (sometimes only in half or less from anterior scutellum); gaster appearing sessile with only a slight dorsal or lateral constriction between mesosoma and metasoma; phragma usually projecting into gaster 28

- Frenum entire; gaster appearing subsessile, with a more definite dorsal and lateral constriction between mesosoma and metasoma, the petiole ring-like; phragma usually not projecting into gaster or only slightly so..33

28(27) Clava entire; protibial spur comb-like...................... ....*Anagrus* **Haliday**

- Clava 2- or 3-segmented; protibial spur not comb-like **29**

29(28) Clava 2-segmented **30**
- Clava 3-segmented **31**

30(29) Ovipositor strongly exserted (for almost entire length of metasoma) beyond apex of metasoma; fore wing narrow and somewhat pointed at apex..................... ....*Omyomymar* **Schauf**

- Ovipositor at most only slightly exserted beyond apex of metasoma; fore wing broad with rounded apex...............................*Schizophragma* **Ogloblin**

31(29) Fore wing narrow; discal blade sharply pointed distally..............................*Platystethynium* **Ogloblin**

- Fore wing with or without distinct round lobe on hind margin behind venation **32**

32(31) Clava compact, with sutures usually oblique; fore wing with distinct rounded lobe on hind margin behind venation ....*Stethynium* **Enock**

- Clava loose, with transverse or only slightly oblique, complete, and distinct sutures; fore wing without distinct rounded lobe on hind margin behind venation.......................*Allanagrus* **Noyes & Valentine**

33(27) Ovipositor much shorter than length of gaster, arising in apical half at about level of gastral tergum 4 (fore wing parallel-sided or almost so, knife-like, at least 8x as long as wide, and posterior margin with weak lobe behind apex of venation).......................... ....*Cleruchus* **Enock**

- Ovipositor about length of gaster, arising near base usually at or before level of gastral tergum 2 ....**34**

34(33) Clava entire..**35**
- Clava 3-segmented (fore wing with curved dark mark behind venation and with relatively long marginal fringe setae).........*Pseudanaphes* **Noyes & Valentine**

35(34) Fore wing venation at least half wing length; ovipositor distinctly exserted, often for at least length of gaster and sheaths with at least a few setae along length ....*Australomymar* **Girault**

- Fore wing venation at most about one-third wing length; ovipositor not or barely exserted and sheaths without setae **36**

36(35) Clava at least as long as four or five preceding funicle segments, widest in basal third and tapering to apex ....*Dorya* **Noyes and Valentine**

- Clava shorter, symmetrical, and more broadly rounded apically, usually appearing oval **37**

37(36) Hypopygium short, inconspicuous; head in lateral view relatively thick, the gena relatively wide so posterior margin of eye separated along entire length from back of head; mandibles normal, crossing medially and with three equal teeth; fore wing with single, socketed seta on posterior margin near apex of retinaculum and basal to the marginal setae; body usually black or dark brown *fuscipennis* group ....*Anaphes* **Haliday**

- Hypopygium well developed, extending almost to apex of gaster; head in lateral view thin, the gena very narrow so posterior margin of eye, at least dorsally, touching back of head; mandibles minute, not meeting medially and apparently without teeth; fore wing without socketed seta as above; body mostly yellow or light brown.......*Erythmelus* **Enock**

MAJOR WORK ON INDIAN FAUNA

The mymarid fauna of the Indian subcontinent was initially studied by Mani (1938), Narayanan et al. (1960), Subba Rao (1966), Mani and Saraswat (1973), and Husain and Agarwal (1981). Subba Rao and Hayat (1983) catalogued the Oriental species of mymarids and provided a key to the genera and recognized 20 genera and 67 species. The other major contributors were Hayat (1992), Hayat and Anis (1999a, 1999b, 1999c), Hayat and Singh (2001), Hayat and Khan (2009), Hayat et al. (2003, 2008), Rehmat et al. (2009), Rehmat and Anis, (2014a, 2014b, 2015, 2016a, 2016b), Anis and Rehmat (2013), Anwar et al. (2014a, 2014b), Zeya and Hayat (1995), Zeya and Khan (2012), Zeya (2011), Zeya et al. (2014), Amer and Zeya (2016, 2018), Amer et al. (2016, 2017)–all from Aligarh Muslim University, largely on the mymarids of North India. The South Indian fauna was mainly studied by researchers from Annamalai University, Tamil Nadu, and the National Bureau of Agricultural Insect Resources, Bengaluru, Karnataka. The major contributors from South India are Manickavasagam et al. (2011, 2017, 2018), Manickavasagam and Rameshkumar (2011, 2012, 2013a, 2013b), Rameshkumar et al. (2011a, 2011b, 2013, 2017), Manickavasagam and Palanivel (2013, 2014, 2015), Singh and Manickavasagam (2014), Palanivel and Manickavasagam (2015, 2016a, 2016b), Palanivel et al. (2015, 2017), Rameshkumar and Manickavasagam (2016), and Gowriprakash and Manickavasagam (2016).

BIODIVERSITY AND DISTRIBUTION PATTERNS IN INDIA

Mymaridae currently includes 116 genera and 1628 species worldwide (Noyes 2018), with 194 species in 38 genera from India (Manickavasagam and Athithya 2018). The genus *Gonatocerus* Nees is highly cosmopolitan and most speciose with over 400 species globally. In India, it is represented by 60 species. However, this genus was reclassified recently by Huber (2015), wherein he split this single genus into 14 genera of which only five [*viz., Cosmocomoidea* Howard (11 species), *Gonatocerus* Nees (10 species), *Lymaenon* Walker (29 species), *Tanyxiphium* Huber (1), and *Zeyanus* Huber (9 species)] are known from India.

The other common genera from India are (number of species in parentheses) *Polynema* (15), *Acomopolynema* (11), *Anagrus* (10), *Alaptus* (8), *Camptoptera* (8), *Mymar* (8), *Erythmelus* (6), *Dicopormorpha* (6), *Omyomymar* (6), *Himopolynema* (5), *Litus* (5), *Palaeoneura* (6), *Ptilomymar* (4), *Cleruchus* (3), *Stephanocampta* (3), *Anaphes* (3), *Eubroncus* (3), *Allanagrus* (2), *Arescon* (2), *Dicopus* (2), *Eofoersteria* (2), *Ooctonus* (2), *Narayanella* (2), *Stethynium* (2), *Callodicopus* (1), *Dorya* (1), *Kikiki* (1), *Schizophragma* (1), *Stephanodes* (1), *Anagroidea* (1), *Australomymar* (1), *Pseudanaphes* (1), and *Platystethynium* (0). Of the 38 Indian genera, *Lymaenon, Polynema, Cosmocomoidea, Acmopolynema, Anagrus, Gonatocerus,* and *Zeyanus* account for 50 percent of the species known from India.

IMMATURE TAXONOMY

Studies on immature stages of mymarids with reference to India are scanty. However, the general morphology of immature stages is known (Clausen 1940). In general, the eggs of mymarids are laid inside the eggs of the host. It is ellipsoidal-, ovoid-, or spindle-shaped with a slender tapering peduncle at the anterior end. They are so minute, varying in size from 0.06 to 0.25 mm. The larvae have not been adequately studied mainly because of their exceedingly minute size and the lack of sclerotized structures. Generally, two types of first instars are recognized. One is sacciform, and the other is mymariform. The former lies in the egg fluid and is incapable of movement. The latter, which is more common, is capable of considerable movement. The tail, along with several rows of spines on the body, is utilized in making movements inside the host egg. This constant agitation lasting for 3–4 days disorganizes the contents of the egg. The number of larval instars remains uncertain, though it is thought to be three. The second instar is distinct and is known as "histriobdellid." In many species of mymarids, host eggs containing advanced stages of larvae and pupae are bright red or yellow in colour which is visible through the delicate chorion. The total life cycle from egg to adult may last for about 7–22 days depending upon the genus and species.

MOLECULAR CHARACTERIZATION AND PHYLOGENY

Work on the molecular characterization and phylogeny of Indian Mymaridae is wanting. A few such studies have been conducted elsewhere. For example, the COI gene of the *Anagrus* Haliday "*atomus*" group was studied from field collected specimens from Italy. Females were morphologically identified as *A. atomus* L. and *A. parvus* Soyka *sensu* Viggiani (= *A. ustulatus sensu* Chiappini). Alignment of COI gene sequences from this study permitted recognition of a total of 34 haplotypes. Phylogenetic and network analyses of molecular data confirmed that *A. atomus* is a species distinct from *A. parvus*. It also suggested that two species may be included within morphologically identified *A. parvus*. Different geographical distributions and frequencies of haplotypes were also evidenced. For males considered in this study, morphometric analysis revealed a character that could be useful to discriminate *A. atomus* from *A. parvus*. Both species were found in vineyards and surrounding vegetation, confirming the role of weeds as a source of parasitoids for leafhopper control in vineyards.

TAXONOMIC PROBLEMS

Among the microhymenoptera, mymarids are among the most difficult to collect, process, and study. However, Mymaridae is presently one of the best defined families in the Chalcidoidea. Care must be taken in collection as the specimens could pass out through the mesh of the net used for collection. Care must also be taken during slide making as the wasps are not only tiny, but also fragile. Without proper care they could become too transparent or remain opaque due to inadequate digestion when treated with KOH. Care must also be taken to see that parts are not lost while dissecting and mounting. Many taxonomic problems in Mymaridae have stemmed from lack of adequately prepared specimens for study.

BIOLOGY

Little is known about the biology of most species, especially those from India. Usually only a single parasitoid emerges per host egg, but at least one species is gregarious, with two or three adults emerging from the same host (Triapitsyn et al. 2003). The best studied species are those that parasitize economically important hosts like leaf and planthoppers on major crops like rice in the Oriental region (Sahad 1982; Sahad and Hirashima 1984).

ECONOMIC IMPORTANCE

Mymarids are economically very important as they parasitize the eggs of agriculturally important insect pests like leafhoppers (Cicadellidae) and planthoppers (Delphacidae) in

TABLE 9.1
Host Associations of Mymaridae

Order Hemiptera

Cicadellidae: *Empoasca vitis, Erythroneura pallifrons, Erymeloides punctata, Eurymela distincta, Erythroneura elegantula, Tettigoniella viridis, Hishimonus sellatus,* and *Typhlocyba froggatti*
Delphacidae: *Delphacodes havwardi, Delphacodes kuscheli, Delphacodes tirginus,* and *Toya propinqua*
Diaspididae: *Quadraspidiotus perniciosus*
Miridae: *Lygus elisus, L. hesperus, L. lineolaris,* and *Helopeltis antonii*
Tingidae: *Stephanitis pyri, Agramma hupehanum,* and *Pontanus puerilis*

Order Lepidoptera

Momphidae: *Pyroderces rileyi*

Order Odonata

Aeschnidae: *Aeschna brevistyla*
Coenagrionidae: *Agrion* sp.
Lestidae: *Lestes* sp.

Order Orthoptera

Tettigoniidae: *Sexava* sp., *S. novaeguineae,* and *S. nubile*

Order Hymenoptera

Cynipidae: *Dryophanta folii*

Order Coleoptera

Dytiscidae: *Agabus* sp., *A. bipustulatus, A. sturmi, Colymbetes* sp., *Dytiscus* sp., *D. marinalis, Hydroporus* sp., *llybius* sp., *l. ater,* and *l fuliginosus.*
Curculionidae: *Gonipterus* sp., *G. gibberus, G. platensis,* and *G. scutellatus*

Secondary Hosts

Rosaceae: *Prunus* sp.
Vitaceae: *Vitis vinifera*
Cyperaceae: *Rhynchospora pubera*
Lythraceae: *Lagerstroemia speciosa*
Myrtaceae: *Eucalyptus* sp., *E. dunnii, E. globosus, E. globules,* and *E. viminalis*

Source: Noyes, J.S., *Universal Chalcidoidea Database.* Worldwide Web electronic Publication, 2018; Ballal, C.R., Other egg parasitoids: Research for utilisation. In *Biological Control of Insect Pests using Egg Parasitoids.* S. Sithanantham, C.R. Ballal, S.K. Jalali, and N. Bakthavatsalam (eds.), 223–270, New Delhi, India, Springer, 2013.

different crop ecosystems (Table 9.1). In the biological control programme, mymarids have been utilized to manage insect pests. The *Eucalyptus* snout-beetle, *Gonipterus scutellatus* Gyllenhal, a serious pest of *Eucalyptus* in South Europe, South Africa, New Zealand, and South America, was managed by the introduction of *Anaphes nitens* (Girault) (=*Patasson nitens* Girault) (Tooke 1955). Another species, *Anagrus armatus* (Ashmead) was used to manage the leafhopper, *Typhlocyba froggatti* Baker, a pest of apple in New Zealand (Dumbleton 1934).

Anagrus species are egg parasitoids, mainly of Cercopidae, Cicadellidae, and Delphacidae, though some species parasitize eggs of Odonata (Bakkendorf 1926; Dozier 1936). It is one of the most important mymarid genera for biological control. Members of this genus are extremely important in paddy fields, where they parasitize eggs of rice planthoppers and leafhoppers (Sahad and Hirashima 1984). Detailed studies on its economic importance in India have not been conducted.

CONSERVATION STATUS

In the absence of any successful mass culturing techniques for mymarids, it is important that we rely on conservation methods to increase the efficiency of the existing mymarids in the field. In general, in any biological control effort, conservation of natural enemies is a critical component. This involves identifying the factor(s) which may limit the effectiveness of a particular natural enemy and modifying them to increase the effectiveness of the beneficial species or by providing resources that natural enemies need in their environment.

To be effective, mymarids may need access to alternate hosts, adult food resources, overwintering habitats (during winter in North India), constant supply of food, and appropriate microclimate (Rabb et al. 1976). In a classic example, Doutt and Nakata (1973) determined that *Anagrus epos* Girault, the principal parasitoid of the grape leafhopper, *Erythroneura elegantula* Osborne in California grape vineyards required an alternate host for overwintering. This host, another leafhopper, only overwintered on blackberry foliage, often quite distant from the vineyards. Vineyards close to natural blackberry stands experienced earlier colonization by the parasitoid in the spring and better biological control. Wilson et al. (1989) found that French prune trees which harbour another overwintering host could be planted upwind of vineyards and effectively conserve *Anagrus epos*.

COLLECTION AND PRESERVATION

COLLECTION METHODS

There are many methods for the collection of mymarids. For details, one can refer to Noyes (1982) or consult the Universal Chalcidoidea Database. However, the common methods normally adopted include the use of yellow pan and Malaise traps (Townes 1972), sweep nets, dipper nets, and the rearing of host eggs.

REARING OF HOST EGGS

Rearing is the most rewarding method of obtaining specimens, since much can be learnt about their biology. But mymarid recovery from host rearing is tedious and laborious as one has to search for the eggs of their leafhopper and planthopper hosts that are partially concealed/embedded in plant tissue.

Host eggs collected along with plant parts should be kept in polythene bags with cut ends embedded in wet cotton and observed under ambient conditions for emergence of parasitoids. Parasitoids that emerge should be aspirated and transferred to 70 percent ethyl alcohol and preserved at −20°C for processing and further study.

PRESERVATION

Three categories of permanent preservation *viz.*, liquid preservation, dry preservation, and slide mounting can be practiced (Noyes 1982).

- **Liquid preservation:** Field collected mymarids should be preserved in airtight vials containing 70 percent ethyl alcohol and stored in a deep freezer (−20°C). Before storing, locality labels (written in pencil) should be placed in the vial.
- **Dry preservation:** Mymarids can be card mounted on rectangular cards and preserved in insect cabinet boxes with camphor.
- **Slide mounting:** As mymarids are minute parasitoids, it is mandatory that they be mounted on slides for morphological study. Slides may be prepared as follows [procedure modified from Triapitsyn (personal communication)].

Remove the wings of a good card mounted specimen with a fine pin and place in a drop of balsam on a slide. Similarly detach antennae, head, thorax, and abdomen and place in 10 percent KOH for 24–48 hours (depending on the extent of sclerotization) at room temperature. Alternatively, the process may be hastened by placing the vial with the specimens in KOH on a hot plate at 40°C, taking care that the KOH solution does not dry up (if necessary, 10 percent KOH may further be added). Take care to ensure that the specimens are not excessively cleared. Once the body parts are clear, about 2/3 KOH by volume is removed and an equal volume of 35 percent acetic acid added. This solution is changed every 5–15 minutes, (the larger specimens requiring more time) in the following order: first add three drops of 35 percent alcohol and keep for 10 minutes; then remove the liquid and add three drops of 70 percent alcohol and keep for 5–15 minutes; remove the liquid and add three drops of 95 percent alcohol; and after another 5–15 minutes remove the liquid and add 100 percent ethanol.

Now add a drop of clove oil every 10–15 minutes, about 2–3 times, until the alcohol is completely removed. Remove excess clove oil, leaving just 1–2 drops. Add a drop of Canada balsam to the clove oil every 30 minutes until the resulting mixture is of the same consistency as Canada balsam. Then transfer the appendages separately onto a glass slide (1.0 mm thick) which has been cleaned thoroughly with absolute alcohol and labelled.

Place the slide in an oven at 40°C for at least 2 weeks until the first layer of balsam is hardened. The slides should be kept in covered petri plates to protect them from dust. Once the Canada balsam dries, a second layer of balsam is added to cover the parts completely. This is allowed to dry for another 2 weeks. A third drop of balsam is then added and a cover slip is carefully placed on each body part. The slide is once again dried for at least 2 more weeks.

USEFUL WEBSITES

www.nhm.ac.uk/entomology/chalcidoids/index.html
http://www.nbair.res.in/IndianMymaridae

CONCLUSION

Indian mymarids have not yet been well documented as compared to the European and American continents. A total of 194 species under 38 genera alone have been documented so far from India with about half of them being described in the last decade. Integrative taxonomic studies have to be initiated and their potential as biological control agents has to be exploited through intensive studies.

REFERENCES

Anis, S. B., and T. Rehmat. 2013. An updated checklist of fairyflies (Hymenoptera-Chalcidoidea-Mymaridae) occurring in India. *Colemania* 36: 1–12.
Amer, F. S. K., and S. B. Zeya. 2016. Two new mymarid wasps of the genus *Gonatocerus* Nees with new records from India. *Oriental Insects* 49: 16–24.
Amer, F. S. K., and S. B. Zeya. 2018. Review of the Indian species of *Palaeoneura* Waterhouse (Hymenoptera: Mymaridae). *Oriental Insects* 1–21.

Amer, F. S. K., S. B. Zeya, and K. Veenakumari. 2016. A review of the Indian species of Mymar Curtis (Hymenoptera: Mymaridae). *Journal of Insect Systematics* 3(1&2): 16–37.

Amer, F. S. K., S. B. Zeya, and M. M. Jamali. 2017. A new species of *Lymaenon* Walker (Hymenoptera: Mymaridae: Gonatocerini) from India with the country checklist and new records. *Oriental Insects* 51(4): 353–369.

Annecke, D. P., and R. L. Doutt. 1961. The genera of the Mymaridae Hymenoptera: Chalcidoidea. *Entomology Memoirs, Department of Agricultural Technical Services, Republic of South Africa* 5: 1–71.

Anwar, P. T., S. B. Zeya, and M. Veenakumari. 2014a. Two new species of *Omyomymar* Schauff (Hymenoptera: Mymaridae) from India. *Journal of Insect Systematics* 1(2): 139–144.

Anwar, P. T., S. B. Zeya, and M. Veenakumari. 2014b. First record of *Stephanocampta* Mathot (Hymenoptera: Mymaridae) from India, with description of a new species. *Journal of Insect Systematics* 1(2): 149–151.

Ashmead, W. H. 1904. Classification of the chalcid flies of the superfamily Chalcidoidea, with descriptions of new species in the Carnegie Museum, collected in South America by Herbert H. Smith. *Memoirs of the Carnegie Museum* 1(4): 225–551.

Ballal, C. R. 2013. Other egg parasitoids: Research for utilisation. In: *Biological Control of Insect Pests Using Egg Parasitoids*. S. Sithanantham, C.R. Ballal, S.K. Jalali, and N. Bakthavatsalam (eds.). pp. 223–270. New Delhi, India, Springer.

Bakkendorf, O. 1926. Recherches sur la biologie de l' *Anagrus incarnates* Haliday. *Annales de Biologie Lacustre* 14: 249–270.

Clausen, C. P. 1940. *Entomophagous Insects*. McGraw-Hill, New York. 688 pp.

DeBauche, H. R. 1949. Mymaridae (Hymenoptera: Chalcidoidea). *Exploration du ParcNational Albert, mission G.F. de Witte* 49: 1–105.

Dozier, H. L. 1936. Several undescribed mymarid egg parasites of the genus *Anagrus* Haliday. *Proceedings of the Hawaiian Entomological Society* 9: 175–178.

Doutt, R. L., and J. Nakata. 1973. The *Rubus* leafhopper and its egg parasitoid: An endemic biotic system useful in grape pest management. *Environmental Entomology* 2: 381–386.

Dumbleton, L. J. 1934. The apple leaf-hopper (*Typhlocyba australis* Frogg.). *New Zealand Journal of Science and Technology* 16: 30–38.

Feitosa, M. C. B., R. B. Ouerino, and N. Hamada. 2016. Association of Anagrus amazonensis Triapitsyn, Querino & Feitosa (Hymenoptera, Mymaridae) with aquatic insects in upland streams and floodplain lakes in Central Amazonia, Brazil. *Revista Brasileria de Entomologia* 60(3): 267–269.

Girault, A. A. 1911. Descriptions of North American Mymaridae with synonymic and other notes on described genera and species. *Transactions of the American Entomological Society* 37: 253–324.

Girault, A. A. 1912. On the occurrence of a European species of Mymaridae in North America. *Canadian Entomologist* 44: 88–89.

Girault, A. A. 1929. North American Hymenoptera Mymaridae. pp. 1–27, Addendum New insects mostly Australian. pp. 28–29. Privately printed. [Reprinted in Gordh, G., A. S. Menke, E. C. Dahms and J. C. Hall. 1979. The privately printed papers of A. A. Girault. *Memoirs of the American Entomological Institute* no. 28, Ann Arbor, Michigan. 400 pp].

Gowriprakash, J., and S. Manickavasagam. 2016. Two new species of *Omyomymar* Schauff (Hymenoptera: Mymaridae) from India with key to Oriental species. *Journal of Insect Biodiversity* 4(20): 1–8.

Hayat, M. 1992. Records of some Mymaridae from India, with notes (Hymenoptera: Chalcidoidea). *Hexapoda* 4: 83–89.

Hayat, M., and S. B. Anis. 1999a. New record of two genera *Ptilomymar* and *Himopolynema* from India, with description of two new species (Hymenoptera: Mymaridae). *Shashpa* 6: 15–22.

Hayat, M., and S. B. Anis. 1999b. The Indian species of *Acmopolynema* with notes on *Acanthomymar* (Hymenoptera: Chalcidoidea: Mymaridae). *Oriental Insects* 33: 297–313.

Hayat, M., and S. B. Anis. 1999c. The Indian species of *Polynema* with notes on *Stephanodes reduvioli* (Hymenoptera: Mymaridae). *Oriental Insects* 33: 315–331.

Hayat, M. and F.R Khan. 2009. First record of *Eubroncus* from India (Hymenoptera: Chalcidoidea: Mymaridae), with description of a new species. *Journal of Threatened Taxa* 1(8): 439–440.

Hayat, M., and S. Singh. 2001. Description of new species of *Polynema* from India with further records of *Himopolynema hishimonus* (Hymenoptera: Chalcidoidea: Mymaridae). *Shashpa* 8: 95–97.

Hayat, M., S. B. Anis, and F. R. Khan. 2008. Descriptions of two new species of Mymaridae (Hymenoptera: Chalcidoidea) from India, with some records. *Oriental Insects* 42: 327–333.

Hayat, M., M. C. Basha, and S. Singh. 2003. Descriptions of three new species of *Himopolynema* from India (Hymenoptera: Chalcidoidea: Mymaridae). *Shashpa* 10 (1): 1–6.

Huber, J. T. 1986. Systematics, biology, and hosts of the Mymaridae and Mymarommatidae (Insecta: Hymenoptera). *Entomography* 4: 185–243.

Hagen, K. S. 1996. Aquatic hymenoptera. In: *An Introduction to the Aquatic Insects of North America*. R. W. Merrit and K. W. Cummins (eds.). Kendall-Hunt Publishing Company, Dubuque, IA. 862p.

Huber, J. T. 1997. Chapter 14. Mymaridae. In: *Annotated Keys to the Genera of Nearctic Chalcidoidea (Hymenoptera)*. Gibson, G. A. P., Huber, J. T. and Woolley, J. B. (eds.). NRC Research Press, Ottawa, Canada. pp. 499–530.

Huber, J. T. 2015. World reclassification of the *Gonatocerus* group of genera (Hymenoptera: Mymaridae). *Zootaxa* 3967(1): 1–184.

Huber, J. T., Z. Mendel, A. Protasov, and J. La Salle. 2006. Two new Australian species of *Stethynium* (Hymenoptera: Mymaridae), larval parasitoids of *Ophelimus maskelli* (Ashmead) (Hymenoptera: Eulophidae) on *Eucalyptus*. *Journal of Natural History* 40: 1909–1921.

Huber, J. T., G. Viggiani, and R. Jesu. 2009. Order Hymenoptera, family Mymaridae. pp. 270–297. In: *Arthropod Fauna of the UAE*. A. van Harten (ed.). Vol. 2, 786 pp.

Husain, T., and M. M. Agarwal. 1981. A new species of the genus *Narayanella* Subba Rao (Hymenoptera: Mymaridae) from India. *Journal of Entomological Research* 5 (2): 118–120.

Jackson, D. J. 1966. Observation on the biology of *Caraphractus cinctus* Walker (Hymenoptera: Mymaridae), a parasitoid of the eggs of Dytiscidae (Coleoptera). *Transactions of the Royal Entomological Society, London* 118 (2): 23–49.

Kryger, J. P. 1950. The European Mymaridae comprising the genera known up to c. 1930. *Entomologiske Meddelelser* 26: 1–97.

Kieffer, J. J. 1902. Description de quelques cynipides nouveaux ou peu connus et deux de leurs parasites (Hymenopteres). *Bulletin de la Societe d'Histoire Naturelle de Metz* 22: 8–9.

Lin, N. Q., J. T. Huber, and J. LaSalle. 2007. The Australian genera of Mymaridae (Hymenoptera: Chalcidoidea). *Zootaxa* 1596: 1–111.

Mani, M. S. 1938. *Catalogue of Indian Insects, Part 23—Chalcidoidea*. Government of India, New Delhi, India. pp. 1–174.

Mani, M. S., and G. G. Saraswat. 1973. On some Chalcidoidea from India. Memoirs of School of Entomology. St. John's College, Agra, 2: 78–125.

Manickavasagam, S., and A. Athithya. 2018. An updated checklist of Mymaridae (Hymenoptera: Chalcidoidea) of India. *Journal of Entomology and Zoology Studies* 6 (4): 1654–1663.

Manickavasagam, S., and S. Palanivel. 2013. First report of two mymarid genera, *Cleruchus* Enock and *Kikiki* Huber and Beardsley (Hymenoptera: Mymaridae) from India. *Journal of Biological Control* 27(2): 81–82.

Manickavasagam, S., and S. Palanivel. 2014. A new mymarid of the genus *Ptilomymar* with new records from India. *Oriental Insects* 48(3&4): 323–324.

Manickavasagam, S., and S. Palanivel. 2015. Description of a new and records of other fairyfly species (Hymenoptera: Mymaridae) from India. *Journal of Insect Systematics* 2(1): 16–17.

Manickavasagam, S., and A. Rameshkumar. 2011. First report of three genera of Fairyflies (Hymenoptera: Mymaridae) from India with description of a new species of *Dicopus* and some other records. *Zootaxa* 3094: 63–68.

Manickavasagam, S., and A. Rameshkumar. 2012. First report of *Callodicopus* Ogloblin (Mymaridae) from India and new records of some Chalcidoidea (Hymenoptera) from Andaman and Nicobar Islands. *Journal of Biological Control* 26(4): 321–328.

Manickavasagam, S., and A. Rameshkumar. 2013a. Four new species of *Gonatocerus* Nees (Hymenoptera: Mymaridae) and a key to the species of *asulcifrons* group from India. *Oriental Insects* 47(1): 86–98.

Manickavasagam, S., and A. Rameshkumar. 2013b. A checklist of Mymaridae (Hymenoptera: Chalcidoidea) of India. *Madras Agricultural Journal* 100(4–6): 562–570.

Manickavasagam, S., A. Rameshkumar, and K. Rajmohana, 2011. First report of four species of fairyflies from India, Key to Indian species of four genera and additional distributional records of mymaridae (Hymenoptera: Chalcidoidea). *Madras Agricultural Journal* 98(10–12): 393–408.

Manickavasagam, S., S. Palanivel, and S. V. Triapitsyn. 2017. Two new species and additional distributional records of *Acmopolynema* Ogloblin (Hymenoptera: Mymaridae) from India. *Journal of Natural History* 51: (33–34), 1971–1987.

Manickavasagam, S., S. V. Triapitsyn, and S. Palanivel. 2018. Five new species of *Cleruchus* from the Oriental region and report of *Anaphes quinquearticulatus* (Hymenoptera: Mymaridae) from India. *Zootaxa* 4387(1): 134–156.

Matheson, R., and C. R. Crosby. 1912. Aquatic Hymenoptera in America. *Annals of Entomological Society of America* 5: 65–71.

Narayanan, E. S. Subba Rao, B. R. and R. B. Kaur. 1960. Studies on Indian Mymaridae II. (Hymenoptera: Chalcidoidea). *Beitrage zur Entomol* 10: 886–891.

Noyes, J. S. 1982. Collecting and preserving chalcid wasps (Hymenoptera: Chalcidoidea). *Journal of Natural History* 16: 315–334.

Noyes, J. S. 2018. Universal Chalcidoidea Database. Worldwide Web electronic Publication. www.nhm.ac.uk/entomology/chalcidoids/index.html (accessed June 27, 2018).

Noyes, J. S., and E. W. Valentine. 1989. Mymaridae (Insecta: Hymenoptera)—Introduction, and review of genera. *Fauna of New Zealand* 17: 1–95.

Palanivel, S., and S. Manickavasagam. 2015. Description of a new species of *Eubroncus* Yoshimoto (Hymenoptera: Mymaridae) from India, with a key to world species. *Journal of Threatened Taxa* 7 (5): 153–155.

Palanivel, S., and S. Manickavasagam. 2016a. Description of a new species of *Dorya* Noyes and Valentine (Hymenoptera: Mymaridae) from Tamil Nadu, India and key to species. *Oriental Insects* 50(1): 3–6.

Palanivel, S., and S. Manickavasagam. 2016b. Description of a new species of *Callodicopus* Ogloblin (Hymenoptera: Mymaridae) from India with key to species. *Oriental Insects* 50(2): 97–101.

Palanivel, S., S. Manickavasagam, and S. V. Triapitsyn. 2015. *Stephanocampta* Mathot (Hymenoptera: Mymaridae) descriptions of two new species and the female of *S. indica* Anwar & Zeya from India with a key to world species. *Zootaxa* 4012(3): 480–482.

Palanivel, S., S. Manickavasagam, and J. T. Huber. 2017. Review of *Allanagrus* Noyes & Valentine (Hymenoptera: Mymaridae) with a key to species. *Zootaxa* 4299(4): 507–520.

Rabb, R. L., R. E. Stinner, and R. Van den Bosch. 1976. Conservation and augmentation of natural enemies. In: *Theory and Practice of Biological Control*. C. B. Huffaker and P. S. Messenger (Eds.). Academic Press, New York. pp. 233–254.

Rameshkumar, A., and S. Manickavasagam. 2016. Descriptions of four new species of *Dicopomorpha* Ogloblin (Hymenoptera: Chalcidoidea: Mymaridae) from India with a key to Indian species. *Journal of Threatened Taxa* 8(1): 8385–8388.

Rameshkumar, A., S. Manickavasagam, and A. Jebanesan. 2011a. Diversity and new distributional records of fairyflies (Hymenoptera: Chalcidoidea: Mymaridae) from the state of Kerala, India. *Plant Archives* 11(2): 769–774.

Rameshkumar, A., S. Manickavasagam, and A. Jebanesan. 2011b. New distributional records of fairyflies (Hymenoptera: Chalcidoidea: Mymaridae) from Pudhucherry, India. *Madras Agricultural Journal* 98(7–9): 279–281.

Rameshkumar, A., S. Manickavasagam, J. Poorani, and C. Malathi. 2013. *Indian Genera of Mymaridae*. World Wide Web Electronic Publication, NBAIR (ICAR), Bangalore, India. http://www.nbair.res.in/IndianMymaridae.

Rameshkumar, A., P., Mohanraj, and K. Veenakumari. 2017. First report of *Dicopomorpha zebra* Huber (Hymenoptera: Chalcidoidea: Mymaridae) for India and distribution records of mymarids from Andaman and Nicobar Islands. *Journal of Entomology and Zoology Studies* 5(4): 228–232.

Rehmat, T., and S. B. Anis. 2014a. Description of a new species of *Eofoersteria* Mathot (Hymenoptera: Mymaridae) from India, with key to world species. *Entomon* 39(3): 129–134.

Rehmat, T., and S. B. Anis, 2014b. Record of some species of Mymaridae from India (Hymenoptera: Chalcidoidea) with description of two new species. *Journal of Insect Systematics* 1(1): 53–54.

Rehmat, T., and S. B. Anis. 2015. Record of genus *Schizophragma* Ogloblin (Hymenoptera: Mymaridae) in Oriental region, with description of a new species from India. *Oriental Insects* 48(3–4): 308–310.

Rehmat, T., and S. B. Anis. 2016a. A review of Indian species of *Polynema* Haliday (Hymenoptera: Mymaridae). *Journal of Insect Systematics* 2(2): 141,153–154.

Rehmat, T., and S. B. Anis. 2016b. Description of a new species of the genus *Litus* Haliday (Hymenoptera: Chalcidoidea: Mymaridae) from India. *Journal of Threatened Taxa* 8(3): 8615–8616.

Rehmat, T., S. B. Anis, and M. Hayat. 2009. Record of the genus *Litus* Haliday (Hymenoptera: Chalcidoidea: Mymaridae) from India, with description of two species. *Journal of Threatened Taxa* 1(7): 370–374.

Sahad, K. A. 1982. Biology and morphology of *Gonatocerus* sp. (Hymenoptera, Mymaridae), an egg parasitoid of the green rice leafhopper, *Nephotettix cincticeps* Uhler (Homoptera, Deltocephalidae).II. Morphology. *Kontyû* 50(3):467–476.

Sahad, K. A., and Y. Hirashima. 1984. Taxonomic studies on the genera *Gonatocerus* Nees and *Anagrus* Haliday of Japan and adjacent regions, with notes on their biology (Hymenoptera: Mymaridae). *Bulletin of the Institute of Tropical Agriculture, Kyushu University* 7: 1–78.

Schauff, M. E. 1984. The Holarctic genera of Mymaridae (Hymenoptera: Chalcidoidea). *Memoirs of the Entomological Society of Washington* 12: 1–67.

Singh, S., and S. Manickavasagam. 2014. Additional Fauna of hymenopterous parasitoids (Insecta: Hymenoptera) from Manipur. *Journal of Biological Control* 28(3): 132–136.

Subba Rao, B. R. 1966. Records of known and new species of mymarid parasites of *Empoasca devastans* from India. *Indian Journal of Entomology* 28(2): 187–196.

Subba Rao, B. R. 1976. *Narayana*, gen. nov. from Burma and some synonyms (Hymenoptera: Mymaridae). *Oriental Insects* 10(1): 88–89.

Subba Rao, B. R., and M. Hayat. 1983. Key to the genera of Oriental Mymaridae, with a preliminary catalog (Hymenoptera: Chalcidoidea). *Contributions of the American Entomological Institute* 20: 125–150.

Townes, H. 1972. A light weight Malaise trap. *Entomological News* 83: 239–247.

Tooke, F. G. C. 1955. The *Eucalyptus* snout beetle *Gonipterus scutellatus* Gyllenhal—a study of its ecology and control by biological control. *Entomological Memoirs, Department of Agriculture, South Africa* 3: 1–282.

Triapitsyn, S. V., and J. T. Huber. 2000. Nadsem Chalcidoidea 51. 51. Sem. Mymaridae. *Opredeliteli nasekomikh dialinego vostoka Rossii* 4(4): 603–614 (Ed: Lera, P.A.) Dalinauka, Vladivostok.

Triapitsyn, S. V., D. J. W. Morgan, M. S. Hoddle, and V. V. Berezovskiy. 2003. Observations on the biology of *Gonatocerus fasciatus* Girault (Hymenoptera: Mymaridae), egg parasitoid of *Homalodisca coagulate* (Say) and *Oncometopia orbona* (Fabricius) (Hemiptera: Clypeorrhyncha: Cicadellidae). *Pan-Pacific Entomologist* 79(1): 75–76.

Triapitsyn, S. V., A. Logarzo., J. H. de Leon, and E. G. Virla. 2008. A new *Gonatocerus* (Hymenoptera: Mymaridae) from Argentina, with taxonomic notes and molecular data on the *G. tuberculifemur* species complex. *Zootaxa* 1949: 1–29.

Viggiani, G. 1989. A preliminary classification of the Mymaridae (Hymenoptera: Chalcidoidea) based on the external genitalic characters. *Bollettino del Laboratorio di Entomologia Agraria "Fillippo Silvestri"* 45: 141–148.

Wilson, L. T., C. H. Pickett, D. L. Flaherty, and T. A. Bates. 1989. French prune trees: Refuge for grape leafhopper parasite. *California Agriculture* 43: 7–8.

Yoshimoto, C. M. 1990. *A Review of the Genera of New World Mymaridae (Hymenoptera: Chalcidoidea)*. Flora and Fauna Handbook no. 7. Sandhill Crane Press, Gainesville, FL. 166 pp.

Yoshimoto, C. M., M. A. Kozlov, and V. A. Trjapitzin. 1972. A new subfamily of Mymaridae (Hymenoptera: Chalcidoidea). *Entomologicheskoe obozrenie* 51(4): 879.

Zeya, S. B. 2011. A new name for *Gonatocerus orientalis* Zeya (Hymenoptera: Mymaridae). *Bionotes* 13: 33.

Zeya, S. B., and M. Hayat. 1995. A revision of the Indian species of *Gonatocerus* Nees (Hymenoptera: Chalcidoidea: Mymaridae). *Oriental Insects* 29: 47–160.

Zeya, S. B., and F. R. Khan. 2012. The genus *Gonatocerus* Nees (Chalcidoidea: Mymaridae) from India with descriptions of two new species. *Oriental Insects* 46: 53–62.

Zeya, S. B., S. U. Usman, and K. Veenakumari. 2014. Record of some Mymaridae from India, with description of a new species of *Gonatocerus* Nees (Hymenoptera: Chalcidoidea). *Journal of Insect Systematics* 1(1): 64–65.

10 Scelionidae and Platygastridae (Hymenoptera: Platygastroidea) of India

K. Rajmohana and Sunita Patra

CONTENTS

Introduction 159
Platygastroidea: Diagnosis and Key Characters 159
    Key to Families of Platygastroidea 160
Diagnosis 160
    Family Scelionidae 160
    Subfamilies of Scelionidae 160
    Key to Subfamilies of Scelionidae 160
    Family Platygastridae 161
    Subfamilies of Platygastridae 162
    Key to Subfamilies of Platygastridae 162
Classification 162
Major Work on Indian Fauna 162
Molecular Characterization and Phylogeny 162
Diversity and Distribution in India 163
Biogeographical Affinities 163
Taxonomic Problems and Integrative Taxonomy 164
Biology and Ecology 164
    Habitats 164
    Phoresy 164
    Sexual Dimorphism 167
Economic Importance 167
Collection and Preservation 167
Useful Websites 167
Conclusion 168
References 168

INTRODUCTION

Platygastroidea (with two families: Platygastridae and Scelionidae) is the third largest of the parasitic hymenopteran superfamilies after Ichneumonoidea and Chalcidoidea represented worldwide by about 6048 described species under 264 genera in 2 families (Johnson 2018). In India, about 490 species under 83 genera are presently known (K. Rajmohana, unpublished data). They are ubiquitous and are encountered in all habitats including meadows, grasslands, cultivated areas, forests, and urban gardens.

Scelionidae are idiobiont endoparasitoids exclusively attacking the eggs of diverse insect groups like Lepidoptera, Heteroptera, Diptera, Coleoptera, Mantodea, Neuroptera, as well as araneomorph spiders. Under Hymenoptera, about 14 families are known to be egg parasitoids of arthropods (Romani et al. 2010). Of these, only three families, Mymaridae and Trichogrammatidae (Chalcidoidea) and Scelionidae (Austin et al. 2005) are true egg parasitoids, as they complete their development within a host egg (Mills 1994), eventually killing the embryo. Platygastridae attack the eggs of Coleoptera, the immature stages of Hemiptera (including planthoppers, whiteflies, aphids, and mealybugs), and the eggs or early larval stages of dipteran cecidomyiid gallflies (Masner 1993). The classification followed here is that of Masner (1976, 1993), Johnson (1992), and Taekul et al. (2014).

PLATYGASTROIDEA: DIAGNOSIS AND KEY CHARACTERS

These wasps are small in size, generally less than 2.5 mm, with a few in the range of 0.5–12 mm. They have a well sclerotized body, with an intricate sculpture and can be recognized easily by the following combination of characters. Antennae in both sexes, rarely exceeding 12 segments and inserted near clypeus, clava bear basiconic sensillae, which are at times paired. Prepectus is usually absent. The metasoma is often acutely margined at sides. Wings have much reduced venation, or at times have hardly any venation at all. Their uniqueness is largely attributed to their metasomal structure and the operating mechanism of the ovipositor (Masner 1993). The ovipositor which is relatively weakly sclerotized, stays fully retracted within metasoma when not in use, but can be extruded along with the telescopic tube. Both Platygastridae and Scelionidae were originally described by Haliday in 1833 and 1839, respectively.

The keys and the diagnosis of the families and subfamilies presented here are modified from Masner (1993).

KEY TO FAMILIES OF PLATYGASTROIDEA

1. Antenna in both sexes usually elbowed after scape and also after fourth or fifth segment and never with more than ten segments in females; clava compact or segmentation distinct....................... **Platygastridae**
- Antenna in both sexes not elbowed and usually not exceeding 12 segments (rarely with 14), if with 6–9 segments, then claval segmentation often indistinct... **Scelionidae**

DIAGNOSIS

FAMILY SCELIONIDAE

Average body size, 1–2.5 mm, though a few are as small as 0.5 mm and as large as 12 mm. They are mostly black, but yellow, brown, or pale forms are also met with, rarely with metallic colours. Body sculpture often prominent. Antenna usually 11- or 12-segmented, but rarely 6- and 14-segmented too. In males, 5th antennal segment usually modified. Fore wing mostly with sub-marginal, marginal, stigmal, and post-marginal veins. The sub-marginal vein in hind wing in most cases complete. Metasoma often moderately to strongly depressed dorsoventrally. Metasomal segments either subequal or variable in length, if segment 2 is the longest, then sub-marginal vein reaching the anterior margin of wing, with all the three veins, marginal, stigmal, and post-marginal indicated. In females, metasomal segment 7 can be either external or internal, with cerci or sensory plates, and may be extruded with the ovipositor during oviposition or remain attached to the sixth tergum.

Body sculpture, number and nature of antennal segments, nature of antennal clava, ocellar and ocular position, pilosity, structure and shape of mandibles, body armature including spines and teeth, proportion of metasomal segments, and wing venation and nature of ovipositor are some of the major characters used in delineating their underlying genera. Brachypterous and apterous forms are also met with.

SUBFAMILIES OF SCELIONIDAE

Scelioninae (Figure 10.1[**1–3**]), with 16 tribes (Masner 1993), is the largest of the three subfamilies, comprising nearly 70% of the species of Indian Scelionidae. They are the most diverse group of parasitic wasps and inhabits even dry deserts (Masner 1993), several species still awaiting discovery. Many new genera like *Neoduta* Rajmohana and Peter (2012), *Neoceratobaeus* Rajmohana (2014), *Chakra* Rajmohana and Veenakumari (2014), *Anokha* Rajmohana and Veenakumari (Rajmohana et al. 2017), and *Indiscelio* Veenakumari (Veenakumari et al. 2018b) have been described in recent times.

Teleasinae (Figure 10.1[**4**]) are distinguished from Scelioninae mainly by the venation of their fore wings. They have a long marginal vein, but very short stigmal vein and almost lack the post-marginal. The anterior pronotal process is distinct, and the ocelli are placed in a compact triangle. The apparent tergite 7 in females is not extruded with the ovipositor during oviposition. The subfamily is represented by six genera in India, including the recently described genus *Dvivarnus* Rajmohana and Veenakumari. Several new species under the most speciose genus *Trimorus* Förster are currently being described.

Telenominae (Figure 10.1[**5 and 6**]) is quite speciose in India, though low in generic diversity (Johnson 2018). In Telenominae, laterosternites are absent, while laterotergites are wide and overlap the sterna. Tergite 2 is the largest of all metasomal tergites. Female antennae have 10 or 11 segments, while male antennae are usually 12-segmented. In females, apparent tergum 7 is external, not extruded with the ovipositor during oviposition, and the cerci are transformed into sensory plates studded with long hairs. The most speciose genus is *Telenomus* Haliday, but its species largely remain undescribed.

KEY TO SUBFAMILIES OF SCELIONIDAE

1. Antenna in females with 10–11 segments, clava distinctly segmented; in males, antenna with 12 segments, two terminal segments never confluent; second metasomal tergite distinctly longest; laterotergites broad, laterosternites and sub-marginal groove absent; and metapostnotum externally demarcated .... **Telenominae**
- Antenna in females with 6–14 segments, clava distinct, either segmented or unsegmented; in males 12 segmented, terminal two segments at times confluent; second or third metasomal tergite longest, if second tergite longest, then laterotergites very narrow, lateral sternites well defined and sub-marginal groove present; and metapostnotum not demarcated externally...................2
2. Fore wings with marginal vein usually more than 3x longer than stigmal vein; post-marginal vein

FIGURE 10.1 Genera of Scelionidae and Platygastridae **1**. *Gryon* sp.; **2**. *Macroteleia* sp.; **3**. *Gryon* sp. (emerging from eggs of coreid bug); **4**. *Trissolcus* sp.; **5**. *Telenomus* sp. **6**. *Xenomerus* sp.; **7**. *Isolia* sp.; **8**. *Gastrotrypes* sp. and Phoresy **9**. *Sceliocerdo viatrix* on *Neorthacris acuticeps* (Acrididae) (Courtesy of Alfred Daniel) **10**. *Trissolcus* sp. on *Cletus* sp. (Coreidae).

rudimentary or absent; and of all tergites of metasoma, 3rd tergite always longest .... **Teleasinae**

- Fore wings with marginal vein usually shorter than stigmal vein; post-marginal vein present or absent; in case marginal vein longer than stigmal, then metasoma elongate and post marginal vein distinct, in case sub-marginal vein absent or rudimentary, then antennal clava unsegmented or postgena and temples with tuft of pilosity; 3rd tergite not always longest of tergites.. **Scelioninae**

FAMILY PLATYGASTRIDAE

Platygastridae is considerably smaller in size, compared to Scelionidae. Usually 1–2 mm in length, 4 mm forms are also met with. Body slender and usually black, rarely yellow, but never metallic. Antenna with 7–10 segments and strongly elbowed. In males, antennal segment 3 or 4 modified. Fore wing without veins, but in a few genera, sub-marginal vein present as a short stub. Both stigmal and post-marginal veins absent. Female metasoma with 2nd segment always the longest and widest and with only six apparent tergites (exceptionally fewer); metasomal tergum 7, often internal, but never extruded with ovipositor. Key morphological characters in distinguishing the genera include body sculpture, number of antennal segments, nature of the female antennal clava, ocellar and ocular position, body pilosity, number and proportion of metasomal segments, the shape of wings, and venation and nature of marginal fringes.

SUBFAMILIES OF PLATYGASTRIDAE

Sceliotrachelinae (Figure 10.1[**7**]) are mostly squat to plump. They have relatively wide laterotergites and the antennal clava in females is usually abrupt, with three clavomeres, which are partially to completely fused; the males often have a subclavate antenna. Their forewing has a tubular and apically knobbed sub-marginal vein. They are idiobionts, attacking mainly the immature stages of Pseudococcidae or Aleyrodidae (Homoptera) (Masner 1993; Manser and Huggert 1989; Vlug 1995).

Compared to Sceliotrachelinae, Platygastrinae (Figure 10.1[**8**]) are slender and elongate. Their metasoma is compact, with narrow laterotergites tightly appressed against the sternites. The antennal clava of females is cylindrical and with four or five well defined clavomeres. Males have a thread-like flagellum. In a few genera, the sub-marginal vein is distinct. Platygastrinae parasitise the early stages of cecidomyiid Diptera and are mostly koinobionts, developing only once their hosts are in pupal/prepupal stages (van Noort 2018).

Platygastrinae is richer in species than Sceliotrachelinae. They are biologically very cohesive, and as koinobionts, attack the eggs or early larval stages of the gall making cecidomyiid Diptera. They can be species specific.

KEY TO SUBFAMILIES OF PLATYGASTRIDAE

1. Female antennal clava moderately abrupt, subcylindrical with five clearly separated clavomeres; body usually cylindrical .... **Platygastrinae**
- Female antennal clava distinctly abrupt, ovoid, and composed of 3–4 subcompact clavomeres; body stocky and short, wider than high................................. ....**Sceliotrachelinae**

CLASSIFICATION

Both the families Scelionidae and Platygastridae were earlier placed within the superfamily Proctotrupoidea, until Masner (1993) designated the superfamily Platygastroidea (Austin et al. 2005). Sharkey (2007), based on the results of a phylogenetic analysis by Murphy et al. (2007), synonymized Scelionidae with Platygastridae. Though this was accepted initially, almost all subsequent workers (Talamas & Buffington 2014; Engel et al. 2016; Talamas et al. 2016; Talamas et al. 2017; Popovici et al. 2017) reverted to the old classification with both Platygastridae and Scelionidae as valid families. The phylogenetic study by Murphy et al. (2007) was found to be wanting as the number of samples of taxa analysed was inadequate. Mckellar and Engel (2012) recognized four families Nixoniidae, Sparasionidae, Scelionidae, and Platygastridae, while Ortega-Blanco et al. (2014) recognized only two families, Scelionidae and Platygastridae. Until the completion of a comprehensive phylogenetic study of the superfamily, confusion will continue to prevail (van Noort 2018).

Classifications of the Scelionidae and Platygastridae exist at the tribal level (Masner 1976; Galloway and Austin 1984). The general trend, however, has been to proceed to the generic level from the subfamilies, bypassing the tribes.

MAJOR WORK ON INDIAN FAUNA

Several new taxa, both species and genera, are regularly being described from India. Some of the new genera described recently from India are *Indiscelio* Veenakumari, Popovici, and Talamas (Veenakumari et al. 2018b), *Neoceratobaeus* Rajmohana (Rajmohana 2014), *Chakra* Rajmohana and Veenakumari (2014), *Dvivarnus* Rajmohana and Veenakumari (Vennakumari et al. 2012), *Narendraniola* (Rajmohana 2012), and *Neoduta* Rajmohana and Peter (2013). Early works on Indian Platygastroidea were by Mani (1936a, 1936b, 1941, 1942, 1973) and Mukerjee (1978, 1979, 1981, 1992, 1993, 1994). Over 150 new species of Platygastroidea were described recently from India (Rajmohana 2007a, 2007b, 2012, 2014; Rajmohana and Narendran 2007a; Rajmohana and Talukdar 2010; Rajmohana and Bijoy 2011; Rajmohana and Abhilash 2012; Rajmohana and Nisha 2013; Rajmohana and Peter 2013; Rajmohana and Anto 2014; Abhilash and Rajmohana 2014; Anjana and Rajmohana 2015; Anjana et al. 2016a, 2016d, 2017; Rajmohana et al. 2013b, 2017; Veenakumari et al. 2011, 2014, 2016, 2017, 2018a, 2018b). Twenty species of *Cremastobaeus* Ashmead (Veenakumari and Mohanraj 2017a) and 40 species of *Leptacis* (Veenakumari et al. 2017–18) were also of late described from India. The diversity of Platygastroidea in different habitats in India was studied by Manoj et al. (2017).

MOLECULAR CHARACTERIZATION AND PHYLOGENY

The past decade has seen significant progress in the molecular characterization and phylogenetic studies of Hymenoptera in general resulting in a revised higher level classification (Aguiar et al. 2013). Sharkey (2007), based on the molecular phylogenetic analysis of Platygastroidea by Murphy et al. (2007), synonymized Scelionidae with Platygastridae. A phylogenetic analysis of subfamily Telenominae by Taekul et al. (2014) using sequence data from multiple genes (18S, 28S, COI, EF-1α), concluded that the nominate tribe Telenomini was not monophyletic. This work led to the generic transfer of *Psix* Kozlov and Le and *Paratelenomus* (Dodd) from Telenominae to Scelioninae, since they formed a monophyletic group which was sister to *Gryon* Haliday. The monophyly of the remaining groups was confirmed. The genus *Phanuromyia* Dodd along with the *Telenomus crassiclava* group, which are egg parasitoids of Fulgoridae, were found to be monophyletic. Hence twenty-nine species under *Telenomus crassiclava* and *aradi* groups were transferred to *Phanuromyia*. The study could not resolve the monophyletic status of two speciose telenomine genera *Telenomus* and *Trissolcus*. Assessing the relationships and the pattern of shifts in host groups, the study also commented upon a clade shifting to parasitism of lepidopteran eggs and a subsequent clade shifting to parasitism of dipteran eggs. With the exception of the generation of DNA barcodes for a few species of *Telenomus* (Rajmohana and Nisha 2013; Rajmohana and Anto 2014; Rajmohana et al. 2013b), molecular studies are lacking on Indian Platygastroidea.

DIVERSITY AND DISTRIBUTION IN INDIA

Scelionidae is represented by 323 species under 64 genera, while Platygastridae has 167 spp. in 19 genera in India (K. Rajmohana, unpublished data). Of the Indian genera of Scelionidae, *Telenomus* Haliday (Telenominae), *Scelio* Latreille, *Calliscelio,* Ashmead, *Gryon* Haliday, *Cremastobaeus* Kieffer, *Ceratobaeus* Ashmead, *Idris* Förster (Scelioninae), and *Trimorus* Förster (Teleasinae) are the most speciose.

Platygastrinae is more common, diverse, and speciose than Sceliotrachelinae. While the former has about 151 species in 14 genera, the latter is known by only 16 species in 5 genera (K. Rajmohana, unpublished data). *Leptacis* Förster, *Platygaster* Latreille, and *Synopeas* Förster (Platygastrinae) are the most speciose genera. Several new species are currently being described from India under Platygastridae. Under *Leptacis* alone, over 40 species were described in 2017 (Veenakumari et al. 2017–2018).

Several taxonomic changes were effected through world revisions of a few scelionid genera (Johnson et al. 2008; Taekul et al. 2010; Talamas et al. 2013; Talamas and Buffington 2014; Chen et al. 2018) and the photographic catalogue (Talamas et al. 2017). The latter work is based on primary types of Indian Platygastroidea based on the work of pioneers like M.S. Mani, S.K. Sharma, and G. Saraswat housed in the National Insect Collection, Smithsonian National Museum of Natural History, Washington, DC, USA. Through the world revisions of *Platyscelio* (Taekul et al. 2010) and *Habroteleia* (Chen et al. 2018), all species originally described under these genera from India lost their taxonomic validity, and they became junior synonyms of other widely distributed species.

As the exploration and inventorization of Indian Platygastroidea is far from complete and the known occurrence patterns are only a reflection of collection effort, nothing definitive can be said about their distribution patterns. Studies on Indian Platygastroidea have largely been confined to S. India, with most species being from the Western Ghats and the Deccan Plateau. Currently, less than 30 species each are known from the highly biodiverse regions of Northeast India, Andaman and Nicobar Islands, and the Himalaya.

BIOGEOGRAPHICAL AFFINITIES

Though several species of Platygastroidea have their distributions restricted to India, endemism at the generic level is extremely low. Of the 83 genera known from India, only two genera, *Mudigere* Johnson and *Sceliocerdo* Muesebeck (both from South India), are endemic to India (Johnson 2018).

The biogeographic affinities of species can best be evaluated only in cases where revisionary work has been published on a global scale or at least regionally, as in a few scelionid taxa like *Paratelenomus* Dodd (Johnson 1996), *Psix* Kozlov and Le (Johnson and Masner 1985), *Nixonia* Masner (Johnson and Masner 2006), *Fusicornia* Risbec (Taekul et al. 2008), *Platyscelio* Kieffer (Taekul et al. 2010), *Heptascelio* Kieffer (Johnson et al. 2008), *Odontacolus* Kieffer (Valerio et al. 2013), *Oxyscelio* Kieffer (Burks et al. 2013), *Scelio pulchripennis* group (Yoder et al. 2009), *Xenomerus* Walker (Miko et al. 2010), *Paridris* Kieffer (Talamas et al. 2013), *Macroteleia* Westwood (Chen et al. 2013), and *Habroteleia* Kieffer (Chen et al. 2018). These have resulted in extensive changes in the nomenclature and taxonomic status of several species in India due to generic transfers and new synonymies. Most Indian species belonging to genera that have undergone world revisions have relatively good distribution in the Indo-Malayan or the Oriental regions, with a few being known from the Ethiopian, Palearctic, and Australasian regions. Of the six species of *Macroteleia* in India, five have an Indo-Malayan distribution and of the five species of *Oxyscelio* in India, two have been documented as occurring widely in Southeast Asian countries. Among the seven species of *Odontacolus* Kieffer in India, four have been reported from Southeast Asia as well. *Paridris taekuli* is a widely distributed species in the Indo-Malayan region including India. It is also known to occur in Melanesia, New Caledonia, Ivory Coast, and Madagascar. *Platyscelio pulchricornis* Kieffer, the only representative of the genus *Platyscelio* in India and *Fusicornia indica* Mani and Sharma (1982) have Indo-Malayan as well as Australasian distributions. *Paratelenomus striativentris* (Risbec), which has been documented in southern India is found in sub-Saharan Africa too, while *Paratelenomus saccharalis* (Dodd) is a widespread species found in southern Palearctic, Africa, tropical Asia, and northern Australia. Except for the two recently described species of *Psix* (Kozlov and Le) from India, four are known from the Oriental region while a few others are known from Australia, Ethiopia, Tanzania, and Malagasy.

The widely distributed tramp species *Calliscelio elegans* (Perkins) is a truly cosmopolitan species (Masner et al. 2009) which is present in India too. In spite of its wide occurrence, the male is known from a single specimen collected from India (Rajmohana et al. 2013a). *Xenomerus ergenna* Walker (Teleasinae), known to occur widely across various biogeographic realms, has also been documented in India. *Nixonia* Masner, the most plesiomorphic genus in Scelionidae, is represented by *Nixonia krombeini* Johnson and Masner in India. It is widely distributed in the Oriental region.

Duta serraticeps (Priesner) from India (Rajmohana 2010) was originally described from Ethiopia. The primitive *Heptascelio striatosternus* Narendran and Ramesh Babu originally described from India was found to occur even in Madagascar. *Scelio poecilopterus* Risbec and *Scelio variegatus* Kozlov and Kononova which are well represented in the Arabian Peninsula have also been found occurring in India. With the exception of *Xenomerus yamagishi* Miko and Masner, not many species of Platygastridae seen in India and the Indo-Malayan region occur in the Palearctic region (Rajmohana et al. 2015).

Except for Scelionidae, there have been very few world revisions of Platygastridae. Over 90% of the species reported from India are restricted in their distribution being known from nowhere else. However, none of the genera in the family are endemic to India.

TAXONOMIC PROBLEMS AND INTEGRATIVE TAXONOMY

In most platygastroid genera, species can only be identified using sound morphological characters of females to the exclusion of males. Males lack distinctive characters for species identification and even generic identification in some cases. As there are only a few identification keys for males, associating the sexes within a species from field-collected samples is quite challenging.

In the case of the highly speciose *Telenomus,* even the females lack distinctive characters. Male genitalia too are not distinctive within species groups. Hence, the genus remains to be worked out taxonomically. Being egg parasitoids of several pestiferous Lepidoptera and Heteroptera, the group is of great significance as biocontrol agents. DNA barcodes may prove to be effective in delimiting species. An integrated taxonomic approach with data on hosts, ecology, morphology, and molecular characteristics may be the method to resolve such issues.

BIOLOGY AND ECOLOGY

Of the subfamilies under Scelionidae, Telenominae have Lepidoptera and Heteroptera as their main host groups, though they also attack Diptera and Neuroptera (Masner 1976). Teleasinae attack the eggs of carabid beetles, while the host groups of Scelioninae are mainly Orthoptera and Heteroptera (Galloway and Austin 1984). Host specificity is very high, at least at the tribal level (Austin et al. 2005), and they show a high degree of host partitioning, with very little overlap between the host groups (Galloway and Austin 1984). Thus, the Scelionini parasitise Acrididae and Baryconini; Platyscelionini are parasitoids of Tettigonidae; Calliscelionini attack both Tettigonidae and Gryllidae; Embidobiini attack Embioptera and Araneae; Gryonini parasitise several families of Heteroptera; Mantibariini attack Mantodea, while species of Idrini and Baeini use araneomorph spiders (Araneae) as their hosts. Thoronini, which are good swimmers (Austin 2005), attack the eggs of aquatic Heteroptera, mostly Gerridae and Nepidae (Masner 1976). Platygastroids in general have a narrow host range and a relatively high degree of specificity of the host stage attacked (Rajmohana and Abhilash 2014). It is interesting to note that unlike Chalcidoidea, a species under Platygastroidea has never been reared from hosts of more than one order (Taekul et al. 2014). Definitive host data of scelionids and platygastrids in India are scanty (Table 10.1) (host data published in taxonomic works only have been considered).

HABITATS

Platygastroidea are encountered commonly in good numbers in all habitats including forests and cultivated landscapes. In rice agroecosystems, they usually outnumber other parasitoid groups like Chalcidoidea and Ichneumonoidea (K. Rajmohana and S. Patra, personal observation). In a study comparing scelionine egg parasitoids in rice ecosystems along elevational ranges in Kerala, it was noted that elevation did not have any major effect on the overall diversity patterns in Platygastroidea, although differences in species assemblages were found (Shweta and Rajmohana 2016a). The results were similar in a study comparing Platygastroidea in organic and conventionally cultivated rice (Gnanakumar et al. 2012).

Brachyptery or microptery, prominent in several genera of Platygastroidea, are adaptations for living in and searching for hosts in confined habitats like soil, leaf litter, etc. Frontal prominence, aptery or brachyptery and protruded and elongate mandibles are sand burrowing adaptations in *Encyrtoscelio* Dodd which attack the eggs of *Cydnus* bugs. A streamlined body and brachyptery or microptery seen in several genera attacking the eggs of spiders like *Idris, Ceratobeaus,* and *Beaus* aid female wasps in penetrating the silken egg sacs of their host spiders. Such modifications are functional specializations seen in groups living in cryptic niches and are assumed to be their habitat dependent morphological modifications (Austin et al. 2005).

PHORESY

Phoresy is the transport of adult insects on the bodies of their hosts for purposes other than parasitization (Lesne 1896). A few species of egg parasitoids belonging to Trichogrammatidae, Scelionidae, and Mymaridae are known to be phoretic on their hosts (Clausen 1976). They use phoresy as a mode of dispersal. The phoretic females usually locate their female hosts, mount their bodies, and ride on them. Soon after the host female oviposits, the wasp dismounts to parasitize the freshly laid host eggs (Naumann and Reid 1990).

Phoresy has been documented multiple times in Scelioninae and Telenominae. In India too, this phenomenon has been reported in a few instances. *Sceliocerdo viatrix* (Brues) was reported to be phoretic on acridid grasshoppers

TABLE 10.1
Platygastroidea and Their Hosts

No.	Subfamily	Parasitoid	Host	References
1.	Platygastridae	*Anectadius bengalensis* Kieffer	Ex. midge gall on *Artemisia* sp. (Diptera: Cecidomyiidae)	Vlug (1995)
2.	Platygastridae	*Anectadius striolatus* Kieffer	*Lasioptera textor* Kieffer (Diptera: Cecidomyiidae); Leaf gall of *Barringtonia acutangula* (Linn.)	Vlug (1995)
3.	Platygastridae	*Inostemma apsyllae* Austin	Ex. psyllid gall-*Apsylla cistellata* (Buckton 1896) (Hemiptera: Aphalaridae) on *Mangifera indica*	Austin (1984)
4.	Platygastridae	*Inostemma indica* Mani	Ex. midge gall-*Neolasioptera cephalandrae* Mani (Diptera: Cecidomyidae)	Mani (1941)
5.	Platygastridae	*Inostemma oculare* Austin	Ex. midge gall-*Procontarinia matteiana* Kieffer & Cecconion (Diptera: Cecidomyiidae) on *Mangifera indica*	Austin (1984)
6.	Platygastridae	*Platygaster luteipes* Buhl	Unknown leaf galls on *Piper nigrum*	Anjana et al. (2016c)
7.	Platygastridae	*Platygaster oryzae*, Camron	Ex. midge gall-*Cecidomyia oryzae* (Diptera: Cecidomyiidae) in paddy	Vlug (1995)
8.	Platygastridae	*Synopeas pauropsylla* Veenakumari and Buhl	Ex. psyllid gall-*Pauropsylla* cf. *depressa* (Hemiptera:Triozidae) on *Ficus benghalensis*	Veenakumari et al. (2018b)
9.	Platygastridae	*Synopeas procon* Austin	Ex. midge gal-*Procontarinia matteiana* Kieffer & Cecconion (Diptera: Cecidomyidae) on *Mangifera indica*	Austin (1984)
10.	Platygastridae	*Synopeas temporale* Austin	Ex. midge gal-*Procontarinia matteiana* Kieffer & Cecconion (Diptera: Cecidomyidae) on *Mangifera indica*	Austin (1984)
11.	Platygastridae	*Trichacoides indicus* Jackson, 1968	Ex. midge gall-*Procontarinia matteiana* Kieffer & Cecconi (Diptera: Cecidomyiidae) on *Mangifera indica* and on *Mimosops elengi* leaves	Anjana et al. (2016b)
12.	Platygastridae	*Amitus aleurolobi* Mani	Ex. early instar *Aleurolobus barodensis* (Maskell) (Hemiptera: Aleurodidae) on *Saccharum officinarum* L	Mani (1939)
13.	Platygastridae	*Amitus japonica* Ashmead	Ex. early instar of *Maconellicoccus hirsutus* (Green) (Hemiptera: Pseudococcidae).	Mukerjee (1978)
14.	Platygastridae	*Amitus vignus* Anjana, Rajmohana.	Ex. early instars of *Zaphanera* sp. (Hemiptera: Aleyrodidae) on *Vigna trilobata*	Anjana et al. (2016d)
15.	Scelionidae	*Ceratobaeus dunensis* Mukerjee	Ex. eggs of *Cheriacanthium* sp. (Araneomorphae: Miturgidae)	Shweta et al. (2014)
16.	Scelionidae	*Gryon diadematis* Mineo	Ex. eggs of Heteroptera on teak leaf	Mineo 1983
17.	Scelionidae	*Gryon fulviventre* (Crawford)	Ex. eggs of *Antestiopsis cruciate* (Hemiptera: Pentatomidae) on Jasmine; Ex. eggs of *Nezara viridula* (Hemiptera: Pentatomidae); Ex eggs of *Clavigralla gibbosa* (Hemiptera: Coreidae)	Mani and Sharma (1982)
18.	Scelionidae	*Gryon ingens* Veenakumari & Rajmohana	Ex. eggs of Reduvidae	Veenakumari et al. (2016)
19.	Scelionidae	*Gryon orestes* (Dodd)	Ex. eggs of Coreidae	K. Rajmohana and S. Patra, personal observation
20.	Scelionidae	*Platyscelio pulchricornis* Kieffer	Ex. orthopteran eggs on paddy leaf	Voucher specimen examined at NBAIR, Bangalore
21.	Scelionidae	*Sceliocerdo viatrix* (Brues)	Phoretic on *Neorthacris acuticeps* (Bolìvar) (Orthoptera: Acrididae)	Veenakumari et al. 2012
22.	Scelionidae	*Eumicrosoma cumaeum* (Nixon)	Ex. eggs of *Macropes excavatus* Distant = *Cavelerius excavatus* (Distant) (Heteroptera: Lygaeidae) on sugarcane	Nixon (1938)
23.	Scelionidae	*Eumicrosoma phaeax* (Nixon)	Ex. eggs of *Macropes excavatus* Distant = *Cavelerius excavatus* (Distant) (Heteroptera: Lygaeidae) on sugarcane	Nixon (1938)
24.	Scelionidae	*Paratelenomus saccharalis* (Dodd)	Phoretic on *Megacopta cribraria* (Hemiptera: Plataspidae:) on *Dolichos lablab*	Rajmohana and Narendran (2001)
25.	Scelionidae	*Protelenomus flavicornis* Kieffer (Coreidae)	Phoretic on *Anoplocnemis phasiana* (Heteroptera: Coreidae)	Veenakumari and Mohanraj (2015)
26.	Scelionidae	*Psix robustus* Rajmohana	Ex. heteropteran Eggs	Rajmohana and Talukdar (2010)
27.	Scelionidae	*Psix sunithae* Singh, Johnson and Ramamurthy	Ex. eggs of Coreidae (Heteroptera) on *Nux vomica* leaves	Singh et al. (2012)

(Continued)

TABLE 10.1 (Continued)
Platygastroidea and Their Hosts

No.	Subfamily	Parasitoid	Host	References
28.	Scelionidae	*Telenomus beneficiens* (Zehntner)	Ex. eggs of *Chilo infuscatellus* (Lepidoptera: Crambidae), *Chilo sacchariphagus* (Lepidoptera: Crambidae) and *Scirpophaga nivella* (Lepidoptera: Crambidae) on sugarcane	Narasimham et al. (1997)
29.	Scelionidae	*Telenomus cuspis* Rajmohana & Srikumar	Ex. eggs of *Helopeltis antonii* (Hemiptera: Miridae) on guava neem and tea	Rajmohana et al. (2013b)
30.	Scelionidae	*Telenomus dignoides* Nixon	Ex. egg mass of *Scirpophaga nivella* (Lepidoptera: Crambidae) on sugarcane, *Erianthus munja*, *Chilo saccariphagus indicus* (Lepidoptera: Crambidae) on sugarcane	Nixon (1937)
31.	Scelionidae	*Telenomus dignus* (Gahan)	Ex. eggs of *Scirpophaga nivella* (Lepidoptera: Crambidae) on sugarcane and *Scirpophaga incertulas* on rice	Gahan (1925)
32.	Scelionidae	*Telenomus dilatus* Rajmohana and Anto	Ex. eggs of *Papilio* sp. (Lepidoptera: Papilionidae)	Rajmohana and Anto (2014)
33.	Scelionidae	*Telenomus euproctiscidis*	Ex. eggs of *Euproctis lunata* (Lepidoptera: Erebidae)	Mani and Sharma (1982)
34.	Scelionidae	*Telenomus globosus* Bin and Johnson	Ex. eggs of *Chilo sacchariphagus* (Lepidoptera: Crambidae) on sugarcane	Bin and Johnson (1982)
35.	Scelionidae	*Telenomus oryzae* Rajmohana and Nisha	Ex. eggs of *Scotinophara* sp. (Heteroptera: Scutellaridae) on *Oryza sativa*	Rajmohana and Nisha (2013)
36.	Scelionidae	*Telenomus otones* Nixon	Ex. eggs of *Tarucus theophrastus* (Lepidoptera: Lycaenidae) on lac insect	Nixon (1940)
37.	Scelionidae	*Telenomus pegasus* Nixon	Ex. eggs of *Tarucus theophrastus* (Lepidoptera: Lycaenidae) on lac insect	Nixon (1940)
38.	Scelionidae	*Telenomus proditor* Nixon	Ex. Eggs of *Eupterote undata* (Lepidoptera: Eupterotidae) on *Gmelina arborea*	Nixon (1937)
39.	Scelionidae	*Telenomus remus* Nixon	Ex. eggs of *Spodoptera litura* (Lepidoptera: Noctuidae)	Narasimham et al. (1997)
40.	Scelionidae	*Telenomus rowani* Gahan	Ex. eggs of *Scirpophaga nivella* (Lepidoptera: Crambidae) on sugarcane and *S. incertulas* (Lepidoptera: Crambidae) on rice	Narasimham et al. (1997)
41.	Scelionidae	*Telenomus samueli* (Mani)	Ex. eggs of *Bagrada cruciferarum* (Hemiptera: Pentatomidae)	Mani and Sharma (1982)
42.	Scelionidae	*Telenomus seychellensis* Kieffer	Ex. eggs of *Antestiopsis cruciata* (Hemiptera: Pentatomidae) on jasmine; and eggs of *Cantheconidea furcellata* (Hemiptera: Pentatomidae)	Mani and Sharma (1982)
43.	Scelionidae	*Telenomus talaus* Nixon	Ex. eggs of *Papilio* sp. (Lepidoptera: Papilionidae)	Nixon (1937)
44.	Scelionidae	*Trissolcus barrowi* (Dodd)	Ex. eggs of sphingid moth (Lepidoptera: Sphingidae)	Mani and Sharma (1982)
45.	Scelionidae	*Trissolcus hyalinipennis* Rajmohana & Narendran	Ex. eggs of *Bagrada cruciferarum* Kirkaldy (Hemiptera: Pentatomidae)	Subba Rao and Chacko (1961); Rajmohana and Narendran (2007b)
46.	Scelionidae	*Trissolcus jatrophae* Rajmohana and Narendran	Ex. eggs of *Scutellara nobilis* Fabricius on *Jatropha* (Hemiptera: Scutelleridae)	Rajmohana et al. (2011)
47.	Scelionidae	*Trissolcus nigrus* Rajmohana & Narendran	Host: Ex. eggs of *Plautia fimbriata* (Hemiptera: Pentatomidae)	Narayanan and Kaur (1959); Rajmohana and Narendran (2007b)
48.	Scelionidae	*Trissolcus orontes* (Nixon)	Ex. eggs of an unidentified insect on Tiliaceae and eggs of *Utetheisa pulchella* (Lepidoptera: Arctiidae)	Nixon (1935)

from India (Brues 1917; Basavanna 1953). Recently on the margins of a rice field in Mandya, Karnataka, six individuals of *S. viatrix* were found attached by their mandibles to the abdominal plates of the wingless grasshopper *Neorthacris acuticeps* (Bolìvar) on *Solanum melongena* (Veenakumari et al. 2012) and also from Coimbatore on *Nerium oleander* (Figure 10.1[9]). A few individuals of *Protelenomus flavicornis* Kieffer were found clinging with their legs to the body and legs of *Anoplocnemis phasiana* (Heteroptera: Coreidae) in Bangalore (Karnataka) (Veenakumari and Mohanraj 2015). A single individual of *Trissolcus* sp. was found on the head of the female of a mating pair of *Cletus* sp. (Heteroptera: Coreidae) (Figure 10.1[10]) in a paddy field in Palghat (Kerala) (Rajmohana and Bijoy 2011a). Further, though the phoretic association was not observed, two species of *Mantibaria*, *M. kerouci* Veenakumari and Rajmohana, and

M. mantis (Dodd) were collected in India (Veenakumari et al. 2012; Veenakumari and Mohanraj 2017a). *Mantibaria* Kirby is known to be a phoretic taxon ectoparasitic on praying mantids (Masner 1976; Galloway and Austin 1984).

Sexual Dimorphism

In Platygastroidea, generally both sexes are very similar in appearance except in the structure of their antennae. *Xenomerus* males have flagellomeres with long bristles, sometimes arranged in a whorl, while males of a few species of *Trimorus* have bristle-like hairs on their antennae. Males have more antennomeres and more tergites and sternites on the metasoma than females. In a few genera like *Baryconus* Förster and *Triteleia* Kieffer, the tip of the last metasomal segment is bidentate or with two small spikes (Masner 1976). However, an instance of very high degree of sexual dimorphism, in colour and metasomal shape, was reported in *Gryon ingens* Veenakumari and Rajmohana (Veenakumari et al. 2016).

In *Baeus* Haliday—egg parasitoids of spiders—all females are apterous, while the males are fully winged. On the other hand in *Idris* Förster and *Ceratobaeus* Ashmead brachyptery is prevalent as a species character. The loss or reduction of wings is an adaptation to ease the entry of the females into the egg cases of spiders for oviposition.

ECONOMIC IMPORTANCE

A major component of the natural enemy complex of insect pests are parasitic Hymenoptera. They are the most common agents introduced for the biological control of insect pests of crops (Ketipearachchi 2002). Platygastroidea are economically significant due to their role as parasitoids, particularly egg parasitoids, killing their hosts before the crops are damaged by even freshly hatched pest species (Mills 2010). They parasitize the eggs of grasshoppers, locusts, crickets, bugs, and moths, which are prominent pests in agriculture and forestry. *Gryon* Haliday is a solitary primary egg parasitoid of several hetropteran bugs, like the rice ear bug (*Leptocorisa* spp.), the pod bug (*Clavigralla* spp.), the rice black bug (*Scotinophara* spp.), etc., which are serious pests of rice. Along with members of *Trissolcus* Ashmead, other scelionids too attack the eggs of several heteropteran families like Pentatomidae, Scutelleridae, and Lygaeidae (K. Rajmohana and S. Patra, personal observation). *Telenomus* spp. also attack a wide range of insect pests across different orders, including several species of *Spodoptera*, shoot and fruit boring moths (Lepidoptera), tea mosquito bugs, and rice black bugs (Heteroptera). Recently, the incidence of the global invasive fall army worm, *Spodoptera fruigiperda* (J. E. Smith) (Lepidoptera: Noctuidae), has been reported from India. Both *Telenomus* spp. and *Trichogramma* spp. have also been found attacking their eggs in India (Shylesha et al. 2018). Among platygastrids, *Amitus* Haldeman attack the early instars of whiteflies, while *Inostemma* Haliday and *Synopeas* Förster are known to attack the cecidomyiids responsible for leaf galls in mango. Species of *Platygaster* Förster have been observed emerging from cecidomyiid galls on paddy and pepper leaves (Table 10.1).

COLLECTION AND PRESERVATION

Platygastroidea can be collected by both active and passive trapping methods. For a good representation of the fauna at a given site, different trapping methods have to be employed, since each trap collects a different community (Kennedy et al. 1961; Hollingsworth et al. 1970).

Active collecting involves net-sweeping, which is the most common strategy employed for general insect collection. Sweep netting is the simplest and best method for collecting parasitic hymenopterans (Narendran 2001). Triangular frame nets are recommended for their collection (Noyes 1982) in preference to round nets for better coverage of swept area. After a few sweeps, the minute wasps from within the nets are sucked into an aspirator, preserved in airtight vials containing 70% or 100% alcohol, and kept under refrigeration.

For gathering definitive host data and related biological information of the parasitoids, potential hosts are collected and reared. Rearing is the only foolproof means to associate the sexes.

Passive collecting mainly involves the use of traps viz., yellow pan traps/Moericke traps, malaise traps, and pitfall traps.

Malaise traps are very efficient in trapping diurnal, swift flying insects. As these traps need to be serviced or emptied only once a week, they can be used for long term studies as a standard collection method. In a study comparing efficiencies of sweep net, yellow pan trap, and Malaise trap in sampling Platygastridae from urban habitats in Kozhikode district, it was seen that the Malaise trap performed better in the collection of Platygastridae over a sweep net or yellow pan trap (Shweta and Rajmohana 2016b).

Pitfall traps can be used to collect Platygastroidea attacking diverse ground dwelling arthropods like ground dwelling spiders and grasshoppers which lay their eggs on or buried in the soil.

The specimens collected are relaxed, suitably spread, and glued onto triangular card tips using a transparent and water soluble glue, with their body tilted slightly on their sides, enabling proper visibility of parts like face and mandibles. Specimens in general do not shrink as their bodies are highly sclerotized. The cards are held using entomological pins. After proper labelling, they are pinned in insect boxes along with 1,2 paradichlorobenzene/naphthalene/camphor for protection from insect attack. The addition of phenol/thymol balls prevents damage to the specimens from fungal attack.

USEFUL WEBSITES

https://hol.osu.edu/index.html?id=195000.
http://www.nbair.res.in/Platygastroidea/.
http://www.waspweb.org/Platygastroidea/.

CONCLUSION

With appreciable species diversity, wide distribution, and common occurrence, Platygastroidea are an integral component of all habitats and ecosystems in India, and their role in maintaining the general heath of an ecosystem cannot be underestimated. As egg parasitoids of a wide range of insect groups, they play a major role in regulating the populations of their hosts. They are of great economic significance being natural enemies of several notorious pests in agriculture and forestry. Taxonomic studies are, however, needed to reveal the diversity of the group to its fullest extent. An integrated taxonomic approach, taking inputs from bioecological and molecular fields, is essential to develop a robust and complete knowledge base of the group.

REFERENCES

Abhilash, P., and K. Rajmohana. 2014. A new species of *Gryon* Haliday (Hymenoptera: Platygastridae) from India, *Journal of Threatened Taxa* 6(14):6711–6714.

Aguiar A. P., A. R. Deans, M. S. Engel et al. 2013. Order Hymenoptera. *Zootaxa* 3703(1):51–62.

Anjana, G., K. Rajmohana, and M. Shweta. 2016a. New species of *Gastrotrypes* Brues (Hymenoptera: Platygastridae) from India along with key to world species. *Oriental Insects* 50(1):9–22.

Anjana, M., and K. Rajmohana. 2015. Three new species of *Amblyaspis* Förster (Hymenoptera: Platygastridae) from India along with a key to Indian species. *Halteres* 6:113–120.

Anjana, M., K. Rajmohana, A. Athira, and A. P. Ranjith. 2016b. New report and host association of *Trichacoides* Dodd (Hymenoptera: Platygastridae), with a key to species from India. *Journal of Insect Systematics* (2)2:81–88.

Anjana, M., K. Rajmohana, and P. A Sinu. 2017. Two new species of *Iphitrachelus* Haliday (Hymenoptera: Platygastridae) from India with a key to oriental species. *Insect Diversity and Taxonomy, T. C. Narendran Com.* Volume November 2017:213–224.

Anjana, M., K. Rajmohana, D. Vimala, and R. Sundararaj. 2016d. On a new species of *Amitus* Haldeman (Hymenoptera: Platygastridae) parasitizing whitefly *Zaphanera* sp. (Aleyrodidae) on *Vigna trilobata* from India. *Halteres* 7:106–111.

Anjana, M., K. Rajmohana, P. N. Buhl, and K. M. Shameem. 2016c. First record of *Platygaster luteipes* Buhl (Hymenoptera: Platygastridae) from leaf galls on black pepper along with first report of the species from India, *International Journal of Environmental Studies* 73(2):186–195. doi:10.1080/00207233.2015.1135569.

Austin, A. D. 1984. New species of Platygastridae (Hymenoptera) from India which parasitise pests of mango, particularly *Procontarinia* spp. (Diptera: Cecidomyiidae). *Bulletin of Entomological Research* 74:549–557.

Austin, A. D., N. F. Johnson, and M. Dowton. 2005. Systematics, evolution, and biology of Scelionid and platygastrid wasps. *Annual Review of Entomology* 50:553–582.

Basavanna, G. P. C. 1953. Phoresy exhibited by *Lepidoscelio viatrix* Brues (Scelionidae, Hymenoptera). *Indian Journal of Entomology* 15:264–266.

Bin, F., and N. F. Johnson. 1982. New species of *Telenomus* (Hymenoptera: Scelionidae), egg parasitoids of tropical pyralid pests (Lepidoptera: Pyralidae). *Redia* 65:229–252.

Brues C. T. 1917. Adult hymenopterous parasites attached to the body of their host. *Proceedings of the National Academy of Sciences* 3:136–140.

Burks, R. A., L. Masner, N. F. Johnson, and A. D. Austin. 2013. Systematics of the parasitic wasp genus *Oxyscelio* Kieffer (Hymenoptera, Platygastridae s.l.), Part I: Indo-Malayan and Palearctic fauna. *ZooKeys* 292:1–263. doi:10.3897/zookeys.292.3867.

Chen, H., E. J. Talamas, L. Masner, and N. F. Johnson. 2018. Revision of the world species of the genus *Habroteleia* Kieffer (Hymenoptera, Platygastridae, Scelioninae). *ZooKeys* 730:87–122. doi:10.3897/zookeys.730.21846.

Chen, H., N. F. Johnson, L. Masner, and Z. Xu. 2013. The genus *Macroteleia* Westwood (Hymenoptera, Platygastridae s.l., Scelioninae) from China. *ZooKeys* 300:1–98. doi:10.3897/zookeys.313.5106.

Clausen, C. P. 1976. Phoresy among entomophagous insects. *Annual Review Entomology* 21:343–367.

Engel, M. S., D. Huang, A. S. Alqarni et al. 2016. An apterous scelionid wasp in mid-Cretaceous Burmese amber (Hymenoptera: Scelionidae). *Comptes Rendus Palevol* 16(1):5–11.

Gahan, A. B. 1925. A second lot of parasitic Hymenoptera from the Philippines. *Philippine Journal of Science* 27:83–109.

Galloway, I. D., and A. D. Austin. 1984. Revision of the Scelioninae (Hymenoptera: Scelionidae) in Australia. *Australian Journal of Zoology*, Supplementary Series No. 99. 138 pp.

Gnanakumar, M., K. Rajmohana, C. Bijoy, D. Balan, and R. Nishi. 2012. Diversity of hymenopteran egg parasitoids in organic and conventional paddy ecosystems. *Tropical Agricultural Research* 23 (4):300–308.

Haliday, A. H. 1833. An essay on the classification of the parasitic Hymenoptera of Britain, which correspond with the *Ichneumones minuti* of Linnaeus. *Entomological Magazine* 1:259–276.

Haliday, A. H. 1839. Hymenopterorum synopsis ad methodum clm. *Fallenii ut Plurimum accommodata* I–IV Belfast with Hymenoptera Britannica, Alysia, London, UK.

Hollingsworth, J. P., A. W. T. Hartstack, and P. D. Lingren. 1970. The spectral response of *Campoletis perdistinctus*. *Journal of Economic Entomology* 63:1758–1761.

Johnson, N. F. 1992. Catalog of world Proctotrupoidea excluding Platygastridae. *Memoirs of the American Entomological Institute* 51:1–825. doi:10.5281/zenodo.23657.

Johnson, N. F. 1996. Revision of world species of *Paratelenomus* (Hymenoptera: Scelionidae). *Canadian Entomologist* 128:273–291.

Johnson, N. F. 2018. Hymenoptera Online (HOL). Available from https://hol.osu.edu/search.html?limit=50&name=platygastroidea. [Accessed October 10, 2018]

Johnson, N. F., and L. Masner. 1985. Revision of the genus *Psix* Kozlov & Lê (Hymenoptera: Scelionidae). *Systematic Entomology* 10:33–58.

Johnson, N. F., and L. Masner. 2006. Revision of world species of the genus *Nixonia* Masner (Hymenoptera: Platygastroidea, Scelionidae). *American Museum Novitates* 3518:1–32.

Johnson, N. F., L, Masner, L. Musetti, S. van Noort et al. 2008. Revision of world species of the genus *Heptascelio* Kieffer (Hymenoptera: Platygastroidea, Platygastridae). *Zootaxa* 1776:1–51.

Kennedy, J. S., C. O. Booth, and W. J. S. Kershaw. 1961. Host finding by aphids in the field. iii Visual attraction. *Annual Applied Biology* 49:1–24.

Ketipearachchi, Y. 2002. Hymenopteran parasitoids and hyperparasitoids of crop pests at Aralaganwila in the North Central Province of Sri Lanka. *Annals of the Sri Lanka Department of Agriculture* 4:293–306.

Lesne, P. 1896. Moeurs de *Limosina sacra* Meig. (Famille Muscidae, tribu Borborenae). Phenomenes de transport mutuel chez les animaux asticales. Origines de parasitisme chez les insectes Dypteres. *Bulletin de la Société entomologique de France*, 162–165.

Mani, M. S. 1936a. Some new and little known parasitic Hymenoptera from India. *Records of Indian Museum* 38:333–340.

Mani, M. S. 1936b. On a collection of parasitic Hymenoptera from the Government Museum, Madras. *Records of Indian Museum* 38:469–472.

Mani, M. S. 1939. Descriptions of new and records of some known chalcidoid and other hymenopterous parasites from India. *Indian Journal of Entomology* 1:69–99.

Mani, M. S. 1941. Studies on Indian parasitic Hymenoptera. II. *Indian Journal of Entomology* 3:25–36.

Mani, M. S. 1942. Studies on Indian parasitic Hymenoptera. II. *Indian Journal of Entomology* 4:153–162.

Mani, M. S. 1973. On a new scelionid parasite (Hymenoptera: Serphoidea). *Oriental Insects* 7:353–354.

Mani, M. S., and S. K. Sharma. 1982. Proctotrupoidea (Hymenoptera) from India. A review. *Oriental Insects* 16:135–258.

Manoj, K., T. P. Rajesh, B. Prashanth et al. 2017. Diversity of Platygastridae in leaf litter and understory layers of tropical rainforests of the Western Ghats biodiversity hotspot, India. *Environmental Entomology* 46 (3):685–692. doi:10.1093/ee/nvx080.

Masner, L. 1976. Revisionary notes and keys to world genera of Scelionidae (Hymenoptera: Proctotrupoidea). *Memoirs of the Entomological Society of Canada* 97:1–87.

Masner, L. 1993. Superfamily Proctotrupoidea, Superfamily Platygastroidea, Superfamily Ceraphronoidea. In *Hymenoptera of the World: An Identification Guide to Families*, eds. H. Goulet and J. Huber, 537–569. Agriculture Canada Research Branch, Ottawa, ON, 1894E.

Masner, L., and L. Huggert. 1989. World review and keys to genera of the subfamily Inostemmatinae with reassignment of the taxa to the Platygastrinae and Sceliotrachelinae (Hymenoptera: Platygastridae). *Memoirs of the Entomological Society of Canada* 147:1–214.

Masner, L., N. F. Johnson, and L. Musetti. 2009. *Calliscelio elegans* (Perkins), a tramp species, and a review of the status of the genus *Caenoteleia* Kieffer (Hymenoptera: Platygastridae). *Zootaxa* 2237:59–66.

McKellar, R. C., and M. S. Engel. 2012. Hymenoptera in Canadian Cretaceous amber (Insecta). *Cretaceous Research* 35:258–279. doi:10.1016/j.cretres.2011.12.009.

Miko, L., L. Masner, and R. Deans. 2010. World revision of *Xenomerus* Walker (Hymenoptera: Platygastroidea, Platygastridae). *Zootaxa* 2708:1–73.

Mills, N. J. 1994. "Parasitoid guilds" defining the structure of the parasitoid communities of endopterygote insect hosts. *Environmental Entomology* 23:1066–1983.

Mills, N. J. 2010. Egg parasitoids in biological control and integrated pest management. In *Egg Parasitoids in Agroecosystems with Emphasis on Trichogramma*, eds. J. R. Postali Parra, F. L. Consoli, and R. A. Zucchi, 384–411. Springer, London, UK.

Mineo, G. 1983. Studies on the Scelionidae (Hymenoptera: Proctotrupoidea) XV111. Revision of the genus *Gryon* Haliday (Ethiopian-Oriental regions) the Charon group. *Phytophaga* 1:11–26.

Mukerjee, M. K. 1978. Descriptions of some new and records of known Platygastridae (Hymenoptera: Proctotrupoidea) from India. *Memoirs of School of Entomology St. John's College Agra* 5:47–66, 80.

Mukerjee, M. K. 1979. On a collection of the genus *Scelio* Latreille (Scelionidae: Proctotrupoidea) from India. *Memoirs of School of Entomology, St. John's College Agra* 7:89–117.

Mukerjee, M. K. 1981. On a collection of Scelionidae and Platygastridae (Hymenoptera: Proctotrupoidea) from India. *Records of Zoological Survey of India* Miscellaneous publication Occasional paper, No. 2. 72 pp.

Mukerjee, M. K. 1992. Three Platygastrid wasps (Hymenoptera: Proctotrupoidea) from Garhwal Himalayas, India. *Hexapoda* 4(2):175–182.

Mukerjee, M. K. 1993. On a collection of Scelionidae (Proctotrupoidea: Hymenoptera) from Garhwal Himalayas, India. *Hexapoda* 5(1):75–105.

Mukerjee, M. K. 1994. Descriptions of some new and records of some known Proctotrupoidea (Hymenoptera) from Garhwal Himalayas India. *Records of Zoological Survey of India* Miscellaneous publication Occasional paper, No. 163. 73 pp.

Murphy, N. P., D. Carey, L. R. Castro, M. Downton, and A. D. Austin. 2007. Phylogeny of the platygastroid wasps (Hymenoptera) based on sequences from the 18SrRNA, 28S rRNA and cytochrome oxidase I genes: Implications for the evolution of the ovipositor system and host relationships. *Biological Journal of the Linnaean Society* 91:653–669.

Narasimham, A. U., B. S. Bhumannavar, S. Ramani, J. Poorani, and S. K. Rajeshwari. 1997. A catalogue of natural enemies and other insects in PDBC reference collection, Technical bulletin no. 17, Project Directorate of Biological Control, Bangalore.

Narayanan, E. S., and R. B. Kaur. 1959. A new species of *Microphanurus* Kieffer. *Proceedings of the Indian Academy of Science* (B)49:136–138.

Narendran, T. C. 2001. *Parasitic Hymenoptera and Biological Control*. Palani, India, Palani Paramount Publications. 190 pp.

Naumann, I. D., and C. A. M. Reid. 1990. *Ausasaphes shzralee* sp.n. (Hymenoptera: Pteromalidae: Asaphinae) a brachypterous wasp phoretic on a flightless chrysomelid beetle (Coleoptera: Chrysomelidae). *Journal of Australian Entomological Society* 29:319–325.

Nixon, G. E. J. 1935. A revision of the African telenominae (Hymenoptera: Proctotrupoidea: Scelionidae). *Proceedings of the Royal Society* 5:131–134.

Nixon, G. E. J. 1937. Some Asiatic Telenominae (Hymenoptera: Proctotrupoidea). *The Annals and Magazine of Natural History* 10(20):444–475.

Nixon, G. E. J. 1938. Asiatic species of *Microphanurus* (Hymenoptera: Proctotrupoideae). *The Annals and Magazine of Natural History* 11(2):122–139.

Nixon, G. E. J. 1940. New species of Proctotrupoidea. *The Annals and Magazine of Natural History* (11)6:497–512.

Noyes, J. S. 1982. Collecting and preserving chalcid wasps (Hymenoptera: Chalcidoidea). *Journal of Natural History* 16:315–334.

Ortega-Blanco, R. C., J. McKellar, and M. S. Engel. 2014. Diverse scelionid wasps in Early Cretaceous amber from Spain (Hymenoptera: Platygastroidea). *Bulletin of Geosciences* 89:553–571. doi:10.3140/bull.geosci.1463.

Popovici, O. A., L. Vilhelmsen, Masner, L. Mikó, and N. Johnson. 2017. Maxillolabial complex in scelionids (Hymenoptera: Platygastroidea): Morphology and phylogenetic implications. *Insect Systematics & Evolution* 48:1–125.

Rajmohana K., and C. Bijoy. 2011. A new species of *Paridris* Kieffer (Hymenoptera: Platygastridae) from India. *Hexapoda* 18(1):9–12.

Rajmohana K., and T. C. Narendran. 2007a. A new species of *Paratelenomus* (Hymenoptera: Scelionidae) from India. *Zoos' Print Journal* 22(1):2522–2523.

Rajmohana K., and T. C. Narendran. 2007b. A systematic note on the genus *Trissolcus* Ashmead (Hymenoptera: Scelionidae) with a key to its species from India. *Records of Zoological Survey of India* 107(3): 101–103.

Rajmohana, K. 2007a. On a new species of *Duta* Nixon (Hymenoptera: Scelionidae) from India. *Records of Zoological Survey of India* 107(4):7–12.

Rajmohana, K. 2007b. Insecta: Scelionidae (Platygastroidea): Hymenoptera Fauna of Kudremukh National Park. *Conservation Area Series* 32:49–69.

Rajmohana, K. 2010. First report and a redescription of *Duta serraticeps* (Priesner) (Hymenoptera: Platygastridae) from India. *Biosystematica* 4(2):51–55.

Rajmohana, K. 2012. Descriptions of new taxa in Thorinini (Scelioninae: Platygastridae: Hymenoptera: Insecta) from India. *Recent Advances in Biodiversity of India*, published by Director, Zoological Survey of India, West Bengal, India, pp. 267–276.

Rajmohana, K. 2014. A systematic inventory of Scelioninae & Teleasinae (Hymenoptera: Platygastridae) in the rice ecosystems of North central Kerala. *Memoirs of Zoological Survey of India* 22:72 pp.

Rajmohana, K. and P. Abhilash. 2012. A new species of *Calliscelio* Ashmead (Platygastridae : Hymenoptera: Insecta) from India: *Records of Zoological Survey of India* (Part -1): 75–79.

Rajmohana, K. and P. Abhilash. 2014 A review of the hymenopteran egg parasitoids of Lepidoptera in India. *Forest Entomology: Emerging Issues and Dimensions*, eds. M. Rahman and M. Anto, Delhi, Narendra Publishing House, 27–47.

Rajmohana, K., and A. Peter. 2012. On a new genus and species of Scelioninae (Hymenoptera: Platygastridae) from India. *Biosystematica* 6(1):19–25.

Rajmohana, K., and A. Peter. 2013. Fauna of Bhadra Wildlife Sanctuary & Tiger Reserve. Insecta: Platygastridae (Hymenoptera: Platygastroidea). *Conservation Area Series* 46:11–23.

Rajmohana, K., and C. Bijoy. 2011a. First report of phoresy of *Trissolcus* Ashmead (Hymenoptera: Platygastridae) from India in *Skaphion*, Platygastroidea. *Planetary Biodiversity Inventory Project Newsletter* 5:49–50.

Rajmohana, K., and K. Veenakumari. 2014. *Chakra*, a new genus of Scelioninae (Hymenoptera: Platygastridae) from India, along with description of a new species. *Zootaxa* 3821(2):285–290.

Rajmohana, K., and M. Anto. 2014. *Telenomus dilatus* sp. n. (Hymenoptera: Platygastridae): An egg parasitoid of swallowtail butterflies from South India. *Halteres* 5:73–78.

Rajmohana, K., and M. S. Nisha, 2013. *Telenomus oryzae* (Hymenoptera: Platygastridae), a new egg parasitoid of the rice black bug, *Scotinophara* (Heteroptera: Pentatomidae) from India. *Halteres* 4:79–86.

Rajmohana, K., and S. Talukdar. 2010. A new species of *Psix* Kozlov & Lê (Hymenoptera: Platygastridae) from India. *Biosystematica* 4(2):57–62.

Rajmohana, K., and T. C. Narendran. 2001. Parasitoid complex of *Coptosoma cribrarium* (Fabricius) (Plataspididae: Hemiptera). *Insect Environment* 6(4):163.

Rajmohana, K., and T. C. Narendran. 2007b. A systematic note on the genus *Trissolcus* Ashmead (Hymenoptera: Scelionidae) with a key to species from India. *Records of the Zoological Survey of India* 107(3):101–103.

Rajmohana, K., K. Veenakumari, C. Bijoy, M. Prashanth, P. A. Sinu, and A. P. Ranjith. 2017. *Anokha* gen. n. (Hymenoptera: Platygastroidea: Scelionidae) and two new species from India. *Halteres* 8:77–84.

Rajmohana, K., P. Abhilash, and T. C. Narendran. 2013a. First record of the male sex of the wide spread *Calliscelio elegans* (Perkins) (Hymenoptera: Platygastridae) along with some taxonomic notes on the species. *Biodiversity Data Journal* 1–13. doi:10.3897/BDJ.1.e983.

Rajmohana, K., P. Abhilash, M. Shweta, and M. Anjana. 2015. Platygastroidea of India- A taxonomic and biogeographic update. *Proceedings of the National Workshop on the identification of Bees, Wasps, Beetles and Bugs*. Department of Zoology, Malabar Christian College, Kozhikode, Kerala.

Rajmohana, K. K. K. Srikumar, P. S. Bhat et al. 2013b. A new species of platygastrid *Telenomus cuspis* sp. nov. (Hymenoptera), egg parasitoid of tea mosquito bug from India, with notes on its bionomics and mtCo1 data. *Oriental Insects* 47(4):226–232.

Rajmohana, K., T. C. Narendran, and T. Manoharan. 2011. A new species of *Trissolcus* (Hymenoptera: Platygastridae): Egg parasitoid of *Scutellera nobilis* Fabricius (Hemiptera: Scutelleridae) on *Jatropha curcas* Linnaeus (Euphorbiacea) in India. *Hexapoda* 18(2):106–110.

Romani, R., N. Isidoro, and E. Bin. 2010. Antennal Structures Used in Communication by Egg Parasitoids. In: *Egg Parasitoids in Agroecosystems with Emphasis on Trichogramma*. ed. J. R. Postali Parra, F. L. Consoli, and R. A. Zucchi., 57-96. Springer, London, UK.

Sharkey, M. 2007. Phylogeny and classification of Hymenoptera. *Zootaxa* 1668:521–548.

Shweta, M., and K. Rajmohana. 2016a. Egg parasitoids from the subfamily Scelioninae (Hymenoptera: Platygastridae) in irrigated rice ecosystems across varied elevational ranges in southern India. *Journal of Threatened Taxa* 8(6):8898–8904.

Shweta, M., and K. Rajmohana. 2016b. A comparison of efficiencies of sweep net, yellow pan trap and malaise trap in sampling Platygastridae (Hymenoptera: Insecta). *Journal of Experimental Zoology* 19(1):393–396.

Shweta, M., K. Rajmohana, and C. Bijoy. 2014. A biosystematic account on Baeini wasps (Hymenoptera: Platygastridae), the little known natural enemy complex of Spiders in India. *Proceedings of 26th Kerala Science Congress* 10(84):3654–3662.

Shylesha, A. N., S. K. Jalali, A. Gupta et al. 2018. Studies on new invasive pest *Spodoptera frugiperda* (J. E. Smith) (Lepidoptera: Noctuidae) and its natural enemies. *Journal of Biological Control* 32(3):1–7. doi:10.18311/jbc/2018/21707.

Singh, L. R., N. F. Johnson, and V. V. Ramamurthy. 2012. New species of the genus *Psix* Kozlov & Lê (Hymenoptera: Platygastridae). *Zootaxa* 3530:35–42.

Subba Rao, B. R., and M. J. Chacko. 1961. Studies on *Allophanurus indicus* n. sp., an egg parasite of *Bagrada cruciferum* Kirkaldy. *Beitraege zur Entomologie* 11:812–824.

Taekul, C., A. A. Valerio, A. D. Austin, H. Klompen, and N. F. Johnson. 2014. Molecular phylogeny of telenomine egg parasitoids (Hymenoptera: Platygastridae s.l.: Telenominae): Evolution of host shifts and implications for classification. *Systematic Entomology* 39:24–35.

Taekul, C., N. F. Johnson, L. Masner, A. Polaszek, and K. Rajmohana. 2010. World species of the genus *Platyscelio* Kieffer (Hymenoptera, Platygastridae). *ZooKeys* 50:97–126.

Taekul, C., N. F. Johnson, L. Masner, K. Rajmohana, and S. P. Chen. 2008. Revision of the world species of the genus *Fusicornia* Risbec (Hymenoptera: Platygastridae, Scelioninae). *Zootaxa* 1966:1–52.

Talamas, E. J., and M. L. Buffington. 2014. Updates to the nomenclature of Platygastroidea in the Zoological Institute of the Russian Academy of Sciences. *Journal of Hymenoptera Research* 39:99–117. doi:10.3897/JHR.39.7698.

Talamas, E. J., I. Mikó, and R. S. Copeland. 2016. Revision of *Dvivarnus* (Scelionidae, Teleasinae). *Journal of Hymenoptera Research* 49:1–23. doi:10.3897/JHR.49.7714.

Talamas, E. J., J. Thompson, A. Cutler et al. 2017. An online photographic catalog of primary types of Platygastroidea (Hymenoptera) in the National Museum of Natural History, Smithsonian Institution. *Journal of Hymenoptera Research* 56:187–224. doi:10.3897/jhr.56.10774.

Talamas, E. J., L. Masner, and N. F. Johnson. 2013. Systematics of *Trichoteleia* Kieffer and *Paridris* Kieffer (Hymenoptera, Platygastroidea, Platygastridae). *Journal of Hymenoptera Research* 34:1–79. doi:10.3897/jhr.34.4714.

Valerio, A. A., A. D. Austin, L. Masner, and N. F. Johnson. 2013. Systematics of Old World *Odontacolus* Kieffer s.l. (Hymenoptera, Platygastridae s.l.): Parasitoids of spider eggs. *ZooKeys* 314:1–151.

van Noort, S. 2018. WaspWeb: Hymenoptera of the Afrotropical region. URL: www.waspweb.org (Accessed on 11/07/2018)

Veenakumari K., K. Rajmohana, P. Mohanraj, and P. Abhilash, 2016. An unusual, new, sexually dimorphic species of *Gryon* Haliday (Hymenoptera: Scelionidae) from India, *Oriental Insects* 50(1):40–49. doi:10.1080/00305316.2016.1142482.

Veenakumari, K., and P. Mohanraj. 2015. Redescription of *Protelenomus flavicornis* Kieffer (Platygastroidea: Platygastridae) from India. *Entomofauna, Zeitschrift für Entomologie* 36:305–312.

Veenakumari, K., and P. Mohanraj. 2017a. First report of *Mantibaria mantis* (Dodd) (Hymenoptera: Scelionidae: Scelioninae) from India and additional descriptors for the species. *Journal of Threatened Taxa* 9(6):10347–10350. doi:10.11609/jott.2936.9.6.10347-10350.

Veenakumari, K., and P. Mohanraj. 2017b. The genus *Cremastobaeus* Ashmead (Hymenotpera: Scelionidae: Cremastobaeini) from India. *Journal of Natural History* 51:1989–2056.

Veenakumari, K., E. J. Talamas, K. Rajmohana, and P. Mohanraj. 2017. Two new species of *Apteroscelio* Kieffer (Hymenoptera: Scelionidae) from India, *Zootaxa* 4277(1):137–143.

Veenakumari, K., F. R. Khan, and P. Mohanraj. 2014. Three new species of Teleasinae (Hymenoptera: Platygastridae) from India. *Entomologist's Monthly Magazine* 150:227–239.

Veenakumari, K., K. Rajmohana, and P. Mohanraj. 2012. Studies on phoretic Scelioninae (Hymenoptera: Platygastridae) from India along with description of a new species of *Mantibaria* Kirby. *Linzer Biologische Beiträge* 44(2):1715–1725.

Veenakumari, K., K. Rajmohana, S. Manickavasagam, and P. Mohanraj. 2011. On a new genus of Teleasinae (Hymenoptera: Platygastridae) from India. *Biosystematica* 5(2):39–46.

Veenakumari, K., O. A. Popovici, E. J. Talamas, and P. Mohanraj. 2018b. *Indiscelio*: A new genus of Scelionidae (Platygastroidea) from India. *Journal Asia-Pacific Entomology* 21:571–577.

Veenakumari, K., P. N. Buhl, and P. Mohanraj. 2018a. A new species of *Synopeas* (Hymenoptera: Platygastridae) parasitizing *Pauropsylla* cf. *depressa* (Psylloidea: Triozidae) in India. *Acta Entomologica* 58(1):137–141. doi:10.2478/aemnp-2018-0011.

Veenakumari, K., P. N. Buhl, P. Mohanraj, and F. R. Khan. 2017–2018. Revision of Indian species of *Leptacis* Förster (Hymenoptera: Platygastroidea, Platygastridae). *Entomologist's Monthly Magazine* (Four parts) 153:205–231, 279–312; 154:21–52.

Vlug, H. J. 1995. Catalogue of the Platygastridae (Platygastroidea) of the world (Insecta: Hymenoptera). *Hymenopterorum Catalogus* 19:1–168. doi:10.5281/zenodo.24358.

Yoder, M. J., A. A. Valerio, A. Polaszek, L. Masner, and N. F. Johnson. 2009. Revision of *Scelio pulchripennis* group species (Hymenoptera: Platygastroidea, Platygastridae). *ZooKeys* 20:53–118.

11 Bumble Bees (Hymenoptera: Apidae: Apinae: Bombini) of India

Martin Streinzer

CONTENTS

Introduction 173
Diagnosis of Tribe and Key Characters 174
    Tribe and Genus Diagnosis 174
    Key Characters for Identification 174
Sub-generic Classification 174
    Key to Subgenera and Species 175
    Simplified Key to the Subgenera of *Bombus* for India 175
        Females 175
        Males 176
Major Work on Indian Fauna 177
Biodiversity and Species Richness 178
Distribution Patterns in Indian Context 179
Biological Characteristics 179
    Biology 179
    Plant Associations 180
Molecular Characterization and Phylogeny 180
Integrative Taxonomy 180
Taxonomic Problems 181
Conservation Status 181
Commercial Use 181
Collection and Preservation 181
Useful Websites 182
Conclusion 183
References 183

INTRODUCTION

Bumble bees are large, colourful, and conspicuous bees that are usually found in abundance throughout the flowering season. They are easily recognized even by non-specialists and amateur entomologists. For these reasons, bumble bees were collected and described by many early naturalists and scientists. Due to their rather monotonous morphology and chromatic variability (Michener 2000), many of the early descriptions are currently considered to be of subspecific or infrasubspecific rank (Williams 1998). The large number of original descriptions further complicates and lengthens the process of a global revision of bumble bee species, despite the fact that they make up only about one percent of all described bee species (Williams 1998).

Bumble bees are important pollinators for many wild plants, but also fruit and vegetable crops, making them not only ecologically, but also economically of great significance. Artificial rearing of bumble bees and their commercialization for indoor pollination have many negative consequences, e.g., pathogen spill-over to wild populations and introduction of alien species to the wild (Aizen et al. 2018; Colla et al. 2006). In recent years, bumble bees also became an important model in science, e.g., behavioural sciences, sociobiology, and neuroscience (Baer 2003; Perry et al. 2017; Plowright and Laverty 1984).

The roughly 260 species of bumble bees worldwide are adapted to cold habitats and are most abundant in temperate regions, mountain ranges, and at high latitudes. In tropical regions, they are confined to higher altitudes, although there are a few exceptions (see Gonzalez et al. 2004; Williams et al. 2008).

In India, bumble bees are only found in the Himalaya and so far have not been recorded from the southern lowland and mountain ranges (Saini et al. 2015; Williams 1991). In the Himalayan mountain range a large variation of suitable bumble bee habitats is found, ranging in altitude from tropical foothills to alpine environments and from the drier east to the wetter west. The climatic gradient is also reflected in the bumble bee species composition (Saini et al. 2015; Williams et al. 2010). India, with its share in the Himalaya over its entire range, is home to rich bumble bee diversity. The most recent checklist contains 48 species (Saini et al. 2015). Mapping of the Indian bumble bee fauna has been going on for a few decades (Saini et al. 2015; Williams 1991), but so far, most of the work has focussed on the West and Central Himalaya (Saini et al. 2015; Williams 1991; Williams et al. 2010). Only a couple of records are available from the eastern end of the Himalaya (Saini et al. 2015; Williams 2004). Recently, fieldwork began to document the species diversity in the Northeast states of India at the eastern end of the Himalayan range (Streinzer et al. 2019). These field surveys, together with recent and on-going taxonomic revisions (e.g., Hines and Williams 2012; Williams et al. 2011, 2012b, 2016; Williams 2018a), will certainly lead to updates to the current checklist in the next couple of years.

This book chapter is intended to serve as an entry point for students and researchers who would like to study Indian bumble bees. One aim is to provide a general overview of bumble bee taxonomy, its progress, and pitfalls. Another is to review the research in context to India and provide the reader with a simple sub-generic key for the Indian bumble bees and references to relevant sources for more in-depth identification, taxonomic, and general information.

DIAGNOSIS OF TRIBE AND KEY CHARACTERS

TRIBE AND GENUS DIAGNOSIS

Bees are a monophyletic lineage in the large order Hymenoptera, where they form a monophyletic clade (Anthophila) within the superfamily Apoidea (Branstetter et al. 2017).

Currently, more than 20,000 bee species are described (Ascher and Pickering 2018). Within Anthophila, bumble bees are placed in the family Apidae and form one (Bombini) of the four monotypic tribes (together with Apini, Euglossini, and Meliponini) of the corbiculate bees within the subfamily Apinae (Cardinal et al. 2010; Michener 2000). The genus *Bombus* is the only extant genus in the tribe Bombini.

Among bees, the long malar space and the corbicula (pollen basket) on the hind legs distinguish females and workers of (the social) bumble bees from all other bees (Michener 2000). Bumble bees are medium- to large-sized bees that are covered with fine dense branched hair (pile). The female castes (queens and workers) are equipped with a pollen collecting structure, the corbicula, which is the flattened outer surface on the hind tibia that is void of hair. The corbicula is fringed with long hair and stiff setae that act as basket to hold the pollen in place during foraging. The obligate parasitic species of the subgenus *Psithyrus* do not possess a corbicula (Michener 2000).

KEY CHARACTERS FOR IDENTIFICATION

Several morphological characters are used in bumble bees for identification and classification at the sub-generic and specific level (Michener 2000; Richards 1968; Williams 1994). In females, many important characters are found on the head, e.g., the shape and sculpturing of the ocello-ocular space, the clypeus and labrum, the length of the malar space, the form and presence of groves, ridges and teeth on the mandible, as well as the relative length of the antennal segments. On the body, the relative length of the pubescence, the sculpturing of the corbicula, and the shape and sculpturing of the abdominal sclerites are often diagnostic. The sclerotized parts of the female sting apparatus are also diagnostic in many cases. In males, the shape of the genital apparatus is the most reliable taxonomic character, due to its relatively small intraspecific variability and large interspecific differences (Michener 2000; Williams 1994). A few species or groups of species are not distinguishable by the genitalia alone (Rasmont et al. 1986). Further taxonomic characters are the pilosity of the body and legs, the shape and sculpturing of the leg segments, relative antennal segment lengths, and the sculpturing of the head.

The colourful pile of bumble bees is a very important character in the identification process, but the large extent of both intraspecific variation and interspecific convergence of pile colour renders its usefulness in species identification often invaluable (Williams 1998, 2007).

Sub-generic identification is usually based on a combination of morphological characters, although in a few cases, a single character may be sufficient [e.g., the absence of a corbicula in females is diagnostic for the subgenus *Psithyrus*; (Williams et al. 2008)].

Larval characters play a very limited role in the history of bumble bee taxonomy and are generally not used for species identification (reviewed in Ito 1985).

SUB-GENERIC CLASSIFICATION

Initially, all described bumble bee species were placed in the genus *Apis* by Linné (Linnaeus 1758). Latreille (1802) later erected the genus *Bombus* for bumble bees. Lepeletier (1832) was the first to consider two genera, *Psithyrus* for the parasitic and *Bombus* for the non-parasitic species. Subgenera were later introduced by Dalla Torre (1880), who discriminated nine subgenera based on colour pattern. Radoszkowski (1884) then used structural characters, mainly the male genitalia, for his classification. Subsequently, many adaptations to the sub-generic classification were made by different authors (e.g., Franklin 1912, 1913; Krüger 1917; Robertson 1903; Skorikov 1914; Vogt 1911), who used various characters for their classifications. Milliron (1961) proposed a very different system with three genera and a few subgenera, based on wing characters, which have otherwise not been used in the taxonomy of bumble bees. This system has not been widely accepted. In a particularly comprehensive account, Richards (1968) revised the subgenera and described 35 subgenera of (non-parasitic) bumble bees, providing descriptions of both sexes and identification keys that have been adopted by different authors (e.g., Michener 2000).

In the sub-generic system of Richards (1968), many of the subgenera contain only one or very few species of which some have been later found to be paraphyletic (Cameron et al. 2007). *Psithyrus*, which had been treated as a separate genus since Lepeletier (1832), was repeatedly suggested to be nested within *Bombus* in recent decades, and this hypothesis was later confirmed by molecular methods (Cameron et al. 2007).

Based on a genus-wide comprehensive molecular phylogeny (Cameron et al. 2007), Williams et al. (2008) suggested a simplified classification comprising 15 subgenera that comply with the criteria of being monophyletic and diagnosable. This system is currently widely accepted by taxonomists.

A comprehensive review of the historic development of the sub-generic classification of bumble bees is presented by Ito (1985).

Key to Subgenera and Species

A number of regional studies and revisions of bumble bees have provided a fair number of different identification keys. Many of the keys cannot be universally used due to their limited regional extent, incomplete species repertoire, the included characters (e.g., only colour) or the included sex [e.g., only workers, see Williams et al. 2010). The value of these keys largely depends on the sampling region and the material that is to be identified, and their use should be carefully evaluated prior to identification.

Whenever a detailed key for the sampling region is available, the reader is directed towards using this as a first reference. For the Indian region, one of most relevant identification keys is for the region Jammu and Kashmir (Williams 1991). The work not only includes a detailed key and many drawings, but also important comparative notes in the species accounts that are relevant to the bumble bees of the entire Himalayan region. A more recent account of the Indian bumble bee fauna is presented by Saini et al. (2015). The authors provide updated keys for males and females for most Indian species, detailed species descriptions, and drawings of important diagnostic characters. A couple of species are missing in the keys, so the general remarks on species identification also apply here.

While the West Himalaya is fairly well covered by the two abovementioned identification keys (Saini et al. 2015; Williams 1991), the East Himalaya has not been surveyed to a similar extent. For the East Indian states, the identification keys for Nepal (Williams et al. 2010) and Sichuan (Williams et al. 2009) are important starting points.

It is important to keep in mind that species composition of the target region may differ and some of the species may not be accounted for in the available keys. Furthermore, recent taxonomic changes may have been implemented after publication of the keys or have not yet been incorporated in the keys (e.g., in Saini et al. 2015). In a few cases, specimens are still unknown to science, e.g., the male of *Bombus (Melanobombus) tanguticus* Morawitz that occurs in West and Central India (Williams 2018a).

In many cases, it is advisable to first identify the subgenus. Williams et al. (2008) revised the sub-generic system and provided usable keys for both sexes, which allow one to identify most specimens to sub-generic rank. In some cases, the keys to the old sub-generic system may further help by narrowing down the possibilities to species-groups (Michener 2000; Richards 1968). A simplified version of the sub-generic key by Williams et al. (2008), including only the subgenera known from India, is presented below.

Simplified Key to the Subgenera of *Bombus* for India

Females

1. Metatibiae with the outer surface broad and flat, without dense hair throughout, and fringed with long stiff hairs that form a pollen basket. Sternum 6 without two ventro-lateral keels..2
- Metatibiae with the outer surface convex and densely covered with moderate to long hair. Sternum 6 with two ventro-lateral keels............. ***Psithyrus*** (8 species)

2(1). Mandible with the anterior keel reaching the distal margin. Proximal posterior process of the metabasitarsus narrow and pointed (or if shorter, then the lateral ocellus separated by twice or more of its diameter from the inner eye margin) 3
- Mandible with the anterior keel not reaching the distal margin. Proximal posterior process of the metabasitarsus broad and short. Outer surface of the metatibiae coarsely sculptured and with long hair in the middle of its proximal half ....
....***Mendacibombus*** (3 species)

3(2). Mandible distally broadly rounded, with two anterior teeth and often a posterior tooth. Outer surface of the metabasitarsus without very long erect hairs............4
- Mandible distally with six large teeth. Outer surface of the metabasitarsus with long erect hairs...
.... ***Alpigenobombus*** (5 species)

4(3). Mesobasitarsus with the distal posterior corner pointed and spine-like, or if borderline, then the ocello-ocular area punctured in more than $^{1}/_{2}$ the distance between the lateral ocellus and the inner eye margin, and the metabasitarsus with dense, plumose hair continuing onto the outer surface of the proximal-posterior process............5
- Mesobasitarsus with the distal posterior corner rounded and not spinosely produced, or if borderline, then the ocello-ocular area punctured only in $^{1}/_{4}$ of the distance between the lateral ocellus and the inner eye margin 7

5(4). Metabasitarsus without dense, plumose hair continuing onto the (bare and shiny) outer surface of the proximal-posterior process. Ocelli lying on the supraorbital line 6

- Metabasitarus with dense, plumose hair continuing onto the outer surface of the proximal-posterior process. Ocelli lying anterior to the supraorbital line.................***Sibiricobombus*** (3 species)
6(5). Clypeus bulbous and with many large punctures. Sternum 6 with a strong medial ridge. Antennal segment 4 shorter than broad ...***Megabombus*** (1 species) Clypeus flattened and predominantly smooth and shining with only widely scattered micro-punctures. Sternum 6 with a raised and often shiny median longitudinal keel in the posterior $1/3$. Antennal segment 4 longer than broad.................***Subterraneobombus*** (3 species)
7(4). Ocello-ocular area with the unpunctured, shining areas medium or small, the area anterior to the three ocelli unpunctured only in the vicinity of the ocelli....... . 8
- Ocello-ocular area with the unpunctured, shining areas large, including most of the area anterior to the ocelli, except for narrow bands of punctures between the ocelli........................***Orientalibombus*** (2 species)
8(7). Mandible without, or just a shallow *incisura lateralis*. Clypeus weakly swollen or flat. Metabasitarsus with the posterior margin convex only in the proximal $1/4$. Malar space nearly as long as or longer than broad...... 9
- Mandible with pronounced *incisura lateralis*. Clypeus centrally swollen and bulging. Metabasitarsus with the posterior margin convex throughout. Malar space just shorter or distinctly shorter than broad....... ....***Bombus*** s. str. (3 species)
9(8). Inner eye margin with a band of large scattered and only few small punctures. Unpunctured area of the ocello-ocular space either large (occupying about $3/4$ of the distance between the eye and the lateral ocelli) or absent (with medium to large punctures throughout)….................... ***Pyrobombus*** (12 species)
- Inner eye margin with a band of intermixed small and large punctures. Unpunctured area of the ocello-ocular space only about $1/2$ of the ocello-ocular space… .... ***Melanobombus*** (9 species)

Males

1. Volsella and gonostylus strongly sclerotized. Gonostylus with the inner proximal process without long branched hairs. Volsella often with a process or hooks on the inner margin 2
- Volsella and gonostylus weakly sclerotized. Gonostylus with the inner proximal process with many long branched hairs. Volsella without a process or hooks on its inner margin ..... ***Psithyrus*** (8 species)
2(1). Antenna medium to long. Penis spatha pointed proximally, penis valve head straight or curved. Compound eyes enlarged or not enlarged relative to the females 3
- Antenna short. Penis spatha rounded proximally, penis valve head straight. Compound eyes strongly enlarged relative to females ....***Mendacibombus*** (3 species)
3(2). Penis valve dorso-ventrally narrowed, at least in its distal $1/3$. Antenna medium to long............................ 4
- Penis valve dorso-ventrally broadened. Antenna medium..............................***Bombus*** s. str. (3 species)
4(3). Penis valve head turned medially (towards the body midline) as broad hook or spoon................................ 6
- Penis valve head either straight or turned laterally (away from the body midline) 5
5(4). Gonostylus with a distinct inner proximal process separate from the distal lobe, strongly recurved away from the body midline. Volsella distally narrow, with the inner hooks pointed and placed close to the distal end........................***Megabombus*** (1 species)
- Gonostylus without a narrow inner proximal process, the inner distal margin simple, flattened and blade-like in cross section, shorter than broad. Volsella broad, with the inner hooks placed more centrally. Penis valve head straight and pointed............***Orientalibombus*** (2 species)
6(4). Gonostylus triangular without an inner proximal process, or if strongly reduced and spine-like, then the penis valve head has an additional outward pointing process. Gonostylus with the inner distal margin at least slightly thickened in cross section with a longitudinal sub-marginal groove... ....***Pyrobombus*** (12 species)
- Gonostylus usually with a distinct inner proximal process. Gonostylus with the inner distal margin simple, flattened and blade-like in cross section without a longitudinal sub-marginal groove............................. 7
7(6). Penis valve shaft with a ventral angle either absent or reduced and broadly rounded 9
- Penis valve shaft with the ventral angle produced as sharp angle or broad transverse process 8
8(7). Penis valve head sickle-shaped. Penis valve shaft with the ventral angle produced as sharp angle. Mandible with two anterior teeth ....***Alpigenobombus*** (5 species)
- Penis valve head spoon-like. Penis valve shaft with the ventral angle produced as broad transverse process. Mandible with one anterior tooth...................... .... ***Subterraneobombus*** (3 species)
9(7). Outer surface of the mesobasitarsus with many long hairs. Antenna medium or long. Gonostylus with the distal lobe longer or shorter than broad. Gonocoxa with the inner proximal process about as long as broad .... ***Melanobombus*** (9 species)
- Mesobasitarsus with no or few long hairs. Gonostylus with the distal lobe longer than broad. Antenna very long, of if shorter than the penis valve hook strongly club-shaped ....***Sibiricobombus*** (3 species)

Recent global revisions that include species-level identification keys have been published for a few subgenera, e.g., *Subterraneobombus* (Williams et al. 2011) and *Mendacibombus* (Williams et al. 2016). For other subgenera, complete or partial revisions have been published without identification keys, e.g., *Alpinobombus* (Williams et al. 2015b),

Bombus s.str. (Williams et al. 2012b), and *Megabombus* (Huang et al. 2015). The taxonomic remarks and distribution maps may still provide enough information to help in species identification. Similar work is ongoing and will soon add to a more complete revision of many other subgenera (see Hines and Williams 2012; Williams 2018a).

It is recommended to critically check the identification using either *confidently* determined material in museum collections, additional identification keys, the original descriptions, or additional identification aids, e.g., DNA barcoding (see below).

MAJOR WORK ON INDIAN FAUNA

Many researchers have contributed directly or indirectly to the study of bumble bees in India. Saini et al. (2015) provide a detailed list of authors and references for the species of the Indian and Oriental fauna, respectively. Many of the species described from material collected in India (e.g., Bingham 1897; Friese 1918; Skorikov 1914; Smith 1852; Tkalcu 1989) are currently treated as valid species and many more as synonyms.

It seems premature to provide an extensive list of species described from Indian material before recent revisions of all subgenera are available. Such thorough revisions (Williams et al. 2011, 2012b, 2016) not only trace the history of all the descriptions that are regarded as valid species and those that are considered synonyms, but also trace the type material and type location. From the available revisions, it becomes clear that: (i) the current Indian checklist of bumble bees is far from being correct and (ii) the actual type location is not always unambiguously attributable to a current country concept or location.

Of the many authors that described species from the Indian region, Bingham (1897) provided the first comprehensive account, listing 23 species from British India and providing identification keys for these. The next comprehensive work was the paper on bumble bees from Jammu and Kashmir by Williams (1991), and the most recent revision of the entire Indian bumble bee fauna is the book by Saini et al. (2015).

One important revisionary work on the Himalayan bumble bee fauna, Tkalcu's *Monographie der Unterfamilie Bombinae des Himalaya (Hymenoptera, Apoidea)*, cited as Tkalcu (1976)-in press in Tkalcu (1974), unfortunately has never been published.

The next step in Indian bumble bee research should be an increased effort to collect in the Eastern Himalaya, which has so far been only poorly investigated. This work will likely add new species to the Indian bumble bee fauna, which occur in close vicinity to the Indian border, but have not been genuinely recorded from the country (Williams 2004; Williams et al. 2016). Recent, current, and future revisionary work on different subgenera will also affect the current checklist. Such proposed changes have already taken effect, but were not incorporated in the most recently published checklist list by Saini et al. (2015) (see taxonomic notes in Table 11.1). An updated checklist that incorporates these changes can be found in Table 11.1.

TABLE 11.1
List of Bumble Bee Species from India

Species	Also Found In
Bombus (Mendacibombus) avinoviellus (Skorikov, 1914)	NPL, PAK
Bombus (Mendacibombus) himalayanus (Skorikov, 1914)	PAK
Bombus (Mendacibombus) waltoni Cockerell, 1910	CHN, NPL
Bombus (Orientalibombus) funerarius Smith, 1852	CHN, MMR, and NPL
Bombus (Orientalibombus) haemorrhoidalis Smith, 1852	BTN, CHN, LAO, MMR, NPL, PAK, THA, and VNM
Bombus (Subterraneobombus) difficillimus Skorikov, 1912[a]	CHN, KGZ, PAK, and TJK
Bombus (Subterraneobombus) melanurus Lepeletier, 1835	AFG, ARM, CHN, IRN, KAZ, KGZ, LBN, MNG, NPL, PAK, RUS, SYR, TJK, TKM, and UZB
Bombus (Subterraneobombus) personatus Smith, 1879	CHN, NPL, and PAK
Bombus (Megabombus) albopleuralis Friese, 1916[b]	CHN, NPL
Bombus (Bombus s.str.) jacobsoni Skorikov, 1912[c]	
Bombus (Bombus s.str.) longipennis Friese, 1918[c]	CHN, NPL
Bombus (Bombus s.str.) tunicatus Smith, 1852	AFG, PAK, and NPL
Bombus (Alpigenobombus) breviceps Smith, 1852	CHN, LAO, MMR, NPL, THA, and VTN
Bombus (Alpigenobombus) genalis Friese, 1918	CHN, MMR
Bombus (Alpigenobombus) grahami (Frison, 1933)	CHN, NPL
Bombus (Alpigenobombus) kashmirensis Friese, 1909	CHN, NPL, and PAK
Bombus (Alpigenobombus) nobilis Friese, 1905	CHN, MMR, and NPL
Bombus (Sibiricobombus) asiaticus Morawitz, 1875	AFG, CHN, KAZ, KGZ, MNG, NPL, PAK, and TJK
Bombus (Sibiricobombus) oberti Morawitz, 1883	CHN, KAZ, KGZ, and TJK
Bombus (Sibiricobombus) sibiricus (Fabricius, 1781)	CHN, KAZ, MNG, and RUS
Bombus (Psithyrus) branickii (Radoszkowski, 1893)	AFG, CHN, KAZ, KGZ, MNG, PAK, RUS, and TJK
Bombus (Psithyrus) cornutus Frison, 1933	CHN, NPL

(*Continued*)

TABLE 11.1 (Continued)
List of Bumble Bee Species from India

Species	Also Found In
Bombus (Psithyrus) ferganicus (Radoszkowski, 1893)	AFG, CHN, KAZ, KGZ, and PAK
Bombus (Psithyrus) morawitzianus (Popov, 1931)	AFG, CHN, KAZ, KGZ, and PAK
Bombus (Psithyrus) novus (Frison, 1933)	NPL, PAK
Bombus (Psithyrus) skorikovi (Popov, 1927)	CHN, NPL, and PAK
Bombus (Psithyrus) tibetanus (Morawitz, 1887)	CHN
Bombus (Psithyrus) turneri (Richards, 1929)	CHN, NPL
Bombus (Pyrobombus) abnormis (Tkalcu, 1968)	NPL
Bombus (Pyrobombus) biroi Vogt, 1911	AFG, CHN, KAZ, KGZ, MGN, PAK, and TJK
Bombus (Pyrobombus) flavescens Smith, 1852	CHN, MMR, MYS, NPL, PHL, THA, TWN, and VTN
Bombus (Pyrobombus) hypnorum s.l. (Linnaeus, 1758)	widespread - Palaearctic + Oriental
Bombus (Pyrobombus) lemniscatus Skorikov, 1912	CHN, NPL
Bombus (Pyrobombus) lepidus Skorikov, 1912	CHN, NPL, and PAK
Bombus (Pyrobombus) luteipes Richards, 1934	BTN, CHN, MMR, and NPL
Bombus (Pyrobombus) mirus (Tkalcu, 1968)	CHN, NPL
Bombus (Pyrobombus) parthenius Richards, 1934	BTN, CHN, and NPL
Bombus (Pyrobombus) pressus (Frison, 1935)	CHN, NPL
Bombus (Pyrobombus) rotundiceps Friese, 1916	CHN, LAO, MMR, NPL, and THA
Bombus (Pyrobombus) subtypicus Skorikov, 1914	CHN, AFG, KAZ, KGZ, PAK, and TJK
Bombus (Melanobombus) eximius Smith, 1852	CHN, MMR, NPL, THA, and TWN
Bombus (Melanobombus) festivus Smith, 1861	CHN, MMR, and NPL
Bombus (Melanobombus) keriensis Morawitz, 1887	AFG, CHN, IRN, KAZ, KGZ, MNG, PAK, TJK, and TUR
Bombus (Melanobombus) ladakhensis Richards, 1928	CHN, NPL
Bombus (Melanobombus) miniatus Bingham, 1897[d]	CHN, NPL, and PAK
Bombus (Melanobombus) rufofasciatus Smith, 1852	CHN, MMR, NPL, and PAK
Bombus (Melanobombus) semenovianus (Skorikov, 1914)	AFG, PAK
Bombus (Melanobombus) simillimus Smith, 1852	PAK
Bombus (Melanobombus) tanguticus Morawitz, 1887	CHN

Taxonomic notes:

[a] Formerly treated as synonym to *B. melanurus* (Saini et al. 2015; Williams 2004), now confirmed to be a separate species by Williams et al. (2011).

[b] Formerly treated as synonym to *B. trifasciatus* Smith (Saini et al. 2015; Williams 2004), but recent studies show that *B. albopleuralis* is a separate species, confined to the Himalaya (Huang et al. 2015; Hines & Williams 2012).

[c] Formerly treated as synonym to *B. lucorum* (Linnaeus) (Saini et al. 2015; Williams 2004), but confirmed to be separate species in recent studies (Williams et al. 2012b).

[d] Also listed as *B. pyrosoma* Morawitz by Saini et al. (2015), which is currently considered a Chinese endemic (P.H. Williams, pers. communication).

Country abbreviations: AFG – Afghanistan; ARM – Armenia; BTN – Bhutan; CHN – China; IRN—Iran; KAZ—Kazakhstan; KGZ—Kyrgyzstan; LAO—Laos; LBN—Lebanon; MMR—Myanmar; MNG—Mongolia; MYS—Malaysia; NPL—Nepal; PAK—Pakistan; PHL—Philippines; RUS—Russia; SYR—Syria; TWN—Taiwan; TJK—Tajikistan; THA—Thailand; TUR—Turkey; TKM—Turkmenistan; UZB—Uzbekistan; VNM—Vietnam

Updated checklist of Indian Bumble bees, based on Saini et al. (2015) and Williams (2004), incorporating new taxonomic and distribution information (Williams et al. 2009, 2011, 2012b, 2016).

Saini et al. (2015) listed for each species in their account the depository of the type material and discussed that much of the material is scattered throughout the world, with no material deposited in Indian collections. For many of the species treated in their account (Saini et al. 2015), reference material had to be procured from the Natural History Museum, London, UK, due to a lack of museum material within the country. Many of these species are considered very rare (Saini et al. 2015; Williams and Osborne 2009), but have recently been recorded in the Eastern Himalaya and are now deposited in the NCBS Research Collection [National Centre for Biological Sciences, Tata Institute of Fundamental Research, Karnataka, Bangalore; (Streinzer et al. 2019)].

BIODIVERSITY AND SPECIES RICHNESS

The roughly 260 species of bumble bees are distributed widely in the Northern Hemisphere and South America, but are naturally absent from sub-Saharan Africa and Australasia (Williams 1998). Biogeographic reconstruction suggests an origin in Central Asia roughly around 30 million years ago (mya) during a global cooling period, with subsequent diversification in several world regions (Hines 2008). Colonialization of the Indian region occurred through different routes from the Central Asian and Oriental region, respectively (Hines 2008). Williams (2004) lists 67 species for the Indian region, and the most recently published checklist for the country of

India contained 48 species (Saini et al. 2015). An updated checklist for the Indian bumble bee species, incorporating recent taxonomic changes (Williams et al. 2011; 2012b), is presented in Table 11.1.

Considering equal size sampling regions, the global biodiversity hotspot of bumble bees is found in Central China (Williams 1998).

DISTRIBUTION PATTERNS IN INDIAN CONTEXT

In India, suitable bumble bee habitats include the northern mountain ranges (Himalaya), but potentially also the southern highlands. So far, bumble bees have only been recorded in the Himalaya (Saini et al. 2015). It is highly unlikely that they naturally occur in the southern mountain ranges where they never had the chance to naturally disperse due to the lack of suitable habitat corridors (for a discussion, see Williams et al. 2018).

In the Himalaya, bumble bees find suitable habitats covering a wide range of altitudinal ranges from the low land tropical region to alpine environments. Bumble bee species usually inhabit a distinct range of altitudes. Both the mean altitude and the altitude range breadth are species specific and range from very wide to very restricted (Saini et al. 2015; Williams 1991; Williams et al. 2009). Bumble bee species also show species-specific climatic preferences. In the Himalaya, a distinct climatic transition from the drier west to the wetter east is found that is reflected in the bumble bee species composition (Saini et al. 2015; Williams et al. 2010, 2015a).

The functional adaptations and proximate mechanisms that lead to the observed distribution patterns are currently not well understood. Further work is urgently needed to improve our knowledge of the pattern of occurrence in the Himalaya, e.g., by model calculations of the climatic niches from known occurrence data (Rasmont et al. 2015; Williams et al. 2015a). While the West Himalayan bumble bee fauna is well documented, occurrence data from the eastern regions are scarce (Saini et al. 2015; Williams 2004) and detailed investigations have recently just commenced (Streinzer et al. 2019).

BIOLOGICAL CHARACTERISTICS

BIOLOGY

Bumble bees are (primitively) eusocial bees. They show a reproductive division of labour, with the queen usually being the only colony member that lays eggs. Workers are usually sterile, but may also produce ovaries and lay (unfertilized) eggs. In Hymenoptera, unfertilized eggs develop into males and a certain number of the male offspring are produced by bumble bee workers. The generations usually do not overlap, and the mated queen enters diapause while the rest of the colony dies during the winter season. The sociobiology of bumble bees was comprehensively reviewed by Plowright and Laverty (1984).

The mated queen emerges after diapause and seeks suitable nesting sites. These are usually cavities in the ground, trees, or manmade objects and (abandoned) mammal or bird nests. A few species nest in the open space, under dense vegetation cover. The choice of the nesting site is to some extent species-specific (Kearns and Thomson 1955). Bumble bee nests are well camouflaged and are very hard to locate in the wild (Saini et al. 2015).

After locating a suitable nesting site, the queen starts to forage for nectar and pollen and constructs a clump of pollen and wax in which the first batch of eggs is deposited. The larvae are then provided with food and heat. After the emergence of the first workers, the queen ceases its foraging activity and stays within the colony for the rest of her life (Kearns and Thomson 1955).

Workers perform indoor tasks (cleaning the nest, feeding, and heating the brood) and outdoor activities (foraging for nectar and pollen) (Plowright and Laverty 1984). They are very variable in body size, which leads to a certain size-dependent division of labour among the colony members (Kapustjanskij et al. 2007; Spaethe and Weidenmüller 2002). Excess nectar and pollen are deposited into wax pots for storage to overcome periods of unsuitable weather conditions. Two major strategies in larval provisioning are distinguished. Pocketmakers feed larvae indirectly through a wax pocket adjacent to larval clumps, while pollen storers feed pollen from storage pots directly to the larvae (Michener 2000). The strategies correlate with the morphology of the mid and hind basitarsi, which are spinosely produced in the pocket makers (Williams et al. 2008). Bumble bee nests vary considerably in size, both between colonies and species. They range from very small nests (Ponchau et al. 2006) to nests that are inhabited by a few thousand individuals, often with several reproductive queens (Garófalo et al. 1986).

Later in the season, the colony starts to produce sexual offspring, i.e., young queens and males. A few days after their emergence, they leave the colony to mate. Mated queens then usually dig into the ground and enter diapause, while males commence with their mating behaviour until they die (Plowright and Laverty 1984).

In a couple of species, a certain percentage of newly mated queens immediately start a new colony. In favourable years, these second generations also manage to rear sexual offspring.

A few tropical species produce very large colonies that persist for several years and continuously produce sexual offspring. They also appear to have multiple reproductive females and show signs of swarming reproduction similar to that in stingless bees and honeybees (Gonzalez et al. 2004).

Bumble bee males usually do not engage in colony work and focus on mating. A number of species-specific mating strategies are described, which differ in the signals and cues used by the males to encounter the young queens. Despite the often conspicuous mating behaviour of males, the process of mate finding and the sexual communication in bumble bees is little understood and actual matings are only rarely observed (Ayasse and Jarau 2014; Baer 2003; Streinzer and Spaethe 2014).

Because of the different life histories, bumble bees are sexually dimorphic. Males are usually (much) smaller than queens (del Castillo and Fairbairn 2011), and they have dimorphic sensory organs (e.g., one additional antennal segment and enlarged compound eyes in many species), reduced mandibles, an additional abdominal segment, and lack a corbicula (Michener 2000). They often also differ in hair colour and are more variable than workers in this trait.

About ten percent of all *Bombus* species have adopted a parasitic lifestyle. All members of the subgenus *Psithyrus* and a few other species are obligate parasites that depend on a eusocial *Bombus* host (Hines and Cameron 2010). They invade young *Bombus* colonies at an early stage and kill or subdue the resident queen. The resident workers then rear the parasite's offspring. Only reproductive females and males are reared, and a worker caste is lacking. *Psithyrus* females are morphologically distinct; they lack a corbicula, are heavily armoured, and have evolved chemical mimicry to trick their hosts (Ayasse and Jarau 2014; Michener 2000). Among the non-*Psithyrus* parasitic species, a reduction of the pollen collecting structure and the lack of a worker caste is observed (Yarrow 1970). A couple of species show facultative nest usurpation of either their own or a closely related species [discussed in Hines and Cameron (2010)].

PLANT ASSOCIATIONS

As eusocial "central place foragers" that provision their brood collaboratively, bumble bees are important visitors and pollinators of flowering plants. Their abundance makes them an important pollinator group for many species, and some flowering plants rely almost completely on pollination performed by bumble bees (Løken 1973; Prys-Jones 1982).

In general, bumble bees can be considered polylectic (i.e., they visit a variety of different plant species for the collection of nectar and pollen), with a few notable exceptions (see Huang et al. 2015; Løken 1973). Bumble bee species differ in their proboscis length, and species with different proboscis lengths visit different groups of plants. Long-tonged species are usually more restricted in their floral choice than the more generalist short-tonged species, a factor that has also been discussed as one reason for recent declines in bumble bee populations (Huang et al. 2015; Williams and Osborne 2009).

Bumble bee-plant associations have been studied in great detail. For India, several studies report on flower visitation of bumble bees (e.g., Deka et al. 2011; Saini and Ghator 2007). A comprehensive account on the flower visitation by Indian bumble bees is presented by Saini et al. (2015).

MOLECULAR CHARACTERIZATION AND PHYLOGENY

Molecular tools are now used in many fields of bumble bee research, e.g., to understand the evolution of sociality, population dynamics, pathogen resistance, systematics, and taxonomy. A detailed review on the genetic and genomic work in bumble bees was recently published by Woodard et al. (2015).

Molecular methods helped to overcome difficulties in reconstructing the phylogeny from morphological characters (Williams 1985, 1994). A robust, almost genus-wide molecular phylogeny based on five genes was published by Cameron et al. (2007). The phylogeny provided the basis for many scientific inquiries, such as biogeography (Hines 2008), susceptibility to health threats (see Woodard et al. 2015), or functional adaptations (Streinzer and Spaethe 2014). The phylogeny was based on single individuals for each species and therefore did not include much of the global variation within species. Since then, many subgenera were treated in detail, based on larger samples and including much of the known variation for each species and in consideration of its global range (e.g., Hines and Williams 2012; Williams et al. 2011, 2012b, 2016).

Near the species-level scale, fast evolving genes, such as the COI region of the mitochondrial genome are used to understand population variation and species level relationships within subgenera (Williams et al. 2012b). COI is further used in DNA barcoding as a fast and reliable species identification tool (Woodard et al. 2015). This approach is particularly important in several complexes of cryptic species that cannot be identified to species level based on their morphology (Ellis et al. 2006; Murray et al. 2008; Woodard et al. 2015).

Multiple genetic markers and microsatellites are further utilized to understand population dynamics and reproductive isolation among members of putative species-complexes to determine whether they should be treated as a complex of separate species or as a single variable species (Duennes et al. 2012; Woodard et al. 2015).

INTEGRATIVE TAXONOMY

Following the period of traditional convention, whereby bumble bees were classified using colour pattern or morphology alone, the advent of molecular methods has enabled a rapid progress in our taxonomic and systematic understanding and uncovered many incorrect systematic treatments. Apart from these methods, a variety of different approaches, including geometric morphometric of the wing venation (Aytekin et al. 2007) or chemotaxonomy based on male marking pheromones (de Meulemeester et al. 2011) were used to study species delimitation and identification. Such approaches have been proven to allow clarification of the species status (e.g., Rasmont et al. 2005). The largest set of different approaches was probably applied to the problematic *Bombus* (*Bombus* s. str.) *lucorum*-complex from Europe, using enzyme electrophoresis, microsatellite analysis, DNA barcoding, chemotaxonomy, morphometric, and artificial breeding experiments (reviewed in Bossert 2015; McKendrick et al. 2017; Williams et al. 2012b).

Recent revisionary work usually includes large samples of bumble bees, sampled across their known distribution range and covering much of the known variation and using a combination of different approaches (e.g., Hines and Williams 2012; Potapov et al. 2018). Indian bumble bee material will play a crucial role in the future revision of many subgenera. For instance, some of the putative members of the *B. trifasciatus*-species group are in close contact in the East Himalaya, but recent DNA barcoded material to confirm their species identity is not available (Hines and Williams 2012). Furthermore, a few species of the subgenus *Pyrobombus* are either narrowly distributed in the East Himalaya (e.g., *B. mirus* (Tkalcu), *B. abnormis* (Tkalcu), see Streinzer et al. 2019) or their species status is still questionable (e.g., *B. hypnorum* s.l. (Linnaeus), see Tkalcu 1974; Williams et al. 2010). Future work in this direction, adding to a complete coverage of the species' distribution range is urgently needed.

TAXONOMIC PROBLEMS

Bumble bees are morphologically relatively monotonous, but chromatically very variable (Michener 2000; Williams 1998). The pile colour of bumble bees often converges among sympatric species in the same region, but may vary over the entire distribution range, which results in correlated patterns of colour similarity (Hines and Williams 2012; Tkalcu 1968; Williams 1998, 2007). Bumble bee colour is in many cases not diagnostic and species identification is often more complicated than anticipated. In a few species-groups, the similarity of both the colour pattern and morphology show such a broad overlap that they cannot be identified to species level without application of molecular methods (Bossert 2015; Williams et al. 2012b).

Recent revisionary work in bumble bees includes material from the entire range of distribution including much of the variation within species using molecular methods to understand species taxonomy and solve identification problems. Recent revisions usually include identification keys (e.g., Williams et al. 2011, 2016), but for a number of species, identification based on morphology or colour is still not reliable (Williams et al. 2012b). Male individuals can be usually more reliably identified from the genital morphology, and one must rely on male individuals for confirmation of the species' occurrence in these cases (Williams 1994).

CONSERVATION STATUS

Bumble bees are generally conceived as important pollinators of wild plants and agricultural crops. Dramatic declines of species worldwide (Williams and Osborne 2009) and local extinction of subspecies (Williams et al. 2013) are alarming signs that there is urgent need for the protection of bumble bee species and their habitats.

The *International Union for Conservation of Nature* (IUCN) installed a bumble bee specialist group whose work aims at implementing a global assessment of all bumble bee species (IUCN SSC Bumblebee Specialist Group 2018). The work is making progress and many species have already received a detailed IUCN assessment. So far, species from the Indian region have not been assessed, mainly due to a paucity of available data. Many rare or confined species that occur in India are currently being investigated (e.g., *B. tanguticus* Morawitz, see Williams 2018a) or have been found in the East Himalaya during recent fieldwork (e.g., *B. abnormis* (Tkalcu), *B. mirus* (Tkalcu), *B. pressus* (Frison), *B. genalis* Friese, see Streinzer et al. 2019). Future field observations, especially of these rare species, are urgently needed before a detailed assessment can be made.

COMMERCIAL USE

Besides their ecological importance as pollinators of wild plant species, bumble bees became an important economic factor in recent years. Development of artificial insemination (Baer and Schmid-Hempel 2000) and rearing of colonies around the year (Röseler 1985) allowed their use for year-round pollination of fruits and vegetables in greenhouses.

In contrast to the Western honey bee, which is widely used as a pollinator of fruit and agricultural crops, the history of domestication of bumble bees is short; see review by Velthuis (2002). Artificial breeding began with the European species *Bombus (Bombus* s. str.*) terrestris* (Linneaus), but several species are bred commercially for various markets (Williams et al. 2012a).

In the Oriental region, two species are currently being evaluated for their suitability for commercial rearing, *B. breviceps* Smith in Vietnam (Thai and Van Toan 2018) and *B. haemorrhoidalis* Smith in India (Chauhan et al. 2014). Both species are important pollinators of Indian fruits and vegetables (Deka et al. 2011).

Commercialization has led to the distribution of pathogens (Colla et al. 2006) and the introduction of alien species in many world regions (Aizen et al. 2018; Williams et al. 2012a) and has been considered the main factor contributing to rapid population declines of many *Bombus* species in the USA (Cameron et al. 2011; Jacobson et al. 2018).

The availability and suitability of bumble bees has made them an important model for neuroscience and cognition research, and the commercially bred species are the main model taxa in these fields (Perry et al. 2017).

COLLECTION AND PRESERVATION

Bumble bees are large conspicuous bees that are abundantly found foraging on flowers. They can be captured either using sweep nets or by scooping them off the flowers using small vials or containers. Catching specimens in flight is also possible after a certain amount of practice. Trapping techniques are less commonly used for bumble bees. Among these, flight interception traps and coloured pan or funnel traps are the most common methods (Stephen and Rao 2005).

Bumble bees are typically killed using toxic fumes (e.g., cyanide, ethyl acetate) or by freezing. The choice of killing agent depends on the scientific aim. Ethyl acetate has been suspected to interfere with DNA sampling and thus other methods should be preferred (Dillon et al. 1996).

Specimens can be stored for a long time in vials containing some tissue paper to absorb moisture and a few drops of ethyl acetate to prevent mould. This facilitates collection during long field trips in humid environments. The specimens can then be stored in a freezer until they are further processed. If specimens are intended to be used in genetic work, it is advisable to remove a leg immediately after killing the specimen and storing the leg separately in ethanol (Dillon et al. 1996).

The usual way of preservation is dry-mounting the bees on insect pins. Bees from liquid collection should also be dry-mounted for identification purposes. Submersed bumble bees are more difficult to identify due to the wet long hair that obstructs the view to the cuticle and usually appears dark or colourless within the liquid.

Mounted specimens receive standard curation treatment, including labelling with collecting information, identification information, and additional relevant details. The entire specimen information may be databased and stored electronically, but it is advisable to attach the most important information to the insect pin.

Bumble bees are pinned through the thorax. The body parts should be placed in a fashion that the relevant details for identification are accessible and well visible. Body parts can be arranged using forceps and pinned in place with insect pins, until the specimen is completely dry. For bumble bees, an unobstructed view of the diagnostic characters is essential (e.g., basitarsi and hind tibia, upper vertex of the head, clypeus, labrum and the mandibles, sculpture of the sclerites). In males, the genital capsule is either removed, glued to a small card and pinned with the specimen, or pulled out, but not detached. In some species, the sclerotized parts of the female sting capsule are diagnostic. The sting apparatus can be treated identical to the male genitalia. Any adhering pollen (except for pollen in the corbicula), debris, or dirt that obstructs the view to relevant parts should be removed during the preparation procedure. When the specimen has particularly matted hair (e.g., from wet storage or regurgitated nectar), washing and blow-drying the specimen before pinning should be considered. To reposition or pin previously dried specimens, they can be softened in a moist chamber for a couple of hours or days or placed in a steam chamber for about 20 minutes (see Droege 2016; Plant and Dubitzky 2008). Pinned specimens should be stored in tightly closed insect boxes in a dark and dry environment after the specimens have completely dried.

In many cases, identification of cryptic species to the species level is only possible using genetic techniques. Usually part of a leg or an entire leg (the front leg, since it does not contain taxonomically relevant characters) is removed and stored dry or in ethanol for DNA extraction.

Directions and techniques for bumble bee collection and specimen treatment are also outlined in great detail by Saini et al. (2015). A very good resource for basic and advanced preparation techniques for bees is the *Very Handy Bee Manual* compiled by Sam Droege, which can be downloaded from the United States Geological Survey (USGS) Native Bee Inventory and Monitoring Lab webpage (Droege 2016).

USEFUL WEBSITES

The most important online reference for bumble bee taxonomy is the *Bombus* webpage maintained by Paul H. Williams (Williams 2018b). It contains an extensive overview and introduction to bumble bee taxonomy. The main and most valuable part of the webpage is the comprehensive checklist of all currently recognized bumble bee species. The online list is a regularly updated version of the original annotated checklist published earlier by Paul H. Williams (1998). Each species account contains taxonomic information (e.g., a reference to the original description, the most important synonyms, and taxonomic notes) and (for most species) a colour photo of the male genitalia. Species accounts are sorted by subgenus and species-group and can be accessed through various routes (e.g., from an alphabetic list of the species names and synonyms, by world region, colour pattern, or via the subgenus). The website further lists an extensive number of relevant references, links to PDF versions of the identification keys for the subgenera, and a lucid matrix version of the key. There is an additional lucid matrix key for the identification of bumble bee workers based on colour pattern. This resource is helpful to some extent, but species level identification is not possible in every case.

A second notable online resource is the *Discover Life bee species guide and world checklist* (Ascher and Pickering 2018). This project currently lists over 20,000 valid bee species with taxonomic information (e.g., authority, synonymy) and usually a world map with occurrence data from museum collection databases. With respect to bumble bees, this resource is useful, as it provides an overview about the global distribution of individual species. The maps can be customized according to the individual preferences, and each individual data point can be accessed to show the available meta-data.

A recent development in biodiversity research is citizen science through online biodiversity portals. These are often aimed at interested non-specialists. Images and observations can be uploaded, and the species identification can be confirmed by specialists. These portals are an important source of information for both parties. The observer is provided with a correct identification for their observation and the scientific community gains access to a larger dataset of species occurrence data. Two important portals that have to be mentioned in this context are iNaturalist (iNaturalist 2018) and the India Biodiversity Portal (Vattakaven et al. 2016). The former is well curated and rather accurate in the taxonomic treatment and identification, while the latter is rather incomplete and contains many incorrect taxonomic treatments and names. In general, both portals currently list only few observations from India.

It must be emphasized that a reliable species identification of bumble bees based on photographs is not always possible, especially when the local bumble bee fauna is not comprehensively understood.

CONCLUSION

Bumble bees are a popular group of insects that received much attention from the general public and entomologists. Despite the long tradition of research, there is still a lack of comprehensive taxonomic understanding in many species, species-groups, and subgenera. Furthermore, a few regions worldwide are still white spots in terms of bumble bee inventory and distribution data. One of these regions is the East Himalaya in the northeast region of India. Fieldwork and taxonomic treatment of bumble bees from that region is urgently needed. This chapter is meant as a starting point for interested students of bumble bees that would like to contribute to our global understanding of bumble bee taxonomy, ecology, and behaviour in the next few decades.

REFERENCES

Aizen, M. A., Smith-Ramírez, C., Morales, C. L., et al. 2018. Coordinated species importation policies are needed to reduce serious invasions globally: The case of alien bumblebees in South America. *Journal of Applied Ecology*. doi: 10.1111/1365-2664.13121.

Ascher, J. S., and J. Pickering. 2018. Discover Life bee species guide and world checklist (Hymenoptera: Apoidea: Anthopila). http://www.discoverlife.org/mp/20q?guide=Apoidea_species (accessed May 23, 2018).

Ayasse, M., and S. Jarau. 2014. Chemical ecology of bumble bees. *Annual Review of Entomology* 59(1): 299–319.

Aytekin, M. A., Terzo, M., Rasmont, P., and N. Çağatay. 2007. Landmark based geometric Morphometric analysis of wing shape in *Sibiricobombus* Vogt (Hymenoptera: Apidae: *Bombus* Latreille). *Annales de La Societe Entomologique de France* 43(1): 95–102.

Baer, B. 2003. Bumblebees as model organisms to study male sexual selection in social insects. *Behavioral Ecology and Sociobiology* 54(6): 521–533.

Baer, B., and P. Schmid-Hempel. 2000. The artificial insemination of bumblebee queens. *Insectes Sociaux* 47(2): 183–187.

Bingham, C. T. 1897. *The Fauna of British India including Ceylon and Burma. Hymenoptera. Vol. I. Wasps and Bees*. London, UK: Taylor & Francis Group.

Bossert, S. 2015. Recognition and identification of bumblebee species in the *Bombus lucorum*-complex (Hymenoptera, Apidae)—A review and outlook. *Deutsche Entomologische Zeitschrift* 62(1): 19–28.

Branstetter, M. G., Danforth, B. N., Pitts, J. P. et al. 2017. Phylogenomic insights into the evolution of stinging wasps and the origins of ants and bees. *Current Biology* 27(7): 1019–1025.

Cameron, S. A., Hines, H. M., and P. H. Williams. 2007. A comprehensive phylogeny of the bumble bees (*Bombus*). *Biological Journal of the Linnean Society* 91(1): 161–188.

Cameron, S. A., Lozier, J. D., Strange, J. P. et al. 2011. Patterns of widespread decline in North American bumble bees. *Proceedings of the National Academy of Sciences* 108(2): 662–667.

Cardinal, S., Straka, J., and B. N. Danforth. 2010. Comprehensive phylogeny of apid bees reveals the evolutionary origins and antiquity of cleptoparasitism. *Proceedings of the National Academy of Sciences* 107(37): 16207–16211.

Chauhan, A., Rana, B. S., and S. Katna. 2014. Successful rearing of bumble bee, *Bombus haemorrhoidalis* Smith year round in Himachal Pradesh in India. *International Journal of Current Research* 6: 10891–10896.

Colla, S. R., Otterstatter, M. C., Gegear, R. J., and J. D. Thomson. 2006. Plight of the bumble bee: Pathogen spillover from commercial to wild populations. *Biological Conservation* 129(4): 461–467.

de Meulemeester, T., Gerbaux, P., Boulvin, M., Coppée, A., and P. Rasmont. 2011. A simplified protocol for bumble bee species identification by cephalic secretion analysis. *Insectes Sociaux* 58(2): 227–236.

Deka, T. N., Sudharshan, M. R., and K. A. Saju. 2011. New record of bumble bee, *Bombus breviceps* Smith as a pollinator of large cardamom. *Current Science* 100(6): 926–928.

del Castillo, R. C., and D. J. Fairbairn. 2011. Macroevolutionary patterns of bumblebee body size: Detecting the interplay between natural and sexual selection. *Ecology and Evolution* 2(1): 46–57.

Dillon, N., Austin, A. D., and E. Bartowsky. 1996. Comparison of preservation techniques for DNA extraction from hymenopterous insects. *Insect Molecular Biology* 5: 21–24.

Droege, S. 2016. Very handy bee manual. https://www.pwrc.usgs.gov/nativebees/ (accessed May 23, 2018).

Duennes, M. A., Lozier, J. D., Hines, H. M., and S. A. Cameron. 2012. Geographical patterns of genetic divergence in the widespread Mesoamerican bumble bee *Bombus ephippiatus* (Hymenoptera: Apidae). *Molecular Phylogenetics and Evolution* 64(1): 219–231.

Ellis, J. S., Knight, M. E., Carvell, C., and D. Goulson. 2006. Cryptic species identification: A simple diagnostic tool for discriminating between two problematic bumblebee species. *Molecular Ecology Notes* 6(2): 540–542.

Franklin, H. J. 1912. The Bombidæ of the New World. *Transactions of the American Entomological Society* 38(3/4): 177–486.

Franklin, H. J. 1913. The Bombidæ of the New World (Continued). *Transactions of the American Entomological Society* 39(2): 73–200.

Friese, H. 1918. Über Hummelformen aus dem Himalaja. *Deutsche Entomologische Zeitschrift* 1/2: 6–12.

Garófalo, C. A., Zucchi, R., Muccillo, G., Garofalo, C. A., Zucchi, R., and G. Muccillo. 1986. Reproductive studies of a Neotropical bumblebee *Bombus atratus* (Hymenoptera, Apidae). *Revista Brasileira de Genética* 9(2): 231–243.

Gonzalez, V. H., Mejia, A., and C. Rasmussen. 2004. Ecology and nesting behavior of *Bombus atratus* Franklin in Andean Highlands (Hymenoptera: Apidae). *Journal of Hymenoptera Research* 13(2): 28–36.

Hines, H. M. 2008. Historical biogeography, divergence times, and diversification patterns of bumble bees (Hymenoptera: Apidae: *Bombus*). *Systematic Biology* 57(1): 58–75.

Hines, H. M., and S. A. Cameron. 2010. The phylogenetic position of the bumble bee inquiline *Bombus inexspectatus* and implications for the evolution of social parasitism. *Insectes Sociaux* 57(4): 379–383.

Hines, H. M., and P. H. Williams. 2012. Mimetic colour pattern evolution in the highly polymorphic *Bombus trifasciatus* (Hymenoptera: Apidae) species complex and its comimics. *Zoological Journal of the Linnean Society* 166(4): 805–826.

Huang, J., An, J., Wu, J., and P. H. Williams. 2015. Extreme food-plant specialisation in *Megabombus* bumblebees as a product of long tongues combined with short nesting seasons. *PLoS ONE* 10(8): e0132358.

iNaturalist. 2018. https://www.inaturalist.org (accessed June 25, 2018).

Ito, M. 1985. Supraspecific classification of bumblebees based on the characters of male genitalia. *Contributions from the Institute of Low Temperature Science* Series B 20: 1–143.

IUCN SSC Bumblebee Specialist Group. 2018. https://bumblebeespecialistgroup.org. (accessed June 22, 2018).

Jacobson, M. M., Tucker, E. M., Mathiasson, M. E., and S. M. Rehan. 2018. Decline of bumble bees in northeastern North America, with special focus on *Bombus terricola*. *Biological Conservation* 217: 437–445.

Kapustjanskij, A., Streinzer, M., Paulus, H. F., and J. Spaethe. 2007. Bigger is better: Implications of body size for flight ability under different light conditions and the evolution of alloethism in bumblebees. *Functional Ecology* 21(6): 1130–1136.

Kearns, C. A., and J. D. Thomson. 1955. *The Natural History of Bumblebees: A Sourcebook for Investigations*. Boulder, CO: University Press of Colorado.

Krüger, E. 1917. Zur Systematik der mitteleuropäischen Hummeln (Hym.). *Entomologische Mitteilungen* 6(1): 55–66.

Latreille, P. A. 1802. *Histoire naturelle des fourmis: et recueil de mémoires et d'observations sur les abeilles, les araignées, les faucheurs, et autres insects*. Paris, France: T. Barrois.

Lepeletier De Saint-Fargeau, A. L. M. 1832. Observations sur l'ouvrage institute: Bombi scandinaviae monographice tractate etc. it Gustav, Dahlbom. *Annales de la Société Entomologique de France* 1: 366–382.

Linnaeus, C. 1758. *Systema naturæ per regna tria naturæ, secundum classes, ordines, genera, species, cum characteribus, differentiis, synonymis, locis*. Holmiae, Sweden: Salvius.

Løken, A. 1973. Studies on Scandinavian bumble bees (Hymenoptera, Apidae). *Norsk Entomologisk Tidsskrift* 20: 1–218.

McKendrick, L., Provan, J., Fitzpatrick, Ú. et al. 2017. Microsatellite analysis supports the existence of three cryptic species within the bumble bee *Bombus lucorum* sensu lato. *Conservation Genetics* 18(3): 573–584.

Michener, C. D. 2000. *The Bees of the World*. Baltimore, MD: Johns Hopkins University Press.

Milliron, H. E. 1961. Revised classification of the bumblebees: A synopsis (Hymenoptera: Apidae). *Journal of the Kansas Entomological Society* 34: 49–61.

Murray, T. E., Fitzpatrick, Ú., Brown, M. J. F., and R. J. Paxton. 2008. Cryptic species diversity in a widespread bumble bee complex revealed using mitochondrial DNA RFLPs. *Conservation Genetics* 9(3): 653–666.

Perry, C. J., Barron, A. B., and L. Chittka. 2017. The frontiers of insect cognition. *Current Opinion in Behavioral Sciences* 16: 111–118.

Plant, J., and A. Dubitzky. 2008. Relaxing bee specimens: A quick and easy method. *Entomologica Austriaca* 15: 41–44.

Plowright, R. C., and T. M. Laverty. 1984. The ecology and sociobiology of bumble bees. *Annual Review of Entomology* 29: 175–199.

Ponchau, O., Iserbyt, S., Verhaeghe, J.-C., and P. Rasmont. 2006. Is the caste-ratio of the oligolectic bumblebee *Bombus gerstaeckeri* Morawitz (Hymenoptera: Apidae) biased to queens? *Annales de La Société Entomologique de France* 42(2): 207–214.

Potapov, G. S., Kondakov, A. V., Spitsyn, V. M. et al. 2018. An integrative taxonomic approach confirms the valid status of *Bombus glacialis*, an endemic bumblebee species of the High Arctic. *Polar Biology* 41(4): 629–642.

Prys-Jones, O. 1982. *Ecological Studies of Foraging and Life History in Bumblebees*. Cambridge, UK: University of Cambridge.

Radoszkowski, O. 1884. Révision des armures copulatrices des mâles *Bombus*. *Bulletin de La Société Impériale Des Naturalistes de Moscou* 60: 51–92.

Rasmont, P., Franzén, M., Lecocq, T. et al. 2015. Climatic risk and distribution atlas of European bumblebees. *BioRisk* 10: 1–236.

Rasmont, P., Scholl, A., Dejonghe, R., Obrecht, E., and A. Adamski. 1986. Identification and variability of males of the genus *Bombus* Latreille sensu-stricto in Western and Central Europe (Hymenoptera, Apidae, Bombinae). *Revue Suisse De Zoologie* 93(3): 661–682.

Rasmont, P., Terzo, M., Aytekin, A. M. et al. 2005. Cephalic secretions of the bumblebee subgenus *Sibiricobombus* Vogt suggest *Bombus niveatus* Kriechbaumer and *Bombus vorticosus* Gerstaecker are conspecific (Hymenoptera, Apidae, *Bombus*). *Apidologie* 36(4): 571–584.

Richards, O. W. 1968. The subgeneric divisions of the genus *Bombus* Latreille (Hymenoptera: Apidae). *Bulletin of the British Museum (Natural History) (Entomology)* 22: 209–276.

Robertson, C. 1903. Synopsis of Megachilidæ and Bombinæ. *Transactions of the American Entomological Society* 29: 163–178.

Röseler, P. F. 1985. A technique for year-round rearing of *Bombus terrestris* (Apidae, Bombini). *Apidologie* 16: 165–169.

Saini, M. S., and H. S. Ghator. 2007. Taxonomy and food plants of some bumble bee species of Lahaul and Spiti Valley of Himachal Pradesh. *Zoos' Print Journal* 22: 2648–2657.

Saini, M. S., Raina, R. H., and H. S. Ghator. 2015. *Indian Bumblebees*. Dehra Dun, India: Bishen Singh Mahendra Pal Singh.

Skorikov, A. C. 1914. Les formes nouvelles des bourdons (Hymenoptera, Bombidae). *Russkoe Éntomologicheskoe Obozrênie* 6(14): 119–129.

Smith, F. 1852. Descriptions of some hymenopterous insects from Northern India. *Transactions of The Royal Entomological Society of London* 2: 45–48.

Spaethe, J., and A. Weidenmüller. 2002. Size variation and foraging rate in bumblebees (*Bombus terrestris*). *Insectes Sociaux* 49(2): 142–146.

Stephen, W. P., and S. Rao. 2005. Unscented color traps for non-*Apis* bees (Hymenoptera: Apiformes). *Journal of the Kansas Entomological Society* 78(4): 373–380.

Streinzer, M., and J. Spaethe. 2014. Functional morphology of the visual system and mating strategies in bumblebees (Hymenoptera, Apidae, *Bombus*). *Zoological Journal of the Linnean Society* 170(4) 735–747.

Streinzer, M., Chakravorty, J., Neumayer, J. et al. 2018. Species composition and elevational distribution of bumble bees (Hymenoptera: Apidae: *Bombus*) in the East Himalaya, Arunachal Pradesh, India. *ZooKeys* 852: 71–89.

Thai, P. H., and T. Van Toan. 2018. Beekeeping in Vietnam. In *Asian Beekeeping in the 21st Century*, eds. P. Chantawannakul, G. Williams, and P. Neumann. Singapore: Springer.

Tkalcu, B. 1968. Revision der vier sympatrischen, homochrome geographische Rassen bildenden Hummelarten SO-Asien (Hymenoptera, Apoidea, Bombinae). *Annotationes Zoologicae at Botanicae* 52: 1–31.

Tkalcu, B. 1974. Eine Hummel-Ausbeute aus dem Nepal-Himalaya (Insecta, Hymenoptera, Apoidea, Bombinae). *Senckenbergiana Biologica* 55: 311–349.

Tkalcu, B. 1989. Neue Taxa asiatischer Hummeln (Hymenoptera, Apoidea). *Acta Entomologica Bohemoslovaca* 86: 39–60.

Vattakaven, T., George, R., Balasubramanian, D. et al. 2016. India Biodiversity Portal: An integrated, interactive and participatory biodiversity informatics platform. *Biodiversity Data Journal* 4: e10279.

Velthuis, H. H. W. 2002. The historical background of the domestication of the bumble-bee, *Bombus terrestris*, and its introduction in agriculture. In *Pollination Bees—The Conservation Link Between Agriculture and Nature,* eds. P. Kevan and V. L. Imperatriz Fonseca. Brasília, Brazil: Ministry of Environment.

Vogt, O. 1911. Studien über das Artproblem. 2. Mitteilung. Über das Variieren der Hummeln. 2. Teil. (Schluss). *Sitzungsberichte der Gesellschaft naturforschender Freunde zu Berlin* 31–74.

von Dalla Torre, K. W. 1880. Unsere Hummel-(*Bombus*) Arten. *Naturhistoriker* 2(4): 30 & 2(5): 40–41.

Williams, P. H. 1985. A preliminary cladistic investigation of relationships among the bumble bees (Hymenoptera, Apidae). *Systematic Entomology* 10(2): 239–255.

Williams, P. H. 1991. The bumblebees of the Kashmir Himalaya (Hymenoptera: Apidae, Bombini). *Bulletin of the British Museum* 60: 1–204.

Williams, P. H. 1994. Phylogenetic relationships among bumble bees (*Bombus* Latr.): A reappraisal of morphological evidence. *Systematic Entomology* 19(4): 327–344.

Williams, P. H. 1998. An annotated checklist of bumble bees with an analysis of patterns of description (Hymenoptera: Apidade, Bombini). *Bulletin of the Natural History Museum(Entomology)* 67: 79–152.

Williams, P. H. 2004. Genus *Bombus* Latreille. In *An Annotated Catalogue of the Bee Species of the Indian Region,* ed. R. K. Gupta. Jodhpur, India: Jai Narain Vyas University.

Williams, P. H. 2007. The distribution of bumblebee colour patterns worldwide: Possible significance for thermoregulation, crypsis, and warning mimicry. *Biological Journal of Linnean Society* 92: 97–118.

Williams, P. H. 2018a. In a group of its own? Rediscovery of one of the world's rarest and highest mountain bumblebees, *Bombus tanguticus. Journal of Natural History* 52(5–6): 305–321.

Williams, P. H. 2018b. *Bombus, Bumble Bees of the World.* Natural History Museum, London, UK. http://www.nhm.ac.uk/research-curation/research/projects/bombus/ (accessed May 25, 2018).

Williams, P. H., An, J., Brown, M. J. F. et al. 2012a. Cryptic bumble-bee species: Consequences for conservation and the trade in greenhouse pollinators. *PLoS ONE* 7(3): e32992.

Williams, P. H., An, J., and J. Huang. 2011. The bumblebees of the subgenus *Subterraneobombus*: Integrating evidence from morphology and DNA barcodes (Hymenoptera, Apidae, *Bombus*). *Zoological Journal of the Linnean Society* 163(3): 813–862.

Williams, P. H., Brown, M. J. F., Carolan, J. C. et al. 2012b. Unveiling cryptic species of the bumblebee subgenus *Bombus* s. str. worldwide with COI barcodes (Hymenoptera: Apidae). *Systematics and Biodiversity* 10(1): 21–56.

Williams, P. H., Bystriakova, N., Huang, J., Miao, Z., and J. An. 2015a. Bumblebees, climate and glaciers across the Tibetan plateau (Apidae: *Bombus* Latreille). *Systematics and Biodiversity* 13(2): 164–181.

Williams, P. H., Byvaltsev, A. M., Cederberg, B. et al. 2015b. Genes suggest ancestral colour polymorphisms are shared across morphologically cryptic species in arctic bumblebees. *PLoS ONE* 10(12): e0144544.

Williams, P. H., Byvaltsev, A., Sheffield, C., and P. Rasmont. 2013. *Bombus cullumanus*—An extinct European bumblebee species? *Apidologie* 44(2): 121–132.

Williams, P. H., Cameron, S., Hines, H. M., Cederberg, B., and P. Rasmont. 2008. A simplified subgeneric classification of the bumblebees (genus *Bombus*). *Apidologie* 39(1): 1–29.

Williams, P. H., Huang, J., Rasmont, P., and J. An. 2016. Early-diverging bumblebees from across the roof of the world: The high-mountain subgenus *Mendacibombus* revised from species' gene coalescents and morphology (Hymenoptera, Apidae). *Zootaxa* 4204(1): 1–72.

Williams, P. H., Ito, M., Matsumura, T., and I. Kudo. 2010. The bumblebees of the Nepal Himalaya (Hymenoptera: Apidae). *Insecta Matsumurana* 66: 115–151.

Williams, P. H., Lobo, J. M., and A. S. Meseguer. 2018. Bumblebees take the high road: Climatically integrative biogeography shows that escape from Tibet, not Tibetan uplift, is associated with divergences of present-day *Mendacibombus. Ecography* 41(3): 461–477.

Williams, P. H., and J. L. Osborne. 2009. Bumblebee vulnerability and conservation world-wide. *Apidologie* 40(3): 367–387.

Williams, P., Tang, Y., Yao, J., and S. Cameron. 2009. The bumblebees of Sichuan (Hymenoptera: Apidae, Bombini). *Systematics and Biodiversity* 7(2): 101–189.

Woodard, S. H., Lozier, J. D., Goulson, D., Williams, P. H., Strange, J. P., and S. Jha. 2015. Molecular tools and bumble bees: Revealing hidden details of ecology and evolution in a model system. *Molecular Ecology* 24(12): 2916–2936.

Yarrow, I. H. H. 1970. Is *Bombus inexspectatus* (Tkalcu) a workerless obligate parasite? (Hym. Apidae). *Insectes Sociaux* 17(2): 95–111.

12 Potter Wasps (Hymenoptera: Vespidae: Eumeninae) of India

P. Girish Kumar, Arati Pannure, and James M. Carpenter

CONTENTS

Introduction 187
Subfamily Diagnosis and Key Characters 187
    Key to Indian Genera of Eumeninae 189
Major Work on Indian Fauna 191
Biodiversity and Species Richness 191
Distribution Patterns in the Indian Context 192
Immature Taxonomy 193
Molecular Characterization and Phylogeny/Barcoding 193
Taxonomic Problems 193
Biology 193
Economic Importance 194
Collection and Preservation 194
Useful Websites 195
Conclusion 195
Acknowledgements 195
References 195

INTRODUCTION

The subfamily Eumeninae, commonly called potter wasps, is the most species rich subfamily among the Vespidae. This cosmopolitan subfamily consists of about 3,794 described species in 205 genera (Pannure et al. 2016; Tan et al. 2018). They are usually solitary and rarely sub-social (Ducke 1914; Bohart and Stange 1965). Adults are small to large and compact to elongate with a sessile to strongly petiolate metasoma. The taxonomy and other aspects of natural history of the Eumeninae are poorly studied in India. For most species nothing is known about their biology, behaviour, prey associations, and hence conservation status. The need for taxonomic work on Eumeninae in India is underlined by the lack of adequate and well-illustrated keys, both at generic and species level (Pannure et al. 2016). This chapter aims to bring together all the fragmentary literature on the subfamily, thereby providing a review that could serve as a probable basis for increasing the knowledge of the group.

SUBFAMILY DIAGNOSIS AND KEY CHARACTERS

Subfamily Diagnosis: Eumeninae can be easily distinguished from other hymenopterans by a combination of following morphological features: usually claws bifid (Figure 12.1, **1**); head with emarginate eyes (**2**); middle tibia usually with one spur (**3**) except in a few genera; tegula emarginate on inner side to receive the parategula (**4**); mandibles long, usually crossing each other at rest (**5**); hind wing with anal lobe (**6**); hind coxa with a longitudinal dorsal carina (**7**); pronotum extending back to tegula (**8**); pronotal lobe usually separated from tegula by a distance equal to or less than length of lobe (**9**), but distance rarely greater; posterior lingual plate longer than wide; fore wing longitudinally folded at rest (**10**), usually reaching at least the posterior margin of metasomal segment 4; solitary or sub-social; making earthen pot-like nests, or nests in burrows or in wood cavities; predatory in nature.

FIGURE 12.1 Morphology of Eumeninae (*Delta pyriforme pyriforme*) **1**. Tarsal claws in apical tarsal segment **2**. Ocular sinus in frontal view **3**. Mid tibial spur **4**. Dorsal view of mesosoma showing tegula and parategula **5**. Mandibles **6**. Anal lobe in hind wing **7**. Carina in hind coxa **8**. Pronotal lobe extension in dorsal view **9**. Lateral view of pronotum **10**. Longitudinal folding in fore wing.

Subfamily Key Characters: Mesoscutum with parategula; posterior lingual plate longer than wide; tarsal claws usually bifid; solitary or sub-social.

Tribes/Genera: Uncertainties in the tribal and generic classification of the subfamily Eumeninae in the past are partly a result of the morphological intricacy of the group (Carpenter and Cumming 1985; Carpenter and Garcete-Barrett 2002; Hermes et al. 2014).

The first generic classification was given by Latreille (1802), who divided species into three genera, namely, *Eumenes, Odynerus,* and *Synagris* based on characters of mandibles, labial and maxillary palpi, and shape of metasoma. Currently, 205 genera are recognized in the world. Hermes et al. (2014) proposed the first tribal division of the subfamily based on cladistic methods, which include Zethini, Eumenini

(= *Eumenes* sensu lato), and Odynerini (= *Odynerus* sensu lato), and this is presently accepted as the classification of the subfamily into tribes. From India, 189 species in 47 genera under all the three tribes have been reported (Table 12.1).

TABLE 12.1
Tribes and Genera of Eumeninae from India

Tribe	Genus	No. of Species
Eumenini	*Delta* de Saussure, 1855	7
	Eumenes Latreille, 1802	18
	Katamenes Meade-Waldo, 1910	2
	Omicroides Giordani Soika, 1934	1
	Oreumenoides Giordani Soika, 1961	1
	Phimenes Giordani Soika, 1992	4
Odynerini	*Alastor* Lepeletier de Saint Fargeau, 1841	4
	Allodynerus Blüthgen, 1937	1
	Allorhynchium van der Vecht, 1963	5
	Ancistrocerus Wesmael, 1836	8
	Antepipona de Saussure, 1855	20
	Anterhynchium de Saussure, 1863	6
	Antodynerus de Saussure, 1855	3
	Apodynerus Giordani Soika, 1993	3
	Chlorodynerus Blüthgen, 1951	3
	Coeleumenes van der Vecht, 1963	3
	Cyrtolabulus van der Vecht, 1969	3
	Ectopioglossa Perkins, 1912	2
	Epsilon de Saussure, 1855	4
	Euodynerus Dalla Torre, 1904	4
	Gribodia Zavattari, 1912	1
	Indodynerus Gusenleitner, 2008	1
	Knemodynerus Blüthgen, 1940	4
	Labus de Saussure, 1867	4
	Leptochilus de Saussure, 1853	2
	Lissodynerus Giordani Soika, 1993	3
	Malayepipona Giordani Soika, 1993	2
	Orancistrocerus van der Vecht, 1963	2
	Paraleptomenes Giordani Soika, 1970	6
	Parancistrocerus Bequaert, 1925	11
	Pararrhynchium de Saussure, 1855	2
	Pareumenes de Saussure, 1855	3
	Pseudepipona de Saussure, 1856	1
	Pseudonortonia Giordani Soika, 1936	4
	Pseudozumia de Saussure, 1875	2
	Pseumenes Giordani Soika, 1935	1
	Rhynchium Spinola, 1806	4
	Stenancistrocerus de Saussure, 1863	1
	Stenodyneriellus Giordani Soika, 1961	6
	Stenodynerus de Saussure, 1863	2
	Subancistrocerus de Saussure, 1855	4
	Symmorphus Wesmael, 1836	9
	Tropidodynerus Blüthgen, 1939	2
	Xenorhynchium van der Vecht, 1963	1
Zethini	*Calligaster* de Saussure, 1852	1
	Discoelius Latreille, 1809	2
	Zethus Fabricius, 1804	6
Total species		**189**

KEY TO INDIAN GENERA OF EUMENINAE

(modified based on keys of Pannure et al. 2016 and Girish Kumar and Sharma 2015b)

(The metasomal terga 1, 2 are denoted as T1, T2, respectively; metasternum 1 denoted as S1; SMC: sub-marginal cell of fore wing)

1. Metasoma petiolate, T1 in dorsal view usually longer, less than half as wide as T2 in dorsal view **2**
- Metasoma not petiolate, T1 in dorsal view usually broader, more than half as wide as T2 in dorsal view **17**
2. Mid tibia with two spurs **3**
- Mid tibia with one spur **5**
3. Propodeal orifice narrow dorsally; sub-marginal carina strongly produced, propodeal valvula elongate and free from sub-marginal carina posteriorly.... ***Zethus* Fabricius**
- Propodeal orifice broad and rounded dorsally; sub-marginal carina weakly produced, propodeal valvula and sub-marginal carina not free posteriorly............. **4**
4. T1 more than twice as long as wide, dorsal surface with longitudinal striae; labial palpus with three palpomeres .... ***Calligaster* de Saussure**
- T1 less than twice as long as wide, dorsal surface without longitudinal striae; labial palpus with four palpomeres .... ***Discoelius* Latreille**
5. Mesosoma globular, as wide as high **6**
- Mesosoma more or less flattened dorso-ventrally, distinctly longer than high **11**
6. Clypeus bluntly angular (female) or flatly convex (male) at apex; temple in dorsal view as long as eye .... ***Katamenes* Meade-Waldo**
- Clypeus weekly to strongly emarginate (female) at apex; temple in dorsal view shorter than eye.............. **7**
7. Pronotum without pretegular carina; T2 with apical lamella... **8**
- Pronotum with pretegular carina; T2 without apical lamella... **9**
8. Mesepisternum with epicnemial carina present; T2 apical lamella strongly and broadly concave medially ....***Omicroides* Giordani Soika**
- Mesepisternum with epicnemial carina absent; T2 apical lamella not concave medially..... ***Eumenes* Latreille**
9. T1 (petiole length) less than 1.25 × length of mesosoma, never shorter than mesosoma .... ***Delta* de Saussure**
- T1 (petiole length) 1.25 × or more than length of mesosoma **10**
10. Fore wing with prestigma longer than pterostigma; F11 of male long and hooked..
....***Phimenes* Giordani Soika**
- Fore wing with prestigma shorter or equals pterostigma; F11 of male short, not hooked...
.... ***Oreumenoides* (Giordani Soika)**
11. Axillary fossa oval; propodeal valvula not fused to sub-marginal carina, sub-marginal carina produced as a pointed process above valvula; SMC 2 basally truncate; body size smaller than 10 mm........................ **12**

- Axillary fossa slit-like; propodeal valvula fused to submarginal carina, sub-marginal carina not produced as a pointed process above valvula; SMC 2 basally acute; body size more than 10 mm **13**
12. Female with a distinct subcircular fovea below median ocellus; metanotum unidentate; propodeum not produced; T1 in dorsal view conspicuously swollen in apical half .... ***Labus* de Saussure**
- Female without fovea below median ocellus; metanotum not unidentate, precipitous; propodeum with extensive horizontal portion, somewhat narrowed apically, behind the postscutellum, abruptly sloping posteriorly; T1 in dorsal view not conspicuously swollen in apical half ....***Cyrtolabulus* van der Vecht**
13. Mesepisternum with epicnemial carina present **14**
- Mesepisternum with epicnemial carina absent **16**
14. TI basally with transverse carina; ventral margins of TI touching each other except for posterior diverging part, thus SI visible only in posterior triangular part ....***Ectopioglossa* Perkins**
- TI without transverse carina; ventral margins of TI basally close to each other, but not touching each other **15**
15. Mesoscutum without notauli; S1 transversely striate posteriorly; fore wing with parastigma shorter than pterostigma .... ***Coeleumenes* van der Vecht**
- Mesoscutum with notauli; S1 irregularly rugose posteriorly; fore wing with parastigma longer than pterostigma .... ***Pseudozumia* de Saussure**
16. Fore wing with prestigma longer than pterostigma; female with cephalic foveae..
.... ***Pareumenes* de Saussure**
- Fore wing with prestigma shorter than pterostigma; female without cephalic foveae.....................................
.... ***Pseumenes* Giordani Soika**
17. SMC 2 petiolate anteriorly; propodeal orifice narrow dorsally............. ***Alastor* Lepeletier de Saint Fargeau**
- SMC 2 not petiolate anteriorly; propodeal orifice broad and rounded dorsally.................................... **18**
18. Anterior face of pronotum with two small, deeply impressed medial pits or foveae; which may be sparse, contiguous, or faint in some species **19**
- Anterior face of pronotum without medial pits or foveae **24**
19. T1 with transverse carina.................................. **20**
- T1 without transverse carina................................. **23**
20. T1 with two transverse carinae close together at crest of declivity; T1 wider than long in dorsal view............
.... ***Subancistrocerus* de Saussure**
- T1 with one transverse carina; T1 variable in dorsal view... **21**
21. Anterior face of pronotum with foveae separated; T2 smooth basally, forming acarinarium beneath apex of T1 that is often full of mites .... ***Parancistrocerus* Bequaert**
- Anterior face of pronotum with foveae coalesced (foveae contiguous); T2 without acarinarium **22**
22. T1 in dorsal view subsessile, longer than wide, T2 much wider than T1 .... ***Pseudonortonia* Giordani Soika**
- T1 sessile; about as wide as T2 ....
.... ***Stenancistrocerus* de Saussure**
23. Anterior face of pronotum with foveae coalesced, punctate laterally, T1 subsessile, in dorsal view usually narrower than T2 .... ***Paraleptomenes* Giordani Soika**
- Anterior face of pronotum usually with foveae separated, smooth laterally; T1 in dorsal view about as broad as T2...................... ***Stenodynerus* de Saussure**
24. T1 with transverse carina..**25**
- T1 without transverse carina...................................**29**
25. TI with broad longitudinal median furrow posterior to transverse carina; male antenna simple apically; notauli present; female cephalic foveae well separated, located midway between posterior ocelli and occipital margin....................***Symmorphus* Wesmael**
- TI without broad longitudinal furrow/groove; mesepisternum without epicnemial carina; male antenna hooked; other characters variable; female cephalic foveae closely spaced, nearer occipital margin than posterior ocelli.. **26**
26. Axillary fossa oval, broader than long; tegula exceeding parategula....................... ***Ancistrocerus* Wesmael**
- Axillary fossa narrower than long, slit-like; tegula not exceeding parategula **27**
27. Propodeal dorsum forming shelf-like area behind metanotum; fore wing with prestigma less than half as long as pterostigma .... ***Pararrhynchium* de Saussure**
- Propodeal dorsum below plane of metanotum, sloping posteroventrally; prestigma more than half as long as pterostigma, measured along posterior part **28**
28. T2 with well-developed apical lamella..........................
.... ***Lissodynerus* Giordani Soika**
- T2 without apical lamella..
.... ***Orancistrocerus* van der Vecht**
29. Tegula evenly rounded posteriorly, not emarginate adjoining parategula and usually not reaching apex of latter .... ***Tropidodynerus* Blüthgen**
- Tegula not evenly rounded posteriorly, emarginate adjoining parategula and often reaching or surpassing apex of latter **30**
30. T1 with transparent or translucent apical border.........**31**
- T1 without transparent or translucent apical border.....
.... **34**
31. Parategula not visible from above; tegulae posteriorly bent inwards .... ***Knemodynerus* Blüthgen**
- Parategula visible from above.................................**32**
32. Metanotum semicircular in shape from above, carinate posteriorly .... ***Antodynerus* de Saussure**
- Metanotum not semicircular in shape from above ... **33**
33. Pronotum with anterior face densely punctate laterally, without dorsal carina .... ***Chlorodynerus* Blüthgen**
- Pronotum with anterior face not densely punctate, with dorsal carina.................... ***Euodynerus* Dalla Torre**
34. Metanotum bidentate **35**

- Metanotum not dentate **36**
35. Clypeus higher than wide; mid-anterior face of pronotum smooth and with short transverse rugae; T1 distinctly narrower than T2; male terminal antennal segment small.................. ***Apodynerus* Giordani Soika**
- Clypeus wider than high; mid-anterior face of pronotum usually densely punctate and with an upper trace of transverse carina; T1 slightly narrower than T2; male terminal antennal segment relatively large............. ....***Antepipona* de Saussure**
36. T2 with apical lamella... **37**
- T2 without lamella **39**
37. T1 depressed subapically, gradually widened with lateral sides divergent in dorsal view; propodeum with sub-marginal carina projecting as rounded lobe above valvula, bilamellate; epicnemial carina absent; axillary fossa oval, broader than long.. ....***Leptochilus* de Saussure**
- T1 not depressed subapically, usually with lateral sides roughly parallel in dorsal view; propodeum with sub-marginal carina not differentiated from valvula, monolamellate (except *Epsilon*); epicnemial carina present; axillary fossa narrower than long, slit-like **38**
38. Palpal formula 5:3; male vertex sometimes with large and deep depression; propodeum with well developed lateral carinae; T1-5 each with apical lamella........ .... ***Gribodia* Zavattari**
- Palpal formula 6:4; male vertex without large and deep depression; propodeum without well developed lateral carina; only T2 with apical lamella.................. .... ***Epsilon* de Saussure**
39. Tegula never exceeds parategula posteriorly; axillary fossa much narrower than long, sometimes slit-like **40**
- Tegula usually exceeds parategula posteriorly, or at least equalling it; axillary fossa oval, at least as wide as long.. **44**
40. Fore wing with prestigma half or less than the length of the pterostigma .... ***Stenodyneriellus* Giordani Soika**
- Fore wing with prestigma more than half the length of the pterostigma, often nearly equal **41**
41. SMC 3 separated from the apex of marginal cell by distance shorter than its minimum width...................... ....***Allorhynchium* van der Vecht**
- SMC 3 separated from the apex of marginal cell by distance longer than its minimum width **42**
42. Scutum posteriorly and scutellum anteriorly smooth and impunctate..............................***Rhynchium* Spinola**
- Scutum and scutellum punctate **43**
43. Mesepisternum without epicnemial carina...................... ....***Indodynerus* Gusenleitner**
- Mesepisternum with epicnemial carina present.............. ....***Anterhynchium* de Saussure**
44. T1 subsessile, in dorsal view narrower than T2, wider apically than basally........***Malayepipona* Giordani Soika**
- T1 sessile, in dorsal view about as wide as T2 .... **45**

45. Metanotum projecting over propodeum; propodeum with sclerotized dorsolateral projections; propodeum with sub-marginal carina not differentiated from valvula; SMC 2 with second recurrent vein far from SMC 3 ....***Xenorhynchium* van der Vecht**
- Metanotum not projecting over propodeum; propodeum without dorsolateral projections; propodeum with sub-marginal carina projecting as rounded lobe above valvula; propodeal valvula free from sub-marginal carina posteriorly.. **46**
46. Tegula narrower and longer, surpassing parategula posteriorly. [female vertex with reniform fovea, about as wide as ocellar triangle; hind coxa with ventral lobes] .... ***Allodynerus* Blüthgen**
- Tegula broad, equal to parategula posteriorly......... ....***Pseudepipona* de Saussure**

MAJOR WORK ON INDIAN FAUNA

The Indian Eumeninae was first comprehensively studied by Bingham (1897). Many taxonomic changes were made after his studies especially at generic level. Most of the earlier genera like *Eumenes*, *Odynerus*, etc. have been split into different genera like *Eumenes* into *Coeleumenes*, *Delta*, *Eumenes* s. str., *Oreumenoides*, *Pareumenes*, *Phimenes*, *Pseumenes*, etc. and *Odymerus* into *Ancistrocerus*, *Antepipona*, *Antodynerus*, *Apodynerus*, *Epsilon*, *Odynerus* s. str., *Paraleptomenes*, *Subancistrocerus*, etc. After Bingham (1897), only scattered papers were published. Cameron (1897, 1902, 1903a, 1903b, 1904, 1907, 1908, 1909, 1913), Nurse (1903, 1914), Rothney (1903), Paiva (1907), Meade-Waldo (1910a, 1910b), Dover (1921, 1925), Dover and Rao (1922), van der Vecht (1937, 1959a, 1959b, 1961, 1963, 1969, 1981), Giordani Soika (1941, 1960, 1966, 1982), Wain (1956), Gusenleitner (1988, 1996, 1998, 2001, 2006, 2007, 2008), Giordani Soika and Khan (1991), and Krombein (1978, 1991) are some of the important work published through the twentieth century containing descriptions of Eumeninae. During the last two decades, Lambert and Narendran (2002), Lambert (2004), Lambert et al. (2007, 2008), Srinivasan and Girish Kumar (2010, 2013), Girish Kumar (2011, 2012a, 2012b, 2012c, 2013a, 2013b, 2013c), Girish Kumar and Lambert (2011, 2012), Girish Kumar and Sharma (2012, 2013, 2014, 2015a, 2015b), Girish Kumar and Carpenter (2013, 2015a, 2015b), Girish Kumar et al. (2013a, 2013b, 2013c, 2013d, 2014a, 2014b, 2014c, 2015a, 2015b, 2016a, 2016b, 2016c, 2016d, 2017a, 2017b, 2017c), Mohammed Shareef et al. (2013), Girish Kumar and Sureshan (2016), Pannure et al. (2016, 2017, 2018), and Selis (2018) have made contributions to the Indian eumenid fauna.

BIODIVERSITY AND SPECIES RICHNESS

India has four biodiversity hotspots, the Western Ghats, the Himalayas, the Indo-Burma region, and Sundaland (which includes the Nicobar group of islands). Most of the faunal elements in India are very significant and peculiar with many endemic species. The diversity and species richness of potter wasps is fairly high in India with 189 species (about 4.8% of the world species) and many additional subspecies under 47 genera in 3 tribes (Table 12.1). Several workers during the past decade have been exploring the potter wasp fauna of the Indian subcontinent mainly in the north and northeastern regions and have described a number of new species and found new records from this region (Girish Kumar 2011, 2012a, 2012c; Girish Kumar and Carpenter 2013, 2015a, 2015b; Girish Kumar and Lambert 2011; Girish Kumar et al. 2013a, 2013b, 2013c, 2013d, 2014a, 2014b, 2015, 2016a, 2016b, 2016c, 2016d, 2017a, 2017b) (Table 12.2). In a series of studies, 17 species in 12 genera have been reported from Rajasthan (Girish Kumar and Sharma 2014), 19 species in 11 genera from Chhattisgarh (Girish Kumar and Sharma 2015b), and 11 species from Arunachal Pradesh (Srinivasan and Girish Kumar 2010). Girish Kumar (2012) recorded the genus *Omicroides* for the first time from NE India. Girish Kumar et al. (2017c) recorded an unreported genus *Pseudepipona* from India and the species *Pseudepipona* (*Pseudepipona*) *vicina* Gusenleitner, 1973 from the northern Himalaya. North and northeastern regions seem to be more diverse than the south Indian region. A total of 72 species and subspecies in 31 genera have been reported from South India, and it appears like the number of species will definitely see an increase (Pannure et al. 2016). Eumenine fauna from most Indian states such as Andhra Pradesh, Bihar, Delhi, Goa, Gujarat, Haryana, Jammu and Kashmir, Jharkhand, Lakshadweep Islands, Madhya Pradesh, Maharashtra, Manipur, Mizoram Nagaland, Odisha, Punjab, Uttar Pradesh, Telangana, and Tripura is very poorly and fragmentarily known. Considering the fact that Western Ghats and Eastern Ghats and Andaman and Lakshadweep Islands are still untapped regions, diversity of potter wasp species in the region should definitely be higher.

DISTRIBUTION PATTERNS IN THE INDIAN CONTEXT

Accounts of the distribution patterns of eumenine wasps within India is lacking. From the geological standpoint, the Indian fauna is comprised of two components, namely, the major Oriental component (represents most of India) and the minor Palearctic component (mainly northern Himalayas). Some genera such as *Allorhynchium, Antepipona, Anterhynchium, Antodynerus, Delta, Eumenes, Paraleptomenes, Phimenes, Rhynchium, Stenodyneriellus,* and *Subancistrocerus* are widely distributed within India (Srinivasan and Girish Kumar 2010; Pannure et al. 2016). Genera such as *Alastor, Cyrtolabulus, Epsilon, Euodynerus, Knemodynerus, Labus, Oreumenoides,* and *Pareumenes* are usually not common, but occur in different parts of India (Pannure et al. 2016). The genus *Xenorhynchium* is widely distributed, but not recorded from northeastern India. Many genera such as *Apodynerus, Coeleumenes, Discoelius, Ectopioglossa, Parancistrocerus, Pseudozumia, Pseumenes,* and *Zethus* show a peculiar distribution pattern with presence in the southern part of India and the north-eastern part of India. The northern part of India has rich biodiversity and Palearctic elements. Many genera such as *Allodynerus, Calligaster, Gribodia, Malayepipona, Omicroides, Orancistrocerus,* and *Pararrhynchium* are restricted to north-east India. The genus *Indodynerus* generally occurs in the southern part of India (Girish Kumar et al. 2013a). The genus *Lissodynerus* has so far been recorded from the Andamans, South India and Sikkim (Girish Kumar et al. 2015b; Selis 2018). The genera *Ancistrocerus, Katamenes,* and *Symmorphus* mainly occur in the Himalayas. The genera *Chlorodynerus, Leptochilus,* and *Stenancistrocerus* have been recorded from western India only. The genus *Tropidodynerus* has been recorded from west, north, and north-eastern India (Girish Kumar et al. 2013d).

TABLE 12.2
Recent Records of Eumeninae from India

	Species	Year	Distribution	Reference
1	*Subancistrocerus venkataramani* Girish Kumar	2013	Jharkhand	Girish Kumar (2013a)
2	*Epsilon manasicum* Girish Kumar & Carpenter	2014	Assam	Girish Kumar et al. (2014a)
3	*Paraleptomenes darugiriensis* Girish Kumar, Carpenter & Sharma	2014	Arunachal Pradesh, Assam, Meghalaya, Sikkim, West Bengal	Girish Kumar et al. (2014b)
4	*Alastor (Alastor) venkataramani* Girish Kumar & Carpenter	2015	Telangana	Girish Kumar and Carpenter (2015a)
5	*Leptochilus (Neoleptochilus) hassani* Girish Kumar & Carpenter	2015	Maharashtra	Girish Kumar and Carpenter (2015b)
6	*Lissodynerus rutlandicus* Girish Kumar, Srinivasan & Carpenter	2015	South Andaman (Rutland Island)	Girish Kumar et al. (2015b)
7	*Allorhynchium tuberculatum* Girish Kumar & Carpenter	2016	Kerala	Girish Kumar et al. (2016a)
8	*Parancistrocerus jaferpaloti* Girish Kumar & Carpenter	2016	Kerala	Girish Kumar et al. (2016c)
9	*Parancistrocerus loharbandensis* Girish Kumar & Carpenter	2016	Assam	
10	*Parancistrocerus turensis* Girish Kumar & Carpenter	2016	Meghalaya	
11	*Discoelius vasukii* Pannure & Carpenter	2017	Tamil Nadu	Pannure et al. (2017)
12	*Pararrhynchium venkataramani* Girish Kumar & Carpenter	2017	Arunachal Pradesh, Meghalaya	Girish Kumar et al. (2017b)
13	*Antepipona tricolorata* Selis	2018	Sikkim	Selis (2018)
14	*Lissodynerus unicus* Selis	2018	Sikkim	
15	*Symmorphus (Symmorphus) incisus* Selis	2018	Sikkim	

The genus *Pseudonortonia* has been recorded from western, southern, and north-eastern India. The genus *Stenodynerus* has been recorded from south, north, and north-eastern India. The genus *Pseudepipona* has been recorded from the northern Himalayas. The genus *Xenorhynchium* is endemic to the Indian subcontinent (Girish Kumar and Kishore 2011; Pannure et al. 2016).

IMMATURE TAXONOMY

The taxonomy of eumenid wasps rests largely on the external morphology of the adults. There is hardly any information available for the immature stages, even though they could be useful. There are only about 2%–3% of known species for which larval descriptions are available. The most important descriptions of immature stages, mainly larvae of Palearctic Eumeninae species, are provided by Tormos et al. (1997a, 1997b, 1997c, 1998, 2005, 2008). Identification of Eumeninae at the generic and specific levels using larval characters is difficult (Grandi 1961; Evans 1977; Tormos et al. 1998). It is therefore imperative to conduct detailed descriptive studies that will help to know characters that define the taxa at genus and species level. It is believed that such studies will contribute greatly to the clarification of the phylogeny and, hence, the systematics of Eumeninae (Tormos et al. 2008). The morphology of immature stages of Oriental species is largely unknown.

MOLECULAR CHARACTERIZATION AND PHYLOGENY/BARCODING

The efforts made to achieve a robust established classification that reflects evolutionary relationships in Eumeninae has been very limited. Hermes et al. (2014) proposed the first formal tribal division of the subfamily based on cladistic methods into Zethini, Eumenini (= *Eumenes* sensu lato), and Odynerini (= *Odynerus* sensu lato). In conflict with this tribal subdivision of Eumeninae, Bank et al. (2017), based solely on a molecular approach, proposed subfamily ranks for the two major clades of "Zethini": Raphiglossinae and Zethinae. Eumeninae, a primary monophyletic lineage of the Vespidae, turned out to be a paraphyletic group based on their phylogenomic approach (Bank et al. 2017), which, however, was based on a limited taxonomic sample. The phylogenetic relationships among the major lineages of Eumeninae are little investigated for two main reasons: few taxonomists are working on the group compared to those who are working on social subfamilies, Polistinae and Vespinae; and the remarkable diversity found among the Eumeninae wasps, which has resulted in a troubled taxonomic history (Hermes et al. 2014). In general, comprehensive investigations of the phylogenetic relationships of Eumeninae are very scarce for the Oriental fauna, compared to the Nearctic (Carpenter and Cumming 1985), Neotropical (Hermes and de Oliveira 2016), and Palearctic Region (Vernier 1997) fauna. A recent study by Hermes et al. (2014) included 18 species of the Oriental fauna. Another effort was made by Nugroho (2015) to study the potter wasps with a petiolate metasoma (Vespidae, Eumeninae) in the Indonesian Archipelago based on a phylogenetic analysis at the generic level. But for these two studies, no efforts have been made to provide a natural classification/phylogenetic analysis of Oriental Eumeninae and nothing at all with reference to the Indian fauna at the generic level. Comprehensive work is required, supplementing the previous work in other regions with an analysis of the Indian fauna.

TAXONOMIC PROBLEMS

Taxonomic studies on Eumeninae have been limited with no intensive research on this group. No revisionary work is available on this subfamily from India, and even faunal studies in the Indian region have received little attention. Major contributions on Indian fauna were done by Giordani Soika (1941, 1960, 1966, 1982, 1991) and Gusenleitner (1988, 1996, 1998, 2001, 2006, 2007, 2008). No updated key to the genera and species of Indian Eumeninae is available. Recently, Srinivasan and Girish Kumar (2010), Girish Kumar and Sharma (2015b), and Pannure et al. (2016) have published generic keys to the fauna of Arunachal Pradesh, Chhattisgarh, and South India, respectively.

BIOLOGY

Nesting: An amazing plasticity of nesting habits is shown by Eumeninae. They owe the common name "potter wasps" to the fact that some of these species build more or less free-standing mud nests looking like earthen pots. Free-standing mud nests may be unicellular or multicellular (Isely 1914). In addition to species that construct their nests with mud or masticated plant material, there are species of Eumeninae that excavate the soil or occupy and modify pre-existing cavities (Iwata 1976; Evans 1966; Cowan 1991; Nugroho et al. 2016). In fact, the primitive and the most common nesting type of wasp is the "renting", nesting in pre-existing cavities. The most commonly used cavities are borings found on decaying wood material, but some species also use other structures such as hollow plant twigs and stems, artificial cavities on man-made structures and ceilings, old mud nests, small cavities on rocks, walls, or concrete slabs, and even in deserted ground burrows of other aculeates (Carpenter and Cumming 1985; Cowan 1991). They easily accept artificially prepared trap nests such as burrows drilled into the wood blocks or bundles of cut hollow reed or plastic tubes, greatly facilitating the study of their biology (Medler and Fye 1956; Parker and Bohart 1966; Krombein 1967; Jayasingh and Freeman 1980; Budrienè et al. 2004; Buschini and Buss 2010; Fateryga 2013a). Several species dig their own ground burrows, and of these some construct a mud tube or "turret" over the nest entrance (e.g., Evans 1956). A few species build free-standing mud nests on the surfaces of rocks or plant twigs (e.g., Evans 1977). Based on the nesting habits, they may be classified into three types: excavators, renters, and builders (Maindron 1882; Iwata 1976). The construction material and form of nests are influenced by the availability of nest sites and construction materials, as well as the ability of particular designs to thwart nest parasites and predators (Cowan 1991; Hermes et al. 2013). For example, the turret of burrowing species functions to protect against ants (Fateryga 2013b). A few species are even polymorphic in nest construction (Cooper 1979; Hermes et al. 2015). Nesting behaviour of most Indian species have not been studied. Jayakar (1963) and Jayakar and Spurway (1967) observed the nest building of *Delta esuriens* (Fabricius). Jayakar and Spurway (1965) observed the nesting behaviour of *Delta conoideum* (Gmelin). Jayakar and Spurway (1971) reported the nesting of *Pareumenes brevirostratus* (de Saussure), involving a primitive form of cooperation. Srinivasan and Girish Kumar (2009) described the nesting behaviour of *Xenorhynchium nitidulum* (Fabricius).

Mass Provisioning vs Progressive Provisioning: Most of the Eumeninae wasps provision their nests with caterpillars or coleopteran larvae. Each cell within a nest will be provisioned with several prey items before the egg hatches ("mass provisioning") so that the hatchlings will be provided with ready-made food in their nest itself. Cells giving rise to female wasps usually receive a greater amount of food than the cells containing males, which usually are smaller (Buck et al. 2008). However, some wasps show primitive signs of social behaviour (pre-social), such as *Xenorhynchium nitidulum, Paraleptomenes miniatus, Eumenes pomiformis,* etc. (Deleurance 1946; Jayakar and Spurway 1966; Krombein 1978; West-Eberhard 1987), which practice progressive provisioning, especially when there is scarcity of prey (Cowan 1991). In a series of studies, Jayakar and co-workers have studied nesting biology and behaviour of a few Indian eumenine wasps (Jayakar 1963, 1966; Jayakar and Spurway 1964, 1965, 1966, 1967, 1968, 1971). Srinivasan and Girish Kumar (2009) also studied some behavioural aspects, such as provisioning of the nest, of the potter wasp *Xenorhynchium nitidulum* (Fabricius).

Sex Determination: Like other hymenopterans, sex determination in Eumeninae is through haplodiploidy. The males are smaller than females with a shorter larval/pupal development time than the females (Buck et al. 2008). In mixed-sex nests of renting wasps, females develop in the inner cells of the nest and emerge after the males, which develop in the outer cells (Buck et al. 2008). Most of the species spend more than one generation per nest. Usually there is only one generation, with the nest then abandoned after emergence. Only where there is progressive provisioning is there overlap of generations (Field 2005).

Mite Association: Association of Eumeninae with some mite species is an interesting phenomenon observed in certain genera (Krombein 1961; Cooper 1954). In some genera, both male and female wasps carry the mites in a specialized region called the acarinarium. In the genus *Parancistrocerus,* this mite chamber (acarinarium) is formed by the transversely depressed base of second tergum which is usually covered by the apical portion of first tergum. These have long been considered as morphological adaptations to securely transfer beneficial mites into nests, and thus are thought to be the product of a mutualistic relationship where mites protect potter wasps against natural enemies (Okabe and Makino 2008, 2011). However, detailed investigations have not been made to understand the association of mites with potter wasps.

Strepsipteran Association: An interesting phenomenon of parasitic association of strepsipteran insects under the metasomal tergum is observed in many species of Eumeninae (e.g., Salt and Bequaert 1929; Girish Kumar and Carpenter 2013).

Prey Associations: Larvae of Coleoptera and Lepidoptera form the majority of the prey of several Eumeninae which provision their nests (Melo et al. 2011), but a few species like *Paragymnomerus signaticollis tauricus* (Kostylev) take sawflies (Fateryga 2018). The knowledge of prey-predator associations in Eumeninae is scanty and fragmentary (Krombein 1967; Itino 1992; Callan 1993).

ECONOMIC IMPORTANCE

They are chiefly predators of many insect larvae of Lepidoptera (Geometridae, Tortricidae) or Coleoptera (Chrysomelidae and Curculionidae), some of which are pests of agricultural importance (Cowan 1991). They can be potentially useful biological control agents for some herbivorous insects such as the spruce budworm (Jennings and Houseweart 1984), other caterpillars (Fye 1965), larvae of alfalfa weevils (Bohart et al. 1982), or leaf beetles (Sears et al. 2001). They may also play a significant role as pollinators in the environment, as their flower preferences are marked over a season (e.g., Cooper 1953). Trap-nesting potter wasps can be used as bioindicators of environmental change (Tscharntke et al. 1998).

COLLECTION AND PRESERVATION

Specimen Collection: Adult specimens for taxonomic studies can be collected by using different methods such as sweep nets, aspirators, malaise traps, yellow pan traps, rearing of adults from the nests, etc. Adult insects can also be collected from the nests by hand picking or using vials.

Preservation: Collected specimens can be put directly from the net into vials of 70% alcohol. Long term preservation in alcohol can be improved by freezing. The alcohol should be changed periodically so as to prevent damage. The adults from alcohol can be mounted or pinned whenever convenient. Usually very small specimens are mounted on cards in such a way that all characters are visible easily. Larger specimens are pinned by using standard entomological pins, passing through the mesosoma from dorsal side. The structures like genitalia are mounted on micro slides.

Dissecting and studying genitalia can be performed by allowing preserved specimens in a relaxing chamber for a few hours till they become flexible for genitalia extraction. The genital capsule can be taken off with a forceps and entomology pins, then immersed in 10% potassium hydroxide, and heated at medium temperature (50°C) for clearing. The genitalia can then be immersed in acetic acid to neutralize the bleaching and either temporary or permanent slides prepared for study.

USEFUL WEBSITES

1. *http://www.zobodat.at* (Gusenleitner 2015). This site is prepared and maintained by Landesmuseum, Zurich and Biologiezentrum, Linz. This is a very useful site for getting information regarding species of Eumeninae.
2. *http://data.gbif.org/search/taxa/* The Global Biodiversity Information Facility (GBIF) is an international network aimed at providing open access to data about all types of life on earth. This site is coordinated through its secretariat in Copenhagen, and India is an associate participant from Asia since 2003. This is also a very useful site for information on species of Eumeninae.
3. *http://www.eol/pages/*Encyclopedia of Life is another very useful site for getting information about life on earth. This website is maintained by the Field Museum, Harvard University, MacArthur Foundation, Marine Biological Laboratory, Missouri Botanical Garden, Sloan Foundation, and Smithsonian Institution. Information about species of Eumeninae is available.

CONCLUSION

Taxonomy and natural history studies on Indian Eumeninae have not been undertaken for many genera, and a comprehensive revision is needed in order to clarify the status of problematic species and subspecies. For example, *Allorhynchium argentatum* (Fabricius, 1804) and *Allorhynchium metallicum* (de Saussure, 1852) are very similar except some minor differences in the intensity of punctures on abdominal tergum. Several subspecies are required to be studied in great detail to confirm the taxonomic status and determine their actual status as subspecies/species. For example, *Anterhynchium (Anterhynchium) abdominale abdominale* (Illiger, 1802) and *Anterhynchium (Anterhynchium) abdominale bengalense* (de Saussure, 1855) show noticeable difference in colour patterns on their abdominal tergum, but intermediate forms are also available. Studies on the taxonomy of potter wasps in India have been very scanty and fragmentary compared to other Asian countries, but it has progressed in very recent years. Earlier workers like Cameron (1897, 1902, 1903a, 1903b, 1904, 1907, 1908, 1909, 1913), Nurse (1903, 1914), Rothney (1903), Paiva (1907), Meade-Waldo (1910ab), Dover (1921, 1925), and Dover and Rao (1922) have published accounts of some species. The studies on the Eumeninae in Asian countries like China and Japan have been impressive as compared to Indian fauna (Yamane 1990; Zhou et al. 2011). Studies on the taxonomy of Eumeninae in India have made progress in the last few years. Many common genera have been reviewed (Girish Kumar 2013a, 2013b, 2013c; Girish Kumar and Sharma 2012, 2013, 2015a; Girish Kumar and Carpenter 2013; Girish Kumar et al. 2014a, 2014b, 2015a, 2016b, 2016c, 2017a, 2017b; Pannure et al. 2016). However, all the regions of India have not been thoroughly surveyed, and such surveys might help in unravelling new distributional records for several species. There is a need for furthering investigations regarding diversity, distribution, zoogeography, ecology, and biology of Indian Eumeninae.

ACKNOWLEDGEMENTS

The authors are grateful to Dr. Kailash Chandra, Director, Zoological Survey of India, Kolkata and to Dr. P. M. Sureshan, Officer-in-Charge, Western Ghat Regional Centre of Zoological Survey of India, Kozhikode, for providing facilities and encouragement. The second author (AP) gratefully acknowledges the University of Agricultural Sciences, Bengaluru for providing facilities and encouragement.

REFERENCES

Bank, S., M. Sann, C. Mayer et al. 2017. Transcriptome and target DNA enrichment sequence data provide new insights into the phylogeny of vespid wasps (Hymenoptera: Aculeata: Vespidae). *Molecular Phylogenetics and Evolution* 116: 213–226.

Bingham, C. T. 1897. *The Fauna of British India, including Ceylon and Burma, Hymenoptera, I. Wasps and Bees*. Taylor & Francis Group, London, UK 579+ i–xxix. doi:10.5962/bhl.title.100738.

Bohart, G. E., F. D. Parker, and V. J. Tepedino. 1982. Notes on the biology of *Odynerus dilectus* (Hym.: Eumenidae), a predator of the alfalfa weevil, *Hypera postica* (Col.: Curculionidae). *Entomophaga* 27: 23–31.

Bohart, R. M., and L. A. Stange. 1965. *A Revision of the Genus Zethus Fabricius in the Western Hemisphere (Hymenoptera: Eumenidae)*. University of California Publications in Entomology 40: 1–208.

Buck, M., T. P. Cobb, J. K. Stahlhut, and R. H. Hanner. 2012. Unravelling cryptic species diversity in eastern Nearctic paper wasps, *Polistes* (*Fuscopolistes*), using male genitalia, morphometrics and DNA barcoding, with descriptions of two new species (Hymenoptera: Vespidae). *Zootaxa* 3502: 1–48.

Buck, M., S. A. Marshall, and D. K. B. Cheung. 2008. Identification atlas of the Vespidae (Hymenoptera, Aculeata) of the northeastern Nearctic region. *Canadian Journal of Arthropod Identification* 5: 1–492.

Budrienè, A., E. Budrys, and Z. Nevronyte. 2004. Solitary Hymenoptera Aculeata inhabiting trap-nests in Lithuania: Nesting cavity choice and niche overlap. *Latvijas Entomologs* 41: 19–31.

Buschini, M. L. T., and C. E. Buss. 2010. Biologic aspects of different species of *Pachodynerus* (Hymenoptera; Vespidae; Eumeninae). *Brazilian Journal of Biology* 70(3): 623–629.

Callan, E. McC. 1993. Nesting behaviour of *Paralastor debilitatus* Perkins (Hymenoptera: Vespidae: Eumeninae) preying on weevil larvae in Australia. *The Entomologist* 112: 95–98.

Cameron, P. 1897. Hymenoptera Orientalia, or contributions to a knowledge of the Hymenoptera of the Oriental Zoological Region. Part V. *Memoirs and Proceedings of the Manchester Literacy & Philosophical Society* 41(4): 1–144.

Cameron, P. 1902. Descriptions of new genera and new species of Hymenoptera collected by Mayor C.S. Nurse at Deesa, Simla and Ferozepore, Part I and II. *Journal of the Bombay Natural History Society* 14: 267–293, 419–449.

Cameron, P. 1903a. Description of nineteen new species of Larridae, Odynerus and Apidae from Barrackpore. *Transactions of the Entomological Society of London* 1903: 117–132.

Cameron, P. 1903b. On some new genera and species of parasitic and fossorial Hymenoptera from the Khasia Hills, Assam. *Annals and Magazine of Natural History* (7)11: 313–336.

Cameron, P. 1904. Descriptions of new genera and new species of Hymenoptera from India. *Zeitschrift für Systematische Hymenopterologie und Dipterologie* 4: 5–15.

Cameron, P. 1907. Description of a new genus and some new species of Hymenoptera captured by Lieut. Col. C.G. Nurse at Deesa, Matheran and Ferozepore. *Journal of the Bombay Natural History Society* 17: 1001–1012.

Cameron, P. 1908. A contribution to the aculeate Hymenoptera of the Bombay Presidency. *Journal of the Bombay Natural History Society* 18(2): 300–311.

Cameron, P. 1909. On some undescribed bees and wasps captured by Lieut-Col C. G. Nurse in India. *Journal of Bombay Natural History Society* 19(1): 129–138.

Cameron, P. 1913. On some new and other species of Hymenoptera in the collections of the Zoological Branch of the Forest Research Institute Dehra Dun. *The Indian Forest Record* 4(2): 26–28.

Carpenter, J. M., and J. M. Cumming. 1985. A character analysis of the North American potter wasps (Hymenoptera: Vespidae: Eumeninae). *Journal of Natural History* 19(5): 877–916. doi:10.1080/00222938500770551.

Carpenter, J. M., and B. R. Garcete-Barrett. 2002. A key to the Neotropical genera of Eumeninae (Hymenoptera: Vespidae). *Boletin Del Museo Nacional de Historia Natural Del Paraguay* 14(1–2): 52–73.

Cooper, K. W. 1953. Biology of eumenine wasps. I. The ecology, predation and competition of *Ancistrocerus antilope* (Panzer). *Transactions of the American Entomological Society* 79: 13–35.

Cooper, K. W. 1954. Venereal transmission of mites by wasps, and some evolutionary problems arising from the remarkable association of *Enslinniella trisetosa* with the wasp *Ancistrocerus antilope*. Biology of Eumenine wasps II. *Transactions of the American Entomological Society* 80: 119–174.

Cooper, K. W. 1979. Plasticity in nesting behavior of a renting wasp, and its evolutionary implications. Studies on eumenine wasps VIII (Hymenoptera, Aculeata). *Journal of the Washington Academy of Sciences* 69: 151–158.

Cowan, D. P. 1991. The solitary and presocial Vespidae. In *The Social Biology of Wasps*, eds. K. G. Ross and R. W. Matthews. New York: Cornell University Press. pp. 33–73

Deleurance, E. P. 1946. Les Eumènes de la region niçoise: Essai de Monographie biologique. *Bulletin de la Societé Zoologique de France* 70: 85–100.

Dover, C. 1921. The wasps and bees of Barkuda Island. *Records of the Indian Museum* 22: 381–391.

Dover, C. 1925 (1924). Further notes on the Indian diplopterous wasps. *The Journal of the Asiatic Society of Bengal, new series* 20: 289–305.

Dover, C., and H. S. Rao. 1922. A note on the diplopterous wasps in the collection of the Indian Museum. *The Journal of the Asiatic Society of Bengal, new series* 18: 235–249.

Ducke, A. 1914. Über phylogenie und classification der sozialen Vespiden. *Zoologische Jahrbücher (Albeitlung für Systematik)* 36: 303–330.

Evans, H. E. 1966. The behaviour patterns of solitary wasps. *Annual Review of Entomology* 11: 123–154. doi:10.1146/annurev.en.11.010166.001011.

Evans, H. E. 1977. Notes on the nesting behavior and immature stages of two species of *Pterocheilus* (Hymenoptera: Eumenidae). *Journal of the Kansas Entomological Society* 50: 329–334. EOL (Encyclopedia of Life) Data Portal Webpage. http://www.eol/pages/.

Evans, H. E. 1956. Notes on the biology of four species of ground-nesting Vespidae (Hymenoptera). *Proceedings of the Entomological Society of Washington* 58(5): 265–270.

Evans, H. E. 1977. Observations on the nests and prey of eumenid wasps (Hymenoptera, Eumenidae). *Psyche: A Journal of Entomology* 84: 255–259.

Fateryga, A. V. 2013a. The nest structure in four species of solitary wasps of the subfamily Eumeninae (Hymenoptera: Vespidae). *Entomological Review* 93(3): 281–292.

Fateryga, A. V. 2013b. Nesting biology of *Odynerus albopictus calcaratus* (Morawitz, 1885) and *Odynerus femoratus* de Saussure, 1856 (Hymenoptera: Vespidae: Eumeninae). *Journal of Insects Article* ID 597583: 1–8.

Fateryga, A. V. 2018. Nesting biology of *Paragymnomerus signaticollis tauricus* (Kostylev, 1940) (Hymenoptera: Vespidae: Eumeninae). *Zootaxa* 4378(3): 429–441. doi:10.11646/Zootaxa.4378.3.10.

Field, J. 2005. The evolution of progressive provisioning. *Behavioral Ecology* 16(4): 770–778.

Fye, R. E. 1965. The biology of the Vespidae, Pompilidae, and Sphecidae (Hymenoptera) from trap nests in northwestern Ontario. *The Canadian Entomologist* 97: 716–744.

GBIF (Global Biodiversity Information Facility) Data Portal Webpage. http://data.gbif.org/search/taxa/.

Giordani Soika, A. 1941. Studi sui Vespidi Solitari. *Bollettino della Società veneziana di storia naturale* 2(3): 130–279.

Giordani Soika, A. 1960. Notulae Vespidologicae XII – Sugli *Eumenes* s. str. dell'India continentale. *Bollettino della Società Entomologica Italiana* 90(9–10): 158–161.

Giordani Soika, A. 1966 (1964). Eumenidi raccolti dalla spedizione Schaefer nel Tibet meridionale e Sikkim. *Bollettino del Museo civico di storia naturale di Venezia* 17: 97–112.

Giordani Soika, A. 1982 (1981). Revisione delle specie orientali del genere *Antepipona* Sauss. *Bollettino del Museo civico di storia naturale di Venezia* 32: 205–257.

Giordani Soika, A., and R. Khan. 1991 (1989). I Rhynchium del Medio Oriente e territori limitrofi (Hymenoptera, Vespidae). *Bollettino del Museo civico di storia naturale di Venezia* 40: 99–106.

Girish Kumar, P. 2011. Redescription and new distributional records of *Oreumenoides edwardsii* (de Saussure) (Hymenoptera: Vespidae: Eumeninae) from India. *Records of Zoological Survey of India* 111(1): 11–16.

Girish Kumar, P. 2012a. New record of the genus *Omicroides* Giordani Soika (Hymenoptera: Vespidae: Eumeninae) from India. *Records of Zoological Survey of India* 112(Part-3): 89–93.

Girish Kumar, P. 2013 (2012b). On *Delta dimidiatipenne* (de Saussure, 1852) (Hymenoptera: Vespidae: Eumeninae) from India. *Journal of Environment and Sociobiology* 9(1): 43–49.

Girish Kumar, P. 2013 (2012c). Redescription and new distributional records of *Delta esuriens* (Fabricius) (Hymenoptera: Vespidae: Eumeninae) from Indian states. *Records of Zoological Survey of India* 112(4): 55–60.

Girish Kumar, P. 2013a. A revision of the genus *Subancistrocerus* de Saussure (Hymenoptera: Vespidae: Eumeninae) from Indian subcontinent. In *Animal Diversity, Natural History and Conservation*, eds. V. K. Gupta, and A. K. Verma, Vol. 2, pp. 53–68. New Delhi, India: Daya Publishing House.

Girish Kumar, P. 2013b. A taxonomic revision of *Phimenes* Giordani Soika (Hymenoptera: Vespidae: Eumeninae) of Indian subcontinent. *Records of Zoological Survey of India* 113(3): 119–135.

Girish Kumar, P. 2014 (2013c). A taxonomic study on the genus *Anterhynchium* de Saussure (Hymenoptera: Vespidae: Eumeninae) from Indian subcontinent. *Records of Zoological Survey of India* 113(4): 139–158.

Girish Kumar, P., and J. M. Carpenter. 2013. A taxonomic review of the genus *Antodynerus* de Saussure, 1855 (Hymenoptera: Vespidae: Eumeninae) from the Indian subcontinent. *Zootaxa* 3731(2): 267–278.

Girish Kumar, P., and J. M. Carpenter. 2015a. Description of a new species of *Alastor (Alastor)* Lepeletier, 1841 (Hymenoptera: Vespidae: Eumeninae) from Telangana, India, with a key and a checklist of Oriental species. *Halteres* 6: 79–84.

Girish Kumar, P., and J. M. Carpenter. 2015b. Description of a new species of *Leptochilus (Neoleptochilus)* Blüthgen, 1961 (Hymenoptera: Vespidae: Eumeninae) from India. *Journal of Threatened Taxa* 7(11): 7786–7790. doi:10.11609/JoTT.04072.7786-90.

Girish Kumar, P., J. M. Carpenter, L. Castro, and P. M. Sureshan. 2017a. A taxonomic review of the Indian species of the genus *Eumenes* Latreille (Hymenoptera: Vespidae: Eumeninae). *Zootaxa* 4317(3): 469–498. doi:10.11646/zootaxa.4317.3.3.

Girish Kumar, P., J. M. Carpenter, and Lambert Kishore. 2014a. A review of the genus *Epsilon* de Saussure (Hymenoptera: Vespidae: Eumeninae) from India. *Journal of Threatened Taxa* 6(1): 5380–5385. doi:10.11609/JoTT.o3626.5380-5.

Girish Kumar, P., J. M. Carpenter, and Lambert Kishore. 2017b. A review of the genus *Pararrhynchium* de Saussure (Hymenoptera: Vespidae: Eumeninae) from India with the description of a new species. *Halteres* 8: 85–91. doi:10.5281/zenodo.894185.

Girish Kumar, P., J. M. Carpenter, Lambert Kishore, and K. P. Mohammed Shareef. 2014c. Additional notes on the genus *Apodynerus* Giordani Soika, 1993 (Hymenoptera: Vespidae: Eumeninae) from the Indian subcontinent. *Halteres* 5: 46–50.

Girish Kumar, P., J. M. Carpenter, and G. Sharma. 2014b. A review of the genus *Paraleptomenes* Giordani Soika, 1970 (Hymenoptera: Vespidae: Eumeninae: Odynerini) from the Indian subcontinent, with the description of a new species from the eastern Himalayas. *Zootaxa* 3802(1): 131–143. doi:10.11646/Zootaxa.3801.2.11.

Girish Kumar, P., J. M. Carpenter, and P. M. Sureshan. 2015a. A taxonomic review of the genus *Cyrtolabulus* van der Vecht, 1969 (Hymenoptera: Vespidae: Eumeninae) from India. In *Animal Diversity, Natural History and Conservation*, eds. V. K. Gupta and A. K. Verma, Vol. 5. Pages 49–55. New Delhi: Daya Publishing House.

Girish Kumar, P., J. M. Carpenter, and P. M. Sureshan. 2016a. Additions to the knowledge of the genus *Allorhynchium* van der Vecht from the Indian subcontinent with the description of a new species from Kerala (Hymenoptera: Vespidae: Eumeninae). *Halteres* 7: 29–34.

Girish Kumar, P., J. M. Carpenter, and P. M. Sureshan. 2016b. A taxonomic review of the genus *Antepipona* de Saussure, 1855 (Hymenoptera: Vespidae: Eumeninae) from India. *Zootaxa* 4150(5): 501–536. doi.org/10.11646/zootaxa.4150.5.1.

Girish Kumar, P., J. M. Carpenter, and P. M. Sureshan. 2016c. A taxonomic review of the genus *Parancistrocerus* Bequaert (Hymenoptera: Vespidae: Eumeninae) from the Indian subcontinent with the description of three new species. *Halteres* 7: 136–156. doi:10.5281/zenodo.192283.

Girish Kumar, P., L. Castro, J. M. Carpenter, and A. H. Sheikh. 2017c. First record of the genus *Pseudepipona* de Saussure, 1856 (Hymenoptera, Vespidae, Eumeninae) from India with the species *Pseudepipona (Pseudepipona) vicina* Gusenleitner, 1973 from the northern Himalaya. *Boletín de la Asociación Española de Entomología* 41(3–4): 347–354.

Girish Kumar, P., and Lambert Kishore. 2011. Redescription and new distributional records of *Xenorhynchium nitidulum* (Fabricius) (Hymenoptera: Vespidae: Eumeninae) from various states of India. *Uttar Pradesh Journal of Zoology* 31(3): 311–316.

Girish Kumar, P., and Lambert Kishore. 2012. On *Delta pyriforme pyriforme* (Fabricius, 1775) (Hymenoptera: Vespidae: Eumeninae) from India. *Uttar Pradesh Journal of Zoology* 32(3): 269–276.

Girish Kumar, P., Lambert Kishore, and K. P. Mohammed Shareef. 2013a. New distributional records of *Indodynerus capitatus* Gusenleitner, 2008 (Hymenoptera: Vespidae: Eumeninae). *Uttar Pradesh Journal of Zoology* 33(1): 81–83.

Girish Kumar, P., K. P. Mohammed Shareef, Lambert Kishore, and J. M. Carpenter. 2013b. A taxonomic study on the Oriental genus *Apodynerus* Giordani Soika (Hymenoptera: Vespidae: Eumeninae) from the Indian subcontinent. *Biosystematica* 7(1): 23–31.

Girish Kumar, P., K. P. Mohammed Shareef, Lambert Kishore, and J. M. Carpenter. 2013c. A taxonomic review of the Oriental genus *Labus* de Saussure (Hymenoptera: Vespidae: Eumeninae) from the Indian subcontinent. *Biosystematica* 7(2): 29–37.

Girish Kumar, P., and G. Sharma. 2014 (2012). A review of the little known genus *Orancistrocerus* van der Vecht (Hymenoptera: Vespidae: Eumeninae) from the Indian subcontinent. *Hexapoda (Insecta indica)* 19(2): 14–22.

Girish Kumar, P., and G. Sharma. 2013. A taxonomic study on the genus *Rhynchium* Spinola (Hymenoptera: Vespidae: Eumeninae) from the Indian subcontinent. *Records of Zoological Survey of India* 113(2): 105–122.

Girish Kumar, P., and G. Sharma. 2014. Taxonomic studies on vespid wasps (Vespidae: Vespoidea: Hymenoptera: Insecta) of Rajasthan, India with six new records from the state. *Journal on New Biological Reports* 3(3): 233–251.

Girish Kumar, P., and G. Sharma. 2015a. A review of the genus *Allorhynchium* van der Vecht, 1963 (Hymenoptera: Vespidae: Eumeninae) from the Indian subcontinent. *Prommalia* 3: 20–34.

Girish Kumar, P., and G. Sharma. 2015b. Taxonomic studies on vespid wasps (Hymenoptera: Vespoidea: Vespidae) of Chhattisgarh. *Journal of Threatened Taxa* 7(14): 8096–8127. doi:10.11609/jott.2426.7.14.8096-8127.

Girish Kumar, P., G. Srinivasan, and J. M. Carpenter. 2013d. A taxonomic study on the genus *Tropidodynerus* Blüthgen (Hymenoptera: Vespidae: Eumeninae) from the Indian subcontinent. *Prommalia* 1: 162–174.

Girish Kumar, P., G. Srinivasan, and J. M. Carpenter. 2015b. A new species of *Lissodynerus* Giordani Soika, 1993 (Hymenoptera: Vespidae: Eumeninae) from Rutland Island, South Andaman, India. *Journal of Threatened Taxa* 7(10): 7664–7667. doi:10.11609/JoTT.o4022.7664-7.

Girish Kumar, P., and P. M. Sureshan. 2016. New record of the species *Ectopioglossa sublaevis* (Smith, 1857) (Hymenoptera: Vespidae: Eumeninae) from the Indian subcontinent. *Records of Zoological Survey of India* 116(4): 489–492.

Girish Kumar, P., P. M. Sureshan, and K. G. Emiliyamma. 2016d. First record of the genus *Coeleumenes* van der Vecht from Peninsular India with the species *C. burmanicus* (Bingham) from Kerala (Hymenoptera: Vespidae: Eumeninae). *Records of Zoological Survey of India* 116 (Part-4): 493–496.

Grandi, G. 1961. Studi di entomologo sugli Imenotteri Superiori. *Bollettino dell'Instituto di Entomologia "Guido Grandi"dell' Università degli Studi di Bologna* 25: 397–400.

Gusenleitner, J. 1988. Uber Eumenidae aus Thailand, mit einer Bestimmungstabelle fur orientalischer *Labus*- Arten (Hymenoptera Vespoidea). *Linzer Biologische Beiträge* 20(1): 173–198.

Gusenleitner, J. 1996. Uber Eumenidae der orientalischen Region (Hymenoptera, Vespoidea). *Linzer biologischen Beiträge* 28(1): 39–56.

Gusenleitner, J. 1998. Uber Faltenwespen aus dem asiatischen Raum (Hymenoptera, Eumenidae, Masaridae). *Linzer biologischen Beiträge* 30(2): 503–513.

Gusenleitner, J. 2001. Beitrag zur Kenntnis von Faltenwespen der orientalischen Region (Hymenoptera, Vespidae, Eumenidae). *Linzer biologischen Beiträge* 33(2): 655–662.

Gusenleitner, J. 2006. Über Aufsammlungen von Faltenwespen in Indien (Hymenoptera: Vespidae). *Linzer biologische Beiträge* 38(1): 677–695.

Gusenleitner, J. 2007. Bemerkenswerte Faltenwespen-Funde aus der orientalischen Region Teil 3 (Hymenoptera: Vespidae, Polistinae, Eumeninae). *Linzer biologische Beiträge* 39(1): 97–104.

Gusenleitner, J. 2008. Bemerkenswerte Faltenwespen-Funde aus der orientalischen region Teil 4. Mit einem Anhang uber eine Art aus Neu-Kaledonien (Hymenoptera, Vespidae, Eumeninae). *Linzer biologische Beiträge* 40(2): 1495–1503.

Gusenleitner, J. 2015. ZOBODAT: Zoological-Botanical Database (Vespoidea) (Version 4.0, October 2011). Available from: http://www.zobodat.at.

Hermes, M. G., G. Araújo, and Y. Antonini. 2015. On the nesting biology of eumenine wasps yet again: *Minixi brasilianum* (de Saussure) is a builder and a renter... at the same time! (Hymenoptera, Vespidae, Eumeninae). *Revista Brasileira de Entomologia* 59(2): 141–142.

Hermes, M. G., G. A. R. Melo, and J. M. Carpenter. 2014. The higher-level phylogenetic relationships of the Eumeninae (Insecta, Hymenoptera, Vespidae), with emphasis on *Eumenes* sensu lato. *Cladistics* 30: 453–484. doi:10.1111/cla.12059.

Hermes, M. G., and L. A. de Oliveira. 2016. Morphological cladistic analysis resolves the generic limits of the Neotropical potter wasp genera *Minixi* Giordani Soika and *Pachyminixi* Giordani Soika (Hymenoptera: Vespidae: Eumeninae). *Invertebrate Systematics* 30(3): 187–200.

Hermes, M. G., A. Somavilla, and B. R. Garcete-Barrett. 2013. On the nesting biology of *Pirhosigma* Giordani Soika (Hymenoptera, Vespidae, Eumeninae), with special reference to the use of vegetable matter. *Revista Brasileira de Entomologia* 57(4): 433–436.

Isely, D. 1914. The biology of some Kansas Eumenidae. *The Kansas University Science Bulletin* 8: 231–309.

Itino, T. 1992. Differential diet breadths and species coexistence in leafroller-hunting eumenid wasps. *Researches on Population Ecology* 34: 203–211. doi:10.1007/BF02513531.

Iwata, K. 1976. *Evolution of Insect: Comparative Ethology of Hymenoptera*. New Delhi, India: Amerind Publishing, 535 pp.

Jayakar, S. D. 1963. "Proterandry" in solitary wasps. *Nature* 198(4876): 208–209.

Jayakar, S. D. 1966. Sexual behavior in solitary eumenid wasps. *Journal of the Bombay Natural History Society* 63(3): 760–763.

Jayakar, S. D., and H. Spurway. 1965 (1964). Winter diapause in the squatter wasps *Antodynerus flavescens* (Fabr.) and *Chalybion bengalense* (Dahlb.) (Vespoidea and Sphecoidea). *Journal of the Bombay Natural History Society* 61: 662–667.

Jayakar, S. D., and H. Spurway. 1965. Normal and abnormal nests of *Eumenes emarginatus conoideus* (Gmelin) including notes on crèpissage in this and other members of the genus (Vespoidea, Hymenoptera). *Journal of the Bombay Natural History Society* 62: 193–200.

Jayakar, S. D., and H. Spurway. 1966. Re-use of cells and brother-sister mating in the Indian species *Stenodynerus miniatus* (Sauss.) (Vespidae: Eumenidae). *Journal of the Bombay Natural History Society* 63: 378–398.

Jayakar, S. D., and H. Spurway. 1967. The nesting activities of the vespoid potter wasp *Eumenes campaniformis esuriens* (Fabr.) compared with the ecologically similar sphecoid *Sceliphron madraspatanum* (Fabr.) (Hymenoptera). *Journal of the Bombay Natural History Society* 64(2): 307–332.

Jayakar, S. D., and H. Spurway. 1968. The nesting activities of the vespoid potter wasp *Eumenes campaniformis esuriens* (Fabr.) compared with the ecologically similar sphecoid *Sceliphron madraspatanum* (Fabr.) (Hymenoptera). *Journal of the Bombay Natural History* 65(1): 148–181.

Jayakar, S. D., and H. Spurway. 1971. The nesting of *Pareumenes brevirostratus* (Saussure), involving a primitive form of cooperation. *Journal of the Bombay Natural History* 68(1): 153–160.

Jayasingh, D. B., and B. E. Freeman. 1980. The comparative populations dynamics of eight solitary bees and wasps (Aculeata: Apocrita: Hymenoptera) trap-nested in Jamaica. *Biotropica* 12(3): 214–219.

Jennings, D. T., and M. W. Houseweart. 1984. Predation by eumenid wasps (Hymenoptera: Eumenidae) on spruce budworm (Lepidoptera: Tortricidae) and other lepidopterous larvae in spruce-fir forests of Maine. *Annals of the Entomological Society of America* 77(1): 39–45.

Krombein, K. V. 1961. Some symbiotic relationships between saproglyphid mites and solitary vespid wasps (Acarina, Saproglyphidae and Hymenoptera, Vespidae). *Journal of the Washington Academy of Sciences* 51: 89–93.

Krombein, K. V. 1967. *Trap-nesting Wasps and Bees: Life Histories, Nests, and Associates*. Washington, WA: Smithsonian Press, 570 pp.

Krombein, K. V. 1978. Biosystematic studies of ceylonese wasps III. Life history, nest and associates of *Paraleptomenes mephitis* (Cameron) (Hymenoptera: Eumenidae). *Journal of the Kansas Entomological Society* 51(4): 721–734.

Krombein, K. V. 1991. Biosystematic studies of ceylonese wasps, xix: Natural history notes in several families (Hymenoptera: Eumenidae, Vespidae, Pompilidae, and Crabronidae). *Smithsonian Contributions to Zoology* 283: 1–41.

Lambert Kishore. 2004. Two new species and a new subspecies of *Antepipona* Saussure (Hymenoptera: Vespidae) from India with a key to species. In *Perspectives on Biosystematics and Biodiversity, Prof. T. C. Narendran Commemoration Volume*, eds. K. Rajmohana, K. Sudheer, P. Girish Kumar, and S. Santhosh, 553–566. Thenjipalam, India: SERSA.

Lambert Kishore, and T. C. Narendran. 2002. A new species of *Antepipona* Saussure (Hymenoptera: Vespidae) from India. *Journal of Zoological Society of Kerala* 10(1&2): 1–4.

Lambert Kishore, P. Girish Kumar, and T. C. Narendran. 2008. A new species of *Antepipona* Saussure (Hymenoptera: Vespidae). *Uttar Pradesh Journal of Zoology* 28(1): 121–123.

Lambert Kishore, T. C. Narendran, and P. Girish Kumar. 2007. A new record of *Ancistrocerus tinctipennis* (Walker) (Hymenoptera: Vespidae) from India. *Journal of Entomological Research* 31(2): 169–171.

Latreille, P. A. 1802–1803. *Histoire Naturelle, Génèrale et Particuliére des Crustacés et des Insectes, Tome 3. Familles Naturelles des Genres*. Paris: F. Dufart, Xii + 13 + 467 pp.

Maindron, M. M. 1882. Histoire des guepes solitaires (Eumeniens) de l'Archipel Indienen et de la Nouvelle-Guinee. 2e Partie (1). *Annales de la Société entomologique de France* ser. 6, 2: 267–286.

Meade-Waldo, G. 1910a. New species of Diploptera in the collection of the British Museum, Part I. *Annals and Magazine of Natural History* (8)5: 30–51. doi:10.1080/00222931008692723.

Meade-Waldo, G. 1910b. New species of Diploptera in the collection of the British Museum, Part II. *Annals and Magazine of Natural History* (8)6: 100–110.

Medler, J. T., and R. E. Fye. 1956. Biology of *Ancistrocerus antilope* (Panzer) (Hymenoptera, Vespidae) in trap-nests in Wisconsin. *Annals of the Entomological Society of America* 49: 97–102.

Melo, G. A. R., M. G. Hermes, and B. R. Garcete-Barrett. 2011. Origin and occurrence of predation among Hymenoptera: A phylogenetic perspective. In *Predation in the Hymenoptera: An Evolutionary Perspectives*, ed. Carlo Polidori, 1–22. Thiruvananthapuram, India: Transworld Research Network.

Mohammed Shareef, K. P., Lambert Kishore, and P. Girish Kumar. 2013. New record of *Pseudozumia indica* (de Saussure, 1855) (Hymenoptera: Vespidae: Eumeninae) from Peninsular India. *Biological Forum – An International Journal* 5(2): 94–97.

Nugroho, H. 2015. *Taxonomic and phylogenetic study of potter wasps with a petiolate metasoma (Vespidae, Eumeninae) in Indonesian Archipelago*. Graduate School of Science and Engineering, Ibaraki University. Pp. 280.

Nugroho, H., R. Ubaidillah, and J. Kojima. 2016. Taxonomy of the Indo-Malayan presocial potter wasp genus *Calligaster* de Saussure (Hymenoptera, Vespidae, Eumeninae). *Journal of Hymenoptera Research* 48: 19–32.

Nurse, C. G. 1903. New species of Indian aculeate Hymenoptera. *Annals and Magazine of Natural History* (7)11: 528–549.

Nurse, C. G. 1914. Zoological results of the Arbor Expedition, 1911-1912. Hymenoptera, V: Fossores, Diploptera, Chrysididae. *Records of the Indian Museum* 8: 443–447.

Okabe, K., and S. Makino. 2008. Parasitic mites as part-time bodyguards of a host wasp. *Proceedings of the Royal Society B*, 275: 2293–2297.

Okabe, K., and S. Makino. 2011. Behavioural observations of the bodyguard mite *Ensliniella parasitica*. *Zoosymposia* 6: 193–199.

Paiva, C. A. 1907. Records of Hemiptera and Hymenoptera from the Himalayas. *Records of the Indian Museum* 1: 13–20.

Pannure, A., V. V. Belavadi, and J. M. Carpenter. 2016. Taxonomic studies of potter wasps (Hymenoptera: Vespidae: Eumeninae) of South India. *Zootaxa* 4171(1): 1–51. doi:10.11646/zootaxa.4171.1.1.

Pannure, A., V. V. Belavadi, and J. M. Carpenter. 2017. A new species of the genus *Discoelius* Latrielle, 1809 (Hymenoptera: Vespidae: Eumeninae) from India. *Zootaxa* 4272(4): 583–586. doi:10.11646/zootaxa.4272.4.7.

Pannure, A., V. V. Belavadi, and J. M. Carpenter. 2018. Taxonomic notes on poorly known species of potter wasps (Vespidae: Eumeninae: Zethini) from India. *Oriental Insects* doi:10.1080/00305316.2018.1521345.

Parker, F. D., and R. M. Bohart. 1966. Host-parasite associations in some twig-nesting Hymenoptera from western North America. *Pan-Pacific Entomologist* 42: 91–98.

Rothney, G. A. J. 1903. The aculeate Hymenoptera of Barrackpore, Bengal. *Transactions of the Royal Entomological Society London* 93–116. doi:10.1111/j.1365-2311.1903.tb01128.x.

Salt, G., and J. Bequaert. 1929. Stylopized Vespidae. *Psyche: A Journal of Entomology* 36: 249–282.

Sears, A. L. W., J. T. Smiley, M. Hilker, F. Muller, and N. E. Rank. 2001. Nesting behaviour and prey use in two geographically separated populations of the specialist wasp *Symmorphus cristatus* (Vespidae: Eumeninae). *The American Midland Naturalist* 145: 233–246.

Selis, M. 2018. Additions to the knowledge of solitary wasps (Hymenoptera: Vespidae: Eumeninae), with description of eight new species. *Zootaxa* 4403(3): 441–468. doi:10.11646/Zootaxa.4403.3.2.

Srinivasan, G., and P. Girish Kumar. 2009. Observations on some behavioural aspects of the potter wasp *Xenorhynchium nitidulum* (Fabricius) (Hymenoptera: Vespidae: Eumeninae). *Records of Zoological Survey of India* 109(4): 77–78.

Srinivasan, G., and P. Girish Kumar. 2010. New records of potter wasps (Hymenoptera: Vespidae: Eumeninae) from Arunachal Pradesh, India: Five genera and ten species. *Journal of Threatened Taxa* 2(12): 1313–1322. doi:10.11609/JoTT.o2468.1313-22.

Srinivasan, G., and P. Girish Kumar. 2013. A study on vespid wasps (Insecta: Hymenoptera: Vespidae) of Itanagar Wildlife Sanctuary, Arunachal Pradesh. *Records of Zoological Survey of India* 113(1): 115–127.

Tan, J. L., J. M. Carpenter, and C. van Achterberg. 2018. An illustrated key to the genera of Eumeninae from China, with a checklist of species (Hymenoptera, Vespidae). *ZooKeys* 740: 109–149.

Tormos, J., J. D. Asis, and S. F. Gayubo. 1997a. Description of the mature larva of *Pterocheilus phaleratus yeguasicus* (Hymenoptera, Vespidae). Fragment. *Entomologist* 29(2): 395–398.

Tormos, J., J. D. Asís, S. F. Gayubo, and F. Torres. 1997b. Description of the mature larvae of *Symmorphus bifasciatus* (L., 1758) and *S.crassicornis crassicornis* (Panzer, 1798) (Hymenoptera, Vespidae, Eumeninae). *Mitteilungen der Münchner Entomologischen Gesellschaft* 87: 23–27.

Tormos, J., J. D. Asis, S. F. Gayubo, and F. Torres. 1997c. Description of the mature larva of *Microdynerus exilis* and *M. timidus* (Hymenoptera: Vespidae). *Entomological News* 108(4): 259–264.

Tormos, J., J. D. Asís, S. F. G. Gayubo, and F. Torres. 1998. Description of the mature larvae of *Ancistrocerus kitcheneri* (Dusmet, 1917), *A. longispinosus* (Saussure, 1885) and redescription of that of *A. trifasciatus* (Müller, 1776). (Hymenoptera, Vespidae). *Nouvelle Revue d' Entomologie* (NS) 15: 31–36.

Tormos, J., R. Boesi, J. D. Asís, and C. Polidori. 2005. Description of the mature larva of *Ancistrocerus sikhimensis* (Hymenoptera: Eumenidae). *Florida Entomologist* 88: 188–190.

Tormos, J., C. Polidori, J. D. Asís, and S. F. Gayubo. 2008. Description of mature larvae of *Allodynerus rossii* (Lepeletier), *Ancistrocerus auctus* (Fabricius), *Euodynerus dantici* (Rossi) and *Symmorphus murarius* (Linnaeus) (Hymenoptera, Vespidae). *Zootaxa* 1946: 42–54.

Tscharntke, T., A. Gathmann, and I. Steffan-Dewenter. 1998. Bioindication using trap-nesting bees and wasps and their natural enemies: Community structure and interactions. *Journal of Applied Ecology* 35: 708–719.

van der Vecht, J. 1937. Descriptions and records of Oriental and Papuan solitary Vespidae. *Treubia* 16: 261–293.

van der Vecht, J. 1959a. On some Fabrician types of Indo-Australian Vespidae (Hymenoptera). *Archives Néerlandaises de Zoologie* 13(1 Suppl): 234–247.

van der Vecht, J. 1959b. On *Eumenes arcuatus* (Fabricius) and some allied Indo-Australian wasps (Hymenoptera, Vespidae). *Zoologische Verhandelingen, Leiden* 41: 1–71.

van der Vecht, J. 1961. Evolution in a group of Indo-Australian *Eumenes* (Hymenoptera, Eumenidae). *Evolution* 15: 468–477.

van der Vecht, J. 1963. Studies on Indo-Australian and East Asiatic Eumenidae (Hymenoptera: Vespoidea). *Zoologische Verhandelingen, Leiden* 42(24): 255–259.

van der Vecht, J. 1969. A new name for the genus *Cyrtolabus* van der Vecht (Hymenoptera, Eumenidae). *Entomologische Berichten, Amsterdam* 29: 2.

van der Vecht, J. 1981. Studies in Indo-Australian Solitary wasps. *Proceedings Koninklijke nederlandse akademie van Wetenschappen* C 84(4): 443–464.

Vernier, R. 1997. Essai d'analyse cladistique des genres d'Eumeninae (Vespidae, Hymenoptera) reprèsèntès en Europe septentrionale, occidentale et centrale. *Bulletin de la Société Neuchâteloise des Sciences Naturelles* 120: 87–98.

Wain, F. L. 1956. Notes on some wasps and Bees (Hymenoptera) of Poona and the Western Ghats. *Journal of the Bombay Natural History Society* 54: 22–36.

West-Eberhard, M. J. 1987. Observations of *Xenorhynchium nitidulum* (Fabricius) (Hymenoptera, Eumeninae), a primitively social wasp. *Psyche* 94(3–4): 317–323.

Yamane, S. K. 1990. A revision of the Japanese Eumenidae (Hymenoptera, Vespoidea). *Insecta matsumurana. Series Entomology, new series* 43: 1–189.

Zhou, X., B. Chen, and T.-J. Li. 2011. The taxonomic research progress of Eumeninae (Hymenoptera: Vespidae). *Journal of Chongqing Normal University (Natural Science)* 28(6): 22–29.

13 Tiger Beetles (Coleoptera: Cicindelidae)
Their History and Future in Indian Biological Studies

David L. Pearson

CONTENTS

Introduction 201
Family Diagnosis and Key Characters 202
An Overview of Indian Genera 202
    Genus *Tricondyla* Latreille 202
    Genus *Neocollyris* Fabricius 203
    Genus *Therates* Latreille 203
    Genus *Prothyma* Hope 203
    Genus *Heptodonta* Hope 203
    Genus *Lophyra* Motschulsky 204
    Genus *Jansenia* Chaudoir 205
    Genus *Apteroessa* Hope 205
Biodiversity and Species Richness 206
Distribution Patterns within India 206
Molecular Characterization and Phylogeny 207
Adult Morphology 207
Adult Ecology, Behaviour, and Physiology 209
Larval/Pupal Biology 209
Integrative Taxonomy 211
Taxonomic Problems 212
Conservation Status 212
Economic Importance 213
Collection and Preservation 213
Future Directions 214
    Occupancy Modelling and Presence/Absence Data 214
    Climate Change 214
    Citizen Scientists 215
    Biomimicry 215
Conclusion 216
Acknowledgements 216
References 217

INTRODUCTION

Tiger beetles are arguably one of the most popular and most studied beetle groups in the world (Pearson 1988; Pearson and Vogler 2001; Pearson 2011). With nearly 2900 species described worldwide, it is a group that presents sufficient diversity to ask interesting questions for scientists and challenge amateurs to take it up as a hobby, but with not so many species that it becomes overwhelming.

Adult and larval tiger beetles are predatory and occur in a wide variety of biomes, from high-elevation alpine forests and high-latitude taiga (boreal) forests to tropical rain forests, from desert washes to ocean beaches. They are found in almost every part of the terrestrial world except Antarctica and isolated oceanic islands. Wherever they reside, however, each species tends to occupy a narrow or highly specialized habitat such as tall termite mounds, edges of mountain streams, forest floors, boulder surfaces, tree trunks, and alkaline flats. Even adults of the most widespread species occupy relatively narrow habitats—sandy or muddy beaches and grasslands. Larvae of each species are generally more restricted to microhabitats than are their adult stages.

Because their taxonomy, morphology, physiology, behaviour, and distribution are so well known in even remote parts of the world, tiger beetles lend themselves both to basic studies as well as applied areas such as crop protection and conservation efforts. This combination of utility, access, and deep background data have made tiger beetles a valuable tool for understanding a broad range of basic and applied questions in India. With 241 described species, India boasts the third highest number of tiger beetle species of any country in the world (Jürgen Wiesner 1992, *pers. comm.*; Pearson and Cassola 1992). There are presently more than 130 published articles and chapters that focus on or include Indian tiger beetles. The goal of this article is to celebrate the generosity, encouragement, and foresight of Professor C. A. Viraktamath during my research years in India. I want to thank him for making me feel welcome in India both as a colleague and a friend with this highlight of the history of tiger beetle studies in India and their use for future studies and understanding of Indian nature.

FAMILY DIAGNOSIS AND KEY CHARACTERS

Tiger beetles form a discrete group of species within the beetle order Coleoptera. Their status as a distinct group (subfamily Cicindelinae, supertribe Cicindelitae, or tribe Cicindelini) within the family of ground beetles called Carabidae, or a distinct family of their own (Cicindelidae) has vacillated back and forth with advancing techniques of analysis. The most recent molecular and morphological analysis places them in their own family (López-López and Vogler 2017).

They share many characters in common with and are most closely related to the predaceous ground beetles (Carabidae), predaceous diving beetles (Dytiscidae), whirligig beetles (Gyrinidae), and crawling water beetles (Haliplidae). These five families and a few others are placed together as a suborder called Adephaga. Twenty-seven genera of tiger beetles are now recognized as occurring in India (Table 13.1), and several characters in combination reliably distinguish tiger beetles from all other groups of Adephaga: (1) long, sickle-shaped mandibles; (2) simple teeth arranged along the inner side of the mandible with a compound (molar-like) tooth on the inner base of each mandible; (3) long, thin antennae with 11 segments and attached to the head between the eye and base of the mandible; (4) long body form with eyes and head together wider than the thorax; (5) long, thin running legs; (6) tunnel-building behaviour of the larvae; and (7) peculiar forward-facing sets of hooks on the backs of the larvae.

Most adult tiger beetle species look remarkably similar in body shape, proportions, and behaviour. They vary primarily in size, colour, and shape of markings. In India, the smallest species is 6.5 mm long, while the largest is more than 25 mm. Some species are dull black or brown, but many species are brilliantly emblazoned with bright green, violet, blue, red, and yellow colours. Most species have streamlined bodies and long, slender legs for fast sprinting across the ground or vegetation.

The prominent eyes of these visual hunters are usually so large that they make the head wider than the relatively narrow thorax. Adults have transparent hind wings that are folded and hidden under the hard front wings, the elytra. In flight, these elytra conveniently open forward to allow the flight wings underneath to unfold and extend out to the sides. The wings are used for short and low escape flights from predators. A few species, however, use their wings for long-range dispersion, while a few others have lost these flight wings and are earthbound.

AN OVERVIEW OF INDIAN GENERA

To more easily understand Indian tiger beetles and their range of form and function, I include here a synopsis of the current knowledge of the natural history and appearance of eight genera (Pearson and Vogler 2001) (Table 13.1).

Genus *Tricondyla* Latreille

These long, narrow beetles (Figure 13.1) are all blackish, flightless with fused elytra, and primarily arboreal and diurnal. They are limited to moist forests and resemble large ants. Some species are remarkably mimicked by late instar grasshoppers

TABLE 13.1

Number of Species and Endemic Species within each of the 27 Genera of Tiger Beetles Found in India

Genus	No. of Species	No. of Endemic Species
Grammognatha	1	
Tricondyla	4	1
Derocrania	4	4
Protocollyris	4	3
Neocollyris	57	34
Collyris	5	2
Therates	10	9
Rhytidophaena	2	1
Prothyma	2	
Heptodonta	2	1
Pronyssa	4	3
Calochroa	11	4
Cicindela	10	7
Calomera	9	2
Cosmodela	6	
Plutacia	2	2
Lophyra	11	2
Chaetodera	2	
Jansenia	34	32
Glomera	2	2
Setinteridenta	1	
Cylindera	43	9
Myriochila	7	
Salpingophora	1	
Hypaetha	3	
Callytron	3	
Apteroessa	1	1
TOTAL	**241**	**119**

Tiger Beetles (Coleoptera: Cicindelidae)

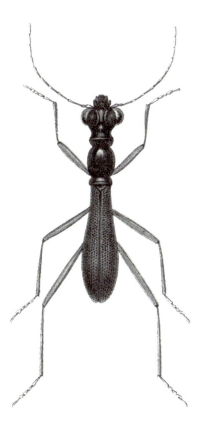

FIGURE 13.1 Genus *Tricondyla*. (From Fowler, W.W., *Fauna of British India including Ceylon and Burma (Coleoptera General Introduction and Cicindelidae and Paussidae)*, Today and Tomorrow's Printers and Publishers, New Delhi, 529 pp, 1912.)

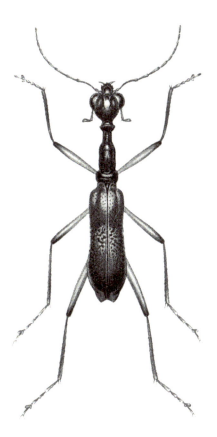

FIGURE 13.2 Genus *Neocollyris*. (From Fowler, W.W., *Fauna of British India including Ceylon and Burma (Coleoptera General Introduction and Cicindelidae and Paussidae)*, Today and Tomorrow's Printers and Publishers, New Delhi, 529 pp, 1912.)

(Shelford 1902). They run up and down the trunks of trees and move, squirrel-like, to the opposite side of the tree when threatened by a predator. When pressed by danger, they will fall to the ground and remain still in the leaf litter. They also move between trees by walking across the ground. Larval burrows have been found in the bark of large trees, and the larvae are active primarily at night. The 48 species of this genus (Fleutiaux 1920; Naviaux and Moravec 2001; Naviaux 2002) occur throughout southeastern Asia to northern Australia. Four species are known from India, one of which is endemic.

Genus *Neocollyris* Fabricius

Adults of these narrow-bodied species (Figure 13.2) run along leaf surfaces, flowers, and small branches of undergrowth bushes and trees to hunt for insect prey and seek mates, but they readily fly from bush to bush to escape danger. Most are dark bluish to black. The larvae make their tunnels in dead and decaying stems of bushes and trees. The more than 250 species in the genus occur throughout southeastern Asia from India to eastern Indonesia. Fifty-eight species are found in India, 34 of which are endemic (Naviaux 1994, 1995, 2003, 2004).

Genus *Therates* Latreille

Adults of this genus range in size from 5 to 23 mm in length (Figure 13.3). They vary in colour from all black to various shades of dark blue with orange or yellow patterns on the elytra. Adults have peculiarly reduced mouth parts except for an immense labrum that hides the closed mandibles. They forage on moist forest floors and quickly fly to leaves of undergrowth bushes to escape danger. The more than 70 species occur from Nepal to the Solomon Islands and Taiwan. Ten species occur in India, all in the northeastern quarter of the country, and all but one of them are endemic (Wiesner 1988, 1996, 2013).

Genus *Prothyma* Hope

This is an ambiguous genus whose tentative species members share only a few characters, such as a lack of white, hair-like setae on the dorsal surface (Figure 13.4). Most are active during the day on open ground and fly short distances to escape danger. However, they are regularly attracted to lights at night. They are found in a wide range of habitats from open savanna to forest floor. Species representing this genus are found from central Africa to Taiwan and the Philippines. As presently defined, there are 40 species in this genus. Only two species occur in India, neither of which is endemic.

Genus *Heptodonta* Hope

Most species of this genus have greenish-olive body colour with few to no hair-like setae on the ventral surface (Figure 13.5).

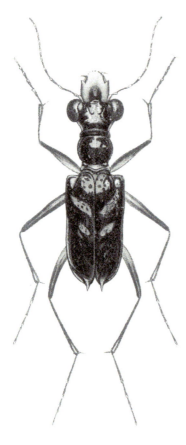

FIGURE 13.3 Genus *Therates*. (From Fowler, W.W., *Fauna of British India including Ceylon and Burma (Coleoptera General Introduction and Cicindelidae and Paussidae)*, Today and Tomorrow's Printers and Publishers, New Delhi, 529 pp, 1912.)

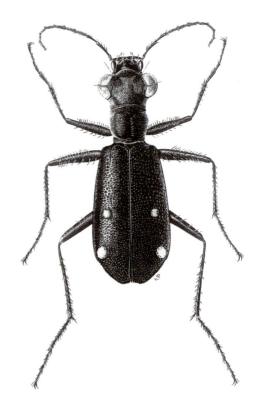

FIGURE 13.4 Genus *Prothyma*. (From Acciavatti, R. E., and Pearson, D. L., 1989. *Ann. Carnegie Mus.*, 58, 77–355, 1989. with permission from Carnegie Museum.)

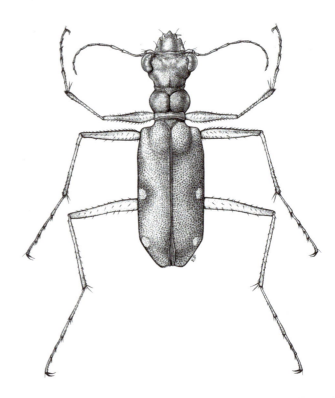

FIGURE 13.5 Genus *Heptodonta*. (From Acciavatti, R. E., and Pearson, D. L., 1989. *Ann. Carnegie Mus.*, 58, 77–355, 1989. with permission from Carnegie Museum.)

They have a characteristic labrum with seven pointed tooth-like projections, and the femora of the middle legs are distinctively expanded. Active during the day, they feed on the ground in moist open forests, but fly to leaves of undergrowth bushes to escape danger. The 11 species in this genus occur from northeastern India to the Philippines. Two species are found in India, one of which is endemic.

Genus *Lophyra* Motschulsky

The genus *Lophyra* is one of more than 50 groups of species that have been variously considered subgenera of *Cicindela* (*sensu lato*) or now more commonly as separate genera within the subtribe Cicindelina (Rivalier 1971). This is the largest subtribe in the family Cicindelidae and includes more than half of the presently known 2841 species. Large eyes, long, thin legs, long, thin antennae with 11 segments, and peculiar tunnelling larvae broadly characterize this entire subtribe. Each genus, such as *Lophyra* (Figure 13.6), has peculiar attributes including male aedeagus shape, hair-like setal patterns, and other subtle characteristics that distinguish them. More recently, molecular differences have proven useful in justifying the status of these genera. In general, they are diurnal on hot soil surfaces that range from water's edge to beaches, dunes, and grasslands. They fly short distances to escape predators. Most of the 80 species in

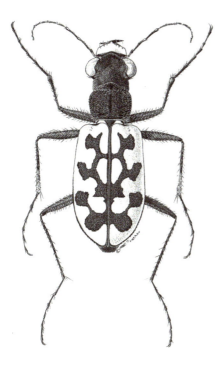

FIGURE 13.6 Genus *Lophyra*. (From Acciavatti, R. E., and Pearson, D. L., 1989. *Ann. Carnegie Mus.*, 58, 77–355, 1989. with permission from Carnegie Museum.)

the genus *Lophyra* are found throughout Africa, but some species are found as far west as central Europe and as far east as Indonesia. Eleven species have been found in India, two of which are endemic (Acciavatti and Pearson 1989).

Genus *Jansenia* Chaudoir

Jansenia was formerly deemed a subgenus within the *Cicindela*, but it is now considered a separate genus (Figure 13.7). Members of *Jansenia* are distinguished by a large and peculiarly tapered aedeagus with a long flagellum, long labrum, thickened segment of labial palpi, and head and pronotum without hair-like setae. Most species are found on the ground in open forest, but some are restricted to faces of large boulders and rocky cliff sides. The 39 species in this genus are confined to the Indian subcontinent. Thirty-two are endemic to India, one occurs only in Myanmar, four are endemic to Sri Lanka, and two Indian species are shared with Bhutan, Nepal, and Bangladesh (Acciavatti and Pearson 1989; Cassola and Werner 2003; Naviaux 2010). The origin of this unique genus is uncertain, but it may involve their ancestors carried on the ancient Indian tectonic plate as it moved across the Tethys Sea from Gondwana towards Laurasia millions of years ago (Pearson and Ghorpade 1989).

Genus *Apteroessa* Hope

The most enigmatic genus in India, *Apteroessa*, has not been recorded since its original discovery in South India in the early nineteenth century. Its stout legs, lack of flight wings, large body, unusually large head, and elytral surfaces covered with hair-like setae (Figure 13.8) provide few clues as to its

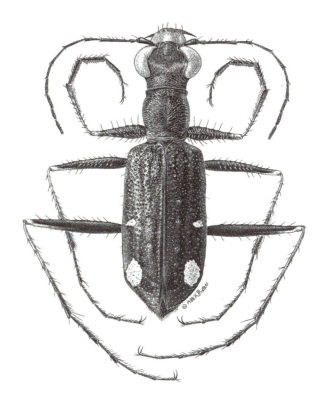

FIGURE 13.7 Genus *Jansenia*. (From Acciavatti, R. E., and Pearson, D. L., 1989. *Ann. Carnegie Mus.*, 58, 77–355, 1989. with permission from Carnegie Museum.)

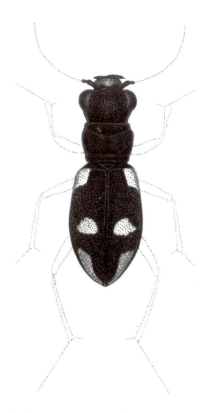

FIGURE 13.8 Genus *Apteroessa*. (From Fowler, W.W., *Fauna of British India including Ceylon and Burma (Coleoptera General Introduction and Cicindelidae and Paussidae)*, Today and Tomorrow's Printers and Publishers, New Delhi, 529 pp, 1912.)

phylogenetic position within the family. Numerous researchers have tried unsuccessfully to rediscover it in areas surrounding the type locality near the town of Ammayanayakanur, in Dindigul district, Tamil Nadu. There is only one species in this genus, *Apteroessa grossa* (Fabricius).

BIODIVERSITY AND SPECIES RICHNESS

The 1970's began a renewed interest in field collection and description of new tiger beetle species from India (Mandl and Wiesner 1975; Mandl 1977, 1981; Werner 1987; Acciavatti and Cassola 1989; Acciavatti and Pearson 1989; Wiesner 1996; Naviaux and Moravec 2001; Cassola and Werner 2003; Naviaux 2003, 2004, 2010; Werner and Wiesner 2008; Cassola 2009; Matalin and Anichtchenko 2012; Matalin 2013). In the last 13 years, 49 new species have been described from India. This rate of new species description is only exceeded by pioneer taxonomists in the 1800s and early 1900s (Pearson and Cassola 2005). Presently, India, with 241 described species of tiger beetles, has the third highest number in the world. Only Indonesia with 301 and Brazil with 257 are higher (Cassola and Pearson 2000; Jürgen Wiesner *pers. comm*).

Published monographs that compile unique characters and behaviour as well as revise nomenclature for entire subgroups of Indian tiger beetles have made identification and diversity patterns within India more accessible. The major genus *Cicindela* (*sensu lato*, subtribe Cicindelina) (Pajni and Bedi 1974; Acciavatti and Pearson 1989), the arboreal genera *Neocollyris*, *Collyris* (Naviaux 1994, 1995), and *Tricondyla* (Naviaux 2002), and the genus *Therates* (Wiesner 1988, 2013) have been detailed in such monographs and together provide taxonomic, distributional, and ecological details for 95% of all the tiger beetle species known from India.

Mathematical modelling is useful for filling in missing data to clarify and broaden incomplete spatial and temporal patterns of tiger beetle species richness (Carroll 1998). However, most statistical procedures, parametric and nonparametric, assume that data points are independent from each other, an often unrealistic assumption in the real world. To overcome this crucial problem, Indian tiger beetles have been tested with models and statistical procedures, such as geostatistics, that assume dependence rather than independence of data (Carroll and Pearson 2000). Exploiting these spatially dependent statistics, several questions of Indian biodiversity patterns, such as spatial scale and sample size, were elucidated with tiger beetles (Pearson and Carroll 1998; Carroll and Pearson 2006).

DISTRIBUTION PATTERNS WITHIN INDIA

Local studies that resulted in lists of tiger beetle species within India have been published for Maharashtra (Jadhavand Sharma 2012; Bharamal et al. 2014), Rajasthan (Kazmi and Ramamurthy 2004), Uttarakhand (Bhardwaj et al. 2008), Chandigarh (Pajni and Bedi 1973; Pajni et al. 1984; Kumar 1999), Siliguri-Darjeeling (Pearson and Ghorpade 1987), Coorg (Fletcher 1914), Kerala (Saha and Halder 1986), Meghalaya (Sawada and Wiesner 1997, 1999, 2006), Himachal Pradesh (Uniyal and Mathur 2000; Uniyal and Bhargav 2007), Tamil Nadu (Thanasingh and Ambrose 2011; Mohan and Padmanaban 2013), Orissa (Gravely 1912), Mizoram (Harit 2013), Kashmir (Mandl 1963), and Andhra Pradesh (Zoological Survey of India 2007). Other surveys have included larger regions of India (Wiesner 1975; Saha and Biswas 1985; Werner 1987; Werner and Wiesner 2008) or subgroups of tiger beetles and their biogeography within India (Wiesner 1996; Mawdsley 2010). Of the 241 tiger beetle species recorded for India, 119 (49%) are endemic to the country. Sixty-six (56%) of these endemic species belong to two genera, *Neocollyris* and *Jansenia* (Jürgen Wiesner *pers. comm.*) (Table 13.1).

Distinct clines in species richness across India reveal the highest diversity in the northeast and southwest (Figure 13.9) (Pearson and Ghorpade 1989; Pearson and Juliano 1993; Aravind et al. 2007). Using clues from rainfall patterns, habitat, mountain range locations, and presumptive regions of origin for subgroups of tiger beetles, biogeographic dispersal routes and speciation centres for tiger beetles were hypothesized to explain the present day clines of species richness across the subcontinent (Figure 13.10) (Reid 1988; Pearson and Ghorpade 1989; Pearson and Juliano 1993).

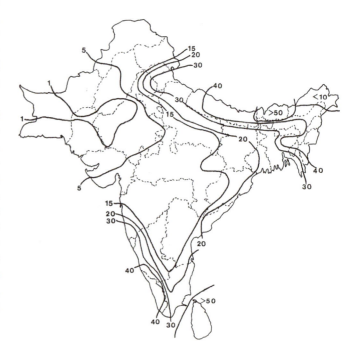

FIGURE 13.9 Isoclines connecting areas with similar numbers of tiger beetle species within a grid of squares (each 3° latitude and 3° longitude) on the Indian subcontinent. (From Pearson, D. L., and Cassola, F.: Worldwide species richness patterns of tiger beetles (Coleoptera: Cicindelidae): indicator taxon for biodiversity and conservation studies. *Conservation Biology*. 1992. 6. 376–391. Copyright Wiley-VCH Verlag GmbH & Co. KGaA. Reproduced with permission.)

Tiger Beetles (Coleoptera: Cicindelidae)

FIGURE 13.10 Theorized dispersal routes used by tiger beetles moving onto and within the Indian subcontinent. Purported ancestors dispersed from Africa (route 1), southwestern Asia (route 2), northern Asia (route 3), and southeastern Asia (route 4), within the subcontinent they moved via forested mountain chains (routes 5a and 5b), land bridges between Sri Lanka and the southern tip of India (route 6), and sandy ocean beaches (routes 7 and 8). (From Pearson, D. L., and Cassola, F.: Worldwide species richness patterns of tiger beetles (Coleoptera: Cicindelidae): indicator taxon for biodiversity and conservation studies. *Conservation Biology*. 1992. 6. 376–391. Copyright Wiley-VCH Verlag GmbH & Co. KGaA. Reproduced with permission.)

MOLECULAR CHARACTERIZATION AND PHYLOGENY

Theories of the phylogeny of tiger beetles started 250 years ago when Linnaeus described the first species. Subsequently, Fabricius, Dejean, Bates, and others began to organize the increasing numbers of described species into groupings and subgroupings of species. Because of the paucity of tiger beetle fossils (Wiesner et al. 2017), Walther Horn used his insight into adult morphological differences to develop a subjective system of classification. He assumed that these differences reflected evolutionary relationships and thus developed the first formal phylogeny of tiger beetles (Horn 1905c, 1915). The rapidly growing discoveries of new tiger beetles from India provided many intermediated morphological characters that reinforced his interpretation of relationships (Schaum1863; Atkinson 1889; Horn 1894, 1897, 1905a, 1905b, 1908, 1924, 1926, 1932; Fleutiaux and Maindron 1903; Maindron and Fleutiaux 1905; Annandale and Horn 1909; Fowler 1912; Fletcher 1914; Fleutiaux 1920; Dover and Ribeiro 1921, 1923; Heynes-Wood and Dover 1928).

The next major advance in tiger beetle phylogeny came from Rivalier (1971), who used the shape and structure of male genitalia as a unique system to group species of tiger beetles into genera and subtribes. His work also relied considerably on Indian species (Rivalier 1950, 1961).

Starting in the 1960s, Indian tiger beetles became instrumental in introducing chromosomes and genes into phylogenetic studies (Dasgupta 1967; Yadav and Karamjeet 1981; Yadav et al. 1985, 1989; Sharma1988; Mittal et al. 1989). Body colour patterns also have been used to test phylogenies that include Indian tiger beetles (Tsuji et al. 2016). Later, the complex array of various numbers of sex chromosomes and their patterns among tiger beetle species groups considerably advanced their phylogeny (Galián et al. 2002). Most recently, sophisticated molecular analysis of rDNA (Zacaro et al. 2004), congruence of mitochondrial and nuclear rDNA (Vogler and Pearson 1996), exon and intron sequences (Pons et al. 2004), and mitogens (López-López et al. 2015) have used Indian tiger beetle species to gain knowledge of a global phylogeny of tiger beetles.

"Mitochondrial genomes successfully resolve deep phylogenetic relationships. Cicindelidae are an independent lineage, separated from Carabidae. Geadephaga and Hydradephaga are reciprocally monophyletic groups. Long Branch lineages and 18S expansion segments distort the phylogenies. These results support a basal split of Geadephaga and Hydradephaga, and reveal Cicindelidae, together with Trachypachidae, as sister to all other Geadephaga, supporting their status as Family. Densely sampled mitogenomes, analysed with site heterogeneous mixture models, support a plausible hypothesis of basal relationships in the Adephaga" (López-López and Vogler 2017).

ADULT MORPHOLOGY

Adult tiger beetles can fly short distances to escape danger, but they spend most of their time on the ground, although a few tropical species patrol tree trunks and leaves. Adults of some species have been clocked running at 2.49 meters per second, making them one of the fastest running arthropods in the world (Kamoun and Hogenhout 1996). They typically use their long, thin legs to run in short, but fast spurts interspersed with brief stops. The stops are necessary because they literally run so fast that they cannot see their prey (Gilbert 1997). During their stops, they search for moving insects. If they see a potential prey item, such as an ant, small spider, or fly, the tiger beetle quickly turns in that direction and waits for another movement. The tiger beetle then runs the prey down and, if successful, grabs it with its long, thin, sickle-shaped mandibles. These mandibles are used to chew the prey into a puree. The beetle's mandibular glands near the base of each mandible release enzymes that begin the digestion process. The fluid flows from the gland to the mandibular tip and teeth via a groove. This chewing tobacco-like substance is also used in defence.

The hard armour-like skin (cuticle) that covers the adult tiger beetle is critical for survival, proving to be useful in identification, too. The outermost layer (epicuticle) has patterns of tiny pits, larger punctures, ridges, and undulations called microsculpture. The differences in these patterns of microsculpture are frequently used to distinguish tiger beetle species and genera. The cuticle is laminated with layers of melanin pigment and translucent waxes that alternately reflect and pass light. The distance between these alternating layers produces a broad range of metallic colours through reflectance and interference. The degree of uniformity of the cuticular reflector determines the purity of reflected colour. Highly sculptured and non-uniform reflectors produce a broad blend of colours of different wavelengths reflected at various angles from different locations. This type of integument gives rise to dull green or brown colours similar to those made by pigments in other insects. In bright iridescent species, the cuticular sublayers are more uniform, and the surface is relatively smooth. In all-black species, like those of the genera *Tricondyla* and *Derocrania*, the melanin is deposited in relatively thick and disorganized patterns that absorb most light. In other species, parts of the integument have no melanin deposited, and these areas are pale yellow or white (Schultz and Rankin 1985).

The most frequently studied anatomical features of adult tiger beetles involve the head (Figure 13.11) where the distinctive characters include long, thread-like (filiform) segmented antennae (colour, distribution of hair-like setae, relative overall length) used primarily as tactile sense organs; mandibles (relative length, number and position of "teeth") used for capturing and processing prey, and in males for grasping females during mating; upper lip or labrum (colour, length-width ratio, number and position of "teeth," number and position of setae) used with the mandibles to help grasp and process prey; and labium (presence and position of setae, relative length and colour of the segments of its finger-like palpi) and maxillae (relative length and colour of the segments of its finger-like palpi) used to manipulate and analyse the quality of food items. Other important parts of the head include: the compound eyes (degree of bulging, relative size); the complexity of surface microsculpturing and depth of grooves (rugae) between the eyes; and other parts of the head as well as the distribution of white, hair-like, or thick and flattened setae that may function as sense organs and/or insulation against hot surfaces.

Various adult eye sizes and shapes produce variable areas of stereoscopic (three-dimensional) vision. Nocturnal adults, such as those of *Grammognatha*, have small, relatively flat eyes compared to the bulbous eyes of genera active during daylight.

On the thorax (Figure 13.12), the most frequent distinctions are found in the proportions of the thorax (rectangular, square, or elongate), its shape as viewed from above (cylindrical, parallel, rounded, trapezoid), the texture (shiny, metallic, dull) and colour of the upper surface (pronotum), and the patterns or absence of setae on its side, lower, or upper surfaces.

In back of the pronotum (Figure 13.12), the hardened elytra (modified front wings) that cover the flight (hind) wings and top of the abdomen are probably the most commonly used identification characters. These elytra are spread and rotated forward in flight where they may function as airfoils, but do not flap. The elytral surface texture or microsculpturing can include large individual punctures (foveae), patterns of small pits (punctation), grooves (rugae), smooth (impunctate) areas or undulations and tiny, saw-like teeth (microserrations), and spines on the rear edge of the elytra. Also important are the shape of the elytra from above (parallel-sided, rounded, or oval), their profile as viewed from the side (domed, flattened), their dark background colour and texture (shiny metallic, dull), and the pattern, colour, and position of spots, lines, and stripes (maculations) or their absence (immaculate).

The flight wings are membranous with a distinctive framework of thickened ridges called veins. Modifications of tiny structures allow for a triple folding so that the flight wings can be stored completely under the elytra. In some flightless species, these flight wings are shrunk or even totally absent and accompanied by permanently fused elytra.

The prominent legs of tiger beetles are thin and long for running fast. Leg colour, positions of setae, and relative size (both to the overall size of the tiger beetle as well as to other segments on the same leg) are sometimes important as identifying characters. Males of most species have white pads of long, curved setae on the feet (tarsi) of the front legs.

The form and shape of the mating structures in tiger beetles are well known and have been used extensively in taxonomic comparisons (Freitag et al. 2001). The extensible penis (Figure 13.13) is located within a hardened (sclerotized) sheath called the aedeagus. The general shape of the aedeagus, the form of its tip, and the shape and position of

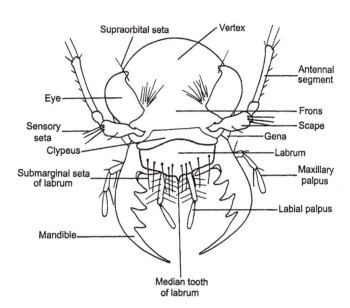

FIGURE 13.11 Dorsal head anatomy of an adult of the genus *Cicindela*. (From From Pearson, D. L., and Vogler, A. P., *Tiger Beetles: the Evolution, Ecology, and Diversity of the Cicindelids*. Ithaca, NY, Cornell University Press, 333 pp, 2001. With permission from Cornell University Press.)

Tiger Beetles (Coleoptera: Cicindelidae)

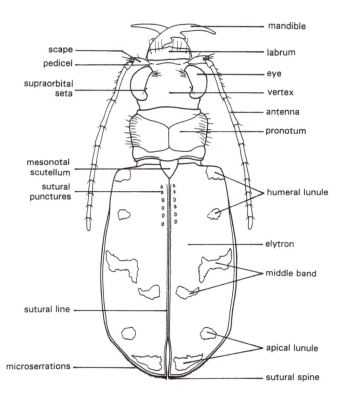

FIGURE 13.12 Dorsal aspect of an adult of the genus *Cicindela*. (From Pearson, D. L., and Vogler, A. P., *Tiger Beetles: The Evolution, Ecology, and Diversity of the Cicindelids*. Ithaca, NY, Cornell University Press, 333 pp, 2001. With permission from Cornell University Press.)

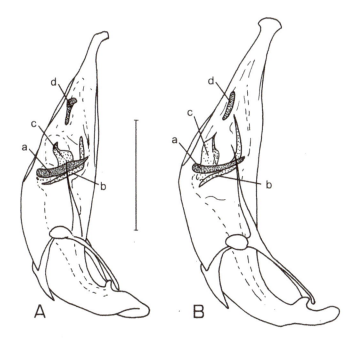

FIGURE 13.13 (A and B) Male genitalia of two closely related tiger beetle species in the genus *Cicindela* as an example of how subtle differences in internal elements and external shape can be used to distinguish species. a, arciform piece; b, stylet; c, shield; and d, stick. Scale bar = 1 mm. (From Pearson, D. L., and Vogler, A. P., *Tiger Beetles: the Evolution, Ecology, and Diversity of the Cicindelids*. Ithaca, NY, Cornell University Press, 333 pp, 2001. With permission from Cornell University Press.)

the sclerotized rings and internal elements are distinctive for many species and give clues not only to species identification, but also to phylogenetic relationships among species. When the aedeagus is not extended, it is maintained internally with the right side down. Upon extension, it rotates 90° clockwise. At the tip of the extended aedeagus, there is a less sclerotized area (internal sac). This internal sac is turned inside out during copulation to extend beyond the tip of the hard aedeagus, and it delivers sperm or packets of sperm (spermatophores) to the female.

The female mating structures also include some specialized characters. The eighth and ninth abdominal segments are modified to form a telescopic ovipositor, which is used to insert eggs, one at a time, into the substrate. Species differ considerably in the form and shape of the ovipositor, especially the terminal portion (gonapophysis). The differences in these characters have only begun to be studied, but offer the potential for many insights into phylogeny, ecology, and behaviour.

ADULT ECOLOGY, BEHAVIOUR, AND PHYSIOLOGY

Only a few ecological studies of Indian tiger beetles have been conducted to date (Pearson 1988). They have centred primarily on the role of competition for food (Ganeshaiah and Belavadi1986; Shivashankar and Veeresh 1987; Pearson and Juliano1991; Ganeshaiah et al. 1999; Satpathi 2000; Sinu et al. 2006; Dangalle and Pallewatt 2015), avoiding predation (Acorn 1988), and habitat restrictions limiting distribution (Bhargav et al. 2009; Edirisinghe et al. 2014). Behavioural studies include communal roosts (Bhargav and Uniyal 2008) and mating strategies in which the male grasps the sides of the female thorax with his mandibles and continues gripping long after copulation is completed. This behaviour, called mate guarding (Figure 13.14), precludes other males from fertilizing the female's eggs as the last sperm deposited are most likely to have contact with the eggs as they pass through to be oviposited (Shivashankar 1990; Shivashankar and Pearson 1994; Bhargav and Uniyal 2008). General reproductive success (Shivashankar and Veeresh 1987) and the impact of activity patterns (Bajpeyi et al. 1997) are other behavioural themes that have been studied among Indian tiger beetles.

Even fewer physiological studies have included Indian tiger beetles. They have ranged from the nervous system and tactile sensing (Ache and Dürr 2015), influence of extreme physical factors on activity (Rensch 1957; Sharma et al. 2015), heart beat rate (Soans and Soans 1968), and hearing (Spangler 1988; Yager et al. 2000).

LARVAL/PUPAL BIOLOGY

The larvae are unique among beetles. They all are designed for life in a narrow burrow in which they go through three instars. As a result, even though the adults may be nocturnal or diurnal, long and thin or short and wide, the larvae are white and grub-like (Figure 13.15), with much of the outer

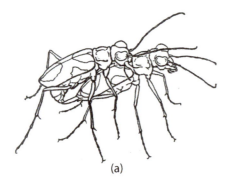

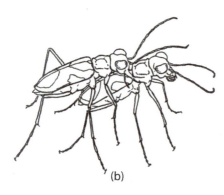

FIGURE 13.14 Schematic drawing of male and female *Cicindela aurofasciata* from peninsular India in coitus (a) and post copulatory amplexus (b), a type of mate guarding. (From Shivashankar, T., and Pearson, D. L.: A comparison of mate guarding among five syntopic tiger beetle species from peninsular India (Coleoptera: Cicindelidae). *Biotropica*. 1994. 26. 436–442. Copyright Wiley-VCH Verlag GmbH & Co. KGaA. Reproduced with permission.)

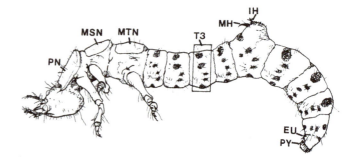

FIGURE 13.15 Lateral aspect of third instar larva of the genus *Cicindela* highlighting morphological characters used to differentiate species. EU, eusternum; IH, inner hook; MH, median hook; MSN, mesonotum; MTN, metanotum; PN, pronotum; PY, pygopod; and T3 third abdominal sclerites. (From Knisley, C. B., and D. L. Pearson. *T. Am. Entomol. Soc.*,110: 465–551, 1984. With permission from American Entomological Society.)

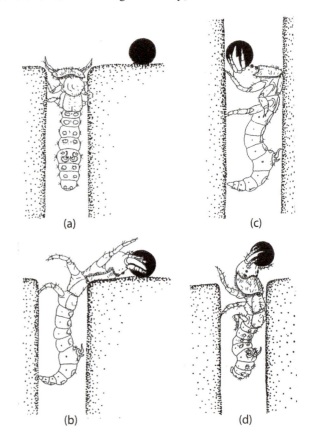

FIGURE 13.16 Larval tiger beetle foraging from its tunnel. (a) waiting at the soil surface entrance hole as prey approaches; (b) extending backward to capture prey as it enters the larva's zone of capture; (c) descending to the bottom of the tunnel to consume digestible parts of prey; and (d) throwing discarded indigestible remains of prey away from tunnel entrance (From Faasch, H. *Zool. Jahrb. Abt. Anat. Ontog. Tiere.*, 95, 477–522, 1968.)

covering of their bodies membranous. A dark-armoured capsule covers the head, and scattered dark plates are especially noticeable on the top of the thorax (pronotum). They have a large head with up to six small eyes on top and formidable mandibles underneath. A particularly striking and unique feature is on the larva's lower back, which includes a prominent hump with two pairs of large hooks that face forward.

The larvae, like adult tiger beetles, are predaceous, but unlike the adults, the larvae wait for prey to come to them. Each larva positions itself at the top of a long burrow with its head and thorax flush to the substrate surface and exactly filling the diameter of the burrow entrance (Figure 13.16). Larval burrows, depending on the species, can be on flat soil, vertical clay banks, forest leaf litter or, for a few tropical taxa, in rotted wood of branches and twigs. When a prey item approaches the burrow entrance closely, the larva extends its body, anchored by the back hooks into the side of the tunnel, and quickly reaches out backward to grab the prey in its powerful mandibles. The larva then pulls the struggling prey down into the depths of its burrow and dispatches it with a few mighty bites.

Unlike most other larval tiger beetles, two Indian species have evolved distinctive differences in their tunnel construction. Larval *Cicindela aurofasciata* construct a low, thick turret over the tunnel entrance above the ground surface. It is bent over to face its opening easterly into the early morning sun (Figure 13.17). This construction apparently enables the larva to quickly warmup on cool mornings, which in turn increases hunting time throughout the day. Increased food accelerates

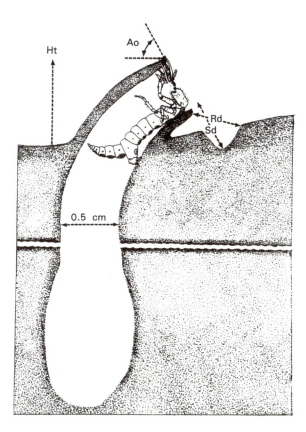

FIGURE 13.17 The curved larval tunnel of *Cicindela aurofasciata* on open forest floor of peninsular India. Ao, angle of tunnel opening from the horizontal; Ht, height of the larval turret; Rd, reaching distance for the larva attacking prey; and Sd, depth of the pit excavated by the larva. (From Shivashankar, T. et al., *Coleopt. Bull.*, 42, 63–68, 1988. With permission from Coleopterists Society.)

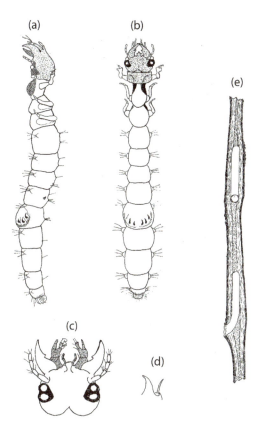

FIGURE 13.18 Larva of the arboreal genus *Neocollyris*. (a) lateral view; (b) dorsal view; (c) head; (d) detail of the one of the hooks on the 8th segment; and (e), excavated pith in the stem of a coffee bush occupied by the larva. (From Shelford, R., *Ecol. Entomol.*, 55: 83–90, 1907.)

growth into the pupal and adult stages (Shivashankar et al. 1988). The extraordinarily thin, elongated larvae of the arboreal genus, *Neocollyris*, burrow into the narrow stems of undergrowth bushes, including coffee plants on plantations. Here, they construct a lengthwise burrow into the pith of the branch and hunt aphids and other prey from the burrow's entrance (Shelford 1907) (Figure 13.18).

Because the head and thorax are usually the same colour and texture as the surrounding soil surface, most larvae are hard to see as they wait at the top of the burrows. The reaction to danger is to retreat immediately down into its tunnel and away from the mouth of their burrows. Often their presence is made obvious only when a black hole suddenly appears where before there was none.

Larval tiger beetles have fewer characters (Figure 13.15) that distinguish species than do the adults. Important distinctions among larvae are found in the shape and relative size of the inner and median hooks on the back of the fifth abdominal segment. The relative size, number, and placement of the simple eyes (stemmata), and the relative lengths of the segments of the short antennae are often useful taxonomically. Size and shape and the presence of ridges on the head and dorsal thoracic plates (nota), mouthparts, and terminal abdominal segment (pygopod) can distinguish taxa. Also, sometimes important are subtle differences in the number and position of hair-like setae on the pronotum and throughout the body (Shivashankar 1990; Putchkov 1994).

The eyes are the most studied organs of larval tiger beetles. The sedentary larvae have less difficulty detecting prey movement than do the mobile adults. Unlike the grub-like larvae of some insects, which only can achieve a coarse visual pattern with their simple eyes, the eyes of tiger beetle larvae have dense photoreceptors that permit detailed focusing and 3-dimensional perception (Toh and Okamura 2001).

The pupal stage of tiger beetles has been largely ignored. Species of only 14 genera worldwide have had this stage described. A recent article (Roza and Mermudes 2017) standardizes the morphological characters and summarizes the species, including some Indian forms that have a published description of the pupal stage.

INTEGRATIVE TAXONOMY

Integrative taxonomy is defined as "the science that aims to delimit the units of life's diversity from multiple and complementary perspectives (phylogeography, comparative morphology, population genetics, ecology, development, behaviour, etc.)" (Dayrat 2005).

A few preliminary studies of Indian tiger beetles have used integrative taxonomy, and one example compared the

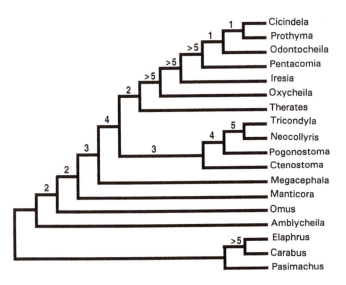

FIGURE 13.19 Alternative phylogeny of tiger beetles derived from integrated use of mitochondrial DNA and larval morphology with support indicated at each branching (Bremer calculations). The higher the number, the greater the support for that node. (From Vogler, A. P., and Barraclough, T. G., Reconstructing shifts in diversification rate during the radiation of Cicindelidae (Coleoptera), In *Phylogeny and Classification of Carboidea*, eds., G. E. Ball, A. Casale and A. Vigna Taglianti, 251–260, Florence, Italy, 1998. With permission from authors.)

phylogenies of Indian tiger beetles derived from molecular DNA and larval morphological characters (Figure 13.19) (Vogler and Barraclough 1998). Also, the occurrence of the defensive chemical, benzaldehyde, in some groups of tiger beetles was tested to see whether it was associated with shared habitats or with phylogenetic relationships (Pearson et al. 1988). Based on a preliminary phylogeny, the more closely related species tended to produce benzaldehyde regardless of habitat. Alternative interpretations agreed or strongly disagree with that interpretation (Altaba 1991). Eventually, when a much more sophisticated phylogeny became available that was based on congruence of mitochondrial and nuclear DNA (Vogler and Pearson 1996), a dataset that included measures of chemical defence, habitat association, and body colouration could be tested with more reliability (Vogler and Kelley 1998), and integrative taxonomy showed its usefulness in answering important questions about Indian tiger beetles.

TAXONOMIC PROBLEMS

The position of tiger beetles as alternatively a tribe, subtribe, or subfamily within the family of ground beetles, Carabidae, or as its own family has been a source of some controversy. With recent molecular analysis, the status as a family appears to be confirmed (López-López and Vogler 2017).

Although locally differentiated groups or subpopulations have an important role in fathoming the diversity of tiger beetles, we do not understand all the factors involved in separating lineages as well as the rate of separation (Pearson and Vogler 2001). For taxa that are extensively studied, these subpopulations are often given names as subspecies. But a taxonomic controversy revolves around the concept of subspecies, a problem not restricted to tiger beetles. More than 70 years ago, a commentary about the futility of subspecies taxonomy ignited a dispute that has continued to flare among taxonomists (Wilson and Brown 1953). Too many subspecies have been described for trivial or inconsistent reasons and with little effort to establish standards that some have argued that the taxonomic rank is arbitrary and ought to be discarded. Many tiger beetle species have been divided into often subjective subspecies as well, and this controversy continues to be a major one for those studying the family (Pearson et al. 2015). These often ambiguous taxonomic distinctions must be defined, and we should develop guidelines and quantitative standards for using genetic data to delimit subspecies (Patten and Remsen 2017). Many of the purported subspecies of tiger beetles, unfortunately, are far from that point.

CONSERVATION STATUS

Largely because of the cooperative efforts between passionate amateurs and a few dedicated professionals over the past two centuries, the taxonomy of tiger beetles is relatively stable, even for species in such remote parts of the world as Sulawesi, New Guinea, and Sudan (Pearson and Carroll 1998; Cassola and Pearson 2000; Pearson and Cassola 2007). These days, it is easier for inexperienced helpers and students to learn to reliably census tiger beetles than it is for them to learn to census most other taxa (Pearson and Cassola 2007).

And the work itself is faster: students of tiger beetles can quite easily census an area during the season of adult activity and reliably find most of the species within a short time, even in such complex and species-rich habitats as tropical forests (Pearson 2011). At one site, ornithologists took almost 5 years of intensive work to document 90 percent of the bird species occurring there, while in the same area, butterfly and dragonfly workers took 2 or 3 years to arrive at this level of knowledge for their respective taxa; those of us looking for tiger beetles found 90 percent of the fauna within the first 55 hours of searching. Their presence and level of abundance become quickly and reliably apparent (Pearson 1984; Fattorini and O'Grady 2013).

Around the world, they are among the few insect groups for which endangered species can be declared with certainty and placed on national red lists. In the United States, Israel, Bolivia, Spain, and Sweden populations and entire species of tiger beetles have officially been declared threatened, primarily due to habitat destruction. Those of us working in conservation cannot afford having to defend false claims of rarity. The reliability of accurately censusing tiger beetles minimizes questions of detectability that haunt conservationists who study many other taxa that are harder to observe and easier to miss. By protecting threatened populations of tiger beetles, we also secure habitats for many other species that need protection—an umbrella effect (Pearson and Cassola 1992).

We have strong evidence that, at large spatial scales, the species richness of tiger beetles is a good predictor of the species richness of other, harder-to-census taxa, such as butterflies and birds (Figure 13.20) (Pearson1994; Gerlach et al. 2013). And because the number of species in a given locale can be so quickly determined, we can census hundreds of hectares for tiger beetles in the time it would take to census one hectare for birds or butterflies. Tiger beetles thus make excellent bioindicators, and they have been used to monitor diversity in several poorly known areas of the world (Pearson and Cassola 1992; Carroll and Pearson 1998).

Bioindicators also play a role in the early detection of habitat degradation and monitoring restoration progress (McGeogh 2007; Langman et al. 2012). Because tiger beetle adults and larvae are so specialized in habitat use, they tend to be highly sensitive to minor changes. They can function as barometers of degradation that might imperil them and their habitats. Collections made long ago are valuable aids in comparing the historic distributions of tiger beetle species with their current geographic ranges; tiger beetle records accumulated over the last century and a half document habitat changes that might not otherwise have been obvious (Pearson and Carroll 1998; Dangalle et al. 2011).

ECONOMIC IMPORTANCE

In search of an economically viable alternative to expensive and often dangerous pesticides, several studies in India have investigated the effectiveness of predators such as tiger beetles as natural pest control agents in crops of cauliflower (Bhati and Srivastava 2016), sugarcane (Butani1960), tea (Das et al. 2010; Muraleedharan and Roy 2016), cereals (Kalaisekar 2003), and rice (Sharma et al. 2015; Mishra et al. 2017; Yadav et al. 2018) in kharif (Naikwadi et al. 2015) and multiple crops (Rahmanet al. 2017; Thanasingh and Ambrose 2011; Harit 2012). These studies, however, only assumed that the presence of predatory insects in these crop fields would have some impact on controlling pests (Rahman 1940). To date, few studies in India have attempted to measure how much the presence of predatory insects actually reduces pest populations. One of these (Sinu et al. 2006), attempted to measure how much tiger beetles control pests in rice paddies of the Western Ghats, but their results were ambiguous. A single study also tested the influence of pesticide residues on tiger beetles and other predatory insects (Agnihotrudu and Mithyanantha 1978).

COLLECTION AND PRESERVATION

Adult tiger beetles readily accommodate to the presence of observers who remain motionless or make very slow, smooth movements. Close-focusing binoculars can help in recording detailed behaviour. Hand nets can be used to capture, mark, and release individuals for further observation. Larval tiger beetles quickly accommodate to nearby observers, even to the point of accepting prey items offered them with forceps. Both adults and larvae will adjust to life in a covered terrarium with native soil, appropriate light, and moisture. Adult feeding, mating behaviour, and oviposition can be observed and tested easily. Larvae can be raised to test such factors as moisture, temperature, and food level influence on larval development, survival, and their fecundity as adults.

Pitfall traps are useful for measuring potential prey abundance, but are only of limited value in capturing adults. Diurnal adult tiger beetles readily avoid them. Nightlights, ultraviolet, or full spectrum, attract many, but not all species of adult tiger beetles, even diurnally active ones. Toy water pistols filled with soapy water can be effective for collecting difficult to capture species on open ground. Some species, especially arboreal forms in dense undergrowth, are best collected by beating the foliage with a stick into a portable sheet or inverted umbrella. Flight interruption traps, such as Malaise traps, are effective in confined habitats along streambeds, paths through forests, or at the base of cliffs.

Larvae make round holes in the soil surface that then open down into their vertical tunnels. Once you learn to recognize these distinctive holes, you can collect individual larvae by inserting a grass stem or other thin, flexible object down the length of the larval tunnel. Then dig carefully down the length of the inserted object until the larva is exposed. Alternatively,

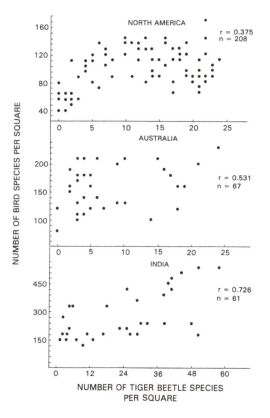

FIGURE 13.20 Correlation of the number of bird species and tiger beetle species in gridded squares across North America, India, and Australia. (From Pearson, D. L., and Cassola, F.: Worldwide species richness patterns of tiger beetles (Coleoptera: Cicindelidae): Indicator taxon for biodiversity and conservation studies. *Conservation Biology*. 1992. 6. 376–391. Copyright Wiley-VCH Verlag GmbH & Co. KGaA. Reproduced with permission.)

larvae will sometimes grasp the intruding grass stem inserted into the tunnel, and they can be carefully extracted as they cling to the object with their mandibles.

Adults can be pinned and dried with labels on insect pins or preserved in vials of 70% alcohol. Legs and heads of dried specimens are extremely fragile and easily damaged. Larvae can be placed into 10% formalin that has been heated to boiling temperature. Then they can be removed, soaked in water for 5 hours, and finally placed into vials of 70% alcohol for long-term preservation. Replicate data labels should be placed both inside the vial and affixed to the outside.

FUTURE DIRECTIONS

Occupancy Modelling and Presence/Absence Data

Assuming spatial dependency in statistical procedures such as geostatistics, we can fill in missing distribution data points along wide swaths of habitat. This technique can generate estimates of the quantitative spatial patterns of tiger beetle species across huge areas such as the Indian subcontinent. These patterns reveal areas of high and low species richness, which in turn can help determine priorities and boundaries for protected areas (Carroll and Pearson 2000). This approach, however, does not incorporate the differences in being unable to readily locate some species that are more cryptic than others, a challenge called detectability.

Occupancy models were developed to solve the problems created by imperfect detectability. Whether a population or species at a site is absent, rare, or just hard to find (detect) can be difficult to distinguish. For instance, few animals are so conspicuous that they are always detected at each survey, making it difficult to determine how long a species can go undetected before it should be considered extinct. This problem is not trivial and makes a great difference in determining many aspects of study including threatened status (Richoux 2001). With simple presence/absence data from repeated observations at each site, these occupancy models can estimate mean detectability.

Unlike most other parametric and non-parametric statistical methods elaborated for inferring the probability of extinction from a sighting record (Solow 2005), occupancy modelling does not assume equal sampling effort among observation periods, an assumption that is rarely met for informally gathered species lists. MacKenzie et al. (2006) proposed a model and maximum-likelihood-based method for estimating occupancy rates when the probability of detection is less than one. Initial occupancy, colonization, local extinction, and detection probability parameters then can be estimated based on repeated visits during multiple seasons or years.

These types of models have begun to be applied to tiger beetles as well (Hudgins et al. 2012). They not only resolve a persistent problem for conservation and ecological studies, but they make informally accumulated species observations and lists for a site extremely valuable and quantitatively rigorous. Thus, it is not necessary to rely solely on elaborate and formal data collection. Citizen scientists also can contribute their notes from as far back as they have them for a site. Occupancy modelling, with some filtering for obvious errors in identification, can then reveal reliable patterns for use in a wide range of studies, temporal and spatial.

Another use for a measure of detectability is in better defining and quantifying classification systems of endangered and threatened species. Presently many of these "Red Lists" include a strong subjective effort (Keith et al. 2015), especially when comparing a definition of endangered or threatened from country to country or across classes of organisms, for example, mollusks and birds. Mean detectability and its standard deviation can be calculated for a species regardless of its systematic position. If a global consensus could be reached for a mean detectability value that constitutes endangered or threatened, any population that meets this standard, regardless of taxon or locality, would be placed on a Red List. The status of this population is thus quantitatively consistent with any other population meeting this criterion.

Climate Change

Another area in which tiger beetles could be a significant player is as vehicles to better understand and anticipate global changes in climate. Evidence of long-term changes in the distribution of global weather patterns has become a major goal over the last few decades. The causes of climate change patterns can range from factors such as carbon cycles and variations in solar radiation received by Earth to plate tectonics, volcanic eruptions, and certain human activities.

Scientists actively work to understand past and future climate by using a wide range of experiments and observations to develop a climate record that extends well into the past. This record can then be used to predict future events and link causes and effects. Besides glacial movements, geological evidence, sea level changes, and past climate data, changes in faunal records and behaviour can be useful for this purpose.

Because of their hard exoskeletons, remains of beetles are common in geological sediments (Elias 2014). Combined with present day occurrence patterns, they can help reveal both long-term and short-term trends in climate change (Vickers and Buckland 2015). Each beetle species tends to be found in areas with specific climatic conditions. With the extensive phylogenetic knowledge of lineages of tiger beetles whose genetic make-up has not changed significantly over eons, knowledge of the present climatic range of the different species, and the age of the sediments in which remains are found, past climatic conditions can be inferred (Coope et al. 1998).

Evidence for shorter term changes in climate change can also be inferred from historical comparisons of changes in distribution, especially latitudinal and altitudinal changes. Because the recent historical records of tiger beetles are so extensive in many areas, scientists can attempt to distinguish the primary causes of these changes in distribution. Are they more likely associated with human destruction of habitat (Staines 2005; Kritsky et al. 2009; Karube 2010; Dangalle et al. 2011; MacRae and Brown 2011), or are they more likely "natural" range contractions or expansions associated with habitat movements caused by climate change (Krotzer 2013; Braud et al. 2016).

These kinds of data can be valuable for documenting ongoing changes and trends in climate change as well as helping determine the cause(s) of these changes. Because the biology, behaviour, physiology, distribution, and genetic make-up of so many tiger beetles are well studied around the world, future studies of climate change will likely involve these beetles more and more extensively (Russell 2014).

CITIZEN SCIENTISTS

Information on taxonomy, reliable identification, descriptions of behaviour, geographical distribution, and population trends are vital for basic studies and conservation efforts. Yet support for professional biologists to pursue these types of data has decreased considerably in the past few decades (Prathapan et al. 2009). Salaries and administrative support have shifted to areas considered more sophisticated, such as molecular genetics, mathematical modelling, and population regulation, all of which, however, still depend on basic natural history data (Pearson and Cassola 2012).

The British social critics, Leadbeater and Miller (2004), identified a rapidly growing involvement of amateurs in science from astronomy to medicine that is not fully recognized or utilized. These citizen scientists are a new breed of largely self-trained experts or professional amateurs (Pro-Ams), who, using modern technology, such as the Internet, fill in information gaps. Many of these citizen scientists are as skilled as professionals in basic descriptive stages of basic and conservation research, and they gather the data at their own cost in time and travel. They freely share results, specimens, and information with other amateurs and professionals by word of mouth, websites, and publications in amateur and professional journals (Pearson and Cassola 2007).

The majority of tiger beetle enthusiasts are among these citizen scientists, and without them, many cutting edge studies would be impossible (Russell 2014). Books on the biology of tiger beetles are part of a program to help educate amateurs as well as provide common ground to share ideas and pertinent data with professional colleagues (Pearson and Vogler 2001, 2017). Websites that introduce tiger beetles to youth and budding amateur enthusiasts at a level they can understand are also important for developing citizen science. Indian tiger beetles are highlighted in such a website sponsored by Arizona State University's Ask-A-Biologist (https://askabiologist.asu.edu/tiger-beetle-anatomy).

Even more importantly, however, are field identification guides for tiger beetles, which have been published for many countries, including Bolivia, Venezuela, Colombia, Thailand, and Madagascar (Pearson and Shetterly 2006) (Figure 13.21). Once these popular field guides are available, they attract more professionals and amateurs to go into the field in the region covered by the book and study the tiger beetles there in greater depth. Their studies then provide more detailed information and data on the species that need to be incorporated into the next and more sophisticated edition of the field guide. Thus, field guides often reflect the stage of development of the study of organisms and influence its activity directly. They also accelerate skills that in turn help basic and applied knowledge grow. Conservation policy and decision makers rely on this growing body of knowledge, and the line separating professionals and amateurs becomes blurred.

BIOMIMICRY

Biomimicry uses patterns and adaptations from Nature to innovate solutions for complex human problems and challenges. The goal is to create new products, processes, and policies that are sustainable and often profitable. Some of the best known examples include observations of bird flight and aerodynamics to develop aeroplanes and microscopic investigations of tiny hooks on burs to create Velcro.

From engineering and medicine to crowd control and nanoparticles, natural systems and species have provided the basis for a growing number of innovative products and systems, many of which we now take for granted. Tiger beetles have already been used to investigate several areas of biomimicry. They include studies of: their visual and tactile senses in hunting prey (Van Dooren and Matthysen 2004) to model human neural pathways (Ache and Dürr 2015); adaptations for avoiding obstacles at high running speeds (Zurek et al. 2014) to develop software that aid unmanned aerial vehicles (UAVs) in collision avoidance (Xin et al. 2015); the pointillistic reflective colouration of their exoskeleton surface (Schultz and Bernard 1989) to manufacture multicolour optical reflectors that are visible at different angles (Yabu et al. 2014) as well as metallic paints (Lenau and Barfoed 2008); and the streamlined body shape of sleek insects like tiger beetles (Zurek and Gilbert 2014; Ache and Dürr 2015) have been used to construct robots that can more efficiently cross densely cluttered terrain (Li et al. 2015).

Studies of tiger beetle adult (Layne et al. 2006; Haselsteiner et al. 2014) and larval (Toh et al. 2003) vision and eye characters, thermoregulation (Edirisinghe et al. 2014), activity patterns (Young 2015), and ability to survive flooding and lack of oxygen (Hoback et al. 2000) are only a few more examples of the potential these insects have in future biomimicry development.

FIGURE 13.21 Published tiger beetle field guides from various countries in North America, South America, Africa, and Asia. (From D.L. Pearson and J. A. Shetterly. *Am. Entomol.*, 52:246–252, 2006. With permission from Entomological Society of America.)

CONCLUSION

When my colleagues and I first published the *Field Guide to the Tiger Beetles of the United States and Canada* (Pearson et al. 2006), there were likely only a hundred or so tiger beetle aficionados in North America, most of them amateurs. Now, just a few years later, largely due to the popularity of the field guide, we could hardly keep up with the flood of new distribution records, natural-history observations, and innovative insights into the study and uses of tiger beetles that we received from hundreds more enthusiasts. A second edition was published to update all the new information (Pearson et al. 2015). Much the same phenomenon occurred following the publication of field guides to tiger beetles for Canada, Colombia, Bolivia, and Thailand.

With growing economies and middle class in China, India, and much of South America, the field guides and web sites that focus on tiger beetles attract a growing number of hobbyists who have the time and money to support their avocation. The future of species distributions, basic natural history, and insect conservation is more and more in the hands of these professional amateurs, whose contributions should help guide future policy decisions and budget planning by professional biologists, politicians, legislators, and policy makers. This passion for tiger beetles illuminates the ways in which insects and their admirers can advance basic science and conservation policy everywhere in our threatened world. One of the most effective ways to attract more attention to and involve additional professionals and citizen scientists in a group like tiger beetles would be to publish a *Field Guide to the Tiger Beetles of India* with colour plates, identification, behaviour notes, and distributional maps. Alternatively, or in addition, an interactive website dedicated to the tiger beetles of India could be made available (Assmann et al. 2018).

ACKNOWLEDGEMENTS

I want to thank Professor C. A. Viraktamath, Professor G. K. Veeresh, Professor H. R. Pajni, and Professor V. P. Uniyal for collaborating with me on research in Bengaluru, Chandigarh, and Dehradun. Many others shared time in the field pursuing tiger beetles, but A. Kumar and R. Sani (Panjab University); K. D. Ghorpade, A. R. V. Kumar, P. Mohanraj,

S. Subramanya, and T. Shivashankar (University of Agricultural Sciences, Bengaluru); and S. C. Goel (Sanatan Dharm College, Muzaffarnagar) stand out as the most diligent and persistent. R. L. Huber, C. B. Knisley, N. B. Pearson, and J. Wiesner critically reviewed early drafts of this article, and their suggestions for changes are greatly appreciated. My travel, local student support, and research in India were supported by funding from the Smithsonian Institution, Washington, DC, USA (PL-480).

REFERENCES

Acciavatti, R. E., and F. Cassola. 1989. A new species of tiger beetle from India (Coleoptera: Cicindelidae), (Studies on Cicindelids 56). *Annals of Carnegie Museum* 58: 71–76.

Acciavatti, R. E., and D. L. Pearson. 1989. The tiger beetles genus *Cicindela* (Coleoptera: Insecta) from the Indian subcontinent. *Annals of the Carnegie Museum* 58: 77–355.

Ache, J. M., and V. Dürr. 2015. A computational model of a descending mechanosensory pathway involved in active tactile sensing. *PLoS Computational Biology* 11: e1004263.

Acorn, J. H. 1988. Mimetic tiger beetles and the puzzle of cicindelid coloration (Coleoptera: Cicindelidae). *The Coleopterists' Bulletin* 42: 28–33.

Agnihotrudu, V., and M. S. Mithyanantha. 1978. *Pesticide Residues: A Review of Indian Work*, Bangalore: Rallis India Limited.

Altaba, C. R. 1991.The importance of ecological and historical factors in the production of benzaldehyde by tiger beetles. *Systematic Zoology* 40: 101–105.

Annandale, N., and W. Horn. 1909. Annotated list of the Asiatic beetles in the collection of the India Museum, Part 1, family Carabidae, subfamily Cicindelinae. *Calcutta*, 1–31, Tafel 1.

Aravind, N. A., B. Tambat, G. Ravikanth, K. N. Ganeshaiah, and R. Uma Shaanker. 2007. Patterns of species discovery in the Western Ghats, a megadiversity hot spot in India. *Journal of Biosciences* 32: 781–790.

Assmann, T., E. Boutaud, J. Buse et al. 2018.The tiger beetles (Coleoptera, Cicindelidae) of the southern Levant and adjacent territories: From cybertaxonomy to conservation biology. *Zookeys* 734.21989: 43–103. doi:10.3897/zookeys.

Atkinson, E. T. 1889. Catalogue of the Insecta of the Oriental region. No. 1. order Coleoptera, family Cicindelidae. *Journal of the Asiatic Society of Bengal, Part II. Natural Science Supplement*, No.1, 24 pp.

Bajpyei, C. M., N. S. Sen, and P. B. Sinha. 1997. Activity pattern and dispersal rate of some cursorial insects in Betla Reserved Forest. *Recent Advances in Ecobiological Research* 2: 462–467.

Bharamal, D. L., Y. J. Koli, and G. P. Bhawane. 2014. An inventory of the Coleopteran fauna of Sindhudurg district, Maharashtra, India. *International Journal of Current Microbiology and Applied Science* 3: 189–193.

Bhardwaj, M., V. K. Bhargav, and V. P. Uniyal. 2008. Occurrence of tiger beetles (Cicindelidae: Coleoptera) in Chilla Wildlife Sanctuary, Rajaji National Park, Uttarakhand. *Indian Forester* 134: 1636–1645.

Bhargav, V. K., and V. P. Uniyal. 2008. Communal roosting of tiger beetles (Cicindelidae: Coleoptera) in the Shivalik hills, Himachal Pradesh, India. *Cicindela* 40: 1–12.

Bhargav, V. K., V. P. Uniyal, and K. Sivakumar. 2009. Distinctive patterns in habitat association and distribution of tiger beetles in the Shivalik landscape of North Western India. *Journal of Insect Conservation* 13: 459–473.

Bhati, D., and M. Srivastava. 2016. A study on entomo-fauna as recorded from cauliflower crop in an agro-ecosystem near Bikaner, Rajasthan, India. *International Journal of Current Microbiology and Applied Sciences* 5: 539–545.

Braud, Y., P. Richoux, E. Sardet, J.-L.Hentz, and F. Rymarczyk. 2016. Actualisation des connaissances sur *Myriochila melancholica* (Fabricius, 1798) en France continentale. *Revue Assocation Roussill d'Entomologie* 25:18–22.

Butani, D. K. 1960. Parasites and predators recorded on sugarcane pests in India. *Indian Journal of Entomology* 20: 270–282.

Carroll, S. S. 1998. Modelling abiotic indicators when obtaining spatial predictions of species richness. *Environmental and Ecological Statistics* 5: 257–276.

Carroll, S. S., and D. L. Pearson. 1998. Spatial modeling of butterfly species richness using tiger beetles (Cicindelidae) as a bioindicator taxon. *Ecological Applications* 8: 531–543.

Carroll, S. S., and D. L. Pearson. 2000. Detecting and modeling spatial and temporal dependence in conservation biology. *Conservation Biology* 14: 1893–1897.

Carroll, S. S., and D. L. Pearson. 2006. The effects of scale and sample size on the accuracy of spatial predictions of tiger beetle (Cicindelidae) species richness. *Ecography* 21: 401–414.

Cassola, F. 2009. Studies of tiger beetles. CLXXXI. A new *Cylindera* (subgenus *Ifasina*) from Andhra Pradesh, Central India (Coleoptera: Cicindelidae). *Zeitschrift Arbeitsgemeinschaft Österreichischer Entomologen* 61: 15–18.

Cassola, F., and D. L. Pearson. 2000. Global patterns of tiger beetle species richness (Coleoptera: Cicindelidae): Their use in conservation planning. *Biological Conservation* 95: 197–208.

Cassola, F., and K. Werner. 2003. Two new *Jansenia* species from South India (Coleoptera: Cicindelidae). *Mitteilungen des InternationalenEntomologischen Vereins* 28: 77–92.

Coope, G. R., G. Lemdahl, J. J. Lowe, and A. Walkling. 1998. Temperature gradients in northern Europe during the last glacial-Holocene transition (14-9 14C kyr BP) interpreted from coleopteran assemblages. *Journal of Quaternary Science* 13: 419–433.

Dangalle, C. D., and N. Pallewatt. 2015. An invertebrate perspective to Hutchinson's ratio using co-occurring tiger beetle (Coleoptera: Cicindelidae) assemblages. *Taprobanica* 7: 224–234.

Dangalle, C. D., N. Pallewatt, and A. P. Vogler. 2011. The current occurrence, habitat and historical change in the distribution range of an endemic tiger beetle species *Cicindela* (*Ifasina*) *willeyi* Horn (Coleoptera: Cicindelidae) of Sri Lanka. *Journal of Threatened Taxa* 3: 1493–1505.

Das, S., S. Roy, and A. Mukhopadhyay. 2010. Diversity of arthropod natural enemies in the tea plantations of North Bengal with emphasis on their association with tea pests. *Current Science* 99: 1457–1463.

Dasgupta, J. 1967. Meiosis in male tiger beetle *Cicindela catena* Fabr. (Cicindelida: Coleoptera). *Science and Culture* 33: 491–493.

Dayrat, B. 2005.Towards integrative taxonomy. *Biological Journal of the Linnean Society* 85: 407–415.

Dover, C., and S. Ribeiro. 1923. A list of the Indian Cicindelidae with localities. *Records of the Indian Museum* 25: 345–363.

Dover, C., and S. Ribeiro. 1921. Records of some Indian Cicindelidae. *Records of the Indian Museum* 22: 721–727.

Edirisinghe, H. M., C. D. Dangalle, and K. Pulasinghe. 2014. Predicting the relationship between body size and habitat type of tiger beetles (Coleoptera, Cicindelidae) using artificial neural networks. *Journal on New Biological Reports* 3: 97–110.

Elias, S. A. 2014. Environmental interpretation of fossil insect assemblages from MIS 5 at Ziegler Reservoir, Snowmass Village, Colorado. *Quaternary Research* 82: 592–603.

Faasch, H. 1968. Beobachtungen zur Biologie und zum Verhalten von *Cicindela hybrida* L. und *Cicindela campestris* L. und experimentelle Analyse ihres Beutefangverhaltens. *Zoologische Jahrbücher: Abteilung für Systematik, Geographie und Biologie der Tier* 95: 477–522.

Fattorini, S., and P. O'Grady. 2013. Regional insect inventories require long time, extensive spatial sampling and good will. *PLoS ONE* 8(4):e62118.

Fletcher, T. B. 1914. Notes on tiger-beetles from Coorg. *Journal of Bombay Natural History Society* 23: 239.

Fleutiaux, E. D. 1920. Tableau pour la determination rapide des *Tricondyla* d'Indo-Chine. *Bulletin de la Société entomologique de France* 1920: 38.

Fleutiaux, E. D., and M. Maindron. 1903. Diagnose d'une espece nouvelle de *Cicindela* [Col]. *Bulletin de la Société entomologique de France* 1903: 1–72.

Fowler, W. W. 1912. *Fauna of British India including Ceylon and Burma (Coleoptera General Introduction and Cicindelidae and Paussidae).* Today and Tomorrow's Printers and Publishers, New Delhi, 529 pp.

Freitag, E., A. Hartwick, and A. Singh. 2001. Flagellar microstructures of male tiger beetles (Coleoptera: Cicindelidae): Implications for systematics and functional morphology. *The Canadian Entomologist* 133: 633–641.

Galián, J., J. E. Hogan, and A. P. Vogler. 2002. The origin of multiple sex chromosomes in tiger beetles. *Molecular Biology and Evolution* 19: 1792–1796.

Ganeshaiah, K. N., A. R. V. Kumar, and K. Chandrashekara. 1999. How much should the Hutchinson ratio be and why? *Oikos* 87: 201–203.

Ganeshaiah, K. N., and V. V. Belavadi. 1986. Habitat segregation in four species of adult tiger beetles (Coleoptera: Cicindelidae). *Ecological Entomology* 11: 147–154.

Gerlach, J., M. Samways, and J. Pryke. 2013. Terrestrial invertebrates as bioindicators: An overview of available taxonomic groups. *Journal of Insect Conservation* 17: 831–850.

Gilbert, C. 1997. Visual control of cursorial prey pursuit by tiger beetles (Cicindelidae). *Journal of Comparative Physiology A* 181: 217–230.

Gravely, F. H. 1912. The habits of some tiger-beetles from Orissa. *Records of the Indian Museum* 7: 207.

Harit, D. N. 2012. Tiger beetles associated with agriculture in Mizoram, North East India. *Annals of Plant Protection Sciences* 20: 344–347.

Harit, D. N. 2013. Tiger beetles (Coleoptera: Cicindelidae) along riverine habitat in Mizoram, north east India. *International Research Journal of Biological Sciences* 2: 30–34.

Haselsteiner, A. F., C. Gilbert, and Z. J. Wang. 2014. Tiger beetles pursue prey using a proportional control law with a delay of one half-stride. *Journal of the Royal Society of London—Interface* 11: 20140216.

Heynes-Wood, M., and C. Dover. 1928. *Catalogue of Indian Insects, Part 13–Cicindelidae.* Calcutta, West Bengal: Government of India Central Publications Branch, 138 pp.

Hoback, W. W., J. E. Podrabsky, L. G. Higley, D. W. Stanley, and S. C. Hand. 2000. Anoxia tolerance of con-familial tiger beetle larvae is associated with differences in energy flow and anaerobiosis. *Journal of Comparative Physiology B* 170: 307–314.

Horn, W. 1894. Beitrag zur Cicindeliden-Fauna von Vorder-Indien. *Deutsche Entomologische Zeitschrift* 2: 169–175.

Horn, W. 1897. Zwei neue Cicindeliden. *Entomologische Nachrichten* 23: 98–99.

Horn, W. 1905a. Eine neue *Derocrania* (*Tricondyla*) aus Vorder-Indien. *Deutsche Entomologische Zeitschrift* 1905: 152.

Horn, W. 1905b. Ein zweiter Beitrag zur Cicindelien-Fauna von Vorder-Indien (incl. Ceylon). *Deutsche Entomologische Zeitschrift* 1905: 59–64.

Horn, W. 1905c. Systematischer Index der Cicindeliden. *Deutsche Entomologische Zeitschrift* 2: 1–56.

Horn, W. 1908. Six new Cicindelinae from the oriental region. *Records of the Indian Museum* 2: 409–412.

Horn, W. 1915. Coleoptera Adephaga (Family Carabidae subfamily Cicindelinae). In *Genera Insectorum* (Fascicles 82 A-C), ed. P. Wytsman. Brussels, Belgium, 484 p.

Horn, W. 1924. On new and old oriental Cicindelidae. *Memoirs of the Department of Agriculture in India, Entomological Series* 8: 89–91.

Horn, W. 1926. Carabidae, Cicindelinae. In *Coleopterorum Catalogus*, ed. W. Schenkling, Pars 86: 1–345. Junk.

Horn, W. 1932. Eine neue Relikt-Form der Gattung *Cicindela* aus dem Süden von Vorder-Indien. *Stylops* 1: 81.

Hudgins, R. M., C. Norment, and M. D. Schlesinger. 2012. Assessing detectability for monitoring of rare species: A case study of the cobblestone tiger beetle (*Cicindela marginipennis* Dejean). *Journal of Insect Conservation* 16: 447–455.

Jadhav, S. S., and R. M. Sharma. 2012. Insecta: Coleoptera: Cicindelidae. *Zoological Survey of India, Fauna of Maharashtra, State Fauna Series* 20: 511–512.

Kalaisekar, A. 2003. *Biosystematic studies on the Coleoptera biodiversity in cereal crops of India.* (PhD thesis) New Delhi: Division of Entomology, Indian Agricultural Research Institute, 118 pp.

Kamoun, S., and S. A. Hogenhout. 1996. Flightlessness and rapid terrestrial locomotion in tiger beetles of the *Cicindela* L. Subgenus *Rivacindela* van Nidek from saline habitats of Australia (Coleoptera: Cicindelidae). *The Coleopterists' Bulletin* 50: 221–230.

Karube, H. 2010. Endemic insects in the Ogasawara Islands: Negative impacts of alien species and a potential mitigation strategy. In *Restoring the Oceanic Island Ecosystem Impact and management of Invasive Alien Species in the Bonin Islands*, eds. K. Kawakami and I. Okochi, 133–137. Japan: Springer.

Kazmi, S. I., and V. V. Ramamurthy. 2004. Coleoptera (Insecta) fauna from the Indian Thar Desert, Rajasthan. *Zoos Print Journal* 19: 1447–1448.

Keith, D. A., J. P. Rodríguez, T. M. Brooks et al. 2015. The IUCN Red List of ecosystems: Motivations, challenges, and applications. *Conservation Letters* 8: 214–226. doi:10.1111/conl.12167.

Knisley, C. B., and D. L. Pearson. 1984. Biosystematics of larval tiger beetles of the Sulphur Springs Valley, Arizona. *Transactions of the American Entomological Society* 110: 465–551.

Kritsky, G., B. Cortright, M. Duennes, and J. Smith. 2009. The status of *Cicindela marginipennis* (Coleoptera: Carabidae) in southeastern Indiana. *Proceedings of the Indiana Academy of Sciences* 118:139–142.

Krotzer, R. S. 2013. New records of *Cicindelidia ocellata rectilatera* (Chaudoir) and *Cicindela formosa* in the southeastern United States. *Cicindela* 45:1–7.

Kumar, A. 1999. *Bionomics of some Indian Tiger Beetles from North West regions Cicindelidae Coleoptera.* (MSc. Thesis) Chandigarh: Punjab University, 54 p.

Langman, P. O. C., J. A. Hale, C. D. Cormack, M. J. Risk, and S. P. Madon. 2012. Developing multimetric indices for monitoring ecological restoration progress in salt marshes. *Marine Pollution Bulletin* 64: 820–835.

Layne, J. E., P. W. Chen, and C. Gilbert. 2006. The role of target elevation in prey selection by tiger beetles (Carabidae: *Cicindela* spp.). *Journal of Experimental Biology* 209: 4295–4303.

Leadbeater, C., and P. Miller. 2004. *The Pro-Am Revolution: How Enthusiasts are Changing Our Society and Economy.* London, UK: Demos, 74 pp.

Lenau, T., and M. Barfoed. 2008. Colours and metallic sheen in beetle shells: A biomimetic search for material structuring principles causing light interference. *Advanced Engineering Materials* 10: 299–314.

Li, C., A. O. Pullin, D. W. Haldane, H. K. Lam, R. S. Fearing, and R. J. Full. 2015. Terradynamically streamlined shapes in animals and robots enhance traversability through densely cluttered terrain. *Bioinspiration and Biomimetics* 10:046003.

López-López, A., A. A. Aziz, and J. Galián. 2015. Molecular phylogeny and divergence time estimation of *Cosmodela* (Coleoptera: Carabidae: Cicindelinae) tiger beetle species from Southeast Asia. *Zoologica Scripta* 44: 437–445. doi:10.1111/zsc.12113.

López-López, A., and A. P. Vogler. 2017.The mitogenome phylogeny of Adephaga (Coleoptera). *Molecular Phylogenetics and Evolution* 114: 166–174.

MacKenzie, D. I., J. D. Nichols, J. A. Royle, K. H. Pollock, L. L. Bailey, and J. E. Hines. 2006. *Occupancy Estimation and Modeling: Inferring Patterns and Dynamics of Species Occurrence.* Boston, MA: Elsevier, 344 pp.

MacRae, T. C., and C. R. Brown. 2011. Historical and contemporary occurrence of *Cylindera* (s. str.) *celeripes* (LeConte) (Coleoptera: Carabidae: Cicindelinae) and implications for its conservation. *The Coleopterists' Bulletin* 65: 230–241.

Maindron, M., and E. D. Fleutiaux. 1905. Voyage de M. Maurice Maindron dans l'Inde meridionale (mai a novembre 1901), 6e memoire, Cicindelides. *Annales de la Société Entomologique de France* 74: 1–19.

Mandl, K. 1963. Zwei neue *Cicindela*-Formen aus dem vorderen Orient. *Mitteilungen der Münchner Entomologischen Gesellschaft* 53: 113–115.

Mandl, K. 1977. Drei neue Collyrini-Formen aus Süd-Indien (Coleoptera, Collyrini). *Zeitschrift der Arbeitsgemeinschaft Österreichischer Entomologen* 29: 113–116.

Mandl, K. 1981. Neun neue Formen aus der Familie Cicindelidae aus fünf Kontinenten (Col.). *Koleopterologische Rundschau* 55: 3–18.

Mandl, K., and J. Wiesner. 1975. Neue Cicindelidae-Formen aus Asien und Afrika: *Cicindela nathanae, Prothyma leprieuri reductesignata* und *Derocrania indica. Zeitschrift der Arbeitsgemeinschaft. Österreichischer Entomologen* 26: 2–4.

Matalin, A. V. 2013. New records of tiger beetles (Coleoptera: Cicindelidae) from different Asian regions. *Russian Entomological Journal* 22: 119–125.

Matalin, A. V., and A. V. Anichtchenko. 2012. New records of tiger beetles (Coleoptera, Cicindelidae) from India and Sri Lanka with description of a new subspecies. *Eurasian Entomological Journal* 11: 151–156.

Mawdsley, J. R. 2010. Cladistic analysis of *Cicindela* Linnaeus 1758, subgenus *Pancallia* Rivalier 1961, a lineage of tiger beetles from southern India (Coleoptera: Cicindelidae). *Tropical Zoology* 23: 195–203.

McGeogh, M. A. 2007.The selection, testing and application of terrestrial insects as bioindicators. *Biological Reviews* 73: 181–201.

Mishra, Y., A. K. Sharma, R. Pachori, and A. Kurmi. 2017. Taxonomic documentation of insect pest fauna of rice collected in light trap at Jabalpur district of Madhya Pradesh. *Journal of Entomology and Zoology Studies* 5: 1212–1218.

Mittal, O. P., T. K. Gill, and S. Chugh. 1989. Chromosome studies on three species of Indian cicindelids (Adephaga: Coleoptera). *Caryologia* 42: 115–120.

Mohan, K., and A. M. Padmanaban. 2013. Diversity and abundance of coleopteran insects in Bhavani Taluk Erode District, Tamil Nadu, India. *International Journal of Innovations in Bio-Sciences* 3: 57–63.

Muraleedharan, N., and S. Roy. 2016. Arthropod pests and natural enemy communities in tea ecosystems of India. In *Economic and Ecological Significance of Arthropods in Diversified Ecosystems*, eds. A. Chakravarthy and S. Sridhara, pp. 36–38. Singapore: Springer.

Naikwadi, B., S. M. Dadmal, and S. Javalage. 2015. Diversity study of predaceous insect fauna in major kharif crop agroecosystem in Akola, Maharashtra (India). *The Bioscan* 10: 1521–1524.

Naviaux, R. 1994. Les *Collyris* (Coleoptera, Cicindelidae): révison de genres et description de nouveaux taxons. *Société linnéenne de Lyon* 63: 106–290.

Naviaux, R. 1995. Les *Collyris* (Coleoptera, Cicindelidae). Révision des genres et description de nouveaux taxons. *Bulletin mensuel de la Société linnéenne de Lyon*, Separatum: 1–332; Periodicum: 63 (4): 106–116 (1994), 63 (5): 133–164 (1994), 63 (6): 185–216 (1994), 63 (7): 233–264 (1994), 63 (8): 273–304 (1994), 64 (1): 9–40 (1995), 64 (2): 57–88 (1995), 64 (3): 105–136 (1995), 64 (4): 153–184 (1995), 64 (5): 201–232 (1995), 64 (6): 259–290 (1995).

Naviaux, R. 2002. Les Tricondylina (Coleoptera, Cicindelidae): Révison de genres *Tricondyla* Latreille et *Derocrania* Chaudoir et descriptions de nouveaux taxons. *Annales de la Société entomologique de France* 5: 1–106.

Naviaux, R. 2003. Diagnoses de trois *Collyris* (*s. lato*) de l'Inde du Sud (Col., Cicindelidae). *Bulletin de la Société entomologique de France* 108: 404.

Naviaux, R. 2004. Les *Collyris* (Coleoptera, Cicindelidae). Complément à la "Révision du genre *Collyris* (sensu lato)" et description de nouveaux taxons. *Bulletin Mensuel de la Société linnéenne de Lyon* 73: 56–142.

Naviaux, R. 2010. *Jansenia biundata*, nouvelle espèce du sud de l'Inde (Coleoptera, Cicindelidae). *Bulletin de la Société entomologique de France* 115: 417–419.

Naviaux, R., and J. Moravec. 2001. *Derocrania dembickyi*, nouvelle especie de l'Inde du Sud (Coleoptera, Cicindelidae). *Bulletin de la Société entomologique de France* 106: 161–162.

Pajni H. R., and S. S. Bedi. 1973. Preliminary survey of the cicindelid fauna of Chandigarh, Punjab, India. *Cicindela* 5: 41–54.

Pajni H. R., and S. S. Bedi. 1974. Revision of the generic status of *Chaetodera albina* and *Chaetodera vigintiguttata* (Col: Cicindelidae). *Annales de la Societe Entomologique de France* 10: 939–941.

Pajni, H. R., A. Kumar, and D. L. Pearson. 1984. Corrections and additions to the tiger beetle fauna (Coleoptera: Cicindelidae) of the Chandigarh area of Northwest India. *Cicindela* 16: 21–38.

Patten, M. A., and J. V. Remsen, Jr. 2017. Complementary roles of phenotype and genotype in subspecies delimitation. *Journal of Heredity* 108: 462–464.

Pearson, D. L. 1984. The tiger beetles (Coleoptera: Cicindelidae) of the Tambopata Reserved Zone, Madre de Dios, Perú. *Revista peruana de Entomologia* 27: 15–24.

Pearson, D. L. 1988. Biology of tiger beetles Coleoptera: Cicindelidae. *Annual Review of Entomology* 33:123–147.

Pearson, D. L. 1994. Selecting indicator taxa for the quantitative assessment of biodiversity. *Philosophical Transactions of the Royal Society B: Biological Sciences* 345: 75–79.

Pearson, D. L., and J. A. Shetterly. 2006. How do published field guides influence interactions between amateurs and professionals in entomology? *American Entomologist* 52: 246–252.

Pearson, D. L., and A. P. Vogler. 2001. *Tiger Beetles: The Evolution, Ecology, and Diversity of the Cicindelids*. Ithaca, NY: Cornell University Press, 333 p.

Pearson, D. L., and A. P. Vogler. 2017. *Tiger Beetles: The Evolution, Ecology, and Diversity of the Cicindelids*, Japanese translation (M. Hori and A. Sato). Ithaca, NY: Cornell University Press, 267 p.

Pearson, D. L., and F. Cassola. 1992. Worldwide species richness patterns of tiger beetles (Coleoptera: Cicindelidae): Indicator taxon for biodiversity and conservation studies. *Conservation Biology* 6: 376–391.

Pearson, D. L., and F. Cassola. 2005. A quantitative analysis of species descriptions of tiger beetles (Coleoptera: Cicindelidae), from 1758 to 2004, and notes about related developments in biodiversity studies. *The Coleopterists Bulletin* 59: 184–193.

Pearson, D. L., and F. Cassola. 2007. Are we doomed to repeat history? A model of the past using tiger beetles (Coleoptera: Cicindelidae) and conservation biology to anticipate the future. *Journal of Insect Conservation* 11: 47–59.

Pearson, D. L., and F. Cassola. 2012. Insect conservation biology: What can we learn from ornithology and birding? In *Insect Conservation: Past, Present and Prospects*, ed. T. R. New, 377–399. New York: Springer.

Pearson, D. L., and K. D. Ghorpade. 1989. Geographical distribution and ecological history of tiger beetles (Coleoptera: Cicindelidae) of the Indian Subcontinent. *Journal of Biogeography* 16: 333–344.

Pearson, D. L., and S. A. Juliano. 1993. Evidence for the influence of historical processes in co-occurrence and diversity of tiger beetles. In *Species Diversity in Ecological Communities. Historical and Geographical Perspectives*, eds. R. E. Ricklefs and D. Schluter, 194–202. Chicago, IL: University of Chicago Press.

Pearson, D. L., and S. S. Carroll. 1998. Global patterns of species richness: Spatial models for conservation planning using bioindicator and precipitation data. *Conservation Biology* 12: 809–821.

Pearson, D. L., C. B. Knisley, and C. J. Kazilek. 2006. *A Field Guide to the Tiger Beetles of the United States and Canada: Identification, Natural History, and Distribution of the Cicindelidae*. New York: Oxford University Press, 227 pp.

Pearson, D. L., C. B. Knisley, D. P. Duran, and C. J. Kazilek. 2015. *A Field Guide to the Tiger Beetles of the United States and Canada: Identification, Natural History, and Distribution of the Cicindelinae*, 2nd edn. New York: Oxford University Press, 251 pp.

Pearson, D. L., M. S. Blum, T. H. Jones, H. M. Fales, E. Gonda, and B. R. White. 1988. Historical perspective and the interpretation of ecological patterns: Defensive compounds of tiger beetles (Coleoptera: Cicindelidae). *American Naturalist* 132: 404–416.

Pearson, D.L. 2011. Six-legged tigers. *Wings: Essays on Invertebrate Conservation* Spring: 19–23.

Pearson, D.L., and K. D. Ghorpade. 1987. Tiger beetles (Coleoptera: Cicindelidae) of the Siliguri-Darjeeling area in India. *Colemania* 4: 1–22.

Pearson, D.L., and S. A. Juliano. 1991. Mandible length ratios as a mechanism for co-occurrence: Evidence from a world-wide comparison of tiger beetle assemblages (Cicindelidae). *Oikos* 60: 223–233.

Pons, J., T. G. Barraclough, K. Theodorides, A. Cardoso, and A. P. Vogler. 2004. Using exon and intron sequences of the gene mp20 to resolve basal relationships in *Cicindela* (Coleoptera: Cicindelidae). *Systematic Biology* 53: 554–570.

Prathapan, K. D., P. D. Rajan, and J. Poorani. 2009. Protectionism and natural history research in India. *Current Science* 97: 1411–1413.

Putchkov, A. V. 1994. State-of-the-art and world perspectives of studies on tiger beetle larvae (Coleoptera, Carabidae, Cicindelinae). In *Carabid Beetles: Ecology and Evolution*, eds. K. Desender, M. Dufrêne, M. Loreau, M. L. Luff, and J. P. Maelfait, pp. 51–54. Series Entomologica, vol. 51. Dordrecht, the Netherlands: Springer.

Rahman, A., M. Bathari, P. Borah, R. R. Taye, and P. Patgiri. 2017. Diversity of insect species along with their host in Assam Agricultural University, Jorhat. *Journal of Entomology and Zoology Studies* 5: 2307–2312.

Rahman, K. A. 1940. Important insect predators of India. *Proceedings of the Indian Academy of Sciences—Section B* 12: 67.

Reid, W. V. 1998. Biodiversity hotspots. *Trends in Ecology and Evolution* 13: 275–280.

Rensch, B. 1957. Aktivitätsphasen von *Cicindela*-Arten in klimatisch stark unterschiedenen Gebieten. *Zoologische Anzeiger* 158: 33–38.

Richoux, P. 2001. Sensibilité de *Cylindera arenaria* aux aménagements fluviaux: l'exemple de la region lyonnaise (Coléoptères Cicindelidae). *Archives du Muséum d'histoire naturelle de Lyon* 2: 63–74.

Rivalier, E. 1950. Démembrement du genre *Cicindela* Linne (Travail préliminaire limité á la faune palearctique). *Revue Française d'Entomologie* 17: 217–244.

Rivalier, E. 1961. Démembrement du genre *Cicindela* L. (suite). IV. Faune indomalaise. *Revue française d'Entomologie* 28: 121–149.

Rivalier, E. 1971. Remarques sur la tribu de Cicindelini (Col. Cicindelidae) et sa subdivision en soustribus. *Nouvelle Revue d'Entomologie* 1: 135–143.

Roza, A. S., and J. R. M. Mermudes. 2017. Tiger beetles' (Coleoptera: Carabidae, Cicindelinae) pupal stage: Current state of knowledge and future perspectives. *Zootaxa* 4226: 348–358.

Russell, S. A. 2014. *Diary of a Citizen Scientist: Chasing Tiger Beetles and Other New Ways of Engaging the World*. Corvallis, OR: Oregon State University Press, 222 pp.

Saha, S. K., and S. Biswas. 1985. Insecta: Coleoptera, Carabidae and Cicindelidae (Part I). *Records of the Zoological Survey of India* 82: 117–127.

Saha, S. K., and S. K. Halder. 1986. Tiger beetles (Coleoptera, Cicindelidae) of Silent Valley (Kerala, India). *Records of the Zoological Survey of India* 84: 1–4.

Satpathi, C. R. 2000. Preys of the carabid beetle *Anthia sexguttata* and tiger beetle, *Cicindela* sp. *Insect Environment* 6: 1–16.

Sawada, H., and J. Wiesner. 1997. Beitrag zur Kenntnis der Cicindelidae von Meghalaya (Nordost-Indien) und Bemerkungen zu *Therates* (Coleoptera), 46. Beitrag zur Kenntnis der Cicindelidae. *Entomologische Zeitschrift* 107: 73–86.

Sawada, H., and J. Wiesner. 1999. Records of tiger beetles collected in North India (Coleoptera: Cicindelidae). *Entomological Review of Japan* 54: 189–195.

Sawada, H., and J. Wiesner. 2006. Records of tiger beetles collected in North India II (Coleoptera: Cicindelidae). 97. Contribution towards the knowledge of Cicindelidae. *Entomologische Zeitschrift* 116: 127–134.

Schaum, H. R. 1863. Contributions to the knowledge of the Cicindelidae of tropical Asia, containing descriptions of new species, a list of those hitherto described, and synonymical notes. *The Journal of Entomology, Descriptive and Geographical, London* 2: 57–73.

Schultz, T. D., and G. D. Bernard. 1989. Pointillistic mixing of interference colors in cryptic tiger beetles. *Nature* 337: 72–73.

Schultz, T. D., and M. A. Rankin.1985. The ultrastructure of the epicuticular interference reflectors of tiger beetles (*Cicindela*). *Journal of Experimental Biology* 117: 87–110.

Sharma, A. K., Y. Muchhala, and R. Pachori. 2015. Impact of abiotic factors on population dynamics of major predatory and parasitic fauna of paddy. *The Ecoscan* 9: 597–600.

Sharma, P. C. 1988. Karyomorphology of three arboreal species of Indian tiger beetles (Coleoptera: Cicindelidae: Collyrinae) from Himachal Pradesh. *International Symposium on Recent Adventures in Cytogenetic Research*, Kurushetra, India, p. 9.

Shelford, R. 1902. Observations on some mimetic insects and spiders from Borneo and Singapore. *Proceedings of the Zoological Society of London* 1902: 230–284.

Shelford, R. 1907. The larva of *Collyris emarginatus*, Dej. *Ecological Entomology* 55: 83–90.

Shivashankar, T. 1990. *Ecology, larval taxonomy and adult behavior of tiger beetles (Coleoptera: Cicindelidae) of peninsular India.* (PhD Thesis) Bangalore,India: Department of Entomology, University of Agricultural Sciences, 123 p.

Shivashankar, T., A. R. V. Kumar, G. K. Veeresh, and D. L. Pearson. 1988. Angular turret building behaviour in a larval tiger beetle from South India (Coleoptera: Cicindelidae). *The Coleopterists' Bulletin* 42: 63–68.

Shivashankar, T., and D. L. Pearson. 1994. A comparison of mate guarding among five syntopic tiger beetle species from peninsular India (Coleoptera: Cicindelidae). *Biotropica* 26: 436–442.

Shivashankar, T., and G. K. Veeresh. 1987. Impact of differential feeding on the reproduction of tiger beetle *Cicindela cancellata* DeJean (Cicindelidae: Coleoptera). *Proceedings Indian Academy of Sciences (Animal Science)* 96: 317–321.

Sinu, P. A., M. Nasser, and P. D. Rajan. 2006. Feeding fauna and foraging habits of tiger beetles found in agro-ecosystems in Western Ghats, India. *Biotropica* 38: 500–507.

Soans, A. B., and J. S. Soans. 1968. Observations on the rate of heart-beat in the various stages of the tiger beetle, *Cicindela cancellata* Dej. *Current Science* 37: 108.

Solow, A. R. 2005. Inferring extinction from a sighting record. *Mathematical Biosciences* 195: 47–55.

Spangler, H. G. 1988. Hearing in tiger beetles (Cicindelidae). *Physiological Entomology* 13: 447–452.

Staines, C. L. 2005. *Cicindela hirticollis* Say (Coleoptera: Cicindelidae) naturally colonizing a restored beach in the Chesapeake Bay, Maryland. *Cicindela* 37: 79–80.

Thanasingh, P. D., and D. P. Ambrose. 2011. Biodiversity and distribution of entomofauna in three ecosystems in Thoothukudi District, Tamil Nadu. In *Insect Pest Management, A Current Scenario*, ed. D. P. Ambrose, 38–57. Palayamkottai, India: Entomology Research Unit, St. Xavier's College, India.

Toh, Y., and J. Y. Okamura. 2001. Behavioural responses of the tiger beetle larva to moving objects: Role of binocular and monocular vision. *Journal of Experimental Biology* 204: 615–625.

Toh, Y., J. Y. Okamura, and Y. Takeda. 2003. The neural basis of early vision distance and size estimation in the tiger beetle larva: Behavioral, morphological, and electrophysiological approaches. In *The Neural Basis of Early Vision, Volume 11 Keio University International Symposia for Life Sciences and Medicine*, ed. A. Kaneko, pp. 80–85. Tokyo, Japan: Springer.

Tsuji, K., M. Hori, M. H. Phyu, H. Liang, and T. Sota. 2016. Colorful patterns indicate common ancestry in diverged tiger beetle taxa: Molecular phylogeny, biogeography, and evolution of elytral coloration of the genus *Cicindela* subgenus *Sophiodela* and its allies. *Molecular Phylogenetics and Evolution* 95: 1–10.

Uniyal, V. P., and P. K. Mathur. 2000. Altitudinal distribution of tiger beetles (Cicindelidae: Coleoptera) in Great Himalayan National Park Conservation Area, Western Himalaya. *Indian Forester* 126: 1141–1143.

Uniyal, V. P., and V. Bhargav. 2007. *Tiger Beetles: A Field Study in the Shivaliks of Himachal Pradesh*. Dehradun, India: Wildlife Institute of India. http://oldwww.wii.gov.in/faculty/vpuniyal/tiger_beetle.

Van Dooren, T. J. M., and E. Matthysen. 2004. Generalized linear models for means and variances applied to movement of tiger beetles along corridor roads. *Journal of Animal Ecology* 73: 261–271.

Vickers, K., and P. I. Buckland. 2015. Predicting island beetle faunas by their climate ranges: The tabula rasa/refugia theory in the North Atlantic. *Journal of Biogeography* 42: 2031–2048.

Vogler, A. P., and D. L. Pearson. 1996. A molecular phylogeny of the tiger beetles (Cicindelidae): Congruence of mitochondrial and nuclear rDNA data sets. *Molecular Phylogenetics and Evolution* 6: 321–338.

Vogler, A. P., and K. C. Kelley. 1998. Covariation of defensive traits in tiger beetles (genus *Cicindela*): A phylogenetic approach using mtDNA. *Evolution* 52: 529–538.

Vogler, A. P., and T. G. Barraclough.1998. Reconstructing shifts in diversification rate during the radiation of Cicindelidae (Coleoptera). In *Phylogeny and Classification of Carboidea*, eds. G. E. Ball, A. Casale, and A. Vigna Taglianti, 251–260, Florence, Italy: XXth International Congress of Entomology (1996).

Werner, K. 1987. Collecting Cicindelidae in West Africa and India (Coleoptera). *Young Entomologists Society Quarterly* 4: 25–30.

Werner, K., and J. Wiesner. 2008. New records of tiger beetles from India with description of new taxa (Coleoptera: Cicindelidae). *Entomologische Zeitschrift* 118: 15–18.

Wiesner, J. 1975. Notes on Cicindelidae of India and Sri Lanka. *Cicindela* 7: 61–70.

Wiesner, J. 1988. Die Gattung *Therates* Latr. und Ihre Arten (15. Beitrag zur Kenntnis der Cicindelidae, Coleoptera). *Mitteilungen der Münchner Entomologischen Gesellschaft* 78: 5–107.

Wiesner, J. 1992. *Checklist of the Tiger Beetles of the World (Coleoptera: Cicindelidae)*. Keltern, Germany: Verlag Erna Bauer, 364 p.

Wiesner, J. 1996. Neues über *Therates* aus Indien und Vietnam (Coleoptera. Cicindelidae) (44. Beitrag zur Kenntnis der Cicindelidae). *Entomologische Zeitschrift* 106: 504–508.

Wiesner, J. 2013. The *chennelli* group of the Genus *Therates* Latreille (Coleoptera: Cicindelidae) 114. Contribution towards the knowledge of Cicindelidae. *Insecta Mundi* 315: 1–86.

Wiesner, J., K. Will, and J. Schmidt. 2017. Two new genera and species of tiger beetles from Baltic amber (Coleoptera: Carabidae: Cicindelinae). *Insecta Mundi* 577: 1–14.

Wilson, E., and W. Brown. 1953. The subspecies concept and its taxonomic application. *Systematic Zoology* 2: 97–111.

Xin, L., W. Xiangke, L. Jie, Z. Guozhong, and S. Lincheng. 2015. Expansion rate based collision avoidance for unmanned aerial vehicles. *Proceedings of the 34th Chinese Control Conference (CCC)*, pp. 8393–8398.

Yabu, H., Y. Saito, and M. Shimomura. 2014. Unique light reflectors that mimic the structural colors of tiger beetles. *Polymer Journal* 46: 212–215.

Yadav J. S., K. Kondal, and A. S. Yadav. 1985. Cytology of *Cicindela* (*Myriochile*) *undulata* and *C.* (*M.*) *fastidiosa* with a summary of chromosomal data on the Cicindelidae. *Cicindela* 17:1–11.

Yadav, J. S., and K. Karamjeet. 1981. Chromosome studies on three species of Cicindelidae (Adephaga: Coleoptera) from Haryana. *Zoologischer Anzeiger* 206: 121–128.

Yadav, J. S., M. R. Burra, and M. P. Dange. 1989. Chromosome number and sex-determining mechanism in 32 species of Indian Coleoptera (Insecta). *National Academy of Sciences and Letters* 12: 93–97.

Yadav, M., R. Prasad, P. Kumari et al. 2018. Potential and prospects of natural enemies in rice ecosystem in Jharkhand. *International Journal of Current Microbiology and Applied Sciences* 7: 3389–3396.

Yager, D. D, A. P. Cook, D. L. Pearson, and H. G. Spangler. 2000. A comparative study of ultrasound-triggered behaviour in tiger beetles (Cicindelidae). *Journal of Zoology* 251: 355–368.

Young, O. P. 2015. Activity patterns, associated environmental conditions, and mortality of the larvae of *Tetracha* (=*Megacephala*) *carolina* (Coleoptera: Cicindelidae). *Annals Entomological Society of America* 108: 130–136.

Zacaro, A. A., S. J. R. Proença, C. Lopes-Andrade, and A. R. M. Serrano. 2004. Cytogenetic analysis of Ctenostomini by C-banding and rDNA localization and its relevance to the knowledge of the evolution of tiger beetles (Coleoptera: Cicindelidae). *Genetica* 122: 261–268.

Zoological Survey of India (ed.). 2007. *Fauna of Andhra Pradesh*. Part 3: Insects. Kolkata, iiv + 544 pp.

Zurek D. B., and C. Gilbert. 2014. Static antennae act as locomotory guides that compensate for visual motion blur in a diurnal, keen-eyed predator. *Proceedings of the Royal Society of London B* 281: 20133072.

Zurek, D. B., Q. Perkins, and C. Gilbert. 2014. Dynamic visual cues induce jaw opening and closing by tiger beetles during pursuit of prey. *Biology Letters* 10: 20140760.

14 Coccinellidae of the Indian Subcontinent

J. Poorani

CONTENTS

Introduction 223
Family Diagnosis and Key Characters 224
    Key Identification Characters 227
    Subfamilies and Tribes 227
Major Work on Indian Fauna 228
Biodiversity and Species Richness 229
Distribution Patterns in the Indian Context 229
Phylogeny of Coccinellidae 229
Taxonomic Problems 230
Biology and Ecology 230
    Habitat 230
    Life Cycle and Voltinism 230
    Searching and Feeding Behaviour 230
    Cannibalism 231
    Natural Enemies 231
        Predators 231
        Parasites 232
        Pathogens 232
    Defence Mechanisms 232
    Polymorphism/Melanism 233
    Hibernation and Mass Assemblages 233
    Prey Associations 234
Immature Taxonomy 234
    Egg 235
    Larva 235
    Pupa 236
Economic Importance 237
    Introductions 238
    Introductions of Coccinellids into India 238
    Augmentation 238
    Conservation 238
    Non-target Effects of Exotic Coccinellids 239
Collection and Preservation 239
Conclusion 240
References 240

INTRODUCTION

Ladybird beetles, also known as ladybugs and lady beetles, are a hugely popular group of insects. They have been revered through the centuries in several European countries, for the ladybird is usually dedicated to the Virgin Mary and thought to have supernatural powers. Popular interest in the ladybird goes back at least to the fifteenth century and probably much farther (Gordon 1985). Aspects of their biology have been incorporated into folklore and cultures of several Western European countries. In many countries, ladybirds are considered as signs of good luck and prosperity. The Republic of Latvia and at least six states of the United States of America have adopted ladybirds as their national insects. Most of the species are beneficial, both as adults and larvae, as predators of crop pests such as aphids, mealybugs, scales, mites, and others. Ladybirds are linked to biological control more often than any other group of predators. The beneficial status of ladybirds has a rich history that is well known and recognized by the public and biological control practitioners alike (Gordon 1985; Obrycki and Kring 1998). The vedalia beetle, *Rodolia cardinalis* (Mulsant), is the most famous introduced beneficial insect in the history of classical biological control. The present status of Coccinellidae of the Indian subcontinent is briefly reviewed here.

FAMILY DIAGNOSIS AND KEY CHARACTERS

The family Coccinellidae was included under the superfamily Cucujoidea as part of the Cerylonid Series of families, a diverse assemblage of highly problematic beetle families, until recently. Robertson et al. (2015) carried out a large-scale phylogenetic study which has drastically changed the phylogenetic framework and higher classification of Cucujoidea. In this study, the Cerylonid Series of Cucujoidea was found to be distinctly monophyletic and only distantly related to the rest of the cucujoid families and hence formally recognized as a separate superfamily Coccinelloidea. "In terms of taxonomic constitution, Coccinelloidea is synonymous with the concept of the erstwhile Cerylonid Series" and accommodates 15 families, with Coccinellidae being the largest (Robertson et al. 2015).

Its closest relatives in the same group were believed to be Corylophidae and Endomychidae (Crowson, 1955; Sasaji 1971) or Alexiidae (=Sphaerosomatidae) and Endomychidae (Ślipiński and Pakaluk 1991). Within the Cerylonid Series, Endomychidae has always been closely associated with Coccinellidae; in fact, some endomychids were originally described as coccinellids and vice versa. Robertson et al. (2015) in their molecular phylogenetic study recognized "two well-supported superfamilial coccinelloid clades, one clade comprising Bothrideridae, Cerylonidae, and Discolomatidae (the bothriderid group) and the second clade including Alexiidae, Akalyptoischiidae, Corylophidae, Coccinellidae, Latridiidae, and multiple endomychid lineages" (the coccinellid group).

Lady beetles vary in size from minute (0.7–1.0 mm long) to large (25–28 mm long). The body outline is usually round or short oval to distinctly elongate. The dorsal surface is weakly to very strongly convex, glabrous and shiny, or with sparse to dense pubescence. The dorsum has fine to coarse punctures, but never has distinct striations and sculptures. The interspaces between punctures are smooth, matt-like, or with microsculpture, which can be observed properly only under a microscope at high magnification. Many species have bright red, orange, or yellow elytra with distinctive spots, stripes, and other markings/patterns (Figure 14.1). A few species are brightly metallic blue, green, or violet. Lady beetles of the

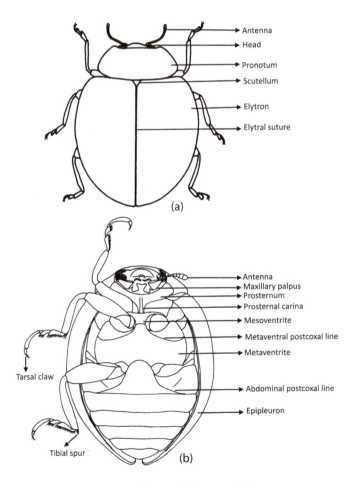

FIGURE 14.1 Coccinellidae, general appearance: (a) dorsal view and (b) ventral view.

Coccinellidae of the Indian Subcontinent

FIGURE 14.2 Coccinellidae—adult beetles of different tribes. a-d, f-i. Coccinellini; e. Noviini; j. Ortaliini; k. Platynaspini; l. Serangiini; m, n. Chilocorini; o. Hyperaspini; p, q, u. Coccidulini (=Scymnini); r. Coccidulini (=Stethorini); s, t, v, x. Epilachnini; w. Sticholotidini.

tribe Coccinellini are considered as true lady beetles and are readily identified by their pretty appearance and striking spots and colour patterns, but the much smaller and obscure forms are more difficult to recognize. Almost all the available keys are based on adult characters. The most important characters for recognition of adult Coccinellidae are: (i) body outline (Figure 14.2) usually round or oval, moderately to distinctly convex dorsally, rarely elongate oval and flattened; (ii) dorsal surface apparently glabrous and polished or sparsely to densely pubescent; (iii) colour bright and striking orange, red, or yellow with spots, bands, and other patterns or duller in various shades of brown, reddish brown, or black; (iv) four-segmented maxillary palpi (Figure 14.3) with a prominent, axe-head shaped (securiform) or otherwise modified terminal segment; (v) presence of postcoxal lines on the metaventrite and the first visible abdominal segment on the underside of the body (Figure 14.4); and (vi) three- to four-segmented tarsi of legs with appendiculate, bifid, or simple claws (Figure 14.5).

Although the above characters are usually diagnostic of the family, all of them are not unique to Coccinellidae. For example, the postcoxal line is also present in some species of the cerylonid group of families such as Cerylonidae, Endomychidae, and Alexiidae (=Sphaerosomatidae). In some coccinellids, the postcoxal line is absent (e.g., *Epiverta* Dieke, some members of *Hippodamia* Chevrolat) (Yu 1994). All the species of Coccinellidae themselves do not have all these characters. For example, in the tribe Sticholotidini, the terminal segment of maxillary palpi is greatly modified and not of the typical axe-head shape. If all or two of these characters are present in a beetle, it can be determined as a

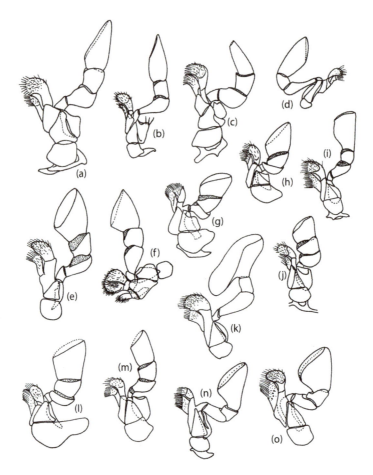

FIGURE 14.3 Maxilla in different tribes of Coccinellidae: (a–d) Sticholotidini; (e) Noviini; (f) Epilachnini; (g) Coccidulini (=Scymnini); (h) Hyperaspini; (i) Ortaliini; (j) Aspidimerini; (k, n, o) Coccinellini; (l) Platynaspini; (m) Chilocorini.

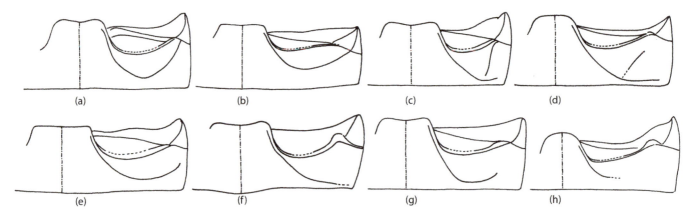

FIGURE 14.4 Abdominal postcoxal line in Coccinellidae: (a, b) complete; (c,d) incomplete, with an associate line; (e–h) incomplete.

member of the Coccinellidae (Pang and Mao 1979). The size of adult coccinellids varies from 0.8 to 25 mm. *Stethorus keralicus* Kapur and *Scotoscymnus popei* (Vazirani), measuring 0.7–0.8 mm, are probably the smallest coccinellids in India. The giant bamboo ladybird, *Synonycha grandis* (Thunberg), *Aiolocaria hexaspilota* (Hope), *Callicaria superba* (Mulsant), and *Palaeoneda auriculata* (Mulsant) rank among the largest coccinellids in the Oriental region.

Some beetles belonging to other beetle families are strikingly similar to coccinellids in external appearance. Members of some Endomychidae such as *Cyclotoma* Mulsant (e.g., *Cyclotoma monticola* Arrow) look remarkably like ladybirds. Some beetles belonging to the families Chrysomelidae (e.g., *Cryptocephalus ovulum* Suffrian, and species of alticine genera such as *Argopistes* Motschulsky and *Chilocoristes* Weise) and Tenebrionidae (e.g., fungus feeding species of *Diaperis*

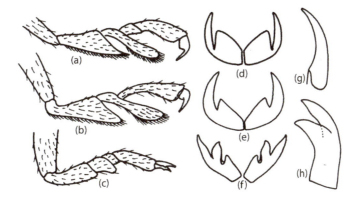

FIGURE 14.5 Tarsi in Coccinellidae: (a) Pseudotrimerous; (b) Trimerous; (c) Tetramerous; (d–h) Tarsal claw: (d, e) appendiculate; (f) bifid with a basal tooth; (g) simple; (h) apically bifid.

Müller with red and black spots, *Leiochrodes* Westwood, and *Leiochrinus* Westwood) are also superficially similar to Coccinellidae, though they are not taxonomically related.

Key Identification Characters

Excellent and exhaustive accounts of coccinellid morphology, with illustrations, given by Ślipiński (2007), Sasaji (1968a, 1971), Hodek (1973), Kovář (1996a), Kuznetsov (1997), and to some extent, Gordon (1985), can be referred to for detailed information. Recently, Ślipiński and Tomaszewska (2010) have provided excellent morphological descriptions of egg, larval, pupal, and adult stages using updated terminology. Websites on lady beetles of the Indian subcontinent (Poorani 2008) and Australia (Ślipiński et al. 2008) also provide accounts of morphological characters with illustrations.

Subfamilies and Tribes

Seven to eight subfamilies of Coccinellidae were recognized until recently, mainly Chilocorinae, Coccidulinae, Coccinellinae, Epilachninae, Ortaliinae, Scymninae, and Sticholotidinae. Chazeau et al. (1989, 1990) summarized the taxonomy of World Coccinellidae and listed the valid genera and subgenera under various subfamilies and tribes. As per the latest classification proposed by Seago et al. (2011) based on a combined morphological and molecular phylogenetic analysis, only two subfamilies, Microweiseinae and Coccidulinae, are recognized. The following subfamilies and tribes are represented in the Indian subcontinent (Tables 14.1 & 14.2).

All these groups are predacious except the Epilachnini, which are exclusively phytophagous. With practice, most coccinellids can be easily placed up to subfamily/tribe level just by sight and further identification can be started at the genus level. At present, satisfactory keys to identify all the known tribes and the genera therein are not available for the Oriental region. Only for the tribe Epilachnini, a key to the world genera is available (Tomaszewska and Szawaryn 2016).

The genera of the Indian region and many common species can be identified with minimal difficulty based on available resources (e.g., Sasaji 1971; Hoang 1982, 1983; Gordon 1985; Pope 1989; Kuznetsov 1997; Ślipiński 2007; Ren et al. 2009), which are by taxonomists working on the fauna of other countries/regions. Reliable identification of species is, however, problematic in view of our poor knowledge of Coccinellidae of this region. Particularly, species belonging to the tribes Sticholotidini, Scymnini, and Ortaliini are difficult to identify in view of their small to minute size and/or very similar external appearance/coloration in scores of species. In the other subfamilies as well, many species remain to be studied in detail.

TABLE 14.1
Subfamilies and Tribes of Coccinellidae Present in India

Family Coccinellidae Latreille 1802

Subfamily Microweiseinae Leng, 1920

1. Tribe Serangiini Pope, 1962
2. Tribe Sukunahikonini Kamiya, 1960

Subfamily Coccinellinae Latreille, 1807

1. Tribe Aspidimerini Mulsant, 1850
2. Tribe Chilocorini Mulsant, 1846
3. Tribe Coccidulini Mulsant, 1846
   - = Exoplectrini Crotch, 1874
   - = Scymnini Mulsant, 1846
   - = Stethorini Dobzhansky, 1924
   - = Tetrabrachini Kapur, 1948a
4. Tribe Coccinellini Latreille, 1807
   - = Halyziini Musant, 1846
   - = Tytthaspidini Crotch, 1847
   - = Bulaeini Savoiskaja, 1969
   - = Singhikaliini Miyatake, 1972
5. Tribe Diomini Gordon, 1999[a]
6. Tribe Epilachnini Mulsant, 1846
7. Tribe Hyperaspini Mulsant, 1846
8. Tribe Noviini Mulsant, 1846
9. Tribe Ortaliini Mulsant, 1850
10. Tribe Platynaspini Mulsant, 1846
11. Tribe Plotinini Miyatake, 1994
12. Tribe Shirozuellini Sasaji, 1967
13. Tribe Sticholotidini Weise, 1901
14. Tribe Telsimiini Casey, 1899

Source: Seago, A. et al., *Mol. Phylogenet. Evol.,* 60, 137–151, 2011.
[a] Present, but without known species.

TABLE 14.2

Subfamilies and Tribes of Coccinellidae Known from the Indian Subcontinent (as per the old classification)

Subfamily	Tribes
Chilocorinae	Chilocorini, Platynaspidini, and Telsimiini
Coccidulinae[a]	Coccidulini, Exoplectrini, and Lithophilini (=Tetrabrachini)
Coccinellinae	Tytthaspidini[a], Coccinellini, Halyziini (=Psylloborini), and Singhikaliini[a]
Epilachninae	Epilachnini
Scymninae	Aspidimerini, Hyperaspidini, Scymnini, and Stethorini
Sticholotidinae	Plotinini[a], Serangiini, Shirozuellini[a], Sticholotidini, and Sukunahikonini[a]
Ortaliinae	Noviini, Ortaliini[a]

[a] Rarely seen or collected in agroecosystems.

MAJOR WORK ON INDIAN FAUNA

Along with other animal groups, the classification of Coccinellidae began with Linnaeus in the middle of the eighteenth century (Gordon 1985). The first Coccinellidae to be described from the Indian region were among those named by Fabricius (1775, 1798) such as *Chilocorus nigrita*, *Menochilus sexmaculatus*, *Micraspis discolor*, and *Harmonia octomaculata*. Hope (1831) described some coccinellids from Nepal. Mulsant in his monumental world monograph "Species des Coléoptères trimères Sécuripalpes" (1850) and later supplements (1853a, 1853b, 1866), described most of the species currently recognized from this region. His work provided the foundation for modern classification of the family. This treatment was so well done that large portions of it remain unaltered by subsequent research (Gordon 1985). After Mulsant, Motschulsky (1858, 1866) described several species from this region, the types of which were later revisited by Iablokoff-Khnzorian (1972). Crotch (1874) revised the world Coccinellidae, mainly based on Mulsant's monograph and described a few genera and species from this region. Subsequently, Sicard, Weise, and Gorham published a series of papers describing several Indian species. Weise was the first coccinellid taxonomist to recognize the importance of male genitalia in species diagnosis (Gordon 1985). Korschefsky (1931, 1932) catalogued the Coccinellidae of the world. This is still considered indispensable for both beginners and experts in coccinellid taxonomy. Mader's monographic works on the fauna of the Palaearctic region (1926–1937) and other papers included several species found in this region. Iablokoff-Khnzorian's (1982) monograph on Coccinellini of the Palaearctic and Oriental regions is a major, but inadequate resource for this tribe.

Sasaji (1968a, 1971) gave a detailed historical account of the higher classification of Coccinellidae. His phylogenetic classification has served as the basis for subfamily and tribal assignments for all recent coccinellid workers and "is a seminal contribution, comparable in its significance to Mulsant's classification of 1850" (Gordon 1985). His "Fauna Japonica" (Sasaji 1971) and several other earlier papers (as H. Kamiya) covered many genera and species found in the Indian subcontinent. Miyatake also contributed a great deal in his studies on the tribes Chilocorini (1970), Platynaspini (1961a), Serangiini (1961b), Hyperaspini (1961c), Plotinini (1969), and Asiatic Sticholotidinae (1994). Bielawski described several species from Sri Lanka (1957), Bhutan (1979), Nepal (1971, 1972), Afghanistan (1963), and the Indian subcontinent in general. Miyatake (1967, 1985) and Canepari (1986, 1997, 2003) described several species from Nepal. Canepari (1997) summarized the fauna of Nepal. In recent times, Booth (1997, 1998) has published important papers on some important genera of coccinellids. Seventy-two years after Korschefsky's work, Jadwiszczak & Węgrzynowicz (2003) published the first part of the *World Catalogue of Coccinellidae* (covering Epilachninae), but the other groups have not been catalogued so far.

Ramakrishna Ayyar was the first Indian taxonomist to describe a new coccinellid species (*Scymnus coccivora*) in 1925. Among coccinellid taxonomists of the Indian subcontinent, A.P. Kapur, former Director of the Zoological Survey of India, contributed the lion's share of the work. He published a series of revisionary papers on major tribes such as Aspidimerini (Kapur 1948b) and Telsimiini (Kapur 1969) and genera such as *Rodolia* (Kapur 1949), *Stethorus* (Kapur, 1948c), and *Jauravia* Motschulsky (1946) and described several new species from this region (e.g., Kapur 1958, 1963, 1967, 1973). His manuscript on the coccinellid fauna of the Indian subcontinent based on the work of his lifetime is believed to be lying unpublished after his demise and efforts to trace it have proved unsuccessful. Anand et al. (1988) provided an annotated checklist of the subfamily Epilachninae, with host plants. Poorani (2002) compiled an annotated checklist of the Coccinellidae of the Indian subcontinent, excluding Epilachninae. Ahmad (1968, 1970, 1973) and Ghorpade (1974, 1976, 1977) described a few species from this region. Ahmad and Ghani (1966, 1971, 1972) studied some economically important species of Pakistan. Irshad (2001) listed the species along with their distribution and hosts in Pakistan. Illustrated/pictorial accounts of Coccinellidae have been brought out for Vietnam (Hoang 1982, 1983), Australia (Ślipiński 2007), and China (Ren et al. 2009), which are noteworthy as they can be used to identify the common genera and many species of this region as well. The Chinese and Vietnamese handbooks on Coccinellidae provide accounts of several species found in India with good illustrations.

Except for a handful of common species, the biology of coccinellids of this region has not been studied in detail. Stebbing (1903) provided brief notes on some common predacious coccinellids of this region. Subramaniam (1924b), Puttarudriah and Channabasavanna (1953, 1955, 1956, 1957), and Channabasavanna and Puttarudriah (1957) gave very good, albeit brief, accounts of common coccinellids of South India and their bioecology, which remain till today the only works of note on the bioecology of common Indian species.

Kapur (1956) provided systematic and biological notes on coccinellids predacious on San Jose scale in Kashmir. Nagarkatti and Ghani (1972) studied in detail the bioecology of coccinellid predators of *Adelges* spp. in the Himalayas. Irshad (2001) summarized the bioecology of important species of coccinellids in Pakistan.

BIODIVERSITY AND SPECIES RICHNESS

The coccinellid fauna of the world consists of about 360 genera and more than 6000 species in two subfamilies and 30 tribes (Ślipiński 2007; Ślipiński and Tomaszewska 2010). The Indian coccinellid fauna is rich and varied, but very poorly known compared to those from other parts of the world. Poorani (2002) provided an annotated checklist of the Indian Coccinellidae excluding Epilachninae, and an updated checklist including Epilachnini is available online (Poorani 2008). From the Indian subcontinent, about 550 species are known at present under 90 genera, 16 tribes, and 2 subfamilies, with scores of undescribed species in major tribes like Scymnini, Sticholotidini, etc.

Compared to this, "only 42 species occur naturally in Great Britain (Majerus 1994), 83 species in the Russian Far East (Kuznetsov 1997), 27 species in the Arctic zone of North America (Belicek 1976), 93 species in Chile (Gonzáles 2008), 481 species from North America (Gordon 1985; Vandenberg 2002), and approximately 500 species from Australia" (as cited in Ślipiński & Tomaszewska 2010). Some of the major catalogue resources on world Coccinellidae include Korschefsky's world catalogues (1931, 1932), Kovár's (2007) catalogue of the Palaearctic fauna (including the northern part of the Oriental region), and the world catalogue of Epilachninae by Jadwiszczak and Węgrzynowicz (2003).

DISTRIBUTION PATTERNS IN THE INDIAN CONTEXT

Zoogeographically, the Indian subcontinent falls in the Oriental region. The Indian mainland lies within the Indomalayan ecozone and is home to three biodiversity hotspots—Eastern Himalayas, Western Ghats, and northeastern India. Ten biogeographic zones are known in India with habitats ranging from tropical rainforests of the Andamans, Western Ghats, and NE region to coniferous forests of the Himalayas, with many climatic extremes in between. The Indian fauna of Coccinellidae is predominantly Oriental, with strong Palaearctic components in the north and northwestern regions. Peninsular India has a high degree of endemism at species level. Genera such as *Macrolasia, Stictobura,* and most of the species of *Jauravia* are endemic to India. The northeastern region is probably the richest in terms of faunal diversity and forms a part of the Eastern Himalayan hotspot. Anand et al. (1989) carried out a preliminary zoogeographical analysis of Indian Coccinellidae and listed as many as 220 species as endemic to India, 105 represented in Oriental region, and 35 in Palaearctic region.

PHYLOGENY OF COCCINELLIDAE

The phylogenetic classification of the subfamilies and tribes of Coccinellidae given by Sasaji (1968a), based on extensive study of comparative morphology of external characters of both adults and larvae, had been almost universally widely accepted and followed until recently. Kovář (1996b) proposed a phylogenetic classification, with minor modifications from Sasaji's, particularly the inclusion of the tribes Ortaliini and Noviini in a separate subfamily Ortaliinae. However, Sasaji's classification is now considered as based on a geographically and taxonomically limited study, with further refinements and additions by many authors (Ślipiński & Tomaszewska 2002), resulting in a proliferation of subfamilies and tribes to suit the needs of the students in their regional perspective (Vandenberg 2002). Based on Sasaji's (1968a) work, six or seven subfamilies of Coccinellidae were recognized until now (Coccinellinae, Coccidulinae, Scymninae, Chilocorinae, Epilachninae, Sticholotidinae, and, sometimes, Ortaliinae), each comprising many tribes (Sasaji 1968a, 1971; Gordon 1985; Kovař 1996b; Vandenberg 2002).

Ślipiński (2007) found that many taxa defined by Sasaji (1968a) and Kovař (1996b) were non-monophyletic assemblages and proposed only two subfamilies Microweiseinae and Coccinellinae, based on the presence of distinct morphological apomorphic characters of adult and larvae. Subsequently, four papers (Giorgi et al. 2009; Aruggoda et al. 2010; Magro et al. 2010; Seago et al. 2011) analysed molecular and/or morphological data and suggested relationships between major taxa similar to those proposed by Ślipiński (2007). In all these studies (except Magro et al. 2010), the Epilachninae were monophyletic in Coccinellinae and as a result, it was reduced to a tribe Epilachnini within a broadly defined Coccinellinae (Ślipiński 2007; Ślipiński and Tomaszewska 2010; Seago et al. 2011). Seago et al. (2011) constructed the phylogeny of Coccinellidae using morphological and molecular characters and came up with a totally revamped system of classification of subfamilies and tribes. This system recognizes only two subfamilies, Microweiseinae and Coccidulinae. At present, this system is widely followed by coccinellid taxonomists, though morphological characters appear to have been grossly overlooked in coming up with certain conclusions. Nedvěd and Kovař (2012) revisited Kovař's (1996b) classification and tried to reconcile it with recently published research by proposing nine subfamilies (Coccinellinae, Coccidulinae, Scymninae, Chilocorinae, Epilachninae, Sticholotidinae, Ortaliinae, Exoplectrinae, and Microweiseinae) and 42 tribes without further discussion. Szawaryn et al. (2015) studied the phylogeny of phytophagous coccinellids (Epilachnini) and proposed a revised generic classification for this tribe with a key to world genera. Despite these developments, a stable classification has not been achieved for the family because published data on partial morphological and molecular phylogenies appear to be incongruent.

TAXONOMIC PROBLEMS

At present, a handbook/identification guide to Indian Coccinellidae on the lines of major faunal volumes on Japanese (Sasaji 1971), Vietnamese (Hoang 1982, 1983), Chinese (Pang and Mao 1979; Chengyi 1992; Pang et al. 2002; Ren et al. 2009; Yu 2009), and Taiwanese (Yu and Wang 1999; Yu 2011) genera and species is not available. Among the more recent works on Coccinellidae, Ślipiński's (2007) conspectus of Australian ladybirds has set the standards for excellence in illustrations. Ren et al. (2009) have brought out an excellent illustrated handbook of Chinese Coccinellidae which has the habitus and genitalia illustrations of several species found in the Indian subcontinent as well. Web resources including a checklist and images of the common species of coccinellids of the Indian subcontinent are, however, available (Poorani 2008).

Unfortunately, systematic studies on Indian Coccinellidae have been hampered due to various problems. The available literature on Indian fauna is fragmented, and most of the Indian works on Coccinellidae are scattered, single species descriptions, and comprehensive revisions of major tribes and genera are lacking. Most of the original, old descriptions of Indian species are based only on external morphology and of limited value. Besides, female types are of limited value in the absence of corresponding male specimens. Primary revisions and value addition to original descriptions of Indian species with information on male genitalia based on examination of types are needed. Type specimens of Indian species are lodged in foreign museums and are difficult to obtain for studies as international collaboration is difficult due to the perceived problems with the Biodiversity Act of India and funds to visit foreign museums are not available for Indian taxonomists.

BIOLOGY AND ECOLOGY

Various aspects of the bioecology of coccinellids have been studied in detail worldwide and represent a significant portion of the international symposia on ecology of aphidophagous insects (Obrycki and Kring 1998). Hagen (1962), Hodek (1967, 1973), Hodek and Honek (1996), Hodek et al. (2012) (an updated version of the former), Majerus (1994, mainly concerning British ladybirds), and numerous other workers (not mentioned here) have also comprehensively reviewed the bioecology of coccinellids. A brief, general account of the bioecology of Coccinellidae with Indian examples is given below.

HABITAT

Coccinellids occur in a wide range of habitats and are found worldwide in almost all climates, except the most arid and the coldest (Hodek 1973). They live in all terrestrial ecosystems: tundra, forest, grassland, agroecosystems, and from the plains to mountains; and at nearly all elevations (Iperti 1999). The most ubiquitous Oriental coccinellid, *Menochilus sexmaculatus* (Fabricius), is a habitat generalist found in a variety of habitats. On the other hand, many species are known to be restricted to a certain, more or less strictly defined, habitat. For example, *Aiolocaria hexaspilota* (Hope) is an arboreal species, found mostly in forest ecosystems of the northwestern region of the Indian subcontinent. Species of *Calvia* Mulsant from the Indian region are more usually associated with trees and shrubs than herbaceous vegetation (Booth 1997). *Scymnus fuscatus* Boheman and *Scymnus hoffmanni* Weise are usually found associated with aphids on aquatic weeds such as *Eichhornia crassipes*, *Pistia stratiotes*, *Salvinia molesta*, and *Trapa* spp., and rarely found in agroecosystems. Species such as *Harmonia expallida* (Weise), *Coccinella luteopicta* (Mulsant), and *Calvia breiti* Mader have a very restricted distribution and are found only at very high altitudes in the Himalayas in north and northwestern India. *Sticholotis ferruginea* (Gorham) and *S. humida* Poorani, species commonly found in the plantations of South India, prefer cryptic and moist niches, and small aggregations are found in wet, moss-covered crevices and nooks on tree trunks. Species of Lithophilini are found under stones, moss, tree bark, in soil, and other cryptic habitats (Kapur 1948a). Deserts contain many phytophagous species (Epilachninae) and some coccidophagous species. Coccinellini, particularly *Hippodamia* spp., are found on mountains (Iperti 1999). Many coccinellids do not remain in a single habitat; particularly aphidophagous coccinellids are known to be highly mobile and move between different habitats depending on the season and prey availability (Sloggett and Majerus 2000).

LIFE CYCLE AND VOLTINISM

Coccinellids have relatively short life cycles. Most of the species have 3–4 weeks long life cycles, which can be prolonged by adverse temperature and shortage of food. The egg period usually lasts 4–5 days. The larvae have usually four, and rarely three or five, instars. The larval period is around 11–15 days, and the pupal stage about 4–5 days.

In temperate climates, predacious coccinellids generally reproduce in spring when their prey is abundant and become quiescent in summer. Some species exhibit renewed activity in autumn, and all coccinellids display varying levels of dormancy in winter (Iperti 1999). All individuals of a coccinellid species do not react similarly in the same geographical area, and a species does not necessarily produce the same number of generations over its entire distribution range (Hagen 1962). In the Indian subcontinent, species such as *Menochilus sexmaculatus* have several generations in a year, whereas species found in temperate regions, such as *Priscibrumus lituratus*, *P. uropygialis*, and *Calvia breiti* Mader, are univoltine, i.e., having one generation only in a year and undergo hibernation during winter months.

SEARCHING AND FEEDING BEHAVIOUR

The larvae and adults of coccinellids search randomly rather than systematically for prey and do not perceive their prey until contact. After encountering prey, the larval search pattern changes from extensive search or rapid movement at random to more intensive search as reflected by more frequent turns, which is an advantage when feeding on colonized prey (Hagen 1962; Kawai 1976). Recent evidence indicates that vision and olfaction may also be used for prey detection. Evans (2003) reviewed the prey searching and oviposition behaviour of aphidophagous ladybirds.

The young coccinellid larvae usually pierce and suck the contents from their prey. Besides sucking body fluids, solid parts are also eaten as evidenced by gut content analysis. The bifid or unidentate apex of the mandibles pierces the prey, rather than chewing it (Samways et al. 1997). The unidentate mandible found in the coccidophagous species is an adaptation for lifting the scale cover by cutting it open in a tin-opener type of way (Samways & Wilson 1988). Based on investigations on the morphology of the mandibles of adults in relation to diet, Samways et al. (1997) concluded that it can be used only to indicate the general nature of the diet such as phytophagous, carnivorous, or mycophagous type.

Later instar larvae develop a chewing action also in addition, enabling consumption of whole prey. In some species, extra-intestinal digestion occurs, i.e., periodically the larvae regurgitate fluid from the gut into the chewed prey and suck back the predigested food (Hodek 1973). The larvae of *Brumoides suturalis* (Fabricius) exhibit this type of alternate sucking and regurgitation; this process is repeated several times until only the chitinous skin of the host is left behind (Kapur 1942). Larval feeding in *Stethorus pauperculus* (Weise) also consists of alternate sucking and regurgitation (Puttaswamy & Channabasavanna 1977). The mandibles in *Stethorus* spp., *Platynaspis* spp., and some Scymnini have a groove to introduce midgut juices into the prey and then suck the digested liquids (Ricci 1979). The plant feeding species (Epilachnini) scrape the parenchyma off the leaves with their multidentate mandibles to ingest the sap, but not the solid plant material. The setal areas of the mouthparts (galea and lacinia) trap the plant juices (Samways et al. 1997). Due to this, the leaf becomes skeletonized, presenting a lace-like pattern or windowing, which is the typical symptom of damage.

Cannibalism

Cannibalism is a common phenomenon in Coccinellidae. Cannibalism in ladybirds can be divided into four categories: cannibalism of eggs by adults; of eggs, larvae, and pupae by unrelated larvae; of larvae, pupae, and adults by adults; and, of eggs by neonate sibling larvae (Majerus & Majerus 1997a). In most coccinellids, especially those that lay eggs in groups, it is of adaptive significance under conditions of prey scarcity. Newly hatched larvae feed on their sibling eggs in the same batch. This can be beneficial at low prey densities and helps larval survival until the first prey is found and also increases searching capacity. Predacious, phytophagous, and fungivorous species commonly eat their own eggs. Egg cannibalism is rare in many Scymnini and Chilocorini, which lay eggs singly that are often hidden. Majerus & Majerus (1997a) discussed in detail different types of cannibalism in ladybirds and their influence on evolution of life history traits.

Natural Enemies

A wide range of parasitoids, predators, and pathogens attacks coccinellids, but natural enemies are not considered to be capable of causing drastic changes in coccinellid populations. Hodek (1973), Ceryngier and Hodek (1996), Kuznetsov (1997), Riddick et al. (2009), and Ceryngier et al. (2012) have reviewed/given detailed accounts of the natural enemies of Coccinellidae. Richerson (1970) listed the natural enemies of Coccinellidae of the world. Schaefer and Semyanov (1992) provided a world list and bibliography of arthropod parasites of the seven-spotted ladybird, *C. septempunctata*.

Predators

In several laboratory studies, most coccinellids were rejected by many mammals, birds, reptiles, and amphibians. However, large-scale predation of some ladybirds in nature by mammals and birds such as swallows, martins, and swifts, particularly during ladybird population explosions, has been observed (Majerus 1994). Hibernating aggregations are eaten by birds and bears (Belicek 1976). Coccinellid aggregations could provide a large food source for birds, mammals, or insect predators, but they are rarely eaten by these (Hagen 1962) due to their general unpalatability.

Nearly all species of ladybirds possess a chemical defence mechanism (Holloway et al. 1993). These defence chemicals, mainly alkaloids, can cause severe damage to avian livers (Marples et al. 1989) and be lethal to bird predators. Majerus & Majerus (1997b) discussed in detail predation of ladybirds by birds in the wild. Predatory lizards are considered to have prevented successful establishment of *Cryptolaemus montrouzieri* Mulsant in Bermuda (Simmonds 1958).

Many species of ants attending honeydew-producing Hemiptera are hostile to natural enemies including coccinellids. Ants are reported to reduce the effectiveness of coccinellid predators and may kill both larvae and adults. For example, non-establishment of *C. montrouzieri* in mealybug-infested citrus orchards in Assam was attributed to the activity of the red tree ant, *Oecophylla smaragdina* (Fabricius) (Narayanan 1957). Jahn (1992) suggested that *Pheidole megacephala* (Fabricius) suppressed the population of *Curinus coeruleus* (Mulsant) preying on *Dysmicoccus neobrevipes* Beardsley on pineapple in Maui. Similarly, interference of *Camponotus compressus* (Fabricius) with the activity of *C. montrouzieri* in sugarcane affecting the control of *Saccharicoccus sacchari* (Cockerell) has been reported (Srikanth et al. 2001). Soldier-like aphids or "samurai" aphids belonging to genera such as *Pseudoregma* Doncaster and *Ceratovacuna* Zehntner also have been reported to display aggressive behaviour towards coccinellids. On the other hand, species such as *Coccinella magnifica* Redtenbacher and many Platynaspini are myrmecophilous and have been recorded in association with ant species. The larvae of the latter are similar in appearance to the larvae of lycaenid butterflies, which are commonly attended by ants.

Parasites

Parasitoids belonging to the families Ichneumonidae, Braconidae, Eulophidae, Encyrtidae, Pteromalidae, Chalcididae, Eupelmidae, Trichogrammatidae, Proctotrupidae, Diapriidae, Scelionidae (Hymenoptera), Tachinidae, and Phoridae (Diptera) have been recorded on various stages of coccinellids. Very few egg parasitoids have been recorded, probably because coccinellid larvae on hatching eat unhatched eggs of the same batch. It is also known that emergence of egg parasitoids takes place only after coccinellid larvae from unparasitized eggs have hatched (Hodek and Honěk 1996).

The braconid, *Dinocampus coccinellae* (Schrank) (=*Perilitus coccinellae* Schrank) is the most extensively studied parasitoid of coccinellids. It is cosmopolitan in distribution and is a solitary endoparasitoid of adults, mostly of the Coccinellinae, preferring larger species in general. It has been recorded on other coccinellid tribes such as Chilocorini and Scymnini. Though Ceryngier et al. (2012) opined that the reports on tribes other than Coccinellini are most probably erroneous, in India, it has been found to parasitize several Coccinellinae and also Chilocorini. At least two Chilocorini, namely, *Priscibrumus uropygialis* (Mulsant) and *P. lituratus* (Gorham), have been recorded as hosts of *D. coccinellae* by Nagarkatti and Ghani (1972) and recently confirmed by Maqbool et al. (2018). Ghorpade and Sasaji (1977) and Ghorpade (1979a) listed the hosts of this parasitoid and discussed phylogenetic implications.

The encyrtid genus *Homalotylus* Mayr is nearly exclusively parasitic on coccinellids. The eulophid, *Pediobius foveolatus* (Crawford), is a major parasitoid of ladybirds in India and has been imported into the USA for controlling the Mexican bean beetle, *Epilachna varivestis* Mulsant. The genus *Nothoserphus* Brues (Hymenoptera: Proctotrupidae) is parasitic on the larvae of Coccinellidae in the Oriental and Palaearctic regions (Lin 1987). In southern India, *Nothoserphus* spp. (*mirabilis*-group) are commonly collected as major late larval-pupal parasitoids of Coccinellinae. Besides these, members of the genera *Cowperia* Girault (Encyrtidae), *Tetrastichus* Haliday, *Oomyzus* Rondani (Eulophidae), and *Phalacrotophora* Enderlein (Diptera: Phoridae) are commonly found parasitising coccinellids in the Indian subcontinent. Several species of tachinids have been recorded in different parts of the world on coccinellids, mainly Epilachninae, but none from India so far.

Mites of the genus *Coccipolipus* Cooreman (Acari: Podapolipidae) are apparently specific parasites of Coccinellidae, especially on those species that tend to aggregate in large numbers. All stages of *Coccipolipus* live under the host's elytra. *Coccipolipus epilachnae* Smiley is specific to coccinellids of the subfamily Epilachninae (Schroder 1979). From southern India, *Coccipolipus synonychae* (Ramaraju and Poorani 2012) has been recorded on the giant bamboo ladybird, *Synonycha grandis*. Trombiculid mites are also known to infest adults and late larval instars (Kuznetsov 1997). Nematodes belonging to Mermithidae and Aphelenchidae are also often recorded from coccinellid adults, larvae, and pupae.

Pathogens

Several pathogens such as fungi, bacteria, and protozoa have been recorded on coccinellids. The commonly used entomopathogenic fungus *Beauveria bassiana* (Balsamo) Vuillemin is reported to infect several coccinellids such as *Cryptolaemus montrouzieri*, *Menochilus sexmaculatus* (Fabricius), *Scymnus coccivora* Ayyar, and *Serangium parcesetosum* Sicard, sometimes adversely affecting their efficacy. Maternally inherited bacteria of the genera *Wolbachia*, *Spiroplasma*, and *Rickettsia* that kill male, but not female hosts, during embryogenesis, resulting in all-female broods occur in many coccinellids. These are cytoplasmically inherited and distort the sex ratio by killing male progeny and giving rise to predominantly or exclusively female progeny. This phenomenon has been recorded in species such as *Coccinella septempunctata*, *Adalia bipunctata* (Linnaeus), *Coleomegilla maculata* (DeGeer), *Menochilus sexmaculatus*, *Hippodamia variegata* (Goeze), and *Harmonia axyridis* (Pallas) (Majerus and Hurst 1997). In view of their widespread occurrence in aphidophagous coccinellids, Majerus and Hurst (1997) proposed ladybirds as a model system for the study of male-killing symbionts. Administration of antibiotics like tetracyclines to the mother beetles has been reported to fully or partially cure this trait and restore normal males (Hodek and Honěk 1996).

DEFENCE MECHANISMS

- **Warning Colouration**: The bright orange, yellow, or red coloration with contrasting patterns in many coccinellids is aposematic in function, i.e., acting as a warning coloration to potential, visually hunting predators, particularly birds. Many ladybirds aggregate into large groups at certain times, which perhaps enhances the visual impact of their bright, contrasting, colour patterns (Majerus and Majerus 1997b).
- **Reflex Bleeding**: When disturbed, coccinellid larvae and adults eject small droplets of glandular secretions from the femoro-tibial joints of legs. This fluid is also exuded from ruptures at the base of hairs that terminate the branches on larval tubercles. This phenomenon is called "reflex bleeding". This liquid is usually orange, yellow, or red, has a strong acrid smell, and tastes bitter. Almost all coccinellids exhibit this phenomenon, which can be readily observed in the species of the genus *Rodolia* and the larger members of the Coccinellinae. The reflex blood consists of the highest concentration of defence chemicals, which are also distributed throughout the body of the beetle (Holloway et al. 1991) and repels potential predators, is even reported to cause allergic reactions or stinging sensation in some humans who come into contact with it, and is sometimes toxic to predators.

Waxy Covering: The immature stages of many coccinellids have dense white waxy covering on their bodies that also may have a protective function. Many species producing wax feed on scale insects and mealybugs, and the wax threads in *Cryptolaemus montrouzieri* are sticky and may entangle the limbs or block the sense organs of potential predators and also reduce parasitism. The waxy covering on their body probably protects coccinellid larvae from molestation by ants attending their coccid and aphid hosts for the honeydew they exude (Pope 1979). Agarwala and Yasuda (2001) reported that vulnerability of the larvae of *Scymnus posticalis* Sicard to predation by syrphid larvae was directly related to the thickness of wax cover, which acted as an effective defence against predation more in later instars than early instars.

Others: Like most beetles, coccinellids also fall to the ground when disturbed and feign death. Adults defend themselves by withdrawing their legs and antennae underneath their dome-shaped bodies, firmly attaching themselves to the substrate (Booth et al. 1990). Aggressive and territorial behaviour by adult males, including guarding of female pupae has been reported in *Leptothea galbula* (Mulsant) (Richards 1980). Pupal guarding appeared to be linked with the production of a sex pheromone in both the pupal and adult stages of the female.

Coccinellid pupae are commonly protected by so-called gin traps, abdominal biting, or pinching devices that they activate by flipping their bodies upward and use as jaws to deter ants (Eisner & Eisner 1992). Pupae of some Epilachninae are densely covered by glandular hairs, each with a tiny droplet of apparently defensive secretions at the tip, containing alkaloids (Attygalle et al. 1993; Schroeder et al. 1998). Similar glandular hairs with defensive droplets are seen in the pupae of some Aspidimerini and Sticholotidini in India.

Polymorphism/Melanism

The colour patterns of many coccinellids, especially the elytral patterns, show a great deal of variability, often within a species. These variations have been studied through classical genetic methods in several species such as *Adalia bipunctata*, *Coccinella septempunctata*, *Harmonia axyridis*, and *Coelophora inaequalis* (Fabricius). The heritability of elytral pattern variations is known to be controlled by multiple allelic genes (Komai 1956). Geographic variations also occur in coccinellid species populations, and the proportion of dark and light pigmentation varies according to the geographic position in a manner which is consistent among many species. Kapur (1957, 1962) studied the geographic variations in the populations of some Indian ladybirds, including those from high altitudes in the Himalayas, and concluded that pigmentation became gradually darker in coccinellid populations towards northern India.

In several instances, different morphs or colour variants of the same species have been described as separate species (e.g., *Harmonia eucharis* (Mulsant) and *Calvia shiva* Kapur), resulting in numerous synonyms. Among the species of this region, many species such as *Coccinella septempunctata*, *Coccinella transversalis* F., *Adalia tetraspilota* (Hope), *Harmonia octomaculata* (Fabricius), *Harmonia eucharis*, *Harmonia expallida*, *Hippodamia variegata*, *Calvia shiva*, *Menochilus sexmaculatus*, *Propylea* spp., *Phrynocaria unicolor* (Fabricius), *P. perrotteti,* and *Coelophora bissellata* Mulsant exhibit moderate to extreme degrees of polymorphism. In *H. eucharis* (Ghani 1962) and *C. sexmaculata* (Subramaniam 1924a), the different morphs were found to freely interbreed and produce still further intermediate forms in breeding experiments. In some cases, species identities have been resolved through classical breeding experiments. For instance, two extremely similar looking species, *Propylea japonica* (Thunberg) and *P. quatuordecimpunctata* (Mulsant), were proved to be reproductively isolated and distinct sibling species through crossing studies (Sasaji et al. 1975). Sasaji and Akamatsu (1979) proved that all the forms described under *Menochilus sexmaculatus* and *C. quadriplagiata* (Swartz) were taxonomically conspecific and hence synonymous.

Hibernation and Mass Assemblages

Mass migrations, aggregations, huge hibernating populations, and population explosions or "plagues" of ladybirds, particularly the convergent ladybird beetle (*Hippodamia convergens* Guérin-Méneville), *Adalia bipunctata* and *Coccinella septempunctata*, are common phenomena in Europe and North America. In many Holarctic temperate countries, huge aggregations of coccinellids undergo hibernation to tide over the very harsh winter conditions. Several billion *H. convergens* are collected annually from overwintering sites in California and sold for augmentative releases (Dreistadt and Flint 1996). The tendency to aggregate is found in species associated with "ephemeral" prey, mostly aphids, and not with the more sessile Hemiptera. Further, aggregating species usually exhibit long dormancy or diapause periods.

Hibernating populations can be hypsostatic, i.e., aggregating in protected places outdoors, in cliffs and crevices, and on mountaintops, or synanthropic, i.e., aggregating in and around houses, buildings, and other man-made structures, causing annoyance to people. The Asian multicoloured ladybird (*Harmonia axyridis*), a major invasive coccinellid in the USA and Europe, has been reported to be a great seasonal nuisance pest due to its habit of hibernating in huge numbers in and around homes, resulting in home invasions, causing serious allergic reactions, and occasionally even nibbling or biting human beings (Mizell 2001).

In the Indian subcontinent, many species such as *Priscibrumus* spp., *H. eucharis*, *H. expallida*, *C. septempunctata,* and *C. luteopicta*, undergo hibernation in winter in north and northwestern regions. Mani (1962) reported mass assemblages of *C. septempunctata* on the glacier beds at the Lakka Pass in the Western Himalayas. Apparently, hibernating mass aggregations of *Afidentula bisquadripunctata* (Gyllenhal) have been reported from Bihar, India (Kapur 1954). Mass assemblages of *Chilocorus nigrita* (Fabricius) on banyan tree (Ketkar 1959) and *Psyllobora bisoctonotata* (Mulsant) on *Dalbergia sissoo* (Kapur 1943) are quite common in the Indian subcontinent. Stebbing (1903) reported gregarious behaviour in *Rodolia sexnotata* (Mulsant), the adults of which passed the heat of the day collected thickly together on the undersurface of large leaves. Froggatt (1907) and Puttarudriah et al. (1952) recorded swarms of *C. montrouzieri* on scale-infested *Araucaria* pines (Christmas tree) in Australia and India, respectively.

Prey Associations

Coccinellids are predacious on homopterous insects such as aphids, psyllids, whiteflies, leaf- and planthoppers, scales, and mealybugs, and early instar larvae of Lepidoptera, Diptera, Coleoptera, Hymenoptera, and Thysanoptera, and mites (Acari). The degree of prey specialization varies from group to group, but as a rule, both adults and larvae feed on the same principal food. Although the entomophagous members of Coccinellidae can all be categorized as predators, the larvae of some minute scale feeding *Hyperaspis* species will complete their development by burrowing into the large egg sac attached to a single female scale and thus approach a parasitic mode of existence (Vandenberg 2002).

In general, members of the tribe Coccinellini are aphidophagous and also feed on other Hemiptera such as psyllids, whiteflies, and leaf- and planthoppers, and early stage larvae of Lepidoptera (Escalona et al. 2017). However, some members of Coccinellini have evolved different host affinities. For example, *Anegleis cardoni* (Weise) and *Phrynocaria perrotteti* (Mulsant) show a propensity to feed on whiteflies in preference to aphids (J. Poorani, personal observation/label data). The genus *Aiolocaria* Crotch and some species of *Calvia* are notable for being predatory on the eggs, larvae, and pupae of chrysomelid beetles (Booth 1997). Members of the genus *Synona* Pope have been recorded as predators of plataspidid bugs (Hemiptera) in India and Australia. *Synona* spp. are well known predators of *Coptosoma ostensum* Distant in India (Poorani et al. 2008). *Bulaea lichatschovi* (Hummel) (Coccinellinae) is reported to feed on pollen and fungal spores and occasionally is even phytophagous. Recently, Escalona et al. (2017) carried out a molecular phylogenetic analysis of Coccinellini and reconstructed food preferences in Coccinellini based on ancestral states. They observed "phylogenetically independent food preference transitions from aphidophagy to other food sources in many Coccinellini".

The members of Chilocorini (particularly *Chilocorus* Leach) and Telsimiini are specific scale predators, especially Diaspididae (Ślipiński 2007). Within Chilocorini, members of the genera *Brumoides*, *Exochomus*, and *Priscibrumus* feed on Coccoidea and aphids. The tribe Sticholotidini includes mainly scale feeders. The tribe Serangiini comprises specific predators of whiteflies, and to a lesser extent, scales. Species of the tribe Scymnini feed on aphids, mealybugs, scales, and whiteflies. Species of *Scymnus* are primarily aphid feeders, though other hosts like mealybugs are often eaten. Members of *Nephus* are usually predatory on mealybugs (e.g., *Nephus regularis* and *N. tagiapatus* in southern India, based on label data). The genus *Axinoscymnus* Kamiya is whitefly-specific. Some *Ortalia* spp. are reported to be myrmecophagous (ant-feeders). Species belonging to Exoplectrini and Noviini feed mainly on margarodid scales (*Icerya* spp. and their relatives). The tribe Stethorini includes exclusively mite feeding species. The tribe Epilachnini is entirely phytophagous. The tribe Halyziini (=Psylloborini) comprises mycophagous species, which commonly feed on lower fungi, especially powdery mildews (Ślipiński & Tomaszewska 2010).

Members of the genus *Coleomegilla* Timberlake probably have the broadest host range as they are not only able to survive on a variety of foods, but to complete development on unusual diets such as mites and plant pollen (Gordon 1985). Some coccinellids indulge in a sort of drinking from plant surfaces, to replenish body fluids after a long hibernation, which has been erroneously reported as plant feeding (Hodek 1973). Members of the genus *Micraspis* Mulsant are known to prefer pollen to insect prey. Coccinellids are at times common and occasionally abundant visitors to extrafloral nectary (EFN)-bearing plants, mostly in the absence of prey as well as ants (Pemberton & Vandenberg 1993).

Schilder and Schilder (1928) listed the prey of Coccinellidae systematically. Agarwala and Ghosh (1988) reviewed and listed the prey records of aphidophagous Coccinellidae, mostly Coccinellinae, from India. Schaefer (1983) provided a catalogue of host plants and natural enemies of Epilachnini of the world. Omkar and Pervez (2004) provided a predator-prey catalogue for the predacious coccinellids of India, which is incomplete with several erroneous records. Most of the host records available in literature are based on casual observations of the predator feeding on a particular host and hence not correct. Thompson (1951) pointed out that various species of ladybirds do not feed on all the host insects with which they seem to be physically associated. Unless the essential prey is taken, oviposition and larval development are affected or prevented even in seemingly polyphagous species. Correct prey records based on actual rearing are therefore important for possible utilization in IPM later.

IMMATURE TAXONOMY

Taxonomic studies on immature stages of coccinellids have been much fewer than those on adults. Gage (1920) was the first to work on larval taxonomy of coccinellids, and his terminology for different structures on the body of larvae is still widely used. Sasaji (1968b), Savoiskaya and Klausnitzer (1973), Savoiskaya (1983), Booth et al. (1990), and Ślipiński (2007) provided detailed accounts of the taxonomy and morphology of coccinellid larvae belonging to different subfamilies. Van Emden (1949) studied the larvae of British coccinellids. Ślipiński (2007) described the larvae of several Australian genera. Pope (1979) described wax production and wax producing structures in coccinellid larvae. Studies on pupal taxonomy are still rarer. Duang and Frederick (1974) described the pupae of some species of North America and discussed the phylogenetic relationships based on pupal taxonomy. The only other notable work on the taxonomy of coccinellid pupae is by Phuoc and Stehr (1974). In India, very few publications on immature stages are available, the only notable one being Kapur's (1950) work on immature stages of some Epilachnini.

Egg (Figure 14.6)

Eggs of coccinellids are usually oval and spindle-shaped and the surface is smooth or with reticulate microsculpture. They are typically orange yellow or bright golden yellow and rarely cream or white. They are laid singly or in groups in the open or in the case of many scale feeders, under the host colonies and attached to the substrate by the narrow end. Unlike most coccinellids, in the tribe Chilocorini (e.g., *Priscibrumus lituratus*, *P. uropygialis*, and some *Chilocorus* spp.), eggs are laid flat on the substrate. The eggs are laid in the vicinity of prey, apparently because adults usually oviposit where they are feeding on essential prey. The most unusual eggs are found in Aspidimerini (e.g., *Pseudaspidimerus*), in which eggs are scale-like and dorsoventrally flattened.

Larva (Figure 14.7)

The larval structure varies among different groups. A summary of the general appearance of the larvae in different subfamilies and tribes is given below.

Coccinellini: General appearance more or less alligator-like (campodeiform), ground colour salty grey to dark chocolate brown or black, with red, orange, yellow, or white spots/patterns, sometimes pruinose/covered with a whitish dust. Larvae of mycophagous species [Tribe Halyziini (=Psylloborini)] are bright lemon yellow or pale greyish with black spots and patterns.

Chilocorini: Body elongate cylindrical or spindle-shaped, with numerous protuberant and branched spines on the dorsal side.

Epilachnini: General appearance similar to that of Chilocorini, more robust and ellipsoidal in outline with branched spines on the dorsal side.

Aspidimerini: Elongate oval or elliptical in outline, remarkably scale-like, dorsoventrally flattened, lacking spines, warts, and other setose protuberances found in other groups.

FIGURE 14.6 Eggs of Coccinellidae: (a, b) Coccinellini; (c, d) Chilocorini; (e) Scymnini; (f) Shirozuellini.

FIGURE 14.7 Larvae of Coccinellidae: (a) Stethorini; (b) Epilachnini; (c) Coccidulini (=Scymnini); (d, e) Aspidimerini; (f) Noviini; (g) Sticholotidini; (h) Shirozuellini; (i, k) Chilocorini; (j, l, o) Coccinellini; (m) Sticholotidini; (n) Serangiini; (p) Hyperaspini; (q) Coccidulini (=Scymnini); (r) Sticholotidini.

Scymnini: Elongate, cylindrical, or spindle-shaped, profusely covered with waxy filaments and extrusions on the dorsal and lateral sides, rarely lacking waxy filaments.

Hyperaspidini: Ellipsoidal in outline, greyish with powdery, white waxy coating.

Noviini: Fusiform or ellipsoidal in outline, fleshy, usually reddish or pink, dorsal surface covered with a thin, powdery wax coating.

Sticholotidini: Form variable, inverted conical/top-like, spindle-shaped, or cylindrical with long, setose projections on the dorsal side.

Pupa (Figure 14.8)

Coccinellid pupae are usually found attached to the substrate by their tail ends. The pupae are exarate, though legs and other appendages are not fully free, with some degree of partial fusion. Species of Sticholotidini and Coccinellini have uncovered pupae, with the last larval skin attached to the caudal end. Pupae of the members of the Coccinellini have spots and other patterns, whereas those of Sticholotidini are densely pubescent, with the last larval skin prominent at the caudal end. Pupae of the tribes Chilocorini, Noviini, and Hyperaspidini are characteristic in appearance, with the last larval skin vertically

FIGURE 14.8 Pupae of common tribes of Coccinellidae: (a, d) Coccinellini; (b, o) Chilocorini; (c) Epilachnini; (e) Coccidulini (=Stethorini); (f) Aspidimerini; (g) Noviini; (h, i, m, p) Coccidulini (=Scymnini); (j) Serangiini; (k, l) Shirozuellini; (n) Hyperaspini.

split along the middle, exposing the pupae. The pupae of some large species of Coccinellini have gin traps, protective devices which occur on the pupal abdomen. They are lateral and obviously toothed structures used like jaws to close on potential predators by rapid movements of the abdomen.

In the tribe Scymnini, pupae are usually covered by the larval skin with its white waxy covering. In *Chilocorus* spp., some species of *Rodolia* and Epilachnini, several larvae usually aggregate in a common site on the lower side of leaves or on the tree trunk for pupation. Some species like *C. nigrita, Jauravia* spp., *Pharoscymnus horni,* and *Sasajiscymnus dwipakalpa* use inhabited or deserted spider nests as pupation sites/shelters (Ghorpade 1979b). The adults on emergence have very soft, pale elytra and the elytra harden and acquire their spots and other patterns only gradually over a few hours or even days.

ECONOMIC IMPORTANCE

Predacious coccinellids have been widely used in biological control for over a century, and the methods used for using these predators have remained virtually unchanged (Obrycki and Kring 1998). The standard tactics of applied biological control, namely introduction, augmentation, and conservation, are employed in the case of coccinellids also. Some of the outstanding examples in biological control involve coccinellids, which in fact launched the concept in the western mind. But barring a few spectacular successes, other attempts at using them in biological control have been mediocre successes or outright failures. Obrycki and Kring (1998) reviewed the use of Coccinellidae in biological control. Semyanov (1980) discussed the ways to employ coccinellids in biological control.

INTRODUCTIONS

The introduction of the eleven-spotted ladybird *Coccinella undecimpunctata* to New Zealand in 1874 has been widely quoted in the history of biological control as the first importation of an insect for biological control in New Zealand and one of the first anywhere (Dumbleton 1936; Doutt 1958; Hagen and Franz 1973; DeBach and Rosen 1991; Dixon 2000; Michaud 2012). However, Galbreath and Cameron (2015) found no evidence that such an introduction was ever made or attempted based on a comprehensive search of historical records. They detailed how a spurious record had arisen and concluded there is no foundation for this record of introduction.

The spectacularly successful introduction of the Vedalia beetle, *Rodolia cardinalis*, in California later in 1888 against the cottony cushion scale, *Icerya purchasi* (Maskell), was the single defining moment of classical biological control and a milestone in applied entomology (Caltagirone and Doutt 1989). The period from 1888 to the early 1900s following the astounding success of the vedalia introduction, was the "ladybird era" or the "ladybird fantasy" period, during which several coccinellid species were introduced in various parts of the world for controlling pests (Hagen 1974). Most of these were not successful. The most successful introductions of coccinellids are in orchards and perennial habitats, as is the case for all introduction programs. The successful introduction of *Chilocorus nigrita* from this region for the control of coconut scale in Seychelles has been well documented by Vesey-Fitzgerald (1941, 1953). *Serangium parcesetosum* Sicard from the Indian region has proved successful in controlling whiteflies in other parts of the world (Legaspi et al. 1996), though it has never been used in an applied context in India so far. Similarly, *Scymnus coccivora* was introduced from India into the Caribbeans as a promising bioagent of the hibiscus mealybug, *Maconellicoccus hirsutus* (Green) (Chong et al. 2015), but it is not mass produced in India for augmentative purposes.

Several successful coccinellid introductions in many historically significant projects involve scale feeders such as *Rodolia cardinalis*, *Cryptognatha nodiceps* (Marshall), *Chilocorus* spp., and *Rhyzobius lophanthae* (Blaisdell). In comparison, relatively fewer aphidophagous species have been established through importations. The tendency of aphidophagous coccinellids to disperse and migrate makes their population build up difficult. Further, in temperate climates, most aphidophagous coccinellids are univoltine, while their aphid prey produce many generations quickly, resulting in unsatisfactory control (Iperti 1999). Coccinellids are considered poor colonizers and are generally ineffective as classical biocontrol agents, with few exceptions (Obrycki and Kring 1998). For instance, the estimated establishment rate for coccinellids into North America (0.10) is lower than a worldwide estimate (9.34) for all biological control programmes (Greathead 1986). The causes for the relatively low rates of establishment of introduced coccinellids have not been examined for most species (Obrycki and Kring 1998).

INTRODUCTIONS OF COCCINELLIDS INTO INDIA

Introductions of coccinellids into India have been few, most of them imported for controlling a single pest, namely, *Melanaspis glomerata* (Green) (Hemiptera: Diaspididae) on sugarcane, during the late 1960s and early 1970s (Poorani 2002). *Cryptolaemus montrouzieri*, introduced in 1898 (Mayne 1953), was perhaps the first natural enemy deliberately introduced in India. Of the 22 species introduced into the Indian region, only three, namely, *Cryptolaemus montrouzieri*, *Curinus coeruleus*, and *Rodolia cardinalis*, have established well, the last two mainly in the southern region, and are providing substantial natural control (Poorani 2002). Jenkins (1946), Rao et al. (1971), Sankaran (1974), Gordon (1985), Booth and Polaszek (1996), Legaspi et al. (1996), and Samways et al. (1999) gave accounts of coccinellid imports into and exports from this region. Gordon's (1985) summary of coccinellid introductions into North America included many from the Indian subcontinent.

AUGMENTATION

In India, very few species including *Menochilus sexmaculatus*, *Coccinella septempunctata*, *Pharoscymnus horni* (Weise), *Pharoscymnus flexibilis* (Mulsant), *Chilocorus nigrita*, *Brumoides suturalis* (Fabricius), and *Scymnus coccivora* among native species and *Cryptolaemus montrouzieri* and *Curinus coeruleus* among introduced species, have been studied well and utilized in biological control programmes to some extent (AICRP-BC 1977–2014). Augmentation of coccinellid predators in agroecosystems has been limited by the lack of effective mass production techniques using artificial diets. *Cryptolaemus montrouzieri*, and to a limited extent, *C. nigrita*, are the only species mass produced to a limited extent in India for augmentation purposes against mealybugs and scales on horticultural and plantation crops. Augmentative releases of *C. montrouzieri* have been found to be particularly effective in horticultural crops such as guava, citrus, and grapevine (Mani & Thontadarya 1988; Mani and Krishnamoorthy 1997; Mani et al. 1990). Attempts have not been made to utilize native species such as *Serangium parcesetosum*, *Scymnus coccivora*, *S. latemaculatus*, and *Nephus regularis* for management of sucking pests by augmentation though they are abundantly found in nature.

CONSERVATION

For most agricultural systems, conservation techniques for coccinellids are lacking, even though they are abundant in these habitats (Obrycki and Kring 1998). Insecticides and fungicides (both biologically and chemically derived) can reduce coccinellid populations and have the greatest effect on the survival of coccinellids in agroecosystems. Several chemical insecticides and bioagents such as *Bacillus thuringienis* (Berliner), entomopathogenic fungi, and entomophilic nematodes (EPNs) have been evaluated for their impact on coccinellids. Use of safe pesticides, strip cutting, providing refuges, etc. can conserve the coccinellid populations. Use of food sprays such as various sugars and/or proteins (e.g., artificial honeydew) has been found to result in increased populations of coccinellids (Evans and Swallow 1993; Hagen 1987). Conservation techniques are needed to enhance the effectiveness of naturally occurring and released coccinellids (Obrycki and Kring 1998).

NON-TARGET EFFECTS OF EXOTIC COCCINELLIDS

Non-target effects due to the introduction of exotic coccinellids have been extensively reviewed and documented, especially in North America (Obrycki et al. 2000). Exotic aphidophagous coccinellids, particularly *Coccinella septempunctata* and *Harmonia axyridis*, have been implicated as potential sources of serious population-level non-target effects on aphids and their natural enemies and also displacement of native coccinellid species in North America and other parts of the world. Elliott et al. (1996) clearly linked the 20-fold or greater decline in *Coccinella transversoguttata richardsoni* Brown and *Adalia bipunctata*, both native coccinellid species, to the invasion of the exotic species *C. septempunctata*. It has also been associated with possible adverse effects on the biological control of the alfalfa weevil, *Hypera postica* (Gyllenhal), in Utah by reducing aphid densities that produce honeydew used by larval parasitoids of the weevil (Evans and England 1996). It also apparently poses a threat to several rare species of lycaenid butterflies by feeding on their eggs as alternative food (Horn 1991). Koch (2003) reviewed the non-target effects due to *Harmonia axyridis*. Possible adverse effects of indigenous coccinellids on introduced aphid parasitoids and exotic coccinellids on indigenous aphid parasitoids have been predicted, though not proved (Obrycki and Kring 1998). Koch et al. (2003) identified *H. axyridis* as a potential hazard to immature monarch butterflies.

The introduction of *Cryptolaemus montrouzieri* was reportedly responsible for population level non-target effects on *Dactylopius opuntiae* (Cockerell) in Mauritius and South Africa, adversely affecting the biocontrol of prickly pear (Goeden and Louda 1976). In South Africa, establishment of *C. montrouzieri* against *Planococcus citri* (Risso) only occurred after the establishment of *Dactylopius* spp. for control of *Opuntia* spp. (Greathead 1977). Similarly, coccinellid predation of phytophagous species feeding on alligator weed and lantana (Goeden and Louda 1976) and purple loosestrife (Nechols et al. 1996) has been reported. Especially in the case of aphidophagous coccinellids, at least some environmental effects due to introduction are predictable due to their broad polyphagy and ability to disperse and establish. It is still not clear as to whether the pest management benefits due to exotic coccinellids outweigh the environmental impacts (Lynch and Thomas 2000).

COLLECTION AND PRESERVATION

Schauff (2003) has described in detail the techniques and tools used in collecting and preserving insects in general, including packing and shipping methods. Booth et al. (1990) gave an elaborate account of methods used for collection, mounting, and labelling of beetles and their larvae. A brief account of the methods used for collecting, processing, and preserving coccinellids is given below.

Collection: The simplest way of collecting coccinellids is by visual searching and hand collecting. This method is, however, time consuming and cumbersome. Most beetles feign death and fall off the vegetation when disturbed; hence, beating the vegetation with a stick and collecting the falling beetles using a tray or white cloth spread underneath is very efficient. Sweeping grasses and low vegetation with an insect net is very useful. Malaise traps, yellow pan interception, and even pitfall traps sometimes yield sizeable numbers of coccinellids, especially the small to minute ones. An aspirator is useful for collecting very small species such as mite feeders. Collecting the host insects also along with the coccinellid will provide vital information on the host associations of the species. After collection, if immediate mounting is not possible after killing, the specimens can be stored dry in butter paper pouches. Another easy method is to use paper cylinders for storage. Slightly thick paper or cardboard can be rolled into cylinders and plugged with cotton at both ends and used to store dry coccinellid specimens, with the collection data written on top.

Killing and Mounting: Adult beetles are best killed using ethyl acetate vapour. Tissue paper or cotton dipped in ethyl acetate can be used to kill the beetles collected in glass tubes or jars. After killing, the specimens should be mounted for long-term storage and/or detailed studies. Preservation of adult coccinellids in alcohol is generally not desirable. But very small specimens and larvae can be preserved in 75%–80% ethyl or isopropyl alcohol. Specimens preserved in 95%–100% ethanol are required for molecular characterization/barcoding. For large specimens (e.g., species of *Synonycha*, *Aiolocaria*, and *Anisolemnia*), direct pinning through the body can be done using stainless/rust-free insect pins. Double mounting is followed for specimens that are not too small or large. However, for most coccinellids, direct pinning should be avoided to prevent serious damage to the specimens. For mounting most of the small to very small coccinellids, card points are usually recommended. Card points can be hand-cut from strips of stiff, white ivory boards/index cards or punched with specially made punches, to ensure uniform size and shape. Insect pins of size 2 or 3 can be used to hold the card points.

Labelling: It is essential to label the specimens after proper mounting, to ensure that they have taxonomic value. The essential data must answer the questions of where, when, and who, in that order and as exactly as feasible (Schauff 2003). In other words, the label should contain at least minimum collection data such as locality, date, collector's name, and wherever possible, other biological information like host insect/associated crop, method of trapping, if any, etc. The date of collection is usually written with the month in Roman numerals and the day and year in Arabic numerals, as in 5.vi.1995. It is advisable to use two or three labels, with separate collection and biological data, as indicated. The word "ex" (Latin for "out of") should mean that the insect was observed feeding on or in or was bred from the mentioned plant or insect host. Insect labels can be easily generated in a computer and laser-printed for use. Printed labels (font size 3.5 or 4) should not be larger than 18 × 8 mm. For specimens in alcohol and other liquid preservatives, a single label large enough to include all data, written with pencil or Indian ink, should be included.

Preparation of Genitalia: For preparation of genitalia for taxonomic studies, the specimens need to be first softened by immersing in warm soapy water for 10–30 minutes, depending on their size and oldness. The abdomen is gently detached by inserting a minuten pin between the metaventrite and the first visible abdominal ventrite and warmed in 10% aqueous solution of potassium hydroxide or sodium hydroxide in a hot air oven for 30–45 minutes. The specimen can also be left overnight in cold KOH or NaOH for clearing the soft tissues. When the specimen is well macerated, it is washed in distilled water, and then the genitalia are dissected in a drop of glycerine under a Stereozoom microscope. The genitalia can be easily pulled out with a pair of minuten pins mounted on spent ballpoint pen refills or wooden sticks. Leaving the specimen in 30% lactic acid overnight allows the siphonal apex to become fully distended to reveal its fine structure (Booth and Pope 1986) (this step, though, can be usually dispensed with). After studies, the genitalia are rinsed in distilled water, mounted permanently on a piece of ivory board with water-soluble gum (particularly in the case of robust or large species), or stored in genitalia vials in a drop of glycerine and pinned with the specimen. Nowadays, dimethyl hydantoin formaldehyde (DMHF), a resin-like substance, is increasingly used for permanent storage of genitalia instead of genitalia microvials. The genitalia can be mounted in a drop of DMHF on a card, which can be pinned with the specimen.

CONCLUSION

The higher classification of Coccinellidae has undergone a major transformation recently due to comprehensive analysis of phylogenetic relationships within Coccinellidae based on molecular and morphology data. Major tribes, genera, and species of Indian Coccinellidae need to be revised in light of the latest classification, and lack of primary revisions has been a major bottleneck in placing the known groups on a sound systematic footing. This problem is further compounded because the Indian mainland in its entirety has not been surveyed for Coccinellidae, and even biodiversity hotspots of Indian mainland (Western Ghats, northeastern region) have not been systematically covered mainly due to lack of funds for faunal surveys, curation, and maintenance of collections and for taxonomic research in general. Besides, inter-institutional collaboration in India is non-existent or difficult due to protectionism which has made access to collections in major institutions very difficult for *bona fide* taxonomists. Networking of collections and individuals working on Coccinellidae systematics in India and other parts/museums of the world and funding to facilitate visits to foreign museums to examine type specimens are essential to promote further research on this important group. This will enable complete documentation of the Coccinellidae of India and facilitate the identification and utilization of promising species in applied biocontrol programmes.

REFERENCES

Agarwala, B. K., and A. K. Ghosh. 1988. Prey records of aphidophagous Coccinellidae in India: A review and bibliography. *Tropical Pest Management* 34:1–14.

Agarwala, B. K., and H. Yasuda. 2001. Larval interactions in aphidophagous predators: Effectiveness of wax cover as defence shield of *Scymnus* larvae against predation from syrphids. *Entomologia Experimentalis et Applicata* 100:101–107.

Ahmad, R. 1968. A new species of *Pseudoscymnus* Chapin (Col.: Coccinellidae) predaceous on scale insects in West Pakistan. *Entomophaga* 13:377–379.

Ahmad, R. 1970. A new species of *Pharoscymnus* Bedel (Coleoptera: Coccinellidae) predaceous on scale insects in Pakistan. *Entomophaga* 15:233–235.

Ahmad, R. 1973. A new tribe of the family Coccinellidae (Coleoptera). *Bulletin of Entomological Research* 62(3):449–452.

Ahmad, R., and M. A. Ghani. 1966. Biology of *Chilocorus infernalis* Muls. (Col.: Coccinellidae). *Technical Bulletin, Commonwealth Institute of Biological Control* 7:101–106.

Ahmad, R., and M. A. Ghani. 1971. The biology of *Sticholotis marginalis* Kapur (Col.: Coccinellidae). *Technical Bulletin, Commonwealth Institute of Biological Control* 14:91–95.

Ahmad, R., and M. A. Ghani. 1972. Coccoidea and their natural enemy complexes in Pakistan. *Technical Bulletin, Commonwealth Institute of Biological Control* 15:59–104.

AICRP-BC. 1977–2014. Annual reports of all India Co-ordinated Research Programme on Biological Control of crop pests and weeds, 1977–2014. Indian Council of Agricultural Research, New Delhi, India.

Anand, R. K., A. K. Gupta, and S. Ghai. 1989. Zoogeographical analysis of Indian Coccinellidae. *Y.E.S. Quarterly* 6(3):13–20.

Anand, R. K., A. K. Gupta, and S. Ghai. 1988. A check-list of Indian Epilachninae (Coccinellidae: Coleoptera) with recorded host plants. *Bulletin of Entomology* 29:121–137.

Aruggoda, A. G. B., R. Shunxiang, and Q. Baoli. 2010. Molecular phylogeny of ladybird beetles (Coccinellidae: Coleoptera) inferred from mitochondria 16S rDNA sequences. *Tropical Agricultural Research* 21:209–217.

Attygalle, A. B., K. D. McCormick, C. L. Blankespoor et al. 1993. Azamacrolides: A family of alkaloids from the pupal defensive secretion of a ladybird beetle (*Epilachna varivestis*). *Proceedings of the National Academy of Sciences, USA* 90:5204–5208.

Belicek, J. 1976. Coccinellidae of Western Canada and Alaska with analyses of the transmontane zoogeographic relationships between the fauna of British Columbia and Alberta (Insecta: Coleoptera: Coccinellidae). *Quaestiones Entomologicae* 12(4):283–409.

Bielawski, R. 1957. Coccinellidae (Coleoptera) von Ceylon. *Verhandlungen der Naturforschenden Gesellschaft in Basel* 68:72–96.

Bielawski, R. 1963. Beiträge zur Kenntnis der Coccinelliden von Afghanistan. III. *Lunds Universitets Årsskrift (N.F.2)* 59:1–21.

Bielawski, R. 1971. Über Coccinellidae (Coleoptera) aus Nepal. Khumbu Himal, Innsbruck-München. *Ergebnisse Forsch Unternehemens Nepal Himalaya* 4:1–9.

Bielawski, R. 1972. Die Marienkäfer (Coleoptera: Coccinellidae) aus Nepal. *Fragmenta Faunistica* 18:283–312.

Bielawski, R. 1979. Ergebnisse der Bhutan-Expedition 1972 des Naturhistorischen Museums in Basel. Coleoptera: Fam. Coccinellidae. *Entomologica Basiliensia* 4:83–125.

Booth, R. G. 1997. A review of the species of *Calvia* (Coleoptera: Coccinellidae) from the Indian subcontinent, with descriptions of two new species. *Journal of Natural History* 31:917–934.

Booth, R. G. 1998. A review of the species resembling *Chilocorus nigrita* (Coleoptera: Coccinellidae): Potential agents for biological control. *Bulletin of Entomological Research* 88:361–367.

Booth, R. G., and A. Polaszek. 1996. The identities of ladybird beetle predators used for whitefly control, with notes on some whitefly parasitoids, in Europe. In *Proceedings of the Brighton Crop Protection Conference-Pests & Diseases, 1996*, pp. 69–74. Brighton, UK: Thornton Heath, British Crop Protection Council.

Booth, R.G. and Pope, R.D. (1986) A review of the genus *Cryptolaemus* (Coleoptera: Coccinellidae) with particular reference to the species resembling *C. montrouzieri* Mulsant. *Bulletin of Entomological Research* 76:701–717.

Booth, R. G., M. L. Cox, and R. B. Madge. 1990. *IIE Guides to Insects of Importance to Man. 3. Coleoptera*. Wallingford, UK: CAB International.

Caltagirone, L. E., and R. L. Doutt. 1989. The history of the vedalia beetle importation to California and its impact on the development of biological control. *Annual Review of Entomology* 34:1–16.

Canepari, C. 1986. Su alcuni Coccinellidi dell'India e Nepal settentrionale del Museo di Storia Naturale di Ginevra (Coleoptera: Coccinellidae). *Revue Suisse de Zoologie* 93:21–36.

Canepari, C. 1997. Coccinellidae (Coleoptera) from the Nepal Himalayas. *Stuttgarter Beiträge zur Naturkunde, Serie A (Biologie)* 565(65):1–65.

Canepari, C. 2003. Coccinellidae (Insecta: Coleoptera) of Nepal from the collection of the Naturkundemuseum Erfurt. In *Biodiversity and Natural Heritage of the Himalaya*, eds. M. Hartmann and H. Baumbach, pp. 261–265. Erfurt, Germany: Verein der Freunde & Förderer des Naturkundemuseums Erfurt e. V., Naturkundemuseum Erfurt.

Ceryngier, P., and I. Hodek. 1996. Enemies of the Coccinellidae. In *Ecology of Coccinellidae*, eds. I. Hodek, and A. Honek, pp. 319–350. Dordrecht, the Netherlands: Kluwer Academic Publishers.

Ceryngier, P., H. E. Roy, and R. L. Poland. 2012. Natural enemies of ladybird beetles. In *Ecology and Behaviour of the Ladybird Beetles (Coccinellidae)*, eds. I. Hodek, H. F. Van Emden, and A. Honek, pp. 375–443. Oxford, UK: Wiley-Blackwell.

Channabasavanna, G. P., and M. Puttarudriah. 1957. Some predators of mites in Mysore. *The Mysore Agricultural Journal* 32(3–4):179–185.

Chazeau, J., H. Fürsch, and H. Sasaji. 1989. Taxonomy of Coccinellidae. *Coccinella* 1:6–8.

Chazeau, J., H. Fürsch, and H. Sasaji. 1990. Taxonomy of Coccinellidae (corrected version). *Coccinella* 2(2):4–6.

Chengyi, C. 1992. *Fauna of Yunnan. Coccinellidae*. Kunming, China: Yunnan Science & Technology Publication.

Chong, J. H., L. F. Aristizábal, and S. P. Arthurs. 2015. Biology and management of *Maconellicoccus hirsutus* (Hemiptera: Pseudococcidae) on ornamental plants. *Journal of Integrated Pest Management* 6(1):1–14.

Crotch, G. R. 1874. *A Revision of the Coleopterous Family Coccinellidae*. London, UK: Jason.

Crowson, R.A. 1955. *The natural classification of the families of Coleoptera*. N. Lloyd, London, 214 p.

DeBach, P., and D. Rosen. 1991. *Biological Control by Natural Enemies*. Cambridge, UK: Cambridge University Press.

Dixon, A. F. G. 2000. *Insect Predator-Prey Dynamics: Ladybird Beetles and Biological Control*. Cambridge, UK: Cambridge University Press.

Doutt, R. L. 1958. Vice, virtue, and the Vedalia. *Bulletin of the Entomological Society of America* 4:119–123.

Dreistadt, S. H., and M. L. Flint. 1996. Melon aphid (Hemiptera: Aphididae) control by inundative convergent lady beetle (Coleoptera: Coccinellidae) release on chrysanthemum. *Environmental Entomology* 25:688–697.

Duang, T. H., and W. S. Frederick. 1974. Morphology and taxonomy of the known pupae of Coccinellidae (Coleoptera) of North America, with a discussion of phylogenetic relationships. *Contributions of the American Entomological Institute* 10(6):1–125.

Dumbleton, L. D. 1936. The biological control of fruit pests in New Zealand. *New Zealand Journal of Science and Technology*, 18:588–592.

Eisner, T., and M. Eisner. 1992. Operation and defensive role of "gin traps" in a Coccinellid pupa (*Cycloneda sanguinea*). *Psyche* 99:265–273.

Elliott, N., R. Kieckhefer, and W. Kauffman. 1996. Effects of an invading coccinellid on native coccinellids in an agricultural landscape. *Oecologia* 105:537–544.

Escalona, H., A. Zwick, H. S. Li et al. 2017. Molecular phylogeny reveals food plasticity in the evolution of true ladybird beetles (Coleoptera: Coccinellidae: Coccinellini). *BMC Evolutionary Biology* 17:151. doi: 10.1186/s12862-017-1002-3.

Evans, E. W. 2003. Searching and reproductive behaviour of female aphidophagous ladybirds (Coleoptera; Coccinellidae): A review. *European Journal of Entomology* 100:1–10.

Evans, E. W., and J. G. Swallow. 1993. Numerical responses of natural enemies to artificial honeydew in Utah alfalfa. *Environmental Entomology* 22:1392–1401.

Evans, E. W., and S. England. 1996. Indirect interactions in biological control of insects: Pests and natural enemies in alfalfa. *Ecological Applications* 6:920–930.

Fabricius, J. C. 1775. *Systema Entomologiae*. Lipsiae, Germany: Officina Libraria Kortii.

Fabricius, J. C. 1798. *Supplementum Entomologiciae Systematicae*. Hafniae [=Copenhagen].

Froggatt, W. W. 1907. *Australian Insects*. Sydney, Australia: Octavo.

Gage, J. H. 1920. The larvae of the Coccinellidae. *Illinois Biological Monographs* 6:1–62.

Galbreath, R. A., and P. J. Cameron. 2015. The introduction of the eleven-spotted ladybird *Coccinella undecimpunctata* L. (Coleoptera: Coccinellidae) to New Zealand in 1874: A spurious record created by cumulative misreporting. *New Zealand Entomologist* 38:1, 7–9. doi: 10.1080/00779962.2014.924467.

Ghani, M. A. 1962. A note on the identity of some species of the genus *Ballia* (Coleoptera: Coccinellidae). *Proceedings of the Royal Entomological Society of London(B)* 31:7–8.

Ghorpade, K. D. 1974. Description of a new *Cryptogonus* Mulsant from Bangalore, Southern India (Coleoptera: Coccinellidae). *Oriental Insects* 8:55–60.

Ghorpade, K. D. 1976. An undescribed species of *Illeis* (Coleoptera: Coccinellidae) from South India. *Oriental Insects* 10:579–585.

Ghorpade, K. D. 1977. A new species of *Pseudoscymnus* (Coleoptera: Coccinellidae) predacious on coconut scale in peninsular India. *Journal of Natural History* 11:465–469.

Ghorpade, K. D. 1979a. Further notes on *Perilitus coccinellae* (Hymenoptera: Braconidae) in India. *Current Research* 8:112–113.

Ghorpade, K. D. 1979b. On the association of some Coccinellidae (Coleoptera) with spider nests. *Current Research* 8:105–106.

Ghorpade, K. D., and H. Sasaji. 1977. On *Perilitus coccinellae* (Schrank) (Hymenoptera: Braconidae), an endoparasite of adult Coccinellidae (Coleoptera), in Karnataka. *Mysore Journal of Agricultural Sciences* 11:55–59.

Giorgi, J. A., N. J. Vandenberg, J. V. McHugh et al. 2009. The evolution of food preferences in Coccinellidae. *Biological Control* 51:215–231.

Goeden, R. D., and S. M. Louda. 1976. Biotic interference with insects imported for weed control. *Annual Review of Entomology* 21:325–342.

González G. F. 2008. Lista y distribución geográfi ca de especies de Coccinellidae (Insecta: Coleoptera) presentes en Chile. *Boletin del Museo Nacional de Historia Natural, Chile* 57:77–107.

Gordon, R. D. 1985. The Coccinellidae of America, north of Mexico. *Journal of the New York Entomological Society* 93:1–912.

Greathead, D. J. 1977. Biological control of mealybugs (HEMIPTERA: Pseudococcidae) with special reference to cassava mealybug (*Phenacoccus manihoti* Mat.-Fer.). In *Proceedings of the International Workshop on the cassava mealybug, Phenacoccus manihoti Mat.-Ferr. (Pseudococcidae)*, eds. K. F. Nwanze and K. Leuschner, pp. 70–80. Mvuazi, Zaire: INERA.

Greathead, D. J. 1986. Parasitoids in classical biological control. In *Insect Parasitoids*, eds. J. Waage and D. J. Greathead, pp. 289–298. New York: Academic Press.

Hagen, K. S. 1962. Biology and ecology of predaceous Coccinellidae. *Annual Review of Entomology* 7:289–326.

Hagen, K. S. 1974. The significance of predaceous Coccinellidae in biological and integrated control of insects. *Entomophaga, Memoires Hors de Series* 7:25–44.

Hagen, K. S. 1987. Nutritional ecology of terrestrial predators. In *Nutritional Ecology of Insects, Mites, Spiders, and Related Invertebrates*, eds. F. Slansky and J. G. Rodriguez, pp. 533–577. New York: John Wiley & Sons.

Hagen, K. S. and J. M. Franz. 1973. A history of biological control. In *History of Entomology* eds. R. F. Smith, T. E. Mittler, and C. N. Smith, pp. 433–476. Palo Alto, CA: Annual Reviews.

Hoang, D. N. 1982. *Bo Rua Coccinellidae o Viet Nam (Insecta, Coleoptera) Tap 1.* Hanoi, Vietnam: Nha Xuat Ban Khoa Hoc Va Ky Thuat. (In Vietnamese with English summary.)

Hoang, D. N. 1983. *Bo Rua Coccinellidae o Viet Nam (Insecta, Coleoptera) Tap 2.* Hanoi, Vietnam: Nha Xuat Ban Khoa Hoc Va Ky Thuat. (In Vietnamese with English summary.)

Hodek, I. 1967. Bionomics and ecology of predaceous Coccinellidae. *Annual Review of Entomology* 12:79–104.

Hodek, I. 1973. *Biology of Coccinellidae*. The Hague, the Netherlands: W. Junk.

Hodek, I., and A. Honěk. 1996. *Ecology of Coccinellidae*. Dordrecht, the Netherlands: Kluwer Academic Publishers.

Hodek, I., H. F. Van Emden, and A. Honěk. 2012. *Ecology and Behaviour of the Ladybird Beetles (Coccinellidae)*. Oxford, UK: Wiley-Blackwell.

Holloway, G. J., P. W. de Jong, and M. Ottenheim. 1993. The genetics and cost of chemical defense in the two-spot ladybird (*Adalia bipunctata* L.). *Evolution* 47:1229–1239.

Holloway, G. J., P. W. de Jong, P. M. Brakefield, et al. 1991. Chemical defence in ladybird beetles (Coccinellidae). I. Distribution of coccinelline and individual variation in defence in 7-spot ladybirds (*Coccinella septempunctata*). *Chemoecology* 2:7–14.

Hope, F. W. 1831. Synopsis of the new species of Nepaul insects in the collection of Major General Hardwicke. In *The Zoological Miscellany*, ed. Gray, J.E., pp. 21–32. London, UK: Treuttel, Wurte and Co.

Horn, D. J. 1991. Potential impact of *Coccinella septempunctata* on endangered Lycaenidae (Lepidoptera) in Northwest Ohio, USA. In *Behaviour and Impact of Aphidophaga*, eds. L. Polgar, R. J. Chambers, A. F. G. Dixon, et al., pp. 159–162. The Hague, the Netherlands: SPB Academic Publishing.

Iablokoff-Khnzorian, S. M. 1972. Les types de Coccinellidae de la collection Motschulsky (Coléoptères Coccinellidae). *Nouvelle Revue d'Entomologie* 2(2):163–184.

Iablokoff-Khnzorian, S. M. 1982. *Les Coccinelles Coléoptères-Coccinellidae. Tribu Coccinellini des régions Paléarctique et Orientale*. Paris, France: Boubée. [In French].

Iperti, G. 1999. Biodiversity of predaceous Coccinellidae in relation to bioindication and economic importance. *Agriculture, Ecosystems & Environment* 74:323–342.

Irshad, M. 2001. Distribution, hosts, ecology and biotic potentials of coccinellids of Pakistan. *Pakistan Journal of Biological Sciences* 4(10):1259–1263.

Jadwiszczak, A., and P. Węgrzynowicz. 2003. *World Coccinellidae. Part 1: Epilachninae*. Olsztyn, Poland: Mantis.

Jahn, G. 1992. The ecological significance of the big headed ant in mealybug wilt disease of pineapple. PhD diss. Manoa, Hawaii: University of Hawaii.

Jenkins, C. F. H. 1946. Biological control in Western Australia. *Journal of the Royal Society of Western Australia* 32(1945–1946):1–17.

Kapur, A. P. 1942. Bionomics of some Coccinellidae predaceous on aphids and coccids in north India. *Indian Journal of Entomology* 4:49–66.

Kapur, A. P. 1943. On the biology and structure of the coccinellid, *Thea bisoctonotata* Muls. in north India. *Indian Journal of Entomology* 5:165–171.

Kapur, A. P. 1946. A revision of the genus *Jauravia* Mots. *Annals and Magazine of Natural History* (11)13:73–93.

Kapur, A. P. 1948a. The genus *Tetrabrachys* (*Lithophilus*) with notes on its biology and a key to the species (Coleoptera: Coccinellidae). *Transactions of the Royal Entomological Society of London* 99:319–339.

Kapur, A. P. 1948b. A revision of the tribe Aspidimerini Weise (Coleoptera: Coccinellidae). *Transactions of the Royal Entomological Society of London* 99:77–128.

Kapur, A. P. 1948c. On the Old World species of the genus *Stethorus* Weise. *Bulletin of Entomological Research* 39:297–320.

Kapur, A. P. 1949. On the Indian species of *Rodolia* Mulsant (Coleoptera: Coccinellidae). *Bulletin of Entomological Research* 39:531–538.

Kapur, A. P. 1950. The biology and external morphology of the larvae of Epilachninae (Coleoptera: Coccinellidae). *Bulletin of Entomological Research* 41(1):161–208.

Kapur, A. P. 1954. Mass assemblage of the coccinellid beetle, *Epilachna bisquadripunctata* (Gyllenhal) in Chota Nagpur. *Current Science* 23:230–231.

Kapur, A. P. 1956. Systematic and biological notes on the ladybird beetles predacious on the San Jose scale in Kashmir with description of a new species (Coleoptera: Coccinellidae). *Records of the Indian Museum* 52 (1954):257–274.

Kapur, A. P. 1957. Variation in the colour pattern of certain ladybird beetles from high altitudes in the Himalayas. *Bulletin of the National Institute of Sciences of India* 9:269–273.

Kapur, A. P. 1958. Coccinellidae of Nepal. *Records of the Indian Museum* 53:309–338.

Kapur, A. P. 1962. Geographical variations in the colour patterns of some Indian Ladybeetles (Coccinellidae: Coleoptera). Part I. *Coccinella septempunctata* Linn., *C. transversalis* Fabr., and *Coelophora bissellata* Muls. *Proceedings of the First All India Congress of Zoology (1959)* 2:479–492.

Kapur, A. P. 1963. The Coccinellidae of the third Mount Everest expedition, 1924 (Coleoptera). *Bulletin of the British Museum (Natural History), Entomology* 14:1–48.

Kapur, A. P. 1967. The Coccinellidae (Coleoptera) of the Andamans. *Proceedings of the National Institute of Sciences of India* 32(B) [1966]:148–189.

Kapur, A. P. 1969. On some Coccinellidae of the tribe Telsimiini with descriptions of new species from India. *Bulletin of Systematic Zoology* 1:45–56.

Kapur, A. P. 1973. The Coccinellidae (Coleoptera) of the Italian expedition to Karakoram and Hindu Kush. *Records of the Zoological Survey of India* 67:373–378.

Kawai, A. 1976. Analysis of the aggregation behaviour in the larvae of *Harmonia axyridis* Pallas (Coleoptera: Coccinellidae) to prey colony. *Researches in Population Ecology* 18:123–134.

Ketkar, S. M. 1959. Mass assemblage of the coccinellid beetle, *Chilocorus nigritus* Fabr. on banyan trees in Poona. *Science & Culture* 25:273.

Koch, R. L. 2003. The multicolored Asian lady beetle, *Harmonia axyridis*: A review of its biology, uses in biological control, and non-target impacts. *Journal of Insect Science* 3:1–16.

Koch, R. L., W. D. Hutchison, R. C. Venette et al. 2003. Susceptibility of immature monarch butterfly, *Danaus plexippus* (Lepidoptera: Nymphalidae: Danainae), to predation by *Harmonia axyridis* (Coleoptera: Coccinellidae). *Biological Control* 28:265–270.

Komai, T. 1956. Genetics of lady-beetles. *Advances in Genetics* 8:155–185.

Korschefsky, R. 1931. *Coleopterorum Catalogus. Pars 118. Coccinellidae I.* Berlin, Germany: W. Junk.

Korschefsky, R. 1932 *Coleopterorum Catalogus. Pars 120. Coccinellidae II.* Berlin, Germany: W. Junk.

Kovář, I. 1996a. Morphology and anatomy. In *Ecology of Coccinellidae*, eds. I. Hodek & A. Honěk, pp. 1–18. Dordrecht, the Netherlands: Kluwer Academic.

Kovář, I. 1996b. Phylogeny. In *Ecology of Coccinellidae*, eds. I. Hodek & A. Honěk, pp. 19–31. Dordrecht, the Netherlands: Kluwer Academic.

Kovář, I. 2007. Family Coccinellidae Latreille, 1807. In *Catalogue of Palaearctic Coleoptera*, Vol. 4, eds. I. Löbl and A. Smetana, pp. 568–631. Stenstrup, Denmark: Apollo Books.

Kuznetsov, V. N. 1997. *Lady Beetles of the Russian Far East. Memoir No. 1.* Gainsville, FL: Centre for Systematic Entomology.

Legaspi, J. C., B. C. Legaspi Jr., R. Meagher, Jr. et al. 1996. Evaluation of *Serangium parcesetosum* (Coccinellidae) as a biological control agent of the silverleaf whitefly. *Environmental Entomology* 25:1421–1427.

Lin, K. S. 1987. On the genus *Nothoserphus* Brues, 1940 (Hymenoptera: Serphidae) from Taiwan. *Taiwan Agricultural Research Institute, Special Publication* 22:51–66.

Lynch, L. D., and M. B. Thomas. 2000. Nontarget effects in the biological control of insects with insects, nematodes and microbial agents: The evidence. *Biocontrol News and Information* 21:117N–130N.

Mader, L. 1926–1937. Evidenz der paläarktischen Coccinelliden und ihrer Aberrationen in Wort und Bild. I. Teil: Epilachnini, Coccinellini, Halyziini, Synonychini. *Zeitschrift des Vereins der Naturbeobachter und Sammler, Wien*, 1934(9):289–336.

Magro, A., E. Lecompte, Magne, F. et al. 2010. Phylogeny of ladybirds (Coleoptera: Coccinellidae): Are the subfamilies monophyletic? *Molecular Phylogenetics and Evolution*, 54:833–848.

Majerus, M. E. N. 1994. *Ladybirds*. London, UK: Harper Collins.

Majerus, M. E. N., and G. D. D. Hurst. 1997. Ladybirds as a model system for the study of male-killing symbionts. *Entomophaga* 42:13–20.

Majerus, M. E. N., and T. M. O. Majerus. 1997a. Cannibalism among ladybirds. *Bulletin of the Amateur Entomologist's Society* 56:235–248.

Majerus, M. E. N., and T. M. O. Majerus. 1997b. Predation of ladybirds by birds in the wild. *Entomologist's Monthly Magazine* 133:55–61.

Mani, M. S. 1962. *Introduction to High Altitude Entomology: Insect Life Above the Timber-line in the North-west Himalaya.* London, UK: Methuen.

Mani, M., A. Krishnamoorthy, and S. P. Singh. 1990. The impact of the predator *Cryptolaemus montrouzieri* Mulsant on pesticide-resistant populations of the striped mealybug *Ferrisia virgata* (Ckll.) on guava in India. *Insect Science and its Application* 11:167–170.

Mani, M., and A. Krishnamoorthy. 1997. Australian ladybird beetle, *Cryptolaemus Montrouzieri*. *Madras Agricultural Journal* 84:237–249.

Mani, M., and T. S. Thontadarya. 1988. Field evaluation of *Cryptolaemus montrouzieri* Muls. in the suppression of grape mealybug, *Maconellicoccus hirsutus* (Green). *Journal of Biological Control* 2:14–16.

Maqbool, A., I. Ahmed, P. Kiełtyk, and P. Ceryngier. 2018. *Dinocampus coccinellae* (Hymenoptera: Braconidae) utilizes both Coccinellini and Chilocorini (Coleoptera: Coccinellidae: Coccinellinae) as hosts in Kashmir Himalayas. *European Journal of Entomology* 115(1):2018.000. doi: 10.14411/eje.2018.033.

Marples, N. M., P. M. Brakefield, and R. J. Cowie. 1989. Differences between the 7-spot and 2-spot ladybird beetles (Coccinellidae) in their toxic effects on a bird predator. *Ecological Entomology* 14:79–84.

Mayne, W. W. 1953. *Cryptolaemus montrouzieri* Mulsant in South India. *Nature* 172:85.

Michaud, J. P. 2012. Coccinellids in biological control. In: *Ecology and Behaviour of Ladybird Beetles (Coccinellidae)*, eds. I. Hodek, H. F. Van Emden, and A. Honek, pp. 488–519. Chichester, UK: John Wiley & Sons.

Miyatake, M. 1961a. The East-Asian coccinellid beetles preserved in the California Academy of Sciences, tribe Platynaspini. *Memoirs of the Ehime University* (6)6:67–86.

Miyatake, M. 1961b. The East-Asian coccinellid beetles preserved in the California Academy of Sciences, tribe Serangiini. *Memoirs of the Ehime University* (6)6:135–146.

Miyatake, M. 1961c. The East-Asian coccinellid beetles preserved in the California Academy of Sciences, tribe Hyperaspini. *Memoirs of the Ehime University* (6)6:147–155.

Miyatake, M. 1967. Notes on some Coccinellidae from Nepal and Darjeeling district of India (Coleoptera). *Transactions of the Shikoku Entomological Society* 9:69–78.

Miyatake, M. 1969. The genus *Plotina* and related genera (Coleoptera: Coccinellidae). *Pacific Insects* 11:197–216.

Miyatake, M. 1970. The East-Asian coccinellid beetles preserved in the California Academy of Sciences. Tribe Chilocorini. *Memoirs of the College of Agriculture, Ehime University* 14(3):303–340.

Miyatake, M. 1985. Coccinellidae collected by the Hokkaido University Expedition to Nepal Himalaya, 1968 (Coleoptera). *Insecta Matsumurana (New Series)* 30:1–33.

Miyatake, M. 1994. Revisional studies on Asian genera of the subfamily Sticholotidinae (Coleoptera: Coccinellidae). *Memoirs of the College of Agriculture, Ehime University* 38:223–294.

Mizell, R. F. 2001. *Harmonia axyridis* Pallas, multicoloured Asian lady beetle. http://creatures.ifas.ufl.edu/beneficial/multicolored_Asian_lady_beetle.htm, September 2012 (accessed September 10, 2018).

Motschulsky, V. 1858. Insectes des Indes Orientales. *Etudes Entomologiques* 7:117–122.

Motschulsky, V. 1866. Essai d'un Catalogue des Insectes de l'ile de Ceylan. Supplement. *Bulletin de la Societe Imperiale des Naturalistes de Moscou* 39:393–446.

Mulsant, E. 1846. *Securipalpes. Histoire Naturelle des Coleopteres de France.* Vol. 4. Paris, France: Maison.

Mulsant, E. 1850. Species des Coléoptères trimères sécuripalpes. *Annales des Sciences Physiques et Naturelles, d'Agriculture et d'Industrie, Lyon* (2)2:1–1104.

Mulsant, E. 1853a. Supplément à la monographie des coléoptères trimères sécuripalpes. *Annales de la Société Linnéenne de Lyon(N.S.)* 1:129–333.

Mulsant, E. 1853b. Supplement a la monographie des Coléoptères trimères sécuripalpes. *Opuscules Entomologiques* 3:1–178.

Mulsant, E. 1866. *Monographie des Coccinellides. 1re partie Coccinelliens.* Paris, France: Savy & Deyrolle.

Nagarkatti, S., and M. A. Ghani. 1972. Ecology and biology of predators. Coleoptera: Coccinellidae. In *Studies on Predators of Adelges spp. in the Himalayas,* eds. V. P. Rao and M.A. Ghani, pp. 58–88. Bangalore, India: Commonwealth Institute of Biological Control.

Narayanan, E. 1957. A note on the performance of *Cryptolaemus montrouzieri* Muls. in citrus orchards at Burnihat (Assam). *Technical Bulletin, Commonwealth Institute of Biological Control* 9:137–138.

Nechols, J. R., J. J. Obrycki, C. A. Tauber et al. 1996. Potential impact of natural enemies on *Galerucella* spp. (Coleoptera: Chrysomelidae) imported for biological control of purple loosestrife: A field evaluation. *Biological Control* 7:60–66.

Nedvěd, O., and I. Kovář. 2012. Appendix: List of genera in tribes and subfamilies. In *Ecology and Behaviour of the Ladybird Beetles (Coccinellidae),* eds. I. Hodek, A. Honěk, and H. F. van Emden, pp. 526–531. Chichester, UK: John Wiley & Sons.

Obrycki, J. J., and T. J. Kring. 1998. Predaceous Coccinellidae in biological control. *Annual Review of Entomology* 43:295–321.

Obrycki, J. J., N. C. Elliott, and K. L. Giles. 2000. Coccinellid introductions: Potential for and evaluation of nontarget effects. In *Nontarget Effects of Biological Control,* eds. P. A. Follett, and J. J. Duan, pp. 127–145. Dordrecht, the Netherlands: Kluwer.

Omkar and A. Pervez. 2004. Predaceous coccinellids in India: Predator-prey catalogue. *Oriental Insects* 38:27–61.

Pang, H., B. Huang, and X. F. Pang. 2002. Coleoptera: Coccinellidae. In: *Fauna of Insects in Fujian Province of China,* Vol. 6, ed. B. K. Huang, pp. 281–357. Fujian, China: Fujian Science Technology Publishing House.

Pang, X. F., and J. L. Mao. 1979. Coleoptera: Coccinellidae, II. *Khongguo Jingi Kunchang Zhi* 14:1–170.

Pemberton, R. W., and N. J. Vandenberg. 1993. Extrafloral nectar feeding by ladybird beetles (Coleoptera: Coccinellidae). *Proceedings of the Entomological Society of Washington* 95:139–151.

Phuoc, D. T., and F. W. Stehr. 1974. Morphology and taxonomy of the known pupae of Coccinellidae (Coleoptera) of North America, with a discussion of phylogenetic relationships. *Contributions of the American Entomological Institute* 10:1–125.

Poorani, J. 2002. An annotated checklist of the Coccinellidae (Coleoptera) (excluding Epilachninae) of the Indian Subregion. *Oriental Insects* 36:307–383.

Poorani, J. 2008. Coccinellidae of the Indian Subcontinent. http://www.angelfire.com/bug2/j_poorani/index.html (accessed September 15, 2018).

Poorani, J., A. Ślipiński, and R. G. Booth. 2008. A revision of *Synona* Pope (Coleoptera: Coccinellidae: Coccinellini). *Annales Zoologici* 58:579–594.

Pope, R. D. 1979. Wax production by coccinellid larvae (Coleoptera). *Systematic Entomology* 4:171–196.

Pope, R. D. 1989. A revision of the Australian Coccinellidae (Coleoptera). Part I. Subfamily Coccinellinae. *Invertebrate Taxonomy* 3:633–735.

Puttarudriah, M., and G. P. Channabasavanna. 1953. Beneficial coccinellids of Mysore-I. *Indian Journal of Entomology* 15:87–96.

Puttarudriah, M., and G. P. Channabasavanna. 1955. Beneficial coccinellids of Mysore-II. *Indian Journal of Entomology* 17:1–5.

Puttarudriah, M., and G. P. Channabasavanna. 1956. Some beneficial coccinellids of Mysore. *Journal of the Bombay Natural History Society* 54:156–159.

Puttarudriah, M., and G. P. Channabasavanna. 1957. Notes on some predators of mealybugs (Coccidae, Hemiptera). *The Mysore Agricultural Journal* 32:4–19.

Puttarudriah, M., G. P. Channabasavanna, and B. Krishnamurti. 1952. Discovery of *Cryptolaemus montrouzieri* Mulsant (Coccinellidae, Coleoptera, Insecta) in Bangalore, South India. *Nature* 169:377.

Puttaswamy, M, and G. P. Channabasavanna. 1977. Biology of *Stethorus pauperculus* Weise (Coleoptera: Coccinellidae), a predator of mites. *Mysore Journal of Agricultural Sciences* 11:81–89.

Ramakrishna Ayyar, T. V. 1925. An undescribed coccinellid beetle of economic importance. *Journal of the Bombay Natural History Society* 30:491–492.

Ramaraju, K., and J. Poorani. 2012. A new species of *Coccipolipus* (Acari: Podapolipidae) parasitic on the giant coccinellid beetle from India. *International Journal of Acarology* 38:260–296.

Rao, V. P., M. A. Ghani, T. Sankaran, et al. 1971. *A Review of the Biological Control of Insects and Other Pests in South-east Asia and the Pacific Region.* Slough: Commonwealth Institute of Biological Control.

Ren, S. X., X. M. Wang, H. Pang et al. 2009. *Colored Pictorial Handbook of Ladybird Beetles in Beijing.* Beijing, China: Science Press [In Chinese].

Ricci, C. 1979. L'apparato boccale pungente succhiante della larva di *Platynaspis luteorubra* Goeze (Col., Coccinellidae). *Bollettino del Laboratorio di Entomologia Agraria 'Filippo Silvestri' di Portici* 36:179–198.

Richards, A. M. 1980. Sexual selection, guarding and sexual conflict in a species of Coccinellidae (Coleoptera). *Journal of the Australian Entomological Society* 19:26.

Richerson, J. V. 1970. A world list of parasites of Coccinellidae. *Journal of the Entomological Society of British Columbia* 67:33–48.

Riddick, E. W., T. E. Cottrell, and K. A. Kidd. 2009. Natural enemies of the Coccinellidae: Parasites, pathogens, and parasitoids. *Biological Control* 51:306–312.

Robertson, J., A. Ślipiński, M. Moulton, et al. 2015. Phylogeny and classification of Cucujoidea and the recognition of a new superfamily Coccinelloidea (Coleoptera: Cucujiformia). *Systematic Entomology* 40:745–778.

Samways, M. J., and S. J. Wilson. 1988. Aspects of the feeding behaviour of *Chilocorus nigritus* (F.) (Col., Coccinellidae) relative to its effectiveness as a biocontrol agent. *Journal of Applied Entomology* 106:177–182.

Samways, M. J., R. Osborn, and T. L. Saunders. 1997. Mandible form relative to main food types in ladybirds (Coleoptera: Coccinellidae). *Biocontrol Science and Technology* 7:275–286.

Samways, M. J., R. Osborn, H. Hastings et al. 1999. Global climate change and accuracy of prediction of species' geographical ranges: Establishment success of introduced ladybirds (Coccinellidae, *Chilocorus* spp.) worldwide. *Journal of Biogeography* 26:795–812.

Sankaran, T. 1974. Natural enemies introduced in recent years for biological control of agricultural pests in India. *Indian Journal of Agricultural Sciences* 44:425–433.

Sasaji, H. 1968a. Phylogeny of the family Coccinellidae (Coleoptera). *Etizenia* 35:1–37.

Sasaji, H. 1968b. Descriptions of the known coccinellid larvae of Japan and the Ryukyus (Coleoptera). *Memoirs of the Faculty of Education, Fukui University, Series II* 18:93–135.

Sasaji, H. 1971. *Fauna Japonica. Coccinellidae (Insecta: Coleoptera).* Tokyo, Japan: Academic Press of Japan.

Sasaji, H., and M. Akamatsu. 1979. Reproductive continuity and genetic relationships in the forms of the genus *Menochilus* (Coleoptera: Coccinellidae). *Memoirs of the Faculty of Education, Fukui University, Series II (Natural Sciences)* 29:1–18.

Sasaji, H., R. Yahara, and M. Saito. 1975. Reproductive isolation and species specificity in two ladybirds of the genus *Propylea* (Coleoptera). *Memoirs of the Faculty of Education, Fukui University, Series II (Natural Sciences)* 25:13–34.

Savoiskaya, G. I. 1983. *Larvae of the coccinellid (Coleoptera: Coccinellidae) fauna of the USSR.* Leningrad, Russia: Nauka. [In Russian.]

Savoiskaya, G. I., and B. Klausnitzer. 1973. Morphology and taxonomy of the larvae with keys for their identification. In: *Biology of the Coccinellidae,* ed. Hodek, I., pp. 36–55. The Hague, the Netherlands: W. Junk.

Schaefer, P. W. 1983. *Natural Enemies and Host Plants of Species in the Epilachninae (Coleoptera: Coccinellidae): A World List.* Newark, NJ: University of Delaware.

Schaefer, P. W., and V. P. Semyanov. 1992. Arthropod parasites of *Coccinella septempunctata* (Coleoptera: Coccinellidae): World parasite list and bibliography. *Entomological News* 103:125–134.

Schauff, M. E. 2003. Collecting and preserving insects and mites: Techniques and tools. https://www.ars.usda.gov/ARSUserFiles/80420580/CollectingandPreservingInsectsandMites/collpres.pdf (accessed October 10, 2018).

Schilder, F. A., and M. Schilder. 1928. Die Nahrung der Coccinelliden und ihre Beziehung zur Verwandtschaft der Arten. *Arbeiten aus der Biologischen Reichsanstalt für Land- und Forstwirtschaft* 16:213–282.

Schroder, R. F. W. 1979. Host specificity tests of *Coccipolipus epilachnae,* a mite parasitic on Mexican bean beetles. *Environmental Entomology* 8:46–47.

Schroeder, F. C., S. R. Smedley, L. K. Gibbons et al. 1998. Polyazamacrolides from ladybird beetles: Ring-size selective oligomerization. *Proceedings of the National Academy of Science, USA* 95:13387–13391.

Seago, A., J. A. Giorgi, and A. Ślipiński. 2011. Phylogeny, classification and evolution of ladybird beetles (Coleoptera: Coccinellidae) based on simultaneous analysis of molecular and morphological data. *Molecular Phylogenetics and Evolution* 60(1):137–151.

Semyanov, V. P. 1980. Ways to employ the Coccinellidae. *Zashchita Rastenii* 8:20–21.

Simmonds, F. J. 1958. The effect of lizards on the biological control of scale insects in Bermuda. *Bulletin of Entomological Research* 49:601–612.

Ślipiński, A. 2007. *Australian Ladybird Beetles (Coleoptera: Coccinellidae): Their Biology and Classification.* Canberra, Australia: ABRS.

Ślipiński, A., A. Hastings, and B. Boyd. 2008. Ladybirds of Australia. http://www.ento.csiro.au/biology/ladybirds/ladybirds.htm (accessed October 10, 2018).

Ślipiński, S.A. and J. Pakaluk. 1991. Problems in the classification of the Cerylonid series of Cucujoidea (Coleoptera). pp. 79–88. In: M. Zunino, X. Belles and M. Blas (eds.), *Advances in Coleopterology.* European Association of Coleopterology, Silvestrelli and Cappelletto, Torino.

Ślipiński, A., and W. Tomaszewska. 2010. Coccinellidae Latreille 1802. In *Handbuch der Zoologie/Handbook of Zoology. Band/Volume IV. Arthropoda: Insecta Teilband/Part 38. Coleoptera, Beetles. Volume 2. Morphology and Systematics (Polyphaga partim),* eds. R. A. B. Leschen, R. G. Beutel, and J. F. Lawrence, pp. 454–472. Berlin, Germany: W. DeGruyter.

Sloggett, J. J., and M. E. N. Majerus. 2000. Habitat preferences and diet in the predatory Coccinellidae (Coleoptera): An evolutionary perspective. *Biological Journal of the Linnean Society* 70:63–88.

Srikanth, J., S. Easwaramoorthy, and N. K. Kurup. 2001. *Camponotus compressus* F. interferes with *Cryptolaemus montrouzieri* Mulsant activity in sugarcane. *Insect Environment* 7:51–52.

Stebbing, E. P. 1903. Coleoptera 2. Notes upon the known predaceous Coccinellidae of the Indian region, Part I. *Indian Museum Notes* VI(1):47–62.

Subramaniam, T. V. 1924a. A note on colour variations in a common ladybird beetle, *Chilomenes sexmaculata* Fb. In *Report of the Proceedings of the Fifth Entomological Meeting, 1923,* ed. T. B. Fletcher, pp. 363–364. Calcutta, India: Government Press.

Subramaniam, T. V. 1924b. Some coccinellids of South India. In *Report of the Proceedings of the Fifth Entomological Meeting, 1923,* ed. T. B. Fletcher, pp. 108–118. Calcutta, India: Government Press.

Szawaryn, K., L. Bocak, A. Ślipiński et al. 2015. Phylogeny and evolution of phytophagous ladybird beetles (Coleoptera: Coccinellinae: Epilachnini), with recognition of new genera. *Systematic Entomology* 40(3):547–569.

Thompson, W. R. 1951. The specificity of host relations in predacious insects. *The Canadian Entomologist* 83:262–269.

Tomaszewska, W., and K. Szawaryn. 2016. Epilachnini (Coleoptera: Coccinellidae)—A revision of the world genera. *Journal of Insect Science* 16(1):101.

Van Emden, F. 1949. *The Larvae of British Beetles: VII. Coccinellidae. Entomologists' Monthly Magazine* 85:265–283.

Vandenberg, N. J. 2002. Family 93. Coccinellidae Latreille 1807. In *American Beetles, Volume 2. Polyphaga: Scarabaeoidea Through Curculionoidea,* eds. R. H. Jr. Arnett, P. E., Thomas, P. E., Skelley, and J. H. Frank, pp. 371–389. Boca Raton, FL: CRC Press.

Vesey-Fitzgerald, D. 1941. Progress of the control of coconut feeding Coccidae in Seychelles. *Bulletin of Entomological Research* 32:161–164.

Vesey-Fitzgerald, D. 1953. Review of the biological control of coccids on coconut palms in Seychelles. *Bulletin of Entomological Research* 44:405–413.

Yu, G., 1994. Cladistic analyses of the Coccinellidae (Coleoptera). *Entomologica Sinica* 1:17–30.

Yu, G. Y. 2009. *Chinese Lady Beetles (The subfamily Coccinellinae).* Beijing, China: Science Press, 180 p.

Yu, G. Y. 2011. *The Coccinellidae of Taiwan.* Beijing, China: Chemical Industry Press, Beijing. [In Chinese with English summary.]

Yu, G. Y., and H. Y. Wang. 1999. *Guidebook to Lady Beetles of Taiwan.* Taipei, Taiwan: Shih Pui Ni. http://www.angelfire.com/bug2/j_poorani/morphology.htm.

15 Flea Beetles of South India (Coleoptera: Chrysomelidae: Galerucinae: Alticini)

K. D. Prathapan

CONTENTS

Introduction 247
Natural History of Flea Beetles in South India 247
Key to the Flea Beetle Genera of South India 249
Checklist of Flea Beetles of South India 267
Flea Beetles as Pests 267
Acknowledgements 271
References 271

INTRODUCTION

Flea beetles constitute Alticini, the largest tribe within Chrysomeloidea, the leaf beetles. This is a polyphyletic tribe of about 9900 valid species classified in 577 valid genera (Konstantinov 2016). They have earned the name "flea beetles" due to the enlarged hind femora and remarkable jumping ability. These are mostly 2–10 mm long leaf beetles occurring in all terrestrial ecosystems. With adults measuring ~2 cm in length, *Podontia lutea* (Olivier), the golden leaf beetle, is the largest flea beetle in the world (Furth 1999). All flea beetles are phytophagous, mostly on higher plants, while a few genera are specialists on ferns or mosses. There are also aberrant ones occurring in leaf litter and humus. Habits of the immature stages of flea beetles vary considerably. Eggs are laid on the plant or in soil. Larvae are root feeders, leaf miners, or feed openly on the leaves. Fruit borers and stem borers are also known among flea beetles. Most larvae of the blepharidine flea beetles feed openly on the leaves and are covered themselves with excreta (Prathapan and Chaboo 2011). In the case of open leaf feeding larvae, such as *Altica* (Figure 15.5) and *Podontia* (Figure 15.3), eggs are deposited in clusters on the leaves (Figures 15.2, 15.4). Pupation, as a rule, occurs in soil, exceptions being rare. Alticine adults feed on the leaves in a fashion characteristic to the species. Consumption of flowers is rare, but pollen feeding is common amongst members of certain genera.

NATURAL HISTORY OF FLEA BEETLES IN SOUTH INDIA

In south India, 192 named species placed in 54 genera are known. Formal studies on flea beetles of south India started with J. S. Baly's description of *Podontia congregata* (Figure 15.1), endemic to the locality, in 1865, though the country of occurrence of the species was unknown to him. This was followed by publication of several species and genera by Martin Jacoby, mostly from the Nilgiri Hills. Maulik's (1926) monograph on Chrysomelinae and Halticinae, under the *Fauna of British India* series, is the first comprehensive study on the flea beetles of the Indian subcontinent. His work is outdated, yet remains the most useful resource on the alticine fauna of south India. The latest monograph on flea beetles of the subcontinent was published by Scherer in German in 1969.

Information on the classification and biology of south Indian alticines is limited. Several genera and species still await discovery as biodiverse areas such as the Eastern Ghats and much of the Western Ghats remain unexplored.

Leaf beetles are specialist herbivores evolved over millions of years of coevolution with their host plants. Host plant information is important in the study of leaf beetles as it provides critical insights into the classification and phylogeny of the beetles as well as their plant hosts. Most flea beetles are monophagous or oligophagous, while generalists being exceptions. Their trophic selections are often confined to a single species or genus of plants. Host plants are known for about 73 species (38%) in 27 genera in south India, due to the recent efforts. However, much of the information published in the last century is inaccurate and needs verification. World over, beetle-plant interactions have earned considerable attention as powerful tools in solving problems in ecology and systematics. Unfortunately, host plant information is not available for the vast majority of the Indian leaf beetles. Jolivet and Hawkeswood (1995), in their world review of the host plants of leaf beetles, have treated India as "*the terra-incognita* for the leaf beetle/host-plant relationship". Their statement "*All of the recent Indian lists are at least 90% inaccurate*" prevails due to lack of leaf beetle systematicians in the country who could incorporate host plant information into the systematics of leaf beetles.

Life histories of most alticines in south India remain unknown. The best known are those of the common *Altica* [probably *A. aenea* (Olivier): see Reid and Beatson 2015], the endemic pepper *pollu* beetle *Lanka ramakrishnai* Prathapan & Viraktamath, and *Podontia congregata* Baly. Appanna and Sastry. (1958) (as *Haltica cyanea*) and Sankaran et al. (1967)

(as *A. caerulea*) worked out the life cycle of *A. ?aenea* in south India on *Jussiaea* sp. and *Jussiaea repens* L., respectively. This flea beetle thrives on members of Onagraceae and is wrongly implicated as a pest of rice, prompting development of chemical control measures (see Christudas et al. 1972). Visalakshy and Jayanth (1998) explored the possibility of its use in the control of *Ludwigia adscendens* (L.) Hara. in south India. The unique biology of *Lanka ramakrishnai*—larva is a fruit borer (Figures 15.8, 15.9)—was reported by Ayyar in 1919. Devasahayam et al. (1998) studied its reproductive system. Biology and host plants of the largest south Indian flea beetle, *Podontia congregata*, was reported only in 2011 (Pathapan and Chaboo 2011). Larva (Figure 15.3), festooned with excreta, is an open leaf feeder on *Garcinia gummi-gutta* (L.) N. Robson (Clusiaceae). The cosmopolitan *Longitarsus* Latreille, with more than 500 named species, is the most speciose genus of the tribe. Most larvae of the genus are root feeders. However, in the case of the south Indian *Longitarsus limnophilae* Prathapan & Viraktamath, the aquatic habitat of the host plant prevents access to soil, making root feeding and pupation in soil impossible. Hence, their eggs are laid on tender leaf buds and the larvae feed on leaves. The full grown larvae bore into the stem and pupate inside the aerenchyma (Figure 15.10). This is the only Indian species that does not pupate in soil (Prathapan and Viraktamath 2011). Based on larval mode of life, south Indian flea beetles can be divided into four major ecological groups: (i) root feeders (most *Longitarsus* and majority of the little-known species); (ii) open leaf feeders (e.g., *Altica*, Figure 15.5; *Podontia*, Figure 15.3; and *Ophrida*, Figure 15.6); (iii) leaf miners (e.g., *Chilocoristes*, Figure 15.7; *Halticorcus, Sphaeroderma*, and *Aphthona nigrilabris* Duvivier); and (iv) fruit borers (*Lanka ramakrishnai*, Figures 15.8, 15.9). Larva of *Demarchus pubipennis* Jacoby, that feeds on the leaves of *Dendrophthoe falcata* (L. f.) Etting., is an intermediate between leaf miners and open feeders as they bore initially into the thick, fleshy leaves of the host plant and later feed openly (K. D. Prathapan, unpublished data). At least two species, the introduced *Chaetocnema confinis* Crotch (Prathapan and Balan 2010) and *Hyphasis sita* (Maulik) (K. D. Prathapan unpublished data) reproduce exclusively through parthenogenesis, as males do not occur here.

FIGURES 15.1–15.6 **1.** *Podontia congregata* (adult); **2.** *Podontia congregata* (egg mass); **3.** *Podontia congregata* (larva); **4.** *Altica ?aenea* (eggs); **5.** *Altica ?aenea* (larva); **6.** *Asiophrida marmorea* (larva).

FIGURES 15.7–15.10 **7.** *Chilocoristes* sp.—larval leaf mine; **8.** *Lanka ramakrishnai* – larva inside berry. (Adapted from Prathapan, K. D., and Viraktamath, C. A., *Zootaxa*, 1681, 1–30, 2008. With permission); **9.** Pollu berries due to infestation of *Lanka ramakrishnai*; **10.** *Longitarsus limnophilae* – pupa inside stem aerenchyma (Adapted from Prathapan K. D., and Viraktamath, C. A., *ZooKeys*, 87, 1–10, 2011. With permission.).

Natural enemies of flea beetles in south India include the eugregarine parasite *Gregarina phygasiae* Prema & Janardanan (Apicomplexa: Cephalina) on *Phygasia silacea* Illiger (Prema and Janardanan 1991), *Bacillus* sp. (Sankaran et al. 1967) and *Microctonus* sp. (Shamalamma et al. 1977) on *Altica ?aenea,* and *Ooencyrtus keralensis* Hayat & Prathapan (Hymenoptera: Encyrtidae) and *Eucanthecona parva* (Distant) (Heteroptera: Pentatomidae) on *Podontia congregata.* Natural enemies recorded on *Lanka ramakrishnai,* the most thoroughly studied south Indian flea beetle, are an unnamed spider (Premkumar 1980), unidentified entomophagous nematode (Mermithidae), and the weaver ant *Oecophylla smaragdina* Fabr. (Formicidae) (Devasahayam and Koya 1994.). In laboratory studies, the entomopathogenic fungi *Beauveria bassiana* (Bals.-Criv.) and *B. brongniartii* (Saccardo) Petch, isolated from coffee berry borer, were pathogenic to adult *pollu* beetle (Devasahayam et al. 2005). Raj and Reddy (1989) reported parasitism by the nematode *Howardula* sp. on *Longitarsus birmanicus* Jacoby (as *L. belgaumensis*).

Alticines in south India exhibit considerable diversity in morphology and adaptations. *Podontia congregata* (Figure 15.1), the largest south Indian flea beetle, measures up to 14.1 mm, while the smallest *Clavicornaltica rileyi* is only 1.5 mm (Doeberl 2003). Apterism is common in the genera *Phaelota* and *Ivalia*. In both the genera, alate species are widely distributed, across altitudinal gradient, while the apterous ones are narrowly distributed and confined to high altitudes. In *Phaelota*, two species groups are discernible. The alate species, which are larger and widely distributed, feed on ferns, while the smaller apterous species occur in moss cushions in altitudes of about 1000 m or more. Wing polymorphism has been recorded in one species, *Longitarsus serrulatus* Prathapan et al. Recently Konstantinov et al. (2018) reported a unique variant of masquerade in flea beetles, including ones from south India, which resemble their feeding damage to escape predation.

KEY TO THE FLEA BEETLE GENERA OF SOUTH INDIA

1 Antenna with nine (Figure 15.14) or ten (Figure 15.15) antennomeres..2
- Antenna with eleven antennomeres (Figure 15.16).....3
2.(1) Antenna with nine antennomeres (Figure 15.14); distal antennomeres flat, triangularly widened; body oval in dorsal view ....***Nonarthra* Baly**
- Antenna with ten antennomeres (Figure 15.15); distal antennomeres narrow, neither flattened nor triangularly widened; body elliptical in dorsal view.........
....***Psylliodes* Latreille**
3.(1) Antenna geniculate and terminal antennomeres forming a stout club (Figure 15.17) ..... ***Clavicornaltica* Scherer**
- Antenna neither geniculate nor terminal antennomeres forming a stout club (Figure 15.16) 4
4.(3) Posterior femur with a long apical spine, exceeding tibia in length (Figure 15.18) ....
....***Aphthonoides* Jacoby**
- Posterior femur with a short apical spine, much shorter than metatibia in length (Figure 15.20) or apical spine absent (Figures 15.21, 15.25) 5
5.(4) Posterior tibia with a broad apical spine (Figure 15.19) ....***Paradibolia* Baly**
- Posterior tibia with apical spine simple (Figure 15.20) or absent (Figures 15.21, 15.25) ... 6
6.(5) Claw tarsomere of hind leg strongly dilated (Figure 15.22) 7
- Claw tarsomere of hind leg not strongly dilated (Figures 15.20, 15.21).. 8
7.(6) Pronotum with antebasal transverse impression; lateral pronotal margin broadly explanated (Figure 15.23) ....***Philopona* Weise**
- Pronotum without antebasal transverse impression; lateral pronotal margin narrow to broad (Figure 15.24) ....***Hyphasis* Harold**
8.(6) Metatibia without apical spine (Figure 21) (spines may be present on either side of apex; Figure 15.25) 9
- Metatibia with apical spine (Figure 15.20) (examine thoroughly as tiny apical spine may be hidden within setae or bristles)....................................... 12
9.(8) Antennal calli posteriorly delimited from vertex by well developed supracallinal sulcus (Figure 15.26); second and third antennomeres subequal, very small, hardly longer than wide; fourth antennomere about twice longer than third (Figure 15.27); body length 4–6 mm........... ***Chalaenosoma* Jacoby**
- Antennal calli posteriorly merge with vertex, supracallinal sulcus absent (Figure 15.28); second antennomere very small, about as wide as long, third antennomere longer than second; fourth subequal to or longer than third, shorter than twice length of third (Figure 15.28); body length 2.7–3.6 mm 10
10.(9) Midfrontal sulcus unusually deep and furrow-like (Figure 15.29); pronotum 1.4 times wider than long, posteriorly slightly narrower than anteriorly (Figure 15.30) .... ***Micraphthona* Jacoby**
- Midfrontal sulcus weak, neither very deep nor furrow-like (Figure 15.28); pronotum 1.9–2.4 times wider than long, posteriorly wider than anteriorly (Figures 15.31, 15.32)... 11
11.(10) Anterolateral corners of pronotum produced forward (Figure 15.31); prosternal intercoxal process as wide as one-third of procoxa ....
....................................***Elytropachys* Motschulsky**
- Anterolateral corners of pronotum almost quadrate (Figure 15.32), not produced forward; prosternal intercoxal process very narrow and hardly visible between procoxae...........***Panilurus* Jacoby**
12.(8) Pronotum neither with a distinct antebasal transverse impression nor a pair of short antebasal longitudinal impressions (Figures 15.33, 15.36) (easily overlooked in *Euphitrea*) 13

- Pronotum with distinct antebasal transverse impression (Figure 15.34) or a pair of short antebasal longitudinal impressions (Figure 15.35) **35**
13.(12) Pronotum and elytra densely pubescent (Figure 15.36)............................... ***Hespera*** **Weise**
- Pronotum and elytra glabrous (Figures 15.33, 15.39, 15.40) **14**
14.(13) Anterior coxal cavities open posteriorly (Figure 15.37) **15**
- Anterior coxal cavities closed posteriorly (Figure 15.38) **31**
15.(14) Body highly convex, hemispherical to ovate in lateral view (Figure 15.39).................... **16**
- Body oblong, neither hemispherical nor extremely convex in lateral view (Figure 15.40) **21**
16.(15) Metatibia produced apically, projection sharp, tibial spine and tarsus inserted subapically (Figure 15.41); first abdominal ventrite medially with a pair of longitudinal ridges (Figure 15.42) ....***Argopistes* Motschulsky**
- Metatibia not produced apically, tibial spine and tarsus inserted at apex (Figure 15.20); first abdominal ventrite without a pair of longitudinal ridges medially (Figure 15.43) **17**
17.(16) Two apical maxillary palpomeres forming a spherical globule (Figure 15.44); feed on *Smilax*............ ....***Chilocoristes* Weise**
- Last maxillary palpomere elongate, pointed (Figure 15.45); does not occur on *Smilax*............ **18**
18.(17) Antenna with first antennomere very long, as long as the next three combined (Figure 15.46); feed on ferns..............................***Halticorcus* Lea**
- Antenna with first antennomere shorter than next three combined (Figure 15.47); feed on moss, dicots, or monocots **19**
19.(18) Hind wings absent or rarely present, antennal calli obsolete (Figure 15.48), labrum deeply emarginate (Figure 15.48); prosternum with a highly raised process along its middle, mesosternum with a raised circular area in middle (Figure 15.49); feed on moss ....***Ivalia* Jacoby**
- Hind wings always present; antennal calli distinct; labrum entire; sternites not modified; feed on monocots or dicots **20**
20.(19) Length 1.7–2.8 mm; frontal ridge narrowly raised between antennal sockets; distance between antennal sockets less than diameter of one socket (Figure 15.45); third tarsomere entire...................... ....***Sphaeroderma* Stephens**
- Length 5–7 mm; frontal ridge broad and flat between antennal sockets (Figure 15.50); distance between antennal sockets more than diameter of a socket (Figure 15.50); third tarsomere bilobed....... .... ***Chabria* Jacoby**
21.(15) Elytra with punctation confused (Figure 15.51) **22**
- Elytra with punctation regularly arranged in eleven longitudinal rows, including a short scutellar row and a marginal one (Figure 15.52)..................... **29**
22.(21) Antennal calli subquadrate (Figure 15.53); posterior tibia with axial excavation extending from apex to basal 1/4th or more (Figure 15.54); body usually ovate and not very convex .... ***Hemipyxis* Dejean**
- Antennal calli not subquadrate (Figures 15.55, 15.60, 15.61, 15.62); posterior tibia without axial excavation or with a short excavation not reaching middle (Figure 15.56); body form oblong, convex **23**
23.(22) First metatarsomere equal to or exceeding half of tibia in length and inserted always on a small callosity at apex of tibia (Figure 15.56)..................... ....***Longitarsus* Latreille**
- First metatarsomere shorter than half of tibia, and never inserted on a callosity at apex of tibia (Figure 15.20)...**24**
24.(23) Prosternum narrow (Figure 15.57), second and third antennomeres always small (Figure 15.47); posterior region of elytra sparsely set with very short setae (Figure 15.58) ...***Luperomorpha* Weise**
- Prosternum not extremely narrow (Figure 15.59), third antennomere longer than second (Figure 15.59); posterior region of elytra not set with sparse short setae... **25**
25.(24) Antennal calli obsolete, sulci delimiting antennal calli very shallow (Figure 15.60); body flat, nearly parallel sided....................... ***Phyllotreta* Stephens**
- Antennal calli prominent, sulci delimiting antennal calli well developed (Figures 15.55, 15.61–15.63); body oblong, moderately convex **26**
26.(25) Antennal calli subovate to subquadrate, not extending to interantennal space (Figures 15.55, 15.63)... ....**27**
- Antennal calli more or less triangular with pointed anteromesal end extending into interantennal space (Figures 15.61, 15.62) **28**
27.(26) Frontal ridge long; anterofrontal ridge flat (Figure 15.63); lateral margin of metatibia with small denticles bearing setae (Figure 15.64)............. ....***Kashmirobia* Konstantinov & Prathapan**
- Frontal ridge short; anterofrontal ridge raised (Figure 15.55); lateral margin of metatibia with or without small denticles (Figure 15.20).................... ....***Aphthona* Chevrolat**
28.(26) Labrum with two pairs of setae arranged in a transverse row; frontal ridge and anterofrontal ridge together form a swollen triangular ridge (Figure 15.61) .... ***Orisaltata* Prathapan & Konstantinov**
- Labrum with three pairs of setae arranged in a transverse row; frontal ridge and anterofrontal ridge together form a flat, triangular ridge (Figure 15.62) .... ***Trachytetra* Sharp**

29.(21) Elytral rows of punctures not strongly impressed, interstices flat; first abdominal ventrite with a pair of subparallel longitudinal ridges along middle (Figure 15.66).. **30**
- Elytral rows of punctures well impressed with raised interstices (Figure 15.67); first abdominal ventrite without a pair of subparallel longitudinal ridges along middle ....
.... ***Brancucciella* Medvedev**
30.(29) Antennal calli unusual, depressed (Figure 15.65); first metatarsomere as long as half of metatibia; feed on *Piper*.....................***Lanka* Maulik**
- Antennal calli normal, as high as vertex, delimited by sulci (Figure 15.68); first metatarsomere shorter than half of metatibia; feed on Euphorbiaceae ....
.................................***Bikasha* Maulik**
31.(14) Size 8.6 mm; claws bifid (Figure 15.69) ....
.... ***Furthia* Medvedev**
- Size less than 5 mm; claws appendiculate (Figure 15.70) **32**
32.(31) Middle and posterior tibiae each with an obtuse tooth behind middle followed by an excavation with a marginal row of stiff bristles (Figure 15.71)
....***Chaetocnema* Stephens**
- Middle and posterior tibiae without such an obtuse teeth and excavation (Figures 15.73, 15.74).......... **33**
33.(32) Distance between antennal sockets subequal to diameter of one eye or more (Figure 15.72); base of elytra not wider than base of pronotum (Figure 15.33); first abdominal ventrite shorter than next two ventrites combined (Figure 15.73)
.... ***Amphimela* Chapuis**
- Distance between antennal sockets less than diameter of one eye; base of elytra wider than base of pronotum; first abdominal ventrite not shorter than next two ventrites combined (Figures 15.74, 15.75)
.... **34**
34.(33) Elytra with punctures confused; width of elytral epipleura less than width of midfemur (Figure 15.74)
.... ***Bimala* Maulik**
- Elytra with punctures regularly arranged in rows; width of elytral epipleura more than width of midfemur (Figure 15.75)........................***Erystus* Jacoby**
35.(12) Pronotum with two short longitudinal impressions near basal margin, other transverse or longitudinal impressions absent (Figure 15.35) **36**
- Pronotum with transverse impression(s), short longitudinal antebasal impressions absent (Figure 15.34); if short longitudinal impressions present, additional impressions present (Figure 15.85)........................**37**
36.(35) Head with vertex longitudinally elevated at disc, sides deeply excavated above eye (Figure 15.76); metasternaum extending anteriorly, covering mesosternum (Figure 15.77); body-form hemispherical
.... ***Euphitrea* Baly**
- Head with vertex evenly convex (Figure 15.78); metasternum not extending anteriorly, mesosternum visible, not covered by metasternum (Figure 15.79); body-form oblong or oblong-oval...
.... ***Podagrica* Foudras**
37.(35) Body massive, length 8–17 mm; posterior tibiae each, in lateral view, with an excavation with a marginal row of stiff bristles (Figure 15.80)........ **38**
- Length less than 8 mm; posterior tibiae without excavation in lateral view **39**
38.(37) Prosternum triangularly excavated to fit mesosternum; mesosternum saddle-like (Figure 15.81); posterior femur angularly dilated on inner edge (Figure 15.82)............................***Podontia* Dalman**
- Prosternum truncate along posterior margin; mesosternum simple (Figure 15.83); posterior femur not angularly dilated (Figure 15.84)
.... ***Asiophrida* Medvedev**
39.(37) Pronotum with longitudinal impressions on anterior and basal margins (Figure 15.85); body rounded ovate ....***Nisotra* Baly**
- Pronotum without short longitudinal impressions; transverse impressions present (Figures 15.34, 15.86, 15.87, 15.89), (transverse impression may be laterally delimited by short longitudinal impressions - Figure 15.88); body oblong..................... **40**
40.(39) Pronotum deeply constricted basally (Figure 15.86).......***Lipromorpha* Chujo & Kimoto**
- Pronotum not deeply constricted basally (Figures 15.87–15.89)... **41**
41(40) Elytra densely pubescent; pronotum bare, with a second transverse depression near anterior margin (Figure 15.87)........................***Demarchus* Jacoby**
- Elytron as well as pronotum bare, pronotum with only a single antebasal transverse impression, anteapical transverse impression absent (Figures 15.88, 15.89) **42**
42.(41) Elytra with punctures arranged in eleven longitudinal rows, including marginal and short scutellar rows (Figure 15.52) **43**
- Elytra with punctures confused (Figure 15.51)... **48**
43.(42) Antebasal transverse impression on pronotum bound laterally by exact longitudinal or curved impressions (Figure 15.88) **44**
- Antebasal transverse impression on pronotum reaching lateral or postero-lateral margin, not bounded laterally by longitudinal impressions (Figure 15.89).. **46**
44.(43) Head with vertex sharply angulate anteriorly (Figure 15.90); antebasal transverse impression on pronotum sinuate (Figure 15.88); feed on Urticaceae or Melastomataceae .... ***Xuthea* Baly**
- Head with vertex not sharply angulate anteriorly (Figure 15.91); antebasal transverse impression not sinuate; do not feed on Urticaceae or Melastomataceae..**45**

45.(44) Distal five antennomeres distinctly thicker, forming an elongate club (Figure 15.93); longitudinal ridge along mesal side of metafemur sexually dimorphic, serrulate in male while smooth in female; feed on ferns or moss........ .... *Phaelota* **Jacoby**
- Distal five antennomeres not distinctly thickened (Figure 15.94); metafemur not sexually dimorphic; feed on Malvaceae .... *Sinocrepis* **Chen**
46.(43) Flightless; humeral calli weakly developed; occur in moss .... *Benedictus* **Scherer**
- Hindwings present; humeral calli prominent (Figure 15.89); do not occur in moss **47**
47.(46) Facial part of head elongate; frons with a carina medially, anterofrontal ridge flat (Figure 15.95); pronotum posteriorly narrower than anteriorly; feed on ferns.............................. *Manobia* **Jacoby**
- Facial part of head short; frons without carina medially, anterofrontal ridge triangular, well developed (Figure 15.92); pronotum anteriorly as broad as posteriorly (Figure 15.89); feed on Euphorbiaceae .... *Lesagealtica* **Doeberl**
48.(42) First abdominal ventrite medially with a pair of subparallel ridges originating in the intercoxal part (Figure 15.66); feed on Piperaceae............... ....*Tegyrius* **Jacoby**
- First abdominal ventrite medially without a pair of subparallel ridges originating in the intercoxal part; do not feed on Piperaceae.......................... **49**
49.(48) Frontal ridge broad between antennal sockets, sulci surrounding antennal calli shallow and poorly developed (Figure 15.96); antebasal transverse impression on pronotum laterally limited by short longitudinal furrows (Figure 15.97); feed on Euphorbiaceae................ *Hermaeophaga* **Foudras**
- Frontal ridge narrowly raised between antennal sockets; antennal calli well developed (Figure 15.98). Antebasal transverse impression complete across base of pronotum, reaching postero-lateral margins or posterior angles (Figure 15.99); do not feed on Euphorbiaceae **50**
50.(49) Supracallinal sulci delimiting antennal calli from vertex poorly developed (Figure 15.96); lateral pronotal margins evenly and strongly curved (Figure 15.99); feed on Asclepiadaceae................ ....*Phygasia* **Dejean**
- Supracallinal sulci delimiting antennal calli from vertex well developed (Figures 15.61, 15.100, 15.101); lateral pronotal margins unevenly and weakly curved (Figure 102); feed on Verbenaceae, Onagraceae, or Aristolochiaceae **51**
51.(50) Length 1.6–2 mm; labrum with four setiferous pores; feed on Aristolochiaceae............................ .... *Orisaltata* **Prathapan & Konstantinov**
- Length 2.7–5.8 mm; labrum with at least six setiferous pores; feed on Onagraceae or Verbenaceae **52**
52.(51) Metallic blue-black; length 4.7–5.8 mm; antennal calli large, more or less rounded, frons distinctly carinate, antennal calli not entering into interantennal space (Figure 15.100); pronotum without shallow anteapical impressions; feed on Onagraceae ....*Altica* **Geoffrey**
- Non-metallic dark brown; length 2.7–3.3 mm; antennal calli longer than wide, anterior ends partially entering into interantennal space; frons raised, not carinate (Figure 15.101); pronotum with shallow anteapical impressions (Figure 15.102); feed on Verbenaceae .... *Yaminia* **Prathapan & Konstantinov**

Konstantinov and Vandenberg (1996) and Konstantinov (1998) described and illustrated the morphological structures of flea beetles. In the key, the morphological terminology, illustrated in Figures 15.11–15.13, follows Konstantinov (1998). The genera *Haemaltica* Chen and *Pseudaphthona* Jacoby are not separated from *Xuthea* (see Medvedev 2004) and *Trachytetra*, respectively, and are not included in the key. Report of the north-east Indian species, *Microcrepis polita* Chen from south India needs confirmation and hence the genus is not included in the key. The non-flea beetle genus *Aulacothorax* Boheman (= *Orthaltica*), hither to treated under "Alticinae", is excluded from the key. *Liprus obscurus* Chen, most probably is a *Lipromorpha* Chŭjô & Kimoto, and hence the genus is not given in the key.

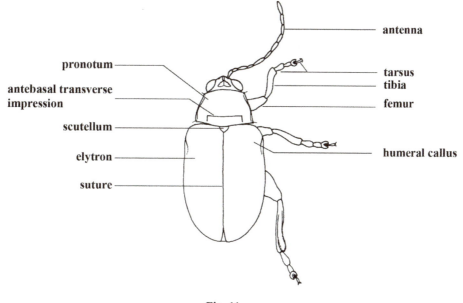
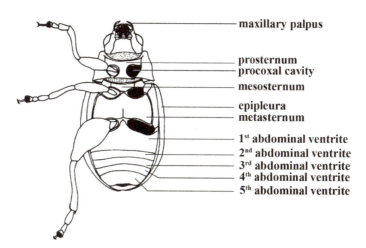
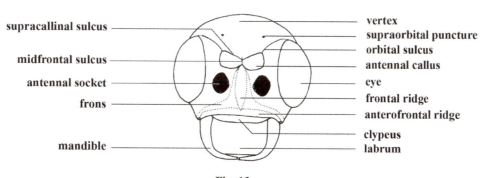

FIGURES 15.11–15.13 Flea beetle (habitus) **11.** Dorsal habitus; **12.** Ventral habitus; **13.** Head.

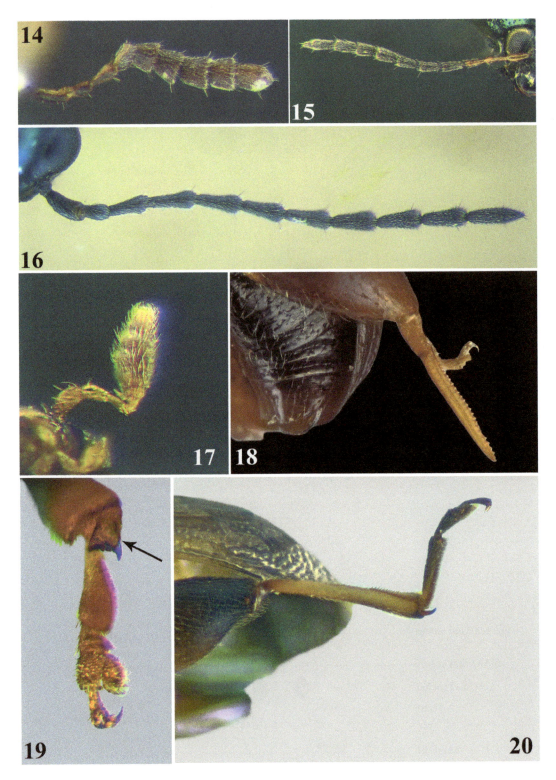

FIGURES 15.14–15.20 **14.** *Nonarthra* (antenna); **15.** *Psylliodes* (antenna); **16.** *Altica ?aenea* (antenna); **17.** *Clavicornaltica* sp. (antenna); **18.** *Aphthonoides* sp. (hind leg); **19.** *Paradibolia nila* (metatibial spine); **20.** *Aphthona* sp. (metatibia and tarsus).

FIGURES 15.21–15.25 21. *Chalaenosoma* sp. (tibia); 22. *Philopona* sp. (metatibia and tarsus); 23. *Philopona* sp. (head and pronotum); 24. *Hyphasis* sp. (head and pronotum); 25. *Elytropachys* sp. (metatibial apex, ventral view).

FIGURE 15.26–15.30   **26.** *Chalaenosoma* sp. (head, frontal view); **27.** *Chalaenosoma* sp. (proximal antennomeres); **28.** *Panilurus nilgiriensis* (head, frontal view); **29.** *Micraphthona fulvipes* (head, frontal view); **30.** *Micraphthona fulvipes* (pronotum).

FIGURES 15.31–15.38 **31.** *Elytropachys* sp. (head and pronotum); **32.** *Panilurus nilgiriensis* (head and pronotum); **33.** *Amphimela picta* (pronotum and base of elytra); **34.** *Altica ?aenea* (pronotum); **35.** *Podagrica* sp. (head and pronotum); **36.** *Hespera* sp. (head, pronotum, and base of elytra); **37.** *Aphthona nigrilabris* (ventral view of prothorax showing open procoxal cavities); **38.** *Chaetocnema* sp. (ventral view of prothorax showing closed procoxal cavities).

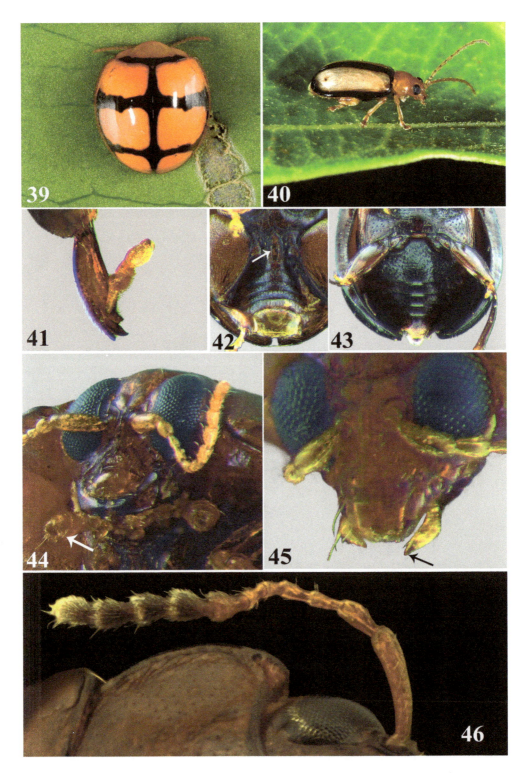

FIGURES 15.39–15.46 39. *Chilocoristes* sp. (dorsal habitus); 40. *Phygasia marginata* (dorso-lateral view); 41. *Argopistes* sp. (metatibia and tarsus); 42. *Argopistes* sp. (abdomen, ventral view); 43. *Sphaeroderma* sp. (abdomen, ventral view); 44. *Chilocoristes* sp. (head); 45. *Sphaeroderma* sp. (head); 46. *Halticorcus* sp. (antenna).

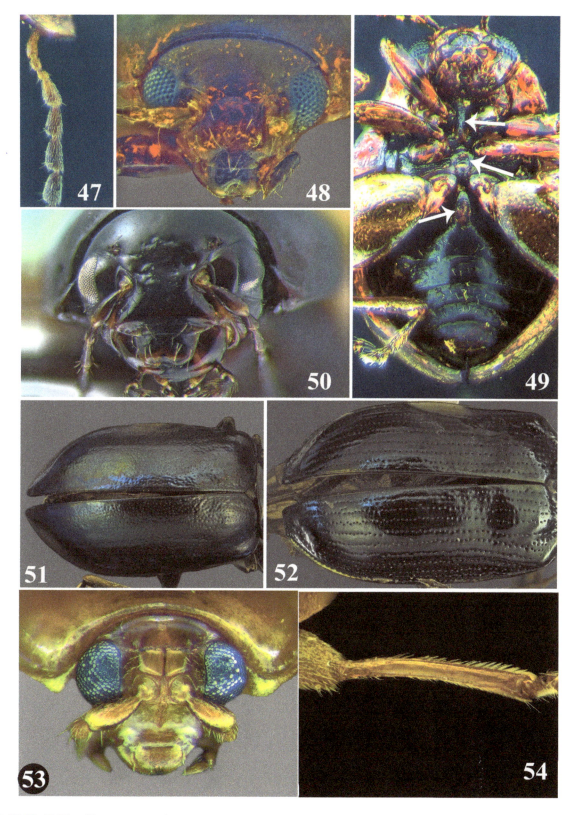

FIGURES 15.47–15.54 **47.** *Luperomorpha vittata* (proximal antennomeres); **48.** *Ivalia* sp. (head); **49.** *Ivalia* sp. (ventral view); **50.** *Chabria decemplagiata* (head, frontal view); **51.** *Altica ?aenea* (elytra); **52.** *Xuthea* sp. (elytra); **53.** *Hemipyxis* sp. (head); **54.** *Hemipyxis* sp. (metatibia, dorsal view).

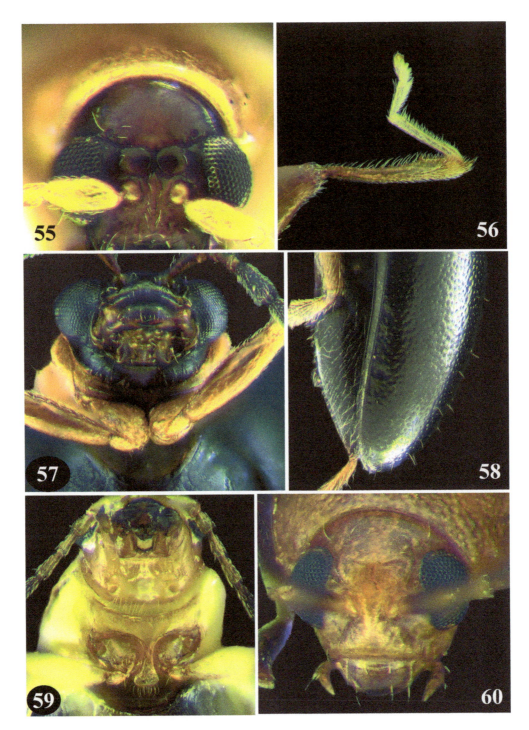

FIGURES 15.55–15.60 **55.** *Aphthona* sp. (head, frontal view); **56.** *Longitarsus* sp. (metatibia and tarsus); **57.** *Luperomorpha vittata* (head and prothorax in ventral view showing procoxae); **58.** *Luperomorpha* sp. (elytral apex); **59.** *Aphthona* sp. (prothorax, ventral view); **60.** *Phyllotreta* sp. (head, frontal view)

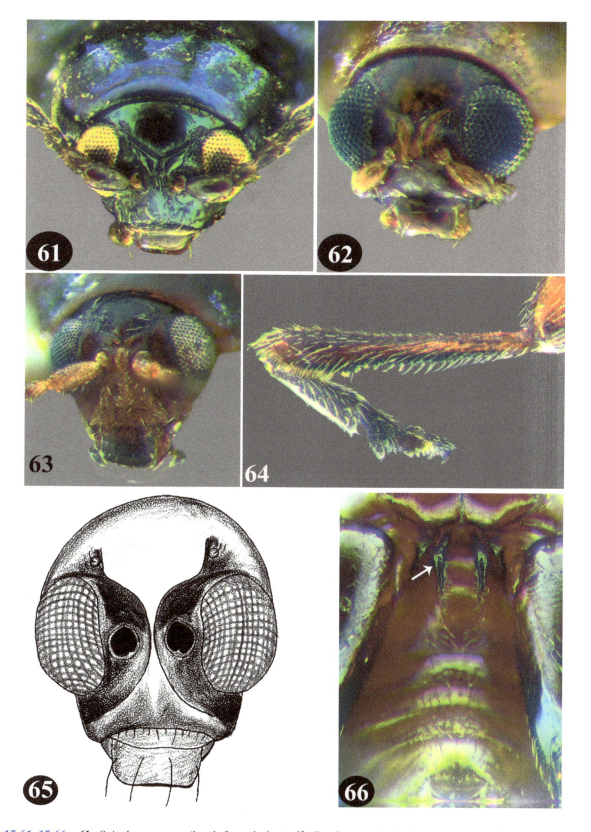

FIGURE 15.61–15.66 **61.** *Orisaltata azurea* (head, frontal view); **62.** *Trachytetra* (head, frontal view); **63.** *Kashmirobia hugeli* (head, frontal view); **64.** *Kashmirobia hugeli* (metatibia and tarsus); **65.** *Lanka ramakrishnai* (head, frontal view). (Adapted from Prathapan, K. D., and Viraktamath, C. A., *Zootaxa*, 1681, 1–30, 2008. With permission); **66.** *Lanka ramakrishnai* – first abdominal ventrite with subparallel ridges.

FIGURE 15.67–15.72 **67.** *Brancucciella kolibaci* (elytra); **68.** *Bikasha* sp. (head, frontal view); **69.** *Asiophrida marmorea* (claw); **70.** *Altica ?aenea* (claw); **71.** *Chaetocnema* sp. (metatibia and tarsus); **72.** *Amphimela picta* (head, frontal view).

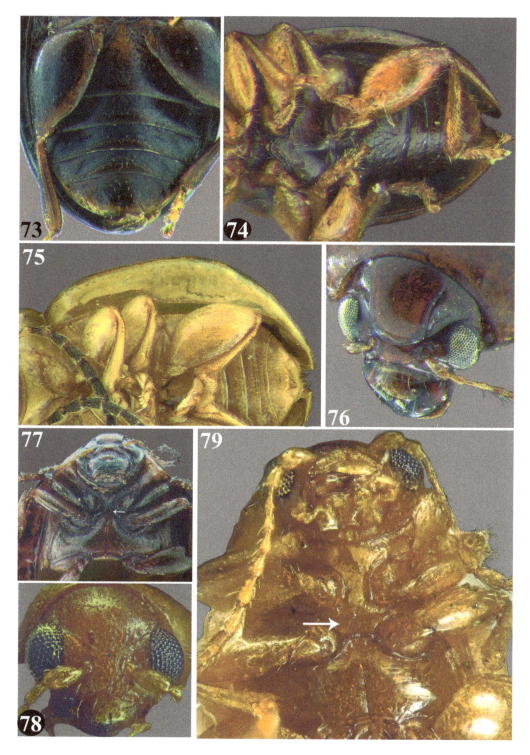

FIGURE 15.73–15.79 **73.** *Amphimela picta* (abdomen and hind legs, ventral view); **74.** *Bimala* sp. (abdomen and thorax, ventral view); **75.** *Erystus andamanica* (abdomen and thorax, ventral view); **76.** *Euphitrea* sp. (head, fronto-lateral view); **77.** *Euphitrea* sp. (thorax, ventral view); **78.** *Podagrica* sp. (head, frontal view); **79.** *Podagrica* sp. (head and thorax, ventral view.)

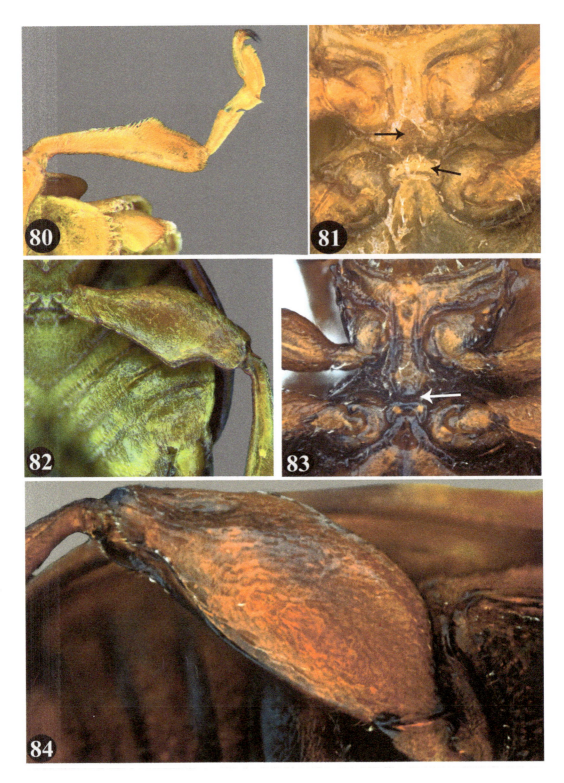

FIGURES 15.80–15.84 **80.** *Podontia congregata* (metatibia and tarsus); **81.** *Podontia congregata* (pro- and mesosternum); **82.** *Podontia congregata* (metafemur); **83.** *Asiophrida marmorea* (thoracic sternites); **84.** *Asiophrida marmorea* (metafemur).

FIGURES 15.85–15.93 **85.** *Nisotra* sp. (pronotum); **86.** *Lipromorpha* sp. (pronotum); **87.** *Demarchus pubipennis* (pronotum); **88.** *Xuthea* sp. (pronotum); **89.** *Lesagealtica* sp. (head, pronotum and elytral base); **90.** *Xuthea* sp. (head, frontal view); **91.** *Phaelota* sp. (head, frontal view); **92.** *Lesagealtica* sp. (head, frontal view); **93.** *Phaelota* sp. (antenna).

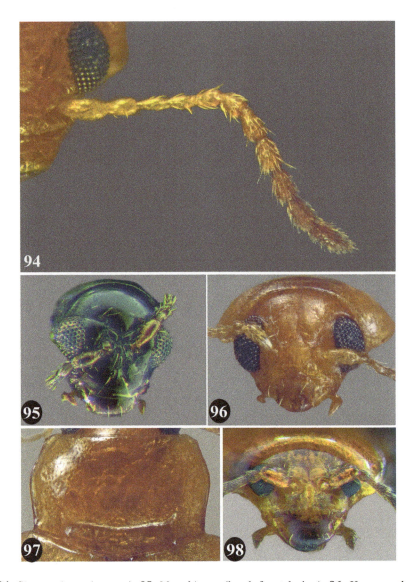

FIGURE 15.94–15.98 **94.** *Sinocrepis* sp. (antenna); **95.** *Manobia* sp. (head, frontal view); **96.** *Hermaeophaga ruficollis* (head, frontal view); **97.** *Hermaeophaga ruficollis* (head and pronotum); **98.** *Phygasia silacea* (head, frontal view).

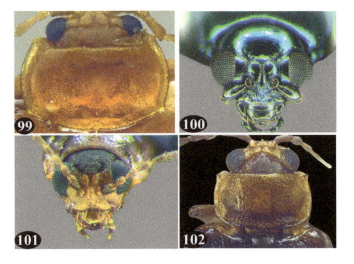

FIGURE 15.99–15.102 **99.** *Phygasia silacea* (head and pronotum) **100.** *Altica ?aenea* (head, frontal view); **101.** *Yaminia gmelini* (head, frontal view); **102.** *Yaminia gmelini* (head and pronotum, dorsal view).

CHECKLIST OF FLEA BEETLES OF SOUTH INDIA

Peninsular India, south of river Tapti, is considered as south India here. The published records of many flea beetles in south India, originally described from elsewhere, is doubtful. Concerted efforts to collect many species recorded by Doeberl (2003) in Kerala and Tamil Nadu have failed. Hence, the material that forms the basis of such records should be studied to ascertain the accuracy of such records. There are records of *Altica brevicosta* (Weise), *A. cyanea* (Weber), and *A. caerulea* (Olivier) from south India. However, two decades of field work has yielded only a single species (identified as *A. cyanea* or *A. caerulea* by authors) that feeds on Onagraceae. Reid and Beatson (2015) have clarified that *A. cyanea* is endemic to the Sundaland (Peninsular Malaysia, Sumatra, Boreneo, and Java) and does not occur in south India. They also indicate that the south Indian species could be *A. aenea* (Olivier). The checklist is based on published records and is provisional, as the reported occurrence of many species need to be confirmed (Table 15.1).

FLEA BEETLES AS PESTS

Being phytophagous both in the larval and adult stages, many species of flea beetles are important crop pests. The pepper *pollu* beetle, *Lanka ramakrishnai* Prathapan & Viraktamath, is the most economically important flea beetle crop pest in south India, that causes 30%–40% yield loss. *Longitarsus birmanicus* Jacoby (=*L. belgaumensis* Jacoby) is a serious pest of the green manure plant *Crotalaria juncea* L. Kumaraswamy (1961) studied its nature of damage, biology, and larval characters. *Amphimela indica* (Jacoby) on *Aegle marmelos* and *Podontia*

TABLE 15.1
Checklist of the Flea Beetles of South India

1. *Altica* Geoffroy
 – ?*aenea* (Olivier)
2. *Amphimela* Chapuis
 – *indica* (Jacoby), – *mouhoti* (Chapuis), – *picta* (Baly)
3. *Aphthona* Chevrolat
 – *atripes* (Motschulsky), – *bombayensis* Scherer, – *chrozophorae* Prathapan & Konstantinov, – *glochidionae* Prathapan & Konstantinov, – *kanaraensis* Jacoby, – *macarangae* Prathapan & Konstantinov, – *mallotae* Prathapan & Konstantinov, – *marataka* Prathapan & Konstantinov, – *nandiensis* Prathapan & Konstantinov, – *nigrilabris* Duvivier, – *opaca* Allard, – *phyllanthae* Prathapan & Konstantinov, – *yercaudensis* Prathapan & Konstantinov
4. *Aphthonoides* Jacoby
 – *keralaensis* Doeberl, – *konstantinovi* Doeberl, – *ovipennis* Heikertinger, – *prathapani* Doeberl, – *rugiceps* Wang
5. *Argopistes* Motschulsky
 – *flavus* Chen, – *lamprotes* Maulik, – *nigristriga* Maulik – *quadrimaculatus* Jacoby
6. *Asiophrida* Medvedev
 – *hirsuta* (Stebbing), – *marmorea* (Wiedemann)
7. *Benedictus* Scherer
 – *robropunctatus* Sprecher-Uebersax, Konstantinov, Prathapan & Döberl
8. *Bimala* Maulik
 – *indica* (Jacoby)
9. *Brancucciella* Medvedev
 – *kolibaci* Medvedev
10. *Chabria* Jacoby
 – *decemplagiata* Maulik
11. *Chaetocnema* Stephens
 – *belli* Jacoby, – *bretinghami* Baly, – *cognata* Baly, – *concinnipennis* Baly, – *confinis* Crotch, – *puncticollis* (Motschulsky), – *gracilis* (Motschulsky), – *harita* Maulik, – *kumaonensis* Scherer, – *longipunctata* Maulik, – *modigliani* Jacoby, – *bella* (Baly), – *nigrica* (Motschulsky), – *pusaensis* Maulik, – *singala* Baly, – *tonkinensis* Chen, – *warchalowskii* Doeberl.
12. *Chalaenosoma* Jacoby
 – *anaimalaiense* Scherer, – *antennatum* Jacoby, – *bulbifera* (Medvedev), – *fulvitarsis* Jacoby, – *hindustanica* (Scherer), – *kolibaci* Medvedev, – *madurensis* Scherer, – *metallicum* Jacoby, – *mimica* Medvedev, – *schereri* Medvedev, – *travancoreense* Scherer, – *viridis* (Jacoby)
13. *Chilocoristes* Weise
 – *bistripunctatus* (Duvivier)
14. *Clavicornaltica* Scherer
 – *rileyi* Döberl

(Continued)

TABLE 15.1 (*Continued*)
Checklist of the Flea Beetles of South India

15. *Demarchus* Jacoby
– *pubipennis* Jacoby
16. *Elytropachys* Motschulsky
– *sandalensis* Bryant, – *viridescens* Motschulsky
17. *Erystus* Jacoby
– *andamanensis* Maulik
18. *Euphitrea* Baly
– *fulva* (Jacoby), – *grossa* Medvedev, – *indica* Jacoby, – *micans* Baly
19. *Haemaltica* Chen
– *indica* Doeberl, – *rubra* Chen
20. *Halticorcus* Lea
– *keralaensis* Basu & Sengupta
21. *Hemipyxis* Dejean
– *brevicollis* (Jacoby), – *fimbriata* (Maulik), – *fulvipennis* (Illiger), – *intermedia* (Jacoby), – *nigricornis* (Baly), – *nigritarsis* (Jacoby), – *pallidicincta* (Jacoby)
22. *Hermaeophaga* Foudras
– *indica* Jacoby, – *ruficollis* (Lucas)
23. *Hespera* Weise
– *dakshina* Maulik, – *lomasa* Maulik, – *sericea* Weise
24. *Hyphasis* Harold
– *anaimalaiensis* Scherer, – *atricorne* Chen, – *discipennis* (Jacoby), – *discoidalis* Jacoby, – *femoralis* Jacoby, – *fulvicornis* Jacoby, – *inconspicua* (Jacoby), – *intermedia* Jacoby, – *nilapita* (Maulik), – *nilgiriensis* Scherer, – *sita* (Maulik), – *submetallica* (Jacoby), – *tenuilimbatus* Jacoby, – *thoracica* Jacoby
25. *Ivalia* Jacoby
– *indica* (Gruev & Askevold), – *korakundah* Prathapan, Konstantinov & Duckett, – *obrieni* (Gruev & Askevold)
26. *Kashmirobia* Konstantinov & Prathapan
– *hugeli* (Jacoby)
27. *Lanka* Maulik
– *aruna* Prathapan & Viraktamath, – *ramakrishnai* Prathapan & Viraktamath, – *sahyadriensis* Prathapan & Viraktamath
28. *Lesagealtica* Doeberl
– *nigripennis* (Ogloblin)
29. *Lipromorpha* Chûjô & Kimoto
– *deformicornis* Medvedev
30. *Liprus* Motschulsky
– *obscurus* Chen
31. *Longitarsus* Latreille
– *birmanicus* Jacoby, – *fumidus* Maulik, – *gilli* Gruev & Askevold, – *hina* Maulik, – *limnophilae* Prathapan & Viraktamath, – *liratus* Maulik, – *longicornis* Jacoby, – *nigronotatus* Jacoby, – *pandura* Maulik, – *rohtangensis* Shukla, – *rufipennis* Jacoby, – *sari* Maulik, – *serrulatus* Prathapan, Faizal & Anith, – *sundara* Maulik, – *suturellus* (Motschulsky)
32. *Luperomorpha* Weise
– *birmanica* (Jacoby), – *bombayensis* (Jacoby), – *nigripennis* Duvivier, – *vittata* Duvivier
33. *Manobia* Jacoby
– *dorsalis* Jacoby, – *sexguttata* Scherer
34. *Micraphthona* Jacoby
– *fulvipes* (Jacoby)
35. *Microcrepis* Chen
– *polita* Chen
36. *Nisotra* Baly
– *apicefulva* (Bryant), – *cardoni* Jacoby, – *madurensis* Jacoby, – *semicoerulea* Jacoby, – *striatipennis* Jacoby, – *viridipennis* Motschulsky
37. *Nonarthra* Baly
– *birmanicum* (Jacoby), – *variabilis* Baly

(*Continued*)

TABLE 15.1 (*Continued*)
Checklist of the Flea Beetles of South India

38. *Orisaltata* Prathapan & Konstantinov
 – *azurea* (Jacoby), – *medvedevi* Prathapan & Konstantinov
39. *Panilurus* Jacoby
 – *agasthyamalaiensis* Prathapan & Viraktamath, – *nilgiriensis* Jacoby, – *ponmudiensis* Prathapan & Viraktamath
40. *Paradibolia* Baly
 – *indica* Baly
41. *Phaelota* Jacoby
 – *jacobyi* Prathapan & Viraktamath, – *kottigehara* Prathapan & Konstantinov, – *maculipennis* Prathapan & Konstantinov, – *mauliki* Prathapan & Konstantinov, – *saluki* Prathapan & Konstantinov, – *sindhoori* Prathapan & Viraktamath, – *vaisakha* Prathapan & Viraktamath, – *viridipennis* Prathapan & Konstantinov
42. *Philopona* Weise
 – *decemmaculata* Maulik, – *indica* Medvedev, – *inornata* (Jacoby), – *nilgiriensis* (Jacoby), – *vibex* (Erichson)
43. *Phygasia* Baly
 – *marginata* Medvedev, – *minuta* Medvedev, – *nigripennis* Jacoby, – *silacea* (Illiger), – *violaceipennis* Jacoby
44. *Phyllotreta* Stephens
 – *birmanica* Harold, – *chotanica* Duvivier, – *depressa* (Chen), – *downesi* Baly, – *indica* Chen
45. *Podagrica* Foudras
 – *ceylonensis* Jacoby
46. *Podontia* Dalman
 – *congregata* Baly
47. *Pseudaphthona* Jacoby
 – *humeralis* Jacoby
48. *Psylliodes* Latreille
 – *brettinghami* Baly, – *chlorophanus* Erichson, – *viridanus* Motschulsky
49. *Sinocrepis* Chen
 – *obscurofasciata* (Jacoby)
50. *Sphaeroderma* Stephens
 – *decemmaculatum* Allard, – *minuta* Chen
51. *Tegyrius* Jacoby
 – *agasthya* Prathapan & Viraktamath, – *dalei* Prathapan & Viraktamath, – *keralaensis* (Doeberl), – *nigrotibialis* Prathapan & Viraktamath, – *pucetibialis* Prathapan & Viraktamath, – *radhikae* Prathapan & Viraktamath, – *tippui* Prathapan & Viraktamath
52. *Trachytetra* Sharp
 – *indica* Doeberl, – *fusca* (Scherer)
53. *Xuthea* Baly
 – *fulvitarsis* Scherer, – *metallica* (Jacoby), – *orientalis* Baly
54. *Yaminia* Prathapan & Konstantinov
 – *gmelini* Prathapan & Konstantinov

congregata on *Garcinia gummi-gutta* are major pests. Species of *Nisotra* are minor pests of malvaceous crops such as hibiscus and bhindi. Cereals and millets, sweet potato, and amaranthus are damaged by species of *Chaetocnema*. A revision of the Oriental species of *Chaetocnema* (Ruan et al. 2019) has brought clarity to the classification of south Indian species of the genus. A list of flea beetle pests in south India is given in Table 15.2. Records, which are evidently wrong, are omitted from the list.

Absence of certain pestiferous species in south India is conspicuous and intriguing. *Phyllotreta striolata* (Fabricious), one of the most widely distributed leaf beetle pests, is common in the sub-Himalayan area and the Andaman Islands. However, it has not yet invaded south India, though the host plants are available and the climate is congenial. *Podontia quatordecimpunctata* (L.) and the citrus leaf mining flea beetle *Podagricomela nigripes* Medvedev (Prathapan 2017) are the other two species likely to invade south India.

Species of *Chaetocnema*, *Phyllotreta*, *Podagrica*, and *Psylliodes* are known as vectors of plant pathogens. However, no flea beetle in south India is known to be associated with plant diseases.

TABLE 15.2
Flea Beetle Pests of South India

Pest	Host and Notes	Source
Amphimela indica (Jacoby) (= *Clitea indica*)	Serious defoliator of bael (*Aegle marmelos* Corr.) throughout southern India. Both adults and larvae feed on the leaves	Anonymous (1954); Ayyar (1963); Wadhi and Batra (1964); Batra (1969); Scherer (1969)
Amphimela picta (Baly) (= *Clitea picta*)	Serious pest of bael (*Aegle marmelos*) in north India, also occurs in Andhra Pradesh	Stebbing (1914); Misra and Fletcher (1919); Maulik (1926); Scherer (1969); Zaka-ur-Rab (1991); Yadav and Rizvi (1994)
Aphthona phyllanthae Prathapan & Konstantinov	Minor pest on Indian gooseberry (*Phyllanthus emblica* L.)	Prathapan and Konstantinov (2003)
Asiophrida hirsuta (Stebbing) (= *Ophrida hirsuta*)	Adults and larvae feed on the leaves of Indian frankincense tree (*Boswellia serrata* Roxb. ex Colebr.)	Stebbing (1914); Beeson (1919, 1941); Maulik (1926); Scherer (1969)
Asiophrida marmorea (Wiedemann) (= *Ophrida marmorea*)	Defoliates *Garuga pinnata* Roxb., which is used as a standard for black pepper in south India	Mathew and Mohandas (1989)
Chaetocnema concinnipennis Baly	Minor pest of rice (*Oryza sativa* Linn.) throughout India	Batra (1969); Scherer (1969); Kulshreshtha and Mishra (1970); Kulshreshtha and Mishra (1970); Nayar et al. (1976); Nair (1986)
Chaetocnema confinis Crotch	Invasive pest of sweet potato (*Ipomoea batatas* Poir.). Occurs throughout India	Prathapan and Balan (2010)
Chaetocnema puncticollis (Motschulsky) (= *C. discreta* (Baly) = *C. kanika* Maulik)	Infest many Amaranthaceae, including cultivated amaranthus (*Amaranthus* spp.)	Scherer (1969); K. D. Prathapan, unpublished data
Chaetocnema gracilis (Motschulsky) (= *C. indica* Weise = *C. minuta* Jacoby)	Minor pest of millets such as finger millet (*Eleusine coracana* Gaertn.) and sorghum (*Sorghum vulgare* Pers.)	Nair (1986); Mote and Ghule (1989a, 1989b); Kishore (1996); Prathapan (unpublished data)
Chaetocnema nigrica (Motschulsky) (= *C. basalis* (Baly))	Minor pest of rice throughout India	Maulik (1926); Krishnaswamy et al. (1982); Garg and Sethi (1983)
Chaetocnema pusaensis Maulik	Recorded on common millet (*Panicum milaceum* L.) and finger millet (*Eleusine coracana*)	Ayyar (1963); Anonymous (1954); Scherer (1969); Jagadish and Ali (1986)
Hermaeophaga ruficollis (Lucas)	Minor pest of castor (*Ricinus communis* L.)	Anonymous (1954); Ayyar (1963); Batra (1969); Nayar et al. (1976); Nair (1986); Satyanarayana et al. (2000)
Hyphasis sita (Maulik)	Infests cultivated jasmine [*Jasminum sambac* (L.) Ait.] and other *Jasminum* species	Prathapan (unpublished data)
Lanka ramakrishnai Prathapan & Viraktamath (*Longitarsus nigripennis* (Motschulsky) *Auctt.*)	Major pest of black pepper (*Piper nigrum* L.) in south India	Prathapan and Viraktamath (2008) and the references there in
Lanka sahyadriensis Prathapan & Viraktamath	Minor pest of black pepper (*Piper nigrum* L.) in high altitudes	Prathapan and Viraktamath (2008) and the references there in
Longitarsus birmanicus Jacoby (= *L. begaumensis* Jacoby)	Serious pest of sunnhemp (*Crotalaria juncea* L.) and other *Crotalaria* species	Anonymous (1954); Kumaraswamy (1961); Ayyar (1963); Batra (1969); Nayar et al. (1976); Nair (1986); Pandit and Pradhan (1995); Prathapan (unpublished data)
Longitarsus serrulatus Prathapan, Faizal & Anith	Larvae feed on the roots and adults scrape the leaves of chinese potato (*Plectranthus rotundifolius* (Poir.) Spreng.)	Prathapan et al. (2005)
Luperomorpha vittata Duvivier (All host records of *L. vittata* and *L. bombayensis* (Jacoby) probably refers to the same species.)	A polyphagous pest on several species of vegetables, ornamental plants, and weeds	Chandrasekhara(1972); Lingappa and Siddappaji (1978); Santhakumari et al. (1979); Rajamma (1982); Srikanth et al. (1991);
Luperomorpha nigripennis Duvivier	Recorded on plants across families. Found feeding on *Clerodendrum inerme* Gaertn. (Verbenaceae) in Bangalore	Stebbing (1903, 1914); Batra (1969); Scherer (1969); Nair (1986); Prathapan (unpublished data)

(Continued)

TABLE 15.2 (Continued)
Flea Beetle Pests of South India

Pest	Host and Notes	Source
Nisotra apicefulva (Bryant),	Minor pest of shoe flower (*Hibiscus rosa-sinensis* L.)	Prathapan (unpublished data)
Nisotra cardoni Jacoby	Minor pest of bhindi (*Abelmoschus esculentus* Moench., *Hibiscus rosa-sinensis* L.)	Prathapan (unpublished data)
Nisotra madurensis Jacoby	Minor pest of gogu (*Hibiscus cannabinus* L.) and shoe flower (*Hibiscus rosa-sinensis*)	Ayyar (1933); Prathapan (unpublished data)
Nisotra semicoerulea Jacoby	Minor pest of shoe flower (*Hibiscus rosa-sinensis*)	Prathapan (unpublished data)
Phyllotreta chotanica Duvivier	Feeds on most crucifers	Chandrasekhara(1972); Prathapan (unpublished data)
Podontia congregata Baly	Serious defoliator of the Malabar tamarind [*Garcinina gummi-gutta* (L.) N. Robson.] Both adults and larvae feed on the leaves	Prathapan and Chaboo (2011)
Psylliodes brettinghami Baly	Recorded on brinjal (*Solanum melongena* L.), however, species identity is uncertain	Maulik (1926); Batra (1969); Singh et al. (1996)
Tegyrius keralaensis (Doeberl)	Minor pest of black pepper (*Piper nigrum* L.) throughout south India	Prathapan and Viraktamath (2009)
Tegyrius radhikae Prathapan & Viraktamath	Minor pest of black pepper (*Piper nigrum*) in Karnataka	Prathapan and Viraktamath (2009)

ACKNOWLEDGEMENTS

S. R. Hiremath prepared the plates. I am indebted to Alex Konstantinov for access to the synoptic collection of Oriental flea beetle genera in the US National Museum of Natural History and for reviewing a draft version of the manuscript. I am thankful to the editors and publishers for permission to use Figures 15.14 and 15.37 in *Zootaxa*, 1681 and Figure 15.15 in ZooKeys, 87.

REFERENCES

Anonymous. 1954. *Memoirs of the Department of Agriculture*. Madras, India: Government Press.

Appanna, M., and K. S. S. S. Sastry. 1958. Some observations on the biology and habits of *Haltica cyanea* Weber. *The Mysore Agricultural Journal* 33(1): 13–16.

Ayyar, T. V. R. 1919. Notes on the life-history of the *pollu* flea beetle (*Longitarsus nigripennis*, Mots.) of pepper. In: *Proceedings of the 3rd Entomological Meeting, Pusa*, February 3–15, 1919, ed. T. B. Fletcher, pp. 925–928. Calcutta, India: Superintendent of Government Printing.

Ayyar, T. V. R. 1933. *Insects Injurious to Vegetable and Flower Plants in south India*. Bulletin No. 34. Madras, India: The Agricultural Department, Government Press.

Ayyar, T. V. R. 1963. *Handbook of Economic Entomology for south India*. Madras, India: Government Press.

Baly, J. S. 1865. Descriptions of new genera and species of Galerucidae. *Annals and Magazine of Natural History: Zoology, Botany, and Geology* 25: 402–410.

Batra, H. N. 1969. Food plants, bionomics and control of flea beetles. *Indian Farming* 19(3): 38–40.

Beeson, C. F. C. 1919. The food plants of Indian forest insects. *Indian Forester* 45(6): 312–323.

Beeson, C. F. C. 1941. *The Ecology and Control of the Forest Insects of India and the Neighbouring Countries*. Dehradun, India: Vasant Press.

Chandrasekhara, K. S. 1972. Faunistic study of flea beetles of Mysore and Biology of *Phyllotreta chotanica* Duvivier (Coleoptera: Chrysomelidae: Halticinae). M. Sc (Ag.) Thesis. Bangalore, India: University of Agricultural Sciences.

Christudas, S. P., S. Chandrika, S. Mathai, and N. M. Das. 1972. Relative toxicity of some newer pesticides to *Haltica cyanea* Web. *Agricultural Research Journal of Kerala* 10(2): 118.

Devasahayam, S., and K. M. A. Koya. 1994. Natural enemies of major insect pests of black pepper. *Journal of Spices and Aromatic Crops* 3: 50–55.

Devasahayam, S., K. M. A. Koya, T. J. Zachariah, and T. K. Jacob. 2005. Biological control of insect pests of spice crops. Final Report of Research Project. Calicut, India: Indian Institute of Spices Research.

Devasahayam, S., P. S. P. V. Vidyasagar, and K. M. A. Koya. 1998. Reproductive system of *pollu* beetle, *Longitarsus nigripennis* Motschulsky (Coleoptera: Chrysomelidae), a major pest of black pepper, *Piper nigrum* Linnaeus. *Journal of Entomological Research* 22(1): 77–82.

Doeberl, M. 2003. Álticinae from India and Pakistan stored in the collection of the Texas Á. & M. University, U.S.Á. (Coleoptera, Chrysomelidae). *Bonner zoologische Beiträge* 51(4): 297–304.

Furth D. G. 1999. Searching for sumacs and flea beetles: From African poison arrows to Mexican poison ivy. *Entomological News* 110: 183.

Garg, A. K., and G. R. Sethi. 1983. New record of beetles in rice fields at Delhi. *Bulletin of Entomology* 24(2): 71–74.

Jagadish, P. S., and T. M. M. Ali. 1986. Insect pests of Millets. In *Souvenir, International Workshop on Small Millets, September 29–October 3, 1986, Bangalore*, ed. K. Krishnamurthy, pp. 46–50. Bangalore, India: University of Agricultural Sciences.

Jolivet P., and T. J. Hawkeswood. 1995. *Host-Plants of Chrysomelidae of the World: An Essay About the Relationships between the Leaf-Beetles and their Food Plants*. Leiden, the Netherlands: Backhuys Publishers.

Kishore, P. 1996. Evolving management strategies for pests of millets in India. *Journal of Entomological Research* 20(4): 287–297.

Konstantinov, A. S. 1998. *Revision of the Palearctic Species of Aphthona Chevrolat and Cladistic Classification of the Aphthonini (Coleoptera: Chrysomelidae: Alticinae)*. Memoirs on Entomology, International. Boca Raton, FL: Associated Publishers.

Konstantinov, A. S. 2016. Possible living fossil in Bolivia: A new genus of flea beetles with modified hind legs (Coleoptera, Chrysomelidae, Galerucinae, Alticini). *ZooKeys* 592: 103–120. doi:10.3897/zookeys.592.8180.

Konstantinov, A. S., and N. J. Vandenberg. 1996. Handbook of Palearctic flea beetles (Coleoptera: Chrysomelidae: Alticinae). *Contributions on Entomology, International* 1(3): 238–439.

Konstantinov, A. S., K. D. Prathapan, and F. V. Vencl. 2018. Hiding in plain sight: Leaf beetles (Chrysomelidae: Galerucinae) use feeding damage as a masquerade decoy. *Biological Journal of the Linnean Society* 123: 311–320.

Krishnaswamy, N., D. P. Chauhan, and R. K. Das 1982. The flea beetle as a rice pest in Assam, India. *International Rice Research Newsletter* 7(6): 16.

Kulshreshtha, J. P., and B. C. Mishra. 1970. Occurrence of *Chaetocnema concinnipennis* Baly (Chrysomelidae: Coleoptera) on rice in the Indian Union. *Indian Journal of Entomology* 32(2): 166–167.

Kumaraswamy, T. 1961. Feeding and biological characteristics of Halticid beetles with special reference to *Longitarsus belgaumensis* Jacoby. Thesis submitted to the Tamil Nadu Agricultural University; Abstract; *Madras Agricultural Journal* 51(10): 441.

Lingappa, S., and C. Siddappaji. 1978. *Luperomorpha vittata* Duvivier (Coleoptera: Chrysomelidae), a pest of ornamental crops. *Current Research* 7(11): 188–189.

Mathew, G., and K. Mohandas. 1989. Insects associated with some forest trees in two types of natural forests in the Western Ghats, Kerala (India). *Entomon* 14(3&4): 325–333.

Maulik S. 1926. Coleoptera. Chrysomelidae (Chrysomelinae and Halticinae). In *The fauna of British India including Ceylon and Burma*, ed. A. E. Shipley. London, UK: Taylor & Francis Group.

Medvedev, L. N. 2004. New and firstly recorded Chrysomelidae from Laos. *Entomologica Basiliensia* 26: 299–323.

Misra C. S., and Fletcher, T. B. 1919. 128. *Clitea picta* as a pest of Bael. In *Second Hundred Notes on Indian Insects*, Bulletin No. 89, ed. T. B. Fletcher, pp. 22–23. Pusa, India: Imperial Agricultural Research Institute.

Mote, U. N., and B. D. Ghule. 1989a. Chemical control of flea beetles infesting *rabi* sorghum. *Journal of Maharashtra Agricultural Universities* 14(2): 161–163.

Mote, U. N., and B. D. Ghule. 1989b. Effects of density of flea beetle on sorghum seedlings. *Journal of Maharashtra Agricultural Universities* 14(3): 375–376.

Nair, M. R. G. K. 1986. *Insects and Mites of Crops in India*. New Delhi, India: Indian Council of Agricultural Research.

Nayar, K. K., T. N. Ananthakrishnan, and B. V. David. 1976. *General and Applied Entomology*. New Delhi, India: Tata Mc Graw-Hill.

Pandit, N. C., and S. K. Pradhan. 1995. Bionomics of the Sunnhemp flea-beetle *Longitarsus belgaumensis* Jac. (Coleoptera: Chrysomelidae). *Indian Journal of Entomology* 57(3): 219–223.

Prathapan K. D., and A. P. Balan. 2010. Report of the occurrence of the sweet potato flea beetle *Chaetocnema confinis* Crotch (Coleoptera: Chrysomelidae) in southern India. *Journal of Root Crops* 36(2): 272–273.

Prathapan K. D., and C. A. Viraktamath. 2008. The flea beetle genus *Lanka* (Coleoptera: Chrysomelidae) of India with descriptions of three new species. *Zootaxa* 1681: 1–30.

Prathapan K. D., and C. A. Viraktamath. 2009. Revision of *Tegyrius* Jacoby (Coleoptera: Chrysomelidae) with descriptions of eight new species. *Zoological Journal of the Linnean Society* 157: 326–358.

Prathapan K. D., and C. A. Viraktamath. 2011. A new species of *Longitarsus* Latreille, 1829 (Coleoptera: Chrysomelidae: Galerucinae) pupating inside stem aerenchyma of the hydrophyte host from the Oriental region. *ZooKeys* 87: 1–10.

Prathapan K. D., and C. S. Chaboo. 2011. Biology of Blepharida-group flea beetles with first notes on natural history of *Podontia congregata* Baly, 1865 an endemic flea beetle from southern India (Coleoptera, Chrysomelidae, Galerucinae, Alticini). *ZooKeys* 157: 95–130.

Prathapan K. D., M. H. Faizal, and K. N. Anith. 2005. A new species of *Longitarsus* (Coleotpera: Chrysomelidae) feeding on Chinese potato, *Plectranthus rotundifolius* (Lamiaceae) in south India. *Zootaxa* 966: 1–8.

Prathapan, K. D., 2017. Identity of the citrus leaf mining flea beetle in northeast India and nomenclatural changes in *Amphimela* (Coleoptera: Chrysomelidae: Galerucinae: Alticini). *Florida Entomologist* 100(2): 276–280.

Prathapan, K. D., and A. S. Konstantinov. 2003. The flea beetle genus *Aphthona* of south India (Coleoptera: Chrysomelidae: Alticinae) with descriptions of seven new species. *Proceedings of the Entomological Society of Washington* 105(1): 154–179.

Prema, S., and K. P. Janardhanan. 1991. Systematics and biology of *Gregarina phygasiae* sp. nov. (Apicomplexa: Cephalina) from *Phygasia silacea* (Illiger). *Journal of Ecobiology* 3(2): 123–127.

Premkumar, T. 1980. *Ecology and control of pepper "pollu" beetle* Longitarsus nigripennis *Motschulsky (Chrysomelidae: Coleoptera)*. Ph. D. Thesis. Kerala, India: Kerala Agricultural University.

Raj, K. D., and Y. N. Reddy. 1989. Parasitism of *Longitarsus belgaumensis* Jacoby (Coleoptera: Chrysomelidae) by Nematode of the genus *Howardula*. *Indian Journal of Nematology* 19(1): 82–84.

Rajamma, P. 1982. New records of insect pests of sweet potato in Kerala. *Agricultural Research Journal of Kerala* 20(1): 84–88.

Reid, C. A. M., and M. Beatson. 2015. Disentangling a taxonomic nightmare: A revision of the Australian, Indomalayan and Pacific species of *Altica* Geoffroy, 1762 (Coleoptera: Chrysomelidae: Galerucinae). *Zootaxa* 3918(4): 503–551.

Ruan, Y., X. Yang, A. S. Konstantinov, K. D. Prathapan, and M. Zhang. 2019. Revision of the Oriental *Chaetocnema* species (Coleoptera, Chrysomelidae, Galerucinae, Alticini). *Zootaxa* (In review).

Sankaran, T., D. Srinath, and K. Krishna. 1967. *Haltica caerulea* Olivier (Col: Halticidae) as a possible agent of biological control of *Jussiaea repens* L. *Technical Bulletin. Commonwealth Institute of Biological Control* 8: 117–138.

Santhakumari, K., T. Nalinakumari, and M. R. G. K. Nair. 1979. New record of a pest of brinjal. *Entomon* 4(2): 215–216.

Satyanarayana, J., T. V. K. Singh, and T. B. Reddy. 2000. Pest complex in Rabi Castor in Mahabubnagar District of Andhra Pradesh. *Insect Environment* 5(4): 152–153.

Scherer G. 1969. Die Alticinae des indischen Subkontinentes (Coleoptera: Chrysomelidae). *Pacific Insects Monograph* 22: 1–251.

Shamalamma, M., A. R. K. Bai, and A. S. Swamy. 1977. Report of the occurrence of a species of *Microctonus* (Hymenoptera: Braconidae) parasitizing the adults of *Haltica caerulea* Olivier (Coleoptera). *Indian Journal of Entomology* 39(4): 387–388.

Singh, B., S. V. Singh, and V. Kumar. 1996. Evaluation of some insecticides against jassids and flea beetles on Brinjal. *Annals of Plant Protection Sciences* 4 (1): 38–41.

Srikanth, J., S. Joshi, and C. A. Viraktamath. 1991. *Luperomorpha vittata* Duvivier: A new association with *Parthenium hysterophorus* L. and other weeds. *Current Science* 60(3): 177–178.

Stebbing, E. P. 1903. Insect pests of fruit trees. *Indian Museum Notes* 5(3): 117–127.

Stebbing, E. P. 1914. *Indian Forest Insects of Economic Importance. Coleoptera*. London, UK: Eyre & Spottiswoode.

Visalakshy, P. N. G., and K. P. Jayanth. 1998. Suppression of *Ludwigia adscendens* by naturally occurring biocontrol agents in Bangalore, India. *Entomon* 23(2): 123–126.

Wadhi, S. R., and H. N. Batra. (1964) Pests of tropical and subtropical fruit trees. In *Entomology in India*, pp. 227–260. New Delhi, India: The Entomological Society of India.

Yadav, L. B., and S. M. A. Rizvi. 1994. Studies on the insect pests of wasteland plantations at Faizabad (India). *Journal of Entomological Research* 18(2): 115–120.

Zaka-ur-Rab. 1991. Leaf mining Coleoptera of the Indian subcontinent. *Journal of Entomological Research* 15(1): 20–30.

16 Taxonomy, Systematics, and Biology of Indian Butterflies in the 21st Century

Krushnamegh Kunte, Dipendra Nath Basu, and G. S. Girish Kumar

CONTENTS

Introduction 275
Indian Butterflies, a Taxonomic Legacy 276
Traditional Tools of Taxonomy 282
    Genitalia 282
    Wing Venation 282
    Colour Patterns 285
Evolutionary Biology, Phylogenetics, and Molecular Systematics as Pillars of Modern Taxonomy ... 286
    What Are Species? 286
    What Are Subspecies? 288
    When Are Subspecies Names Inappropriate? 289
        Linear Clinal Variation 289
        Phenotypic Plasticity 289
    Reconstructing and Interpreting Phylogenies for Systematic and Taxonomic Studies 289
Recent Insights from Molecular Systematic Studies into the Evolution and Higher Classification of Butterflies ... 291
    Butterflies Are Moths, Butterfly-Moths Are Butterflies, Swallowtails Are Older than Skippers, and Other Startling Observations ... 291
    Family- and Subfamily-Level Changes 292
    Resolutions at Generic and Species Levels 292
Indian Butterflies, a Taxonomic Impediment 292
    The Genus *Baracus* 294
    The *homolea* Species-Group of *Halpe* 294
    The *Euploea core* Species-Group 295
    More Examples 296
A Modern *Mantra* for Butterfly Biology in the 21st Century India 296
    Collection of Field Data 296
    Developing Museum Resources 296
    Scientific Research 297
Conclusion 298
Acknowledgements 299
References 299

INTRODUCTION

India is a large landmass with a unique geological history, which is placed at a critical meeting point of several biogeographic regions and subregions of the world. It provides diverse habitats and environmental gradients in its oceans and mountains, and encompasses some conspicuous biogeographic breaks, which together form complex land- and seascapes that support remarkable biodiversity. Indeed, India is one of the most biodiverse countries of the world, hosting four globally recognized biodiversity hotspots that contribute to the densest cluster of hotspots in the world (Figure 16.1) (Marchese 2015; Myers et al. 2000). As a result, India is one of the most critical centres for the exploration and conservation of tropical biodiversity. Yet, India has not been able to keep pace in documenting species diversity, from the basic taxonomic characterization of species to deeper understanding of their biology, compared to the progress that has been made by foreign researchers in this area. This stunted growth of biological research has especially affected studies of exceptionally biodiverse groups such as insects, plants, and microbes, which still remain poorly inventoried in India's biodiversity hotspots. As India makes progress in building scientific institutions and infrastructure in its ambition of becoming a scientific and technological entity, it is crucial to keep a firm focus on its unique and invaluable biological heritage—its biodiversity—that can place Indian science in a unique leadership.

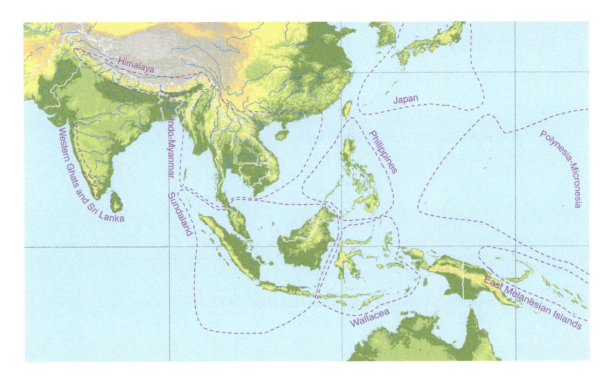

FIGURE 16.1 India's four globally recognized biodiversity hotspots, which are part of the densest cluster of hotspots placed in the Indo-Australian Region. Indian biodiversity needs to be understood in the larger context of biogeography and diversification in the Oriental, Afrotropical, and Palearctic Zones. The map of biodiversity hotspots is based on (Marchese, C., *Glob. Ecol. Conserv.*, 3, 297–309, 2015; Myers, N. et al. *Nature*, 403, 853–858, 2000) and the Conservation International map of biodiversity hotspots.

Butterflies are charismatic, so they are often considered to be a flagship group for insect biology and conservation. In India, the state of butterfly biology instead effectively represents the problems that have dogged insect biodiversity research: (a) taxonomy of butterflies is outdated, (b) there is widespread misunderstanding of species and subspecies names, synonyms, and taxonomic rules, (c) competent taxonomists are largely missing, and those who exist largely lack easy access to historically most important European museum collections, (d) knowledge of current distributions and populations was virtually absent until 10 years ago, and it is still sketchy, (e) detailed morphological characterization, which is very important for taxonomic works, has been done poorly, (f) information on basic natural history such as habitat use, early stages, larval host plants, nectar plants, etc., is just beginning to accumulate, and (g) advanced knowledge of population and community ecology, plant-butterfly associations and other inter-species interactions, trait evolution, biogeography, population genetics, speciation patterns, etc., is almost completely lacking. This situation needs to change rapidly.

The primary goals of this chapter are to: (a) provide a historical account of the development of taxonomy and biology of Indian butterflies, including a brief review of trends in taxonomic, natural historic, and scientific research on Indian butterflies, (b) point out recent developments in the fields of evolutionary biology, phylogenetics, molecular systematics, and biogeography, which have a strong bearing on current taxonomic developments and trends, (c) highlight problems with current taxonomic practices in India, and (d) offer a vision for growth of modern taxonomy, systematics, and butterfly biology, in the 21st century India. It is meant to be a primer on historical and modern methods in taxonomy, molecular systematics, and museum sciences in general, for students of entomology, who will benefit from using butterflies as a case study to understand insect systematics and biology. Thus, this chapter will present an array of illustrations and examples that have a special significance in the Indian context.

INDIAN BUTTERFLIES, A TAXONOMIC LEGACY

Historical accounts and literature on Indian butterflies, natural history, and naturalists, are extensive. It is not possible to include an exhaustive account here, although such an account would be an excellent academic pursuit, and it needs to be written for posterity. However, the short account given below will serve as a basic introduction that will get students initiated. It will also highlight some important milestones that will be relevant for the remainder of the chapter.

Indian butterflies started to be systematically studied and formally named with the publication of Carl Linnaeus's *Systema Naturae* in 1758 in which he established the binomial system of naming species. This was a novel way of assigning unique scientific names to any species of organism on earth based on a genus name followed by a species name, which would ease communication among scientists. This proved especially timely since the natural world around Linnaeus was rapidly expanding as European empires reached far corners of the world, bringing in previously unfamiliar species from distant lands. Linnaeus's binomial system was an instant success: tens of thousands of species had been named using this system, and its acceptance spread far and wide in Europe, within decades. The system has subsequently become the core of taxonomic nomenclature for all life on earth. Linnaeus's work was soon followed by two sets of volumes in which dozens more Indian butterfly species were described, one written by Linnaeus's student Johan Christian Fabricius and the other by a Dutch merchant and amateur entomologist, Pieter Cramer. Together, these three taxonomists described nearly 350 butterfly taxa from the Indian region (Figure 16.2). Most of the butterfly type specimens used by Linnaeus, which have been intensively inventoried, are now housed in the Linnean Society of London, Museum Ludovicae Ulricae (Uppsala University), the Clerck and De Geer collections in the Naturhistoriska Riksmuseet (Stockholm), and the James Petiver Collection in the Sir Hans Sloane Collection at The Natural History Museum, London (Honey and Scoble 2001). The Linnean type specimens have also been photographed by The Natural History Museum, London (NHMUK, previously the British Museum of Natural History, or BMNH) (http://www.nhm.ac.uk/research-curation/research/projects/linntypes/), and by the GART project (Globales Ahgister Tagfalter), i.e., The Global Species Register Butterflies (http://www.naturkundemuseum-bw.de/sites/default/files/forschung/user_122/gart_biolog-status_2001.pdf). Many of Cramer's type specimens are also in NHMUK (Chainey 2005). The Fabricius Collection is largely in the Natural History Museum of Denmark in the University of Copenhagen and in the Zoological Museum of the University of Kiel (Germany) (https://samlinger.snm.ku.dk/en/dry-and-wet-collections/zoology/entomology/fabricius-collection/). These are significant type collections since Fabricius alone described nearly 10,000 species of insects, including many butterflies, for which he is considered one of the founders of systematic entomology (see the last link).

This initial flurry of species descriptions was followed by a relative lull in the early 1800s. What might be considered a golden period in the discovery and naming of Indian butterflies started from the 1820s with the publication of Thomas Horsfield and Frederic Moore's *A Catalogue of the Lepidopterous Insects in the Museum of the Honorary East-India Company*. Both Horsfield and Moore had very long careers, and they together covered nearly 90 years of studies of Indian butterflies, describing over 500 taxa. Indeed, Moore holds a record for describing more butterfly taxa from the Indian region than any other butterfly taxonomist in history (Figure 16.2C), although some of his species descriptions were published as collaborations, and therefore their authorships were shared with other taxonomists. The golden period of taxonomic discovery of butterflies on the Indian subcontinent peaked from ca 1840 to 1900, when the majority of the species and subspecies that are currently considered taxonomically valid were described (Figure 16.2B). Apart from Moore, the bulk of this work was done by entomologists of the British Raj, such as Hewitson, Doubleday, Horsfield, Westwood, and de Nicéville, although continental European entomologists such as Kollar, C. Felder, and R. Felder also made important contributions (Figure 16.2).

Details of their active periods on Indian butterflies, landmark publications, and the number of taxa described by them from the Indian subcontinent, are given in Figure 16.2B,C. The majority of the type specimens used by these British entomologists are housed in NHMUK, except many of de Nicéville's types that are in the Zoological Survey of India in Kolkata (ZSI-K). Types described by the continental European entomologists are scattered, and some of them have not been properly catalogued and therefore not easily accessible. Many of the Indian butterfly types housed in NHMUK, and the majority of the de Nicéville types in ZSI-K, have recently been photographed by one of us (KK) (e.g., Box 16.1). These are being made publicly available on the Butterflies of India website, at http://www.ifoundbutterflies.org/ (Kunte et al. 2018).

The 1880s–1910s saw the publication of two critically important compilations of Indian butterflies. The first was a 3-volume series, *The Butterflies of India, Burmah and Ceylon*, by Marshall and de Nicéville (Marshall and de Nicéville 1882; de Nicéville 1886b, 1890b). The second set was a 10-volume series started by Moore and finished by Swinhoe, called *Lepidoptera Indica* (Moore 1892, 1896, 1899, 1900, 1903, 1905; Swinhoe 1910, 1911, 1912, 1913). Both these sets of volumes compiled all the available information on species descriptions, distributional ranges, and larval host plants. The ten volumes of *Lepidoptera Indica* especially still stand strong in the history of Indian butterfly research as the only set of volumes that extensively reviewed and illustrated every known species known at the time.

The 1890s–1910s also saw two other revolutions in taxonomic practices as applied to Indian butterflies. The first revolution was the application of subspecies names, which led to the trinomial system composed of genus, species, and subspecies names. By this time, there was a widespread recognition that different, often spatially widely separated populations of the same species varied in morphological features such as wing colouration and body size. Biologists had by then realized the importance of such variation in understanding morphological diversification, evolution, and natural selection in isolated populations. This led to the widespread establishment of the subspecies concept, which cast a long shadow on taxonomic practices especially in well-collected and extensively studied groups such as butterflies, birds, and mammals. Thus, both taxonomists and evolutionary biologists embraced the subspecies concept and its use in trinomial names as a means of formally referring to geographically structured, morphologically distinct populations of the same species. Three people led this practice in butterfly taxonomy, especially as relevant to Indian butterflies: Walter Rothschild, Karl Jordan (Mallet 2007; Riley 1960), and Hans Fruhstorfer. Rothschild and Jordan's contributions were particularly valuable because they bolstered the conceptual growth of the subspecies concept and widespread acceptance of trinomial names in butterfly taxonomy. The extensive use of subspecies by Fruhstorfer, and later by W. H. Evans, has survived to this day. The taxonomic and biological implications of this are discussed in later sections of this chapter.

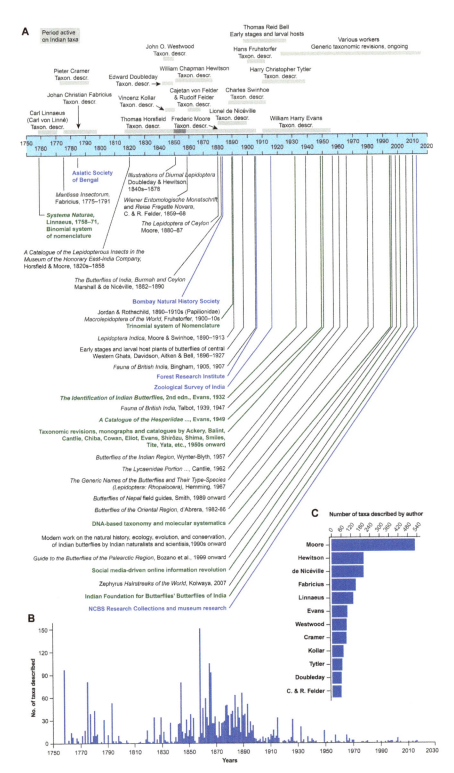

FIGURE 16.2 History of taxonomy and biology of Indian butterflies. **Panel A:** Major milestones and prominent figures along a timeline from Linnaeus's *Systema Naturae* that laid the foundation of modern taxonomy in the mid-1700s to the present time. Entries in green are highlighted milestones that contributed towards the growth of butterfly taxonomy and biology in India. Entries in blue show establishment of major societies, institutions, and infrastructures that have played critical roles in the development of scientific research and publications on Indian butterflies. Names above the timeline represent taxonomists who contributed in major ways to the development of butterfly taxonomy and biology in India, although the list is naturally somewhat selective. Only the years in which these lepidopterists were active on Indian butterflies are marked in this panel. The beginning years of multi-year works are marked on the timeline, the actual ranges in years are given in the text. **Panel B:** A timeline of taxonomic descriptions of valid species and subspecies, as recognized at present, of butterflies of the Indian region. **Panel C:** The number of taxa described by major authors of Indian butterflies, i.e., authors who published over 50 valid taxonomic names (species and subspecies) of butterflies of the Indian region. Based on data from K. Kunte, unpublished.

BOX 16.1 A CLASSIFICATION OF TYPE SPECIMENS USED COMMONLY IN TAXONOMIC LITERATURE

The quoted definitions given below are from the latest (4th) edition of the International Code of Zoological Nomenclature, or "The Code" as it is popularly called, and its latest online version (International Commission on Zoological Nomenclature (ICZN) 1999). Further explanations provided for each definition are also largely based on The Code. ICZN is a world authority on nomenclatural matters of zoological nature, which rules on nomenclatural acts through its bulletin and other publications.

Holotype: "The single specimen (except in the case of a hapantotype, q.v.) designated or otherwise fixed as the name-bearing type of a nominal species or subspecies when the nominal taxon is established" (ICZN 1999). Holotype is thus taxonomically the most important specimen of a species or subspecies. E.g., holotype of *Hypolycaena narada* Kunte, 2015, in National Centre for Biological Sciences (NCBS).

Paratype: "Each specimen of a type series other than the holotype [Recommendation 73D]" (ICZN 1999). Paratypes cannot be treated as syntypes and used for lectotype selection if the holotype is lost or destroyed; however, they are eligible for neotype selection. E.g., paratype of *Hypolycaena narada* Kunte, 2015, in NCBS.

Syntype: "Each specimen of a type series (q.v.) from which neither a holotype nor a lectotype has been designated [Arts. 72.1.2, 73.2, 74]. The syntypes collectively constitute the name-bearing type." (ICZN 1999). For a nominal species-group taxon established before 2000, all the specimens of the type series are treated as syntypes if neither a holotype nor a lectotype had been fixed. E.g., syntype of *Lethe tristigmata* Elwes, 1887, in NHMUK.

Lectotype: "A syntype designated as the single name-bearing type specimen subsequent to the establishment of a nominal species or subspecies [Art. 74]" (ICZN 1999). In taxonomic value, lectotypes are equivalent to holotypes. E.g., lectotype of *Charaxes dolon magniplagus* (Fruhstorfer, 1904), in NHMUK.

(Continued)

BOX 16.1 (Continued) A CLASSIFICATION OF TYPE SPECIMENS USED COMMONLY IN TAXONOMIC LITERATURE

Neotype: "The single specimen designated as the name-bearing type of a nominal species or subspecies when there is a need to define the nominal taxon objectively and no name-bearing type is believed to be extant" (ICZN 1999). See under Paratype above.

Paralectotype: "Each specimen of a former syntype series remaining after the designation of a lectotype [Art. 72.1.3, Recommendation 74F]" (ICZN 1999). In taxonomic value, paralectotypes are equivalent to paratypes. E.g., paralectotype of *Charaxes dolon magniplagus* (Fruhstorfer, 1904), in NHMUK.

Cotype: "A term not recognized by The Code, formerly used for either syntype or paratype, but that should not now be used in zoological nomenclature [Recommendation 73E]" (ICZN 1999). E.g., cotype of *Hyponephele (pulchra) astorica* (Tytler, 1926), in NHMUK.

Allotype: "A term, not regulated by The Code, for a designated specimen of opposite sex to the holotype [Recommendation 72A]" (ICZN 1999). For example, if a species is described from a male holotype, then a female type may be designated as an allotype, and vice versa. E.g., allotype (female) of *Chrysozephyrus tytleri tytleri* (Howarth, 1957), in NHMUK.

The second revolution of the very early twentieth century was the use of differences in the male genitalia in making taxonomic decisions. It was realized by then that sister species of insects often have distinctly different male genitalia. With the availability of microscopes, the use of genitalia dissections became a gold standard in taxonomic studies of butterflies. Early studies that described male genitalia of Indian butterflies were performed by Bethune-Baker and Swinhoe (Bethune-Baker 1918; Swinhoe 1910, 1911, 1912, 1913). The structures of male genitalia continue to be used as an important morphological dimension in making taxonomic decisions to this day. The use of female genitalia has also proved useful in resolving species relationships in some genera, although taxonomists have tended to rely on evidence from female genitalia to a lesser degree. Some examples of male and female genitalia will be discussed later in this chapter.

Brigadier William Harry Evans entered the Indian butterfly scene in early 1900s and made a deep impact with his extensive work that spanned nearly seven decades. Evans secured his position in the annals of Indian butterfly taxonomy not only with his dozens of species descriptions, but also with his generic revisions (e.g., Evans 1954, 1957), an influential catalogue and identification key of Hesperiidae of Asia, Europe, and Australia in the BMNH collections (Evans 1949), and an identification key to all the Indian butterfly species and subspecies that were known in his time (Evans 1932), which is followed by butterfly watchers to this day. Evans did some of this work as a serious amateur like so many other British lepidopterists of his generation, while serving as a military engineer. His military postings offered excellent opportunities to collect butterflies far and wide, from Balochistan and Chitral to Simla, Jabalpur and Kodaikanal. Based on this growing experience, he initially made some preliminary attempts to list Indian butterflies (Evans 1912). After World War I, however, Evans started working on a more comprehensive catalogue and identification key to butterflies of the Indian subcontinent, which was published in multiple volumes in the *Journal of the Bombay Natural History Society*. This series later contributed to Evans's most well-known identification guide, *The Identification of Indian Butterflies* (Evans 1932). This book was remarkable in that it provided the first synonymic catalogue of Indian butterflies in a trinomial system, describing several dozen new subspecies in the process. After retirement in 1931, Evans moved to London where he spent the remainder of his life devoted to the taxonomic listings and in-depth studies of butterflies in the BMNH, taking advantage of the millions of butterfly specimens deposited there. He dissected and illustrated male genitalia of thousands of hesperiid and lycaenid butterflies, which informed his taxonomic decisions. A large part of this effort resulted in his most prominent works—catalogues of Hesperiidae of the world, including that of Asia (Evans 1949), and very close to his death in 1956, of *Arhopala* (Lycaenidae) (Evans 1957). In all, Evans published dozens of papers and comprehensive catalogues in what might be considered by any standards a very productive lifetime (Evans and Bellinger 1956; Riley 1956). Evans's specimens were largely deposited in the BMNH, where his study materials of world Hesperiidae and *Arhopala* are maintained separately as reference collections to this day, which one of us (KK) has extensively photographed.

Two of Evans's contemporaries need a particular mention: Harry Christopher Tytler, also a British army officer, and George Talbot. Tytler collected extensively while on duty in Chitral in the western Himalaya and the Naga-Manipur-Chin Hills in north-eastern India and northern Myanmar. He described several dozen new species from this material (e.g., Tytler 1911, 1914, 1915, 1926, 1940) whose type specimens were deposited in the BMNH. Some of his species have not been recorded from India since their descriptions. The areas where Tytler collected are just beginning to be properly explored once again, so many of the butterfly populations that Tytler reported on may be rediscovered in coming years. On the other hand, Talbot's main contribution was not his new taxonomic descriptions, but his extensive notes on original species descriptions, natural history, and distributional ranges of butterflies of the Indian subcontinent that he published in two volumes of the *Fauna of British India* (Talbot 1939, 1947). His *Fauna* volumes also provided illustrations of male genitalia of many species. For these reasons, Talbot's *Fauna* volumes remain among the most comprehensive taxonomic works on Indian butterflies.

Scientific publications on Indian butterflies until the 1880s were largely species descriptions and/or catalogues of specimens in museum and private collections. However, natural history papers with relevance to the then newly minted theory of evolution by natural selection and other papers of interest to modern ecologists and evolutionary biologists started to appear from the 1870s (Aitken 1897; Davidson and Aitken 1890; Dudgeon 1895; Forsayeth 1884; Fryer 1914; Punnett 1908). From the 1880s, comprehensive regional checklists and other compilations of Indian butterflies started to appear since there was so much information and specimens available by then (e.g., Aitken 1886; Betham 1890a, 1890b, 1891, 1894; Cantlie 1952, 1956; de Nicéville 1885a, 1885b, 1886a, 1886b, 1890a, 1890b, 1883; Davidson et al. 1896, 1897; Doherty 1886; Elwes and de Nicéville 1886; Elwes and Möller 1888; Marshall and de Nicéville 1882; Mackinnon 1898; Mackinnon and de Nicéville 1898; Parsons and Cantlie 1948; Swinhoe 1886). Butterfly surveys in the Nilgiris and Palnis in southern Western Ghats, which had been frequented by the British tea planters and missionary school teachers, culminated in a series of papers over several decades, although the efforts still continue (Larsen 1987; Mathew and Kumar 2003; Rufus and Sabarinathan 2007; Wynter-Blyth 1944a, 1944b, 1945, 1947; Yates 1946). Studies of early stages and larval host plants of Indian butterflies received a major uplift with James Davidson, Edward Hamilton Aitken, and Thomas Reid Bell's series of papers on butterflies of the Karwar area (reported as North Kanara in the erstwhile Bombay Presidency) (e.g., Bell 1909, 1927; Davidson et al. 1896). Bell's papers, which were published over nearly 20 years in the *Journal of the Bombay Natural History Society*, offered unmatched detailed descriptions of larval and pupal morphology and behaviour of butterflies of the Western Ghats. Some of this information has subsequently contributed to a deeper understanding of larval host plant specialization that can shed light on speciation and diversification.

Demise of Tytler in the decade preceding India's independence, and that of Bell, Talbot, and Evans in the decade following, drew the direct British engagement in field expeditions and the butterfly taxonomy and biology in the Indian subcontinent more or less to a close. From World War I, attention and priorities of British officers, including that of professional and amateur lepidopterists, had already started to wander elsewhere. Shortly after World War II and India's independence in 1947, the British enterprise of natural historic and taxonomic studies largely wrapped up from the Indian subcontinent. Most of the works published after this period, including last pieces of work by Evans and Talbot, were based exclusively on specimens that had accumulated in BMNH in the preceding 125 years or so. Major highlights and expansive works after independence were few and far in between. Only a few new species have been described from India since independence, almost exclusively from the eastern Himalaya and NE India (e.g., Cantlie 1958; Cantlie and Norman 1959, 1960; Koiwaya 2002; Kunte 2015; Roy 2013). Several more species and subspecies have been described, but their taxonomic validity needs to be verified after making adequate comparisons with related species and improving morphological diagnosis (Kumar et al. 2009; Sharma 2013a, 2013b; Singh 2007; Smetacek 2004, 2011, 2012). Many more species and subspecies have been described from the neighbouring Nepal and Tibet, some of which likely occur in India (Fujioka 1970; Huang 1998, 2000, 2001, 2002, 2003; Huang and Xue 2004). However, the most important taxonomic works on Indian butterflies in the second half of the 20th century involved revisions and catalogues rather than species discovery and descriptions. Apart from Evans's *Hesperiidae Catalogue* (Evans 1949), an updated list and identification key of Indian Lycaenidae was published by Cantlie (1962). An important generic catalogue of the world butterflies that also included all the Indian genera known at the time still remains valuable today (Cowan 1970; Hemming 1967). The bulk of the progress in this area, however, involved a number of tribal and generic revisions and monographs by a large number of taxonomists (Figure 16.2) (e.g., Cowan 1966, 1967, 1974; Chiba 2009; Chiba and Eliot 1991; Chiba and Tsukiyama 1993, 1994; Eliot and Kawazoé 1983; Eliot 1963, 1967, 1969; Eliot 1973, 1986, 1990; Evans 1957; Fan et al. 2010; Howarth 1957; Masui 2004; Shirôzu and Shima 1979; Smiles 1982; Tite 1963, 1966; Tsukiyama and Chiba 1994; Yata 1989, 1991, 1992, 1994, 1995 Yata et al. 2010). A glance at these publications reveals a striking pattern: the racial dominance in taxonomic expertise and scientific output on Indian butterflies has shifted from early to late 1900s from predominantly white Europeans to Japanese Asians. However, Indians continue to be under-represented in biodiversity-related studies—from species discoveries to taxonomic revisions—in the Indian region itself. This has reinforced a "taxonomic impediment" that continues to cause trouble for Indian taxonomy and systematics (see further sections). In any case, it is very likely that traditional taxonomic works will continue in large and historically strong museums abroad with or without Indian participation. However, most of the critically important revisions in the future will be based on molecular phylogenies that are likely to stabilize tribal and generic classification and illuminate species- and subspecies-level relationships of butterflies. This is an area where Indian biologists may perhaps be able to make relatively rapid progress. Several notable recent developments in this area will be reviewed in the section below, "Recent Insights from Molecular Systematic Studies ...."

This historical legacy has had a profound influence on the current state of butterfly taxonomy in India. Before reviewing these influences and current practices, we will briefly review traditional tools of butterfly taxonomy. We will view these in light of some recent trends in evolutionary biology, phylogenetics, and molecular systematics, which form pillars of modern taxonomy. This perspective is critical for the vision of future growth in this field.

TRADITIONAL TOOLS OF TAXONOMY
GENITALIA

From Linnaeus's time, butterfly taxonomists have relied to a great extent on wing colouration to define species. Body form and size, wing shape, structure of legs, labial palpi, structure of scales, etc., have also been used in various species groups, but to a lesser degree. The historical summary of work on Indian butterflies above shows that from the late 1800s, the use of male genitalia became widespread in insect taxonomy. In the case of butterflies, it became particularly common from 1910s onward, and by the 1920s and 1930s, differences in the structures of male genitalia had become one of the most predominant ways in which taxonomists tried to distinguish between closely related species in many difficult species groups. Studies of male genitalia, and in some species groups female genitalia, continue to be one of the prominent axes of evidence on which taxonomic decisions are based. However, the nomenclature of genital parts has changed to a degree in the past century. Comprehensive, well-marked reference diagrams of male and female genitalia are rarely accessible, so we provide detailed diagrams and nomenclature of butterfly genitalia in Figure 16.3. For male genitalia, structures of the uncus, cuiller, clasp, and aedeagus, are considered taxonomically especially important since they often differ prominently even among sister groups. However, not all structures are found in all butterflies, and there is considerable variation in the presence and shapes of different genital components in different groups of butterflies (Figure 16.4). The cartoon shown in Figure 16.3B is more complex than average male genitalia in any species because this is a composite of various structures that may be found in different butterfly groups and have therefore been shown together for reference. For female genitalia, structures of signum and bursa copulatrix are considered taxonomically important, although they are less frequently used for species delineation compared to structures of the male genitalia. The illustrations in Figure 16.3 are based on (Cordero and Baixeras 2015; Evans 1949; Mehta 1933).

WING VENATION

Wings of butterflies and moths are transparent, membranous structures that are crossed by veins through which haemolymph, which is somewhat equivalent to vertebrate blood, is circulated (see below for scales and colouration).

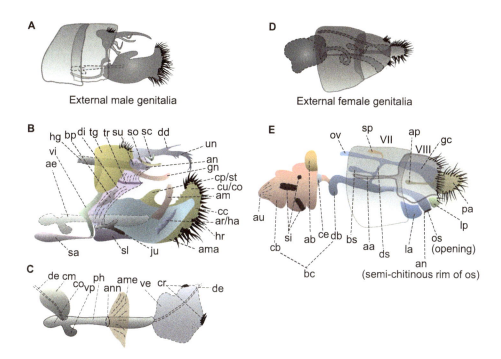

FIGURE 16.3 Idealized structures of the male and female genitalia of butterflies, which have been used extensively in the past for taxonomic studies. **Panel A:** External male genitalia, including the last two abdominal segments that envelop the genitalia. **Panel B:** Detailed structures of the male genitalia, excluding the aedeagus (see Panel C). **ae** aedeagus, **am** ampullary process, **an** anal opening, **ama** anal margin, **ar/ha** ampullary ridge/harpe, **bp** basal process, **cc** cucculus, **cp/st** costal process/style, **cu/co** cuiller/corona, **dd** dorsal dentate process, **di** diaphragma, **gn** gnathos, **hg** hind-gut, **hr** hairy sensilla, **ju** juxta, **sa** saccus, **sc** scaphium, **sl** sacculus, **so** socius, **su** subuncus, **tg** tegumen, **tr** transtilla, **un** uncus, and **vi** vinculum. Among these, ama, ar/ha, am, bp, cc, cp, cu/co, hr, and sl, together form the clasps or valvae (singular "valve"), with which male butterflies hold females during copulation. **Panel C:** Structure of aedeagus: **ann** annulus, **ame** annular membrane, **cm** coecal membrane, **co** coecum, **cr** cornuti, **de** ductus ejaculatorius, **ph** phallus, **ve** vesica, and **vp** ventral process. **Panel D:** External female genitalia, including the last two abdominal segments that envelop the genitalia. **Panel E:** Female genitalia: **aa** apophysis anterior, **ab** accessory bursa, **an** antrum, **ap** apophysis posterior, **au** appendix bursa, **bc** bursa copulatrix, **bs** bulla seminalis, **cb** corpus bursa, **ce** cervix, **db** ductus bursa, **ds** ductus seminalis, **gc** genital chamber, **la** lamella antevaginalis, **lp** lamella postvaginalis, **os** ostium, **ov** oviduct, **pa** anal papilla, **si** signum, and **sp** spermatheca.

Veins are stiff, and stronger than wing membranes, giving the wings sufficiently strong support that they sustain sometimes extremely high-speed, strenuous flights. Thus, veins are prominent wing structures that also have group-specific spatial arrangements. Therefore, venation is widely used in defining genera and other higher taxonomic categories of butterflies, although there may sometimes be differences even between sister species. For these reasons, wing venation has been characterized in a broad range of butterflies. Specific veins may be missing, distinctly arranged, or prominently modified, in some butterfly families, subfamilies, or tribes. Additionally, wing colour patterns develop during late larval and pupal development in relation to veins, so veins typically define boundaries between various colour patches, spots, and bands on butterfly wings (Nijhout 1991). These colour markings are commonly used in identification keys (see the next section). Thus, wing venation has great utility in studying butterflies whether from a taxonomic or developmental perspective.

Wing venation of butterflies is described with two commonly used systems (Figure 16.5). The Comstock-Needham system (Figure 16.5A) is universal, i.e., it describes wing venation across insects, because it relies on the origins of veins in relation to the cells (cell is an area of the wing that is enclosed by veins on all sides. However, open cells, i.e., cells that are not enclosed on all sides by veins, are also known in some butterflies, e.g., groups of nymphalids). This system is commonly used all over the world in entomological and other scientific literature.

The numerical system simply numbers veins from 1a through 12 on the forewing, and 1a through 8 on the hindwing, of butterflies and moths (Figure 16.5B). It is popular because of the simplicity of numbering veins. It is commonly used in identification keys of Asian butterflies, including in the extensive keys developed by Evans (see above for Evans's works) and used subsequently. In both the systems, wing areas between the numbered veins are also numbered. These numbered areas are often used to localize spots, bands, and other colour markings, as well as brands, specialized scales, and other distinctive features, while describing butterflies. Thus, it is important to familiarize oneself with wing venation before diving into serious

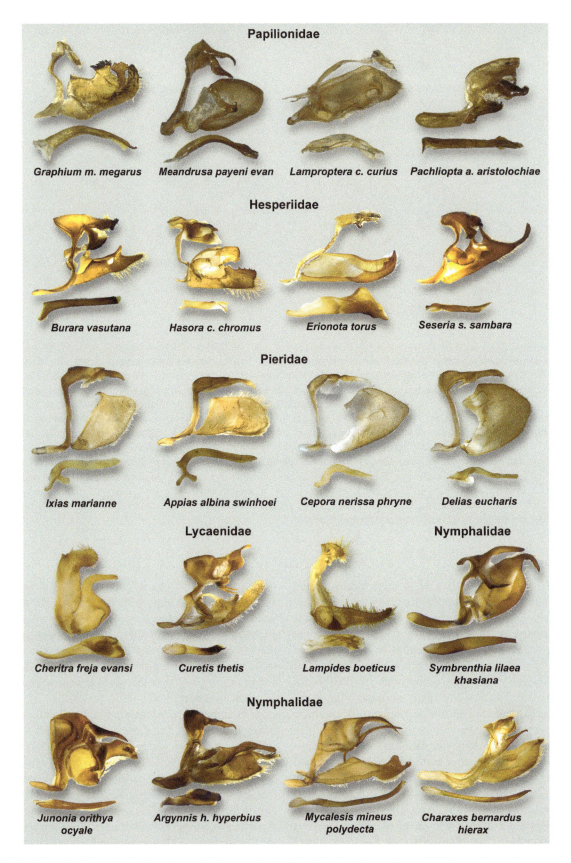

FIGURE 16.4 Diversity and variation in male genitalia across butterflies. Note that the male genitalia for each species are usually simpler than the cartoon shown in Figure 16.3. Also note that genitalia structures in Lycaenidae are fused in some groups. All genitalia dissections are of specimens deposited in the NCBS Research Collections.

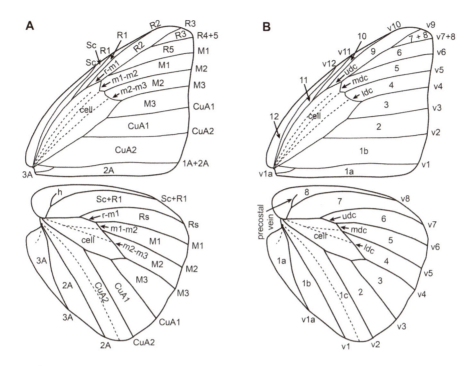

FIGURE 16.5 Wing venation of *Appias paulina* (Pieridae) as a representative butterfly species, with the relationship shown between the two most commonly used systems of classifying veins. **Panel A:** The Comstock-Needham system is universal, i.e., used across insects, because it relies on the origins of veins in relation to the cells. **Sc** subcosta (see costa, the leading edge of the wings, in Fig. 16.6), **R** radius, **M** media (middle), **CuA** as **Cu** cubitus (elbow), **A** anal veins (indicating their position near dorsum, the trailing edge of the wings, see Fig. 16.6), and **h** precostal. Some of these veins are linked with cross-veins, which are named after the veins that they link. **Panel B:** The numerical system simply numbers veins from dorsum to costa. Cross-veins in the numerical system are named after **dc** discocellular veins, as **udc** (upper discocellular vein), **mdc** (middle discocellular vein), and **ldc** (lower discocellular vein). In both panels, wing vein numbers are shown at the ends of the veins, i.e., outside the wings, except in case of veins around the cells. Numbers on the wings represent wing areas defined in relation to veins. Based and redrawn from (Sondhi, S. and Kunte, K., *Butterflies and Moths of Pakke Tiger Reserve*, 2nd edn., Titli Trust (Dehradun), National Centre for Biological Sciences (Bengaluru), and Indian Foundation for Butterflies (Bengaluru), India, p. 242, 2018, with information based on Yata, O. et al., *Syst. Entomol.*, 35, 764–800, 2010; Miller, L.D., *J. Res. Lepidoptera.*, 8, 37–48, 1970.)

butterfly literature, including identification keys. Further details of venation systems, and conversions between the Comstock-Needham and numerical systems, may be found in a number of scientific papers (e.g., de Jong 2004; Miller 1970; Yata et al. 2010).

Colour Patterns

Both wing surfaces of butterflies and moths are covered with rows of scales, which are dead cells. This has earned their order the scientific name Lepidoptera, which means "scale-winged." Scales are often pigmented, which give them colour, and some have characteristic hollow spaces and nanostructures that give them other spectral properties such as iridescence and fluorescence. Butterflies that have mostly transparent wings (e.g., some Neotropical ithomiine and *Cithaeria* butterflies), or transparent wing patches (e.g., Indian *Kallima*), are still covered by scale cells that may be highly underdeveloped, differently shaped, and/or sparsely placed, on the wing membranes.

Many butterflies are incredibly colourful and prominently patterned. Since even closely related species have distinct colour patterns, however subtly different, wing colour patterns of butterflies are commonly used to distinguish between species. One must closely inspect spots, bands, and other colour patches in relation to wing venation and wing areas (Figure 16.5) to distinguish between closely related and subtly different species. For this purpose, different wing margins and areas have been defined in detail, although their names are the same on forewing and hindwing (Figure 16.6). The leading edge of the wing is called costa, the outer edge is termen, and the trailing edge is dorsum. The wing tip is called apex, and the lower corner of the wing between termen and dorsum is called tornus. The area immediately beyond the cell is called discal area, the area just inside of the wing edge is called marginal, and before that, sub-marginal. Various colour patches and markings on butterfly wings are defined in relation to these wing margins and areas (Figure 16.6B) (Kunte and Tiple 2009).

Butterflies have also been classified based on other morphological characters such as legs, labial palpi, and mouthparts (Chapman 1982).

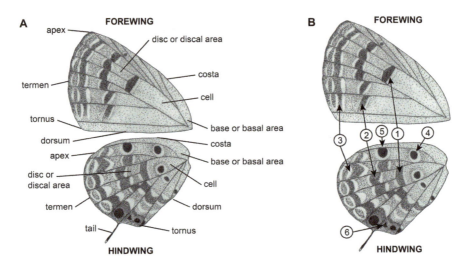

FIGURE 16.6 Wing areas of butterflies, which feature prominently in identification keys of butterflies in conjunction with wing venation (**FIGURE 16.5**), as illustrated from the wing pattern of *Chilades pandava* (Lycaenidae). **Panel A:** Nomenclature of wing margins and areas used to describe markings on butterfly wings. **Panel B:** Nomenclature of markings used commonly to describe wing colour patterns of butterflies. (1) cell-end bars, (2) discal bands, (3) sub-marginal bands, (4) sub-basal spot in space 7, followed by three sub-basal spots in the cell and spaces 1c and 1a, (5) sub-costal spot in space 7, and (6) tornal orange-crowned black spots. (Based on Kunte, K. and Tiple, A., *News Lepid. Soc.*, 51, 86–88, 109, 2009; Sondhi, S. and Kunte, K., *Butterflies and Moths of Pakke Tiger Reserve*, 2nd edn., Titli Trust (Dehradun), National Centre for Biological Sciences (Bengaluru), and Indian Foundation for Butterflies (Bengaluru), India, p. 242, 2018.)

EVOLUTIONARY BIOLOGY, PHYLOGENETICS, AND MOLECULAR SYSTEMATICS AS PILLARS OF MODERN TAXONOMY

Taxonomy and evolutionary biology grew in large part as distinct, independent fields with no academic connections between them. Linnaeus believed in a world created by a divine power, and he invented the binomial system of naming species simply as a formal way to give unique names to each species to facilitate scientific communication. The practice of binomial classification flourished in the entire century prior to the Darwin-Wallace theory of evolution by natural selection. On his part, although Darwin wrote extensively about evolutionary divergence and speciation, he did not define species in an evolutionary sense, and he was largely silent on applying evolutionary thinking to the science of recognizing and naming species (Darwin 1859). Evolutionary biology, the concept of species, and the practice of systematics grew somewhat in parallel in the decades following (Mallet 2004, 2007). Eventually, building up from a number of significant works by German systematists, Willi Hennig's efforts and influential book, *Phylogenetic Systematics*, brought a fundamental shift in integrating evolutionary biology and phylogenetics with taxonomy (Hennig 1966) (this book was first published in German in 1950). Conceptual, computational, and empirical developments in phylogenetic systematics have since made these three areas inseparable. The processes of population divergence, sub-speciation, and speciation have been intensively studied in the past few decades using rigorous mathematical models and computer simulations. As a result, there is a general consensus about the conceptual framework to think about these evolutionary processes. These developments have provided clear directions for systematic studies and taxonomic practices. This has already resulted in significant advances in the higher classification of butterflies and other organisms (see below). Students are encouraged to refer to recent excellent books on systematics and phylogenetics to gain a deeper understanding of the issues discussed below. However, the following briefs (based on Hennig 1966; Wiley and Lieberman 2011) will suffice for the objectives of this chapter. The concepts of species and subspecies will be briefly reviewed before discussing phylogenetic principles that are useful to understand Indian butterflies in the historical and current context. This will also establish some definitions and concepts that are used in the discussion towards the end of this chapter.

WHAT ARE SPECIES?

Species are considered fundamental units in biology. However, opinions on the matter of species are diverse in terms of philosophy (what constitutes a species; are species real entities; and are they kinds, sets, or individuals?) as well as practice (how to delimit and distinguish between species?). Thus, dealing with species is more complicated than an average non-biologist might expect. For practical reasons, most biologists work with the assumption that species are real, identifiable biological entities. However, which species concepts they use in practice varies considerably (Box 16.2). While the biological species concept has been adopted very widely in principle, in practice, its use is limited because of the lack of knowledge about reproductive isolation. Phylogenetic species concept is increasingly more popular because of a flood of phylogenetic studies and phylogeny-based species delimitation algorithms that

BOX 16.2 SPECIES CONCEPTS IN BIOLOGY

Given below are definitions and brief notes about the most popular species concepts. See the main text for further discussion.

Biological Species Concept: "... groups of actually or potentially interbreeding natural populations which are reproductively isolated from other such groups" (Mayr 1965, quoting earlier work). This is by far the most popular species definition, and it has been used widely. However, it cannot be used for asexually reproducing plants, bacteria, and many other groups of organisms and under varied ecological circumstances. The issues of reproductive isolation, allopatry, etc., can also be problematic under this definition.

Phylogenetic Species Concept: "... the smallest aggregation of populations (sexual) or lineages (asexual) diagnosable by a unique combination of character states in comparable individuals..." (Nixon and Wheeler 1990). This definition has gained substantial popularity in recent literature on phylogenetics and systematics. This definition is also practical in the sense that it may be used to delineate species even when there is no relevant information on reproductive isolation, which is the case for most species. The widespread use of phylogenetic inference has also contributed to its recent popularity.

Evolutionary Species Concept: "... a species is a lineage of ancestral descendant populations which maintains its identity from other such lineages and which has its own evolutionary tendencies and historical fate" (Wiley 1978). This definition combines several useful elements of the above two species concepts, for which it has become well-known among evolutionary biologists.

Genealogical Species Concept: Genealogical species are "... basal, exclusive groups of organisms, where exclusive groups are ones whose members are all more closely related to each other than to any organisms outside the group" (Baum and Shaw 1995).

Unified Species Concept: This concept interprets "... the common fundamental idea of being a separately evolving lineage segment as the only necessary property of species and viewing the various secondary properties either as lines of evidence relevant to assessing lineage separation or as properties that define different subcategories of the species category (e.g., reproductively isolated species, monophyletic species, diagnosable species)" (de Queiroz 2005a). This species concept has recently been discussed extensively because it argues for integrating various kinds of evidence—from natural history to genetic to phylogenetic—in defining species. This approach is likely to gain greater favour in the future as more information accumulates on the biology of species and their phylogenetic relationships, without necessarily relying directly or exclusively on the knowledge of reproductive isolation.

depend on extensive molecular evidence. Molecular datasets and phylogenetic studies have rapidly outpaced the generation of knowledge about the biology and reproductive isolation among species. For the discussion below, species are treated as a unified concept, i.e., species are separately evolving lineages with diagnosable characters and distinctive biology, whose reproductive isolation from related lineages may be inferred from a phylogenetically cohesive population structure (de Queiroz 1998, 2005a, 2005b, 2007). In practice, in the absence of data on reproductive isolation, species biology, or phylogenetic cohesiveness, most taxonomists treat species as diagnosable entities that appear different along several morphological, ecological, and behavioural trait axes. For butterflies, differences in morphology may include traits such as wing patterns, genitalia, venation, labial palps, tibial claws, and larval and pupal characters. Ecological and behavioural traits may include the use of larval host plants/habitats, flight periods, and mating preferences.

Species can either be sympatric (i.e., existing in the same area, or in other words, with overlapping distributions), parapatric (occurring in neighbouring regions with little distributional overlap), or allopatric (existing in different areas, with widely non-overlapping distributions) (Figure 16.7). If populations are allopatric, then the amount of morphological and ecological/behavioural differences between them, along with the nature of geographical separation between them with respect to well-known biogeographical breaks, may be particularly important

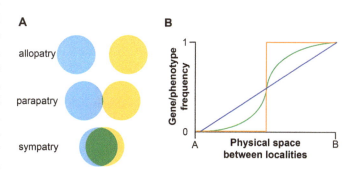

FIGURE 16.7 Evolutionary aspects of species, speciation, and population divergence that affect taxonomic interpretation and application of names to taxa. **Panel A:** Three common kinds of distributions that determine the degree to which populations may be isolated in space. Yellow and blue circles represent two populations, and green areas represent geographical overlap between the populations. **Panel B:** Different types of clines. Gene/phenotype frequencies may change in space linearly (dark blue) or in a step-like manner (green and orange).

in determining their species status (see Sections 'What are Subspecies?', and 'When are Subspecies Names Inappropriate?'). However, classifying populations as sympatric, parapatric, or allopatric may not always be easy: how much geographical separation is enough separation (Figure 16.7A)? One solution to this problem is to focus on gene flow between the populations rather than geographical distance, since the amount of gene flow is more directly relevant to reproductive isolation. This, however, is by no means easy to measure, although modern sequencing technologies have made this feasible.

WHAT ARE SUBSPECIES?

As mentioned above, by the mid-1800s, many naturalists and collectors had started to appreciate the value of recognizing geographically structured variations within species of well-collected groups such as butterflies, birds, and mammals. Before the concept of subspecies was formalized, these geographical variations were often referred to as varieties or local forms. Taxonomists of the time dealt with the problem of distinguishing between species and subspecies (varieties) thus: "... we have a definition [of species] which will compel us to neglect altogether the *amount* of difference between any two forms, and to consider only whether the differences that present themselves are *permanent*. The rule, therefore, I have endeavoured to adopt is, that when the difference between two forms inhabiting separate areas seems quite constant, when it can be defined in words, and when it is not confined to a single peculiarity only, I have considered such forms to be species. When, however, the individuals of each locality vary among themselves, so as to cause the distinctions between the two forms to become inconsiderable and indefinite, or where the differences, though constant, are confined to one particular only, such as size, tint, or a single point of difference in marking or in outline, I class one of the forms as a variety of the other" (Wallace 1865). This reasonable practice has changed little in the past over 150 years. Only recently have phylogenetic methods and molecular datasets begun to address the problem of distinguishing between species and subspecies. The solution is to sample molecular variation in multiple individuals from several populations of each putative species. The general expectation is that the genetic (and morphological) divergence within interbreeding populations/subspecies of the same species will be relatively smaller than divergence across reproductively isolated sister species. Thus, populations/subspecies of the same species should form a cohesive phylogenetic cluster that is separated from other cohesive clusters that represent related species. Cohesive clusters thus obtained may form an objective basis for delineating species and subspecies.

In addition to considering how to distinguish between species and subspecies, it is important to understand why and how populations diverge and subsequently form subspecies and species. A tremendous amount of scientific literature exists on the subject (Coyne and Orr 2004; Dobzhansky 1951; Futuyma 1998; Howard and Berlocher 1998; Mayr 1942, 1965, 1982; Nosil 2012; Simpson 1961). The following is a very brief, simplified summary of the general understanding of population divergence from this evolutionary biology literature.

Populations have a tendency to diverge in space over time. This may be because space is not uniform: abiotic (e.g., climate and soil) and biotic factors (e.g., intra- and interspecific interactions) may vary even over short distances. Thus, geographically isolated or separated populations may be under ecological selection for local adaptation. For example, in butterflies: (a) body size and the amount of wing melanization may change in response to thermal envelopes across elevational gradients, and (b) ovipositing females, and subsequently caterpillars, may prefer or avoid certain host plants based on locally variable secondary compounds in plants or pressure from competing species, predators, and parasitoids. Such local adaptation will cause inter-populational divergence in relevant morphological and behavioural characters. However, populations will diverge even in absence of any local adaptation if they are isolated for a sufficiently long time. In isolated populations, phenotypes and/or genotypes will diverge even under neutral processes such as random genetic drift. Many isolated butterfly populations (e.g., island populations) show differences in the intensity or hue of wing colours, or the presence/absence/extent of spots, bands, and other colour patches. The functional significance of such visual differences are largely unknown at present, but many may involve random genetic drift. Thus, over a sufficiently large number of generations, geographically structured populations with little gene flow between them will diverge in their phenotypic and genotypic composition because of ecological selection or genetic drift. Such geographically structured variations among populations may be formally described at the subspecies level.

Based on this summary, one might ask whether subspecies are on their way to becoming species, which appears to be a popular belief. If populations have a tendency to diverge and they have accumulated many phenotypic and genetic differences (enough to be recognized as subspecies), then it is possible that they will continue to diverge and, given sufficient amount of time, turn into reproductively isolated species. However, this is only one of the expected evolutionary outcomes. It is also possible that slightly diverged populations/subspecies may homogenize once again if gene flow increases. It is also possible that some of the populations/subspecies will go extinct. These three outcomes might occur because of changes in ecological selection, changes in distributional ranges, and breakdown of geographical barriers, among a number of reasons (Mayr 1942; Simpson 1961). The taxonomic and evolutionary value of named subspecies must be judged with these outcomes in mind.

When Are Subspecies Names Inappropriate?

It is taxonomically inappropriate to name all variations within a species at subspecies level. Two common cases where this applies are considered below.

Linear Clinal Variation

Spatial divergence across populations might take place along elevational, temperature, rainfall, and other environmental and resource gradients such as those found across the Himalaya and the Western Ghats. As a result, the mean of a character and/or allele frequency gradually change across a geographic transect (Figure 16.7). This is called a cline (Futuyma 1998; Huxley 1939; Simpson 1961). The ends of a gradual, linear cline may appear distinct, but they are connected by intermediate variations across adjacent populations. Hence, gradual, linear clinal variations should not be given separate subspecies names, they should really be recognized as a cline (Huxley 1939; Simpson 1961). It is interesting to note that a population may belong to several distinct clines going in different directions and along multiple phenotypic axes, but it can only get a single subspecies name; hence, taxonomists should avoid assigning subspecific names to such clinal variations (Huxley 1939; Mayr 1942; Simpson 1961). Clinal variations must abound in the Indian region because of large gradients in elevation, temperature, rainfall, forest types, and other climatic and biotic regimes in the hills and the plains. It is known that size and colouration of organisms varies with rainfall, e.g., individuals occurring in wetter areas tend to be darker than individuals occurring in drier areas. The Western Ghats-endemic *Idea malabarica* provides a good example of clinal variation across the south-north rainfall gradient. There is a prominent decline in rainfall from the southern to northern Western Ghats, so forests in the south are much wetter overall. Southern populations associated with wetter forests (e.g., Shendurney WLS, Kerala) are darker and larger, whereas populations from the drier north (e.g., Goa and Maharashtra) are brighter white and smaller. Although these south and north ends of the cline are easily distinguished, darkness and size of butterflies is quite variable from the Nilgiris to the Karwar-Goa border depending on the season and wetness of forests. Because of these rainfall gradient-associated darkness and size clines, the two described subspecies of *Idea malabarica*—the darker-larger *malabarica* and the paler-smaller *kanarensis*—should really be considered infrasubspecific, clinal variations. Similarly, western Himalayan populations of many butterflies are paler compared to their eastern Himalayan darker counterparts. Since the amount of rainfall declines from the east to the west, it is natural that butterfly populations in the drier western Himalaya are paler compared to butterfly populations in the wetter eastern Himalaya, but one also expects to find intermediate, more or less continuously variable forms in areas between. Such gradual, nearly continuously varying clinal variations have been given subspecies names in the Indian region. These subspecific names will need to be synonymized as evolutionary biologists and taxonomists gain greater insights into clines that are relevant for Indian butterflies.

On the other hand, sharp step clines (Figure 16.7B) may form based on the nature of the environmental gradient, the nature of selection, and any developmental genetic thresholds that may alter phenotypes non-linearly. Following Wallace's logic, if a sharp step cline occurs along a single phenotype/genotype axis, then the resultant variations may be treated as subspecies. If sharp step clines overlap along a number of independent phenotype/genotype axes such that the populations on two sides of the step differ sharply by sets of independent characters, then these populations likely represent distinct, parapatric species.

The nature and distribution of clines need to be carefully studied because of the geological and climatic heterogeneity apparent in the Indian subcontinent and because of the implications of clinal variation for taxonomy and biology of species. However, clines are greatly understudied, and indeed neglected by Indian taxonomists and other biologists.

Phenotypic Plasticity

Inter-population phenotypic differences are not exclusively controlled by genotypic variation, they may also be environmentally induced and occur in absence of genetic differentiation. Indeed, phenotypic differences are often a product of interaction between the genotype and the environment. Two classic examples illustrate this point. First, many butterfly species show seasonally changing forms, often called dry or wet season forms, spring or summer forms, etc., that are produced with the same genomic information, but in different environments (Nijhout 1991). Switches between seasonal forms are caused by non-genetic, external factors such as climatic parameters, food resources, and substrate texture. Second, many insects in colder areas have dark stripes and patches on wings, which help them warm up faster in cold climates (Clusella Trullas et al. 2007). This thermal melanism is in many cases purely environmentally induced and produced during development because of differential gene expression profiles rather than due to gene sequence variation. Similarly, animals in colder climates tend to be larger in size (Bergmann's rule) (Futuyma 1998). If inter-population differences are purely because of environment-induced phenotypic plasticity, then it is inappropriate to name subspecies based on these differences. In this case, populations may have morphological differences, but they are *not constant*, e.g., when early stages are raised under different climatic conditions.

Reconstructing and Interpreting Phylogenies for Systematic and Taxonomic Studies

All life on earth has descended from a common ancestor, and all the species—from bacteria and viruses to highly multicellular and complex eukaryotes—are a product of evolutionary processes involving ancestor-descendant relationships between species. A natural extension of this understanding is

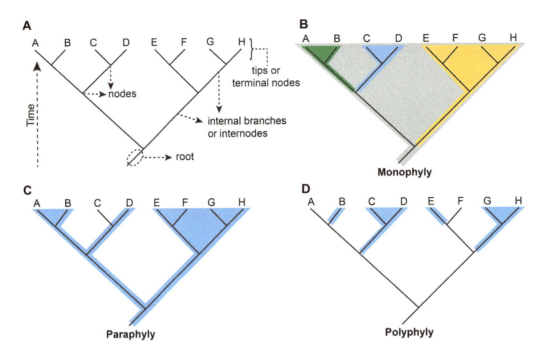

FIGURE 16.8 A simplified cartoon of a phylogeny (in this case, a species tree) on which phylogenetic concepts of monophyly, paraphyly, and polyphyly, which affect taxonomic grouping from subspecies and species to family level and above, are illustrated. **Panel A:** Terminology used in describing various components of a phylogeny. Terminal nodes A to H represent eight extant species with phylogenetic relationships as illustrated. **Panels B–D:** Representation of monophyly, paraphyly, and polyphyly. In each panel, similarly coloured species with their common ancestors represent distinct taxonomic rankings that might have been in use. See Section 'Reconstructing and Interpreting Phylogenies ...' for further details.

that species show evolutionary relationships with each other, which may be represented as a species tree with a root, stems, and branches (Figure 16.8). Such a tree is called a species phylogeny, in which the most recent common ancestor (MRCA) of all the species in the phylogeny is represented as a root (Figure 16.8A). Branching events, when one ancestral species splits into two descendent species, are represented by nodes. Branches between the nodes are called internal branches or internodes, which represent ancestral species. The tips of the species tree, also called terminal nodes, represent extant species. Phylogenies are reconstructed in reverse, i.e., data from extant species are tested against models of character evolution to reconstruct a relative topology of ancestral nodes. For example, variation in molecular sequences of sampled genes and/or morphological characters of a group of extant species can be used to reconstruct relationships between those species through to the phylogenetic root. The following ideas are fundamental to interpreting and using phylogenies.

A primary motive behind reconstructing phylogenies is to discover monophyletic groups. Monophyly, in which all the extant species along with all their ancestors down to their MRCA are included, defines a clade. Such monophyletic groups or clades is an inclusive, taxonomically desirable arrangement. However, monophyly can be applied at different levels, so exactly where a clade is delineated may depend on accessory information and/or preference of a taxonomist. For example, in Figure 16.8B, species A-B and their MRCA (green) represent a monophyletic group or a clade, as do species C-D and their MRCA (blue), species E-F and their MRCA (not shown), species G-H and their MRCA (not shown), species A-D and their MRCA (not shown), species E-H and their MRCA (dark yellow), or species A-H and their MRCA (grey). A genus may be defined at any of these levels. Many of the currently accepted genera of world butterflies are likely monophyletic. However, this hypothesis needs to be rigorously tested in a molecular phylogenetic framework before genus- and species-level arrangements of butterflies become stabilized.

A paraphyletic group includes most of the species and their MRCA, but some embedded groups are excluded (Figure 16.8C). The clade (species A-H) is said to be paraphyletic with respect to species C, which is excluded from the group in a taxonomic arrangement. One of the most well-regarded examples of paraphyly is a clade that includes turtles, crocodilians, dinosaurs, and related reptiles, from which birds are excluded and classified into their own class. Thus, Class Reptilia is paraphyletic with respect to Class Aves. An example from Lepidoptera will be explained below.

A polyphyletic group is composed of species that are not related to each other through a common ancestor, but which are classified into a single taxonomic unit (Figure 16.8D). Species B, C, D, E, G, and H in Figure 16.8D, if classified as a single genus, will represent a polyphyletic group. Paraphyly and polyphyly are taxonomically undesirable outcomes as they are not inclusive groups in a phylogenetic sense. The idea that taxonomic groups must represent clades, and therefore classification of organisms should reflect their evolutionary relationships, was one of the most critical contributions of Henning to the field of systematics.

Taxonomy, Systematics, and Biology of Indian Butterflies in the 21st Century

Identification of monophyletic groups across the tree of life will stabilize classification and nomenclature in the long term. Detailed discussions may be found elsewhere (Futuyma 1998; Hennig 1966; Wiley and Lieberman 2011).

RECENT INSIGHTS FROM MOLECULAR SYSTEMATIC STUDIES INTO THE EVOLUTION AND HIGHER CLASSIFICATION OF BUTTERFLIES

Since the 1990s, there has been a strong emphasis on using molecular data and phylogenetic methods to gain insights into organic evolution and to resolve taxonomic problems, in a field that is now known as molecular systematics. Several studies have emphasized the importance of combining morphological and molecular data in resolving issues in Lepidoptera classification and taxonomy (Aduse-Poku et al. 2016; Huang et al. 2018; Wahlberg et al. 2005). This was especially important when a small number of molecular markers was used in generating phylogenies. However, modern, cheap sequencing and phylogenomic methods have now nearly obviated a need for using more labour intensive but limited morphological data into phylogenetic analysis. Standardized, large molecular (nuclear) marker sets ranging from ca 10 to over 400 genes have now been developed that are being used in most large-scale butterfly taxonomic studies (Espeland et al. 2018; Kawahara et al. 2018; Wahlberg and Wheat 2008). Such a large molecular marker set promises to offer a robust phylogenetic framework that was nearly inaccessible using morphological data. Indeed, most recent butterfly phylogenies are reconstructed exclusively using molecular phylogenetic approaches (see below). These molecular phylogenetic analyses have fundamentally changed the understanding of butterfly evolution and higher classification, as summarized below.

BUTTERFLIES ARE MOTHS, BUTTERFLY-MOTHS ARE BUTTERFLIES, SWALLOWTAILS ARE OLDER THAN SKIPPERS, AND OTHER STARTLING OBSERVATIONS

For a long time, butterflies have been assumed to be sister to, but somewhat distinct from moths. Indeed, this thinking has pervaded popular understanding of butterflies to the extent that a chart of differences between moths and butterflies is included in nearly every book and article introducing butterflies. However, butterflies have been classified in several taxonomic categories that are no longer tenable as a result of insights provided by recent molecular phylogenetic studies. One of the most startling findings is that butterflies are indeed moths. It has now been established from robust molecular phylogenetic analyses that the superfamily Papilionoidea that encompasses all butterflies (see below) is monophyletic, but completely embedded among other moth superfamilies (Figure 16.9) (Kawahara and Breinholt

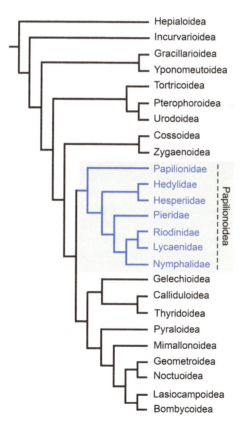

FIGURE 16.9 Lepidoptera phylogeny showing relationships of superfamilies of moths that are closely related to the butterfly superfamily, Papilionoidea. Note that: (a) butterflies are a monophyletic group (i.e., a clade), but it includes Hedylidae ("butterfly- moths"), (b) Papilionidae is the most basal family of butterflies, not Hesperiidae, as is still widely believed among Indian lepidopterists. Hedylidae and Hesperiidae are sister families. Superfamily and family relationships are redrawn from (Kawahara, A.Y. et al. *Mol. Phylogenet. Evol.*, 127, 600–605, 2018; Kawahara, A.Y. and Breinholt, J.W., *Proc. R. Soc. Lond. B Biol. Sci.*, 281, 20140970, 2014; Mutanen, M. et al. *Proc. R Soc. B*, 277, 2839–2848, 2010).

2014; Mutanen et al. 2010). Thus, moths are paraphyletic with respect to butterflies, if one wants to treat butterflies and moths as somehow different. In other words, butterflies are specialized and monophyletic, but they are only a subgroup of moths, without any exclusive morphological characters that distinguish all butterflies from all the other moths.

Another startling finding is that in spite of a long-held belief, Papilionidae (swallowtails) are the most basal, i.e., the oldest, butterflies, not Hesperiidae (skippers) (Figure 16.9) (Espeland et al. 2018; Heikkilä et al. 2012). Additionally, the South American Hedylidae (butterfly-moths; earlier classified under superfamily Hedyloidea) are sister to Hesperiidae (earlier classified under superfamily Hesperioidea), and both are embedded within true butterflies (superfamily Papilionoidea) (Figure 16.9) (Espeland et al. 2018; Heikkilä et al. 2012). Thus, superfamilies Hedyloidea and Hesperioidea are synonymized with superfamily Papilionoidea. According to this recent, but broadly accepted classification, the monophyletic butterfly superfamily Papilionoidea contains seven families: Papilionidae (swallowtail butterflies), Hesperiidae (skippers), Hedylidae (butterfly-moths), Pieridae (whites and yellows), Lycaenidae (blues, hairstreaks), Riodinidae (metalmark butterflies), and Nymphalidae (brush-footed butterflies), with the family-level phylogenetic relationships as given in Figure 16.9.

FAMILY- AND SUBFAMILY-LEVEL CHANGES

Molecular systematic findings are shaking other long-held taxonomic frameworks as well. For example, family Lycaenidae has long been classified among two large subfamilies: Theclinae ("Strong Blues") and Polyommatinae ("Weak Blues"), among a host of other smaller subfamilies. Although recent phylogenetic work supports monophyly of the smaller subfamilies, it shows Theclinae to be paraphyletic with respect to Polyommatinae, i.e., Polyommatinae is embedded within Theclinae (Espeland et al. 2018). This finding, which is complicated by the fact that Polyommatinae is an older name compared to Theclinae, prompts a reclassification of a significant fraction of the tribes and a major subfamily under Lycaenidae. Another finding that prompts a subfamily-level synonymy is that Neotropical Ithomiinae is embedded within Danainae, and must therefore be subsumed as a tribe under Danainae (Espeland et al. 2018). This has long-reaching implications for our understanding of the evolution and biology of Danainae. On the other hand, Riodinidae is now normally treated as a sister family of Lycaenidae, rather than its subfamily, an arrangement that is phylogenetically well supported (Espeland et al. 2015, 2018; Wahlberg et al. 2005). Similar re-examinations of higher classification of butterflies are likely in the next decade or two as molecular systematics advances significantly and as a large fraction of butterfly species are sequenced and put in a phylogenetic context.

RESOLUTIONS AT GENERIC AND SPECIES LEVELS

The standardized nuclear and mitochondrial markers that are now commonly used in butterfly phylogenetics also provide resolution at genus and species levels. After preliminary phylogenetic analysis (Kodandaramaiah et al. 2010), recent comprehensive studies split the nymphalid genus *Heteropsis* in three well-supported and geographically subdivided genera: Asian *Telinga*, Malagasy *Heteropsis*, and African *Brakefieldia* (Aduse-Poku et al. 2016). Another recent molecular phylogenetic analysis found many of the genera under the tribe Aeromachini (Hesperiidae) to be either paraphyletic, prompting synonymy of some genera, or polyphyletic, prompting description of two new genera and significant movement of species across the newly delineated genera in that tribe (Huang et al. 2019). As the world's butterflies are sequenced and analysed in coming years, generic placements of many butterflies will alter and species-level classifications will undergo prominent changes. These taxonomic changes will be especially prominent in the Asian, African, and American tropics as cryptic species are discovered and specific relationships are clarified (Hebert et al. 2004; Huang et al. 2018; Kawahara 2013; Kawahara et al. 2018; Toussaint et al. 2015; Yata et al. 2010). In the Asian context, this will certainly affect a considerable number of genera under Nymphalidae, Lycaenidae, and Hesperiidae, which have historically been created with poor taxonomic characterization, and they continue to be commonly used without sufficient modern scientific investigations and revisions, e.g., the *Euthalia-Bassarona-Tanaecia-Dophla-Symphaedra* group (Nymphalidae), the *Zephyrus* hairstreaks, and many other genera under the traditional subfamilies Theclinae and Polyommatinae (Lycaenidae), and genera under the tribes Tagiadini, Aeromachini, and Baorini (Hesperiidae).

Genomic sequencing has become reasonably cheap and widely accessible even in developing tropical countries such as India. The next-generation sequencing platforms are providing large amounts of molecular data that have the power to reveal cryptic genetic variation and reproductive isolation that was difficult to decipher using traditional morphological data. Rapidly growing computational infrastructures and establishment of research groups in developing countries themselves are also likely to drive taxonomic discovery and resolution in the world's super-biodiverse tropics in the next few decades. Since taxonomic studies have hardly taken place in India in the past few decades, Indian butterfly taxonomy will no doubt see a flood of taxonomic rearrangements and other kinds of updates based on molecular systematic studies.

INDIAN BUTTERFLIES, A TAXONOMIC IMPEDIMENT

In Section "Indian butterflies, a taxonomic legacy", a glimpse was offered into how taxonomic studies on Indian butterflies have progressed through time, and how this field has historically been dominated by non-native taxonomic experts. At the same time, natural history museums and museum-based sciences have been neglected in India. Indeed, historically significant collections in India's natural history museums have hardly grown in the past few decades, with the museums, research staff, and field surveys grossly under-funded. The quality of academic training leaves much room for improvement. There are very few large-scale international collaborations, which has bred a culture of insularity. Naturally, the volume of work produced from India has been very small and the quality alarmingly poor. Restrictive laws and permitting procedures have also burdened taxonomic studies (Prathapan et al. 2006, 2008). Along with taxonomy, complimentary fields such as natural history, ecology, evolutionary biology, systematics and phylogenetics, and biogeography, which generate information on species that is critical in advancing taxonomic studies, have also been neglected. The lack of native expertise and uninterrupted taxonomic works has had a profound impact on the scientific development and current practices in this field. One of the unfortunate consequences of this stunted growth is that taxonomists working on Indian taxa continue to rely on outdated taxonomic arrangements generated at the time when concepts of species and subspecies were not properly elucidated in systematics and evolutionary biology, and in a phylogenetic context. In many cases, evidence exists that contradicts old arrangements and current practices. Some of these problems are easily addressed if taxonomists consulted extensive type collections, museum specimens, catalogues, and records in European research collections, where most of the type specimens of Indian species are deposited. It is true that these museums openly welcome foreign and especially native scientists from countries where the type specimens were collected a long time ago. However, funding for scientists from developing countries to make such research visits is scarce at a time when visiting these European countries is prohibitively expensive for most practicing taxonomists. Therefore, most of the collections are practically inaccessible, especially to Indian taxonomists. Nonetheless, several attempts have recently been made to update the taxonomy of Indian butterflies (Varshney 1993, 2010; Varshney and Smetacek 2015). Although these compilations are useful in the overall vacuum that exists in India, their scientific value is compromised because of: (a) inadequate notes and annotations, (b) glaring omissions resulting from neglect of recent taxonomic and phylogenetic literature, (c) lack of museum work especially including inspection of type and other specimens as well as genitalia dissections, (d) a critical review and reassessment of available evidence, (e) heavy reliance on old Indian literature that is vastly outdated, and (f) poor understanding of modern trends in taxonomy and systematics. This has given rise to an alarmingly flawed culture of "copy-and-paste," where successive generations have carried taxonomic arrangements forward even as the academic landscape in systematics and taxonomy was transformed elsewhere in the world. The specific examples given below and illustrated in Figures 16.10 through 16.12 offer detailed case studies of taxonomic problems pertaining to Indian butterflies.

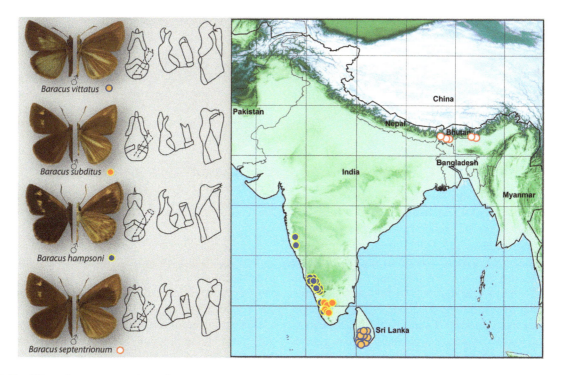

FIGURE 16.10 Wing phenotypes, male genital morphology, and distributional ranges of Indian *Baracus* (Hesperiidae). For each species, images of adult butterflies show dorsal (left) and ventral (right) views, and drawings of the male genitalia illustrate ventral view of uncus (left), lateral view of uncus with aedeagus (centre), and inside of valve with distal end facing up (right). Illustrations of male genitalia are redrawn from (Evans 1949). Distributional maps are generated from published spot records (Evans 1949; van der Poorten and van der Poorten 2016), museum records from NHMUK and NCBS (K. Kunte, unpublished data), and spot records on the *Butterflies of India* website (From Kunte, K. et al. *Butterflies India*, v. 2.56, 2018.)

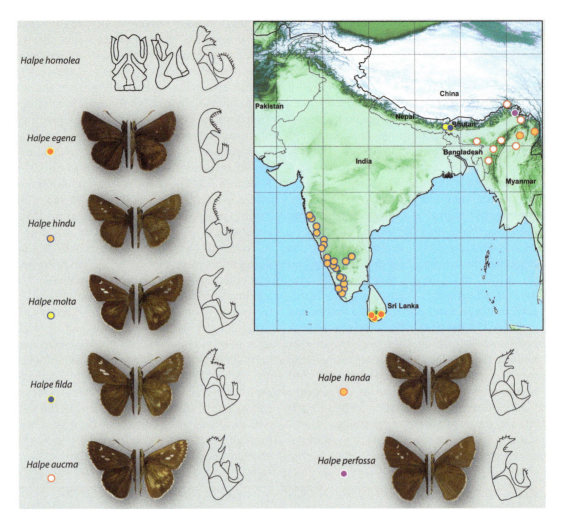

FIGURE 16.11 Wing phenotypes, male genital morphology, and distributional ranges of S. Asian *Halpe* (Hesperiidae). For *Halpe homolea*, ventral view of uncus (left), lateral view of uncus with aedeagus (centre), and inside of valve with distal end facing up (right) are illustrated. For all other *Halpe*, only the inside of valve, which is characteristic, with distal end facing up is illustrated. Illustrations of male genitalia are redrawn from Evans, W.H., *A Catalogue of the Hesperiidae from Europe, Asia and Australia in the British Museum (Natural History)*, British Museum (Natural History), London, UK, p. 502, 1949. See a note about distributional maps in the legend of Figure 16.10.

The Genus *Baracus*

Evans listed two species under the hesperiid genus *Baracus*: (a) *vittatus* from Sri Lanka and India, including four subspecies: *vittatus*, *hampsoni*, *subditus*, and *septentrionum* (Figure 16.10), and (b) *plumbeola* from Luzon, the Philippines (Evans 1932). Later, he moved *plumbeola* under the genus *Aeromachus*, treating *Baracus* as a monobasic genus (Evans 1949). He continued to treat the four Sri Lankan and Indian taxa as subspecies of *vittatus* and described one more subspecies, *gotha*, from the Anaimalai (Evans 1949) that appears to be an aberration and has not been seen since the description. His treatment of the Sri Lankan and Indian taxa as subspecies of *vittatus* was surprising given that his genitalia dissections showed these four taxa to have distinct structures (specifically, clasps or valvae; Figure 16.10) (Evans 1949). This evidence should have prompted their treatment as allopatric species by Evans himself. Recent, presumably updated checklists have continued to treat these taxa as subspecies following Evans (Varshney 2010; Varshney and Smetacek 2015), without reassessing available evidence and its current taxonomic implications.

The *homolea* Species-Group of *Halpe*

Another glaring example of taxonomic lumping by Evans and subsequent authors is the *homolea* species-group of South Asian *Halpe* (Hesperiidae). In this species-group, Evans once again found considerable differences in the structure of clasps or valvae of the male genitalia (Figure 16.11), but he still listed all the S. Asian, S. Chinese, and Indo-Chinese taxa as subspecies of the Singaporean *H. homolea* (Evans 1949). The differences in the male genitalia are so prominent that any modern taxonomist who sees this evidence would treat these taxa as distinct species. Indeed, the S. Chinese and Indo-Chinese species have recently been treated as distinct species (Huang 2003; Inayoshi 2018). However, Indian taxa that were not reassessed by these authors have been continued

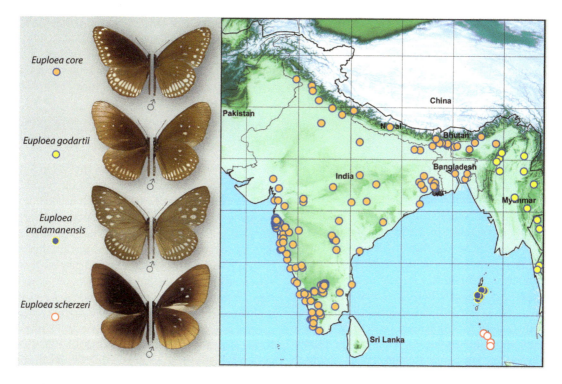

FIGURE 16.12 Wing morphology, sex-brands, and distributional ranges of some Indian *Euploea* (Nymphalidae). *godartii*, *andamanensis*, and *scherzeri* have for a long time been treated as subspecies of *Euploea core*, although each has characteristic androconia (sex-brands) on forewing and hindwing, and a distinctive forewing shape. Distributional maps are generated from museum records from NHMUK, NCBS, and ZSI Port Blair (K. Kunte, unpublished data), and spot records on the *Butterflies of India* website. (From Kunte, K. et al. *Butterflies of India*, v. 2.56, 2018.)

to be treated as subspecies of *H. homolea* by Indian authors. This has led to completely untenable biogeographic and taxonomic arrangements: the Singaporean *homolea* is replaced in Indo-China and S. China by various species (Huang 2003; Inayoshi 2018), but it is supposed to reappear in NE India, Western Ghats, and Sri Lanka in several subspecific forms—each with highly distinctive male genitalia (Figure 16.11)—according to recent Indian authors (Varshney 2010; Varshney and Smetacek 2015). Considering the available evidence of prominent differences in the male genitalia, all these *Halpe* should really be considered distinct species, some as allopatric (e.g., Sri Lankan *egena* and the Western Ghats *hindu*) and others as sympatric (e.g., *aucma*, *perfossa*, and *handa* in NE India and northern Myanmar; Figure 16.11).

The examples of *Baracus* and the *homolea* species-group of *Halpe* suffice to illustrate that in spite of his stellar work and a highly productive career, Evans had a flawed understanding of subspecies. This compelled him to treat many good species as subspecies, even though the evidence from genitalia structures indicated otherwise. Looking back, it is perhaps easy to see the roots of this error: most of Evans's work on Indian butterflies took place prior to 1950, just as the concepts of species and subspecies were being clarified in evolutionary and systematic literature. Evans certainly did not appear to have been cognizant of these developments based on his writings and taxonomic treatments. In any case, he misapplied the concept of subspecies far and wide, creating dozens of cases where taxa were inaccurately treated as subspecies. Many of his generic placements were also flawed. Some of these errors have been corrected in other parts of the world, but they are largely carried forward in India from the very old literature.

THE *EUPLOEA CORE* SPECIES-GROUP

Euploea core is a complex species-group which has been taxonomically tossed around for a long time. Evans treated *core*, *godartii*, and *andamanensis* as distinct species and *scherzeri* as a subspecies of *climena* (Evans 1932). Without any justification, Talbot lumped all these taxa as subspecies of *core* (Talbot 1947). Talbot's arrangement was followed by Ackery and Vane-Wright, again without giving justification for the lumping (Ackery and Vane-Wright 1984). Vane-Wright, however, later realized that there are differences in the male genitalia and androconia that show these taxa (except *godartii*, which he did not consider) to belong to potentially different species-groups and revised their taxonomic status to be semispecies (Vane-Wright 1993). However, the presence and/or size of androconia (also known as sex-brands) on both the wings, and the forewing shape, of males of all these taxa are distinctly different (Figure 16.12). Given that the scent produced by androconia is one of the most important sexual traits that strongly influence chemical communication and mate preference in the Danaini (Ackery and Vane-Wright 1984), and the fact that the male genitalia as well as androconia of these taxa are different (Vane-Wright 1993), these taxa are reasonably treated as distinct species that have parapatric and allopatric distributions. In summary, there is no justification for lumping these taxa as subspecies of *core*, but the available evidence supports treating them as distinct species. And yet, Indians continue to treat them as subspecies of *core* without any justification (Varshney and Smetacek 2015), and most seem unaware of the available literature on this matter.

More Examples

Apart from the three examples just discussed in some detail and illustrated in Figures 16.10 through 16.12, dozens more taxonomic treatments of Indian butterflies readily point to the neglect of taxonomic works in India, to the culture of "copy-and-paste" that is evident in the Indian literature, and the lack of new, scholarly works that are required to truly update the taxonomy of Indian butterflies. Indeed, the culture of academic insularity in India has long affected access to scientific literature and international collaborative networks and research projects. As a result, Indian taxonomists, ecologists, and naturalists seem to frequently miss developments in scientific research and publications. For example, the following three findings affect taxonomic treatments and names of Indian butterflies, but they are largely neglected in India, as judged from lack of citation to the recent research papers as well as continued use of older arrangements in recent Indian literature. First, the southern Indian *lepitoides* has since the 1930s been listed as a subspecies of either *Libythea lepita* or *L. celtis* (d'Abrera 1985; Evans 1932). A recent monograph of Libytheinae moved it under *Libythea laius* (Kawahara 2013). Second, *galba* has historically been listed as a subspecies of *Appias nero* (d'Abrera 1982; Evans 1932; Talbot 1939), but was recently shown to be a distinct species (Yata et al. 2010). Third, a subclade of Indian Mycalesina is now placed in the resurrected genus, *Telinga* (Aduse-Poku et al. 2016), which affects name combinations of approx. 10 Indian species.

Many of the problems highlighted above appear to stem from inaccessibility of resources in India, whether scientific literature or museum resources, and inadequate training. It is important to sort out these taxonomic problems not only for academic reasons (e.g., an enriched understanding of diversity of life on the planet), but also for practical reasons. National legislation such as the Wildlife (Protection) Act (WPA) and National Biodiversity Act depend on reliable and updated biological information. Many butterflies are protected in India either at subspecies or species levels under various Schedules of Wildlife (Protection) Act. There should be periodic taxonomic updates and reassessments of these scheduled species, but the lists of scheduled butterflies have never been properly reassessed. It is unfortunate that such an important piece of conservation legislation continues to rely on woefully inadequate biological information and reassessment framework. The biological information is critical in other ways, too. While conservationists may want to protect every population, it may be feasible to protect only a subset of populations. Under these circumstances, taxonomically distinctive, phylogenetically unique, and biologically specialized species may be given greater priority over some populations that show minor phenotypic and genetic differences. Thus, addressing the taxonomic impediment is one of the most significant challenges while rejuvenating taxonomic and systematic studies in India, which will have wide-ranging practical outcomes as well.

A MODERN *MANTRA* FOR BUTTERFLY BIOLOGY IN THE 21ST CENTURY INDIA

Biodiversity sciences are undergoing a long-awaited renaissance due to a strong and very broad interest in exploring and conserving biodiversity that is increasingly threatened with habitat loss and climate change, among other threats. Scientific discovery of species and their biology is no longer an arcane activity only for taxonomists in the world's museums, but it is a central goal for any modern evolutionary biologists, ecologists, systematists, conservation biologists, and even citizen scientists. In India, amateur and professional scientists are increasingly forming closely knit teams to discover and describe new species as well as study natural history (Karmakar et al. 2018; Kunte et al. 2012, 2018; Nitin et al. 2018; Rai et al. 2012; Sondhi et al. 2016; Sondhi and Kunte 2016). It is widely recognized that organismal biology and taxonomy have to grow side-by-side. Integrative taxonomy indeed relies on a deep understanding of ecology, behaviour, and morphological diversity of butterfly populations and species, that is used in conjunction with application of concepts and methods of species delimitation, biogeography, and molecular systematics (de Queiroz 2005a, 2005b). These modern scientific explorations critically depend on museums that hold deep-frozen tissue libraries and georeferenced data that can assist in understanding the evolutionary history of species and their potential for future survival in a changing world under the influence of climate change and other human impacts on climate and habitats. In this context, Indian scientific community needs to step up in several major ways (Figure 16.13).

Collection of Field Data

Very little is known about the natural history of Indian butterflies. Extensive field observations are needed to study climatic envelopes in which species occur, seasonal population dynamics and phenology, community dynamics and inter-specific interactions, foraging and host use, reproductive behaviour and early stages, host plant use, etc. (Figure 16.13). Although some good natural history publications exist on these subjects [see Section 'Scientific Research' below], very few scientific, high-quality studies have emerged on species biology from India. Any other work on Indian butterflies needs a solid foundation of the fundamental understanding of butterfly biology.

Developing Museum Resources

The Natural History Museum, London, has an estimated 11 million Lepidoptera specimens (Geoff Martin and Blanca Huertas, personal communication with KK), along with many

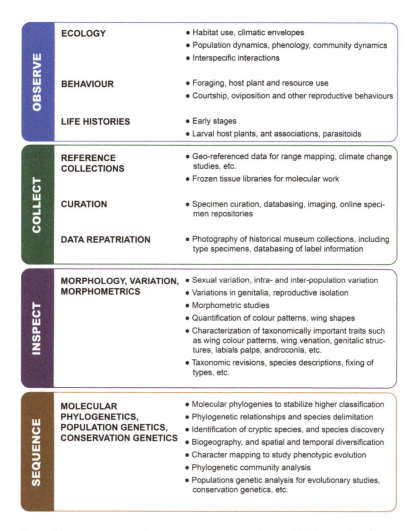

FIGURE 16.13 A roadmap for making progress on the taxonomy, systematics, and biology of Indian butterflies in a truly integrative manner. The modern *mantra* of "Observe-Collect-Inspect-Sequence" depends on intensive field work, strong academic institutions, rich museum resources, and joint efforts by professional and amateur scientists.

more million specimens in dozens of other museums in Europe and North America. These collections are at the heart of hundreds of scientific publications every year by taxonomists and butterfly/moth biologists from all over the world. The largest Lepidoptera collections in India are in the Zoological Survey of India among its various regional centres and the head office in Kolkata, and the Forest Research Institute in Dehradun, among dozens of collections spread in smaller institutions all over the country. However, the museum resources in India are tiny by any comparison. Only a strong culture of museum-based research will build strengths in species discovery and systematic revisions that are necessary to advance the understanding of Indian butterflies. Note that almost every leading university and research centre in the world engaged in research on organismal biology has an entomological collection. Therefore, Indian institutions and researchers need to start building significant research collections with modern strengths: geo-referenced data, deep-frozen tissue libraries, online or otherwise electronic databases, and high-throughput sequencing facilities. Specimen curation should include high-quality imaging and online image repositories. At the same time, we should recognize that most of the type specimens of currently known Indian species are in European museums. It is unlikely that these specimens will be repatriated to countries of origin, but it should be easy to repatriate label data and high-quality images of type specimens. Fixing of type specimens (Box 16.2) is also important because of its implications for taxonomic stability. Such museum-based research alone will facilitate a large number of studies of Indian butterflies. This work has already been initiated at NCBS, with imaging of nearly 15,000 butterfly specimens, including dozens of type specimens, completed in major international museums and other repositories. The NCBS research collections are also making a significant impact on the availability of modern infrastructures for taxonomic, systematic, evolutionary and conservation studies by providing state-of-the-art museum facilities.

SCIENTIFIC RESEARCH

Although taxonomic discovery of Indian butterflies has slowed down, natural historic, evolutionary ecological, biogeographic, and phylogenetic information has grown in leaps and bounds in the past two decades. Modern Indian naturalists, entomologists, and biologists are better trained to do natural history and scientific studies, and they are also better provisioned. Many Indians have received training in biology in some of the best research institutions and museums in the world. Their return to India to take up academic positions, in what might be termed "reverse brain-drain," has made a deep impact on the direction of scientific research in India. The rise of institution-building in the era of reverse brain-drain and citizen science has also increased the pace at which new discoveries are made and published. Numerous checklists and other records of butterflies from the Garo Hills, Eastern Himalaya, Assam Valley, and NE India, have reported many species rediscoveries as well as species new to India (e.g., Balaji et al. 2018; Gogoi 2012, 2013, 2015; Karimbumkara et al. 2016; Kunte 2009, 2010; Kunte et al. 2012; Rai et al. 2012; ; Sondhi and Kunte 2016). Since Bell's time, early stages and larval host plants have been reported to some extent (e.g., Kalesh and Prakash 2007; Karmakar et al. 2018; Kunte 2006; Kunte et al. 2018; Nitin et al. 2018; Pant and Chatterjee 1949; Sevastopulo 1973), although rigorous scientific studies are still lacking. Recent studies have investigated microbial associations of adult butterflies and caterpillars, effect of host plants and environmental factors on pupal polymorphism, and diversification of butterflies following host shifts (Ankola et al. 2012; Mayekar and Kodandaramaiah 2017; Phalnikar et al. 2018; Sahoo et al. 2017; Sahoo and Kodandaramaiah 2018). There is considerable promise in similar studies in the Indian context because of the rich diversity of butterflies, plants, and microbial communities. Building up on historic work, the southern Indian butterfly migrations are being studied (e.g., Bharos 2000; Bhaumik and Kunte 2017; Briscoe 1952; Chaturvedi and Satheesan 1979; Chaturvedi 1992; French 1943; Kunte 2005; Larsen 1978; Williams 1938). Flight morphology and other aspects have provided interesting insights into dispersal (Sekar 2012; Sekar and Karanth 2013, 2015). Recent work on mimicry has provided insights into morphological evolution and community assembly (Joshi et al. 2017; Su et al. 2015). Some preliminary work on population dynamics has also been published (Tiple et al. 2009). Taxonomy, phylogenetic relationships, biogeography, and related fields are gaining greater ground (Aduse-Poku et al. 2016; Huang et al. 2019; Kunte 2015; Sahoo et al. 2016, 2018; Toussaint et al. 2015). The overall picture that emerges from this summary is that butterfly biology is growing strong in India. This growth will be stronger in the future if long-term research programmes are built by specific labs and research centres around themes of particular significance for Indian butterfly biology and conservation.

There are other promising developments that will facilitate rapid growth of butterfly biology in India. The information age, the internet, and social media, have given easy access to vast literature, museum resources, and a globally interactive pool of subject experts whether they are amateur or professional. Butterfly-watching has become popular, with parallel growth in the volume of photographic records of Indian butterflies. At the same time, several citizen science projects have taken off, which are accumulating an unparalleled amount of data on the natural history and biology of Indian butterflies. The most successful citizen science project on Indian butterflies is *Butterflies of India* (http://www.ifoundbutterflies.org/), which has created a powerful online platform for aggregating spot records and reference images of butterflies (Kunte et al. 2018). These records are largely contributed by amateur butterfly-watchers and peer-reviewed before publication by advanced amateurs and professional biologists. Thus, this is one of the best-integrated professional-amateur scientific communities that aggregates big data on Indian butterfly diversity with the goal of studying ecological trends. This web-project has already accumulated over 55,000 reference images representing detailed distributional ranges, seasonal occurrence, early stages, and larval and nectar plants of Indian butterflies. It has also resulted in over a dozen scientific publications. Similar citizen science projects have a tremendous potential for future growth.

The modern mantra of "Observe-Collect-Inspect-Sequence" has significant potential to impact the growth of systematics and biology of butterflies in India. Figure 16.13 illustrates a roadmap for future growth.

CONCLUSION

India faces a major taxonomic impediment due to many decades of neglect of modern systematic research, the inaccessibility of reference materials of taxonomic importance, and a culture of academic isolation that has led to scarcity of in-house taxonomic expertise. This is compounded by a lack of vision that taxonomic advancement is central to understanding basic evolutionary and ecological processes. In this chapter, it was shown how these four problems have impeded scientific growth in India, specifically as applied to the taxonomy, systematics, biogeography, evolution, and ecology of butterflies. It was also shown that butterfly taxonomy currently used in India largely dates back to the early and mid-1900s, before the concepts of species and subspecies were clarified. This was also a time when the importance of biogeographic barriers, allopatry, vicariance, and dispersal in (sub)speciation process was just beginning to be explored in evolutionary biology. As a result, our understanding of Indian butterflies is vastly incomplete and to a large degree outdated. It is common to see the use of species and subspecies concepts/delimitations that confuse between clinal variation, environmentally induced variation, individual variation, and distinct geographic variation as applied to subspecies. Misapplications of the concepts of sympatry and parapatry to define species are also commonly observed. On the whole, the taxonomy and systematics of Indian butterflies has largely lagged behind the work on other Asian butterflies. These various problems were outlined in this chapter with examples. As a way forward, a roadmap was proposed to modernize taxonomy and species discovery of Indian butterflies. This should be put in the larger context of Asian butterflies and Oriental biogeography, with a combination of: (a) intensive taxonomic work with types and other important specimens housed in the research collections of European museums and, more recently, Japanese collections, (b) collecting new geo-referenced specimens, especially around critical climatic transitions and biogeographic barriers that have influenced diversification and endemism in the Oriental Region, and (c) using morphometric, molecular, and phylogenetic methods to distinguish between polymorphisms, subspecific, and specific variations to define taxa in modern evolutionary and phylogenetic frameworks with a strong emphasis on systematics (Figure 16.13). This will require considerable new work in the field and also substantial collaborations across museums, universities, and governments; leading to a truly international scientific enterprise.

ACKNOWLEDGEMENTS

We thank the editors of this volume for inviting us to contribute this chapter to a book that celebrates a lifetime of work by C. A. Viraktamath, a living legend of insect taxonomy in India. We also thank: (a) Blanca Huertas, David Lees and Geoff Martin, Lepidoptera curators at the Natural History Museum, London (NHMUK), for permission to inspect and photograph the historical type and other specimens and materials deposited in the museum, (b) Gaurav Agavekar for assistance in taking images at NHMUK, (c) Tarun Karmakar for assistance in curation of specimens in the Research Collections of the National Centre for Biological Sciences (NCBS), (d) Anuradha Joglekar for preparation of the map of Indian Subcontinent used in Figures 16.1 and 16.10 through 16.12, and (e) Mark Sterling and David Lees for comments on Figure 16.3. All the digitally enhanced illustrations of butterflies used in Figures 16.10 through 16.12 were based on original images taken at the NHMUK, except that images of types in Box 16.1 have not been digitally altered, and the images in that text box of *Hypolycaena narada* are from NCBS Research Collections (images by Krushnamegh Kunte, copyright of NCBS), the remaining from NHMUK (images by Krushnamegh Kunte, Gaurav Agavekar, and Dipendra Nath Basu, copyright of NHMUK, used with permission). Specimens used in genitalia dissections in Figure 16.4 were collected under research and collection permits to KK, issued by the state forest departments in Kerala [permit no. WL 10-3781/2012 dated 18/12/2012, and GO (RT) No. 376/2012/F and WLD dated 26/07/2012], Karnataka (permit no. 227/2014-2015 dated 2015/04/16), Goa (permit no. 2/21/GEN/WL and ET(S)/2013-14/387 dated 2013/06/20), Nagaland (permit no. CWL/GEN/240/522-39, dated 14/08/2012), Meghalaya (permit no. FWC/G/173/Pt-II/474-83, dated 27/05/2014), Arunachal Pradesh (permit no. CWL/G/13(95)/2011-12/Pt-III/2466-70, dated 16/02/2015), Sikkim (dated 21/03/2011), and West Bengal [permit no. 2115(9)/WL/4K-1/13/BL41, dated 06/11/2013], for which we thank the Principal Chief Conservators of Forest, Deputy Conservators of Forest, Wildlife Wardens, and field officers of those states. This work was funded by a Ramanujan Fellowship (Dept. of Science and Technology, Govt. of India) and a research grant from NCBS to KK and a CSIR-UGC Research Fellowship to DNB.

REFERENCES

Ackery, P. R., and R. I. Vane-Wright. 1984. *Milkweed Butterflies: Their Cladistics and Biology*. 1st ed. New York: Cornell University Press. p. 425.

Aduse-Poku, K., D. C. Lees, O. Brattström, U. Kodandaramaiah, S. C. Collins, N. Wahlberg, and P. M. Brakefield. 2016. Molecular phylogeny and generic-level taxonomy of the widespread palaeotropical "*Heteropsis* clade" (Nymphalidae: Satyrinae: Mycalesina). *Systematic Entomology*, 41:717–731.

Aitken, E. H. 1886. A list of the butterflies of the Bombay Presidency in the society's collections. *Journal of the Bombay Natural History Society*, 1:215–218.

Aitken, E. H. 1897. The migration of butterflies. *Journal of Bombay Natural History Society*, 11:336–337.

Ankola, K., D. Brueckner, and H. P. Puttaraju. 2012. *Wolbachia* endosymbiont infection in two Indian butterflies and female-biased sex ratio in the Red Pierrot, *Talicada nyseus*. *Journal of Biosciences*, 36:845–850.

Balaji, P. B., K. Kunte, and H. Chiba. 2018. *Erynnis pelias* Leech, 1891: Frosted Duskywing. Kunte, K., S. Sondhi, and P. Roy (Chief Editors). *Butterflies of India, v. 2.56*. Indian Foundation for Butterflies and National Centre for Biological Sciences, Bengaluru. http://www.ifoundbutterflies.org/sp/3170/Erynnis-pelias. Accessed November 2018.

Baum, D. A., and L. Shaw. 1995. Genealogical perspectives on the species problem in *Experimental and Molecular Approaches to Plant Biosystematics*. St. Louis, MO: Missouri Botanical Garden. pp. 289–303.

Bell, T. R. 1909. The common butterflies of the plains of India (including those met with the hill stations of the Bombay Presidency). *Journal of Bombay Natural History Society*, 19:16–58.

Bell, T. R. 1927. The common butterflies of the plains of India (including those met with the hill stations of the Bombay Presidency). *Journal of the Bombay Natural History Society*, 31:951–974.

Betham, J. A. 1890a. The butterflies of the central provinces. *Journal of the Bombay Natural History Society*, 5:279–286.

Betham, J. A. 1890b. The butterflies of the central provinces. *Journal of the Bombay Natural History Society*, 6:19–28.

Betham, J. A. 1891. The butterflies of the central provinces. *Journal of the Bombay Natural History Society*, 6:175–183.

Betham, J. A. 1894. Note on some of the butterflies of Matheran. *Journal of the Bombay Natural History Society*, 8:421–423.

Bethune-Baker, G. T. 1918. A revision of the genus *Tarucus*. *Transactions of the Entomological Society of London*, 1918:269–296.

Bharos, A. K. M. 2000. Large scale emergence and migration of the common Emigrant butterflies *Catopsilia pomona* (Family-Pieridae). *Journal of Bombay Natural History Society*, 97:301.

Bhaumik, V., and K. Kunte. 2017. Female butterflies modulate investment in reproduction and flight in response to monsoon-driven migrations. *Oikos*, 127:285–296.

Briscoe, V. 1952. Butterfly migration in the Nilgiris. *Journal of Bombay Natural History Society*, 50:417–418.

Cantlie, K. 1952. More butterflies of the Khasi and Jaintia Hills, Assam. *Journal of the Bombay Natural History Society*, 51:42–60.

Cantlie, K. 1956. Hesperiidae of Khasi and Jaintea Hills. *Journal of the Bombay Natural History Society*, 54:212–215.

Cantlie, K. 1958. A new butterfly from Assam. *Journal of the Bombay Natural History Society*, 55:180–181.

Cantlie, K. 1962. *The Lycaenidae Portion (Except the* Arhopala *Group), of Brigadier Evans' The Identification of Indian Butterflies 1932 (India, Pakistan, Ceylon, Burma)*. Mumbai, India: Bombay Natural History Society. p. 159.

Cantlie, K., and T. Norman. 1959. A new butterfly from Assam. *Journal of the Bombay Natural History Society*, 56:357–358.

Cantlie, K., and T. Norman. 1960. Four new butterflies from Assam. *Journal of the Bombay Natural History Society*, 57:424–429.

Chainey, J. E. 2005. The species of Papilionidae and Pieridae (Lepidoptera) described by Cramer and Stoll and their putative type material in the Natural History Museum in London. *Zoological Journal of the Linnean Society*, 145:283–337.

Chapman, R. F. 1982. *The Insects: Structure and Function*. Cambridge, UK: Harvard University Press. p. 919.

Chaturvedi, N. 1992. Northward migration of the common Indian crow butterfly *Euploea core* (Cramer) in and around Bombay. *Journal of Bombay Natural History Society*, 90:115–116.

Chaturvedi, N., and S. M. Satheesan. 1979. Southward migration of *Euploea core* at Khandala, Western Ghats. *Journal of the Bombay Natural History Society*, 76:534.

Chiba, H. 2009. A revision of the subfamily Coeliadinae (Lepidoptera: Hesperiidae). *Bulletin of the Kitakyushu Museum of Natural History and Human History, Serial A*, 7:1–102.

Chiba, H., and H. Tsukiyama. 1993. A revision of the genus *Pirdana* Distant (Lepidoptera: Hesperiidae). *Butterflies*, 6:19–25.

Chiba, H., and H. Tsukiyama. 1994. A revisional note on the genus *Pirdana* Distant (Lepidoptera: Hesperiidae). *Butterflies*, 7:56.

Chiba, H., and J. N. Eliot. 1991. A review of the genus *Parnara* (Lepidoptera: Hesperiidae) with special reference to the Asian species. *Tyô to Ga*, 42:179–194.

Clusella Trullas, S., J. H. van Wyk, and J. R. Spotila. 2007. Thermal melanism in ectotherms. *Journal of Thermal Biology*, 32:235–245.

Cordero, C., and J. Baixeras. 2015. Sexual selection within the female genitalia in Lepidoptera. in *Cryptic Female Choice in Arthropods: Patterns, Mechanisms and Prospects*, eds. A. V. Peretti and A. Aisenberg. Cham, Switzerland: Springer International Publishing. pp. 325–350.

Cowan, C. F. 1966. The Indo-Oriental Horagini (Lepidoptera: Lycaenidae). *Bulletin of the British Museum, Natural History (Entomology)*, 18:103–141.

Cowan, C. F. 1967. The Indo-Oriental tribe Cheritrini (Lepidoptera: Lycaenidae). *Bulletin of the British Museum, Natural History (Entomology)*, 20:75–103.

Cowan, C. F. 1970. *Annotationes Rhopalocerologicae*. Herts, UK: Clunbury Press. p. 70.

Cowan, C. F. 1974. The Indo-Oriental genus *Drupadia* Moore (Lepidoptera: Lycaenidae). *Bulletin of the British Museum, Natural History (Entomology)*, 29:281–356.

Coyne, J. A., and H. A. Orr. 2004. *Speciation*. Sunderland, MA: Sinauer Associates. p. 545.

d'Abrera, B. 1982. *Butterflies of the Oriental Region. Part I: Papilionidae, Pieridae & Danaidae*. Victoria, New York: Hill House Publishing. pp. 1–244.

d'Abrera, B. 1985. *Butterflies of the Oriental Region. Part II: Nymphalidae, Satyridae & Amathusiidae*. Melbourne, Australia: Hill House. pp. 246–534.

Darwin, C. 1859. *On the Origin of Species by Means of Natural Selection, or the Preservation of Favoured Races in the Struggle for Life*. London, UK: John Murray. p. 502.

Davidson, J., and E. H. Aitken. 1890. Notes on the larvae and pupae of some of the butterflies of the Bombay presidency. *Journal of the Bombay Natural History Society*, 5:349–375.

Davidson, J., T. R. Bell, and E. H. Aitken. 1896. The butterflies of the North Canara district of the Bombay Presidency. Part I. *Journal of the Bombay Natural History Society*, 10:22–63.

Davidson, J., T. R. Bell, and E. H. Aitken. 1897. The butterflies of the North Canara district of the Bombay Presidency. Part II. *Journal of the Bombay Natural History Society*, 10:372–393.

de Jong, R. 2004. Phylogeny and biogeography of the genus *Taractrocera* Butler, 1870 (Lepidoptera: Hesperiidae), an example of Southeast Asian-Australian interchange. *Zoologische Mededelingen*, 78:383–415.

de Nicéville, L. 1883. Third list of butterflies taken in Sikkim in October, 1883, with notes on habits, &c. *Journal of the Asiatic Society of Bengal (II)*, 52:92–100.

de Nicéville, L. 1885a. Fourth list of butterflies taken in Sikkim in October, 1884, with notes on habits, &c. *Journal of the Asiatic Society of Bengal (II)*, 54:1–5.

de Nicéville, L. 1885b. List of the butterflies of Calcutta and its neighbourhood, with notes on habits, food-plants, &c. *Journal of the Asiatic Society of Bengal (II)*, 54:39–54.

de Nicéville, L. 1886a. On the life-history of certain Calcutta species of Satyrinae, with special reference to the seasonal dimorphism alleged to occur in them. *Journal of the Asiatic Society of Bengal (II)*, 55:229–238.

de Nicéville, L. 1886b. *The Butterflies of India, Burmah and Ceylon. Volume II. Nymphalidae, Nymphalinae. Lemoniidae, Libythaeinae, Nemeobiinae*. Calcutta (Kolkata), India: The Calcutta Central Press Co. p. 332.

de Nicéville, L. 1890a. List of Chin-Lushai butterflies. *Journal of the Bombay Natural History Society*, 5:295–298.

de Nicéville, L. 1890b. *The Butterflies of India, Burmah and Ceylon. Volume III. Lycaenidae*. Calcutta (Kolkata), India: The Calcutta Central Press Co. p. 503.

de Queiroz, K. 1998. The general lineage concept of species, species criteria, and the process of speciation: A conceptual unification and terminological recommendations in *Endless Forms: Species and Speciation*, edited by Howard DJ and Berlocher SH. Oxford, UK: Oxford University Press. pp. 57–75.

de Queiroz, K. 2005a. A unified concept of species and its consequences for the future of taxonomy. *Proceedings of the California Academy of Sciences*, 56 Suppl.:196–221.

de Queiroz, K. 2005b. Different species problems and their resolution. *BioEssays*, 27:1263–1269.

de Queiroz, K. 2007. Species concepts and species delimitation. *Systematic Biology*, 56:879–886.

Dobzhansky, T. 1951. *Genetics and the Origin of Species* (3rd edn.). New York: Columbia University Press. p. 364.

Doherty, W. 1886. A list of butterflies taken in Kumaon. *Journal of the Asiatic Society of Bengal (II)*, 55:103–140.

Dudgeon, G. C. 1895. Life history of *Kallima inachus* Biosduval, a Nymphaline butterfly. *Journal of the Bombay Natural History Society*, 9:342–343.

Eliot, J. N. 1963. The *Heliophorus epicles* (Godart, 1823) (Lepidoptera: Lycaenidae) complex. *The Entomologist*, 96:169–180.

Eliot, J. N. 1967. The *sakra* Moore, 1857, section of the satyrid genus *Ypthima* Hübner. *The Entomologist*, 100:49–61.

Eliot, J. N. 1969. An analysis of the Eurasian and Australian Neptini (Lepidoptera: Nymphalidae). *Bulletin of the British Museum, Natural History (Entomology)*, Suppl. 15:3–155.

Eliot, J. N. 1973. The higher classification of the Lycaenidae: A tentative arrangement. *Bulletin of the British Museum, Natural History (Entomology)*, 28:371–505.

Eliot, J. N. 1986. A review of the Miletini (Lepidoptera: Lycaenidae). *Bulletin of the British Museum, Natural History (Entomology)*, 53:1–105.

Eliot, J. N. 1990. Notes on the genus *Curetis* Hübner (Lepidoptera, Lycaenidae). *Tyô to Ga*, 41:201–225.

Eliot, J. N., and A. Kawazoé. 1983. *Blue Butterflies of the Lycaenopsis Group*. London, UK: British Museum (Natural History). p. 309.

Elwes, H. J., and L. de Nicéville. 1886. List of the lepidopterous insects collected in Tavoy and in Siam during 1884–1885 by the Indian Museum collector under C. E. Pitman, Esq., C. I. E., Chief Superintendent of Telegraphs. Part II, Rhopalocera. *Journal of the Asiatic Society of Bengal (II)*, 55:413–442.

Elwes, H. J., and O. Möller. 1888. A catalogue of the Lepidoptera of Sikkim; with additions, corrections, and notes on seasonal and local distribution. *Transactions of the Entomological Society, London*, 1888:269–464.

Espeland, M., J. Breinholt, K. R. Willmott et al. 2018. A comprehensive and dated phylogenomic analysis of butterflies. *Current Biology*, 28:770–778.e5.

Espeland, M., J. P. W. Hall, P. J. DeVries et al. 2015. Ancient Neotropical origin and recent recolonisation: Phylogeny, biogeography and diversification of the Riodinidae (Lepidoptera: Papilionoidea). *Molecular Phylogenetics and Evolution*, 93:296–306.

Evans, W. H. 1912. A list of Indian butterflies. *Journal of the Bombay Natural History Society*, 21:553–584.

Evans, W. H. 1932. *The Identification of Indian Butterflies* (2nd edn.). Mumbai, Inida: Bombay Natural History Society. p. 454.

Evans, W. H. 1949. *A Catalogue of the Hesperiidae from Europe, Asia and Australia in the British Museum (Natural History)*. London, UK: British Museum (Natural History). p. 502.

Evans, W. H. 1954. A revision of the genus *Curetis* (Lepidoptera: Lycaenidae). *The Entomologist*, 87:190–194, 212–216, 241–247.

Evans, W. H. 1957. A revision of the *Arhopala* group of Oriental Lycaenidae (Lepidoptera: Rhopalocera). *Bulletin of the British Museum, Natural History (Entomology)*, 5:85–137.

Evans, W. H., and P. F. Bellinger. 1956. Lepidoptera publications by Brigadier W. H. Evans. *Lepidopterists' News*, 10:197–199.

Fan, X.-L., H. Chiba, and M. Wang. 2010. The genus *Scobura* Elwes and Edwards, 1897 from China, with the description of two new species (Lepidoptera: Hesperiidae). *Zootaxa*, 2490:1–15.

Forsayeth, R. W. 1884. Life history of sixty species of Lepidoptera observed in Mhow, Central India. *Transactions of the Entomological Society of London*, 3:377–419.

French, W. L. 1943. Butterfly migration (*Danais melissa dravidarum* and *Euploea c. core*). *Journal of the Bombay Natural History Society*, 44:310.

Fryer, J. C. F. 1914. An investigation by pedigree breeding into the polymorphism of *Papilio polytes*, Linn. *Philosophical Transactions of the Royal Society of London, B*, 204:227–254.

Fujioka, T. 1970. Butterflies collected by the lepidopterological research expedition to Nepal Himalaya, 1963. Part I Papilionoidea. *Special Bulletin of the Lepidopterists' Society of Japan*, 4:1–125.

Futuyma, D. J. 1998. *Evolutionary Biology*. 3rd edn. Sunderland, MA: Sinauer Associates. p. 763.

Gogoi, M. J. 2012. Butterflies (Lepidoptera) of Dibang Valley, Mishmi Hills, Arunachal Pradesh, India. *Journal of Threatened Taxa*, 4:3137–3160.

Gogoi, M. J. 2013. A preliminary checklist of butterflies recorded from Jeypore-Dehing forest, eastern Assam, India. *Journal of Threatened Taxa*, 5:3684–3696.

Gogoi, M. J. 2015. Observations on lycaenid butterflies from Panbari Reserve Forest and adjoining areas, Kaziranga, Assam, northeastern India. *Journal of Threatened Taxa*, 7:8259–8271.

Hebert, P. D. N., E. H. Penton, J. M. Burns, D. H. Janzen, and W. Hallwachs. 2004. Ten species in one: DNA barcoding reveals cryptic species in the neotropical skipper butterfly Astraptes fulgerator. *Proceedings of the National Academy of Sciences, USA*, 101:14812–14817.

Heikkilä, M., L. Kaila, M. Mutanen et al. 2012. Cretaceous origin and repeated tertiary diversification of the redefined butterflies. *Proceedings of the Royal Society B*, 279:1093–1099.

Hemming, F. 1967. The generic names of the butterflies and their type-species (Lepidoptera: Rhopalocera). *Bulletin of the British Museum, Natural History (Entomology)*, Supplement 9:1–509.

Hennig, W. 1966. *Phylogenetic Systematics. Translated by D. Dwight Davis and Rainer Zangerl. Illinois Reissue 1999.* Urbana, IL: University of Illinois Press. p. 263.

Honey, M. R., and M. J. Scoble. 2001. Linnaeus's butterflies (Lepidoptera: Papilionoidea and Hesperioidea). *Zoological Journal of the Linnean Society*, 132:277–399.

Howard, D. J., and S. H. Berlocher (ed.). 1998. *Endless Forms: Species and Speciation*. New York: Oxford University Press. p. 496.

Howarth, T. G. 1957. A revision of the genus *Neozephyrus* Sibatani and Ito (Lepidoptera: Lycaenidae). *Bulletin of the British Museum, Natural History (Entomology)*, 5:235–285.

Huang, H. 1998. Research on the butterflies of the Namjagbarwa Region, S. E. Tibet. *Neue Entomologische Nachrichten*, 41:207–263.

Huang, H. 2000. A list of butterflies collected from tibet during 1993–1996, with new descriptions, revisional notes and discussion on zoogeography - 1 (Lepidoptera: Rhopalocera). *Lambillionea*, 100:141–158.

Huang, H. 2001. Report of H. Huang's 2000 expedition to SE. Tibet for Rhopalocera (Insecta, Lepidoptera). *Neue Entomologische Nachrichten*, 51:65–151.

Huang, H. 2002. Some new butterflies from China - 2 (Lepidoptera, Hesperiidae). *Atalanta*, 33:109–122, 226–229.

Huang, Z., H. Chiba, J. Jin, Athulya Girish K., M. Wang, K. Kunte, and X. Fan. 2019. A multilocus phylogenetics framework of the tribe Aeromachini (Lepidoptera: Hesperiidae: Hesperiinae), with implications for taxonomy and historical biogeography. *Systematic Entomology*, 44:163–178.

Huang, H., and Y.-P. Xue. 2004. Notes on some Chinese butterflies. *Neue Entomologische Nachrichten*, 57:171–177.

Huang, Z., H. Chiba, J. Jin et al. 2018. A multilocus phylogenetics framework of the tribe Aeromachini (Lepidoptera: Hesperiidae: Hesperiinae), with implications for taxonomy and historical biogeography. *Systematic Entomology*, in press.

Huxley, J. S. 1939. Clines: An auxiliary method in taxonomy. *Bijdragen Tot de Dierkunde*, 27:491–520.

Inayoshi, Y. 2018. *A Check List of Butterflies in Indo-China (chiefly from Thailand, Laos and Vietnam)*. http://yutaka.it-n.jp/. Accessed November 2018.

International Commission on Zoological Nomenclature (ICZN). 1999. *International Code of Zoological Nomenclature*, 4th edn. London, UK: The International Trust for Zoological Nomenclature. p. 364.

Joshi, J., A. Prakash, and K. Kunte. 2017. Evolutionary assembly of communities in butterfly mimicry rings. *The American Naturalist*, 189:E58–E76.

Kalesh, S., and S. K. Prakash. 2007. Additions to larval host plants of butterflies of the Western Ghats, Kerala, southern India (Rhopalocera, Lepidoptera): Part 1. *Journal of the Bombay Natural History Society*, 104:237–240.

Karimbumkara, S. N., R. Goswami, and P. Roy. 2016. A report of False Tibetan Cupid *Tongeia pseudozuthus* Huang, 2001 (Lepidoptera: Lycaenidae) from the Upper Dibang Valley, Arunachal Pradesh: An addition to the Indian butterfly fauna. *Journal of Threatened Taxa*, 8:8927–8929.

Karmakar, T., R. Nitin, V. Sarkar et al. 2018. Early stages and larval host plants of some northeastern Indian butterflies. *Journal of Threatened Taxa*, 10:11780–11799.

Kawahara, A. Y. 2013. Systematic revision and review of the extant and fossil snout butterflies (Lepidoptera: Nymphalidae: Libytheinae). *Zootaxa*, 3631:1–74.

Kawahara, A. Y. and J. W. Breinholt. 2014. Phylogenomics provides strong evidence for relationships of butterflies and moths. *Proceedings of the Royal Society B*, 281:20140970.

Kawahara, A. Y., J. W. Breinholt, M. Espeland et al. 2018. Phylogenetics of moth-like butterflies (Papilionoidea: Hedylidae) based on a new 13-locus target capture probe set. *Molecular Phylogenetics and Evolution*, 127:600–605.

Kodandaramaiah, U., D. C. Lees, C. J. Müller, E. Torres, K. P. Karanth, and N. Wahlberg. 2010. Phylogenetics and biogeography of a spectacular Old World radiation of butterflies: The subtribe Mycalesina (Lepidoptera: Nymphalidae: Satyrini). *BMC Evolutionary Biology*, 10:172.

Koiwaya, S. 2002. Descriptions of five new species and a new subspecies of Theclini (Lycaenidae) from China, Myanmar and India. *Gekkan-Mushi*, 377:2–8.

Kumar, R., G. S. Arora, and V. V. Ramamurthy. 2009. A new subspecies of *Talicada nyseus* (Guerin) (Lepidoptera: Lycaenidae) from Delhi, India. *Oriental Insects*, 43:297–307.

Kunte, K. 2005. Species composition, sex-ratios and movement patterns in danaine butterfly migrations in southern India. *Journal of the Bombay Natural History Society*, 102:280–286.

Kunte, K. 2006. Additions to known larval host plants of Indian butterflies. *Journal of the Bombay Natural History Society*, 103:119–122.

Kunte, K. 2009. Occurrence of *Elymnias obnubila* Marshall and de Nicéville, 1883 (Lepidoptera: Nymphalidae: Satyrinae) in southern Mizoram: Range extension of the species and an addition to the Indian butterfly fauna. *Journal of Threatened Taxa*, 1:567–568.

Kunte, K. 2010. Rediscovery of the federally protected Scarce Jester butterfly *Symbrenthia silana* de Nicéville, 1885 (Nymphalidae: Nymphalinae) from the Eastern Himalaya and Garo Hills, northeastern India. *Journal of Threatened Taxa*, 2:858–866.

Kunte, K. 2015. A new species of *Hypolycaena* (Lepidoptera: Lycaenidae) from Arunachal Pradesh, north-eastern India. *The Journal of Research on the Lepidoptera*, 48:21–27.

Kunte, K. and A. Tiple. 2009. The Polyommatine wing pattern elements and seasonal polyphenism of the Indian *Chilades pandava* butterfly (Lepidoptera: Lycaenidae). *News of the Lepidopterists' Society*, 51:86–88 and 109.

Kunte, K., S. Sondhi, and P. Roy. 2018. *Butterflies of India. v. 2.56.* http://ifoundbutterflies.org/. Accessed November 2018.

Kunte, K., S. Sondhi, B. M. Sangma, R. Lovalekar, K. Tokekar, and G. Agavekar. 2012. Butterflies of the Garo Hills of Meghalaya, northeastern India: Their diversity and conservation. *Journal of Threatened Taxa*, 4:2933–2992.

Larsen, T. B. 1978. Butterfly migrations in the Nilgiri Hills of South India. *Journal of Bombay Natural History Society*, 74:546–549.

Larsen, T. B. 1987. The butterflies of the Nilgiri Mountains of Southern India (Lepidoptera: Rhopalocera). *Journal of the Bombay Natural History Society*, 84:26–54, 291–316, 560–584; 85:26–43.

Mackinnon, P. W. 1898. A list of the butterflies of Mussoorie in the Western Himalayas and neighboring regions. *Journal of the Bombay Natural History Society*, 11:368–389.

Mackinnon, P. W., and L. de Nicéville. 1898. A list of the butterflies of Mussoorie in the Western Himalayas and neighbouring regions. Part III. *Journal of Bombay Natural History Society*, 11:585–605.

Mallet, J. 2004. Poulton, Wallace and Jordan: How discoveries in *Papilio* butterflies led to a new species concept 100 years ago. *Systematics and Biodiversity*, 1:441–452.

Mallet, J. 2007. Subspecies, semispecies, superspecies in *Encyclopedia of Biodiversity*, edited by S. A. Levin. Amsterdam, the Netherlands: Elsevier Academic Press. pp. 1–5.

Marchese, C. 2015. Biodiversity hotspots: A shortcut for a more complicated concept. *Global Ecology and Conservation*, 3:297–309.

Marshall, G. F. L., and L. de Nicéville. 1882. *The Butterflies of India, Burmah and Ceylon. Volume I. Nymphalidae. Danainae, Satyrinae, Elymniinae, Morphinae, Acraeinae.* Calcutta (Kolkata), India: The Calcutta Central Press Co. p. 327.

Masui, A. 2004. A revision on *Chitoria sordida* and *C. naga* (Lepidoptera, Nymphalidae). *Tyô to Ga*, 55:243–250.

Mathew, G., and M. M. Kumar. 2003. State of the art knowledge on the butterflies of Nilgiri Biosphere Reserve, India. *Envis (Wildlife and Protection Areas): Conservation of Rainforests in India*, 4:115–120.

Mayekar, H. V., and U. Kodandaramaiah. 2017. Pupal colour plasticity in a tropical butterfly, *Mycalesis mineus* (Nymphalidae: Satyrinae). *PLoS ONE*, 12:e0171482.

Mayr, E. 1942. *Systematics and the Origin of Species*. New York: Columbia University Press. p. 334.

Mayr, E. 1965. *Animal Species and Evolution*. Cambridge, UK: Harvard University Press. p. 797.

Mayr, E. 1982. *The Growth of Biological Thought: Diversity, Evolution, and Inheritance*. Cambridge, UK: Harvard University Press. p. 974.

Mehta, D. R. 1933. Comparative morphology of the male genitalia in Lepidoptera. *Records of the Indian Museum (A Journal of Indian Zoology)*, 35:197–266.

Miller, L. D. 1970. Nomenclature of wing veins and cells. *Journal of Research on the Lepidoptera*, 8:37–48.

Moore, F. 1892. *Lepidoptera Indica. Vol. I. Rhopalocera. Family Nymphalidae. Sub-Families Euploeinae and Satyrinae [1890–1892].* London, UK: Reeve & Co. p. 317.

Moore, F. 1896. *Lepidoptera Indica. Vol. II. Rhopalocera. Family Nymphalidae. Sub-Families Satyrinae (Continued), Elymniinae, Amathusiinae, Nymphalinae (Group Charaxina) [1893–1896].* London, UK: Reeve & Co. p. 274.

Moore, F. 1899. *Lepidoptera Indica. Vol. III. Rhopalocera. Family Nymphalidae. Sub-Families Nymphalinae (Continued), Groups Potamina, Euthaliina, Limenitina [1896–1899].* London, UK: Reeve & Co. p. 254.

Moore, F. 1900. *Lepidoptera Indica. Vol. IV. Rhopalocera. Family Nymphalidae. Sub-Families Nymphalinae (Continued), Groups Limenitina, Nymphalina, and Argynnina [1899–1900].* London, UK: Reeve & Co. p. 260.

Moore, F. 1903. *Lepidoptera Indica. Vol. V. Rhopalocera. Family Nymphalidae. Sub-Family Nymphalinae (Continued), Groups Melitaeina and Eurytelina. Sub-Families Acraeinae, Pseudergolinae, Calinaginae, and Libytheinae. Family Riodinidae. Sub-Family Nemeobiinae. Family Papilionidae.* London, UK: Reeve & Co. p. 248.

Moore, F. 1905. *Lepidoptera Indica. Vol. VI. Rhopalocera. Family Papilionidae. Sub-Family Papilioninae (Continued). Family Pieridae. Sub-Family Pierinae [1903–1905].* London, UK: Reeve & Co. p. 240.

Mutanen, M., N. Wahlberg, and L. Kaila. 2010. Comprehensive gene and taxon coverage elucidates radiation patterns in moths and butterflies. *Proceedings of the Royal Society B*, 277:2839–2848.

Myers, N., R. A. Mittermeier, C. G. Mittermeier, G. A. B. da Fonseca, and J. Kent. 2000. Biodiversity hotspots for conservation priorities. *Nature*, 403:853–858.

Nijhout, H. F. 1991. *The Development and Evolution of Butterfly Wing Patterns*. Washington, DC: Smithsonian Institution Press. p. 297.

Nitin, R., V. C. Balakrishnan, P. V. Churi, S. Kalesh, S. Prakash, and K. Kunte. 2018. Larval host plants of the butterflies of the Western Ghats, India. *Journal of Threatened Taxa*, 10:11495–11550.

Nixon, K. C., and Q. D. Wheeler. 1990. An amplification of the phylogenetic species concept. *Cladistics*, 6:211–223.

Nosil, P. 2012. *Ecological Speciation*. Oxford, UK: Oxford University Press. p. 304.

Pant, G. D., and N. C. Chatterjee. 1949. A list of described immature stages of Indian Lepidoptera, Part I: Rhopalocera. *Indian Forest Records (New Series): Entomology*, 7:213–255.

Parsons, R. E., and K. Cantlie. 1948. The butterflies of the Khasia and Jaintia hills, Assam. *Journal of the Bombay Natural History Society*, 47:498–522.

Phalnikar, K., K. Kunte, and D. Agashe. 2018. Dietary and developmental shifts in butterfly-associated bacterial communities. *Royal Society Open Science*, 5:171559.

Prathapan, K. D., P. Dharma Rajan, T. C. Narendran, C. A. Viraktamath, N. A. Aravind, and J. Poorani. 2008. Death sentence on taxonomy in India. *Current Science*, 94:170–171.

Prathapan, K. D., P. Dharma Rajan, T. C. Narendran et al. 2006. Biological Diversity Act, 2002: Shadow of permit-raj over research. *Current Science*, 91:1006–1007.

Punnett, R. C. 1908. "Mimicry" in Ceylon butterflies, with a suggestion as to the nature of polymorphism. *Spolia Zeylanica*, 5:1–24.

Rai, S., K. D. Bhutia, and K. Kunte. 2012. Recent sightings of two very rare butterflies, *Lethe margaritae* Elwes, 1882 and *Neptis nycteus* de Nicéville, 1890, from Sikkim, eastern Himalaya, India. *Journal of Threatened Taxa*, 4:3319–3326.

Riley, N. D. 1956. Obituary: William Harry Evans. *Lepidopterists' News*, 10:193–196.

Riley, N. D. 1960. Heinrich Ernst Karl Jordan. *Biographical Memoirs of Fellows of the Royal Society*, 6:106–133.

Roy, P. K. 2013. *Callerebia dibangensis* (Lepidoptera: Nymphalidae: Satyrinae), a new butterfly species from the eastern Himalaya, India. *Journal of Threatened Taxa*, 5:4725–4733.

Rufus, K., and S. P. Sabarinathan. 2007. A checklist of butterflies of Thengumarahada in the Nilgiris, Southern India. *Zoos' Print Journal*, 22:2818–2837.

Sahoo, R. K., A. D. Warren, N. Wahlberg, A. V. Z. Brower, V. A. Lukhtanov, and U. Kodandaramaiah. 2016. Ten genes and two topologies: An exploration of higher relationships in skipper butterflies (Hesperiidae). *PeerJ*, 4:e2653.

Sahoo, R. K., A. D. Warren, S. C. Collins, and U. Kodandaramaiah. 2017. Hostplant change and paleoclimatic events explain diversification shifts in skipper butterflies (Family: Hesperiidae). *BMC Evolutionary Biology*, 17:174.

Sahoo, R. K., and U. Kodandaramaiah. 2018. Local host plant abundance explains negative association between larval performance and female oviposition preference in a butterfly. *Biological Journal of the Linnean Society*, 125:333–343.

Sahoo, R. K., D. J. Lohman, N. Wahlberg et al. 2018. Evolution of *Hypolimnas* butterflies (Nymphalidae): Out-of-Africa origin and *Wolbachia*-mediated introgression. *Molecular Phylogenetics and Evolution*, 123:50–58.

Sekar, S. 2012. A meta-analysis of the traits affecting dispersal ability in butterflies: Can wingspan be used as a proxy? *Journal of Animal Ecology*, 81:174–184.

Sekar, S., and P. Karanth. 2015. Does size matter? Comparative population genetics of two butterflies with different wingspans. *Organisms, Diversity and Evolution*, 15:567–575.

Sekar, S., and P. Karanth. 2013. Flying between sky islands: The effect of naturally fragmented habitat on butterfly population structure. *PloS ONE*, 8:e71573.

Sevastopulo, D. G. 1973. The food-plants of Indian Rhopalocera. *Journal of the Bombay Natural History Society*, 70:156–183.

Sharma, N. 2013a. Two new species of the genus *Ypthima* Hübner (Lepidoptera: Papilionoidea: Satyridae) from India and Myanmar. *Records of the Zoological Survey of India*, 113:1–10.

Sharma, N. 2013b. *Ypthima kedarnathensis* Singh: New synonym of *Ypthima sakra* Moore (Nymphalidae: Satyrinae) from Garhwal Himalaya, India. *Journal of Bombay Natural History Society*, 110:230–232.

Shirôzu, T. and H. Shima. 1979. On the natural groups and their phylogenetic relationships of the genus *Ypthima* Hübner mainly from Asia (Lepidoptera: Satyridae). *Sieboldia*, 4:231–295.

Simpson, G. G. 1961. *Principles of Animal Taxonomy*. New York: Columbia University Press. p. 247.

Singh, A. P. 2007. A new butterfly species of the genus *Ypthima* Hübner (Nymphalidae: Satyrinae) from Garhwal Himalaya, India. *Journal of the Bombay Natural History Society*, 104:191–194.

Smetacek, P. 2004. Descriptions of new Lepidoptera from the Kumaon Himalaya. *Journal of the Bombay Natural History Society*, 101:269–276.

Smetacek, P. 2011. A review of West Himalayan Neptini (Nymphalidae). *Journal of the Lepidopterists' Society*, 65:153–161.

Smetacek, P. 2012. A new subspecies of *Mycalesis suaveolens* Wood-Mason & de Nicéville 1883 from the western Himalaya, India (Lepidoptera, Nymphalidae, Satyrinae). *Nachrichten Des Entomologischen Vereins Apollo*, 32:105–108.

Smiles, R. L. 1982. The taxonomy and phylogeny of the genus *Polyura* Billberg (Lepidoptera: Nymphalidae). *Bulletin of the British Museum, Natural History (Entomology)*, 44:115–237.

Sondhi, S., and K. Kunte. 2016. Butterflies (Lepidoptera) of the Kameng Protected Area Complex, western Arunachal Pradesh, India. *Journal of Threatened Taxa*, 8:9053–9124.

Sondhi, S., and K. Kunte. 2018. *Butterflies and Moths of Pakke Tiger Reserve*, 2nd edn. Bengaluru, India: Titli Trust (Dehradun), National Centre for Biological Sciences (Bengaluru), and Indian Foundation for Butterflies (Bengaluru). p. 242.

Sondhi, S., T. Karmakar, Y. Sondhi, R. Jhaveri, and K. Kunte. 2016. Re-discovery of *Calinaga aborica* Tytler, 1915 (Lepidoptera: Nymphalidae: Calinaginae) from Arunachal Pradesh, India. *Journal of Threatened Taxa*, 8:8618–8622.

Su, S., M. Lim, and K. Kunte. 2015. Prey from the eyes of predators: Color discriminability of aposematic and mimetic butterflies from an avian visual perspective. *Evolution*, 69:2985–2994.

Swinhoe, C. 1886. On the Lepidoptera of Mhow in Central India. *Proceedings of the Zoological Society, London*, 1886: 421–465.

Swinhoe, C. 1910. *Lepidoptera Indica. Vol. VII. Rhopalocera. Family Pieridae [Printed in Error: "Family Papilionidae"]. Sub-Family Pierinae (Continued). Family Lycaenidae. Sub-Families Gerydinae, Lycaenopsinae and Everinae [1905–1910]*. London, UK: Reeve & Co. p. 286.

Swinhoe, C. 1911. *Lepidoptera Indica. Vol. VIII. Rhopalocera. Family Lycaenidae. Sub-Families Lycaeninae, Plebeinae, Lampidinae, Chrysophaninae, Poritiinae, Amblypodiinae, Curetinae, Liphyrinae, Ruralinae [1910–1911]*. London, UK: Reeve & Co. p. 293.

Swinhoe, C. 1912. *Lepidoptera Indica. Vol. IX. Rhopalocera. Family Lycaenidae (Continued). Sub-Families Horaginae, Deudorixinae, Hypolycaeninae, Zesiusinae, Aphnaeinae, Biduandinae, Cheritrinae, Loxurinae. Family Hesperiidae. Sub-Families Ismeneinae, Achalarinae [1911–1912].* London, UK: Reeve & Co. p. 278.

Swinhoe, C. 1913. *Lepidoptera Indica. Vol. X. Rhopalocera. Family Hesperiidae (Concluded). Sub-Families Celaenorrhinae, Hesperiinae, Pamphilinae, Astictopterinae, Suastinae, Erionotinae, Matapinae, Notocryptinae, Plastingiinae, Erynninae [1912–1913].* London, UK: Reeve & Co. p. 364.

Talbot, G. 1939. *The Fauna of British India, Including Ceylon and Burma: Butterflies, Vol. 1*. London, UK: Taylor & Francis Group. p. 600.

Talbot, G. 1947. *The Fauna of British India, Including Ceylon and Burma: Butterflies, Vol. 2*. London, UK: Taylor & Francis Group. p. 506.

Tiple, A., D. Agashe, A. M. Khurad, and K. Kunte. 2009. Population dynamics and seasonal polyphenism of *Chilades pandava* butterfly (Lycaenidae) in central India. *Current Science*, 97:1774–1779.

Tite, G. E. 1963. A synonymic list of the genus *Nacaduba* and allied genera (Lepidoptera: Lycaenidae). *Bulletin of the British Museum, Natural History (Entomology)*, 13:69–116.

Tite, G. E. 1966. A revision of the genus *Anthene* from the Oriental Region (Lepidoptera: Lycaenidae). *Bulletin of the British Museum, Natural History (Entomology)*, 18:253–275.

Toussaint, E. F. A., J. Morinière, C. J. Müller et al. 2015. Comparative molecular species delimitation in the charismatic Nawab butterflies (Nymphalidae, Charaxinae, *Polyura*). *Molecular Phylogenetics and Evolution*, 91:194–209.

Tsukiyama, H., and H. Chiba. 1994. A review of the genus *Odina* Mabille, 1891 (Lepidoptera: Hesperiidae). *Butterflies*, 8:30–33.

Tytler, H. C. 1911. Notes on butterflies from the Naga Hills. Part II. *Journal of the Bombay Natural History Society*, 21:588–606.

Tytler, H. C. 1914. Notes on some new and interesting butterflies from Manipur and the Naga Hills. Part I. *Journal of the Bombay Natural History Society*, 23:216–229.

Tytler, H. C. 1915. Notes on some new and interesting butterflies from Manipur and the Naga Hills. *Journal of the Bombay Natural History Society*, 23:502–515.

Tytler, H. C. 1926. Notes on some new and interesting butterflies from India and Burma. Part II. *Journal of the Bombay Natural History Society*, 31:579–590.

Tytler, H. C. 1940. Notes on some new and interesting butterflies chiefly from Burma. *Journal of the Bombay Natural History Society*, 42:109–123.

van der Poorten, N. E. and G. M. van der Poorten. 2016. *The Butterfly Fauna of Sri Lanka*. Toronto, Canada: Lepodon Books. p. 418.

Vane-Wright, R. I. 1993. Milkweed butterflies (Lepidoptera: Danainae) and conservation priorities in the Andaman and Nicobar islands, India. *Butterflies (Teinopalpus)*, 4:21–33.

Varshney, R. K. 1993. Index Rhopalocera Indica. Part III. Genera of butterflies from India and neighbouring countries (Lepidoptera: (A) Papilionidae, Pieridae and Danaidae). *Oriental Insects*, 27:151–198.

Varshney, R. K. 2010. *Genera of Indian Butterflies*. New Delhi, India: Nature Books India. p. 186.

Varshney, R. K., and P. Smetacek (eds.). 2015. *A Synoptic Catalogue of the Butterflies of India*. New Delhi, India: Butterfly Research Centre, Bhimtal and Indinov Publishing. p. 261.

Wahlberg, N., and C. W. Wheat. 2008. Genomic outposts serve the phylogenomic pioneers: Designing novel nuclear markers for genomic DNA extractions of Lepidoptera. *Systematic Biology*, 57:231–242.

Wahlberg, N., M. F. Braby, A. V. Z. Brower, et al. 2005. Synergistic effects of combining morphological and molecular data in resolving the phylogeny of butterflies and skippers. *Proceedings of the Royal Society B*, 272:1577–1586.

Wallace, A. R. 1865. On the phenomena of variation and geographical distribution as illustrated by the Papilionidae of the Malayan Region. *Transactions of the Linnean Society of London*, 25:1–71.

Wiley, E. O. 1978. The evolutionary species concept reconsidered. *Systematic Zoology*, 27:17.

Wiley, E. O., and B. S. Lieberman. 2011. *Phylogenetics: Theory and Practice of Phylogenetic Systematics*. 2nd edn. Hoboken, NJ: John Wiley & Sons. p. 406.

Williams, C. B. 1938. The migration of butterflies in India. *Journal of Bombay Natural History Society*, 40:439–457.

Wynter-Blyth, M. A. 1944a. The butterflies of the Nilgiris (part I). *Journal of the Bombay Natural History Society*, 44:536–549.

Wynter-Blyth, M. A. 1944b. The butterflies of the Nilgiris (part II). *Journal of the Bombay Natural History Society*, 45:47–61.

Wynter-Blyth, M. A. 1945. Addenda and corrigenda to "The butterflies of the Nilgiris" published in vols. XLIV and XLV of the journal. *Journal of the Bombay Natural History Society*, 45:613–615.

Wynter-Blyth, M. A. 1947. Additions to "The butterflies of the Nilgiris" published in vol. 44, no. 4 and vol. 45, no. 1. *Journal of the Bombay Natural History Society*, 46:736.

Yata, O. 1989. A revision of the Old World species of the genus *Eurema* Hübner (Lepidoptera; Pieridae). Part I. Phylogeny and zoogeography of the subgenus *Eurema* Hübner. *Bulletin of the Kitakyushu Museum of Natural History*, 9:1–103.

Yata, O. 1991. A revision of the Old World species of the genus *Eurema* Hübner (Lepidoptera; Pieridae). Part II. Description of the *smilax*, the *hapale*, the *ada* and the *sari* (part) groups. *Bulletin of the Kitakyushu Museum of Natural History*, 10:1–51.

Yata, O. 1992. A revision of the Old World species of the genus *Eurema* Hübner (Lepidoptera; Pieridae). Part III. Description of the *sari* group (part). *Bulletin of the Kitakyushu Museum of Natural History*, 11:1–77.

Yata, O. 1994. A revision of the Old World species of the genus *Eurema* Hübner (Lepidoptera; Pieridae). Part IV. Description of the *hecabe* group (part). *Bulletin of the Kitakyushu Museum of Natural History*, 13:59–105.

Yata, O. 1995. A revision of the Old World species of the genus *Eurema* Hübner (Lepidoptera; Pieridae). Part V. Description of the *hecabe* group (part). *Bulletin of the Kitakyushu Museum of Natural History*, 14:1–54.

Yata, O., J. E. Chainey, and R. I. Vane-Wright. 2010. The golden and mariana albatrosses, new species of pierid butterflies, with a review of subgenus *Appias* (Catophaga) (Lepidoptera). *Systematic Entomology*, 35:764–800.

Yates, J. A. 1946. The butterflies of the Nilgiris: A supplementary note. *Journal of the Bombay Natural History Society*, 46:197–198.

17 Taxonomy and Diversity of Indian Fruit Flies (Diptera: Tephritidae)

K. J. David, S. Ramani, and S. K. Singh

CONTENTS

Introduction	306
Family Diagnosis and Key Characters	306
Key Characters at Subfamily Level	306
Key Characters at Tribal Level	306
Subfamily and Tribes	306
Subfamily Tachiniscinae	307
Subfamily Blepharoneurinae	307
Subfamily Phytalmiinae	307
Subfamily Dacinae	307
Subfamily Tephritinae	307
Subfamily Trypetinae	307
Key to Subfamilies, Tribes, Genera, and Species in India	308
Key to Subfamilies of Tephritidae	308
Key to Tribes of Dacinae	308
Key to Genera under Dacini	308
Key to Important Subgenera and Species in *Bactrocera* Macquart	308
Key to Some Subgenera and Species in *Zeugodacus* Hendel	309
Key to Important Subgenera and Species in *Dacus* Fabricius	309
Key to Some Genera of Tribe Gastrozonini	309
Key to Some Common Genera of Phytalmiinae	309
Key to Some Common Genera of Trypetinae	310
Key to Some Common Genera of Tephritinae	310
Major Work on Indian Fauna	314
Biodiversity and Species Richness	314
Immature Taxonomy	315
Molecular Characterisation and Phylogeny	316
Integrative Taxonomy	317
Taxonomic Problems	317
Biology and Behaviour	317
Host Plant Associations	318
Economic Importance	320
Collection and Preservation	320
Collection of Fruit Flies	320
Processing and Preservation	320
Dissection of Adult Females	320
Processing of Larvae	321
Useful Websites	321
Conclusion	321
References	321

INTRODUCTION

Fruit flies (Diptera: Tephritidae) with nearly 5000 described species in 500 genera and six subfamilies, have immense economic importance, both due to the large number of pestiferous species, and, conversely, because of their many actual and potential weed biocontrol agents. Several species in this group are major pests of quarantine concern across the world mainly of the genera *Bactrocera*, *Dacus*, *Zeugodacus*, *Ceratitis,* and *Anastrepha*. The family is extremely diverse in many aspects, such as life history, mode of infestation, behaviour, and morphological and anatomical structure. Most of its member species, even the tiniest, are ornamented by brightly coloured body or wing patterns and are aesthetically pleasing (Freidberg 2006). Besides 35% of the species attacking soft fruits, larvae of 40% of the species develop in flowers of Asteraceae, and most of the remaining species are associated with flowers or their larvae are miners of leaf, stem, or root tissues (White and Elson-Harris 1992). They are characterised by a well developed frontal setae, costal break, incomplete subcostal vein (Sc), and extended cell cup. Of the six subfamilies known, Dacinae and Trypetinae are predominantly frugivorous and are pests of various horticultural crops, whereas Tephritinae are potential biocontrol agents for weeds except a few pestiferous species like *Acanthiophilus helianthi*. Nearly 20 species of fruit flies have been used as biocontrol agents of invasive weeds. In addition, they have been employed in genetic research for (Mediterranean fruit fly, *Ceratitis capitata*) and also in speciation studies (Apple maggot, *Rhagoletis pomonella*). In this chapter, apart from diagnostic characters and keys, biodiversity, zoogeography, biology, behaviour, and phylogeny of Tephritidae are also dealt with.

FAMILY DIAGNOSIS AND KEY CHARACTERS

Tephritidae are characterized by well developed frontal plates, frontal setae always much longer than surrounding setulae developed on the frontal plate (Figure 17.1), costal vein with a deep constriction or break before the apex of subcostal vein, two or three spines guarding such break, vein Sc abruptly bent forward at nearly 90°, weakened beyond bend and ending abruptly without joining the costa at subcostal break, cell bcu with an acute extension (Figure 17.53), dorsal side of vein R_1 with setulae, and wings usually patterned by coloured bands (White and Elson-Harris 1992; Korneyev 1999). Platystomatidae, Pyrgotidae, and Ulidiidae are often confused with Tephritidae. Pyrgotidae can be easily distinguished from Tephritidae by the presence of porrect antenna and tubular oviscape, whereas Ulidiidae can be separated by the presence of completely developed vein Sc and lack of well developed frontal setae. In Platystomatidae, cell bcu is closed by straight vein Cu_2 that is perpendicular to Cu.

Key Characters at Subfamily Level

- Scapular seta: It is present in all subfamilies of Tephritidae except Tephritrinae. Absence of scapular seta is one of the diagnostic characters of this subfamily
- Number of scutellar setae: Phytalmiinae, Tachiniscinae, and Blepharoneurinae possess six scutellar setae, whereas Dacinae, Trypetinae, and Tephritinae possess two to four scutellar setae
- Number of spermathecae: Dacinae and Trypetinae possess two spermathecae, whereas Trypetinae, Phytalmiinae, and Tachiniscinae possess three spermathecae
- Type of aculeus: Tactile type of aculeus with elongate sensory setae are found in Phytalmiinae, whereas cutting aculeus with acute apex are seen in other subfamilies
- Number of postpronotal lobe setae: All subfamilies of Tephritidae possess 0–1 postpronotal lobe seta, whereas Tachiniscinae and Blepharoneurinae possess two pairs of postpronotal lobe setae
- Anepisternal setae: In Blepharoneurinae, anepisternal seta anterior to phragma is present
- Epandrium: Shape of epandrium and lateral surstyli is significant in delineating subfamilies. In Phytalmiinae, epandrium and surstyli are often bar-shaped without distinct differentiation, whereas in Dacinae, epandrium is quadrate with a distinct surstylus. Epandrial shape (posterior view) is oval or circular without distinctly developed surstyli in Tephritinae, which is one of the diagnostic feature of the subfamily.

Key Characters at Tribal Level

- Depth of basal medial cell (bm): Basal medial cell (bm) is deeper than basal cubital cell (bcu) in the tribe Dacini compared to other tribes
- Extension of cell bcu: Extension of cell bcu is much longer than the length of cell bcu
- Shape of spermathecae: Spermatheca in the tribe Dacini are typically mulberry-shaped.

SUBFAMILY AND TRIBES

Hardy (1973, 1974) listed four subfamilies, namely, Dacinae, Schistopterinae, Tephritinae, and Trypetinae. Freidberg and Kugler (1989) included eight subfamilies in Tephritidae, namely, Aciurnae, Dacinae, Myopitinae, Oedaspidinae, Schistopterinae, Tephritinae, Terrellinae, and Trypetinae. Norrbom et al. (1999) listed only three subfamilies in his checklist, which includes Phytalmiinae, Trypetinae, and

Tephritinae. Korneyev (1999) included six subfamilies in Tephritidae based on phylogenetic analysis using several morphological markers, which is followed presently. The six subfamilies are Blepharoneurinae, Dacinae, Phytalmiinae, Tachiniscinae, Tephritinae, and Trypetinae.

SUBFAMILY TACHINISCINAE

The subfamily is characterised by the presence of: (i) a short-pubescent arista; (ii) 1–3 (usually 2) postpronotal setae; (iii) 1–2 postsutural supra-alar setae; (iv) oviscape opened posterodorsally; (v) eversible membrane of the ovipositor devoid of taeniae, but with basoventral area of dense, very dark, larger scales; and (vi) aculeus short and stiletto-like (Korneyev 1999). Members of this subfamily are represented in the Afrotropical, Oriental, Australasian, Palaearctic, and Neotropical Regions of the world. Two tribes, Tachiniscini Kertész and Ortalotrypetini Ito, are recognised, each with five genera. It is represented by a single species, *Ortalotrypeta isshiki* (Matsumura) in India (David and Hancock 2013)

SUBFAMILY BLEPHARONEURINAE

They can be distinguished by the presence of a single large setae just anterior to the phragma of anepisternum, three pairs of scutellar setae, two or three pairs of post pronotal setae, vanes of the phallapodeme broadly separate, scales on the anterior half of eversible membrane well developed, but mostly tri or bidentate with a large central tooth, and aculeus cutting or lobed (Condon and Norrbom 1994; 1999). It comprises of 5 genera and 34 species and is phytophagous. No member of this subfamily has been encountered in India.

SUBFAMILY PHYTALMIINAE

Phytalminae comprises species which are saprophytic in habit with a primitive tactile type of aculeus. Korneyev (1999) considered this as a monophyletic group in light of the following characters; in profile, epandrium is elongated in the dorsoventral direction, often bar-like, with lateral surstylus, hypandrium with lateral sclerites rudimentary, vanes of the phallapodeme articulated with anterior end of hypandrium, and abdominal sterna 4–6 without anterior apodeme. It is divided into 4 tribes and 100 genera with 369 species (Norrbom et al. 1999), of which a single tribe Acanthonevrini with 18 species and 8 genera are reported from India (Agarwal and Sueyoshi 2005). *Acanthonevra*, *Diarrhegma*, *Rioxa* and *Themara* are the genera reported from India.

SUBFAMILY DACINAE

It consists of three tribes, Gastrozonini, Dacini, and Ceratitidini, 47 genera and 1267 species in the world (Norrbom et al. 1999; Kovac et al. 2006; Doorenweerd et al. 2018).

Monophyly of the Dacinae is supported by the large size of male proctiger, which is usually larger than the epandrium, and only two spermathecae, larvae with numerous oral ridges, long sickle-like mouth hooks, and well developed creeping welts (Korneyev 1999). In India, 18 genera and 127 species under Gastrozonini and Dacini were reported (Agarwal and Sueyoshi 2005; David and Ramani 2011; David et al. 2014, 2016, 2017). *Acroceratitis*, *Anoplomus*, *Chaetellipsis*, *Dietheria*, *Gastrozona*, *Bactrocera*, *Zeugodacus*, and *Dacus* are some of the genera reported from India. The three pestiferous genera, *Ceratitis*, *Bactrocera*, and *Dacus*, are the members of this subfamily (White and Elson-Harris 1992).

SUBFAMILY TEPHRITINAE

They are mainly associated with flower heads of Asteraceae and are found in all world regions with 219 genera and 1847 species (Norrbom et al. 1999). They are characterized by dense grey microtrichose bodies and spotted wings, absence of scapular setae, presence of fine tomentosum on the anepisternum, which makes the suture obscure, and two spermathecae in females (Hardy 1973, 1974; Hardy and Drew 1996). Korneyev (1999) strongly believes that this subfamily is monophyletic based on the above morphological characters as well as the biology of the group, as all of them breed in plants of the family Asteraceae as flower feeders, gall formers, or stem miners. It is an economically important group with pestiferous species; *Acanthiophilus helianthi* (Rossi) on safflower as well as biocontrol agents, *Cecidochares connexa* Macquart on Siam weed, *Chromolaena odorata*, and *Procedidochares utilis* Stone on *Eupatorium adenophorum* that cause stem galls (White and Elson-Harris 1992). In India, it is well represented with 34 genera and 70 species (Agarwal and Sueyoshi 2005).

SUBFAMILY TRYPETINAE

They are predominantly frugivorous in their feeding habit except a few which are leaf and stem miners. They are characterized by wing with basal medial cell (bm) about the same width as posterior cubital cell or anal cell (cup) narrowed distinctly towards the base, apical extension of posterior cubital cell or anal cell (cup) broad and acute, wing often brown with or without hyaline spots and indentations, presence of scapular setae, and anepisternum with visible suture (Permkam and Hancock 1995). Korneyev (1999) suggests that the subfamily Trypetinae may be paraphyletic. Many species are economically important pests of commercial fruit crops such as *Anastrepha*, *Carpomya*, and *Rhagoletis* (Robinson and Hooper 1989; White and Elson-Harris 1992). There are 969 species under 229 genera and 9 tribes in the world, of which, 49 species under 19 genera are reported from India (Agarwal and Sueyoshi 2005). *Adrama*, *Euphranta*, and *Carpomya* are some of the genera reported.

KEY TO SUBFAMILIES, TRIBES, GENERA, AND SPECIES IN INDIA

KEY TO SUBFAMILIES OF TEPHRITIDAE

1. Scutum with two postpronotal lobe setae, two postsutural supra-alar setae (Figures 17.20, 17.32), oviscape opened posterodorsally ....**Tachiniscinae**
- Scutum with single (Figure 17.19) or no postpronotal lobe seta (Figures 17.11–17.13), single postsutural supra-alar seta (Figures 17.11–17.13), oviscape not as above **2**
2. Scutum without scapular seta, dorsocentral seta usually placed anterior to postsutural supra-alar seta/anterior supra-alar seta (Figure 17.23) ....**Tephritinae**
- Scutum with scapular seta (Figures 17.11–17.13), dorsocentral seta present or absent, if present, then aligned with or posterior to postsutural supra-alar seta (Figure 17.18)..**3**
3. Dorsocentral seta usually absent (Figures 17.11–17.13), wing predominantly hyaline with a costal band and narrow anal streak (Figures 17.54–17.60), if dorsocentral, seta present, then aligned with postsutural supra-alar seta (Figures 17.18, 19), wing variously marked, with longitudinal and transverse bands (Figure 17.53); scutellar setae one or two; female with two spermathece ....**Dacinae**
- Dorsocentral seta present, scutellar setae range from two to five (Figures 17.22 through 17.24), female usually with three spermathecae **4**
4. Three to five scutellar setae (Figures 17.22, 17.24); wing predominantly dark with hyaline spots and wedges (Figures 17.63, 17.64); tip of aculeus tactile with long bristles (Figure 17.51) [associated generally with dead and decaying plant material]......................................
....**Phytalmiinae (Acanthonevrini)**
- Two scutellar setae (Figure 17.21); wing predominantly hyaline with various dark markings (Figure 17.63); tip of aculeus pointed and of piercing type (Figures 17.47– 17.50) ....**Trypetinae**

KEY TO TRIBES OF DACINAE

1. Wing with cell bm deeper than cell bcu, extension of cell bcu longer than cell bcu, predominantly hyaline with costal band [dark band along costal margin extending from subcostal cell (sc) to wing apex] and anal streak (Figures 17.54–17.60); dorsocentral seta absent...**Dacini**
- Wing with cell bm as deep as cell bcu, extension of cell bcu short, wing variously marked with longitudinal and transverse bands (Figure 17.53); dorsocentral seta present; arista plumose ....
.... **Gastrozonini**

KEY TO GENERA UNDER DACINI

1. Abdominal tergites separate (tergites overlapping in lateral view) (Figures 17.38, 17.42).................................... 2
- Abdominal tergites fused (tergites not overlapping in lateral view) (Figures 17.40, 41) ....**Dacus Fabricius**
2. Medial postsutural vitta often present (Figures 17.25–17.29), sternite V of male with shallow posterior emargination (Figure 17.44)...................**Zeugodacus Hendel**
- Medial postsutural vitta not present (Figures 17.11–17.13), sternite V of male with deep concavity (Figure 17.43)............................**Bactrocera Macquart**

KEY TO IMPORTANT SUBGENERA AND SPECIES IN *BACTROCERA* MACQUART

1. Oviscape bottle-shaped, rounded in transverse section....................................subgenus *Tetradacus* Miyake.......abdomen petiolate (Figure 17.40), costal band broad slightly overlapping vein R_{4+5} ....
....*minax* **(Enderlein)**
- Oviscape dorsoventrally flattened (Figures 17.37, 17.41), not with bulbous base abdomen not petiolate............subgenus *Bactrocera*. 2
2. Face entirely black (Figure 17.3) ....
...................................*nigrofemoralis* **Tsuruta & White**
- Face with two separate black spots or transverse line or fulvous (Figures 17.2, 17.4, 17.8, 17.9)..........................3
3. Costal band continuous, extends from cell sc to wing apex (Figures 17.55–17.59)...4
- Costal band discontinuous (Figure 17.54)................... 10
4. Postsutural supra-alar seta present, aculeus tip pointed or trilobed (Figures 17.49, 17.50).................................5
- Postsutural supra-alar seta absent, aculeus tip forked (Figure 17.48) ....*digressa* **Radhakrishnan**
5. Prescutellar acrostichal seta absent, costal band overlapping vein R_{2+3} and expanded apically (Figure 17.56) ....*affinis* **(Hardy)**
- Prescutellar acrostichal seta present, costal band confluent or slightly overlapping vein R_{2+3} (Figure 17.55) 6
6. Scutum reddish-brown, scutellum with an apical black spot (Figure 17.12)............................*versicolor* **(Bezzi)**
- Scutum black-brown, scutellum without apical black spot (Figure 17.11) 7
7. All femora yellow (Figure 17.31)..................................8
- All or at least forefemur with black spots/markings (Figure 17.30) 9
8. Scutum with lateral postsutural vitta broad (>0.15 mm) (Figure 17.11), abdominal tergites III–V with black

lateral markings (Figure 17.37).......................
....*syzygii* **White & Tsuruta**
- Scutum with lateral postsutural vitta narrow (<0.15 mm) (Figure 17.21), abdominal tergites III–V without black lateral markings (Figure 17.36) ....
....*dorsalis* **(Hendel)**
9. Anepisternal stripe broad reaching notopleuron (Figure 17.33), aculeus tip trilobed in females (Figure 17.49), costal band overlapping vein R_{2+3}, breed on *Solanum* spp.................................*latifrons* **(Hendel)**
- Anepisternal stripe narrow (Figure 17.30), not reaching notopleuron, costal band confluent with R_{2+3}, aculeus tip not trilobed (Figure 17.50)..............*caryeae* **(Kapoor)**
10. Scutum black, face with two black spots coalesces to form a transverse line (Figure 17.2) .... *correcta* **(Bezzi)**
- Scutum reddish-brown, face with two separate black spots (Figure 17.29)..........................*zonata* **(Saunders)**

KEY TO SOME SUBGENERA AND SPECIES IN *ZEUGODACUS* HENDEL

1. Postsutural supra-alar seta absent (Figure 17.25) subgenus *Javadacus* **Hardy** face fulvous without any markings (Figure 17.8)............................*trilineatus* **(Hardy)**
- Postsutural supra-alar seta present (Figures 17.28, 17.29) 2
2. Pecten absent on tergite III of males (Figure 17.38) subgenus *Hemigymnodacus* **Hardy** face fulvous without any markings in males, females with black transverse band on face ....*diversus* **(Coquillett)**
- Pecten present on tergite III of males (Figure 17.36) subgenus *Zeugodacus* **Hendel**......................................3
3. Scutum and abdomen predominantly black, scutellum with an apical black spot (Figure 17.26), costal band narrow constricted beyond cell r_1 and expanded slightly towards apex..................................... *scutellaris* **(Bezzi)**
- Scutum black/reddish-brown, abdomen not black, reddish-brown or fulvous with various markings; costal band usually broad confluent with R_{4+5}, if narrow and confluent with R_{2+3}, not constricted beyond cell r_1........ 4
4. Costal band expanded into an apical spot, wing with subapical band and radial-medial band (Figure 17.59) ....*cucurbitae* **(Coquillett)**
- Costal band expanded into an apical spot, wing without subapical and radial-medial band (Figure 17.58 ....5)
5. Face with transverse black band (Figure 17.2) ....
.... *caudatus* **(Fabricius)**
- Face with two separate black spots (Figure 17.9).......... 6
6. Scutellum with an apical black spot (Figure 17.28), aculeus with preapical indentations (Figure 17.47) ....
.... *watersi* **(Hardy)**
- Scutellum without an apical black spot (Figure 17.27), aculeus pointed..7
7. Anepisternal stripe broad, inverted L-shaped (Figure 17.35); costal band confluent with vein R_{2+3} ....*gavisus* **(Munro)**
- Anepisternal stripe not inverted L-shaped (Figure 17.34); costal band overlapping vein R_{2+3} expanded into an apical spot .... *tau* **(Walker)**

KEY TO IMPORTANT SUBGENERA AND SPECIES IN *DACUS* FABRICIUS

1. Combined length of scape and pedicel longer than vertical length of face (Figure 17.5), wing with broad costal band and radial-medial band (Figure 17.60); abdomen petiolate or club-shaped ....
......................*D.* (*Mellesis*) *ramanii* **Drew & Hancock**
- Combined length of scape and pedicel as long as or shorter than face (Figure 17.4), wing with narrow costal band, without radial medial band; abdomen oval 2
2. Medial postsutural vitta present; scutum grey-brown to black (Figure 17.16), elongate oviscape (2.7 mm) with a bulbous base and narrow neck, aculeus elongate (3.5–4.0 mm)..............*D.* (*Leptoxyda*) *persicus* **Hendel**
- Medial postsutural yellow vitta absent; scutum reddish-brown (Figure 17.15), oviscape short (1.2 mm) triangular, dorsoventrally flattened, aculeus short (1.4 mm) ....*D.* (*Didacus*) *ciliatus* **(Loew)**

KEY TO SOME GENERA OF TRIBE GASTROZONINI

1. First flagellomere dorsoapically pointed (Figure 17.1), aculeus tip, pointed......................*Acroceratitis* **Hendel**
- First flagellomere not dorsoapically pointed (Figure 17.6), aculeus tip bifid or trilobed....................2
2. Scutum yellow with longitudinal black bands, 1 scutellar seta (Figure 17.18); aculeus tip bifid (Figure 17.52) ....*Dietheria* **Hardy**
- Scutum black/reddish-brown without black bands, 2 scutellar setae (Figure 17.19), aculeus tip often trilobed...*Gastrozona* **Bezzi**

KEY TO SOME COMMON GENERA OF PHYTALMIINAE

1. Scutum predominantly dark brown with yellow triangular patch from dorsocentral seta to posterior margin; scutellum yellow (Figure 17.17) ....
....*Diarrhegma* **Bezzi**
- Scutum reddish-brown or fulvous with black or brown longitudinal vittae; scutellum not completely yellow (Figures 17.22, 17.24) 2

2. Scutum with two longitudinal brown vittae extending up to the apex of scutellum (Figure 17.22); wing predominantly dark brown with hyaline spots in apical and subapical margin (Figure 17.64).............***Rioxa* Walker**
- Scutal characters and wing markings variable 3
3. Wing with vein M and Cu$_1$ setose, males with broadened eyes (Figure 17.7)......................***Themara* Walker**
- Wing with vein M and Cu$_1$ not setose, male eyes not broadened.............................***Rioxoptilona* Hendel**

Key to Some Common Genera of Trypetinae

1. Scutum black with or without medial vittae (Figure 17.21); scutum devoid of black spots; wing hyaline with black bands (Figure 17.65).............................2
- Scutum yellowish-brown with black spots laterally; scutellum with black marginal spots (Figure 17.14); wing hyaline with yellowish bands..............................***Carpomya* Costa**
2. Scutum often with a broad medial yellow postsutural vitta; without presutural setae, anatergite with fine erect hairs, wing usually hyaline with longitudinal/transverse bands (Figure 17.65)........................... ***Euphranta* Loew**
- Scutum black without yellow vitta (Figure 17.21), presutural setae well developed; 0.5 of wing black with hyaline apex (Figure 17.66); lateral surstylus with broadly flattened anterior lobe and elongate posterior lobe..................
....***Philophylla* Rondani**

Key to Some Common Genera of Tephritinae

1. Scutum glossy black; wing predominantly black with hyaline wedges (Figures 17.61, 17.62) 2
- Scutum grey pollinose not black (Figure 17.23); wing mostly hyaline with dark bands/spots especially towards apex (Figures 17.67, 17.68) 3
2. Wing with broad transverse hyaline band from anterior to posterior margin distal of cross veins (Figure 17.62); single orbital seta .... ***Procecidochares* Hendel**
- Wing without a broad transverse hyaline band distal of cross veins (Figure 17.61); two orbital setae .... ***Cecidochares* Bezzi**
3. Wing with few brown bands and hyaline spots towards apex (Figure 17.67), three pairs of frontal setae, proboscis not geniculate..................... ***Acanthiophilus* Becker**
- Wing with dark bands and hyaline spots almost for entire posterior half portion (Figure 17.68), two pairs of frontal setae, proboscis geniculate..........................***Dioxyna* Frey**

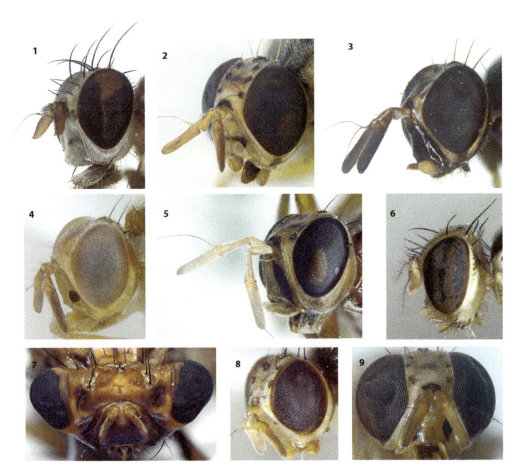

FIGURES 17.1–17.9 Heads of Tephritidae **1**. *Acroceratitis tomentosa*; **2**. *Bactrocera correcta*; **3**. *B. nigrofemoralis*; **4**. *Dacus ciliatus*; **5**. *D. sphaeroidalis*; **6**. *Gastrozona fasciventris*; **7**. *Themara yunnana*; **8**. *Zeugodacus trilineata*; **9**. *Z. cucurbitae*. (Figure 17.7 from David, K. J., and Ramani, S., *Zootaxa*, 3021, 1–31, 2011. with permission from Zootaxa www.mapress.com/j/zt.)

FIGURES 17.10–17.29 Thorax (dorsal view) **10**. *Acroceratitis tomentosa*; **11**. *Bactrocera syzygii*; **12**. *B. versicolor*; **13**. *B. dorsalis*; **14**. *Carpomya vesuviana*; **15**. *Dacus ciliatus*; **16**. *D. persicus*; **17**. *Diarrhegma modestum*; **18**. *Dietheria fasciata*; **19**. *Gastrozona fasciventris*; **20**. *Ortalotrypeta isshikii*; **21**. *Philophylla indica*; **22**. *Rioxa sexmaculata*; **23**. *Scedella spiloptera*; **24**. *Themara yunnana*; **25**. *Zeugodacus trilineatus*; **26**. *Z. scutellarius*; **27**. *Z. tau*; **28**. *Z. watersi*; **29**. *Z. caudatus*. (Figure 17.20 from David, K.J. and Hancock, D.L., *Aust. Entomol.*, 40, 131–135, 2013. With permission from Queensland Entomological Society; Figures 17.22, 17.24 from David, K. J., and Ramani, S., *Zootaxa*, 3021, 1–31, 2011. With permission from Zootaxa www.mapress.com/j/zt.)

FIGURES 17.30–17.42 Thorax (lateral view) **30**. *Bactrocera caryeae*; **31**. *Zeugodacus gavisus*; **32**. *Ortalotrypeta isshikii*; **33**. *B. latifrons*; **34**. *Z. tau*; **35**. *Z. gavisus*. Abdomen (dorsal view) **36**. *Bactrocera dorsalis*; **37**. *B. syzygii*; **38**. *Zeugodacus diversus*; **39**. *Dacus discophorus*; **40**. *B. minax*; **41**. *D. ciliatus*; **42**. *Z. havelockiae*. (Figure 17.32 from David, K.J. and Hancock, D.L., *Aust. Entomol.*, 40, 131–135, 2013. With permission Queensland Entomological Society.)

Taxonomy and Diversity of Indian Fruit Flies (Diptera: Tephritidae)

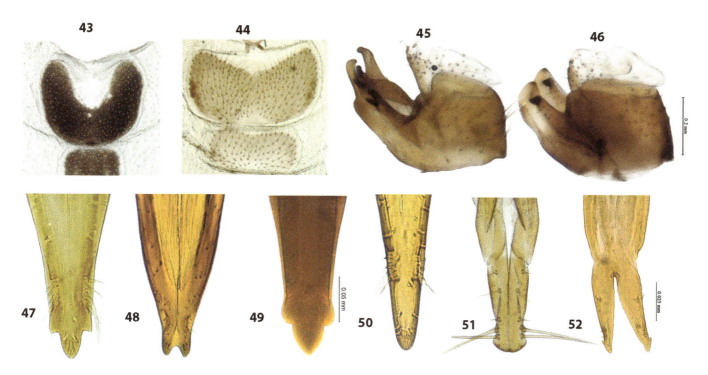

FIGURES 17.43–17.52 Sternite V of male **43**. *Bactrocera aethriobasis* and **44**. *Z. cucurbitae*. Epandrium and surstyli of Dacini **45**. *Z. brevipunctatus* **46**. *B. affinis*. Aculeus tip **47**. *Zeugodacus watersi*; **48**. *B. digressa*; **49**. *B. latifrons*; **50**. *B. caryeae*; **51**. *Tritaeniopteron punctatipleura*; **52**. *Dietheria fasciata*.

FIGURES 17.53–17.60 Wings **53**. *Acroceratitis ceratitina*; **54**. *Bactrocera correcta*; **55**. *B. dorsalis*; **56**. *B. affinis*; **57**. *B. latifrons*; **58**. *Zeugodacus tau*; **59**. *Z. cucurbitae*; **60**. *Dacus ramanii*.

FIGURES 17.61–17.68 Wings **61**. *Cecidochares connexa*; **62**. *Procecidochares utilis*; **63**. *Rioxoptilona* sp.; **64**. *Rioxa sexmaculata*; **65**. *Euphranta crux*; **66**. *Philophylla indica*; **67**. *Acanthiophilus helianthi*; **68**. *Dioxyna sororcula*. (Figures 61, 62, & 67 from David, K. J., and Ramani, S., *Zootaxa*, 3021, 1–31, 2011. With permission from Zootaxa www.mapress.com/j/zt.)

MAJOR WORK ON INDIAN FAUNA

Bezzi (1913, 1915, 1916), Senior-White (1921, 1922, 1924), Munro (1935, 1938, 1939), Perkins (1938), Hering (1938, 1941, 1956), Hardy (1971), Kapoor (1971), Kapoor et al. (1980), White and Hancock (1997), and Hancock and Drew (2004, 2005) studied the tephritid fauna of India. Kapoor (1993) provided a checklist and key for species from India. Drew and Raghu (2002) recorded 21 species of dacines from Western Ghats and provided a key to 56 species of *Bactrocera* from the Indian subcontinent. Agarwal and Sueyoshi (2005) published a catalogue for 243 species of fruit flies in 74 genera and 18 tribes from India. An illustrated key and a checklist for 126 species of fruit flies from peninsular India and Andaman and Nicobar Islands were published by David and Ramani (2011). Seventeen species of fruit flies from India have been described and 14 species were recorded for the first time from India by David et al. (2013, 2014a, 2014b, 2015, 2016, 2017), David and Hancock (2013, 2017), David and Singh (2015), Drew and Romig (2013) described ten new species and gave distributional records for several species from India. A key to fruit flies of the tribe Dacini from Southeast Asia by Drew and Romig (2016) contains several Indian species and is very useful for identifying species from the country. The Pest Fruit Fly Identification System developed by Carroll et al. (2002, 2004) includes interactive keys for adults and for larvae. The adult key includes data and images to identify nearly 190 economically important species of fruit flies from throughout the world including several species from India. Virgilio et al. (2015) elevated *Zeugodacus* to generic status based on phylogenetic analysis using molecular markers. Norrbom et al. (1999) provided a world checklist of Tephritidae, whereas Doorenweerd et al. (2018) published a world checklist for the tribe Dacini, which includes several species from India. It is very evident from Table 17.1 that major faunal contributions in India were made by Bezzi, Drew and Hancock, Hering, Hardy, and Kapoor.

BIODIVERSITY AND SPECIES RICHNESS

Species diversity over the major zoogeographic regions indicates that diversity is greatest in the tropics (Table 17.2). The Oriental region appears to have the largest diversity in both genera and species (975 species, 155 genera); the Afrotropical region is a close second (929 species, 154 genera) and may eventually prove to have more genera and species. In terms of number of genera, the Palearctic, Australasian, Neotropical, and Nearctic regions are similar, except that the Palearctic has fewer genera than the Australasian. The Nearctic fauna is by far the least diverse, in both genera and species, followed by the Neotropical, although it has almost as many species as the Australasian (Norrbom 2004).

Nearly one-third of fruit fly species in the Oriental region are in India, with 284 described species in 88 genera and 5 subfamilies (Table 17.3). It is as diverse as the Nearctic region which has nearly 365 species in 62 genera. Of the six subfamilies, Dacinae is more predominant in India

TABLE 17.1
Taxonomists and Their Contribution to the Indian Fauna of Tephritidae

Period	Taxonomist	Species	Cumulative No. of Species
1758–1800	Linnaeus	1	
	Fabricius	6	
			7
1801–1850	Wiedemann	9	
	Macquart	4	
			13
1851–1900	Walker	9	
	Coquillett	2	
	Loew	2	
	Wulp	4	
	Schiner	7	
	Rondani	3	
			27
1901–1950	Bezzi	46	
	Hendel	14	
	Hering	25	
	Munro	12	
	Zia	5	
	Meijere	5	
	Enderlein	4	
	Senior-White	4	
	Brunetti	3	
	Shiraki	2	
	Perkin and May	2	
			124
1951–2000	Hardy	23	
	Drew and Hancock	27	
	Kapoor and co-workers	17	
	Radhakrishnan	2	
	Han	2	
			71
2001–Present	Tsuruta and White	5	
	David and co-workers	16	
	Drew and Raghu	8	
	Drew and Romig	7	
	Hancock and Whitmore	1	37

with 119 described species (nearly 42 percent of Indian fauna), followed by Tephritinae (69 species), and Trypetinae (65 species). The subfamily Phytalmiinae is represented by 18 species in 10 genera in a single tribe, Acanthonevrini. The subfamily Tachiniscinae is represented by a single species in India (David and Hancock 2013). The zoogeographic distribution indicates that the species diversity is almost similar in the eastern (67 species) and southern (67 species) regions, followed by the northern (44 species) region. The north-east-south region and north-east region showed a roughly identical, maximum percentage of overlapping species diversity (6%), followed by the north-south (5%) and east-south (4%) regions (Table 17.4).

TABLE 17.2
Genera and Species of Tephritidae by Zoogeographic Regions

Region	Genera		Species[b]		Species/Genus
	Valid	Native	Valid	Native	
World[a]	481	481	4,352	4,352	9.05
Afrotropical	154	151	929	923	6.11
Australasian	145	139	785	772	5.55
Nearctic	62	60	361	345	5.75
Neotropical	72	70	747	745	10.64
Oriental	157	155	975	970	6.26
Palearctic	137	135	882	875	6.48

Source: Norrbom, A.L., Fruit fly (Diptera: Tephritidae) Classification & Diversity. Fruit fly taxonomy pages, 2004.

[a] World totals are not the sum of the regional totals as some genera and species occur in more than one region.

[b] The average species per genus for the regions are based on native taxa only.

TABLE 17.3
Diversity of Tephritidae in India

Subfamilies	Tribe	No. of Genera	No. of Species
Dacinae	Dacini	3	89
	Gastrozonini	15	40
Phytalmiinae	Acanthonevrini	10	18
Trypetinae	Adramini	9	30
	Carpomyini	2	3
	Callistomyiini	1	1
	Trypetini	11	31
Tephritinae	Cecidocharini	2	2
	Dithrycini	2	6
	Eutretini	1	1
	Myopitini	1	1
	Noeetini	1	1
	Pliomelaenini	4	5
	Schistopterini	3	4
	Tephrellini	6	10
	Tephritini	12	33
	Terellini	3	5
	Xyphosini	1	1
Tachiniscinae	Ortalotrypetini	1	1
Total		**88**	**284**

IMMATURE TAXONOMY

In the case of fruit flies, morphological characters of third instar larva and egg chorionic sculpturing are often used. In the third instar larva, characters like facial mask, number of oral ridges, number of segments of antennal lobes, maxillary organ, preoral organ, chaetotaxy of creeping welt, shape of anal lobes, cephalopharyngeal skeleton, and anterior posterior spiracular characters are often considered important in segregating species. In the cephalopharyngeal skeleton, shape of mouth hook, presence/absence of preapical tooth, dental sclerite, hypopharyngeal sclerite, labial sclerite, length of parastomal bar, presence/absence of anterior sclerite, dorsal arch, and dorsal and ventral cornua are the other characters which show variations among subfamilies/tribes/genera/species. In spite of the economic importance of the family, immature stages of only 7% of the species and 17% of the genera of Tephritidae have been described (Norrbom et al. 1999). Knowledge of the immature stages of five most economically important genera is limited; *Anastrepha* (10%), *Bactrocera* (9%), *Ceratitis* (13%), *Dacus* (4%), and *Rhagoletis* (51%). White and Elson-Harris (1992) illustrated and provided keys to larvae of 12 genera of phytophagous fruit flies, 8 species of *Anastrepha*, 22 species of *Bactrocera*, 2 species of *Carpomya*, 8 species of *Ceratitis*, 3 species of *Dacus,* and 13 species of *Rhagoletis*. The Pest Fruit Fly Identification System developed by Carroll et al. (2004) includes interactive keys for larvae. The website contains illustrations, descriptions of larval morphology, and keys to third instar larvae of 80 species of fruit flies in two subfamilies, namely, Dacinae and Trypetinae, of which, 13 species are found in India. David (2016) described, illustrated, and provided keys to 12 species of Dacinae. Chorionic microsculpturing of four species of the Dacini were illustrated and described by Singh and Ramamurthy (2010). In the case of Phytalmiinae, larvae of only four species have been described. Best studied groups include Carpomyini, Tephritini, and Terellini. In Gastrozonini, *Ichneumonopsis burmensis*, *Acrotaeniostola spiralis,* and five species *Acroceratitis* have been illustrated and described by Kovac et al. (2013, 2017) and Schneider et al. (2018), respectively.

TABLE 17.4
Region Wise Distribution of Species of Tephritidae in India

Regions[a]	% of Overlap	Percentage	No. of Species
All four	4.38	4	12
Unknown	3.85	4	10
Northern, Eastern, and Southern	5.84	6	16
Northern, Eastern, and Central	0.36	0	1
Northern, Central, and Southern	2.55	3	7
Eastern and Southern	4.38	4	12
Northern and Eastern	5.47	6	15
Northern and Southern	5.11	5	14
Northern and Central	0.73	1	2
Central and Southern	1.46	2	4
Northern	16.06	16	44
Eastern	24.45	24	67
Central	1.09	1	3
Southern	24.45	24	67

[a] **Northern region**: Geographically, North Himalaya, Indo-Gangetic Plain, Aravalli Range and politically Bihar, Haryana, Himachal Pradesh, Jammu and Kashmir, Jharkhand, Punjab, Uttar Pradesh (including Uttarakhand), Chandigarh, and National Capital Territory of Delhi
Southern region: Geographically, Deccan Plateau left and right coastal ranges, Western Ghats and Eastern Ghats and politically Andhra Pradesh, Goa, Tamil Nadu, Kerala, Karnataka, Andaman and Nicobar Islands, Lakshadweep, and Puducherry
Central region: Geographically, Satpura, Bindha, Thar Desert, left and right coastal ranges and politically Maharashtra, Gujarat, Madhya Pradesh, Chhattisgarh, Rajasthan, Orissa, Dadra and Nagar Haveli, and Daman and Diu
Eastern region: Geographically, North Eastern Himalaya, Garo Range, lower Indo-Gangetic and Bramhaputra Plane and politically Arunachal Pradesh, Assam, Manipur, Meghalaya, Mizoram, Nagaland, Sikkim, Tripura, and West Bengal.

MOLECULAR CHARACTERISATION AND PHYLOGENY

Han and Ro (2016) analysed 53 tephritoid species belonging to 10 families and 30 outgroup (14 families) species using four mitochondrial genes, namely, COI, COII, 16s rRNA, and 12s rRNA. Their study revealed that all families of Tephritoidea are monophyletic except Tephritidae. Tephritidae branched along with Tachiniscinae, Ctenostylidae, and Pyrgotidae. Korneyev (1999) analysed 149 terminal taxa and 112 characters of the major subgroups of the family Tephritidae. His studies could not satisfactorily resolve the relationships among the subfamilies and tribes of Tephritidae. Condon and Norrbom (1999) investigated phylogenetic relationships between 5 genera of Blepharoneurinae using 22 morphological characters with 8 outgroup species. Smith and Bush (1999) analysed the relationships between 87 taxa in 8 genera of Carpomyini using 77 morphological features. Han (1999) studied the phylogenetic relationships of 35 genera of the tribe Trypetini using both morphological and molecular markers. White (1999) analysed phylogenetic relationships of 51 pest species of Dacini using 38 morphological characters. His studies revealed *Monacrostichus* to be a sister group of *Bactrocera* and *Dacus*. The subgenus *Tetradacus* Miyake was found to be a sister group of *Bactrocera*, while the subgenus *Callantra* of *Dacus* was closely allied to *Monacrostichus* owing to the wasp-waisted abdomen. The *Bactrocera* group of subgenera were paraphyletic to *Dacus* and *Zeugodacus* groups of subgenera in his study. Phylogenetic studies on Dacini using molecular markers were explored by several workers; studies using 34 species of Tephritidae by Smith et al. (2002) confirmed the monophyly of the genus *Dacus* and *Bactrocera*, Smith et al. (2003) also reported monophyly of the *Bactrocera* subgenus group and stated the paraphyletic status of the subgenus *Zeugodacus* in their analysis. Jamnongluk et al. (2003) and Nakahara and Muraji (2008) reported separate branching of *Bactrocera* and *Zeugodacus* subgenus groups in their studies, and polyphyletic branching of species of the subgenus *Zeugodacus* was reported by Jamnongluk et al. (2003). Han and Ro (2009) reconfirmed monophyly of the subfamily Dacinae using 12s, 16s, and COII gene fragments. Asokan et al. (2011) and Yong et al. (2015) revealed monophyly of *Bactrocera* using COI and 13 protein coding genes, respectively. *Bactrocera* (*Bactrocera*) and *Bactrocera* (*Zeugodacus*) were paraphyletic in the trees generated by Zhang et al. (2010) using COI and 16s RNA. The phylogenetic tree of the tribe Dacini generated by Krosch et al. (2012) using molecular markers, namely, 16S, COI, COII, and white eye genes, had two major clades: *Bactrocera* sensu stricto and the *Zeugodacus* group of subgenera; the *Zeugodacus* clade branched as sister group to *Dacus*, not *Bactrocera*. White (2006) also clearly stated the sister group relationship between *Zeugodacus* group and the genus *Dacus*. The phylogenetic studies on the subfamily Dacinae by Virgilio et al. (2015), using four mitochondrial and one nuclear gene fragment, proved the monophyly of tribes Dacini and Ceratitidini. *Dacus*, *Zeugodacus* subgenus of *Bactrocera,* and all other subgenera of *Bactrocera* formed separate clades in their studies; hence the subgenus *Zeugodacus* was raised to generic level. San Jose et al. (2018) could confirm monophyly of *Bactrocera*, *Dacus*, and *Zeugodacus* based on 7 genes in 167 species of the tribe Dacini. Phylogenetic revision of *Acanthiophilus* and *Tephritomyia* were done by Morgulis et al. (2015, 2016). Barcodes for 28 species of Dacinae from India were generated by David (2016).

INTEGRATIVE TAXONOMY

In fruit flies, a classic example in this regard would be the work by Schutze et al. (2015), where they have integrated a series of traits (morphological, molecular, chemecology, cytological, and sexual compatibility) to study the variations between several species of *Bactrocera dorsalis* complex. Through their studies they were able to confirm that *Bactrocera dorsalis*, *B. invadens*, and *B. papayae* are one and the same, i.e., *B. dorsalis*. Thus, *B. invadens* and *B. papayae* were synonymised with *B. dorsalis*. De Meyer et al. (2015) adopted an integrative approach employing morphology, cuticular hydrocarbons, pheromones, microsatellites, developmental physiology, and geographical distribution to resolve the cryptic species complex within *Ceratitis* FAR, i.e., *C. fasciventris*, *C. anonae,* and *C. rosae*. In *Anastrepha fraterculus* species complex, six species have been identified/delineated based on morphological, molecular, cytological, sexual behaviour, post zygotic compatibility, pheromones, cuticular hydrocarbons, distribution, and host range (Hendrichs et al. 2015)

TAXONOMIC PROBLEMS

Higher level classification in Tephritidae is not fully resolved as per Korneyev (1999). His studies using morphological markers could not resolve the relationships among subfamilies and tribes of Tephritidae. This was mainly due to the higher level of homoplasy associated with characters selected and lack of information regarding the postabdominal structures and immature stages. Even though six subfamilies have been recognized by Korneyev (1999) based on cladistic analysis, the basic branching of Tephritidae could not be resolved, although the subfamilies Tachiniscinae, Blepharoneurinae, and Tephritinae, and most tribes considered were well supported monophyletic groups. In fruit flies, identification problems have always been centred on species complexes of the tribe Dacini, which are notorious pests of quarantine concern. Of the several species complexes, *Bactrocera dorsalis* complex, is of great significance owing to its economic importance and difficulty in identification. It has been known since early this century that the Oriental fruit fly was more than just a single species. Its greatest diversity is seen in the islands of the Indonesian Archipelago with a decline in diversity to the east towards Australia and west towards India (Clarke et al. 2005). As many as 83 species are in this complex (Clarke et al. 2005; Ramani et al. 2008; Drew and Romig 2013, 2016), of which, 14 species were recorded from India (David and Ramani 2011; David et al. 2017). Five of these species have become adventive outside these regions (*B. carambolae* in Suriname, French Guyana, and northern Brazil, *B. dorsalis* in Hawaii and Mauritius, *B. papayae* in Irian Jaya, Papua New Guinea, and northern Queensland, Australia, *B. philippinensis* in Australia, and recently *B. invadens* from Sri Lanka that has invaded several parts of Africa) (Hancock and Drew 1994; Drew and Raghu 2002; Clarke et al. 2005; Drew et al. 2005). Still there is no consensus regarding the identity of *Bactrocera dorsalis* and *B. invadens*. Drew and Romig (2016) treat them as separate valid species even after synonymisation of the *B. invadens* with *B. dorsalis* by Schutze et al. (2015). Same is the case with acceptance of *Zeugodacus* as a separate genus; Krosch et al. (2012), Virgilio et al. (2015), and San Jose et al. (2018) proved that *Zeugodacus* is a separate genus based on molecular methods, but Hancock and Drew (2018) consider it as a subgenus of *Bactrocera*.

BIOLOGY AND BEHAVIOUR

The larvae of most species of fruit flies develop in the seed bearing organs of plants, and about 35% of species attack soft fruits, including many commercial fruits. Besides attacking soft fruit, the larvae of about 40% of species develop in the flowers of Asteraceae, and most of the remaining species are associated with flowers of other families, or their larvae are miners in leaf, stem, or root tissue (White and Elson-Harris 1992). Fruit flies are synovigenic, i.e., female flies require proteinaceous diet for egg maturation and oviposition after emergence. It has got three larval instars and a puparial stage. Most of the pestiferous species are multivoltine (10–12 generations per year), except *B. minax*, which is univoltine.

> **Mimicry**: Fruit flies do mimic wasps and jumping spiders of the family Salticidae. The papaya fruit fly, *Toxotrypana curvicauda*, appears to mimic social wasps and may gain some protection from predators by this mimicry. They mimic species of *Polistes* and *Ischocyttarus* in Florida and other species in Central

America (Landolt 1984). It was noted by Mather and Roitberg (1987) and Greene et al. (1987) that the tephritids, *R. zephyria* and *Z. vittigera,* resemble salticid jumping spiders.

- **Mating trophallaxis**: This is relatively rare and has been reported for only about 20 species. Premating trophallaxis is the most widespread kind of mating trophallaxis in Tephritidae. Freidberg (1981) studied this in several species of Tephritinae, namely, *Shistopterum moebiusi, Stenopa vulnerata* (Loew), *Icrerica seriata* (Loew), *Eutreta novaeboracensis* (Fitch), and *Dirioxa pornia* (Walker). Post mating trophallaxis is reported in *Spathulina sicula* (Sivinski et al. 1999).
- **Lek behaviour**: It is a type of non-resource based aggregation behaviour exhibited by fruit flies to encounter mates especially in polyphagous species. Usually lekking will be on non-host plants in order to avoid predation except in the case of *Rhagoletis* (Landolt 1999).
- **Sexual dimorphism**: Fruit flies exhibit variations in morphological characters between sexes ranging from facial marking to widely protruded eye lobes. In the tribe Dacini, it has been noted in few species, for example, *Zeugodacus (Hemigymnodacus) diversa* (Coquillett), in which males possess a fulvous face devoid of any markings, whereas females have a black band. In males of the species of *Themara* Walker (Phytalmiinae), the head is hammer-shaped due to the protrusion of eye-stalks (Hancock 2013). This has been recorded in species of *Vidalia, Stemonocera,* and *Phytalmia* also.

HOST PLANT ASSOCIATIONS

The majority of fruit flies are highly host-specific except polyphagous pestiferous taxa and saprophytes. Species of the tribe Dacini are predominantly frugivorous with 932 described species in 5 genera from around the world (Doorenweerd et al. 2018), of which, 85 species in three genera are known from India (Agarwal and Sueyoshi 2005; David and Ramani 2011; Drew and Romig 2013; David et al. 2016; David et al. 2017). Of the other two tribes of the subfamily Dacinae, Ceratitidini is not found in India, whereas Gastrozonini are well represented in India, which are associated with plants of the family Poaceae. *Bactrocera, Dacus,* and *Zeugodacus* are the predominant genera in tropical regions. Host plant details of these three genera are presented in Table 17.5. The data reveal that in the genus *Bactrocera, dorsalis, correcta, carambolae,* and *zonata,* are highly polyphagous with respect to host plant families and number of species, whereas the polyphagous species in *Zeugodacus, cucurbitae,* and *tau,* are mainly associated with several host plants of Cucurbitaceae. Among the pestiferous species, *B. minax* is highly specific to host plants of Rutaceae. *Dacus ciliatus* and *D. longicornis* are the two pest species in the genus *Dacus* specific to Cucurbitaceae, whereas several monophagous species are associated with Asclepediaceae.

In Dacinae, only the tribe Gastrozonini has acquired bamboo as its host. The Gastrozonini are considered to be a monophyletic tribe with 141 described species placed in 25 genera (Wang and Chen 2002; Kovac et al. 2006; De Meyer 2006; Dohm et al. 2014). Larvae of Gastrozonini mainly feed on the shoot meristem or other soft tissue of living or dead shoots. Most species of Phytalmiinae appear to be saprophagous, although Hardy (1986) and Permkam and Hancock (1995) reported that species of *Clusiosoma, Clusiosomina, Cheesmanomyia,* and *Rabaulia* are known to infest fruit of *Ficus* spp. *Dirioxa pornia* (Walker) has been reared from fruits of a wide variety of damaged or decaying fruits and also from fallen *Araucaria* cones (Permkam and Hancock 1995). Species of *Rioxoptilona, Felderimyia, Polyara,* and *Ptilona* have been reared from damaged bamboo shoots, and species of various other genera have been collected on bamboo and may breed in it (Hardy 1986; Hancock and Drew 1995). Larvae of *Austronevra, Dacopsis, Diarrhegma, Diarrhegmoides, Lumirioxa,* and *Phytalmia* develop in decomposing tree trunks or rotting parts of trees, although individual species of at least *Phytalmia* are fairly specific to certain types of trees (Hardy 1986; Permkam and Hancock 1995). Few species of *Afrocneros* and *Ocnerioxa* have been found under the bark of living trees (Munro 1967). *Territorioxa termitoxena* (Bezzi) breeds in termite galleries in tree trunks (Hill 1921). Trypetinae is the most diverse subfamily in terms of larval feeding habits. Adramini, which are restricted to the Old World except for two species of *Euphranta* from North America, breed in fruits, seeds, flower buds, or stems (Hardy 1986; White and Elson-Harris 1992; Hancock and Drew 1994), although one species, *E. toxoneura* (Loew), is a predator of sawfly larvae within their galls (Kopelke 1984). The tribe Carpomyini, breed exclusively in fruits, and most species are highly host specific (White and Elson-Harris 1992). The tribe Toxotrypanini, which are restricted to the Neotropical region, also breed almost exclusively in fruits or in the seeds within; some species are generalists, but many are specialized on specific plant groups such as *Passiflora* or latex-bearing families such as Sapotaceae, Moraceae, Apocynaceae, Caricaceae, or Asclepiadaceae. The tribe Rivelliomimini, which occur only in tropical Africa, Asia, and the Pacific, are poorly known biologically; only one species appears to have been reared, from *Cycas* spp. (White and Elson-Harris 1992). Trypetini is most diverse in the Oriental and Palaearctic regions. Species of this tribe are mostly leaf or stem miners or fruit breeders, although the small subtribe Acidoxanthina which is sometimes included in the Trypetini includes flower breeders. The two species of Zaceratini, one Afrotropical and one Palearctic, are stem borers (White and Elson-Harris 1992). The larvae of Tephritinae predominantly infest flower heads of the Asteraceae, the largest and the most advanced and widespread family of the angiosperms.

TABLE 17.5
Host Plants of Important Species in Tribe Dacini

Fruit Fly Species[a]	Host Plant Families/Species
B. albistrigata (6 families 11 species)	Anacardiaceae, Apocyanaceae, Moraceae, Combretaceae, Myrtaceae, and Verbenaceae
B. carambolae (26 families 75 species)	Alangiaceae, Anacardiaceae, Annonaceae, Apocyanaceae, Arecaceae, Clusiaceae, Combretaceae, Euphorbiaceae, Lauraceae, Loganiaceae, Meliaceae, Moraceae, Myristacaceae, Myrtaceae, Olacaceae, Oxalidaceae, Polygalaceae, Punicaceae, Rhamnaceae, Rhizophoraceae, Rutaceae, Sapindaceae, Sapotaceae, Simaroubaceae, Solanaceae, and Symplocaceae
B. caryeae (6 families 8 species)	Anacardiaceae, Lecythidaceae, Malphigiaceae, Myrtaceae, Rutaceae, and Sapotaceae
B. correcta (28 families 58 species)	Anacardiaceae, Annonaceae, Apocyanaceae, Capparaceae, Combretaceae, Cucurbitaceae, Dipterocarpaceae, Elaeocarpaceae, Euphorbiaceae, Flacourtiaceae, Lecythidaceae, Malphigiaceae, Meliaceae, Moraceae, Musaceae, Myristicaceae, Myrtaceae, Olacaceae, Oxalidaceae, Rhamnaceae, Rosaceae, Rutaceae, Sapindaceae, Sapotaceae, and Simaroubaceae
B. dorsalis (40 families 116 species)	Alangiaceae, Anacardiaceae, Annonnaceae, Apocyanaceae, Arecaceae, Burseraceae, Capparaceae, Caprifoliaceae, Caricaceae, Celastraceae, Chrysobalanceae, Clusiaceae, Combretaceae, Convolvulaceae, Cucurbitaceae, Ebenaceae, Elaeocarpaceae, Euphorbiaceae, Fabaceae, Flacourtiaceae, Lauraceae, Lecythidaceae, Malpighiaceae, Meliaceae, Moraceae, Musaceae, Myrtaceae, Olacaceae, Oleaceae, Oxalidaceae, Polygalaceae, Rhamnaceae, Rosaceae, Rubiaceae, Rutaceae, Sapindaceae, Sapotaceae, Simaroubaceae, and Solanaceae
B. latifrons (10 families 28 species)	Lythraceae, Myrtaceae, Oleaceae, Passifloraceae, Punicaceae, Rahmnaceae, Rutaceae, Sapindaceae, Solanaceae, and Verbenaceae
B. limbifera (2 families 2 species)	Anacardiaceae (*Dracontomelon dao*), Meliaceae (*Aglaia sp.*)
B. melastomatos (1 family 2 species)	Melastomataceae (*Melastoma malabathrica, Melastoma sanguineum*)
B. nigrofemoralis (1 family 1 species)	Myrtaceae (*Psidium guajava*)
B. tuberculata (8 families 11 species)	Anacardiaceae, Caricaceae, Euphorbiaceae, Lecythidaceae, Myrtaceae, Polygalaceae, Rosaceae, and Sapotaceae
B. versicolor (1 family 1 species)	Sapotaceae (*Manilkara zapota*)
B. zonata (15 families 20 species)	Anacardiaceae, Annonaceae, Arecaceae, Caricaceae, Combretaceae, Cucurbitaceae, Fabaceae, Lecythidaceae, Malpighiaceae, Malvaceae, Myrtaceae, Punicaceae, Rosaceae, Rutaceae, and Tiliaceae
Z. diversus (1 family 9 species)	Cucurbitaceae
Z. trilineatus (1 family 1 species)	Cucurbitaceae (*Coccinia grandis*)
B. garciniae (1 family 4 species)	Clusiaceae
B. minax (1 family 8 species)	Rutaceae
Z. caudatus (1 family 1 species)	Cucurbitaceae (*Cucurbita moschata*)
Z. cucurbitae (12 families 42 species)	Agavaceae, Capparidaceae, Cucurbitaceae, Fabaceae, Malvaceae, Moraceae, Myrtaceae, Rhamnaceae, Rutaceae, Sapotaceae, Solanaceae, and Vitaceae
Z. scutellaris (1 family 4 species)	Cucurbitaceae (*Cucurbita maxima, Cucurbita moschata, Lagenaria siceraria, and Melothria wallichii*)
Z. tau (9 family 34 species)	Arecaceae, Cucurbitaceae, Fabaceae, Loganiaceae, Moraceae, Myrtaceae, Oleaceae, Sapotaceae, and Vitaceae
D. polistiformis (1 family 1 species)	Asclepiadaceae (*Oxystelma esculentum*)
D. sphaeroidalis (1 family 1 species)	Asclepiadaceae (*Telosma cordata*)
D. ciliatus (1 family 6 species)	Cucurbitaceae
D. longicornis (1 family 4 species)	Cucurbitaceae (*Luffa acutangula, Luffa aegyptiaca, Melothria wallichii, and Trichosanthes cucumerina*)
D. persicus (1 family 2 species)	Asclepiadaceae (*Calotropis gigantea, Calotropis procera*)

Source: Allwood, A.J. et al., *Raff. Bull. Zool. Suppl.*, 7, 1–92, 1999.

[a] Number of host plant families and plant species attacked is given in parentheses.

Acanthiophilus, *Cecidochares*, and *Procecidochares* are the few economically important genera. They are useful as biocontrol agents of weeds, *Cecidochares connexa* for *Chromolaena odorata* and *Procecidochares utilis* for *Eupatorium adenophorum*. *Acanthiophilus helianthi* (capsule fly) is a pest of safflower. Blepharoneurinae are mostly a tropical group of Tephritidae that includes five genera: *Ceratodacus* Hendel, *Problepharoneura* Norrbom and Condon, *Blepharoneura* Loew, *Baryglossa* Bezzi, and *Hexaptilona* Hering. Although the biology of the species of *Ceratodacus* and *Problepharoneura* is unknown, the species of *Blepharoneura*, *Baryglossa*, and *Hexaptilona* are known, or suspected, to breed in plants of the family Cucurbitaceae.

ECONOMIC IMPORTANCE

Fruit flies rank top among the pests of quarantine concern across the continents due to their concealed feeding and invasive potential. Fruit flies belonging to the subfamilies Dacinae and Trypetinae are predominantly frugivorous and form the lion's share of the economically important species. Economic effects of pest species include not only direct loss of yield and increased control costs, but also the loss of export markets and/or the cost of constructing and maintaining fruit treatment and eradication facilities. In many countries, the exportation of most commercial fruits is severely restricted by quarantine laws to prevent the spread of fruit fly species. Nearly 40 species of Tephritidae have been spread intentionally or accidentally by man beyond their natural ranges of which *Bactrocera dorsalis* (Hendel), *B. carambolae* Drew and Hancock, *B. papayae* Drew and Hancock, *B. latifrons* (Hendel), *Z. cucurbitae* (Coquillett), and *B. zonata* (Saunders) have been introduced to other parts from the Oriental region, whereas *B. oleae* (Rossi) got introduced to India from the Mediterranean region (Norrbom et al. 1999). More than 20 species have been used as biological control agents for adventive weedy species of Asteraceae. Most of the species that have been introduced belong to the subfamily Tephritinae. Several species of *Urophora* have been introduced into North America for the management of knapweeds and thistles (White and Elson-Harris 1992). Of significance to India are *Cecidochares connexa* Macquart on Siam weed, *Chromolaena odorata* and *Procedidochares utilis* Stone on *Eupatorium adenophorum* that cause stem galls, and these have been introduced for the biological control of these weeds (Bhumannavar et al. 2004; 2007; Ramani 2004). Some tephritids are key subjects for studying basic biology (*Ceratitis capitata* in genetics and molecular biology) and as models for testing evolutionary or ecological theories (*Rhagoletis* spp. for sympatric speciation and *C. capitata* for demographic research) (Norrbom 2004).

COLLECTION AND PRESERVATION

COLLECTION OF FRUIT FLIES

Fruit fly collections can be made in the field by the following methods:

a. Sweep netting: Sweep netting on vegetation is one of the general methods of collection. Members of the subfamily Tephritinae are mostly collected by sweep netting on host plants of the families Acanthaceae and Asteraceae as tephritines are flower-head borers/stem gall markers of aforementioned families. In case of fruit infesting tephritids, sweeping on fruiting trees and shrubs can be employed to collect them
b. Host plant rearing: Infested fruits, leaves, stems, and flower heads of various plants have to be collected and to be reared to adult stage. For the collection of species of the subfamily Phytalmiinae, rotten wood, bark of trees, and trunk of stem can be examined
c. Traps: For the sake of collection of fruit flies of the tribe Dacini, parapheromone traps impregnated with methyl eugenol and cuelure along with insecticide can be employed. Steiner traps and McPhail traps are widely used. In the case of fruit flies infesting bamboo shoots, Gastrozonini, chopped tender shoots of bamboo can be used to lure the flies. The same shoots will serve as a source to trap saprophytic species of fruit flies of the subfamily Phytalmiinae which are associated with bamboo.

PROCESSING AND PRESERVATION

Direct pinning: Fruit flies can be directly pinned on the thorax, right of the scutum just below the transverse suture. Several times, important setae and other characters on the thorax needed for identification are lost in the process and so direct pinning through the top of the thorax should be avoided.

Double mounting: Fruit flies can be double mounted, using a minuten on foam using an insect pin.
Double mounting on side: Specimens may be pinned through the side of the thorax with a micropin (D2 or B2 or A1 size) by inserting the pin just behind the wing base at such an angle that it emerges just in front of the wing base on the other side of the specimen.
Pointing: Fruit flies can be glued to a paper point on right side of the scutum.
Preservation: Specimens can be preserved after drying by pinning them in insect boxes with camphor or paradichlorobenzene (PDB) as fumigants. For doing molecular studies, specimens can be preserved in absolute alcohol and stored in deep freezer.

DISSECTION OF ADULT FEMALES

The abdomen has to be separated carefully using fine forceps. Dissected abdomen has to be boiled in 10% KOH for softening the cuticle. The digested abdomen has to transferred to glacial acetic acid for 2 minutes. Further, the specimen can be transferred to 70%–80% and absolute ethanol for dehydration. The processed abdomen can be now dissected in glycerol under a Stereozoom microscope. The ovipositor can be easily pulled out using a fine needle from the rest of the abdomen by applying gentle pressure near the base of oviscape while holding the abdomen firmly with the help of a needle. Gently insert a fine needle into the oviscape so that the apex of aculeus is visible. Holding the oviscape with a bent fine needle, gently pull out the aculeus and eversible membrane. The dissected ovipositor can be mounted in Canada balsam for permanent mounts or in glycerol for temporary use.

Processing of Larvae

Larvae can be collected from the infested plant part (fruit, leaf mine, flower heads, stem galls) and washed thoroughly in clean cold water. They can then be killed by immersing in hot water (just off the boil) for 2–3 minutes. After allowing the water to cool to room temperature, the larvae can be transferred to a serial dilution of 30%–50% ethanol and finally stored in 70% ethanol. Specimens to be slide-mounted can be slit and macerated overnight in a 10% solution of cold potassium hydroxide. After the digestion of specimens, transfer it to acetic acid for 2 minutes followed by serial dehydration in 70%, 80%, and absolute alcohol. For clearing, the digested larvae can be transferred to clove oil, further, it can be cleared by removing remaining musculature and internal contents through the slits. Place the larvae length wise to observe the anterior spiracles and cephalopharyngeal skeleton. The posterior spiracles can be better viewed if placed dorsoventrally. The cuticle, cephalopharyngeal skeleton, and the anterior and posterior spiracles of larvae have to be dissected and mounted (Frias et al. 2006).

USEFUL WEBSITES

Global fruit fly keys and information—Fruit Flies (Diptera: Tephritidae)
http://www.sel.barc.usda.gov/Diptera/tephriti/tephriti.htm

The Diptera Site—The Biosystematic Database of World Diptera
http://www.sel.barc.usda.gov/Diptera/biosys.htm

Pest Fruit Flies of the World—Adults
L. E. Carroll, I.M. White, A. Freidberg, A.L. Norrbom, M.J. Dallwitz, and F.C. Thompson
http://delta-intkey.com/ffa/

Pest Fruit Flies of the World—Larvae
L. E. Carroll, A.L. Norrbom, M.J. Dallwitz, and F.C. Thompson
http://delta-intkey.com/ffl/

Identification key: African Tephritids
https://fruitflykeys.africamuseum.be/wrapper-african-tephritids.html

Handbook for the Identification of Fruit Flies
http://www.planthealthaustralia.com.au/national-programs/fruit-fly/handbook-for-the-identification-of-fruit-fly/

Fruit Fly Identification Australia
http://fruitflyidentification.org.au/

World Diptera Systematists-Directory of Tephritidae workers
http://hbs.bishopmuseum.org/dipterists/

Dipterists Forum—The Society for the study of flies (Diptera)
http://www.dipteristsforum.org.uk/t601-Some-Diptera-Links.html

Diptera Info
http://www.diptera.info/

Parasitoids of Fruit-Infesting Tephritidae
http://www.paroffit.org/public/site/paroffit/home

Background information on morphology of cyclorrhaphous Diptera—Australian National Insect Collection (ANIC) anatomical atlas of flies
http://www.csiro.au/resources/ps252.html

Fighting Fruit Flies-Videos from Access Agriculture
Integrated approach against fruit flies
http://www.accessagriculture.org/integrated-approach-against-fruit-flies

Collecting fallen fruit against fruit flies
http://www.accessagriculture.org/collecting-fallen-fruit-against-fruit-flies

Mass trapping of fruit flies
http://www.accessagriculture.org/mass-trapping-fruit-flies

Killing fruit flies with food baits
http://www.accessagriculture.org/killing-fruit-flies-food-baits

CONCLUSION

Fruit flies have received immense attention in research and development across the world owing to their economical significance as pests of various horticultural crops and as biocontrol agent for several invasive weeds. It is the most diverse and speciose family in Tephritoidea compared to the other nine families. Even though it is a well studied group, higher level phylogeny of the family and subfamilies have not been fully resolved neither using morphological nor molecular characters which demands a comprehensive study of the group across the zoogeographic regions integrating holomorphological and molecular markers. As it is a pest of quarantine concern, taxonomy of immature stages of all the economically important species needs to be studied intensively and documented properly. Moreover, it is the need of the hour for both traditional taxonomists and molecular biologists to come to a consensus in defining species and generic limits; examples in this case would be the identity of *Bactrocera dorsalis* and *B. invadens* and generic status of *Zeugodacus*, respectively. Apart from the major economically important genera of fruit flies, the host plants, biology, and ecology of several taxa of Tephritidae are still enigmatic with reference to the fauna of India. Oriented research in this regard is required to have a comprehensive knowledge about this group.

REFERENCES

Agarwal, M. L., and M. Sueyoshi. 2005. Catalogue of Indian fruit flies (Diptera: Tephritidae). *Oriental Insects* 39: 371–433.

Allwood, A. J., A. Chinajariyawong, R. A. I. Drew et al. 1999. Host plant records for fruit flies (Diptera: Tephritidae) in South East Asia. *Raffles Bulletin of Zoology Supplement* 7: 1–92.

Asokan, R., K. B. Rebijith, S. K. Singh et al. 2011. Molecular identification and phylogeny of *Bactrocera* species (Diptera: Tephritidae). *Florida Entomologist* 94 (4): 1026–1035.

Bezzi, M. 1913. Indian trypaneids (fruit-flies) in the collection of the Indian Museum, Calcutta. *Memoirs of Indian Museum* 3: 53–175.

Bezzi, M. 1915. Two new species of fruit flies from southern India. *Bulletin of Entomological Research* 5: 153–154.

Bezzi, M. 1916. On the fruit-flies of the genus *Dacus* (s.l.) occurring in India, Burma, and Ceylon. *Bulletin of Entomological Research* 7: 99–121.

Bhumannavar, B. S., S. Ramani, S. K. Rajeswari et al. 2007. Field release and impact of *Cecidochares connexa* (Macquart) (Diptera: Tephritidae) on *Chromolaena odorata*. *Journal of Biological Control* 21(1): 59–64.

Bhumannavar, B. S., S. Ramani, S. K. Rajeswari et al. 2004. Host specificity and biology of *Cecidochares connexa* (Macquart) (Diptera: Tephritidae) introduced into India for the biological suppression of *Chromolaena odorata* (Linnaeus) King & Robinson. *Journal of Biological Control* 18(2): 111–120.

Carroll, L. E, I. M. White, A. Freidberg et al. 2002 onwards. Pest fruit flies of the world. Version: 13th September 2018. delta-intkey.com.

Carroll, L. E., A. L. Norrbom, M. J. Dallwitz et al. 2004 onwards. Pest fruit flies of the world—larvae. Version: 13th September 2018. delta-intkey.com.

Clarke, A. R., K. F. Armstrong, A. E. Carmichael et al. 2005. Invasive Phytophagous pests arising through a recent tropical evolutionary radiation: The *Bactrocera dorsalis* complex of fruit flies. *Annual Review of Entomology* 50: 293–319.

Condon, M. A., and A. L. Norrbom. 1994. Three sympatric species of *Blepharoneura* (Diptera: Tephritidae) on a single species of host (*Gurania spinulosa*, Cucurbitaceae): New species and taxonomic methods. *Systematic Entomology* 19: 279–304.

Condon, M. A., and A. L. Norrbom. 1999. Phylogeny of the subfamily Blepharoneurinae. In *Fruit Flies (Tephritidae): Phylogeny and Evolution of Behavior,* eds. M. Aluja and A. L. Norrbom, pp. 135–154. Boca Raton, FL: CRC Press.

David, K. J. 2016. Systematic studies on fruit flies of subfamily Dacinae (Diptera: Tephritidae) with special emphasis to peninsular India. PhD diss., Thesis. Bangalore, India: University of Agricultural Sciences.

David K. J., and D. L. Hancock. 2013. The first record of *Ortalotrypeta isshikii* (Matsumura) and subfamily Tachiniscinae (Diptera: Tephritidae) from India, with redescription of the species. *Australian Entomologist* 40 (3): 131–135.

David K. J., D. L. Hancock, A. Freidberg et al. 2013. New species and records of *Euphranta* Loew and other Adramini (Diptera: Tephritidae: Trypetinae) from south and southeast Asia. *Zootaxa* 3635 (4): 439–458.

David, K. J., and D. L. Hancock. 2017. A new species of *Gastrozona* Bezzi (Diptera: Tephritidae: Dacinae: Gastrozonini) with an updated key to species from India. *Zootaxa* 4216 (1): 55–64.

David, K. J., and S. K. Singh. 2015. Two new species of *Euphranta* Loew (Diptera: Tephritidae: Trypetinae) and an updated key for the species from India. *Zootaxa* 3914 (1): 064–070.

David, K. J., and S. Ramani. 2011. An illustrated key to fruit flies (Diptera: Tephritidae) from Peninsular India and the Andaman and Nicobar Islands. *Zootaxa* 3021: 1–31.

David, K. J., D. L. Hancock, and S. Ramani. 2014a. Two new species of *Acroceratitis* Hendel (Diptera: Tephritidae) and an updated key for the species from India. *Zootaxa* 3895 (3): 411–418.

David, K. J., D. L. Hancock, S. K. Singh et al. 2017. New species, new records and updated subgeneric key of *Bactrocera* Macquart (Diptera: Tephritidae: Dacinae: Dacini) from India. *Zootaxa* 4272 (3): 386–400.

David, K. J., S. K. Singh, and S. Ramani. 2014b. New species and records of Trypetinae (Diptera: Tephritidae) from India. *Zootaxa* 3795(2): 126–134.

David, K. J., S. Ramani, D. Whitmore et al. 2016. Two new species and a new record of *Bactrocera* Macquart from India. *Zootaxa* 4103 (1): 25–34.

De Meyer, M. 2006. Systematic revision of the fruit fly genus *Carpophthoromyia* Austen (Diptera, Tephritidae). *Zootaxa* 1235: 1–48.

De Meyer, M., H. Delatte, S. Ekesi et al. 2015. An integrative approach to unravel the Ceratitis FAR (Diptera: Tephritidae) cryptic species complex: A review. *Zookeys* 540: 405–427.

Dohm, P., D. Kovac, A. Freidberg et al. 2014. Basic biology and host use patterns of Tephritid flies (Phytalmiinae: Acanthonevrini, Dacinae: Gastrozonini) breeding in Bamboo (Poaceae: Bambusoideae). *Annals of Entomological Society of America* 107(1): 184–203.

Doorenweerd, C., L. Leblanc, A. L. Norrbom et al. 2018. A global checklist of the 932 fruit fly species in the tribe Dacini (Diptera: Tephritidae). *Zookeys* 730: 17–54.

Drew, R. A. I., and M. C. Romig. 2013. *Tropical Fruit Flies (Tephritidae: Dacinae) of South-East Asia*. Wallingford, UK: CAB International.

Drew, R. A. I., and M. C. Romig. 2016. *Keys to The Tropical Fruit Flies (Tephritidae: Dacinae) of South-East Asia*. Wallingford, UK: CAB International.

Drew, R. A. I., and S. Raghu. 2002. The fruit fly fauna (Diptera: Tephritidae: Dacinae) of the rainforest habitat of the Western Ghats, India. *Raffles Bulletin of Zoology* 50: 327–352.

Drew, R. A. I., K. Tsuruta, and I. M. White. 2005. A new species of pest fruit fly (Diptera: Tephritidae: Dacinae) from Sri Lanka and Africa. *African Entomology* 13: 149–154.

Freidberg, A. 1981. Mating behavior of *Schistopterum moebiusi* Becker (Diptera: Tephritidae). *Israel Journal of Entomology* 15: 89–95.

Freidberg, A. 2006. Biotaxonomy of Tephritoidea. *Israel Journal of Entomology* 35–36: 1–597.

Freidberg, A., and J. Kugler. 1989. *Fauna Palestina. Insecta IV-Diptera: Tephritidae*. Jerusalem, Israel: Israel Academy of Sciences and Humanities.

Frías, D., V. Hernández-Ortiz, N. C. Vaccaro et al. 2006. Comparative Morphology of Immature Stages of Some Frugivorous Species of Fruit Flies Diptera: Tephritidae, In *Biotaxonomy of Tephritoidea*, ed. A. Freidberg, pp. 459–475. Entomological Society of Israel, Israel.

Greene, E., L. J. Orsack, and D. W. Whitman. 1987. A tephritid fly mimics the territorial displays of its jumping spider predators. *Science* 236: 310–312.

Han, H.-Y. 1999. Phylogeny and behaviour of flies in tribe Trypetini (Trypetinae). In *Fruit Flies (Tephritidae): Phylogeny and Evolution of Behavior,* eds. M. Aluja and A. L. Norrbom, pp. 253–297. Boca Raton, FL: CRC Press.

Han, H.-Y., and K.-E. Ro. 2016. Molecular phylogeny of the superfamily Tephritoidea (Insecta: Diptera) reanalysed based on expanded taxon sampling and sequence data. *Journal of Zoological Systematics and Evolutionary Research* 54 (4): 276–288.

Hancock, D. L. 2013. *Themara maculipennis* (Westwood) and *Themara hirtipes* Rondani (Diptera: Tephritidae: Acanthonevrini): A case of confused synonymies. *Australian Entomologist* 40 (2): 93–98.

Hancock, D. L., and R. A. I. Drew. 1994. New species and records of Asian Trypetinae (Diptera: Tephritidae). *Raffles Bulletin of Zoology* 42: 555–591.

Hancock, D. L., and R. A. I. Drew. 1995. New genus, species and synonyms of Asian Trypetinae (Diptera: Tephritidae). *Malaysian Journal of Science* 16A: 45–59.

Hancock, D. L., and R. A. I. Drew. 2004. Notes on the genus *Euphranta* Loew (Diptera: Tephritidae), with description of four new species. *Australian Entomologist* 31(4): 151–168.

Hancock, D. L., and R. A. I. Drew. 2005. New genera, species and records of Adramini (Diptera: Tephritidae: Trypetinae) from the South Pacific and southern Asia. *Australian Entomologist* 32 (1): 5–16.

Hancock, D. L., and R. A. I. Drew. 2018. A review of the Indo-Australian subgenera *Apodacus* Perkins, *Hemizeugodacus* Hardy, *Neozeugodacus* May, stat. rev., *Semicallantra* Drew and *Tetradacus* Miyake of *Bactrocera* Macquart (Diptera: Tephritidae: Dacinae). *Australian Entomologist* 45 (1): 105–132.

Hardy, D. E. 1974. The fruit flies of the Philippines (Diptera-Tephritidae). *Pacific Insects Monograph* 32: 1–266.

Hardy, D. E. 1986. Fruit flies of the subtribe Acanthonevrina of Indonesia, New Guinea and the Bismarck and Solomon Islands (Diptera: Tephritidae: Trypetinae: Acanthronevrini). *Pacific Insects Monograph* 42: 1–42.

Hardy, D. E. 1971. Diptera: Tephritidae from Ceylon. *Entomologia Scandinavica Supplement* 1: 287–292.

Hardy, D. E. 1973. The fruit flies (Tephritidae: Diptera) of Thailand and bordering countries. *Pacific Insects Monograph* 31:1–353.

Hardy, D. E., and R. A. I. Drew. 1996. Revision of the Australian Tephritini (Diptera: Tephritidae). *Invertebrate Taxonomy* 10: 213–405.

Hendrichs, J., T. Vera, M. De Meyer et al. 2015. Resolving cryptic species complexes of major tephritid pests. *Zookeys* 540: 5–39.

Hering, E. M. 1938. Entomological results from the Swedish expedition 1934 to Burma and British India. Diptera: Fam. Trypetidae. (23. Beitrag zur Kenntnis der Trypetidae). *Arkiv för zoologi* 30A (25): 1–56.

Hering, E. M. 1941. Entomological results from the Swedish expedition to Burma and British India. Diptera: Trypetidae. Nachtrag [30. Beitrag zur Kenntnis der Trypetidae]. *Arkiv för zoologi* 33B (11):1–7.

Hering, E. M. 1956. Trypetidae (Dipt.) von Ceylon (53. Beitrag zur Kenntnis der Trypetidae). *Verhandlungen der Naturforschenden Gesellschaft in Basel* 67: 62–74.

Hill, G. F. 1921. Notes on some Diptera found in association with termites. *Proceedings of Linnaean Society of New South Wales* 46: 216–220.

Jamnongluk, W., V. Baimai, and P. Kittayapong. 2003. Molecular evolution of tephritid fruit flies in the genus *Bactrocera* based on the cytochrome oxidase I gene. *Genetica* 119 (1): 19–25.

Kapoor, V. C. 1971. Four new species of fruit flies (Tephritidae) in India. *Oriental Insects* 5: 477–482.

Kapoor, V. C. 1993. *Indian fruit flies (Insecta: Diptera: Tephritidae)*. New Delhi, India: Oxford & IBH.

Kapoor, V. C., D. E. Hardy, M. L. Agarwal et al. 1980. *Fruit fly (Diptera: Tephritidae) Systematics of the Indian Subcontinent*. Jalandhar, India: Export India Publications.

Kopelke, J. P. 1984. Der erste Nachweis eines Brutparasiten unter den Bohrfliegen. *Nat. Mus. Frankfurt* 114: 1–28.

Korneyev, V. A. 1999. Phylogenetic relationships among higher groups of Tephritidae. In *Fruit flies (Tephritidae): Phylogeny and Evolution of Behavior*, eds. M. Aluja and A. L. Norrbom, pp. 73–113. Boca Raton, FL: CRC Press.

Kovac, D., P. Dohm, A. Freidberg et al. 2006. Catalog and revised classification of the Gastrozonini (Diptera: Tephritidae: Dacinae). In *Biotaxonomy of Tephritoidea*, ed. A. Freidberg, pp. 163–196. Jerusalem, Israel: Entomological Society of Israel.

Kovac, D., A. Freidberg, and G. J. Steck. 2013. Biology and description of the third instar larva and puparium of *Ichneumonopsis burmensis* Hardy (Diptera: Tephritidae: Dacinae: Gastrozonini), a bamboo-breeding fruit fly from the Oriental region. *Raffles Bulletin of Zoology* 61 (1): 117–132.

Kovac, D., A. Schneider, A. Freidberg et al. 2017. Life history and description of the larva of *Acrotaeniostola spiralis* (Diptera: Tephritidae: Dacinae: Gastrozonini), an Oriental fruit fly inhabiting bamboo twigs. *Raffles Bulletin of Zoology* 65: 154–167.

Krosch, M. N., M. K. Schutze, K. F. Armstrong et al. 2012. A molecular phylogeny for the tribe Dacini (Diptera: Tephritidae): Systematic and biogeographical implications. *Molecular Phylogenetics and Evolution* 64: 513–523.

Landolt, P. J. 1984. Behavior of the papaya fruit fly *Toxotrypana curvicauda* Gerstaecker, in relation to its host plant. *Carica papaya* L. *Folia Entomologica Mexicana* 61: 215–224.

Landolt, P. J. 1999. Behaviour of flies in the genus *Toxotrypana* (Trypetinae: Toxotrypanini). In *Fruit Flies (Tephritidae): Phylogeny and Evolution of Behavior*, eds. M. Aluja and A. L. Norrbom, pp. 363–372. Boca Raton, FL: CRC Press.

Mather, M. H., and B. D. Roitberg. 1987. A sheep in wolf's clothing: Tephritid flies mimic spider predators. *Science* 236: 308–310.

Morgulis, E., A. Freidberg, and N. Dorchin. 2015. Phylogenetic revision of *Acanthiophilus* (Diptera: Tephritidae) with a description of three new species and a discussion of zoogeography. *Annals of Entomological Society of America* 108 (6):1–28.

Morgulis, E., A. Freidberg, and N. Dorchin. 2016. Phylogenetic revision of *Tephritomyia* Hendel (Diptera: Tephritidae) with description of 14 new species. *Annals of Entomological Society of America* 109 (4): 595–628.

Munro, H. K. 1935. Records of Indian Trypetidae (Diptera) with descriptions of some apparently new species. *Records of Indian Museum* 37: 15–27.

Munro, H. K. 1938. Studies on Indian Trypetidae (Diptera). *Records of Indian Museum* 40: 21–37.

Munro, H. K. 1939. The fruit fly, *Dacus ferrugineus* Fabr., and its variety *dorsalis* Hendel in north west India. *Indian Journal of Entomology* 1: 101–105.

Munro, H. K. 1967. Fruit flies allied to species of *Afrocneros* and *Ocnerioxa* that infest *Cussonia*, the umbrella tree or kiepersol (Araliaceae) (Diptera: Trypetidae). *Annals of the Natal Museum* 18: 571–594.

Nakahara, S., and M. Muraji. 2008. Phylogenetic analyses of *Bactrocera* fruit flies (Diptera: Tephritidae) based on nucleotide sequences of the mitochondrial. *Research Bulletin of Plant Protection Service Japan* 44: 1–12.

Norrbom A. L., L. E. Carroll, F. C. Thompson et al. 1999. Systematic database of names. In *Fruit Fly Expert Identification System and Systematic Information Database*, ed. F. C. Thompson pp. 65–251. Leiden, the Netherlands: Backhuys publishers.

Norrbom, A. L., 2004. Fruit fly (Diptera: Tephritidae) classification & diversity. Fruit fly taxonomy pages. http://www.sel.barc.usda.gov/diptera/tephriti/TephClas.htm (Accessed January 2010.)

Perkins, F. A. 1938. Studies in Oriental and Australian Trypaneidae-Part II. Adraminae and Dacinae from India, Ceylon, Malaya, Sumatra, Java, Borneo, Philippine Islands and Formosa. *Proceedings of Royal Society of Queensland* 49: 120–144.

Permkam, S., and D. L. Hancock. 1995. Australian Trypetinae (Diptera: Tephritidae). *Invertebrate Taxonomy* 9: 1047–1209.

Ramani, S. 2004. Classical biological control attempts in India. In *Quarantine Procedures and Facilities for Biological Control Agents*, eds. S. Ramani, B. S. Bhumannavar, and R. J. Rabindra, pp. 17–22. Bangalore, India: Project Directorate of Biological Control.

Ramani, S., K. J. David, C. A. Viraktamath et al. 2008. Identity and distribution of *Bactrocera caryeae* (Kapoor) (Insecta: Diptera: Tephritidae)-A species under the *Bactrocera dorsalis* complex in India. *Biosystematica* 2 (1): 49–57.

Robinson, A. S., and G. Hooper. 1989. *Fruit Flies: Their Biology, Natural Enemies and Control*. In *World Crop Pests*, ed. W. Helle, vol. 3A, 372 pp. and vol. 3B, 447 pp. Elsevier Science Publishers, Amsterdam.

San Jose, M., C. Doorenweerd, L. Leblanc et al. 2018. Incongruence between molecules and morphology: A seven-gene phylogeny of Dacini fruit flies paves the way for reclassification (Diptera: Tephritidae). *Molecular Phylogenetics and Evolution* 121: 139–149. doi:10.1016/j.ympev.2017.12.001.

Schneider, A., D. Kovac, G. J. Steck et al. 2018. Larval descriptions of five Oriental bamboo-inhabiting *Acroceratitis* species (Diptera: Tephritidae: Dacinae) with notes on their biology. *European Journal of Entomology* 115: 535-561.

Schutze, M. K., N. Aketarawong, W. Amornsak et al. 2015. Synonymization of key pest species within the *Bactrocera dorsalis* species complex (Diptera: Tephritidae): Taxonomic changes based on a review of 20 years of integrative morphological, molecular, cytogenetic, behavioural and chemoecological data. *Systematic Entomology* 40 (2): 456–471.

Senior-White, R. A. 1921. New Ceylon Diptera. *Spolia Zeylan*11: 381–396.

Senior-White, R. A. 1922. Notes on Indian Diptera 1. Diptera from the Khasia Hills. 2. Tabanidae in the collection of the forest zoologist. 3. New species of Diptera from the Indian Region. *Memoirs of the Department of Agriculture in India Entomological Series* 7: 83–170.

Senior-White, R. A. 1924. *Catalogue of Indian Insects. Part 4-Trypetidae (Trypaneidae)*. Calcutta, India: Central Publication Branch. 33 p.

Singh, S. K., and V. V. Ramamurthy. 2010. Chorionic microsculpturing as a diagnostic character for species identification of fruit flies (Diptera: Tephritidae). *Entomological News* 121 (1): 91–94.

Singh, S. K., K. J. David, D. Kumar et al. 2015. A new species of *Magnimyiolia* Shiraki (Diptera: Tephritidae: Trypetinae) and new records of Acanthonevrini from India. *Zootaxa* 3949 (1):129–134.

Sivinski, J., M. Aluja, G. Dodson et al. 1999. Topics in the evolution of sexual behaviour in the Tephritidae. In *Fruit flies (Tephritidae): phylogeny and evolution of behavior*, eds. M. Aluja and A. L. Norrbom, pp. 751–786. Boca Raton, FL: CRC Press.

Smith, J. J., and G. L. Bush. 1999. Phylogeny of the subtribe Carpomyina (Trypetinae), emphasizing relationships of the genus *Rhagoletis*. In *Fruit Flies (Tephritidae): Phylogeny and Evolution of Behavior*, eds. M. Aluja and A. L. Norrbom, pp. 187–215. Boca Raton, FL: CRC Press.

Smith, P. T., B. A. McPheron, and S. Kambhampati. 2002. Phylogenetic analysis of mitochondrial DNA supports the monophyly of Dacini fruit flies (Diptera: Tephritidae). *Annals of Entomological Society of America* 95 (6): 658–664.

Smith, P. T., S. Kambhampati, and B. A. Mc Pheron. 2003. Phylogenetic relationships among *Bactrocera* species (Diptera: Tephritidae) inferred from mitochondrial DNA sequences. *Molecular Phylogenetics and Evolution* 26: 8–17.

Virgilio, M., K. Jordaens, C. Verwimp et al. 2015. Higher phylogeny of frugivorous flies (Diptera: Tephritidae: Dacini): Localised partition conflicts and a novel generic classification. *Molecular Phylogenetics and Evolution* 85: 171–179.

Wang, X.-J., and X.-L. Chen. 2002. A revision of the genus *Gastrozona* Bezzi from China (Diptera: Tephritidae). *Acta Entomologica Sinica* 45 (4): 507–515.

White, I. M. 1999. Morphological features of the tribe Dacini (Dacinae): Their significance to behavior and classification. In *Fruit Flies (Tephritidae): Phylogeny and Evolution of Behavior*, eds. M. Aluja and A. L. Norrbom, pp. 505–534. Boca Raton, FL: CRC Press.

White, I. M. 2006. Taxonomy of the Dacina (Diptera: Tephritidae) of Africa and the Middle East. *African Entomology, Memoir* 2: 1–156.

White, I. M., and D. L. Hancock. 1997. *CABIKEY to the Dacini (Diptera, Tephritidae) of the Asia-Pacific-Australasian Regions*. Wallingford, UK: CAB International, Windows CD.

White, I. M., and M. M. Elson-Harris. 1992. *Fruit flies of Economic Significance: Their Identification and Bionomics*. Wallingford, UK: CAB International.

Yong, H. S., S. L. Song, P. E. Lim et al. 2015. Complete mitochondrial genome of *Bactrocera arecae* (Insecta: Tephritidae) by next-generation sequencing and molecular phylogeny of Dacini tribe. *Scientific Reports* 5: 15155. doi:10.1038/srep15155.

18 Hover-Flies (Diptera: Syrphidae) Recorded from "Dravidia," or Central and Peninsular India and Sri Lanka

An Annotated Checklist and Bibliography[1]

Kumar Ghorpadé

CONTENTS

Introduction 325
Systematic Annotated Checklist 329
    Subfamily MICRODONTINAE 329
        Tribe Microdontini 329
    Subfamily ERISTALINAE 332
        Tribe Brachyopini 332
Acknowledgements 382
References 382

INTRODUCTION

This is the second, after my first paper (Ghorpadé 2015a) that documented all that has been published on the Syrphidae of the north-western borderlands of our subcontinent, which area holds all the quasi-Palaearctic and some Afrotropical elements that are found in the Indian subregion of the Oriental region. Three hundred and forty species were included there, of the total of 493 species recorded from the Indian subcontinent (subregion; Ghorpadé 2014b), or around 60% of the entire Indian subcontinental fauna. It is recommended that readers go through that first paper, especially the introductory portion (pp. 1–8), to comprehend this subject totally.

As mentioned in that paper, the three papers planned are a summation of my pre-doctoral, doctoral, post-doctoral, and later research, carried out from 1973 until now and ongoing. These are here published ahead of my planned further generic reviews and hopefully a revision of Brunetti's (1923) Syrphidae Fauna volume, to present data that co-specialists working on Oriental—Papuan (and Palaearctic) Syrphidae could make use of until these papers of mine (with or without collaborators), or of others, are finally published in the near future. My checklist (Ghorpadé 2014b) is the basic database (*q.v.*).

For preparing the checklist given below, almost all published work on the hover-flies of central and southern India (see list in Ghorpadé et al. 2011: 79) was consulted (Delfinado and Hardy, 1973, 1975, 1977). Species and genera (I have treated all currently recognized "subgenera" as full genera here, like earlier) are cited as are now nomenclaturally valid. Subfamilies are listed in presently researched phylogenetic sequence, but their tribes, genera, and species are subsequently cited in alphabetical order, so also are the Indian states, for convenience; the countries are listed from north to south. Only the original reference citations and regional synonymies are given, chronologically, and distribution of each species listed, by country, giving records of occurrence in this southern Indian area, the involved Indian states are listed alphabetically. For some species, undetected and possible synonymy, or misidentification, are also indicated for each species in my "NOTES" appended, for information and further research. My unpublished specimen data (as "Ghorpadé, unpubl.") are also indicated for some species when felt necessary and relevant here and this includes new species of some genera which will be named and described in forthcoming papers.

This chapter covers the hover-fly fauna of "Dravidia," or central and peninsular India and Sri Lanka (Table 18.1), this geographical area made up of land that is most ancient here (primal India), lying south of the recently formed Indo-Gangetic Plains, the Himalayas, and the north-eastern, "Assam" areas of this Indian sub-continent. "Dravidia" was a portion of the "Lesser (not Greater!) Indian Plate," which was once part of the prehistoric Southern Hemisphere supercontinent, Gondwanaland. The age and character of "Dravidia" was most aptly described by the celebrated British Indian

---

[1] This paper is dedicated to my friend and senior colleague, Dr C. A. Viraktamath, UAS, Bangalore, who helped me with my early research on Syrphidae, and Diptera, and was the main taxonomist authority to consult then, after the retirement of Drs. M. Puttarudriah and G.P. Channabasavanna, my Professor teachers during my B.Sc., M.Sc., and also Ph.D. degree programmes from 1966 to 1981 at the Agricultural University, Bangalore. "CAV," as we all called him, was also an interesting and enthusiastic member on many field trips we undertook all over "Dravidia" and other parts of India, from the 1970s to recently.

TABLE 18.1
Number of Genera and Species (before and after hyphen respectively) of Each Subfamily Recorded from Central and Peninsular India and Sri Lanka

Subfamily	Kerala	Tamil Nadu	Pondicherry	Karnataka	Goa
SYRPHINAE	17–26	20–33	2–3	17–36	1–1
ERISTALINAE	9–20	16–33	4–6	14–27	2–2
MICRODONTINAE	0	1—1	0	1—1	0
TOTAL	26—46	39—67	6—9	32—64	3—3
Subfamily	Andhra Pradesh	Telangana	Maharashtra	Orissa	Chattisgarh
SYRPHINAE	8–13	1–1	9–11	3–4	8–10
ERISTALINAE	6–13	1–1	6–11	4–6	7–16
MICRODONTINAE	1–1	0	0	1–1	0
TOTAL	15—27	2—2	15—22	8—11	15—26
Subfamily	Sri Lanka	Madhya Pradesh	Jharkhand (± Bihar)	Rajasthan	Gujarat (+ DD, DN)
SYRPHINAE	20(1)–37(1)	8–9	12–16	3–4	6–8
ERISTALINAE	16–33	6–9	10–19	5–6	5–9
MICRODONTINAE	5–7(1)	0	0	0	1–1
TOTAL	41(1)—77 (2)	14—15	22—35	8—10	12—18

writer of the Raj era, E.M. Forster (1924; introductory lines of Chapter 12, "Caves"), in his novel "*A Passage to India*," where he himself had coined the term "Dravidia," and wrote:

"But India is really far older. In the days of the prehistoric ocean the southern part of the peninsula already existed, and the high places of "Dravidia" have been land since land began, and have seen on the one side the sinking of a continent that joined them to Africa, and on the other the upheaval of the Himalayas from a sea. They are older than anything in the world. No water has ever covered them, and the sun who has watched them for countless aeons may still discern in their outlines forms that were his before our globe was torn from his bosom. If flesh of the sun's flesh is to be touched anywhere, it is here, among the incredible antiquity of these hills."

See Christophers (1921), Gressitt (1961), Gupta (1962), Croizat (1968), Fairchild (1969), Darlington (1970), Holloway (1974), Mani (1968, 1974), Legris & Meher-Homji (1977), Illies (1983), Meher-Homji (1983), Erwin (1985), Pearson & Ghorpadé (1989), Briggs (1989, 2003), Cox (1990), Pearson & Cassola (1992), Humphries (2000), Lomolino (2001), Ghorpadé (2014a), and Gaonkar (*in prep.*) for helpful biogeographical notes.

In recent years, molecular biology studies have been applied to understand phylogeny better and the following papers on Syrphidae phylogeny can be consulted—Hull (1949), Goffe (1952), Vockeroth (1969), Darlington (1970), Hippa (1978, 1990), Erwin (1985), Wiegmann (1986), Rotheray & Gilbert (1989), Rotheray & Gilbert (1999), Skevington & Yeates (2000), Ståhls et al. (2003), Lyneborg & Barkemeyer (2005), Vujic et al. (2008), Mengual et al. (2008a, b, 2015), Ghorpadé (2012), Reemer & Ståhls (2013b), and Mengual et al. (2015).

For preparing the checklist given below, almost all published work on hover-flies of central and peninsular India and Sri Lanka was consulted. Species and genera (I have treated all currently recognized "subgenera" as full genera here) are cited as are now nomenclaturally valid. Subfamilies are listed in presently researched phylogenetic sequence, but their tribes, genera, and species are subsequently cited in alphabetical order, so also are the Indian states, for convenience, the countries are listed from north to south. Only the original reference citations and regional synonymy are given, chronologically, and distribution of each species is listed by country, giving records of occurrence in this area, these are listed alphabetically. For India, the concerned and geographically relevant, 16 central and southern states and Union territories are also indicated in alphabetical sequence, using their acronyms in the checklist below, *viz.,* AH (Andhra), AN (Andaman and Nicobar islands), BI (Bihar), CG (Chattisgarh), DD (Daman and Diu), DN (Dadra and Nagar Haveli), GJ (part of Gujarat, less Cutch and Little Rann), GO (Goa), JH (Jharkhand), KL (Kerala), KN (Karnataka), MH (Maharashtra), MP (Madhya Pradesh), OR (Orissa), PO (Pondicherry), RJ (Rajasthan, eastern part), TG (Telengana), and TN (Tamil Nadu). For some species, undetected and possible synonymy, or misidentification, are also indicated for each species in my "NOTES" appended, for information and further research. My unpublished specimen data (as "Ghorpadé, unpubl.") are also indicated for some species when felt necessary and relevant here.

Recently, a 30 page paper was published (Thompson et al. 2017) on what was called the PSWNG (Philippines, Solomon Is., Wallacea and New Guinea) region (see Map 1 on its p. 518 which shows range extent of this area) where "currently 278 (286) species in 69 (73) genera and subgenera are recognized (numbers in brackets represent these tallies including known but undescribed species and genera." This fine paper contributes a comprehensive introduction covering diagnosis, biology, phylogenetic relationships, classification, fossils, and identification of these Syrphidae which is an updated and all-round in-depth treatment useful for any student of Syrphidae, with adequate supporting illustrations of Syrphidae (head, body, wing, thorax, and larvae) and fine colour figures (images) (Plates 167–172) of the head, metasternum and metaleg, dorsal and lateral species habitus, and wings, with a useful key (pp. 504–508) to the identification of Philippine, Solomon Islands, Wallacea, and New Guinea flower fly groups. I introduce their notes on the synopsis of that fauna in each taxon treated by me below which should assist the reader/user adequately.

For the states and Union territories of the "Dravidia" area, now existing as political divisions (some 16 in all), the following additional information is also provided here.

AH = Andhra Pradesh, including Rayalaseema, the Andhra Carnatic, Northern Circars, the "Jeypore Agency," and Yanam, the last being a part of erstwhile French Pondicherry. Gaonkar (*in prep.*) wrote: "The present Andhra Pradesh earlier included the former Hyderabad Princely State of the Nizam and some districts of the former Madras Presidency. It now has just the southern Rayalseema area and the Andhra Carnatic. Two great rivers, the Godavari and Krishna, both originating in the Western Ghats cross this state from west to east and empty into the Bay of Bengal. Biogeographically, the mountainous land north of the River Godavari is different from what is found in the south. The Visakhapatnam district in the northeast, with hill ranges extending into Orissa, has the same butterfly fauna as Orissa. Many northern species find their southernmost distribution limits here. See Ghorpadé et al. (2011) for a faunistic list.

- AN = Andaman and Nicobar Islands in the Bay of Bengal, a Union territory.
- BI = Bihar, now only the northern portion of the former larger state, largely the floodplains of the rivers Ganga (Ganges) and Son. Pusa is in Samastipur district, where the British had founded and initiated the Indian Agricultural Research Institute. This is territory of the ancient Magadh and Maithila (Videha) kingdoms of ancient Hindostan, B.C.
- CG = Chhattisgarh, the eastern portion of former Madhya Pradesh, including many small former princely states, of which the largest were Bastar and Surguja.
- DD = Daman and Diu, now Union territories, but formerly Portuguese colonies, located in present Gujarat (*q.v.*).
- DN = Dadra and Nagar Haveli, Union territory and former Portuguese colony, currently located in present southern Gujarat (*q.v.*).
- GJ = Part of Gujarat, formerly part of the Sind and Bombay Presidency, including former Baroda (princely) state of the erstwhile H.H. Gaekwar, and Kathiawar and Saurashtra (with erstwhile Bhavnagar, Gondal, Junagadh, and Rajkot princely states of British India), and the Surat Dangs. Also includes the Union territories of Daman & Diu and Dadra & Nagar Haveli (*q.v.*) within its land area. Excludes Cutch (Kutch), and the Little Rann areas of Gujarat. Gaonkar (*in prep.*) wrote: "The Western Ghats are located in its southeastern part, and locally known as the Surat Dangs. The westernmost part of the Satpura Range is located north of the river Tapti and south of the river Narmada. The southernmost part of the Aravalli Hills is also situated in the north-eastern part of Gujarat. Forest cover exists on the hilly areas, and Gujarat has some interesting Afrotropical species that have not been found south of this State."
- GO = Goa, a former Portuguese colony, annexed and now a state of India. Gaonkar (*in prep.*) wrote: "The tiny State of Goa was a Portuguese colony until 1961. Until quite recently Goa had significant areas under forests (now 34% of total area), except for the coastal plains. However, due mainly to recent mining activity, most of the area has been degraded beyond recovery. But the eastern and southern (Canacona) parts still contain some magnificent remains of the evergreen and semi-evergreen forests of the Western Ghats. The southern province of Canacona has much insect diversity. Most of the central Western Ghats is found here, where most endemics also have their northern distributional limit. T.R. Bell worked here, and in the area just south of it, North Kanara."
- JH = Jharkhand, the southern bifurcated hilly portion of the earlier larger Bihar state, including all of Chhota Nagpur, and much of this state lying south of the rivers Ganges (Ganga) and Son. Gaonkar (*in prep.*) wrote: "Some forests are left along the eastern Vindhyas where some diversity exists."
- KL = Kerala, including former Malabar in the north, the erstwhile Travancore, and Cochin (princely) states and Mahé, part of the former French territory of Pondicherry (*q.v.*). Gaonkar (*in prep.*) wrote: "Kerala was formerly divided into three regions. The state of Travancore in the south, the state of Cochin in the centre, and the Malabar area in the north. These are the names that are typically used in older literature. Apart from Goa, Kerala is the only state where the Western Ghats are the dominating feature of topography and landscape. However, the coastal landscape of Kerala was quite different in van Rheede's time. The entire lowlands have now become an extended township from north to south. The forests of the midlands, which were intact only fifty years ago, have also given way to agriculture and plantations. It is only along the slopes of the Western Ghats that Kerala is well endowed with dense tropical rain forests, supporting a varied flora and fauna. However, the forests in the Palghat Gap (or Pass), which served as a corridor for lowland species between the southern and central Western Ghats, have completely disappeared. S.N. Ward, a 19th-century naturalist of some repute, collected many endemic species of butterflies near Calicut in Malabar (see original descriptions of these by Moore and Westwood). Nevertheless, the richest butterfly diversity in all of peninsular India is found in Idukki district, south of the Palghat Gap. About the time of independence from the British, Kerala had more

than 50% of its area under forest, but now hardly 27% remains, much of this being plantation areas!"

KN = Karnataka, including all of former Mysore princely state of the erstwhile H.H. Wodeyar; also Coorg (Kodagu), North and South Canara (Kanara), as well as Karnatak with some northern parts of the earlier Madras Presidency like Bellary district with the former Sandur princely state of erstwhile H.H. Ghorpade. Also southern portions of the earlier Bombay province (Maharashtra, *q.v.*), which were "Bombay Karnatak," and parts of current Andhra Pradesh (now to be Telengana) and the former Nizam's Dominions of Hyderabad—the "Hyderabad Karnatak." Gaonkar (*in prep.*) wrote: "Formerly the princely state of Mysore, with the additions of four northern districts from the Bombay Presidency and from the Hyderabad state, Karnataka still has some of the finest forested areas in southern India. The Western Ghats are situated mainly in the western districts of Hassan, Shimoga, Chikmagalur, North Kanara, South Kanara, and Coorg. The Malnaad districts, east of the crest of the Ghats, which used to have quite extensive forests until recently, have been transformed into commercial crop areas of plantations such as coffee and oranges. In the south, bordering Tamil Nadu, the Biligirirangan Hills run from west to east. Up to the bend of the river Cauvery, these hills are biogeographically a part of the Western Ghats complex. The hills east of the river Cauvery are the Eastern Ghats. With the exception of strict endemic species of the Western Ghats, the Eastern Ghats in Tamil Nadu have the same kind of flora and fauna as the Western Ghats [This is erroneous theory, see my papers on biogeography—Kumar Ghorpadé]. Many endemics on the Western Ghats also occur on the Eastern Ghats [= Eastern Droogs—Kumar Ghorpadé]. The North Kanara district (formerly in the old Bombay Presidency) is one of the best studied areas for butterfly fauna anywhere in India [hardly so for many other insect groups, sadly—Kumar Ghorpadé]. This is primarily due to three "Canara" naturalists, Aitken, Davidson, and Bell (see Ghorpadé et al. 2013: 2). The diversity is concentrated mainly in the three Western Ghats districts of North Kanara, South Kanara, and Coorg and about 17% of the state is under forest cover today."

MH = Maharashtra, formerly part of erstwhile "Bombay Province" and most of the earlier Bombay & Sind Presidency, now including Marathwada, Vidarbha, Khandesh, and Konkan, and many erstwhile princely states of former British India, especially Kolhapur of H.H. Bhosle Chhatrapati. Gaonkar (*in prep.*) wrote: "Formerly a part of the Bombay Presidency, which included some northern districts of Karnataka, Andhra Pradesh, and Gujarat. With a varied topography and vegetation, Maharashtra also has very varied ecological zones. However, the highest diversity is concentrated along the Western Ghats. Biogeographically, most of the northern Western Ghats are situated in this state. However, in the southernmost District of Sindhudurg there are patches of real evergreen forests still left, and this is where the endemics are found in this state. In most early literature, the word Bombay was used to designate the entire Presidency and not only the city." See Ghorpadé (2017) for a faunistic list.

MP = Madhya Pradesh, now the larger western portion of the erstwhile bigger state, but without Chhattisgarh. Including former Central India, Central Provinces, Madhya Bharat, Vindhya Pradesh, Bundelkhand, Baghelkhand, Malwa, Bhilistan, and the erstwhile princely states of Berar, Bhopal, Indore, Rewa, and Gwalior, the last of H.H. Scindia. Gaonkar (*in prep.*) wrote: "Madhya Pradesh was known in earlier literature as the Central Provinces or Central India. It is the heart of India, and the country's largest state. It also has the largest cover of forests, about 30% of the total area, and also includes a number of the largest sanctuaries and national parks in the country. In the north, there is the southernmost part of the Ganga Plains, in the central plateau, the Satpura and Vindhya ranges stretch from west to east." See Evans (1932) for early literature and Tiple & Ghorpadé (2012) for an updated list.

OR = Orissa (Odisha, Kalinga), now a separate state, the southern part of former "Bihar and Orissa" province of British India, and including the northern edges of the Northern Circars, and the "Jeypore Agency" of British India, including several erstwhile princely states, like Dhenkanal, Kalahandi, Keonjhar, and Mayurbhanj. Gaonkar (*in prep.*) wrote: "The relatively small state of Orissa still retains much of the original forest cover. The northernmost part of the Eastern Ghats are located in this state [This is incorrect, the river Godavari is the northern limit and Orissa has the eastern central highlands—Kumar Ghorpadé]. There are also large tracts of forest running into Madhya Pradesh and Andhra Pradesh. Famous for its Sal tree (*Shorea robusta*), the forests are mainly evergreen (some areas) and deciduous in other places. However, Orissa also has a coastline, with SOME mangrove vegetation. Many species found in Orissa are not found in the rest of peninsular India. The estimated forest cover is about 30% of the total area."

PO = Pondicherry (now Puducherry), a former French colony, now a state located within the land area of Tamil Nadu (*q.v.*), with further territories like Karaikal (just south of the famous Tranquebar, now in Tamil Nadu), Mahé (in Kerala, *q.v.*), and Yanam (in Andhra Pradesh, *q.v.*).

RJ = Rajasthan, eastern part, or Mewar, including the Aravalli Hills, made up mostly of erstwhile princely states, some of the larger ones being Ajmer, Jaipur, and Udaipur. Gaonkar (*in prep.*) wrote: "The Aravalli Hills cross the state in the south and east. The fauna of the state is predominantly Oriental with a few Afrotropical elements and has not been completely documented. Rajasthan has a forest cover of about 4% of the total area only."

TG = Telengana, including most of former Hyderabad (princely) state of the erstwhile Nizam, (approximating to much of the old Hyderabad). See notes under Andhra Pradesh.

TN = Tamil Nadu, part of the former extensive Madras Presidency in southern India, including erstwhile Madura and the Carnatic (Kongunad and Tamil Carnatic) and all of the famous Nilgiri Mountains. Included a few small erstwhile princely states like Arcot and Pudukkottai and with Pondicherry (*q.v.*) within its land area. Gaonkar (*in prep.*) wrote: "Formerly a part of the old Madras Presidency, which also included parts of Kerala, Andhra Pradesh, and Karnataka. Dominated by the fertile Cauvery river valley, has substantial portions of the Western Ghats (eastern slopes), the southern eastern Ghats portions of the Shevaroy Hills, the Bodi Hills, and the Kolli Hills. The latter hills are poor in diversity compared to the Western Ghats and have not been surveyed as comprehensively. The Nilgiris (north of the Palghat Gap), the Palnis, and the Anaimalais (south of the Palghat Gap) have been explored for over a hundred years. The Nilgiris district is perhaps the best known area in all of butterfly literature in India (see Larsen 1987–1988), and the Palnis were covered and updated by Ghorpadé and Kunte (2010). Most of the Western Ghats endemics fly here and Tamil Nadu has about 14% of its total as forest area."

SL = Sri Lanka, or the previously called island of Ceylon, is a very interesting part of "Dravidia" for its biodiversity, including many strict endemics, especially at genus level. It can be recognized as having the southernmost section of the Western Ghats, its high range. Its Syrphidae were very well sampled and documented by Keiser (1958) of the Naturhistorisches Museum Basel (Switzerland), but the paper was in German ! Keiser went to Ceylon (as it was then called) during 1953–1954, and collected some 1504 specimens (831♂, 673♀) during that trip. He went on to document 57 species (+1 undetermined) and 3 varieties in that paper, all that were by then recorded from that island. He named and described 2 new genera, 11 new species, and 1 new variety from Ceylon in his paper. Earlier, only scattered descriptions of species or records from Sri Lanka were given by de Meijere, Wiedemann, Austen, Brunetti, Curran, Hull, Shiraki, and Stuckenberg. Besides Keiser's landmark work, there were other expeditions made to Sri Lanka, by the Lund Museum (Sweden) and by an extensive Smithsonian Institution multi-specialist team over many years, coordinated by Dr. Karl Krombein and published in Smithsonian publications. See Keiser (1958) for a faunistic paper. Gaonkar (*in prep.*) also wrote: "Its fauna is also closely related to that of the southern Western Ghats, many closely related species (sister-species) being found in both places. However, most of the Dipte. species, including the endemics, are now concentrated (and isolated) in the southwestern montane part of the island. This area is now regarded as a biodiversity "hot spot" area with very high diversity, needing immediate conservation efforts. The fauna is almost entirely Oriental, with weak representation from the Afrotropical region. There is only one species of Palaearctic affinity on the mountains of Sri Lanka.

SYSTEMATIC ANNOTATED CHECKLIST

Subfamily MICRODONTINAE

Tribe Microdontini

Archimicrodon lanka (Keiser, 1958)

Microdon lanka Keiser, 1958, Revue Suisse Zool., 65: 213 (♀; "Carney, Ratnapura, Ceylon") [NM, Basel]

Archimicrodon lanka: Reemer, *in* Reemer & Ståhls, 2013a; Zookeys, 288:129 (as *comb. n.*)

Sri Lanka (Keiser 1958; Knutson et al. 1975: 370; Cheng & Thompson 2008; Reemer & Ståhls 2013a; Ghorpadé 2014b: 6; Thompson et al. 2017: 509.)

Note: Cheng & Thompson (2008: 30) wrote: "*Archimicrodon* was proposed for those Australian species of *Microdon* which have antennae short, "quite short and about as long as in some species of *Syrphus*," which were considered a primitive character by Hull (1949). Reemer & Ståhls (2013a) completed a generic revision and species classification of all Microdontinae, based on morphological and molecular data (454 spp.), and I follow their conclusions made in this paper. However, they break up species-groups into some 67 groups, 43 of them as genera, but 7 as "subgenera" and 17 synonyms. Their Introduction needs to be read and become updated with this the smallest of the earlier recognized families of Syrphidae, and being the plesiomorphic taxon among this family. Now the Pipizinae are also considered a fourth subfamily, the smallest. Reemer & Ståhls (2013a: 19–21, 127–130, figs. 7–26) gave 45 described species of this genus which they split into three "subgenera"—also *Hovamicrodon* Keiser, 1971 and a "leftover group, which they called *Archimicrodon s.l.* Keiser (1958: 213) named this new species, after Ceylon also known as lanka, as "*Microdon lanka*" in his paper documenting specimens taken from Carney, near Ratnapura, 9.x.1953, taken from plants in a rubber estate.

I wish to clarify my objections to two aspects of current taxonomists' logic. First, as I wrote in my updated checklist also (Ghorpadé 2014b: 3) all "sub" taxa now recognized are ignored by me as they complicate nomenclatural matters and lumping genus-level taxa seems more advisable than splitting them into doubtful subgenera and giving them burdenous nomenclatural validity. Again, biogeography gives better leads to phylogeny (see also Ghorpadé 2007). The action of Reemer & Ståhls (2013a: 129), placing *simplex* Shiraki, 1930 treated as a "variety" of the Oriental *caeruleus* Brunetti, 1908 (Margherita, Arunachal Pradesh) as the only "Palaearctic" species of *Archimicrodon* is flawed, basing this on a single ♀ specimen taken also in Japan other than others found largely in Taiwan ! Thompson et al. (2017: 509) provide a synopsis of this "Old World genus of 39 species, 11 of them found in the PSWNG area."

Heliodon elisabeth (Keiser, 1958)

Microdon elisabeth Keiser, 1958, Revue Suisse Zool., 65: 211 (♂; "Kandy, Lady Hortons," Ceylon') [NM, Basel]

Archimicrodon elisabeth: Reemer, *in* Reemer & Ståhls, 2013a; Zookeys, 288: 131 (as *comb. n.*)

Sri Lanka (Keiser 1958; Knutson et al. 1975: 370; Reemer & Ståhls 2013a, b; Ghorpadé 2014b: 6; Thompson et al. 2017: 515.)

Note: Reemer & Stahls (2013a: 32) erected this new genus for eight Oriental species, diagnostically separate from those species retained in *Microdon s. str.* Keiser (1958: 211) named this new species, in honour of his wife and for her help and collaboration, as "*Microdon elisabeth*" in his paper documenting specimens taken from Lady Horton's, Kandy, 7.ix.1953, taken from flowers of bushes. Knutson et al. (1975: 370) listed it as a *Microdon* (*Microdon*) in their Oriental catalogue. Ghorpadé (2014b: 6) included this as the only species of this genus in our subcontinent. Thompson et al. (2017: 515) provide a synopsis of this Indo-Malayan genus of eight species, only one of them found in Wallacea of the PSWNG area.

Indascia gracilis Keiser, 1958

Microdon lanka Keiser, 1958, Revue Suisse Zool., 65: 223 (♂; "Bot. Gardens, Peradeniya, Ceylon") [NM, Basel]

Indascia gracilis: Reemer, *in* Reemer & Ståhls 2013a, Zookeys, 288: 132.

India: KN, TN; Sri Lanka (Keiser 1958; Knutson et al. 1975: 368; Cheng & Thompson 2008: 39; Reemer & Ståhls 2013a: 132; Ghorpadé 2014b: 6, unpubl.; Thompson et al. 2017: 515).

Note: Cheng and Thompson (2008: 30) wrote: "*Indascia* was originally placed in the tribe Spheginini, subfamily Cheilosiinae [= Ersitalinae]. The genus, however, clearly belongs to the subfamily Microdontinae and is very similar to *Paramicrodon*." Keiser (1958: 223–224, fig. 4) included this as "*Indascia gracilis*" in his paper on Sri Lanka, documenting specimens taken from Peradeniya Botanical Gardens, 10.vi.1953, on monsoon flowers on bushes, in Central Province. Reemer & Stahls (2013a: 34–35, 132) recognized four Oriental species (two new), with more undescribed species known to Reemer from India and Pakistan also. Ghorpadé (2014b: 6) listed two species from our sub-continent, the type-species *brachystoma* Wiedemann, 1824 with unknown locality ("Ind. Or."). I have seen specimens (Ghorpadé, unpubl.) of what is probably this species, from Pakistan, and from localities in Karnataka, Tamil Nadu, and Delhi in India that are still to be documented in publications. Thompson et al. (2017: 515) provide a synopsis of this genus of four described species, five others undescribed from the PSWNG area.

Metadon auricinctus (Brunetti, 1908)

Microdon auricinctus Brunetti, 1908, Rec. Indian Mus., 2:93 (♀; "Kandy, Ceylon") [British Museum (Natural History) (BMNH), London]

Metadon auricinctus: Reemer, *in* Reemer & Ståhls, 2013a; Zookeys 288:134 (as *comb. n.*)

Sri Lanka (Brunetti 1908, 1910, 1923, 1925; Shiraki 1930; Keiser 1958; Gokulpure 1972; Knutson et al. 1975: 369; Reemer & Ståhls 2013a: 41–43, 109–110, 133–135, figs. 163–175; Ghorpadé 2014b: 6; Mitra et al. 2015; Thompson et al. 2017: 516.)

Note: Reemer & Ståhls (2013a: 34–35, 132) recognized around 40 species (22 Oriental, about one-half) of this new genus, providing a diagnosis and discussion. On p. 134, however, note that they misspelled *pendleburyi* Curran, 1931b! Brunetti (1923: 309–310), in his key gave its abdomen as short and rounded or elongate-conical, scutellum oblong or rounded, and not produced into teeth, large, non-metallic species, 13 mm long, surface of wing without trace of scales, with abdomen in considerable part red and thorax black. Brunetti (1910: 171) wrote "the type (male) [*sic*] is from Kandy, X.07, whilst I possess females from Kandy, IV., V., and IX." He added abdomen "brown with gold pubescence in *flavipes*." Gokulpure (1972) recorded it from Damoh in Madhya Pradesh. Knutson et al. (1975: 369), Ghorpadé (2014b: 8, 2015: 17), and Mitra et al. (2015: 62) listed it. Thompson et al. (2017: 516) provide a synopsis of this Old World genus of 42 species, 6 of which occur in the PSWNG area.

Metadon montis (Keiser, 1958)

Microdon montis Keiser, 1958, Revue Suisse Zool., 65: 214 (♂; "Pidrutalagala, Ceylon") [NM, Basel]

Metadon montis: Reemer, *in* Reemer & Ståhls 2013a; Zookeys 288: 134 (as *comb. n.*)

Sri Lanka (Keiser 1958; Knutson et al. 1975: 371; Reemer & Ståhls 2013a: 41–43, 109–110, 133–135, figs. 163–117; Ghorpadé 2014b: 6).

Note: Reemer & Ståhls (2013a: 42, 134) gave these diagnostic characters that separate their *Metadon* from *Microdon*: anepisternum (almost) entirely pilose; phallus projecting not or only little beyond apex of hypandrium; aedeagus furcated in apical half. Keiser (1958: 214) named this new species, after montane Sri Lanka, as "*Microdon montis*" in his paper documenting specimens from Pidrutalagala, 2100–12,300 m, 30.v.1953, taken from flowers of a herb/shrub, *Osbeckia wightiana* (Melastomaceae).

Metadon taprobanicus (Keiser, 1958)

Microdon taprobanicus Keiser, 1958, Revue Suisse Zool., 65: 212 (♂; "Kandy, Hantana, Ceylon") [NM, Basel]

Metadon taprobanicus: Reemer, *in* Reemer & Ståhls 2013a; Zookeys 288: 135 (as *comb. n.*)

Sri Lanka (Keiser 1958; Knutson et al. 1975: 372; Reemer & Ståhls 2013a: 41–43, 109–110, 133–135, figs. 163–175; Ghorpadé 2014b: 6).

Note: Keiser (1958: 211) named this new species, after Ceylon, also known as taprobana, as "*Microdon taprobanicus*" in his paper documenting specimens from Hantana, Kandy, 23.xii.1953, taken from grasses on the path in a tea estate. Reemer and Ståhls (2013a: 42, 134) gave these diagnostic characters that separate their *Metadon* from *Microdon*: anepisternum (almost) entirely pilose; phallus projecting not or only little beyond apex of hypandrium; aedeagus furcated in apical half.

[Species Incertae Sedis:

Metadon flavipes Brunetti, 1908

Microdon flavipes Brunetti, 1908, Rec. Indian Mus., 2: 92 (♀♀; "Mergui, Lower Burma") [Zoological Survey of India, Calcuttta—examined]

Metadon flavipes: Reemer, *in* Reemer & Ståhls 2013a; Zookeys 288: 134 (as *comb. n.*)

Sri Lanka ? (Brunetti 1908, 1910, 1923; Knutson et al. 1975: 370; Reemer & Ståhls 2013a: 134; Ghorpade (2014b); Mitra et al. 2015: 62).

Note: Brunetti (1910: 171) gave this species with a query, noting "A large female from 'Nitre Cave,' Ceylon, VII. 02, is apparently this species, which was described (Rec. Ind. Mus., II., 92) from specimens in the Indian Museum." Nothing further has been published on this "large female," so it remains in doubt. See also under *auricinctus*. This species was transferred to *Metadon* by Reemer and Ståhls (2013a). Knutson et al. (1975: 370), Ghorpadé (2014b: 6), and Mitra et al. (2015: 6) listed it.

Microdon fulvopubescens Brunetti, 1923

Microdon fulvopubescens Brunetti, 1923, *Fauna Br. India*, Dipt., 3: 313 (♀; "Ceylon") [BMNH, London]

Microdon fulvopubescens: Reemer, *in* Reemer & Ståhls, 2013a; Zookeys 288: 139.

Sri Lanka (Brunetti 1923; Knutson et al. 1975: 370; Reemer & Ståhls 2013a: 47–52, 110–114, 137–142, figs. 196–214; Ghorpadé 2014b: 6, 2015b: 9. Thompson et al. 2017: 516.)

Note: Reemer & Ståhls (2013a: 48) diagnosed and described Microdon s. str. (and s. lat. and two other "subgenera"), recognizing 126 species of Microdon (62 s. str., besides 9 s. str., and 4 s. str., s. lat., all Oriental). Brunetti (1923: 309) in his key gave its abdomen as short and rounded or elongate-conical, scutellum with hind corners produced into an obvious blunt point or tooth, large, non-metallic species, 13 mm long, and with abdomen being much broader than thorax. I (Ghorpade 2015b: 9) wrote about this species as an "incertae sedis" referring to an incorrect record from Pakistan. Thompson et al. (2017: 516) provide a synopsis of this practically cosmopolitan genus of 126 species, three of them occurring in the PSWNG area.

Microdon unicolor Brunetti, 1915

Microdon unicolor Brunetti, 1915, Rec. Indian Mus., 11: 255 (♂; "Puri, Orissa") [ZSI, Calcutta—examined]

Microdon (s.lat.) unicolor: Reemer, *in* Reemer & Ståhls, 2013a, Zookeys, 288: 142.

India: AH ?, OR (Brunetti 1915, 1923; Knutson et al. 1975: 372; Ghorpadé et al. 2011; Reemer & Ståhls 2013a: 47–52, 110–114, 137–142, figs. 196–214; Ghorpadé 2014b: 6; Mitra et al. 2015:62).

Note: Reemer & Ståhls (2013a: 142) treated this as an unplaced species of *Microdon s.l.* In March 1981, I found the holotype male in ZSI, Calcutta, but with head lost. Brunetti (1915: 255) described this "perfect ♂" taken from near Puri in Orissa in November 1912 and collected by Gravely, and stated, "the only other violet black species from the East is *sumatranus*, Wulp, which is punctuated freely on the body and legs with white hair spots." In the *Fauna* volume (Brunetti 1923: 318), the original page number was mistakenly printed as "225" for 255, and in his key to Indian *Microdon* (p. 309–310), he gave its abdomen as short and rounded or elongate-conical, scutellum oblong or rounded and not produced into teeth, large, non-metallic species, 10–11 mm long, wing surface without traces of scales, abdomen in no part red, and thorax being dark shining violet. Ghorpadé et al. (2011: 83) mentioned this described from near Puri (Orissa) and as may be occurring in the Andhra Carnatic. Knutson et al. (1975: 372), Ghorpadé (2014b: 6), and Mitra et al. (2015: 62) listed it.

Paramixogaster contractus (Brunetti 1923)

Microdon contractus Brunetti, 1923, *Fauna Br. India*, Dipt., 3: 31o (♀; "Deesa, Bombay Presidency") [BMNH, London]

Paramixogaster contractus: Reemer, *in* Reemer & Ståhls 2013a; Zookeys 288: 145. (as *comb. n.*)

India: GJ (Brunetti 1915, 1923; Knutson et al. 1975: 370, 373; Reemer & Ståhls 2013a: 58–59, 114–115, 144–145, figs. 272–275; Ghorpadé 2014b: 6, 2015b: 9–10; Mitra et al. 2015: 62; Thompson et al. 2017: 517.)

Note: Reemer & Ståhls (2013a: 142) listed 12 Oriental and another nine "Australian/Oceanian" species of this genus and wrote that *Paramixogaster contractus* (Brunetti 1923), and *P. conveniens* (Brunetti 1923) are aberrant from all other known species of *Paramixogaster* in their complete transverse suture, a bare postpronotum, and in presence of pile on metaepisternum. The "unique ♀" was collected by C.G. Nurse from Deesa in northern Gujarat in March 1897. Diagnostics given were second abdominal segment distinctly, but not greatly contracted in middle, making the abdomen appear subclavate, with a small rounded scutellum, clear wings with fourth vein recurrent at a right angle, and orange legs. Brunetti (1923: 309–311, 319–321) had placed this and *conveniens* in *Microdon* though erecting *Paramixogaster* as a new genus with *vespiformis* Brunetti as the type-species, having diagnostic wing venation and peculiar antennae. Ghorpadé (2015b: 9–10) included this in his N.W. Frontier Syrphidae paper. Knutson et al. (1975: 372), Ghorpadé (2014b: 6), and Mitra et al. (2015: 62) listed it. Thompson et al. (2017: 517) provide a synopsis of this Old World tropical genus of 26 species, four of them found in the PSWNG area.

Subfamily ERISTALINAE

Tribe Brachyopini

Myolepta sp.

> India: KN (Brunetti 1915, 1923; Sack 1932; Thompson 1971, 1974; Knutson et al. 1975: 337; Wiegmann 1986; Thompson & Rotheray 1998; Ghorpadé 2014b: 7, 2015b: 12; Mitra et al. 2015: 62; Thompson et al. 2017: 517.)

Note: I have a single ♀ specimen that I recovered while curating flies from my extensive papered material of Diptera, including Syrphidae. This was labelled "India: Karnataka: Gulbarga, 454m, 21.ix.1980, A.R.V. Kumar coll. No. 204 [KGC]." I was surprised to find this to be a *Myolepta*, this being the first of this genus sampled from anywhere outside of the Himalayas! The type-locality, Gulbarga, is bang on the dry Deccan Plateau (in Karnataka, S. India), and one wonders what its immature stages are up to there, a habitat so diametrically different from the mountainous, humid, forested Himalaya range? This Gulbarga female specimen has an elongate flagellum and goes to the *luteola* group in Thompson (1974: 326). It has no facial tubercle and so cannot be a *Lepidomyia* Loew and remains the fourth Oriental *Myolepta* species with elongate antennae (Thompson 1971: 343) pending description and a new name. See Ghorpadé (2015a: 12) for information about *M. himalayana* Brunetti, the first *Myolepta* known from the Indian subcontinent, taken on the Simla Hills in north-west Himalaya. The other two *Myolepta* newly described by Thompson (1971) were taken in the forest in the north-west Chiangmai province of Thailand. This Gulbarga female is closer to *orientalis* Thompson with an oval not petiolate abdomen and the dark scutellum with a distinct yellow posterior margin. All the African species (Thompson 1974) were also taken in the forest, and it remains to be discovered if A.R.V. Kumar had actually taken it in once remnant ancient forest near Gulbarga, but his information was that it was taken well inside the small town it was then (1980), now built up, and nothing like an undisturbed forest! On my visit to Gulbarga, I saw the apparent "type-locality" and was disappointed that it was located inside town! See also Wiegmann (1986). Thompson et al. (2017: 517) provide a synopsis of this almost cosmopolitan (except Australian) genus of 42 species, none so far found in the PSWNG area. This specimen will be sent to Dr. F.C. Thompson, U.S.A. for naming and description,

Tribe Cerioidini

Ceriana ornatifrons (Brunetti 1915)

> *Ceria ornatifrons* Brunetti, 1915, Rec. Indian Mus., 11: 252, Pl. XIII, fig. 22 (♀; "Kumdhik, Nepal Hmalaya") [ZSI, Calcutta—examined]
> Sri Lanka (Brunetti 1915, 1923; Shannon 1927; Sack 1932; Keiser, 1958; Knutson et al. 1975: 344; Thompson & Rotheray 1998; Ghorpadé 2014b: 8, 2015b: 17; Mitra et al. 2015; Thompson et al. 2017: 513.)

Note: Brunetti (1915: 252, 1923: 331) described this, as a *Ceria*, from the base of the Nepal Himalayas, 22.iii.1909, and also mentioned specimens from Singla, 1500 ft, Darjeeling district, iv.1913, taken by Lord Carmichael's collector. The holotype was seen by me in Box 17 at the ZSI, Calcutta in 1981. Shannon (1927: 51) listed it from Sri Lanka. Keiser (1958: 227) included this as "*Tenthredomyia ornatifrons*" in his paper on Sri Lanka, documenting specimens taken from north-western and western provinces. See also Sack (1932: 339) and Thompson and Rotheray (1998: 109) for discussion and key to this genus in the Palaearctic. Knutson et al. (1975: 344), Ghorpadé (2014b: 8, 2015b: 17), and Mitra et al. (2015: 63) listed it. Thompson et al. (2017: 513) provide a synopsis of this almost cosmopolitan (except Neotropics) genus of 50–60 species, three so far found in the PSWNG area.

Monoceromyia eumenioides (Saunders 1842)

> *Ceria eumenioides* Saunders, 1842, Trans. ent. Soc. Lond., 3: 60 (♀; "northern India") [BMNH, London]
> *Ceria eumenoides*: Brunetti 1908, Rec. Indian Mus., 2: 96; Bhatia 1931, Indian J. Agric. Sci., 1(4): 503 [*lapsus calami*]
> *Ceria apicata* Brunetti, 1908, Rec. Indian Mus., 2: 95 (♂; "Ganjam, Madras Presidency") [ZSI, Calcutta—examined]
> *Ceria apicata* Bigot, *nomen nudum*: Brunetti 1908, Rec. Indian Mus., 2: 96.
> India: BI, KN, MH, OR, TN (Brunetti 1923; Shannon 1927; Bhatia 1931; Beeson 1953; Knutson et al. 1975: 344; Ghorpadé 2014b: 8, 2015b: 17, 2017, unpubl.; Mitra et al. 2015; Thompson et al. 2017: 517.)

Note: Brunetti (1908: 96) mentioned two female specimens in the ZSI, Calcutta collection. In the *Fauna* volume (Brunetti 1923: 337–338, P. VI, fig. 8), he gave localities: "Kohat, North-West Frontier Province, 7.v.1916 (*Fletcher*), Matheran, Bombay Presidency, iii. & v.1899 (*Nurse*), Chapra, Bengal [*sic*] (*Mackenzie*), Calcutta, Abbottabad, Hazara District, 21.v.1915 (*Fletcher*), and northern Bengal (Saunders Collection)." Also from "Ganjam (Madras Presidency) [= Orissa] for *apicata*. Ghorpadé (unpubl.) collected it in Karnataka and in Coimbatore (Tamil Nadu). Bhatia (1931: 503–508, Pl. LVI) gave details of its life history at Pusa (Bihar) where a gravid female was caught in April on flowing sap of a siris (*Albizia lebbeck*) tree. Beeson (1953: 339) gave this as breeding in sap of fermenting *Albizia lebbeck* and quoted Bhatia (1931). I included this from Maharashtra (Ghorpadé 2017) based on two ♀♀ from Matheran documented by Brunetti (1923: 38) taken by Nurse at Matheran in March and May 1899. I saw 2 ♀♀ in the Tamil Nadu Agricultural University (TNAU), Coimbatore museum taken from tree holes in Coimbatore in March 1933 (K. Ghorpade unpubl.). Shannon (1927: 52), Knutson et al. (1975: 344), Ghorpadé (2014b: 8, 2015b: 17), and Mitra et al. (2015: 63) listed it. Thompson et al. (2017: 517) provide a synopsis of this cosmopolitan genus of 75 species, three so far found in the PSWNG area.

Monoceromyia javana (Wiedemann 1824)

Ceria javana Wiedemann, 1824, Analecta Ent., p. 32 (♀; "Java") [Universitetes Zoologiske Museum, Copenhagen ?]

Ceria vittigera Bigot, *nomen nudum*: Brunetti, 1908, Rec. Indian Mus., 2: 96.

India: KN ? (Brunetti 1908, 1915, 1923; Shannon 1927; Knutson et al. 1975: 344; Ghorpadé 2014b: 8, 2015b: 17; Mitra et al. 2015)

Note: Brunetti (1908: 96) wrote, "A specimen from the Naga Hills in the collection [ZSI, Calcutta], labelled *C. vittigera*, Big., is only *C. javana*, Wied., the former name being a *nomen nudum*." Brunetti (1915: 251) then wrote "I have seen several specimens of *C. javana*, W., ♀" [in his own collection]. In the *Fauna* volume (Brunetti 1923: 339–340, fig. 71), he gave a full description and a figure based on several specimens "in the British and Indian Museums. Darjeeling district, 1000–3000 ft., v. 1912 (*Lord Carmichael*); Sukna, 500 ft., 2.vii.1908 (*Annandale*); Sidapur, Coorg, S. India, 3000 ft., 29.iv.1917; Margherita, Assam." I suspect that the material from Coorg (Karnataka) may be misidentified? I have seen material from Karnataka (K. Ghorpadé unpubl.). Shannon (1927: 52), Knutson et al. (1975: 344), Ghorpadé (2014b: 8, 2015b: 17), and Mitra et al. (2015: 63) listed it.

Tribe Eristalini

Digulia kochi de Meijere, 1913

Digulia kochi de Meijere, 1913, Nova Guinea, 9: 357 (♀; "Digul River, New Guinea") [ZM, Amsterdam]

Sri Lanka (Keiser 1958; Ghorpadé, 2014b: 8; Knutson et al. 1975: 347; Thompson 2003; Thompson et al. 2017: 514.)

Note: Knutson et al. (1975: 347) first pointed out the presence of this genus and species from the Indian subregion, based on Keiser (1958). Keiser (1958: 236) included this as "*Digulia kochi*" in his paper on Sri Lanka, documenting specimens taken from Elephant Pass and Kanniyai in the Northern and Eastern Provinces. Thompson (2003) keyed out this genus among others of Eristalina. Knutson et al. (1975: 347), and Ghorpadé (2014b: 8) listed it. Thompson et al. (2017: 517) provide a synopsis of this small genus of a single species restricted to New Guinea.

Dolichomerus crassa (Fabricius 1787)

Syrphus crassus Fabricius, 1787, Mantissa Insect., 2: 334 (sex?; "Tranquebar, Madras, India") [?]

India: AH, BI, GO, JH, KL, KN, MH, MP, TN; Sri Lanka (Brunetti 1908, 1910, 1913, 1915, 1923; Cherian 1934; Keiser 1958; Gokulpure 1972; Ghorpadé 1973b, 2014b: 8, 2015b: 19, 2017; Knutson et al. 1975: 358; Datta & Chakraborti 1986; Joseph & Parui 1986; Ghorpadé et al. 2011; Mitra et al. 2015; Thompson et al. 2017: 514.)

Note: Brunetti (1908: 71–74) gave a key to Oriental species of *Megaspis* including this (as *crassus*, F.) and remarked, "This is a good genus, but the roughness of the frons, which distinguishes it, is not always easily visible." He mentioned differences of this from *sculptata* van der Wulp [as "*sculptatus*," now known only from Timor, Indonesia]. Brunetti (1910: 171) later noted this (as *Megaspis*) being "generally common throughout the summer" in Sri Lanka. Cherian (1934: 699) listed this (as *Megaspis crassus*) from Coimbatore, Saidapet (Chingleput) [Tamil Nadu], Kurnool, Amalapuram (Godavari) [Andhra Pradesh], noting flies were collected resting on safflower and that "a tooth on the hind femora is an important mark of the species." Keiser (1958: 237) included this as "*Megaspis (Dolichomerus) crassus*" in his paper on Sri Lanka, documenting specimens taken from Sabaragamuva and North-West, North-Central, Western, and Central Provinces. Gokulpure (1972) recorded it from Damoh (Madhya Pradesh). In my masters thesis (Ghorpadé 1973b, 1974), I had written, "This striking jet black species was rather rare around Bangalore, taken only once on niger flowers around Bangalore." Datta & Chakraborti (1986: 63, fig. 5) mentioned a male taken at Idamalaya in the Ernakulam district (Kerala) in December 1979, illustrated its male terminalia, and gave southern states where it has been reported, as Andhra Pradesh, Bihar, Goa, Karnataka, Kerala, and Tamil Nadu. Joseph & Parui (1986: 161) gave it from Kumattan Thodu in Kerala in December 1980 and stated it is well distributed in India. Ghorpadé et al. (2011: 81) recorded it from the Andhra Carnatic and gave notes. I have also taken it at Coimbatore (K. Ghorpadé unpubl.). I also included it from Maharashtra (Ghorpadé 2017) based on a ♀ taken in Bombay [=Mumbai] in September 1928 by P.F. Gomes (Ghorpade 2017: 107). Knutson et al. (1975: 358), Ghorpadé (2014b: 8, 2015b: 19), and Mitra et al. (2015) listed it. Thompson et al. (2017: 514) provide a synopsis of this small Indo-Malayan genus of a single species, found in the PSWNG area.

Eoseristalis cerealis (Fabricius 1805)

Eristalis cerealis Fabricius, 1805, Syst. Antliat., p. 232 (♀; "China") [UZM, Copenhagen]

Eristalis solitus Walker, 1849, List Dipt. Colln Br. Mus., 3: 619 (♂; "Nepal") [BMNH, London]

Eristalis incisuralis Loew 1858, Wien. Ent. Mschr., 2: 108 (♂♀; Japan) [?]

Eristalis barbata Bigot 1880, Annls Soc. Ent. Fr., (5) 10: 214 (♂; "Indostan") ["Bigot Colln"?BMNH, London]

Eristalis lunar Nayar 1968b, Agra Univ. J. Res. (Sci.), 16(3): 27 (♂; "Kalatop") [ZSI, Calcutta—examined]; Ghorpadé 2014c, Colemania, 46: 2 (as *n. comb., n. syn.*)

Eristalis (Eoseristalis) cerealis Fabricius: Knutson et al. 1975, Cat. Dipt. Orient. Reg., p. 351 (Catalog).

India: CG, TN ? (Brunetti 1908, 1915, 1923, 1925; Hervé-Bazin 1914, Knutson et al. 1975: 351; Thompson 2003; Ghorpadé 2014b: 8, 2015b: 21, unpubl.]

Note: Brunetti (1908: 71) wrote of his identification of this species as *solitus* being corroborated by Mr. Austen, and that a "considerable series were seen in the Indian Museum collection." Knutson et al. (1975: 351) gave its distribution as "throughout Oriental region," but specimens from southern high altitude areas need to be confirmed as conspecific with this chiefly north Indian and Himalayan species. I (K. Ghorpadé unpubl.) have specimens from the Nilgiris, but these need to be properly determined. Thompson (2003) may be consulted for information on what he recognized as *Eristalis* (*Eoseristalis*) and provided a key to this genus (subgenus?).

Eoseristalis curvipes (Schiner 1868)

Eristalis curvipes Schiner, 1868, Reise der österreichischen Fregatte Novara, p. 363 (♂♀; "Ceylon") [Naturhistorisches Museum, Vienna]

Eristalis (*Eoseristalis*) *curvipes*: Knutson et al. 1975, Cat. Dipt. Orient. Reg., p. 351 (Catalog).

India: CG ?, TN; Sri Lanka (Brunetti 1923, 1925; Keiser 1958; Knutson et al. 1975: 351; Ghorpadé 2014b: 8, unpubl.)

Note: Brunetti (1923: 192, 412) copied the description of Schiner, and then made his own, based on specimens he saw from Ceylon. He wrote, "From the grey transverse cross-band on the suture [on mesonotum] it appears to be allied to *latus* and *suturalis*." He expanded on its diagnostic characters thus: "the brown band of pubescence on the eyes, the extremely strongly produced, snout-like lower part of the head and the strikingly bent hind tibiae which is less noticeable in the female." I have taken specimens (K. Ghorpadé unpubl.) at Ootacamund (7500 ft), Nilgiri Hills, in December and at Yercaud, Shevaroy Hills, in May, October, and December. See also notes under *simulata* (Brunetti) below.

Eoseristalis simulata (Brunetti 1923)

Eristalis simulatus Brunetti 1923, *Fauna Br. India*, Dipt., 3: 31o, Pl. IV fig. 1 (♂; "Ootacamund, 7500 ft.") [BMNH, London]; Knutson et al. 1975, Cat. Dipt. Orient. Reg., p. 353 (Catalog).

India: TN (Brunetti 1923, 1925; Knutson et al. 1975: 353; Mahalingam 1988; Ghorpadé 2014b: 9, unpubl.)

Note: Brunetti (1923: 177–179) named and described this from "some ♂♂ and ♀♀ in the Pusa collection from Ootacamund, 7500 ft., 24–31.xii.1913 (Fletcher)." He wrote "the Type ♂ and ♀ was sent to the British Museum and that it closely resembled *E. angustimarginalis* [Brunetti]." But *simulatus* has "a conspicuously conical epistome; and the hind tibiae are rather abruptly bent in the middle, rather less so in the ♀." Curiously, no mention was made of its resemblance to *curvipes* (Schiner) from Ceylon (!) both included by him in his *Fauna* volume. Mahalingam (1988: 161) gave it as recorded from the Nilgiri Hills among ten species recorded by Brunetti (1923). Thompson (*in litt.*) considers this a synonym of *curvipes* (Schiner), but this needs to be investigated by comparing male terminalia and morphology of these two species with modified hind legs, as they are biogeographically allopatric.

Eristalinus aeneus (Scopoli 1763)

Conops aeneus Scopoli 1763, Ent. Carniolica, p. 356 (sex?; "Idria," = Idrija, Yugoslavia) [type destroyed ?]

Eristalis taphicus Wiedemann 1830, Aussereurop. Zweifl. Insekt., 2: 191 (♀; "Egypten") [UZM, Copenhagen?]

Eristalis ridens Walker 1849, List Dipt. Brit. Mus., 3: 610 (♀; "Egypt") [ZM, Frankfurt]

Lathyrophthalmus aeneus var. *nigrolineatus* Hervé-Bazin, 1923b, Annls Sci. Nat., (10) 6: 134 (♂♀; "Kurrachee") [Muséum national d'histoire naturelle, Paris]

India: AN, CG, GJ, KN, MH, (Verrall 1901; Brunetti 1915, 1923; Hervé-Bazin 1923b, 1924; Gokulpure 1972; Knutson et al. 1975: 347; Peck 1988; Parui et al. 2002; Thompson 2003; Ghorpadé 2014b, c, 2015b: 26–27, 2017; Thompson et al. 2017: 514–515.)

Note: Brunetti (1915: 229) wrote of *taphicus*: "A few in the Indian Museum from Karachi, both sexes." However, Brunetti (1923: 163) concluded that: "Though regarded by many authors as a variety of *aeneus*, Scop., *E. taphicus* appears to be a valid species." He then goes on to give the differences and puts *ridens* Walker as a synonym of *taphicus* Wiedemann. Note that Hervé-Bazin (1923b: 141) gave *ridens* Walker as a synonym of *quinquelineatus* Fabricius which the latter is an African species! These synonymies still remain to be checked by study of terminalia, especially, and based on a larger series of specimens. Hervé-Bazin (1924: 293) treated this (as "*taphicus*") which he mentioned is a synonym of *aeneus*, like *ridens* is. Gokulpure (1972: 848) recorded this (as "*Eristalis*") from Damoh in Madhya Pradesh. Parui et al. (2002) recorded a male of what they determined as *Eristalinus aeneus* var. *taphicus* (Wiedemann) from Casuarina Bay, Great Nicobar in 1966 taken by A. Daniel and H.K. Bhowmick, which is another southernmost record of a species of *Eristalinus*. I included this from Maharashtra (Ghorpadé 2017) based on "*Eristalis taphicus*" documented by Brunetti (1923: 163) from Bombay on 2 and 21 March 1905 on seaweed. Knutson et al. (1975: 347), Ghorpadé (2014b: 9), and Mitra et al. (2015: 63) listed it. Thompson (2003: 4) included it in his useful key to genera of the "subtribe" Eristalina, as "*Eristalinus* (*Eristalinus*)." Thompson et al. (2017: 514–515) provide a synopsis of this worldwide genus of 87 species, with 17 (of "subgenera" *Eristalodes* and *Lathyrophthalmus*) found in the PSWNG area. This genus *Eristalinus* in the Oriental—Papuan area requires detailed revision and many synonymies remain to be discovered. The taxonomy of this genus ("subgenus" ?) is in a terrible state and only study of type-specimens and a large collection can solve the problem.

Eristalinus arvorum (Fabricius 1787)

Syrphus arvorum Fabricius 1787, Mantissa Insectorum, 2: 335 (♂♀; China) [?]

Syrphus quadrilineatus Fabricius 1787, Mantissa Insectorum, 2: 336 (♂?; Tranquebar) [UZM, Copenhagen?]

Musca tranquebarica Gmelin, 1790, Syst. Nat., 5: 2870 (unjustified *nom. nov.* for *quadrilineatus* Fabricius).

Eristalis fulvipes Macquart, 1846, Dipt. Exot., Suppl., 1: 256 (128) (♂; "Nouvelle-Hollande" = Queensland, Australia) [MNHN, Paris?]

Eristalis anicetus Walker, 1849, List Dipt. Colln Br. Mus., 3: 624 (♂; unknown) [BMNH, London]

Eristalis antidotus Walker, 1849, List Dipt. Colln Br. Mus., 3: 626 (♂; China) [BMNH, London]

Eristalomyia fo Bigot, 1880, Annls Soc. Ent. Fr., (5) 10: 220 (♂; Amoy = Xiamen in Fujian, China) [MNHN, Paris]

Eristalomyia eunotata Bigot, 1891, Nouv. Archs Mus. Hist. nat. Paris, (3) 2: 208 (♂; Laos) [MNHN, Paris]

Eristalis okinawensis Matsumura, 1916, Thous. Insects Japan, Add. 2, p. 261 (♂; Okinawa, Ryukyu Is) [HU, Sapporo]

Eristalis (*Lathyrophthalmus*) *haileyburyi* Nayar, 1968, Pan-Pacific Ent., 44: 121 (♀; "St. John's College, Agra, India") [ZSI, Calcutta—examined]; Ghorpadé, 2014c, Colemania, 46: 4. (as *n. syn.*)

India: peregrine in subcontinent ?; Sri Lanka (Brunetti 1915, 1923, 1925, Hervé-Bazin 1923b, c, 1924, 1926; Cherian 1934; Keiser 1958; Baid 1959; Joshee 1968; Nayar 1968c; Gokulpure 1972; Ghorpade 1973b, 2014b: 9, 2015b: 27–28, 2017; Knutson et al. 1975: 347; Joseph & Parui 1977, 1986; Panchabhavi & Rao 1978; Peck 1988; Ghorpade et al. 2011; Mitra et al. 2015).

Note: Brunetti (1915: 228) wrote, "The species is the commonest of the Indian ones (of "*Eristalis*") and occurs apparently all over the country from the Himalayas to the south. It has been found by Dr. Annandale breeding in rotting seaweed in brackish water at Lake Chilka, Orissa, in February and November." Brunetti (1923: 183) wrote "It is very common apparently all over India, in hills and plains, and occurs probably in all parts of the Orient. The actual dates run from March to October," and then continued with its synonymy. He also made some corrective notes (Brunetti 1925: 76–77). Hervé-Bazin (1923b: 145–146, figs. 20–21) gave two nice figures of both sexes along with some notes. Hervé-Bazin (1923c: 174) mentioned a misspelling "*acervorum* Macquart" and gave "Bengale and Bombay" as localities from where this was known. Hervé-Bazin (1924: 294) gave *fo* Bigot and *eunotata* Bigot as synonyms, in addition to Brunetti's (1923: 181) synonymy. Cherian (1934: 698) listed it from Coimbatore (Tamil Nadu), Hadagalle, Bellary (Karnataka), Devakonda, Amalapuram, and Madanapalle (Andhra Pradesh). Keiser (1958: 194) included this as "*Asarcina* (*Dideopsis*) *aegrota*" in his paper on Sri Lanka, documenting specimens taken from Uva, Sabaragamuva, and all other, Northern, Eastern, North-Western, Western, North Central, and Central Provinces. Keiser (1958: 194) included this as "*Eristalis* (*Lathyrophthalmus*) *arvorum*" in his paper documenting almost 200 specimens taken mostly from plains areas on flowers of *Euphorbia antiquorum* and *Mimosa* sp. Baid (1959) gave a "*Eristalis* sp." flying from September to March around the Sambhar Lake in Rajasthan, which could be this common species. Joshee (1968) found rat-tailed maggots of doubtful name (as *Eristalis tenax*? *q.v.*) from bull frog stomachs in the Greater Bombay area; which could be this widespread species. Gokulpure (1972: 848) recorded it from Damoh (central M.P.). I examined the sole holotype ♀ of *haileyburyi* Nayar (1968c: 121, figs. 2, 4, 6) in the collection of ZSI, Calcutta, labelled "Haileybury House/St. John's College, Agra, India/3.iii.1960/J.L. Nayar/*Eristalis* (*Lathyrophthalmus*) *haileyburyi* Nayar, Holotype." It is nothing but *L. arvorum* (Fabricius) and was synonymized by me (Ghorpadé 2014c: 4). I have also taken specimens (K. Ghorpade, unpubl.) from Tamil Nadu at Coimbatore and Yercaud. In my masters thesis (Ghorpade 1973b, 1974), I had written "This species was rather uncommon around Bangalore, adult flies were secured on niger and mango flowers and in wheat fields, similar to *obliquus*, male face in profile with prominent tubercle as in that species; hairs on frons denser and black, hind femora wholly yellowish or orange-yellow, never black." Joseph & Parui (1977: 230) listed specimens taken at Basia in Ranchi district (Jharkhand) in December 1967, mentioning it is common all over India. Joseph & Parui (1986: 161) listed it from "towards Mukkali" in December 1980 and wrote it is very common all over India. Panchabhavi & Rao (1978: 254) reported it visiting flowers of niger (*Guizotia abyssinica*) in northern Karnataka. Ghorpadé et al. (2011: 81) recorded it from the Andhra Carnatic and gave detailed notes on this species from southern India. I included this from Maharashtra (Ghorpadé 2017: 108) based on Brunetti (1923: 183) stating, "It is very common apparently all over India, in hills and plains, and occurs probably in all parts of the Orient. Knutson et al. (1975: 347), Ghorpadé (2014b: 9), and Mitra et al. (2015) listed it.

Eristalinus aurulans (Wiedemann 1824)

Eristalis aurulans Wiedemann 1824, Analecta Ent., p. 37 (sex ?; "Trevancour," India) [type destroyed ?]

Eristalis kochi de Meijere 1908, Tijdschr Ent., 51: 255 (♂♀; "Etna-Bai, Meranke," New Guinea) [ZM, Amsterdam?]

Lathyrophthalmus diffidens Curran 1947, Amer. Mus. Novitates, 1364: 16 (♂; "Guadalcanal," Solomon Is.) [American Museum of Natural History, New York]

India: KL (Brunetti 1915: 230; Hervé-Bazin 1923b, 1924; Knutson et al. 1975: 348; Thompson & Vockeroth 1989; Thompson 2003; Ghorpadé 2014b: 9; Mitra et al. 2015).

Note: Wiedemann named and described this from what was labelled as Travancore, S., Kerala, which could have been mislabelled ? Brunetti (1915: 230) listed *kochi* de Meijere, that was a synonym, from New Guinea. Curran described another synonym as *diffidens*, from the Solomon Islands." Knutson et al. (1975: 348), Thompson & Vockeroth (1989: 449), Ghorpadé (2014b: 9), and Mitra et al. (2015) listed it.

Eristalinus invirgulatus (Keiser, 1958)

Lathyrophthalmus invirgulatus Keiser 1958, Revue Suisse Zool., 65: 231 (♂; "Kandy, Hantana, Ceylon") [NM, Basel]

Eristalinus invirgulatus: Knutson et al. 1975, Cat. Dipt. Orient. Reg., p. 348 (Catalog) (as *n. comb.*)

Sri Lanka (Keiser 1958; Knutson et al. 1975: 348; Thompson 2003; Ghorpadé 2014b: 9; Mitra et al. 2015).

Note: Keiser (1958: 23194) named and described this as "*Eristalis (Lathyrophthalmus) invirgulatus*" in his paper on Sri Lanka, documenting one ♂ taken from Hantana, Kandy, 14.x.1953, taken on flowers on plants in a tea estate in the Central Province. Ghorpadé (2014b: 9) and Mitra et al. (2015: 63) listed it.

Eristalinus laetus (Wiedemann 1830)

Eristalis laetus Wiedemann 1830, Aussereurop. Zweifl. Insekt., 2: 192 (♂; "China") [UZM, Copenhagen]

Eristalis pallinevris Macquart 1842, Dipt. Exot., 2(2): 46 (♂; "Bengal, India") [MNHN, Paris]

Eristalis quinquefasciatus Schiner, 1849: *nomen nudum*

Lathyrophthalmus ishigakiensis Shiraki, 1968, *Fauna Japonica*, Syrphidae, 3: 177 (♂; "Ishigaki Is., Japan") [National Institute of Agrobiological Sciences (NIAS), Tsukuba]

Eristalis laetus: Knutson et al. 1975, Cat. Dipt. Orient. Reg., p. 348 (Catalog) (as *n. comb*)

India: AH, BI, CG, GJ ?, KN ?, MH, RJ; Sri Lanka (Brunetti 1923; Hervé-Bazin 1924; Cherian 1934; Rahman 1940; Usman & Puttarudriah 1955; Baid 1959; Mahalingam 1988; Peck 1988.)

Note: Brunetti (1923: 165–167) perhaps misidentified this, noting it was "Originally described from China" and listed specimens from Nepal, Mt Abu (Rajasthan), and Shillong (Meghalaya), as well as from South India at Ootacamund and Mysore. This species is treated (F.C. Thompson, pers.comm.) as a synonym of *megacephalus* (Rossi 1794), *loc. cit.*, but which the latter is probably a southern European and African species and not Oriental? Hervé-Bazin (1923e: 141, 1924: 293) treated that as a synonym of *quinquestriatus* Fabricius, *q.v.* (!) and clarified that *laetus* of Brunetti (1923) was *ocularius* Coquillett and questioned his inclusion of South Indian localities. As noted under *L. aeneus* above (*op. cit.*, pp. 8–9), this complex of species requires study of types and adequate material to clear synonymy and establish species validity. Brunetti (1923: 160–161), in his discussion of the genus *Eristalis* Latreille, placed *pallineuris* [sic] and *laetus* in two of his different groups and noted "I have seen no specimen of this species (*E. pallineuris*, Macq.), nor can I glean any information as to which subgenus it should be referred." Cherian (1934: 699) listed it from "Bababudin hills, Mysore," and was quoted by Usman and Puttarudriah (1955: 150). Baid (1959) gave a "*Eristalis* sp." flying from September to March around the Sambhar Lake in Rajasthan, which may be this species. Mahalingam (1988: 161) gave it as recorded from the Nilgiri Hills among ten species recorded by Brunetti (1923).

Dr. F.C. Thompson (*pers. comm.*) treated this species, *pallinevris* Macquart, and *obscuritarsis* de Meijere as synonyms of *megacephalus* Rossi (*q.v., op. cit.*), and this needs to be carefully researched to arrive at the correct interpretation, especially as the type is lost. I (K. Ghorpadé, unpubl.) have specimens from Palni Hills (Tamil Nadu) and perhaps from Amboli (Maharashtra). Knutson et al. (1975: 348), Ghorpadé (2014b: 9), and Mitra et al. (2015: 63) listed it.

Eristalinus lucilia (de Meijere 1911)

Eristalis lucilia de Meijere 1911, Tijdschr. Ent., 54: 341 (♂; "Semarang, Java") [ZM, Amsterdam]

Eristalinus lucilia: Knutson et al. 1975, Cat. Dipt. Orient. Reg., p. 348 (Catalog) (as *n. comb.*)

Sri Lanka (de Meijere 1911; Keiser 1958; Knutson et al. 1975: 348; Thompson 2003; Ghorpadé 2014b: 9.)

Note: Known from the peculiarly frequent southern Asian islands species, ranging from Sri Lanka east through the Malay and Indonesian archipelagoes. I saw one blue specimen loaned to me by the Lund University Museum (Sweden). Dr. F.C. Thompson (*pers. comm.*) thinks it deserves a different genus. Keiser (1958: 233) included this as "*Eristalis (Lathyrophthalmus) lucilia*" in his paper on Sri Lanka, documenting specimens taken from Haragama and Nugawela in December and January in the Central Province. He wrote: "Under the species of the subgenus *Lathyrophthalmus*, there is a striking, bright metallic, blue green *lucilia* de Meij., which cannot be confused with any other species in Ceylon. I observed these flies in large numbers on plain, small flowers of shrubs, their strong odour reminded me of fresh excrement. Great multitudes of muscids, tachinids, and calliphorids were found frequenting these flowers, and under the presence of these flies, this species was seen next to the *Lucilia*s. On more close observation, it was seen that the resemblance was not so great, as its gait and flight were essentially different. I know no similarly coloured *Lathyrophthalmus* from the south Asian region; the identity with *lucilia* could be fixed based on a determined example by de Meijere from Sumatra." Knutson et al. (1975: 348), and Ghorpadé (2014b: 9) listed it.

Eristalinus megacephalus (Rossi, 1794)

Syrphus megacephalus Rossi, 1794, Mantissa Insectorum, 2: 63 (♂: lost; "Etruria" = Toscana, Italy) [lost]

Eristalis obscuritarsis de Meijere 1908, Tijdschr. Ent., 51: 250 (♂♀; "Java, Singapore, Malaya, Bombay") [ZM, Amsterdam]

Eristalis (Lathyrophthalmus) lalitai Nayar 1968, Pan-Pacific Ent., 44: 119 (♀; "St. John's College, Agra, India") [ZSI, Calcutta—examined]; Ghorpadé 2014d, Colemania, 46: 5. (as *n. syn.*)

India: AH, BI, CG, GJ, KL, KN, MH, TN (Brunetti 1923; Hervé-Bazin 1914, 1923a, b, 1924; Cherian 1934; Ghorpadé 1973b, 2014b, c, 2017, unpubl.; Knutson et al. 1975; Datta & Chakraborti 1986; Mitra et al. 2015).

Note: I noticed this extralimital (?) species name in a manuscript on Oriental—Papuan Syrphidae Conspectus sent to me by Dr. Chris Thompson. As noted under *laetus* above, it remains to be determined if this is actually the same as the truly Oriental species, and their synonyms, like *ishigakiensis, laetus, obscuritarsis, pallinevris,* and *quinquefasciatus*. Hervé-Bazin (1923b: 141) treated it as a synonym of *quinquelineatus,* but which later Knutson et al. (1975: 349) had clearly noted was "Not Oriental"! Brunetti (1923: 187) had placed it as a synonym of *quinquestriatus* and noted, "*Syrphus megacephalus* Rossi, is evidently synonymous with *E. quinquestriatus,* and was described in the same year. I have accepted the latter name, as Dr. de Meijere has adopted it." But see Hervé-Bazin's (1923b: 141) synonymy of this under *quinquelineatus* as stated above !

E. obscuritarsis was generally known by this name in our subcontinent, but is treated in synonymy now. However, this is yet unconfirmed and is uncorroborated by accurate taxonomic research and study of types. See notes under *aeneus* also above (*op. cit.,* pp. 25–27) for more details. The Oriental Catalog (Knutson et al. 1975: 348) placed it as a new synonym of *laetus,* but I am not aware if a lectotype of *obscuritarsis* was designated and types checked and confirmed as conspecific. Hervé-Bazin (1923b: 127) mentioned that *obscuritarsis, nigroscutatus,* and *tristriatus,* all from Java, were a distinct "species-group," but then did not treat *obscuritarsis* at all, neither in any detail, nor with any figures. I believe Hervé-Bazin had misidentified *obscuritarsis,* but this can only be proven after a lectotype of this species is selected and so designated, since the syntypes of de Meijere (1908) could be mixed and polytypic, hailing from Java, Singapore, Malay Peninsula, and even Bombay [now Mumbai] in western India. Brunetti (1923: 191–192) had given detailed notes on this and *quinquestriatus* which he considered distinct: "*E. obscuritarsis,* de Meij., is very closely allied to *E. quinquestriatus,* Fabr., yet it is a perfectly distinct species, as is quite obvious when several specimens of each sex of both forms are placed side by side." He goes on to list the diagnostic characters in some detail. Again, Hervé-Bazin (1924: 297) stated that this is *quinquelineatus* F., but see notes under *paria* below (*op. cit.,* p. 24). Brunetti (1915: 230) listed it. Bhatia (1931: 508–519, Pl. LVII) described its life history from larvae "observed in large numbers floating in every dirty drain during the last week of April and the whole of the month of May 1930 at Pusa [Bihar.]" But this was misidentified as *quinquestriatus* Fabricius (*q.v.*). Cherian (1934: 699) listed it as *E. obscuritarsis* from Kurnool (Andhra). I examined the sole holotype ♀ of *lalitai* Nayar (1968c: 119) in the collection of ZSI, Calcutta, labelled "St. John's College, Agra, India/10.xii.1962/Lalita Taneja/*Eristalis* (*Lathyrophthalmus*) *lalitai* Nayar, Holotype." It is nothing but *L. obscuritarsis* (de Meijere) and was synonymized as such by Ghorpadé (2014c: 5). In my masters thesis (Ghorpade 1973b), I had written, "This species was fairly common around Bangalore in winter when it was collected from November to January on mango and niger flowers, in herbaceous weeds and in ragi [= finger millet] fields." It was keyed out as "tarsi black, at most pale at base; abdomen comparatively longer, more elongate-conical." Datta and Chakraborti (1986: 59) listed it from Parambikulam, Palghat district, Kerala and from Chalakudy, Trichur district, Kerala, both in December 1979. Ghorpadé et al. (2011: 82) gave notes on this species. I included this from Maharashtra (Ghorpadé 2017) based on a ♂ taken in Bombay in August 1911 by N.B. Kinnear and Brunetti (1923: 230) giving "Bombay (Biro)." Kinnear also took a ♂ in Mercara (Coorg, Karnataka) in October 1918 (Ghorpadé 2017: 107). Knutson et al. (1975: 348; as synonym of *laetus* Wied.), Ghorpadé (2014c: 9, 2015: 5), and Mitra et al. (2015: 64) listed it.]

Eristalinus obliquus (Wiedemann, 1824)

Eristalis obliquus Wiedemann, 1824 Analecta Ent., p. 38 (♂♀; "Bengal") [UZM, Copenhagen]

Lathyrophthalmus connectens Hervé-Bazin 1923b, Annls Sci. nat. Zool., 6: 148 (♂♀; "Saigon, Java") [MNHN, Paris]

Eristalinus connectens: Knutson et al. 1975 Cat. Dipt. Orient. Reg., p. 348 (Catalog). (as *n. comb.*)

India: AH, AN, KL, KN, MH, PO ?, TN; Sri Lanka (Brunetti 1915, 1923, 1925; Hervé-Bazin 1923b, 1926; Curran 1930; Keiser 1958; Ghorpadé 1973b, 2014b: 9, 2015b: 29, 2017; Knutson et al. 1975: 348, 349; Mitra et al. 2015: 63, 64.)

Note: Brunetti (1915: 28) gave New Guinea and Java as distributional records and added "It is closely allied to *arvorum,* F." In his *Fauna* volume Brunetti (1923: 164–165, Pl. IV, fig. 4) gave a description "from a few indifferent examples in the Indian Museum" from West Bengal and Tamil Nadu states. Then (Brunetti, 1925: 76) he wrote about *E.* (*L.*) *connectens* Hervé-Bazin that was "omitted in error from the *Fauna* volume." *L. connectens* was cited from "Trichinopoly, Inde méridionale (ma collection)" also by Hervé-Bazin (1923b: 148), but not from "Ceylon" as Brunetti (1925: 28) wrongly indicated. Hervé-Bazin (1926: 86) gave "Cochinchine, Java, Inde méridionale" as localities for his *connectens;* the fine figures of *obliquus* female and *connectens* male are both of this same species. He also gave comparisons (Hervé-Bazin 1923b: 151) of this species with *arvorum* F., *ferrugineus* de Meij., and *connectens* H.-B., and I find weak arguments for sustaining *connectens* as a separate species, as Hervé-Bazin himself wrote, and so synonymized it (Ghorpadé 2015b: 29). Curran (1930) gave *connectens* as a doubtful synonym of *obliquus,* and included this in a key to Oriental *Lathyrophthalmus,* with species from Australia also. Keiser (1958: 233) included this as "*Eristalis* (*Lathyrophthalmus*) *obliquus*" in his paper on Sri Lanka, documenting specimens taken from Sabaragamuva and North-western, Western, Southern, and Central Provinces. See comments in Ghorpadé et al. (2011: 81) which establishes this as a common and widespread species in southern India. In my master's thesis (Ghorpadé, 1973b), I had written, "This species was found to be rather abundant around Bangalore in winter from September to February. Adult flies were secured on flowers of mango, cashew, niger and *Tridax procumbens* and in fields of wheat, radish and ragi [= finger millet]." It was keyed out as "tergite 4 with a pair of obliquely placed white pollinose spots; some black hairs in centre of scutellum." I included this from Maharashtra (Ghorpadé 2017) based on Ghorpadé (2014b: 9, 2015b: 29). I have much material, and several States listed above are based on my collections (K. Ghorpadé, unpubl.). Knutson et al. (1975: 348, 349; gave "southern Oriental region; New Guinea"). Ghorpadé (2014c: 9, 2015b: 5) and Mitra et al. (2015: 63, 64) listed it.

Eristalinus quadristriatus (Macquart 1846)

Eristalis quadristriatus Macquart 1846, Dipt. Exot. Suppl., 1: 127 (♀; "Inde") [BMNH, London]

India; Sri Lanka (Brunetti 1923, 1925; Knutson et al. 1975: 349; Ghorpadé 2014b: 9, c, 2015; Mitra et al. 2105: 64)

Note: Brunetti (1923: 176) just mentioned that his *basifemoratus* "might be regarded as the ♀ of *quadristriatus*, Macq. The 3rd abdominal segment is obviously discoloured in the unique *type*, and may or may not resemble that of *quadristriatus*, Maq." But Brunetti (1925: 76–77) mentioned that, "This species, described from India, was inadvertently omitted [in his *Fauna* volume, 1923]." He gave Macquart's description and commented, "It must be rather closely allied to *arvorum* but differences in the frons, abdominal markings and the colour of the femora may make it distinct." It was described based on a single female in Bigot's collection from India, with no specific locality cited. As with *basifemoratus*, I could not find the holotype female of *quadristriatus* in ZSI, Calcutta type collection when I visited there in March 1981. Coe (1964: 276) listed specimens (as "*Eristalis*") taken in east Nepal, adding, "Represented in the Brit. Mus. (Nat. Hist.) collection by a single ♀ labelled "ex Bigot Coll. Pres. By G.H. Verrall. B.M. 1894–234." I am satisfied that my above series from Nepal is the same species. There is a dark spot at *both* extremities of the stigma in all the material, an unusual feature in the genus." Kapoor et al. (1979: 64) listed it from Nepal, India, and Sri Lanka in this subcontinent. Ghorpadé (2014b: 9, 2015b: 5) and Mitra et al. (2105: 64) listed it.

Eristalinus quinquestriatus (Fabricius 1794)

Syrphus quinquestriatus Fabricius, 1794, Ent. Syst., 4: 289 (sex?; "India orientali") [UZM, Copenhagen ?]

Eristalis aesepus Walker 1849, List Dipt. Coll. Br. Mus., 3: 625 (♂♀; "China") [BMNH, London]

? *Eristalis quinquevittatus* Macquart, *in* Lucas 1849, Explor. Scient. Algerie, Zool., 3: 465 (sex ?; "d"Ain-Drean, aux environs du cercle de Lacalle" = nr La Calle, Algeria) [?]

Eristalomyia picta Bigot 1880, Annls Soc. Ent. Fr., (5) 10: 219 (♂; "Indostan") [?]

Lathyrophthalmus basalis Shiraki, 1968, *Fauna Japonica*, Syrphidae, 3: 175 (♂; "Iriomote I., Ryukyu Is., Japan") [NIAS, Tsukuba]

India: AH, BI, KN, MP, OR, TN (Brunetti 1915, 1923; Hervé-Bazin 1923b, 1924; Bhatia 1931; Cherian 1934; Beeson 1953; Usman & Puttarudriah 1955; Keiser 1958; Ghorpadé 1973b, 2014b, 2015b; Datta & Chakraborti 1986; Peck 1988; Thompson 2003; Mitra et al. 2015).

Note: See notes under *aeneus*, *obscuritarsis*, and *megacephalus* above for more details and the situation with the identity and correct names of some of these *Eristalinus* (or *Lathyrophthalmus*) species here. Brunetti (1915: 228) listed it, quoting de Meijere. Brunetti (1923: 187–189, Pl. IV, figs. 11–14) described it "from several of each sex in the Indian Museum and Pusa collections" from Kausanie, Kumaon district, and Kathmandu, Nepal besides other localities in West Bengal, Madhya Pradesh, Karnataka, Orissa, and "Assam." He was not sure of the true identity of *quinquevittatus* Macquart, 1849, being either a synonym of this species or of *quinquelineatus* (F., 1794). Hervé-Bazin (1923b: 141–143, figs. 15 and 16, 1924: 293) noted that *laetus* Wiedmann [sic] was also synonymous! Bhatia (1931: 508–519, Pl. LVII) described its life history from larvae "observed in large numbers floating in every dirty drain during the last week of April and the whole of the month of May 1930 at Pusa [Bihar.]" But this was misidentified and actually was *obscuritarsis* [= *megacephalus*] (q.v.). Beeson (1953: 339) gave this as breeding in stagnant water and quoted Bhatia (1931). Cherian (1934: 699) recorded it from Bababudan hills (Mysore) and was quoted by Usman & Puttarudriah (1955: 150). Keiser (1958) recorded it from Sri Lanka. Keiser (1958: 194) included this as "*Eristalis* (*Lathyrophthalmus*) *quinquestriatus*" in his paper on Sri Lanka, documenting around 150 specimens taken from localities of over 1900 m in the provinces. In my master's thesis (Ghorpadé 1973: 73–74, 129–143, figs. 57–64, 79–82). I had included and discussed four *Eristalinus* (or *Lathyrophthalmus*—as "*Eristalis*" then) species: *arvorum* (F.), *obliquus* (Wied.), *obscuritarsis* (de Meij.), and *quinquestriatus* (F.). In the key provided, *arvorum* was separated from the other three in having its hind femur wholly yellow or orange-yellow, never black or brownish-black as in the other three. *L. obliquus* was then separated by having a pair of white pollinose spots, placed obliquely on abdominal tergum 4 and not with any arcuate white fascia on it. *L. obscuritarsis* had black tarsi and a comparatively longer, elongate-conical abdomen, unlike *quinquestriatus* with a shorter, more ovate-conical abdomen and yellowish-white tarsi. Male genitalia were also dissected and illustrated, being diagnostic (figs. 79–82). I had also written, "This was not a very abundant species around Bangalore, secured only from October to February on flowers of niger, mango and *Tridax procumbens*, also in wheat and in herbaceous vegetation. It could perhaps be mistaken for *E. obscuritarsis* de Meij in the field but can be distinguished from that species by the characters given in the key, *viz*., tarsi yellowish-white nearly to tips, abdomen comparatively shorter, more ovate-conical." Datta & Chakraborti (1986: 14, fig. 7) illustrated male terminalia again. See also Hervé-Bazin (1923b: 139–143, 145–146, 147–148, figs. 13–16, 20–21, 23) for descriptions and figures, but note that his *"quinquelineatus* F." was misidentified, and is my *obscuritarsis* de Meij, as above. Gokulpure (1972: 848) mentioned it from Damoh, Madhya Pradesh. I saw specimens (K. Ghorpadé unpubl.) from Coimbatore (Tamil Nadu), Panyam, Kurnool District (Andhra), Goorghalli Estate, 3300 ft, and Puttur, S. Canara (Karnataka). Peck (1988: 184), Ghorpadé (2014b: 9), and Mitra et al. (2015: 64) listed it.

Eristalinus tristriatus (de Meijere, 1911)

Eristalis tristriatus de Meijere 1911, Tijdschr. Ent., 54: 341 (♂; "Semarang and Batavia, Djakarta, Muara Angke, Java") [ZM, Amsterdam]

Eristalis polychromatus Brunetti 1923, *Fauna Br. India*, Dipt., 3: 180, Pl. IV fig. 5 (♂; "Calcutta.") [BMNH, London]

India: BI, KL, OR (Brunetti 1923; Hervé-Bazin 1923; Datta & Chakraborti 1986: 60; Parui & Datta 1987: 245; Knutson et al. 1975: 349, 353; Ghorpadé 2014b: 9; Mitra et al. 2015: 64.)

Note: Brunetti (1923: 180, Pl. IV, fig. 5) described and illustrated it in his *Fauna* volume. The female of *L. tristriatus* (de Meijere) was illustrated by Hervé-Bazin (1923: 144–145, fig. 19). Ghorpade & Anooj (2016: 6) synonymized *polychromatus* (Hervé-Bazin 1923; Brunetti 1923) under the older *tristriatus* (de Meijere 1911) for the first time. When describing *Eristalis polychromatus* as new, Brunetti (1923: 181) had called it "an unsatisfactory species, of which the limits are by no means clear." Brunetti's male type was from "Calcutta" and he wrote of two other males in the Bigot collection, now in the ZSI, Calcutta, plus a probable female from Katihar, Purneah district, Bihar, taken on 15.vi.1907 by Paiva, also in the same collection, agreeing with Brunetti's description (1923: 181). The type locality of *Eristalis tristriatus* de Meijere was "Semarang and Batavia (Djakarta), Muara Angke, Java" (see Knutson et al. 1975: 349). Note that Parui & Datta (1987: 245) listed this species also from Orissa (Sambalpur) and Datta & Chakraborti (1986: 60, fig. 2) had also reported a male from "Konnakuzi, chalakudy [sic], Trichur Dist., Kerala, 22.xii.1979, Coll. A.N.T. Joseph," recording it from Kerala, in south India, and illustrating its male terminalia as well! Mr Mohd. Asghar Hassan, a graduate student from Pakistan, has sent me photos of this from northern Pakistan where he says it is frequent! If this is correct, *tristriatus* could be widespread also in the Indian subcontinent, but rarely collected. The Oriental Syrphidae Catalog (Knutson et al. 1975: 349) listed it only from Java and the Negros Island of the Philippines, so it is apparently poorly sampled in southeast Asia as well. The known distribution of this species should now read—Pakistan, India, Burma, Singapore, Java, Philippines. Knutson et al. 1975: 349, 353; Ghorpadé 2014b: 9; and Mitra et al. 2015: 64 listed it.

[Species Incertae Sedis:

Eristalis canocincta Brunetti, nomen nudum ??

India: TN ? (Brunetti, *in litt.*; K. Ghorpadé, unpubl.)

Note: Brunetti (*in litt.*, specimen notes) had mentioned *Eristalis canocincta* on specimens I found in the Indian Agricultural Research Institute (IARI), New Delhi which has not been formally named and described (*nomen nudum*). These specimens were from Kodaikanal (Tamil Nadu) from where I also took some flies that resemble those present in the IARI collection. In the Pusa collection (IARI, New Delhi), I found a ♂ labelled Pusa, Bengal [sic], 24.ix.1907, H.N. Sharma, B8595, *Eristalis canocincta* Brunetti, named by E. Brunetti 1919, with head lost, but with black thorax, and with abdomen terga 2–4 with curved, white pollinose fasciae and those on terga 3 and 4 with curved fasciae meeting at a point. Also found in IARI 4♂♀ from E. India which I noted as "different from Kody, with spotted eyes and black, with white narrow bands on abdomen.

Eristalodes paria (Bigot, 1880)

Eristalomyia paria Bigot 1880, Annls Soc. Ent. Fr., (5) 10: 218 (♂♀; "Ceylon") [BMNH, London]

Eristalomyia zebrina Bigot 1880, Annls Soc. Ent. Fr., (5) 10: 222 (♂; "Ternate, Moluccas") [MNHN, Paris ?]

Eristalis kobusi de Meijere 1908, Tijdschr. Ent., 51: 252 (♂♀; "Tosari, W. Java") [ZM, Amsterdam ?]

Eristalis arisanus Matsumura 1916, Thousand Insects Japan, 2: 264 (♂; "Arisan, Formosa" = Taiwan) [HU, Sapporo ?]

Eristalis quinquelineatus var. *orientalis* Brunetti, 1923, *Fauna Brit. India,* Dipt., 3: 183 (♂♀; several localities for syntypes; lectotype not designated) [ZSI, Calcutta and BMNH, London]

Eristalis santoshi Nayar 1968, Agra Univ. J. Res. (Sci.), 16(3): 28 (♂; "Kalatop, 8000 ft.") [ZSI, Calcutta— examined]; Ghorpade 2014d, Colemania, 46: 3 (as *n. comb., n. syn.*)

India: CG, KN, TN; Sri Lanka (Brunetti 1923; Hervé-Bazin 1924, 1926; Cherian 1934; Keiser 1958; Knutson et al. 1975: 349–350; Mahalingam 1988; Ghorpadé 2014b: 9, 2015b: 25, unpubl.; Mitra et al. 2015: 63; Thompson et al. 2017: 514.)

Note: Brunetti (1923: 183–186) treated this as *Eristalis quinquelineatus* var. *orientalis* nov., which he described "from a short series of both sexes in the Indian Museum and elsewhere." He mentioned seeing specimens from the following localities: Kasauli, Theog, Phagu, Simla, Naini Tal, Kousanie, Kurseong, Sureil, Bijnor, as well as from southern Indian locales like Mysore, Ootacamund, and Yercaud. But he did not specify a holotype ("type"), so all these must be treated as syntypes and a lectotype selected later. He also gave extensive notes and indicated a wide range, from "South Europe, Africa, India, Ceylon and Java." He also mentioned that this species "appears to be variable in the relative extent of black and yellow on the abdomen. The arcuate whitish band (usually present in several Indian species [*sic*]) on the

2nd, 3rd and 4th abdominal segments is apparently normally absent on the 2nd segment in African specimens, though occasionally it is more or less indistinctly discernible, but it is nearly always present in Indian examples. The eye-stripes are normally three or four in number, in addition to the usually dark inner and hind margins, but the number is variable; I have seen a specimen with five complete dark stripes in addition to the inner and hind margins." Brunetti concluded that *paria* Bigot is "synonymous with *Eristalis quinquelineatus.*" But Knutson et al. (1975: 349) clearly note that *quinquelineatus* (Fabricius) is "Not Oriental." Hervé-Bazin (1924: 294–296) gave a lengthy discussion about Brunetti's *quinquelineatus orientalis* and presented a synonymy of this, *quinquestriatus* and *paria*, which would be useful in any future revisionary work on this group. Hervé-Bazin (1926: 82–85, fig. 11) gave a lengthy discussion about this and *taeniops*, with figures of dorsal views of heads? Cherian (1934: 699) recorded a *E. quinquelineatus* from Yercaud (Salem) [Tamil Nadu] which must be this species. Keiser (1958: 194) included this as "*Eristalis (Lathyrophthalmus) quinquelineatus*" in his paper on Sri Lanka, documenting specimens taken from Northern, Eastern, North-western, and Central Provinces. Nayar (1968b: 28, 30) gave a female from the Kalatop-Lakkarmandi bridal path near Dalhousie, and described as new *santoshi* from Kalatop which I synonymized with *paria* (Ghorpade 2014c: 3). I studied the male holotype of *lunar* Nayar in the Z.S.I. collection at Calcutta, with the following labels: "HOLOTYPE [red label]/*Eristals santoshi* Nayar sp. nov., Det. J.L. Nayar/Kalatop—L. Mandi, 29.IX."61, Coll. J.L. Nayar/3731/H6/*Eristalodes paria* (Bigot) det. Ghorpade 1981," and found it synonymous with *paria* (Bigot), n. comb., n. syn (see Ghorpade 2014d: 3). Knutson et al. (1975: 349), Ghorpadé (2014b: 9, 2015b: 25), and Mitra et al. (2015: 63) listed it. Mahalingam (1988: 161) gave it (*Eristalis quinquelineatus*) as recorded from the Nilgiri Hills among ten species recorded by Brunetti (1923). Thompson et al. (2017: 517) provide a synopsis of this "subgenus" of *Eristalinus q.v.* with a single species in the PSWNG area.

Mallota curvigaster (Macquart, 1842)

Helophilus curvigaster Macquart 1842, Dipt. Exot., 2(2): 62 (♂; "Java") [MNHN, Paris ?]; Bhatia 1931, Indian J. Agric. Sci., 1: 510, Pl. LVIII (life history) [MNHN, Paris]

Merodon interveniens Walker 1860, J. Linn. Soc. Lond.,4: 120 (♂; "Makasar, Celebes) [BMNH, London]

Tigridemyia pictipes Bigot 1882, Annls Soc. Ent. Fr., Bull., (6) 2: cxxi (♂; "Java") [UM, Oxford]

Polydonta (?) orientalis Brunetti 1908, Rec. Indian Mus., 2: 74 (♂; "Inde") [ZSI, Calcutta—examined]

Teuchomerus orientalis (Brunetti): Sack 1922, Arch. Naturg., Abt. A, p. 265–266.

India: BI, KN, RJ, TN; Sri Lanka (Brunetti 1908, 1915, 1923, 1925; Hervé-Bazin 1924; Sack 1922; Bhatia 1931; Beeson 1953; Keiser 1958; Knutson et al. 1975: 354; Ghorpadé 2014b: 10; Mitra et al. 2015: 64; Thompson et al. 2017: 516.)

Note: Brunetti (1908: 74–76) formally named and described (?) *Polydonta orientalis* "from a single ♂ in fair condition in the Indian Museum collection, bearing no data, but marked "Inde" in Bigot's handwriting." Brunetti was unsure of its generic placement, noting "knowing of no other [genus] in which to place the Oriental species, I leave it here, where Bigot placed it." Curiously, Brunetti also included *Helophilus curvigaster* Macquart [*sic*!] in that same paper, mentioning "the transverse bands of pubescence on the thoracic dorsum in *curvigaster*," but then not realizing that *curvigaster* and *orientalis* were the same species! I observed the single male type of *Polydonta orientalis* in the ZSI, Calcutta, when I had then checked Brunetti's Diptera types housed there, in March 1981, and then making a note that it was *Merodon intervenius* Wlk. Brunetti (1915: 231) mentioned de Meijere's notes on this species (as *Helophilus*!). Brunetti (1923: 211–218) had also mixed up his *Merodon* species, some of which were *Mallota* and some *Mesembrius* as well! Only *pallidus* Macquart (= *pruni* Rossi) and *albifasciatus* Macquart were true *Merodon* as included in the *Fauna* volume. In fact, in the Appendix to that book, Brunetti (1923: 414–415) had accepted the synonymy of *Merodon interveniens* Walker and *Teuchomerus orientalis* Brunetti with *Helophilus curvigaster* Macquart, stating, "The species is undoubtedly a *Merodon*." He also gave a locality in Sri Lanka, where he stated that it was rare. Brunetti (1925: 77) gave notes on *Merodon interveniens*, discussing its generic relationship. Bhatia (1931: 510, PL. LVIII) described its life history in tree-holes full of decomposing vegetable matter in August at Pusa (Bihar), but misidentified it as *Helophilus curvigaster* Macquart. Beeson (1953: 340) quoted the above paper and wrote that its (as "*Helophilus*") larvae "breed in holes in trees where wet vegetable matter is putrifying." I saw a ♀ taken from a tree hole in Coimbatore in July 1933 (K. Ghorpadé unpubl.). Sack (1922: 265) erected *Teuchomerus* for *Polydonta orientalis* Brunetti, but that genus is now a synonym of *Mallota* Meigen, as also are *Imatisma* Macquart, *Tigridemyia* Bigot, and *Paramallota* Shiraki. Knutson et al. (1975: 354), Ghorpadé (2014b: 1o), and Mitra et al. (2015: 64) listed it. Thompson et al. (2017: 516) provide a synopsis of what they call "a paraphyletic group in need of taxonomic revision." with at least two species in two subgenera in the PSWNG area.

Generic concepts in the Syrphidae have altered radically in present times and, as noted in the Abstract above, the higher classification and phylogeny is also in need of professional analysis, involving molecular biology as well.]

Mallota vilis (Wiedemann, 1830)

Eristalis vilis Wiedemann 1830, Aussereurop. zweifl. Insekt., 2: 164 (♂; "Java") [?]

Mallota eristaloides Curran 1928, J. fed. Malay St. Mus., 14: 293 (♂; "Khao Luang, Nakhon si Thammarat, Thailand) [BMNH, London]

Mallota malayana Curran 1931b, J. fed. Malay St. Mus., 16: 328 (nom. nov. for *eristaloides* Curran not Loew).

India; Sri Lanka (Brunetti 1923; Keiser 1958; Knutson et al. 1975: 355; Ghorpadé 2014b: 10; Mitra et al. 2015: 64.)

Note: This is one of the rarely noticed syrphid genera in the field whose species were taxonomically confused earlier, but now are better known. There should be more species in this area, and it will need extensive trapping to sample them. Keiser (1958: 229) included this as "*Mesembrius (Tigridomyia) vilis*" in his paper on Sri Lanka, documenting specimens taken from Central Province, at Lady Horton's and Haragama, Kandy, 7.ix and 30.xii.1953.

Merodonoides multifarius (Walker, 1852)

Eristalis multifarius Walker 1852, Insecta Saundersiana, 1: 248 (♀; "East Indies") [BMNH, London]

Merodonoides circularis Curran 1931b, J. Fed. Malay St. Mus., 16: 333 (♂; Kedah Peak, 3300 ft, Malaya) [BMNH, London]

Merodonoides minutus Hull 1944, Ann. Mag. Nat. Hist., (11) 11: 43 (♂; "Jubblepore, Central India") [BMNH. London]

Merodonoides czernyi Hull 1944, J. Wash. Acad. Sci., 34: 400 (♂; "Tonkin, Montes Mauson, Vietnam") [NM, Vienna]

Eristalis (Merodonoides) kandyensis Keiser 1958, Revue Suisse de Zoologie, 65(1): 234 (♂; "Kandy, Asgiriya") [NHM, Basel]

Eristalis yamunanagarensis Awtar et al. 1986b, J. Bombay nat. Hist. Soc., 83(2): 395 (♂; "Yamunanagar, Haryana, N. India" [IARI, New Delhi ?]; Ghorpade 2014b, Colemania, 44: 10 (as *n. syn.*)

Pseudomeromacrus setipenitus Li 1994, Entomologia Sinica, 1(2): 146 (♂; "Guangzhou, China") [SCAU, Guangzhou]

India: GJ, RJ, TN; Sri Lanka (Brunetti 1923; Hervé-Bazin 1924; Curran 1931a; Cherian 1934; Hull 1944a, b; Keiser 1958; Coe 1964; Knutson et al. 1975: 350; Datta & Chakraborti 1986; Awtar et al. 1986b; Ghorpadé 2014b: 10, 2015b: 33; Mitra et al. 2015: 64; Thompson et al. 2017: 516.)

Note: This could probably be a polytypic species, certainly morphologically diverse, and it remains to be examined if all six synonyms indicated above are genuine and valid ? Walker's type from "East Indies" is apparently a A.R. Wallace collection from one of the Indonesian islands or from the Malay Peninsula. But Brunetti (1923: 195) mentioned "the type ♀ in the British Museum from India and another ♀ from the Lower ranges, N. Khasi Hills, Assam, 1878 (Chennell)." It is distributed all over the Indian subcontinent, from Haryana in the north to Sri Lanka down south, and I myself have collected many specimens of this species from Karnataka and Tamil Nadu on both the Western Ghats and the Eastern Droogs. It also occurs on the Indochinese Peninsula and in China. Careful study of all this widespread material may and could reveal more than a single species. Curran (1931b: 333) supposed "*Merodon tuberculatus* Brunetti may also belong to this genus [*Merodonoides*]," but that species is currently placed in *Mesembrius*, which genus also requires more careful study. Curran also likened his new *Merodonoides* to *Tigridemyia* Bigot, but it was "at once distinguished by the petiolate marginal cell." Hervé-Bazin (1924: 297) felt this was an "*Eristalodes*." Hull (1944a: 43) separated his *minutus* from *circularis* Curran "by the absence of a fascia upon the posterior portion of the third and fourth abdominal segments, the wholly reddish hind femora, the smaller size and different eye-pattern [four conspicuous brown stripes]." Cherian (1934: 698) listed it from Coimbatore. Hull (1944b: 400) noted his *czernyi* was "[r]elated to *circularis* Curran, [but] this species is distinguished by the chiefly reddish femur and a different pattern of eye stripes." Keiser (1958: 234–236) included this as "*Eristalis (Merodonoides) kandyensis*" in his paper on Sri Lanka, documenting one ♂ specimen taken from Asgiriya, Kandy, 10.i.1954, in grass taken in the Central Province. He (Keiser, 1958: 234, fig. 7) gave differences of his *kandyensis* from *minutus* and *circularis*, but stated better material would be required to come to a correct conclusion; his was based on a single male holotype. Coe (1964: 277) retained *multifarius* in the large genus *Eristalis* agreeing with Hull (1949: 397) that characters separating *Merodonoides, Eristaloides* [sic !] etc., were deserving "only of minor group value." Coe then synonymized *circularis* Curran and *minutus* Hull after studying holotypes, noting that the latter "is only a pale variety of *multifarius*." Interestingly, Brunetti (1923: 195) had opined: "I do not think that this species [*multifarius*] should remain in *Eristalis*, on account of the greatly thickened hind femora; the dipped 2nd vein is an additional abnormality." Thompson (2003: 4) separated this genus from others in his "subtribe" Eristalina. Datta & Chakraborti (1986: 59) recorded it from Parambikulam, Palghat district, Kerala in December 1979. Awtar et al. (1986a: 192) listed an "*Eristalis* sp. nov." from Yamunanagar taken in November in wild weeds, and later (Awtar Singh et al. 1986b: 395) described this as a new species *yamunanagarensis* which I (Ghorpadé 2014b: 10) then synonymized under *Merodonoides multifarius* (Walker). I saw a ♀ taken in Coimbatore in January 1913 (K. Ghorpadé unpubl.). Knutson et al. (1975: 350), Ghorpadé (2014b: 10, 2014c: 5, 2015: 5), and Mitra et al. (2015: 64) listed it. Thompson et al. (2017: 516) provide a synopsis of this, treated as another "subgenus" of *Eristalinus*, a large worldwide genus (87 spp.) with few New World species introduced from the Old World, and the genus divided into five subgenera of which two, *Eristalodes* and *Lathyrophthalmus* [treated now as *Eristalinus*] are found in the PSWNG area.

Mesembrius bengalensis (Wiedemann, 1819)

Eristalis bengalensis Wiedemann 1819, Zool. Mag. (Wied.), 1: 16 (♂♀; "Bengal, India") [NM, Vienna & UZM, Copenhagen]

Eumerosyrphus indianus Bigot 1882, Annls Soc. Ent. Fr., Bull., (6) 2: cxxviii (♂; "India") [UM, Oxford ?]

Eumerosyrphus indicus Bigot 1883, Annls Soc. Ent. Fr. Bull., (6) 3: 349 [*lapsus calami* ?]

India: AH, BI, CG, GJ, KN, TN, Sri Lanka (Brunetti 1907b, 1908, 1923; Bhatia & Shaffi 1933; Beeson 1953; Keiser 1958; Knutson et al. 1975: 356; Datta & Chakraborti 1986; Ghorpadé et al. 2011; Ghorpadé 2014b: 10, c, 2015b: 34; Mitra et al. 2015: 64; Thompson et al. 2017: 516.)

Note: Brunetti (1907a: 379, figs. 4–6, 1908: 69–70) wrote in some detail on this species and mentioned diagnostics of markings on abdominal tergum 4. The male middle femur has a tooth that is diagnostic of this species. In the *Fauna* volume, Brunetti (1923: 209–210, Pl. V, figs. 103) gave a full description and cited specimens from Calcutta, Pusa, Kathmandu (Nepal), Deesa, Bangalore, and "Assam," stating it "is apparently widely distributed throughout India." Bhatia & Shaffi (1933: 567–569, Pl. LXVII) described its life history using field collected gravid females found "hovering over wild plants near the river bank in Pusa [Bihar]." Beeson (1953: 340) quoted the above paper and wrote that its (as "*Helophilus*") larvae "breed in wet fermenting woody pulp." Keiser (1958: 228) included this as "*Mesembrius (Eumerosyrphus) bengalensis*" in his paper on Sri Lanka, documenting specimens taken from Sabaragamuva and Northern, North-Western, Western, Southern, and Central Provinces. Datta and Chakraborti (1986: 60, fig. 3) recorded it from Mullayarm (Tamil Nadu) in June 1981 and figured its male terminalia. Ghorpade et al. (2011: 82) recorded it from the Andhra Carnatic and gave some notes. Knutson et al. (1975: 356), Ghorpadé (2014b: 10, 2015b: 34), and Mitra et al. (2015: 64) listed it. Thompson et al. (2017: 516) provide a synopsis of this Old World tropical group with 56 species, 10 of which occur in the PSWNG area.

Mesembrius quadrivittatus (Wiedemann, 1819)

Eristalis quadrivittatus Wiedemann 1819, Zool. Mag. (Wied.), 1: 17 (♂♀; "Bengal, India") [NM, Vienna & UZM, Copenhagen]

Helophilus quadrivittatus Wiedemann: Brunetti 1923, *Fauna Brit. India*, Dipt., 3: 210.

Mesembrius quadrivittatus Wiedemann: Knutson et al. 1975, *Cat. Dipt. Orient. Reg.*, p. 356 (Catalog).

Merodon brunetti [sic] Sodhi & Awtar 1991, Acta zool Cracov., 34: 315 (♂; Morinda) [IARI, New Delhi—examined]; Ghorpade 2014b, Colemania, 41: 9 (as *n. comb., n. syn.*)

India: AH, BI, CG, GJ, KN, MP, OR, TN (Brunetti 1907b, 1915, 1923; Ghorpadé 1973b, 1974, 2014b, c, 2015b; Knutson et al. 1975; Ghorpadé et al. 2011: 82; Mitra et al. 2015).

Note: Brunetti (1907b: 379, figs. 1–3, 1908: 69–70) wrote in some detail of this species and of diagnostic markings on abdominal tergum 4. The male middle femur lacks the tooth that is diagnostic of *M. bengalensis*. Brunetti (1915: 231) mentioned de Meijere's redescription of this species. Brunetti (1923: 210–211, Pl. V, figs. 4–5) gave a description, comparison with *bengalensis*, and cited specimens seen from Pusa, Deesa, Calcutta, Port Canning, Puri, Katihar, Jabalpur, and Siliguri, and remarked "[a]pparently widely distributed in the plains of India." In my master's thesis (Ghorpade 1973b, 1974), I had written of this species (as "*Helophilus*") with marginal cell open, "This species was decidedly rare around Bangalore, a single male taken in a wheat field in the Agricultural University campus in February." Sodhi & Awtar (1991: 315–319, figs. 1–4) named and described a new species, *Merodon brunetti*, which I synonymized under this species (Ghorpadé 2014b: 9). Knutson et al. (1975: 356), Ghorpadé (2014b: 10, 2015b: 34), and Mitra et al. (2015: 64) listed it.

Phytomia argyrocephala (Macquart, 1842)

Eristalis argyrocephala Macquart 1842, Dipt. Exot., 2(2) 45 (♂♀; "Indes orientales") [MNHN, Paris]

Megaspis transversus Brunetti 1908, Rec. Indian Mus., 2: 73, figure (♂♀; "Bangalore and Calcutta, India") [ZSI, Calcutta—examined]

India: AH, BI, GJ, JH, KL, KN, MH, MP, TN (Brunetti 1908, 1923; Cherian 1934; Gokulpure 1972; Ghorpadé 1973b, 1974, 2014b: 10, 2015b: 36, 2017: 107; Knutson et al. 1975; Joseph & Parui 1977; Mahalingam 1988; Thompson 2003; Ghorpade et al. 2011; Mitra et al. 2015; Thompson et al. 2017: 519.)

Note: Species of *Phytomia* (earlier *Megaspis*) are large and stocky eristalines, dominantly tropical in range and generally avoiding cold habitats, both latitudinal or altitudinal, unlike species of *Eristalis* (*s. str.*), *Eoseristalis*, and *Eristalodes* which prefer temperate climes. Brunetti (1908: 73–74) described *transversus* as new based on eight specimens from Bangalore and Calcutta, giving diagnostics from *errans* F. I confirmed the holotype specimen in ZSI, Calcutta in 1981. Brunetti also gave a key to six species of "*Megaspis*." Brunetti (1915: 231) quoted de Meijere, synonymizing his *transversus* with this species and adding a table (= key), and commented, "Mr. Austen writes me that *Megaspis* is antedated by *Phytomia*, Guer. (1833), but I do not like to change the name after it has stood so long." Brunetti (1923: 201–203, Pl. IV, fig. 20) synonymized his *transversus* with this earlier name and listed its occurrence all over the subcontinent, specifically from Deesa, Parasnath (Chhota Nagpur), Ranchi, Calcutta, Pusa (Bihar), Bangalore, Talewadi (N. Kanara District, Karnataka), Travancore, Belgaum, Coonoor, Purneah, Calcutta, and Burma. Cherian (1934: 699) recorded it from Coimbatore (Tamil Nadu). Gokulpure (1972: 848) recorded it from Damoh, Madhya Pradesh. In my master's thesis (Ghorpade 1973b, 1974), I had written, "This species was rather common in the winter months of October to December around Bangalore, taken in fields of niger, wheat, sannhemp and ragi [= finger millet] feeding on the flowers." It was keyed out as "arista microscopically haired on basal half; hind femora wholly black." Joseph & Parui (1977: 230) recorded it from Mandar and Basia in the Ranchi district in December 1967. Mahalingam (1988: 161) gave it as recorded from the Nilgiri Hills among ten species recorded by Brunetti (1923). Thompson (2003: 4) included this genus in a key to others of his "subtribe" Eristalina. See also Ghorpadé et al. (2011: 82) for records from the Andhra Carnatic and more notes. I included this (Ghorpadé 2017: 107) from Maharashtra based on a ♀ specimen taken in that state found in the BNHS, Bombay collection by "P.G." [= Gomes ??]. Knutson et al. (1975: 357), Ghorpadé (2014b: 10, 2015b: 36) and Mitra et al. (2015: 64) listed it. Thompson et al. (2017: 519) provide a synopsis of this Old World tropical group, divided into two subgenera (the other is *Dolichomerus q.v.*) with 19 species, two of which occur in the PSWNG area.

Phytomia errans (Fabricius, 1787)

Syrphus errans Fabricius 1787, Mantissa Insect., 2: 337 (♀; "China") [UZM, Copenhagen ?]
Eristalis varipes Macquart 1842, Dipt. Exot., 2(2): 46 (♂♀; "Indes orientales and China") [MNHN, Paris ?]
Eristalis amphicrates Walker 1849, List Dipt. Colln Br. Mus., 3: 623 (♂♀; "N. Bengal, Java, East Indies, China" [BMNH, London]
Eristalis plistoanax Walker 1849, List Dipt. Colln Br. Mus., 3: 628 (♂; "Philippine Is." [BMNH, London]
Eristalis agyrus Walker 1849, List Dipt. Colln Br. Mus., 3: 629 (♂; "Philippine Is." [BMNM London]
Eristalis babytace Walker 1849, List Dipt. Colln Br. Mus., 3: 629 (♂; "Philippine Is." [BMNH, London]
Eristalis macquartii Doleschall 1856, Natuurk. Tijdschr. Ned.-Indië, 10: 410 (o ?; "Java") [?]
India: AH, BI, KL, KN, MH, TN; Sri Lanka (Brunetti 1908, 1915, 1923; Hervé-Bazin 1923e, 1924; Cherian 1934; Keiser 1958; Ghorpadé 1973b, 2014b: 10, 2015b: 36, 2017: 107; Knutson et al. 1975: 357; Joseph & Parui 1977; Datta & Chakraborti 1986; Peck 1988; Ghorpade et al. 2011; Mitra et al. 2015).

Note: Brunetti (1908: 72–73) included this in a key to six species of this genus and made notes on it. Then he listed (Brunetti 1915: 231) a record of a female of this species "from Cochin State, 1700–3200 ft., 16—24-ix-14 [*Gravely*]," and cited de Meijere's paper recording this and others of this genus from Java, with a key. Brunetti (1923: 199–200, 414) noted that, "This species and *argyrocephalus* are very closely allied, but the colour of the hind femora and the presence or absence of pubescence on the arista will separate them with certainty. Very common throughout the East." He listed localities of Sukna (West Bengal), Pusa (Bihar), Trivandrum (Travancore), Parambikulam (Cochin State) (both in present Kerala), Bangalore, Coorg (Karnataka), Hadagalli (Ceylon), Margherita (Arunachhal), Myingyan (Burma), Maymyo (Upper Burma), and Sibsagar (Assam) in the Indian subcontinent. Hervé-Bazin (1923e: 253, 1924: 297) gave a discussion and another possible synonym of this species, *albifrons* Macquart from "Bengale, M. Duvaucel." Cherian (1934: 699) recorded it from Coimbatore (Tamil Nadu), Hadagalle, Bellary (Karnataka), Amalapuram, Godavari, Devakonda, Kurnool, Madanapalle, and Chittoor (Andhra Pradesh), where flies were found resting on manure heap, safflower, cholam and sugarcane. Keiser (1958: 237) included this as "*Megaspis (Megaspis) erans*" in his paper on Sri Lanka, documenting specimens taken from Uva, Sabaragamuva and Northern, North-western, Western, Southern, and Central Provinces. In my master's thesis (Ghorpade 1973b, 1974), I had written, "Another species encountered only once around Bangalore when two males were collected from niger flowers on my farm." It was keyed out as "arista bare; basal half of hind femora orange." Joseph & Parui (1977: 230) recorded it from Basia, Ranchi district (Bihar) in December 1967, one specimen having mesonotum all black with brown hairs except for some whitish hairs on either side of the suture. Datta and Chakraborti (1986: 61, fig. 4) recorded it from Top Slip, Coimbatore district (Tamil Nadu), Idamalaya, Ernakulam district, Kumily, Idukki district (both Kerala), and figured its male terminalia. See also Ghorpadé et al. (2011: 82) for records from the Andhra Carnatic and more notes. I included this (Ghorpadé, 2017: 107) from Maharashtra based on a ♀ taken in Bombay in October 1909 by N.B. Kinnear and found in the BNHS, Bombay collection. Knutson et al. (1975: 357), Ghorpadé (2014b: 10, 2015b: 36), and Mitra et al. (2015: 64) listed it.

Phytomia zonata (Fabricius, 1787)

Syrphus zonatus Fabricius, 1787 Mantissa Insect., 2: 337 (♀; "China") [UZM, Copenhagen]
Syrphus zonalis Fabricius, 1794 Ent. Syst., 4; 294 (Unjustified nom. nov. for zonata F., 1787); Brunetti, 1908, et seq. [*lapsus calami*?]
Musca sinensis Gmelin 1790, Syst. Nat., 5: 2872 (Unjustified nom. nov. for *zonata* F., 1787)
Eristalis rufitarsis Macquart, 1842, Mem. Soc. R. Sci. Agric. Arts, Lille, Pl. 10, p. 58 (♀; "Patrie inconnue") [MNHN, Paris ?]
Eristalis lata Macquart, 1842, Dipt. Exot., 2(2): 35 (♀; "Patrie inconnue") [MNHN, Paris ?]
Eristalis andraemon Walker, 1849, List Dipt. Colln Brit. Mus., 3: 627 (♀; "Sylhet and Sikkim") [BMNH, London]
Eristalis datamus Walker, 1849, List Dipt. Colln Brit. Mus., 3: 628 (♀; "?") [BMNH, London]
Eristalis babytace Walker, 1849, List Dipt. Colln Brit. Mus., 3: 629 (♀; "Philippine Islands") [BMNH, London]
Eristalis flavofasciatus Macquart, 1850, Dipt. Exot., Suppl., 4: 136 (sex ? "Java") [MNHN, Paris ?]
Eristalis exterus Walker, 1852, *Insecta Saundersiana*, 3: 248 (♀; "East Indies") [BMNH, London]
Megaspis cingulata Snellen van Vollenhoven, 1863, Versl. Meded. K. Akad. Wet. Amst., 15: 12 (sex ?; "Japan") [ZM, Amsterdam ?]
India: KL, KN, TN; Sri Lanka (Brunetti 1908, 1910, 1913b, 1915, 1923; Hervé-Bazin 1914, 1923e, 1924; Sack 1932; Keiser 1958; Knutson et al. 1975; Datta & Chakraborti 1986; Mahalingam 1988; Peck 1988; Thompson 1998; Ghorpadé 2014b: 120, 2015b: 37; Mitra et al. 2015)

Note: Brunetti (1908: 72–73) included this in a key to six species of this genus, and (Brunetti 1910: 171) noted this (as *Megaspis*) "generally common throughout the summer" in Sri Lanka. Then he mentioned Darjeeling, 1000–3000 ft. (Brunetti 1913b: 272). Brunetti (1915: 231) seeing specimen(s) of this species (as "*zonalis*") from Darjeeling, cited de Meijere's paper recording this and others of this genus from Java, with a key, and also mentioning "*Eristalis externus* [sic], Walk.," stating "a ♂ and ♀ under this name exist in the Indian Museum collection. They were identified by Bigot but, I think, incorrectly, owing to discrepancies in the size, the length of the abdomen and marks of the latter," In his *Fauna* volume (Brunetti 1923: 203–204), he gave a full description and mentioned specimens from north-east India; a key to six species of "*Megaspis*" (pp. 196–197) was also given. Hervé-Bazin (1914: 151, 1923e: 252–253, 1924:297) gave some notes. I (K. Ghorpadé unpubl.) took specimens on the Nilgiris (3000 ft.) and near Yercaud (Shevaroy Hills). Keiser (1958: 237) included this as "*Megaspis (Megaspis) zonatus*" in his paper on Sri Lanka, documenting specimens taken from Uva, Sabaragamuva, and Central Provinces. Mahalingam (1988:161) gave it as recorded from the Nilgiri Hills among ten species recorded by Brunetti (1923). Knutson et al. (1975: 357), Peck (1988: 193), Ghorpadé (2014c: 10, 2015b: 5), and Mitra et al. (2015: 64) listed it. See also Sack (1932: 250) and Thompson & Rotheray (1998: 113) for discussion and key to this genus in the Palaearctic.

Kertesziomyia nigra (Wiedemann, 1824) ?

 Eristalis niger Wiedemann 1824, Analecta Ent., p, 38 (♀; "Java") [UZM, Copenhagen]
 Eristalis bomboides Walker 1860, J. Linn. Soc. Lond., 4: 119 (♂♀; "Makasar, Celebes") [BMNH, London]
 Eristalis obscurata Walker 1860, J. Linn. Soc. Lond., 5: 239 (♂♀; "Dorey, New Guinea") [BMNH, London]
 Eristalist tortuosa Walker 1861, J. Linn. Soc. Lond., 5: 266 (♂; "Tond, Celebes") [BMNH, London]
 Pseuderistalis nigra: Knutson et al. 1975, Cat. Dipt. Orient. Reg., p. 358 (Catalog).
 India: AN; Sri Lanka (Brunetti 1915, 1923; Keiser 1958; Knutson et al. 1975: 358; Thompson & Rotheray 1998; Parui et al. 2002; Ghorpadé 2014b: 10, 2015b: 38; Mitra et al. 2015: 64; Thompson et al. 2017: 515.)

Note: Brunetti (1915: 229) listed this as *Eristalis niger*, Wied., and wrote "The ♂ redescribed by Meijere from Sukabumi [sic], Java [*Kramer*]. A ♀, without data, is under this name in the Indian Museum, identified by Bigot, but I cannot be sure that it is this species." In his *Fauna* volume (Brunetti 1923: 163–164, 414), he redescribed it "mainly from a large ♀ in good condition in the British Museum from Singapore, and a ♂ and ♀ in the Indian Museum. The type ♂♀ of *bomboides*, type ♀ of *obscurata*, and type ♂ of *tortuosa* are all in the British Museum and are certainly conspecific. The synonymy is by Major Austen." Keiser (1958: 194) included this as "*Eristalis (Eristalomyia) niger*" in his paper on Sri Lanka documenting specimens taken from Kitulgala, 9.iv.1929 in Sabaragamuva Province. Parui et al. (2002: 135) interestingly recorded a male specimen of what they determined as "*Pseuderistalis fascipennis* Thompson" from Manarghat, South Andaman, taken in April 1964 by B.S. Lamba. The species identity of that ♂ needs to be confirmed.

This is exactly the peculiar biogeographic range of South Asian fauna which are found on the island of Sri Lanka, skip the Indian mainland (or are extinct there in "Dravidia"), and occur also in the islands of Malaysia and Indonesia. "Dravidia" has been a source area for many taxa but is now terribly impoverished through invasion of human populations over the last millennia. What if these extreme S. Asian fauna are still holed up in rare undisturbed tracts in "Dravidia" needing future workers to sort out through intensive field collecting and trapping. Knutson et al. (1975: 358), Ghorpadé (2014b: 10, 2015b: 38), and Mitra et al. (2015: 64) listed it. See Thompson and Rotheray (1998: 112) for a key to this genus in the Palaearctic. Thompson et al. (2017: 515) provide a synopsis of this "broad genus with three subgenera" and four species, none of *Kertesziomyia* s. str. currently known from the PSWNG area. I found two specimens loaned to me by the Lund University Museum (Sweden) in the late 1970s which was determined as *Pseuderistalis ? nigra* (Wied.).]

Tribe Merodontini

Eumerus aeneithorax Brunetti, 1915

 Eumerus aeneithorax Brunetti 1915, Rec. Indian Mus., 11: 244 (♂; "Simla") [ZSI, Calcutta—examined]
 India: CG ? (Brunetti 1915, 1923; Knutson et al. 1975: 340; Ghorpadé 2014b: 11, 2015b: 39; Mitra et al. 2015: 84; Thompson et al. 2017: 513).

Note: Brunetti (1915: 244–245) described this as new based on "a single perfect male taken by Capt. Evans, R.E., at Simla in August 1914, and generously presented by him, with other diptera, to the Indian Museum." Brunetti (1923: 257) then noted that he saw "Further specimens of each sex from Simla, ix.1898 (*Nurse*)." I consider it worth mentioning that Brigadier W. Harry Evans, a leading worker on Indian subregion butterflies (see Ghorpadé & Kunte 2010) also picked up small sized Syrphidae a hundred years ago while generally hunting larger butterflies of his fancy ! This was the quintessential "character" of the typical erstwhile British naturalist of days gone by who enthusiastically sampled and made known the insect fauna of India. This character and habit is far removed from current professional (are there really any amateurs here now? The amateur "collector" naturalist is almost extinct here now!— see Ghorpadé, 1997) entomologists who only concentrate on their special select taxon and ignore additional sampling of any other insect families, just through lack of interest (and laziness?), and possession of that human "character," not of time and opportunity. I examined the holotype present in ZSI, Calcutta in March 1981 when I visited there to study types. Knutson et al. (1975: 340), Ghorpadé (2014b: 11) and Mitra et al. (2015: 64) listed it. It is interesting that Misra & Verma (1975) recorded larvae of an unidentified species of this genus infesting potato tubers in Simla. It would be useful to consult staff of the Central Potato Research Institute at Simla if syrphid damage to tubers is frequent? Thompson et al. (2017: 513) provide a synopsis of this large Old World genus (266 spp.) "with several economic pests introduced into the New World, nine are known from the PSWNG area."

Eumerus albifrons Walker, 1852

Eumerus albifrons Walker 1852, Insecta Saundersiana, 1: 224 (♂; "East Indies") [BMNH, London]

Eumerus halictiformis Brunetti 1915, Rec, Indian Mus., 11: 241 (♂♀; "Puri, Orissa Coast") [ZSI, Calcutta—examined]

India: OR, KN, TN (Brunetti 1915, 1923; Cherian 1934; Knutson et al. 1975: 341; Patnaik & Bhagat 1976; Patnaik et al. 1977; Sathiamma 1979; Ghorpade et al. 2011; Ghorpadé 2014b: 11; Mitra et al. 2015: 64).

Note: Brunetti (1923: 258) noted that the type of his *halictiformis* "agrees exactly with Walker's type" [of *albifrons*] "in British Museum, without data." He also gave Coimbatore and Kangra Valley as other locations for this species. Cherian (1934: 699) listed this [correct identity ?—Kumar Ghorpadé] from Salem and Coimbatore, adults resting on castor and watermelon leaves. I noted two type specimens of *halictiformis* present in ZSI, Calcutta in March 1981 when I visited there to study types. Brunetti (1915: 241–242) had described *halictiformis* as new based on a male and female taken at Puri on the Orissa coast in August by Annandale. Patnaik & Bhagat (1976) and Patnaik et al. (1977) recorded this from Puri district, Orissa, but that was misidentified for a *Paragus* species as the photograph in the former paper amply testifies to. It was assumed to be a "predator" of the sorghum aphid which was another doubtful claim! Sathiamma (1979) recorded its larvae infesting ginger. I saw (K. Ghorpadé unpubl.) ♀♀ from S. Canara (Karnataka) on rotting ginger. Ghorpadé et al. (2011; 82) mentioned it being described from Puri (Orissa), and that this could occur in the Andhra Carnatic. Ghorpadé (2014b: 11) and Mitra et al. (2015: 64) listed it.

Eumerus argentipes Walker, 1861

Eumerus argentipes Walker 1861, J. Linn. Soc. Lond., 5: 284 (♀; "Batjan, Moluccas") [BMNH, London]

Eumerus argyropus Doleschall 1857, Natuurk. Tijdschr. Ned.-Indië, 14: 410 (sex ?; "Amboina, Moluccas") [?]

Eumerus doleschalli Shiraki 1930, Mem. Fac. Agric. Taihako imp. Univ., 1: 97 (*nom. nov.*)

Sri Lanka (Brunetti 1908, 1923; Knutson et al. 1975: 341; Ghorpadé 2014b: 11; Mitra et al. 2015: 64).

Note: Brunetti (1908: 76) listed this as *E. argyropus* Dol. (*E. argentipes*, Wlk.) of which he found 3♂ and 1♀ "from Assam in the Indian Museum collection which I have identified with this species." Then in Brunetti (1923: 415) in his appendix, he wrote of this species, "A form which is either this species, or very close to it, is "rare in Ceylon". I have not been able to decide the status of this form, and have therefore omitted it from the text of this volume [Fauna]. The genus requires thorough revision on abundant and fresh material. *E. argentipes*, Walk., is recorded (Kertész, *Cat. Dipt.* vii, p. 313, 1910) from Batchian, Amboina, and New Guinea." Knutson et al. (1975: 341), Ghorpadé (2014b: 11), and Mitra et al. (2015: 64) listed it.

Eumerus aurifrons (Wiedemann, 1824)

Pipiza aurifrons Wiedemann 1824, Analecta Ent., p. 32 (♂; "Ostindien") [UZM, Copenhagen]

Eumerus aurifrons var. *similis* Keiser, 1958, Revue Suisse de Zoologie, 65(1): 216 (♀; "Deiyannewela, Kandy," Sri Lanka) [NM, Basel]

India: BI, CG ?, KN, MH, TN; Sri Lanka (Brunetti 1915, 1923; Curran 1926; Beeson 1953; Keiser 1958; Knutson et al. 1975; Ghorpadé 2014b: 11, 2015b: 40, 2017: 108; Mitra et al. 2015: 64.)

Note: Brunetti (1915: 240) had noted: "Dr. Meijere makes *splendens*, W., a synonym of this [*aurifrons* Wied.] This may be the species described by me as *nepalensis*" (see below, *op. cit., q.v.*). Brunetti (1923: 252–253) described this from several of each sex in the Pusa collection and gave locations as Pusa (Bihar), Chapra (Bengal) [*sic*], Bombay, and Ceylon, besides from Java and the Philippines, the flies found on gum-saturated earth, under mango bark, and on stable wall. Curran (1926: 115) studied the type and gave a key to Oriental "forms" related to *aurifrons* Wied. Beeson (1953: 340) wrote its larvae breed in fermenting sap of *Albizia lebbek* and other forest trees. Keiser (1958: 215) included this as "*Eumerus aurifrons*" in his paper documenting Sri Lankan specimens taken from Uva and Central Provinces, flying among grass over soil, rarely on flowers on shrubs. He named and described a new variety, *similis* (p. 216) which differed only in colour of legs. Knutson et al. (1975: 341), Ghorpadé (2014b: 11), and Mitra et al. (2015: 64) listed it.

Eumerus coeruleifrons Keiser, 1958

Eumerus coerulifrons Keiser, 1958, Revue suisse Zool., 65: 218 (♂; "Peradeniya, Expt. Stn, Ceylon") [NM, Basel]

Sri Lanka (Keiser 1958; Knutson et al. 1975; Ghorpadé 2014b: 11.)

Note: Keiser (1958: 218) included this as "*Eumerus coeruleifrons*" in his paper documenting specimens in Sri Lanka taken from Peradeniya Exp. Stn, 15.viii.1953, in Central Province, on flowers of plants. He also commented it is "medium-sized, almost black species, with shining blue forehead and face, without thoracic stripes, fine abdominal spots, thickened hind femur and brownish wing base. It has thin thoracic white stripes, wing hyaline, with femur not much swollen." Knutson et al. (1975: 341), Ghorpadé (2014b: 11), and Mitra et al. (2015: 64) listed it.

Eumerus figurans Walker, 1859

Eumerus figurans Walker 1859, J. Linn. Soc. Lond., 4: 121 (♀; "Makassar, Celebes") [BMNH, London]

Eumerus marginatus Grimshaw, 1902, in Grimshaw & Speiser, Fauna hawaiiensis, Diptera, Suppl.3: 82 (♂; "Honolulu, Hawaii") [USNM/BPBM]

India: KL; Sri Lanka (Keiser 1958; Knutson et al. 1975; Misra & Verma 1975; Ghorpadé 2014b: 11; Mitra et al. 2015; Sandhya et al. 2016.)

Note: Keiser (1958: 194) included this as *"Eumerus figurans"* in his paper documenting specimens taken from Uva and Western and Central Provinces of Sri Lanka. He also commented "Among the Ceylonese and Indonesian material at my disposal, I can recognize no more distinct one. In size, coloration, stripes and spots on thorax and abdomen and the remaining morphological characters, this material and the description of *E. flavicinctus* de Meijere (1908) seems to be identical in comparison. It has a preference for grass; occasionally also found on flowers of shrubs."

I was sent and requested to determine a syrphid whose larvae were found to infest ginger (Zinziber officinale Rosc.) rhizomes in Kerala which I found to be this species. Sandhya et al. (2016) published on this where a Micropezidae was the major pest affecting ginger during 2013–2014 in the Trichur and Palghat districts of Kerala. Earlier papers recorded a syrphid affecting ginger rhizomes, but identifications were incorrect, as Eumerus albifrons Walker or E. pulcherrimus Brunetti (q.v.). E. figurans is the actual species that infests ginger here in Kerala and most probably in other parts of India as well and has also been recorded earlier in the Hawaiian islands as a minor pest of ginger, as the current synonym E. marginatus Grimshaw. Sandhya sent me ten specimens from Trichur taken from 1–16. ix. 2014. This species is an Oriental, Papuan, and Pacific taxon, widespread, distributed from Sri Lanka through the Malay archipelago from Sumatra and Java to the Lesser Sunda Islands, Taiwan, Philippines, and across the Pacific to Hawaii. Misra & Verma (1975) reported a *Eumerus* damaging potato. Knutson et al. (1975: 341), Ghorpadé (2014b: 11), and Mitra et al. (2015: 64) listed it.

Eumerus nepalensis Brunetti, 1908

Eumerus nepalensis Brunetti 1908, Rec. Indian Mus., 2: 76 (♀; Chonebal, Nepal) [ZSI, Calcutta—examined]
India: CG ? (Brunetti 1908, 1915, 1923; Knutson et al. 1975: 342; Ghorpadé 2014b; 11, 2015: 41; Mitra et al. 2015: 64.)

Note: Brunetti (1908: 76–77) had noted: "Described from the one type-specimen in the Indian Museum collection. It is near *argyropus*, Dol., but distinct by the wholly clear wing." I confirmed the type in the ZSI, Calcutta collection when I visited there in 1981. *E. argyropus* Doleschall, 1857 was preoccupied by Loew, 1848 and thus Shiraki published a new name *doleschalli* in 1930. But these are now synonyms of *argentipes* Walker, 1861, all collected in the Moluccas in Indonesia. About *nepalensis* Brunetti (1915: 239) wrote: "It is probable that my *nepalensis* will sink to synonymy, but it is not certain which species it is identical with, as three or four appear very closely allied if allowances for variation are made. These are *macrocerus*, W., *aurifrons*, W. (*splendens*, W.), *nicobarensis*, Sch., and *niveipes*, Meij. Specimens agreeing with the description of *nepalensis* are in the Indian Museum from Mergui, Margherita, Pallode and Travancore and from Mergui, Nepal (the type specimen of *nepalensis*) [this is incorrect, note Brunetti's (1923: 252–253) correction], and Sibu, Sarawak." Brunetti (1915: 240) ultimately mentioned the chances in favour of *aurifrons*, W., being the senior synonym. Brunetti's (1915: 239) comments on the difficulty in *Eumerus* species taxonomy deserve mention here. He had written, "I had anticipated drawing up a table of oriental species in this genus, but from the descriptions only this is quite impracticable, the species being very closely allied, whilst the few characters that appear mostly useful taxonomically, *viz.*, the width and shape of the frons, the structure of the hind tarsi and the degree of pubescence or bareness of the eyes, are ignored by all the older writers. The presence or absence of a infuscation at the wing tip, the intensity or entire absence of the pale stripes on the thorax, and the proportion of tawny colour in the legs are all characters subject to considerable variation." Knutson et al. (1975: 342), Ghorpadé (2014b: 11, 2015b: 41), and Mitra et al. 2015: 64) listed it.

Eumerus nicobarensis Schiner, 1868

Eumerus nicobarensis Schiner, 1868, *in: Reise der österreichischen Fregatte Novara*, Dipt., p. 368 (♀; "Nicobar is.") [NM, Vienna]
Eumerus nepalensis Brunetti, 1915, Rec. Indian Mus., 11: 240 (♀; "Sibu, Sarawak, Borneo") [ZSI, Calcutta]
India: AN, BI TN ?; Sri Lanka (Brunetti, 1915, 1923; Keiser, 1958; Knutson et al. 1975; Ghorpadé, 2014b, 2015b; Mitra et al. 2015).

Note: Brunetti (1923: 251–252) gave a literal translation of Schiner's original description; and then went on to add, giving differences from *aurifrons*, Wied.: "Four ♂ ♂ and two ♀ ♀ (one headless) in the Indian Museum agree very well with Schiner's description. Mergui; Margherita; Pallade [*sic*], Travancore. These specimens were incorrectly attributed by me firstly to *argyropus*, Dol., and subsequently to my *nepalensis*, of which the only specimen known to me now is the type. The Sarawak specimen (Sibu, Sarawak, 2.vii.1910 (*Beebe*)) referred by me to *nepalensis* and redescribed (Brunetti 1915: 240) is also *nicobarensis* ♀. I have subsequently seen further specimens of this species from Pusa, 14.viii.1907, 8.iii.1908, and 17.iii.1908, "under mango bark"; Chapra, Bengal [*sic*]." I did not find the type of *nepalensis* Brunetti, 1915 in ZSI, Calcutta. This species is evidently a tropical rain forest dweller known from NE. India and the southernmost Western Ghats. Its record from NW India is suspicious and mostly based on misidentification(s). Keiser (1958: 216) included this as "*Eumerus nicobarensis*" in his paper documenting specimens taken from Ambacotta and Teldeniya in Central Province of Sri Lanka, flying among grass over soil. Its diagnostics are spotted wings, and large spots on the second tergum, and being a large fly. Knutson et al. (1975: 342), Ghorpadé (2014b: 11, 2015b: 42), and Mitra et al. (2015: 64) listed it.

Eumerus pulverulentus Brunetti, 1923

Eumerus pulverulentus Brunetti 1923, *Fauna Brit. India*, Dipt., 3: 258 (♂; "Pusa") [BMNH, London]
India: BI, RJ (Brunetti 1923; Knutson et al. 1975; Ghorpadé 2014b: 11, 2015b: 42; Mitra et al. 2015: 64.)

Note: Brunetti (1923: 258–259) had noted, "Described from two ♂ ♂ and nine ♀ ♀ from Pusa; all bred, 30.iv.1908, in stem of *Euphorbia* sp., "C. No. 696," and 5.vi.1907, in stem of "Sig." [Fig ?], "C. No. 535." Type ♂ and ♀ sent to the British Museum, cotype ♂ and ♀♀ in the Pusa collection. One ♀ from Abu (*Nurse*). This species is strikingly distinct from all others by the very elongate narrow antennae and by the median stripe on the 4th abdominal segment. The yellow-margined scutellum also separates it from the majority of the Oriental species." In October 2012, I found one ♂ and four ♀ "co-types" in the IARI, New Delhi collection labelled "Pusa, Bengal, stem of *Euphorbia* sp., 30.iv.1908, C. No. 696 No. L, *E. pulverulentus* Brun. Cotype, Brunetti det." Also a label with D 9011–D 9015 on each of the five specimens. Knutson et al. (1975: 342), Ghorpadé (2014b: 11), and Mitra et al. (2015: 64) listed it.

Eumerus rufoscutellatus Brunetti, 1913

Eumerus rufoscutellatus Brunetti, 1913, Rec. Indian Mus., 9: 269 (♂; "Singla, Darjeeling District") [ZSI, Calcutta—examined]

India: CG ? (Brunetti 1913b, 1923; Knutson et al. 1975; Ghorpadé 2014b; Mitra et al. 2015).

Note: Brunetti (1913b: 269–270, Pl. xiv, fig. 13) described it from a male in inferior condition from Singla, Darjeeling District. Brunetti (1923: 255) noted "in inferior condition through immersion in spirit, from Singla, Darjeeling district. A species conspicuous by its densely yellow-haired scutellum and its large size [12 mm]." I had seen the single type in ZSI, Calcutta in 1981. Knutson et al. (1975: 342), Ghorpadé (2014b: 11), and Mitra et al. (2015: 64) listed it.

Eumerus singhalensis Keiser, 1958

Eumerus singhalensis Keiser, 1958, Revue suisse Zool., 65: 219 (♂♀; "Kalpitiya, Ceylon") [NM, Basel]

Sri Lanka (Keiser, 1958; Knutson et al. 1975: 342; Ghorpadé 2014b: 11.)

Note: Keiser (1958: 219) included this as "*Eumerus singhalensis*" in his paper documenting specimens taken from Kalpitiya, 24.i.1954, in grass and herbs, in North-western and North Central Provinces of Sri Lanka. He also commented "species is close to *E. aurifrons*, and distinguished by the characters—short eye-hairs, contiguous eyes, and yellow hind tarsi." Knutson et al. (1975: 342), Ghorpadé (2014b: 11), and Mitra et al. (2015: 64) listed it.

Eumerus sita Keiser, 1958

Eumerus coerulifrons Keiser, 1958, Revue suisse Zool., 65: 217 (♀; "Kandy, Hantana, Ceylon") [NM, Basel]

Sri Lanka (Keiser 1958; Knutson et al. 1975: 342, Ghorpadé 2014b: 11.)

Note: Keiser (1958: 217–218) included this as "*Eumerus sita*" in his paper documenting specimens taken from Hantana, Kandy, 23.xii.1953, in grass in tea estate. He commented that it is a "small species, characterized by the broad, shining-silvery forehead and face, blackish scutellum, a large pair of spots on the second abdominal segment and the white-haired last segment of the hind tarsus." Taken in the Central Province of Sri Lanka. Knutson et al. (1975: 342), Ghorpadé (2014b: 11), and Mitra et al. (2015: 64) listed it.

Tribe Milesiini

Calcaretropidia triangulifera (Keiser, 1958)

Syritta triangulifera Keiser, 1958, Revue suisse Zool., 65: 226 (♂♀; "Belihul Oya, Ceylon") [NM, Basel]

Calcaretropidia triangulifera (Keiser), 2005. The genus *Syritta*, pp. 213–216. (as *comb.n.*)

India: south ?; Sri Lanka (Keiser 1958; Knutson et al. 1975: 365; Ghorpadé 2014b: 12; Mitra et al. 2015; Thompson et al. 2017.)

Note: Keiser (1958: 226–227, fig. 5) included this as "*Syritta triangulifera*" new species, in his paper documenting specimens taken from Sabaragamuva and Central Provinces, he took in December and February. Lyneborg & Barkemeyer (2005) transferred it to *Calcaretropidia*, a new genus erected by Keiser from a species he described in Madagascar. Thompson et al. (2017: 513) provide a synopsis of this small genus with six species ranging from tropical Africa to New Guinea, with a single undescribed species from the PSWNG area (Thompson et al. 2017). Knutson et al. (1975: 342), Ghorpadé (2014b: 11), and Mitra et al. (2015: 65) listed it.

Korinchia rufa Hervé-Bazin, 1922

Korinchia rufa Hervé-Bazin, 1922, Bull. Soc. Ent. Fr., 27; 122 (♂♀; "Kodaikanal") [MNHN, Paris]

India: TN (Hervé-Bazin 1922; Brunetti 1923; Curran 1928, 1929, 1930, 1931a; Shiraki 1930; Thompson 1975; Knutson et al. 1975; van Steenis & Hippa 2012; Ghorpade 2014b; Mitra et al. 2015.)

Note: Hervé-Bazin (1922) named and described it from a ♂ and 2 ♀ from Kodaikanal (Tamil Nadu) taken by him ("mai 1913"). Then Brunetti (1923: 224) gave a description and figures of these same specimens (types) in his *Fauna* volume. Curran (1928, 1929, 1930, 1931b) keyed it out or listed it in his papers on Malay Peninsula Syrphidae. Van Steenis & Hippa (2012) revised *Korinchia* and discussed its phylogeny. For this species, *rufa*, they gave additional specimens which were taken by me, Ghorpade, a ♂ from Yercaud, 1370 m (Tamil Nadu), 18.ix.1978, A673, and a ♀ from Kodaikanal, 2150 m (Tamil Nadu), 28.ix.1985, B421. Besides these couple of specimens sent by me to USNM, Washington, DC, I have another two (♂♀) which I took visiting flowers in the garden and in the room of a bungalow at Kodaikanal (2150 m) on 28.ix.1985, B419, which will be sent to the CNC, Ottawa (Canada). It may be useful to mention that my collection no. B421 was done in Bear Sholah near Kodaikanal (2160 m), on September 28, 1985. Knutson et al. (1975: 361), Ghorpadé (2014b: 12), and Mitra et al. (2015: 65) listed it. Thompson et al. (2017) omitted this otherwise dominantly Oriental genus from their paper, which ranges all over (76 spp.) the world, except in the Australian region, but surprisingly no species are yet recorded from the PSWNG area.

Milesia caesarea Hippa, 1990

Milesia caesarea Hippa, 1990, Acta Zool. Fenn., 187: 93, figs. 56C–56E (♂; "Cinchona, 3500 ft, Anaimalai Hills") [ZM, Amsterdam]

India: KL, KN, TN, MP ? (Hippa 1990; Ghorpadé 2014b; Mitra et al. 2015; Thompson et al. 2017.)

Note: Hippa (1990) named and described it from some 30+ ♂ and around 10 ♀ from the Anaimalai Hills (Cinchona, Top Slip), 2♂ 3♀ from Ponmudi range (Kerala), and 4♂ from Agumbe ghat (Karnataka), plus, doubtfully, 1♂ 2♀ from Pachmarhi on the Satpura Hills (Madhya Pradesh). Thompson et al. (2017: 516) provide a synopsis of this large Indomalayan group with a few species in the north temperate region and in the northern Neotropics, with 11 species recorded from the PSWNG area. Ghorpadé (2014b: 13) and Mitra et al. (2015: 65) listed it.

I believe Hippa erred in clubbing the specimens from Central India with this southern Indian species. The material from Pachmarhi (3500 m ft) are distinct as he himself noted (Hippa 1990: 94) even though this is an otherwise variable species as to tergum and abdominal pattern. These central Indian highlands specimens should be named and described as a separate, new species by future worker(s).

Milesia cinnamomea Hippa, 1990

Milesia cinnamomea Hippa, 1990, Acta Zool. Fenn., 187: 167, 106, 109 A, B, 110C, D, 113F (♂; "Agumbe ghat, 2000 ft") [ZM, Amsterdam]

India: KL, KN, TN (Hippa 1990; Ghorpadé 2014b; Mitra et al. 2015.)

Note: Hippa (1990) named and described this from south-west Indian Ghats (7 ♂ 5 ♀) taken from April to June at Agumbe (Shimoga district—correct spelling!), Mercara (Coorg), both in Karnataka, and at Chembra Peak and Ponmudi range in Kerala and at Top Slip, Anaimalai Hills (Tamil Nadu). Ghorpadé (2014b: 13) and Mitra et al. (2015: 65) listed it.

Milesia mima Hippa, 1990

Milesia cinnamomea Hippa, 1990, Acta Zool. Fenn., 187: 85, figs. 47, 48B, 49 (♂; "Cinchona, 3500 ft, Anaimalai Hills") [ZM, Amsterdam]

India: KL, TN (Hippa 1990; Ghorpadé 2014b; Mitra et al. 2015.)

Note: Hippa (1990) named and described this from Anaimalai Hills and from Ponmudi range in south India, taken from April to May. Ghorpadé (2014b: 13) and Mitra et al. (2015: 65) listed it.

Milesia querini Ghorpadé, 2014

Milesia querini Ghorpadé, 2014, Colemania, 44: 13 (♂; "cote du Coromandel") [lost ?]

Milesia gigas Querin: Hippa, 1990, Acta Zool. Fenn., 187: 109 (??) [?]

India: TN, AH ?, PO ? (Hippa 1990; Ghorpadé et al. 2011; Ghorpadé 2014b; Mitra et al. 2015.)

Note: Hippa (1990: 109) mentioned this species which was named and described based on a single male specimen taken on the "côte du Coromandel," with no exact locality specified. Hippa misinterpreted this as being on the "west coast of Indian Peninsula." The original publication by Querin also included "an excellent coloured illustration in natural size." But this type specimen has never been located nor have any other specimens from this area been collected. *M. gigas* Querin is preoccupied by *M. gigas* Macquart, 1834 from Java (and by *gigas* Rossi, 1790), so Hippa (1990: 102) proposed *M. gigantea* as a new name for the Javan species (q.v.). This Coromandel *gigas* (new name also required, as it is unlikely to be conspecific with the Javan species) was presumably a relic, inhabiting the dry evergreen forest on this coast which has now been almost wholly obliterated by human agency. I have included this rare species in Ghorpade et al. (2011: 83) only to bring attention to this fact and hoping that specimens can be found flying in surviving and restored (e.g., near Pondicherry) forested habitats on this coast. Ghorpadé (2014b: 13 proposed a new name (*nom. nov.*) and listed it.

Milesia sexmaculata Brunetti, 1915

Milesia sexmaculata Brunetti, 1915, Rec. Indian Mus., 11: 248 (♂; "Trivandrum, Travancore State") [ZSI, Calcutta—examined]

Milesia sexmaculata Hippa, 1990, Acta Zool. Fenn., 187: 94, figs. 54A,C, 56A,B ♂;"Cinchona, 3500 ft, Anaimalai Hills") [ZM, Amsterdam]

India: KL, TN (Brunetti 1915, 1923, 1925; Hervé-Bazin 1926; Hippa 1990; Ghorpadé 2014b, 2015b; Mitra et al. 2015.)

Note: Brunetti (1915: 246) named and described this from a single ♂ from Trivandrum, Travancore State (Kerala) received by the Indian Museum (now ZSI) from the Trivandrum Museum. Then (Brunetti, 1923 265, Figure 51) he described and figured it again in his *Fauna* volume. Brunetti (1925: 77) mentioned a ♀ received from "Coimbatore, South India, sent by Mr. Bainbrigge Fletcher. It agrees very well with the ♂ except that the abdominal spots are rather smaller and narrower." Hippa (1990: 106–107, figs. 50, 54, 56) described and illustrated it and cited specimens seen from Kerala and Tamil Nadu in southern India. The record from "Kurseong" in West Bengal is certainly incorrect, possibly because of mislabelling. Note (Hippa, 1990: 94) that P. Caius who is named as the collector also collected in Kodaikanal in May 1921. The specimen in the Paris Museum needs to be checked for its label and mislabelling. See also my notes under *Allograpta bouvieri* below. Hervé-Bazin (1926: 96) included this in his long list of 34 species from Asia and Oceania. In recent years (2007–2010), I have taken some ten specimens of this species in Koolak Sholah (2000–2200 m) in deep forest in March and May, and these large flies were investigating insides of hollowed out trunk bases. Ghorpadé (2014b: 13, 2015b: 51), and Mitra et al. (2015: 65) listed it.

Syritta fasciata (Wiedemann, 1830)

Xylota fasciata Wiedemann, 1830, Aussereurop. Zweifl. Insekt., 2: 103 (♀; "Nubien" = Sudan) [ZMHU, Berlin]

India: GJ (Lyneborg & Barkemeyer 2005; Ghorpadé 2014b, 2015b; Mitra et al. 2015; Thompson et al. 2017).

Note: Lyneborg & Barkemeyer (2005: 98–104) wrote that this "is the *Syritta* species which is most often misidentified. Belonging to the subgroup of 4 species with a very thin to nearly obsolete spurious vein, it is easily separated by the different curvature of the apical marginal vein (M1). The slender body–shape distinguishes it from *S. pipiens*, but a dissection of the genitalia is often necessary for a safe identification. The fourth species in the subgroup, *S. stylata* new species has a [*sic*] almost obsolete spurious vein, and its genitalia have larger cerci and surstyli." Thompson et al. (2017: 516) provide a synopsis of this "large and now worldwide genus of 60 described species (with several more undescribed known); those found in the New World were introduced from the Old World, nine species occur in [the] PSWNG." area. Ghorpadé (2014: 13, 2015b: 52) and Mitra et al. (2015: 65) listed it.

Syritta indica (Wiedemann, 1824)

Eumerus indica Wiedemann, 1824, Analecta Ent., p. 33 (♂♀; "India Or.") [ZMUC, Copenhagen]

Syritta rufifacies Bigot, 1883, Annls Soc. Ent. Fr., (6) 3: 538 (♂; "Pondicherry, India" [UZM, Oxford ?]

Syritta femorata Sack, 1913, Ent. Mitt., 2: 8 (♂; "Tainan, Formosa") [DEI, Eberswalde]

India: AH, BI ?, KL, KN ?, MH, MP ?, PO, TN; Sri Lanka (Cherian 1934; Keiser 1958; Gokulpure 1972; Ghorpadé 1973b, 1974, 2014b: 13, 2015b: 52–53; Knutson et al. 1975; Datta & Chakraborti 1986; Lyneborg & Barkemeyer 2005; Ghorpadé et al. 2011; Mitra et al. 2015).

Note: Cherian (1934: 69) listed it as *Syritta rufifacies* from Coimbatore where its maggots were found feeding on aphids on mango flowers (?). Keiser (1958) listed this from Sri Lanka. Lyneborg & Barkemeyer (2005: 127–129) placed this in their *indica* species-group, characterized by spurious vein well sclerotized, hind femur without a subbasal posteroventral spina, but with a number of setula-bearing tubercles and whitish setae, hind femur unmodified, i.e., without a lamina on or excavation into the ventral surface. They distinguished *indica* from its sister-species *proximata* by the profile of the terminal abdominal segment, "in *S. indica* with a hump on tergum 4 located at one-sixth of tergal length from the posterior margin, and a continuous brown posterior fascia on tergum 4." Brunetti (1908: 77) gave an interesting discussion of this genus and wrote, "In a subsequent paper I hope to deal with this genus." He was unsure if there were only three or eight or ten good species in the Orient. In a later paper (Brunetti, 1915: 237–239) he gave more notes and took this as a good species with *orientalis* Macq., *lutescens* Dol., *illucida* Wlk., and *laticincta* Big., *nom. nud.*, as synonyms. He also recognized *pipiens* L., *amboinensis* Dol., and *rufifacies* Big., which he thought "possibly synonymous with *orientalis*. In his *Fauna* volume (Brunetti 1923: 244–248) he recognized only three species from the Indian subcontinent: *pipiens* L., *orientalis* Macq., and *rufifacies* Big., gave a key to separate them and full descriptions and figures. This species he listed as a synonym of *pipiens* L., noting (pp. 245–246, Pl. V, figs. 14 and 15) "This species is common and generally distributed in both hills and plains in India... A widely distributed species: from North America and Europe to Asia and Africa." Of flies of this genus he wrote, "the insects occurring almost anywhere amongst flowers and leaves, in hedges and fields; the ♂♂ are frequently seen hovering." Of *rufifacies* Big. (pp. 247–248, Pl. V, fig. 18) he wrote it "differs from *pipiens* and *orientalis* by the hind femora being wholly orange or brownish for from one-third to two-thirds of their length from the base, while the rest is black." He mentioned several specimens in good condition in the Indian Museum from most parts of India including Satara District and United Provinces [= Uttar Pradesh]. Gokulpure (1972) mentioned this (?) as "*Syritta pipiens*" from Damoh (Madhya Pradesh). I (Ghorpadé, 2017) included it from Maharashtra based on Brunetti (1923: 246) who stated, "This species is common and generally distributed in both hills and plains in India." The species identity needs to be confirmed (*indica* ?) as Bombay specimens will not be *S. pipiens*. Brunetti (1923: 248) noted as "*Syritta rufifacies* Big." "based on several in good condition in the Indian Museum from... Satara District... The species identity also needs to be confirmed." Keiser (1958: 225) included this as "*Syritta orientalis* var. *rufifacies*" in his paper documenting specimens taken from Northern and Central Provinces of Sri Lanka. In my master's thesis, I (Ghorpade 1973b, 1974) mentioned this being recorded from Bangalore, but not finding it myself. Datta & Chakraborti (1986: 64) mentioned a ♂ from Chalakudy, Trichur District (Kerala), taken in December 1979. Ghorpadé et al. (2011: 82–83) mentioned it being taken at Rayudupalem and near Rajahmundry in the Andhra Carnatic and gave some notes. Knutson et al. (1975: 365), Ghorpadé (2014b: 13, 2015b: 52), and Mitra et al. (2015: 65) listed it.

Syritta orientalis Macquart, 1842

Syritta orientalis Macquart, 1842, Dipt. Exot., 2(2): 76 (♂; "Pondicheri") [MNHN, Paris]

Senogaster ? lutescens Doleschall, 1856, Natuurk. Tijdschr. Ned.-Indië, 10: 410 (sex ?; "Djokjokarta, Java") [?]

Syritta ? amboinensis Doleschall, 1859, Natuurk. Tijdschr. Ned.-Indië, 17: 97 (sex ?; "Amboina, Moluccas") [?]

Syritta illucida Walker 1859, J. Linn. Soc. Lond., 4: 121 (♀; "Mak." = Makasar, Celebes, Indonesia) [NHM, London]

Syritta laticincta "Bigot": Brunetti, 1915, Rec. Indian Mus., 11: 238 (sex ?; Karachi and Calcutta) [ZSI, Calcutta] *nom. nud.*

Spheginobaccha christiani Sodhi & Awtar, 1991, Acta zool., Cracov, 34: 319 (♂; "Nainital") [DZPU, Chandigarh—lost ? or in IARI, New Delhi ?]; Ghorpadé, 2014b, Colemania, 41: 9 (as *n. comb., n. syn.*)

India: BI ? CG, PO, TN; Sri Lanka (Brunetti 1908, 1915; Keiser 1958; Ghorpadé 1973b, 1974, 2014b, 2015b, unpubl.; Knutson et al. 1975; Datta & Chakraborti 1986; Sodhi & Awtar 1991; Mitra et al. 2015.)

Note: Brunetti (1908: 77, 1915: 237–239) gave notes on Oriental species of this genus; see notes under *indica* above. About *orientalis* Macq. (pp. 246–247, Pl. V, figs. 16 and 17) he placed *illucida* Wlk., *lutescens* Dol., and doubtfully *amboinensis* Dol., as synonyms, giving extended notes and noting specimens seen from Pusa in Bihar and from localities in Bengal and Burma. Lyneborg & Barkemeyer (2005: 157–159) wrote: "*S. orientalis* is a variable species, especially with regard to colour of the hind femora and of tergum 2. It is a widely distributed species ranging from India and Sri Lanka through Indonesia, continental SE Asia and the Philippines to Australia in the south and the Hawaiian islands in the east, can be distinguished by a combination of the presence of a well sclerotized spurious vein and of a cone-shaped subbasal spina on the posteroventral surface of the hind femur; however, in the female this spina is minute." Brunetti (1908: 77) gave notes (see under *indica* above) on this and others of this genus here. Keiser (1958: 225) included this as "*Syritta orientalis*" in his paper documenting specimens taken from Uva, Sabaragamuva and Northern, Eastern, North Central, and Central Provinces of Sri Lanka. Datta & Chakraborti (1986: 65) mentioned a ♀ from Top Slip (Tamil Nadu) taken in December 1979. I (Ghorpadé 2014b: 9) had corrected Sodhi & Awtar's (1991) misidentification of their *christiani* as a *Spheginobaccha* and synonymized that under *Syritta orientalis* based on the illustrations provided. The holotype was not found in the Punjabi University, Chandigarh, nor in the IARI, New Delhi when I visited these institutions in late 2012. In my master's thesis, I (Ghorpadé 1973b, 1974) wrote of taking this (as "*pipiens*") in Bangalore from November to February, and April, flies taken from flowers of mango, wheat, niger, *Tridax procumbens*, *Bidens pilosa*, and *Sonchus oleraceus*, and mentioning that, "The status of the Indian species of *Syritta* needs investigation and the material under study cannot be determined specifically without doubt." Knutson et al. (1975: 365), Ghorpadé (2014b: 13, 2015b: 53), and Mitra et al. (2015: 65) listed it.

Syritta pipiens (Linnaeus, 1758)

Musca pipiens Linnaeus, 1758, Syst. Nat., Ed. 10, p. 594 (sex ?; "Europe" = Sweden vide Thompson et al. 1982: 159) [NHM, London]

India: CG? (Brunetti 1908, 1915. 1923; Knutson et al. 1975; Thompson et al. 1982; Thompson et al. 1990; Lyneborg & Barkemeyer 2005; Ghorpadé 2014b, 2015b, unpubl.; Mitra et al. 2015.)

Note: Brunetti (1908: 77) gave notes on this and other species of this genus here. Brunetti (1915: 237–239) gave more notes on other Oriental species; see my notes under *indica* above. *q.v.*). Thompson et al. (1982: 159) gave information on the types. Thompson et al. (1990) gave a diagnostic key and illustrations to this species to separate it from the Afrotropical *S. flaviventris* Macquart, newly recorded as an immigrant, a synanthrophic flower fly, in Texas state (U.S.A.) and in Mexico. Knutson et al. (1975: 365), Ghorpadé (2014b: 13, 2015b: 53–54), and Mitra et al. (2015: 65) listed it.

Syritta proximata Lyneborg & Barkemeyer, 2005

Syritta proximata Lyneborg & Barkemeyer, 2005, The genus *Syritta*, p. 129 (♂; "Bangalore") [USNM, Washington, DC]

India: AH, KL, KN, MP, PO, RJ, TN; Sri Lanka (Cherian 1934; Lyneborg & Barkemeyer 2005; Ghorpade et al. 2011; Ghorpadé 1973b, 1974, 2014b, 2015b; Mitra et al. 2015).

Note: Cherian (1934: 699) probably listed this as *Syritta pipiens* from Coimbatore, Bellary, and Madanapalle (Chittoor). Lyneborg & Barkemeyer (2005: 95–98) named and described this "remarkable, and in India, widely distributed [*sic*] species. It was present in collections under the names of *S. orientalis* Macquart or *S. rufifacies* Bigot... *S. proximata* seems to be restricted to India and Sri Lanka." Incidentally, the holotype selected was collected by myself labelled "India, Karnataka, Bangalore, 916m, 5.x.1980, leg Ghorpade [A 899]." The 27 ♂♀ paratypes listed were taken in the Indian states of Andhra Pradesh, Delhi (?), Kerala, Karnataka, Madhya Pradesh, Pondicherry, Rajasthan, Tamil Nadu, and West Bengal (?), besides Sri Lanka as well. The Andhra locality "Fayndupalem" is an error for Rayudupalem. The Rajasthan locality "Udsipur" is correctly Udaipur. The Central Indian Satpura Hills locality "Pachmarti" is actually Pachmarhi, and the "Anaimalai Hills, Chinchona" location should correctly be spelled Anaimalai Hills, Cinchona. The Delhi and Rajasthan specimens need to be checked and confirmed, especially as the "India, N.E. [!], Delhi, XI.1966, leg Jermyn (BMNH)" labelling is incorrect—see Ghorpadé (2007: 5). So this species may not really be a NW Frontier resident ? In my master's thesis, I (Ghorpadé 1973b, 1974) included most probably this (as *pipiens*) from Bangalore. Ghorpadé et al. (2011: 83) mentioned taking this at Rayudupalem, Potunuru, and Naguldevpadu near Eluru in the Andhra Carnatic and gave some notes. Ghorpadé (2014b: 13, 2015b: 54) and Mitra et al. (2015: 65) listed it.

Syritta stylata Lyneborg & Barkemeyer, 2005

Syritta stylata Lyneborg & Barkemeyer, 2005, the genus *Syritta*, p. 104 (♂; "Yercaud, 4500 ft, Shevaroy Hills") [RMNH, Leiden]

India: KL, KN, MP, TN (Lyneborg & Barkemeyer 2005; Ghorpadé 2014b; Mitra et al. 2015).

Note: Lyneborg & Barkemeyer (2005: 104–106) named and described this species. They wrote, "The discovery of this undescribed species in southern India is of special interest, because it is a member of the otherwise non-Oriental *S. pipiens* species-group and one of several examples of parallelism in the genus *Syritta* regards to the reduction of the spurious vein. The species has been collected mainly in hilly country in southern India, but there is also an isolated record from Punjab in northern India [this is an obvious mislabelling as Susai Nathan mainly collected in south India and the "C. Punjab, XII.1955" label must be actually Central Provinces? At Jabalpur?—Kumar Ghorpadé] *S. stylata* is a larger and more broadly built species than *S. fasciata* (Wiedemann) which we can record from the foothills of the Himalayas some 1,500 km further north. It can be confused with *S. pipiens* (Linnaeus) which occurs in NW India. Both *S. fasciata* and *S. pipiens* have a thin but almost complete spurious vein, and there are distinct differences in the male genitalia (see the illustrations)." The types (10 ♂ 4 ♀) were from Shevaroy Hills, Nilgiri Hills, Anaimalai Hills (Tamil Nadu), Chembra Peak (Kerala), and Mudigere (Karnataka). The last (♂) was collected by me from "Mudigere, 970m, 8.iv.1975, A191, Ghorpade [USNM]." My collection diary adds that that area was an old neglected coffee plantation to the right of the Regional Research Station, Mudigere, in thinned forest with much leaf litter, but not many herbaceous weeds, and with a nullah (stream) with little water flowing through at lowest spot. Syrphids were taken on flowers of *Terminalia* sp. Ghorpadé (2014b: 13), and Mitra et al. (2015: 65) listed it.

Diagnostic characters given were as follows: *S. stylata* is distinguishable from other species of the pipiens species-group [five Afrotropical, plus pipiens, fasciata, and *stylata*— Kumar Ghorpadé] by the indistinct spurious vein. In the male genitalia, the dorsal lobe of the surstylus is a slender and almost uniformly wide rod, without any hump; the cerci are conspicuous, with an almost straight median margin and concave lateral margin, apex broadly rounded.

Three other Syritta species occur in southern India: *S. orientalis* Macquart, *S. indica* (Wiedemann), and *S. proximata* L & B. All three species possess a well developed spurious vein. *S. orientalis* has a ventral spina at the base of the hind femur, and the two other species do not possess a distinct black median vitta on tergum 2 as does *S. stylata*.

Xylota atroparva Hippa, 1974

Xylota atroparva Hippa, 1974, Ann. Ent. Fenn., 40(2): 58, figs. 3, 5, 7, 10 (♂; "Haycock Hill, Ceylon") [BMNH, London]

Sri Lanka (Hippa 1974, 1978; Ghorpadé 2014b; Mitra et al. 2015; Thompson et al. 2017).

Note: Hippa (1974: 58–59) named and described this as a new species from Haycock Hill, Kandy, Kitulgala, and Labugama, in Sri Lanka, taken in April, May, and September. He compared it with *X. carbonaria* Brunetti, 1923 which was taken from the Shevaroy Hills in Tamil Nadu, India. He also mentioned another possible other new species, from Sri Lanka, like *carbonaria,* but distinct from the Shevaroys types, which he named as "*Xyota* sp. 5" (Hippa, 1978: 70) in a larger revision of Xylotini (see below). Thompson et al. (2017: 516) provide a synopsis of this "largely north temperate group (132 species), six species occur in [the] PSWNG." area. Ghorpadé (2014b: 13) and Mitra et al. (2015: 65) listed it.

Xylota bistriata Brunetti, 1915

Xylota bistriata Brunetti, 1915, Rec. Indian Mus., 11: 235 (♂♀; "Parambikulam, Cochin") [ZSI, Calcutta—examined]

India: KN, KL (Brunetti 1915, 1923; Cherian 1934; Knutson et al. 1975; Hippa 1978; Ghorpadé 2014b; Mitra et al. 2015).

Note: Brunetti (1915: 235) named and described this from 3 ♂ 3 ♀ "in perfect condition" from "Parambikulam, Cochin, 1700–3200 ft., 17–24.ix.[19]14, Gravely." In his *Fauna* volume, Brunetti (1923: 238, Pl. V, fig. 13) gave a full description and figure of the male hind leg, and added more specimens examined from Talewadi, 3–10.x.1916 and Castle Rock, 11–26.x.1916, Kemp, in N. Kanara district (Karnataka); Kollur Ghat, 3000 ft., S. Kanara district, 18–21.ix.1913; N. Coorg, S. India, 29.v.1918, on coffee flowers, I.R. No. 64, Newcome. I saw the last (Coorg) ♂ specimen in the TNAU, Coimbatore collect ion and that specimen has these labels: "Mysore, N. Coorg, Somvarpet, Cowcoody Estate, 29.v.1915, on coffee flowers, Newcome, 3565." Cherian (1934: 699) listed it from North Coorg, adults collected resting on coffee flowers. Hippa (1978: 69) placed it in his *Xylota pendleburyi* group and incorrectly gave "Assam" as its locality, instead of Cochin. Knutson et al. (1975: 366), Ghorpadé (2014b: 13), and Mitra et al. (2015: 65) listed it.

Xylota carbonaria Brunetti, 1923

Xylota carbonaria Brunetti, 1923, *Fauna Brit. India*, Dipt., 3: 240 (♂; "Yercaud, Shevaroys") [TNAU, Coimbatore—examined, lectotype designated here]

India: KN ?, TN (Brunetti 1923; Cherian 1934; Knutson et al. 1975; Hippa 1974, 1978; Ghorpadé 2014b; Mitra et al. 2015).

Note: Brunetti (1923: 240) named and described this from 2 ♂ from "Shevaroys, Yercaud, 23.iv-4.v.1913 (Fletcher)" which Brunetti (1923: 232) diagnosed as having head not descending below eyes in profile, body black, thorax with two obvious dorsal stripes of pale pubescence, always visible from behind, femora mainly or wholly black. He gave "*Type* in British Museum" *which is in error* as I found 2♂ with labels "Madras, Shevaroys—Yercaud, 4300 ft, 23.iv-4.v.1914, Fletcher" in the TNAU, Coimbatore Museum, on my last visit there in August 2016. At that time I realized that the cotypes were in the TNAU collection, and I designated one ♂ labelled "Shevaroys—Yercaud, 4500 ft., 23 Apr—4 May 13, Fletcher coll., 0003563, *Xylota carbonaria* Brunetti, Ghorpade det. 2016, LECTOTYPE." as the LECTOTYPE of *Xylota carbonaria* Brunetti, 1923. Of its legs, only the left legs remain as well as the right hind leg; its mesonotum had its anterior edges eaten by dermestid. The other ♂ with the same labels was designated the PARALECTOTYPE of *Xylota carbonaria* Brun. Cotype ♂ and this specimen had only its left legs remaining, all femora, but only middle tibia existing—its sternum was mostly eaten up by dermestid. These types were hand given to the curator there for safekeeping. Cherian (1934: 699) listed it from Yercaud (Salem). Hippa (1978: 70) placed it in his *Xylota carbonaria* group and mentioned "For ♂ genitalia, see Hippa (1974)." Knutson et al. (1975: 367), Ghorpadé (2014b), and Mitra et al. (2015: 65) listed it.

Xylota sp. 5 Hippa, 1978

Xylota sp. 5 Hippa, 1978, Acta Zool. Fenn., 156: 70 (♂; "Rakwana, Morningride, 4000 ft., 8.v.1929, Ceylon") [CNM, Colombo; BMNH, London]
Sri Lanka (Hippa, 1974, 1978; Ghorpadé, 2014b.)

Note: Hippa (1974: 57) named this apparently new species as "sp. 5" and listed it (Hippa 1978: 70) in his *Xylota carbonaria* group, writing about its resemblance to *X. carbonaria* and *X. atroparva*, but mentioning distinctive characters. Future workers must write of this, give it a new name, and other diagnostic characters. Ghorpadé (2014b: 14) listed it.

Tribe Volucellini

Graptomyza brevirostris Wiedemann, 1820

Graptomyza brevirostris Wiedemann, 1820, Nova Dipt. Gen., p. 17 (♀; "Batavia, Djakarta, Java") [UZM, Copenhagen]
India: AN, CG, KL, KN, TN; Sri Lanka (Brunetti 1915, 1923; Kertész 1910, 1913; Curran 1942; Keiser 1958; Knutson et al. 1975; Ghorpadé 2014b, 2015b, unpubl.; Mahalingam 1988; Mitra et al. 2015; Thompson et al. 2017.)

Note: Brunetti (1915: 226) wrote that this species "also occurs in the Nilgiri Hills." In the *Fauna* volume, Brunetti (1923: 138–139) gave a detailed description of the female and listed specimens from parts of NE. and S. India, Burma, Sri Lanka, and the Nicobar Islands, and gave a key to seven Indian species, all from these above areas only. Kertész (1914: 76–77) gave a key to separate a dozen species present in the Hungarian National Museum, including this species. Curran (1942: 4–5) presented a key to Oriental species of this genus, which included this species. Keiser (1958: 224) included this as "*Graptomyza brevirostris*" in his paper documenting specimens taken from Kandy, Hantana, Horton's, and Peradeniya in Central Province. Mahalingam (1988: 161) gave it as recorded from the Nilgiri Hills among ten species recorded by Brunetti (1923). Thompson et al. (2017: 515) provide a synopsis of this widespread genus (90 species) in the Old World tropics and in the Pacific, with 18 species recorded from the PSWNG area. Knutson et al. (1975: 333) listed it. I saw (K. Ghorpadé unpubl.) a damaged ♀ from Taliparamba, N. Malabar (Kerala) taken in December 1924 and preserved in the TNAU, Coimbatore collection. Also saw a couple taken *in copula* on *Bambusa vulgaris* at Thallapalem, Erattupatta, Kottayam district (Kerala) in October 2014 sampled by K.J. David and preserved in the NBAIR (ICAR) collection, Bangalore. With European colleagues I am revising *Graptomyza* in the Oriental region, and I probably have another undescribed species in south India from the lower Palni Hills. Ghorpadé (2014b: 15, 2015b: 62) and Mitra et al. (2015: 66) listed it.

Volucella sp. nov.

India: KL, KN, TN (Brunetti 1923: 144–153; Knutson et al. 1975: 335–337; Thompson & Rotheray 1998; Ghorpadé 2014b: 15, 2015b:63–64; Mitra et al. 2015: 66; Thompson et al. 2017: 520.)

Note: I have more than ten specimens of a *Volucella* from southern India that is undoubtedly undescribed and needs a new name. I took it first at Kemmangundi on the Bababudan Hills in Karnataka during my postgraduate studies and have taken more on other hills in Karnataka, Tamil Nadu, and Kerala since then. This is the first species of this genus found in southern India, and it is close to *Volucella trifasciata* Wiedemann, 1830, which occurs in north-east India. Thompson et al. (2017: 520) provide a synopsis of this mainly north temperate group occurring marginally in the Orient, but absent from the Afrotropical, Australian, and Neotropical regions; just a couple being found in the Philippines of the PSWNG area. Knutson et al. (1975: 335–337), Ghorpadé (2014b: 15, 2015b: 63–64), and Mitra et al. (2015: 66) listed species of *Volucella*.

Subfamily SYRPHINAE
Tribe Bacchini

Baccha sp.

India: KN, TN (Brunetti 1908, 1915, 1917, 1923; Curran 1928; Coe 1964; Knutson et al. 1975; Peck 1988; Ghorpadé 2014b, 2015b; Mitra et al. 2015; Thompson et al. 2017: 513.)

Note: I found "*Baccha maculata*" to actually be a "lumped" species. In his *Fauna* volume, Brunetti (1923: 119–120, fig. 22) gave a description and figure of the wing of *Baccha maculata* Walker, 1852, and mentioned several specimens from the Simla Hills and Mussoorie, besides NE India, and made his *tinctipennis* a synonym, of which he was certain (but see below). Brunetti (1923: 113–127) had also dealt with a dozen species of "*Baccha*" which also included what are now some *Allobaccha* or *Asiobaccha* and also a *Spheginobaccha* of the Microdontinae! Curran (1928: 244–256; see for notes) also dealt with some nine Malayan species, but was the first to separate *Allobaccha* (p. 251) as a new "subgenus." He placed only *virtuosa* Curran, *eronis* Curran, *nigricoxa* Curran, and di*spar* Walker in the "subgenus" *Baccha* Fabricius, with bare humeri. In the Oriental Syrphidae Catalog (Knutson et al. 1975), only *virtuosa* was retained in *Baccha s str.*, and *eronis* made a synonym of *maculata* Walker. "*Baccha*" *dispar* and *nigricoxa* were placed in Curran's new "subgenus" *Allobaccha* (with pubescent humeri), but as maybe needing revision. In those times, predacious syrphids with a petiolate abdomen were all lumped as *Baccha*, but are now separated correctly in different genera. See Vockeroth (1969: 11–16) for a discussion of previous classifications and his classical revision of the tribe Syrphini. This work had revolutionized Syrphini classification and nomenclature and put me myself on the path to my own careful revisionary work on Indian subcontinent and other east Oriental Syrphini for my doctoral research, with Vockeroth's cooperative specialist guidance and that of Chris Thompson and Lloyd Knutson as well, ever since I had begun work as a postgraduate researcher in 1973 (see also Ghorpadé 1994, and other papers of mine cited in the references below). Knutson et al. (1975: 323) placed *tinctipennis* Brunetti, 1907, *austeni* de Meijjere, 1908, *tenera* de Meijere, 1910, and *eronis* Curran, 1928 as synonyms, which need clarification and confirmation, after comparing types. These synonymized Indian species are not conspecific, as discussed below, *q.v.* Thompson et al. (2017: 513) provided a synopsis of this "small north temperate genus (13 species) of predaceous flies. Only one species (*maculata* Walker, 1852) occurs in PSWNG area." In fact *maculata* barely enters the Oriental region and this statement is incorrect. Knutson et al. (1975: 323), Peck (1988: 55), Ghorpadé (2014b: 15, 2015b: 65–66), and Mitra et al. (2015: 66) listed the species of *Baccha* Fabricius *s. str.*

The southern Indian specimens are also not *maculata* Walker, but need to be named and described as a separate, new species, which I propose to do in the future. During my doctoral research, I also worked on this Bacchini genus and kept notes which I will place here below for information. I wrote: "The Oriental Diptera Catalog (Knutson et al. 1975) listed only two species, *maculata* Walker (with four synonyms) and *virtuosa* Curran. I examined specimens of the latter from Malaya, determined by Curran and Vockeroth [American Museum of Natural History, New York] and noted that it was an *Asiobaccha* (see also Mengual et al. 2015). Baccha s. str., of which I had studied almost 200 specimens from the Orient, appeared to be a complex of 4–5 species, so far identified as *maculata*. This also Oriental species seemed distinct from both *elongata* Fabricius and *obscuripennis* Meigen, especially the latter, from the Palaearctic region. The extent of white pile on the frons easily separated these two from each other, and the totally black first tergum separated these two from *maculata*. *Baccha maculata* s. str. is widely distributed along the Himalaya in India, from Kashmir to Burma and also in southern China (Fukien Province). The new species from southern India can be separated from *maculata* in having the white pile on male frons restricted only to the upper third or quarter, in having the first tergum entirely white and in tergum 2 being white on anterior margins. So far I have specimens of this new species from Coorg (Karnataka), Kaikatty (Kerala), Nilgiri Hills, Longwood sholah, and Shevaroy Hills near Yercaud (Tamil Nadu).

Allobaccha amphithoe (Walker, 1849)

Baccha amphithoe Walker, 1849, List Dipt. Br. Mus., 3: 549 (sex ?; "Mulmein, E. Ind.") [BMNH, London]

Baccha pedicellata Doleschall, 1856, Natuurk. Tijdschr. Ned.-Indie, 10: 411 (sex?; "Java") [ZM, Amsterdam ??]

Baccha bicincta de Meijere, 1910, Tijdschr. Ent., 53: 104 (♂♀; "Batavia, Djakarta, Tandjong Priok, Bekassi, Java, Krakatau") [ZM, Amsterdam]

Baccha flavopunctata Brunetti, 1913, Rec. Indian Mus., 8: 165 (♀; Dibrugarh, Assam, India") [ZSI, Calcutta—examined]

Baccha fulvicostalis Matsumura, 1916, Thousand Insects Japan, 2: 226 (♀; "Formosa") [HU, Sapporo]

India: AN, BI ?, KL, KN; Sri Lanka (Austen 1893; Kertész 1910; Brunetti 1913a, 1915, 1923; Shiraki 1930; Keiser 1958; Curran 1930, 1931a, b; Knutson et al. 1975, Ghorpadé 1981a, b, 1994, 2014c, unpubl.; Peck 1988; Mitra et al. 2015; Thompson et al. 2017.)

Note: Austen (1893: 142–144) in his paper giving notes on species of Syrphidae described by the late Francis Walker saw the type ["Mulmein" is in Burma] of *amphithoe* and wrote: "Walker's type, however, is a mere fragment, the sex of which is impossible to determine, and was minus its head when Walker described it." Austen described it based on three specimens received from Ceylon, gifted by Lieut.-Col. Yerbury. The label data of these were "nr Trincomali, Ceylon." Two ♂♂ from Kanthalai, 8.iii.1892, and 31.i.1891, plus a ♀ from Kottawa, 24.iv.1892. Yerbury took two other specimens, a ♂ from Bentota, 6.vi.1890 and a ♀ from Huldamulla, 10.vi.1892. Yerbury stated that this species is very rare. Brunetti (1923: 126–127, 414, figs. 25–26, Pl. III, fig. 12) synonymized his *flavopunctata* with *amphithoe* in the Fauna. He gave several localities in NE. India and Sri Lanka. Kertész (1910: 156), van der Wulp (1896: 121), and Shiraki (1930) mentioned it from Sri Lanka, and Keiser (1958) included it in his paper on Sri Lanka Syrphidae. I saw the type♀ of *flavopunctata* in ZSI, Calcutta labelled "*Baccha amphithoe* Wlk (*flavopunctata* Brun Type ♀, det Brun. 1923 [*Baccha flavopunctata* Brun. Typ ♀ Dibrugarh, N.E. Assam, Abor Exped. 17–19-xi-11, Kemp/type/1968/HI." Brunetti (1913: 165) named and described it as a new species *flavopunctata* from Dibrugarh, Assam, 17–19.xi.1911. Brunetti (1915: 219) wrote that "Of true *flavopunctata* further specimens have been acquired [from Bengal and Assam] and all are females." He then compared it with few other "*Baccha*" species and concluded that he should include all under one species! In his *Fauna* volume, Brunetti (1923: 126–127, 414, figs. 25, 26, Pl. III, fig. 12) gave an "amalgamated" description of his *flavopunctata* and Austen's *amphithoe* Walker, and figures, placing it as a synonym of *amphithoe*, and gave localities from Bengal, Assam, and NE. India, from the type—locality Moulmein (Burma), taken by Gravely, Kemp, and Yerbury, as well as from Ceylon (uncommon), all taken by Yerbury. Curran's (1930, 1931a, b) set of papers were very important documentations of Syrphidae from the adjacent Malay Peninsula to the Indian subcontinent from which latter Brunetti gave his own valuable writings. Curran (1930: 244–246, 251–256) erected a subgenus *Allobaccha*, separating it by its species having humeri with a row of hairs behind or almost half hairy. He went on to comment, "It will be impossible accurately to place most of the described species in this genus [*Allobaccha*] until they have been checked over with attention to the following points: position of ocellar triangle, hairiness of humeri, presence of "collar," colour of squamae, presence of squamal fringe, presence of scutellar fringe, and long abdominal pile. These characters are much more reliable than the colour of the wings and presence or absence of abdominal markings in dealing with certain groups of Oriental species." Curran (1931a: 321–323) gave a key to 18 species of Malayan *Baccha* in which he included *amphithoe* Walker, keying this out as having abdomen reddish with three broad, transverse black fasciae. Keiser (1958: 201) included this as "*Baccha amphithoë*" in his paper documenting specimens taken from the Central Province, Ceylon. Thompson et al. (2017: 508) provide a synopsis of this Old World tropical group occurring in the Afrotropics, Indomalaya, and Australia with extension to Japan (86 species), with 26 species from the PSWNG area. Knutson et al. (1975: 321), Peck (1988: 53), Ghorpadé (2014b: 17, 2015b: 84), and Mitra et al. (2015: 67) listed it. See also Dirickz (2010) for Madagascar *Allobaccha* species. I saw specimens (K. Ghorpadé, unpubl.) from Mandya, ex. *Megatrioza hirsutus* on *Terminalia* sp., and from Bangalore ex psyllids on *Bauhinia*.

Allobaccha apicalis (Loew, 1858)

Baccha apicalis Loew, 1858, Wien. Ent. Mschr., 2: 106 (♀; "Japan") [MCZ, Cambridge]

? *Baccha pulchrifrons* Austen, 1893, Proc. Zool. Soc., London, p. 139 (♂♀; "Hot Wells, Trincomali, Ceylon" [BMNH, London—examined; also see below]

? *Baccha nigricosta* Brunetti, 1907b, Rec. Indian Mus., 1: Pl. XI, fig. 5, 1908: 50 (♂; "Bhim Tal, Kumaon, India") [ZSI, Calcutta—examined]

Baccha apicenotata Brunetti, 1915, Rec. Indian Mus., 11: 221 (♀; "Bhowali, 5700 ft") [ZSI, Calcutta—examined]

India: AN, BI, GJ, GO, KL, KN, TN; Sri Lanka (Austen 1893; van der Wulp 1896; Kertész 1910, 1913; Brunetti 1915, 1923; Shiraki 1930, 1963; Curran 1930; Bhatia & Shaffi 1933; Keiser 1958; Knutson et al. 1975; Ghorpadé 1981a, b, 1994, 2014b, 2015b, unpubl.; Mahalingam 1988; Peck 1988; Mitra et al. 2015).

Note: Brunetti (1915: 221) described his *apicenotata* based on "a single ♀ from Bhowali, 5700 ft., vii-09 [*Imms*]" and mentioned two other specimens from "Jungle at base of Dawna Hills, 1-iii-08" and "Cherrapunji, Assam, 4400 ft., 2—8-x-14 [Kemp]." Brunetti (1923: 122–124) later placed his *apicenotata* as a synonym of *pulchrifrons* Austen and this synonymy needs to be studied and determined and also that of *apicalis* Loew. Curran (1930: 325) wrote: "In his original description Major Austen called attention to the similarity of his species [*pulchrifrons*] and *apicalis* Loew. In view of the fact that *pulchrifrons* occurs commonly in Japan, whence Loew's damaged specimen came, it seems probable that Loew's name should be applied to this species." Bhatia & Shaffi (1930: 549–555, Pl. LX) described and illustrated this syrphid and its life history (as "*Baccha pulchrifrons*") on the psyllid *Ctenophalara elongata* on the red silk cotton tree *Bombax malabaricum* at Pusa in Bihar. I had not included this in my key (Ghorpadé, 1994: 6–7) to Indian subregion *Allobaccha*, but later noted that "it could be a junior synonym (?) of *Allobaccha apicalis* (Loew, 1858) which is probably the same as *pulchrifrons* Austen, 1893 and so synonymous with *apicenotata* Brunetti, 1915? Shiraki (1930, 1963) listed this species. Keiser (1958: 201) included this as "*Baccha pulchrifrons*" in his paper documenting specimens taken from Western and Central Provinces. Knutson et al. (1975: 321) placed *pulchrifrons* and *apicenotata* as synonyms of *apicalis* Loew. Ghorpadé (1981b: 66) listed recorded prey of this species from our subcontinent. Mahalingam (1988: 161) gave it as recorded from the Nilgiri Hills among ten species recorded by Brunetti (1923), but misidentified it as "*Baccha dispar* Walk." I studied the type of *Baccha apicenotata* in the ZSI, Calcutta, a ♀ labelled "near Bhowali, Kumaon, 5,700 ft., July 1909, A.D. Imms/Baccha apicenotata Brun. Typ ♀/ TYPE/1977/HI." I had then noted that, "In FBI key (p. 114–115) it goes to *pulchrifrons*." Brunetti (1923: 124) incorrectly gave a specimen from Ceylon as "type ♀ of *apicenotata*," under his full description of *pulchrifrons* and a figure of its wing. Ghorpadé (1994: 6–7) included this species in his key to *Allobaccha*. Van der Wulp (1896: 122), Knutson et al. (1975: 321), Peck (1988: 53), Ghorpadé (2014b: 17, 2015b: 84), and Mitra et al. (2015: 67) listed it. I saw specimens (K. Ghorpadé, unpubl.) from Mandya (Karnataka) reared from *Megatrioza hirsutus* on *Terminalia* sp. and from Bangalore ex psyllids on *Bauhinia* sp. in NBAIR, Bangalore.

Allobaccha elegans (Brunetti, 1915)

Baccha elegans Brunetti, 1915, Rec. Indian Mus., 11: 220 (♂; "Sukna, base of Darjeeling Hills, India") [ZSI, Calcutta—examined]

India: KN; Sri Lanka ? (Austen 1893; van der Wulp 1896; Kertész 1910, 1913; Brunetti 1915, 1923; Shiraki 1930, 1963; Curran 1930; Bhatia & Shaffi 1933; Knutson et al. 1975; Ghorpadé 1981a, b, 1994, 2014b, 2015b, unpubl;, Peck 1988; Mitra et al. 2015.)

Note: See Brunetti (1915: 220) who described this new species based on several ♂♂ in the Indian Museum from Sukna, 500 ft., in July 1908 and from jungle at base of Dawna Hills (Burma) in March 1908, and also from Sikkim, 1400 ft., in September 1909. Then, in his *Fauna* volume, Brunetti (1923: 124–126, 414) synonymized it under *triangulifera* Austen, 1893 and created confusion, mixing up species ! In my doctoral thesis (Ghorpadé 1981a) I had omitted *Allobaccha* as a non-Syrphini genus, but in my later paper (Ghorpadé 1994: 7), done after my Smithsonian postdoc, I solved the problem. Austen (1893: 138–139, Pl. IV, Figure 5) described *Baccha triangulifera* as a new species based on a pair *in copula* taken in Huldamulla, *ca* 400 ft., Ceylon, collected in June 1892 by Lt-Col. Yerbury. He mentioned that "the sharply-defined yellow triangle on the third abdominal segment distinguishes the species from any other known to me." Dr. Chris Thompson (*in litt*.) believes this (*triangulifera*) requires a new genus and will publish this shortly. My Indian Syrphinae diagnostics paper (Ghorpadé 1994: 7) and the later update on Indian Syrphidae (Ghorpadé 2014b: 17–18) has cleared up the situation with this taxon. The checklist gave *elegans* from India (KN, SI, WB), Nepal, and Burma, but *triangulifera* from India (KN, WB), Nepal, and Sri Lanka, which was again erroneous. The diagnostics paper, however, gave a key where *elegans* and *triangulifera* were separated by diagnostic characters, and the former recognized from India (KN, WB), Nepal, and Burma (?), while *triangulifera* was confirmed only from Sri Lanka. With pencilled notes in my *Fauna* volume copy, I had written that *elegans* type was from Sukna, and from Dawna Hills, Burma; while the Rungpo, Sikkim specimens were new, and I named it *binghami* Ghorpadé. The "other specimens... from Kollur Ghat, S. Kanara district, 3000 ft., taken in September 1913 were *triangulifera*", of which I found another ♀ in the IARI collection in New Delhi! I (Ghorpadé 1994: 7) showed that that species was distinct from *elegans*, and that *triangulifera* was restricted to the island of Sri Lanka and that *elegans* was found in NE India, Nepal, and peninsular India, but material being scarce. I studied the holotype male of *elegans* in ZSI, Calcutta in 1981, which was labelled "Sukna, 500 ft., E. Himalayas, 1-VII-08, N.A./Baccha elegans Brun Type ♂/TYPE/1972/HI/Baccha triangulifera Aust./elegans Brun. Type ♂, det. Brun. 1923." I had then noted, "Goes to *Allobaccha*. In FBI key to *triangulifera*, but no yellow spot on tip of T 3." Van der Wulp (1896: 122), Kertész (1910), Curran (1930), Shiraki (1930, 1963), Knutson et al. (1975: 323), Peck (1988), Ghorpadé (2014b: 17, 2015b: 84), and Mitra et al. (2015: 67) listed it. See also comments under *triangulifera* (Austen) below.

I also wrote: "The genus Allobaccha in the Orient is quite speciose and requires careful study and revision, based on good many specimens. I had recognized and keyed out eight species (Ghorpadé, 1994: 7) in the Indian subcontinent, but have seen more than 300 specimens from other parts of the Orient and am in the process of revising Oriental—Papuan species of this genus. Curran (1931a: 325–327) gave a key to Malayan species of his *Allobaccha*, which is a workable base."

Allobaccha fallax (Austen, 1893)

Baccha fallax Austen, 1893, Proc. Zool. Soc., London, p. 142 (♂; "Haycock Hill, nr Galle, Ceylon" [BMNH, London]

Sri Lanka (Austen 1893; Kertész 1910; Brunetti 1923; Keiser 1958; Curran 1931a; Knutson et al. 1975; Ghorpadé 1994, 2014b; Mitra et al. 2015.)

Note: Austen (1893: 142, Pl. IV, fig. 12) described this based on a ♂ taken at Haycock Hill, near Galle, Sri Lanka, in April 1892. One other ♂ was collected at Kandy in May 1892, both by Lt-Col. Yerbury. In his *Fauna* volume, Brunetti (1923: 117–118, 413) reproduced Austen's description and keyed it out as "wings wholly pale brown except basally nearly clear". Keiser (1958: 194) included this as "*Baccha fallax*" in his paper documenting specimens taken from North-western and Central Provinces. Curran (1931a: 323) keyed it out as having scutellum wholly blackish, sides of face broadly yellowish in ground colour on more than the upper half, alula distinctly narrowed, and yellow of sides of thorax extending onto the surrounding parts in front of the mesonotal suture. I keyed it out (Ghorpadé 1994: 7) from *apicalis* using most of Curran's characters and recognizing it only from Sri Lanka, doubtfully from India. Kertész (1910), Knutson et al. (1975: 322), Peck (1988), Ghorpadé (2014b: 17), and Mitra et al. (2015: 67) listed it.

Allobaccha oldroydi Ghorpadé, 1994

Allobaccha oldroydi Ghorpadé, 1994, Colemania, 3: 7 (♂; "Nadungayam, Kerala") [USNM, Washington, DC]

India: KL; Sri Lanka (Ghorpadé 1994, 2014b; Mitra et al. 2015.)

Note: I named and described this new species that I found in borrowed material from the British Museum, London taken by their B.M-C.M. Expedition to South India, from Nadungayam (Kerala), 16–22.ix.1938. Listed it in my checklist (Ghorpadé, 2014b: 17) as did Mitra et al. (2015: 67).

Allobaccha sapphirina (Wiedemann, 1830)

Baccha sapphirina Wiedemann, 1830, Aussereurop. Zweifl. Insekt., 2: 96 (♀; "Ind. Or., East Indies") [BMNH, London—examined]

Baccha umbrosa Brunetti, 1923, 1923, *Fauna Brit. India*, Dipt., 3: 119 (♂; "Abu, Rajputana, India") [BMNH, London—examined]

India: GJ, KN, MH, MP, RJ, TN; Sri Lanka (Kertész 1910, 1913; Brunetti 1923; Shiraki 1930; Curran 1931a; Cherian 1934; Beeson 1953; Gokulpure 1972; Ghosh 1974; Knutson et al. 1975; Ghorpadé 1981b, 1994, 2014b, 2015, 2017; Peck 1988; Dirickx 2010; Mitra et al. 2015.)

Note: *Baccha umbrosa* Brunetti (1923) is sometimes treated as a synonym (*q.v., loc. cit.*), but this needs further study and verification. Kertész (1913: 279–281, fig. 7) described material from Taiwan and figured wing of both sexes. Brunetti (1923: 122) described it from Indian (Mt Abu in Rajasthan and Deesa in Gujarat) and African specimens, but the African distribution needs to be verified for species identity. Cherian (1934: 698) recorded this from Coimbatore being "predaceous on psyllid bugs." The species treated as *sapphirina* by Dirickx (2010: 215, 218–220, 232, figs. 2 and 3) from Madagascar appears to me to be misidentified. Dirickx (2010: 229–232) also gave notes on the status of *Allobaccha*. Beeson (1953: 339) gave this as being predacious on Psyllidae. Gokulpure (1972) gave it from Damoh, Madhya Pradesh. Ghosh (1974: 196–197) recorded *Pterochlorus persicae* as prey of this species (as "*Bacca ? sapphirinia*" [sic]). Knutson et al. (1975: 322) gave a wide distribution, from Africa to New Guinea which requires confirmation by studying specimens. I (Ghorpadé 1981b: 66) listed its aphid and psyllid prey and presented a key (Ghorpadé 1994: 6). In the Indian subregion checklist (Ghorpadé 2014b: 17), I gave its range which included Delhi also from NW. India. I examined the holotype ♀ in NHM, London, labelled "TYPE [red]/B: n: fp: u (?) Ind. Or., Baccha sapphirina Wied./Allobaccha sapphirina." It is interesting that Curran (1931: 321–323) did not include it in his key to *Baccha s. lat.* from Malaya. Kertész (1910), Knutson et al. (1975: 322), Peck (1988: 54) Ghorpadé (2014b: 17, 2015: 85), and Mitra et al. (2015: 67) listed it. I saw specimens (K. Ghorpadé unpubl.) from Hebbal (Karnataka) and Coimbatore Botanical garden, Yercaud (Tamil Nadu) in collections.

Allobaccha triangulifera (Austen, 1893)

Baccha triangulifera Austen, 1893, Proc. zool. Soc. Lond., p. 138 (LT ♂; "Huldamulla, Ceylon," here designated) [NHM, London—examined]

India: KN, TN; Sri Lanka (Austen 1893; Kertész 1910; Brunetti 1923; Curran 1931a; Cherian 1934; Knutson et al. 1975; Ghorpadé 1981a, 1994, 2014b, 2015b; Mitra et al. 2015)

Note: This peculiar species needs comment on its misidentification and incorrect range concept. Austen (1893: 138, Pl IV, fig. 5) described this from a pair taken at "Huldamulla, Ceylon, *circa* 400 ft. (*Lieut.-Colonel Yerbury*), a pair taken *in copulâ*, June 10, 1892." Cherian (1934: 698) recorded it from South Canara. I formally designated (Ghorpadé 2015: 87) a Lectotype and Paralectotype [both on same pin, *in copula*] of *triangulifera*, a male in the NHM, London, labelled as follows: "Type ♂, ♀ [circular white label with red submarginal ring]/Huldamulla., 10.VI.92, circa 3000 ft., Ceylon., Col. Yerbury., 92.—192./*Baccha triangulifera*, Aust., Type ♂, ♀/ LECTOTYPE, *Baccha triangulifera* Austen ♂, Ghorpade des. 1983," and "Paralectotype, *Baccha triangulifera* Austen ♀, Ghorpade des. 1983." I found some nine specimens of this species in then Commonwealth Institute of Biological Control, India (CIBCI), Bangalore [now specimens in National Bureau of Agricultural Insect Resources (NBAIR), Bengaluru], taken in December and January, whose larvae were found feeding on fulgorids on *Casuarina equisetifolia* trees there. Ghorpadé (1994: 6) included it in his key to Indian subregion *Allobaccha* species. Kertész (1910), Knutson et al. (1975: 323), Ghorpadé (2014b: 18, 2015b: 87), and Mitra et al. (2015: 67) listed it. See also my comments under *Allobacha elegans* above. I saw specimens (K. Ghorpadé unpubl.) from Chidambaram (Tamil Nadu), Kanakapura, Hebbal, Mandya, Mudigere, Vaderahalli, and Nagody, 2500 ft., S. Canara District, Gandhi Krishi Vignana Kendra (GKVK) Campus and Lal Bagh, Bangalore (Karnataka). Some were reared from *Bauhinia* psyllids, *Megatrioza hirsutus* on *Terminalia* sp. galls and also found feeding on leaf hoppers (Homoptera: Cicadellidae). Specimens were also taken in Namchi (Sikkim) from a "log hive" (?), but were in poor shape and could be the north-eastern *elegans* (Brunetti) q.v. .

Melanostoma ceylonense (de Meijere, 1911)

Melanostoma ceylonense de Meijere, 1911, Tijdschr. Ent., 54: 348; 51: 312 (♂♀; "Pattipola, 2000m, Ceylon") [ZMA. Amsterdam]

Sri Lanka (de Meijere 1908, 1911; Keiser 1958; Knutson et al. 1975; Ghorpadé 2014b.)

Note: de Meijere named this *ceylonense* in 1911 after placing it as *orientale* Wiedemann earlier in 1908. Chris Thompson (*in litt.*) in a manuscript of our Oriental—Papuan Syrphidae Conspectus (Ghorpadé, *in prep.*) listed *ceylonense* as a synonym of *scalare* (F.), but this requires more study and confirmation or correction. Keiser (1958: 194) included this as "*Melanostoma ceylonense*" in his paper documenting specimens taken from Uva and the Central Province. Knutson et al. (1975: 325) listed it as a synonym of *scalare* (Fabricius 1794). In a recent paper, Thompson & Skevington (2014) presented the latest understanding of the phylogeny of what they recognized as Tribe Bacchini, recognized "as a working and possibly paraphyletic group with Melanostomini as a well supported monophyletic group." See also Vockeroth (1969: 14). Thompson et al. (2017: 520) provide a synopsis of this genus which is found in all regions (56 species), four species occurring in the PSWNG area. Knutson et al. (1975: 325; as synonym), and Ghorpade (2014b: 15) listed it. See also Thompson & Skevington (2014), Haarto & Ståhls (2014), and Fluke (1950, 1958).

Melanostoma melanoides Lambeck & Kiauta, 1973

Melanostoma orientale (Wiedemann, 1824) form *melanoides* Lambeck & Kiauta, 1973, Ent. Bericht., 33: 74 (♀; "Lughla, 2800m, Nepal") [NCABR, Utrecht]

India: TN; Sri Lanka ? (Lambeck & Kiauta 1973; Lambeck & van Brink 1973; Ghorpadé 2014b, 2015, unpubl.)

Note: Lambeck & Kiauta (1973: 74–75) found a female in Lughla (Nepal), "captured in the vicinity of running water," which looked melanistic and had deviant characters from *orientale* which made them give this a new name, even if just as a "form" of *orientale* Wied. In the same year, Lambeck & van Brink (1973: 92) reported on a couple of males from Kashmir which they wrote "are apparently melanistic and completely devoid of yellow markings on the abdominal tergites, thus resembling the female specimen from Nepal referred to as form *melanoides*" by Lambeck & Kiauta (1973). They also reported some other specimens in their collection "from southern India (Anaimalai [sic] Hills near Coimbatore, 1050 m) which have even darker legs; those specimens have tergite 2 completely black, but have distinct pairs of yellow spots on tergites 3 and 4 and may belong to *M. ceylonense* de Meijere, 1908 recorded from mountainous areas in Ceylon..."

I have similar specimens of a "melanistic" *Melanostoma* which I collected on the Nilgiri Hills in Tamil Nadu at Longwood Sholah (ca 2000 m) and in my forthcoming revision of Oriental *Melanostoma* species will give it a name, either already published or a new species. In the Z.S.I. Chennai Centre I found (K. Ghorpadé, unpubl.) 3♂ 6♀ of what I determined as this species (or variety?) taken from Masinagudi in December 2012. Lambeck & van Brink (1973: 92) had concluded, "The oriental species of *Melanostoma* are in need of a revision. However in order to solve the pending problems concerning the identity of some taxa, long series of material from different localities will have to be taken into consideration." Ghorpadé (2014b: 15, 2015b: 66) listed it.]

Melanostoma orientale (Wiedemann, 1824)

Syrphus orientale Wiedemann, 1824, Analecta Ent., p. 36 (♂; "Ind. Or.") [UZM, Copenhagen ?]

India: AH, CG, KL. KN, TN; Sri Lanka (Brunetti 1913a, 1915, 1917, 1923; Hervé-Bazin 1924, 1926; Shiraki 1930; Curran 1930, 1931b; Lambeck & van Brink

1973, 1975; Ghorpadé 1973b, 2014b, 2015, unpubl.; Ghosh 1974; Knutson et al. 1975; Datta & Chakraborti 1986; Joseph & Parui 1986; Mitra et al. 2015).

Note: This again appears to be a "widespread," but "lumped" species. It requires careful study of material from the entire area of its supposed distribution. It is a higher altitude and northern latitudes species. Brunetti (1913a: 164) listed it in his Abor Hills paper (as "*orientalis*") and gave diagnostic characters from specimens in the "Indian Museum, as determined by me, from Bangalore and Mergui [Burma]." Brunetti (1915: 207–208) wrote, "However, it seems to me highly probable that *orientale* is not specifically distinct from *mellinum*, a species it is more akin to than *scalare*. The principal alleged difference is the grey-dusted frons and face in *orientale*, but numerous specimens occur in which this is much less conspicuous than usual, thereby closely approximating to *mellinum*... The females in *orientale* are more easily recognized by the dust spots on the frons being more closely approximate, so that the vertex and the lower part of the frons are more clearly demarcated, but a near approximation to this is not infrequently met with in *mellinum* ♀." Brunetti (1917: 85) wrote, "The specimens originally referred by me to *scalare* are certainly *orientale*, Wied., and in my second paper on Oriental Syrphidae (1915: 207), I have suggested that Wiedemann's species is only a variety of *mellinum*." In 1923 (Fauna, p. 50) he wrote, "Described from a good number of specimens from various sources [including "Peshawur" in Pakistan, "various locs." in Simla and in Naini Tal]... Apparently the commonest and most widely distributed species of the genus in the East, occurring throughout all the warm weather in the plains [?] and hills. I am still undecided whether *orientale* is specifically distinct from the common European *mellinum*..." Hervé-Bazin (1924: 289) mentioned Brunetti's omission of data in de Meijere's (1908, 1911) papers, and then he (Hervé-Bazin, 1926: 62–63) recorded it from Tonkin and Laos and gave notes on its differences from *scalare, mellinum, unittavitatum* [*sic*]*, pedium,* and *ceylonense*. Shiraki (1930) noted it was known from "Ceylon" [Sri Lanka]. Curran (1930: 257–258, 1931b: 359) recognized and treated species of *Melanostoma* from the Malay Peninsula and north Borneo (Sarawak) [= Malaysia] and provided a key to separate these—*algens* Curran, *gedehensis* de Meijere, *normalis* Curran, *orientale* Wiedemann, *quadrifasciatum* Curran, *talamaui* de Meijere, and *univitattum* Wiedemann. All of these were listed by Knutson et al. (1975: 325). Curran (1931b: 358) wrote, "I have attempted to prepare a key for the separation of the known species of *Melanostoma* from India and Malaya. However, I have no representatives of two of the species described by de Meijere and the resultant table may prove inadequate in the last two or three couplets." Earlier Curran (1930: 256) had also written "The females of many of the species can only be placed in their proper genera [*Melanostoma* or *Platycheirus*] through familiarity with the species." Lambeck & van Brink (1973: 92) took it at Nagin Lake near Srinagar in Kashmir (India) and made some notes on its identity as being this species "for having the frons and face distinctly pollinose," which, as according to Brunetti (1915), "this feature seems to be the most obvious distinction between this species and the palearctic [*sic*] *M. mellinum* (Linnaeus, 1758)." They mentioned that their specimens were melanistic and could be *melanoides* (*q.v.*), and that other specimens they have from the Anaimalai Hills in southern India could be *M. ceylonense* de Meijere recorded from the Sri Lanka mountains (*q.v.*). They concluded—"The oriental species of *Melanostoma* are in need of a revision [there are 10 recognized Oriental species (Thompson & Skevington, 2014: 106), but I had put down 11 species + 1 Palaearctic in my Oriental—Papuan Conspectus photocopy]. However in order to solve the pending problems concerning the identity of some taxa, long series of material from different localities will have to be taken into consideration." In my master's thesis, I did not find it around Bangalore and stated, "The Indian species of this genus require revision." I have more than 1,500 specimens of *Melanostoma* in my collection now and this revision is possible in the near future. Ghosh (1974: 196) recorded *Lipaphis erysimi* as prey of this species. Lambeck & van Brink (1975: 8–10) described karyotypes from specimens taken at Srinagar in Indian Kashmir and concluded, "Thus, the specific controversy *orientale* versus *mellinum* seems not to be solved by our karyological information. Datta & Chakraborti (1986) listed specimens from "Nopokiola" [Napoklu], Mercara, Muthodi, Chikmagalur, Belgaum district (all Karnataka), and from Top Slip, Coimbatore (Tamil Nadu) and stated it is reported also from Sri Lanka. Joseph & Parui (1986: 161) gave one specimen taken on "road to Cardamom Estate" in December 1980 and stated this was "the first record of the species from Kerala" [which appears untrue!]..." Knutson et al. (1975: 325) gave its distribution as "throughout Oriental region" in their catalog. Ghorpadé (2014b: 15, 2015b: 67–68) and Mitra et al. (2015: 66) also listed it. See also my notes under *melanoides* (*loc. cit.*). I have seen specimens (K. Ghorpadé unpubl.) from Kallar 1250 m, Nilgiris, Kookal (Tamil Nadu), Sadahalli, Hessaraghatta, GKVK, Bangalore, Jamboti, Khanapur, Belgaum dist., Thadiyandamol 1087 m (Coorg), Yellapur 446 m, Bellamani 478 m, Mudigere, Hogalkere, Nandidrug 1467 m, Sahasralinga 446 m, Sirsi, Bidar 584 m, Chettalli, Sidlaghata, Tirthalli, Agumbe (Karnataka), and Nellore (Andhra).

Melanostoma pedium (Walker, 1852)

Syrphus pedius Walker, 1852, *Insecta Saundersiana*, Dipt., 3: 234 (♂; "East Indies") [BMNH, London]
Syrphus cothonea 1852, *Insecta Saundersiana*, Dipt., 3: 235 (♀; "East Indies") [BMNH, London]
India: KN ? (Brunetti 1923; Knutson et al. 1975; Ghorpadé 1973b, 2014b, 2015b; Mitra et al. 2015).

Note: Brunetti (1923: 52, 410) listed this as a separate species, but wrote: "A species requiring confirmation as to its validity." He also treated *cothonea* Walker as a synonym. He also wrote: "As there are altogether twelve specimens, all ♀♀, there can be no doubt as to the validity of this form." The specimens are in the Indian Museum, from the Simla and Darjeeling districts, the United Provinces, Bengal, and Bangalore... A revision needs to decide the status of this name and even though I (Ghorpadé, 2014b: 15, 2015b: 68) recognized it and listed it from Himachal Pradesh, Uttarakhand,

West Bengal, and Karnataka (!), these were from published data and cannot be considered infallible. In my master's thesis (Ghorpade 1973b), I listed it and wrote: "This species was held to be a valid one by Brunetti (1923), but it was not found around Bangalore in the present study." The type locality is "East Indies" and could mean either the Indonesian archipelago or even NE India. Knutson et al. (1975: 325) listed it as "*pedius*" in their catalogue. Ghorpadé (2014b: 15, 2015b: 68) and Mitra et al. (2015: 66) listed it.

Melanostoma scalare (Fabricius, 1794)

Syrphus scalare Fabricius, 1794, Ent. Syst., 4: 308 (sex ?; "Kiliae" = Kiel, Denmark) [UZM, Copenhagen]

Sri Lanka ? (Brunetti 1907a, 1915, 1917; Sack 1932; Knutson et al. 1975; Peck 1988; Thompson & Rotheray 1998; Ghorpadé 2014b, 2015b; Mitra et al. 2015).

Note: Brunetti (1907a: 168) had written, "A series of thirteen females from Simla, Theog and Matiana also appear to be the true *scalare*." But he later (Brunetti 1915: 207) corrected himself—"examples referred to *scalare* are only *orientale*..." Brunetti (1917: 85) then wrote, "The specimens originally referred by me to *scalare* are certainly *orientale*, Wied." He gave Theog, Simla, and Matiana as localities where this was taken, in April. Peck (1988: 67) also gave Afghanistan and the Oriental region (!). Sack (1932: 29, fig. 2) reported specimens from Lombok and Flores (Indonesia), gave a brief description and a figure of the female abdomen. Knutson (1975: 325), Ghorpadé (2014b: 15, 2015b: 68), and Mitra et al. (2015: 66) also listed it. See also my notes under *ceylonense* above. See also Sack (1932: 157) and Thompson & Rotheray (1998: 105) for discussion and key to this genus in the Palaearctic. See also Dirickx (2001) for notes on this genus.

Melanostoma univittatum (Wiedemann, 1824)

Syrphus univittatus Wiedemann, 1824, Analecta Ent., p, 36 (♂; "Ind. Or.") [UZM, Copenhagen]

Syrphus planifacies Macquart, 1848, Dipt. Exot., Suppl., 3: 43 (♀; "Java") [MNHN, Paris]

Syrphus cyathifer Walker, 1856, J. Linn. Soc. Lond., 1: 125 (♀; "Sarawak, Borneo") [NHM, London]

India: AH, BI, CG, JH, KL, KN, TN; Sri Lanka (Brunetti 1913, 1915, 1923; Curran 1931b; Keiser 1958; Ghorpadé 1973b, 1974, 2014b, 2015b, unpubl.; Knutson et al. 1975; Joseph & Parui 1977; Datta & Chakraborti 1986; Joseph & Parui 1986; Thompson & Rotheray 1998; Ghorpadé et al. 2011: 80; Mitra et al. 2015).

Note: Brunetti (1915: 208–209) wrote in some detail about this species which Wiedemann had described based only on a male, since no females were described until then. He went on, "As regards *planifacies*, Macq. I think it may also be regarded as the ♀ of *univittatum*. The sole disagreement in Macquart's description is the colour of the thorax and frons, which he says is greenish black." Specimens in the "Indian Museum" [= ZSI, Calcutta] had them aeneous black and some exhibited "a distinctly greenish tinge." The males came from many localities including Kathmandu and the females from Bhim Tal. So he ended, "the localities of both sexes thus supporting the view that they are the same species. Its range of distribution is evidently very wide." He also wrote: "The ♂ *univittatum* specimens in the Indian Museum come from Darjeeling, Katmandu, Dibrugarh, the Assam-Bhutan Frontier, Mergui, Travancore, Bangalore, and Coromandel; whilst the ♀♀ hail from Bhim, Tal, the Assam-Bhutan Frontier, Sadiya, Travancore, Bangalore, Coromandel and Sarawak, the localities of both sexes thus supporting the view that they are the same species. Its range of distribution is evidently very wide." Brunetti (1923: 50–51) gave a description and figure of head profile, listing it from many localities including "Katmandu, Nepal, Bhim Tal 4450 ft., Pusa, not uncommon, and three rather small specimens of *planifacies*, Macq., from Dehra Dun." Curran (1931a) gave localities for this species in Malaya and then (Curran 1931b: 358–359) gave a key to *Melanostoma* species from "India and Malaya" and separated *univittatum* as having a "Face with only a trace of a tubercle." Shiraki (1930) and Keiser (1958) listed it from "Ceylon" [Sri Lanka]. Ghorpadé (1973a, 1974) recorded it in his master's thesis from Bangalore and wrote; "This species was fairly common around Bangalore during the winter months of November to February and was collected from fields of maize, wheat, paddy, radish and from grassy and herbaceous borders. It was taken on flowers of niger, *Guizotia abyssinica*." Keiser (1958: 207) included this as "*Melanostoma univitatum*" in his paper documenting specimens taken from Uva, Sabaragamuva, and the North-western, North Central, and Central Provinces. Knutson et al. (1975: 325) gave its distribution as "throughout S.E. Asia." Joseph & Parui (1977: 228–229) listed one specimen from Manoharpur, Chaibasa district (Jharkhand) and stated that it is recorded from Coorg, Nedumangad, Coromandel, and Sri Lanka, etc. Datta & Chakraborti (1986: 56) documented specimens from Belgaum district, "Mudigore" [= Mudigere], Chikmagalur, "Nopokula" [= Napoklu], Mercara (all Karnataka), and Kumily, Idukki district, Takkachi and Eluppara, Kottayam district (Kerala). Joseph & Parui (1986: 161) gave one specimen from "towards Mukkali" in December 1980 and stated, "This species enjoys wide distribution all over India [*sic*]." In Ghorpadé et al. (2011: 80), it was mentioned from "Coromandel," but not taken then in the Andhra Carnatic. Ghorpadé (2014b: 15, 2015b: 69), and Mitra et al. (2015: 66) listed it. I have seen specimens (K. Ghorpadé unpubl.) from Chidambaram, Dindigul, Coimbatore, Yercaud, Kotagiri (Nilgiris) (Tamil Nadu), Magadi, Hebbal, GKVK, Attur, Kunigal, Chettali, Cherampand, Appangala & Arvathakkuli (Coorg), Poodhipadugai, Chamarajnagar dist., Balakola (Mysore dist.), Seringapatam, Madiyahalli (Mysore dist.), Nanjangud, Chintamani, Hoskote (Karnataka), Neliampathi Hills 1110 m (Kerala), Araku Valley 934 m, and Srisailam (Andhra).

Melanostoma sp.

India: KN (Brunetti, 1915; Ghorpadé, 2014b: 16).

Note: Brunetti (1915: 208, pl. xiii, fig. 3) wrote in some detail about "an apparently undescribed form... with a facial profile intermediate between *orientale* and *univitttatum*, in which the central bump though distinct is much less conspicuous than in *orientale*..." and "by the 1st pair of abdominal spots being larger than in *orientale*, oval, and carried over the side of the 2nd segment below the base. Also the hind femora are all yellow, the hind tibiae bearing only an indistinct median dark band which is frequently absent." Ghorpadé (2014b: 16) listed it.

Xanthandrus ceylonicus Keiser, 1958

Xanthandrus ceylonicus Keiser, 1958, Revue Suisse Zool., 65: 205 (♂; "Pidrutalagala, Ceylon") [NM, Basel]

Sri Lanka (Keiser 1958; Knutson et al. 1975: 326; Ghorpadé 2014b: 16; Mitra et al. 2015; Thompson et al. 2017)

Note: Keiser (1958: 205–206) named and described it from Sri Lanka and included it in his paper documenting specimens taken from Pidrutalagala 2200–2400 m, 30.v.1953, in the Central Province. Thompson et al. (2017: 520) provide a synopsis of this nearly cosmopolitan genus (except New Zealand) of 29 species, with five described species from the PSWNG area and two undescribed ones from the Philippines and New Guinea. Knutson et al. (1975: 326), and Ghorpadé (2014b: 16) listed it.

Xanthandrus indicus Curran 1933

Xanthandrus indicus Curran, 1933, Stylops, 2: 46 (♂♀; "Rahatgaon, Hoshangabad, India") [FRI, Dehra Dun—examined]

India: AH ?, MP (Curran 1933; Beeson 1953; Knutson et al. 1975: 326; Ghorpadé et al. 2011: 80; Ghorpadé 2014b: 16, 2015b: 75; Mitra et al. 2015: 67)

Note: I examined the holotype male of this species in Forest Research Institute (FRI) (Dehra Dun). Beeson (1953: 340) wrote, "The larva of this hoverfly feeds on young caterpillars of *Hyblaea puera* [pyralid moth] which it discovers in the folded edges of the leaves of teak [*Tectona grandis*] or *Vitex negundo*... The puparium is formed on the leaf and the fly emerges in seven days." The maggot eats 2–3 caterpillars each day. *X. indicus* was taken at Rahatgaon near Hoshangabad (Madhya Pradesh) in August 1926, reared from larvae found "predacious on larvae of *Hyblaea puera*" (Curran, 1933: 46), the Pyralid moth pest of teak. I have seen the holotype male in FRI, Dehra Dun, but the allotype female is in the American Museum of Natural History, New York and needs to be studied and compared with this Amalapuram female to determine correct identity. This is an exciting find: A female specimen from Amalapuram, collected by Ramakrishna Ayyar (♀, Amalapuram, Godavari district, 16–18.ix.1921), which may be a new species. See Ghorpadé et al. (2011: 80) for notes on this specimen. Species of *Xanthandrus* are very rare in our subcontinent, only this and *X. ceylonicus* Keiser from Sri Lanka being known so far.

I caught another, probably undescribed, species in the Mawphlang Sacred Grove near Shillong (Meghalaya) in May 1985, and recently took a pair of an undetermined *Xanthandrus* at Sangmein, Upper Shillong, 1800 m on 13.iii.2015. Knutson et al. (1975: 326), Ghorpadé (2014b: 16, 2015b: 75), and Mitra et al. (2015: 67) listed it. The tribal name was Melanostomatini (or Melanostomini) earlier.]

Pandasyopthalmus rufocinctus (Brunetti, 1908)

Pipizella rufocincta Brunetti, 1908, Rec. Indian Mus., 2: 53 (♂; "Rangoon, Burma") [NHM, London]

India: BI, CG, JH, KL, KN, TN; Sri Lanka (Brunetti 1908, 1913, 1915, 1923; Kertész 1910; Stuckenberg 1954a; Keiser 1958; Coe 1964; Knutson et al. 1975; Joseph & Parui 1977; Ghorpadé 1981b, 2014b, 2015b, unpubl.; Datta & Chakraborti 1986; Thompson & Ghorpadé 1992; Vujić et al. 2008; Mitra et al. 2015; Thompson et al. 2017.)

Note: Brunetti (1908: 53) described this (and his *indica*) as a *Pipizella*, perhaps recognizing the distinctness from *Paragus s. str.* and antedating Stuckenberg's (1954a) concept of *Pandasyopthalmus* as a distinct genus-group. *P. rufocincta* was described from Rangoon in Burma and from "Umballa, N.-W. India" [= Ambala, now in Haryana] noting it was different from *Pipizella* species in having a marked (red) abdomen, and that it was "apparently widely distributed." In his Abor Expedition paper (Brunetti 1913: 157), he described as new *rufiventris* from the Assam area in the north-east, but was clearly confused and did "lumping," as other specimens he wrote came from Dhikala, Gharwal [sic] district, base of Western Himalayas and from Bijrani, Naini Tal district, as well as from Sri Lanka! And his new *rufiventris* was a *politus* ! He mentions this *rufiventris* in his next paper (Brunetti 1915: 201), again lumping it, giving specimens from "Assam, the Western Himalayas and Ceylon." In the *Fauna* volume, Brunetti (1923: 37) gave a description and listed it (as "*Pipizella rufocincta*") also from "Umballa (N.W. India, altitude 900 ft.), 8–13.v.1905." Kertész (1910) listed *P. politus* from Sri Lanka, but that was a misidentification. Keiser (1958: 210) included this as "*Paragus tibialis* and var. *rufiventris*" in his paper documenting specimens taken from Uva and Central Provinces. Coe (1964: 256) wrote: "This form of *tibialis* occurs in India and Ceylon," for his "*Paragus tibialis rufiventris*." Joseph & Parui (1977: 228) listed one specimen from Tholkabad, Chaibasa district (Jharkhand) taken in February 1955, but misidentified as "*rufiventris*." Ghorpade (1981b: 65) listed prey of *tibialis* recorded in India. Datta & Chakraborti (1986: 57) again mis-named (?), misidentified (?) it as "*rufiventris*" [which is a *Betasyrphus* ! see below], listing one ♂ from Chalakudy, Trichur district (Kerala) taken in December 1979. They also lumped it like Brunetti, giving a wide distribution. Thompson & Ghorpadé (1992: 7–8, figs. 7–8) gave details and listed it from Tangmarg 2200 m and Srinagar 1893 m, both from JK, from Manali 1828 m, and Simla 2133 m (HP), and gave prey and flower records. Vujić et al. (2008: 535) placed it in their *tibialis*-group. Thompson et al. (2017: 517) provide a synopsis of this primarily Old World group (30 species), considered a subgenus of *Paragus*, with three species from the PSWNG area. Knutson et al. (1975: 328), Ghorpadé (2014c: 16, 2015: 77), and Mitra et al. 2015: 67) listed it. I saw four specimens borrowed from the Lund University, Sweden collection in the late 1970s.

Serratoparagus auritus (Stuckenberg, 1954b)

Paragus auritus Stuckenberg, 1954, Trans R. ent. Soc. Lond., 105: 418 (♂; "Kandy, Ceylon") [NHM, London]

India: AH, KN, MH, OR, TG, TN; Sri Lanka (Brunetti 1908; Cherian 1934; Stuckenberg 1954b; Keiser 1958; Coe 1964; Ghorpadé 1973b, 1974, 1981b, 2014b, 2015b, unpubl.; Knutson et al. 1975; Musthak Ali & Sharatchandra 1985; Datta & Chakraborti 1986; Joseph & Parui 1986; Thompson & Ghorpadé 1992; Vujić et al. 2008; Ghorpadé et al. 2011; Mitra et al. 2015.)

Note: Brunetti (1908: 52) listed a *serratus* [*s. lat.*] as a "common species variable both in size, and coloration of the abdomen," and distributed "from Calcutta to Nepal and reaching as far west as Karachi and as far south as Bangalore." Similarly, he (Brunetti, 1915: 201) gave *serratus* as a "common and widely distributed species" extending to Assam and having in his collection specimens from Cawnpore [Kanpur], Calcutta, and Rangoon, and stating it is "common at Pusa in Bihar." He included a description of it in the *Fauna* volume (Brunetti, 1923: 31, 413) based on material in the Indian Museum [ZSI], Pusa [IARI], and other collections. He gave locality records from Nepal, N. India, NE. India, then Bombay Presidency, and from Bangalore, Travancore, and Goa ("Mormugao"), also Bengal, Assam, Burma, Java, Sarawak, and Papua, writing it was "immediately recognized by its serrated scutellum." These serrated scutelli individuals were found to be polyphyletic by Stuckenberg (1954b) and separated into many species in Africa and the Indian subregion. Stuckenberg (1954b: 418–420, figs. 30–33) described this from specimens taken in Sri Lanka, India, and, curiously, also from Kenya (E. Africa). The holotype male was collected at "Ceylon: Kandy, 29.vi.1892 (*Lt.-Col. Yerbury*)" and the Indian specimen was a male taken at "Calcutta, 1–17.xii.1908, ex. Coll. Brunetti." Stuckenberg (1954b: 418) illustrated the terminalia and abdomens and wrote "This is a large species with elongated antennae. The abdomen is usually broad in relation to the thorax, the females especially with greatly distended abdomens." One male from Kenya was taken at the Lumi River, Teita Hills, Kenya, in December [Coryndon Museum, Nairobi], but this African specimen showed some differences from the Indian sub-continent ones. I believe this species does not occur on the African continent and this Kenyan specimen is of a separate, probably undescribed, species. Cherian (1934: 697) documented a "*serratus*" giving life history, habits, food, distribution based on TNAU, Coimbatore specimens from Coimbatore, Ramnad (Tamil Nadu), Hadagalle (Karnataka), and Samalkota (Andhra), mentioning that, "The presence of the serrated scutellum is one of the most important characters of the fly," lumping all species of *Serratoparagus*. Keiser (1958: 207) included this as "*Paragus auritus*" in his paper documenting specimens taken from Uva all Northern, North Central, and Central Provinces. Coe (1964: 257) reported specimens taken in eastern Nepal "from blooms of *Guizotia abyssinica*," but misspelt it as "*auratus*." Ghorpade (1973b, 1974) did not find it then around Bangalore or list prey (Ghorpade 1981b) and Mitra et al. (2015) curiously omitted it from their review of hover-flies from India! Musthak Ali & Sharatchandra (1985: 18) reported the root aphid *Forda orientalis* as prey of this species in Karnataka. Datta & Chakraborti (1986: 57) and also Joseph & Parui (1986: 162) again misidentified their "*serratus*" which was mainly *crenulatus* (see below), taken in Kerala. Thompson & Ghorpadé (1992: 10–11, map 5) gave details, reported specimens from southern and eastern India, gave a map and a key, as well as prey and flower records. See also Ghorpadé et al. (2011: 79) where this species was taken in the Andhra Carnatic and stated to have larvae that feed on root aphids. Vujić et al. (2008: 536) placed this in *Paragus* (*Serratoparagus*), a new subgenus they erected for a group of mainly Oriental and Afrotropical species with striped eyes and scutellum with conspicuous teeth on posterior margin. Thompson et al. (2017: 519) provide a synopsis of this usually accepted subgenus of *Paragus* restricted to the Afrotropics and the Indomalayan region (?? species), with only one species from the PSWNG area. Knutson et al. (1975: 327) and Ghorpadé (2014b: 17, 2015: 80) listed it. Curiously, Shah et al. (2013) and Mitra et al. (2015) omitted this species in their checklists. I have seen specimens (K. Ghorpadé unpubl.) taken in Salvan forest, Kolhapur (Maharashtra), Bangalore, Chintamani, Sadahalli 906 m, Kunigal, Attur, Hebbal (Karnataka), Nizamabad (Telengana), and Coimbatore (Tamil Nadu), and I saw one specimen borrowed from the Lund University, Sweden collection in the late 1970s.

Serratoparagus crenulatus (Thomson, 1869)

Paragus crenulatus Thomson, 1869, *in*: K. svenska fregatten Eugenies resa, Zool., Dipt., p. 503 (♂♀; "China") [NRS, Stockholm]

India: BI, KL, KN, TN; Sri Lanka (Brunetti 1908, 1913, 1915, 1923; Stuckenberg 1954b; Keiser 1958; Coe 1964; Knutson et al. 1975; Datta & Chakraborti 1986; Joseph & Parui 1986; Peck 1988; Thompson & Ghorpadé 1992; Vujić et al. 2008; Ghorpadé 2014b, 2015b, unpubl.; Mitra et al. 2015)

Note: See notes given below by Brunetti (1908, 1913, 1915, and 1923) under *S. serratus*. Stuckenberg (1954b: 408–413, Figure 17–20) wrote that this species seems to "differ from *P. serratus* as then understood in having extensively dark femora, longish antennae, the mesonotal stripes narrowed behind, and the wings yellowish at the base... There is considerable variation within the species. Each of the larger islands [of the Malay Archipelago] has a slightly different form, and there are small differences between specimens from Malaya, Siam [= Thailand], India, and Hong Kong. *P. crenulatus* could probably be divided into several subspecies; I lack sufficient material to attempt this." About variation, differences between sections of this population ranging widely, from India to probably Australia, he wrote, "This is one of the most widely distributed species of the [*serratus-*]complex. It ranges over the whole of the Oriental Region and Austro-Malayan subregion... The hypopygium of the male from

Ceylon differs in certain respects from that of the Chinese specimens... three specimens from Calcutta are rather dark... This species can usually be recognized by its suffused wings, and by the extensively dark posterior femora. It may be confused with *P. serratus* (Fabricius) when the wings are clear, but the longer mesonotal pile and darker posterior femora distinguish it from that species. The shape of the abdomen is often characteristic." Keiser (1958: 208) included this as "*Paragus crenulatus*" in his paper documenting specimens taken from Uva, Sabaragamuva, and the North-western, Western, North Central, and Central Provinces. Coe (1964: 256–257) gave some notes confirming Stuckenberg's paper. He mentioned a "dark form" also from eastern Nepal, with a more coarsely punctuate thorax than in the typical form, and with "dull violaceous and a few cupreous reflections." Datta & Chakraborti (1986: 57) and Joseph & Parui (1986: 162) gave "*serratus*" which were misidentifications for *crenulatus,* several specimens documented from Ernakulam, Parambikulam, Palghat district, Chalakudy, Trichur district, Idamalaya, Ernakulam district (all in Kerala), and from Top Slip, Coimbatore district (Tamil Nadu), and "Mustor" [?], Chitradurga district (Karnataka). Thompson & Ghorpadé (1992: 13–14, map 4) cited more than 200 specimens which they studied from India (Bihar), Sri Lanka, Nepal, Burma, to Taiwan, and doubtfully Australia. They gave a key, notes on diagnostics and prey, and flower records as well. They wrote that this species "is restricted to the humid forested areas, unlike *serratus*, which clearly favours an open, dry, plains habitat." And that "dark specimens can be distinguished from *P. yerburiensis* in never being wholly black behind the transverse ridge on the first tergum..." Perhaps more extensive material from most of the Malay Archipelago islands and more critical study of variations, along with molecular biology data will result in confirming *crenulatus* as a "lumped" species (?), and is a study worth waiting for, though neither Thompson & Ghorpadé (1992) nor any other workers on Paragini have made any revelations after the fundamental revision by Stuckenberg (1954b) done for a M.Sc. degree at Rhodes University, South Africa. Vujić et al. (2008: 536) placed this in *Paragus* (*Serratoparagus*), their new subgenus. Knutson et al. (1975: 327), Peck (1988: 81), Ghorpadé (2014b: 17, 2015b: 80), and Mitra et al. (2015: 67) listed it. I have seen four specimens borrowed from the Lund University, Sweden collection in the late 1970s. I have also seen specimens (K. Ghorpade unpubl.) from Eruvadi, T.K. district (Tamil Nadu), Navile 588m (Shimoga), Vittal, Sagar, Mudigere 930 m, Kanakapura, Hebbal (Karnataka), and Nelliampathi Hills 1110m (Kerala).

Serratoparagus serratus (Fabricius, 1805)

Mulio serratus Fabricius, 1805, Syst. Antliat., p. 186 (♀; "Tranquebar, India") [UZM, Copenhagen]

Paragus serratus (Fabricius): Stuckenberg, 1954, Trans. R. ent. Soc. Lond., 105: 413, figs. 21–25.

Paragus femoratus Kohli et al., 1988, J. Insect Sci., 1(2): 120 (♂; "Pantnagar, U.P.") [IARI, New Delhi—examined]; Ghorpadé 2014d: 17 (as *n. syn.*)

Serratoparagus serratus Fabricius: Vujić & Radenković 2008, Zool. J. Linn. Soc., 152: 536 (as *n. comb.*)

India: AH, BI, CG, GJ, KN, MH, MP, PO, TN (Brunetti 1908, 1915, 1923; Kertész 1910; Shiraki 1930; Bhatia & Shaffi 1933; Rahman 1940; Beeson 1953; Stuckenberg 1954b; Ghosh 1974; Patel et al. 1975; Knutson et al. 1975; Ghorpadé 1981a, b, 2014b, d, 2015b, 2017, unpubl.; Agarwala et al. 1984; Peck 1988; Kohli et al. 1988; Thompson & Ghorpadé 1992; Vujić et al. 2008; Ghorpadé et al. 2011; Mitra et al. 2015.)

Note: Brunetti (1908: 52) wrote of this species "with serrated scutellum" and "variable both in size, and coloration of the abdomen." He mentioned having seen specimens in his own collection and in the Indian Museum [= ZSI, Calcutta] taken "from Calcutta to Nepal and reaching as far west as Karachi and as far south as Bangalore." Brunetti (1915: 201) mentioned seeing specimens from the Assam area and he himself having collected it at "Cawnpore, 29-xi-04, Calcutta 1-ii-07 and Rangoon 9-ii-06. It is common at Pusa in Bihar." In his *Fauna* volume (Brunetti 1923: 32), he mentioned a "good series in the Indian Museum [ZSI, Calcutta], Pusa [IARI, New Delhi] and other collections. A variable species, especially in the extent of pale colour in the abdomen." Bhatia & Shaffi (1933: 555–556, Pl. LXI: f) described and figured its immature stages. Rahman (1940: 72) noted it as a "widely distributed syrphid in Orient," larvae feeding on aphids on red gram, watermelon, lablab, cotton, mustard, and sugarcane. Beeson (1953: 340) gave it as predaceous on the psyllid *Euphalerus vittatus*, and aphids, quoting Bhatia & Shaffi (1933). Stuckenberg (1954b: 413–415, figs. 21–25) revised what he then called the "*Paragus serratus* Complex" (now *Serratoparagus*) and mentioned specimens in the BMNH (London) from "Jubblepore, North-West India" [= Jabalpur, M.P.] and Deesa [Gujarat], as well as from Delhi, North-East India [! in error, based on mislabelled museum specimens of T. Jermyn; see Ghorpadé 2007: 5], Poona, Coimbatore, Mysore, Bangalore, and "Hasi" a place unknown to me, a misprint, probably, for Itarsi in north-central India (?). He mentioned its "scutellum black, with extensive yellow on the rim, saw-like with many little teeth. Abdomen black on the first segment, red on the rest, with a whitish edge, clear, unspotted wings; rust-red tarsi." He also ended "It is interesting to note that a Danish factory was opened at Tranquebar [= Tharangambadi in Tamil Nadu now], the type locality in 1620. Danish influence continued until 1845, except for a brief period of British occupation from 1801 to 1814... Fabricius probably came by his material of *P. serratus* through the efforts of these settlers. The species seems to be confined to India." It has not been recorded from Sri Lanka as he rightfully stated, but incorrectly assumed so by Kertész (1910), Brunetti (1923), and Shiraki (1930). Ghosh (1974) recorded several aphid prey of this species (see Ghorpadé, 1981b: 64–65). Patel et al. (1975) listed a *Syrphophagus* sp. (Encyrtidae) larval parasitoid in Gujarat. Ghorpadé (1981b: 64–65) listed all its aphid prey, and a psyllid, from India and Pakistan. Agarwala et al. (1983: 391) gave its aphid prey from India. Agarwala et al. (1984: 18) listed aphid prey from India. Peck (1988: 82) gave its distribution as Afghanistan and the Oriental Region. Kohli et al. (1988: 120–121, figs. 30–35) described a *Paragus (Paragus) femoratus* as new from Pantnagar (Uttarakhand) which type specimen I examined in IARI, New Delhi (see Ghorpadé 2014d: 17) and found it to be *serratus* (F.), a new synonym. Thompson & Ghorpadé (1992: 14–16) revised Oriental Paragini and can be consulted for details on this species, the "smallest and palest species of the *serratus*-complex and confined to the hot, dry plains of the subcontinent... From small specimens of *auritus*, it can be separated by its clear wings and brownish black base of the fore femur." Prey and flower records were also given. I included it from Maharashtra based on Stuckenberg (1954b: 413) record from "Poona" and Brunetti (1923: 32) record from "Igatpuri, Bombay," but this specimen needs careful confirmation, if available. See also Thompson & Ghorpadé (1992: 15) and Ghorpadé (2015b: 81–82). Vujić & Radenković (in Vujić et al. 2008b) erected a new "subgenus" *Serratoparagus* with *Paragus pusillus* Stuckenberg, 1954b as the type species and included *serratus* and six other Afrotropical and Oriental species under it (Stuckenberg, 1954b had included seven species in his *serratus*-complex then). I raise this "subgenus" to genus status here. Knutson et al. (1975: 327), Peck (1988: 82), Ghorpadé (2011: 79), Ghorpadé (1981a, 2014b: 17, 2015b: 81, 2017: 108), and Mitra et al. (2015: 67) listed it. I have seen specimens (K. Ghorpadé, unpubl.) from Tamil Nadu (Chidambaram, Vallavannur, Villupuram, Coimbatore), Karnataka (Hadagalli, Hagari, Chikmagalur, Dodballapaura, Bangalore, Chethalli, Uppinakakere, nr Maddur, Karkala, Hoskote, CPCRI, Kidu (S. Canara), Kanakapura, Kunigal, Hebbal, Attur, GKVK, Bangalore, Chikballapur 594m, Hessaraghatta, Haveri, Ranebennur, Doddaballapura), Andhra (Guvvvalacheruvu, nr Cuddapah), Rajasthan (Udaipur), and from Gujarat (Navsari).

Serratoparagus yerburiensis Stuckenberg, 1954

Paragus yerburiensis Stuckenberg, 1954, Trans, R. ent. Soc. Lond., 105: 415, figs. 26–29 (♂; "Velverry, Ceylon" = Sri Lanka) [BMNH, London]

India: AH, BI, KL, KN, MP, OR, PO, TN; Sri Lanka (Stuckenberg 1954b; Keiser 1958; Coe 1964; Rao 1969; Gokulpure 1972; Ghorpadé 1973b, 1974, 1981a, b, 2014b, 2015b, unpubl.; Ghosh, 1974; Knutson et al. 1975; Patnaik & Bhagat 1976; Thompson & Ghorpadé 1992; Vujić et al. 2008; Ghorpadé et al. 2011; Mitra et al. 2015).

Note: Before Stuckenberg (1954b: 415–418, figs. 26–29), this was "lumped" with *serratus* (Fabricius). The epithet *–ensis* in honour of Lt-Col. J.W. Yerbury is incorrect and unfortunately chosen, as this epithet should only be added to a geographical name. A male patronym ends with an "*-i*" and a female one with "*-ae*." Stuckenberg's type material was mainly from Sri Lanka and from "Jubblepore" [= Jabalpur, M.P.] in central India. This is a distinct species with tergum 1 all black and fore femur without any black markings. Stuckenberg (1954b) further wrote, "This species has a transverse band of black across most of the first abdominal segment. The posterior corners of the abdomen are very much produced, and the abdomen appears to be truncated in the males... it seems that *P. yerburiensis* is a rather isolated species within the complex. There seems to be considerable variation in size between individuals of this species." Keiser (1958: 209) included this as "*Paragus yerburiensis*" in his paper documenting specimens taken from Northern, Northwestern, North Central, and Central Provinces. Coe (1964: 257) gave specimens taken in eastern Nepal from blooms of *Guizotia abyssinica* and stated "described from India and Ceylon" which was incorrect! Rao (1969) gave *Aphis gossypii, A. spiraecola*, and *Myzus persicae* as larval prey for this species (see Ghorpadé, 1981b: 65). Gokulpure (1972: 848) recorded it from Madhya Pradesh. See also Ghorpade (1981b: 65). Ghorpade (1973b, 1974) in his master's thesis wrote: "this species, first recorded from Bangalore and Karnataka, was as common as *serratus* around Bangalore and was collected from August to March; also several prey records were given (see also Ghorpadé 1981b: 65). Patnaik & Bhagat (1976: Pl. 2(d)) misidentified this as *Eumerus* sp. nr *albifrons* Walker, as their illustration of the adult fly indicates. It was reared from several aphid species listed found in the Bhubaneswar area and in other parts of Puri district in Orissa. Thompson & Ghorpadé (1992: 16–18, map 7) and Ghorpadé et al. (2011: 79–80) may be consulted for details, also of prey and flower records. The former paper listed more than 350 specimens studied from many parts of India and Sri Lanka and gave a psyllid as prey as well. Vujić et al. (2008: 536) placed this in *Paragus (Serratoparagus)*, a new subgenus. Knutson et al. (1975: 328), Ghorpadé (2011: 79–80), Ghorpadé (1981a, 2014b: 17, 2014e: 9), and Mitra et al. (2015: 67) listed it. See also Sack (1932: 131) and Thompson & Rotheray (1998: 101) for discussion and key to *Paragus s. lat.* in the Palaearctic. I have seen (k. Ghorpadé, unpubl.) specimens from Tamil Nadu (Coimbatore, Salem, Yercaud, Nathampatti, Ramnad district, Tindivanam, S. Arcot, Chidambaram), Karnataka (Hunsur, Chintamani, Bijapur, Bangalore, Hebbal, Hessaraghatta, Hoskote ex cowpea aphid, Chetthalli, Kanakapura, Hebbal, Kaimara (Chikballapur District), Attur, Hebbal), and Andhra (Bapatla, Samalkot).

Allograpta bouvieri (Hervé-Bazin, 1923)

Xanthogramma bouvieri Hervé-Bazin, 1923, Bull. ent. Soc. Fr., p. 26 (LT ♂ "Trichinopoly, Coonoor, Nilgiris, India; Laos) [MNHN, Paris—examined]

India: KL, TN (Hervé-Bazin 1924; Knutson 1975; Ghorpadé 1994, 2014b; Mitra et al. 2015.)

Note: In my doctoral thesis (Ghorpade 1981a), I had written, "This species was described from a series of specimens collected at Trichinopoly [= Tiruchirapalli] and Coonoor in south India, and in Laos (syntypes). Vockeroth (*in litt.*) examined 4♂ and 1♀ syntypes ("cotypes") in the Paris Museum and felt, like Hervé-Bazin (1924) that they represented *maculipleura* (Brunetti). However, I consider *bouvieri* a distinct species as distinguished in the key (see Ghorpadé 1994: 7)." Only males are known so far, and the female syntype (from Laos) almost certainly belongs to a different species, probably *maculipleura*. I have designated a lectotype ♂ (here designated) labelled "Inde Meridionale, Trichinopoly, F. Caius 1911/Museum Paris, Coll. J. Hervé-Bazin, 1923/*Xanthogramma bouvieri* H.-B. ♂ Cotype (blue label) [MNHN, Paris] and so labelled it." My 50 specimens from Munnar (Kerala) labelled "India: 12km N. Munnar, 2000 m" were all taken in a "swarm" of males hovering in shade by the roadside, in forest. Tiruchirapalli being on the plains below, I suspect the specimens were actually collected on the Kodai Hills (Palni Hills), where the species appears to be fairly common during March and April. I have seen a ♂ also at Kodaikanal 2250 m, and another ♂ at Shembaganur 1800 m a little below in California Academy of Sciences (CAS), San Francisco, and Canadian National Collection of Insects, Arachnids and Nematodes (CNC), Ottawa, museums borrowed on loan. Knutson et al. (1975: 308), Ghorpadé (2014c: 18), and Mitra et al. (2015: 67) listed it. Thompson et al. (2017: 509) provide a synopsis of this cosmopolitan genus (73 species), with four species from the PSWNG area. I have seen specimens (K. Ghorpadé unpubl.) from Kodaikanal, Singara, Nilgiri Hills, Maruthamalai, 1600 ft., Kallar 1250 ft., Nilgiri Hills (Tamil Nadu), Hunsur, Chintamani, Vijayapura, Sadahalli, Hebbal, Bangalore (Karnataka), and Munnar 200 m (Kerala), in collections.

Allograpta dravida Ghorpadé, 1994

Allograpta dravida Ghorpade, 1994, Colemania, 3: 7 (♂; "Bangalore") [USNM, Washington, DC]

India: KN (Ghorpadé, 1994, 2014b.; Mitra et al. 2015.)

Note: In my doctoral thesis (Ghorpade 1981a), I had written: "This new species is differentiated from the others by characters given in the key [see Ghorpadé 1994: 7] and by the distinctive male terminalia. The holotype ♂ was caught while hovering under a tall tree, in shade, in scrub forest on the outskirts of Bangalore (GKVK, UASB campus). The name "*dravida*," meaning "southern," in Sanskrit, alludes to its distribution in southern India." Ghorpadé (2014b: 18) and Mitra et al. (2015: 67) listed it.

Allograpta javana (Wiedemann, 1824)

Syrphus javanus Wiedemann, 1824, Analecta Ent., p. 34 (LT ♂; "Java") [Naturhistorisches Museum Wien (NMW), Vienna—examined]

Xanthogramma nakamurae Matsumura, 1918, J. Coll. Agric, Hokaido Imp. Univ., 8: 9 (sex ?; "Japan") [?]

Miogramma iavanus: Frey, 1946, Notul. Ent., 25: 165 (as n. gen.)

Helenomyia javana: Bańkowska, Bull. L'Acad. Pol. Sci., 10: 311 (as n. gen.)

Paraxanthogramma nakamurae Tao & Chiu, 1971, Taiwan Agril. Res. Inst., Spl. Publ., 10: 74 (as n. gen., *nom. nud*.)

India: AH, BI, KL, KN, MP, OR, TN; Sri Lanka (Kertész 1910; Brunetti 1915, 1917, 1923; Hervé-Bazin 1923, 1924, 1926; Shiraki 1930; Curran 1928, 1931a, b; Bhatia & Shaffi 1933, Cherian 1934; Beeson 1953; Keiser 1958; Bańkowska 1962; Rawat & Modi 1968; Rao 1969; Vockeroth 1969; Vockeroth 1971; Knutson et al. 1975; Patnaik et al. 1977; Ghorpadé 1973a, b, 1981a, b, 1994, 2014c, 2015b, unpubl.; Peck 1988; Mengual et al. 2008b, 2009; Mitra et al. 2015).

Note: In my doctoral thesis (Ghorpade 1981a) I had written that this species "is widespread in south and southeast Asia (recorded from Japan, Australia, and even Fiji in the literature)... It is very variable in colour markings, especially the extent of the black facial vitta. The form of the male terminalia, especially the dorsal margin of the minis (see Hippa 1968: Figure 4, mi) but the surstylus, is however characteristic and constant. It is fairly common in the Indian plains and foothills and its larvae have been reared on several prey species (see Ghorpadé 1973a, 1981b: 72–73)." Wiedemann (1824) described this species from an unspecified number of specimens. What remains of the primary type material (see Osten Sacken 1878: XV-XVI, and Zimsen 1954, for the types of Wiedemann) is a single syntype ♂ in NMW, Vienna, and the other syntypes in UZM, Copenhagen are lost. Dr. Leif Lyneborg informed me (*pers. comm.*) that there are no identified specimens of "*iavanus*" in UZM. Wiedemann (1824: 35) indicated the type material as being "In museo nostro; a Westermannio ex Insula Iava allatus," meaning that they were in "nostro" [= our] museum [NMV] and in Killiae (= UZM, Copenhagen), brought by Westermann from the island of Java. A male syntype, labelled "Java, Coll. Winthem" exists in Vienna and is designated lectotype here. *Allograpta distincta* (Kertész) and *A. medanensis* (de Meijere) are probably distinct species and not synonyms or varieties of *javana* as previously treated. The former has all yellow antennae and no black facial vitta and was described from New Guinea (see also Bezzi 1928: 73 and van Doesburg Sr 1966: 64). While some specimens of *javana* lack the facial vitta (especially males), or have it extremely reduced, *distincta* is not accepted as a synonym of *javana* (see Knutson et al. 1975: 308) until fresh specimens are examined from the type locality (Erima, Astrolabe Bay, New Guinea). The type(s) [in Hungarian Natural History Museum, Budapest] were destroyed. *A. medanensis*, described from Medan (Sumatra) as a variety of *javana*, is a distinct species, as is apparent from its distinct male terminalia and in the black facial vitta reaching the oral cavity and in its fore femur being black pilose on its apical one-third. Mengual et al. (2009) presented a comprehensive "conspectus" of this genus *Allograpta* with a checklist of world species and a key to its "subgenera," and their phylogenetic relationships. I (Ghorpadé 1981b: 72) listed its prey species and also gave a key (Ghorpadé 1994: 7) to species in this subcontinent. I studied the syntypes in NMW, Vienna and designated a ♂ Lectotype labelled "Java, Coll. Winthem [square pink card]/*javanus*, det. Wiedem./*javanus* Wied., Java/LECTOTYPE, *Syrphus javanus* Wiedemann, Ghorpade des. 1983. In his *Fauna* volume (Brunetti 1923: 100), he gave a full description, but still retained it in *Sphaerophoria* (with *indiana* Bigot and *viridaenea* Brunetti) in the key given to these species. He also noted it from Coorg in south India. Hervé-Bazin (1924: 291, 1926: 70–71) debated its placement in *Sphaerophoria* and stated it to be a *Xanthogramma*. Curran (1928: 241–243, 1931a: 321, 1931b: 355–356) retained it in

Sphaerophoria and reported it from Malaya, and gave a key. Bhatia & Shaffi (1933: 557–559, Pl. LXII) described and illustrated its life-history in Pusa (Bihar). Cherian (1934: 698) recorded it from Maruthamalai (Coimbatore). It is fairly common in the Indian plains and foothills, and its larvae have been reared on several prey species (see Ghorpadé 1973a, 1981b: 72–73). Wiedemann (1824) described this species from an unspecified number of specimens. *Allograpta distincta* (Kertész) and *A. medanensis* (de Meijere) are probably distinct species and not synonyms or varieties of *javana* as previously treated. The former has all yellow antennae and no black facial vitta, and was described from New Guinea (see also Bezzi 1928: 73 and van Doesburg Sr 1966: 64). While some specimens of *javana* lack the facial vitta (especially males), or have it extremely reduced, *distincta* is not accepted as a synonym of *javana* (see Knutson et al. 1975: 308) until fresh specimens are examined from the type locality (Erima, Astrolabe Bay, New Guinea). The type(s) [in Hungarian Natural History Museum, Budapest] were destroyed. *A. medanensis*, described from Medan (Sumatra) as a variety of *javana*, is a distinct species, as is apparent from its distinct male terminalia and in the black facial vitta reaching the oral cavity and in its fore femur being black pilose on its apical one-third." Mengual et al. (2009) presented a comprehensive "conspectus" of this genus *Allograpta* with a checklist of world species and a key to its "subgenera," and their phylogenetic relationships. I (Ghorpadé 1981b: 72) listed its prey species and also gave a key (Ghorpadé, 1994: 7) to species in this subcontinent. I studied the syntypes in NMW, Vienna and designated a ♂ Lectotype labelled "Java, Coll. Winthem [square pink card]/javanus, det. Wiedem./javanus Wied., Java/ LECTOTYPE, *Syrphus javanus* Wiedemann, Ghorpade des. 1983." Brunetti acknowledged this "clerical error" in his later paper (Brunetti, 1915: 214, 217) and listed this species from more localities in NE India and the Burma-Siam [= Thailand] Frontier and separated it in a key to Oriental species and the "forms" of *Sphaerophoria*. Then (Brunetti 1917: 85) he mentioned it as a *Sphaerophoria*, common on the Simla Hills. In his *Fauna* volume (Brunetti 1923: 100), he gave a full description, but still retained it in *Sphaerophoria* (with *indiana* Bigot and *viridaenea* Brunetti) in the key given to these species. He also noted it was from Coorg in south India. Hervé-Bazin (1924: 291, 1926: 70–71) debated its placement in *Sphaerophoria* and stated it to be a *Xanthogramma*. Curran (1928: 241–243, 1931a: 321, 1931b: 355–356) retained it in *Sphaerophoria* and reported it was from Malaya, and gave a key. Bhatia & Shaffi (1933: 557–559, Pl. LXII) described and illustrated its life-history in Pusa (Bihar). Beeson (1953: 340) wrote it feeds on *Phylloplecta* sp., "a psyllid species making pit-galls on the leaf of *Shorea robusta*, and *Psylla* sp. on *Bauhinia variegata*, and *Tenaphalara acutipennis*," quoting Bhatia & Shaffi (1933). Bańkowska (1962: 311–314, figs. 1, 5–12) differed from placing this species either in *Sphaerophoria* or in *Ischiodon,* and erected a new genus *Helenomyia* for it, with *javana* as the type species. She gave a description of the genus and illustrations of *javana*: head, pleura, abdomen, terminalia. She mentioned it as a "predator of Psyllidae," and gave its range as India to Australia, with note of a male from Bhim Tal [UK] in November, and "N.E. India" coll. T. Jermyn [this is in error, as noted elsewhere in this paper, *q.v.*]. Rawat & Modi (1968) recorded its larvae feeding on early nymphs of *Ferrisia virgata* at Jabalpur (M.P.). Vockeroth (1969: 126–130) was the first to put it in the correct genus *Allograpta* in his excellent review of world Syrphini. Then he gave diagnostics and figures of head profile and surstylus (Vockeroth 1971: 1628). I (Ghorpadé, 1973a) gave an Araeopid as prey, listed its other prey from India, and included it in my master's thesis (Ghorpadé, 1973b) on Syrphidae of the Bangalore area. Knutson et al. (1975: 308) listed it from India to Japan, Fiji, and Guadalcanal. Rao (1969: 787) gave *Aphis gossypii* and *Myzus persicae* as larval prey for this species (see Ghorpadé 1981b: 65). Patnaik et al. (1977: 585) recorded it from the Puri district of Orissa and gave sorghum aphid as prey there. I (Ghorpadé 1981b: 72–73) listed prey from the Indian sub-continent. Ghorpadé (1994: 7) gave a key to this and three other species of *Allograpta* from the Indian subcontinent and noted *javana* to be the correct original spelling, not *"iavana"* as used by some authors. Mengual et al. (2008b) wrote on the molecular phylogeny of *Allograpta* and then a detailed conspectus of this genus (Mengual et al. 2009). In the species listed in their *obliqua* species-group, they listed this species with *Xanthogramma nakamurae* Matsumura, 1918 as its synonym. Ghorpadé (2014b: 18) listed it from India to the Solomon Islands and even Australia. Kertész (1910), Knutson et al. (1975: 308), Peck (1988: 12), and Mitra et al. (2015: 67) listed it. See also Thompson & Rotheray (1998: 99) for a key to this genus in the Palaearctic. I have seen specimens (K. Ghorpadé unpubl.) from Maruthamalai, Coimbatore, and Nilgiri Hills (Tamil Nadu), Mudigere ARS 930 m, and Yalagere, Chikballapur (Karnataka) in collections.

Allograpta sp.

India: KN (K. Ghorpadé unpubl.)

Note: I have taken a specimen on the Bababudan Hills in Karnataka recently which looks to be yet another undescribed species of Indian *Allograpta*. This was sent to Dr. X. Mengual (Bonn, Germany), and we will publish a name and description soon.

Asarkina ayyari Ghorpadé, 1994

Asarkina ayyari Ghorpade, 1994, Colemania, 3: 7 (♂; "Yercaud, Shevaroy Hills") [USNM, Washington, DC]

India: AH, KL, KN, TN; Sri Lanka (Keiser 1958; Ghorpadé 1981a, 1994, 2014b; Mitra et al. 2015.)

Note: In my doctoral thesis (Ghorpade 1981a), I had written: "This species is easily distinguished by its strongly produced face and distinct male terminalia. Other differences were given in my key (Ghorpadé 1994: 7). It is distributed in southern India and Sri Lanka and appears to be the commonest *Asarkina* on the hills here. It was misidentified as "*porcina porcina* (Coquillett)" by Keiser (1958) from Sri Lanka; a ♂ and two ♀♀ in the Nietner Collection [ZMHU, Berlin] were determined as *ericetorum* var. *kelantanensis* Bezzi (♂) and as *macquarti* Doleschall (♀♀)." Keiser (1958: 194) included this as "*Asarcina* (*Asarcina*) *porcina*" in his paper documenting specimens taken from Uva, Sabaragamuva and North-western, and Central Provinces. Perhaps Vockeroth's (1969) "*consequens*" from "S. India" actually represented the present new species ? It is named after the late Dr. T.V. Ramakrishna Ayyar, a pioneering Indian entomologist who was also the earliest Indian systematist of stature, and who built up a fine collection of south Indian insects at the Agricultural College, Coimbatore, following up on work done by T. B. Fletcher. My paratypes were 4 ♂ 14 ♀ taken from Nandi Hills 1467 m, Mudigere 970 m (Karnataka), Yercaud 1370 m, Shevaroy Hills, Ootacamund 2350 m, Nilgiri Hills, Anaimalai Hills (Tamil Nadu), Meppadi 690 m, Wynaad (Kerala), and from Sri Lanka: N. Pundaloya, Labugama reservoir jungle, Laksapana 823 m, Kanda-ela reservoir, 9 km W. Nuwara Eliya 1890 m, and Rangala. Mitra et al. (2015: 67) listed it. Thompson et al. (2017: 509) provide a synopsis of this Old World genus of tropical distribution (47 species), with nine species from the PSWNG area. I have seen specimens (K. Ghorpadé unpubl.) from Santhanathod 250 m, Malabar District (Kerala), and from Amalapuram, Godavari District (Andhra).

Asarkina belli Ghorpadé, 1994

Asarkina belli Ghorpade, 1994, Colemania, 3: 8 (♂; "35km W. Jog Falls") [USNM, Washington, DC]

India: AH, KN; Sri Lanka (Keiser, 1958; Ghorpadé, 1981a, 1994, 2014b; Mitra et al. 2015.)

Note: In my doctoral thesis (Ghorpade, 1981a) I had written: "This species, along with other south Indian ones other than *incisuralis* have been confused with *consequens* (Walker) in existing literature (see also Vockeroth 1969). It is a member of the group having a well developed tubercle on the face as distinct from those with a carinate face. It is closest to *hema* Ghorpadé, sp. nov., but its third antennal segment is much longer, facial tubercle more pronounced, and male terminalia different. It is named after the late T.R.D. Bell, a well-known naturalist recognized for his detailed studies of life-histories and immatures of the butterflies of North Canara and adjacent areas" (see Ghorpadé et al. 2013). See also diagnostic key in Ghorpadé (1994: 7–8). I have seen specimens (K. Ghorpadé unpubl.) from Dandeli 481 m, and Belgaum (Karnataka).

Asarkina hema Ghorpadé, 1994

Asarkina hema Ghorpade, 1994, Colemania, 3: 8 (♂; "Bannerghatta Park") [USNM, Washington, DC]

India: KL, KN, (Ghorpadé, 1981a, 1994, 2014b; Mitra et al. 2015.)

Note: In my doctoral thesis (Ghorpade 1981a) I had written: "This species resembles *belli* sp. nov., but is easily separated by the very short third antennal segment, different male terminalia, and other characters (see key in Ghorpadé 1994: 8). The abdomen is very broad, almost rectangular and light in colour. This species seems widely distributed in the foothills of the Western Ghats and on the Eastern Droogs. The species name is taken from Sanskrit and means "golden," alluding to the general colour and appearance of the species." The single ♀ from Thekkady 884 m, Walayar forest (Kerala) is apparently conspecific, though different from the ♂♂ in shape of facial tubercle and in other respects. My paratype males were all taken from Bannerghatta forest 900 m near Bangalore, the species flying in March and September. Mitra et al. (2015: 67) listed it. I have also seen specimens from Ootacamund (K. Ghorpadé unpubl.).

Asarkina incisuralis (Macquart, 1855)

Syrphus incisuralis Macquart, 1855, Dipt. Exot., Suppl., 5: 94 (LT ♂; "Inde") [UM, Oxford]

Asarcina ericetorum var. *formosae* Bezzi, 1908, Annls hist.-nat. Mus. Natn. Hung., 6: 499 (♂♀; "Takao, Formosa") [MNHN, Budapest, lost ?]; Ghorpadé, Colemania, 15: 8 (as *n. syn.*)

Asarcina ericetorum (Fabricius), *A. salviae* (F.) *auct.*: misident.

India: AH, BI, CG, KL, KN, MH, PO, TN; Sri Lanka (Bezzi 1908; Sack 1913, 1922, 1932; Brunetti 1910, 1913, 1915, 1917, 1923; Hervé-Bazin 1926; Curran 1928, 1931a; Cherian 1934; Keiser 1958; Coe 1964; Knutson et al. 1975; Ghorpadé 1973a, 1974, 1981a, 1994, 2009, 2014b: 18, 2015b: 90, 2017: 108, unpubl.; Datta & Chakraborti 1984, 1986; Mitra & Parui 2002; Ghorpadé et al. 2011; Mitra et al. 2015).

Note: In my doctoral thesis (Ghorpade 1981a) I had written: "This is one of the most widely distributed syrphids in the Orient. All earlier workers have misidentified it either as *ericetorum* (F.) or *salviae* (F.) which are distinct African species. The oldest available name for this Indian *Asarkina* is *incisuralis* (Macquart) which has priority over *formosae* Bezzi which latter becomes a new synonym (see above). I have seen examples of this species in the FRI (Dehra Dun) collection labelled as *incisuralis* by Curran. It is a distinct species with a carinate face, frons with yellow hairs (completely in male, lower half in female), and narrow black bands on each tergum. It apparently occurs all over India except in the dry north-western region and on the higher elevations on the hills." Sack (1913: 3, fig. 2) noted specimens taken in Taiwan and gave a figure of the head profile which shows a carinate face rather like *incisuralis* (Macquart). Later (Sack 1922: 3) he mentioned more specimens of what he named as "var. *orientalis* Bezzi" which he noted could be a synonym of *incisuralis* but required study. Cherian (1934: 698) listed an "*Asarcina ericetorum*" from "South Canara, Coimbatore, Bellary, Godavari, and Chepauk (Madras)." Keiser (1958: 192) included this as "*Asarcina* (*Asarcina*) *ericetorum formosae*" in his paper documenting specimens taken from Sabaragamuwa, and Northern, Western, Central, and Southern Provinces. Knutson et al. (1975: 310) treated it as a synonym of *ericetorum* (Fabricius). Then Sack (1932: 230) gave "*ericetorum*" from the Indonesian Lesser Sunda Islands of Lombok and Flores. Brunetti (1910: 171, 1913: 164, 1917: 84) noted that *Syrphus salviae*, W. [sic] was "generally common during summer... now relegated to *Asarcina*" in Sri Lanka." Later, Brunetti (1915: 210) curiously stated that *Asarcina* [sic] "is not a good genus," I wonder why? Then he mentioned under *Syrphus* (*Asarcina*) *ericetorum*, F. that "*S. salviae*, W., is identical with *ericetorum*, F., described originally from Africa, and the latter name will have to be used for it." He then listed localities he had seen specimens from, ranging from Simla and Sri Lanka through Assam to Java. Then Brunetti (1917: 84) listed "*salviae*" from Simla, taken in August by Capt. Evans, the famous butterfly specialist (see Ghorpadé & Kunte 2010). In his *Fauna* volume, Brunetti (1923: 63–64) described and illustrated this (as "*Asarcina ericetorum*, Fabr.") and treated *salviae* Fabr. and *incisuralis* Macq. as synonyms, noting that it is, "Widely distributed in India and the East, in both hills and plains, throughout the greater part of the year." No doubt this note stems from a misunderstanding that all of the several species of *Asarkina* flying here were one and the same species, which I had separated recently (Ghorpadé 1994: 7–8). Hervé-Bazin (1926: 64) mentioned a single female taken in Tonkin (as "*ericetorum*") and gave a brief description. Curran (1928: 232, fig. 13, 1931a: 320) treated this as *salviae* Fabricius and reported it from the Malay Peninsula. His "*salviae*" was possibly *incisuralis* based on the narrow black fasciae of its figured abdominal terga. Keiser (1958: 195–196) interestingly named his Ceylonese specimens, of what evidently were *incisuralis,* as *ericetorum* var. *formosae*, and probably some others as "*porcina*" that were *ayyari* Ghorpadé. Coe (1964: 265) treated it as "*Asarcina ericetorum* (Fabricius)" when reporting specimens from E. Nepal and, like Brunetti, mistook *Asarkina* here as a single species stating, "This species with its numerous named varieties is common and widespread in the Ethiopian, Oriental and Australian regions." In my master's thesis (Ghorpadé 1973a, 1974: 636), I reported it from the Bangalore area and continued to use *ericetorum* (Fabr.) then. Knutson et al. (1975: 309–310) listed more than a dozen *Asarkina* (*s. str.*) species from the Oriental region and again wrongly included both *ericetorum* and *salviae* from the Oriental region ! Macquart's *incisuralis* was treated as a synonym of *ericetorum* Fabricius by them in their Oriental catalog. Datta & Chakraborti (1984: 237, fig. 1) illustrated its male terminalia and recorded it (again as *ericetorum*) from Jammu & Udhampur (but not from the Kashmir Valley) on flowers of a few plants. Datta & Chakraborti (1986: 53) listed "*Asarkina* (*Asarkina*) *ericetorum*" from Top Slip, Coimbatore, taken in December 1979, which is apparently this species. I recognized and keyed seven species of *Asarkina* from the Indian subcontinent (Ghorpadé, 1994: 3, 7–8) and for the first time treated this species as *incisuralis* (Macquart). This was based on a revision of Indian Syrphini for my doctoral thesis research (Ghorpade, 1981a) where I wrote, "This is one of the most widely distributed syrphids in the Orient. All earlier workers have misidentified it either as *ericetorum* (F.) or as *salviae* (F.), which are distinct African species, based on my examination of types. The oldest name available for it is *incisuralis* (Macquart) which has priority over *formosae* Bezzi which becomes a new synonym. I have seen examples of this species in the FRI, Dehra Dun, labelled as "*incisuralis*" by Curran... It apparently occurs all over India except in the dry north-western region (?) and on the higher elevations on hills and mountains." Mitra & Parui (2002: 45) recorded "*ericetorum*" from N. Gujarat visiting flowers of *Cassia tora* and *Commelina* sp. in the Jessore and Balaram-Ambaji Wildlife Sanctuary (WLS) there. I clarified the correct name for this species and gave synonymy (Ghorpadé, 2009: 8). See notes in Ghorpadé et al. (2011: 80) also. Mitra et al. (2015: 67) listed it. My doctoral thesis had specimens from Dehra Dun (Uttarakhand), Kanpur (Uttar Pradesh), Pusa (Bihar), Kalimpong 1275 m, Lalmanirhat, Calcutta, Suryaberia Is. (West Bengal), Kobo, Abhoypur forest, Kohora, nr Kaziranga Sanctuary (Assam), nr Eluru (Andhra), 20 km N. Yelburga, Ramandrug 990 m, Sandur, Tarikere area 900 m, Devarayadurga 1187 m, Mudigere 970 m, Nandi Hills 1467 m, Bangalore 916 m, Bannerghatta Park, Bandipur Sanctuary 1110 m (Karnataka), Yercaud 1370 m, 12 km N. Salem, Coimbatore, Nedungadu (Tamil Nadu), Kaikatty 937 m, Maraiyur 1066 m (Kerala), Kurumbagaram, and nr Karaikal (Pondicherry), Lothar, nr Birganj (Nepal), Baraiyadhala Forest Reserve (Bangladesh), Kalatuwawa, Labugama Reservoir, Ratmalana, Padaviya, Katagamuwa, Koite, Kandy, Nugegoda, and Colombo (Sri Lanka).

I studied syntypes of *incisuralis* in the Macquart Collection in the UM, Oxford and designated a Lectotype ♂ labelled "*Syrphus incisuralis* Macquart, 1855/*Asarkina incisuralis* Macquart, Ghorpade det. 1994." Other syntypes, two males and a female, were from "China," but are actually *A. porcina* (Coquillett)! Knutson et al. (1975: 309) gave the type as a ♀ from "Inde" which was incorrect. I also studied syntypes of *Syrphus ericetorum* Fabricius in the NHM, London, present in the Banks Collection. These were two ♀ without any labels (!), but the BMNH Curator gave labels which read, "Syntypes of Syrphus ericetorum Fabr. From Banks' Collection." These were abundantly distinct from *incisuralis* with different facial profile and other characters. When I visited the CNC, Ottawa in August 1983, I examined a female of this species taken at Lothar, 450" near Birganj, in Nepal in August. Knutson et al. (1975: 309) and Mitra et al. (2015: 67) listed it. I have seen specimens from Chidambaram, Coimbatore, Chepauk, Chennai (Tamil Nadu), Mandagadde, Shimoga, Malaya Marutha, Mudigere 913 m, Chikmagalur, Nagody 250 m, S. Canara District, Sadahalli 913 m, Bangalore (Karnataka), and Pathanamthitta (Kerala).

Asarkina pitambara Ghorpadé, 1994

Asarkina pitambara Ghorpade, 1994, Colemania, 3: 8 (♂; "Jog Falls") [USNM, Washington, DC]
India: KL, KN; Sri Lanka (Keiser 1958; Ghorpadé 1981a, 1994, 2014b; Mitra et al. 2015.)

Note: In my doctoral thesis (Ghorpade 1981a) I had written: "This species is intermediate between *incisuralis* (carinate face) and the other species (tuberculate faces). The face is carinate but the male terminalia, hairing, etc., are similar to the majority of species in India. The species name, "*pitambara*" means clothed in yellow in Sanskrit, and alludes to its overall yellow coloration." Only the single holotye ♂ is known, taken in November 1976 by me at Jog Falls 534 m (Karnataka). See also my diagnostic key in Ghorpadé (1994: 7–8). I have seen specimens (K. Ghorpadé unpubl.) from Taliparamba, N. Malabar (Kerala).

Asiobaccha nubilipennis (Austen, 1893)

Baccha nubilipennis Austen, 1893, Proc. zool. Soc. Lond., p. 136 (LT ♂; "Kandy, 1700 ft., Ceylon," here designated) [NHM, London]

India: KL, KN, TN; Sri Lanka (Austen 1893; Kertész 1910; Brunetti 1910, 1923; Matsumura 1916; Keiser 1958; Knutson et al. 1975; Ghorpadé 1981a, 1994, 2014b, 2015b; Mahalingam 1988; Thompson & Rotheray 1998; Dirickx 2010; Mitra et al. 2015; Thompson et al. 2017.)

Note: Austen (1893: 136–137, Pl. IV, figs. 7,9, Pl. V, fig. 14) described this based on specimens from Kandy, ca 1800 ft., in Sri Lanka, taken in May-June by Lieut.-Colonel Yerbury there and presented by Brunetti to the British Museum in 1927. Brunetti (1910: 170–171) noted four males seen from Kandy, taken in October and November in Sri Lanka, "by that indefatigable collector Mr. E.E. Green." Then, in the *Fauna* volume, Brunetti (1923: 116–117) gave a detailed description and mentioned specimens seen from Sri Lanka as well as from several localities in southern and north-eastern India and mentioned Austen's diagnosis of this by its different abdominal markings and "the sharply defined facial and antenniferous tubercles when viewed in profile." Matsumura described it as new again with the same binomen (preoccupied) from Okinawa, Ryukyu Is. Keiser (1958: 201) included this as "*Baccha nubilipennis*" in his paper documenting specimens taken from Sabaragamuva, North-western and Central Provinces. Knutson et al. (1975: 322) listed this as *Baccha* (*Allobaccha*) from Sri Lanka, India, Nepal, Taiwan, and even Japan. Mahalingam (1988: 161) gave it as recorded from the Nilgiri Hills among ten species recorded by Brunetti (1923). In my doctoral thesis (Ghorpadé, 1981a) I had written, "A Lectotype ♂ is here designated labelled "Type ♂ [red bordered circular label]/Kandy, Ceylon, 28.VI.92, *circa* 1,700 ft., Col. Yerbury, 92–192/*Baccha nubilipennis*, Aust. Type ♂/LECTOTYPE, *Baccha nubilipennis* Austen, Ghorpade des. 1983/*Asiobaccha nubilipennis* (Austen) ♂, Ghorpade det. 1983." [NHM, London]. A Paralectotype ♀ was also designated, labelled as "Type ♀ [red bordered circular label]/Kandy, Ceylon, 25.V.92, *circa* 1,700 ft., Col. Yerbury, 92–192/*Baccha nubilipennis*, Aust. Type ♀/PARALECTOTYPE, *Baccha nubilipennis* Austen ♀, Ghorpade des. 1983/*Asiobaccha nubilipennis* (Austen) ♀, Ghorpade det. 1983." [NHM, London]." I saw several specimens from Sri Lanka and India (KL, KN, TN, WB) and Burma, besides also from Thailand, Java, Taiwan, Philippines, Ryukyu Is, and Malaysia. I separated this in a key to Indian subcontinent genera (Ghorpadé 1994: 4). Dirickx (2010: 231) gave notes on this species, *q.v.* Mengual (2015) recently wrote on the systematics and phylogeny of *Asiobaccha*, reviving its status to a good genus, with seven known species in the Oriental—Papuan region. He concluded, "Consequently, the combination of pilose anterior anepisternum and sclerotized maculae on posterior wing margin seems to define the "*Episyrphus* clade," with *Asiobaccha*, *Episyrphus*, and *Meliscaeva*." He then also revised *Asiobaccha* in a separate paper (Mengual 2016). Kertész (1910), Knutson et al. (1975: 322), Ghorpadé (2014b: 18, 2015b: 92), and Mitra et al. (2015: 67) listed it. See Thompson & Rotheray (1998: 97) for a key to this genus in the Palaearctic. Thompson et al. (2017: 509) provide a synopsis of this group restricted to the Indomalayan and Australian regions (19 species), with ten species from the PSWNG area. I have seen other specimens (K. Ghorpadé unpubl.) from Nagody 2500 m, S. Canara District (Karnataka).

Betasyrphus fletcheri Ghorpadé, 1994

Betasyrphus fletcheri Ghorpade, 1994, Colemania, 3: 8 (♂; "Ootacamund") [USNM, Washington, DC]

India: KL, KN, TN; Sri Lanka (Kertész 1910; Brunetti 1915, 1923; Shiraki 1930; Keiser 1958; Knutson et al. 1975; Ghorpadé 1981a, 1994, 2014b; Mahalingam 1988; Mitra et al. 2015; Thompson et al. 2017.)

Note: In my doctoral thesis (Ghorpade 1981a) I had written: "South Indian and Ceylonese specimens of *Betasyrphus* with narrow grey or yellow tergal bands, hitherto misidentified as *serarius* (Wiedemann) (e.g., Keiser 1958; Brunetti 1915, 1923), turned out to be an undescribed species, distinct from the similar north Indian ones (*aeneifrons, isaaci*), and is described here. The species name alludes to the late T. Bainbrigge Fletcher, Imperial Entomologist, who was primarily responsible for the development of Insect Systematics in India. He was himself a specialist on the Microlepidoptera and accumulated a vast amount of insect material from India and adjacent countries. This new species is dedicated to him as a token of my appreciation of his efforts, especially as his written works provided me with an adequate stimulus and zeal for taxonomic research on Indian insects." Keiser (1958: 192) included this as "*Metasyrphus serarius*" in his paper documenting specimens taken from Uva and Central Provinces. This species is restricted to southernmost India and Sri Lanka, being common on the hilly tracts in this zoogeographically distinct area, which Wallace had termed the "Ceylonese Subregion." My holotype is from Ootacamund (see Ghorpadé 1994: 8). My paratypes came from Yercaud 1370 m, Ootacamund 2350 m, Shembaganur 1800 m, Kodaikanal 1981 & 2250 m, and Kodaikanal grade 1600 m (Tamil Nadu), Manantoddy [= Manantavadi] 840 m, Kaikatty 937 m, Munnar 1900 & 2000 m (Kerala), and from Namunukuli (Sri Lanka), all taken from January to April and August to November. I have seen specimens (K. Ghorpadé unpubl.) from Mudigere ZARS, Attur, nr Bangalore! (Karnataka), and Yercaud, Salem district, Kallar 1250 m, Nilgiri Hills (Tamil Nadu). Ghorpade (1981b: 68) listed prey from southern India, but as of "*serarius.*" Thompson et al. (2017: 509) provide a synopsis of this small, but widespread genus in the Old World (19 species), with a single species in the PSWNG area.

Betasyrphus linga Ghorpadé, 1994

Betasyrphus linga Ghorpade, 1994, Colemania, 3: 8 (♂; "Nandi Hills") [USNM, Washington, DC]

India: KL, KN, TN (Brunetti 1915, 1923; Keiser 1958; Knutson et al. 1975; Ghorpadé 1981a, 1994, 2009, 2014b; Mitra et al. 2015.)

Note: In my doctoral thesis (Ghorpade, 1981a) I had written: "Along with *fletcheri* sp. nov. this is the only species of *Betasyrphus* occurring in southern India and Sri Lanka (?). Specimens determined as "*transversus*" in the BMNH (London) and those mentioned in De Silva (1961) are probably *linga*—see also remarks under *bazini* (Ghorpadé, 2009: 5, 2015b: 94). This species is close to *bazini*, but males can be easily separated on the basis of terminalia, which are very atypical in *linga*, being grotesquely enlarged (this is apparent even in undissected specimens) and asymmetrical. The species name "*linga*" refers to the male phallus in Sanscrit and is to be treated as a noun in apposition to the genus name. The females are very similar to *bazini*, but have the hind femur almost wholly black, and the hairs below the facial tubercle are black, not white as in *bazini*." See also key characters in Ghorpadé (1994: 8). It should occur also in Sri Lanka, but there are no positive records (specimens) as of now from that island. I have seen specimens (K. Ghorpadé unpubl.) from Biligirirangan WLS, Chamrajnagar district, and Hebbal, Bangalore (Karnataka).

Chrysotoxum baphyrum Walker, 1849

Chrysotoxum baphyrus Walker, 1849, List. Dipt. Br. Mus., 3: 542 (♂; "north Bengal, India") [NHM, London]

Chrysotoxum indicum Walker, 1852, *Insecta Saundersiana*, 3: 218 (♂; "East Indies") [NHM, London—examined]

Chrysotoxum sexfasciatum Brunetti, 1907, Rec. Indian Mus., 1: Pl. XIII, fig. 9; Brunetti, 1908, *ibid.*, 8: 89 (♀; "Rampore Chaka, Bijnor district, United Provinces") [ZSI, Calcutta—examined]

Chrysotoxum citronellum Brunetti, 1908, Rec. Indian Mus., 8: 90 (♂; "Kandy, Ceylon") [NHM, London]

Chrysotoxum testaceum Sack, 1913, Ent. Mitt., 2: 9 (ST ♂♀; "Yama and Tappani, Taiwan" [SMF, Frankfurt]; Ghorpadé 2012, Colemania, 32: 2. (as *n. syn.*)

Chrysotoxum mundulum Hervé-Bazin, 1923, Bull. Soc. Ent. Fr., p. 27 (♂; "Cochinchine") [MNHN, Paris—examined]

Chrysotoxum fasciatus Kohli et al. 1988, J. Insect Sci., 1(2): 115 (♀; "Panjab Univ., Chandigarh") [IARI, New Delhi—examined]; Ghorpadé, 2012, Colemania, 32: 2. (as *n. syn.*)

India: CG, KN, MH, TN; Sri Lanka (Brunetti 1907b, 1908, 1910, 1923; Sack 1913; Hervé-Bazin 1923, 1924; Cherian 1934; Keiser 1958; Violovitsh 1974; Knutson et al. 1975; Kohli et al. 1988; Ghorpadé 1981a, 1994, 2012, 2014b, 2015b, 2017, unpubl.; Mahalingam 1988; Patil et al. 2013; Mitra et al. 2015; Thompson et al. 2017.)

Note: See Ghorpadé (2012) for a detailed review of this genus in the Indian subcontinent. I had written "*C. baphyrum* is extremely variable in markings on the face, and in relative lengths of antennal segments, the opaque yellow scutellar patches, black markings on terga 2–5, posterior anepisternum (which may also be yellow in some specimens), black sternal fasciae (especially on sternum 4), facial black vitta, infuscation on wing, etc. The extreme bases of first and second basal cells may be narrowly bare, and the barette and anterior anepisternum may be black in some specimens." Types seen by me are as follows: holotype of *sexfasciatum* Brunetti is labelled "Simla, W. Himalayas, 7000 ft., 9-V-10, Annandale/*Chrysotoxum 6 fasciatum* Brun. Type ♂/TYPE/2159/HI/*Chrysotoxum 6 fasciatum = baphyrus* Walk., det. Brun. 1923/*Chrysotoxum baphyrus* Walker ♂, Ghorpade det. 1981." Holotype of *indicum* Walker is labelled "*Chrysotoxum* Type *indicum* Walk. [circular white green bordered label]/India/Ind./*Chrysotoxum baphyrum* Walker ♂, Ghorpade det. 1983." Holotype ♂ of *mundulum* Hervé-Bazin is labelled "COCHINCHINE, Chuachan le 26.I.1921, R. Vitalis de Salvaza/*Chrysotoxum mundulum* H.-B. ♂ Type [blue label, handwritten]/*Chrysotoxum baphyrum* Walker ♂. Ghorpade det. 1983." I also examined two ♂ Paratypes of *mundulum* in MNHN (Paris). I also saw a Homotype ♂ of *testaceum* Sack (in CNC, Ottawa; compared by Vockeroth; head lost), labelled "Chisan Park, Koashung, TAIWAN, 19.VII.1968, J.W. Boyes/Boyes Cytolog. Coll. #956, To remain in C.N.C. [yellow label]/comp. with STs *Chrysotoxum testaceum* Sack, Vockeroth"69 [blue label]/*Chrysotoxum baphyrum* WALKER ♂, Ghorpade det. 1983." Brunetti (1907b: Pl. XIII, fig. 9, 1908: 89–90) described and illustrated his new *sexfasciatum* from a female "taken 23—31-xi-07, at Rampore Chaka, Bijnor district, United Provinces." Brunetti (1908: 90–91, figure) also described and illustrated his new *citronellum* "described from a unique ♂ in my own collection, sent by Mr. E. Green; taken at Kandy (Ceylon), December 1907" (with missing antennae). He then wrote, "At first I thought it was the ♂ of my *sexfasciatum*, but am now convinced it is quite distinct. The three colours in the abdomen stand out very clearly." He mentioned this in his Notes on Ceylon Diptera (Brunetti, 1910: 171). In his *Fauna* volume (Brunetti, 1923: 296–298, figs. 62 and 63), he synonymized his *sexfasciatum* and *citronellum* along with Walker's *indicum* with this species giving it a wide range from NW. to NE. India, peninsular India, and Sri Lanka. Sack (1913: 9–10) described his *testaceum* from "Formosa" [= Taiwan] which Knutson et al. (1975: 326–327) kept as valid, and listed *baphyrum* (as "*baphyrus*"] and its synonyms. I (Ghorpadé 2012: 2) synonymized it with *baphyrum* after examining types. Hervé-Bazin (1923: 27) described his *mundulum* from "Cochinchine" mentioning its likeness to *sexfasciatum* Brun. and *indicum* Wlk., and this was also synonymized by Knutson et al. (1975: 327). Hervé-Bazin (1924: 299) gave notes on this (as "*Baphyrus*") and noted that his *mundulum* was a synonym. Keiser (1958: 210) included this as "*Chrysotoxum baphyrus*" in his paper documenting specimens taken from Uva and Central Provinces. Violovitsh (1974) gave a key to Palaearctic species of this genus and many illustrations of abdomen and antennae, but did not include *baphyrum*. Kohli et al. (1988: 115, 18, figs. 8 and 9) described it as new *fasciatus* from the Panjab Univ., Chandigarh, which I synonymized after examining the holotype female (Ghorpadé 2012: 2). Mahalingam (1988: 161) gave it as recorded from the Nilgiri Hills among ten species recorded by Brunetti (1923). Ghorpadé (1994: 8–9) keyed ten species of *Chrysotoxum* recorded in this subcontinent. Recently, Patil et al. (2013) recorded the sugarcane root aphid, *Tetraneura javensis* as its prey in northern Karnataka, India. Ghorpadé (2014b: 18, 2015b: 96) and Mitra et al. (2015: 68) listed it. Curiously this genus, which is speciose in the Palaearctic, does not occur in the PSWNG area and only reaches Taiwan and Sumatra, other than the Indian and Indochinese peninsulae [*cf* Thompson et al. (2017: 509)]. I have seen specimens (K. Ghorpadé unpubl.) from Maddur, Mandya district, UAS Hebbal, and GKVK, nr Bangalore, Mudigere ZARS 915 m, Chikmagalur district, Gunjanur, Holalkere, Balehonnur CCRI, Bangalore (Karnataka), and Moulvie estate, Yercaud, Shevaroy Hills, Kookal, and Palni Hills (Tamil Nadu). I also experienced a large number of flies active at Kunjappanai, Nilgiris, in May 2011, moving near ground level, but did not find any prey they were hunting for.

Citrogramma amarilla Mengual, 2012

Citrogramma amarilla Mengual, 2012, Zool. J. Linn Soc., 164: 106 (♂; "Kadamparai, Anaimalai Hills") [NNM, Leiden]

India: TN (Mengual 2012; Wyatt 1991; Ghorpadé 1994, 2014b, 2015b; Mitra et al. 2015)

Note: The type material of this new species erected by Mengual makes me suspicious of its identity. Only the holotype ♂ is from S. India, while all other paratypes come from Nepal, NE. India, Thailand, Malaysia, Indonesia, Laos, and the Philippines ! This kind of pan-Indian distribution is otherwise non-existent in non-palaeogene (common) species, and Mengual's analysis is certainly suspect. Only the holotype ♂ from Kadamparai 3500 ft. on the Anaimalai Hills deserves to be called *amarilla*, and all others from the north-east and the Indochinese, Indomalayan areas invariably are not *amarilla* and belong to other species. I would like to examine all other S. Indian *Ciitrogramma—flavigenum* Wyatt, 1991 (= *chola* Ghorpadé), *frederici* Mengual & Ghorpadé, 2012, and perhaps *henryi* Ghorpadé, 1994 species and confirm the true validity of *amarilla* Mengual. Ghorpadé (2014b: 18) and Mitra et al. (2015: 68) listed it. Thompson et al. (2017: 509) provide a synopsis of this group restricted to the Indomalayan and Australian regions (42 species), with 19 species in the PSWNG area.

Citrogramma flavigenum Wyatt, 1991

Citrogramma flavigena Wyatt, 1991, Oriental Insects, 25: 159 (♂; "Kodaikanal 2135 m") [BMNH, London]
Citrogramma chola Ghorpadé, 1994, Colemania, 3: 9 (♂; "Nandi Hills") [USNM, Washington, DC] syn. nov.
India: KL, KN, TN (Herve-Bazin 1924; Wyatt 1991; Mengual 2012; Ghorpadé 1994, 2014b, Mitra et al 2015)

Note: This species was described by Wyatt based on a single holotype ♂ from Kodaikanal 2135 m on the Palni Hills, S. India, taken in March 1936 by the B. M.-C.M. Expedition to South India in 1936. I had later named and described *chola* from Nandi Hills and Sidapur (Karnataka), Kaikatty and Munnar (Kerala), and Kodaikanal and Ootacamund (Tamil Nadu), and a ♂ "paratype" (of *citrinum* Brunetti) labelled "Sidapur, Coorg, 8000" [this altitude is wrong as Coorg does not have as high a mountain peak anywhere!], 15.ii.1917, T.R.N. "*Xanthogramma citrinum* Brun. Cotype ♂/D/8303" [IARI]. The "Cotype" of *citrinum* (Brunetti) labelled "Sidapur, Coorg, 8000," 15.iii.1917 (T.R.N.)" "*Xanthogramma citrinum* Brun. Cotype ♂," "D/8303" which I examined in IARI, New Delhi is not *citrinum*, and then appeared to belong to my *chola*, except that it had the yellow fascia on tergum 5 narrowly separated in the centre and the lunule was not brownish black [See also notes below on *C. frederici*]. Now that more species of *Citrogramma* have been discovered and described from peninsular India, this Coorg (Karnataka) specimen needs to be re-examined for proper placement. Most flies taken by me on November 19th, 2006 were caught while visiting flowers of *Polygonum chinense* L. (Polygonaceae) in the morning hours, coming in large numbers. The material of this species that I have examined gives its range as Western Ghats from Coorg south to Munnar (High Range), on the Nilgiri and Palni Hills, at altitudes over 2000 m. It was also found on the summit of the Nandi Hills (Nandidrug 1467 m) and should also fly on the other Eastern Droogs ("Ghats") like the Shevaroy Hills in particular (Yercaud area) from where no specimens have so far surfaced. It flies all around the year except no specimens were seen taken in February, June, and December. Most flies taken on 19th November 2006 were caught while visiting flowers of *Polygonum chinense* L. (Cross) (Polygonaceae) near Kookal in the morning hours, coming in large numbers. Males were also taken hovering in forest shade, but no prey association has so far been encountered!

Hervé-Bazin (1924: 290) wrote about *Xanthogramma citrinum* Brun., compared it to *Olbiosyrphus* Mik, and discussed *javanum* Wied. (an *Allograpta*), *obscuricorne* de Meij. (also an *Allograpta*), and *maculipleura* Brun. (another *Allograpta*!) and confused it with *citrinum* Brun. Only Vockeroth (1969: 92–95) noticed this was different and erected *Citrogramma* for it. Ghorpadé (1994) named and described *chola* from part of examined specimens from Nandi Hills, Ootacamund, Kaikatty, Munnar, and Kodaikanal. Wyatt (1991) had earlier described his "*flavigena*" from a single ♂ from "Kodaikanal, 2135 m, 23.iii.1936 (B.M.-C.M. Expedition to south India 1936)," and he separated it from my *chola* (named by Wyatt 1991: 166, as "sp. A") by the presence of metasternal hairs which he claimed sp. A (*chola*) did not possess. My re-examination of the entire large series I now have resulted in the discovery that all specimens had metasternal hairs (though some had just a pair at the posterior edge), and that body coloration was variable even in a series taken at the same locality and date. Mengual's examination of Wyatt's material led him to also believe that these two were distinct species, but that is evidently based on sparse comparative material in the BMNH, London. I therefore made chola a junior synonym of flavigenum. The species name "chola" was based on the Chola dynasty which, in historical times, ruled the area now known as northern Tamil Nadu, between the 9th and 13th centuries. Ghorpadé (1994: 9) also commented on "flavigena" in a footnote in that paper on Indian Syrphini. The material examined gives the distributional range of this southern Indian *Citrogramma* species as occurring on the Western Ghats from Coorg (?) or the Nilgiri Hills to Kodaikanal and Munnar on the Palni Hills, flying in some abundance above 2000 m. It was also found on the peak of the Nandi Hills (at Nandidrug 1467 m altitude) and so should also fly on the other higher Eastern Droogs ("ghats"), like the Shevaroys in particular, from where no specimen has so far been procured. This species flies all round the year, though no specimens have so far been caught in February, June, or December. Male flies were also taken hovering singly in forest shade in Longwood sholah, near Kookal, and on the Nandi Hills. Otherwise, flies were caught while visiting garden flowering plants especially in the botanical gardens in Ooty and Kody. However, no prey have yet been found nor any immatures of this species. I have seen specimens from Kodaikanal 2135 & 2250 m, Sowrikadu 1000 m, Kookal 2000 m, all Palni Hills, Naduvattam, Glenmorgan 2000 m, Kotagiri, Longwood sholah 1900 m, and Ootacamund 2350 m, all on the Nilgiri Hills,

Citrogramma frederici Mengual & Ghorpadé, in Mengual, 2012

Citrogramma frederici Mengual & Ghorpadé *in* Mengual, 2012, Zool. J. Linn. Soc., 164: 141 (♂; "Agumbe Ghat, 2000 ft.") [ZM, Amsterdam]

India: KN (Wyatt 1991; Mengual 2012; Ghorpadé 1994, 2014b; Mitra et al. 2015.)

Note: The specific epithet refers to Frederic Christian Thompson (USNM, Smithsonian Institution). Since his retirement in 2009, the government addresses him by his first name, Frederic, but he is hailed as Chris by his friends and colleagues. This new species is named, by Ximo Mengual and myself, after Chris in his honour, for dedicating his life to the study of the family Syrphidae and other Diptera, and for his guidance and advice to all of his students, especially us, his only postdocs at the Smithsonian Institution, Washington, DC, U.S.A. (X. Mengual and K. Ghorpadé), in pursuing our own research on this family of flies.

This species is very distinct and large, and only known from southern India. *Citrogramma frederici* keys out with *C. shirakii*, but they differ in male genitalia and wing microtrichia, being bare basally in *C. frederici*. A female specimen is very similar to *C. clarum*; a dark orange macula on the frons, wing partially bare, and dark protarsus separate this female from the ones of *C. clarum*. The species is only known from the holotype male and paratype female. This new species was originally based on a single male by Mengual. Later, I (Ghorpadé) sent my female specimen to Mengual, which was identified by me as a new species too. Thus both of us appear as co-authors of this new name. The holotype male is deposited in the Zoölogisch Museum, Amsterdam (Universiteit van Amsterdam, the Netherlands) and is labelled: "SOUTH INDIA/Mysore State/Shimoga Dist./Agumbe Ghat/2000 ft. –V -1981/T.R.Susai Nathan" "HOLOTYPE/*Citrogramma/frederici*/det. X. Mengual 2009" [red, second and third lines handwritten]." The type locality of the paratype ♀ is "INDIA: Karnataka, 4 km N. Agumbe, 621 m, 13°30′N, 75°05′E. 22.xi.1984, K. Ghorpadé B201 [1 ♀, USNM]. I found another ♂ specimen of *frederici* in the Pusa collection, IARI, New Delhi when I visited again in October 2012, with these labels—"Sidapur, Coorg, 8000," 15.iii.1917, D8303, T.R.N., *Xanthogramma citrina* Brun. ♂, Brunetti det. 1920 COTYPE."

Citrogramma henryi Ghorpadé, 1994

Citrogramma henryi Ghorpadé, 1994, Colemania, 3: 9 (♂; "Kanda-ela, Sri Lanka") [USNM, Washington, DC]

Sri Lanka (Brunetti 1915, 1923; Keiser 1958; Vockeroth 1969; Knutson et al. 1975; Ghorpadé 1981a, 1994, 2009, 2014b; Mitra et al. 2015.)

Note: In my doctoral thesis (Ghorpade 1981a) I had written: "This species is endemic to Sri Lanka and has a distinctive black "vitta" on the lower face adjacent to the gena. Keiser (1958) and Vockeroth (1969) misidentified their Ceylonese *Citrogramma* material as *citrinum* (Brunettti). However, Curran (1928, 1931a, b, 1942) had interpreted *citrinum* correctly. This species name is dedicated to G.M. Henry, an orthopterist and ornithologist, formerly curator at the Colombo National Museum. My interest in natural history (initially on birds, later on insects) germinated through the gift of Henry's book "*Birds of Ceylon*," in 1955, by M. Krishnan, himself a noted Indian naturalist and photographer/writer, and an authority on the Indian Elephant. Besides being a specialist on birds and grasshoppers of Sri Lanka, Henry was a most accomplished artist, and his works on these groups include excellent pen and ink sketches and colour paintings." He was on the B.M.-C.M. Expedition to South India in 1936 and wrote about all the locations the team visited and collected in. Keiser (1958: 193) included this as "*Xanthogramma citrinum*" in his paper documenting specimens taken from Uva and Central Provinces.

Dasysyrphus rossi Ghorpadé, 1994

Dasysyrphus rossi Ghorpadé, 1994, Colemania, 3: 9 (♂; "12km NE. Munnar") [CAS, San Francisco]

India: KL, TN; Sri Lanka (Brunetti 1915, 1923; Keiser 1958; Vockeroth 1969; Knutson et al. 1975; Ghorpadé 1981a, 1994, 2009, 2014b; Mitra et al. 2015.)

Note: In my doctoral thesis (Ghorpade, 1981a) I had written: "This species is very distinct and is the first *Dasysyrphus* known from southern India. The record of *orsua* (Walker) by

Keiser (1958) from Sri Lanka "Pidurutalagala" was perhaps of this new species, or an allied and yet undescribed one. It is named for late Dr. Edward S. Ross, California Academy of Sciences, one of its collectors and a world authority on the Embioptera. He has travelled almost the entire globe collecting insects, and besides being an excellent photographer, has done much to popularize insect study as an invigorating hobby. My holotype ♂ was from Kerala, 12 km NE. Munnar, 20.iii.1962, Ross & Cavagnaro; and a Paratype ♀ also with labels –"Sri Lanka, Bad. Dist., Haputale, 4.vi.1975, S.L. Wood & J.L. Petty, *Dasysyrphus serarioides* (de Meijere) det. F.C. Thompson, 1976" and my paratype label. See my diagnostic keys paper (Ghorpadé 1994: 9) also. Knutson et al. (1975: 312) gives *D. orsua* (Walker) also from "Ceylon," but that was undoubtedly an error for my new species. Thompson et al. (2017: 509) provide a synopsis of this north temperate genus with small extensions into the Neotropical and Oriental regions (50 species), with a single species in the PSWNG area.

Dideopsis aegrota (Fabricius, 1805)

Eristalis aegrotus Fabricius, 1805, Syst. Antliat., p. 243 (♀; "China") [UZM, Copenhagen—examined]

Syrphus fascipennis Macquart, 1834, Hist. nat. Ins. Dipt., 1: 537 (♂; "Java") [MNHN, Paris]

Didea ellenriederi Doleschall, 1857, Natuurk. Tijdschr. Ned.-Indië, 14: 407 (sex ?; "Amboina, Moluccas") [ZMA, Amsterdam ?]

Syrphus infirmus Rondani, 1875, Annali Mus. Civ. Stor. Nat. Giacomo Doria, 7: 423 (♀; "Sarawak, Borneo") [?]

Asarkina pura Curran, 1928, J. Fed. Malay St. Mus., 14: 230 (♀; "Ampang F.R., 600 ft., Selangor") [NHM, London—examined]; Ghorpadé, 2009, Colemania, 15: 9. (n. syn.)

Dideopsis hemipennis Hull, 1945, Ent. News, 56: 212 (♂; New Georgia") [CNC, Ottawa—examined] ?

India: AN, KL, KN, MH, MP, TN; Sri Lanka (Brunetti 1910, 1913, 1915, 1923; Cherian 1934; Beeson 1953; Keiser 1958; De Silva 1961; Coe 1964; Vockeroth 1969; Rao 1969; Gokulpure 1972; Knutson et al. 1975; Ghorpadé 1973a, 1974, 1981a, b, 1994, 2009, 2014b, 2015b, 2017; Datta & Chakraborti 1986; Puttannavar et al. 2005; Ramegowda et al. 2006; Mitra et al. 2015.)

Note: In my doctoral thesis (Ghorpade, 1981a) I had written that, "This is a very widespread Oriental species and very distinctive, on account of the black-banded wings. Besides the types of *aegrota*, *pura*, and *hemipennis*, I have seen over 400 specimens from all over the Oriental Region, including two specimens from Australia. I am convinced that there is only one, variable, species throughout this area (confirmed by dissections of male terminalia) and that *pura* and *hemipennis* are synonyms." But, being restricted to Guadalcanal, *hemipennis* Hull could be a distinct species. Being a typically "Oriental" species, restricted to forested habitats and probably Indo-Malayan in origin, it is noticeably absent from the arid parts of north-western India, Pakistan, and Afghanistan. In my nomenclature paper (Ghorpadé 2009: 9) I wrote: "This is one of the largest species of Syrphinae and very distinctive owing to its bright orange-yellow and black coloration and the broadly black-banded wings. After examining hundreds of specimens from all over the Oriental region, and even two specimens from Australia, which is a new regional record here, and doing dissections of terminalia, I remain convinced it is a single species with some irregular variation which is not linked to geographical areas." The species *fascipennis* Macquart (Java), *ellenrieder*i Doleschall (Amboina, Moluccas) and *infirmus* Rondani (Sarawak, Borneo) vide Knutson et al. (1975: 313), and *pura* Curran (Malaya) are all synonyms. Hull's *hemipennis* ("type" examined in CNC, Ottawa) is probably a *nomen nudum* (??) (see Curran 1947: 4), but also seems to be a synonym. In some specimens the black wing cross-band begins from the base of the wing hazily, but this is variable, as Brunetti (1913: 164, 1915: 210) also noticed. In this subregion, it has been recorded from all countries except Pakistan and Bhutan.

The holotype female of aegrota is labelled "E. aegrotus, e China "Sflueg"/*Dideopsis aegrotus* Fabr., det. FC Thompson, 1974/*Dideopsis aegrota* (Fabricius) ♀, Ghorpade det. 1980/ HOLOTYPE [red label]" [in UZM, Copenhagen]. The holotype female of pura is labelled "Type [circular red label]/ Pres. By Fed. Malay States Museum, B.M. 1934–74/Oct 1920, H.C. Abraham/Ex. Coll. F.M.S. Museum/*Dideopsis aegrota* (Fabricius), Ghorpade det. 1979" [in NHM, London]. The holotype male of hemipennis is labelled "New Georgia, 1944, C.O. Berg/Syrph. C./HOLOTYPE, *Dideopsis hemipennis* Hull [red label]/*Dideopsis aegrota* (Fabricius), Ghorpade det. 1982" [in CNC. Ottawa] Brunetti (1910: 171) wrote "*Syrphus aegrotus*, F. generally common during summer [in Sri Lanka], relegated to *Asarcina*." Then Brunetti (1913a: 164) wrote for *Asarkina aegrotus*, "One ♂, Sadiya, 28-xi-11. The wings are infuscated on the entire basal half instead of bearing, as in typical forms, a broad cross band, but other specimens in the Indian Museum have only the shortest possible clear space at the wing base." Then, for *Asarcina aegrotus*, F. he wrote (Brunetti, 1913b: 267), "One from Darjeeling, 1000–3000 ft. The head being crushed, the sex is indeterminable." This was at a time when the male terminalia were not even looked at as a character, hence his confusion. Nowadays, besides the holoptic head in males (for most species) the distinct male terminalia is also a pointer to the male sex. Females have dichoptic eyes and just a pointed abdominal tip, and no distinct genitalia which indicates a female. Brunetti (1915: 209) under "*Melanostoma cingulatum*, Big." which is an *Asarkina*, mentioned, "In the Indian Museum are two specimens marked "Melanostoma, hemiptera, Big." In that author's handwriting which are merely the common *Syrphus* (*Asarcina*) *aegrotus* F." In the *Fauna* volume, Brunetti (1923: 64–65) named it "*Asarcina aegrota*, Fabr." gave a description and stated, "As widely distributed as A. ericetorum in India and the East, both from hills and plains, at almost all seasons of the year." Cherian (1934: 698) recorded it from Coimbatore. Keiser (1958: 194) included this as "*Asarcina* (*Dideopsis*) *aegrota*" in his paper documenting specimens taken from Uva, Sabaragamuva, and all other, Northern, Eastern, Northwestern, Western, North Central, and Central Provinces. Coe (1964: 265) mentioned one male taken in east Nepal in August of "*Asarcina aegrota* (Fabricius)" "Common throughout the Oriental Region. Recorded from NORTH AUSTRALIA." Vockeroth (1969: 13–114, fig. 70, map 20) did not include northern Australia in his map for this genus, placed pura newly in this genus, and wrote, "No other genus in the Syrphini has a broad dark band covering ⅓ or more of the wing." Beeson (1953: 339) gave this as "commonly predacious on Aphidae" in Indian forests. Rao (1969: 787) listed *Aphis spiraecola* on citrus as prey. Gokulpure (1972: 848) recorded it from Madhya Pradesh. In my master's thesis (Ghorpadé, 1973a, 1974: 636), I reported it from the Bangalore area, and wrote: "This species was collected in winter (September to November) every year from 1970 to 1972 around Bangalore, but is decidedly uncommon here. The larvae were found feeding on the following aphids in nature: *Toxoptera citricidus* (Kirk.) on *Citrus* spp., *Toxoptera aurantii* (B.d. F.) on coffee, *Toxoptera odinae* on cashew, *Aphis craccivora* on avare (= field bean), and *Aphis gossypii* Glov. on bhendi or lady's finger." See also prey records in Ghorpade (1981b: 70–71). Datta & Chakraborti (1986: 54) recorded it from Idamalaya, Ernakulam district, and Konnakuzhi, Chalakudy, Trichur district (both Kerala) taken in December 1979. Ghorpadé (1994: 4) keyed out several species in a key and also gave notes on synonymy (Ghorpadé 2012: 9). Puttannavar et al. (2005: 44) and Ramegowda et al. (2006: 22) recorded this species preying on the sugarcane woolly aphid, *Ceratovacuna lanigera* in Karnataka, S. India. The larvae of this syrphid are large and broadly flattened (not narrow and cylindrical) and coloured black and white and very noticeable on aphid colonies on the plant. Knutson et al. (1975: 313), Ghorpadé (2014b: 19, 2015b: 103), and Mitra et al. (2015: 68) listed it. Ghorpade (2017: 108) listed it from Maharashtra, citing Brunetti (1923: 65). Thompson et al. (2017: 509) provide a synopsis of this small genus of two species, with one or two species in the PSWNG area. I have seen specimens (K. Ghorpadé, unpubl.) from Chidambaram, Maruthamalai, Coimbatore, Pannai kadu (Tamil Nadu), Hebbal, Attur, nr Bangalore, Jog Falls, Charmadi Ghat, Dandeli 481 m, Subramanya, and Mudigere (Karnataka).]

Eosphaerophoria dentiscutellata (Keiser, 1958)

Tambavanna dentiscutellata Keiser, 1958, Revue Suisse Zool., 65: 202 (♂; "Kandy, Deiyannewela, Ceylon") [NM, Basel]
Eosphaerophoria dentiscutellata: Knutson et al.: Catalog. Orient. Reg., 3: 313
Sri Lanka (Keiser 1958; Knutson et al. 1975: 313; Mengual & Ghorpadé 2010: 55: Mengual 2013: 388)

Note: In my doctoral thesis (Ghorpade, 1981a) I had written that, "Vockeroth (*in litt.*) has examined the single holotype ♂ and wrote as follows "Holotype in Basle. [*sic*] Apex of sternum 3 flat with about 15 short black setae in 2 or 3 rows in each side of mid-line [about 10 short black setae in a single row. Apex of tergite 4 flat] Femur 3 with a row of short, weak but distinct anteroventral and posteroventral spines on apical half. Wing bare on about apical 2/3 anteriorly. Scutellum flat basally with only spine slightly upcurved." This species was described by Keiser (1958: 202–204, fig. 3) from Kandy, Deiyannewela (Central Province), in Sri Lanka and was placed in the Tribe Bacchini. His record of *dentiscutellata* from Vietnam needs confirmation. I think *Eosphaerophoria* is an aberrant relative of *Citrogramma*; females being very similar to some species of the latter genus. The male terminalia are asymmetrical. Knutson et al. (1975: 313) listed it. I have examined a couple of other specimens of *Eosphaerophoria* species from the Oriental region. See Mengual & Ghorpadé (2010: 55) and Mengual (2013: 388) also. The latter paper claims to have seen a female specimen taken from S. India, but see in next species notes below. Thompson et al. (2017: 514) provide a synopsis of this group restricted to the Indomalayan region and New Guinea (11 species), with six species in the PSWNG area.

Eosphaerophoria sp. nov.

Eosphaerophoria dentiscutellata (Keiser, 1958): Mengual, 2013: 393: (misident.) (♀; "Cinchona," 150m, Kerala [sic]) [NNM, Leiden]
India: TN (Mengual, 2013)

Note: In Mengual (2013: 393, figs 11–13), it is claimed that the female specimen is of *dentiscutellata* from Sri Lanka. In my opinion, this is an incorrect assumption, and the S. Indian specimen belongs to an undescribed species and not to *dentiscutellata*, as Mengual himself comments about differences in his remarks (p. 394) and also based on the allopatry of the two populations. This will shortly be named and described in future. Mengual (*in litt.*) sent me a photocopy of the labels on the S. Indian specimen, and his published data (p. 393) needs correction to "India: Tamil Nadu: Anaimalai Hills, Cinchona 1050m, xi.1959, P. Susai Nathan."

Episyrphus arcifer (Sack, 1927)

Syrphus arcifer Sack, 1927, Stettin ent. Ztg., 88: 306 (♂♀; "Kankau and Fuhosho, Formosa") [SMF, Frankfurt]
Sri Lanka (Keiser 1958; Knutson et al. 1975: 313; Mengual & Ghorpadé 2010: 55; Mengual 2013: 388)

Note: Ghorpadé (1994: 10, 2014b: 19, 2015: 106) listed it from Sri Lanka. I had also written in my doctoral thesis (Ghorpadé, 1981a), "This species is fairly large for the genus *Episyrphus* and the pattern of markings on terga is peculiar. One other syrphid, *Milesia macularis* Wiedemann, from India (*op. cit.*) has an almost identical tergal pattern (see Brunetti, 1923: 266–268, fig. 52), and the wasp model must be a sympatric species of aculeate Hymenoptera.

The presence of a mesonotal hair collar, otherwise found only in *Asarkina, Dideopsis, Asiobaccha,* and some *Allobacha,* in this *Episyrphus* species is interesting." The extensive notes given under *balteatus* below contain much information that could apply to this Oriental, tropical species and should also be consulted. Thompson et al. (2017: 514) provide a synopsis of this wide-ranging group in the Old World (22 species), with four species in the PSWNG area.

Episyrphus balteatus (De Geer, 1776)

Musca balteata De Geer, 1776, Mem. pour serv. Hist. Ins., 6: 116 (♂♀; "Europe" = Sweden) [NRS, Stockholm]

Syrphus nectarea Fabricius, 1787, Mantissa insect., 2: 341 (LT♂, here designated; "Dania") [UZM, Copenhagen]

Syrphus pleuralis Thomson, 1869, in: K. svenska Fregatten Eugenies resa, 2(1): 497 (♂♀; "China") [NRS, Stockholm]

Episyrphus fallaciosus Matsumura, 1917, Ent. Mag., Kyoto, 2(4): pl. VI, fig. 13; 1917, ibid., 3(1): 18 (♀; "Honshū, Kiushū, Japan") [NIAS, Tsukuba ?]

India: CG, GJ ? (Brunetti 1907a, 1908, 1913, 1917, 1923; de Meijere 1908; Matsumura & Adachi 1917; Hervé-Bazin 1926; Curran 1926, 1928; Misra 1932; Bhatia & Shaffi 1933; Rahman 1940; Beeson 1953; Batra 1960; Coe 1964; Nayar 1964a, b, c, 1965a, b, 1966a, b, 1968a; Nayar & Nayar 1965; Rao 1969; Patel & Patel 1969a; Gokulpure 1972; Knutson et al. 1975; Patnaik & Bhagat 1976; Patnaik et al. 1977; Roy & Basu 1978; Peck 1988; Ghorpadé 1981a, b, c, 1994, 2009, 2014b, c, d, 2015, unpubl.; Claussen & Weipert 2003; Ghorpadé et al. 2011; Mitra et al. 2015).

Note: In my doctoral thesis (Ghorpadé 1981a) I had written, "This species was earlier thought to be widespread in the Palaearctic, Afrotropical and Oriental regions, even reaching Australia. However, the present study of a vast amount of material... has revealed that *balteatus, s. str.,* is confined to the Palaearctic, entering the Oriental only in NW. India and Pakistan. The markings on the sterna readily identify this species in both sexes." The many papers cited above may be consulted for more information, and the Indian states this has been recorded from are cited above. Papers dealing with localities outside of Pakistan and NW India (Himalayas) should correctly refer to the next species *viridaureus*; true *balteatus* (s. str.) occurs here only on the high W. Himalaya and in Pakistan and Afghanistan. I here formally designate the Lectotype male of *Syrphus nectarea* Fabricius (with terminalia in microvial) labelled "TYPE [red label]/Syntype *M. nectarea* F., Lyneborg det. 1979/LECTOTYPE, *Musca nectarea* Fab., Ghorpade des. 1980 [red label]/*Episyrphus balteatus* (De Geer) ♂, Ghorpade det. 1980" [UZM, Copenhagen]. A female Paralectotype is labelled "TYPE [red label]/S. nectareus/PARALECTOTYPE, *Musca nectarea* Fab., Ghorpade des. 1980 [yellow label]/*Episyrphus balteatus* (De Geer) ♀, Ghorpade det. 1980" [UZM, Copenhagen].

Brunetti (1907a: 169, 1908: 57) mentioned specimens taken at Simla, Theog, and Matiana on the Simla Hills, and also from localities in NE. India (Brunetti, 1913a: 159). He later wrote of "*balteatus*, DeGeer" that it was, "Very common in the Himalayas and also in the plains of India and Assam, extending to Java, China and Japan." But true *balteatus* has now been established only to fly in temperate climes on the high Himalayas, and that which is found on the plains and in NE. India and eastwards is *viridaureus* Wied., q.v. In his paper on Simla Hills Diptera, Brunetti (1917: 83–84) gave "Simla, 26-iv-07 (Capt. Evans), Theog, 27-iv-07; Valley of Sutlej River, 6-v-10," and repeated his erroneous view of it being widely distributed. In the *Fauna* volume, Brunetti (1923: 82–84, Pl. I, figs. 19 and 20) gave a description "partly based on Verrall's description of British, or at least European, specimens, and is partly from Indian examples." Therefore being mixed! Brunetti also wrote, "Specimens from Shanghai and Hankow, China, more nearly approach the European form," and mentioned de Meijere's "interesting notes on the typical form and the two varieties, *nectarinus* and *alternans*." See my notes on *viridaureus* below. Brunetti agreed that "*S. balteatus* is distinctly variable" and wrote that "I have generally adopted the principle that anything that looks like *balteatus*, is *balteatus*." He gave notes on variation and mentioned specimens from "Peshawur (19.iii.1913, Howlett), and from Pusa, 29.viii.1912 with different abdominal bands." De Meijere (1908: 297–299) should be seen for his comments on the varieties *nectarinus* Wied., and *alternans* Macq. which he separated in a key with the "typische Form." For *nectarinus*, he mentioned Matheran 800 m (Biro) [in Maharashtra], and also named *triligatus* Walk., and *viridaureus* Wied. very briefly without any useful notes on them. See also Curran (1926: 112) for notes on the Wiedemann types of *nectarinus* and *viridaureus* (= *alternans*), the former known only from China and the latter widely distributed in the Orient. Like true *balteatus* in Europe, this Indian (mostly *viridaureus*) species is exceedingly abundant and widespread and much applied entomology work has been done on its occurrence, prey records, and structure (see citations of authors listed above, q.v., loc. cit.). Without checking specimens identified as *balteatus* by each author, all such reports of this species in the papers mentioned below must be approached with caution; many could actually be *viridaureus*, especially in the plains areas! Misra (1932) saw flies hovering near aphid infested mango trees at Tirhut (Bihar). Bhatia & Shaffi (1933: 561–564, Pl. LXIV) described its life-history in detail and illustrated the fly and immature stages (but these are of *viridaureus*). They wrote that this "is one of the commonest species of the genus *Syrphus* and has been reported from all over India. In Pusa the fly is fairly common and is available during most of the months in the year. From January to March the flies of this species [also *balteatus* in these cold season months?] are seen in sufficient number in the fields and can be collected at any time in the day. They are seen hovering over flowers in search of food which they find in honey [sic] stored in the nectaries of the flowers." Besides feeding on cotton aphids, the larvae also "were found mainly feeding on young nymphs of coccids which attack the shoots of the cotton plant." Matsumura & Adachi (1917: 18) described *Episyrphus fallaciosus* Matsumura as new and wrote "Somewhat resembles to *E. balteatus* Deg." See also Ghorpadé (1981c: 89–91). Curran (1928: 198–200) included it in a key to the many Malayan "*Syrphus*" species. Rahman (1940: 71–72) noted it to be "commonest of Indian syrphids both in hills and plains; most abundant during January-March at Pusa." Beeson (1953: 340) wrote it is "a common hoverfly throughout India, predaceous on aphids and coccids," and quoted Bhatia & Shaffi (1933). Nayar (1964a, b, c, 1965a, b, 1966a, b) in a series of papers on "*Syrphus balteatus*" wrote on many aspects, morphology, anatomy, etc., of this fly, but his male terminalia (Nayar, 1965a: figs. 19 and 20) suggest it to be *viridaureus*, not *balteatus*, which should be noted and so corrected. Coe (1964: 162) wrote of *balteatus* and a "curious variety" from Nepal, the latter from "blooms of *Guizotia abyssinica*," and having "a grayish black longitudinal stripe" on terga which was "very noticeable" in the field. Nayar & Nayar (1965: 241) recorded it (as *Epistrophe*) from the Agra environs, but this also is probably misidentified for *viridaureus*. Nayar (1968a: 123–124) listed specimens taken near Dalhousie in September; specimens need to be examined to confirm "balteatus" or correct it. Rao (1969: 787) gave *Aphis gossypii*, *A. spiraecola*, and *Myzus persicae* as larval prey for this species (see Ghorpadé, 1981b: 65). Patel & Patel (1969b: 86) gave cabbage aphid as prey for balteatus at Anand in Gujarat, but this needs confirmation of species identity. Gokulpure (1972: 848) reported "*Syrphus balteatus*" from Damoh in Madhya Pradesh, but identity needs confirmation and is probably *viridaureus*. Ghosh (1974) recorded several aphid prey of this species (as "*balteatus*"). Patnaik & Bhagat (1976: 44) and Patnaik et al. (1977: 585) recorded *Episyrphus balteatus* from Puri district in Orissa, but this must certainly be *viridaureus*! Roy & Basu (1978: 165) gave bionomics of this species (as "*Syrphus*") from Kalyani in West Bengal on mustard aphids, but this again is certainly a misidentification for *viridaureus*. In their paper on Nepal Syrphidae, Claussen & Weipert (2003: 354–356) listed specimens taken of what purportedly were assumed to be *alternans*, *balteatus*, and *nectarinus* separately, treating these three as distinct species and separating them in a key. See Ghorpadé (2009: 9–10) for synonymy and a clear analysis of *viridaureus* and clinal variation in that species, "where larger and darker specimens (*nectarinus*) fly in the northern colder, wetter areas, and smaller, lighter [paler] specimens (*alternans*) in the hotter areas in the south, mainly in the more open, dry Indian peninsula." Knutson et al. (1975: 314–315), Peck (1988: 22), Ghorpadé (2014b: 19, 2015: 106), and Mitra et al. (2015: 68) listed it.]

Episyrphus divertens (Walker, 1856)

    Syrphus divertens Walker, 1856, J. Linn. Soc. Lond., 1: 1214 (♀; "Sarawak, Borneo") [BMNH, London]

    Syrphus claviger Sack, 1927, Stettin ent. Ztg, 88: 308 (♀; Fuhosho, Formosa) [SMF, Frankfurt]

    Sri Lanka (Knutson et al. 1975; Ghorpadé 2014b, unpubl.)

Note: I have seen specimens (K. Ghorpadé unpubl.) of *divertens* from Sri Lanka through Ms. Sujatha Mayadunagge taken in the course of her research work there, and this is a first record of that species in Sri Lanka (*q.v.*). It has also been recorded from Borneo, Taiwan, and Thailand (Knutson et al. 1975: 315). Ghorpadé (2014b: 20) listed it.

Episyrphus viridaureus (Wiedemann, 1824)

    Syrphus viridaureus Wiedemann, 1824, Analecta Ent., p. 35 (♂; "Batavia, Java") [UZM, Copenhagen—examined]

    Syrphus nectarinus Wiedemann, 1830, Aussereurop zweifl. Insekt, 2: 128 (LT ♀; "China") [UZM, Copenhagen—examined]; Ghorpadé 2009, Colemania, 15: 10. (as *stat. rev.*)

    Syrphus alternans Macquart, 1842, Dipt. Exot., 2(2): 89 (♂♀; "Coromandel, India") [MNHN, Paris]; Ghorpadé 2009, Colemania, 15: 10. (as *stat. rev.*)

    Syrphus triligatus Walker, 1857, J. Linn. Soc. Lond., 1: 19 (♂; "Mt Ophir, Malaya") [NHM, London]; Ghorpadé 2009, Colemania, 15: 10. (as *n. syn.*)

    Syrphus heterogaster Thomson, 1869, in: K. svenska Fregatten Eugenies resa, 2(1): 498 (♀; "China") [NRS, Stockholm—examined]; Ghorpadé 2009, Colemania, 15: 10. (as *n. syn.*)

    Syrphus balteatus var. *formosae* Sack, 1913, Ent. Mitt., 2: 5 (♂; "Formosa" = Taiwan) [SMF, Frankfurt ?]; Ghorpadé 2009, Colemania, 15: 10. (as *n. syn.*)

    Syrphus graptus Hull, 1944, Psyche, Camb., 51: 22 (♂; "Sozan, Formosa) [MCZ, Cambridge, MA]; Ghorpadé 2009, Colemania, 15: 10. (as *n. syn.*)

    Baccha (Baccha) bistriatus Kohli et al., 1988, J. Insect Sci., 1: 113, figs. 1–7 (♀; "Chandigarh") [IARI, New Delhi—examined]; Ghorpadé 2014, Colemania, 44: 20. (as *n. syn.*)

    India: AH, BI, CG, GJ, KL, KN, MH, MP, TN; Sri Lanka (Curran 1926, 1928; Misra 1932; Coe 1964; Patel & Patel 1969a; Vockeroth 1969; Lambeck & van Brink 1973; Knutson et al. 1975; Ghorpadé 1981a, b, c, 1994, 2009, 2014b, c, d, 2017, unpubl.; Kohli et al. 1988; Peck 1988; Mitra et al. 2015; Claussen & Weipert 2003; Wright & Skevington 2013).

Note: The extensive notes given under *balteatus* above contain much information that could apply to this Oriental, tropical species and should also be consulted. I examined the holotype male of *viridaureus* Wiedemann labelled "♂/Mus. Westerm./TYPE [red label]/*S. viridaureus* Wied., Batavia, Aug. 1815/HOLOTYPE [red label]/*Episyrphus viridaureus* Wied., Det. FCThompson, 74/*Episyrphus viridaureus* (Wiedemann) ♂, Ghorpade det. 1980" in UZM, Copenhagen. I here formally designate the Lectotype female of *Syrphus nectarinus* Wiedemann Fabricius labelled "♀/Mus. Westerm./TYPE [red label]/*S. nectarinus* Wied., China, Trentepohl/*Episyrphus nectarinus* Wiedemann, Det. FCThompson, 1974/*Episyrphus viridaureus* (Wiedemann) ♀, Ghorpade det. 1980" in UZM, Copenhagen. Curran (1926: 112) noted that *nectarinus* Wiedemann "apparently is a good species," but placed alternans as a synonym of *viridaureus* Wiedemann after studying types. See Ghorpadé (1981c, 2009) for detailed notes on the identity of this species. Curran (1928: 198–200) included alternans and nectarinus separately in a key to the many Malayan *Syrphus* species.

In my doctoral thesis (Ghorpadé 1981a) I had written, "There has always been a controversy regarding the status of *balteatus* and other closely related "species" like *alternans*, *nectarinus*, *viridaureus*, etc., in the Palaearctic region. Brunetti (1923: 82–84) felt that only one species, *balteatus*, was involved, and that the others were only variations. Curran (1928, 1931a, b) felt careful studies of larvae and morphology of adults along with data on their life cycles, etc., would show that several "cryptic" species were involved. The present study has resulted in an "in-between" situation. Only one widespread and highly variable species occurs in the Oriental part of the range, it being replaced in the Palaearctic, temperate areas, by *balteatus*. The Oriental species is to be called viridaureus by priority and many nominal species, described for extreme variant specimens, are to be synonymized as given above. The Indochinese Peninsula and the Malay Archipelago contain other, more distinct species of *Episyrphus*, like *arcifer*, *contax*, *divertens*, *obligatus*, etc., but, except for *arcifer* and *divertens*, none of these occur on the Indian subcontinent. I have seen specimens of *divertens* from Sri Lanka through Ms. Sujatha Mayadunagge taken in the course of her research work there, and this is a first record of that species in Sri Lanka (q.v.). The typically Palaearcic *balteatus* enters the Indian area only in Kashmir and in the higher Himalaya in Himachal Pradesh and Uttar Pradesh [= Uttarakhand now], and of course, in northern Pakistan and Afghanistan. However, it is easily separated from the more widespread Oriental *viridaureus* by diagnostic markings on the sterna (Ghorpadé 1994: 10). See Wright & Skevington (2013) for Australian records.

The many papers cited above may be consulted for more information, and the Indian states this species has been recorded from are cited above. Before my revisionary work, not many papers (authors) have even referred to *viridaureus*, this name being overlooked (!), and just *nectarinus* and *alternans* sometimes being used for variant specimens, as detailed in my notes to *balteatus* above (q.v.). My key (Ghorpadé 1994: 10) will help authors to finally approach reality and find correct names easily. Knutson et al. (1975: 315), Peck (1988: 22–23), Ghorpadé (2014c: 20, 2015b: 108), and Mitra et al. (2015: 68) listed it. See Thompson & Rotheray (1998: 95) for a key to this genus in the Palaearctic. I also saw specimens (K. Ghorpadé unpubl.) from Chintamani, Kolar Dist., Nandi Hills, GKVK and Hebbal, Hesaraghatta, Bangalore, Balehonnur 867 m, Skandagiri Hills 1288 m, Chikballapur, Someshwar 2000," and Nagody 2500," S. Canara Dist., Ramanagaram, Dharwad, Chamrajnagar (all in Karnataka), Thandigudi 1299 m, Maruthamalai 1000," Coimbatore, Yercaud 4500," Shevaroys, Chidambaram (in Tamil Nadu), Nelliampathi Hills 1110 m (Kerala), and in Udaipur (Rajasthan),]

Ischiodon scutellaris (Fabricius, 1805)

Scaeva scutellaris Fabricius, 1805, Syst. Antliat., p. 252 (LT ♂; "Tranquebar, India") [UZM, Copenhagen—examined]

Syrphus coromandelensis Macquart, 1842, Dipt. exot., 2(2): 80 (♂; "Cote de Coromandel") [MNHN, Paris]

Sphaerophoria annulipes Macquart, 1855, Mem. Soc. Sci. Agric. Arts, Lille, (2) 1: 96 (LT ♀; "Marquesas Is.") [UM, Oxford—examined]

Sijrphus splendens Doleschall, 1856, Natuurk. Tijdschr. Ned.-Indië, 10: 410 (LT sex ?; "Java") [ZM, Amsterdam]

Syrphus erythropygus Bigot, 1884, Annls Soc. Ent. Fr., (6) 4: 87 (LT ♂; "Indes") [UM, Oxford—examined]

Melithreptus novaeguineae Kertész, 1899, Természetr. Füz., 22: 178 (♂; "Friedrich-Wilhelmshafen and Erima, New Guinea") [MNH, Berlin]

Ischiodon trochanterica Sack, 1913, Ent. Mitt., 2: 6 (♂; "Kanshizei, Polishe, Suihenkyaku, Tainan, and Takao, Formosa") [SM, Frankfurt]

Melithreptus ogasawarensis Matsumura, 1916, *Thousand Insects of Japan*, p. 23 (♂; "Bonn Is.) [NIAS, Tsukuba ?]

Ischiodon platychiroides Sack, 1913, Ent. Mitt., 2: 6 (LT ♂; Antimonan, Philippines") [ZMUH, Helsinki—examined]

Epistrophe magnicornis Shiraki, 1963, *Insects Micronesia*, 13: 141 (♀; "Wena, Truk, Caroline Is.") [USNM, Washington, DC—examined]

Sphaerophoria macquarti van der Goot, 1964, Beaufortia, 10: 220 (nom.nov. for *annulipes* Macquart, 1855 not 1842)

India: AH, AN, BI, CG, GJ, KL, KN, MH, MP, OR, RJ, TN; Sri Lanka (Maxwell-Lefroy 1909; Kertész 1910, 1913; Sack 1913; Fletcher 1914, 1916; Brunetti 1915, 1923, 1925; Hervé-Bazin 1924, 1926; Curran 1928, 1931a; Bhatia & Shaffi 1933; Beeson 1953; Lal & Gupta 1953; Lal & Haque 1956; Deoras 1957; Bindra & Saxena 1958; Upadhyaya & Soares 1964; Rao 1969; Patel & Patel 1969a, b; Kalyanam 1970; Gokulpure 1972; Joshi & Sharma 1973; Ghorpadé 1973b, 1974, 1981a, b, 1994, 2014b, c, d, 2017, unpubl.; Ghosh 1974; Knutson et al. 1975; Patel et al. 1976; Patnaik & Bhagat 1976; Patnaik et al. 1977; et al. 1987; Peck 1988; Mitra & Parui 2002; Puttannavar et al. 2005; Ramegowda et al. 2006; Láska et al. 2006; Ghorpadé et al. 2011.)

Note: The first few synonyms given above under this species, of names proposed by Fabricius, Macquart, Doleschall, and Bigot and the museums where the types have been deposited (in Denmark, France, Great Britain, and Holland), show the first countries that began trading and colonial activities in India, and the people associated with sampling early specimens of insects in this subcontinent. In my doctoral thesis (Ghorpadé 1981a) I had written,

"This is one of the most widely distributed and commonly encountered species of the Syrphini in the Indian subcontinent, especially at lower elevations. The larvae feed on a wide variety of prey (Ghorpade 1981b: 69–70), and the adult flies are invariably present wherever plants are infested with aphids. It has been referred to as *Xanthogramma* and *Sphaerophoria*, species of which genera also possess a striking yellow lateral mesonotal margin and black and yellow abdominal markings." After showing that *aegyptius* (Wiedemann) was distinct and confined to S. Europe and Africa (Vockeroth 1969: 105–106), he wrote an excellent paper (Vockeroth 1971: 1635) elucidating the relationships and identities of Old World *"Sphaerophoria"* species, synonymizing many species (of Frey, 1946, and Shiraki, 1963, for example) based on his examination of types. Yerbury and Brunetti both stated (see Brunetti 1923: 314) that this species was never seen by them on the wing, which is strange. The male fly never hovers in swarms and females are adept at seeking out prey, flying in between low-growing plants and are thus not generally visible easily to the collector, as they tend to fly very low, like species of Paragini, close to the ground (males also) and always in the immediate proximity of grass or low vegetation or crop plants, and this habit may be responsible for their "invisibility" to the uninformed, inexperienced collector. The many papers cited above and noted below may be consulted for more information and also for the Indian states this species has so far been recorded from (see above). I may also mention, that as noted under *Agnisyrphus angara* (*q.v.*, *op. cit.*), one female specimen of *I. scutellaris*, taken in Delhi [NHM, London], was found to also carry the incorrect labels "N.E. INDIA, Delhi, iii.47, T. Jermyn, B.M. 1949–53/'B.'" So even Delhi was labelled as "N.E. India" (!), and it is transparently clear that the NHM, London museum staff had then erroneously labelled T. Jermyn and R.C. Jermyn collections as "N.E. India." This error requires to be remembered and corrected by researchers working with the Jermyn Indian material accessed in the BMNH, London museum in 1949.

I have examined the following types and also designated some Lectotypes during my doctoral and postdoctoral research, and now validate the latter: of *scutellaris* Sack, labelled "TYPE [red label]/*S. scutellaris*, e Tranqueb. Daldorf/LECTOTYPE, *Scaeva scutellaris* Fab., Ghorpade des. 1980 [red label]/*Ischiodon scutellaris* (Fabricius) ♂, Ghorpade det. 1980" [UZM, Copenhagen]; of *platychiroides* Sack, labelled "Philipp, Antimonan, XI.1915/Spec. typ. [pink label]/Mus. Zool. H:fors, Spec. typ. No 14038, *Ischiodon platychiroides* Frey/*Ischiodon scutellaris* (Fabricius) ♂, Ghorpade det. 1980" [ZMUH, Helsinki]; of *magnicornis* Shiraki, labelled "Truk Atoll, Moen Is, V-31–46, H Townes 419/on *Sorghum vulgare*/Holotype [red label]/*Epistrophe magnicornis* sp. n., det. T. Shiraki/*Ischiodon scutellaris* (Fabricius) ♀ Ghorpade det. 1983 [USNM, Washington, DC]; of *annulipes* Macquart, labelled "*Sphaerophoria annulipes*. ♀. Macq. n. sp./*Sphaerophoria annulipes*. LectoTYPE Macq. designated Vockeroth"71 [red label]/*Ischiodon scutellaris* (Fabricius) ♀, Ghorpade det. 1983" [UM, Oxford]; and of *erythropygus* Bigot, labelled "*S. erythropygus* Big = *Ischiodon scutellaris* F./*S. erythropygus*. ♂., Indes, J. Bigot [black bordered rectangular white label]/LECTOTYPE, *Syrphus erythropygus* BIG., Ghorpade des. 1983 [red label]/*Ischiodon scutellaris* (Fabricius) ♂, Ghorpade det. 1983" [UM, Oxford].

Howlett (in Maxwell-Lefroy 1909: 611, Pl. LXIV) made mention of this abundant syrphid on the Indian plains and illustrated it in that classic work *Indian Insect Life*. In his world catalogue, Kertész (1910) gave "Coromandel" [= coast] as locality record in India. Fletcher (1914, 1916) cited this species in his economic entomology works of that early period in British India. Sack (1913: 5–7, figs. 3–4) erected a new genus *Ischiodon* for his new species *trochanterica* from Formosa [= Taiwan] which is now a synonym of *scutellaris* Fabricius. Kertész (1913: 273–274) recorded it also from Taiwan and confirmed the correct assignment of *scutellaris* Fabr. and *aegyptius* Wied. to *Ischiodon* Sack. Brunetti (1915: 217) was probably unaware of this new genus and cited "*Sphaerophoria scuttellaris*, F." [sic] specimens in the Indian Museum, and gave localities "from Maho, base of Nepalese Himalayas; 17-iii-09; Ferozepore, 28-iv-05; Agra, 3-iv-05 [both Brunetti]... Bhanwar, 26-ii-07; Bettiah, Champaran, 8-iii-08; Dharampur, 24-ii-07... Kulti Sitarampore, 10-viii-09 [Lord]. I also took it myself at many places in India and the East but exact data are not available." Earlier he had cited this species (Brunetti 1913a: 164) from NE India, but later (Brunetti 1915: 217) mentioned this "clerical error" he had made for *S. javana* Wied., while reporting on the Diptera of the Abor Expedition. Brunetti (1917: 85) briefly stated that, "the species in this genus [*Sphaerophoria*] offer exceptional difficulties, beyond the two common ones, *scutellaris*, F. and *javana*, Wied." In his *Fauna* volume (Brunetti, 1923: 97–99, fig. 17), he cited it as "*Ischiodon scutellaris*, Fabr.," gave the extensive synonymy and a detailed description and figure of its habitus. He wrote, "The species is common in many parts of India and Assam practically all the year round, and is one of the most widely distributed Syrphids in the East... The species has been bred more than once from larvae predaceous on Aphidae on chrysanthemums and watermelons." Later (Brunetti 1925: 76) he corrected his earlier statement "that the stick-like processes are present on the trochanters in both sexes is incorrect, they being present in the ♂ only, as stated by Sack." Hervé-Bazin (1924: 290–291) wrote on this genus and corrected Brunetti's (1923) synonymy, and recognized only two species of *Ischiodon*. He went on to discuss the synonymy (Hervé-Bazin 1926: 69–70) and gave China and Hanoi [= Vietnam] as the then easternmost known provenances of this species in the Orient. Curran (1928: 243–244, 1931a: 321) gave Malay records, a description and differences of *Ischiodon* from *Sphaerophoria*, and the two known species. Bhatia & Shaffi (1933: 548) gave aphids on Solanum, Chrysanthemum, Calotropis, watermelon, cotton, cabbage, wheat, sissoo, and mustard as prey at Pusa (Bihar). Rakshpal (1945: 235) reported it from Gwalior, Madhya Pradesh. Beeson (1953: 339) gave this as "commonly predacious on Aphidae" in Indian forests. Deoras (1957: 306) listed it as *Xanthogramma (Ischiodon) scutellare* Fb., from the then suburban Kurla area of Bombay [= Mumbai]. Bindra & Saxena (1958) gave larvae of this species feeding on mustard aphid at Gwalior (M.P.). Upadhyaya & Soares (1964) wrote on the vena spuria in the wing of this species as a diagnostic character of Syrphidae. Rao (1969) gave *Aphis gossypii* and *A. spiraecola* as larval prey for this species (see Ghorpadé 1981b: 65). Patel & Patel (1969a, b) gave the bionomics of *Xanthogramma scutellare*, prey records, and larval parasitoids studied at Anand in Gujarat. Kalyanam (1970) gave prey, but as for "*Syrphus latifasciatus,*" in error. Gokulpure (1972: 848) recorded it from Damoh in Madhya Pradesh." Joshi & Sharma (1973) gave it from Udaipur and occurring throughout Rajasthan, and gave notes on toxicity of insecticides. Patel et al. (1976) listed *Aphis craccivora* as prey in Kapadwanj taluk in central Gujarat as prey. Patnaik & Bhagat (1976: 43) and Patnaik et al. (1977: 585) gave prey from Puri district in Orissa. Ghosh (1974) recorded several aphid prey of this species. I (Ghorpadé 1981b: 69–70) gave a long list of prey recorded in the Indian subcontinent. This species has also been taken in the hot and arid desert of Thar. It was found visiting flowers of *Anogeissus* sp. and *Cassia tora* in the Jessore and Balaram-Ambaji WLS in northern Gujarat (Mitra & Parui 2002: 45). Puttannavar et al. (2005: 44) and Ramegowda et al. (2006: 22) recorded its larvae feeding on the sugarcane woolly aphid, *Ceratovacuna lanigera* in northern Karnataka. Knutson et al. (1975: 315), Peck (1988: 23–24), Ghorpadé (2014b: 20, 2015b: 7), and Mitra et al. (2015: 68) listed it.

Láska et al. (2006: 651) in a recent paper transferred *scutellaris* to the genus *Simosyrphus* Bigot, 1882 along with *grandicornis* (Macquart, 1842) and *aegyptius* (Wiedemann, 1830), also studying and treating what they announced as similar genera, i.e., *Scaeva (Semiscaeva)* Kuznetzov, 1985, with *S. (S.) selenitica* (Meigen, 1822), *S. (S.) mecogramma* (Bigot, 1860), and *S. (S.) dignota* (Rondani, 1857), and also *Scaeva* (s. str.) *pyrastri* (Linnaeus, 1758), *S.* (s.str.) *albomaculata* (Macquart, 1842), and *S.* (s. str.) *latimaculata* (Brunetti, 1923), and studying molecular and immature stages characters of all. These revolutionary changes in phylogeny require to be carefully checked and corroborated before acceptance by me and perhaps other specialists. Use of minutely differing characters to "split" generic taxa is dangerous as my training has been to look at similarities among genera and differences between species. There is no argument in accepting that *Scaeva, Metasyrphus, Eupeodes, Ischiodon,* and *Simosyrphus* are close phylogenetically (see Vockeroth 1969 and Ghorpadé 2007), but "lumping" *Ischiodon* and *Simosyrphus* and splitting subgroups of *Scaeva* subgenerically is to my mind risky and confusing, not clear. There still is a debate, in my mind at least, about *Metasyrphus* and *Eupeodes* and other genus groups of that clade so I prefer to let taxonomy remain as it was ("let it be") and not upset long followed genus groupings, which Vockeroth (1969: 62–68, 70–72, 102–106) masterfully demonstrated in that excellent generic revision of his of the tribe Syrphini. Readers may note that my own *Scaeva*—group (Ghorpadé 2007: 16) contained *Eupeodes, Ischiodon, Lapposyrphus, Macrosyrphus, Metasyrphus, Scaeva,* and *Simosyrphus*! Should these be "lumped," "split," or left as they are with their characteristic differences as acknowledged so far? However, occurrence of this species even in the temperate, cold, high plateau of Afghanistan, where *Metasyrphus* and *Scaeva* rule the sky, biodiverse in the temperate regions, but disappearing in the hot, steamy tropics, where *Ischiodon* rules, is notable. Also that *scutellaris* was first named and described as a *Scaeva* by that master, J.-C. Fabricius, the celebrated student of Carolus Linnaeus, may suggest true relationships recognized by these legends, in early times, and justify Láska et al.'s new proposition after all? Name change notwithstanding, being irritable to economic entomologists and bibliographers, who have no clue of the principles and workings of taxonomy and who also do not consider, or understand, the real benefits of good, correct taxonomy to all of applied science.

Lastly, in my own recent papers (Ghorpadé et al. 2011: 80–81, Ghorpadé 2014d: 10, 2014e: 11), more records and notes have been given which should be consulted. See also papers of other authors cited below the synonymy, but not expanded on above. Knutson et al. (1975: 315), Peck (1988: 23), Ghorpadé (2014b: 20, 2015b:110–113, 2017: 108), and Mitra et al. (2015: 68) listed it. See also Sack (1932b: 204) and Thompson & Rotheray (1998: 99) for discussion and key to this genus in the Palaearctic. Thompson et al. (2017: 509) provide a synopsis of this common widespread group of two species found throughout the Afrotropics and the Indomalayan region with extensions to the south Palaearctic, and a single species known in the PSWNG area. I also saw specimens (K. Ghorpadé unpubl.) from Attur farm, Sadahalli, GKVK, Hebbal, Nandigrama, Bangalore, Nandi Hills, Shimoga, Hagari, Koppal, Gulbarga 474 m, Raichur 388 m, Dharwad 700 m, Mudigere 915 m (Karnataka), Coimbatore, Kodaikanal, Thandikudi, Yercaud, Salem dist., Villupuram, S., Arcot dist., Vandalur, Chingleput (Tamil Nadu), Samalkota, Bapatla (Andhra), Naloti, Ira Dharan, Kondhwal (Maharashtra), Udaipur (Rajasthan), and from Jirkatang, S. Andaman (A&N Islands)]

Macrosyrphus confrater (Wiedemann, 1830)

Syrphus confrater Wiedemann, 1830, Aussereurop. Zweifl. Insekt., 2: 120(♀; "China") [UZM, Copenhagen—examined]

Syrphus mundulus Walker, 1852, *Insecta Saundersiana,* 3: 23o (♂; "East Indies") [NHM, London]

Syrphus cranapes Walker, 1852, Insecta Saundersiana, 3: 231 (LT♀; "East Indies") [NHM, London]; Ghorpadé 2009, Colemania, 15: 10. (*LT designated*)

Syrphus macropterus Thomson, 1869, in: K. Svenska fregatten Eugenies resa, Zool., Dipt., 2(1): 498 (♀; "China") [NRS, Stockholm]; Ghorpadé 2009, Colemania 15: 10. (as *n. syn.*)

Syrphus trilimbatus Bigot, 1884, Annls Soc. Ent. Fr., (6) 4: 86 (♂; "Indes") [UM, Oxford]

Syrphus torvoides de Meijere, 1914, Tijdschr. Ent., 57: 155 (♀; "Nongkodjadjar, Java") [ZM, Amsterdam—examined]; Ghorpadé 2009, Colemania 15: 11. (as *n. syn.*)

Syrphus (Metasyrphus) okinawae Matsumura in Matsumura & Adachi, 1917, Ent. Mag., Kyoto, 2: Pl. VI, fig. 16; ibid., 3: 23 (LT ♂ des. Vockeroth, 1973, "Kumamota, Okinawa I.") [NIAS, Taihoku or EIHU, Sapporo ?]; Ghorpadé 2009, Colemania 15: 11. (as n. syn.)

India: BI, CG, GJ, KL ?, KN, MH, RJ, TN; Sri Lanka (Sack 1913; Brunetti 1923; Hervé-Bazin 1924, 1926; Curran 1928, 1931a; Bhatia & Shaffi 1933; Keiser 1958; Beeson 1953; Nayar 1968a; Rao 1969; Vockeroth 1969, 1973; Ghorpadé 1973b, 1981a, 1994, 2009, 2014b, c,d, 2017, unpubl.; Ghosh 1974; Patel et al. 1975; Knutson et al. 1975; Diller 1977; Mitra et al. 2015).

Note: In my doctoral thesis (Ghorpade 1981a) I had written, "A large species, widely distributed in the Oriental region. Several "species" have been described, based on extreme variant specimens, but my studies were of a large material from India and SE. Asia, including holotype of *confrater*. Vockeroth's notes (*in litt.*) on the types of *cranapes, macropterus, mundus, torvoides,* and *trilimbatus* tend to support the treatment of *confrater* as a single, highly variable and widely distributed species. Hence I have synonymized all other species in Vockeroth's (1969: 65) list under *confrater* Wiedemann. The type-species of the "subgenus" *Macrosyrphus* (of *Syrphus*) Matsumura, *okinawae* Matsumura from the Ryukyu Islands, is certainly another synonym; Vockeroth (1973: 1075–1076) was unable to find differences in the male terminalia and stated character differences within the variability of *confrater*. This species is an important predator of *Adelges* spp. (Homoptera: Adelgidae) and of *Eriosoma lanigerum* (Homoptera: Eriosomatidae) attacking conifers and apples, respectively, and is mainly distributed on the Himalayas and adjacent mountains. Its occurrence in peninsular India and other southern climes may possibly be by accidental transportation there along with its prey on the host plant ?" I have seen and taken many large flies of this species on the Ghats which appeared to be in their "home territory." At the CNC, Ottawa, I examined two ♀ labelled "B65–101(1), India, 9/viii/95/*Myzus* sp. ? *ornatus*/69.-5816/5815." See also my notes (Ghorpadé 2009: 10–11) on synonymy, and a list of its recorded prey (Ghorpadé 1981b: 67–68). The many papers cited above may be consulted for more information, and the Indian states this has been recorded from are cited above. Some of them are mentioned here: Sack (1913: 2) recorded it from Taiwan as a *Didea*! Brunetti (1923: 92–94; Pl. II, fig. 17) described and illustrated this species (as *Syrphus*) and mentioned some apparently alien characters of some specimens. Hervé-Bazin (1924: 290, 1926: 66) gave notes on Brunetti's Fauna and separated it from *nitidicollis* Meigen, which is now placed in *Epistrophe*. Curran (1928: 203, 1931a: 315) gave a description and included it in a key (as "*confrator*") to Malayan *Syrphus*. He also wrote, "This species and the preceding [*chrysotoxoides* Curran], in the strict sense, belong to the genus *Dideoides* Brun. This genus [*Dideoides*] cannot stand on the characters proposed by Brunetti, although it is quite valid if we limit the genus *Syrphus* to those species with the lower lobe of the squamae pilose. If that is done *Epistrophe* of Walker becomes the next available name, but this might easily be limited to those species without any trace of a raised abdominal margin." Bhatia & Shaffi (1933: 565–566, Pl. LXV) gave its life history and recorded prey species, these from most parts of India. Beeson (1953: 340) noted this (as "*Syrphus*") feeding on aphids on the hills and the plains, quoting Bhatia & Shaffi (1933). Keiser (1958: 191) included this as *Syrphus confrater* from Asgiriya, Kandy, in the Central Province. Vockeroth (1973: 1075–1076, fig. 1) mentioned, "It belongs to the *confrater* group; the name *Macrosyrphus* is available should this group be given generic or subgeneric status. As first reviser I synonymize *Macrosyrphus* with *Metasyrphus*." He illustrated the abdomen of *okinawae* Matsumura and kept it separate from *confrater*, but I synonymized it and raised *Macrosyrphus* to the status of a good genus (Ghorpadé 2009: 11). Diller (1977) recorded its *Diplazon* sp. (Hymenoptera: Ichneumonidae) parasitoids from India. Rao (1969) gave *Aphis gossypii* and *A. spiraecola* as larval prey for this species (see Ghorpadé 1981b: 65). Patel et al. (1975: 40) listed *Aphis craccivora* as prey in Gujarat. Puttannavar et al. (2005: 44) and Ramegowda et al. (2006: 22) recorded it as a predator of the sugarcane woolly aphid, *Ceratovacuna lanigera* in northern Karnataka. I examined the holotype of *confrater* Wiedemann labelled "♀/Mus. Westerm./TYPE [red label]/*S. confrator* [sic] Wied., China, Trentepohl/*Metasyrphus confrater* Wiedemann. Det. FCThompson, 1974/*Macrosyrphus confrater* (Wiedemann) ♀, Ghorpade, det. 1980" [UZM, Copenhagen]. Also of *torvoides* de Meijere labelled "E. Jacobson, Nongkodjadjar, JAVA, Jan 1911/*Syrphus torvoides* det. de Meijere, Type/TYPE [red label]/*Macrosyrphus confrater* (Wiedemann) ♀, Ghorpade, det. 1982" [ZMA. Amsterdam]. My studies of extensive material from India and SE Asia, and holotypes of the synonyms listed above, tend to support the treatment of *confrater* as a single, highly variable and widely distributed, peregrine, species. Even male terminalia are subject to variation as was observed in three series of specimens of populations from Dalhousie, Mussoorie, and China. Knutson et al. (1975: 317), Peck (1988: 32), Ghorpadé (2014b: 20, 2015b: 114, 2017: 108), and Mitra et al. (2015: 68) listed it. Thompson et al. (2017: 515) provide a synopsis of this group, treated as a sungenus of *Eupeodes* Osten Sacken, a mainly north temperate group with limited extensions into the tropics, with a single species known in the PSWNG area. I have seen specimens (K. Ghorpadé unpubl.) from Coonoor (Tamil Nadu), Shimoga ZARS, and Dharwad (Karnataka), their larvae feeding on woolly aphids.

Meliscaeva ceylonica Keiser, 1958

Meliscaeva ceylonica Keiser, 1958, Revue Suisse Zool., 65: 198 (♂; "Pidrutalagala, Ceylon") [NM, Basel]

India: TN; Sri Lanka (Keiser 1958; Ghorpadé 1981a, 1994, 2014b; Peck 1988, Mitra et al. 2015).

Note: In my doctoral thesis (Ghorpade 1981a) I had written, "This species was described from the mountains of Sri Lanka by Keiser (1958: 198–199, fig. 2) as a *Metepistrophe*. It was transferred to *Meliscaeva* and recorded from "S. india" by

Vockeroth (1969). My specimens were taken while feeding on cultivated annual flowers in beds in the Botanical Garden at Ootacamund 2350 m. See Ghorpadé (1994: 12) for diagnostics in a key. This is apparently restricted to the extreme high altitude regions in southern India and Sri Lanka and is easily identified by its produced lower face." Thompson et al. (2017: 509) provide a synopsis of this north temperate group (28 species), with two species known in the PSWNG area.

Meliscaeva mathisi Ghorpadé, 1994

Meliscaeva mathisi Ghorpadé, 1994, Colemania, No. 3, p. 11 (♂; "Yercaud") [USNM, Washington, DC—examined]

India: TN, KL (Brunetti 1923, 1925; Ghorpadé 1981a, 1994, 2014b, unpubl.; Mitra et al. 2015).

Note: In my doctoral thesis (Ghorpadé 1981a) I had written, "A rather distinct, stouter built *Meliscaeva*, related to *strigifrons* (de Meijere) from Malaya, Java and Formosa. Apparently confined to southern India and may turn up in Sri Lanka as well. The ♂♀ were collected by me at Yercaud, 20.ix.1978, on the Shevaroy hills and were taken on separate occasions in tropical forest on the hill slopes. It is named in honour of Dr. Wayne Mathis, my then advisor during my postdoctoral fellowship tenure at the Smithsonian Institution, Washington, DC, and who has been a valuable help on Diptera." Mitra et al. (2015: 69) listed it.

Meliscaeva monticola (de Meijere, 1914)

Syrphus monticola de Meijere, 1914, Tijdschr Ent., 57: 159 (LT♀; "Gunung gede, Tosari, Java") [ZM, Amsterdam—examined]

Sri Lanka ? (Brunetti 1923, 1925; Shiraki 1930, Knutson et al. 1975; Ghorpadé 1981a, 1994, 2014b; Peck 1988).

Note: In my doctoral thesis (Ghorpadé 1981a) I had written, "This species was reported from Sri Lanka by Shiraki (1930) but that was apparently a misidentification, as a study of the lectotype revealed. The female lectotype [in ZM, Amsterdam] I examined is labelled "E. Jacobson, Goenoeng Gedeh, Java Maart 1911"/*Syrphus monticola* det. de Meijere, Type"/ TYPE [red]/Lectotype, *Syrphus monticola* de Meij., Desig. Thompson 1974 [Yellow]. None of the *Meliscaeva* species seen by me from Sri Lanka are conspecific with the lectotype of *monticola*, and Shiraki obviously determined the material from Sri Lanka incorrectly as this species.

Meliscaeva strigifrons (de Meijere, 1914)

Syrphus cinctellus var. *strigifrons* de Meijere, 1914, Tijdschr Ent., 57: 158 (♂♀; "Gede, Nongkodjadjar, Java") [ZM, Amsterdam]

Sri Lanka (Ghorpadé 1981a, 1994, 2014b).

Note: In my doctoral thesis (Ghorpadé 1981a), I had cited this as recorded from Sri Lanka, but further details are unavailable. Ghorpadé (2014b: 20) listed it, but Knutson et al. (1975: 317) gave only Formosa, Java, and Malaya as its Oriental range, and had not mentioned Sri Lanka.

Rhinobaccha gracilis de Meijere, 1908

Rhinnobacha gracilis de Meijere, 1908, Tijdschr Ent., 51: 315 (♂♀; "Patipola, Ceylon") [ZM, Amsterdam]

India: Sri Lanka (Brunetti 1915, 1923; Keiser 1958; Knutson et al. 1975; Ghorpadé 1981a, 1994, 2014b.)

Note: This species is endemic to Sri Lanka and S. India and an important taxon here. In my doctoral thesis (Ghorpadé 1981a) I had written, "Besides being taken in Malaise Traps, specimens have also been taken "at blacklight trap." Keiser (1958: 204–205) collected more than 50 specimens at Ohiya 1900 m (Uva Province) and at Urugala, Kandapola 1900 m, Mousakanda, Gammaduwa, Nanu Oya, Hakgala 180–1900 m, Nuwara Eliya 1900 m, Pidrutalagala 2250–2400 m and Mt. Kikilimana in January, April-May, June-September, and November-December. See my diagnostic key (Ghorpadé 1994: 12). I was loaned five specimens of this by the Lund University, in 1978.

Rhinobaccha krishna Ghorpadé, 1994

Rhinobaccha krishna Ghorpadé, 1994, Colemania, No. 3, p. 12 (♂; "Shembaganur") [USNM, Washington, DC]

India: KL, TN (Ghorpadé 1981a, 1994, 2014b; Mitra et al. 2015).

Note: In my doctoral thesis (Ghorpadé 1981a) I had written, "This is a very distinctive all-black species; the holotype and the only known female paratype were taken "*in copula*" by net sweeping in primary evergreen montane forest. The Munnar specimen was also taken in identical habitat. When flying, it behaved like a jassid-fly (Pipunculidae), with a rapid, erratic, zigzag flight. It is apparently restricted to the area around the Palni Hills and the High Wavy Mountains adjacent to them. The species name *krishna* means "black" in Sanscrit and alludes its general coloration." The three specimens known were taken in March and in October. The types were swept by Dr. C.A. Viraktamath, to whom this chapter in this *Festschrift* is written for. Mitra et al. (2015: 69) listed it. See also my diagnostic key (Ghorpadé 1994: 12).

Rhinobaccha peterseni Ghorpadé, 1994

Meliscaeva peterseni Ghorpadé, 1994, Colemania, No. 3, p. 12 (♂; "Kemmangundi") [ZM, Copenhagen—examined]

India: KN, TN (Ghorpadé 1981a, 1994, 2014b, unpubl.; Mitra et al. 2015).

Note: In my doctoral thesis (Ghorpadé 1981a) I had written "This is the northernmost record for this endemic genus. It seems restricted to primary forest habitat above 1200m, or even higher, on the Western Ghats." I have taken more specimens (K. Ghorpadé unpubl.) recently at Kotagiri and at the Coonoor Botanical Garden (Nilgiri Hills) visiting flowers of extremely low growing herbs. It is named in honour of Dr. Børge Petersen, formerly Director of the Zoologisk Museum, Copenhagen. He had led the expedition to southern India in November 1977 that collected this material on the Bababudan Hills in Karnataka and which was my first expedition along with foreign entomologists that I assisted and helped collect this material, taken in November then, and later by me in Coonoor in August (K. Ghorpadé unpubl.). See my diagnostic key (Ghorpadé 1994: 12).

Scaeva latimaculata (Brunetti, 1923)

Lasiopticus latimaculatus Brunetti, 1923, *Fauna Brit. India*, Dipt., 3: 68 (♂♀; "Allahabad, Peshawur, Ferozepore, Abu") [BMNH, London—examined]

Xanthogramma pruthii Deoras, 1943: 217 (♀; "Delhi") [IARI, New Delhi—examined]

Xanthogramma indica Nayar, 1968: 129 (♀; "Kalatop 2440m, Dalhousie, Himachal Pradesh") [ZSI, Calcutta—examined]

Scaeva montana Violovitsh, 1975: 177 (♂; "Taqkob River 1800m, Hissar Range, Tadzikistan, USSR") [ZIAS, Leningrad]

India: RJ (Brunetti 1923, 1925; Ghorpadé 1981a, 1994, 2014b; Mitra et al. 2015).

Note: In my doctoral thesis (Ghorpadé 1981a) I had written, "This species is often mistaken for *albomaculata* (Macquart) but is distinctly smaller in size and is much more common in northern India. I studied the holotype ♀ (without antennae) and allotype ♂ of *Xanthogramma pruthii* Deoras and discovered its synonymy. Nayar (1968b) was also mistaken by the yellow mesonotal margins and described his *indica* as a *Xanthogramma* as well. My examination of the type ♀ in ZSI (Calcutta) confirms its present synonymy. Some specimens of this species I examined in the BMNH (London) and CIBCI (now NBAIR, Bangalore), had also been labelled as *albomaculata* (Macquart), which, perhaps, does not enter Indian limits except in the higher reaches of the Kashmir area (see also Ghorpadé 2014b: 21). I have also examined specimens of *latimaculata* from Iran. Violovitsh (1975), perhaps not aware of Brunetti's species, described *montana,* and his figures and descriptions indicate that it is a synonym of the present species."

Sphaerophoria bengalensis Macquart, 1842

Sphaerophoria bengalensis Macquart, Dipt. Exot., 2(2): 104 (LT♂; "Bengale") [MNHN, Paris]

Sphaerophoria flavoabdominalis Brunetti, 1915, Rec. Indian Mus., 11: 214 (LT♂; "Dharampur, Simla Hills"; as "form 1") [ZSI, Calcutta—examined]; Ghorpadé, 2009, Colemania, 15: 11. (as *n. syn.*)

Sphaerophoria turkmenica Bankowska, 1964, Annales Zoologici, 22(15): 345 (♂; "Berg Siunt, West Kopet Dag, Turkmenistan" [ZIAS, Leningrad]; Ghorpadé 2009, Colemania, 15: 11. (as *n. syn.*)

India: BI, MH ? (Kertész 1910; Brunetti 1915, 1917, 1923, 1925; Joseph 1968; Vockeroth 1969, 1971; Knutson 1973; Lambeck & van Brink 1973, 1975; Knutson et al. 1975; Skufjin 1982; Peck 1988; Ghorpadé 1981a,b, 1994, 2009, 2014b, 2015b, 2017: 108.)

Note: In my doctoral thesis (Ghorpadé 1981a) I had written, "A widely distributed species, from Iran and Turkmen SSR along the Himalayas to Bengal. The occurrence of this species as far south as Deolali (east of the Western Ghats near Nasik, north of Poona) is noteworthy. Vockeroth's (1969: 134) record from "S. India" is questionable. Some specimens may be difficult to separate from *indiana* Bigot, but the male terminalia and scutellar hair colour, as well as other characters given in the key (Ghorpadé 1994: 13) should make determination easy. I have seen syntypes of Brunetti's "form *flavoabdominalis*" in the ZSI (Calcutta) and have designated a Lectotype male (with distal portion of abdomen along with terminalia cut and not with the specimen on pin), and a Paralectotype (with head lost). The description and figures, also of male terminalia of *turkmenica* (Bankowska 1964: 345–347, figs. 166–175; Knutson 1973: fig. 95) show it also to be identical with *bengalensis*, and a junior synonym. This species belongs to the *scripta*-group, *sensu* Knutson (1973)." Brunetti (1915: 212–215) gave fairly long notes on this genus which would repay consultation. He first separated this as "Form I, *flavoabdominalis*" based on males and females taken in "Baluchistan, Persia, Simla, Nepal, Punjab, Bushire," and gave exact locations as Katmandu, Dharampur, Agra, Ferozepore, and Purneah, mentioning that he "took this form in abundance at both Agra and Ferozepore... in fields of dry grass, stubble and general vegetation." This form, was taken at Simla, 6–8-v-07 (Brunetti, 1917: 85). In his *Fauna* volume (Brunetti, 1923: 100), he omitted to include this species, writing, "The only other species described from the East is Macquart's *bengalensis*, but as he definitely allies his species with *menthastri*, L. ("taeniata, Meig."), the preference is given to Bigot's name." See also Vockeroth (1963, 1971), Bańkowska (1964), Speight (1973), and Skufjin (1982). Brunetti (1925: 76) then wrote—"My reference to this species [= *bengalensis*] (Fauna, p. 100) is not clear. The commonest Indian species of the genus is what I have regarded as *indiana* Big. and which may be a form of the equally common European *scripta* L. The other Indian species may be Macquart's *bengalensis*, which may be synonymous with, or a variety of the common European *menthrastri* L. [*sic*] (*taeniatus* Mg.)." Bankowska (1964: 345) named and described this from Turkmenistan and other parts of the USSR, and then reported it from Afghanistan (Bańkowska 1967: 194, 1968: 201, and 1969: 283 (both as "*turkmenica*"). Vockeroth (1969: 134) cited *turkmenica* in his treatment of *Sphaerophoria*. Vockeroth (1971: 1628–1629, 1634) wrote an excellent paper on Old World species of *Sphaerophoria*, and designated a lectotype male in MNHN, Paris. He treated *flavoabdominalis* as a tentative synonym owing to Joseph's (1968) errors. Lambeck & Kiauta (1973: 74) made longish notes on this species and *indiana* and mentioned papers by Vockeroth (1971) and

Joseph (1968). Knutson (1973: fig. 95) figured its terminalia. Knutson et al. (1975: 318) listed this with *flavoabdominalis* as a synonym from many states in India and from Pakistan and Nepal. Skufjin (1982: 140–141, figs. 13, 24) wrote of Russian specimens, gave illustrations, and listed it from north Caucasus. Ghorpadé (1981b: 73) listed prey [under "*menthastri* (L.)]" for this species. I gave a key to separate Indian subregion species of this most species diverse genus of Indian Syrphini (Ghorpadé 1994: 13) and gave notes on its synonymy (Ghorpadé 2009: 11). Lambeck & van Brink (1973: 91, 1975a: 8) reported it from the Kashmir Valley and hinted at a synonymy with *bengalensis* Macquart. Joseph (1968: 248) listed this, with terminalia figures, but wrongly as *indiana* Bigot. I had designated a Lectotype male of *flavoabdominalis* Brunetti (Ghorpadé 1981a) which is labelled "Dharampur, c. 5000 ft., Simla Hills, 6–8.v.07, N.A./4/9745/H2/Sph. Form 1 ♂ Brun., figure in Fauna II/A/9745/H2/LECTOTYPE, Sphaerophoria flavoabdominalis Brunetti ♂, Ghorpade des. 1981/Sphaerophoria bengalensis Macquart ♂, Ghorpade det. 1981" [ZSI, Calcutta]. Also designated a Paralectotype female from Agra, India in ZSI, Calcutta. Knutson et al. (1975: 318), Peck (1988: 42), Ghorpadé (2014b: 21), and Mitra et al. (2015: 69) listed it. Thompson et al. (2017: 509) provide a synopsis of this largely north temperate genus (77 species) with limited extensions into the southern regions, with a single species known in the PSWNG area.

Sphaerophoria knutsoni Ghorpadé, 1994

Sphaerophoria knutsoni Ghorpadé, 1994, Colemania, No. 3, p. 13 (♂; "Kodaikanal") [USNM, Washington, DC]

India: KL, TN (Brunetti 1923, 1925; Ghorpadé 1981a, 1994, 2014b, unpubl.; Peck 1988; Mitra et al. 2015).

Note: In my doctoral thesis (Ghorpadé 1981a) I had written, "This species belongs to the *scripta*-group and seems most closely related to *indiana* Bigot and *bengalensis* Macquart. It is differentiated from them by the distinctive male terminalia, especially the shape of the surstylus, and by the distinctly produced lower face. The wholly black-haired scutellum and fore femur are further diagnostic characters (see Knutson 1973: 13). It is apparently endemic to the higher hills of southern India (Nilgiris, Palnis, High Range). It is named in honour of now late Dr. Lloyd Knutson, formerly Director, IIBIII, USDA, Beltsville, Maryland, U.S.A., in recognition of his detailed revision of Western Hemisphere *Sphaerophoria*" (Knutson 1973) and especially his valuable encouragement and help to me, and his guidance on Oriental Diptera (also especially with the Sciomyzidae, see Ghorpade 2015a) from 1973.

Sphaerophoria macrogaster (Thomson, 1869)

Syrphus macrogaster Thomson, 1869, K. Svenska fregatten Eugenies resa, Zool., Dipt., p. 501 (LT ♂; "Sydney, Australia") [NRS, Stockholm]

Sphaerophoria koreana Bańkowska, 1964, Annales Zoologici, 22(15): 339 (♂; "Dephun ad Kujang, Korea") [IZ, Warsaw]

Sphaerophoria poonaensis Joseph, 1967, Bull. Ent., 8(2): 79 (♀; "Botanical Garden, nr R. Bhima, Poona") [ZSI, Calcutta—examined]; Ghorpadé, 2009, Colemania, 15: 11. (as *n. syn.*)

India: AH, BI ?, KL, KN, MH, MP, TN; Sri Lanka (Joseph 1967, 1968; Vockeroth 1971; Knutson et al. 1975; Bhumannavar 1977; Skufjin 1982; Ghorpadé 1973b, 1981a, 1994, 2009, 2014b, 2015b, 2017, unpubl.; Mitra et al. 2015: 69.)

Note: This abundantly common species of the Indian subcontinent plains (see Ghorpadé et al. 2011: 81) was not recognized by this species name until Vockeroth (1971: 1630–1632, fig. 3) saw Thomson's types in the Stockholm museum and studied specimens (including terminalia) from a wide range of locations, from the Indian sub-continent to Australia. It was perhaps previously generally misidentified as *S. indica* Bigot by Brunetti and later authors (see Joseph 1970) who worked mainly on northern Indian Syrphidae. Curiously Joseph (1968) also did not mention it! Vockeroth (1971) also made *koreana* Bankowska a synonym, as well as three other names from far eastern Oriental and Australian regions. Bańkowska (1964: 339–342, figs. 143–152) described *koreana* as new from Korea and China, but her illustrations are of *macrogaster* ! Knutson et al. (1975: 318) included it in their Oriental Catalog from Ceylon, China, India [Madras], Nepal; through the eastern USSR, Japan, Korea, Manchuria, New Caledonia, and New Guinea to Australia. That catalogue did much to update the then outdated nomenclature of Oriental Syrphidae, and along with Vockeroth's (1969: 132–134) magnificent revision of the genera of World Syrphini, had formed the foundation, basis, of my own research on Indian Syrphini and other Syrphidae from the early 1970s. Joseph (1968) attempted to decipher Brunetti's four "forms" of *Sphaerophoria,* whose specimens were in the ZSI at Calcutta. But he remained "stuck" with old nomenclature and inexperience with Palaearctic species, even if then adding terminalia characters, and so his work resulted in wrong names for those "forms," species—*brunettii* Joseph (= *scripta* L.), *nigritarsis* Brunetti (= *indiana* Bigot), *viridaenea* Brunetti, and *indiana* Bigot (= *bengalensis* Brunetti). In my doctoral thesis (Ghorpade 1981a) I had written of *macrogaster*— "Widely distributed in the Palaearctic (eastern), Oriental (S. & NE. India to China, Taiwan and New Guinea) and Australian (Australia) regions, and the only species (besides *knutsoni* sp. nov.) occurring in southern India and Sri Lanka. It is very closely related to *vockerothi* Joseph but has sternum 9 and surstylus of different shape, the hind femur with more black hairs, and the face in profile distinctive. Smaller in size than the other species of this genus, it seems to belong to the *scripta*-group." I examined the female holotype (with head glued on to pith) of *poonaensis* Joseph (1967: 79–80, figure) in the ZSI, Calcutta, and discovered its synonymy with *macrogaster* (Thomson). This holotype was labelled "HOLOTYPE/INDIA, Botanical Garden, near Bhima River, Poona, 8.II.1962, K.V.L. Narayana/ Z.S.I. Western and Southern India Survey, 1962/3937/H6/ *Sphaerophoria poonaensis* Joseph/*Sphaerophoria macrogaster* (Thomson) ♀, Ghorpade det. 1981." One specimen seen (in

TNAU, Coimbatore) had a label reading that it was "reared from *Ludwigia parviflora* fruits" at Coimbatore; maybe these fruits were covered with aphids. See also notes and figure in Skufjin (1982: 141, fig. 15) and synonymical notes in Ghorpade (2009: 11–12). I included it in a key to Indian *Sphaerophoria* species (Ghorpadé 1994: 13). Records from the plains of northern India await publication (but see Ghorpadé 2015: 130), though it has been reported from all southern states of India and Sri Lanka (Ghorpadé 1994: 13, Ghorpadé et al. 2011: 81). Ghorpadé (2014e: 14) mentioned a male taken at Jullundur in Punjab on mustard aphid in December that was misidentified by Dr. R.W. Crosskey as "*indiana*" [CIBCI, Bangalore]. Knutson et al. (1975: 318), Peck (1988: 43), Ghorpadé (2014b: 21), and Mitra et al. (2105: 69) listed it. I have seen specimens (K. Ghorpade unpubl.) from Valparai 1062 m (United Planters' Association of Southern India), Thandigudi 1311 m, Kotagiri, Cheyyur 707 m, Tiruvannamalai (Tamil Nadu), Araku Valley (Andhra), Tirthalli, Agumbe, Galibeedu 1074 m, Coorg, Chettalli 1001 m, Unchalli Falls 510 m, Sirsi, Mudigere 970 m., Mandya, Tumkur, Kolar, Kadadhanahgalli 857 m, Masthi, Rampura 847 m, GKVK, Hebbal, Attur farm, Bangalore 916 m, Ramanagaram, Ranganathapura, Basarenahalli, Beerasandra, Kunigal, Hessaraghatta, Hoskote, Chittahalli (Mercara), Chamarajanagara, and Arvathakakkulli (Mercara Dist) (all Karnatak).

Sphaerophoria scripta (Linnaeus, 1758)

Musca scripta Linnaeus, 1758, Syst. Nat. ed. 10, p. 584 (sex?; "Uppsala, Sweden") [BMNH, London—examined]

Sphaerophoria brunettii Joseph, 1968, Orient. Insects, 1: 243 (♂; "Srinagar 1800m, Kashmir, India") [ZSI, Calcutta—examined]

India: GJ (Brunetti 1923, 1925; Ghorpadé 1981a, 1994, 2014b, 2015b: 131, unpubl.; Peck 1988; Mitra et al. 2015).

Note: In my doctoral thesis (Ghorpadé 1981a) I had written, "This species occurs over all of the Palaearctic (except the eastern ?) including Iceland, in the Nearctic only in Greenland, in the northern Afrotropical, and enters the Oriental only in Kashmir. It is easily distinguished by the black hind femoral spicules (or long hairs in females) and the elongate abdomen. The holotype of *brunetti* Joseph [in ZSI, Calcutta] has the abdomen along with terminalia mounted between two very large round cover slips attached to the pin. One paratype ♂ has its antennae missing and have their abdomens cut terminally. Knutson et al. (1975: 318), Peck (1988: 43), Ghorpadé (2014b: 21), and Mitra et al. (2105: 69) listed it.

[Species Incertae Sedis:

Syrphus neitneri Schiner, 1868, nom. nud.

Syrphus nietneri Nietner, 1861: 17 [?? nomen nudum ?] nom. nud.

India: Sri Lanka (Nietner 1861: 17; Cotes 1893: 164; Barlow 1899: 218; Stebbing 1908: 148; Ghorpadé 2014b: 22.)

Note: The description of this species by Schiner (assigned to this author by Cotes, Barlow and Stebbing) is not traceable. But Nietner (1861) gave notes on the adult and larva which may validate the naming of *Syrphus nietneri* Nietner. The larvae were stated to be good predators of the coffee aphid, "*Aphis coffeae*" [= *Toxoptera aurantii* (Boyer de Fonscolombe)] in Sri Lanka. From the scanty descriptive notes, more of the larva than of the adult fly, which was said to be similar to *Syrphus splendens* Doleschall, 1856 [= *Ischiodon scutellaris* (Fabr.), *nietneri* seems to me to be a species of *Betasyrphus*, which could either be *fletcheri* sp. nov., or a different species. This would then be *Betasyrphus nietneri* (Nietner). It is certainly not a true *Syrphus*, which is a north temperate fly. However, until the original description (if there is one by Doleschall or Schiner), other than that by Nietner, is traced, or the types are found, I consider *nietneri* as a generically unassigned name, best probably to treat it as a nom.nud.]

ACKNOWLEDGEMENTS

I wish to say thank you to all my co-specialists on the Syrphidae, especially those working abroad, for all kinds of help, especially in providing me with some relevant recent papers which were inaccessible to me and not in my library. Also to field entomologists all over India who helped in my researches all these years. I want to say a special word of thanks to my classmates and colleagues at the University of Agricultural Sciences, Bangalore, besides Dr. C.A. Viraktamath, especially Mr. K. Durga Prasad, C. Siddappaji, Drs. B. Mallik, A.R.V. Kumar, C.S. Wesley, V.V. Belavadi, S. Ramani, Prashant Mohanraj, and V. Shashidhar. And thanks also to others who helped me in field work or by providing syrphid specimens, like Mss. M.V. Rama, Shubali Rai, Nandini Belamkar, Sandhya Surendran, Shweta Basnett, and Urbashi Pradhan, and Drs. V.K. Gupta, J.S. Bhatti, K. Gunathilagaraj, Chitra Narayanaswamy, S. Manickavasagam, Girish Chandra, S.S. Anooj, Ashish Tiple, K.J. David, Sabbithi Pavan, Sunil Dhaman, and Bipul Rabha, and Messrs. Rojeet Thangjam, Sachin Thapa, Raaj Jadhav, H. Sankararaman, and Shaikh Riyazuddin.

REFERENCES

Agarwala, B. K., P. Laska, and D. N. Raychaudhuri. 1984. Prey records of aphidophagous syrphid flies from India (Diptera, Syrphidae). *Acta ent. Bohemoslov.* 81: 15–21.

Agarwala, B. K., and J. L. Saha. 1986. Larval voracity, development and relative abundance of predators of *Aphis gossypii* on cotton in India. *Series Entomol. (Dordrecht)*, 35: 339–344.

Anand, R. K. 1986. Records of syrphids predating on aphids in Delhi. pp. 197–200. In: *Recent Trends in Aphidological Studies*. ed. S.P. Kurl. Modinagar, India.

Ashwani, K., V. C. Kapoor, and P. Láska. 1987. Immature stages of some aphidophagous syrphid flies of India (Insecta, Diptera, Syrphidae). *Zoologica Scripta* 16(1): 83–88.

Austen, E. E. 1893. Descriptions of new species of Dipterous insects of the family syrphidae in the collection of the British Museum, with notes on species described by the late Francis Walker. Part I. Bacchini and Brachyopini. *Proc. zool. Soc. Lond.* 61: 132–164, Pls IV & V.

Awtar, S., V. Gupta, and N. S. Sodhi. 1986a. External male genitalia of two species of genus *Paragus* Latreille (Diptera: Syrphidae). *Uttar Pradesh J. Zool.* 5(2): 145–148.

Awtar, S., N. S. Sodhi, and V. Gupta. 1986b. A new species of genus *Eristalis* Latreille (Syrphidae: Diptera). *J. Bombay nat. Hist. Soc.* 83: 395–398.

Baid, I. C. 1959. Some preliminary notes on the insect life in Sambhar Lake. *J. Bombay nat. Hist. Soc.* 56: 361–361.

Bańkowska, R. 1962. Studies on the family Syrphidae (Diptera) *Helenomyia* gen. nov. *Bull. l'Acad. Polon. Sci.* 10(8): 311–314.

Bańkowska, R. 1964. Studien über die paläarktischen Arten der Gattung *Sphaerophoria* St. Farg. et Serv. (*Diptera, Syrphidae*). *Annls. Zool. Warszawa.* 22(15): 285–353.

Bańkowska, R. 1967. Die *Sphaerophoria*-arten der Afghanistan-Expedition (1952–1953) J. Klapperichs (Diptera, Syrphidae). *Acta faun. ent. Mus. Nat. Pragae.* 12: 193–195.

Barlow, E. 1899. Notes on insect-pests from the Entomological section, Indian Museum: VI—Some beneficial insects in India. II. Some Diptera (Two-winged Flies) known to be destructive to insect-pests. Indian Some beneficial insects in India. II. Some Diptera (Two-winged Flies) known to be destructive to insect-pests. *Indian Museum Notes* 4(4): 217–219.

Batra, H. N. 1960. The Cabbage Aphis (*Brevicoryne brassicae* L.) and its control schedule. Indian *J. Hort.* 17: 74–80.

Beeson, C. F. C. 1953. *The Ecology and Control of the Forest Insects of India and Its Neighbouring Countries.* 2nd edn. vi+ii+1007 pp. Vasant Press, Dehra Dun, Uttarakhand. [Syrphidae, pp. 339–340].

Bezzi, M. 1908. Secondo contributo alla conoscenza del genere *Asarcina*. *Annls Hist.Nat. Mus. Natn. Hung.* 6: 495–504.

Bezzi, M. 1928. *Diptera Brachycera and Athericera of the Fiji Islands Based on Material in the British Museum (Nat. Hist.), London.* viii+220 pp. British Museum of Natural History, London, UK.

Bhatia, H. L. 1931. Studies in the life-histories of three Indian Syrphidae. *Indian J. Agric. Sci.* 1(4): 503–513.

Batia, H. L., and M. Shaffi. 1933. Life-histories of some Indian Syrphidae. *Indian J. Agric. Sci.* 2(6): 543–570.

Bhumannavar, B. S. 1977. Studies on the bionomics of the Safflower Aphid, *Dactynotus compositae* Theobald, its role on safflower crop loss and its control. [Abstract]. *Mysore J. Agric. Sci.* 11: 261.

Bigot, J.-M.-F. 1884. Dipteres nouveaux ou peu connus. 24e partie, XXXII: Syrphidi (2e partie). Especes nouvelles. No. III. *Ann. Soc. Entomol. Fr.* (6) 4: 73–116.

Bigot, J. M. F. 1891. Catalogue of the diptera of the oriental region. I. *J. Asiat. Soc. Beng.* 40: 250–282.

Bindra, O. S., and D. K. Saxena, 1958. The mustard aphid and its control. *Proc. 45th Indian Sci. Congr.* 3: 462.

Briggs, J. C. 2003. The biogeographic and tectonic history of India. *J. Biogeogr.* 30: 381–388.

Briggs, J. C. 1989. The historic biogeography of India: Isolation or contact?. *Syst. Zool.* 38(4): 322–332.

Brunetti, E. 1907a. Notes on oriental diptera. II.—Preliminary Report on a collection from Simla made in April and May 1907. *Rec. Indian Mus.* 1: 166–170.

Brunetti, E. 1907b. Notes on the Oriental Syrphidae. Part I. *Rec. Indian Mus.* 1: 379–380.

Brunetti, E. 1908. Notes on Oriental Syrphidae with descriptions of new species. Part I. *Rec. Indian Mus.* 2: 49–96 [and] figs. in Brunetti, E. 1907. *Rec. Indian Mus.* 1: 379–380.

Brunetti, E. 1910. Notes on Ceylon Diptera. *Spolia Zeylanica* 6: 170–172.

Brunetti, E. 1913a. Zoological results of the Abor Expedition, 1911–12. XI. Diptera. *Rec. Indian Mus.* 8(2): 149–190.

Brunetti, E. 1913b. New and interesting Diptera from the Eastern Himalayas. *Rec. Indian Mus.* 9(5): 255–277.

Brunetti, E. 1915. Notes on Oriental Syrphidae: With descriptions of new species. Part II. *Rec. Indian Mus.* 11: 201–256.

Brunetti, E. 1917. Diptera of the Simla District. *Rec. Indian Mus.* 13(2): 59–101.

Brunetti, E. 1923. *The Fauna of British India, including Ceylon and Burma. Diptera.* Vol. 3. Pipunculidae, Syrphidae, Conopidae, Oestridae. xii+424 pp., 85 figs., 7 pls. Taylor & Francis Groups, London, UK.

Brunetti, E. 1925. Some notes on Indian Syrphidae, Conopidae, and Oestridae. *Rec. Indian Mus.* 27: 75–79.

Cheng, X. Y., and F. C. Thompson. 2008. A generic conspectus of the Microdontinae (Diptera: Syrphidae) with the description of two new genera from Africa and China. *Zootaxa* 1879: 21–48.

Cherian, M. C. 1934. Notes on some south Indian Syrphids. *J. Bombay nat. Hist. Soc.* 37: 697–699.

Christophers, S. R. 1921. The distribution of mosquitoes in relation to the zoogeographical areas of the Indian Empire. *Rep. Proc. 4th Entom. Mtg, Pusa* 205–215.

Claussen, C., and J. Weipert. 2003. Zur Schwebfliegenfauna Nepals (Insecta: Diptera: Syrphidae) unter besonderer Berucksichtigung Westnepals. In: Hartmann, M., and M. Baumbach [Eds.] *Biodiversitat und Naturausstattung im Himalaya.* Verein der Freunde und Forderer des Naturkundemuseums Erfurt e. V., Erfurt, pp. 343–380, 120 figs., 2 Tables, pls IX–XVI.

Coe, R. L. 1964. Diptera from Nepal. Syrphidae. *Bull. Br. Mus. (Nat. Hist.), Ent.* 15(8): 255–290.

Cotes, E. C. 1893. A conspectus of the insects which affect crops in India. *Indian Mus. Notes.* 2: 145–176.

Cox, C. B. 1990. New geological theories and old biogeographical problems. *J. Biogeogr.* 17: 117–130.

Croizat, L. 1968. The biogeography of India. A note on some of its fundamentals. In: Misra, R., and B. Gopal [Eds.] *Proceedings of the Symposium on Recent Advances in Tropical Ecology* Part II. Varanasi, India.

Curran, C. H. 1926. Notes on Wiedemann's types of Syrphidae (Dipt.). *Canadian Entomologist* 58: 111–115.

Curran, C. H. 1928. The Syrphidae of the Malay Peninsula. *Journal of the Federated Malay States Museums* 14(2): 141–324.

Curran, C. H. 1929. New Syrphidae and Tachinidae. *Ann. Ent. Soc. Amer.* 22: 489–510.

Curran, C. H. 1930. New species of Eristalinae with notes (Syrphidae, Diptera). *Amer. Mus. Novit* 411: 1–27.

Curran, C. H. 1931a. Additional records and descriptions of Syrphidae from the Malay Peninsula. *J. Fed. Malay States. Mus.* 16(3 + 4): 290–338.

Curran, C. H. 1931b. Records and descriptions of Syrphidae from North Borneo, including Mt. Kinabalu. *Fed. Malay States. Mus.* 16(3 + 4): 339–376.

Curran, C. H. 1933. Three new Diptera from India. *Stylops* 2(2): 45–48.

Curran, C. H. 1947. The Syrphidae of Guadalcanal, with notes on related species. *Amer. Mus. Novit.* 1364: 1–17.

Curran, C. H. 1942. Syrphidae from Sarawak and the Malay Peninsula (Diptera). *Amer. Mus. Novit.* 1216: 1–8.

Darlington, P. J. Jr. 1970. A practical criticism of the Hennig-Brundin "Phylogenetic [sic] systematics" and Antarctic Biogeography. *Syst. Zool.* 19: 1–18.

Datta, M., and M. Chakraborti. 1986. On collections of Flower flies (Diptera: Syrphidae) from south India. *Rec. Zool. Surv. India.* 83: 53–67, 6 figs.

De Meijere, J. C. H. 1908. Studien über Südostasiatische Dipteren. III. *Tijdschr. Entomol.* 51: 191–332.

De Meijere, J. C. H. 1911. Studien über Südostasiatische Dipteren. VI. *Tijdschr. Entomol.* 54: 258–432.

De Silva, M. D. 1961. A preliminary list of the native parasites and predators of insect pests in Ceylon. *Trop. Agric. Peradeniya.* 117: 115–141.

Delfinado, M., and D. E. Hardy. 1973, 1975, 1977. *A Catalog of the Diptera of the Oriental Region.* Vol. I, vii+618 pp., Vol. II, vii+459 pp., Vol. III, x+854 pp. The University Press of Hawaii, Honolulu, Hawaii.

Deoras, P. J. 1943. Description of and biological notes on a new species of Syrphidae from India. *Indian J. Entomol.* 4(2): 217–219.

Deoras, P. J. 1957. Notes on some Insects of medical importance from a suburban area of Bombay. *Indian J. Entomol.* 18(3): 305–307.

Diller, E. H. 1977. Die in Indien vorkommenden Taxa der Gattung *Diplazon* Nees 1818 (Hymenoptera, Ichneumonidae, Diplazontinae). *Mitt. Munch. Entomol. Ges.* 66: 21–28.

Dirickx, H. G. 2010. Notes sur le genre *Allobaccha* Curran, 1928 (Diptera, Syrphidae) à Madagascar avec descriptions de cinq nouvelles espèces. *Rev. Suisse Zool.* 117(2): 213–233.

Erwin, T. L. 1985. The nature of taxon pulses and their impact on the distribution of organisms in space and time. pp. 437–472. *In:* G. E. Bal [Ed.] *Taxonomy, Phylogeny and Zoogeography of Beetles and Ants.* Dr W. Junk, Dordrecht, the Netherlands.

Evans, W. H. 1932. *The identification of Indian Butterflies.* The Bombay Natural History Society, Madras: The Diocesan Press. 454 pp.

Fairchild, G. B. 1969. Climate and the phylogeny and distribution of Tabanidae. *Bull. Soc. Ent. Am.* 15: 7–11.

Fletcher, T. B. 1914. *Some South Indian Insects, and Other Animals of Importance Considered Especially from An Economic Point of View.* xxii+565 pp. Government Press, Madras, Tamil Nadu.

Fletcher, T. B. 1916. One hundred Notes on Indian Insects. *Bull. Agric. Res. Inst., Pusa.* 59: v+39 pp., Supdt. Govt. Printing, Calcutta, India

Fluke, C. L. 1958. A study of the male genitalia of the Melanostomini (Diptera-Syrphidae). *Trans. Wisconsin Acad. Sci. Arts and Lett.* 46: 261–279.

Forster, E. M. 1924. *A Passage to India.* pp. 1–317. Penguin Books Ltd, Middlesex, UK.

Frey, R. 1946. Ubersicht der Gattungen der Syrphiden-Unterfamilie Syrphinae (Syrphine+ Bacchinae). *Notulae Entomologicae,* 25: 152–172.

Gaonkar, H. (in prep.). *The Butterflies of the Indian Region, Including Bangladesh, Bhutan, Burma, India, Maldive Islands, Nepal, Pakistan and Sri Lanka.* Volume 1. Biogeography, History, Classification and an Annotated Bibliography of their Natural History and Biodiversity. [Manuscript 458 p.].

Ghorpade, K.D. 1973a. The hover-fly *Allograpta javana* (Wiedmann) [sic], predacious on the jowar shoot-bug *Peregrinus maidis* (Ashmead) together with its recorded hosts [sic] from India. *Sci. Cult.* 39: 400–401.

Ghorpade, K.D. 1973b. *A faunistic study of the Hover-flies (Diptera: Syrphidae) of Bangalore, Southern India.* x+208 pp., 82 figs. M.Sc. (Agri.) thesis, Bangalore, India: University of Agricultural Sciences.

Ghorpade, K.D. 1974. *A faunistic study of the Hover-Flies (Diptera: Syrphidae) of Bangalore, Southern India. Mysore J. agric. Sci.* 8(4): 636.

Ghorpade, K.D. 1981a. *A Taxonomic Revision of Syrphini (Diptera: Syrphidae) from the Indian Subcontinent.* 381 pp. Ph.D. thesis. Bangalore, India: University of Agricultural Sciences.

Ghorpade, K.D. 1981b. Insect prey of Syrphidae (Diptera) from India and neighbouring countries: a review and bibliography. *Trop. Pest Manag.* 27: 62–82.

Ghorpade, K.D. 1981c. An anomalous new *Episyrphus* (Diptera: Syrphidae) from Madagascar. *Colemania,* 1: 89–94.

Ghorpade, K. 1994. Diagnostic keys to new and known genera and species of Indian subcontinent Syrphini (Diptera: Syrphidae). *Colemania,* No. 3: 1–15.

Ghorpadé, K. 1997a. The extinction of the classical naturalist. *Indian J. Biodiversity,* 1(1+2): 193–196.

Ghorpadé, K. 2007. The genus *Agnisyrphus* Ghorpadé (Diptera—Syrphidae), peculiar to the Oriental Region, with notes on phylogeny, evohistory and panbiogeography. *Colemania,* No. 14: 1–35.

Ghorpadé, K. 2009. Some nomenclatural notes on Indian Subregion Syrphini (Diptera—Syrphidae). *Colemania,* No. 15: 3–13.

Ghorpadé, K. 2012. Notes on nomenclature, taxonomy and phylogeny of the genus *Chrysotoxum* Meigen (Diptera—Syrphidae) in the Oriental region. *Colemania,* No. 32: 1–4.

Ghorpadé, K. 2013. Notes on Hawk-moths (Lepidoptera—Sphingidae) in the Karwar-Dharwar transect, peninsular India: a tribute to T.R.D. Bell (1863–1948). *Colemania,* No. 33: 1–16.

Ghorpadé, K. 2014a. The Diptera of the Western Ghats, peninsular India, with knowledge status and richness estimates of families of true flies in the Indian subregion. *Colemania,* No. 40: 3–29.

Ghorpadé, K. 2014b. An updated Check-list of the Hover-flies (Diptera—Syrphidae) recorded in the Indian sub-continent. *Colemania,* No. 44: 1–30.

Ghorpadé, K. 2014c. On the hover-flies (Diptera—Syrphidae) preserved in the collection of the Panjab University, Chandigarh, and further notes on those from the Indian Punjab and NW. India. *Colemania,* No. 46: 1–17.

Ghorpadé, K. 2015a. A Conspectus of Oriental—Papuan Sciomyzidae (Diptera—Acalyptratae) with keys to genera and an updated Check-List. *Colemania,* No. 47: 3–14.

Ghorpadé, K. 2015b. Hover-flies (Diptera—Syrphidae) documented from the Northwest Frontier of the Indian sub-continent: A circumstantial history and inclusive bibliography. *Colemania,* No. 50: 1–151.

Ghorpadé, K. 2017. Hover-flies (Diptera Syrphidae) in the Bombay Natural History Society Collection, with an annotated checklist of those recorded from Maharashtra, India. *J. Bombay Nat.* 112(2): 106–110 [2015].

Ghorpadé, K. and Anooj, S. S. 2016. Hover-flies (Diptera—Syrphidae) of Tripura, north-eastern India. *Colemania,* 51: 3–12.

Ghorpadé, K., Durga Prasad, K. & Pavan, S. 2011. Hover-flies (Diptera: Syrphidae) of the Coromandel Coast in Andhra Carnatic, peninsular India. *Bionotes (Aligarh),* 13(2): 78–86.

Ghorpadé, K. 1994. Diagnostic keys to new and known genera and species of Indian subcontinent Syrphini (Diptera: Syrphidae). *Colemania* 3: 1–15.

Ghorpade, K. and Kunte, K. 2010. Butterflies of the Palani Hills, Southern Western Ghats in Peninsular India. *Colemania,* 23: 1–19.

Ghosh, A. K. 1974. Aphids (Homoptera: Insecta) of economic importance in India. *Indian Agric.* 18(2): 81–214.

Goffe, E. R. 1952. An outline of a revised classification of the Syrphidae (Diptera) on phylogenetic lines. *Trans. Soc. Brit. Ent.* 11(4): 97–124.

Gokulpure, R. S. 1972. Further collection of the Syrphidae (Diptera) from central India. *J. Bombay Nat. Hist. Soc.* 68(3): 848.

Gressitt, J. L. 1961. Problems in the zoogeography of pacific and Antarctic Insects. *Pacific Ins. Monogr.* 2: 1–94.

Gupta, V. K. 1962. Taxonomy, zoogeography and evolution of Indo-Australian Theronia (Hymenoptera: Ichneumonidae). *Pac. Ins. Monogr.* 4: 1–142.

Haarto, A., and G. Ståhls. 2014. When mtDNA COI is misleading: Congruent signal of ITS2 molecular marker and morphology for North European *Melanostoma* Schiner, 1860 (Diptera, Syrphidae). *Zookeys*, 431: 923–134.

Hervé-Bazin, J. 1914. Note sur quelques Syrphides (Diptera) provenant de Java et de l'Inde. Avec la description d'un genre nouveau. *Insecta, Revue Illustrée d'Entomologie*, 41: 149–154.

Hervé-Bazin, J. 1922. Description d'une nouvele espece de *Korinchia* de l'Inde (Dipt. Syrphidae). *Bull. Soc. Entomol. Fr.* 27: 122–124.

Hervé-Bazin, J. 1923a. Diagnoses de Syrphides [Dipt.] nouveaux du Laos (Indo-Chine francaise). *Bull. Soc. Entomol. Fr.* 28: 25–28.

Hervé-Bazin, J. 1923b. Etudes sur les "*Lathyrophthalmus*" (Diptera, Syrphidae) d'extreme-Orient. *Ann. Sci. Nat. Zool.* 6: 125–152.

Hervé-Bazin, J. 1923c. Remarques surl'ouvrage de M. Th. Becker: « Neue Dipteren meiner sammlung », Part. I, Syrphidae, paru dans: Mitteil. Aus dem zoolog. Mus. Berlin, X [1921], pp. 1–93. *Bull. Soc. Entomol. Fr.* 29: 129–131.

Hervé-Bazin, J. 1923e. Premiere note sur les Syrphides (Diptera) de la Collection du Museum National de Paris. Les genres *Megaspis* et *Volucella* en Asie. *Bull. Soc. Entomol. Fr.* 29: 252–259.

Hervé-Bazin, J. 1924. Note sur les Syrphides de I'nde. Remarques et notes synonymiques sur l'ouvrage de M. E. Brunetti: « The Fauna of British India, III. Syrphidae, etc. » *Ann. Soc. Entomol. Fr.* 92: 289–299.

Hervé-Bazin, J. 1926. Syrphides de l'Indo-Chine Francaise. *Encycl. Entomol. (B), Diptera*, 3(2): 61–110.

Hippa, H. 1974. On the taxonomy of the Ceylonese and southern Indian species of the genus *Xylota* Meigen (Dipt, Syrphidae). *Ann. Entomol. Fennici.* 40(2): 56–60.

Hippa, H. 1978. Classification of Xylotini (Diptera, Syrphidae). *Acta Zool. Fennica.* 156: 1–153.

Hippa, H. 1990. The genus *Milesia* Latreille (Diptera, Syrphidae). *Acta Zool. Fennica.* 187: 1–226.

Holloway, J. D. 1974. The biogeography of Indian butterflies. pp. 473–499, figs. 81–94. In: M. S. Mani [Ed.] *Ecology and Biogeography in India.* xix+773 pp. Dr W. Junk, B. V., The Hague, the Netherlands.

Hull, F. M. 1944a. Some flies of the family Syrphidae in the British Museum (Natural History). *Ann. Mag. Nat. Hist.* 11(11): 21–61.

Hull, F. M. 1944b. Studies on flower flies (Syrphidae) in the Vienna Museum of Natural History. *J. Wash. Acad. Sci.* 34: 398–404.

Hull, F. M. 1949. The morphology and inter-relationship of the Genera of Syrphid Flies, recent and fossil. *Trans. Zool. Soc. Lond.* 26(4): 257–408.

Humphries, C. J. 2000. Vicariance biogeography. pp. 767–779. In: S. A. Levin [Ed.] *Encyclopedia of Biodiversity.* Vol. 5, San Diego, Academic Press.

Illies, J. 1983. Changing concepts in Biogeography. *Annu. Rev. Entomol.* 28: 391–406.

Joseph, A. N. T. 1967. A new species of *Sphaerophoria* St. Fargeau and Serville, 1825 (Diptera: Syrphidae). *Bull. Entomol.* 8(2): 79–80.

Joseph, A. N. T. 1968. On the "forms" of *Sphaerophoria* St. Fargeau and Serville (Diptera: Syrphidae) described by Brunetti from India. *Oriental Insects* 1(3 + 4): 243–248.

Joseph, A. N. T. 1970. Two new and two known species of *Sphaerophoria* St. Fargeau and Serville, 1828 (Dipt. Syrphidae). *Eos, Madrid* 45: 165–172.

Joseph, A. N. T. and P. Parui. 1977. On a small collection of Diptera from Chota Nagpur, Bihar. *Rec. Zool. Surv. India.* 72: 227–238.

Joseph, A. N. T. and P. Parui. 1986. Diptera from silent valley. *Rec. Zool. Surv. India.* 84: 157–164.

Joshee, A. K. 1968. Food habits of the Bull Frog *Rana tigerina* (Daud.). *J. Bombay nat. Hist. Soc.* 65: 498–501.

Joshi, F. L., and J. C. Sharma. 1973. Toxicity of certain insecticides to *Menochilus sexmaculatus* F. and *Xanthogramma* (*Ischiodon*) *scutellare* F. predating on mustard aphid, *Lipaphis erysimi* Kalt. JNKVV *Res. J.* 7(2): 112–113.

Kalyanam, N. P. 1970. Occurrence of the aphidophagous syrphid fly *Syrphus latifasciatus,* Macq. (Syrphidae: Diptera) in south India. *Curr Sci.* 39: 66–67.

Kapoor, V.C., Y.K. Malla and Y. Rajbhandari. 1979. Syrphid flies (Diptera: Syrphidae) from Kathmandu valley, Nepal with a checklist of syrphids of Nepal. *J. Nat. Hist. Mus.* 3(1–4): 51–68.

Keiser, F. 1958. Bietrag zur Kenntnis der Syrphidenfauna von Ceylon (Dipt.). *Revue Suisse Zool.* 65(1): 185–239.

Kertész, K. 1910. *Catalogus Dipterorum Hucusque Descriptorum.* Vol. VII. Syrphidae, Dorylaidae, Phoridae, Clythiidae. 470 p.

Kertész, K. 1913. H. Sauter's Formosa-Ausbeute. Syrphidae. [Dipt.]. *Annls. Mus. Nat. Hung.* 11: 273–285.

Kertész, K. 1914. H. Sauter's Formosa-Ausbeute. Syrphidae II. *Annls. Mus. Nat. Hung.* 12: 73–87.

Knutson, L. V. 1973. Taxonomic revision of the aphid-killing flies of the genus *Sphaerophoria* in the Western Hemisphere (Syrphidae). *Misc. Publ. Ent. Soc. Am.* 9(1): 1–50.

Knutson, L. V., Thompson, F. C. and J. R. Vockeroth. 1975. Family Syrphidae. pp. 307–374. In: Delfinado, M. D., and D. E. Hardy [Eds.] *A Catalog of the Diptera of the Oriental Region.* Volume 2, x+459 pp. University Press of Hawaii, Honolulu, Hawaii.

Kohli, V. K., V. C. Kapoor, and S. K. Gupta. 1988. On one new genus and nine species of syrphid flies (Diptera: Syrphidae) from India. *J. Insect Sci.* 1(2): 113–127.

Lambeck, H. J. P., and B. Kiauta. 1973. On a small collection of syrphid flies (Diptrera: Syrphidae) from the Kathmandu Valley and the Khumbu Himal region (Nepal). *Entomol.* 33: 70–78.

Lambeck, H. J. P., and J. M. van Brink. 1973. Contribution to the knowledge of the taxonomy, faunal composition and cytology of the syrphid flies (Diptera: Syrphidae) of Kashmir (India). II. Karyotypes of eleven species. *Genen Phaenen* 18: 1–15.

Lambeck, H. J. P., and J. M. van Brink. 1975. Contribution to the knowledge of the taxonomy, faunal composition and cytology of the syrphid flies (Diptera: Syrphidae) of Kashmir (India). I. Taxonomic account and faunal composition. *Genen Phaenen* 16: 87–100.

Láska, P., C. Pérez-Banón, L. Mazánek. et al. 2006. Taxonomy of the genera *Scaeva, Simosyrphus* and *Ischiodon* (Diptera: Syrphidae): Descriptions of immature stages and status of taxa. *Eur. J. Entomol.* 103: 637–655.

Legris, P., and V. M. Meher-Homji. 1977. Phytogeographic outlines of the hill ranges of peninsular India. *Trop. Ecol.* 18: 10–24.

Lomolino, M. V. 2000. Biogeography, Overview. pp. 455–469. In S. A. Levin [Ed.] *Encyclopedia of Biodiversity.* Vol. 1, Academic Press, San Diego.

Lyneborg, L., and W. Barkemeyer. 2005. *The Genus Syritta. A World Revision of the Genu Syritta Le Peletier & Serville, 1828 (Diptera: Syrphidae).* 224 pp., 224 figs. Apollo Books, Denmark.

Mahalingam, M. 1988. A preliminary report on the aphidophagous syrphids from Ootacamund area, Nilgiris. *Indian Zool.* 12(1 + 2): 161–163.

Mani, M. S. 1968. *Ecology and Biogeography of High Altitude Insects.* xiv+527 pp. W. Junk, B. V., The Hague, the Netherlands.

Mani, M. S. 1974. Physical features. pp. 11–60, Biogeography of the Himalaya. pp. 664–681, Biogeography of the western borderlands. pp. 682–688. In: M. S. Mani [Ed.] *Ecology and Biogeography in India.* xix+773 pp. Dr W. Junk, by, The Hague, the Netherlands.

Matsumura, S., and J. Adachi. 1917. Synopsis of the economic Syrphidae of Japan. (Pt. III). *Entomol. Mag. Kyoto.* 3(1): 14–46.

Maxwell-Lefroy, H. 1909. *Indian Insect Life: A Manual of the Insects of the Plains (Tropical India).* xii+786 pp. Thacker & Spink, Calcutta, India. [Diptera chapter pp. 545–657 by F. M. Howlett].

Mener-Homji, V. M. 1983. On the Indo-Malaysian and Indo-African elements in India. *Feddes Repert.* 94: 407–424.

Mengual, X. 2012. The flower fly genus *Citrogramma* Vockeroth (Diptera: Syrphidae): Illustrated revision with descriptions of new species. *Zool. J. Linn. Soc.* 164: 99–172.

Mengual, X. 2013. Notes on *Citrogramma* Vockeroth and *Eosphaerophoria* Frey (Diptera: Syrphidae). *Zootaxa* 3745 (3): 388–396.

Mengual, X. 2015. The systematic position and phylogenetic relationships of *Asiobaccha* Violovitsh (Diptera, Syrphidae). *J. Asia-Pacific Entomol.* 18: 397–408.

Mengual, X. 2016. A taxonomic revision of the genus *Asiobaccha* Violovitsh (Diptera: Syrphidae). *J. Nat. Hist.* 50: 2585–2645.

Mengual, X., and K. Ghorpadé. 2010. The flower fly genus *Eosphaerophoria* Frey (Diptera, Syrphidae), *Zookeys* 33: 39–80.

Mengual, X., C. Ruiz, S. Rojo, G. Stahls, and F. C. Thompson. 2009. A conspectus of the flower fly genus *Allograpta* (Diptera: Syrphidae) with description of a new subgenus and species. *Zootaxa* 2214: 1–28.

Mengual, X., Ståhls, G., and S. Rojo. 2008a. First Phylogeny of predatory Flower flies (Diptera, Syrphidae, Syrphinae) using mitochondrial COI and nuclear 28S rRNA genes: Conflict and congruence with the current tribal classification. *Cladistics* 23: 1–20.

Mengual, X., Ståhls, G., and S. Rojo. 2008b. Molecular Phylogeny of Allograpta (Diptera, Syrphidae) reveals diversity of lineages and non-monophyly of phytophagous taxa. *Mol. Phylogenet. Evol.* 49: 715–727.

Mengual, X., Ståhls, G., and S. Rojo. 2015. Phylogenetic relationships and taxonomic ranking of Pipizine flower flies (Diptera: Syrphidae) with implications for the evolution of aphidophagy. *Cladistics* 31: 491–508.

Misra, C. S. 1932. The green peach-aphis (*Myzus persicae* Sulz.) and a new pyralid mango defoliator (*Orthaga mangiferae* n. sp.). Indian *J. Agric. Sci.* 2: 536–541.

Misra, S. S., and K. D. Verma. 1975. Record of lunate fly, *Eumerus* sp. (Syrphidae: Diptera) on potatoes from India. *Curr. Sci.* 44(22): 827.

Mitra, B., and P. Parui. 2002. Dipteran flower visitors in Jessore Sloth Bear and Balaram Ambaji Wildlife Sanctuaries, north Gujarat. *Bionotes* 4(2): 45.

Mitra, B., S. Roy, I. Imam, and M. Ghosh. 2015. A review of the hover flies (Syrphidae: Diptera) from India. *Internatl. J. Fauna Biol. Stud.* 2(3): 67–73.

Musthak Ali, and H. C. Sharatchandra. 1985. A new record of *Paragus auritus* (Syrphidae: Diptera) on root aphid (*Forda orientalis*). *Milwai Newsletter* 4: 18.

Nayar, J. L. 1964a. A note on the incidence and flower colour preference in *Syrphus balteatus* (De Geer), (Syrphidae: Diptera). *Agra Univ. J. Res. (Sci).,* 12(3): 79–85.

Nayar, J. L. 1964b. External morphology of head capsule of *Syrphus balteatus* De Geer (Syrphidae: Diptera). *Indian J. Entomol.* 26(2): 135–151.

Nayar, J. L. 1964c. Thoracic morphology of *Syrphus balteatus* De Geer (Syrphidae: Diptera). *Indian J. Entomol.* 26(3): 255–267.

Nayar, J. L. 1965a. Reproductive system and external genitalia of *Syrphus balteatus* De Geer (Diptera: Syrphidae). *Indian J. Entomol.* 27(1): 31–45.

Nayar, J. L. 1965b. A study of the respiratory system of *Syrphus balteatus* De Geer (Syrphidae: Diptera). *J. Anim. Morph. Physiol.* 12(1): 17–31.

Nayar, J. L. 1966a. Musculature studies on *Syrphus balteatus* De Geer, (Syrphidae: Diptera). *Agra Univ. J. Res. (Sci.).* 14(3): 29–46.

Nayar, J. L. 1966b. Circulatory system and associated tissues of *Syrphus balteatus* De Geer (Syrphidae: Diptera). *Agra Univ. J. Res. (Sci.).* 15(1): 91–98.

Nayar, J. L. 1968a. A contribution to our knowledge of high altitude Syrphidae (Cyclorrhapha: Diptera) from N. W. Himalaya Part I—Subfamily Syrphinae. *Agra Univ. J. Res. (Sci.).* 16(2): 121–131.

Nayar, J. L. 1968b. A contribution to our knowledge of high altitude Syrphidae (Cyclorrhapha: Diptera) from N. W. Himalaya. Part II—Subfamily Eristalinae. *Agra Univ. J. Res. (Sci.).* 16(3): 27–31.

Nayar, J. L. 1968c. Two new species of *Eristalis* from India (Diptera: Syrphidae). *Pan-Pacif. Entomol.* 44: 119–122.

Nayar, J. L., and L. Nayar. 1965. Some interesting syrphids of Agra. *Indian J. Entomol.* 27(2): 240–241.

Nietner, J. 1861. Observations on the enemies of the coffee tree in Ceylon. *Ceylon Times,* pp. 16–17.

Panchabhavi, K. S., and K. J. Rao. 1978. Note on the effect of mixed cropping of niger on the activities of insect pollinators and seed-filling of sunflower in Karnataka. *Indian J. Agric. Sci.* 48(4): 254–255.

Pape, T., and N. L. Evenhuis. 2013. *Systema Dipterorum, Version [1.5].* "http://www.diptera.org/"http://www.diptera.org/[Accessed: 15 December 2015]

Parui, P., and M. Datta. 1987. Insecta: Diptera. pp. 241–255. In: Director, Z. S. I. [ed.] *State Fauna Series 1. Fauna of Orissa.*Part 1 (Insects), Zoological Survey of India, Calcutta, India. iv+390pp.

Parui, P., B. Mitra, M. Mukherjee, and R. S. Mridha. 2002. Further contribution on the Diptera (Insecta) fauna of Andaman and Nicobar Islands. *J. Bombay Nat. Hist. Soc.* 99(1): 135–137.

Patel, H. K., J. R. Patel, and S. N. Patel. 1975. Records of predators and their parasites from Gujarat. *Entomologists' Newsl.* 5: 40.

Patel, J. R., and H. K. Patel. 1969a. Some syrphids of Gujarat and their hymenopterous parasites. *Indian J. Entomol.* 31(1): 86–88.

Patel, J. R., and H. K. Patel. 1969b. The bionomics of the Syrphid fly, *Xanthogramma scutellare* [sic] Fab. (Syrphidae: Diptera). *Madras Agric. J.* 56(8): 516–522.

Patel, R. C., D. N. Yadav, and J. R. Patel. 1976. Natural control of ground-nut aphid, *Aphis craccivora* Koch in central Gujarat. *Curr. Sci.* 45(1): 34–35.

Patil, R. R., K. Ghorpadé, P. S. Tippanavar, and M. K. Chandaragi. 2013. New record of syrphid, *Chrysotoxum baphyrum* Walker (Diptera: Syrphidae) on the Sugarcane root aphid, *Tetraneura javensis* (van der Goot) in peninsular India. *J. Exp. Zool. India* 16(2): 557–560.

Patnaik, N. C., and K. C. Bhagat. 1976. Observations on the syrphid fauna of Puri district (Orissa) and bio-ecology of *Ischiodonscutellaris* Fabr., and *Eumerus* sp., nr. *albifrons* Walk. (Diptera: Syrphidae). *Prakruti-Utkal Univ. J. Sci.* 13(1 + 2): 43–50.

Patnaik, N. C., J. M. Satpathy, and K. C. Bhagat. 1977. Note on the occurrence of aphidophagous insect predators in Puri district (Orissa) and their predation on the sorghum aphid, *Longuinguis sacchari* (Zhnt.). *Indian J. Agric. Sci.* 47(11): 585–586.

Pearson, D. L., and F. Cassola. 1992. World-wide species richness patterns of Tiger Beetles (Coleoptera: Cicindelidae): Indicator taxon for biodiversity and conservation studies. *Cons. Biol.* 6(3): 376–391.

Pearson, D. L., and K. Ghorpadé. 1989. Geographical distribution and ecological history of Tiger Beetles (Coleoptera: Cicindelidae) of the Indian subcontinent. *J. Biogeogr.* 16: 333–344, 15 maps.

Peck, L. V. 1988. Family Syrphidae. pp. 11–230. In: Á. Soós. & L. Papp. *Catalogue of Palaearctic Diptera.* Vol. 8 Syrphidae—Conopidae. Akadémiai Kiadó, Budapest. 363 pp.

Puttannavar, M. S., R. K. Patil, G. K. Ramegowda, V. Mulimani, Shekarappa, and S. Lingappa. 2005. Survey of Sugarcane Woolly Aphid, *Ceratovacuna lanigera* Zehntner (Homoptera: Aphididae) incidence and its natural enemies in sugarcane growing areas of northern Karnataka. *Indian Sugar*, 54: 41–46.

Rahman, K. A. 1940. Important insect predators of India. *Proc. Indian Acad. Sci. (Ser. B).* 12: 67–74.

Rakshpal, R. 1945. Notes on a syrphid (*Sphaerophoria scutellaris* Fab.) fly predator on mustard-aphid (*Rhopalosiphum pseudobrassicae* Davis). *Curr. Sci.* 14: 235–236.

Ramegowda, G. K., R. K. Patil, V. Mulimani, M. S. Puttannavar, and S. Lingappa. 2006. Incidence and spread of Sugarcane Woolly Aphid, *Ceratovacuna lanigera* Zehntner (Homoptera: Karnataka Aphididae) in Karnataka. *J. Aphidol.* 20(2): 19–23.

Rao, V. P. 1969. India as a source of natural enemies of pests of Citrus. *Proc. 1st Intl Citrus Symp.* 2: 785–792.

Rawat, R. R., and B. N. Modi. 1968. A record of natural enemies of *Ferrisia virgata* Ckll. In Madhya Pradesh (India). *Mysore J. Agric. Sci.* 2: 51–53.

Reemer, M., and G. Ståhls. 2013a. Generic revision and species classification of the Microdontinae (Diptera, Syrphidae). *Zookeys* 288: 1–213.

Reemer, M., and G. Ståhls. 2013b. Phylogenetic relationships of Microdontinae (Diptera: Syrphidae) based on molecular and morphological characters. *Syst. Entomol.* 38: 661–688.

Rotheray, G. E., and F. S. Gilbert. 1999. Phylogeny of Palaearctic Syrphidae (Diptera): Evidence from larval stages. *Zool. J. Linn. Soc.* 127: 1–112.

Rotheray, G. E., and F. S. Gilbert. 1989. The phylogeny and systematics of European predacious Syrphidae (Diptera) based on larval and puparial stages. *Zool. J. Linn. Soc.* 95: 29–70.

Roy, P., and S. K. Basu. 1978. Bionomics of aphidophagous syrphid flies. *Indian J. Entomol.* 39(2): 165–174.

Sack, P. 1913. H. Sauter's Formosa-Ausbeute: Syrphidae (Dipt.). *Entomol. Mitt.*, 2(1): 1–10.

Sack, P. 1922. H. Sauter's Formosa-Ausbeute: Syrphiden II. (Dipt.). *Archiv für Naturgeschichte* (A) 87(11): 258–275.

Sack, P. 1930. Syrphidae. In: E. Lindner [Ed.] *Die Fliegen der Palaearktischen Region.* Schweizerbart, Stuttgart, Germany, Bd. 4, pt. 6, pp. 145–240.

Sack, P. 1932. Syrphidae. In: E. Lindner [Ed.] *Die Fliegen der Palaearktischen Region.* Schweizerbart, Stuttgart, Germany, Bd. 4, pt. 6, No. 31, 451 pp.

Sandhya, P. T., M. Subramanian, and K. Ghorpadé. 2016. Biology of ginger rhizome fly, *Mimegralla* sp. nr *coeruleifrons* (Diptera: Micropezidae). *Entomon* 41(3): 171–182.

Sathiamma, B. 1979. Occurrence of maggor pests on ginger. *Bull. Entomol.* 20: 143–144.

Shannon, R. C. 1927. Notes on and descriptions of syrphid flies of the subfamily Cerioidinae. *J. Wash. Acad. Sci.* 17(2): 38–53.

Shiraki, T. 1930. Die syrphiden der Japanischen Kaiserreichs, mit berucksichtigung benachbartergebiete. *Mem. Fac. Sci. Agric.* 1: 1–446

Shiraki, T. 1963. Insects of Micronesia. Diptera: Syrphidae. *Insects of Micronesia* 13(5): 129–187. 23 figs. Bernice P. Bishop Museum, Honolulu, Hawaii.

Skevington, J. H., and D. K. Yeates. 2000. Phylogeny of the Syrphoidea (Diptera) inferred from mtDNA sequences and morphology with particular references to classification of the Pipunculidae (Diptera). *Mol. Phylogenetics Evol.* 16(2): 212–224.

Skufjin, K. V. 1982. A review of the genus *Sphaerophoria* Lepeletier et Serville (Diptera, Syrphidae) in the fauna of the USSR. *Entomol. Rev.* 59(4): 134–142, 28 figs.

Sodhi, A. S., and S. Awtar. 1991. Three new species of family Syrphidae (Diptera) from India. *Acta Zool. Cracov.* 34(1): 315–322.

Stahls, G., H. Hippa, G. Rotheray, J. Muona, and F. Gilbert. 2003. Phylogeny of Syrphidae (Diptera) inferred from combined analysis of molecular and morphological characters. *Syst. Ent.* 28: 433–450.

Stebbing, E. P. 1908. *A Manual of Elementary Forest Zoology for India.* xxiii+229+xxiv pp. (Repr. Edn 1977). Superintendent Government Printing, Calcutta, India.

Stuckenberg, B. R. 1954a. Studies on *Paragus,* with descriptions of new species (Diptera: Syrphidae). *Rev. Zool. Bot. Afr.* 49: 97–139.

Stuckenberg, B. R. 1954b. The *Paragus serratus* Complex, with descriptions of new species (Diptera: Syrphidae). *Trans. R. Entomol. Soc., Lond.* 105(17): 393–422.

Thompson, F. C. 1971. Two new Oriental species of the genus *Myolepta* Newman. *Proc. Entomol. Soc. Wash.* 73(3): 343–347.

Thompson, F. C. 1974. Descriptions of the first known Ethiopian *Myolepta* species, with a review of the subgeneric classification of *Myolepta* (Diptera: Syrphidae), *Ann. Natal. Mus.* 22(1): 325–334.

Thompson, F. C. 2003. *Austalis,* a new genus of Flower flies (Diptera: Syrphidae) with revisionary notes on related genera. *Zootaxa* 246: 1–19.

Thompson, F. C. 2004. *Biosystematic Database of World Diptera.*http://www.sel.barc.usda.gov/Diptera/biosys.htm [Accessed: 1 November 2018].

Thompson, F. C. 2010. Syrphidae. *Systema Dipterorum*, Version 1.0. http://www.diptera.org/ [Accessed: 1 November 2018].

Thompson, F. C. 2017. First *Calcaretropidia* flower fly (Diptera: Syrphidae) known from New Guinea: Description of a new species. *Entomol. News* 127: 93–99.

Thompson, F. C., and K. Ghorpadé. 1992. A new coffee aphid predator, with notes on other Oriental species of *Paragus* (Diptera: Syrphidae). *Colemania* 5: 1–24.

Thompson, F. C., and J. H. Skevington. 2014. Afrotropical flower flies (Diptera: Syrphidae). A new genus and species from Kenya, with a review of the melanostomine group of genera. *Zootaxa* 3847(1): 97–114.

Thompson, F. C., F. D. Fee, and L. G. Berzark. 1990. Two immigrant synanthrophic flower flies (Diptera: Syrphidae) new to North America. *Entomol. News* 101(2): 69–74.

Thompson, F. C., X. Mengual, A. D. Young, and J. H. Skevington. 2017. Flower flies (Diptera: Syrphidae) of Philippines, Solomon Islands, Wallacea and New Guinea. pp. 501–524, pls 167–172. In: Telnov, D., M. V. L. Barclay and O. S. G. Pauwels [Eds.], *Biodiversity, Biogeography and Nature Conservation in Wallacea and New Guinea III*. The Entomological Society of Latvia, Riga.

Thompson, F. C., J. R. Vockeroth, and M. C. D. Speight. 1982. The Linnaean species of Flower Flies (Diptera: Syrphidae). *Mem. Ent. Soc. Wash.* 10: 150–165.

Thompson, F. C., and G. Rotheray. 1998. Family Syrphidae. pp. 81–139. In: L. Papp & B. Darvas [Eds.], *Contributions to a Manual of Palaearctic Diptera*. Vol. 3. Science Herald, Budapest.

Thompson, F. C., and J. R. Vockeroth. 1989. Family SYRPHIDAE. Pp. 437–458. In: N. L. Evenhuis [Ed.] *Catalog of the Diptera of the Australasian and Oceanian Regions*. Bishop Museum Press, the Netherlands, 1155 p.

Tiple, A. D., and K. Ghorpade. 2012. Butterflies (Lepidoptera-Rhopalocera) of the Achanakmar-Amarkantak Biosphere Reserve, Chhattisgarh and Madhya Pradesh, with a synopsis of the recorded butterfly fauna of the eastern Central Highlands in India. *Colemania*. 26: 1–38.

Upadhyaya, V. N., and T. Soares. 1964. A characteristic feature of wing venation of *Ischiodon scutellaris* (Fabr.) (Diptera: Syrphidae). *Indian J. Entomol.* 26: 209–210.

Usman, S., and M. Puttarudriah. 1955. A list of the insects of Mysore including the mites. vi+194 pp. *Dept Agric. Mysore State, Entomol. Series Bull.* 16, Bangalore, India.

Van der Wulp, F. M. 1896. *Catalogue of the Described Diptera from South Asia*. vii+220 pp. *Martinus Nijhoff for the Dutch Entomological Society*, The Hague, the Netherlands.

Van Steenis, J., and H. Hippa. 2012. Revision and phylogeny of the Oriental hoverfly genus *Korinchia* Edwards (Diptera: Syrphidae). *Tijdschr. Entomol.*, 155: 209–268, 216 figs.

Verrall, G. H. 1901. *Platypezidae, Pipunculidae, and Syrphidae of Great Britain*. vi+691 + 121 pp. Taylor & Francis Groups, London, UK. [Reprinted 1969 by E. W. Classey Ltd, Hampton, Middlesex, UK].

Violovitsh, N. A. 1974. A review of the Palaearctic species of the genus *Chrysotoxum* Mg. (Diptera, Syrphidae). *Entomol. Obozr.* 53: 196–217.

Violovitsh, N. A. 1975. A revision of the Palaearctic species of the genus *Scaeva* Fabricius, 1805 (Diptera, Syrphidae). *Ent. Obozr.* 54: 176–179.

Vockeroth, J. R. 1969. A revision of the genera of the Syrphini (Diptera: Syrphidae). *Mem. Entomol. Soc. Can.* 62, 176 pp.

Vockeroth, J. R. 1971. Some changes in the use of generic names in the Tribe Cerioidini (Diptera: Syrphidae). *Can. Entomol.* 103(2): 282–283.

Vockeroth, J. R. 1973. The identity of some genera of Syrphini (Diptera: Syrphidae) described by Matsumura. *Can. Entomol.* 105: 1075–1079.

Vujić, A., G. Ståhls, S. Rojo, S. Radenković, and Š. Simić. 2008. Systematics and phylogeny of the Tribe Paragini (Diptera: Syrphidae) based on molecular and morphological characters. *Zool. J. Linn. Soc.* 152: 507–536.

Wiedemann, C. R. W. 1824. Munus rectoris in Academia Christiana Albertina aditurus analecta entomologica ex Museo Regio Havniensi maxime congesta profert iconibusque illustrat. Kiliae [= Kiel], 60 pp.

Wiegmann, B. M. 1986. A new species of *Myolepta* (Diptera: Syrphidae) from Nepal, with its phylogenetic placement and a key to Oriental species. *J. N. Y. Entomol. Soc.* 94(3): 377–382.

Wright, S. G., and J. H. Skevington. 2013. Revision of the subgenus *Episyrphus* (*Episyrphus*) Matsumura (Diptera: Syrphidae) in Australia. *Zootaxa* 3683(1): 51–64.

Wyatt, N. P. 1991. Notes on *Citrogramma* Vockeroth (Diptera: Syrphidae) with descriptions of five new species. *Orient. Insects* 25: 155–169.

Zimsen, E. 1954. The insect types of C. R. W. Wiedemann in the zoological museum in Copenhagen. *Spol. Zool. Mus. Hauniensis* 14: 1–43.

19 A Comparative Study of Antennal Mechanosensors in Insects

Harshada H. Sant and Sanjay P. Sane

CONTENTS

Introduction 389
Antennal Mechanosensory-Motor Circuits in Diverse Insects 390
    Böhm's Bristles in Diverse Insects 392
    Johnston's Organs in Diverse Insects 394
    Additional Antennal Mechanosensors in Insects 395
Antennal Mechanosensory Structures in Other Arthropods 395
Conclusion 396
Acknowledgements 397
References 397

INTRODUCTION

Insects are remarkably diverse, with more than a million described species spanning 450 million years of evolution and an estimated number of 5.5 million (Stork et al. 2015). They occupy nearly every niche in terrestrial ecology, and their forms have evolved myriad adaptations to survive within these niches. They vary in size by three orders of magnitude, from 0.3 mm (e.g., fairyfly wasp, *Kikiki huna*) to 300 mm (e.g., birdwing moth, *Thysania agrippina*), and their lifestyles are diverse ranging from parasitoidism in small wasps, endoparasitism in Strepsiptera, to ferocious predation in praying mantises, or mutualism in ants and aphids (Resh and Cardé 2003). Their morphologies have been crafted by the selective pressures of their distinct habitats. As even an amateur insect collector recognizes, the color patterns of butterflies serve the role of sexual signals or the thickened elytra of beetles protect their hindwings from wear and tear. However, less recognized is the fact that, like these surface structures, insect behaviors are just as diverse and their evolutionary complexity also results from natural and sexual selection. Because behavioral responses are ultimately the output of neural circuits consisting of sensory organs, central circuits, and muscle responses, these components also display similar diversity.

Unlike morphological characteristics however, behavioral or neurobiological traits do not survive fossilization, and it is impossible to infer these traits from genetic sequences. To gain any insights into the evolution of behaviors requires us to adopt the comparative approach, in which we compare and contrast the neurobiological and behavioral traits across extant insect species, and use them to determine the generality of behavioral traits or underlying neural circuits across diverse groups. Such an approach is especially rewarding when synapomorphic traits—both behavioral and neuroanatomical—can be identified from studies on extant taxa. This approach, called "*neural cladistics*," has gained some ground in recent years—most prominently through the detailed comparative analyses of neuroanatomical data to help add resolution to relationships between insects and crustaceans (Strausfeld and Andrew 2011). The *neural cladistics* approach, like its morphological counterpart, adopts the Hennigian method (Hennig and Pont 1981) using neuroanatomical traits as data points for comparison (Strausfeld 2012). This method bases itself on the fundamental assumption that neural structures—whether at the level of ganglionic organization or neural circuitry—are conserved through the process of natural or sexual selection, and diversify in a manner that allows adaptation of their behavioral output to the requirements within their ecological niche. Moreover, these comparisons may not be restricted only to the adult nervous system, but can also be extended to various developmental stages of the insects (Katz et al. 1983; Angelini and Kaufman 2005; Moczek et al. 2006), thereby vastly enhancing the resolution of such derived relationships. Importantly, it synthesizes insights about the evolution of neural circuits and morphology thereby providing a more holistic view, and bringing behavioral data into the taxonomic realm.

The research described in this review was inspired largely by the approach described above, but we extend it to include behavioral output that emerges from neural circuits. Using an established and well-resolved phylogeny (Misof et al. 2014), we compared the behavior and the underlying neural machinery of extant insects. As a point of reference, we chose to focus on the *antennal positioning behavior* during flight as our study system because it offers several advantages. First, it is ubiquitous; to the best of our knowledge, all insects position their antennae at the onset of flight and thereafter maintain their antennal position during flight. Second, the underlying sensory and motor components are well studied in a few systems and appear to be generally conserved at least at the level of their organization. Third, the antennal positioning behavior is of key importance to insects; disrupting it causes loss of flight ability. Fourth, perhaps because the antenna is a critically important mechanosensory and olfactory organ, *antennal positioning behavior* and its underlying circuitry remains strongly conserved across all insects thus readily enabling such comparisons.

The precise functional role of *antennal positioning* during locomotor activities is poorly understood. Data from diverse systems largely point to correlations between antennal activity and sensory input or locomotor kinematics. There are sharp differences in antennal movements during flight vs. walking. During flight, insects precisely position their antennae, and this position is a function of diverse sensory inputs including airflow, optic flow, and perhaps also olfactory input. For example, in hawkmoths, antennal positioning precedes flight initiation, and is maintained and modulated during flight (Krishnan et al. 2012). In honeybees and hawkmoths, antennal positioning relative to speed of the body is regulated during flight in relation to airflow (Heran 1957) and optic flow stimuli (Roy Khurana and Sane 2016; Natesan et al. 2016). In locusts also, antennal positioning is well documented in response to frontal airflow (Heinzel and Gewecke 1987) In *Drosophila melanogaster*, passive or active antennal positioning is associated with visually guided turns (Mamiya et al. 2011). In addition to its putative role in proper acquisition of mechanosensory feedback relevant to flight control (Sane et al. 2007), the antennal positioning behavior has also been implicated in ensuring more efficient olfactory feedback (Duistermars et al. 2009), and it has also been found to be modulated by visual feedback in addition to mechanosensory feedback (Krishnan and Sane 2014; Roy Khurana and Sane 2016).

In contrast to flight, walking insects tend to move their antennae a great deal. Insects such as crickets (Honegger 1981), locusts (Saager and Gewecke 1989), and honeybees (Maronde 1991) actively move their antennae in a manner that can be visually targeted. Cockroaches constantly move their antennae to acquire mechanosensory inputs, which are critical for their ability to follow walls and boundaries, negotiate obstacles, and avoid collisions (Cowan et al. 2006; Harley et al. 2009; Baba et al. 2010). Likewise, stick insects also actively explore their environments using their antennae during walking (Schutz and Durr 2011; reviewed in Staudacher et al. 2005).

The above examples provide a strong motivation for studying antennal mechanosensors from a comparative viewpoint. The behavioral studies in conjunction with neuroanatomical insights enable a better understanding of how these circuits have evolved over the course of insect diversification.

ANTENNAL MECHANOSENSORY-MOTOR CIRCUITS IN DIVERSE INSECTS

The functional importance of antennae for insects can hardly be overstated. Antennae are present in all insects ranging from microscopic parasitoid wasps and flies to giant locusts and birdwings. Antennal morphology across insects is highly variable, reflecting the diversity of insect forms and functions (Krishnan and Sane 2015). The external appearance of the antenna ranges from very small and setaceous in dragonflies, moniliform in termites, filiform in cockroaches, geniculate in bees and ants, to capitate or plumose in butterflies and moths, and funicular-aristate (pouch with bristle) in flies (Figure 19.1a). In hyper-diverse orders like Coleoptera, we see representatives of almost all types of antennal morphologies. The evolution of antennal forms also goes hand in hand with the evolution of insect forms, as exemplified by findings in fossils of different insect orders from early cretaceous period with ramified antennae (Gao et al. 2016). These myriad shapes and sizes influence olfactory sampling by antennae because they determine the physical volume of the world that is sampled. They are also the key features for the mechanosensory modality because their weight and mechanics determine the nature of the mechanical stimulus received by the antennal mechanosensors. Indeed, as argued in his influential review by Schneider (1964), the estimated density of olfactory sensillae on the elaborately plumose silk moths *Bombyx mori* is not high, and hence it is likely that its exaggerated form serves an important role from the mechanosensory rather than an olfactory perspective.

The olfactory and mechanosensory receptors on the antenna mediate a large number of behaviors including olfaction, tactile orientation, and obstacle avoidance responses in cockroaches and stick insects (Okada and Toh 2000; Harley et al. 2009; Schütz and Durr 2011). In social insects including honeybees and ants, they are used for tactile communication between nest mates (Tautz and Rohrseitz 1999; Ettorre et al. 2004). In flies and other insects, antennal mechanosensors mediate audition, flow sensing, and graviception (Manning 1967; Gewecke 1970; Bennet-Clark 1971; Ewing 1978; Göpfert and Robert 2000; Kamikouchi et al. 2009; Yorozu et al. 2009), and their role in flight stabilization is well-documented in moths and bees (Sane et al. 2007; Roy Khurana and Sane 2016). Yet, despite this functional versatility, the basic organization of antennal sensory-motor apparatus is largely conserved across insects.

For the purposes of this review, we equate Ectognatha (which includes bristletails, silverfish, and pterygotes) with insects (Hennig and Pont 1981; Trautwein et al. 2012; Misof et al. 2014). All insects possess a tripartite antenna comprised of three segments—the basal scape, the medial pedicel, and multi-annular flagellum. In this type of antenna, muscle attachments are found only in the head capsule-scape and scape-pedicel joints, and, respectively, called the extrinsic and intrinsic muscles (Chapman 1982). The scape and pedicel of the antenna also possess the principal mechanosensory organs of the antennae. In contrast, the flagellum primarily houses olfactory receptors, in addition to sparsely distributed mechanoreceptors. A key question from the evolutionary perspective is whether the mechanosensory function of the antenna evolved early in insect evolution or whether it was an adaptation that evolved at a later stage.

A Comparative Study of Antennal Mechanosensors in Insects

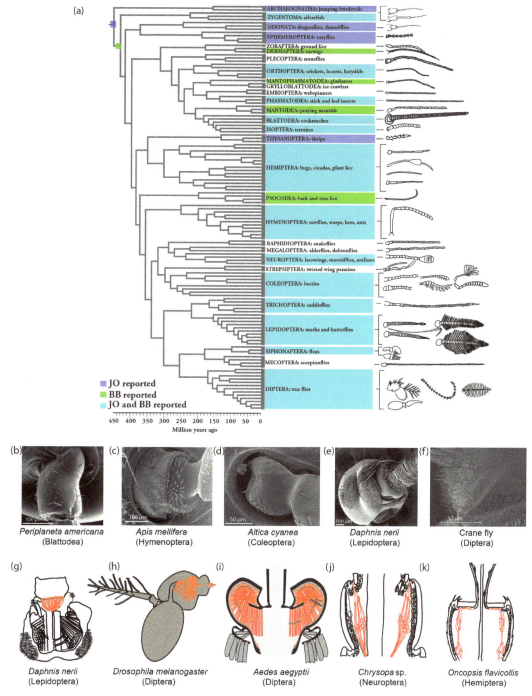

FIGURE 19.1 Antennal mechanosensors in diverse insect orders. (a) Antennal morphologies of different insect orders showcase the wide diversity of forms (right panel). On this phylogeny (Adapted from Misof, B. et al., *Science*, 346, 763–768, 2014.) are marked the insect orders in which both Johnston's organs (JO) and Böhm's bristles (or hair plates) are reported (cyan), or only JO (blue), and or only Böhm's bristles (green). These data were derived from Field and Matheson (1998); Mißbach et al. (2011); Mittmann and Scholtz (2001); Tillyard (1917); Brodskiy (1973); Drilling and Klass (2010); Dürr (2016); Toh (1981); Howse (1965); Bode (1986); Wang et al. (2016); Jeram and Cokl (1996); Hu et al. (2009); Melnitsky et al. (2018); Amrine and Lewis (1978). (Adapted from Krishnan, A. and Sane, S. P., *Adv. Insect Physiol.*, 49, 59–99, 2015.) Examples of Böhm's bristles (b–f) and Johnston's organs (g–k) in diverse insect orders. Böhm's bristles are shown in (b) Blattodea (*Periplaneta americana*), (c) Hymenoptera (*Apis mellifera*), (d) Coleoptera (*Altica cyanea*), (e) Lepidoptera (*Daphnis nerii*), and (f) Diptera (Crane fly). Johnston's organs are shown in (g) Lepidoptera (*Daphnis nerii*); (h) Diptera (*Drosophila melanogaster*); (i) Diptera (*Aedes aegypti*); (j) Neuroptera (*Chrysopa* sp.); and (k) Hemiptera (*Oncopsis flavicollis*). (The schematics are adapted from Sane, S. P. et al., *Science*, 315, 863–866, 2007; Jarman, A. P., Development of the auditory organ (Johnston's organ) in *Drosophila*, in: *Development of Auditory and Vestibular Systems*, Romand, R. and Varela-Nieto V. S., (Eds.), Academic Press, San Diego, CA, pp. 31–61, 2014; Sane, S. P. and McHenry M. J., *Int. Com. Biol.*, 49(6), i8–i23, 2009; Schmidt, K., *Zeitschrift Für Zellforschung Und Mikroskopische Anatomie*, 99, 357–388, 1969; Howse, P. E., *Proc. R.Entomol. Soc. Lond.*, 40, 137–146, 1965.) These data allow us to make preliminary inferences about last common ancestor that possessed JO (marked blue on the phylogeny) and Böhm's bristles (green), respectively. Insect orders for which data were not available were left unmarked.

Moreover, how have these structures evolved over the course of insect diversification? To address this question, we try to derive insights from a comparison of antennal mechanosensors in diverse insects.

Böhm's Bristles in Diverse Insects

Antennal movements are mediated by mechanosensory feedback from a set of bristles that monitor the relative movements of scape and pedicel. These are called *Böhm's bristles*, and they are homologous to hair plates in the legs hence also often termed as *antennal hair plates* (Krishnan and Sane 2015). Böhm's bristles are organized as distinct or continuous fields on the head-scape and scape-pedicel joints on antenna of diverse insects (Figure 19.1b–f). Feedback from Böhm's bristles is involved in active movements of the antennae during walking, as well as the maintenance and control of their position during flight. For example, in cockroaches, different fields of Böhm's bristles are directionally selective and important in guiding antennal movements during object detection and tactile localization responses (Okada and Toh 2001). In hawkmoths (Krishnan et al. 2012) and honeybees (Roy Khurana and Sane 2016), their ablation causes complete loss of the ability to position antennae and severely disables both flight and walking. In hawkmoths, the axonal projections of Böhm's bristles terminate in the antennal mechanosensory and motor centre (AMMC) region of the deutocerebrum, co-localize with the dendrites of antennal motor neurons, and very likely form monosynaptic connections with them (Krishnan et al. 2012; Sant and Sane 2018). This may be modelled as a negative-feedback circuit, in which mechanosensory inputs from the Böhm's bristles directly activate the antennal muscles to ensure that moths can reflexively maintain and modulate their antennal position during flight or other locomotory activities (Krishnan et al. 2012; Krishnan and Sane 2014).

Böhm's bristles mediate control of antennal positioning during locomotion in nearly all insects. The ubiquity of this *antennal positioning response* across insect orders suggests that its underlying sensory-motor components may be evolutionarily conserved. To gain insights into the evolution of these sensors and insect antenna, it is therefore essential to compare the morphology of these mechanosensory structures and their underlying neural circuitry in diverse insects. Such studies may also highlight how the underlying neural circuits are shaped by natural selection. Comparison of homologous circuit structures has been informative in a number of invertebrate systems and used effectively to explore how behaviors have arisen and diversified (e.g., Katz and Harris-Warrick 1999; Katz 2011; Strausfeld 2012).

Using scanning electron microscopy, we systematically explored and established the presence of Böhm's bristles in diverse insect orders including Zygentoma (silverfish), Dermaptera (earwig), Orthoptera (locust, cricket), Mantodea (praying mantis), Blattodea (cockroach), Isoptera (termites), and all holometabolous insects, *albeit* with a few exceptions (Figure 19.1b–f). In holometabolous insects, Böhm's bristles are typically arranged in discrete or continuous fields on the scape and pedicel. The scapal fields vary in number and gross morphology in the different insect orders. For instance, in Hymenoptera, hair plates are arranged as a continuous field of bristles on the scapal surface and two distinct fields on the pedicel. In lacewings (Neuroptera), however, one continuous field of bristles is present on the pedicel. In orthopterans such as crickets, scapal fields are located orthogonal to each other at the hinge-like head-scape joint and with the bristles arranged in two columns, unlike the circular fields in other insects. In Dermaptera, scapal fields contain only 3–4 bristles in each field, substantially lower than most other insects. Interestingly, these structures are also present in miniaturized insects of sub-millimetre sizes. For example, fields of Böhm's bristles are present in the parasitoid wasp, *Trichogramma* sp. on the head-scape joint (Figure 19.2a,b), which differ in their organization as compared to the continuous arrangement seen on the scape of honeybees, which are approximately two orders of magnitude larger (Figure 19.1c). This may not be a general rule, however, because small lepidopterans such as *Plodia interpunctella* (Figure 19.2c,d) contain a similar bristle arrangement as larger moths (Figure 19.1e). In both cases, the number of sensillae in the Böhm's bristles fields are severely reduced, as is the case with sensory units of all other sensory organs.

The organization of Böhm's bristles in insects largely depends on the joint morphologies of the head capsule-scape-pedicel segments, which are variable across insects (reviewed in Staudacher et al. 2005). For instance, ball-and-socket type head capsule-scape joints in cockroaches, beetles, honeybees, etc. usually contain Böhm's bristle fields that cover the entire scape either as discrete or continuous fields. In contrast, antennae with a hinge joint in crickets, stick insects, etc. possess localized bristle fields, which are stimulated when these joints flex. In both cases, the placement of bristles reflects their role in sensing joint movement.

The conserved patterns of Böhm's bristles and their role in antennal positioning are also reflected in the underlying neural circuitry. Insects from phylogenetically distinct orders with diverse antennal morphology and function nevertheless show similar patterns of axonal projections of Böhm's bristles relative to the antennal motor neurons. These projections always terminate in the ipsilateral AMMC where their axonal arbors typically co-localize with the dendritic arbors of the antennal motor neurons (Krishnan et al. 2012; Sant and Sane 2018) (Figure 19.3b,d,f,h). In insects such as crickets and honeybees, the Böhm's bristle arbors extend to the sub-oesophageal zone (SEZ) (Figure 19.3b,f). Similar projection patterns may exist in cockroaches also (Figure 19.3d).

These data indicate that Böhm's bristles were present in the common ancestor of all insects, except Archaeognatha in which it is not clear if these were secondarily lost or if these are only a synapomorphic feature of non-Archaeognathan insects. They also appear to be absent in paleoneopterans such as dragonflies (Odonata), mayflies (Ephemeroptera), and in some species of Condylognatha [Hemiptera (aphids) and Thysanoptera (thrips)]. In many extant insects, including several species of Diptera, Böhm's bristles are not visible and

A Comparative Study of Antennal Mechanosensors in Insects

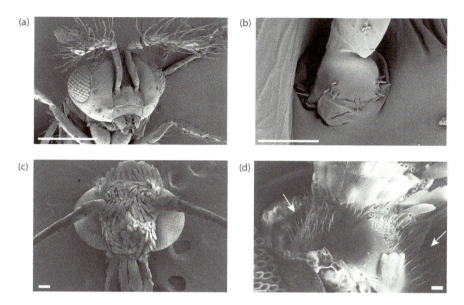

FIGURE 19.2 Böhm's bristles in miniature or small insects. Scanning electron microscopy images of representatives of two different insect orders are shown for (a) the miniature parasitic wasp (*Trichogramma* sp., Hymenoptera) head and (b) the associated Böhm's bristles, which are arranged in discrete fields. Also shown are the head of the Indian meal moth (*Plodia interpunctella*, Lepidoptera), (c) and the associated Böhm's bristles (d), which too are organized as discrete fields (white arrows). Scale bars in (a,c) represent 100 μm and those in (b,d) represent 10 μm.

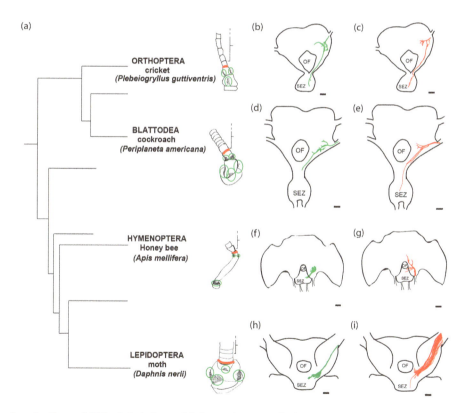

FIGURE 19.3 Axonal projections of Böhm's bristles and Johnston's organs in insects. (a) Phylogenetically distinct species were chosen to compare central projection patterns of antennal mechanosensors. Schematic adjacent to species indicates morphology of Böhm's bristles (green) and Johnston's organs (red). The projection patterns in (b,c) cricket, *Plebeiogryllus guttiventris*, (d,e) cockroach, *Periplaneta americana*, (f,g) honeybee, *Apis mellifera*, and (h,i) moth, *Daphnis nerii* are shown. Axonal projections of Böhm's bristles in crickets (b), honeybees (f) project to AMMC and SEZ. In cockroach too, the antennal mechanosensors project into the AMMC and towards SEZ (d) unlike projections in moths which terminate in AMMC (h). Scolopidial projections in all species project to both AMMC and SEZ (c,e,g,i). All mechanosensory projections are ipsilateral. However, finer arborization patterns of mechanosensory axonal projections differ amongst species. Scale bar represents 100 μm. OF—Oesophageal Foramen, AMMC—antennal mechanosensory and motor centre, SEZ—sub-oesophageal zone. Phylogeny adapted from Krishnan, A. and Sane, S. P., *Adv. Insect Physiol.*, 49, 59–99, 2015.

may have been secondarily lost. Böhm's bristles are absent in certain, but not all, clades of Diptera. Crane flies, which belong to the most basal taxa, Tipulomorpha (Wiegmann et al. 2011) possess scapal hair plates, but these are absent in mosquitoes, although they are closely related to crane flies. Brachyceran flies appear to have lost Böhm's bristles, although they do retain the ability to position their antennae during flight (Fuller et al. 2014). Diptera thus present a fascinating and extreme example of antennal modifications within a clade. More studies are required to fully resolve their evolution and loss in Diptera. It would be interesting to study the consequences of such loss, especially because their role is clearly defined in the insects in which they are present. One testable prediction is that, in these insects, the joint movement is either reduced or absent which could render the presence of such bristles unnecessary. Alternatively, as is the case with Diptera (Gewecke and Schlegel 1970), the role of detecting joint movement may be performed by internal mechanosensors, such as chordotonal organs.

Johnston's Organs in Diverse Insects

Whereas Böhm's bristles sense slower and coarser movements of the antennal scape and pedicel, they cannot sense the subtler movements of the flagellum relative to the pedicel. The movement of this joint is passive in all insects, due to lack of any muscles that can actively move the flagellum relative to the pedicel. Nevertheless, the flagellum does experience small-amplitude subtle vibrations, which encode a range of very critical mechanical stimuli. These vibrations are sensed by a set of highly sensitive scolopidial units that span the pedicel-flagellar joint. Together, these scolopidial units make up the Johnston's organ, which is a characteristic feature of insects with annulate antenna, but are not reported in segmented antenna (Imms 1940). Johnston's organs are extraordinarily sensitive (e.g., Dieudonné et al. 2014) and structurally and functionally varied across insects. Given their extreme sensitivity and the diverse lifestyles of various insect orders, it is not surprising that Johnston's organs take on highly specialized forms across insects. Even within an individual insect, Johnston's organs detect diverse kinds of mechanical stimuli.

The structural diversity of Johnston's organs is reflected in the varying numbers and innervation of scolopidial units in diverse insect orders (for example, see Figure 19.1g–k). These numbers range from 140 scolopidia in Oleander hawk moth, *Daphnis nerii*, 650 scolopidia in *Manduca sexta* (Lepidoptera) (Figure 19.1g) (Vande Berg 1971; Sant and Sane 2018), 150 scolopidia in the American cockroach *Periplaneta americana* (Blattodea) (Toh 1981) to 720 scolopidia in the European honeybee, *Apis mellifera* (Hymenoptera) (Ai et al. 2007). In Diptera, Johnston's organs are widely variable; compare, for instance, *Drosophila melanogaster*, with 477 scolopidial neurons that are focally attached to the base of the funiculus (Figure 19.1h) (Kamikouchi et al. 2006), with 15,000 in males of mosquitoes (*Aedes aegypti*) in which the scolopidia, in addition to being much more numerous, are densely packed in the pedicel (Figure 19.1i) (Boo and Richards 1975).

In *Drosophila*, each scolopidial unit is innervated by two or three bipolar neurons (Kamikouchi et al. 2006), but in the hawk moths, *Daphnis nerii* and *Manduca sexta*, a single sensory neuron is associated with each scolopidial unit (Vande Berg 1971; Sane et al. 2007; Sant and Sane 2018). In the beetle, *Paussus favieri*, nine groups of chordotonal organs are lined along the circumference of the antennal segment with two to five scolopidia per group, each innervated by three sensory neurons (Di Giulio et al. 2012). Similarly, in Thysanoptera, its 32 scolopidia are arranged as five groups of five scolopidia and one group of six scolopidia. However, unlike *Paussus favieri*, ten of these scolopidia have two sensory neurons innervating them (Bode 1986).

The diversity of Johnston's organs also extends to the structure of scolopidia in different insects. Scolopidia are categorized as mononematic if scolopidia have sub-epidermal caps, and amphinematic if there is no cap, but a thin thread drawn from the sheath surrounding the distal end of cilium which joins the cuticle or terminates just beneath it (Field and Matheson 1998). Johnston's organs of lacewings (Neuroptera) are relatively sparsely populated (Figure 19.1j) and have amphinematic scolopidia containing one enveloping cell, one scolopale cell, three sense cells, and one accessory cell (Schmidt 1969). Johnston's organs of leaf-hopper *Oncopsis flavicollis* (Hemiptera) are even sparser (Figure 19.1k) and have three dendrites with terminal ciliary processes (Howse 1965). In contrast, in mayfly, (Ephemeroptera) Johnston's organs are composed of mononematic scolopidia and innervated by two sensory cells except the innermost scolopidia which have three sensory cells (Schmidt 1974). The scolopidia in the beetle *Paussus favieri* (Coleoptera) are also mononematic (Di Giulio et al. 2012). Interestingly, *Thrips validus, Aeolothrips intermedius,* and *Hapfothrips aculeatus* (Thysanoptera) possess both amphinematic and mononematic scolopidia (Bode 1986). These comparisons reveal the sheer variety of Johnston's organ designs.

Comparing Johnston's organs in different insects reveal that despite differences in the number of scolopidial neurons, axonal projections terminate mainly in the AMMC in the deutocerebrum and the SEZ. There are, however, some differences across insects: in *Drosophila melanogaster*, the sensory projections of the Johnston's organs are segregated into five zones—four in the AMMC and one in the ventrolateral protocerebrum and the SEZ, in addition to the AMMC. Scolopidial neurons may project contralaterally in flies (Kamikouchi et al. 2006), but they are strictly ipsilateral in hawk moths (Sant and Sane 2018). We conducted such a comparison in four representative insects (Figure 19.3a) for the central projections of both Böhm's bristles and Johnston's organs (Figure 19.3a–i). In crickets (Figure 19.3b,c) and cockroaches (Figure 19.3d,e), Böhm's bristles and scolopidial projections terminate in AMMC and towards SEZ. Böhm's bristles in honeybees also show a similar projection pattern (Figure 19.3f). Additionally, sensory afferents of Johnston's organs have elaborate projections in the dorsal lobe, medial protocerebrum, as well as dorsal SEZ (Ai et al. 2007) (Figure 19.3g). However, although scolopidial neurons in the Oleander hawkmoth,

Daphnis nerii, terminate in the ipsilateral AMMC, with a diffused arbor with few projections in the SEZ (Figure 19.3i), projections of Böhm's bristles terminate in AMMC (Sant and Sane 2018) (Figure 19.3h). Moreover, axonal arbors are layered in the AMMC, which may indicate a tonotopic arrangement similar to that observed in flies (Kamikouchi et al. 2009; Yorozu et al. 2009; Matsuo et al. 2014; Sant and Sane 2018).

One consequence of the structural diversity of the Johnston's organs is its vast functional versatility within and across insects. They are range-fractionated, which means that individual scolopidia are differently tuned such that they can encode a large range of frequencies of mechanical vibrations. In hawk moths, for example, different scolopidia are tuned to frequency range from 0 to 100 Hz or more (Sane et al. 2007; Ai et al. 2007; Kamikouchi et al. 2009; Yorozu et al. 2009; Dieudonné et al. 2014). Thus, Johnston's organs encode a wide spectrum of mechanosensory stimuli ranging from low-frequency tactile or airflow cues, to high-frequency auditory or antennal vibration cues. Studies in diverse insects have recently begun to document the staggering functional diversity of Johnston's organs (Krishnan and Sane 2015). For example, in the hawk moth *Manduca sexta*, they detect high frequency, small-amplitude deflections of the antenna which help in stabilizing flight (Sane et al. 2007). In *Drosophila melanogaster*, they are activated by passive movements of the funiculus-pedicel joint to modulate stroke amplitude of the contralateral wing (Mamiya et al. 2011) and also the small vibrations due to the sounds from the song generated by the courting male (Göpfert et al. 1999). Perhaps the most elaborate structure of the Johnston's organs appears in mosquitoes, where the scolopidial numbers are enhanced by nearly hundredfold, thereby making the basal segments of the antennae appear doughnut shaped to accommodate them (Boo and Richards 1975). They are responsible for the exquisite sensitivity of the mosquitoes to sounds of conspecific mates, which must occur under low light conditions. In addition, Johnston's organs subserve a wide range of sensory functions in diverse insects including tactile sensing, graviception, anemosensing, and audition (reviewed in Staudacher et al. 2005; Matsuo and Kamikouchi 2013).

Thus, although the central organization of Johnston's organs is conserved, its structure and function can be highly variable and versatile between diverse insects. From an evolutionary standpoint, the Johnston's organ is likely homologous to the leg chordotonal organs (Krishnan and Sane 2015) and hence was probably a key feature also in the common ancestor of all modern insects (Figure 19.1a). These data suggest that the ancestors of modern insects probably had the same basic organization of antennal mechanosensors as extant insects.

ADDITIONAL ANTENNAL MECHANOSENSORS IN INSECTS

In addition to Böhm's bristles and Johnston's organs, insect antennae also possess chordotonal organs in the base of their antennae and mechanosensory structures all along the flagellum. The basal chordotonal organs are likely involved in tracking the finer movements of the antennae (Gewecke 1972) and have been classified into various types based on their innervation and structure (reviewed in Field and Matheson 1998). In addition to Johnston's organs, some insects also possess other antennal chordotonal organs, like Janet's organs in hymenopterans (Janet 1911). Antennal mechanosensors in arthropods include muscle receptor organ, different kinds of sensory setae, unique mechanosensory hair associated with a chordotonal organ, etc. (Strickler and Bal 1973; Vedel and Monnier 1983; Weatherby et al. 1994; Kouyama and Shimozawa 1982). In contrast, the flagellar mechanosensors are thought to be involved in some aspects of tactile sensation, such as antenna-mediated collision avoidance or wall following in cockroaches (Harley et al. 2009; Baba et al. 2010). Due to the paucity of studies on their function, we have excluded these from our review.

In insects, mechanosensors range from long hair-like sensilla trichodea, to shorter hair with characteristic base, sensilla chaetica, to thick-walled campaniform sensilla, and so on. All of them are innervated by bipolar sensory neurons. Based on their location and structural properties, they detect diverse mechanical stimuli to the antenna (reviewed in Schneider 1964). Arachnids and some chelicerates such as ticks have mechanosensory hairs similar to those in insects. However, they have more than one sensory neuron innervating them (three neurons in spiders and two in ticks). In copepods, antennal hairs are not set into a socket, but are constricted basally and innervated by bipolar neurons (McIver 1975). Insects and crustaceans also have internal mechanoreceptors called chordotonal organs that are usually not associated with exoskeletal structures like hairs, bristles, etc. First described in insects, they are found at every skeletal joint providing proprioceptive feedback to the animal. They are comprised of many scolopidia innervated by one or more bipolar sensory neurons with a ciliated dendrite enveloped by scolopale and attachment cells (also see Keil 1998; Field and Matheson 1998; Yack 2004; Staudacher et al. 2005).

ANTENNAL MECHANOSENSORY STRUCTURES IN OTHER ARTHROPODS

Antennae are key features of arthropods (other than insects) including Collembola, Diplura, Crustacea, and Myriapoda (Figure 19.4a). They are also present in ancient arthropods such as trilobites (Strausfeld 2012). Of the extant taxa, Chelicerates and Protura lack antennae (Figure19.4). Crustaceans have two sets of antennae—first pair is homologous to insect antennae and has neural projections in the deutocerebrum. Insects have one pair of antennae, with projections in the deutocerebrum and have lost the second pair (Rospars 1988; Staudacher et al. 2005). Insects have annulate antennae, with muscles in only the basal segments of the antenna (Figure 19.4g,h). Unlike insects, Myriapoda (Figure 19.4b), Collembola (Figure 19.4e), and Diplura (Figure 19.4f) have segmented antennae with musculature in all segments. In Crustacea, some members have segmented antennae (e.g., Ostracoda) (Figure 19.4d) and some have annulate antennae (e.g., Malacostraca) (Figure 19.4c) (Imms 1939; Schneider 1964). The morphology of crustacean

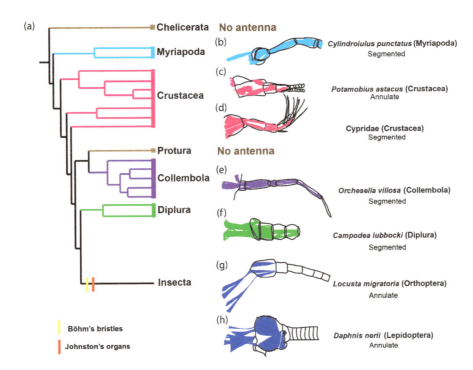

FIGURE 19.4 Antennal structures of other arthropods. (a) Antennae are essential features of all arthropods, with the possible secondary loss or modifications seen in Protura and Chelicerata. Segmented antennae, with muscles in all segments (different colours) are present in (b) Myriapoda (*Cylindroiulus punctatus*; light blue), (e) Collembola (*Orchesella villosa*; purple), and (f) Diplura (*Campodea lubbocki*; green). Annulate antennae have muscles only in the basal segments of antenna and possess a passively moving flagellum. Crustaceans have both segmented antennae as seen in (d) Cypridae (Ostracod), as well as annulate antennae as seen in (c) crayfish, *Potamobius astacus* (both shown in magenta). All insects possess annulate antennae, as seen in (g) Orthoptera (*Locusta migratoria*) and (h) Lepidoptera (*Daphnis nerii*) (both shown in blue). Böhm's bristles (yellow) and Johnston's organs (red) are present in most insects, but have not yet been reported in the other arthropods. Chordotonal organs, on the other hand, are present in antennae of insects and crustaceans. Further studies are required to investigate presence of homologous structures in crustaceans. (Schematics adapted from Imms, A. D., *Q. J. Microsc. Sci.*, 2, 273–320, 1939; Sant, H. H. and Sane S. P., *J. Comp. Neurol.*, 526, 2215–2230, 2018; phylogeny adapted from Misof, B. et al., *Science*, 346, 763–768, 2014.)

annulate antennae resembles insect antennae. For example, in the rock lobster (*Palinurus vulgaris*), the antenna is composed of a long distal flagellum and three basal segments named S1, S2, and S3 articulated by the joints J1 (S1–S2 joint), J2 (S2–S3 joint), and J3 (S3-flagellum joint). J0 joint articulates the basal most segment with the cephalothorax (Vedel 1980). They have proprioceptive chordotonal organs in antennal segments, aiding in a variety of antennal movements like resistance reflexes (muscle activation caused by resisting antennal movements) (Clarac et al. 1976; Barnes and Neil 1982). Similar proprioceptive organs associated with muscles are also found in antennal segments of the Collembolan, *Allacma fusca* (Altner 1988).

The basal segments of *Palinurus* antennae have hair plate-like sensory structures, with striking similarity to Böhm's bristles of insects. These also provide information on joint position (Clarac et al. 1976). Antennae of crayfish and lobster show wide ranges of antennal movements (Sandeman and Wilkens 1983; Barnes and Neil 1982). Considering the diversity of crustacean forms, it will be interesting to determine if crustacea also show presence of hair plates aiding antennal positioning movements (Figure 19.4). Studies have shown that inputs from legs drive the ipsilateral movements of the antennae with movement at the coxo-basal leg joint as a major source of sensory input (Clarac et al. 1976). This sensory-motor system is homologous to insects (reviewed in Krishnan and Sane 2015). In arthropods like Arachnids, which lack antennae, proprioceptive hair plates have been described in legs. These hair plates, which typically comprise of 25–70 hairs, are situated on the ventrolateral leg coxae. Movement of the coxal joint deflects them by the bulging joint membrane. Each plate hair is innervated by one sensory cell (Seyfarth et al. 1990). These are similar to hair plates observed in legs as well as antennae in insects (reviewed in Krishnan and Sane 2015). It will be interesting to investigate homologous structures in crustacean annulate antennae, which closely resemble insect antennae. Comparative studies will also provide insights into the evolution of these mechanosensors in diverse arthropods.

CONCLUSION

Böhm's bristles and Johnston's organs are essential in monitoring the coarse and fine movements of the scape, pedicel, and flagellar segments of the antennae. Because antennal movements arise from a myriad range of mechanical stimuli, antennal mechanosensors are central to the complex role of antennae as a sensory probe in all insects. Despite the vast differences in the antennal morphology and the brain structure across insects, the underlying neural circuitry and behavioral function of antennal mechanosensors is highly conserved, underscoring the importance of the role of antennae as mechanosensory structures, that inform insects about their sensory environment.

ACKNOWLEDGEMENTS

Comparative analyses such as the research showcased in this chapter, rests on the shoulders of careful taxonomic research by natural historians. The writing of this chapter was inspired by the extraordinary research career of our esteemed colleague Prof. C. A. Viraktamath, Gandhi Krishi Vignan Kendra (GKVK), Bangalore, India. At a time when the focus on "modern biology" comes at the cost of detailed taxonomic expertise, we hope that the above chapter highlights the importance of natural history work for gaining insights into the evolution of sensory-motor systems in animals. We wish to thank Prof. Viraktamath for his lifelong service to the natural history of insects and for his inspirational role as an educator for a large number of Indian entomologists. Many thanks also to Prof. K. Chandrashekara, GKVK for his numerous suggestions and help throughout the course of this research. Funding for this work was provided by intramural grants from the National Center for Biological Sciences, Tata Institute of Fundamental Research and extramural grants from the US Air Force Office of Scientific Research (AFOSR; FA2386-11-1-4057 and FA9550-16-1-0155) to SPS.

REFERENCES

Ai, H., H. Nishino, and T. Itoh. 2007. Topographic organization of sensory afferents of Johnston's organ in the honeybee brain. *The Journal of Comparative Neurology* 502(6): 1030–1046. doi:10.1002/cne.21341.

Altner, H. 1988. The scolopidial organs in the first antennal segment in *Allacma fusca* (Collembola, Sminthuridae). *Zoomorphology* 108(3): 173–181. doi:10.1007/BF00363934.

Amrine, J. W., and R. E. Lewis. 1978. The topography of the exoskeleton of *Cediopsylla simplex* (Baker 1895) (Siphonaptera: Pulicidae). I. The head and its appendages. *The Journal of Parasitology* 64(2): 343–358.

Angelini, D. R., and T. C. Kaufman. 2005. Insect appendages and comparative ontogenetics. *Developmental Biology* 286(1): 57–77. doi:10.1016/j.ydbio.2005.07.006.

Baba, Y., A. Tsukada, and C. M. Comer. 2010. Collision avoidance by running insects: Antennal guidance in cockroaches. *The Journal of Experimental Biology* 213(13): 2294–2302. doi:10.1242/jeb.036996.

Barnes, W. J., and D. Neil. 1982. Reflex antennal movements in the spiny lobster, *Palinurus elephas* II. Feedback and motor control. *Journal of Comparative Physiology* 147: 269–280.

Bennet-Clark, H. C. 1971. Acoustics of insect song. *Nature* 232(5313): 655–657. doi:10.1038/234255a0.

Bode, W. 1986. Fine structure of the scolopophorous organs in the pedicel of three species of Thysanoptera (Insecta). *International Journal of Insect Morphology and Embryology* 15(3): 139–154.

Boo, K. S., and A. G. Richards. 1975. Fine structure of the scolopidia in the Johnston's organ of male *Aedes aegypti* (L.) (Diptera: Culicidae). *International Journal of Insect Morphology and Embryology* 4(6): 549–566. doi:10.1016/0020-7322(75):90031-8.

Brodskiy, A. K. 1973. The swarming behavior of mayflies (Ephemeroptera). *Entomological Review* 52: 32–39.

Chapman, R. F. 1982. *The Insects: Structure and Function.* Harvard University Press, Cambridge, MA. Retrieved from https://books.google.co.in/books?id=xZjwAAAAMAAJ.

Clarac, F., D. M. Neil, and J. P. Vedel. 1976. The control of antennal movements by leg proprioceptors in the rock lobster, *Palinurus vulgaris*. *Journal of Comparative Physiology A* 107(3): 275–292. doi:10.1007/BF00656738.

Cowan, N. J., J. Lee, and R. J. Full. 2006. Task-level control of rapid wall following in the American cockroach. *Journal of Experimental Biology* 209(15): 3043–3043. doi:10.1242/jeb.02433.

Di Giulio, A., E. Maurizi, M. V. Rossi Stacconi, and R. Romani. 2012. Functional structure of antennal sensilla in the myrmecophilous beetle *Paussus favieri* (Coleoptera, Carabidae, Paussini). *Micron* 43(6): 705–719. doi:10.1016/j.micron.2011.10.013.

Dieudonné, A., T. L. Daniel, and S. P. Sane. 2014. Encoding properties of the mechanosensory neurons in the Johnston's organ of the hawk moth, *Manduca sexta*. *The Journal of Experimental Biology* 217: 3045–3056. doi:10.1242/jeb.101568.

Drilling, K., and K. D. Klass. 2010. Surface structures of the antenna of Mantophasmatodea (Insecta). *Zoologischer Anzeiger* 249(3–4): 121–137. doi:10.1016/j.jcz.2010.07.001.

Duistermars, B. J., D. M. Chow, and M. A. Frye. 2009. Flies require bilateral sensory input to track odor gradients in flight. *Current Biology* 19(15): 1301–1307. doi:10.1016/j.cub.2009.06.022.

Dürr, V. 2016. Stick insect antennae. In *Scholarpedia of Touch* pp. 45–63. doi:10.2991/978-94-6239-133-8.

Ettorre, P. D., J. Heinze, C. Schulz, W. Francke, and M. Ayasse. 2004. Does she smell like a queen? Chemoreception of a cuticular hydrocarbon signal in the ant *Pachycondyla inversa*. *Journal of Experimental Biology* 207: 1085–1091. doi:10.1242/jeb.00865.

Ewing, A. W. 1978. The antenna of *Drosophila* as a "love song" receptor. *Physiological Entomology* 3(1): 33–36. doi:10.1111/j.1365-3032.1978.tb00129.x.

Field, L. H., and T. Matheson. 1998. Chordotonal organs of insects. *Advances in Insect Physiology* 27(C). doi:10.1016/S0065-2806(08)60013-2.

Fuller, S. B., A. D. Straw, M. Y. Peek, R. M. Murray, and M. H. Dickinson. 2014. Flying *Drosophila* stabilize their vision-based velocity controller by sensing wind with their antennae. *Proceedings of the National Academy of Sciences* 111(13): E1182–E1191. doi:10.1073/pnas.1323529111.

Gao, T., C. Shih, C. C. Labandeira, J. A. Santiago-Blay, Y. Yao, and D. Ren. 2016. Convergent evolution of ramified antennae in insect lineages from the Early Cretaceous of Northeastern China. *Proceedings of the Royal Society B: Biological Sciences*, 283(1839): 20161448. doi:10.1098/rspb.2016.1448.

Gewecke, M. 1970. Antennae: Another wind-sensitive receptor in locusts. *Nature* 225(5239): 1263–1264. doi: 10.1038/2251263a0.

Gewecke, M. 1972. Bewegungsmechanismus und gelenkrezeptoren der Antennen von *Locusta migratoria* L. (Insecta, orthoptera). *Zeitschrift Für Morphologie Der Tiere*. doi:10.1007/BF00298573.

Gewecke, M., and P. Schlegel. 1970. Die Schwingungen der Antenne und ihre Bedeutung far die Flugsteuerung bei *Calliphora erythrocephala*. *Z. Vergl. Physiologie* 67: 325–362.

Göpfert, M. C., H. Briegel, and D. Robert. 1999. Mosquito hearing: Sound induced antennal vibrations male and female *Aedes aegypti*. *The Journal of Experimental Biology* 202: 2727–2738.

Göpfert, M. C., and D. Robert. 2000. Nanometre-range acoustic sensitivity in male and female mosquitoes. *Proceedings of the Royal Society B: Biological Sciences* 267(1442): 453–457. doi:10.1098/rspb.2000.1021.

Harley, C. M., B. A. English, and R. E. Ritzmann. 2009. Characterization of obstacle negotiation behaviors in the cockroach, *Blaberus discoidalis*. *The Journal of Experimental Biology*. 212: 1463–1476. doi:10.1242/jeb.028381.

Heinzel, H. G., and M. Gewecke. 1987. Aerodynamic and mechanical properties of the antennae as air-current sense organs in *Locusta migratoria*. *Journal of Comparative Physiology A* 161: 671–680.

Hennig, W., and A. C. Pont. 1981. *Insect Phylogeny*. New York: John Wiley & Sons.

Heran, H. 1957. Die Bienenantenne als Meflorgan der Fiugeigengeschwindigkeit. *Kurze Originalmitteilungen* 44: 475.

Honegger, H. W. 1981. A preliminary note on a new optomotor response in crickets: Antennal tracking of moving targets. *Journal of Comparative Physiology A* 142(3): 419–421. doi:10.1007/BF00605454.

Howse, P. E. 1965. The structure of the subgenual organ and certain other mechanoreceptors of the termite. *Proceedings of the Royal Entomological Society of London* 40: 137–146.

Hu, F., G. N. Zhang, and J.-J. Wang. 2009. Antennal sensillae of five stored-product psocids pests (Psocoptera: Liposcelididae). *Micron* 40(5): 628–634. doi:10.1016/j.micron.2009.02.006.

Imms, A. D. 1939. On the antennal musculature in insects and other arthropods. *Quarterly Journal of Microscopical Science* 2(322): 273–320.

Imms, A. D. 1940. On growth processes in the antennae of insects. *Quarterly Journal of Microscopical Science* 81: 585–593.

Janet, C. 1911. *Sur l'existence d'un organe chordotonal et d'une vésicule pulsatile antennaires chez l'Abeille et sur la morphologie de la tête de cette espèce* (Vol. 22). Limoges, France: Ducourtieux et Gout.

Jarman, A. P. 2014. Development of the auditory organ (Johnston's organ) in *Drosophila*. In *Development of Auditory and Vestibular Systems*. Romand, R. and V. S. Varela-Nieto (Eds.). San Diego, CA: Academic Press, pp. 31–61. doi:10.1016/B978-0-12-408088-1.00002-6.

Jeram, S., and A. Cokl. 1996. Mechanoreceptors in insects: Johnston's organ in *Nezara viridula* (L) (Pentatomidae, Heteroptera). *European Journal of Physiology*, 431: R281–R282. doi:10.1007/BF02346378.

Kamikouchi, A., H. K. Inagaki, T. Effertz et al. 2009. The neural basis of *Drosophila* gravity-sensing and hearing. *Nature* 458(7235): 165–71. doi:10.1038/nature07810.

Kamikouchi, A., T. Shimada, and K. Ito. 2006. Comprehensive classification of the auditory sensory projections in the brain of the fruit fly *Drosophila melanogaster*. *The Journal of Comparative Neurology* 499: 317–356. doi:10.1002/cne.

Katz, M. J., R. J. Lasek, and J. Silver. 1983. Ontophyletics of the nervous system: Development of the corpus callosum and evolution of axon tracts. *Proceedings of the National Academy of Sciences* 80: 5936–5940.

Katz, P. S. 2011. Neural mechanisms underlying the evolvability of behaviour. *Philosophical Transactions of the Royal Society B: Biological Sciences* 366(1574): 2086–2099. doi:10.1098/rstb.2010.0336.

Katz, P. S., and R. M. Harris-Warrick. 1999. The evolution of neuronal circuits underlying species-specific behavior. *Current Opinion in Neurobiology* 9(5): 628–633. doi:10.1016/S0959-4388(99)00012-4.

Keil, T. A. 1998. Functional morphology of insect mechanoreceptors. *Microscopy Research and Technique* 39: 506–531.

Kouyama, N., and T. Shimozawa. 1982. The structure of a hair mechanoreceptor in the antennule of crayfish (Crustacea). *Cell and Tissue Research* 226(3): 565–578. doi:10.1007/BF00214785.

Krishnan, A., S. Prabhakar, S. Sudarsan, and S. P. Sane. 2012. The neural mechanisms of antennal positioning in flying moths. *The Journal of Experimental Biology* 215: 3096–4105. doi:10.1242/jeb.071704.

Krishnan, A., and S. P. Sane. 2014. Visual feedback influences antennal positioning in flying hawk moths. *The Journal of Experimental Biology* 217: 908–917. doi:10.1242/jeb.094276.

Krishnan, A., and S. P. Sane. 2015. Antennal mechanosensors and their evolutionary antecedents. *Advances in Insect Physiology* 49: 59–99. doi:10.1016/bs.aiip.2015.06.003.

Mamiya, A., A. D. Straw, E. Tómasson, and M. H. Dickinson. 2011. Active and passive antennal movements during visually guided steering in flying *Drosophila*. *The Journal of Neuroscience: The Official Journal of the Society for Neuroscience* 31(18): 6900–6914. doi:10.1523/JNEUROSCI.0498-11.2011.

Manning, A. 1967. Antennae and sexual receptivity in *Drosophila melanogaster* females. *Science* 158(3797): 136–137. doi:10.1126/science.158.3797.137.

Maronde, U. 1991. Common projection areas of antennal and visual pathways in the honeybee brain, *Apis mellifera*. *The Journal of Comparative Neurology* 309(3): 328–340. doi:10.1002/cne.903090304.

Matsuo, E., and A. 2013. Neuronal encoding of sound, gravity, and wind in the fruit fly. *Journal of Comparative Physiology A: Neuroethology, Sensory, Neural, and Behavioral Physiology* 199(4): 253–262. doi:10.1007/s00359-013-0806-x.

Matsuo, E., D. Yamada, Y. Ishikawa, T. Asai, H. Ishimoto, and A. Kamikouchi. 2014. Identification of novel vibration- and deflection-sensitive neuronal subgroups in Johnston's organ of the fruit fly. *Frontiers in Physiology* 5: 1–13. doi:10.3389/fphys.2014.00179.

McIver, S. B. 1975. Structure of cuticular mechanoreceptors of arthropods. *Annual Review of Entomology* 20(15): 381–97. doi:10.1146/annurev.en.20.010175.002121.

Melnitsky, S. I., V. D. Ivanov, M. Y. Valuyskiy, L. V. Zueva, and M. I. Zhukovskaya. 2018. Comparison of sensory structures on the antenna of different species of Philopotamidae (Insecta: Trichoptera). *Arthropod Structure and Development* 47(1): 45–55. doi:10.1016/j.asd.2017.12.003.

Misof, B., S. Liu, K. Meusemann, R. S. Peters, T. Flouri, and R. G. Beutel et al. 2014. Phylogenomics resolves the timing and pattern of insect evolution. *Science* 346(6210): 763–768. doi:10.1126/science.1257570.

Mißbach, C., S. Harzsch, and B. S. Hansson. 2011. New insights into an ancient insect nose: The olfactory pathway of *Lepismachilis y-signata* (Archaeognatha: Machilidae). *Arthropod Structure and Development* 40(4): 317–333. doi:10.1016/j.asd.2011.03.004.

Mittmann, B., and G. Scholtz. 2001. Distal-less expression in embryos of *Limulus polyphemus* (Chelicerata, Xiphosura) and *Lepisma saccharina* (Insecta, Zygentoma) suggests a role in the development of mechanoreceptors, chemoreceptors, and the CNS, *Development Genes and Evolution* 211(5):232–243. doi:10.1007/s004270100150.

Moczek, A. P., T. E. Cruickshank, and A. Shelby. 2006. When ontogeny reveals what phylogeny hides: Gain and loss of horns during development and evolution of horned beetles. *Evolution* 60(11): 2329–2341. doi:10.1111/j.0014-3820.2006.tb01868.x.

Natesan, D., N. Saxena, Ö. Ekeberg, and S. P. Sane. 2016. Airflow mediated antennal positioning in flying hawk moths. *SICB 2016 Annual Meeting Abstracts* 159.

Okada, J., and Y. Toh. 2000. The role of antennal hair plates in object-guided tactile orientation of the cockroach (*Periplaneta americana*). *Journal of Comparative Physiology—A Sensory, Neural, and Behavioral Physiology* 186(9): 849–857. doi:10.1007/s003590000137.

Okada, J., and Y. Toh. 2001. Peripheral representation of antennal orientation by the scapal hair plate of the cockroach *Periplaneta americana*. *The Journal of Experimental Biology* 204(24): 4301–4309. Retrieved from http://www.ncbi.nlm.nih.gov/pubmed/11815654.

Resh, V. H., and R. T. Cardé, (Eds.). 2003. *Encyclopedia of Insects*. Cambridge, MA: Academic Press. (2003rd ed., Vol. 136). pp. 35, 535–539, 676–677, 1094–1095.

Rospars, J. P. 1988. Structure and development of the insect antennodeutocerebral system. *Int. Journal of Insect Morphology & Embryology* 17(3): 243–294.

Roy Khurana, T., and S. P. Sane. 2016. Airflow and optic flow mediates antennal positioning in flying honeybees. *ELife* 5: e14449: 1–20. doi:10.7554/eLife.14449.

Saager, F., and M. Gewecke. 1989. Antennal reflexes in the desert locust *Schistocerca gregaria*. *Journal of Experimental Biology* 147(1): 519–532.

Sandeman, D. and L. Wilkens. 1983. Motor control of movements of the antennal flagellum in the Australian crayfish. *Euastacus armatus*. *Journal of Experimental Biology* 273: 253–273. Retrieved from http://jeb.biologists.org/content/105/1/253.short.

Sane, S. P., A. Dieudonné, M. A. Willis, and T. L. Daniel. 2007. Antennal mechanosensors mediate flight control in moths. *Science* 315(5813): 863–866. doi:10.1126/science.1133598.

Sane, S. P., and M. J. McHenry. 2009. The biomechanics of sensory organs. *Integrative and Comparative Biology* 49(6): i8–i23. doi:10.1093/icb/icp112.

Sant, H. H., and S. P. Sane. 2018. The mechanosensory-motor apparatus of antennae in the oleander hawk moth (*Daphnis nerii*, Lepidoptera). *Journal of Comparative Neurology* 526(14): 2215–2230 doi:10.1002/cne.24477.

Schmidt, K. (1969). Der Feinbau der stiftführenden Sinnesorgane im Pedicellus der Florfliege *Chrysopa* Leach (Chrysopidae, Planipennia). *Zeitschrift Für Zellforschung Und Mikroskopische Anatomie* 99(3): 357–388. doi:10.1007/BF00337608.

Schmidt, K. 1974. Die mechanorezeptoren im pedicellus der eintagsfliegen (insecta, ephemeroptera). *Zeitschrift Für Morphologie Der Tiere* 78(3): 193–220. doi:10.1007/BF00375742.

Schneider, D. 1964. Insect antennae. *Annual Review of Entomology* 9: 103–122. doi:10.1146/annurev.en.09.010164.000535.

Schütz, C., and V. Durr. 2011. Active tactile exploration for adaptive locomotion in the stick insect. *Philosophical Transactions of the Royal Society of London: Series B, Biological Sciences* 366(1581): 2996–3005. doi:10.1098/rstb.2011.0126.

Seyfarth, E. A., W. Gnatzy, and K. Hammer. 1990. Coxal hair plates in spiders: Physiology, fine structure, and specific central projections. *Journal of Comparative Physiology A* 166(5): 633–642. doi:10.1007/BF00240013.

Staudacher, E. M., M. Gebhardt, and V. Dürr. 2005. Antennal movements and mechanoreception: Neurobiology of active tactile sensors. *Advances in Insect Physiology* 32. doi:10.1016/S0065-2806(05)32002-9.

Strausfeld, N. J. 2012. *Arthropod Brains: Evolution, Functional Elegance, and Historical Significance*. Cambridge, MA: Harvard University Press. Retrieved from https://books.google.co.in/books?id=SCqauAAACAAJ.

Strausfeld, N. J., and D. R. Andrew 2011. Arthropod structure & development a new view of insect-crustacean relationships I. Inferences from neural cladistics and comparative neuroanatomy. *Arthropod Structure and Development* 40(3): 276–288. doi:10.1016/j.asd.2011.02.002.

Strickler, J. R., and A. K. Bal. 1973. Setae of the first antennae of the Copepod *Cyclops scutifer* (Sars): Their structure and importance. *Proceedings of the National Academy of Sciences of the United States of America* 70(9): 2656–2659. doi:10.1073/pnas.70.9.2656.

Tautz, K., and J. Rohrseitz. 1999. Honey bee dance communication: Waggle run direction coded in antennal contacts? *Journal of Comparative Physiology A* 184: 463–470.

Tillyard, R. 1917. *The Biology of Dragonflies*. (A. E. Shipley, Ed.). Cambridge, MA: Cambridge University Press.

Toh, Y. 1981. Fine structure of sense organs on the antennal pedicel and scape of the male cockroach, *Periplaneta americana*. *Journal of Ultrastructure Research* 77(2): 119–132. doi:10.1016/S0022-5320(81)80036-6.

Trautwein, M. D., B. M. Wiegmann, R. Beutel, K. M. Kjer, and D. K. Yeates. 2012. Advances in insect phylogeny at the dawn of the postgenomic era. *Annual Review of Entomology* 57: 449–68. doi:10.1146/annurev-ento-120710-100538.

Vande Berg, J. S. 1971. Fine structural studies of Johnston's organ in the tobacco hornworm moth, *Manduca sexta* (Johannson). *Journal of Morphology* (133): 439–455. doi:10.1002/jmor.1051330407.

Vedel, J. P. 1980. The antennal motor system of the rock lobster (*Palinurus vulgaris*): Competitive occurrence of resistance and assistance reflex patterns originating from the same proprioceptor. *Journal of Experimental Biology* 87: 1–22.

Vedel, J. P., and S. Monnier. 1983. A new muscle receptor organ in the antenna of the rock lobster *Palinurus vulgaris*: Mechanical, muscular and proprioceptive organization of the two proximal joints J0 and J1. *Proceedings of the Royal Society of London. B, Biological Sciences* 218(1210): 95–110.

Wang, X., Y. Xie, Y. Zhang, W. Liu, and J. Wu. 2016. The structure and morphogenic changes of antennae of *Matsucoccus matsumurae* (Hemiptera: Coccoidea: Matsucoccidae) in different instars. *Arthropod Structure and Development* 45(3): 281–293. doi:10.1016/j.asd.2016.01.010.

Weatherby, T. M., K. K. Wong, and P. H. Lenz. 1994. Fine structure of the distal sensory setae on the first antennae of *Pleuromamma Xiphias* Giesbrecht (Copepoda). *Journal of Crustacean Biology* 14(4): 670–685. doi:10.1163/193724094X00641.

Wiegmann, B. M., M. D. Trautwein, I. S. Winkler et al. 2011. Episodic radiations in the fly tree of life. *Proceedings of the National Academy of Sciences* 108: 5690–5695. doi:10.1073/pnas.1012675108.

Yack, J. E. 2004. The structure and function of auditory chordotonal organs in insects. *Microscopy Research and Technique* 63(6): 315–337. doi:10.1002/jemt.20051.

Yorozu, S., A. Wong, B. J. Fischer et al. 2009. Distinct sensory representations of wind and near-field sound in the *Drosophila* brain. *Nature* 458(7235): 201–205. doi:10.1038/nature07843.

20 Cross-Kingdom Interactions in Natural Microcosms

The Worlds Within Fig Syconia and Ant-Plant Domatia

Renee M. Borges, Joyshree Chanam, Mahua Ghara, Anusha Krishnan, Yuvaraj Ranganathan, Megha Shenoy, Vignesh Venkateswaran, and Pratibha Yadav

CONTENTS

Value of Microcosms as Model Systems 401
Natural Microcosms 402
Exemplar Microcosms and Their Ecological and Evolutionary Processes 402
    Fig Syconia (Closed Microcosms) 402
        Origin 402
        Internal Environments 403
        Entry Restrictions 404
        Trophic Networks 405
    Ant-Plant Domatia (More Open Microcosms) 406
        Origin 406
        Entry Restrictions 406
        Development and Maintenance 407
Microcosm Studies and the Future 408
Acknowledgements 408
References 408

Dedication: This paper is dedicated to Professor C. A. Viraktamath who has patiently and willingly identified many of the inhabitants of our microcosms and has inspired us with his enthusiasm for the documentation of life's diversity.

> *[I]n a fractal universe, a single mote can mirror the cosmos, giving literal meaning to Blake's famous image of the "world in a grain of sand, and a heaven in a wild flower".*
> (Stephen Jay Gould 2003)

VALUE OF MICROCOSMS AS MODEL SYSTEMS

At human scales, there is likely to be a consensus on what is defined as a microcosm, a mesocosm, or a macrocosm; however, at other scales this definition will surely change (Odum 1996). Moving from a macrocosm to a microcosm, it appears that processes at smaller scales are better characterized at smaller scales, and this is likely the outcome of a perception that there are fewer variables at these levels. While this may be true, it does not make modelling processes within a microcosm, e.g., the metabolome of the cell, any easier. At the ecosystem level, deconstructing a microcosm that is constituted by fewer organisms or types of organisms is likely easier than the challenge of examining processes where hundreds or millions of taxa are involved. This is why microcosms have been considered useful model systems in ecology (Srivastava et al. 2004), especially since ecology encompasses interactions between organisms and their biotic and abiotic environments.

To exploit the benefits of working with simpler ecosystems with fewer variables, ecologists or population biologists have traditionally constructed artificial microcosms, where they have attempted to manipulate conditions periodically or have started with fixed initial conditions and have allowed the system to run its course. Most notable experiments with microcosms have used organisms with short generation times such as bacteria, plankton, protists, and insects (Drake et al. 1996; Jessup et al. 2004; Altermatt et al. 2015). Notable examples of experimental artificial mesocosms on a slightly larger scale have been Ecotrons (Lawton 1995, 1996; Lähteenmäki et al. 2015; Eisenhauer and Türke 2018), Biosphere 2 (Walter and Lambrecht 2004; Osmond 2014), Limnotrons (Verschoor et al. 2003), Planktotrons (Gall et al. 2017), and, more recently, the underground Earthtrons (Kamitani and Kaneko 2006). While artificially assembled ecosystems can provide useful insights into ecosystem functioning, they suffer from biases that are inherent to their unnatural construction (Huston 1999).

NATURAL MICROCOSMS

Natural microcosms are those whose boundaries are well defined, as in artificial microcosms, and that maintain internal environments within which whole communities of organisms at different trophic levels exist. Natural microcosms may occur fortuitously, e.g., tree holes filled with water (Montero et al. 2010); others are often by-products of an organism, e.g., vertebrate gut lumens (Shapira 2016) or water-filled bases of bromeliads (Talaga et al. 2015); some are specifically constructed by organisms, e.g., pitchers of carnivorous plants (Thorogood et al. 2018a); while yet others may be constructed by organisms using exogenous materials, e.g., soil nest mounds of fungus-growing termites (Zachariah et al. 2017). Another category of a microcosm is when two partners may interact to construct a microcosm as occurs in gall construction when plants and gallers interact, or when the bioluminescent light organ of a squid is co-constructed by the host squid and *Vibrio* bacteria (Borges 2017, 2018).

Changes in developmental trajectories of regular plant or animal organs can result in microcosm formation. For example, it is believed that in the evolution of pitcher microcosms, plants with glandular leaves may have undergone epiascidiation in which the leaf lamina curls and fuses at the margins, forming a tubular pitcher such that the upper surface (bearing the glands) forms the inner wall of the pitcher (Owen and Lennon 1999; Heubl et al. 2006). Diversity in pitcher morphology is likely to have been shaped by co-evolution with local faunal assemblages (Thorogood et al. 2018a). For example, ants in *Nepenthes* inhabit hollow pitcher tendrils, are able to swim in pitcher fluid without difficulty or any negative impact, and their wastes that drop into the pitcher contribute significantly to the nitrogen budget of the plant; ants thus engage in a nutritional mutualism with pitcher plants (Bazile et al. 2012). While pitchers are an example of a new organ that develops to create a microcosm, enclosed spaces within existing morphological structures can also serve as microcosms, e.g., nectar-filled spaces within flowers. These harbour predictable communities of yeasts and bacteria and are proving to be excellent model systems in community ecology (Chappell and Fukami 2018).

EXEMPLAR MICROCOSMS AND THEIR ECOLOGICAL AND EVOLUTIONARY PROCESSES

Among the important questions in community ecology are the following: What is the species composition of ecosystems? Are these stable over time? If so, what contributes to their stability? How are communities assembled? What are the trophic interactions within communities? Are communities saturated? Many of these questions are difficult to answer in natural ecosystems because most natural ecosystems have unbounded species memberships, wherein there is a flux, even over ecological time, in community membership. This makes "true" community dynamics difficult to capture within the lifetimes of individual researchers and may require long-term investigations to make sense of seemingly chaotic patterns (Condit et al. 2012). The still unresolved debates over tropical forest diversity, which include issues as fundamental as how many species co-occur (Chase 2014; Huston 2014; Munoz and Huneman 2016; Chao et al. 2017; LaManna et al. 2017) are testimony to the complexity of the patterns and their underlying processes in community ecology. Most of the objections to proofs of general theories in ecology revolve around the lack of good empirical data on underlying processes to make the theories and their assumptions testable; these objections are often sought to be resolved by assembling artificial ecosystems (Huston 1999).

Natural microcosms, as mentioned earlier, exhibit the advantages of more defined ecosystem boundaries, as well as smaller species compositions, which makes for easier and more precise measurements of community and ecosystem parameters. Two such microcosms are the enclosed, globose inflorescences or syconia of *Ficus* species and the enclosed spaces called domatia generated by ant-plants to house ants (Figure 20.1). These have proven to be valuable systems to investigate ecosystem and evolutionary processes at small scales. In this paper, we give a brief overview of these two microcosms, with a strong focus on the data generated by our investigations into both these microcosms.

FIG SYCONIA (CLOSED MICROCOSMS)

Origin

The genus *Ficus* (Moraceae: Tribe Ficeae) is species-rich (over 750 species), mostly tropical, with some representatives in the sub-tropical regions of the world (Datwyler and Weiblen 2004). The inflorescence of *Ficus* is called a syconium and is an enclosed globular or urn-shaped structure with a single opening termed the ostiole; the flowers are therefore internally presented and line the lumen of the syconium in single or many layers. This enclosed syconium functions as a microcosm. In Dorstenieae, a related tribe in the Moraceae, the unisexual flowers are presented on a flat open receptacle that does not have bracts (Datwyler and Weiblen 2004; Thorogood et al. 2018b) and are pollinated by flies that also oviposit into this receptacle (Araújo et al. 2017). In another closely related tribe, Castilleae, the bracts form an involucre surrounding the flowers (Datwyler and Weiblen 2004), and in a dioecious *Antiaropsis* species within this tribe, pollinating thrips feed on pollen from the staminate inflorescences, and oviposit in the pistillate inflorescences (Zerega et al. 2004). In *Ficus*, the involucral bracts completely enclose the unisexual flowers to form the ostiolar opening (Datwyler and Weiblen 2004). It appears that the transition to the syconium may have occurred from the flat, open *Dorstenia* stage through the more enclosed Castilleae stage, followed by complete enclosure in *Ficus* (Thorogood et al. 2018b). Since *Ficus* is pollinated by

FIGURE 20.1 Microcosms and associated features. (a) Trunk of *Ficus racemosa* with pendant syconia. (b) Close-up of the syconial microcosm showing ostiole and an ovipositing wasp (OW). (c) Section of a syconium. Ostiolar bracts (OB) form a lining around the ostiole. The florets inside the syconium host the developing fig wasps and develop into galls. Some florets develop into seeds. Florets are present along the wall of the syconium (SW). The interior of the syconium is a hollow lumen. (d) Inflorescence of *Humboldtia brunonis*. (e) Ants patrolling young leaves of *H. brunonis* and feeding on extrafloral nectaries. A cauline domatium of *H. brunonis* that develops as a hollow internode. (f) Domatium opening (DO). (g) A longitudinal section of a domatium of *H. brunonis* showing earthworms *Perionyx pullus* (EW) and the domatium wall (DW). (Courtesy of Borges Lab.)

agaonid fig wasps that oviposit into female flowers within what is termed a brood-site pollination mutualism, it is interesting that the pollinators of Dorstenieae and Castilleae also oviposit within the inflorescence. It is speculated that, within the Moraceae, *Dorstenia* and *Antiaropsis* may serve as transitions from the wind-pollinated systems of *Morus* (mulberry) to the closed and obligate brood-site pollination mutualism of *Ficus* (Misiewicz and Zerega 2012).

Internal Environments

The ostiole formed by the involucral bracts can serve to seal the syconial microcosm and thereby modulate internal temperatures and gas fluxes. In experiments with 11 species of figs, Patiño et al. (1994) showed that temperatures within syconia that contained developing wasps could reach 2–3°C above ambient conditions; however, if transpiration from the syconia was prevented, internal temperatures increased to 3–8°C above ambient conditions, resulting in temperatures that were lethal to developing wasps. The heat gains and losses due to radiation and diffusion differed with syconial size; large syconia would need to transpire more to maintain suitable internal temperatures compared to small syconia in which heat loss due to diffusion is greater (Patiño et al. 1994). This can explain why even within a fig species, syconia produced in the summer are smaller than those produced in cooler seasons (Krishnan et al. 2014). As compared to the winter months, summer crops of figs exhibited a two-fold increase in seed production and reduced fig wasp production; this variation may be attributed to higher temperatures even though the seasonal temperature increases were only between 3–5°C (Krishnan et al. 2014). Fig species that occur in arid regions usually have small syconia (McLeish et al. 2010, 2011) possibly due to the same thermal constraints. Heat effects have also been documented in other brood-site pollination microcosms; e.g., in yucca flowers into which the pollinating yucca moths lay eggs, desiccation effects can cause egg and presumably also larval mortality (Segraves 2003). The enclosed environments of galls can have similar effects on developing gallers. The microclimate of galls induced on the stems of goldenrod is quite variable, and tephritid flies must exhibit cold and desiccation tolerance to survive overwintering as larvae within these galls (Layne 1993; Nelson and Lee 2004). Thermal ecology must, therefore, be an important consideration in microcosm effects (Pincebourde and Casas 2006; Kierat et al. 2017). This consideration of thermal effects must also extend to the impact of global warming on successful dispersal of species between microcosms as found for fig wasps inhabiting syconia (Jevanandam et al. 2013).

Besides heat fluxes, microcosms also demonstrate gas fluxes or internal gas accumulation. Endophagous insects, e.g., those living within microcosms, such as galler or leaf miner larvae or adults, must have mechanisms to tolerate the low oxygen (hypoxic) and/or high CO_2 (hypercarbic) conditions (Pincebourde and Casas 2016). Changes in the CO_2 microclimate may signal important events to microcosm dwellers. Conditions during development of fig wasps within the syconia of *Ficus religiosa* are hypercarbic at 10% CO_2; male wasps are active at these hypercarbic levels; however, when the syconial wall is breached by male activity in order to release female wasps, the equilibration of CO_2 with outside ambient conditions depresses male activity, a phenomenon that has been shown to be due to gas levels (Galil et al. 1973), suggesting that male fig wasps are adapted to hypercarbia. Other changes in microcosm environments can induce physiological changes in inhabitants, e.g., desiccation changes in goldenrod stems induce the overwintering larvae in stem galls to upregulate desiccation tolerance measures (Williams and Lee 2005). Since microcosms such as fig syconia also progress into fruit, whether these fruit are climacteric (with a sudden burst of fruit ripening hormones such as ethylene) or non-climacteric, in which case the ripening process is slow (Borges et al. 2011), can also change the internal environments of the microcosms at varying rates.

The syconium is a relatively sterile environment; some of the antimicrobial effects are attributed to ficin, a non-specific protease present in fig latex (Baidamshina et al. 2017). Yet, latex is viscous, and while laticifers abound in fig syconia, they are usually restricted to the ostiolar bracts, outer walls of the syconium receptacle, and the pedicels of flowers (Marinho et al. 2018). This distribution of latex ensures that the entry routes into the syconium are well protected, leaving latex-free zones in the syconium for development of legitimate syconium inhabitants. Some syconia also contain less viscous liquid of unknown origin, especially during late developmental phases, and microcosm inhabitants must have adaptations to prevent drowning and to enable respiration within these environments. In certain fig syconia especially those in section Sycomorus, the syconial lumen is filled with liquid when the males eclose; the abdominal spiracles of these males are surrounded by filamentous peritremes. These structures probably function as air channels that capture gas bubbles from within the galls that males can use in respiration while they attempt to release females from their galls and copulate with them (Ramirez 1987, 1996; Compton and McFarlen 1989). The hairy hydrophobic hind legs of *Ceratosolen* male wasps could also serve as traps for air bubbles and enable males to eclose early within the fluid-filled lumens to acquire access to virgin females (Rodriguez et al. 2017). Inhabitants of microcosms also have to deal with efficient waste management; e.g., gall-forming social aphids that excrete honeydew, but cannot induce ants to remove it from the closed environment that they live in, have galls the inner surfaces of which absorb the honeydew and transfer it to the plant vascular system (Kutsukake et al. 2012).

Entry Restrictions

The syconium ostiole can serve as a barrier to entry into the microcosm, thus limiting the entry of fig wasps, including potential pollinators, into this microcosm via this route. This sole opening into the syconium, that is also lined by inward facing bracts, has placed a strong selection pressure on the head morphology of female pollinating wasps, resulting in a laterally compressed and elongated head that is capable of penetrating through the ostiolar channel (Noort and Compton 1996). However, those fig wasps that are unable to enter the syconial microcosm through the ostiole gain membership into the microcosm by ovipositing into the microcosm from the exterior. The ovipositors of these wasps show specific adaptations based on whether these wasps oviposit into an early, less firm syconium compared to a large, harder, and more mature syconium. Ghara et al. (2011) demonstrated that the force required to penetrate the syconium wall for egg-laying increases with syconium development. Therefore, early arriving fig wasp parasites have weaker and less sclerotized ovipositors compared to those arriving later. The ovipositors of the late-arrivers show considerable sclerotization and are also equipped with more armature in terms of cutting teeth presumably to facilitate easier penetration through the syconial wall towards the egg-laying sites. Ghara et al. (2011) also found that the complexity of the sensory structures or sensilla on the ovipositors increased with timing of oviposition into the microcosm, implying that a more complex repertoire of sensory structures is required to determine egg-laying sites as the microcosm begins to get more chemically crowded along its ontogeny. Elias et al. (2018) similarly found ovipositor morphology to be correlated with galler versus parasitoid lifestyles. Pollinating wasps that do not use their ovipositors to penetrate into the syconium, but only insert them down the style of female flowers had the least complex ovipositor with the fewest sensilla (Ghara et al. 2011). For pollinators, ovipositor length must match the style length of female flowers since the egg needs to be placed near the ovule; mismatches of style length and ovipositor length is therefore an additional filter that constrains community membership and facilitates fig–pollinator species fidelity (Souto-Vilarós et al. 2018). For pollinating gallers, the ability to gall the uniovulate flowers is yet another filter that serves to regulate fig species specificity (Ghana et al. 2015). Some non-pollinating gallers, that do not enter syconia, can still insert their ovipositors down the style of flowers rather than via the easier route of access through the base of the flowers, by tracing a complicated path from the exterior of the syconium with long, flexible ovipositors (Elias et al. 2012).

The syconium is the breeding site of a multitude of fig wasps; the development of the brood-site, however, is usually conditional upon the entry of the pollinating fig wasps into the syconium and their delivery of pollination services. Failure of pollinator arrival, or of excessive oviposition relative to pollination, leads to syconium abortion; inadequate pollination could also result in reduction in the size of wasp offspring (Jandér and Herre 2016; Jandér et al. 2016). These sanctions are expressed at the level of the syconium and not at the level of individual flowers within the syconia (Jandér et al. 2012), making data at the syconium level relevant, and emphasizing the role of the syconium as a microcosm. Pollinator wasps are attracted to pollen-receptive syconia by scents produced by osmophores within the syconia (Souza et al. 2015); these scents are highly fig species-specific and usually only attract species-specific female pollinators (Souto-Vilarós et al. 2018); in rare cases, when there is cross-species attraction, additional filters act to maintain species-specificity as mentioned before. Besides scents produced specifically for the purpose of attracting pollinators, fig syconial scents are attractive to non-pollinating gallers and parasitoid wasps (Borges 2016); syconial scents change with the arrival of the first gallers (even prior to the arrival of the pollinators), and the largely diurnal scent signatures include herbivore-induced plant volatiles that indicate internal syconium tissue damage as a result of feeding by galler larvae (Borges et al. 2013). In dioecious figs, only male syconia serve as breeding sites for pollinating wasps, while syconia on female trees do not; consequently, pollinator wasps entering syconia on female trees are doomed to have zero reproductive success while ensuring female fitness via seed production for the host plant (Janzen 1979). Owing to chemical mimicry between the scents of the receptive stages of syconia on male and female trees, pollinator wasps are manipulated to enter female syconia despite their certain reproductive suicide (Hossaert-McKey et al. 2016).

Once fig wasps have arrived at the syconial microcosms using long-range chemical signals and cues, they make additional decisions about entering or ovipositing into syconia based on short-range cues. Pollinator wasps may be deterred from entering ostioles that are blocked by many pollinator wings (pollinators lose their wings while forcing themselves through the ostiole) (Ramya et al. 2011), as this may serve as an indicator of internal competition levels since several foundress females are already present within the syconium. Contact cues on the syconial surface may also serve to maintain species specificity (Wang et al. 2012). Avoiding competition and also ensuring adequate mates for offspring is an important feature within any microcosm. Since fig wasp larvae (whether gallers or parasitoids) are immobile and are confined within their galls which are densely packed within the syconium (Yadav and Borges 2018a), competition must be avoided since larvae are unable to breach the gall and feed on extra-gall material. Furthermore, since offspring of fig wasps (the F1 generation) can usually only mate within the syconium, the syconial microcosm is also a mating arena. It is, therefore, imperative, especially for non-pollinating fig wasps that do not enter the syconium, but only oviposit from the exterior, to know which wasps and how many wasps have already oviposited into the microcosm. Wasps are able to obtain this information on the basis of chemical footprints left by earlier visitors that leave "signatures" of their presence and activity (Yadav and Borges 2018b; Yadav et al. 2018). Wasps avoid ovipositing into those syconia into which too many conspecifics have earlier deposited eggs or those in which there was no prior egg-laying by conspecifics. However, wasps nearing the end of their lifespans are not so choosy (Yadav and Borges 2018b). Wasps also sample the internal chemical environment of the syconium with their volatile-sensing ovipositors and could use the presence of CO_2 and syconial stage-specific volatiles to make decisions about the quality of internal egg-laying sites (Yadav and Borges 2017).

Trophic Networks

Fig microcosms have fig wasp members that occupy a range of trophic levels. Microcosm members include gallers (herbivores), seed predators (herbivores), cleptoparasites/inquilines (herbivores), and parasitoids (predators) (Kerdelhué et al. 2000; Kjellberg et al. 2005; Wang et al. 2014). Cross-continental comparisons of communities (South America, Africa, Australasia) indicate similarity in community compositions with most fig wasp communities consisting of early-arriving large gallers that initiate large galls, followed by pollinators that are also gallers, followed by later-arriving gallers, and a set of small early-arriving and larger late-arriving parasitoids (Segar et al. 2013; Borges 2016). The total number of these fig wasps in the various groups can range from a few species to up to 30 in varying combinations (Borges 2016). It is also possible that the total number of species that can be hosted at any time within a microcosm may depend on microcosm size, especially, since as mentioned earlier, the syconium serves as a mating arena for members of a species. Members of the community divide up the microcosm temporally and spatially based on time of arrival for oviposition as well as the tissues that they use for galling; access to these tissues from the exterior is constrained by ovipositor length and may sometimes result in predator-free spaces, i.e., areas within the microcosm that parasitoids cannot access with their ovipositors (Al-Beidh et al. 2012, but see Ghara et al. 2014). The syconium, as mentioned earlier, can have its own energy budget and may be viewed by the plant as an independent unit; indeed, members of the syconium can influence its development, with late-arriving parasitoids delaying the development of the syconium and early-arriving gallers hastening its development (Krishnan et al. 2014).

Owing to the extreme species specificity of community membership, members of the community must be able to find microcosms on fig plants in which they can breed and where their offspring can mate. In order to prevent self-pollination resulting from pollinator wasps leaving natal microcosms, and entering breeding microcosms on the same tree, syconia on any given fig plant are usually all in the same developmental phase. Therefore, individual fig plants generally exhibit reproductive synchrony (Janzen 1979). Since figs are sparsely distributed in the landscape, this has necessitated fig wasps having to travel long distances to reach suitable breeding microcosms from their natal trees. While the dispersal of the very tiny fig wasps (often <2 mm length) has been thought to be largely wind-assisted (Compton 2002), and the largest distance that pollen has been known to travel via fig wasps is 160 km in Africa (Ahmed et al. 2009), recent work also suggests that the intrinsic flight abilities of wasps within syconial communities vary with life history strategies (Venkateswaran et al. 2017). The fig wasp community comprises members that vary in life history consisting of those that exhibit fast-paced to those that show slow-paced life histories (Ghara et al. 2011; Borges 2016). Pollinators and other gallers tend to be fast-paced, eclosing with a full complement of mature eggs that are ready to be deposited into breeding sites within non-natal syconia. Pollinators, however, will deposit their entire egg complement into a single syconium that they enter, while other parasitic gallers will deposit their eggs into multiple syconia from the exterior. Pollinators and many gallers have short lives (24–48 hours), and generally do not feed as adults, while other gallers can live for longer, and may be able to acquire some nutrition during this time. Consequently, pollinators and most gallers eclose with all the fuel needed to find their breeding syconia, and have greater intrinsic flight capacities as measured by the total time spent in tethered flight, total amount of somatic lipid, and resting metabolic rates (Venkateswaran et al. 2017); parasitoids on the other hand, eclose with only a few eggs matured, have much longer lifespans, are capable of feeding during their lifetime (Ghara et al. 2011), and demonstrate much lower intrinsic flight capacities (Venkateswaran et al. 2017). It is believed that the spatio-temporal dispersion of the microcosms exerts a strong selective force on the dispersal capacities of fig wasps and may constrain their membership in fig species communities (Venkateswaran et al. 2018).

Phoretic organisms can take advantage of organisms dispersing between microcosms and hitch a ride with them, e.g., tank bromeliad tree frogs serving as vehicles for ostracods (Sabagh and Rocha 2014) or flower mites travelling on birds or bats (Reynolds et al. 2014). Fig syconial microcosms harbour plant- and animal-parasitic nematodes (Krishnan et al. 2010; Borges et al. 2018; Van Goor et al. 2018); these nematodes are phoretic on female pollinator wasps and use the chemical cues of wasp cuticular hydrocarbon and volatiles to identify their vehicles within the cocktail of cues of other male and female wasps inhabiting the microcosm (Krishnan et al. 2010). These nematodes appear highly species-specific to the fig microcosms especially in the plant-parasitic genera (Giblin-Davis et al. 2003). The presence of parasitic nematodes in these fig microcosms adds a level of complexity to factors contributing to the stability of these microcosms (Van Goor et al. 2018; S. Gupta and R.M. Borges, unpublished).

At the interface of the syconial microcosm with the exterior, predation effects, especially by ants, can be important and can affect the colonization success of pollinating and non-pollinating fig wasps (Ranganathan et al. 2010). In flower microcosms also, the presence of ants alters microbial community compositions vectored by insect visitors (Vannette et al. 2017).

Ant-Plant Domatia (More Open Microcosms)

Origin

The interaction between ants and plants is ancient and has evolved independently many times. For example, the interaction between ants and rewarding extrafloral nectaries (EFNs) is believed to have evolved at least 450 times (Weber and Keeler 2012), with the earliest EFNs recorded in the Eocene, ca 35–40 million years ago (mya) (Chomicki and Renner 2017). Some plants have taken the association with ants even further and have developed specialized nesting spaces for ants that are called domatia (Davidson and McKey 1993; Heil and McKey 2003; Mayer et al. 2014; González-Teuber and Heil 2015). Ant-plants that bear domatia are called myrmecophytes, while those that offer only EFNs or other food rewards are termed myrmecophiles. As with EFNs, domatia are believed to have at least 158 independent origins and appeared later than EFNs in the Miocene (Chomicki and Renner 2015). The Miocene origin of domatia possibly indicates the benefits to ants and also to plants of having sheltered nesting sites at times when the earth was experiencing arid conditions due to large-scale cooling (Chomicki and Renner 2015). That domatia can serve as microcosms is, therefore, of great significance in the origin and evolution of this trait.

Domatia can be formed from various plant parts that include stems, leaf pouches, leaf rachis, stipules (modified into swollen hollow thorns), and root tubers; of these, the number of species with stem or cauline domatia are the most abundant and constitute >50% of the recorded ant-plant species (Chomicki and Renner 2015), with the least numerous being the root tuber domatium type. While much attention has been paid to the nutrient rewards given by plants to ants in terms of EFNs and food bodies, as well as to protection services received from ants, there has been insufficient attention paid to domatia in terms of development, anatomy, and functionality (González-Teuber and Heil 2015). Furthermore, the role of domatia as microcosms has been scarcely investigated.

Entry Restrictions

Ant–plant interactions centred around domatia may be highly specialized with a single or a few ant species associated with a plant, e.g., the *Vachellia* (formerly *Acacia*) and *Pseudomyrmex* ant interactions (Heil at al. 2005), or facultative with several unspecialized ants associated with domatia (Alonso 1998). The nature of the domatia can also contribute to the specificity of the ant occupants. For example, some cauline domatia have a prostoma, i.e., a relatively unlignified region, which must be chewed through by founding queen ants in order to access the domatium interior; the size of this prostoma serves to filter out parasitic ants whose head width is larger than the domatium opening (Brouat et al. 2001). Sometimes cauline domatia have a self-opening slit; this not only allows ants, but also a variety of invertebrates to enter and take up residence within the domatia (Shenoy and Borges 2010; Chanam and Borges 2017). Winged founding queens may find their domatia-bearing host plants using plant volatiles, e.g., leaf volatiles, even from long distances (Grangier et al. 2009), and also at the seedling stage (Torres and Sanchez 2017); however, after arriving on the plant surface, short-range cues may also be important in colonization decisions (Jürgens et al. 2006; Blatrix and Mayer 2010). Founding queens may also use plant volatiles that provide information about the status of the host plant and prefer volatiles from undamaged and higher quality plants (Razo-Belman et al. 2018).

Founding queens may serve as phoretic vehicles for Hemiptera that might become important food sources for the colony within the domatia (Gaume et al. 2000); however, whether this is a general pattern is debated (Moog et al. 2005; Sanchez 2016). It is also suggested that founding queens collect wind-dispersed coccid nymphs from leaves and carry them into the domatia (Handa et al. 2012). Within the domatia, the identity of the trophobiont can determine colony fitness (Gaume and McKey 2002), suggesting that queens and workers must select their third-partner mutualists judiciously. Within myrmecophytic trees occupied by coccid-tending ants, there is also spatial variation in the location of domatia containing coccids, ant brood, and refuse piles, many of which were infected with rhabditid nematodes (Houadria et al. 2018). The bacteriovorous rhabditid nematodes in this system probably consume bacteria within these refuse piles and may be an important newly discovered component of the domatium microcosm whose exact roles are yet to be discovered (Maschwitz et al. 2016; Morera et al. 2018).

Development and Maintenance

Since domatia of various types have evolved independently many times, and stem domatia are the most frequent, it is possible that domatia have evolved from opportune exaptations (Chomicki and Renner 2015), e.g., thick stems that subsequently were hollowed out, and whose thickening appeared precociously during stem ontogeny (Brouat and McKey 2000) to facilitate early development of domatia even when plants are at the sapling stage. That the relationship between leaf area and the diameter of the domatium stem attached to the leaf is allometric suggests that domatia are costly to produce especially early in development (Brouat and McKey 2001). How these costs are recovered and how such stem domatia could confer a selective advantage before they are colonized by protective ants are questions that need to be answered. Addressing such questions requires comparative analysis between domatia-bearing and non-domatia-bearing individuals (possibly mutants) of the same myrmecophyte species. Since extant myrmecophytes bear domatia constitutively, answering such questions was difficult since the removal of domatia is experimentally difficult and destructive, and individuals that do not bear domatia were not known. There are also only a few examples of ants being able to induce domatia on their host plants (Edwards et al. 2009). Therefore, the discovery of natural intra-population variation in the presence of domatia in a primitive leguminous tree *Humboldtia brunonis* (Fabaceae) that is endemic to the Western Ghats of India was very important in this field and is providing perspectives on the evolution and maintenance of domatia. The stem domatia of *H. brunonis* are self-opening slits; this allows a motley set of invertebrates to reside within the domatia, including several species of ants most of which do not provide protection services, and also others such as arboreal earthworms, pseudoscorpions, millipedes, and resident pollinating *Braunsapis* bees (Gaume et al. 2006; Shenoy and Borges 2008, 2010). Despite the presence of resident ants that are potentially harmful to the plant, i.e., a flower-castrating ant species (Gaume et al. 2005a), it was observed that individual trees that had domatia had greater fruit set than those without domatia (Gaume et al. 2005b). This suggested that there were fitness advantages to bearing domatia that were independent of the presence of protective ants. Furthermore, many domatia were dominated by the presence of arboreal earthworms that had an antagonistic interaction with ants (Gaume et al. 2006).

Since trophic mutualisms, in which nutrients especially nitrogen are absorbed from ant wastes into the plant, are known to exist between protective ants and their plants (Fischer et al. 2003), could such trophic mutualisms also exist between plants bearing domatia and their motley non-protective domatia inhabitants? It was found that not only the protective ants, but even the earthworms and non-protective ants contribute significant amounts of nitrogen to the plant tissues adjacent to and distant from the domatia (Chanam et al. 2014b). This therefore suggests that during the evolution of domatia, a trophic mutualism between non-protective inhabitants could pay the cost of domatia formation well before an interaction between protective resident ants is set up. Therefore, a trophic mutualism could precede a protective mutualism in the evolution of the domatium trait (Chanam et al. 2014b). Indeed, the inner walls of *H. brunonis* domatia also show evidence of specialized structures for the absorption of domatia-derived nutrients (Chanam and Borges 2017). That microcosm availability may become a premium for some domatia inhabitants was also evidenced in the geographical variation in the *H. brunonis* protective mutualism along a latitudinal gradient (Shenoy and Borges 2010). Arboreal earthworms are sensitive to desiccation and shelter within domatia during the dry season, leaving the domatia only in the monsoon in order to mate (M. Shenoy, J. Chanam, and R. M. Borges, pers. observ.); the feeding biology of these earthworms is unknown, but they have a reduced gizzard, and presumably consume plant material within the domatium. In the northern part of the plant's range, the dry season is much longer than in the southern portion and, correspondingly, these earthworms are more dominant in the domatia in the north compared to the south (Chanam et al. 2014a), resulting in the ant protective mutualism mostly in the south. In some cases, ants and earthworms co-exist within the same domatium (Gaume et al. 2006); when this occurs, the ants build a carton partition between themselves and the earthworms, and the ants occupy that portion of the domatium closest to the opening.

As in other microcosms, e.g., nest mounds of fungus-farming termites, where food crops are cultivated by the inhabitants, ant inhabitants of stem domatia also cultivate fungi (Defossez et al. 2009, 2011). These fungi are likely important in recycling nitrogen for the ants and converting nitrogenous wastes into a form that ants can assimilate (Defossez et al. 2011). These fungi largely belong to the ascomycete order Chaetothyriales, i.e., black yeasts (Voglmayr et al. 2011; Nepel et al. 2016; Vasse et al. 2017). Such fungi are associated with the production of ant cartons (the cellulose-fibre rich material that ants use to build structures such as nest galleries) (Nepel et al. 2014). Foundress ants that cultivate these fungi in the domatia bring in the cultures from their natal nests

(Mayer et al. 2018). The ant-inhabited domatia of *H. brunonis* also have fungi in their inner walls; domatia occupied by earthworms are devoid of these fungi (Chanam and Borges 2017). These fungi are dominantly Chaetothyriales; additionally, the carton-discs built by ants to separate themselves from the earthworms is also built largely by a framework of chaetothyrialean hyphae (A. Vishnu and R.M. Borges, unpub.). It is, therefore, interesting that throughout the tropical world, these black yeast ascomyetes are an important feature of the microcosm of domatia, and likely have nutrititive and domatium-remodelling functions.

Since a domatium is a microcosm, microcosm inhabitants must protect it from invasion by interlopers. Entry into the self-opening domatium of *H. brunonis* is relatively unrestricted; however, once occupied, residents may modify the opening to prevent access to other taxa, e.g., the tiny ant *Tapinoma indicum* narrows the opening using carton structures (Chanam and Borges 2017); still other ants may use their shield/wedge-shaped heads to block the entrance in a form of defence called phragmosis (Debout et al. 2003; R. M. Borges, pers. observ.).

Domatia are clearly microcosms; while they are constructed by plants for fitness benefits, they may also be exploited by parasites. For example, the domatia-inhabiting *Braunsapis* pollinator of *H. brunonis* is parasitised by a cuckoo bee (Michener et al. 2003). However, a supposed parasite may also pay its dues by engaging in exchange of nutrients for shelter, and these currencies must be determined. Since there is no research on the microclimates of domatia or other physical attributes of this structure, this is another area that is ripe for research. Several ecological processes occur within the domatia microcosms about which much remains to be investigated. Does the host plant have control over these processes within its lifetime? Can errant residents be sanctioned? As of now, there is some evidence for partner choice and screening by host plants. *Acacia* plants are sometimes co-colonized by high and low quality queens that occupy the domatia resulting in high and low quality colonies; if protection conferred by the better colonies leads to lower losses due to herbivory, then those modules of the plant occupied by the better colonies achieve higher productivity compared to the unproductive modules whose domatia are occupied by the less aggressive colonies (Heil 2013). Over time, the more productive modules become dominant and outgrow the less productive ones. These are examples of self-organized screening mechanisms that help to stabilize insect–plant mutualisms (Borges 2015b). The ultimate expression of microcosm inhabitants providing mutualistic services is when aggressive ants inhabiting domatia are effective at deterring even megaherbivores such as elephants and thereby affect landscape compositions of African savannas (Goheen and Palmer 2010; Riginos et al. 2015).

MICROCOSM STUDIES AND THE FUTURE

The usefulness of microcosms in understanding ecological and evolutionary processes cannot be overestimated. As more natural microcosms are being investigated, there appear to be convergences in processes that mimic those occurring in larger ecosystems (Bittleston et al. 2016). Furthermore, microcosms can be used to successfully evaluate impacts of climate change on ecosystems (Bradshaw and Holzapfel 2001). Microcosms are not only valuable in examining patterns and processes such as keystone predation, trait space, niche evolution, and nutrient stoichiometry, they also help in understanding the range of functional adaptations within a community (Mouquet et al. 2008; Peterson et al. 2008; Miller et al. 2014; Céréghino et al. 2018). Microcosms can also be used to examine how trade-offs in life history parameters coupled with dispersal abilities to reach ephemeral microcosms can help to predict community composition and the co-existence of species (Duthie et al. 2014). Microcosms such as fig syconia are also valuable as model systems for examining theories on sex ratios under local mate competition (Fellowes et al. 1999; Greeff 2002) and the evolution of male conflict (Murray 1987), since mating in pollinating wasps usually is confined within the natal syconium. It is also possible to invoke island biogeography theory to investigate species assemblages on hosts when considered as islands (Janzen 1968).

Studying evolution in the laboratory using simple constructs has long been extolled as a valuable exercise that enables an understanding of the limits to functional traits and other processes; but, as with any artificial construct, there are always doubts about the balance between simplicity and realism (Huey and Rosenzweig 2009). Studying evolution in natural microcosms should now gain centre focus, since microcosms are where simplicity and realism converge (Zuk and Travisano 2018).

ACKNOWLEDGEMENTS

Work reported from our laboratory in this paper has been generously supported by the Department of Science and Technology (DST), DST-FIST, Department of Biotechnology, and Ministry of Environment, Forests & Climate Change. Sunitha Murray, G. Yettiraj, Srinivasan Kasinathan, Satish Desireddy, and Anusha Kumble have provided invaluable logistic support within our laboratory microcosm. The inputs from Lakshya Katariya, Nikita Zachariah, and Satyajeet Gupta are also gratefully acknowledged.

REFERENCES

Ahmed, S., S. G. Compton, R. K. Butlin, and P. M. Gilmartin. 2009. Wind-borne insects mediate directional pollen transfer between desert fig trees 160 kilometers apart. *Proceedings of the National Academy of Sciences USA* 106:20342–20347.

Al-Beidh, S., D. W. Dunn, S. A. Power, and, J. M. Cook. 2012. Parasites and mutualism function: Measuring enemy-free space in a fig–pollinator symbiosis. *Oikos* 121:1833–1839.

Alonso, L. E. 1998. Spatial and temporal variation in the ant occupants of a facultative ant-plant. *Biotropica* 30:201–213.

Altermatt, F., E. A. Fronhofer, A. Garnier et al. 2015. Big answers from small worlds: A user's guide for protist microcosms as a model system in ecology and evolution. *Methods in Ecology and Evolution* 6:218–231.

Araújo, L. M., A. Valentin-Silva, G. W. Fernandes, and M. F. Vieira. 2017. From anthesis to diaspore dispersal: Reproductive mechanisms of rare herbaceous Moraceae species endemic to Brazil. *Darwiniana* 5:83–92.

Baidamshina, D. R., E. Y. Trizna, M. G. Holyavka et al. 2017. Targeting microbial biofilms using Ficin, a nonspecific plant protease. *Scientific Reports* 7: 46068.

Bazile, V., J. A. Moran, G. Le Moguédec, D. J. Marshall, and L. Gaume. 2012. A carnivorous plant fed by its ant symbiont: A unique multi-faceted nutritional mutualism. *PLOS ONE* 7: e36179.

Bittleston, L. S., N. E. Pierce, A. M. Ellison, and A. Pringle. 2016. Convergence in multispecies interactions. *Trends in Ecology & Evolution* 31:269–280.

Blatrix, R., and V. Mayer. 2010. Communication in ant–plant symbioses. In *Plant Communication from an Ecological Perspective* (F. Baluška, V. Ninkovic, eds.), pp. 127–158. Springer, Berlin, Germany.

Borges, R. M. 2015a. How to be a fig wasp parasite on the fig–fig wasp mutualism. *Current Opinion in Insect Science* 8:34–40.

Borges, R. M. 2015b. How mutualisms between plants and insects are stabilized. *Current Science* 108:1862–1868.

Borges, R. M. 2016. On the air: Broadcasting and reception of volatile messages in brood-site pollination mutualisms. In *Deciphering Chemical Language of Plant Communication* (J. D. Blande, R. Glinwood, eds.), pp. 227–255. Springer International Publishing, Cham, Switzerland.

Borges, R. M. 2017. Co-niche construction between hosts and symbionts: Ideas and evidence. *Journal of Genetics* 96:483–489.

Borges, R. M. 2018. The galling truth: Limited knowledge of gall-associated volatiles in multitrophic interactions. *Frontiers in Plant Science* 9:1139.

Borges, R. M., J.-M. Bessière, and Y. Ranganathan. 2013. Diel variation in fig volatiles across syconium development: Making sense of scents. *Journal of Chemical Ecology* 39:630–642.

Borges, R. M., S. G. Compton, and F. Kjellberg. 2018. Fifty years later, figs and their associated communities. *Acta Oecologica* 90:1–3.

Borges, R. M., Y. Ranganathan, A. Krishnan, M. Ghara, and G. Pramanik. 2011. When should fig fruit produce volatiles? Pattern in a ripening process. *Acta Oecologica* 37:611–618.

Bradshaw, W. E., and C. M. Holzapfel. 2001. Genetic shift in photoperiodic response correlated with global warming. *Proceedings of the National Academy of Sciences USA* 98:14509–14511.

Brouat, C., and D. McKey. 2000. Origin of caulinary ant domatia and timing of their onset in plant ontogeny: Evolution of a key trait in horizontally transmitted ant-plant symbioses. *Biological Journal of the Linnean Society* 71:801–819.

Brouat, C., and D. McKey. 2001. Leaf-stem allometry, hollow stems, and the evolution of caulinary domatia in myrmecophytes. *New Phytologist* 151:391–406.

Brouat, C., N. Garcia, C. Andary, and D. McKey. 2001. Plant lock and ant key: Pairwise coevolution of an exclusion filter in an ant–plant mutualism. *Proceedings of the Royal Society of London Series B* 268:2131–2141.

Céréghino, R., V. D. Pillar, D. S. Srivastava et al. 2018. Constraints on the functional trait space of aquatic invertebrates in bromeliads. *Functional Ecology* 32:2435–2447.

Chanam, J., and R. M. Borges. 2017. Cauline domatia of the ant-plant *Humboldtia brunonis* (Fabaceae). *Flora* 236:58–66.

Chanam, J., S. Kasinathan, G. K. Pramanik, A. Jagdeesh, K. A. Joshi, and R. M. Borges. 2014a. Context dependency of rewards and services in an Indian ant-plant interaction: Southern sites favour the mutualism between plants and ants. *Journal of Tropical Ecology* 30:219–229.

Chanam, J., M. S. Sheshshayee, S. Kasinathan, A. Jagdeesh, K. A. Joshi, and, R. M. Borges. 2014b. Nutritional benefits from domatia inhabitants in an ant–plant interaction: Interlopers do pay the rent. *Functional Ecology* 28:1107–1116.

Chao, A., C. H. Chiu, R. K. Colwell, L. F. S. Magnago, R. L. Chazdon, and N. J. Gotelli. 2017. Deciphering the enigma of undetected species, phylogenetic, and functional diversity based on Good-Turing theory. *Ecology* 98:2914–2929.

Chappell, C., and T. Fukami. 2018. Nectar yeasts: A natural microcosm for ecology. *Yeast* 35:417–423.

Chase, J. M. 2014. Spatial scale resolves the niche versus neutral theory debate. *Journal of Vegetation Science* 25:319–322.

Chomicki, G., and S. S. Renner. 2015. Phylogenetics and molecular clocks reveal the repeated evolution of ant-plants after the late Miocene in Africa and the early Miocene in Australasia and the Neotropics. *New Phytologist* 207:411–424.

Chomicki, G., and S. S. Renner. 2017. The interactions of ants with their biotic environment. *Proceedings of the Royal Society of London Series B* 284:20170013.

Compton, S. G. 2002. Sailing with the wind: Dispersal by small flying insects. In *Dispersal Ecology*, (J. M. Bullock, R. E. Kenward and R. S. Hails, eds.), pp. 113–133. Blackwell Publishing, Oxford, UK.

Compton, S. G., and F. A. C. McFarlen. 1989. Respiratory adaptations in some male fig wasps. *Proceedings of the Koninklijke Nederlandse Akademie van Wetenschappen: Series C: Biological and Medical Sciences* 92:57–71.

Condit, R., R. A. Chisholm, and S. P. Hubbell. 2012. Thirty years of forest census at Barro Colorado and the importance of immigration in maintaining diversity. *PLOS ONE* 7:e49826.

Datwyler, S. L., and G. D. Weiblen. 2004. On the origin of the fig: Phylogenetic relationships of Moraceae from *ndhF* sequences. *American Journal of Botany* 91:767–777.

Davidson, D. W., and D. McKey. 1993. The evolutionary ecology of symbiotic ant/plant relationships. *Journal of Hymenoptera Research* 2:13–83.

Debout, G., E. Provost, M. Renucci, A. Tirard, B. Schatz, and D. McKey. 2003. Colony structure in a plant-ant: Behavioural, chemical and genetic study of polydomy in *Cataulacus mckeyi* (Myrmicinae). *Oecologia* 137:195–204.

Defossez, E., C. Djiéto-Lordon, D. McKey, M. A. Selosse, and Blatrix, R. 2011. Plant-ants feed their host plant, but above all a fungal symbiont to recycle nitrogen. *Proceedings of the Royal Society of London Series B* 278:1419–1426.

Defossez, E., M. A. Selosse, M. P. Dubois et al. 2009. Ant-plants and fungi: A new threeway symbiosis. *New Phytologist* 182:942–949.

Drake, J. A., G. R. Huxel, and C. L. Hewitt. 1996. Microcosms as models for generating and testing community theory. *Ecology* 77:670–677.

Duthie, A. B., K. C. Abbott, and J. D. Nason. 2014. Trade-offs and coexistence: A lottery model applied to fig wasp communities. *American Naturalist* 183:826–841.

Edwards, D. P., M. E. Frederickson, G. H. Shepard, and D. W. Yu. 2009. A plant needs ants like a dog needs fleas: *Myrmelachista schumanni* ants gall many tree species to create housing. *American Naturalist* 174:734–740.

Eisenhauer, N., and M. Türke. 2018. From climate chambers to biodiversity chambers. *Frontiers in Ecology and the Environment* 16:136–137.

Elias, L. G., F. Kjellberg, F. H. A. Farache et al. 2018. Ovipositor morphology correlates with life history evolution in agaonid fig wasps. *Acta Oecologica* 90:109–116.

Elias, L. G., S. P. Teixeira, F. Kjellberg, and R. A. S. Pereira. 2012. Diversification in the use of resources by *Idarnes* species: Bypassing functional constraints in the fig–fig wasp interaction. *Biological Journal of the Linnean Society* 106:114–122.

Fellowes, M. D., S. G. Compton, and J. M. Cook. 1999. Sex allocation and local mate competition in Old World non-pollinating fig wasps. *Behavioral Ecology and Sociobiology* 46:95–102.

Fischer, R. C., W. Wanek, A. Richter, and V. Mayer. 2003. Do ants feed plants? A ^{15}N labelling study of nitrogen fluxes from ants to plants in the mutualism of *Pheidole* and *Piper*. *Journal of Ecology* 91:126–134.

Galil, J., M. Zeroni, and D. B. Shalom. 1973. Carbon dioxide and ethylene effects in the co-ordination between the pollinator *Blastophaga quadraticeps* and the syconium in *Ficus religiosa*. *New Phytologist* 72:1113–1127.

Gall, A., U. Uebel, U. Ebensen et al. 2017. Planktotrons: A novel indoor mesocosm facility for aquatic biodiversity and food web research. *Limnology and Oceanography: Methods* 15:663–677.

Gaume, L., and D. McKey. 2002. How identity of the homopteran trophobiont affects sex allocation in a symbiotic plant-ant: The proximate role of food. *Behavioral Ecology and Sociobiology* 51:197–205.

Gaume, L., D. Matile-Ferrero, and D. McKey. 2000. Colony foundation and acquisition of coccoid trophobionts by *Aphomomyrmex afer* (Formicinae): Co-dispersal of queens and phoretic mealybugs in an ant-plant-homopteran mutualism? *Insectes Sociaux* 47:84–91.

Gaume, L., M. Shenoy, M. Zacharias, and R. M. Borges. 2006. Co-existence of ants and an arboreal earthworm in a myrmecophyte of the Indian Western Ghats: Anti-predation effect of the earthworm mucus. *Journal of Tropical Ecology* 22:341–344.

Gaume, L., M. Zacharias, and R. M. Borges. 2005a. Ant–plant conflicts and a novel case of castration parasitism in a myrmecophyte. *Evolutionary Ecology Research* 7:435–452.

Gaume, L., M. Zacharias, V. Grosbois, and R. M. Borges. 2005b. The fitness consequences of bearing domatia and having the right ant partner: Experiments with protective and non-protective ants in a semi-myrmecophyte. *Oecologia* 145:76–86.

Ghana, S., N. Suleman, and S. G. Compton. 2015. Ability to gall: The ultimate basis of host specificity in fig wasps? *Ecological Entomology* 40:280–291.

Ghara, M., L. Kundanati, and R. M. Borges. 2011. Nature's Swiss army knives: Ovipositor structure mirrors ecology in a multitrophic fig wasp community. *PLOS ONE* 6:e23642.

Ghara, M., Y. Ranganathan, A. Krishnan, V. Gowda, and R. M. Borges. 2014. Divvying up an incubator: How parasitic and mutualistic fig wasps use space within their nursery microcosm. *Arthropod-Plant Interactions* 8:191–203.

Giblin-Davis, R. M., K. A. Davies, K. Morris, and W. K. Thomas. 2003. Evolution of parasitism in insect-transmitted plant nematodes. *Journal of Nematology* 35:133–141.

Goheen, J. R., and T. M. Palmer. 2010. Defensive plant-ants stabilize megaherbivore-driven landscape change in an African savanna. *Current Biology* 20:1768–1772.

González-Teuber, M., and M. Heil. 2015. Comparative anatomy and physiology of myrmecophytes: Ecological and evolutionary perspectives. *Research and Reports in Biodiversity Studies* 4:21–32.

Gould, S. J. 2003. The pre-Adamite in a nutshell. In *I have Landed. Splashes and Reflections in Natural History*. pp. 130–146. Vintage, Random House, UK.

Grangier, J., A. Dejean, P. J. G. Male, P. J. Solano, and J. Orivel. 2009. Mechanisms driving the specificity of a myrmecophyte–ant association. *Biological Journal of the Linnean Society* 97:90–97.

Greeff, J. M. 2002. Mating system and sex ratios of a pollinating fig wasp with dispersing males. *Proceedings of the Royal Society of London Series B* 269:2317–2323.

Grison-Pigé, L., J. L. Salager, M. Hossaert-McKey, and J. Roy. 2001. Carbon allocation to volatiles and other reproductive components in male *Ficus carica* (Moraceae). *American Journal of Botany* 88:2214–2220.

Handa, C. 2012. How do scale insects settle into the nests of plant-ants on *Macaranga* myrmecophytes? Dispersal by wind and selection by plant-ants. *Sociobiology* 59:435–446.

Heil, M. 2013. Let the best one stay: Screening of ant defenders by *Acacia* host plants functions independently of partner choice or host sanctions. *Journal of Ecology* 101:684–688.

Heil, M., and D. McKey. 2003. Protective ant-plant interactions as model systems in ecological and evolutionary research. *Annual Review of Ecology, Evolution, and Systematics* 34:425–453.

Heil, M., J. Rattke, and W. Boland. 2005. Postsecretory hydrolysis of nectar sucrose and specialization in ant/plant mutualism. *Science* 308:560–563.

Heubl, G., G. Bringmann, and H. Meimberg. 2006. Molecular phylogeny and character evolution of carnivorous plant families in Caryophyllales—revisited. *Plant Biology* 8:821–830.

Hossaert-McKey, M., M. Proffit, C. C. L. Soler et al. 2016. How to be a dioecious fig: Chemical mimicry between sexes matters only when both sexes flower synchronously. *Scientific Reports* 6:21236.

Houadria, M. Y., P. Klimes, T. M. Fayle, and P. J. Gullan. 2018. Host-plant dissections reveal contrasting distributions of *Crematogaster* ants and their symbionts in two myrmecophytic *Macaranga* species. *Ecological Entomology* 43:601–611.

Huey, R. B., and F. Rosenzweig. 2009. Laboratory evolution meets Catch-22: Balancing simplicity and realism. In *Experimental Evolution: Concepts, Methods, and Applications of Selection Experiments* (T. Garland Jr. and M. R. Rose, eds.). pp. 671–701. University of California Press, Berkeley, CA.

Huston, M. A. 1999. Microcosm experiments have limited relevance for community and ecosystem ecology: Synthesis of comments. *Ecology* 80:1088–1089.

Huston, M. A. 2014. Disturbance, productivity, and species diversity: Empiricism vs. logic in ecological theory. *Ecology* 95:2382–2396.

Jandér, K. C, E. A. Herre, and E. L. Simms. 2012. Precision of host sanctions in the fig tree–fig wasp mutualism: Consequences for uncooperative symbionts. *Ecology Letters* 15:1362–1369.

Jandér, K. C., A. Dafoe, and E. A. Herre. 2016. Fitness reduction for uncooperative fig wasps through reduced offspring size: A third component of host sanctions. *Ecology* 97:2491–2500.

Jandér, K. C., and E. A. Herre. 2016. Host sanctions in Panamanian *Ficus* are likely based on selective resource allocation. *American Journal of Botany* 103:1753–1762.

Janzen, D. H. 1968. Host plants as islands in evolutionary and contemporary time. *American Naturalist* 102:592–595.

Janzen, D. H. 1979. How to be a fig. *Annual Review of Ecology and Systematics* 10:13–51.

Jessup, C. M., R. Kassen, S. E. Forde et al. 2004. Big questions, small worlds: Microbial model systems in ecology. *Trends in Ecology & Evolution* 19:189–197.

Jevanandam, N., A. G. Goh, and R. T. Corlett. 2013. Climate warming and the potential extinction of fig wasps, the obligate pollinators of figs. *Biology Letters* 9:20130041.

Jürgens, A., H. Feldhaar, B. Feldmeyer, and B. Fiala. 2006. Chemical composition of leaf volatiles in *Macaranga* species (Euphorbiaceae) and their potential role as olfactory cues in host-localization of foundress queens of specific ant partners. *Biochemical Systematics and Ecology* 34:97–113.

Kamitani, T., and N. Kaneko. 2006. The Earthtron facility for belowground manipulation study. *Ecological Research* 21:483–487.

Kerdelhué, C., J. P. Rossi, and J.-Y. Rasplus. 2000. Comparative community ecology studies on Old World figs and fig wasps. *Ecology* 81:2832–2849.

Kierat, J., H. Szentgyörgyi, M. Czarnoleski, and M. Woyciechowski. 2017. The thermal environment of the nest affects body and cell size in the solitary red mason bee (*Osmia bicornis* L.). *Journal of Thermal Biology* 68:39–44.

Kjellberg, F., E. Jousselin, M. Hossaert-McKey, and J.-Y. Rasplus. 2005. Biology, ecology, and evolution of fig-pollinating wasps (Chalcidoidea, Agaonidae). In *Biology, Ecology and Evolution of Gall-inducing Arthropods* Vol. 2 (A. Raman, C. W. Shaefer, T. M. Withers, eds.), pp. 539–572. Science Publishers Inc., Enfield, New Hampshire.

Krishnan, A., and R. M. Borges. 2014. Parasites exert conflicting selection pressures to affect reproductive asynchrony of their host plant in an obligate pollination mutualism. *Journal of Ecology* 102:1329–1340.

Krishnan, A., G. K. Pramanik, S. V. Revadi, V. Venkateswaran, and R. M. Borges. 2014. High temperatures result in smaller nurseries which lower reproduction of pollinators and parasites in a brood site pollination mutualism. *PLOS ONE* 9: e115118.

Krishnan, A., S. Muralidharan, L. Sharma, and R. M. Borges. 2010. A hitchhiker's guide to a crowded syconium: How do fig nematodes find the right ride? *Functional Ecology* 24:741–749.

Kutsukake, M., X. Y. Meng, N. Katayama, N. Nikoh, H. Shibao, and T. Fukatsu. 2012. An insect-induced novel plant phenotype for sustaining social life in a closed system. *Nature Communications* 3:1187.

Lähteenmäki, S., E. M. Slade, B. Hardwick et al. 2015. MESOCLOSURES–increasing realism in mesocosm studies of ecosystem functioning. *Methods in Ecology and Evolution* 6:916–924.

LaManna, J. A., S. A. Mangan, A. Alonso et al. 2017. Plant diversity increases with the strength of negative density dependence at the global scale. *Science* 356:1389–1392.

Lawton, J. H. 1995. Ecological experiments with model systems. *Science* 269:328–331.

Lawton, J. H. 1996. The Ecotron facility at Silwood Park: The value of "big bottle" experiments. *Ecology* 77:665–669.

Layne Jr, J. R. 1993. Winter microclimate of goldenrod spherical galls and its effects on the gall inhabitant *Eurosta solidaginis* (Diptera: Tephritidae). *Journal of Thermal Biology* 18:125–130.

Marinho, C. R., R. A. S. Pereira, Y. Q. Peng, and S. P. Teixeira. 2018. Laticifer distribution in fig inflorescence and its potential role in the fig-fig wasp mutualism. *Acta Oecologica* 90:160–167.

Maschwitz, U., B. Fiala, K. Dumpert, R. bin Hashim, and W. Sudhaus. 2016. Nematode associates and bacteria in ant-tree symbioses. *Symbiosis* 69:1–7.

Mayer, V. E., M. E. Frederickson, D. McKey, and, R. Blatrix. 2014. Current issues in the evolutionary ecology of ant–plant symbioses. *New Phytologist* 202:749–764.

Mayer, V. E., M. Nepel, R. Blatrix et al. 2018. Transmission of fungal partners to incipient *Cecropia*-tree ant colonies. *PLOS ONE* 13:e0192207.

McLeish, M. J., S. van Noort, and K. A. Tolley. 2010. African parasitoid fig wasp diversification is a function of *Ficus* species ranges. *Molecular Phylogenetics and Evolution* 57:122–134.

McLeish, M., D. Guo, S. van Noort, and G. Midgley. 2011. Life on the edge: Rare and restricted episodes of a pan-tropical mutualism adapting to drier climates. *New Phytologist* 191:210–222.

Michener, C. D., R. M. Borges, M. Zacharias, and M. Shenoy. 2003. A new parasitic bee of the genus *Braunsapis* from India (Hymenoptera: Apidae: Allodapini). *Journal of the Kansas Entomological Society* 76:518–522.

Miller, T. E., E. R. Moran, and C. P. terHorst. 2014. Rethinking niche evolution: Experiments with natural communities of protozoa in pitcher plants. *American Naturalist* 184:277–283.

Misiewicz, T. M., and N. C. Zerega. 2012. Phylogeny, biogeography and character evolution of *Dorstenia* (Moraceae). *Edinburgh Journal of Botany* 69:413–440.

Montero, G., C. Feruglio, and I. M. Barberis. 2010. The phytotelmata and foliage macrofauna assemblages of a bromeliad species in different habitats and seasons. *Insect Conservation and Diversity* 3:92–102.

Moog, J., L. G. Saw, R. Hashim, and U. Maschwitz. 2005. The triple alliance: How a plant-ant, living in an ant-plant, acquires the third partner, a scale insect. *Insectes Sociaux* 52:169–176.

Morera, J., G. Mora-Pineda, A. Esquivel, P. Hanson, and A. A. Pinto-Tomás. 2018. Detection, ultrastructure and phylogeny of *Sclerorhabditis neotropicalis* (Nematoda: Rhabditidae) nematodes associated with the *Azteca* ant-*Cecropia* tree symbiosis. *Revista de Biología Tropical* 66:368–380.

Mouquet, N., T. Daufresne, S. M. Gray, and T. E. Miller. 2008. Modelling the relationship between a pitcher plant (*Sarracenia purpurea*) and its phytotelma community: Mutualism or parasitism? *Functional Ecology* 22:728–737.

Munoz, F., and P. Huneman. 2016. From the neutral theory to a comprehensive and multiscale theory of ecological equivalence. *Quarterly Review of Biology* 91:321–342.

Murray, M. G. 1987. The closed environment of the fig receptacle and its influence on male conflict in the Old World fig wasp, *Philotrypesis pilosa*. *Animal Behaviour* 35:488–506.

Nelson, D. R., and R. E. Lee Jr. 2004. Cuticular lipids and desiccation resistance in overwintering larvae of the goldenrod gall fly, *Eurosta solidaginis* (Diptera: Tephritidae). *Comparative Biochemistry and Physiology Part B: Biochemistry and Molecular Biology* 138:313–320.

Nepel, M., H. Voglmayr, J. Schönenberger, and V. E. Mayer. 2014. High diversity and low specificity of Chaetothyrialean fungi in carton galleries in a Neotropical ant–plant association. *PLOS ONE* 9:e112756.

Nepel, M., H. Voglmayr, R. Blatrix et al. 2016. Ant-cultivated Chaetothyriales in hollow stems of myrmecophytic *Cecropia* sp. trees–diversity and patterns. *Fungal Ecology* 23:131–140.

Noort, S. V., and S. G. Compton. 1996. Convergent evolution of agaonine and sycoecine (Agaonidae, Chalcidoidea) head shape in response to the constraints of host fig morphology. *Journal of Biogeography* 23:415–424.

Odum, H. T. 1996. Scales of ecological engineering. *Ecological Engineering* 6:7–19.

Osmond, B. 2014. Our eclectic adventures in the slower eras of photosynthesis: From New England down under to Biosphere 2 and beyond. *Annual Review of Plant Biology* 65:1–32.

Owen, T. P., and K. A. Lennon. 1999. Structure and development of the pitchers from the carnivorous plant *Nepenthes alata* (Nepenthaceae). *American Journal of Botany* 86:1382–1390.

Patiño, S., E. A. Herre, and M. T. Tyree. 1994. Physiological determinants of *Ficus* fruit temperature and implications for survival of pollinator wasp species: Comparative physiology through an energy budget approach. *Oecologia* 100:13–20.

Peterson, C. N., S. Day, B. E. Wolfe, A. M. Ellison, R. Kolter, and A. Pringle. 2008. A keystone predator controls bacterial diversity in the pitcher-plant (*Sarracenia purpurea*) microecosystem. *Environmental Microbiology* 10:2257–2266.

Pincebourde, S., and J. Casas. 2006. Multitrophic biophysical budgets: Thermal ecology of an intimate herbivore insect–plant interaction. *Ecological Monographs* 76:175–194.

Pincebourde, S., and J. Casas. 2015. Warming tolerance across insect ontogeny: Influence of joint shifts in microclimates and thermal limits. *Ecology* 96:986–997.

Pincebourde, S., and J. Casas. 2016. Hypoxia and hypercarbia in endophagous insects: Larval position in the plant gas exchange network is key. *Journal of Insect Physiology* 84:137–153.

Ramírez, W. 1987. Biological analogies between some fig-wasps (Hymenoptera: Agaonidae and Toryrnidae: Sycophaginae) and *Varroa jacobsoni* (Acari: Varroidae). *Revista de Biología Tropical* 35:209–214.

Ramírez, W. 1996. Breathing adaptations of males in fig gall flowers (Hymenoptera: Agaonidae). *Revista de Biología Tropical* 44:277–282.

Ramya, K. T., R. A. Fiyaz, R. Uma Shaanker, and K. N. Ganeshaiah. 2011. Pollinators for a syconium: How do wasps choose among syconia? *Current Science* 101:520–527.

Ranganathan, Y., J.-M. Bessière, and R. M. Borges. 2015. A coat of many scents: Cuticular hydrocarbons in multitrophic interactions of fig wasps with ants. *Acta Oecologica* 67:24–33.

Ranganathan, Y., M. Ghara, and R. M. Borges. 2010. Temporal associations in fig–wasp–ant interactions: Diel and phenological patterns. *Entomologia Experimentalis et Applicata*, 137:50–61.

Razo-Belman, R., J. Molina-Torres, O. Martínez, and M. Heil. 2018. Plant-ants use resistance-related plant odours to assess host quality before colony founding. *Journal of Ecology* 106:379–390.

Reynolds, D. R., A. M. Reynolds, and J. W. Chapman. 2014. Non-volant modes of migration in terrestrial arthropods. *Animal Migration* 2:8–28.

Riginos, C., M. A. Karande, D. I. Rubenstein, and T. M. Palmer. 2015. Disruption of a protective ant–plant mutualism by an invasive ant increases elephant damage to savanna trees. *Ecology* 96:654–661.

Rodriguez, L. J., F. Young, J.-Y. Rasplus, F. Kjellberg, and S. G. Compton. 2017. Constraints on convergence: Hydrophobic hind legs allow some male pollinator fig wasps early access to submerged females. *Journal of Natural History* 51:761–782.

Sabagh, L. T., and C. F. Rocha. 2014. Bromeliad treefrogs as phoretic hosts of ostracods. *Naturwissenschaften* 101:493–497.

Sanchez, A. 2016. Establishing an ant-plant mutualism: Foundress queen mortality and acquiring the third partner. *Insectes Sociaux* 63:155–162.

Segar, S. T., R. A. S. Pereira, S. G. Compton, and J. M. Cook. 2013. Convergent structure of multitrophic communities over three continents. *Ecology Letters* 16:1436–1445.

Segraves, K. A. 2003. Understanding stability in mutualisms: Can extrinsic factors balance the yucca–yucca moth interaction? *Ecology* 84:2943–2951.

Shapira, M. 2016. Gut microbiotas and host evolution: Scaling up symbiosis. *Trends in Ecology & Evolution* 31:539–549.

Shenoy, M., and R. M. Borges. 2008. A novel mutualism between an ant-plant and its resident pollinator. *Naturwissenschaften* 95:61–65.

Shenoy, M., and R. M. Borges. 2010. Geographical variation in an ant–plant interaction correlates with domatia occupancy, local ant diversity, and interlopers. *Biological Journal of the Linnean Society* 100:538–551.

Shenoy, M., V. Radhika, S. Satish, and R. M. Borges. 2012. Composition of extrafloral nectar influences interactions between the myrmecophyte *Humboldtia brunonis* and its ant associates. *Journal of Chemical Ecology* 38:88–99.

Souto-Vilarós, D., M. Proffit, B. Buatois et al. 2018. Pollination along an elevational gradient mediated both by floral scent and pollinator compatibility in the fig and fig-wasp mutualism. *Journal of Ecology* 106:2256–2273.

Souza, C. D., R. A. S. Pereira, C. R. Marinho, F. Kjellberg, and S. P. Teixeira. 2015. Diversity of fig glands is associated with nursery mutualism in fig trees. *American Journal of Botany* 102:1564–1577.

Srivastava, D. S., J. Kolasa, J. Bengtsson et al. 2004. Are natural microcosms useful model systems for ecology? *Trends in Ecology & Evolution* 19:379–384.

Talaga, S., O. Dézerald, A. Carteron et al. 2015. Tank bromeliads as natural microcosms: A facultative association with ants influences the aquatic invertebrate community structure. *Comptes Rendus Biologies* 338:696–700.

Thorogood, C. J., U. Bauer, and S. J. Hiscock. 2018a. Convergent and divergent evolution in carnivorous pitcher plant traps. *New Phytologist* 217:1035–1041.

Thorogood, C., N. Dalton, A. Irvine, and S. Hiscock. 2018b. The reproductive biology of two poorly known relatives of the fig (*Ficus*) and insights into the evolution of the fig syconium. *Nordic Journal of Botany* 36:njb-01832.

Torres, M. F., and A. Sanchez. 2017. Neotropical ant-plant *Triplaris americana* attracts *Pseudomyrmex mordax* ant queens during seedling stages. *Insectes Sociaux* 64:255–261.

Van Goor, J., F. Piatscheck, D. D. Houston, and J. D. Nason. 2018. Figs, pollinators, and parasites: A longitudinal study of the effects of nematode infection on fig wasp fitness. *Acta Oecologica* 90:140–150.

Vannette, R. L., P. Bichier, and S. M. Philpott. 2017. The presence of aggressive ants is associated with fewer insect visits to and altered microbe communities in coffee flowers. *Basic and Applied Ecology* 20:62–74.

Vasse, M., H. Voglmayr, V. Mayer et al. 2017. A phylogenetic perspective on the association between ants (Hymenoptera: Formicidae) and black yeasts (Ascomycota: Chaetothyriales). *Proceedings of the Royal Society of London Series B* 284:20162519.

Venkateswaran, V., A. L. Kumble, and R. M. Borges. 2018. Resource dispersion influences dispersal evolution of highly insulated insect communities. *Biology Letters* 14:20180111.

Venkateswaran, V., A. Shrivastava, A. L. Kumble, and R. M. Borges. 2017. Life-history strategy, resource dispersion and phylogenetic associations shape dispersal of a fig wasp community. *Movement Ecology* 5:25.

Verschoor, A. M., J. Takken, B. Massieux, and J. Vijverberg. 2003. The Limnotrons: A facility for experimental community and food web research. *Hydrobiologia* 491:357–377.

Voglmayr, H., V. Mayer, U. Maschwitz, J. Moog, C. Djieto-Lordon, and R. Blatrix. 2011. The diversity of ant-associated black yeasts: Insights into a newly discovered world of symbiotic interactions. *Fungal Biology* 115:1077–1091.

Walter, A., and S. C. Lambrecht. 2004. Biosphere 2 Center as a unique tool for environmental studies. *Journal of Environmental Monitoring* 6:267–277.

Wang, G., S. G. Compton, and J. Chen. 2012. The mechanism of pollinator specificity between two sympatric fig varieties: A combination of olfactory signals and contact cues. *Annals of Botany* 111:173–181.

Wang, R., A. Matthews, J. Ratcliffe et al. 2014. First record of an apparently rare fig wasp feeding strategy: Obligate seed predation. *Ecological Entomology* 39:492–500.

Weber, M. G., and K. H. Keeler. 2012. The phylogenetic distribution of extrafloral nectaries in plants. *Annals of Botany* 111:1251–1261.

Williams, J. B., and R. E. Lee. 2005. Plant senescence cues entry into diapause in the gall fly *Eurosta solidaginis*: Resulting metabolic depression is critical for water conservation. *Journal of Experimental Biology* 208:4437–4444.

Yadav, P., and R. M. Borges. 2017. The insect ovipositor as a volatile sensor within a closed microcosm. *Journal of Experimental Biology* 220:1554–1557.

Yadav, P., and R. M. Borges. 2018a. Host–parasitoid development and survival strategies in a non-pollinating fig wasp community. *Acta Oecologica* 90:60–68.

Yadav, P., and R. M. Borges. 2018b. Why resource history matters: Age and oviposition history affect oviposition behaviour in exploiters of a mutualism. *Ecological Entomology* 43:473–482.

Yadav, P., S. Desireddy, S. Kasinathan, J-M. Bessière, and R. M. Borges 2018. History matters: Oviposition resource acceptance in an exploiter of a nursery pollination mutualism. *Journal of Chemical Ecology* 44:18–28.

Zachariah, N., A. Das, T. G. Murthy, and R. M. Borges. 2017. Building mud castles: A perspective from brick-laying termites. *Scientific Reports* 7:4692.

Zerega, N. J., L. A. Mound, and G. D. Weiblen. 2004. Pollination in the New Guinea endemic *Antiaropsis decipiens* (Moraceae) is mediated by a new species of thrips, *Thrips antiaropsidis* sp. nov. (Thysanoptera: Thripidae). *International Journal of Plant Sciences* 165:1017–1026.

Zuk, M., and M. Travisano. 2018. Models on the runway: How do we make replicas of the world? *American Naturalist* 192:1–9.

Annexure I: Revisions and Reviews of Taxa by Professor C. A. Viraktamath

Note: Entries are in chronological order. The current classification of subfamilies and tribes are mentioned which may be different from those in the original papers. See Annexure IV for the references.

REVISIONS

1. Genus *Austroagallia* (Megaphthalminae: Agalliini) (Viraktamath and Sohi, 1980)
2. Genus *Grammacephalus* (Deltocephalinae: Scaphoideini) (Viraktamath, 1981)
3. Indian Macropsinae (Viraktamath, 1980, 1981)
4. Genus *Hishimonoides* (Deltocephalinae: Opsiini) (Viraktamath, Anantha Murthy, and Viraktamath, 1987)
5. Tribes Nirvanini and Balbillini (Evacanthinae) of the Indian subcontinent (Viraktamath and Wesley, 1988)
6. Genus *Traiguma* (Hylicinae: Hylicini) (Viraktamath and Webb, 1991)
7. Grass feeding genus *Leofa* (Deltocephalinae: Chiasmini) (Viraktamath and Viraktamath, 1992)
8. Macropsinae of Taiwan (Huang, Kun-wei, and Viraktamath, 1993)
9. Oriental Macropsinae (Viraktamath, 1996)
10. Genera *Glossocratus* and *Hecalus* (Deltocephalinae: Hecalini) (Dash and Viraktamath, 1997)
11. Genus *Amritodus* (Idiocerinae: Idiocerini) (Viraktamath, 1998)
12. Genus *Deltocephalus* (Deltocephalinae: Deltocephalini) of India and Nepal (Dash and Viraktamath, 1998)
13. Tribe Drabescini (Deltocephalinae) of the Indian subcontinent (Viraktamath, 1998)
14. Tribe Vartini and Scaphoideini (Deltocephalinae) from India and Nepal (Viraktamath and Anantha Murthy, 1999)
15. Genus *Acacimenus* (Deltocephalinae: Athysanini) (Viraktamath, 1999)
16. Genus *Scaphoideus* (Deltocephalinae: Scaphoideini) from the Indian subcontinent (Viraktamath and Mohan, 2004)
17. Genus *Durgades* (Megaphthalminae: Agalliini) (Viraktamath, 2004)
18. *Varta-Stymphalus* generic complex (Deltocephalinae: Vartini, Scaphoideini, Opsiini) (Viraktamath, 2004)
19. Genus *Sophonia* (Nirvaninae: Nirvanini) (Webb and Viraktamath, 2004)
20. Tribe Krisnini (Iassinae) of the Indian subcontinent (Viraktamath, 2006)
21. Genus *Pythamus* (Evacanthinae: Evacanthini) (Viraktamath and Webb, 2007)
22. Genus *Platyretus* (Deltocephalinae: Scaphoideini) (Viraktamath, Webb, Dai and Zhang, 2008)
23. *Deltocephalus* genus group and other Deltocephalinae (Webb and Viraktamath, 2009)
24. Genus *Idioceroides* (Idiocerinae: Idiocerini) (Zhang and Viraktamath, 2009)
25. Genus *Scaphotettix* (Deltocephalinae: Mukariini) (Dai, Viraktamath, Zhang and Webb, 2009)
26. Genus *Goniagnathus* (Deltocephalinae: Goniagnathini) (Viraktamath and Gnaneswaran, 2009)
27. Genus *Hishimonoides* (Deltocephalinae: Opsiini) (Dai, Viraktamath, and Zhang, 2010)
28. Oriental and Australian tribe Agalliini (Megophthalminae) (Viraktamath, 2011)
29. Tribe Agalliini (Megophthalminae) from China (Viraktamath, Dai and Zhang, 2012)
30. Oriental genus *Parallygus* (Deltocephalinae: Scaphoideini) (Dai, Viraktamath, Webb and Zhang, 2012)
31. Genus *Tambocerus* (Deltocephalinae: Athysanini) (Viraktamath, 2012)
32. Afrotropical and Oriental Signoretiinae (Takiya, Dietrich, and Viraktamath, 2013)
33. Madagascaran Agalliini (Megophthalminae) (Viraktamath and Gonçalves, 2013)
34. Genera *Hishimonus* and *Litura* Deltocephalinae: Opsiini) (Viraktamath and Anantha Murthy, 2014)
35. Genus *Nicolaus* (Deltocephalinae: Paralimini) (Viraktamath and Webb, 2014)
36. Genus *Gurawa* (Deltocephalinae: Chiasmini) (Viraktamath and Gnaneswaran, 2015)
37. Genus *Signoretia* (Signoretiinae: Signoretiini) (Viraktamath and Webb, 2016)
38. Genus *Allophleps* (Deltocephalinae: Athysanini) (Webb and Viraktamath, 2016)
39. Tribe Phlogisini (Sigoretinae) (Viraktamath and Deitrich, 2016)
40. Tribe Adelungiini (Megophthaliminae) (Viraktamath, 2017)
41. Tribe Evacanthini (Evacanthinae) of the Indian subcontinent (Viraktamath and Webb, 2018)
42. Genus *Penthimia* (Deltocephalinae: Penthimiini) of the Indian subcontinent (Shobharani, Viraktamath, and Webb, 2018)

43. Genus *Tambila* (Deltocephalinae: Penthimiini) of the Indian subcontinent (Shobharani, Viraktamath, and Webb, 2018)
44. Bamboo leafhopper tribe Mukariini (Deltocephalinae) of the Indian subcontinent (Viraktamath and Webb, 2019)
45. Genera *Daimachus, Radhades and Ulopsina* (Ulopinae: Ulopini) of the Indian subcontinent (Viraktamath and Webb, 2019)
46. Tribe Coelidiini (Coelidiinae) of the Indian subcontinent (Viraktamath and Meshram, 2019)

REVIEWS

1. Species of subfamily Agallinae described by S. Matsumura (Viraktamath, 1973)
2. Species of subfamily Iassinae described by S. Matsumura (Viraktamath, 1979)
3. Species of *Idioscopus* (Idiocerinae: Idiocerini) described by H. S. Pruthi (Viraktamath, 1980)
4. Redescriptions of *Allectus, Divitiacus,* and *Lampridius* (Deltocephalinae: Stenometopiini) described by W. L. Distant (Viraktamath and Viraktamth, 1980)
5. Auchenorrhyncha associated with mango (Viraktamath, 1989)
6. C. F. Baker's species of Oriental Nirvaninae (Viraktamath, 1992)
7. Genus *Idiocerus* (Idiocerinae: Idiocerini) from India (Viraktamath and Sohi, 1994)
8. Genus *Cofana* (Cicadellinae: Cicadellini) from India (Sindhu and Viraktamath, 2008)
9. Genus *Jilinga* Deltocephalinae: Paralimnini) (Xing, Viraktamath, Dai, and Li, 2012)
10. New and little known Deltocephalinae (Viraktamath and Yeshwanth, 2017)
11. Genus *Mahellus* (Coelidiinae: Coelidiini) (Viraktamath and Meshram, 2017)

Annexure II: New Tribe and Genera Described by Professor C. A. Viraktamath

DELTOCEPHALINAE

Deltocephalini
***Miradeltaphus* Dash & Viraktamath 1995**
Type species: *Miradeltaphus mirabilis* Das & Viraktamath 1995
Drabescini
***Canopyana* Viraktamath & Srinivasa 2006**
Type species: *Canopyana vateriae* Viraktamath & Srinivasa 2006
***Indokutara* Viraktamath 1998**
Type species: *Indokutara conica* Viraktamath 1998
***Kotabala* Viraktamath 1998**
Type species: *Kotabala adiveyyai* Viraktamath 1998
Goniagnathini
***Goniagnathus* (*Tropicognathus*) Viraktamath & Gnaneswaran 2009**
Type species: *Goniagnathus* (*Tropicognathus*) *fumosus* Distant 1908
Koebeliini
***Shivapona* Ghauri & Viraktamath 1987**
Type species: *Shivapona shivai* Ghauri & Viraktamath 1987
***Sohipona* Ghauri & Viraktamath 1987**
Type species: *Sohipona sohii* Ghauri & Viraktamath 1987
***Pinopona* Viraktamath & Sohi 1998**
Type species: *Pinopona minuta* Viraktamath & Sohi 1998
Mukariini
***Aalinga* Viraktamath & Webb 2019**
Type species: *Aalinga brunoflava* Viraktamath & Webb 2019
***Mukariella* Viraktamath & Webb 2019**
Type species: *Mukariella daii* Viraktamath & Webb 2019
*Punctulini Dai, Zahniser, Viraktamath & Webb 2017
Type genus: *Punctulus* Dai, Zahniser, & Viraktamath 2017
***Punctulus* Dai, Zahniser, Viraktamath 2017**
Type species: *Punctulus tumidifrons* Dai, Zahniser, Viraktamath, & Webb 2017
***Hirsutula* Dai, Zahniser, Viraktamath, & Webb 2017**
Type species: *Hirsutula rubrifrons* Dai, Zahniser, Viraktamath, & Webb 2017
***Taveunius* Dai, Zahniser, Viraktamath, & Webb 2017**
Type species *Taveunius megapunctatus* Dai, Zahniser, Viraktamath, & Webb 2017

Scaphoideini
***Scaphodhara* Viraktamath & Mohan 1994**
Type species: *Scaphodhara sahyadrica* Viraktamath & Mohan 1994
***Scaphomonus* Viraktamath, 2009**
Type species: *Scaphotettix freytagi* Viraktamath & Mohan 1993
***Sikhamani* Viraktamath & Webb in Dai et al. 2006**
Type species: *Sikhamani delicatula* Viraktamath & Webb 2006
***Sudhamruta* Viraktamath & Anantha Murthy 1999**
Type species: *Sudhamruta wesleyi* Viraktamath & Anantha Murthy 1999
***Thryaksha* Viraktamath & Anantha Murthy 1999**
Type species: *Thryaksha recurvatus* Viraktamath & Anantha Murthy 1999
***Univagris* Viraktamath & Anantha Murthy 1999**
Type species: *Univagris pallida* Viraktamath & Anantha Murthy 1999
Stenometopiini
***Cymbopogonella* Viraktamath 1976**
Type species: *Cymbopongonella longivertex* Viraktamath 1976
Vartiini
***Curvimonus* Viraktamath & Anantha Murthy 1999**
Type species: *Curvimonus gajadantha* Viraktamath & Anantha Murthy 1999
***Vartalapa* Viraktamath 2004**
Type species: *Vartalapa robusta* Viraktamath 2004
***Vartatopa* Viraktamath 2004**
Type species: *Vartatopa bifurcata* Viraktamath 2004
***Xenovarta* Viraktamath 2004**
Type species: *Xenovarta acuta* Viraktamath 2004

EVACANTHINAE

Evacanthini
***Onukindia* Viraktamath & Webb 2017**
Type species: *Platyretus connexus* Distant 1908

IDIOCERINAE

Idiocerini
***Burmascopus* Viraktamath 2007**
Type species: *Idiocerus fasciolatus* Distant 1908
***Ceylonoscopus* Viraktamath 2007**
Type species: *Ceylonoscopus quadripunctatus* Viraktamath 2007

Jogocerus **Viraktamath 1979**
Type species: *Jogocerus freytagi* Viraktamath 1979
Lankacerus **Viraktamath 2007**
Type species: *Lankacerus rotundus* Viraktamath 2007
Neoscopus **Viraktamath 2007**
Type species: *Neoscopus ceylonensis* Viraktamath 2007
Nilgiriscopus **Viraktamath 2007**
Type species: *Nilgiriscopus transversus* Viraktamath 2007
Periacerus **Viraktamath & Parvathi 2002**
Type species: *Idioscopus lalithae* Viraktamath 1979

MACROPSINAE

Macropsini
Reticopsella **Viraktamath 1996**
Type species: *Reticopsella orientalis* Viraktamath 1996

MEGOPHTHALMINAE

Agalliini
Agallidwipa **Viraktamath & Gonçalves 2013**
Type-species: *Agallidwipa biramosa* Viraktamath & Gonçalves 2013
Formallia **Viraktamath 2011**
Type species: *Formallia truncata* Viraktamath 2011
Hemagallia **Viraktamath 2011**
Type species: *Agallia plotina* Distant 1908
Humpatagallia **Linnavuori & Viraktamath 1973**
Type species: *Humpatagallia scutellaris* Linnavuori & Viraktamath 1973
Ianagallia **Viraktamath 2011**
Type species: *Austroagallia bifurcata* Sawai Singh & Gill 1973
Nandigallia **Viraktamath 2011**
Type species: *Nandigallia nandiensis* Viraktamath 2011
Paulagallia **Viraktamath 2011**
Type species: *Paulagallia punctata* Viraktamath 2011
Purvigallia **Viraktamath, Dai, & Zhang 2012**
Type species: *Purvigallia maculata* Viraktamath, Dai & Zhang 2012
Sangeeta **Viraktamath 2011**
Type species: *Sangeeta sadongensis* Viraktamath 2011
Sinoagallia **Viraktamath, Dai, & Zhang 2012**
Type species: *Sinoagallia serrata* Viraktamath, Dai, & Zhang 2012
Skandagallia **Viraktamath, Dai, & Zhang 2012**
Type species: *Skandagallia dietrichi* Viraktamath, Dai, & Zhang 2012
Sungallia **Viraktamath, Dai, & Zhang 2012**
Type species: *Sungallia truncata* Viraktamath, Dai, & Zhang 2012

SIGNORETIINAE

Phlogisini
Aloka **Viraktamath & Dietrich 2016**
Type species: *Aloka depressa* Viraktamath & Dietrich 2016

TYPHLOCYBINAE

Dikraneurini
Kalkiana **Sohi, Viraktamath & Dworakowska 1980**
Type species: *Kalkiana bambusa* Sohi, Viraktamath & Dworakowska 1980
Sweta **Viraktamath & Dietrich 2011**
Type species: *Sweta hallucinata* Viraktamath & Dietrich 2011
Erythroneurini
Anaka **Dworakowska & Viraktamath 1975**
Type species: *Anaka colorata* Dworakowska & Viraktamath 1975
Mandola **Dworakowska & Viraktamath 1975**
Type species: *Zygina quadrinotata* Ahmed 1969
Meremra **Dworakowska & Viraktamath 1979**
Type species: *Meremra puncta* Dworakowska & Viraktamath 1979
Takama **Dworakowska & Viraktamath 1975**
Type species: *Takama magna* Dworakowska & Viraktamath 1975
Tuzinka **Dworakowska & Viraktamath 1979**
Type species: *Tuzinka acuta* Dworakowska & Viraktamath 1979
Ziginellini
Borulla **Dworakowska, Sohi, & Viraktamath 1980**
Type species: *Borulla gracilis* Dworakowska, Sohi, & Viraktamath 1980

ULOPINAE

Ulopini
Ulopsina **Dai, Viraktamath Zhang 2012**
Type species: *Ulopsina sinica* Dai, Viraktamath, & Zhang 2012
Ulopsina (indoulopa) **Viraktamath & Webb 2019**
Type species: *Ulopsina (Indoulopa) himalayana* Viraktamath & Webb 2019

* New tribe
** New subgenus

Annexure III: New Species Described by Professor C. A. Viraktamath

HEMIPTERA
CICADELLIDAE

CICADELLINAE

    Cicadellini
        ***Anatkina* Young 1986**
            jogensis Krishnankutty & Viraktamath 2011
            multilineata Krishnankutty & Viraktamath 2011
        ***Cofana* Melichar 1926**
            bidentata Krishnankutty & Viraktamath 2008
            spiculata Krishnankutty & Viraktamath 2008

COELIDIINAE

    Coelidiini
        ***Calodia* Nielson 1982**
            deergha Viraktamath & Meshram 2019
            keralica Viraktamath & Meshram 2019
            kumari Viraktamath & Meshram 2019
            neofusca Viraktamath & Meshram 2019
            periyari Viraktamath & Meshram 2019
            tridenta Viraktamath & Meshram 2019
        ***Glaberana* Nielson 2015**
            acuta Viraktamath & Meshram 2019
            purva Viraktamath & Meshram 2019
        ***Mahellus* Nielson 1982**
            cardoni Viraktamath & Meshram 2017
            ungulatus Viraktamath & Meshram 2017
        ***Olidiana* Mckamey 2006**
            lanceolata Viraktamath & Meshram 2019
            flectheri Viraktamath & Meshram 2019
            umroensis Viraktamath & Meshram 2019
            unidenta Viraktamath & Meshram 2019
        ***Singillatus* Nielson 2015**
            parapectitus Viraktamath & Meshram 2019
            serratispatulatus Viraktamath & Meshram 2019
        ***Trinoridia* Nielson 2015**
            dialata Viraktamath & Meshram 2019
            ochrocephala Viraktamath & Meshram 2019
            piperica Viraktamath & Meshram 2019
            ramamurthyi Viraktamath & Meshram 2019
            saraikela Viraktamath & Meshram 2019
            timlivana Viraktamath & Meshram 2019
        ***Webbolidia* Nielson 2015**
            andamana Viraktamath & Meshram 2019
            burmanica Viraktamath & Meshram 2019
        ***Zhangolidia* Nielson2015**
            weicongi Viraktamath & Meshram 2019

DELTOCEPHALINAE

    Athysanini
        ***Acacimenus* Dlabola 1979**
            deccanensis Viraktamath 1999
            inequalis Viraktamath 1999
            maheshai Viraktamath 1999
            variabilis Viraktamath 1999
        ***Allophleps* Bergroth 1920**
            linnavuorii Webb & Viraktamath 2016
        ***Neoreticulum* Dai 2009**
            manipurensis Viraktamath & Yeshwanth 2017
        ***Tambocerus* Zhang & Webb 1996**
            acutus Viraktamath 2012
            cholas Viraktamath 2012
            daii Viraktamath 2012
            furcostylus Viraktamath 2012
            krameri Viraktamath 2012
            nilgiris Viraktamath 2012
            zahniseri Viraktamath 2012
    Cicadulini
        ***Cicadula* Zetterstedt 1840**
            simlaensis Viraktamath & Yeshwanth 2017
        ***Neoreticulum* Dai 2009**
            shillongensis Viraktamath & Yeshwanth 2017
        ***Yuanamia* Zhang & Duan 2006**
            rajiae Viraktamath & Yeshwanth 2017
    Chiasmini
        ***Gurawa* Distant 1918**
            constricta Viraktamath & Gnaneswaran 2013
            ceylonica Viraktamath & Gnaneswaran 2013
        ***Leofa* Distant 1918**
            naga Viraktamath & Viraktamath 1992
            neela Viraktamath & Viraktamath 1992
            robusta Viraktamath & Viraktamath 1992
            truncata Viraktamath & Viraktamath 1992
    Deltocephalini
        ***Deltocephalus* Burmeister 1838**
            vulgaris (Dash & Viraktamath 1998)
        ***Maiestas* Distant 1917**
            acuminatus (Dash & Viraktamath 1998)
            agroecus (Dash & Viraktamath 1998)
            albomaculatus (Dash & Viraktamath 1998)
            aridus (Dash & Viraktamath 1998)
            belonus (Dash & Viraktamath 1998)
            bengalensis (Dash & Viraktamath 1998)
            bilineatus (Dash & Viraktamath 1998)
            bispinosus (Dash & Viraktamath 1998)
            breviculus (Dash & Viraktamath 1998)

brevis (Dash & Viraktamath 1998)
canaraicus (Dash & Viraktamath 1998)
cuculatus (Dash & Viraktamath 1998)
dashi Webb & Viraktamath 2009
hastatus (Dash & Viraktamath 1998)
jagannathi (Dash & Viraktamath 1995)
jogensis (Dash & Viraktamath 1998)
knighti Webb & Viraktamath 2009
raoi (Dash & Viraktamath 1998)
semilimax (Dash & Viraktamath 1998)
spiculatus (Dash & Viraktamath 1998)
systenos (Dash & Viraktamath 1998)
tareni (Dash & Viraktamath 1995)
trispinosus (Dash & Viraktamath 1998)
trisuli (Dash & Viraktamath 1998)
truncatus (Dash & Viraktamath 1998)
variabilis (Dash & Viraktamath 1998)
xanthocephalus (Dash & Viraktamath 1998)
Miradeltaphus **Dash & Viraktamath 1995**
   mirabilis Dash & Viraktamath 1995
Paramesodes **Ishihara 1953**
   iraniensis Webb & Viraktamath 2009

Drabescini
  Bhatia **Distant 1908**
    radhamaniae Viraktamath & Murthy 2009
    serrata Viraktamath & Murthy 2009
  Canopyana **Viraktamath & Srinivasa 2006**
    vateriae Viraktamath & Srinivasa 2006
  Carvaka **Distant 1918**
    clava Viraktamath 1998
    compressa Viraktamath 1998
    confusa Viraktamath 1998
    elongata Viraktamath 1998
    girijae Viraktamath 1998
    kumari Viraktamath 1998
    nielsoni Viraktamath 1998
    pruthii Viraktamath 1998
    sinuata Viraktamath 1998
    synavei Viraktamath 1998
    wellingtoni Viraktamath 1998
  Divus **Distant 1908**
    samanus Viraktamath 1998
  Indokutara **Viraktamath 1998**
    conica Viraktamath 1998
  Kotabala **Viraktamath 1998**
    adiveyyai Viraktamath 1998
  Kutara **Distant 1908**
    breviplata Viraktamath 1998
    crypta Viraktamath 1998
    striata Viraktamath 1998
    trifida Viraktamath 1998
  Megabyzus **Distant 1908**
    ganeshai Viraktamath 1998
    indicus Viraktamath 1998
    jogensis Viraktamath 1998

Eupelicini
  Chloropelix **Lindberg 1936**
    indica Viraktamath & Viraktamath 1989
  Paradorydium **Kirkaldy 1901**
    deccani Viraktamath & Viraktamath 1989
    dharwarensis Viraktamath 1976
    khasiana Viraktamath & Viraktamath 1989
    omani Viraktamath 1976

Goniagnathini
  Goniagnathus (*Tropicognathus*) **Viraktamath & Gnaneswaran 2009**
    anufrievi Viraktamath & Gnaneswaran 2009
    concavus Dash & Viraktamath 2001
    nepalicus Viraktamath & Gnaneswaran 2009
    quadripinnatus Dash & Viraktamath 2001
    symphysis Dash & Viraktamath 2001
    syncerus Dash & Viraktamath 2001
    zeylanicus Viraktamath & Gnaneswaran 2009

Hecalini
  Glossocratus **Fieber 1866**
    ramakrishnai Dash & Viraktamath 1997
  Hecalus **Stal**
    caudatus Dash & Viraktamath 1997
    compressus (Dash & Viraktamath 1997)
    dentatus (Dash & Viraktamath 1997)
    tuberculatus (Dash & Viraktamath 1997)
  Memnonia **Ball 1900**
    bifida (Dash & Viraktamath 1997)

Koebeliini
  Pinopona **Viraktamath & Sohi 1998**
    minuta Viraktamath & Sohi 1998
  Shivapona **Ghauri & Viraktamath 1987**
    shabnami Ghauri & Viraktamath 1987
    shivai Ghauri & Viraktamath 1987
  Sohipona **Ghauri & Viraktamath 1987**
    habibi Ghauri & Viraktamath 1987
    sohii Ghauri & Viraktamath 1987
    thapai Ghauri & Viraktamath 1987
    webbi Ghauri & Viraktamath 1987

Mukariini
  Aalinga **Viraktamath & Webb 2019**
    brunoflava Viraktamath & Webb 2019
  Buloria **Distant 1908**
    indica Viraktamath & Webb 2019
    zeylanica Viraktamath & Webb 2019
  Flatfronta **Chen & Li 1997**
    bella Viraktamath & Webb 2019
  Mohunia **Distant 1908**
    bifurcata Viraktamath & Webb 2019
  Mukaria **Distant 1908**
    omani Viraktamath & Webb 2019
    vakra Viraktamath & Webb 2019
  Mukariella **Viraktamath & Webb 2019**
    daii Viraktamath & Webb 2019
  Myittana (*Benglebra*) **Mohmood & Ahmed 1969**
    cornuta Viraktamath & Webb 2019
  Myittana (*Myittana*) **Distant 1908**
    distincta Viraktamath & Webb 2019
  Myittana (*Savasa*) **Viraktamath & Webb 2019**

constricta Viraktamath & Webb 2019
Scaphotettix Matsumura 1914
acutus Viraktamath & Webb 2018
Opsiini
Cestius Distant 1908
terminaliae Viraktamath 1978
Hishimonoides Ishihara 1965
arbudae Viraktamath, Anantha Murthy & Viraktamath 1987
bougainvilleae Viraktamath, Anantha Murthy & Viraktamath 1987
curvatus Dai, Viraktamath &Yalin 2010
orientalis Mahmood 1975
similis Dai, Viraktamath &Yalin 2010
spinosus Viraktamath, Anantha Murthy & Viraktamath 1987
Hishimonus Ishihara 1953
acuminatus Viraktamath & Anantha Murthy 2014
distinctus Viraktamath & Anantha Murthy 2014
dwipae Viraktamath & Anantha Murthy 2014
longisetosus Viraktamath & Anantha Murthy 2014
spicans Viraktamath & Anantha Murthy 2014
thapai Viraktamath & Anantha Murthy 2014
zeylanicus Viraktamath & Anantha Murthy 2014
Japananus Ball 1931
nepalica Viraktamath & Anantha Murthy 1999
Litura Knight 1970
triangula Viraktamath & Anantha Murthy 2014
Paralimini
Jilinga Ghauri 1974
asymmetrica Xing, Viraktamath, Dai & Li 2012
linzhiensis Xing, Viraktamath, Dai & Li 2012
Nicolaus Linderg 1958
abuensis Viraktamath & Webb 2014
bidentatus Viraktamath & Webb 2014
cornutus Viraktamath & Webb 2014
serratus Viraktamath & Webb 2014
Penthimini
Penthimia Germer 1821
curvata Shobharani, Viraktamath & Webb 2018
meghalayensis Shobharani, Viraktamath & Webb 2018
neoattenuata Shobharani, Viraktamath & Webb 2018
ribhoi Shobharani, Viraktamath & Webb 2018
sahyadrica Shobharani, Viraktamath & Webb 2018
spiculata Shobharani, Viraktamath & Webb 2018
tumida Shobharani, Viraktamath & Webb 2018
Punctulini
Hirsitus Dai, Zahniser, Viraktamath & Webb 2017
rubrifrons Dai, Zahniser, Viraktamath & Webb 2017
Punctulus Dai, Zahniser, Viraktamath & Webb 2017
tumidifrons Dai, Zahniser, Viraktamath & Webb 2017
manipurensis Dai, Zahniser, Viraktamath & Webb 2017
lobatus Dai, Zahniser, Viraktamath & Webb 2017
Taveunius Dai, Zahniser, Viraktamath & Webb 2017
megapunctatus Dai, Zahniser, Viraktamath & Webb 2017
Scaphoideini
Grammacephalus Haupt 1929
indicus Viraktamath & Anantha Murthy 1999
raunoi Viraktamath 1981
Parallygus Melichar 1903
burmindicus Viraktamath & Webb 2012
rameshi Viraktamath & Webb 2012
Platyretus Melichar 1903
gangeticus Viraktamath 2008
javanicus Viraktamath 2008
sudindicus Viraktamath 2008
Scaphodhara Viraktamath & Mohan 1994
biloba Viraktamath & Mohan 1994
neela Viraktamath & Mohan 1994
periyari Viraktamath & Mohan 1994
raoi Viraktamath & Mohan 1994
sahyadrica Viraktamath & Mohan 1994
Scaphoideus Ulher 1889
asymmetricus Viraktamath & Mohan 2004
bicoloratus Viraktamath & Mohan 2004
bifidus Viraktamath & Mohan 2004
dellagiustinai Webb & Viraktamath 2007
hirsutus Viraktamath & Mohan 2004
inequalis Viraktamath & Mohan 2004
jogensis Viraktamath & Mohan 2004
kirti Viraktamath & Mohan 2004
lamellaris Viraktamath & Mohan 2004
malaisei Viraktamath & Mohan 2004
quangtriensis Webb & Viraktamath 2007
sculptellus Viraktamath & Mohan 2004
sculptus Viraktamath & Mohan 2004
spiculatus Viraktamath & Mohan 2004
subsculptus Meshram, Viraktamath & Ramamurthy 2012
trilobatus Viraktamath & Mohan 2004
unimaculatus Webb & Viraktamath 2007
varna Viraktamath & Mohan 2004
vaticus Viraktamath & Mohan 2004
zhangi Viraktamath & Mohan 2004
Scaphomonus Viraktamath in Dai et al. 2006
quadrifidus (Viraktamath & Mohan 1993)
agumbensis (Viraktamath & Mohan) 1993
arcuatus (Viraktamath & Mohan 1993)
freytagi (Viraktamath & Mohan 1993)
malnadicus (Viraktamath & Mohan 1993)
vateriae Viraktamath in Dai et al. 2006

Sikhamani **Viraktamath & Webb in Dai et al. 2006**
    delicatula Viraktamath & Webb in Dai et al. 2006

Sudhamruta **Viraktamath & Anantha Murthy 1999**
    wesleyi Viraktamath & Anantha Murthy 1999

Thryaksha **Viraktamath & Anantha Murthy 1999**
    recurvatus Viraktamath & Anantha Murthy 1999

Univagris **Viraktamath & Anantha Murthy 1999**
    pallida Viraktamath & Anantha Murthy 1999

Stenometopiini
  Doratulina (***Cymbopogonella***) **Viraktamath 1976**
    longivertex Viraktamath 1976

Vartini
  Curvimonus **Viraktamath & Anantha Murthy 1999**
    gajadantha Viraktamath & Anantha Murthy 1999
  Shivania **Viraktamath 2004**
    serrata Virktamath 2004
  Varta **Distant 1908**
    bifida Viraktamath 2004
    japonica Viraktamath 2004
    longula Viraktamath 2004
    sympatrica Viraktamath 2004
  Vartalapa **Viraktamath 2004**
    curvata Viraktamath 2004
    malayana Viraktamath 2004
    robusta Viraktamath 2004
    bifurcata Viraktamath 2004
  Xenovarta **Viraktamath 2004**
    acuta Viraktamath 2004
    ankusha Viraktamath 2004
    compressa Viraktamath 2004
    cylindrica Viraktamath 2004
    harpago Viraktamath 2004

EVACANTHINAE

Balbillini
  Balbillus **Distant 1908**
    indicus Viraktamath & Wesley 1988
  Stenotortor **Baker 1923**
    acuta Viraktamath & Wilson 2018
    subhimalaya Viraktamath & Wesley 1988

Evacanthini
  Apphia **Distant 1918**
    flava Viraktamath & Webb 2018
  Bundera **Distant 1908**
    acutivertex Viraktamath & Webb 2018
  Concavocorana **Wang & Zhang 2014**
    kambaiti Viraktamath & Webb 2018
  Evacanthus **Le Peletier & Serville 1825**
    albipennis Viraktamath & Webb 2018
    convolutes Viraktamath & Webb 2018
    distinctus Viraktamath & Webb 2018
    manaliensis Viraktamath & Webb 2018
    yeshwanthi Viraktamath & Webb 2018

Onukia **Matsumura 1912**
    echina Viraktamath & Webb 2018

Onukindia **Viraktamath & Webb 2018**
    aruna Viraktamath & Webb 2018
    compressa Viraktamath & Webb 2018

Pythamus **Melichar 1903**
    biramosus Webb & Viraktamath 2007
    bispinosus Webb & Viraktamath 2007
    montanus Webb & Viraktamath 2007

Shortcrowna **Li & Li 2014**
    rubrostriata Viraktamath & Webb 2018

Striatanus **Li & Wang 1914**
    delta Viraktamath & Webb 2018

Taperus **Li & Wang 1994**
    indicus Viraktamath & Webb 2018

Nirvanini
  Kana **Distant 1908**
    bispinosa Viraktamath & Wesley 1988
    nigropicta Viraktamath & Wesley 1988
  Nirvana **Kirkaldy 1900**
    peculiaris Viraktamath & Wesley 1988
    striata Viraktamath & Wesley 1988
  Sophonia **Walker 1870**
    bakeri Viraktamath & Wesley 1988
    bifida Viraktamath & Wesley 1988
    complexa Viraktamath & Wesley 1988
    complicata Viraktamath & Wesley 1988
    keralica Viraktamath & Wesley 1988
    picta Viraktamath & Webb 2018
    similis Viraktamath & Wilson 2018

HYLICINAE

Traiguma **Distant 1908**
    nielsoni Viraktamath & Webb 1991

IASSINAE

Krisnini
  Krisna **Kirkaldy 1900**
    bakeri Viraktamath 2006
    burmanica Viraktamath 2006
    delta Viraktamath 2006
    megha Viraktamath 2006
    raja Viraktamath 2006
    varia Viraktamath 2006
    veni Viraktamath 2006
    walayari Viraktamath 2006

IDIOCERINAE

Idiocerini
  Amritodus **Anufriev 1970**
    brevis Viraktamath 1997
    brevistylus Viraktamath 1976

***Busoniomimus* Maldonado-Capriles 1977**
    manjunathi Viraktamath & Viraktamath 1985
    mudigerensis (Viraktamath 1976)
    setulistylus Viraktamath & Murphy 1980
***Ceylonoscopus* Viraktamath 2007**
    quadripunctatus Viraktamath 2007
***Idioceroides* Matsumura 1912**
    sichunensis Zhang & Viraktamath 2009
***Idiocerus* Lewis 1834**
    cedarae Viraktamath & Sohi 1994
    deodarae Viraktamath & Sohi 1994
    sharmai Viraktamath & Sohi 1994
***Idioscopus* Baker 1915**
    anasuyae Viraktamath & Viraktamath 1985
    bellus Viraktamath 1979
    capriliana Viraktamath & Murphy 1980
    decoratus Viraktamath 1976
    dworakowskae Viraktamath 1979
    indicus Viraktamath 1979
    irenae Viraktamath 1980
    jayashriae Viraktamath & Viraktamath 1985
    pretiosus Viraktamath 1979
    robustipennis Viraktamath 1979
    shillongenssis Viraktamath 1976
    spectabilis Viraktamath 1979
    thapai Viraktamath & Hongsaprug 1989
    virescens Viraktamath 1979
    webbi Viraktamath 1979
***Jogocerus* Viraktamath 1979**
    freytagi Viraktamath 1979
***Lankacerus* Viraktamath 2007**
    rotundus Viraktamath 2007
***Neoscopus* Viraktamath 2007**
    ceylonensis Viraktamath 2007
***Nilgiriscopus* Viraktamath 2007**
    transversus Viraktamath 2007
***Paraidioscopus* Viraktamath 1973**
    harrisi Viraktamath 1973
***Periacerus* Viraktamath & Parvathi 2002**
    bidentatus Viraktamath & Parvathi 2002
    lalithae (Viraktamath 1979)
    lankaensis Viraktamath & Parvathi 2002

MACROPSINAE

Macropsini
    ***Macropsis* Lewis 1836**
        ater Huang & Viraktamath 1993
        brunomaculata Huang & Viraktamath 1993
        campbelli Viraktamath 1981
        ceylonica Viraktamath 1981
        ganeshai Viraktamath 1996
        hamiltoni Viraktamath 1980
        irenae Viraktamath 1981
        kanakapurensis Viraktamath 1980
        karnatakana Viraktamath 1980

        kolarensis Viraktamath 1980
        leucasasperae Viraktamath 1980
        linnavuorii Viraktamath 1981
        meifengensis Huang & Viraktamath 1993
        nepalica Viraktamath 1996
        nigrolineata Viraktamath 1980
        shrideviae Viraktamath 1996
        smitae Viraktamath 1996
        sohii Viraktamath 1980
        sundara Viraktamath 1981
        sympatrica Viraktamath 1996
        vagdeviae Viraktamath 1996
        warburgii Huang & Viraktamath 1993
        ziziphii Viraktamath 1980
    ***Oncopsis* Burmeister 1838**
        kuluensis Viraktamath 1996
    ***Pedionis* Hamilton 1980**
        cherraensis Viraktamath 1996
        clypellalta Huang & Viraktamath 1993
        curvata Viraktamath 1981
        lii Zhang & Viraktamath 2010
        palniensis Viraktamath 1981
        rufoscutellata Huang & Viraktamath 1993
        serrata Viraktamath 1981
        spinata Zhang & Viraktamath 2010
        sumatrana Viraktamath 1996
        yunnana Zhang & Viraktamath 2010
    ***Pediopsoides* (*Pediopsoides*) Matsumura 1912**
        kodaina Viraktamath 1996
    ***Pediopsoides* (*Celopsis*) Hamilton 1980**
        pectinata Viraktamath 1996
    ***Pediopsoides* (*Sispocnis*) Anufriev 1977**
        sharmai Viraktamath 1981
    ***Reticopsella* Viraktamath 1996**
        orientalis Viraktamath 1996
    ***Varicopsella* Hamilton 1980**
        elegans Viraktamath 1981

MEGOPHTHALMINAE

Agalliini
    ***Agallia* Curtis 1833**
        distincta Viraktamath 1980
        evansi Viraktamath 1980
        furcostyli Viraktamath 1980
        stenagalloides Viraktamath 1980
    ***Agallidwipa* Viraktamath & Gonçalves 2013**
        biramosa Viraktamath & Gonçalves 2013
        bispinosa Viraktamath & Gonçalves 2013
        webbi Viraktamath & Gonçalves 2013
    ***Austroagallia* Evans 1935**
        arrhenonigra Viraktamath 1972
        balii Viraktamath 2011
        bisinuata Viraktamath & Viraktamath 1981
        distanti Viraktamath 2011
        prachuabensis Viraktamath 2011

Durgades **Distant 1912**
   aviana Viraktamath 2004
   confusa Viraktamath 2004
   dunchensis Viraktamath 2004
   saura Viraktamath 2004
   spatulata Viraktamath 2004
   sympatrica Viraktamath 2004
Formallia **Viraktamath 2011**
   ishiharai Viraktamath 2011
   longipenis Viraktamath 2011
   longistyla Viraktamath 2011
   rugosa Viraktamath 2011
   truncata Viraktamath 2011
Humpatagallia **Linnavuori & Viraktamath 1973**
   scutellaris Linnavuori & Viraktamath 1973
Igerna **Kirkaldy 1903**
   aurora Viraktamath 2011
   channa Viraktamath, Dai & Zhang 2012
   darjeelingensis Viraktamath 2011
   delineata Viraktamath 2011
   delta Viraktamath & Gonçalves 2013
   flavocosta Viraktamath & Gonçalves 2013
   himalayensis Viraktamath 2011
   keyae Viraktamath 2011
   malagasica Viraktamath & Gonçalves 2013
   priyankae Viraktamath 2011
   quinlani Viraktamath 2011
   sikkima Viraktamath 2011
   wilsoni Viraktamath 2011
Japanagallia **Ishihara 1955**
   asymmetrica Viraktamath 2011
   curvata Viraktamath 2011
   curvipenis Viraktamath, Dai & Zhang 2012
   decliva Viraktamath, Dai & Zhang 2012
   javana Viraktamath 2011
   longipenis Viraktamath 2011
   malaisei Viraktamath 2011
   multispina Viraktamath, Dai & Zhang 2012
   mussooriensis Viraktamath 2011
   neotappana Viraktamath 2011
   nepalensis Viraktamath 2011
   palmata Viraktamath, Dai & Zhang 2012
   peculiaris Viraktamath 2011
   sclerotica Viraktamath, Dai & Zhang 2012
   sumatrana Viraktamath 2011
   yoshimoto Viraktamath 2011
Nandigallia **Viraktamath 2011**
   matai Viraktamath 2011
   nandiensis Viraktamath 2011
Paulagallia **Viraktamath 2011**
   maai Viraktamath 2011
   punctata Viraktamath 2011
Purvigallia **Viraktamath, Dai & Zhang 2012**
   maculata Viraktamath, Dai & Zhang 2012
Sangeeta **Viraktamath 2011**
   dentata Viraktamath 2011
   dlabolai Viraktamath 2011
   fyanensis Viraktamath 2011
   linnavuorii Viraktamath 2011
   nigra Viraktamath 2011
   quadriloba Viraktamath 2011
   sadongensis Viraktamath 2011
   sarawakensis Viraktamath *2011*
Sinoagallia **Viraktamath, Dai & Zhang 2012**
   serrata Viraktamath, Dai & Zhang 2012
Skandagallia **Viraktamath, Dai & Zhang 2012**
   dietrichi Viraktamath, Dai & Zhang 2012
Sungallia **Viraktamath, Dai & Zhang 2012**
   truncata Viraktamath, Dai & Zhang 2012

SIGNORETIINAE

Phlogisini
  Aloka **Viraktamath & Dietrich 2016**
    depressa Viraktamath & Dietrich 2016
Signoretiini
  Signoretia **Stål 1859**
    dulitensis Viraktamath & Webb 2016
    lunglei Viraktamath & Webb 2016
    mishmiensis Viraktamath & Webb 2016
    quoinensis Viraktamath & Webb 2016
    rubra Viraktamath & Webb 2016
    sahyadrica Viraktamath & Webb 2016
    similaris Viraktamath & Webb 2016
    sinuata Viraktamath & Webb 2016
    takiyae Viraktamath & Webb 2016

TYPHLOCYBINAE

Dikraneurini
  Kalkiana **Sohi, Viraktamath & Dworakowska 1980**
    bambusa Viraktamath & Dworakowska 1980
Erythroneurini
  Ahmedra **Dworakowska & Viraktamath 1979**
    distincta Dworakowska & Viraktamath 1979
  Anaka **Dworakowska & Viraktamath 1975**
    colorata Dworakowska & Viraktamath 1975
  Anufrievia **Dworakowska 1970**
    bauhinicola Dworakowska & Viraktamath 1978
  Arboridia (*Arboridia*) **Zachvatkin 1946**
    tertina Dworakowska & Viraktamath 1978
  Dayus **Mahmood 1967**
    formosus Dworakowska & Viraktamath 1978
  Empoasca (*Livasca*) **Dworakowska & Viraktamath 1978**
    malliki Dworakowska & Viraktamath 1978
  Empoasca **Walsh 1862**
    spirosa Dworakowska & Viraktamath 1979
  Meremra **Dworakowska & Viraktamath 1979**
    puncta Dworakowska & Viraktamath 1979
  Optya **Dworakowska 1974**
    barbistyla Dworakowska & Viraktamath 1978
  Sweta **Viraktamath & Dietrich 2011**
    hallucinata Viraktamath & Dietrich 2011

  ***Takama* Dworakowska & Viraktamath 1975**
    magna Dworakowska & Viraktamath 1975
  ***Thaia* (*Nlunga*) Dworakowska 1974**
    australis Dworakowska & Viraktamath 1979
    obtusa Dworakowska & Viraktamath 1979
    septima Dworakowska & Viraktamath 1979
  ***Tuzinka* Dworakowska &Viraktamath 1979**
    acuta Dworakowska & Viraktamath 1979
 Ziginellini
  ***Borulla* Dworakowska, Sohi & Viraktamath 1980**
    gracilis Dworakowska, Sohi & Viraktamath 1980
  ***Singapora* Mahmood 1967**
    karnatakana Viraktamath & Dworakowska 1979
  ***Zyginella* Low 1885**
    loewi Dworakowska, Sohi & Viraktamath 1980

ULOPINAE

 Ulopsini
  ***Daimachus* Distant 1916**
    matheranensis Viraktamath & Webb 2019
    robustus Viraktamath & Webb 2019
    sirsiensis Viraktamath & Webb 2019
    sudindicus Viraktamath & Webb 2019
  ***Ulopsina* Dai, Viraktamath & Zhang 2012**
    sinica Dai, Viraktamath & Zhang 2012
    szwedoi Dai, Viraktamath & Zhang 2012
    Ulopsina (*Indoulopa*) Viraktamath & Webb 2019
    himalayana Viraktamath & Webb 2019

CALISCELIDAE

 ***Formiscurra* Gnezdilov & Viraktamath 2011**
   indicus Gnezdilov & Viraktamath 2011

MEGALOPTERA

CORYDALIDAE

 ***Nevromus* Rambur 1842**
   austroindicus Liu & Viraktamath 2012

COLEOPTERA

CHRYSOMELIDAE

 ***Lanka* Maulik 1926**
   ramakrishnai Prathapan & Viraktamath 2008
   sahyadriensis Prathapan & Viraktamath 2008
 ***Longitarsus* Latreille 1825**
   limnophilae Prathapan & Viraktamath 2011
 ***Phaelota* Jacoby 1887**
   jacobyi Prathapan & Viraktamath 2004
   sindhoori Prathapan & Viraktamath 2004
   vaishakha Prathapan & Viraktamath 2004
 ***Tegyrius* Jacoby 1887**
   agasthya Prathapan & Viraktamath 2009
   anupama Prathapan & Viraktamath 2009
   buddhai Prathapan & Viraktamath 2009
   dalei Prathapan & Viraktamath 2009
   nigrotibialis Prathapan & Viraktamath 2009
   pucetibialis Prathapan & Viraktamath 2009
   radhikae Prathapan & Viraktamath 2009
   tippui Prathapan & Viraktamath 2009

Annexure IV: Research Publications of Professor C. A. Viraktamath

Note: Not sequenced chronologically. In each category, all papers independently authored by Prof. CAV are listed first, followed by those in which he is the second author and so on.

TAXONOMY OF LEAFHOPPERS

REVISIONS

1. Viraktamath, C. A. 1980. Indian Macropsinae (Homoptera: Cicadellidae) I. New species of *Macropsis* from south India. *Journal of Natural History* 14: 319–329.
2. Viraktamath, C. A. 1981. Indian species of *Grammacephalus* (Homoptera: Cicadellidae). *Colemania* 1: 7–12.
3. Viraktamath, C. A. 1981. Indian Macropsinae (Homoptera: Cicadellidae). II. Species described by W.L. Distant and description of new species from the Indian subcontinent. *Entomologica Scandinavica* 12: 295–310.
4. Viraktamath, C. A. 1996. New Oriental Macropsinae with a key to species of the Indian subcontinent (Insecta: Auchenorrhyncha: Cicadellidae). *Entomologische Abhandlungen, Stätliches Museum für Tierkunde in Dresden* 57(7): 183–200.
5. Viraktamath, C. A. 1998. A revision of the Idiocerine leafhopper genus *Amritodus* (Hemiptera: Cicadellidae) breeding on mango. *Entomon* 22(1997): 111–117.
6. Viraktamath, C. A. 1998. A revision of the leafhopper tribe Paraboloponini (Hemiptera: Cicadellidae: Selenocephalinae) in the Indian subcontinent. *The Bulletin of the British Museum of Natural History (Ent. Series)* 67(2): 153–207.
7. Viraktamath, C. A. 1999. New species of deltocephaline genus *Acacimenus* (Hemiptera: Cicadellidae) from India and Sri Lanka. *Journal of the Bombay Natural History Society* 96(2): 297–305.
8. Viraktamath, C. A. 2004. Revision of the Agalliinae leafhopper genus *Durgades* (Hemiptera: Cicadellidae) along with description of six new species. pp. 363–380. In: Rajamohana, K., Sudheer, K., Girish Kumar, P. and Santhosh, S. (Eds) 2004. *Perspectives on Biosystematics and Biodiversity, Prof. T.C. Narendran Commemoration Volume*, Sersa Publisher, Calicut, India, xviii+666 pp.
9. Viraktamath, C. A. 2004. A revision of the *Varta-Stymphalus* generic complex of the tribe Scaphytopiini (Hemiptera: Cicadellidae) from the Old World. *Zootaxa* 713: 1–47.
10. Viraktamath, C. A. 2006. Revision of the leafhopper tribe Krisnini (Hemiptera: Cicadellidae: Iassinae) of the Indian subcontinent. *Zootaxa* 1338: 1–32.
11. Viraktamath, C. A. 2011. Revision of the Oriental and Australian Agalliini (Hemiptera: Cicadellidae: Megophthalminae). *Zootaxa* 2844: 1–118.
12. Viraktamath, C. A. 2012. Seven new species of the leafhopper genus *Tambocerus* (Hemiptera: Cicadellidae) from the Indian subcontinent. *Zootaxa* 3385: 43–61.
13. Viraktamath, C. A. 2017. Review of the leafhopper tribe Adelungiini (Hemiptera: Cicadellidae: Megophthalminae) from the Indian subcontinent. *Entomon* 42(1): 47–58.
14. Viraktamath, C. A. and Anantha Murthy, H. V. 1999. A revision of the leafhopper tribe Scaphytopiini from India and Nepal (Insecta, Hemiptera, Cicadellidae, Deltocephalinae). *Senckenbergiana Biologica* 79: 39–55.
15. Viraktamath, C. A. and Anantha Murthy, H.V. 2014. Review of the genera *Hishimonus* Ishihara and *Litura* Knight (Hemiptera: Cicadellidae) from the Indian subcontinent with description of new species. *Zootaxa* 3785(2): 101–138. doi:10.11646/zootaxa.3785.2.1.
16. Viraktamath, C. A. and Dietrich, C. H. 2017. New genus and species of the leafhopper tribe Phlogisini from India with description of male *Phlogis mirabilis* Linnavuori from Africa (Hemiptera: Auchenorrhyncha: Cicdellidae: Signoretiinae). *Entomologica Americana* 122(3): 451–460. doi:10.1664/1947-5144-122.3.451.
17. Viraktamath, C. A. and Gnaneswaran, R. 2009. Three new species of *Goniagnathus* (Hemiptera: Cicadellidae) from the Indian subcontinent with description of a new subgenus. *Zootaxa* 2224: 51–59.
18. Viraktamath, C. A. and Gnaneswaran, R. 2015. Review of the grass feeding genus *Gurawa* Distant (Hemiptera: Cicadellidae: Deltocephalinae) from the Indian subcontinent with description of two new species. *Entomon* [2013] 38(4): 193–212.
19. Viraktamath, C. A. and Gonçalves, A. C. 2013. Review of Madagascaran Agalliini (Hemiptera: Cicadellidae: Megophthalminae) with descriptions of a new genus and six new species. *Zootaxa* 3616(1): 1–21.

20. Viraktamath, C. A. and Meshram, N. 2019. Leafhopper tribe Coelidiini (Hemiptera: Cicadellidae: Coelidiinae) of the Indian subcontinent. *Zootaxa* 4653(1): 001–091.
21. Viraktamath, C. A. and Mohan, G. S. 2004. A revision of the deltocephaline leafhopper genus *Scaphoideus* (Hemiptera: Cicadellidae) from the Indian subcontinent. *Zootaxa* 578: 1–48.
22. Viraktamath, C. A. and Sohi, A. S. 1980. Notes on the Indian species of *Austroagallia* (Homoptera: Cicadellidae). *Oriental Insects* 14: 283–289.
23. Viraktamath, C. A. and Viraktamath, S. 1992. A revision of the deltocephaline leafhoppers of the grass feeding genus *Leofa* Distant (Insecta, Homoptera, Auchenorrhyncha, Cicadellidae). *Entomologische Abhandlungen, Stätliches Museum für Tierkunde in Dresden* 55: 1–12.
24. Viraktamath, C. A. and Webb, M. D. 1991. A review of the Indian hylicine leafhoppers of the genus *Traiguma* Dist., with description of a new species (Insecta, Homoptera, Auchenorrhyncha, Cicadellidae). *Reichenbachia* 28(3): 123–133.
25. Viraktamath, C. A. and Webb, M. D. 2007. Review of the leafhopper genus *Pythamus* Melichar (Hemiptera: Cicadellidae: Evacanthinae) in the Indian subcontinent. *Zootaxa* 1546: 51–61.
26. Viraktamath, C. A. and Webb, M. D. 2014. Four new species of the grass feeding leafhopper genus *Nicolaus* Lindberg (Hemiptera: Cicadellidae: Deltocephalinae) from the Indian subcontinent. *Zootaxa* 3784(5): 528–538. doi:10.11646/zootaxa.3784.5.2.
27. Viraktamath, C. A. and Webb, M. D. 2016. Review of the genus *Signoretia* (Hemiptera: Cicadellidae: Signoretiinae) of the Oriental region with description of nine new species. *Zootaxa* 4193(3): 486–516. doi:10.11646/zootaxa.4193.3.3.
28. Viraktamath, C. A. and Webb, M. D. 2018. Revision of the evacanthine leafhoppers Hemiptera: Cicadellidae: Evacanthinae) of the Indian subcontinent. *Zootaxa* 4386: 1–78. doi:10.11646/zootaxa4386.1.1.
29. Viraktamath, C. A. and Webb, M. D. 2019. Revision of the bamboo leafhopper tribe Mukariini (Hemiptera: Cicadellidae: Deltocephalinae) from the Indian subcontinent with description of new genera and species. *Zootaxa* 4547(1): 1–69.
30. Viraktamath, C. A. and Webb, M. D. 2019. Revision of the Ulopinae leafhoppers (Hemiptera: Cicadellidae) of the Indian subcontinent, I. Ulopini genera: *Daimachus, Radhades* and *Ulopsina*. *Zootaxa* 4613(3): 557–577.
31. Viraktamath, C. A. and Wesley, C. S. 1988. Revision of the Nirvaninae (Homoptera: Cicadellidae) of the Indian subcontinent. *In Research in the Auchenorrhyncha Homoptera: A tribute to Paul W. Oman. Great Basin Naturalist Memoirs No.* 12: 182–223.
32. Viraktamath, C. A., Anantha Murthy, H. V. and Viraktamath, S. 1987. The deltocephaline leafhopper genus *Hishimonoides* in the Indian subcontinent (Homoptera). *Journal of Natural History* 21: 1225–1236.
33. Viraktamath, C. A., Dai, W. and Zhang, Y. 2012. Taxonomic revision of the leafhopper tribe Agalliini (Hemiptera: Cicadellidae: Megophthalminae) from China, with description of new taxa. *Zootaxa* 3430: 1–49.
34. Viraktamath, C. A., Webb, M. D., Dai, W. and Zhang, Y. 2008. A review of the grassland leafhopper genus *Platyretus* Melichar (Hemiptera: Cicadellidae: Deltocephalinae). *Zootaxa* 1696: 37–47.
35. Dash, P. C. and Viraktamath, C. A. 1997. New species of grass feeding hecaline leafhopper genera *Glossocratus* and *Hecalus* (Hemiptera: Cicadellidae) from India. *Journal of the Bombay Natural History Society* 94: 127–138.
36. Dash, P. C. and Viraktamath, C. A. 1998. A review of the Indian and Nepalese grass feeding leafhopper genus *Deltocephalus* (Homoptera: Cicadellidae) with description of new species. *Hexapoda* 10: 1–59.
37. Huang, K.-W. and Viraktamath, C. A. 1993. The macropsine leafhoppers (Homoptera: Cicadellidae) of Taiwan. *Chinese Journal of Entomology* 13: 361–373.
38. Webb, M. D. and Viraktamath, C. A. 2004. On the identity of an invasive leafhopper on Hawaii (Hemiptera, Cicadellidae, Nirvaninae). *Zootaxa* 692: 1–6.
39. Webb, M. D. and Viraktamath, C. A. 2009. Annotated check-list, generic key and new species of Old World Deltocephalini leafhoppers with nomenclatorial changes in the *Deltocephalus* group and other Deltocephalinae (Hemiptra, Auchenorrhyncha, Cicadellide). *Zootaxa* 2163: 1–64.
40. Webb, M. D. and Viraktamath, C. A. 2017. Nomenclatorial changes in the African leafhopper genus *Allophleps* (Hemiptera: Auchenorrhyncha: Cicadellidae: Deltocephalinae). *Entomologica Americana* 122(3): 326–332. doi:10.1664/1947-5144-122.3.326.
41. Zhang, B. and Viraktamath, C. A. 2009. New placement of the leafhopper genus *Idioceroides* Matsumura (Hemiptera: Cicadellidae: Idiocerinae), with description of a new species. *Zootaxa* 2242: 64–68.
42. Dai, W., Viraktamath, C. A. and Zhang, Y. 2010. A review of the leafhopper genus *Hishimonoides* Ishihara (Hemiptera: Cicadellidae: Deltocephalinae). *Zoological Science* 27: 771–781.
43. Shobharani, M., Viraktamath, C. A. and Webb, M. D. 2018. Review of the leafhopper genus *Penthimia* Germar (Hemiptera: Cicadellidae: Deltocephalinae) from the Indian subcontinent with description of seven new species. *Zootaxa* 4369(1): 1–45. doi:10.11646/zootaxa.4369.1.1.

44. Shobharani, M., Viraktamath, C. A. and Webb, M. D. 2018. Revision of the penthimiine leafhopper genus *Tambila* Distant (Hemiptera: Cicadellidae: Deltocephalinae) of the Indian subcontinent. *Zootaxa* 4514(4): 501–515. doi:10.11646/zootaxa.4514.4.4.
45. Dai, W., Viraktamath, C. A., Zhang, Y. and Webb, M. D. 2009. A review of the leafhopper genus *Scaphotettix* Matsumura (Hemiptera: Cicadellidae: Deltocephalinae), with description of a new genus. *Zoological Science* 26: 656–663.
46. Dai, W., Viraktamath, C. A., Webb, M. D. and Zhang Y. 2012. Revision of the Oriental leafhopper genus *Parallygus* Melichar (Hemiptera: Cicadellidae: Deltocephalinae) with description of new species. *Zootaxa* 3157: 41–53.
47. Takiya, D. M., Dietrich, C. H. and Viraktamath, C. A. 2013. The unusual Afrotropical and Oriental leafhopper subfamily Signoretiinae (Hemiptera, Cicadellidae): Taxonomic notes, new distributional records, and description of two new *Signoretia* species. *Zookeys* 319: 303–323. doi:10.3897/zookeys.319.4326.

Reviews

48. Viraktamath, C. A. 1973. Some species of Agalliinae (Cicadellidae: Homoptera) described by Dr. S. Matsumura. *Kontŷu* 41: 307–311.
49. Viraktamath, C. A. 1979. Studies on the Iassinae (Homoptera: Cicadellidae) described by Dr. S. Matsumura. *Oriental Insects* 13: 93–108.
50. Viraktamath, C. A. 1980. Notes on *Idioscopus* species (Homoptera: Cicadellidae) described by Dr. H.S. Pruthi, with description of a new species from Meghalaya, India. *Entomon* 5: 227–231.
51. Viraktamath, C. A. 1989. Auchenorrhyncha associated with mango, *Mangifera indica* L. *Tropical Pest Management* 35: 431–434.
52. Viraktamath, C. A. 1992. Oriental nirvanine leafhoppers (Homoptera: Cicadellidae): A review of C.F. Baker's species and keys to the genera and species from Singapore, Borneo and the Philippines. *Entomologca Scandinavica* 23: 249–273.
53. Viraktamath, C. A. and Meshram, N. 2017. A review of the coelidiine leafhopper genus *Mahellus* (Hemiptera: Cicadellidae: Coelidiinae) with description of two new species. *Zootaxa* 4258(3): 271–280. doi:10.11646/zootaxa.4258.3.4.
54. Viraktamath, C. A. and Sohi, A. S. 1994. Three new species of *Idiocerus* (Hemiptera: Cicadellidae) from north India. *Entomon* 19: 23–28.
55. Viraktamath, C. A. and Yeshwanth, H. M. 2017. New and little known deltocephaline leafhoppers (Hemiptera: Cicadellidae) from the Indian subcontinent. pp. 39–74. In *Insect Diversity and Taxonomy.* Santhosh S., M. Nasser and K. Sudheer (Eds) TCN commemorative volume. Prof. T. C. Narendran Trust for Animal Taxonomy, Calicut, India. [published February 2018].
56. Sindhu, M. K., and Viraktamath, C. A. 2008. Notes on the leafhopper genus *Cofana* Melichar (Hemiptera: Cicadellidae: Cicadellinae) from India with description of two new species. *Zootaxa* 1874: 35–49.
57. Viraktamath, S. and Viraktamath, C. A. 1980. Redescriptions of *Allectus, Divitiacus* and *Lampridius* (Homoptera: Cicadellidae) described by W.L. Distant. *Entomon* 5: 135–140.
58. Xing, J. C., Viraktamath, C. A., Dai, R. H. and Li, Z. Z. 2012. Review of the leafhopper genus *Jilinga* Ghauri (Hemiptera: Cicadellidae: Deltocephalinae: Paralimnini), with description of two new species from China. *Zootaxa* 3164: 49–56.

Descriptions of New Taxa and Other Taxonomic Papers

59. Viraktamath, C. A. 1972. A new species of *Austroagallia* Evans from Galapagos islands (Homoptera: Cicadellidae). *Occasional Papers of the California Academy of Sciences* No. 101, 4 pp.
60. Viraktamath, C. A. 1973. A new species of Idiocerinae (Cicadellidae: Homoptera) on *Semecarpus anacardium* L. *Oriental Insects* 7: 133–135.
61. Viraktamath, C. A. 1976. New species of *Doratulina* and *Bumizana* (Homoptera: Cicadellidae) from Karnataka. *Oriental Insects* 10: 79–86.
62. Viraktamath, C. A. 1976. Four new species of idiocerine leafhoppers from India with a note on male *Balocha astuta* (Melichar) (Homoptera: Cicadellidae: Idiocerinae). *Mysore Journal of Agricultural Sciences* 10: 234–244.
63. Viraktamath, C. A. 1978. A new species of *Cestius* Distant (Homoptera: Cicadellidae) from southern India. *Journal of Natural History* 12: 241–244.
64. Viraktamath, C. A. 1979. *Jogocerus* gen. nov. and new species of idiocerine leafhoppers from southern India (Homoptera: Cicadellidae). *Entomon* 4: 17–26.
65. Viraktamath, C. A. 1979. Four new species of *Idioscopus* (Homoptera: Cicadellidae) from southern India. *Entomon* 4: 173–181.
66. Viraktamath, C. A. 1979. Oriental Agalliinae, Macropsinae, Iassinae and Idiocerinae – A review. *Workshop on Advances in Insect Taxonomy in India and the Orient*. Manali, India, October 9–12. pp. 78–79. (Abstract only)
67. Viraktamath, C. A. 1980. Four new species of Agalliinae (Homoptera: Cicadellidae) from Juan Fernandez. *Journal of Natural History* 14: 621–628.
68. Viraktamath, C. A. 1983. Genera to be revised on a priority basis. The need for keys and illustrations of economic species of leafhoppers and preservation of voucher specimens in recognised institutions. In Knight, W. J., Pant, N.

C., Robertson, T. S. and Wilson, M. R. (Eds) *Proceedings of the 1st International Workshop on Biotaxonomy, Classification and Biology of Leafhoppers and Planthoppers (Auchenorrhyncha) of Economic Importance.* pp. 471–492. October 4–7. Commonwealth Institute of Entomology, London, UK, 500 p.

69. Viraktamath, C. A. 2005. Key to the subfamilies and tribes of leafhoppers (Hemiptera: Cicadellidae) of the Indian subcontinent. *Bionotes* 7(1): 20–24; 7(2): 44–49.

70. Viraktamath, C. A. 2006. Biodiversity of leafhoppers (Hemiptera: Cicadellidae) of the Indian subcontinent. *In* Jain, P.C. and Bhargava, M. C. (Eds) *Entomology: Novel Approaches, 2007.* New India Publishing Agency, New Delhi, India. pp 477–497.

71. Viraktamath, C. A. 2007. New genera and species of idiocerine leafhoppers (Hemiptera: Cicadellidae) from India, Sri Lanka and Myanmar. *Biosystematica* 1: 21–30.

72. Viraktamath, C. A. 2014. Leafhoppers (Hemiptera: Auchenorrhyncha: Cicadellidae) of Indian subcontinent: Systematic and species diversity. Book Chapter. pp. 65–84. *In* Rahman, M. and Anto, M. (Eds) *Forest Entomology: Emerging Issues and Dimensions.* Narendra Publishing House, New Delhi, India.

73. Viraktamath, C. A. and Dash, P. C. 2001. Taxonomic position of the Indian species of grass feeding deltocephaline leafhoppers assigned to the genus *Allophleps* (Hemiptera: Cicadellidae). *Journal of the Bombay Natural History Society* 98 (1): 47–52.

74. Viraktamath, C. A. and Dietrich, C. H. 2011. A remarkable new genus of Dikraneurini (Hemiptera: Cicadomorpha: Cicadellidae: Typhlocybinae) from Southeast Asia. *Zootaxa*, 2931: 1–7.

75. Viraktamath, C. A. and Dworakowska, I. 1979. Indian species of *Singapora* (Homoptera: Cicadellidae: Typhlocybinae). *Oriental Insects* 13: 87–91.

76. Viraktamath, C. A. and Hongsaprug, W. 1989–1990. Two new species of *Idioscopus* (Homoptera: Cicadellidae) from Nepal and Thailand. *Zoogica Orientalis* 6 & 7: 1–5.

77. Viraktamath, C. A. and Mohan, G. S. 1994. Indian species of the deltocephaline leafhopper genus *Scaphotettix* Matsumura (Hemiptera: Cicadellidae). *Journal of Bombay natural History Society* 90(1993): 463–474.

78. Viraktamath, C. A. and Mohan, G. S. 1994. Description of *Scaphodhara* a new genus related to *Scaphoideus* (Homoptera: Cicadellidae) and five new species from south India. *Entomon* 19: 13–22.

79. Viraktamath, C. A. and Murphy, D. H. 1981. Description of two new species with notes on some oriental Idiocerinae (Homoptera: Cicadellidae). *Journal of Entomological Research* 4(1980): 83–90.

80. Viraktamath, C. A. and Parvathi, C. 1993. Utility of ovipositor and associated structures in the higher classification of Cicadellidae with special reference to Indian Idiocerinae. *In Proceedings of the 8th Auchenorrhyncha Congress*, Drosopoulos, S., Petrakis, P. V., Claridge, M. F., and de Vrijer, P. W. F. (Eds), Delphi, Greece, August 9–13. Extended abstract, pp. 33–34.

81. Viraktamath, C. A. and Parvathi, C. 2002. Description of a new idiocerine genus *Periacerus* (Hemiptera: Cicadellidae) and two new species from India and Sri Lanka. *Journal of the Bombay Natural History Society* 99(3): 488–494.

82. Viraktamath, C. A. and Shankara Murthy, M. 2009. Two new species of the genus *Bhatia* (Hemiptera: Cicadellidae) from south India with new locality records of other species. *Zootaxa* 2245: 47–53.

83. Viraktamath, C. A. and Sohi, A. S. 1998. A new grypotine leafhopper genus and species from the Indian subcontinent (Hemiptera: Cicadellidae) *Journal of Insect Science* 11(2): 114–116.

84. Viraktamath, C. A. and Srinivasa, Y. B. 2006. *Canopyana vateriae* gen. nov. and sp. nov. – A leafhopper breeding on *Vateria indica* and a new record of *Bhatia distanti* (Hemiptera: Cicadellidae: Selenocephlinae) from south India. *Zootaxa* 1307: 35–39.

85. Viraktamath, C. A. and Viraktamath, S. 1989. New species of *Bumizana* Dist. and *Chloropelix* Lindb. from Indian subcontinent (Insecta: Homoptera: Auchenorrhyncha, Cicadellidae). *Reichenbachia* 27(4): 19–26.

86. Viraktamath, C. A. and Wilson, M. R. 2018. New species of the leafhopper genera *Sophonia* and *Stenotortor* (Hemiptera: Cicadellidae: Evacanthinae) from the Oriental region. *Zootaxa* 4378(3): 356–366. doi:10.11646/zootaxa.4378.3.8.

87. Dash, P. C. and Viraktamath, C. A. 1996. Description of a new grass feeding deltocephaline leafhopper genus *Miradeltaphus* gen. nov. with notes on the genus *Pruthiorosius* (Homoptera: Cicadellidae). *Hexapoda* 6: 37–44.

88. Dash, P. C. and Viraktamath, C. A. 1996. Two new species of grass feeding leafhopper genus *Deltocephalus* (*Recilia*) (Homoptera: Cicadellidae) from Orissa, India. *Hexapoda* 6: 71–78.

89. Dash, P. C. and Viraktamath, C. A. 2001. Deltocephaline leafhopper genus *Goniagnathus* (Hemiptera: Cicadellidae) in the Indian subcontinent with description of four new species. *Journal of the Bombay Natural History Society* 98(1): 62–79.

90. Dworakowska, I. and Viraktamath, C. A. 1975. On some Typhlocybinae from India (Auchenorrhyncha, Cicadellidae). *Bulletin de l'Académie polonaise des sciences. Série des sciences biologiques II* 23: 521–530.

91. Dworakowska, I. and Viraktamath, C. A. 1978. On some Indian Typhlocybinae (Auchenorrhyncha, Cicadellidae). *Bulletin de l'Académie Polonaise des Sciences. Série des sciences biologiques* 26: 539–548.

92. Dworakowska, I. and Viraktamath, C. A. 1979. On some Indian Erythroneurini (Auchenorrhyncha, Cicadellidae, Typhlocybinae). *Bulletin de l'Académie Polonaise des Sciences. Série des sciences biologiques* 27: 49–59.
93. Ghauri, M. S. K. and Viraktamath, C. A. 1987. New Paraboloponinae from the Sub Himalayan region (Insecta, Homoptera, Cicadelloidea, Iassidae). *Reichenbachia* 25(12): 47–58.
94. Krishnankutty, S. M. and Viraktamath, C. A. 2011. Description of two new species of leafhopper genus *Anatkina* Young (Hemiptera: Cicadellinae) from India. *Biosystematica* 5: 5–11.
95. Linnavuori, R. and Viraktamath, C. A. 1973. A new cicadellid genus from Africa (Homoptera, Cicadellidae, Agalliinae). *Revue de Zoologie et de Botanique Africaines* 87: 485–492.
96. Viraktamath, S. and Viraktamath, C. A. 1996. The leafhoppers (Homoptera: Cicadellidae) and their host plants in Karnataka. *Karnataka Journal of Agricultural Sciences* 8(1995): 249–255.
97. Viraktamath, S. and Viraktamath, C. A. 1981. Biology of two species of *Austroagallia* (Homoptera: Cicadellidae) from India with description of one new species. *Colemania* 1(2): 79–87.
98. Viraktamath, S. and Viraktamath, C. A. 1982. Biology of *Agallia campbelli* (Homoptera: Cicadellidae) in south India. *Colemania* 1(3): 155–162.
99. Viraktamath, S. and Viraktamath, C. A. 1985. New species of *Busoniomimus* and *Idioscopus* (Homoptera: Cicadellidae: Idiocerinae) breeding on mango in south India. *Entomon* 10: 305–311.
100. Webb, M. D. and Viraktamath, C. A. 2007. Three new Old World species of the leafhopper genus *Scaphoideus* Uhler (Hemiptera: Auchenorrhyncha: Cicadellidae). *Zootaxa* 1457: 49–55.
101. Zhang, B. and Viraktamath, C. A. 2010. New species of macropsine leafhopper genus *Pedionis* Hamilton (Hemiptera: Cicadellidae) from China, with a key to Chinese species. *Zootaxa* 2484: 53–60.
102. Dai, W., Viraktamath, C. A. and Zhang, Y. 2012. *Ulopsina*, a remarkable new ulopine leafhopper genus from China. *Journal of Insect Science* 12(70): 1–9.
103. Meshram, N. M., Viraktamath, C. A. and Ramamurthy, V. V. 2012. A new deltocephaline *Scaphoideus subsculptus* sp. nov. (Hemiptera: Cicadellidae) with redescription of *S. harlani* from India. *Oriental Insects* 46: 153–162.
104. Nielson, M. W., Viraktamath, C. A. and Lann van der, R. 2018. Case 3582—*Scaris* Le Peletier & Andinet-Serville, 1828 and *Scarides* (Scaridae) Amyot & Andinet-Serville, 1843 (Insecta, Hemiptera): Proposed suppression to remove homonymy of the latter with Scarini (Currently Scaridae) Rafinesque, 1810 (Osteichthyes). *The Bulletin of Zoological Nomenclature* 75(1): 6–11. doi:10.21805/bzn.v75.a005.
105. Shobharani, M., Viraktamath, C. A. and Webb, M. D. 2018. Review of the leafhopper genus *Penthimia* Germar (Hemiptera: Cicadellidae: Deltocephalinae) from the Indian subcontinent with description of seven new species. *Zootaxa* 4369(1): 1–45. doi:10.11646/zootaxa.4369.1.1.
106. Sohi, A. S., Viraktamath, C. A. and Dworakowska, I. 1980. *Kalkiana bambusa* gen. et sp. nov. (Homoptera: Cicadellidae) a dikraneurine leafhopper breeding on bamboo in northern India. *Oriental Insects* 14: 279–281.
107. Sohi, A. S., Viraktamath, C. A. and Pathania. 2008. Taxonomy of leafhoppers (Hemiptera: Cicadellidae) of India—An insight. *Journal of Insect Science* 21(4): 321–343.
108. Wang, Y., Viraktamath, C. A. and Zhang, Y. L. 2015. *Mediporus*, a new genus of the leafhopper subfamily Evacanthinae (Hemiptera: Cicadellidae), with a key to genera of the Evacanthini. *Zootaxa* 3964(3): 379–385. doi:10.11646/zootaxa.3964.3.7.
109. Xue, Q. Q., Viraktamath, C. A. and Zhang, Y. 2016. Checklist to Chinese Idiocerinae leafhoppers, key to genera and description of a new species of *Anidiocerus* (Hemiptera: Auchenorrhyncha: Cicadellidae). *Entomologica Americana* 122(3): 405–417. doi:10.1664/1947-5144-122.3.405.
110. Gnaneswaran, R. Viraktamath, C. A., Hemachandra, K. S., Ahangama, D., Wijayagundasekara, H. N. P. and Wahundeniya, I. 2008. Typhlocybinae leafhoppers (Hemiptera: Auchenorrhyncha: Cicadellidae) associated with horticultural crops in Sri Lanka. *Tropical Agricultural Research* 20: 1–11.
111. Dworakowska, I., Sohi, A. S. and Viraktamath, C. A. 1980. One new genus and two new species of Zyginellini (Cicadellidae: Typhlocybinae) from India. *Oriental Insects* 14: 271–277.
112. Dai, W., Zhang, Y., Viraktamath, C. A. and Webb, M. D. 2006. Two new Asian Sacphytopiini leafhoppers (Hemiptera: Cicadellidae: Deltocephalinae) with description of a new genus. *Zootaxa* 1309: 37–44.
113. Dai, W., Zahniser, J. N., Viraktamath, C. A. and Webb, M. D. 2017. Punctulini (Hemiptera: Cicadellidae) Deltocephalinae), a new leafhopper tribe from the Oriental region and pacific Islands. *Zootaxa* 4226(2): 229–248. doi:10.11646/zootaxa.4226.2.4.
114. Gnaneswaran, R. Hemachandra, K. S., Viraktamath, C. A., Ahangama, D., Wijayagunasekara, H. N. P. and Wahundeniya, I. 2007. *Idioscopus nagpurensis* (Pruthi) (Hemiptera: Cicadellidae: Idiocerinae): A new member of mango leafhopper complex in Sri Lanka. *Tropical Agricultural Research* 19: 78–90.

TAXONOMY OF OTHER GROUPS

115. Gnezdilov, V. M. and Viraktamath, C. A. 2011. A new genus and new species of the tribe Caliscelini Amyot & Serville (Hemiptera, Fulgoroidea,

116. Kumar, P. and Viraktamath, C. A. 1992. Illustrated keys for identification of common species of short-horned grasshoppers (Orthoptera: Acridoidea) of Karnataka and notes on their ecology and behaviour. *Hexapoda* 3: 53–70.
117. Kumar, P. and Viraktamath, C. A. 1992. Taxonomic significance of the male genitalia (Epiphallus) of some species of short-horned grasshoppers (Orthoptera: Acridoidea). *Journal of Bombay Natural History Society* 88: 200–209.
118. Prathapan, K. D. and Viraktamath, C. A. 2004. Revision of *Phaelota* Jacoby (Coleoptera: Chrysomelidae) with descriptions of three new species. *Zootaxa* 447: 1–18.
119. Prathapan, K. D. and Viraktamath, C. A. 2008. The flea beetle genus *Lanka* (Coleoptera: Chrysomelidae) in India with descriptions of three new species and notes on the identity of the *pollu* beetle infesting black pepper, *Piper nigrum. Zootaxa* 1681: 1–30.
120. Prathapan, K. D. and C. A. Viraktamath, 2011. A new species of *Longitarsus* Latreille, 1829 (Coleoptera: Chrysomelidae) pupating inside stem aerenchyma of the hydrophyte host from the Oriental region. *Zookeys* 87: 1–10.
121. Salini, S. and Viraktamath, C. A. 2015. Genera of Pentatomidae (Hemiptera: Pentatomoidea) from south India—An illustrated key to genera and checklist of species. *Zootaxa* 3924(1): 1–76. doi:10.11646/zootaxa.3924.1.1.
122. Ghate, H. V., Viraktamath, C. A. and Sundararaj, R. 2011. First report of a cerambycid beetle (*Capnolymma cingalensis*) from India. *Taprobanica* 3(2): 104–106.
123. Kalleswaraswamy, C. M., Manjunatha, M. and Viraktamath, C.A. 2011. Diversity of termites (Isoptera: Insecta) in India with special reference to Karnataka. pp. 175–189. *In* Hosetti, B. B. and Chakravarthy, A. K. (Eds) *Biodiversity Conservation*. Avishkar Publishers Distributors, Jaipur, India.
124. Kalleshwaraswamy, C. M., Nagaraju, D. K. and Viraktamath, C.A. 2013. Illustrated identification key to common termite (Isoptera) genera of south India. *Biosystematica* 7(1): 11–21.
125. Liu, X., Hayashi, F., Viraktamath, C. A. and Yang, D. 2012. Systematics and biogeography of the dobsonfly genus *Nevromus* Rambur (Megaloptera: Corydalidae: Corydalinae) from the Oriental realm. *Systematic Entomology* 37: 657–669.
126. Ramani, S., David, K. J., Viraktamath, C. A. and Kumar, A. R. V. 2008. Identity and distribution of *Bactrocera caryeae* (Kapoor) (Insecta: Diptera: Tephritidae)—A species under the *Bactrocera dorsalis* complex in India. *Biosystematica* 2(1): 59–62.

BIOLOGY AND ECOLOGY OF CROP PESTS AND NATURAL ENEMIES

127. Viraktamath, C. A. and Ramakrishna, B. V. 1989. Influence of host plants on the development of legume leafhopper, *Empoasca (Distantasca) terminalis* Distant) (Homoptera: Cicadellidae). *Proceedings of the 76th Indian Science Congress Madurai Part II.* Abstracts, p. 129.
128. Viraktamath, C. A., Bhumannavar, B. S. and Patel, V. N. 2004. Biology and ecology of *Zygogramma bicolorata* Pallister, 1953. In: *New Developments in the Biology of Chrysomelidae*. pp. 767–777. Jolivet, P., Santago-Blay, J. A. and Schmitt, M. (Eds), SPB Academic Publishing bv. The Hague, the Netherlands.
129. Viraktamath, C. A., Puttarudriah, M. and Channa-Basavanna, G. P. 1974. Studies on the biology of rice case-worm, *Nymphula depunctalis* Guenee (Pyraustidae: Lepidoptera). *Mysore Journal of Agricultural Sciences* 8: 234–241.
130. Bhaskar, H. and Viraktamath, C. A. 2001. Prevalence and predator prey relationship of coccinellids in red gram ecosystem. *National Symposium on Pulses and Oilseeds for Sustainable Agriculture*. July 29–31, Tamil Nadu Agricultural University, Coimbatore, India, p. 182.
131. Bhaskar, H. and Viraktamath, C. A. 2002. Spatial distribution of mealy bugs and *Cryptolaemus montrouzieri* Mulsant (Coccinellidae: Coleoptera) in mango orchard. *Insect Environment* **7**: 184.
132. Bhaskar, H. and Viraktamath, C. A. 2002. Diversity and abundance of aphidophagous coccinellids in cabbage field. *Insect Environment* 8: 31.
133. Bhumannavar, B. S. and Viraktamath, C. A. 2000. Biology and behaviour of *Euplectrus maternus* Bhatnagar (Hymenoptera: Eulophidae), an ectoparasitoid of *Othreis* spp. (Lepidoptera: Noctuidae). *Pest Management in Horticultural Ecosystems* 6: 1–14.
134. Bhumannavar, B. S. and Viraktamath, C. A. 2001. Proboscis morphology and nature of fruit damage in different fruit piercing moths (Lepidoptera: Noctuidae). *Pest Management in Horticultural Ecosystems* 7(1): 28–40.
135. Bhumannavar, B. S. and Viraktamath, C. A. 2001. Seasonal incidence and extent of parasitization of fruit piercing moths of the genus *Othreis* (Lepidoptera: Noctuidae). *Journal of Biological Control* 15: 31–38.
136. Bhumannavar, B. S. and Viraktamath, C. A. 2001. Larval host specificity, biology and adult feeding and oviposition preference of *Othreis homaena* Hübner (Lepidoptera; Noctuidae) on different Menispermaceae. *Journal of Entomological Research* 25(3): 165–181.
137. Bhumannavar, B. S. and Viraktamath, C. A. 2001. Rearing techniques for three species of *Othreis*

(Lepidoptera: Noctuidae) and their ectoparasitoid, *Euplectrus maternus* Bhatnagar (Hymenoptera: Eulophidae). *Journal of Biological Control* 15(2): 189–192.
138. Bhumannavar, B. S. and Viraktamath, C. A. 2002. Biology and adult feeding and oviposition preference and seasonal incidence of *Othreis materna* (Linnaeus) (Lepidoptera: Noctuidae). *Entomon* 27(1): 63–77.
139. Bhumannavar, B. S. and Viraktamath, C. A. 2004. Larval host specificity, biology on different Menispermaceae and adult feeding preference of *Othreis fullonia* (Clerck) (Noctuidae) in southern India. *Journal of Insect Science* 17(1): 28–34.
140. Diraviam, J. and Viraktamath, C. A. 1991. Population dynamics of the introduced ladybird beetle, *Curinus coeruleus* Mulsant in relation to its psyllid prey, *Heteropsylla cubana* Crawford in Bangalore. *Journal of Biological Control* 4(2): 99–104.
141. Diraviam, J. and Viraktamath, C. A. 1991. Biology of the introduced ladybird beetle *Curinus coeruleus* Mulsant (Coleoptera: Coccinellidae). *Journal of Biological Control* 5(1): 14–17.
142. Diraviam, J. and Viraktamath, C. A. 1992. Predatory potential and functional response of the introduced ladybird beetle, *Curinus coeruleus* (Mulsant) (Coleoptera: Coccinellidae). *Journal of Biological Control* 5(1991): 78–90.
143. Diraviam, J. and Viraktamath, C. A. 1993. Laboratory studies on the host range of *Curinus coeruleus* Mulsant, an exotic predator of subabul psyllid. *Journal of Biological Control* 6(1992): 42–43.
144. Hiremath, I. G. and Viraktamath, C. A. 1992. Biology of the sorghum earhead bug, *Calocoris angustatus* (Hemiptera: Miridae) with description of various stages. *Insect Science and Application* 13: 447–457.
145. Jagannatha, R. and Viraktamath, C. A. 1997. Host preference by the female serpentine leafminer, *Liriomyza trifolii* (Diptera: Agromyzidae). *Insect Environment* 2(4): 137.
146. Joshi, S. and Viraktamath, C. A. 2004. The sugarcane woolly aphid, *Ceratovacuna lanigera* Zehntner (Hemiptera: Aphididae): Its biology, pest status and control. *Current Science* 87(3): 101–111.
147. Muniyappa, V. and Viraktamath, C. A. 1981. Transmission of yellow dwarf with the blue race of *Nephotettix virescens*. *International Rice Research Newsletter* 6(2): 15.
148. Muniappan, R. and Viraktamath, C. A. 2006. The Asian Cycad Scale *Aulocaspis yasumatsui*. A threat to native cycads in India. *Current Science* 91(7): 868–870.
149. Viswanath, B. N. and Viraktamath, C. A. 1969. A new bug pest (Hemiptera: Miridae) on *Cucumis melo* L. in Mysore state. *Mysore Journal of Agricultural Sciences* 3(4): 475–476.
150. Kulkarni, K. A., Viraktamath, C. A. and Puttaswamy. 1971. Note on the incidence of the brinjal budworm *Phthorimaea blapsigona* Mayerick (Gelechiidae: Lepidoptera) and its natural parasite *Microbracon lefroyi* Dudgeon and Gough (Braconidae: Hymenoptera). *Science and Culture* 37: 211–214.
151. Madhura, H. S., Viraktamath, C. A. and Thippaiah, M. 2002. Population dynamics of fruit flies (Diptera: Tephritidae) in relation to fruiting period and abiotic factors. In Sanjayan, K.P., Mahalingam, V. and Muralirangan, M.C. (Eds). *Vistas of Entomological Research for the New Millennium*, pp. 132–136.
152. Mallik, B., Viraktamath, C. A. and Puttaswamy. 1996. Damage by the slug caterpillar, *Macroplectra nararia* (Lepidoptera: Limacodidae) to coconut around Bangalore, south India. *Indian Coconut Journal* 27(2): 2–4.
153. Nagaraju, D. K., Viraktamath, C. A. and Krishna Kumar, N. K. 2001. Studies on the insect galls of bell pepper (*Capsicum annuum* var. *grossum*) and brinjal (*Solanum melongena* L.), p. 65. In: Verghese A. and Reddy, P. P. (Eds.) *Proceedings of the Second National Symposium on Integrated Pest Management (IPM) in Horticultural Crops: New Molecules, Biopesticides & Environment*. October 17–19, Bangalore, India.
154. Nagaraju, D. K., Viraktamath, C. A. and Krishna Kumar, N. K. 2003. Effect of gall insects on nutritive status of the bell pepper fruits. *Insect Environment* 9: 84–85.
155. Nagaraju, D. K., Viraktamath, C. A. and Krishna Kumar, N. K. 2002. Crop loss caused by gall insects in bell pepper in south India. *Pest Management in Horticultural Ecosystems* 8(1): 20–26.
156. Nagaraju, D. K., Viraktamath, C. A. and Krishna Kumar, N. K. 2002. Species composition and abundance of gall insects on bell pepper (*Capsicum annuum* L.) and brinjal (*Solanum melongena* L.) flowers. In Sanjayan, K. P., Mahalingam, V. and Muralirangan, M. C. (Eds). *Vistas of Entomological Research for the New Millennium*, pp. 5–10.
157. Nagaraju, D. K., Viraktamath, C. A. and Krishna Kumar, N. K. 2004. New report of two phytophagous hymenopterans on flowers of bell pepper and brinjal from South India. *Insect Environment* 10(1): 25–26.
158. Prabhuraj, A., Viraktamath, C. A. and Kumar, A. R. V. 2000. Modified technique for the isolation of insect parasitic nematodes. *Journal of Biological Control* 14(2): 83–85.
159. Rajagopal, B. K., Viraktamath, C. A., and Nachegowda, V. 1998. Incidence of ant associated mealy bug, *Xenococcus annandalei* (Homoptera: Pseudococcidae) on grapes in south India. *Entomon* 22(1997): 165–166.
160. Rangaswamy, H. R., Viraktamath, C. A. and Vijaya Lakshmi, S. 1974. The male reproductive pattern in

Chondriomorpha severini (Silvestri) (Diplopoda: Arthropoda). *Current Research* 3: 6–7.

161. Srinivasa, M. V., Viraktamath, C. A. and Reddy, C. 1999. A new parasitoid of the spiraling white fly, *Aleurodicus dispersus* Russell (Hemiptera: Aleyrodidae) in south India. *Pest Management in Horticultural Ecososystems* 5: 59–61.

162. Srinivasa, N., Viraktamath, C. A. and Thontadarya, T. S. 1990. Influence of moon phase and weather on the light trap catches of insect pests of rice. *Oryza* 27: 183–190.

163. Subba Rao, M., Viraktamath, C. A. and Muniyappa, V. 1988. Incidence of leafhoppers treehoppers and froghopper (Homoptera) in sandal forests of Karnataka in relation to sandal spike disease. *Annals of Entomology* 6(6): 25–34.

164. Suryanarayana, V., Viraktamath, C. A. and Reddy, H. R. 1998. Oviposition hosts of the leafhopper, *Orosius albicinctus* Distant – A vector of phytoplasma diseases of crop and weed plants. *Pest Management in Horticultural Ecosystems* 4: 11–15.

165. Ravi, K. C., Viraktamath, C. A., Puttaswamy and Mallik, B. 1997. *Diaphania indica* (Pyralidae: Lepidoptera)—A serious pest of gherkins. *Insect Environment* 3(3): 81.

166. Ravi, K. C., Viraktamath, C. A., Puttaswamy and Mallik, B. 1998. Seasonal incidence of insect pests of gherkins *Cucumis anguria* L. pp. 132–134. In: *Proceedings of 1st National Symposium on Pest Management in Horticultural Crops: Environmental Implications and Thrusts: Advances in IPM for Horticultural Crops*. P. Parvatha Reddy, N. K. Krishna Kumar and Abraham Verghese (Eds), Association for Advancement of Pest Management in Horticultural Ecosystems, Bangalore, vii+363pp.

167. Srinivasan, K., Viraktamath, C. A., Gupta, M. and Tewari, G. C. 1996. Geographical distribution, host range and parasitoids of serpentine leaf miner, *Liriomyza trifolii* (Burgess) in south India. *Pest Management in Horticultural Ecosystems* 1(2): 93–100.

168. Muniappan, R., Sundaramurthy, V. T. and Viraktamath, C. A. 1990. Distribution of *Chromolaena odorata* (Asteraceae) and bionomics and consumption and utilization of food by *Pareuchaetes pseudoinsulana* (Lepidoptera: Arctiidae) in India. pp. 401–409. In Delfosse, E. S. (Ed.) *Proceedings of the VII International Symposium in Biological Control of Weeds*, March 6–11, 1988, Rome, Italy. 1st Sper. Patol. Veg. (MAF).

169. Narayana Swamy T., Visweswara Gowda, B. L. and Viraktamath, C. A. 1989. A note on the incidence of the lycaenid butterfly on the ornamental plant, *Kalanche blossfeldiana* V. (Crassulaceae). *Current Research* 18: 81–82.

170. Pandurange Gowda, K. T., Reddy, H. R. and Viraktamath, C. A. 1982. Population fluctuation of aphids in chilli field in relation to some meteorological factors. *Mysore Journal of Agricultural Sciences* 16: 160–165.

171. Rao, N. S., Veeresh G. K. and Viraktamath, C. A. 1991. Association of crazy ant, *Anoplolepis longipes* (Jerdon) with different fauna and flora. *Indian Journal of Ecology* 16(2): 205–208.

172. Rao, N. S., Veeresh G. K. and Viraktamath, C. A. 1991. Dispersal and spread of crazy ant, *Anoplolepis longipes* (Jerdon) (Hymenoptera: Formicidae). *Environment and Ecology* 9(3): 682–686.

173. Srikant, J., Joshi, S. and Viraktamath, C. A. 1991. *Luperomorpha vittata* Duvivier: A new association with *Parthenium hysterophorus* L. and other weeds. *Current Science* 60: 177–178.

174. Srikantaiah, M., Joshi, S. and Viraktamath, C. A. 1998. New host records of the almond moth, *Ephestia cautella* (Walker) (Lepidoptera: Phycitidae) in south India. *Entomon* 22(1997): 251–253.

175. Suryanarayan, V., Muniyappa, V. and Viraktamath, C. A. 1989. Transmission, host range and electron microscopy of sunnhemp phyllody. Abstract of paper presented at *Indian Science Congress*, January 7, Madurai, India.

176. Venkateshalu, G. and Viraktamath, C. A. 1998. Conservation of spiders in rice ecosystem. *Entomon* 23: 147–149.

177. Viraktamath, S., Hiremath, S. C. and Viraktamath, C. A. 1996. Varietal influence on the seasonal incidence of mango leafhoppers in Raichur, Karnataka. *Karnataka Journal of Agricultural Sciences* 9: 40–46.

178. Srinivasa, M. V., Viraktamath, C. A. and Reddy, C. 2002. Spiraling white fly, *Aleurodicus dispersus* Russell (Homoptera: Aleyrodidae)-Spatial distribution, sampling method and effect of weather parameters. In Sanjayan, K. P., Mahalingam, V. and Muralirangan, M. C. (Eds). *Vistas of Entomological Research for the New Millennium*, pp. 137–140.

179. Ambika, T., Sheshashayee, Viraktamath, C. A. and Udayakumar, M. 2005. Identifying the dietary source of polyphagous *Helicoverpa armigera* (Hübner) using carbon isotope signatures. *Current Science* 89(12): 1982–1984.

180. Hiremath, I. G., Lingappa, S., Viraktamath, C. A. and Musthak Ali, T. M. 1986. Occurrence of the ground beetle, *Gonocephalum hofmannseggi* (Coleoptera: Tenebrionidae) on finger millet (*Eleusine coracana* (L.) Gaertn.). *Journal of Soil Biology and Ecology* 6: 62–66.

181. Nalawadi, U. G., Viswanath, B. N., Viraktamath, C. A. and Kalolgi, S. D. 1976. Observations on the resistance of mango varieties to the gall fly (*Dasyneura mangiferae* Felt., Cecidomyiidae: Diptera). *Lalbaugh* 21: 5–6.

182. Ravi, K. C., Puttaswamy, Viraktamath, C. A., Mallik, B. and Ambika, T. 1998. Influence of host

plants on the development of *Diaphania indica* (Saunders) (Lepidoptera: Pyralidae). pp. 135–136. In: *Proceedings of 1st National Symposium on Pest Management in Horticultural Crops: Environmental Implications and Thrusts: Advances in IPM for Horticultural Crops*. Parvatha Reddy, P., Krishna Kumar, N. K. and Verghese, A. (Eds), Association for Advancement of Pest Management in Horticultural Ecosystems, Bangalore, India, vii+363pp.

183. Manjunatha, M., Bhat, N. S., Raju, G. T. T. and Viraktamath, C. A. 1988. A new predatory ant, *Aenictus pachycerus* Smith (Hymenoptera: Formicidae) on wireworm (*Agriotes* sp.) (Coleoptera: Elateridae). *Current Science* 57: 74.

184. Maragal, S. M., Hiremath, I. G., Bhuti, S. G. and Viraktamath, C. A. 1992. Armyworm on barbadense cotton. *Current Research* 21: 71.

185. Suryanarayana, V., Muniyappa, V., Singh, S. J., Viraktamath, C. A. and Reddy, H. R. 1997. Phytoplasma—Vector relationships of pigeon pea phyllody disease. *Pest Management in Horticultural Ecosystems* 3: 1–6.

IPM OF INSECT PESTS AND WEEDS

186. Viraktamath, C. A. 1995. Management of horticultural crop pests. pp. 63–84. In: Veeresh, G. K. and Viraktamath, C. A. (Eds) *A Century of Plant Protection Research in Karnataka*. xv+282 pp.

187. Viraktamath, C. A. 2002. Alien invasive insect and mite pests and weeds in India and their management. *Micronesica Supplement* 6: 67–83.

188. Viraktamath, C. A. 2002. "Augmentative Biocontrol"-Proceedings, ICAR-CABI workshop—Book review. *Journal of Biological Control* 16: 94.

189. Viraktamath, C. A. and Muniappan, R. 1992. New records of insects on *Chromolaena odorata* in India. *Chromolaena Newsletter No.* 5, 1&4.

190. Viraktamath, C. A. and Patel, V. N. 1998. Insect pests of grapes and their management (In Kannada). In a bulletin published during *Seminar on Grapes*, March 21, Department of Horticulture, Bangalore, India, pp. 11–16.

191. Viraktamath, C. A. and Bhumannavar, B. S. 2001. Biology, ecology and management of *Diaphorina citri* Kuwayama (Hemiptera: Psyllidae). *Pest Management in Horticultural Ecosystems* 7(1): 1–27.

192. Viraktamath, C. A. and Jagannatha, R. 2000. Serpentine leafminer, *Liriomyza trifolii* (Burgess) (Diptera: Agromyzidae) and its management. In Upadhyay, R. K., Mukerji, K. G. and Dubey, O. P. (Eds) *IPM System in Agriculture, Animal Pests*, Vol. 7.

193. Viraktamath, C. A., Tewari, G. C., Srinivasan, K. and Gupta, M. 1993. American Serpentine leaf miner is a new threat to crops. *Indian Farming* 43(2): 10–12.

194. Diraviam, J. and Viraktamath, C. A. 1993. Toxicity of some insecticides to *Curinus coeruleus* Mulsant (Coleoptera: Coccinellidae) an introduced predator of the subabul psyllid. *Entomon* 18: 77–79.

195. Muniappan, R. and Viraktamath, C. A. 1986. Status of biological control of the weed, *Lantana camera* in India. *Tropical Pest Management* 32: 40–42.

196. Muniappan, R. and Viraktamath, C. A. 1986. Insects and mites associated with *Chromolaena odorata* (L.) R.M. King and Robinson (Asteraceae) in Karnataka and Tamil Nadu. *Entomon* 11: 285–287.

197. Muniappan, R. and Viraktamath, C. A. 1993. Invasive alien weeds in the Western Ghats. *Current Science* 64: 555–558.

198. Muniappan, R. and Viraktamath, C. A. 1994. Biological control programmes in India—A review in retrospect. *Current Science* 65: 899–901.

199. Sannaveerappanavar, V. T. and Viraktamath, C. A. 1997. Management of insecticide resistant diamondback moth, *Plutella xylostella* L. (Lepidoptera: Yponomeutidae) on cabbage using some novel insecticides. *The Mysore Journal of Agricultural Sciences* 31(3): 230–235.

200. Sannaveerappanavar, V. T. and Viraktamath, C. A. 1997. The IGR activity of aqueous neem seed kernel extract against diamondback moth, *Plutella xylostella* L. larvae. *The Mysore Journal of Agricultural Sciences* 31(3): 241–243.

201. Madhura, H. S. and Viraktamath, C. A. 2001. Selective response of fruit flies (Diptera: Tephritidae) to traps at different heights. P27. In Verghese A. and Reddy, P. P. (Eds) *Proceedings of the Second National Symposium on Integrated Pest Management (IPM) in Horticultural Crops: New Molecules, Biopesticides & Environment*. October 17–19, Bangalore, India.

202. Madhura, H. S. and Viraktamath, C. A. 2003. Efficacy of different traps in attracting fruit flies (Diptera: Tephritidae). *Pest Management in Horticultural Ecosystems* 9: 153–154.

203. Sannaveerappanavar, V. T. Viraktamath, C. A. and Puttaswamy 1997. Management of insecticide resistant diamondback moth, *Plutella xylostella* L. on cabbage using neem seed kernel extract and other novel insecticides. *The Karnataka Journal of Agricultural Sciences* 10(3): 886–888.

204. Nagaraju, D. K., Viraktamath, C. A. and Krishna Kumar, N. K. 2002. Screening of bell pepper accessions against gall insects and their chemical control. *Pest Management in Horticultural Ecosystems* 8(1): 12–19.

205. Prabhuraj, A. Viraktamath, C. A. and Kumar, A. R. V. 2000. Field evaluation of entomopathogenic nematodes against brinjal root weevil, *Myllocerus discolor* Boh. *Pest Management in Horticultural Ecosystems* 6(2): 149–151.

206. Madhura, H. S., Viraktamath, C. A. Verghese, A. and Nagaraju, D. K. 2003. Attraction of braconid parasitoid, *Psytallia* sp. to coloured traps. *Pest Management in Horticultural Ecosystems* 10(1): 73–75.

207. Hussain, M. A., Puttaswamy, and Viraktamath, C. A. 1996. Management of citrus mealy bug, *Planococcus citri* Risso on guava using botanical oils. *Insect Environment* 2(3): 73–74.
208. Hussain, M. A., Puttaswamy, and Viraktamath, C. A. 1996. Effect of botanical oils on lantana bug, *Orthezia insignis* Browne infesting crossandra. *Insect Environment* 2(3): 85–86.
209. Venkatesh, J. M., Muniappa, V., Ravi, K. S., Viraktamath, C. A. and Krishnaprasad, P. R. 1996. Management of chilli leaf curl complex disease. Abstract of the paper presented during the *Annual Meeting and Symposium on Epidemiology and Management of Plant Diseases*, November 21–22. Indian Phytopathological Society, Southern Chapter, Dharwad, India. Abstract Page No. 41.

BOOKS

210. Channa-Basavanna, G. P. and Viraktamath, C. A. (Eds). 1989. *Progress in Acarology*. Volume I. xv+525 pp. Oxford and IBH Publishing Co Pvt Ltd.
211. Channa-Basavanna, G. P. and Viraktamath, C. A. (Eds). 1989. *Progress in Acarology*. Volume II. xv+484 pp. Oxford and IBH Publishing Co Pvt Ltd.
212. Veeresh, G. K., Mallik, B. and Viraktamath, C. A. (Eds). 1990. *Social Insects and the Environment – Proceedings of the 11th International Congress of IUSSI*. 1990. xxxi+ 765 pp. Oxford and IBH Publishing Co Pvt Ltd.
213. Veeresh, G. K., Rajagopal, D. and Viraktamath, C. A. (Eds). 1991. *Advances in Management and Conservation of Soil Fauna*. Xv+925 pp. Oxford and IBH Publishing Co Pvt Ltd.
214. Veeresh, G. K. and Viraktamath, C. A. (Eds). 1995. *A Century of Plant Protection Research in Karnataka*. Xv+282 pp. Dr. M. Puttarudriah Memorial Endowment, Institution of Agricultural Technologists, Bangalore, India.
215. Viraktamath, C. A., Mallik, B., Chandrashekar, S. C., Ramakrishna, B. V. and Praveen, H. M. 2003. *Insect Pests and Diseases of Gherkins and Their Management*. UAS, APEDA, KAPPEC, Bangalore, India. 23 pp.

Annexure V: Taxa Named in Honour of Professor C. A. Viraktamath

HEMIPTERA

Cicadellidae

Pythochandra Wei & Webb 2014
  Type species: *Pythamus melichari* Baker
Chandra Meshram 2017
  Type species: *Chandra dehradunensis* Meshram 2017

COLEOPTERA

Curculionidae

Viraktamathia Devi, Ray & Ramamurthy 2014
  Type species: *Viraktamathia srinivasa* Devi, Ray & Ramamurthy 2014

HEMIPTERA

Cicadellidae

Agallia viraktamathi Dlabola 1972
Balclutha viraktamathi Webb & Vilbaste 1994
Cretacoelidia viraktamathi Wang, Dietrich & Zhang 2018 (amber)
Discolopeus viraktamathi Stiller 2019
Drabescus viraktamathi Xu & Zhang 2018
Eleazara viraktamathi Huang & Zhang 2018
Gannia viraktamathi Zahniser & Dietrich 2013
Goniagnathus (*Tropicognathus*) *viraktamathi* Duan & Zhang 2012
Hishimonus viraktamathi Knight 1973
Japanagallia viraktamathi Li, Dai & Li 2014
Jogocerus viraktamathi Xue & Zhang 2017
Kadrabia viraktamathi Dworakowska & Sohi 1978
Kalasha viraktamathi Tang & Zhang 2019
Krisna viraktamathi Yalin Zhang, Xinmin Zhang & Wu Dai 2008
Kusala (*Kusala*) *viraktamathi* Cao, Dmitriev & Zhang 2018
Longistyla viraktamathi Zhang & Webb 2019
Maiestas chandrai Fletcher & Dai 2018
Maiestas viraktamathi Zahniser, McKamey & Dmitriev 2012
Makilingia viraktamathi Dietrich & Zahniser 2019
Orosius viraktamathi El-Sonbati & Wilson 2019
Processina chandrai He, Yang & Yu 2018
Sangeeta viraktamathi Zhang 2104
Scaphoidella viraktamathi Dai & Dietrich 2011
Scaphoideus viraktamathi Meshram 2014
Singillatus viraktamathi (Nielson 1990)
Sophonia chandrai Meshram & Ramamurthy 2013
Tambocerus viraktmathi Rao 1996
Thagria viraktamathi Nielson 2013
Xenovarta viraktamathi Meshram, Stuti & Hashmi 2018

Aleyrodidae

Aleuroclava viraktamathi Pushpa & Sundararaj 2010

COLEOPTERA

Cerambycidae

Notomulciber (*Micromulciber*) *viraktamathi* Hiremath 2018

Mordellidae

Brodskyella viraktamathi Ruizzer and Yeshwanth, 2019

Nitidulidae

Epuraea (*Micruria*) *viraktamathi* Dasgupta, Pal & Hegde 2016

HYMENOPTERA

Apidae

Isotrigona chandrai Shashidhar Viraktamath & Sejan Jose 2017

Dryinidae

Anteon viraktamathi Olmi 1987
Gonatopus viraktamathi Olmi 1987

Encyrtidae

Eugahania viraktamathi Manickavasagam & Ayyamperumal 2018

Eulophidae

Oomyzus viraktamathi Narendran 2007

Platygasteridae

Fidiobia virakthamati Veenakumari 2018

LEPIDOPTERA

Lecithoceridae

Frisilia chandrai Park & Shashank 2018

Annexure VI: Courses Offered by Professor C. A. Viraktamath

POST GRADUATE

Insect Systematics I	1975–2009
Insect Systematics II/Advanced Insect Systematics	1979–2016
Insect Morphology	1976–1991, 2001–2004
Insect Anatomy	1976–1985, 2000–2003
Insect Vectors of Plant Pathogens	1977–2016
Immature Insects	1978–2003
Bio-ecology of Crop Pests	1988–1989, 1992–1993
Literature and Techniques in Entomology	1999–2000

UNDER GRADUATE

Introduction to Entomology I & II	1969–1995
Insect Taxonomy	1969–1980
Economic Entomology I	1969–1977
Integrated Pest Management	1969–1970, 1972–1973, 1986–1987
Veterinary and Medical Entomology	1972–1973
Household and Stored Product Pests	1972–1973
Insecticides and Appliances	1972–1973

Annexure VII: Theses Submitted Under the Guidance of Professor C. A. Viraktamath (1980–2019)

TAXONOMY

LEAFHOPPERS

1. Ramakrishna, B. V. 1980. Leafhopper fauna of pulse crops and biology of *Empoasca* (*Distantasca*) *terminalis* Distant (Homoptera: Cicadellidae). MSc (Ag.) thesis, UAS, Bangalore, India.
2. Kumar, A. R. V. 1980. A revision of Indian *Batracomorphus* Lewis (Homoptera: Cicadellidae: Iassinae). MSc (Ag.) thesis, UAS, Bangalore, India.
3. *Wesley, C. S. 1980. Taxonomic studies on Nirvaninae (Homoptera: Cicadellidae) of the Indian subcontinent. MSc (Ag.) thesis, UAS, Bangalore, India.
4. Ananthamurthy, H. V. 1984. Taxonomic studies on Opsiini and Scaphytopiini (Homoptera: Cicadellidae: Deltocephalinae) of the Indian subcontinent with special reference to Karnataka. MSc (Ag.) thesis, UAS, Bangalore, India.
5. Mohan, G. S. 1986. Taxonomic revision of the species of *Scaphoideus* Uhler and *Scaphotettix* Matsumura (Homoptera: Cicadellidae: Deltocephalinae) of the Indian subcontinent. MSc (Ag.) thesis, UAS, Bangalore, India.
6. Parvathi, C. 1986. Taxonomic revision of idiocerine leafhoppers (Homoptera: Cicadellidae) of Indian subcontinent based on female genitalia. MSc (Ag.) thesis, UAS, Bangalore, India.
7. Dash, P. C. 1993. Systematic studies on the deltocephaline leafhopper tribes Goniagnathini and Deltocephalini (Homoptera: Cicadellidae). PhD thesis, UAS, Bangalore, India.
8. Sindhu, M. K. 2004. Taxonomic studies on Cicadellinae (Hemiptera: Cicadellidae) of the Indian subcontinent. MSc (Ag.) thesis, UAS, Bangalore, India.
9. *Jyothi, R. 2019. Erythroneurini leafhoppers (Hemiptera: Cicadellidae: Typhlocybinae) associated with economically important plants in Karnataka. MSc (Ag.) thesis, UAS, Bangalore, India.

OTHER INSECT TAXA

10. Musthak Ali, T. M. 1981. Ant fauna (Hymenoptera: Formicidae) of Bangalore with observations on their nesting and foraging habits. MSc (Ag.) thesis, UAS, Bangalore, India.
11. Kumar, P. 1986. Faunistic studies on short-horned grasshoppers (Orthoptera: Acridoidea) of Bangalore district. MSc (Ag.) thesis, UAS, Bangalore, India.
12. Ganesh Bhat, U. 1989. Faunistic and some ecological studies on fruit flies (Diptera: Tephritidae) of Bangalore and Kodagu districts. MSc (Ag.) thesis, UAS, Bangalore, India.
13. Bhaskar, H. 1992. Systematic and ecological studies on predaceous coccinellid beetles (Coleoptera: Coccinellidae). MSc (Ag.) thesis, UAS, Bangalore, India.
14. Vinayakachari, M. L. 1996. Systematic and biological studies on arctiid hairy caterpillar pests (Lepidoptera: Arctiidae) with special reference to red headed hairy caterpillar, *Amsacta albistriga* (Walker). MSc (Ag.) thesis, UAS, Bangalore, India.
15. *Jayaram, K. R. 1997. Faunistic studies on the genus *Apogonia* Kirby (Coleoptera: Scarabaeidae: Mololonthinae) of Karnataka and adjoining states and bioecology of *Apogonia brevis* sp. nov.). PhD thesis, UAS, Bangalore, India.
16. Ramani, S. 1997. Biosystematic studies on fruit flies (Diptera: Tephritidae) with special reference to the fauna of Karnataka, and Andaman & Nicobar Islands. PhD thesis, UAS, Bangalore, India.
17. Bhumannavar, B. S. 2000. Studies on fruit piercing moths (Lepidoptera: Noctuidae)—Species composition, biology and natural enemies. PhD thesis, UAS, Bangalore, India.
18. *Prathapan, K. D. 2002. Systematic studies on flea beetles of south India (Coleoptera: Chrysomelidae: Alticinae). PhD thesis, UAS, Bangalore, India.
19. Joshi, S. 2005. Faunistic studies on Aphididae (Hemiptera) of Karnataka and bioecology of the aphid parasitoid, *Diaeretiella rapae* (M'intosh) (Hymenoptera: Braconidae). PhD thesis, UAS, Bangalore, India.
20. *David, K. J. 2005. Diversity and abundance of fruit flies (Diptera: Tephritidae) of Kerala with special reference to Western Ghats. MSc (Ag.) thesis, UAS, Bangalore, India.
21. *Suma, S. 2006. Studies on bee faunal diversity of Bangalore region. MSc (Ag.) thesis, UAS, Bangalore, India.
22. *Salini, S. 2006. Faunistic studies on Pentatomidae (Hemiptera: Pentatomidae) in Karnataka. MSc (Ag.) thesis, UAS, Bangalore, India.

23. *Nayana, E. D. 2008. Studies on non-*Apis* bee fauna of Western Ghats of Karnataka. MSc (Ag.) thesis, UAS, Bangalore, India.
24. *Nimisha, K. K. 2007. Studies on planthoppers (Hemiptera: Delphacidae) occurring in different agroecosystems in Karnataka. MSc (Ag.) thesis, UAS, Bangalore, India.
25. *Dhanyavathi, P. N. 2009. Studies of non-*Apis* bee faunal diversity of Mysore District. MSc (Ag.) thesis, UAS, Bangalore, India.
26. *Yeshwanth, H. M. 2014. Taxonomy of mirid bugs (Hemiptera: Miridae) of south India. PhD thesis, UAS, Bangalore, India.
27. *Salini, S. 2015. Systematic studies on Pentatomidae (Hemiptera: Pentatomidae) of south India. PhD thesis, UAS, Bangalore, India.
28. *Hiremath, S. R. 2015. Cerambycidae fauna of plantation fruit crop ecosystems of Western Ghats in Karnataka. MSc (Hort.) thesis. UAHS, Shimoga, India.
29. *Bhagyashree, S. N. 2017. Taxonomic studies on assassin bugs (Hemiptera: Reduviidae) of south India. PhD thesis, UAS, Bangalore, India.

Entomopathogenic Nematodes

30. Prabhuraj, A. 1998. Faunistic studies on entomopathogenic nematode families Steinernematidae and Heterorhabditidae and their potential as biocontrol agents of white grubs and caterpillars. PhD thesis, UAS, Bangalore, India.

BIO-ECOLOGY AND MANAGEMENT

31. Krishnamurthy, H. S. 1981. Biology and morphology of *Diocalandra frumenti* (Fabricius) (Coleoptera: Curculionidae) and *in vitro* evaluation of four insecticides on adults. MSc (Ag.) thesis, UAS, Bangalore, India.
32. Ramadas Rai, N. 1983. Biology, some aspects of morphology and chemical control of cowpea pod borer, *Maruca testulalis* (Geyer) (Lepidoptera: Pyralidae). MSc (Ag.) thesis, UAS, Bangalore, India.
33. Diraviam, J. 1990. Biology, population dynamics of the introduced lady bird beetle, *Curinus coeruleus* Mulsant (Coleoptera: Coccinellidae) and its susceptibility to some insecticides and *Beavuveria bassiana* (Balsamo) Vuill. MSc (Ag.) thesis, UAS, Bangalore, India.
34. Tulasi Jyothi, D. 1991. Influence of host plants on *Helicoverpa armigera* (Hübner) (Lepidoptera: Noctuidae) with an objective of identifying biotypes and its oviposition preference to pigeon pea genotypes. PhD thesis, UAS, Bangalore, India.
35. Jagannatha, R. 1994. Comparative biology, ecology and management of American serpentine leaf miner, *Liriomyza trifolii* (Burgess) (Diptera: Agromyzidae). MSc (Ag.) thesis, UAS, Bangalore, India.
36. Patel, V. N. 2001. Ecological and behavioural studies on the Mexican beetle *Zygogramma bilocorata* Pallister (Coleoptera: Chrysomelidae) introduced for the biological control of *Parthenium hysterophorus* L. PhD thesis, UAS, Bangalore, India.
37. Nagaraju, D. K. 2001. Biology, ecology and management of the capsicum gall midge, *Asphondylia capparis* Rübsaamen (Diptera: Cecidomyiidae) and other insects associated with galls on bell pepper. MSc (Ag.) thesis, UAS, Bangalore, India.
38. Madhura, H. S. 2001. Management of fruit flies (Diptera: Tephritidae) using physical and chemical attractants. MSc (Ag.) thesis, UAS, Bangalore, India.
39. Subbireddy, S. N. 2002. Bio-ecology and management of whitefly, *Bemisia tabaci* (Gennadius) (Homoptera: Aleyrodidae) in tomato. MSc (Ag.) thesis, UAS, Bangalore, India.
40. **Srinivasa, M. V. 2003. Bio-ecology and management of spiralling white fly *Aleurodicus dispersus* Russell (Hemiptera: Aleyrodidae) PhD thesis, Bangalore University, Bangalore, India.

POLLINATION

41. Girish, P. P. 1981. Role of bees, *Apis cerana* Fabricius (Hymenoptera: Apidae) in the pollination of summer squash (*Cucurbita pepo*). MSc (Ag.) thesis, UAS, Bangalore, India.

INSECTICIDE TOXICOLOGY

42. Sannaveerappanavar, V. T. 1995. Studies on insecticide resistance in the diamondback moth, *Plutella xylostella* (L.) (Lepidoptera: Yponomeutidae) and strategies for its management. PhD thesis, UAS, Bangalore, India.
43. Revannavar, R. 2010. Evaluation of local plants for insecticidal properties. PhD thesis, UAS, Bangalore, India.

LIGHT TRAP STUDIES

44. Jayaramagupta, P. S. 1981. Preliminary studies on the light trap catches of selected insect pests with special reference to the moon phases and some weather factors. MSc (Ag.) thesis, UAS, Bangalore, India.

BIOLOGICAL CONTROL

45. Shirazi, J. 2004. Studies on different aspects of quality control in *Trichogramma chilonis* Ishii (Hymenoptera: Trichogrammatidae). PhD thesis, UAS, Bangalore, India.

FORENSIC ENTOMOLOGY

46. **Shashikanth Naik, C. R. 2015. Study of insects associated with decomposed bodies and their forensic application. MD in Forensic Medicine thesis, RGUHS, Bangalore, India.

* Guidance provided as an Advisory Committee member
** Co-guide

Annexure VIII: Research Projects Operated by Professor C. A. Viraktamath

S. No.	Project Title	Sponsor
1	Biosystematic studies on Asiatic leafhoppers with particular reference to Macropsinae, Iassinae, and Agalliinae	Public Law 480/Department of Science and Technology (PL480/DST), Ministry of Science and Technology, Government of India
2	Status of American leaf miner, *Liriomyza trifolii* in south India	Indian Council of Agricultural Research (ICAR), Ministry of Agriculture and Farmers' Welfare, Government of India
3	Management of red headed hairy caterpillar, *Amsacta albistriga* in Karnataka	Government of Karnataka
4	Studies on pest status of Mexican beetle, *Zygogramma bicolorata* Pallister introduced for biological control of *Parthenium hysterophorus* L.	ICAR
5	Management of insect pests of gherkins	Oceania Peninsula Pvt. Ltd., Bangalore
6	Utilization of non-edible oils for development of insect control agents	Ministry of Environment and Forests (MoEF), Government of India
7	Management of insect pests and diseases of gherkins	Agricultural & Processed Food Products Export Development Authority (APEDA), Ministry of Commerce & Industry, Government of India
8	Taxonomic studies on the economically important leafhoppers (Hemiptera: Cicadellidae) of the Indian subcontinent	ICAR (Emeritus Scientist's Scheme)
9	Network Project on Insect Biosystematics	ICAR

Index

Note: Page numbers in italic and bold refer to figures and tables, respectively.

A

Acanthiophilus helianthi, 313
Acrididae
    classification, 63
    collection, 65
    diagnosis, 60–61
    economic importance, 65
    identification characters, 61–63
    preservation, 65–66
    subfamilies of, 63–64
    taxonomic problems, 64
    work on Indian fauna, 64
Acroceratitis ceratitina, 313
Acroceratitis tomentosa, 310, 311
Adephaga, 202
adult(s)
    Anisoptera, 31
    Cordulegasteroidea, 31
    Ephemeroptera, 21
    females fruit fly, dissection of, 320
    Odonates, 38
    Plecoptera, 48
    Zygoptera, 30
aedeagus, 208–209
Agonoscelis nubilis, 125
Agriocnemis pygmaea, 36
Allobaccha amphithoe, 353–354
Allobaccha apicalis, 354
Allobaccha elegans, 354–355
Allobaccha fallax, 355
Allobaccha oldroydi, 355
Allobaccha sapphirina, 355
Allobaccha triangulifera, 356
Allograpta bouvieri, 362–363
Allograpta dravida, 363
Allograpta javana, 363
Allograpta sp., 364
allotype, 280
Altica ?aenea, 248
    antenna, 254
    elytra, 259
    head/frontal view, 266
    pronotum, 257
Ameletidae, 15, *15*
Amphimela picta
    abdomen/hind legs/ventral view, 263
    head/frontal view, 261, 262
    pronotum/base of elytra, 257
Amyotea malabarica, 123, 135
anal cleft, 73, 74
anal plates, 73, 74
anal ring, 73–74, 75
anal sclerotisation, 74, 75
Anax immaculifrons, 35
Andaman and Nicobar Islands, 327
Andhra Pradesh, **326**, 327
Andrallus spinidens, 123
Anepisternal setae, 306
Anisoptera
    adults, 31
    larvae, 32
anogenital fold, 74–75, 75

antenna
    of mymarid, 148, *148*
    structure, *81*
    types of, *81*
antennal hair plates, 392
antennal mechanosensors
    -motor circuits in diverse insects, 390–395, *391*
    overview, 389–390
    structures in arthropods, 395–396, *396*
antennal mechanosensory and motor centre (AMMC), 392
antennal torulus, 148
Antestiopsis cruciata, 125, *133*, *134*
ant-plant domatia
    development and maintenance, 407–408
    entry restrictions, 406–407
    origin, 406
Aphthona nigrilabris, 257
Aphthona sp., 254, 260
Aphthonoides sp., 254
Apis mellifera, 393
Apteroessa, 205, 205–206
aquatic mymarids, 147
Archimicrodon lanka, 329
Argopistes sp., 258
Asarkina ayyari, 364–365
Asarkina belli, 365
Asarkina hema, 365
Asarkina incisuralis, 365–366
Asarkina pitambara, 366
Asiobaccha nubilipennis, 367
Asiophrida marmorea, 248, 262, 264
Asopinae, *123*, 123–125, *133*
augmentation, 238
austro-oriental region, 93
autocatalytic feedback process, 5

B

Baccha sp., 352–353
Bactrocera aethriobasis, 313
Bactrocera affinis, 313
Bactrocera caryeae, 312, 313
Bactrocera correcta, 310, 313
Bactrocera digressa, 313
Bactrocera dorsalis, 311, 312, *313*, 317
Bactrocera minax, 312
Bactrocera nigrofemoralis, 310
Bactrocera syzygii, 311, 312
Bactrocera versicolor, 311
Baetidae, 15, *15*, 16
Bagrada hilaris, 125, *134*
Baracus, 293, 294
basal chordotonal organs, 395
beetles
    diversity, 2–3
    quips, 1, 5
    richness, 1–2
Bemisia tabaci, 114–115
Betasyrphus fletcheri, 367
Betasyrphus linga, 367–368
Bihar, 327

Bikasha sp., 262
Bimala sp., 263
biodiversity and species richness
    bumble bees, 178–179
    Coccinellidae, 229
    Ephemeroptera, 18–19
    Pentatomidae, 132
    potter wasps, 191–192
    tiger beetles, 206
biogeographical affinities, species, 163–164
bioindicator, 213
biological species, 287
biology
    Ephemeroptera, 20–22
    larval/pupal, 209–211, *210*
    Mymaridae, 152
    Pentatomidae, 135–137
    potter wasps, 193–194
    scale insects, 94, **95**
biology and ecology
    Coccinellidae, 230
    habitats, 164
    phoresy, 164, 166–167
    sexual dimorphism, 167
biomimicry, 215
biomonitoring tool, 21
Bison latifrons
    male, sternite V, *313*
    thorax, *312*
    wings, *313*
black bugs, 122
Blanchia ducalis, 123
Blepharoneurinae, 307
Böhm's bristles, 392, 394
Bombus, 175–177
Brachycerocoris camelus, 131
Brachycoris tralucidus, 125
Brancucciella kolibaci, 262
brood-site pollination mutualism, 403
bumble bees (Hymenoptera: Apidae: Apinae: Bombini), 173–174
    biodiversity and species richness, 178–179
    biological characteristics, 179–180
    collection and preservation, 181–182
    commercial use, 181
    conservation status, 181
    distribution patterns, 179
    integrative taxonomy, 180–181
    molecular characterization and phylogeny, 180
    species, **177–178**
    sub-generic classification, 174–177
    taxonomic problems, 181
    tribe and characters diagnosis, 174
    work on Indian fauna, 177–178

C

Caenoidea, 11–12, *12*
Calcaretropidia triangulifera, 347
Caliscelidae, 425
Canada balsam, 154
cannibalism, 231
Capniidae, 49

445

Cappaea taprobanensis, 125
Carapacea, 9
Carbula biguttata, 125
Carbula scutellata, 133
Cardiococcinae Hodgson, 86
Carpomya vesuviana, 311
Castilleae, 402
Catacanthus incarnates, 125
Catacanthus incarnatus, 133
cauline domatia, 406
Cazira verrucosa, 123
Cecidochares connexa, 313
Cecyrina platyrhinoides, 123
Ceriana ornatifrons, 332
Cerioidini tribe, 332–333
Ceroplastinae Atkinson, 86
Chabria decemplagiata, 260
Chaetocnema sp., 257, 262
Chalaenosoma sp., 255, 256
Chhattisgarh, 327
Chilocoristes sp., 248, 258
Chloroperlids, 50
Chrysomelidae, 425
Chrysotoxum baphyrum, 368–369
Cicadellinae, 419
Cicindela
    dorsal aspect, 209
    dorsal head anatomy, 208
    lateral aspect, 210
    male genitalia, 209
Cicindela aurofasciata, 210
    larval tunnel, 211
    male and female, 210
Cicindelidae, *see* tiger beetles
Cissococcinae Brain, 86
Citrogramma amarilla, 369
Citrogramma flavigenum, 369–370
Citrogramma frederici, 370
Citrogramma henryi, 370
cladogram, 91
Clavicornaltica sp., 254
Cletus sp., 161
coccidology, 70
Coccinae Fallen, 86–87
Coccinellidae, 223–224
    abdominal postcoxal line in, 226
    biodiversity and species richness, 229
    biology and ecology, 230
    cannibalism, 231
    collection and preservation, 239–240
    defence mechanisms, 232–233
    distribution patterns in, 229
    dorsal and ventral view, 224
    economic importance, 237–239
    eggs, 235, 235
    family diagnosis and characters, 224–227
    habitats, 230
    hibernation and mass assemblages, 233–234
    immature taxonomy, 234–237
    key identification characters, 227
    larvae, 235–236, 236
    life cycle and voltinism, 230
    maxilla in, 226
    natural enemies, 231–232
    parasites, 232
    pathogens/phylogeny, 229, 232
    polymorphism/melanism, 233
    predators, 231
    prey associations, 234
    pupa, 236–327, 237
    searching and feeding behaviour, 230–231
    subfamilies and tribes, 227, **227**, **228**
    tarsi in, 227
    taxonomic problems, 230
    tribes, 225
    work on Indian fauna, 228–229
coccinellids, 238
Coccipolipus, 232
Coelidiinae, 419
cohesive clusters, 288
COI gene, *Anagrus*, 152
Coleoptera, *391*, 437
collection and preservation
    Acrididae, 65
    bumble bees, 181–182
    Coccinellidae, 239–240
    Ephemeroptera, 22–23
    fruit flies, 320
    Odonata, 40
    Pentatomidae, 141–142
    Platygastroidea, 167
    Plecoptera, 53
    potter wasps, 194–195
    tiger beetles, 213–214
colouration, 232
colour patterns, butterflies/moths, 285, 286
Common Stoneflies, 50
Comstock-Needham system, 283
conservation techniques, 238–239
corbicula, 174
Cordulegasteroidea, 31
Corydalidae, 425
costa, 285
cotype, 280
Cresphontes monsoni, 125
Cressona valida, 131
Cyphococcinae Hodgson, 87

D

Dacinae, 307
Dacus ciliatus, 310, 311, 312
Dacus discophorus, 312
Dacus persicus, 311
Dacus ramanii, 313
Dacus sphaeroidalis, 310
Dadra and Nagar Haveli, 327
Dalpada complex, 134
Daman and Diu, 327
Daphnis nerii, 393
Dasysyrphus rossi, 370–371
Degonetus serratus, 125, 135
Deltocephalinae, 417, 419–422
Demarchus pubipennis, 265
dermaptera, 392
derm areolations, 75, 76
detectability, 214
Diarrhegma modestum, 311
Dideopsis aegrota, 371–372
Dietheria fasciata, 311, 313
Digulia kochi, 333
dimethyl hydantoin formaldehyde (DMHF), 240
Dioxyna sororcula, 313
Diptera (Crane fly), 391
direct pinning, 320
Disparoneura quadrimaculata, 36
diverse insects
    Böhm's bristles, 392, 394
    Johnston's organs, 393, 394–395
"diversity begets diversity," 3
diversity profiles, 21
DNA barcoding, 19

Dolichomerus crassa, 333
Dolycoris indicus, 126, 133
domatia, 406
domatium opening (DO), 403
dorsal abdominal glands (DAG), 122, 136
dorsal pores, 76, 77
dorsal setae, 76, 77
dorsal tubercles, 77, 78
Dorstenieae, 402
dragonflies and damselflies, *see* Odonata
Drosophila melanogaster (Diptera), 390
dry preservation
    mymarids, 154
    soft scale, 96–97
Dunnius bellus, 126

E

economic importance, 320
    Acrididae, 65
    Coccinellidae, 237–239
    Ephemeroptera, 22
    mymarids, 152–153
    Odonata, 39
    Pentatomidae, 141
    Plecoptera, 53
    potter wasps, 194
    Scelionidae/Platygastridae, 167
    soft scales, 96
    tiger beetles, 213
    whiteflies, 114–115
egg laying stage, Odonates, 39
eggs, Odonates, 37–38
Elytropachys sp., 255, 257
Eocanthecona furcellata, 123, 133
Eoseristalis cerealis, 333–334
Eoseristalis curvipes, 334
Eoseristalis simulata, 334
Eosphaerophoria dentiscutellata, 372
Eosphaerophoria sp., 372
epandrium, 306
Ephemerellidae, 12, 13
Ephemerelloidea, 12–14, 13, 14
Ephemeridae, 10, 11
Ephemeroidea, 10–11, 11
Ephemeroptera, 7–8
    adult and larvae, 8, 9
    biodiversity and species richness, 18–19
    biology, 20–22
    Carapacea, 9
    collection and preservation, 22–23
    conservation status, 22
    distribution patterns, 19
    economic importance, 22
    families of, 16–18
    Furcatergalia, 9–14
    higher classification, 8
    immature taxonomy, 19
    integrative taxonomy, 19–20
    molecular characterization, 19
    phylogeny, 19
    Pisciforma, 15–16
    Setisura, 14
    taxonomic problems, 20
    work on Indian fauna, 18
Episyrphus arcifer, 372–373
Episyrphus balteatus, 373–374
Episyrphus divertens, 374
Episyrphus viridaureus, 374–375
Epophthalmia vittata, 35
Eriopeltinae Sulc, 87

Index

Eristalini tribe, 333–344
Eristalinus aeneus, 334
Eristalinus arvorum, 334–335
Eristalinus aurulans, 335
Eristalinus invirgulatus, 336
Eristalinus laetus, 336
Eristalinus lucilia, 336
Eristalinus megacephalus, 336–337
Eristalinus obliquus, 337–338
Eristalinus quadristriatus, 338
Eristalinus quinquestriatus, 338–339
Eristalinus tristriatus, 339
Eristalis canocincta Brunetti, 339
Eristalodes paria, 339
Erystus andamanica, 263
Ethiopian, 93
Eulecaniinae Koteja, 87
Eumeninae; *see also* potter wasps (Hymenoptera: Vespidae: Eumeninae)
    Indian genera of, 189–191
    morphology, *188*
    nesting, 193–194
    records, **192**
    subfamily diagnosis, 187
    taxonomic studies on, 193
    tribes and genera, **189**
Eumerus aeneithorax, 344
Eumerus albifrons, 345
Eumerus argentipes, 345
Eumerus aurifrons, 345
Eumerus coeruleifrons, 345
Eumerus figurans, 345–346
Eumerus nepalensis, 346
Eumerus nicobarensis, 346
Eumerus pulverulentus, 346–347
Eumerus rufoscutellatus, 347
Eumerus singhalensis, 347
Eumerus sita, 347
Euphaea fraseri, 36
Euphitrea sp., 263
Euphranta crux, 313
Euploea core, 295–296
Eurydema pulchra, 126
Eurysaspis flavescens, 126
Evacanthinae, 417, 422
evisceration, 65
evolutionary species, 287, *287*
exemplar microcosms, 402
    fig syconia, 402–406
exotic coccinellids, 239
extrafloral nectaries (EFNs), 406
Eysarcoris montivagus, 126
Eysarcoris ventralis, 134

F

fairyflies, *see* Mymaridae (Hymenoptera: Chalcidoidea)
family diagnosis and key characters
    Coccinellidae, 224–227
    tiger beetles, 202
    whiteflies, 103–104
feeding stage, Odonates, 38
female scale insects, diagnostic characters of, 70–71
Ficus, 402
Ficus racemosa, *403*
field data collection, 296, *297*
fig syconia
    entry restrictions, 404–405
    internal environments, 403–404
    origin, 402–403
    trophic networks, 405–406
Filippiinae Bodenheimer, 87–88
flea beetles (Coleoptera: Chrysomelidae: Galerucinae: Alticini)
    checklist, 267, **267–269**
    genera, 249–252
    habitus, *253*
    natural history, 247–249
    as pests, 267, 269, **270–271**
flying skills, Odonates, 38
founding queens, 406–407
fruit borers, 248
Furcatergalia
    Caenoidea, 11–12, *12*
    Ephemerelloidea, 12–14, *13*, *14*
    Ephemeroidea, 10–11, *11*
    Leptophlebioidea, 9, *10*

G

Gastrotrypes sp., *161*
Gastrozona fasciventris, *310*, *311*
Gellia nigripennis, *131*, *133*
genealogical species, 287
genitalia, 282, *283*
    preparation, 240
Global Biodiversity Information Facility (GBIF), 195
Goa, **326**, 327
Gonopsis pallescens, *131*
Graptomyza brevirostris, 352
grasshoppers, *see* Acrididae
Gryon sp., *161*
Gujarat, 327

H

Halpe (Hesperiidae), *294*, 294–295
Halticorcus sp., *258*
Halyomorpha picus, *126*, *133*, *134*, *135*
Halys serrigera, *126*, *133*, *135*
Heliocypha bisignata, 36
Heliodon elisabeth, 330
Hemiptera, 437
Hemipyxis sp., *259*
Hennigian method, 389
Heptageniidae, 14, *14*
Heptodonta, 203–204, *204*
Hermaeophaga ruficollis, 266
Hespera sp., *257*
hibernation, 233–234
histriobdellid, 152
holometabolous insects, 392
holotype, 279
homolea, 294–295
Hoplistodera recurva, *126*
Horn, W., 207
host associations, Mymaridae, **153**
host eggs (mymarids), rearing of, 154
host plant associations
    Coccidae, 95
    tribe Dacini, 318–319, **319**
host plant rearing, 320
hover-flies (Diptera: Syrphidae)
    Eristalinae, 332–344
    Microdontinae, 329–331
    overview, 325–329, **326**
    Syrphinae, 352–382
Humboldtia brunonis, *403*, 407
Hylicinae, 422
Hymenoptera, 179, *391*, 392, 437
Hyphasis sp., 255

I

Iassinae, 422
Ictinogomphus rapax, 35
Idea malabarica, 289
Idiocerinae, 417–418, 422–423
immature stages, mymarids, 152
immature taxonomy, 315
    Coccinellidae, 234–237
    Ephemeroptera, 19
Indascia gracilis, 330
Indian butterflies, *276*
    evolutionary biology/phylogenetics/ molecular systematics, 286–291
    modern mantra, 296–298
    molecular systematic studies, 291–292
    overview, 275–276
    taxonomic impediment, 292–296
    taxonomic legacy, 276–282, *278*
    taxonomy, traditional tools, 282–285
Indian Odonata, *see* Odonata
Indolestes davenporti, 36
integrative taxonomy, 19–20, 317
    bumble bees, 180–181
    Ephemeroptera, 19–20
    Scelionidae/Platygastridae, 164
    tiger beetles, 211–212
internal branches, 290
International Commission on Zoological Nomenclature (ICZN), 279
Ischiodon scutellaris, 375–377
Isolia sp., *161*
Ivalia sp., 260

J

Jansenia, 205, *205*
Jharkhand, 327
Johnston's organs, *391*, 394–395

K

Karnataka, **326**, 328
Kashmirobia hugeli, 261
Kerala, **326**, 327
Kertesziomyia nigra, 344
Klugephlebia kodai, 9
Korinchia rufa, 347

L

ladybird beetles (Coccinellidae), 223; *see also* Coccinellidae
Lanka ramakrishnai, 248, 261
larvae
    Anisoptera, 32
    Plecoptera, 48
    processing, 321
    Zygoptera, 31–32
larval characters, 174
larval/pupal biology, 209–211, *210*
larval stage
    Ephemeroptera, 7, 20–21
    Odonates, 38
leaf miners, 248
lectotype, 279
lek behaviour, 318
Lepidoptera, *396*, 437
    phylogeny, *291*

Lepidoptera Indica, 277
Leptophlebiidae, 9, *10*
Leptophlebioidea, 9, *10*
Lesagealtica sp., 265
Lestoidea, 30
Libelluloidea, 31
life cycle
    Ephemeroptera, 20–21
    Odonata, 37–39
    Plecoptera, 52–53
linear clinal variation, *287*, 289
Linnaeus's binomial system, 276–277
Lipromorpha sp., 265
liquid preservation, mymarids, 154
longevity
    mayflies, 21
    Odonates, 39
Longitarsus limnophilae, 248
Longitarsus sp., 260
Lophyra, 204–205, *205*
Luperomorpha sp., 260
Luperomorpha vittata, 259, 260

M

Macromedia donaldi, *35*
Macropsinae, 418, 423
Macrosyrphus confrater, 377–378
Macroteleia sp., *161*
Madhya Pradesh, 328
Maharashtra, **326**, 328
malaise traps, 167
Mallota curvigaster, 340
Mallota vilis, 340–341
Manobia sp., 266
marginal setae, 78–79, *79*
mass assemblages, 233–234
mass *versus* progressive provisioning, 194
mate guarding, 209
mating, 20
    trophallaxis, 318
mayflies, *see* Ephemeroptera
Megarrhamphus sp., *131*
Megophthalminae, 418, 423–424
melanism, 233
Melanostoma ceylonense, 356
Melanostoma melanoides, 356
Melanostoma orientale, 356–357
Melanostoma pedium, 357–358
Melanostoma scalare, 358
Melanostoma sp., 358–359
Melanostoma univittatum, 358
Meliscaeva ceylonica, 378–379
Meliscaeva mathisi, 379
Meliscaeva monticola, 379
Meliscaeva strigifrons, 379
Menida versicolor, *126*
Merodonoides multifarius, 341
Merodontini tribe, 344–347
Mesembrius bengalensis, 341–342
Mesembrius quadrivittatus, 342
mesocosms, 401
mesosoma, *148*
Metadon auricinctus, 330
Metadon flavipes, 331
Metadon montis, 330
Metadon taprobanicus, 330–331
Micraphthona fulvipes, 256
microcosms, *403*
    ant-plant domatia, 406–408
    exemplar, *see* exemplar microcosms

as model systems, 401–402
natural, 402
pitcher, 402
Microdon fulvopubescens, 331
Microdontini tribe, 329–331
Microdon unicolor, 331
microsculpture, 207–208
migration, Odonates, 38
Milesia caesarea, 348
Milesia cinnamomea, 348
Milesia mima, 348
Milesia querini, 348
Milesia sexmaculata, 348
Milesiini tribe, 347–352
mimicry, 317–318
minnow mayflies, *see* Baetidae
Miocene, 406
mite association, 194
molecular characterization and phylogeny, 316–317
    bumble bees, 180
    Pentatomidae, 132–134
    Scelionidae/Platygastridae, 162–163
    tiger beetles, 207
    whiteflies, 108, 114
molecular phylogeny
    Ephemeroptera, 19
    Odonata, 33
    potter wasps, 193
molecular systematics, 291
Monoceromyia eumenioides, 332
Monoceromyia javana, 333
monophyletic, 9, 11, 15
museum resources developing, 296–297
Mymaridae (Hymenoptera: Chalcidoidea), 147–148
    biodiversity and distribution, 152
    biology, 152
    collection methods, 154
    conservation status, 153–154
    economic importance, 152–153
    family diagnosis, 148
    females, 149–151
    immature taxonomy, 152
    molecular characterization and phylogeny, 152
    preservation, 154
    subfamilies and tribes, 148–149
    taxonomic problems, 152
    work on Indian fauna, 151
Myolepta sp., 332
myrmecophiles, 406
myrmecophytes, 406
Myzolecaniinae Hodgson, 88

N

natural microcosms, 402
Nearctic region, 93
Nemouridae, 50, *51*
Nemouroidea, 49–50, *51*
Neocollyris, 203, *203*
    larva, *211*
Neoephemeridae, 11–12, *12*
Neohalys serricollis, *127*
Neorthacris acuticeps, *161*
Neotropics, 93
neotype, 280
nesting, 193–194
neural cladistics, 389
Neurobasis chinensis, *35*

Nezara viridula, *127*, *134*
Nisotra sp., 265
Nonarthra, 254

O

occupancy models, 214
Odonata, 29
    adults, 30–31, *35*, *36*
    biodiversity, 33, **34**
    biology, 37–39
    classification of, 30
    collection and preservation, 40
    conservation status, 39
    economic importance, 39
    endemic species, distribution of, **37**
    larvae, 31–32
    molecular phylogeny, 33
    odonatological studies, 32–33
    species richness, 33
    taxonomic problems, 36
    threatened species of, **39**
open leaf feeders, 248
Oriental region, 93
Orisaltata azurea, 261
Orissa, **326**, 328
Ortalotrypeta isshikii, *311*, *312*
Orthetrum chrysis, *35*
Orthoptera, 57–58; *see also* Acrididae
    classification, 58, **58**
    in Indian subcontinent, **59**, 59–60
    Locusta migratoria, *396*
orthopterans, 392
ostiole, 402–403
oviposition, 20
ovipositors, 320, 404

P

Pacific region, 93
Palaearctic region, 92
Pandasyopthalmus rufocinctus, 359
Panilurus nilgiriensis, 256, *257*
Paradibolia nila, 254
paralectotype, 280
Paramixogaster contractus, 331
parasites, 232
parasitoids, 154
paratype, 279
pathogens, 232
Patterned Stoneflies/Springflies, 50
Paussus favieri, 394
Peltoperlidae, 50–51, *51*
Pentatomidae (Hemiptera: Heteroptera: Pentatomoidea)
    Asopinae, *123*, 123–125
    biodiversity and species richness, 132
    biology, 135–137
    classification, 122–123
    collection and preservation, 141–142
    diagnosis, 122
    economic importance, 141
    female and male genitalia, *136*, 142
    male genital capsule, *135*
    molecular characterization and phylogeny, 132–134
    Pentatominae, *125*, 125–129, *126*, *127*
    Phyllocephalinae, 129–130, *131*
    Podopinae, 130–131, *131*
    taxonomic problems, 134–135
    works on Indian fauna, 131–132

Index

Pentatominae, *125*, 125–129, *126*, *127*, *133*
    external scent efferent system, *134*
    genera and species in, **132**
    head-dorsal, *133*
    suprageneric classification, **128**
Perillus bioculatus, 123
Periplaneta americana, 393
Perlidae, *51*
perlids, 50
Perlodidae, 50
Perloidea, 50, *51*
Petersula courtallensis, 9
Phaelota sp., 265
phenotypic plasticity, 289
Philophylla indica, *311*, 313
Philopona sp., 255
Phoresy, *161*, 164, 166–167
Phricodus hystrix, *127*
Phygasia marginata, 258
Phygasia silacea, 266
Phyllocephalinae, 129–130, *131*, *133*
Phyllotreta sp., 260
Phyllotreta striolata, 269
phylogenetic species, 287
phylogenetic systematics, 286
phylogeny
    monophyly/paraphyly and polyphyly, *290*
    scale insects, 91–92
Phytalminae, 307
Phytomia argyrocephala, 342–343
Phytomia errans, 343
Phytomia zonata, 343–344
Phytophagous Pentatomidae, host plants, 137, **137–141**
Picromerus sp., 123
Piezodorus hybneri, *127*
Pisciforma, 15, *15*, *16*
pitcher microcosms, 402
Placosternum sp., *127*, *133*
Platygastridae; *see also* Scelionidae/Platygastridae (Hymenoptera: Platygastroidea)
    family, 161–162
    subfamilies, 162
Platygastrinae, 162–163
Platygastroidea, 159–160; *see also* Scelionidae/Platygastridae (Hymenoptera: Platygastroidea)
    collection and preservation, 167
    families, 160
    habitats, 164
    and hosts, **165–166**
    sexual dimorphism, 167
Plautia crossota, *127*, *134*
Plebeiogryllus guttiventris, 393
Plecoptera, 47–48
    biogeography, 52
    biology, 52–53
    biomonitoring potential, 53
    classification overview, 48
    climate change impacts, 53
    collection and preservation, 53
    conservation, 53
    diagnostic features, 51–52
    diversity, 48, **49**
    economic importance, 53
    historical review, 48
    life cycle, 52–53
    Nemouroidea, 49–50, *51*
    Perloidea, 50, *51*
    phylogeny, 52

Pteronarcyoidea, 50–51, *51*
    taxonomic problems, 52
pocket-like sclerotisations, 77–78, *78*
Podagrica sp., 257, 263
Podontia congregata, 248, 264
Podontia lutea, 247
Podopinae, 122, 130–131, *131*
Polya-Urn probability model, 4
Polymitarcyidae, 10–11, *11*
polymorphism, 233
polyphyletic group, 290
polyphyletic tribe, 247
Pondicherry, **326**, 328
positive feedback process, 3–5
Potamanthidae, 11, *11*
potter wasps (Hymenoptera: Vespidae: Eumeninae), 187
    biodiversity and species richness, 191–192
    biology, 193–194
    collection and preservation, 194–195
    distribution patterns in, 192–193
    economic importance, 194
    immature taxonomy, 193
    molecular characterization and phylogeny/barcoding, 193
    subfamily diagnosis and key characters, 187–191
    taxonomic problems, 193
    work on Indian fauna, 191
pre-antennal pores, 84
predators, 137, 231
pregenital disc-pores, *83*, 83–84
preservation, *see* collection and preservation
preservation and storage
    dry, 96–97
    wet, 96
prey associations, 194, 234
Procecidochares utilis, 313
Prosopistomatidae, 9, *10*
prostoma, 406
Prothyma, 203, *204*
Protosticta gravelyi, 36
Psithyrus, 175
Psylliodes, 254
Pteronarcyoidea, 50–51, *51*
puparium, 103–104
pyrgotidae, 306

R

Rajasthan, **326**, 328
rearing, 22
reflex bleeding, 232
reproduction stage, Odonates, 38–39
resolution, generic/species levels, 292
reverse brain-drain, 298
Rhinobaccha gracilis, 379
Rhinobaccha krishna, 379
Rhinobaccha peterseni, 379–380
Rhynchocoris humeralis, *127*
Rhynchocoris plagiatus, *133*
Rioxa sexmaculata, *311*, 313
Rioxoptilona sp., 313
Rodolia cardinalis, 224
Rolled-Winged Stoneflies, 49
root feeders, 248

S

Salvianus sp., *131*
Scaeva latimaculata, 380

scale insects (Hemiptera: Coccidae), 70
    biology of, 94, **95**
    checklist of, 89, **89–90**
    classification, 86–88
    collection, 96
    distribution patterns, 88–89
    economic importance, 96
    family diagnosis, 70–86
    host plant association, 95
    immature taxonomy, 93
    molecular characterisation, 93–94
    phylogeny, 91–92
    preservation and storage, 96–97
    slide preparation, 97, **97**
    taxonomic problems, 94
    work on Indian fauna, 88
    zoogeography of, 92–93
scapular seta, 306
Scedella spiloptera, 311
Sceliocerdo viatrix, 161
Scelionidae/Platygastridae (Hymenoptera: Platygastroidea), 159
    biogeographical affinities, 163–164
    biology and ecology, 164–167
    classification, 162
    diversity and distribution, 163
    economic importance, 167
    family, 160
    molecular characterization and phylogeny, 162–163
    subfamilies, 160–161
    taxonomic problems/integrative taxonomy, 164
    work on Indian fauna, 162
Sceliotrachelinae, 162
scent efferent system, 122, 136
Sciocoris indicus, *127*, *133*
scolopidia, 394
Scotinophara sp., *131*
Serratoparagus auritus, 360
Serratoparagus crenulatus, 360–361
Serratoparagus serratus, 361–362
Serratoparagus yerburiensis, 362
Setisura, 14, *14*
sex determination, 194
sexual dimorphism, 136, 167, 318
Signoretiinae, 418, 424
Sinocrepis sp., 266
slide mounting, mymarids, 154
species, 286–288
    phylogeny, 290
    radiation as positive feedback, 3–4
    simulation, 4–5
species richness
    Ephemeroptera, **18**, 18–19
    on global data, 2
    on Indian insect data, 2
    Odonata, 33
    and rank order, *3*
specimen preparation, Ephemeroptera, 22
Sphaeroderma sp., 258
Sphaerophoria, 376
Sphaerophoria bengalensis, 380–381
Sphaerophoria knutsoni, 381
Sphaerophoria macrogaster, 381–382
Sphaerophoria scripta, 382
spiny crawlers, 12
spiracles, 82, *83*
spiracular disc-pores, 84
Spring/Brown Stoneflies, 50
Sri Lanka, **326**, 329

stigmatic clefts, 79
stigmatic spines, 79, *80*
stink bugs, 121; *see also* Pentatomidae (Hemiptera: Heteroptera: Pentatomoidea)
stoneflies, *see* Plecoptera
Storthecoris nigriceps, *131*
strepsipteran association, 194
stridulation, 57
subimago, 8
subspecies, 288
Surat Dangs, 327
swarming, 20
sweep netting, 167, 320
Sycomorus, 404
syconium, 402, *403*, 404
syntype, 279
Syritta fasciata, 349
Syritta indica, 349
Syritta orientalis, 349–350
Syritta pipiens, 350
Syritta proximata, 350
Syritta stylata, 350–351
Syrphus neitneri, 382

T

Tachiniscinae, 307
Taeniopterygidae, 50
Talbot's arrangement, 295
Tamil Nadu, **326**, 329
taprobana, 330
taxonomic characters, adult female scale insects, 72–73
    dorsal structures, 73–78
    marginal structures, 78–79
    ventral pores, 83–86
    ventral structures, 79–83
taxonomic characters, adult male scale insects, 71
    abdomen, 72
    dermal structure, 72
    head, 71
    legs, 71
    thorax, 71
    wings, 71
taxonomic problems
    bumble bees, 181
    Coccinellidae, 230
    Pentatomidae, 134–135
    potter wasps, 193
    Scelionidae/Platygastridae, 164
    tiger beetles, 212
Teleasinae, 160
Telengana, **326**, 329
Telenominae, 160
Telenomus sp., *161*
Teloganodidae, 12–13, *13*
Tephritidae
    biodiversity/species richness, 314, **315**
    biology/behaviour, 317–318
    collection/preservation, 320–321
    economic importance, 320
    heads, *310*
    host plant associations, 318–319, **319**
    immature taxonomy, 315
    integrative taxonomy, 317
    molecular characterisation/phylogeny, 316–317
    region wise distribution, **316**
    subfamily/tribes, 306–308
    taxonomic problems, 317
    thorax, *311*
    work on Indian fauna, 314, **314**
Tephritinae, 307, 310
terminal nodes, 290
Tetroda histeroides, *131*, *133*
Themara yunnana, *310*, *311*
Therates, 203, *204*
tiger beetles (Coleoptera: Cicindelidae), 201–202
    adult morphology, 207–209
    alternative phylogeny, *212*
    Apteroessa, *205*, 205–206
    biodiversity and species richness, 206
    bird species and, *213*
    citizen scientists, 215
    climate change, 214–215
    collection and preservation, 213–214
    conservation status, 212–213
    dispersal routes by, *207*
    distribution patterns, 206–207
    ecology, behaviour, and physiology, 209
    economic importance, 213
    family diagnosis and key characters, 202
    genera, **202**
    guides from countries, *216*
    Heptodonta, 203–204, *204*
    integrative taxonomy, 211–212
    isoclines connecting areas, *206*
    Jansenia, *205*, 205
    larval/pupal biology, 209–211, *210*
    Lophyra, 204–205, *205*
    molecular characterization and phylogeny, 207
    Neocollyris, 203, *203*
    occupancy modelling and data, 214
    Prothyma, 203, *204*
    taxonomic problems, 212
    Therates, 203, *204*
    Tricondyla, 202–203, *203*
tornus, 285
Trachytetra, 261
traps, 320
tribal level, 306
Tricondyla, 202–203, *203*
Tricorythidae, 13–14, *14*
Trissolcus sp., *161*
Tritaeniopteron punctatipleura, *313*
trophic categorization, 21
trophic mutualisms, 407
Trypetinae, 307
Typhlocybinae, 418, 424–425

U

Udonga montana, *127*
Ulopinae, 418, 425
unified species, 287

V

vasiform orifice, 104
ventral microduct, 84
ventral pores, 83–86
Volucella sp., 352
Volucellini tribe, 352

W

waxy covering, 233
wet preservation, 96
whiteflies (Hemiptera: Aleyrodidae), 103
    Aleyrodidae, 105
    Aleyrodinae, 105–108
    checklist, **109–114**
    economic importance, 114–115
    family diagnosis, 103–104
    key characters, 104
    molecular characterization and phylogeny, 108, 114
    puparium, *104*
    species diversity, 108
    works on Indian fauna, 108
Wildlife Protection Act (WPA), 296
wing venation, 282, *285*
Winter Stoneflies, 49, 50

X

Xanthandrus ceylonicus, 359
Xanthandrus indicus, 359
Xanthogramma, 376
Xenomerus sp., *161*
Xuthea sp., *259*, *265*
xwing venation, 282
Xylota atroparva, 351
Xylota bistriata, 351
Xylota carbonaria, 351–352
Xylota sp., 352

Y

Yaminia gmelini, *266*

Z

Zeugodacus brevipunctatus, *313*
Zeugodacus caudatus, *311*
Zeugodacus cucurbitae
    heads, *310*
    male, sternite V, *313*
    wings, *313*
Zeugodacus diversus, 312
Zeugodacus gavisus, 312
Zeugodacus havelockiae, 312
Zeugodacus scutellarius, *311*
Zeugodacus tau, *311*, *313*
Zeugodacus trilineata, 310
Zeugodacus trilineatus, 311
Zeugodacus watersi, *311*, *313*
Zicrona caerulea, 123
Zygoptera
    adults, 30
    larvae, 31–32